THERMAL–FLUID SCIENCES
An Integrated Approach

Thermal-Fluid Sciences is a truly integrated textbook for an engineering course covering thermodynamics, heat transfer, and fluid mechanics. This integration is based on the fundamental conservation principles of mass, energy, and momentum; a hierarchical grouping of related topics; and the early introduction and revisiting of practical device examples and applications.

As with all textbooks the focus is on accuracy and pedagogy. To enhance the learning experience *Thermal-Fluid Sciences* features full-color illustrations. The robust pedagogy includes chapter learning objectives, overviews, historical vignettes, and numerous examples that follow a consistent problem-solving format, enhanced by innovative self tests and color coding to highlight significant equations and advanced topics. Each chapter concludes with a brief summary and a unique checklist of key concepts and definitions. Integrated tutorials show the student how to use modern software including the NIST database (included on the in-text CD) to obtain thermodynamic and transport properties.

Stephen R. Turns has been a Professor of Mechanical Engineering at The Pennsylvania State University since completing his Ph.D. at the University of Wisconsin in 1979. Before that Steve spent five years in the Engine Research Department of General Motors Research Laboratories in Warren, Michigan. His active research interests include the study of pollutant formation and control in combustion systems, combustion engines, combustion instrumentation, slurry fuel combustion, energy conversion, and energy policy. He has published numerous referreed journal articles on many of these topics. Steve is a member of the ASME and many other professional organizations and has been an ABET Program Evaluator since 1994. He is also a dedicated teacher, for which he has won numerous awards including the Penn State Teaching and Learning Consortium, Hall of Fame Faculty Award; Penn State's Milton S. Eisenhower Award for Distinguished Teaching; the Premier Teaching Award, Penn State Engineering Society; and the Outstanding Teaching Award, Penn State Engineering Society. Steve's talent as a teacher is also reflected in his best-selling advanced undergraduate textbook *Introduction to Combustion: Concepts and Applications*, 2nd ed. Steve's commitment to students and teaching is shown in the innovative approach and design of *Thermal-Fluid Sciences: An Integrated Approach* and its companion volume *Thermodynamics*, also published by Cambridge University Press.

In lecturing on any subject, it seems to be the natural course to begin with a clear explanation

of the nature, purpose, and scope of the subject. But in answer to the question

"What is thermo-dynamics?" I feel tempted to reply

"It is a very difficult subject, nearly, if not quite, unfit for a lecture."

Osborne Reynolds
On the General Theory of Thermo-dynamics
November 1883

THERMAL–FLUID SCIENCES

AN INTEGRATED APPROACH

Stephen R. Turns

The Pennsylvania State University

CAMBRIDGE
UNIVERSITY PRESS

CAMBRIDGE UNIVERSITY PRESS

Cambridge, New York, Melbourne, Madrid, Cape Town, Singapore, São Paulo

CAMBRIDGE UNIVERSITY PRESS

40 West 20th Street, New York, NY 10011–4211, USA

www.cambridge.org

Information on this title:

First published 2006

Printed in Hong Kong by Golden Cup Printing Co. Ltd.

A catalog record for this publication is available from the British Library.

Library of Congress Cataloging-in-Publication Data

Pages 1157–1158 constitute a continuation of the copyright page.

Turns, Stephen R.
Thermal-fluid sciences : an integrated approach / Stephen R. Turns.
p. cm.
Includes bibliographical references and index.
ISBN-13: 978-0-521-85043-8 (hardback : alk. paper)
ISBN-10: 0-521-85043-6 (hardback : alk. paper)
1. Thermodynamics. 2. Gas flow. 3. Heat–Transmission. I. Title.

QC311.T86 2005
536'.7–dc22

2005026545

ISBN-13 978 0 521 85043 8 hardback
ISBN-10 0 521 85043 6 hardback

*This book is dedicated to
Mike, Matt, Sara, and Bryan*

Contents

PART ONE: FUNDAMENTALS 1

Chapter 1 • BEGINNINGS 2

Chapter 2 • **THERMODYNAMIC PROPERTIES, PROPERTY RELATIONSHIPS, AND PROCESSES** *46*

Chapter 3 • **CONSERVATION OF MASS** *172*

Chapter 6 • **CONSERVATION OF MOMENTUM** *406*

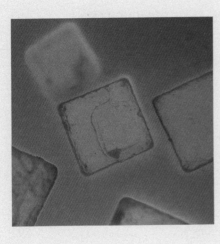

Chapter 7 • **SECOND LAW OF THERMODYNAMICS
AND SOME OF ITS CONSEQUENCES**   *512*

Chapter 8 • **SIMILITUDE AND DIMENSIONLESS PARAMETERS** *582*

PART TWO: **BEYOND THE FUNDAMENTALS** *627*

Chapter 9 • EXTERNAL FLOWS: FRICTION, DRAG,
AND HEAT TRANSFER   *628*

Chapter 11 • **THERMAL-FLUID ANALYSIS OF STEADY-FLOW DEVICES** *820*

Chapter 12 • **SYSTEMS FOR POWER PRODUCTION, PROPULSION, AND HEATING AND COOLING** *948*

APPENDIX C   THERMODYNAMIC AND THERMO-PHYSICAL PROPERTIES OF AIR   *1087*

APPENDIX D   THERMODYNAMIC PROPERTIES OF H₂O   *1092*

APPENDIX E   VARIOUS THERMODYNAMIC DATA   *1113*

APPENDIX F   THERMO-PHYSICAL PROPERTIES OF SELECTED GASES AT 1 ATM   *1114*

APPENDIX G   THERMO-PHYSICAL PROPERTIES OF SELECTED LIQUIDS   *1120*

APPENDIX H   THERMO-PHYSICAL PROPERTIES OF HYDROCARBON FUELS   *1126*

Sample Syllabi

Sample Syllabus–One-Semester Survey Course*

Period, Topic(s) followed by Textbook Reference

1 Preliminaries & course introduction ➤ *1.1*

Introductory Topics & Groundwork

2 Thermal-science applications, systems & control volumes ➤ *1.2–1.3*

3 General conservation principles & key concepts ➤ *1.4*

4 Properties, states, processes, cycles, & equilibrium concepts ➤ *1.5*

5 Real flows, dimensions & units, problem-solving method ➤ *1.6–1.8*

Thermodynamic Properties & Equations of State

6 Common properties related to 1st law & equation of state ➤ *2.1–2.2a*

7 State principle, ideal-gas equation of state, P–v–T space ➤ *2.3, 2.4a–2.4c*

8 Multiphase substances: phase boundaries, x, tables, & NIST software ➤ *2.6*

9 Multiphase substances (continued) ➤ *2.6*

Conservation of Mass

10 Conservation of mass: systems & control volumes, flow rates ➤ *3.1–3.3c*

11 Conservation of mass: control volumes, steady state, steady flow ➤ *3.3d–3.3e*

Calorific Properties & Calorific Equations of State

12 Internal energy, enthalpy, & specific heats ➤ *2.3b*

13 Calorific equation of state: ideal gases ➤ *2.4d*

Groundwork for Energy Conservation

14 Energy: macroscopic & microscopic ➤ *4.1–4.2*

15 Identifying heat & work interactions ➤ *4.3–4.4*

* A semester of forty-five class periods is assumed with two periods used for examinations or other activities.

16 Identifying heat & work interactions (continued) ➤ *4.3–4.4*

A Closer Look at Heat Transfer

17 Heat transfer modes: conduction & convection ➤ *4.5a–4.5b*

18 Heat transfer modes: radiation ➤ *4.5c*

Energy Conservation—Thermodynamic Systems

19 Energy conservation for a system & applications ➤ *5.1a–5.2a*

20 Energy conservation for a system & applications (continued) ➤ *5.2a*

Application of Energy Conservation to Conduction Heat Transfer

21 Conduction analysis: integral (lumped) formulations ➤ *5.2c*

22 1-D conduction (planar) & electrical analog ➤ *5.2c*

23 1-D conduction (cylindrical) & electrical analog ➤ *5.2c*

Energy Conservation—Control Volumes

24 Energy conservation for a control volume ➤ *5.3a–5.3b*

25 Steady-flow processes & devices ➤ *5.3a–5.3b, 11.1*

Second Law of Thermodynamics & Related Topics

26 2nd law: statement, consequences, & prerequisite concepts ➤ *7.1–7.4a*

27 Carnot efficiency, reversibility, & entropy ➤ *7.4b–7.6b*

28 2nd-law property relationships ➤ *2.2c, 2.3d, 2.4e–2.4h*

29 Isentropic efficiency ➤ *7.6e*

Conservation of Momentum & Fluid Statics

30 Conservation of linear momentum—forces ➤ *6.1–6.2*

31 Conservation of linear momentum—fluid statics ➤ *6.3a*

32 Fluid statics: manometry & forces on submerged surfaces ➤ *6.3a*

Momentum Conservation—Control Volumes

33 Momentum flows & conservation of linear momentum: integral CVs ➤ *6.4*

34 Momentum flows & conservation of linear momentum: integral CVs ➤ *6.5a–6.5b*

35 Mechanical energy equation, Bernoulli equation ➤ *6.6–6.7*

Similitude & Dimensionless Parameters

36 Similitude & dimensionless parameters ➤ *8.1–8.5*

Conservation Principles Applied to Internal & External Flows

37 External flows: friction and heat transfer; basic flow patterns & physics ➤ *9.1–9.2*

38 Forced flows—flat plate ➤ *9.3–9.4*

39 Forced flows—other geometries ➤ *9.5*

40 Internal flows: friction & heat transfer; integral analyses ➤ *10.2a–10.2d*

41 Internal flows: friction & heat transfer; integral analyses (continued) ➤ *10.2e*

42 Fully developed laminar flows ➤ *10.3*

43 Fully developed turbulent flows ➤ *10.4*

Sample Syllabus–Thermal-Fluid Sciences 1*

Period, Topic(s) followed by Textbook Reference

1 Preliminaries & course introduction ➤ *1.1–1.2*

Introductory Topics & Groundwork

2 Physical frameworks & introduction to conservation principles ➤ *1.3–1.4*

3 Key concepts & definitions ➤ *1.5*

4 Key concepts & definitions (continued) ➤ *1.5*

5 Real flows, dimensions & units, problem-solving method ➤ *1.6–1.9*

Thermodynamic Properties

6 Motivation for study of properties, common thermodynamic properties ➤ *2.1–2.2*

7 Properties related to first & second laws of thermodynamics ➤ *2.1–2.2*

8 State principle, state relationships, ideal-gas state relationships ➤ *2.3, 2.4a–2.4c*

9 Calorific equation of state; $P–v$, $T–v$, $u–T$, $h–T$ plots for ideal gases ➤ *2.4d, 2.4g*

10 Nonideal gases: van der Waals equation of state & generalized compressibility ➤ *2.5*

11 Multiphase substances: phase boundaries, quality, $T–v$ diagrams ➤ *2.6a*

12 Multiphase substances: tabular data, NIST database, log P–log v diagrams ➤ *2.6b*

13 Compressed liquids & solids ➤ *2.7–2.8*

Conservation of Mass

14 Conservation of mass: systems ➤ *3.1–3.2*

15 Conservation of mass: flow rates & average velocities ➤ *3.3a–3.3c*

16 Conservation of mass for integral control volumes: steady state & steady flow ➤ *3.3e*

17 Conservation of mass for integral control volumes: unsteady flow ➤ *3.3e*

Groundwork for Energy Conservation

18 Energy storage, heat & work interactions at boundaries ➤ *4.1–4.3*

19 Identifying heat & work interactions ➤ *4.3–4.4*

A Closer Look at Heat Transfer

20 Rate laws for heat transfer: conduction & convection ➤ *4.5a–4.5b*

21 Rate laws for heat transfer: radiation ➤ *4.5c*

Energy Conservation—Thermodynamic Systems

22 Energy conservation for a system: finite processes ➤ *5.1a–5.2a*

23 Energy conservation for a system: at an instant ➤ *5.2a*

24 Energy conservation for a system: examples & applications ➤ *5.2a*

Application of Energy Conservation to Conduction Heat transfer

25 Conduction heat transfer: Integral (lumped) analysis ➤ *5.2c*

* A semester of forty-five class periods is assumed with two periods used for examinations or other activities.

Sample Syllabus–Thermal-Fluid Sciences II*

Period, Topic(s) followed by Textbook Reference

* A semester of forty-five class periods is assumed with two periods used for examinations or other activities

Nozzles & Diffusers—Application of Conservation Principles & Property Relations

34 General analysis ➤ *11.2a–11.2c*

35 Compressible flow introduction ➤ *11.2d*

36 Choked flow; converging–diverging nozzles ➤ *11.2d*

37 Choked flow; converging–diverging nozzles (continued) ➤ *11.2d*

Turbojet Engines—Application of Fundamentals to a Complex System

38 Turbojet components & integral control volume analysis ➤ *12.2a–12.2b*

39 Integral mass, energy, & momentum analyses (continued) ➤ *12.2a–12.2b*

40 Air-standard turbojet cycle analysis & performance measures ➤ *12.2c–12.2d*

41 Air-standard turbojet cycle analysis & performance measures (continued) ➤ *12.2c– 12.2d*

42 Combustor analysis ➤ *12.2f*

43 Combustor analysis (continued) ➤ *12.2f*

Sample Syllabus–Thermal-Fluid Sciences III*

Period, Topic(s) followed by Textbook Reference

1 Preliminaries & course introduction

Differential Forms of the Conservation Principles

2 Mass & energy conservation: differential forms ➤ *3.3f, 5.3e*

3 Momentum conservation: differential form ➤ *6.4d*

Turbulence Review

4 Tubulent flows & Reynolds decomposition ➤ *1.6, 3.4a*

5 Time-averaged integral & differential mass & momentum equations ➤ *3.4b, 6.8*

Application of Dimensional Analysis to Thermal-Fluids Sciences

6 Parametric testing & similitude ➤ *8.1–8.4*

7 Dimensionless parameters: origins & applications ➤ *8.5*

8 Dimensionless parameters (continued) ➤ *8.5*

Introduction to External Flows

9 Basic flow patterns & concept of boundary layers ➤ *9.1–9.2*

Forced Laminar Flow over Flat Plates

10 Velocity & temperature profiles in boundary layers ➤ *9.3a–9.3b*

11 Friction & drag solutions ➤ *9.3c*

12 Friction & drag solutions (continued) ➤ *9.3c*

13 Heat transfer solutions ➤ *9.3c–9.3e*

14 Heat transfer solutions (continued) ➤ *9.3c–9.3e*

*A semester of forty-five class periods is assumed with two periods used for examinations or other activities.

Forced Turbulent Flow over Flat Plates

15 Laminar–turbulent transition & boundary-layer growth ➤ *9.4a–9.4b*

16 Friction & drag ➤ *9.4c*

17 Heat transfer ➤ *9.4d–9.4e*

Forced Flow over Cylinders, Spheres, & Other Geometries

18 Friction & form drag ➤ *9.5a*

19 Empirical correlations for drag & heat transfer ➤ *9.5b*

20 Applications & examples ➤ *9.1–9.5*

Free Convection

21 Vertical flat plate—physical & mathematical description ➤ *9.6a*

22 Vertical flat plate—dimensionless parameters, solutions, & correlations ➤ *9.6a*

23 Other geometries, applications, & examples ➤ *9.6b*

Introduction to Internal Flows—Integral Analyses

24 Mass, momentum, & mechanical energy conservation ➤ *10.1–10.2c*

25 Mass, momentum, & mechanical energy conservation (continued) ➤ *10.2d*

26 Energy conservation: isothermal & nonisothermal flows with uniform heat flux ➤ *10.2e*

27 Energy conservation: nonisothermal flows with uniform wall temperature ➤ *10.2e*

Fully Developed Laminar Flows

28 Laminar flow criterion, differential conservation equations, & solutions ➤ *10.3a–10.3b*

29 Laminar velocity distribution: average velocity, wall shear stress, pressure drop ➤ *10.3c*

30 Laminar velocity distribution: head loss & friction factor ➤ *10.3c*

31 Temperature distributions: heat-transfer coefficients ➤ *10.3d*

32 Applications & examples ➤ *10.1–10.3*

Fully Developed Turbulent Flows

33 Velocity distributions & wall friction: theory & experiment ➤ *10.4a*

34 Average velocity, friction factor, rough tubes, & pipes ➤ *10.4a*

35 Heat-transfer relationships ➤ *10.4b*

Internal Flows—Additional Considerations

36 Developing flows, ducts of noncircular cross section, & minor losses ➤ *10.5–10.7*

37 Developing flows, ducts of noncircular cross section, & minor losses (continued) ➤ *10.5–10.7*

Applications to Heat Exchangers

38 Heat exchanger classifications & applications; integral analysis ➤ *11.6*

39 Overall heat-transfer coefficient ➤ *11.6*

40 Log-mean temperature difference method for heat exchanger design ➤ *11.6*

41 NTU–effectiveness method for heat exchanger design & analysis ➤ *11.6*

42 NTU–effectiveness method for heat exchanger design & analysis (continued) ➤ *11.6*

43 Review and synthesis

Sample Syllabus–Traditional One-Semester Thermodynamics Course*

Period, Topic(s) followed by Textbook Reference

1 Preliminaries & course introduction ➤ *1.1–1.2*

Introductory Topics & Groundwork

2 Physical frameworks & introduction to conservation principles ➤ *1.3–1.4*

3 Key concepts & definitions ➤ *1.5*

4 Key concepts & definitions (continued) ➤ *1.5*

5 Real flows, dimensions & units, problem-solving method ➤ *1.6–1.9*

Thermodynamic Properties & State Relationships

6 Motivation for study of properties, common thermodynamic properties ➤ *2.1–2.2*

7 Properties related to first & second laws of thermodynamics ➤ *2.2*

8 State principle, state relationships, ideal-gas state relationships ➤ *2.3–2.4*

9 Calorific equation of state; P–v, T–v, u–T, h–T plots for ideal gases ➤ *2.4*

10 Nonideal gases: van der Waals equation of state & generalized compressibility ➤ *2.5*

11 Multiphase substances: phase boundaries, quality, T–v diagrams ➤ *2.6*

12 Multiphase substances: tabular data, NIST database, log P–log v diagrams ➤ *2.6*

13 Compressed liquids & solids ➤ *2.7–2.8*

Conservation of Mass

14 Conservation of mass: systems; flow rates ➤ *3.1–3.3b*

15 Conservation of mass: control volumes ➤ *3.3d–3.3e*

Groundwork for Energy Conservation

16 Energy storage, heat & work interactions at boundaries ➤ *4.1–4.3*

17 Identifying heat & work interactions ➤ *4.3*

Energy Conservation—Thermodynamic Systems

18 Energy conservation for a system: finite processes ➤ *5.1–5.2a*

19 Energy conservation for a system: at an instant ➤ *5.2a*

20 Energy conservation for a system: examples & applications ➤ *5.2a*

Energy Conservation—Control Volumes & Some Applications

21 Energy conservation for a control volume: introduction ➤ *5.3a–5.3b*

22 Steady-flow processes & devices ➤ *5.3a, 11.1–11.2a*

23 Steady-flow devices: nozzles, diffusers, & throttles ➤ *11.2c, 11.3*

24 Steady-flow devices: pumps, compressors, fans, & turbines ➤ *11.4–11.5*

25 Steady-flow devices: heat exchangers ➤ *11.6*

26 Steam power plants & jet engines revisited ➤ *1.2a, 1.2d, 12.1a, 12.2a*

* A semester of forty-five class periods is assumed with three periods used for examinations or other activities.

Second Law of Thermodynamics

27 2nd law of thermodynamics: overview, Kelvin–Planck statement, consequences ➤ *7.1–7.4a*

28 Carnot cycle & Carnot efficiency, definition of entropy ➤ *7.4b–7.6a*

29 Entropy-based statement of 2nd law, entropy balances, other 2nd-law statements ➤ *7.6b–7.6c*

Second-Law Properties, Property Relationships, & Efficiencies

30 2nd-law property relationships ➤ *2.2c, 2.3d, 2.4e–2.4h*

31 *T–s* relationships for ideal gases, air tables, isentropic relationships ➤ *2.4e–2.4h*

32 Isentropic & polytropic processes, *T–s* & *P–v* diagrams ➤ *2.4e–2.4h*

33 Isentropic efficiencies ➤ *7.6e, 11.4b, 11.5b*

Steam Power Plant—Application of the 1st & 2nd Laws

34 Steam power plant: Rankine cycle ➤ *12.1a*

35 Steam power plant: superheat & reheat ➤ *12.1b*

36 Steam power plant: regeneration ➤ *12.1c*

37 Steam power plant (continued) ➤ *12.1*

Turbojet Engine—Application of the 1st & 2nd Laws

38 Jet engines: overall integral control volume analysis ➤ *12.2a–12.2b*

39 Turbojet engine cycle analysis ➤ *12.2c*

40 Turbojet engine cycle analysis (continued) ➤ *12.2d–12.2e*

Other Applications of Thermodynamics & Conservation Principles

41 Selected topics ➤ *Chapter 12*

42 Selected topics ➤ *Chapter 12*

Preface

THIS BOOK WAS CONCEIVED to address the needs of instructors desiring an integrated approach to teaching the thermal-fluid sciences. Traditionally, the thermal-fluid sciences are taught in a sequence of three or more separate courses treating individually the disciplines of thermodynamics, fluid mechanics, and heat transfer. Although this traditional grouping makes considerable sense, many engineering educators believe that a more effective and efficient treatment of these subjects can be achieved by integrating topics. The author hopes that this book will appeal to these educators.

The following organizational principles undergird this book:

- The fundamental conservation principles of mass, energy, and momentum are used as the primary integrating device.
- Related topics are grouped together hierarchically.
- Many examples revisit a few particular practical devices or applications.

As an understanding of these principles is important to the use of this book, some elaboration is helpful.

The use of the fundamental conservation principles (mass, energy, and momentum) is a natural choice as an integrating device for the thermal-fluid sciences and should be a comfortable choice for many engineering educators. The conservation principles are introduced in Chapter 1 in two general forms: a form associated with a process occurring over a finite time interval and a form expressing the conservation principle at an instant. These general formulations are then elaborated for mass conservation in Chapter 3, for energy conservation in Chapter 5, and momentum conservation in Chapter 6. Entropy balances introduced in Chapter 7 also follow the same general formulation, although entropy is not conserved in the same sense as are mass, energy, and momentum. Showing that all conserved quantities obey a single simple accounting (*in* minus *out* plus *generated* equals *stored*) provides an even higher level of integration. Also, stressing that the many ways these conservation principles are expressed within various subject domains all have their origins in just three basic statements should help engineering students structure their knowledge in useful ways. In a sense, this helps to establish the conservation principles as bedrock in the hierarchy of engineering science.

The second organizing principle, the hierarchical arrangement of subject matter, is perhaps best illustrated in Chapter 2. In this chapter, essentially all material related to thermodynamic and thermophysical properties is grouped together. In this way, we are able to show clearly the hierarchy of thermodynamic state relationships starting with the basic equation of state involving P, v, and T; adding first-law-based calorific equations of state involving u, h, P, T, and v; and ending with the second-law-based state relationships involving s, T, P, and v, for example. Such arrangement requires that Chapter 2 be revisited at appropriate places in the study of later chapters. In this sense, Chapter 2 is a resource that is to be returned to many times. Chapter 5 is another example of a hierarchical arrangement. Here the conservation of energy principle is applied to simple systems and then extended to the most complicated forms typically presented in undergraduate fluid mechanics and heat-transfer textbooks. For example, both simple conduction heat transfer and combustion are integrated within the context of conservation of energy. Therefore, Chapter 5, like Chapter 2, can be revisited and does not have to be studied from beginning to end.

What purpose is served by such an arrangement? First, it provides an important structure for a beginning learner. Experts who have mastered and work within a discipline organize material this way in their minds, whereas novices tend to treat concepts in an undifferentiated way as a collection of seemingly unrelated topics.[1] It is hoped that providing a useful hierarchy from the start may speed learning and aid in retention. A second reason for a hierarchical arrangement is flexibility. In general, the book has been designed to permit an instructor to select topics from within a chapter and combine them with material from other chapters in a relatively seamless manner. This flexibility allows the book to be used in many ways depending on the educational goals of a particular course or a sequence of courses. The several syllabi that follow the table of contents suggest some arrangements.

The third device used to promote the integration of the thermal-fluid sciences is the selection of several topics that are revisited in various examples used throughout the book. Motivation for the particular topics chosen is elaborated in Chapter 1.

With this philosophical understanding behind us, we now examine the specific structure of the book. The book is divided into two parts: Chapters 1–8 comprise the part designated as *Fundamentals*; Chapters 9–12 are denoted *Beyond the Fundamentals*. This division is a natural one in that all of the fundamental concepts are developed in Chapters 1–8, that is, property relationships in Chapter 2; the three conservation principles in Chapters 3, 5, and 6; the second law of thermodynamics in Chapter 7; and dimensional analysis in Chapter 8. Chapters 9–12 then see the application of these fundamentals to a variety of mechanical engineering topics. Chapter 9 treats external flows, combining the problems of friction and drag and heat transfer, topics usually treated separately in traditional fluid mechanics and heat-transfer courses. In a similar way, internal flows are examined in Chapter 10. Both chapters emphasize conservation principles in both integral and differential form. Chapter 11 focuses on steady-flow devices. Portions of this chapter can and should be used earlier than their placement in a later chapter suggests. For example, entry points into Chapter 11 are indicated in Chapters 5, 6, and 7. Coverage of thermal-fluids systems is grouped in Chapter 12. Topics from this chapter can be selected to integrate many of the fundamental concepts developed in earlier chapters. For example, the section on jet

---

[1]Larkin, J., McDermott, J., Simon, D. P., and Simon, H. A., "Expert and Novice Performance in Solving Physics Problems," *Science*, 208: 1335–1342 (1980).

engines combines all three conservation principles. This section also utilizes advanced ways of dealing with properties and the first and second laws of thermodynamics when the air-standard cycle is modified to include a constant-pressure combustion process.

In addition to structure, many other pedagogical devices are employed in this book. These include the following:

- An abundance of color photographs and images illustrate important concepts and emphasize practical applications;
- Each chapter begins with a list of learning objectives, a chapter overview, and a brief historical perspective, where appropriate, and concludes with a brief summary;
- Each chapter contains many examples that follow a standard problem-solving format;
- Self tests follow most examples;
- Key equations are denoted by colored backgrounds;
- Each chapter concludes with a checklist of key concepts and definitions linked to specific end-of-chapter questions and problems;
- The National Institute of Science and Technology (NIST) database for thermodynamic and transport properties (included in the NIST12 v.5.2 software provided with the book) is used extensively.

All of these features are intended to enhance student motivation and learning and to make teaching easier for the instructor. For example, the many color photographs make connections to real-world devices, a strong motivator for undergraduate students. Also, the learning objectives and checklists are particularly useful. For the instructor, they aid in the selection of homework problems and the creation of quizzes and exams, or other instructional tools. For students, they can be used as self tests of comprehension and can monitor progress. The checklists also cite topic-specific questions and problems. In his use of the book, the author utilizes the learning objectives to guide reviews of the material prior to examinations. Having well-defined learning objectives is also useful in meeting engineering accreditation requirements. Many questions and problems are included at the end of each chapter. The purpose of the questions is to reinforce conceptual understanding of the material and to provide an outlet for students to articulate such understanding. Throughout the book, students are encouraged to use the National Institute of Science and Technology (NIST) databases to obtain thermodynamic and transport properties. The online NIST property database is easily accessible and is a powerful resource. It is a tool that will always be up to date. The NIST12 v.5.2 software included with this book has features not available online. This user-friendly software provides extensive property data for eighteen fluids and has an easy-to-use plotting capability. This invaluable resource makes dealing with properties easy and can be used to enhance student understanding.

TO THE INSTRUCTOR

Two specific uses of this book are envisioned: 1. as the primary textbook for a multisemester sequence of thermal-fluid science courses for mechanical engineering majors and 2. as a textbook for a one-semester survey course for engineering students in disciplines outside of mechanical engineering. (A similar one-semester survey course for mechanical engineering majors may also be useful in a modern curriculum.) Because of the inherent flexibility in the organization of the materials, this book can meet the needs of a course sequence, or a survey course, in a variety of ways. To assist in selecting topics, the text distinguishes three levels: level 1 (basic) material is unmarked, level 2 (intermediate) material appears with a blue background and a blue edge stripe, and level 3 (advanced) material is denoted with a light red background and a red edge stripe. Instructors can therefore choose from the numerous topics presented to create courses that meet their specific educational objectives. To show how this might be done, sample syllabi preceding this preface illustrate a three-semester sequence of courses and a one-semester survey course. Both of these examples emphasize topics traditionally included in thermodynamics courses, reflecting the author's view on teaching the thermal-fluid sciences. A three-semester sequence was chosen to be compatible with the present trend in mechanical engineering curricula to limit the core thermal-fluid sciences to three 3-credit courses, or their equivalent. In fact, the repackaging of the thermal-fluids core from typically 12 to 9 credits was a strong motivator for the creation of this book.

Also presented in the preceding is a syllabus showing how this book can be used for a traditional single-semester first course in thermodynamics. This syllabus is presented for two reasons: to illustrate the flexibility of the book in creating courses to meet specific needs and to provide an option for those instructors who might want to use the book in a traditional way before trying a more integrated approach.

Feedback from instructors who use this book is most welcome.

About the Author

Stephen R. Turns has been a Professor of Mechanical Engineering at The Pennsylvania State University since completing his Ph.D. at the University of Wisconsin in 1979. Before that Steve spent five years in the Engine Research Department of General Motors Research Laboratories in Warren, Michigan. His active research interests include the study of pollutant formation and control in combustion systems, combustion engines, combustion instrumentation, slurry fuel combustion, energy conversion, and energy policy. He has published numerous referreed journal articles on many of these topics. Steve is a member of the ASME and many other professional organizations and has been an ABET Program Evaluator since 1994. He is also a dedicated teacher, for which he has won numerous awards including the Penn State Teaching and Learning Consortium, Hall of Fame Faculty Award; Penn State's Milton S. Eisenhower Award for Distinguished Teaching; the Premier Teaching Award, Penn State Engineering Society; and the Outstanding Teaching Award, Penn State Engineering Society. Steve's talent as a teacher is also reflected in his best-selling advanced undergraduate textbook *Introduction to Combustion: Concepts and Applications*, 2nd ed. Steve's commitment to students and teaching is shown in the innovative approach and design of *Thermal-Fluid Sciences: An Integrated Approach* and its companion volume *Thermodynamics*, also published by Cambridge University Press.

Acknowledgments

THIS BOOK HAS BEEN A LONG TIME COMING and many people have contributed along the way. First I would like to thank the many reviewers and students, too many to name here, all of whom contributed both mightily, and subtly. Without their candid comments and careful reading this project would not have been possible.

I am indebted to Peter Gordon at Cambridge University Press for his vision of a richly illustrated and colorful book and the managers at the Press for supporting our shared vision. To this end, I am happy to acknowledge Rick Medvitz and Jared Ahern of the Applied Research Laboratory at Penn State for their creation of the many exciting computational fluid dynamics illustrations sprinkled throughout the text. Thanks also are owed to Joel Peltier and Eric Paterson at ARL for their support of the CFD effort. Regina Brooks and Anne Wells at Serendipity Literary Agency worked hard to find photographs, and the cooperation of AGE fotostock is gratefully acknowledged. Jessica Cepalak and Michelle Lin at Cambridge were indispensable in many ways throughout the project. The stunning book design was the effort of José Fonfrias. Thank you, José.

Special thanks go to Chris Mordaunt for his creation of the self tests and his meticulous reading of the manuscript and insightful comments. Thanks also go to the members of the solutions manual team: Jacob Stenzler and Dave Kraige, leaders of the effort, and Justin Sabourin, Yoni Malchi, and Shankar Narayanan. For nearly a decade, Mary Newby deciphered my pencil scrawls to create a word-processed manuscript. I thank Mary for her invaluable efforts.

I would like to acknowledge the production team at Cambridge—Alan Gold (Senior Production Controller) and Pauline Ireland (Director, Production and New Media Development)—who broke a great number of old rules to make a very new book. I especially want to thank Anoop Chaturvedi and his production team at TechBooks for their careful attention to detail. I also thank fellow textbook authors Dwight Look, Jr., and Harry Sauer, Jr.; Glen Myers; Alan Chapman; David Pnueli and Chaim Gutfinger; and Gertrude Shepherd, wife of deceased author Dennis Shepherd, for their permission to use selected problems from their works. Thanks also are owed to Eric Lemmon at NIST for assembling the software provided with this book and to Joan Sauerwein for making the agreement.

For the hospitality shown during a sabbatical year spent writing, I thank Allan Kirkpatrick and Charles Mitchell at Colorado State University and Taewoo Lee and Don Evans at Arizona State University. I would also like to thank good friends Kathy and Dan Wendland for opening their home to us for an extended stay in Fort Collins. Thanks also are extended to Nancy and Dave Pearson, more good friends, for their help and companionship in Tempe.

The sales and marketing team at Cambridge are a joy to work with. Thanks go to Liza Murphy, Kerry Cahill, Liz Scarpelli, Robin Silverman, Ted Guerney, Catherine Friedl, and Valerie Yaw, along with their counterparts in the UK, Rohan Seery, Ben Ashcroft, Gurdeep Pannu, and Cherrill Richardson. I would also like to thank Jae Hong for his contributions to this project.

Moral support has come from many fronts, especially from the crowd at Saints Cafe and from Bob Santoro. Three people, however, deserve my heartfelt thanks. The first is Peter Gordon at Cambridge. Peter came to the rescue in trying times and breathed new life into this project. Second is Dick Benson, friend and confidant. Without Dick's enthusiastic support, this book would not have been possible. Third, but hardly last, is Joan, my wife. I cannot thank her enough for her help, patience, and support. Thank you, Joan.

PART ONE

FUNDAMENTALS

CHAPTER
ONE

BEGINNINGS

CHAPTER
TWO

THERMODYNAMIC
PROPERTIES, PROPERTY
RELATIONSHIPS,
& PROCESSES

CHAPTER
THREE

CONSERVATION
OF MASS

CHAPTER
FOUR

ENERGY AND
ENERGY
TRANSFER

CHAPTER
FIVE

CONSERVATION OF
ENERGY

CHAPTER
SIX

CONSERVATION
OF MOMENTUM

CHAPTER
SEVEN

SECOND LAW OF
THERMODYNAMICS AND
SOME OF ITS
CONSEQUENCES

CHAPTER
EIGHT

SIMILITUDE AND
DIMENSIONLESS
PARAMETERS

BEGINNINGS

After studying Chapter 1, you should:

- *Have a basic idea of what the thermal-fluid sciences are and the kinds of engineering problems to which they apply.*

- *Be able to distinguish between a system and a control volume.*

- *Understand the differences between integral systems or control volumes and differential systems or control volumes.*

- *Have an understanding of and be able to state the formal definitions of thermodynamic properties, states, processes, and cycles.*

- *Understand the concept of thermodynamic equilibrium and its requirement of simultaneously satisfying thermal, mechanical, phase, and chemical equilibria.*

- *Be able to explain the meaning of a quasi-equilibrium process and the concept of local equilibrium.*

- *Appreciate the general characteristics of laminar and turbulent flows.*

- *Understand the distinction between primary dimensions and derived dimensions, and the distinction between dimensions and units.*

- *Be able to convert SI units of force, mass, energy, and power to U.S. customary units, and vice versa.*

Chapter 1 Overview

IN THIS CHAPTER, we introduce and define the subdisciplines comprising the thermal-fluid sciences: thermodynamics, heat transfer, and fluid dynamics. We also introduce five fairly complex, practical applications of our study of the thermal-fluid sciences: the fossil-fueled steam power plant, solar-heated buildings, jet engines, the spark-ignition reciprocating engine, and biological systems. To set the stage for more detailed developments later in the book, several of the most important concepts and definitions are presented here: These include the concepts of thermodynamic systems and control volumes; the fundamental conservation principles; integral and differential analyses; thermodynamic properties, states, and cycles; equilibrium and quasi-equilibrium processes; local equilibrium; and the differences between laminar and turbulent flows. The chapter concludes with some ideas of how you might optimize the use of this textbook based on your particular educational objectives.

1.1 WHAT ARE THE THERMAL-FLUID SCIENCES?

The three disciplines—**thermodynamics**, **heat transfer**, and **fluid dynamics**—are collectively known as the **thermal-fluid sciences**, or sometimes just the **thermal sciences**. Traditionally, these three subjects are taught as separate entities, and there are many textbooks devoted to each discipline. References [1–9] indicate just a few. To understand what the thermal-fluid sciences involve, we begin with a dictionary definition [10] of thermodynamics:

> **Thermodynamics is the science that deals with the relationship of heat and mechanical energy and conversion of one into the other.**

The Greek roots, *therme*, meaning heat, and *dynamis*, meaning power or strength, suggest a more elegant definition: *the power of heat*. In its common

Examples of energy conversion systems: Fuel cells convert energy stored in chemical bonds to electricity to power a low-pollution bus (left); solar concentrators collect radiant energy from the sun (middle); wind turbines, Tehachapi, California produce electricity (right).

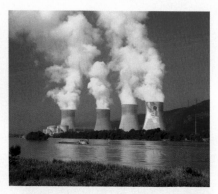

Applications of heat transfer: brake systems (left); steel heat treating and forming (middle); cooling towers (right).

usage in engineering, thermodynamics has come to mean the broad study of energy and its various interconversions from one form to another. The concepts developed in Chapters 2, 4, 5, and 7 are at the heart of thermodynamics.

The second component of the thermal-fluid sciences, heat transfer, deals explicitly with the transfer of energy that results when either a difference in temperature or a temperature gradient exists. Examples of heat transfer are numerous. The design of engine pistons, computer cooling systems, automotive and aircraft braking systems, and boilers and other heat exchangers is dominated by heat-transfer considerations. Keeping warm on a cold day is a very practical example of a heat-transfer problem. Heat-transfer concepts are introduced and developed in Chapters 4, 5, 9, and 10; a portion of Chapter 11 considers the design and performance of heat-exchange equipment.

Fluid dynamics, the last component of our thermal-fluid trilogy, is concerned with the motion of fluids and the effects that such motion produces. The connection of fluid dynamics to thermodynamics and heat transfer is strong and obvious: Moving fluids move energy from place to place. Again, engineering examples abound. Fans, pumps, turbines, rockets, internal combustion engines, and plumbing systems of all types all involve moving fluids. In this book, we explore the interconnections between the forces produced on or by fluids and the work that such forces may produce. Chapters 3 and 6 focus on key principles of fluid dynamics, which are further developed in Chapters 9 and 10.

Throughout this book we emphasize the fundamental principles of conservation of mass, conservation of energy, and conservation of momentum. A

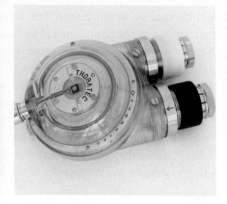

Applications of fluid dynamics: aircraft design and flight (top left); the dive of this Kingfisher (left); portable heart pump (middle); freshwater pumping station (right).

chapter is explicitly devoted to each. In varying degrees, the three thermal-fluid science disciplines all build upon these three fundamental principles; thus, the conservation laws comprise an important organizing thread throughout this book. Also emphasized throughout are fundamental definitions. To this end, the most important definitions are presented in bold type at their first appearance. Following the presentation of some applications, we begin this precedent.

1.2 SOME APPLICATIONS

A major objective of this book is to present an integrated view of the thermal-fluid sciences. The following five applications are sufficiently complex that a comprehensive understanding of each requires concepts from each of the core thermal-fluid disciplines (thermodynamics, heat transfer, and fluid mechanics):

- Fossil-fueled steam power plants,
- Solar-heated buildings,
- Spark-ignition engines,
- Jet engines, and
- Biological systems.

Approximately 20% of the examples presented in subsequent chapters revisit these specific applications, as do many of the end-of-chapter problems. The intent here is to use these applications to provide an additional modest layer of integration to the subject matter. Where these particular examples appear, a note reminds the reader that the example relates to one of these five applications.

Coal-fired power plant in North Rhine-Westphalia, Germany.

1.2a Fossil-Fueled Steam Power Plants

There are many reasons to choose the fossil-fueled steam power plant as an integrating application. First, and foremost, is the overwhelming importance of such power plants to our daily existence. Imagine how your life would be changed if you did not have access to electrical power, or less severely, if electricity had to be rationed so that it would be available to you only a few hours each day! It is easy to forget the blessings of essentially limitless electrical power available to residents of the United States. That the source of our electricity is dominated by the combustion of fossil fuels is evident from an examination of Table 1.1. Here we see that approximately 71% of the electricity produced in the United States in 2002 had its origin in the combustion of a fossil fuel, that is, coal, gas, or oil. A second reason for our choice of steam power plants as an integrating application is the historical significance of steam power. The science of thermodynamics was born, in part, from a desire to understand and improve the earliest steam engines. John Newcomen's first coal-fired steam engine (1712) predates the discovery of the fundamental principles of thermodynamics by more than a hundred years! The idea later to be known as the second law of thermodynamics was published by Sadi Carnot in 1824; the conservation of energy principle, or the first law of thermodynamics, was first presented by Julius Mayer in 1842.[1]

In this chapter we present the basic steam power plant cycle and illustrate some of the hardware used to accomplish this cycle. In subsequent chapters, we will add devices and complexity to the basic cycle. Figure 1.1 shows the basic

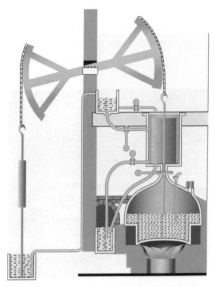

John Newcomen's steam engine was used to pump water from coal mines.

---

[1] A timeline of important people and events in the history of the thermal sciences is presented in Appendix A.

Table 1.1 Electricity Generation in the United States for 2002 [11]

Source		Billion kW·hr	%
Fossil Fuels			
Coal		1926.4	50.2
Petroleum		89.9	2.3
Natural Gas		685.8	17.9
Other Gases		12.1	0.3
	Subtotal	2714.2	70.7
Nuclear		780.2	20.3
Hydro Pumped Storage		−8.8	−0.2
Renewables			
Hydro		263.6	6.9
Wood		36.5	1.0
Waste		22.9	0.6
Geothermal		13.4	0.3
Solar		0.5	—
Wind		10.5	0.3
Other		5.6	0.1
	Subtotal	353.0	9.2
TOTAL		3838.6	100.0

FIGURE 1.1

This basic steam power cycle is also known as the Rankine cycle.

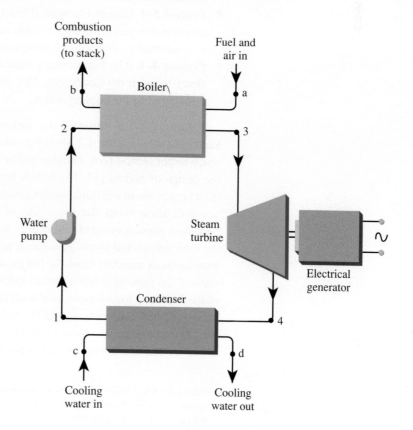

FIGURE 1.2
Cutaway view of boiler showing gas- or oil-fired burners on the right wall. Liquid water flowing through the tubes is heated by the hot combustion products. The steam produced resides in the steam drum (tank) at the top left. The boiler has a nominal 8-m width, 12-m height, and 10-m depth. Adapted from Ref. [12] with permission.

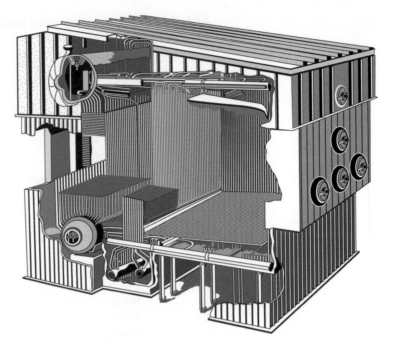

steam power cycle, or Rankine[2] cycle. Water (liquid and vapor) is the working fluid in the closed loop 1–2–3–4–1. The water undergoes four processes:

- **Process 1–2** A pump boosts the pressure of the liquid water prior to entering the boiler. To operate the pump, an input of energy is required.
- **Process 2–3** Energy is added to the water in the boiler, resulting, first, in an increase in the water temperature and, second, in a phase change. The hot products of combustion provide this energy. The working fluid is all liquid at state 2 and all vapor (steam) at state 3.
- **Process 3–4** Energy is removed from the high-temperature, high-pressure steam as it expands through a steam turbine. The output shaft of the turbine is connected to an electrical generator for the production of electricity.
- **Process 4–1** The low-pressure steam is returned to the liquid state as it flows through the condenser. The energy from the condensing steam is transferred to the cooling water.

Figures 1.2–1.6 illustrate the various generic components used in the Rankine cycle. Figure 1.2 shows a cutaway view of an industrial boiler; a much larger central power station utility boiler is shown in Fig. 1.3. Although the design of boilers [14, 15] is well beyond the scope of this book, the text offers much about the fundamental principles of their operation. For example, you will learn about the properties of water and steam in Chapter 2; the necessary aspects of mass and energy conservation needed to deal with both the combustion and steam generation processes in Chapters 3–5; and energy transfer (heat transfer) from the hot products of combustion through the tube walls to the flowing water or steam in Chapters 4, 5, 9, and 10. Consideration of the frictional losses associated with flow through the tubes and tube bends is covered in Chapters 6, 8, and 10. Similar statements could be made about the other components of the cycle (Figs. 1.4, 1.5, and 1.6). For example, a major section of Chapter 11 is devoted to heat exchangers.

---

[2] William Rankine (1820–1872), a Scottish engineer, was the author of the *Manual of the Steam Engine and Other Prime Movers* (1859) and made significant contributions to the fields of civil and mechanical engineering.

FIGURE 1.3
Boilers for public utility central power generation can be quite large, as are these natural gas–fired units. **Photograph courtesy of Florida Power and Light Company.**

Westinghouse Turbine Rotor, 1925
© **Smithsonian Institution**

The "Heart" of the Huge Westinghouse Turbine – An unusual detailed picture showing the maze of minutely fashioned blades – approximately five thousand—of the Westinghouse turbine rotor, or "spindle". Though only twenty-five feet in length this piece of machinery weighs one hundred and fifteen thousand pounds. At full speed the outside diameter of the spindle, on the left, is running nearly ten miles per minute, or a little less than 600 miles per hour. The problem of excessive heat resulting from such tremendous speed has been overcome by working the bearings under forced lubrication, about two barrels of oil being circulated through the bearings every sixty seconds to lubricate and carry away the heat generated by the rotation. The motor is that of the 45000 H.P. generating unit built by the South Philadelphia Works, Westinghouse Electric & Mfg. Co. for the Los Angeles Gas and Electric Company.

Original Caption by Science Service
© Westinghouse

FIGURE 1.4
Steam turbine for power generation. **Photograph and original caption reproduced with permission of the Smithsonian Institution.**

FIGURE 1.5
Cutaway view of shell-and-tube heat exchanger. Energy is transferred from the hot fluid passing through the shell to the cold fluid flowing through the tubes.

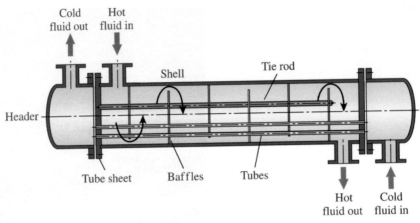

Smog in Lujiazui financial district in Shanghai, China.

Pollution controls are important components of fossil-fueled power plants.

As we begin our study, we emphasize the importance of safety in both the design and operation of power generation equipment. Fluids at high pressure contain enormous quantities of energy, as do spinning turbine rotors. Figure 1.7 shows the results of a catastrophic boiler explosion. Similarly, environmental concerns are extremely important in power generation. Examples here are the emission of potential air pollutants from the combustion process and thermal interactions with the environment associated with steam condensation. You can find entire textbooks devoted to these topics [16, 17].

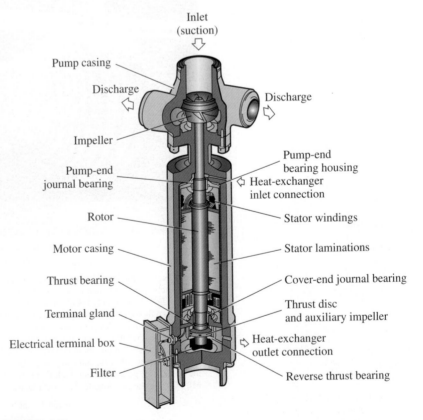

FIGURE 1.6
Typical electric-motor-driven feedwater pump. Larger pumps may be driven from auxiliary steam turbines. Adapted from Ref. [13] with permission.

FIGURE 1.7
Safety is of paramount importance in the design and operation of power generation equipment. This photograph shows the catastrophic results of a boiler explosion at the Courbevoie power station (France).

1.2b Solar-Heated Buildings

Our second application is solar-heated buildings. An example is shown in Fig. 1.8. Unlike the steam power plant, a solar house relies primarily on the essentially inexhaustible energy of the sun, rather than energy stored in ultimately exhaustible fossil fuels. Solar homes, however, may need auxiliary conventional energy supplies to supplement the solar energy, as both weather and climate affect the solar energy available at a particular time at a particular site.

Depending on the degree to which solar energy meets the total heating demands, the solar heating system can be relatively simple or quite complex. The simplest systems are *passive*, with no moving parts or controls. More complex systems are known as *active* systems. The flow schematic in Fig. 1.9 illustrates such a system, which we now explore to set the stage for later chapters.

Radiant energy from the sun passes through the glass cover plates of the solar collector (see Fig. 1.10) and is absorbed at the absorber plate. This energy is then transferred to either a liquid stream or air flow. The system in Fig. 1.9 is an air-based system. The hot air then passes through a heat exchanger (Fig. 1.9) where it loses some of its energy to heat tap water. Depending on the positions of the dampers, the still-hot air is delivered to heat the house, if needed, or routed to the pebble bed. A typical pebble bed may be 1.5 m on a side and 2 m long, and the pebbles that fill it are about 20–30 mm in diameter. The hot air gives up its energy by heating the pebbles, which can stay hot for a relatively long time, thus performing an energy storage role. When solar energy is not available, as on a rainy day or at night, the flow through the pebble bed can be reversed and the cool return air from the house is heated as it passes through the pebble bed.

Much of the apparatus involved in this example is heat-transfer equipment. The design and operation of the solar collector, the pebble bed, and various heat exchangers depend critically on all three thermal-fluid sciences disciplines—thermodynamics, heat transfer, and fluid dynamics. Various mass and energy balances associated with these components are discussed in Chapters 3 and 5, respectively, whereas more specific information concerning

Examples of passive solar heating: Transparent insulation allows buildings to use the rays of the sun to provide direct heat and lighting (top); Sun's rays penetrate simple greenhouse cover (bottom).

FIGURE 1.8
*Some buildings in the Colorado State
University Solar Village utilize a bank
of solar collectors as an integral part
of the roof design. Photograph
courtesy of Colorado State University
Photographic Archives.*

heat-transfer aspects is presented in Chapters 4 and 9–11. Chapters 6, 10, and
11 provide the concepts needed to design the piping system and to calculate
fan power requirements. All of these concerns will be revisited as you proceed
through the text. For readers desiring more detailed information on solar
heating design, Refs. [18, 19] are recommended.

1.2c Spark-Ignition Engines

We choose the spark-ignition engine as one of our applications to revisit
because there are so many of them (approximately 200 million are installed
in automobiles and light-duty trucks in the United States alone) and
because many students are particularly interested in engines. Owing to
these factors, and others, many schools offer entire courses dealing with

FIGURE 1.9
*Schematic diagram of flow paths for a
solar-heated house. Energy is used to
heat domestic water as well as to heat
the house. Energy is stored in a pebble
bed to be used as needed. Adapted
from Ref. [18] with permission.*

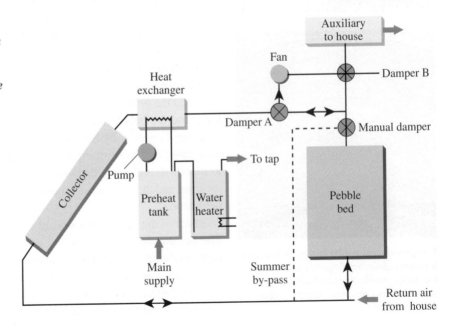

FIGURE 1.10
Cutaway views of solar collectors. The collector at the top of the figure heats a liquid working fluid that flows through integral channels in the absorber plate (Ref. [18] 1st ed.). The collector at the bottom of the figure heats air. Adapted from Ref. [18] with permission.

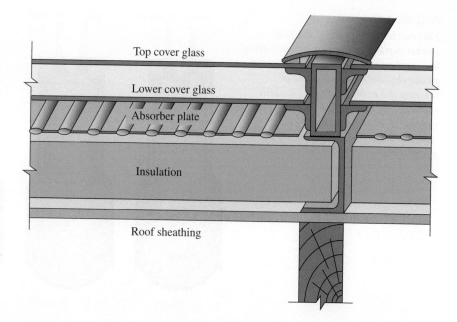

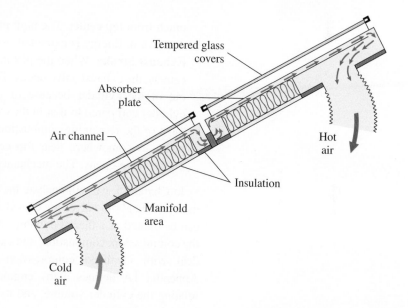

internal combustion engines, and many books are devoted to this subject, among them Refs. [20–23].

Although you may be familiar with the four-stroke engine cycle, we present it here to make sure that all readers have the same understanding. Figure 1.11 illustrates the following sequence of events:

- **Intake Stroke** The inlet valve is open and a fresh fuel–air mixture is pulled into the cylinder by the downward motion of the piston. At some point near the bottom of the stroke, the intake valve closes.
- **Compression Stroke** The piston moves upward, compressing the mixture. The temperature and pressure increase. Prior to the piston reaching the top of its travel (i.e., top center), the spark plug ignites the mixture and a flame begins to propagate across the combustion chamber. Pressure rises above that owing to compression alone.
- **Expansion Stroke** The flame continues its travel across the combustion chamber, ideally burning all of the mixture before the piston descends

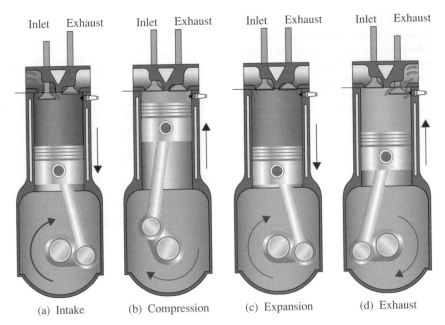

(a) Intake   (b) Compression   (c) Expansion   (d) Exhaust

much from top center. The high pressure in the cylinder pushes the piston downward. Energy is extracted from the burned gases in the process.

- **Exhaust Stroke** When the piston is near the bottom of its travel (bottom center), the exhaust valve opens. The hot combustion products flow rapidly out of the cylinder because of the relatively high pressure within the cylinder compared to that in the exhaust port. The piston ascends, pushing most of the remaining combustion products out of the cylinder. When the piston is somewhere near top center, the exhaust valve closes and the intake valve opens. The mechanical cycle now repeats.

In Chapter 5, we will analyze the processes that occur during the times in the cycle that both valves are closed and the gas contained within the cylinder can be treated as a thermodynamic system. With this analysis, we can model the compression, combustion, and expansion processes. Chapters 2 and 7 also deal with idealized compression and expansion processes. Note that Appendix 1A defines many engine-related terms and provides equations relating the cylinder volume, and its time derivative, to crank-angle position and other parameters.

1.2d Jet Engines

Air travel is becoming more and more popular each year as suggested by the number of passenger miles flown in the United States increasing from 433 billion in 1989 to 651 billion in 1999. Since you are likely to entrust your life from time to time to the successful performance of jet engines, you may find learning about these engines interesting. Figure 1.12 schematically illustrates the two general types of aircraft engines.

The schematic at the top shows a pure turbojet engine in which all of the thrust is generated by the jet of combustion products passing through the exhaust nozzle. This type of engine powered the supersonic J4 Phantom military jet fighter and the recently retired Concorde supersonic transport aircraft. In the turbojet, a multistage compressor boosts the pressure of the entering air. A portion of the high-pressure air enters the combustor, where fuel is added and burned, while the remaining air is used to cool the

FIGURE 1.12
Schematic drawings of a single-shaft turbojet engine (top) and a two-shaft high-bypass turbofan engine (bottom). Adapted from Ref. [24] with permission.

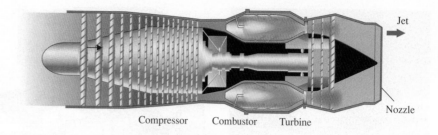

Jet

Nozzle

Compressor   Combustor   Turbine

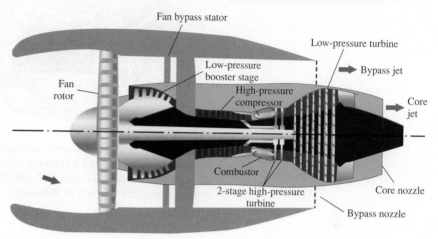

Fan bypass stator

Low-pressure turbine

Low-pressure
booster stage

Bypass jet

Fan
rotor

High-pressure
compressor

Core
jet

Combustor

Core nozzle

2-stage high-pressure
turbine

Bypass nozzle

combustion chamber. The hot products of combustion then mix with the cooling air, and these gases expand through a multistage turbine. In the final process, the gases accelerate through a nozzle and exit to the atmosphere to produce a high-velocity propulsive jet.[3] The compressor and turbine are rotary machines with spinning wheels of blades. Rotational speeds vary over a wide range but are of the order of 10,000–20,000 rpm. Other than that needed to drive accessories, all of the power delivered by the turbine is used to drive the compressor in the pure turbojet engine.

The second major type of jet engine is the turbofan engine (Fig. 1.12 bottom). This is the engine of choice for commercial aircraft. (See Fig. 1.13.) In the turbofan, a bypass air jet generates a significant proportion of the engine thrust. The large fan shown at the front of the engine creates this jet. A portion of the total air entering the engine bypasses the core of the engine containing the compressor and turbine, while the remainder passes through the core. The turbines drive the fan and the core compressors, generally using separate shafts for each. In the turbofan configuration, the exiting jets from both the bypass flow and the core flow provide the propulsive force.

To appreciate the physical size and performance of a typical turbojet engine, consider the GE F103 engine. These engines power the Airbus A300B, the DC-10-30, and the Boeing 747. The F103 engine has a nominal diameter of 2.7 m (9 ft) and a length of 4.8 m (16 ft), produces a maximum thrust of 125 kN (28,000 lb$_f$), and weighs 37 kN (8,325 lb$_f$). Typical core and fan speeds are 14,500 and 8000 rpm, respectively [25].

Jet engines afford many opportunities to apply the theoretical concepts developed throughout this book. For example, we will apply the conservation of energy and momentum principles developed in Chapters 5 and 6, respectively, to estimate the thrust from such engines. Analyzing a turbojet

---

[3] The basic principle here is similar to that of a toy balloon that is propelled by a jet of escaping air.

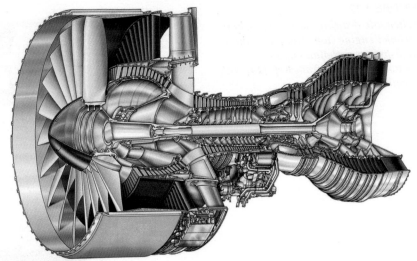

FIGURE 1.13
Cutaway view of PW4000-series turbofan engine. PW4084 turbofan engines (see also Fig. 12.13) power the Airbus A310-300 and A300-600 aircraft and Boeing 747-400, 767-200/300, and MD-11 aircraft. Cutaway drawing courtesy of Pratt & Whitney.

engine cycle is a major topic in Chapter 12, and individual components are treated in Chapter 11.

1.2e Biological Systems

This application is more diffuse than the others in that many different biological systems and processes can be used to illustrate applications of the thermal-fluid sciences (Fig. 1.14). The human body makes an excellent example of a complex system with many thermal-fluid interactions. A specific application is how heat-transfer considerations govern human comfort. We will study the principles involved in insulating the body with clothing and how wind chill relates to convection heat transfer.

FIGURE 1.14
Chemical energy stored in food is released to provide the energy needed for a human body to function. Here some of that energy is also used to produce work (left); fur and clothing provide insulating layers that reduce heat transfer from the body (second from left and third from left); many seeds depend on aerodynamic lift and drag forces for dispersal (right).

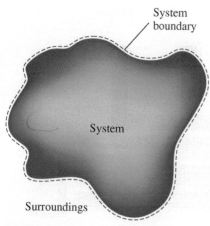

FIGURE 1.15

The system boundary separates a fixed mass, the system, from its surroundings.

1.3 PHYSICAL FRAMEWORKS FOR ANALYSIS

In this section, we define a **thermodynamic system** and a **control volume**. These concepts are central to almost any analysis of a thermal-fluid problem. We further extend these ideas to their applications in **integral** and **differential** analyses.

1.3a Systems

In a generic sense, a **system** is anything that we wish to analyze and distinguish from its **surroundings** or **environment**. To denote a system, all one needs to do is create a **boundary** between the system of interest and everything else, that is, the surroundings. The boundary may be a real surface or an imaginary construct indicated by a dashed line on a sketch. Figure 1.15 illustrates the separation of a system from its surroundings by a boundary.

A more specific definition of a system used in the thermal sciences is the following:

> A *system* is a specifically identified fixed mass of material separated from its surroundings by a real or imaginary boundary.

To emphasize that the mass of the system is always unchanged regardless of what other changes might occur, the word **system** is frequently modified by adjectives, for example, **fixed-mass system** or **closed system**. In this book, we usually use the word *system* without modifiers to indicate a fixed mass.

The boundaries of a system need not be fixed in space but may, out of necessity, move. For example, consider the gas as the system of interest in the piston–cylinder arrangement shown Fig. 1.16. As the gas is compressed, the system boundary shrinks to always enclose the same mass.

Depending upon one's objectives, a system may be simple or complex, homogeneous or nonhomogenous. Our example of the gas enclosed in an engine cylinder (Fig. 1.16) is a relatively simple system. There is only one substance comprising the system: the fuel–air mixture. To further simplify an analysis of this particular system, we might assume that the fuel–air mixture has a uniform temperature, although in an operating engine the temperature will vary throughout the system. The matter comprising the system need not be a gas. Liquids and solids, of course, can be the whole system or a part of it. Again, the key distinguishing feature of a system is that it contains a fixed quantity of matter. No mass can cross the system boundary.

To further illustrate the thermodynamic concept of a system, consider the computer chip module schematically illustrated in Fig. 1.17. One might perform a thermal analysis of this device to ensure that the chip stays sufficiently cool. Considering the complexity of this module, a host of possible systems exist. For example, one might choose a system boundary surrounding the entire device and cutting through the connecting wires, as indicated as system 1. Alternatively, one might choose the chip itself (system 2) to be a thermodynamic system.

Choosing system boundaries is critical to any thermal-fluid analysis. One of the goals of this book is to help you develop the skills required to define and analyze thermodynamic systems.

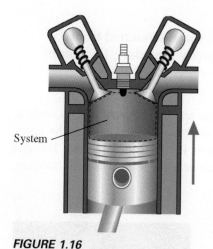

FIGURE 1.16

The gas sealed within the cylinder of a spark-ignition engine constitutes a system, provided there is no gas leakage past the valves or the piston rings. The boundaries of this system deform as the piston moves.

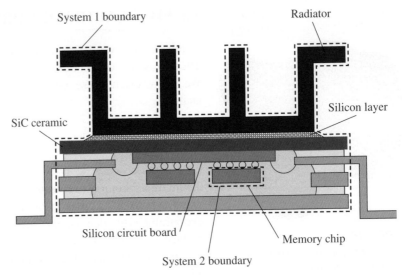

FIGURE 1.17

A computer chip is housed in a module designed to keep the chip sufficiently cool. Various systems can be defined for thermal analyses of this relatively complex device. Two such choices are shown. Basic module sketch courtesy of Mechanical Engineering *magazine, Vol. 108/No. 3, March, 1986, page 41;* © Mechanical Engineering *magazine (The American Society of Mechanical Engineers International).*

1.3b Control Volumes

In contrast to a system, mass may enter and/or exit a **control volume** through a **control surface**. We formally define these as follows:

> A *control volume* is a region in space separated from its surroundings by a real or imaginary boundary, the *control surface*, across which mass may pass.

Figure 1.18 illustrates a control volume and its attendant control surface. Here, mass in the form of water vapor or water droplets crosses the upper boundary as a result of evaporation from the hot liquid coffee. For this example, we chose a fixed control surface near the top of the cup; however,

FIGURE 1.18

The indicated control surface surrounds a control volume containing hot liquid coffee, air, and moisture. Water vapor or droplets exit through the upper portion of the control surface.

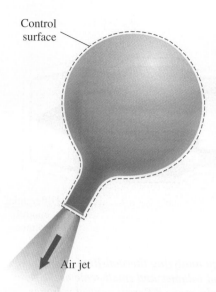

Control
surface

Air jet

FIGURE 1.19

A rubber balloon and the air it contains constitute a control volume. The control volume moves through space and shrinks as the air escapes.

we could have chosen the regressing liquid surface to be the upper boundary. A choice of a moving control surface may or may not simplify an analysis, depending on the particular situation.

Control volumes may be simple or complex. Fixing the control surface in space yields the simplest control volume, whereas moving control volumes with deforming control surfaces are the most complex. Figure 1.19 illustrates the latter, where an inflated balloon is propelled by the exiting jet of air. In this example, the control volume both moves with respect to a fixed observer and shrinks with time. Figure 1.20 illustrates a simple, fixed control volume associated with an analysis of a water pump. Note that the control surface cuts through flanged connections at the inlet and outlet of the pump. The particular choice of a control volume and its control surface frequently is of overwhelming importance to an analysis. A wise choice can make an analysis simple, whereas a poor choice can make the analysis more difficult, or perhaps, impossible. Examples presented throughout this book provide guidance in selecting control volumes.

1.3c Integral versus Differential Analyses

In our development and application of the conservation principles for mass, energy, and momentum, we employ either integral systems/control volumes or differential systems/control volumes (Fig. 1.21). The **integral system or integral control volume** is typically of large scale and macroscopic. All of the systems and control volumes shown so far (see Figs. 1.16–1.20) are of the integral type. One usually selects an integral system or control volume when the details within the system/control volume are not important to the analysis. For example, in a system in which the temperature and all other properties are uniform, an integral system suffices; however, if one is interested in the temperature distribution within an integral system, one begins with an analysis of a differential system within the larger system. Many examples later in this book illustrate these ideas.

Figure 1.22 shows differential systems for one-dimensional planar and three-dimensional Cartesian geometries. In the one-dimensional (1-D) case, the differential system has a length dx and extends indefinitely in the y- and

FIGURE 1.20

A sliding-vane pump and the fluid it contains constitute a simple control volume. Mass crosses the control surface at both the pump inlet and outlet. **Adapted from Ref. [3] with permission.**

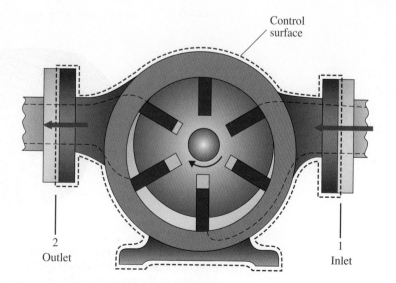

Control
surface

2
Outlet

1
Inlet

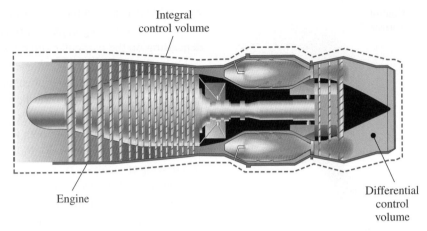

Integral
control volume

Engine

Differential
control
volume

FIGURE 1.21
*Two types of control volumes are used in analyzing thermal-fluids
problems: large-scale or integral control volumes and small-scale or
differential control volumes. For example, many different control volumes,
integral and differential, are used in the detailed design, analysis and
testing of rocket engines (left).* **Photograph (bottom) courtesy of NASA.**

z-directions, whereas the three-dimensional (3-D) system has a length dx,
width dz, and height dy. A differential system or control volume is
sufficiently small that any of its properties can be characterized by single
values at any instant; furthermore, the differential system is sufficiently large
to contain enough atoms or molecules such that the fluid properties exhibit
no statistical fluctuations. For example, the density of a control volume that
contains only a few molecules will vary significantly when a single molecule

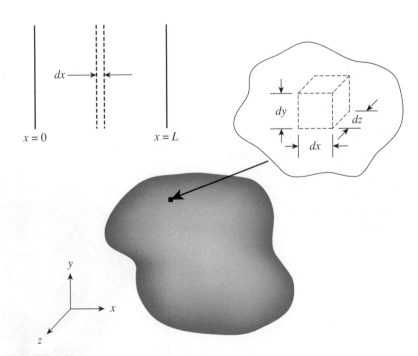

FIGURE 1.22
*Differential (microscopic) systems are shown as dashed lines imbedded within
integral (macroscopic) systems defined by the bold lines. A planar 1-D system
is shown at the top and a 3-D system at the bottom.*

exits without an incoming molecule to compensate for its loss. The importance of these considerations is illuminated in the following sections and in Chapter 2.

The size of the system or control volume also determines the mathematical complexity of an analysis. Integral systems typically yield algebraic equations or, if time is a variable, ordinary differential equations. Differential systems or control volumes, in contrast, yield ordinary differential equations at their lowest level of complexity (one dimensional, time independent) and partial differential equations as either time or more than one spatial dimension is added.

1.4 PREVIEW OF CONSERVATION PRINCIPLES

In your study of physics, you have already encountered conservation of mass, conservation of energy, and conservation of momentum. These fundamental principles form the bedrock of the thermal-fluid sciences. A wise first step in analyzing any new problem addresses the following questions: Do any fundamental conservation principles apply here? If so, which ones? Because of the overriding importance of these principles to the thermal-fluid sciences, we offer a preview here in this first chapter. Later chapters elaborate and apply these principles.

1.4a Generalized Formulation

A moving train possesses large quantities of mass, momentum, and energy.

It is quite useful to cast all three conservation principles in a common form. We do this for two reasons: first, to show the unity among the three conservation principles and, second, to provide an easy-to-remember single relationship that can be converted to more specific statements as needed. To do this, we define X to be the conserved quantity of interest, that is,

$$X \equiv \begin{cases} \textit{Mass or} \\ \textit{Energy or} \\ \textit{Momentum} \end{cases}$$

We now express the general conservation law as follows:
For a finite time interval, $\Delta t = t_2 - t_1$,

$$X_{\text{in}} \quad - \quad X_{\text{out}} \quad + \quad X_{\text{generated}} \quad = \quad \Delta X_{\text{stored}} \equiv X(t_2) - X(t_1). \quad (1.1)$$

| Quantity of X crossing boundary and passing into system or control volume | Quantity of X crossing boundary and passing out of system or control volume | Quantity of X generated within system or control volume | Change of quantity of X stored within system or control volume during time interval |

Instantaneously, we have

$$\dot{X}_{\text{in}} \quad - \quad \dot{X}_{\text{out}} \quad + \quad \dot{X}_{\text{generated}} \quad = \quad \dot{X}_{\text{stored}}. \quad (1.2)$$

| Time rate of X crossing boundary and passing into system or control volume | Time rate of X crossing boundary and passing out of system or control volume | Time rate of generation of X within system or control volume | Time rate of storage of X within system or control volume |

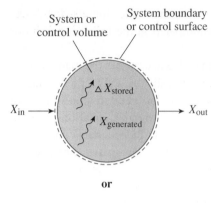

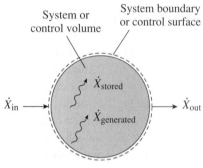

FIGURE 1.23
Schematic representation of conservation principles for a finite time interval (top) and for an instant (bottom).

To interpret and understand these relationships, consider Fig. 1.23. The top drawing of Fig. 1.23 illustrates the situation for a system or control volume that experiences various exchanges of a conserved quantity during a finite time period. To make our discussion less abstract, let us assume for the time being, first, that X is mass and, second, that we are dealing with a control volume, not a system. During the time interval of interest, a certain quantity of mass, X_{in}, crosses the control surface into the control volume. Also during this interval, another quantity of mass, X_{out}, leaves the control volume. Straight arrows represent these two quantities. Note that these arrows are drawn external to the control volume and that they begin or end at the control surface without crossing the boundary. (Attention to such details helps distinguish interactions at the boundary from what is happening within the control volume.) Furthermore, if we allow for nuclear transformations, some mass can be generated (or destroyed). This mass is represented as the squiggly arrow within the control volume labeled $X_{generated}$. If more mass enters or is generated ($X_{in} + X_{generated}$) than exits (X_{out}), the control volume will experience an increase in mass. This mass increase is also represented by a squiggly arrow within the control volume labeled ΔX_{stored}. The subscript *stored* refers to the idea that during the time interval of interest additional mass is stored within the control volume. Of course, X_{out} could be greater than $X_{in} + X_{generated}$, in which case the storage term is negative. This accounting procedure involving quantities *in*, *out*, *generated*, and *stored* is intuitively satisfying and is much like keeping track of your account in a checkbook.[4]

This same general accounting applies to an instant where time rates replace actual quantities (Fig. 1.23 bottom). Continuing with our example of mass, $\dot{X}_{in}$ represents the mass flow rate[5] into the control volume and has units of kilograms per second. The storage term, $\dot{X}_{stored}$, now represents the time rate of change of mass within the control volume rather than the difference in mass over a finite time interval. Again, the ideas expressed in Eq. 1.2 are intuitively appealing and easy to grasp.

The challenge in the application of Eqs. 1.1 and 1.2 lies in the correct identification of all of the terms, making sure that every important term is included in an analysis and that no term is doubly counted. For example, in our study of energy conservation (Chapter 5), the *in* and *out* terms represent energy transfers as work and heat, whereas the *generation* term may result from a conversion of electrical energy to thermal energy. Regardless of the complexity of the mathematical statement of any conservation principle, Eqs. 1.1 and 1.2 are at its heart; that is, *in* minus *out* plus *generated* equals *stored*.

1.4b Motivation to Study Properties

To apply these basic conservation principles to engineering problems requires a good understanding of thermodynamic properties, their interrelationships, and how they relate to various processes. Although mass conservation may require only a few thermodynamic properties—usually, the mass density ρ and its relationship to pressure P and temperature T suffice—successful application of the energy conservation principle relies heavily on thermodynamic properties. To illustrate this, we preview Chapter 5 by applying the generic conservation principle (Eq. 1.1) to a particular conservation of energy statement as follows:

$$Q_{net, in} - W_{net, out} = U_2 - U_1. \qquad (1.3)$$

---

[4] For all purposes of this book we will not be concerned with nuclear reactions; hence, $X_{generated}$ will always be zero for mass conservation.

[5] Flow rates are discussed at length in Chapter 3.

Here, $Q_{net, in}$ is the energy that entered the system as a result of a heat-transfer process, $W_{net, out}$ is the energy that left the system as a result of work being performed, and U_1 and U_2 are the internal energies of the system at the initial state 1 and at the final state 2, respectively.

This first-law statement (Eq. 1.3) immediately presents us with an important thermodynamic property, the internal energy U. Although you likely encountered this property in a study of physics, a much deeper understanding is required in our study of engineering thermodynamics. Defining internal energy and relating it to other, more common, system properties, such as temperature T and pressure P, are important topics considered in Chapter 2. We also need to know how to deal with this property when dealing with two-phase mixtures such as water and steam.

To further illustrate the importance of Chapter 2, we note that the work term in our conservation of energy statement (Eq. 1.3) also relates to system thermodynamic properties. For the special case of a reversible expansion, the work can be related to the system pressure P and volume V, both thermodynamic properties. Specifically, to obtain the work requires the evaluation of the integral of $Pd V$ from state 1 to state 2. To perform this integration, we need to know how P and V relate for the particular process that takes the system from state 1 to state 2. Much of Chapter 2 deals with, first, defining the relationships among various thermodynamic properties and, second, showing how properties relate during various specific processes. For example, the ideal-gas law, $PV = NR_u T$ (where N is the number of moles contained in the system and R_u is the universal gas constant), expresses a familiar relationship among the thermodynamic properties P, V, and T. From this relationship among properties, we see that the product of the system pressure and volume is constant for a constant-temperature process. In addition to exploring the ideal-gas equation of state, Chapter 2 shows how to deal with the many substances that do not behave as ideal gases. Common examples here include water, water vapor, and various refrigerants.

1.5 KEY CONCEPTS AND DEFINITIONS

In addition to systems, control volumes, and surroundings, several other basic concepts permeate our study of the thermal-fluid sciences and are listed in Table 1.2. We introduce these concepts in this section, recognizing that they will be revisited again, perhaps several times, in later chapters.

Table 1.2 Some Fundamental Thermodynamic Concepts

System (integral and differential)
Control volume (integral and differential)
Surroundings
Property
State
Process
Flow process
Cycle
Equilibrium
Quasi-equilibrium
Local equilibrium

1.5a Properties

Before we can begin a study of thermodynamic properties understood in their most restricted sense, we define what is meant by a **property** in general:

> A *property* is a quantifiable macroscopic characteristic of a system.

Examples of system properties include mass, volume, density, pressure, temperature, height, width, and color, among others. Not all system properties, however, are thermodynamic properties. Thermodynamic properties all relate in some way to the energy of a system. In our list here, height, width, and color, for example, do not qualify as thermodynamic properties, although the others do. A precise definition of thermodynamic properties often depends on restricting the system to which they apply in subtle ways. Chapter 2 is devoted to thermodynamic properties and their interrelationships.

Properties are frequently combined to create new ones. For example, a spinning baseball in flight not only has the properties of mass and velocity but also kinetic energy and angular momentum. Thus, any system may possess numerous properties.

1.5b States

Another fundamental concept in thermodynamics is that of a **state**:

> A thermodynamic *state* of a system is defined by the values of all of the system thermodynamic properties.

When the value of any one of a system's properties changes, the system undergoes a change in state. For example, hot coffee in a thermos bottle undergoes a continual change of state as it slowly cools. Not all thermodynamic properties necessarily change when the state of a system changes; the mass of the coffee in a sealed thermos remains constant while the temperature falls. An important skill in solving problems in thermodynamics is to identify which properties remain fixed and which properties change during a change in state.

1.5c Processes

One goal of studying thermal-fluid sciences is to develop an understanding of how various devices convert one form of energy to another. For example, how does the burning coal in a power plant result in the electricity supplied to your home? In analyzing such energy transformations, we formally introduce the idea of a thermodynamic **process**:

> A *process* occurs whenever a system changes from one state to another state.

Figure 1.24a illustrates a process in which the gas contained in a cylinder–piston arrangement at state 1 is compressed to state 2. As is obvious from the sketch, one property, the volume, changes in going from state 1 to state 2. Without

FIGURE 1.24

(a) A system undergoes a process that results in a change of the system state from state 1 to state 2. (b) In a steady-flow process, fluid enters the control volume at state 1 and exits the control volume at state 2.

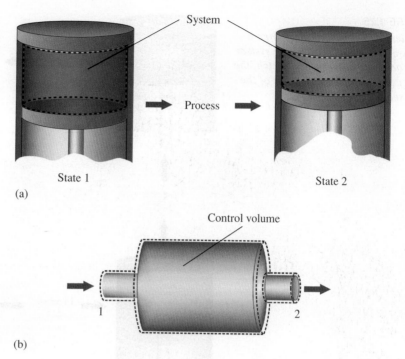

(a)

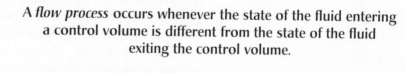

(b)

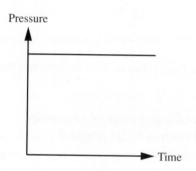

knowing more details of the process, we cannot know what other properties may have changed as well. In many thermodynamic analyses, a single property—for example, temperature, pressure, or entropy[6]—is held constant during the process. In these particular cases, we refer to the processes as a constant-temperature (isothermal) process, constant-pressure (isobaric) process, or a constant-entropy (isentropic) process, respectively.

Although formal definitions of process refer to systems, the terminology is also applied to control volumes, in particular, steady flows in which no properties change with time. We thus define a **flow process** as follows:

> A *flow process* occurs whenever the state of the fluid entering a control volume is different from the state of the fluid exiting the control volume.

Figure 1.24b schematically illustrates a flow process. An example of a steady-flow process is constant-pressure combustion. In this process, a cold fuel–air mixture (state 1) enters the control volume (combustor) and exits as hot products of combustion (state 2). Various flow processes underlie the operation of myriad practical devices, many of which we will study in subsequent chapters.

This jet engine test exemplifies a steady-flow process. Photograph courtesy of U.S. Air Force.

1.5d Cycles

In many energy-conversion devices, the **working fluid** undergoes a thermodynamic **cycle**. Since the word *cycle* is used in many ways, we need a

---

[6] Entropy is an important thermodynamic property and is discussed in detail in Chapters 2 and 7. We only introduce it here as an example of a property.

FIGURE 1.25

A system undergoes a cycle when a series of processes returns the system to its original state. In this sketch, the cycle consists of the repetition of the state sequence 1–2–3–4–1.

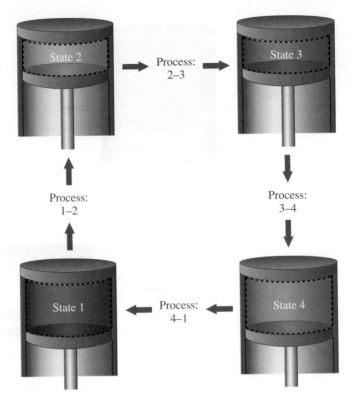

precise definition for our study of thermal-fluid sciences. Our definition is the following:

> **A thermodynamic *cycle* consists of a sequence of processes in which the working fluid returns to its original thermodynamic state.**

An example of a cycle applied to a fixed-mass system is presented in Fig. 1.25. Here a gas trapped in a piston–cylinder assembly undergoes four processes. A cycle can be repeated any number of times following the same sequence of processes. Although it is tempting to consider reciprocating internal combustion engines as operating in a cycle, the products of combustion never undergo a transformation back to fuel and air, as would be required for our definition of a cycle. There exist, however, types of reciprocating engines that do operate on thermodynamic cycles. A prime example of this is the Stirling engine [26]. The working fluid in Stirling engines typically is hydrogen or helium. All combustion takes place outside of the cylinder. Figure 1.26 shows a modern Stirling engine.

Thermodynamic cycles are most frequently executed by a series of flow processes, as illustrated in Fig. 1.27. As the working fluid flows through the loop, it experiences many changes in state as it passes through various devices such as pumps, boilers, and heat exchangers, but ultimately it returns to its initial state, arbitrarily chosen here to be state 1. As we have already seen, the heart of a fossil-fueled steam power plant operates on a thermodynamic cycle of the type illustrated in Fig. 1.27 (cf. Fig. 1.1). Refrigerators and air conditioners are everyday examples of devices operating on thermodynamic cycles. In Chapter 12, we analyze various cycles for power production, propulsion, and heating and cooling.

FIGURE 1.26

A Stirling engine provides the motive power to an electrical generator set. Photograph courtesy of Stirling Technology Company.

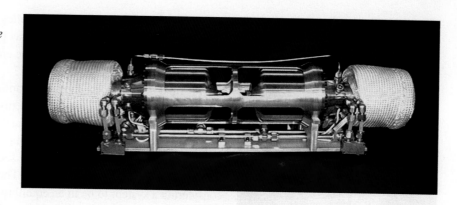

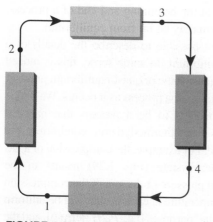

FIGURE 1.27

A cycle can also consist of a sequence of flow processes in which the flowing fluid is returned to its original state: 1–2–3–4–1.

FIGURE 1.28

The flow of energy from a region of higher temperature to a region of lower temperature drives a system toward thermal equilibrium. The temperature is uniform in a system at equilibrium.

The thermodynamic cycle is central to many statements of the second law of thermodynamics and concepts of thermal efficiency (Chapter 7).

1.5e Equilibrium and the Quasi-Equilibrium Process

The science of thermodynamics builds upon the concept of equilibrium states. For example, the common thermodynamic property temperature has meaning only for a system in equilibrium. In a thermodynamic sense, then, what do we mean when we say that a system is in equilibrium? In general, there must be no unbalanced potentials or drivers that promote a change of state. The state of a system in equilibrium remains unchanged for all time. For a system to be in equilibrium, we require that the system be simultaneously in thermal equilibrium, mechanical equilibrium, phase equilibrium, and chemical equilibrium. Other considerations may exist, but they are beyond our scope. We now consider each of these components of thermodynamic equilibrium.

Thermal equilibrium is achieved when a system both has a uniform temperature and is at the same temperature as its surroundings. If, for example, the surroundings are hotter than the system under consideration, energy flows across the system boundary from the surroundings to the system. This exchange results in an increase in the temperature of the system. Equilibrium thus does not prevail. Similarly, if temperature gradients exist within the system, the initially hotter regions become cooler, and the initially cooler regions become hotter, as time passes (Fig. 1.28 top). Given sufficient time the temperature within the system becomes uniform, provided the ultimate temperature of the system is identical to that of the surroundings (Fig. 1.28 bottom). Prior to achieving the final uniform temperature, the system is not in thermal equilibrium.

Mechanical equilibrium is achieved when the pressure throughout the system is uniform and there are no unbalanced forces at the system boundaries. An exception to the condition of uniform pressure exists when a system is under the influence of a gravitational field. For example, pressure increases with depth in a fluid such that pressure forces balance the weight of the fluid above. We will explore this idea in Chapter 6. In many systems, however, the assumption of uniform pressure is reasonable.

Phase equilibrium relates to conditions in which a substance can exist in more than one physical state; that is, any combination of vapor, liquid, and solid. For example, you are quite familiar with the three states of H_2O: water vapor (steam), liquid water, and ice. Phase equilibrium requires that the amount of a substance in any one phase not change with time; for example,

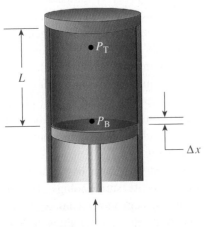

FIGURE 1.29

In the quasi-equilibrium compression of a gas, the pressure is essentially uniform throughout the gas system, that is, $P_B \approx P_T$.

liquid–vapor phase equilibrium implies that the rate at which molecules escape the liquid phase to enter the gas phase is exactly balanced by the rate at which molecules from the gas phase enter the liquid phase. We will discuss these ideas further in Chapters 2 and 7.

Our final condition for thermodynamic equilibrium requires the system to be in **chemical equilibrium**. For systems incapable of chemical reaction, this condition is trivial; however, for reacting systems this constraint is quite important. A significant portion of Chapter 7 is devoted to this topic.

From the foregoing discussion, we see that an equilibrium state is a pretty boring proposition. Nothing happens. The system just sits there. Nevertheless, considering a system to be in an equilibrium state at the beginning of a process and at the end of the process is indeed useful. In fact, it is this idea that motivates our discussion. Our development of the conservation of energy principle relies on the idea that equilibrium states exist at the beginning and end of a process, even though during the process the system may be far from equilibrium.

This discussion suggests that it is not possible to describe the details of a process because of departures from equilibrium. In some sense, this is indeed true. The often-invoked assumption of a quasi-static or quasi-equilibrium process, however, does permit an idealized description of a process as it occurs. We define a **quasi-static** or **quasi-equilibrium process** to be a process that happens sufficiently slowly such that departures from thermodynamic equilibrium are always so small that they can be neglected. For example, the compression of a gas in a perfectly insulated piston–cylinder system (Fig. 1.29) results in the simultaneous increase in temperature and pressure of the gas. If the compression is performed slowly, the pressure and temperature at each instant will be uniform throughout the gas system for all intents and purposes [i.e., $P_B(t) = P_T(t)$], and the system can be considered to be in an equilibrium state.[7] In contrast, if the piston moves rapidly, the pressure in the gas at the piston face (P_B) would be greater than at the far end of the cylinder (P_T) and equilibrium would not be achieved. The formal requirement for a quasi-equilibrium process is that the time for a system to reach equilibrium after some change is small compared to the time scale of the process. For our example of the gas compressed by the piston, the time for the pressure to equilibrate is determined by the speed of sound, that is, the propagation speed of a pressure disturbance. For room-temperature air, the sound speed is approximately 340 m/s. If we assume that at a particular instant the piston is 100 mm ($= L$) from the closed end of the cylinder, the change in the pressure due to the piston motion will be communicated to the closed end in a time equal to 2.9×10^{-4} s [$= (0.100\text{ m})/(340\text{ m/s})$] or 0.29 ms. If during 0.29 ms the piston moves very little (Δx), the pressure can be assumed to be uniform throughout the system for all practical purposes, and the compression can be considered a quasi-equilibrium process. Alternatively, we can say that the speed of the pressure waves that communicate changes in pressure is much faster than the speed of the piston that creates the increasing pressure.[8] Similar arguments involving time scales can be used to ascertain whether or not thermal and chemical equilibrium are approximated for any real processes. Our purposes here, however, are not to define the exact conditions for which a quasi-equilibrium process might be assumed, but rather to acquaint the reader with this commonly invoked assumption. Later in the book we use the quasi-equilibrium process as a standard to which real processes are compared.

---

[7] For the pressure to be truly uniform requires that we neglect the effect of gravity. In Chapter 6, we deal with the influence of gravity on the pressure distribution in a confined fluid.

[8] Pressure waves travel at the speed of sound.

1.5f Local Equilibrium

Were we to consider only thermodynamics as our subject, the aforementioned considerations of equilibrium would be sufficient. Our desire to include heat transfer and fluid dynamics in our study, however, requires us to consider the concept of **local equilibrium**. The idea of local equilibrium is closely related to our previous discussion of differential systems (see Fig. 1.22). To define local equilibrium, we consider a small hypothetical temperature sensor located within a small region of a larger system such as depicted in the 3-D system of Fig. 1.22. For local equilibrium to exist within the volume $d\mathcal{V}$ ($\equiv dx \cdot dy \cdot dz$), the following three conditions must be met [27, 28]:

- The region $d\mathcal{V}$ is sufficiently small that variations in mean properties are small within $d\mathcal{V}$.
- The region contains a large number of molecules that interact with the sensor. Furthermore, the molecules interact with the sensor much more rapidly than any macroscopic changes occur.
- The mean free path of molecules is small compared to the dimensions of $d\mathcal{V}$ so that any molecule interacting with the sensor had its last collision within our region of interest.

Fortunately, most engineering applications allow these three conditions to be met. This circumstance is exceedingly important in that it allows us to use the definitions of thermodynamic properties and property relationships, which are based on true equilibrium, in analyses of systems in which properties vary with both location and time. For example, the theory of conduction heat transfer relies on local equilibrium to describe the transfer of energy within real systems. In Chapter 5, we revisit these ideas.

1.6 SOME GENERAL CHARACTERISTICS OF REAL FLOWS

Although the thermodynamic concepts developed in this book deal with equilibrium systems in the absence of motion, many engineering applications deal with flowing fluids. As suggested in the preceding section, the idea that departures from equilibrium are small in a local sense allows us to use the thermodynamic relationships developed in Chapter 2 to describe real engineering flows. Another, and less easily dismissed, complication is that most flows of engineering interest are turbulent. In this brief section, we introduce this complication. Here we define what is meant by a turbulent flow; in other chapters, we focus on some of the consequences of **turbulence**.[9] The topic of turbulence is thus distributed throughout this book.

Turbulent flow results when instabilities in a flow are not sufficiently damped by viscous action; the fluid velocity at each point in the flow then exhibits random fluctuations. Figure 1.30 illustrates such a flow created by a jet injected into a quiescent reservoir of fluid. Osborne Reynolds, a remarkable engineer and fluid mechanician, conducted a now famous experiment [30] in which dye was injected at the centerline of a flow through a long pipe. Figure 1.31 shows Reynolds's apparatus. Reynolds found that three flow regimes exist, the occurrence of which depends on the flow velocity, tube size, and the fluid density and viscosity. Figure 1.32 is taken from Reynolds's original 1883 publication and illustrates the **laminar** regime (Fig. 1.32a) and

FIGURE 1.30
A dye jet issues into a quiescent reservoir of fluid.

---

[9] Many of these developments can be considered advanced topics and omitted for a single-semester survey course.

FIGURE 1.31
Drawing of Osborne Reynolds's original dye-streak apparatus. Reprinted from Ref. [30].

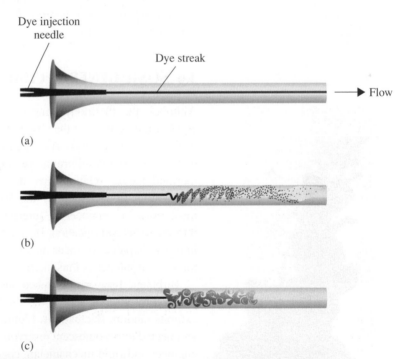

FIGURE 1.32
Sketches from Reynolds's 1883 dye-streak experiments [30] illustrating (a) laminar flow and (b and c) turbulent flow. In Reynolds's own words: "When the velocities were sufficiently low, the streak of colour extended in a beautiful straight line through the tube (a) . . . As the velocity was increased by small stages, at some point in the tube, always at a considerable distance from the trumpet or intake, the colour band would all at once mix up with the surrounding water, and fill the rest of the tube with a mass of coloured water, as in (b) . . . On viewing the tube by the light of an electric spark, the mass of colour resolved itself into a mass of more or less distinct curls, showing eddies, as in (c)."

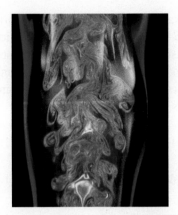

Large- and small-scale eddies can be seen in this turbulent flame. Image from Ref. [29] and used with permission.

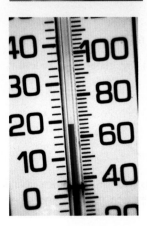

a fully turbulent regime (Figs. 1.32b and 1.32c). In the laminar regime, the flow is steady at every point. In the turbulent regime, the flow is highly unsteady; in particular, the velocity fluctuates greatly with time at any location. The random unsteadiness associated with various flow properties is the hallmark of a turbulent flow. In later chapters, we show how such complex flows can be understood and characterized for engineering purposes.

What is the physical nature of a turbulent flow? Figure 1.30 gives a partial answer to this question. In this figure, we see fluid blobs and filaments of fluid intertwining. A common notion in fluid mechanics is the idea of a fluid **eddy**. An eddy is considered to be a macroscopic fluid element in which the microscopic elements composing the eddy behave in some ways as a unit. A turbulent flow comprises many eddies with a multitude of sizes and angular velocities. A number of smaller eddies may be embedded in a larger eddy. A characteristic of a fully turbulent flow is the existence of a wide range of length scales, or eddy sizes.

The rapid intertwining of fluid elements is a characteristic that distinguishes turbulent flow from laminar flow. The turbulent motion of fluid elements allows momentum, species, and energy to be transported in the cross-stream direction much more rapidly than is possible by the molecular diffusion processes controlling transport in laminar flows. For example, most practical combustion devices employ turbulent flows to enable rapid mixing and heat release in relatively small volumes. Specific examples include steam power plant boilers, spark-ignition engines, and rocket engines.

1.7 DIMENSIONS AND UNITS

The primary dimensions used in this book are *mass*, *length*, *time*, *temperature*, *electric current*, and *amount of substance*. All other dimensions, such as force, energy, and power, are derived from these primary dimensions. Furthermore, we employ almost exclusively the International System of Units, or le Système Internationale (SI) d'unités. The primary dimensions thus have the following associated units:[10]

Mass [=] kilogram (kg),

Length [=] meter (m),

Time [=] second (s),

Temperature [=] kelvin (K),

Electric Current [=] ampere (A), and

Amount of Substance [=] mole (mol).

The derived dimensions most frequently used in this book are defined as follows:

$$\text{Force} \equiv 1 \ \text{kg} \cdot \text{m/s}^2 = 1 \ \text{newton (N)}, \tag{1.4}$$

$$\text{Energy} \equiv 1 \ (\text{kg} \cdot \text{m/s}^2) \cdot \text{m} = 1 \ \text{N} \cdot \text{m or } 1 \ \text{joule (J), and} \tag{1.5}$$

$$\text{Power} \equiv 1 \frac{(\text{kg} \cdot \text{m/s}^2) \cdot \text{m}}{\text{s}} = 1 \ \text{J/s or } 1 \ \text{watt (W).} \tag{1.6}$$

Unfortunately, a wide variety of non-SI units are used customarily in the United States, many of which are industry-specific. Conversion factors from SI units to the most common non-SI units are presented on the inside covers of this book.

[10] The symbol [=] is used to express *"has units of"* and is used throughout this book.

We refer to certain non-SI units in several examples to provide some familiarity with these important non-SI units. To have a single location for their definition, the customary units most important to our study of the thermal-fluid sciences are presented here. In terms of the four primary dimensions, mass ($[=]$ pound-mass or lb_m), length ($[=]$ foot), time ($[=]$ second), and temperature ($[=]$ degree Rankine or R), common derived dimensions and units are as follows:

$$\text{Force} \equiv 32.174 \frac{lb_m \cdot ft}{s^2} = 1 \text{ pound-force or } lb_f, \tag{1.7}$$

$$\text{Energy} \equiv 1 \text{ ft} \cdot lb_f = \frac{1}{778.17} \text{ British thermal unit or Btu, and} \tag{1.8}$$

$$\text{Power} \equiv 1 \frac{ft \cdot lb_f}{s} = \frac{1}{550} \text{ horsepower or hp.} \tag{1.9}$$

In some applications, the unit associated with mass is the slug. Using this unit to define force yields

$$\text{Force} \equiv 1 \frac{slug \cdot ft}{s^2} = 1 \text{ pound-force or } lb_f. \tag{1.10}$$

Further elaboration of units is found in the discussion of thermodynamic properties in Chapter 2.

The use of the units pound-force (lb_f) and pound-mass (lb_m) can cause some confusion to the casual user.[11] The following example clarifies their usage.

Example 1.1

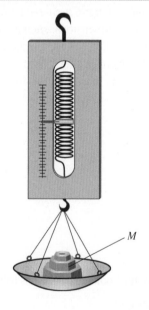

A mass of 1 lb_m is placed on a spring scale calibrated to read in pounds-force. Assuming an earth-standard gravitational acceleration of 32.174 ft/s², what is the scale reading? What is the equivalent reading in newtons? How do these results change if the measurement is conducted on the surface of the moon where the gravitational acceleration is 5.32 ft/s²?

Solution

Known $M = 1 \text{ lb}_m$, $g_{earth} = 32.174$ ft/s²

Find   Force exerted on scale

Analysis   The force exerted on the scale is the weight, $F = W$, which equals the product of the mass and the gravitational acceleration, that is,

$$F = M g_{earth}.$$

Thus,

$$F = (1 \text{ lb}_m) (32.174 \text{ ft/s}^2).$$

Using the definition of pounds-force (Eq. 1.7), we generate the identity

$$1 \equiv \left[\frac{1 \dfrac{lb_f}{ft/s^2}}{32.174 \text{ lb}_m} \right].$$

---

[11] A safe rule of thumb to avoid any confusion with units is to convert all given quantities in a problem to SI units, solve the problem using SI units, and then convert the final answer to any desired customary units.

Using this identity in the previous expression for the force yields

$$F = 1 \text{ lb}_m \left[\frac{1 \frac{\text{lb}_f}{\text{ft/s}^2}}{32.174 \text{ lb}_m} \right] 32.174 \text{ ft/s}^2,$$

or

$$F = 1 \text{ lb}_f.$$

To convert this result to SI units, we use the conversion factor found at the front of the book:

$$F = 1 \text{ lb}_f \left[\frac{1 \text{ N}}{0.224809 \text{ lb}_f} \right] = 4.45 \text{ N}.$$

On the moon, the force is

$$F = M g_{moon}$$

$$= (1 \text{ lb}_m) (5.32 \text{ ft/s}^2)$$

$$= 1 \text{ lb}_m \left[\frac{1 \frac{\text{lb}_f}{\text{ft/s}^2}}{32.174 \text{ lb}_m} \right] 5.32 \text{ ft/s}^2$$

$$F = 0.165 \text{ lb}_f,$$

or

$$F = 0.736 \text{ N}.$$

Comments From this example, we see that the definition of the U.S. customary unit for force was chosen so that a one-pound mass produces a force of exactly one pound-force under conditions of standard earth gravity. In the SI system, numerical values for mass and force are *not* identical for conditions of standard gravity. For example, consider the weight associated with a 1-kg mass:

$$F = Mg$$

$$= (1 \text{ kg})(9.807 \text{ m/s}^2)$$

$$= 1 \text{ kg} \left(\frac{1 \text{ N}}{\text{kg} \cdot \text{m/s}^2} \right) (9.807 \text{ m/s}^2)$$

$$= 9.807 \text{ N}.$$

Regardless of the system of units, care must be exercised in any unit conversions. Checking to see that the units associated with any calculated result are correct should be part of every problem solution. Such practice also helps you to spot gross errors and can save time.

Self Test
1.1

☑ **Repeat Example 1.1 for a given mass of 1 slug.**

(*Answer: 32.174 lb$_f$, 143.12 N, 5.32 lb$_f$, 23.66 N*)

1.8 PROBLEM-SOLVING METHOD

For you to consistently solve engineering problems successfully (both textbook and real problems) depends on your developing a procedure that aids thought processes, facilitates identification of errors along the way, allows others to easily check your work, and provides a reality check at completion. The following general procedure has these attributes and is recommended for your consideration for most problems. Nearly all of the examples throughout the text will follow this procedure.

1. State what is known in a simple manner without rewriting the problem statement.

2. Indicate what quantities you want to find.

3. Draw and label useful sketches whenever possible. (What is useful generally depends on the context of the problem. Suggestions are provided at appropriate locations in the text.)

4. List your initial assumptions and add others to the list as you proceed with your solution.

5. Analyze the problem and identify the important definitions and principles that apply to your solution.

6. Develop a symbolic or algebraic solution to your problem, delaying substitution of numerical values as late as possible in the process.

7. Substitute numerical values as appropriate and indicate the source of all physical data as you proceed. (The appendices of this book contain much useful data.)

8. Check the units associated with each calculation. The factor-label method is an efficient way to do this.

9. Examine your answer critically. Does it appear to be reasonable and consistent with your expectations and/or experience?

10. Write out one or more comments using step 9 as your guide. What did you learn as a result of solving the problem? Are your assumptions justified?

1.9 HOW TO USE THIS BOOK

You may find the organization of this textbook different from others used previously in your study of mathematics, the fundamental sciences, and engineering. To a large degree, the organization here is hierarchical; that is, all related topics are grouped within a single chapter, with the most elementary topics appearing first and additional levels of complexity added as the chapter progresses. The purpose of this particular organization is to show in as strong a fashion as possible that various topics are actually related and not disjoint concepts [31]. Therefore, in the actual use of the book, you will not necessarily proceed linearly, going from the beginning to the end of each chapter, chapter by chapter. In particular, the following chapter dealing with thermodynamic properties provides much more information than you will need initially. You should be prepared to continually revisit Chapter 2 as you need particular information on properties. For example, a study of entropy and related properties should be concurrent with your study of the second law of thermodynamics in Chapter 7. Another example of nonlinear organization is the placement of numerous practical applications in the final chapters. This arrangement does not mean that these applications should be dealt with last; rather, they are separated to provide flexibility in coverage. Do not be surprised to have simultaneous assignments from Chapters 5 and 11, for example. It is the author's hope that this structure will make your learning the thermal-fluid sciences both more efficient and more interesting than more traditional presentations.

SUMMARY

In this chapter, we introduced the three thermal sciences, thermodynamics, heat transfer, and fluid mechanics, and presented the following five practical applications: fossil-fuel steam power plants, solar-heated buildings, spark-ignition engines, jet engines, and biological systems. From these discussions, you should have a reasonably clear idea of what this book is about. Also presented in this chapter are many concepts and definitions upon which we will build in subsequent chapters. To make later study easier, you should have a firm understanding of these concepts and definitions. A review of the learning objectives presented at the start of this chapter may be helpful in that regard. Finally, it is the author's hope that your interest has been piqued in the far-ranging subject matter of this book.

Chapter 1
Key Concepts & Definitions Checklist[12]

1.1 What are the thermal fluid sciences? ➤ *1.1*

❑ Thermodynamics

❑ Heat transfer

❑ Fluid dynamics

1.2 Some applications ➤ *1.3–1.6*

❑ Basic Rankine cycle

❑ Rankine cycle components

❑ Solar collectors

❑ Spark-ignition engines

❑ Four-stroke cycle

❑ Turbojet and turbofan engines

❑ Human comfort and thermal-fluid sciences

1.3 Physical frameworks for analysis

❑ Thermodynamic (or closed) system ➤ *1.7*

❑ Control volume (or open system) ➤ *1.10*

❑ Boundaries ➤ *1.8*

❑ Surroundings ➤ *1.11*

❑ Integral and differential systems or control volumes ➤ *1.12*

1.4 Preview of conservation principles

❑ General conservation law for time interval (Eq. 1.1) ➤ *1.13, 1.17*

❑ General conservation law at an instant (Eq. 1.2) ➤ *1.19*

❑ The need for thermodynamic properties

1.5 Key concepts and definitions

❑ Property ➤ *1.22*

❑ State ➤ *1.22*

❑ Process ➤ *1.22*

❑ Flow process ➤ *1.24*

❑ Cycle ➤ *1.23*

❑ Equilibrium ➤ *1.25*

❑ Quasi-equilibrium ➤ *1.26*

❑ Local equilibrium ➤ *1.27*

1.6 Some general characteristics of real flows ➤ *1.28*

❑ Laminar flows

❑ Turbulent flows

❑ Eddies

❑ Osborne Reynolds's contributions

1.7 Dimensions and units ➤ *1.35, 1.40, 1.47*

❑ The difference between dimensions and units

❑ Primary SI dimensions

❑ Derived dimensions

❑ U.S. customary units

❑ Pounds-force and pounds-mass

1.8 Problem-solving method

❑ Key steps in solving thermal-fluid problems

---

[12] Numbers following arrows below refer to Questions and Problems at the end of the chapter.

REFERENCES

1. Moran, M. J., and Shapiro, H. N., *Fundamentals of Engineering Thermodynamics*, 3rd ed., Wiley, New York, 1995.

2. Wark, K. Jr., and Richards, R. E., *Thermodynamics*, 6th ed., McGraw-Hill, New York, 1999.

3. Obert, E. F., *Thermodynamics*, McGraw-Hill, New York, 1948.

4. White, F. M., *Fluid Mechanics*, 3rd ed., McGraw-Hill, New York, 1994.

5. Fox, R. W., and McDonald, A. T., *Introduction to Fluid Mechanics*, 3rd ed., Wiley, New York, 1985.

6. Shepherd, D. G., *Elements of Fluid Mechanics*, Harcourt, Brace & World, New York, 1965.

7. Incropera, F. P., and DeWitt, D. P., *Fundamentals of Heat and Mass Transfer*, 4th ed., Wiley, New York, 1996.

8. Myers, G. E., *Analytical Methods in Conduction Heat Transfer*, McGraw-Hill, New York, 1971.

9. Holman, J. P., *Heat Transfer*, 8th ed., McGraw-Hill, New York, 1997.

10. *Webster's New Twentieth Century Dictionary*, J. L. McKechnie (Ed.), Collins World, Cleveland, 1978.

11. Energy Information Agency, U.S. Department of Energy, "Annual Energy Review 2002," http://www.eia.doe.gov/emeu/aer/contents.html, posted Oct. 24, 2003.

12. *Steam: Its Generation and Use*, 39th ed., Babcock & Wilcox, New York, 1978.

13. Singer, J. G. (Ed.), *Combustion Fossil Power: A Reference Book on Fuel Burning and Steam Generation*, 4th ed., Combustion Engineering, Windsor, CT, 1991.

14. Basu, P., Kefa, C., and Jestin, L., *Boilers and Burners: Design and Theory*, Springer, New York, 2000.

15. Goodall, P. M., *The Efficient Use of Steam*, IPC Science and Technology Press, Surrey, England, 1980.

16. Flagan, R. C., and Seinfeld, J. H., *Fundamentals of Air Pollution Engineering*, Prentice Hall, Englewood Cliffs, NJ, 1988.

17. Schetz, J. A. (Ed.), *Thermal Pollution Analysis*, Progress in Astronautics and Aeronautics, Vol. 36, AIAA, New York, 1975.

18. Duffie, J. A., and Beckman, W. A., *Solar Engineering of Thermal Processes*, Wiley, New York, 1980. See also 2nd ed., 1991.

19. Harrell, J. J. Jr., *Solar Heating and Cooling of Buildings*, Van Nostrand Reinhold, New York, 1982.

20. Heywood, J. B., *Internal Combustion Engine Fundamentals*, McGraw-Hill, New York, 1988.

21. Obert, E. F., *Internal Combustion Engines and Air Pollution*, Harper & Row, New York, 1973.

22. Ferguson, C. F., *Internal Combustion Engines: Applied Thermosciences*, Wiley, New York, 1986.

23. Campbell, A. S., *Thermodynamic Analysis of Combustion Engines*, Wiley, New York, 1979.

24. Cumpsty, N., *Jet Propulsion*, Cambridge University Press, New York, 1997.

25. St. Peter, J., *History of Aircraft Gas Turbine Engine Development in the United States: A Tradition of Excellence*, International Gas Turbine Institute of the American Society of Mechanical Engineers, Atlanta, 1999.

26. White, M. A., Colendbrander, K., Olan, R. W., and Penswick, L. B., "Generators That Won't Wear Out," *Mechanical Engineering*, 118:92–6 (1996).

27. Carey, V. P., *Statistical Thermodynamics and Microscale Thermophysics*, Cambridge University Press, Cambridge, England, 1999.

28. Brzustowski, T. A., *Introduction to the Principles of Engineering Thermodynamics*, Addison-Wesley, Reading, MA, 1969.

29. Roquemore, W. M., et al., "Jet Diffusion Flame Transition to Turbulence," in *A Gallery of Fluid Motion*, Samimy, M., Breuer, K. S., Leal, L. G., and Steen, P. H. (Eds.), p. 83, Cambridge University Press, Cambridge, England, 2003.

30. Reynolds, O., "An Experimental Investigation of the Circumstances Which Determine Whether the Motion of Water Shall Be Direct or Sinuous, and of the Law of Resistance in Parallel Channels," *Philosophical Transactions of the Royal Society of London*, 174:935–82 (1883).

31. Reif, F., "Scientific Approaches to Science Education," *Physics Today*, November 1986, pp. 48–54.

32. Lide, D. R. (Ed.), *Handbook of Chemistry and Physics*, 77th ed., CRC Press, Boca Raton, FL, 1996.

Some end-of-chapter problems were adapted with permission from the following:

33. Look, D. C. Jr., and Sauer, H. J. Jr., *Engineering Thermodynamics*, PWS, Boston, 1986.

34. Myers, G. E., *Engineering Thermodynamics*, Prentice Hall, Englewood Cliffs, NJ, 1989.

35. Pnueli, D., and Gutfinger, C., *Fluid Mechanics*, Cambridge University Press, Cambridge, England, 1992.

Chapter 1 Question and Problem Subject Areas

QUESTIONS AND PROBLEMS

*Indicates computer required for solution.

1.1 List the three disciplines that comprise the thermal-fluid sciences. Discuss the principal subjects of each.

1.2 Make a list of practical devices for which the design was made possible by the application of the thermal-fluid sciences.

1.3 Draw a schematic diagram of a simple steam power plant. Discuss the processes associated with each major piece of equipment.

1.4 Distinguish between *active* and *passive* solar heating.

1.5 Sketch and discuss the four strokes associated with the four-stroke cycle spark-ignition engine.

1.6 Distinguish between a turbojet engine and a turbofan engine. Which type is most commonly used for commercial airline service?

1.7 Create a list of the factors that differentiate a thermodynamic system from a control volume.

1.8 Consider a conventional toaster—the kind used to toast bread. Sketch two different boundaries associated with an operating toaster: one in which a thermodynamic system is defined, and a second one in which a control volume is defined. Be sophisticated in your analyses. Discuss.

1.9 Discuss what would have to be done to transform the control volume shown in Fig. 1.18 to a thermodynamic system. Ignore the boundary shown in Fig. 1.18. What thermodynamic systems can be defined for the situation illustrated?

1.10 Consider a conventional house in the northeastern part of the United States. Isolate the house from the surroundings by drawing a control volume. Identify all of the locations where mass enters or exits your control volume.

1.11 Consider an automobile, containing a driver and several passengers, traveling along a country road. Isolate the automobile and its contents from the surroundings by drawing an appropriate boundary. Does your boundary enclose a thermodynamic system or a control volume? How would you have to modify your boundary to convert from one to the other? What assumptions, if any, do you have to make to perform this conversion?

1.12 Distinguish between differential (microscopic) and integral (macroscopic) thermodynamic systems. Mathematically, how do they differ?

1.13 Write out in words a general expression of the conservation of energy principle applied to a finite time interval. (See Eq. 1.1.)

1.14 Write out in words a general expression of the conservation of energy principle applied to an instant, that is, an infinitesimal time interval. (See Eq. 1.2.)

1.15 Repeat Questions 1.13 and 1.14, but for conservation of mass.

1.16 On Wednesday morning a truck delivers a load of 100 perfect ceramic flowerpots to a store. Each of these perfect flowerpots is valued at $3 and has a mass of 2

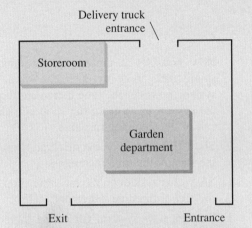

Delivery truck entrance

Storeroom

Garden department

Exit   Entrance

lb_m. Half of the flowerpots are immediately put in the storeroom, while the other half are sent to the garden department to be sold. Wednesday afternoon, five customers enter the store, buy 1 flowerpot apiece, and leave the store. Another customer enters, buys 3 flowerpots, and leaves the store. Wednesday evening, three particularly clumsy customers knock 14 flowerpots on the floor and damage them. Their damaged value is $1 each. For the entire store as the system, fill in the following table:

Quantity (Units)	Mass (lb_m)	Perfect (#)	Damaged (#)	Value ($)
Inflow				
Produced				
Outflow				
Stored				
Destroyed				

1.17 One shop of a roller-bearing plant produces 50-g roller bearings at the rate of 10,000 per day from steel that enters the plant during the day. Of this total production, 3,000 are judged "precision" (value = $5 each), 5,000 are judged "standard" (value = $3 each), and the remainder are judged to be "substandard" (value = 0). The shop sells 1,500 precision bearings per day and 2,000 standard bearings per day to another company. The remaining precision and standard bearings are stored within the shop. The substandard bearings are removed from the shop as waste along with the 20 kg/day of steel that is ground off during the production of the bearings. For the shop as the system, fill the following table for a one-day time increment:

Quantity (Units)	Mass (kg)	Precision (#)	Standard (#)	Substandard (#)	Value (k$)
Inflow					
Produced					
Outflow					
Stored					
Destroyed					

1.18 A chemical plant produces the chemical furfural (FFL). The raw materials are primarily corncobs and oat hulls. One pound of corncobs will make 0.5 lb_m of FFL and 0.5 lb_m of residue. One pound of oat hulls will make 0.75 lb_m of FFL and 0.25 lb_m of residue. Corncobs are delivered to the plant at a rate of 1,000 lb_m/hr and a cost of 10 cents/ lb_m. Oat hulls are delivered to the plant at a rate of 200 lb_m/hr at a cost of 20 cents/lb_m. FFL is sold at a rate of 350 lb_m/hr for a price of 50 cents/lb_m. FFL is stored within the plant at a rate of 100 lb_m/hr. All residue produced is sold for 25 cents/lb_m. At the beginning of the 8:00 A.M. shift, there are 350 lb_m of oat hulls in the plant. One hour later the supply of oat hulls in the plant is down to 250 lb_m.

A. For the chemical plant as the system, complete the following table:

Quantity (Units)	Corncobs (lb_m/hr)	Oat hulls (lb_m/hr)	FFL (lb_m/hr)	Value ($/hr)	Money ($/hr)
Inflow					
Produced					
Outflow					
Stored					
Destroyed					

B. Using a closed system and balance principles, determine the change in value associated with making 1 lb_m of corncobs into FFL. Show and label your system clearly.

1.19 The Foster-Davis Juice Company steadily takes in 1,000 oranges per hour and ships out 150 cans of juice per hour. The average orange has a mass of 0.5 lb_m, costs 25 cents, and is processed into 0.4 lb_m of juice and 0.1 lb_m of peels and pulp (P&P). Each can of juice contains 2 lb_m of juice and sells for 1 dollar. The mass of the can itself may be neglected. The P&P are sold to University Food Service for 20 cents/lb_m at the rate of 50 lb_m/hr. Oranges are put into cold storage in the plant at a rate of 300 oranges per hour. The remaining oranges are processed into juice and P&P. The Foster-Davis Juice Company has a "pay as you go" policy. Initially there are stockpiles of oranges, cans of juice, P&P, and money within the plant.

A. Considering the Foster-Davis plant as a control volume (open system), complete the following table:

Quantity (Units)	Mass (lb_m/hr)	Oranges (#/hr)	Juice (cans/hr)	P&P (lb_m/hr)	Money ($/hr)
Inflow					
Produced					
Outflow					
Stored					
Destroyed					

B. Can the plant go on operating in this way forever? Explain.

C. Can the process within the plant be reversed? Explain.

1.20 Consider a machine shop that stamps washers out of metal disks as a control volume (open system). Disks enter the shop at the rate of 1000 disks per second and each disk has a mass of 2 g. The stamping machine punches out the center of each disk to make a 1.5-g washer and a 0.5-g center. The stamping machine makes washers at the rate of 800 washers per second. A conveyor belt transports 1000 washers per second from the shop. Initially there are 3000 disks and 5000 washers stockpiled in the shop but no centers. The shop then runs for a 1-s time increment.

A. Sketch the machine shop. Be sure to include the control-volume boundaries in your sketch.

B. Fill in the following table for the I-S increment:

Quantity (Units)	Disks (#)	Washers (#)	Centers (#)	Mass (g)
Inflow				
Produced				
Outflow				
Stored				
Destroyed				

Now consider a closed system containing a single disk about to enter the shop. For this system, consider the time increment required for the system to go from the shop entrance to the exit of the stamping machine.

C. Show this system on the machine-shop sketch at the start of the time increment.

D. Show this system on the machine-shop sketch at the *end* of the time increment.

E. For this new system, fill in the table given in Part B.

1.21 A cook (200 lb_m) brings 100 apples (0.5 lb_m each) into the kitchen, stores 36 of the apples in the refrigerator, and puts the rest on the table to make applesauce. While the cook is mashing the apples into sauce, three young friends (70 lb_m each) enter the kitchen. The cook gives them 6 apples each from the table, and they take them outdoors. After finishing the sauce, the cook leaves the kitchen. Considering the kitchen as a control volume (open system), complete the following table:

Quantity (Units)	Mass (lb_m)	Apples (#)	Sauce (lb_m)
Inflow			
Produced			
Outflow			
Stored			
Destroyed			

1.22 Write out the formal definitions of the following terms: property, state, and process. Discuss the interrelationships of these three terms.

1.23 Can you identify any devices that operate in a thermodynamic cycle? If so, list them. Also identify the working fluids, if known.

1.24 Distinguish between a system process and a flow process.

1.25 List the conditions that must be established for thermodynamic equilibrium to prevail (i.e., list the subtypes of equilibrium).

1.26 Define a quasi-static or quasi-equilibrium process and give an example of such a process. Also give an example of a non–quasi-equilibrium process.

1.27 Discuss what is meant by local equilibrium. Why is this concept of importance to engineering?

1.28 Discuss the difference between laminar flow and turbulent flow.

1.29 What is your approximate weight in pounds-force? In newtons? What is your mass in pounds-mass? In kilograms? In slugs?

1.30 Repeat Problem 1.29 assuming you are now on the surface of Mars where the gravitational acceleration is 3.71 m/s^2.

1.31 A residential natural gas–fired furnace has an output of 86,000 Btu/hr. What is the output in kilowatts? How many 100-W incandescent light bulbs would be required to provide the same output as the furnace?

1.32 The 1964 Pontiac GTO was one of the first so-called muscle cars produced in the United States in the 1960s and 1970s. The 389-in^3 displacement V-8

engine in the GTO delivered a maximum power of 348 hp and a maximum torque of 428 lb$_f$·ft. From a standing start, the GTO traveled 1/4 mile in 14.8 s, accelerating to 95 mph. Convert all of the U.S. customary units in these specifications to SI units.

1.33 Coal-burning power plants convert the sulfur in the coal to sulfur dioxide (SO_2), a regulated air pollutant. The maximum allowable SO_2 emission for new power plants (i.e., the maximum allowed ratio of the mass of SO_2 emitted per input of fuel energy) is 0.80 lb$_m$/(million Btu). Convert this emission factor to units of grams per joule.

1.34 The National Ambient Air Quality Standard (NAAQS) for lead (Pb) in the air is 1.5 μg/m^3. Convert this standard to U.S. customary units of lb$_m$/ft^3.

1.35 The astronauts of *Apollo 17*, the final U.S. manned mission to the moon, collected 741 lunar rock and soil samples having a total mass of 111 kg. Determine the weight of samples in SI and U.S. customary units (a) on the lunar surface (g_{moon} = 1.62 m/s^2) and (b) on the earth (g_{earth} = 9.807 m/s^2).

Also determine (c) the mass of the samples in U.S. customary units.

*1.36 The gravitational acceleration on the earth varies with latitude and altitude as follows [32]:

$$g \ (m/s^2) = 9.780356 \ [1 + 0.0052885 \sin^2 \theta - 0.0000059 \sin^2 (2\theta)] - 0.003086 \ z,$$

where θ is the latitude and z is the altitude in kilometers. Use this information to determine the weight (in newtons and pounds-force) of a 54-kg mountain climber at the following locations:

A. Summit of Kilimanjaro in Tanzania (3.07° S and 5895 m)

B. Summit of Cerro Aconcagua in Argentina (32° S and 6962 m)

C. Summit of Denali in Alaska (63° N and 6,194 m)

How do these values compare with the mountain climber's weight at 45° N latitude at sea level? Use spreadsheet software to perform all calculations.

1.37 The gravitational acceleration at the surfaces of Jupiter, Pluto, and the sun are 23.12 m/s², 0.72 m/s², and 273.98 m/s², respectively. Determine your weight at each of these locations in both SI and U.S. customary units. Assume no loss of mass results from the extreme conditions.

1.38 A coal-burning power plant consumes fuel energy at a rate of 4,845 million Btu/hr and produces 500 MW

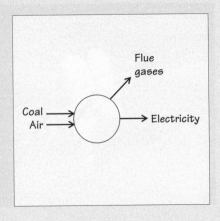

of net electrical power. Determine the overall efficiency of the power plant (i.e., the dimensionless ratio of the net output and input energy rates).

1.39 Derive a conversion factor relating pressures in pascals (N/m^2) and psi (lb_f/in^2). Compare your result with the conversion factor provided inside the front cover of this book.

1.40 The Space Shuttle main engine had the following specifications ca. 1987. Convert these values to SI units.

Maximum Thrust	
At sea level	408,750 lb_f
In vacuum	512,300 lb_f
Pressures	
Hydrogen pump discharge	6872 psi
Oxygen pump discharge	7936 psi
Combustion chamber	3277 psi
Flow Rates	
Hydrogen	160 lb_m/s
Oxygen	970 lb_m/s
Power	
High-pressure H_2 turbopump	74,928 hp
High-pressure O_2 turbopump	28,229 hp
Weight	7000 lb_f
Length	14 ft
Diameter	7.5 ft

1.41 Determine the acceleration of gravity for which the weight of an object will be numerically equal to its mass in the SI unit system.

1.42 Determine the mass in lb_m of an object that weighs 200 lb_f on a planet where $g = 50$ ft/s².

1.43 Determine the mass in kg of an object that weighs 1,000 N on a planet where $g = 15$ m/s².

1.44 You are given the job of setting up an experiment station on another planet. While on the surface of this planet, you notice that an object with a known mass of 50 kg has a weight (when measured on a spring scale calibrated in lb_f on earth at standard gravity) of 50 lb_f. Determine the local acceleration of gravity in m/s² on the planet.

1.45 In an environmental test chamber, an artificial gravity of 1.676 m/s^2 is produced. How much would a 92.99-kg man weigh inside the chamber?

1.46 Compare the acceleration of a 5-lb$_m$ stone acted on by 20 lb$_f$ vertically upward and vertically downward at a place where $g = 30$ ft/s^2. Ignore any frictional effects.

1.47 Use the conversion factors found on the inside covers of the book to convert the flowing quantities to SI units:

Density, $\rho = 120$ lb$_m$/ft^3

Thermal conductivity, $k = 170$ Btu/(hr·ft·F)

Convective heat-transfer coefficient, $h_{conv} = 211$ Btu/(hr·ft^2·F)

Specific heat, $c_p = 175$ Btu/(lb$_m$·F)

Viscosity, $\mu = 20$ centipoise

Viscosity, $\mu = 77$ lb$_f$·s/ft^2

Kinematic viscosity, $\nu = 3.0$ ft^2/s

Stefan–Boltzmann constant, $\sigma = 0.1713 \times 10^{-8}$ Btu/(ft^2·hr·R^4)

Acceleration, $a = 12.0$ ft/s^2

*1.48 Given the following geometric parameters for a spark-ignition engine, plot the instantaneous combustion chamber volume and its time rate of change for a complete cycle (see Appendix 1A):

Bore, 80 mm

Stroke, 70 mm

Connecting-rod length to crank radius ratio, 3.4

Compression ratio, 7.5

Engine speed, 2000 rpm

Appendix 1A
Spark-Ignition Engines

Since we will be revisiting the SI engine, it is useful to define a few engine-related geometrical terms at this point, rather than repeating them in a variety of locations. These are:

B: The **bore** is the diameter of the cylinder.

S: The **stroke** is the distance traveled by the piston in moving from top center to bottom center, or vice versa.

V_c (or V_{TC}):   The **clearance volume** is the combustion chamber volume when the piston is at top center.

V_d (or V_{disp}):   The **displacement** is the difference in the volume at bottom center and at top center, that is,

$$V_{disp} = V_{BC} - V_{TC} \qquad (1A.1)$$

The displacement is also equal to the product of the stroke and the cross-sectional area of the cylinder:

$$V_{disp} = S \, \pi B^2 / 4. \qquad (1A.2)$$

CR: The **compression ratio** is the ratio of the volume at bottom center to the volume at top center, that is,

$$CR = \frac{V_{BC}}{V_{TC}} = \frac{V_{disp}}{V_{TC}} + 1. \qquad (1A.3)$$

The following geometrical parameters associated with a reciprocating engine are illustrated in Fig. 1A.1:

$B \equiv$ bore

$S \equiv$ stroke

$\ell \equiv$ connecting-rod length

$a \equiv$ crank radius

$\theta \equiv$ crank angle

$V_{TC} \equiv$ volume at top center

$V_{BC} \equiv$ volume at bottom center

$CR \equiv$ compression ratio ($= V_{BC}/V_{TC}$)

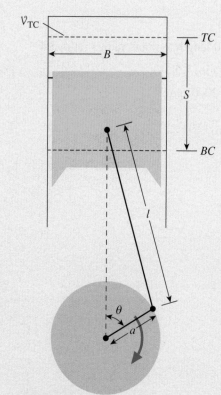

FIGURE 1A.1
Definition of geometrical parameters for reciprocating engines. After Ref. [20].

From geometric and kinematic analyses [20], the instantaneous volume, $\mathcal{V}(\theta)$, and its time derivative, $d\mathcal{V}(\theta)/dt$, are given by

$$\mathcal{V}(\theta) = \mathcal{V}_{TC}\left\{1 + \frac{1}{2}(CR - 1)\left[\frac{\ell}{a} + 1 - \cos\theta - \left(\frac{\ell^2}{a^2} - \sin^2\theta\right)^{1/2}\right]\right\} \quad (1A.4)$$

and

$$\frac{d\mathcal{V}(\theta)}{dt} = SN\left(\frac{\pi B^2}{4}\right)\pi\sin\theta\left[1 + \frac{\cos\theta}{\left(\frac{\ell^2}{a^2} - \sin^2\theta\right)^{1/2}}\right], \quad (1A.5)$$

where N is the crank rotational speed in rev/s.

THERMODYNAMIC PROPERTIES, PROPERTY RELATIONSHIPS, AND PROCESSES

After studying Chapter 2, you should:

- Be able to explain the meaning of the continuum limit and its importance to thermodynamics.

- Be familiar with the following basic thermodynamic properties: pressure, temperature, specific volume and density, specific internal energy and enthalpy, constant-pressure and constant-volume specific heats, entropy, and Gibbs free energy.

- Understand the relationships among absolute, gage, and vacuum pressures.

- Know the four common temperature scales and be proficient at conversions among all four.

- Know how many independent intensive properties are required to determine the thermodynamic state of a simple compressible substance.

- Be able to indicate what properties are involved in the following state relationships: equation of state, calorific equation of state, and the Gibbs (or T–ds) relationships.

- Explain the fundamental assumptions used to describe the molecular behavior of an ideal gas and under what conditions these assumptions break down.

- Be able to write one or more forms of the ideal gas equation of state and from this derive all other (mass, molar, mass-specific, and molar-specific) forms.

- Be proficient at obtaining thermodynamic properties for liquids and gases from NIST software or online databases and from printed tables.

- Be able to explain in words and write out mathematically the meaning of the thermodynamic property quality.

- Be able to draw and identify the following on T–v, P–v, and T–s diagrams: saturated liquid line, saturated vapor line, critical point, compressed liquid region, liquid–vapor region, and the superheated vapor region.

- Be able to draw an isobar on a T–v diagram, an isotherm on a P–v diagram, and both isobars and isotherms on a T–s diagram.

- Be proficient at plotting simple isobaric, isochoric, isothermal, and isentropic thermodynamic processes on T–v, P–v, and T–s coordinates.

- Be able to explain the approximations used to estimate properties for liquids and solids.

- Be able to derive all of the isentropic process relationships for an ideal gas given that $Pv^\gamma = constant$.

- Be able to explain the meaning of a polytropic process and write the general state relationship expressing a polytropic process.

- Be able to explain the principle of corresponding states and the use of generalized compressibility charts.

- Be able to express the composition of a gas mixture using both mole and mass fractions.

- Be able to calculate the thermodynamic properties of an ideal-gas mixture knowing the mixture composition and the properties of the constituent gases.

- Understand the concept of standardized properties, in particular, standardized enthalpies, and their application to ideal-gas systems involving chemical reaction.

- Be able to apply the concepts and skills developed in this chapter throughout this book.

Chapter 2 Overview

THIS IS ONE OF THE LONGER chapters in this book and should be revisited many times at various levels. To cover in detail the subject of the properties of gases and liquids would take an entire book. Reference [1] is a classic example of such a book. We begin our study of properties by defining a few basic terms and concepts. This is followed by a treatment of ideal-gas properties that originate from the equation of state, calorific equations of state, and the second law of thermodynamics. Various approaches for obtaining properties of nonideal gases, liquids, and solids follow. The properties of substances that have coexisting liquid and vapor phases are emphasized. The concept of illustrating processes graphically using thermodynamic property coordinates (i.e., T–v, P–v, and T–s coordinates) is developed.

In addition to dealing with pure substances, we treat the thermodynamic properties of nonreacting and reacting ideal-gas mixtures. The chapter concludes with a brief introduction to the transport properties encountered in this book.

2.1 KEY DEFINITIONS

To start our study of thermodynamic properties, you should review the definitions of **properties**, **states**, and **processes** presented in Chapter 1, as these concepts are at the heart of the present chapter.

The focus of this chapter is not on the properties of the generic system discussed in Chapter 1—a system that may consist of many clearly identifiable subsystems—but rather on the properties of a pure substance and mixtures of pure substances. We formally define a **pure substance** as follows:

> A *pure substance* is a substance that has a homogeneous and unchanging chemical composition.

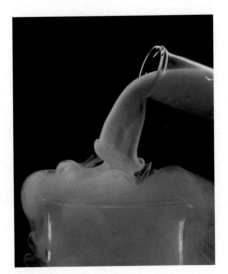

Each element of the periodic table is a pure substance. Compounds, such as CO_2 and H_2O, are also pure substances. Note that pure substances may exist in various physical phases: **vapor**, **liquid**, and **solid**. You are well acquainted with these phases of H_2O (i.e., steam, water, and ice). Carbon dioxide also readily exhibits all three phases. In a compressed gas cylinder at room temperature, liquid CO_2 is present with a vapor phase above it at a pressure of 56.5 atmospheres. Solid CO_2 or "dry ice" exists at temperatures below $-78.5°C$ at a pressure of 1 atmosphere.

In most of our discussions of properties, we further restrict our attention to those pure substances that may be classified as **simple compressible substances**.

The high pressure beneath the blade of an ice skate results in a thin layer of liquid between the blade and the solid ice (top). Surface tension complicates specifying the thermodynamic state of very small droplets (bottom).

> A *simple compressible substance* is one in which the effects of the following are negligible: motion, fluid shear, surface tension, gravity, and magnetic and electrical fields.[1]

The adjectives *simple* and *compressible* greatly simplify the description of the state of a substance. Many systems of interest to engineering closely approximate simple compressible substances. Although we consider motion, fluid shear, and gravity in our study of fluid flow (Chapters 6, 9, and 10), their effects on local thermodynamic properties are quite small; thus, fluid properties are very well approximated as those of a simple compressible substance. Free surface, magnetic, and electrical effects are not considered in this book.

We conclude this section by distinguishing between **extensive properties** and **intensive properties**:

> An *extensive property* depends on how much of the substance is present or the "extent" of the system under consideration.

Examples of extensive properties are volume V and energy E. Clearly, numerical values for V and E depend on the size, or mass, of the system. In contrast, intensive properties do not depend on the extent of the system:

> An *intensive property* is independent of the mass of the substance or system under consideration.

Two familiar thermodynamic properties, temperature T and pressure P, are intensive properties. Numerical values for T and P are independent of the mass or the amount of substance in the system.

Intensive properties are generally designated using lowercase symbols, although both temperature and pressure violate the rule in this text. Any extensive property can be converted to an intensive one simply by dividing by the mass M. For example, the **specific volume** v and the **specific energy** e can be defined, respectively, by

$$v \equiv V/M \quad [=] \, \text{m}^3/\text{kg} \tag{2.1a}$$

and

$$e \equiv E/M \quad [=] \, \text{J/kg.} \tag{2.1b}$$

Intensive properties can also be based on the number of moles present rather than the mass. Thus, the properties defined in Eqs. 2.1 would be more accurately designated as **mass-specific** properties, whereas the corresponding **molar-specific** properties would be defined by

$$\bar{v} \equiv V/N \quad [=] \, \text{m}^3/\text{kmol} \tag{2.2a}$$

and

$$\bar{e} \equiv E/N \quad [=] \, \text{J/kmol,} \tag{2.2b}$$

---

[1] A more rigorous definition of a simple compressible substance is that, in a system comprising such a substance, the only reversible work mode is that associated with compression or expansion, work, that is, $P–dV$ work. To appreciate this definition, however, requires an understanding of what is meant by *reversible* and by $P–dV$ *work*. These concepts are developed at length in Chapters 4 and 7.

where N is the number of moles under consideration. Molar-specific properties are designated using lowercase symbols with an overbar, as shown in Eqs. 2.2. Conversions between mass-specific properties and molar-specific properties are accomplished using the following simple relationships:

$$\bar{z} = z\mathcal{M} \tag{2.3}$$

and

$$z = \bar{z}/\mathcal{M} \tag{2.4}$$

where z (or $\bar{z}$) is any intensive property and $\mathcal{M}$ ($[=]$ kg/kmol) is the molecular weight of the substance (see also Eq. 2.6).

2.2 FREQUENTLY USED THERMODYNAMIC PROPERTIES

One of the principal objectives of this chapter is to see how various thermodynamic properties relate to one another, expressed 1. by an **equation of state**, 2. by a **calorific equation of state**, and 3. by **temperature–entropy, or Gibbs, relationships**. Before we do that, however, we list in Table 2.1 the most common properties so that you might have an overview of the scope of our study. You may be familiar with many of these properties, although others will be new and may seem strange. As we proceed in our study, this strangeness should disappear as you work with these new properties. We begin with a discussion of three extensive properties: mass, number of moles, and volume.

2.2a Properties Related to the Equation of State
Mass

As one of our fundamental dimensions (see Chapter 1), **mass**, like time, cannot be defined in terms of other dimensions. Much of our intuition of what mass is follows from its role in Newton's second law of motion

$$F = Ma. \tag{2.5}$$

In this relationship, the force F required to produce a certain acceleration a of a particular body is proportional to its mass M. The SI mass standard is a platinum–iridium cylinder, defined to be one kilogram, which is kept at the International Bureau of Weights and Measure near Paris.

Number of Moles

In some applications, such as reacting systems, the number of moles N comprising the system is more useful than the mass. The **mole** is formally defined as the amount of substance in a system that contains as many elementary entities as there are in exactly 0.012 kg of carbon 12 (^{12}C). The elementary entities may be atoms, molecules, ions, electrons, etc. The abbreviation for the SI unit for a mole is *mol*, and *kmol* refers to 10^3 *mol*.

The number of moles in a system is related to the system mass through the **atomic** or **molecular weight**[2] $\mathcal{M}$; that is,

$$M = N\mathcal{M}, \tag{2.6}$$

---

[2] Strictly, the atomic weight is not a weight at all but is the relative atomic mass.

Table 2.1 Common Thermodynamic Properties of Single-Phase Pure Substances

Property	Symbolic Designation		Units		Relation to Other Properties	Classification
	Extensive	Intensive*	Extensive	Intensive*		
Mass	M	—	kg	—	—	Fundamental property appearing in equation of state
Number of moles	N	—	kmol	—	—	Fundamental property appearing in equation of state
Volume and specific volume	$\mathcal{V}$	v	m³	m³/kg	—	Fundamental property appearing in equation of state
Pressure	—	P	—	Pa or N/m²	—	Fundamental property appearing in equation of state
Temperature	—	T	—	K	—	Fundamental property appearing in equation of state
Density	—	ρ	—	kg/m³	—	Fundamental property appearing in equation of state
Internal energy	U	u	J	J/kg	—	Based on first law and calculated from calorific equation of state
Enthalpy	H	h	J	J/kg	$U + P\mathcal{V}$	Based on first law and calculated from calorific equation of state
Constant-volume specific heat	—	c_v	—	J/kg·K	$(\partial u/\partial T)_v$	Appears in calorific equation of state
Constant-pressure specific heat	—	c_p	—	J/kg·K	$(\partial h/\partial T)_p$	Appears in calorific equation of state
Specific-heat ratio	—	γ	—	Dimensionless	c_p/c_v	—
Entropy	S	s	J/K	J/kg·K	—	Based on second law of thermodynamics
Gibbs free energy (or Gibbs function)	G	g	J	J/kg	$H - TS$	Based on second law of thermodynamics
Helmholtz free energy (or Helmholtz function)	A	a	J	J/kg	$U - TS$	Based on second law of thermodynamics

* Molar intensive properties are obtained by substituting the number of moles, N, for the mass and changing the units accordingly. For example, $s \equiv S/M$ [=] J/kg·K becomes $\bar{s} \equiv S/N$ [=] J/kmol·K.

where $\mathcal{M}$ has units of g/mol or kg/kmol. Thus, the atomic weight of carbon 12 is exactly 12. The **Avogadro constant** $\mathcal{N}_{AV}$ is used to express the number of particles (atoms, molecules, etc.) in a mole:

$$\mathcal{N}_{AV} \equiv \begin{cases} 6.02214199 \times 10^{23} \text{ particles/mol} \\ 6.02214199 \times 10^{26} \text{ particles/kmol.} \end{cases} \tag{2.7}$$

For example, we can use the Avogadro constant to determine the mass of a single ^{12}C atom:

$$M_{^{12}C} = \frac{M(1 \text{ mol } ^{12}C)}{\mathcal{N}_{AV} N_{^{12}C}}$$

$$= \frac{0.012}{6.02214199 \times 10^{23}(1)} = 1.9926465 \times 10^{-26}$$

$$[=] \frac{\text{kg}}{(\text{atom/mol})\text{mol}} = \frac{\text{kg}}{\text{atom}}.$$

Although not an SI unit, one-twelfth of the mass of a single ^{12}C atom is sometimes used as a mass standard and is referred to as the **unified atomic mass unit**, defined as

$$1 \, m_u \equiv (1/12) \, M_{^{12}C} = 1.66053873 \times 10^{-27} \text{ kg.}$$

Volume

The familiar property, **volume**, is formally defined as the amount of space occupied in three-dimensional space. The SI unit of volume is cubic meters (m^3).

Density

Consider the small volume $\Delta V \, (= \Delta x \Delta y \Delta z)$ as shown in Fig. 2.1. We formally define the density to be the ratio of the mass of this element to the volume of the element, under the condition that the size of the element shrinks to the continuum limit, that is,

$$\rho \equiv \lim_{\Delta V \to V_{continuum}} \frac{\Delta M}{\Delta V} \quad [=] \text{ kg/m}^3. \tag{2.8}$$

What is meant by the **continuum limit** is that the volume is very small, but yet sufficiently large so that the number of molecules within the volume is essentially constant and unaffected by any statistical fluctuations. For a volume smaller than the continuum limit, the number of molecules within the

FIGURE 2.1

A finite volume element shrinks to the continuum limit to define macroscopic properties. Volumes smaller than the continuum limit experience statistical fluctuations in properties as molecules enter and exit the volume.

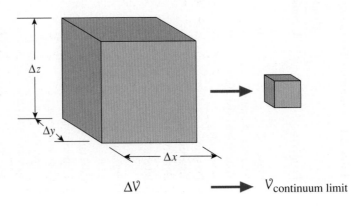

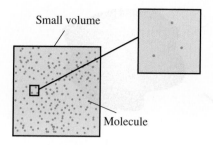

Small volume

Molecule

Table 2.2 Size of Gas Volume Containing n Molecules at 25°C and 1 atm

Number of Molecules (n)	Volume (mm³)	Size of Equivalent Cube (mm)
10^5	4.06×10^{-12}	0.00016
10^6	4.06×10^{-11}	0.0003
10^{10}	4.06×10^{-7}	0.0074
10^{12}	4.06×10^{-5}	0.0344

volume fluctuates with time as molecules randomly enter and exit the volume. As an example, consider a volume such that, on average, only two molecules are present. Because the volume is so small, the number of molecules within may fluctuate wildly, with three or more molecules present at some times, and one or none at other times. In this situation, the volume is below the continuum limit. In most practical systems, however, the continuum limit is quite small, and the density, and other thermodynamic properties, can be considered to be smooth functions in space and time. Table 2.2 provides a quantitative basis for this statement.

Specific Volume

The **specific volume** is the inverse of the density, that is,

$$v \equiv \frac{1}{\rho} \quad [=] \text{ m}^3/\text{kg}. \tag{2.9}$$

Physically it is interpreted as the amount of volume per unit mass associated with a volume at the continuum limit. The specific volume is most frequently used in thermodynamic applications, whereas the density is more commonly used in fluid mechanics and heat transfer. You should be comfortable with both properties and immediately recognize their inverse relationship.

Pressure

For a fluid (liquid or gaseous) system, the pressure is defined as the normal force exerted by the fluid on a solid surface or a neighboring fluid element, per unit area, as the area shrinks to the continuum limit, that is,

$$P \equiv \lim_{\Delta A \to A_{continuum}} \frac{F_{normal}}{\Delta A}. \tag{2.10}$$

This definition assumes that the fluid is in a state of equilibrium at rest. It is important to note that the pressure is a scalar quantity, having no direction associated with it. Figure 2.2 illustrates this concept showing that the pressure at a point[3] is independent of orientation. That the pressure force is always directed normal to a surface (real or imaginary) is a consequence of a fluid being unable to sustain any tangential force without movement. If a tangential force is present, the fluid layers simply slip over one another. Chapter 6 discusses pressure forces and their important role in understanding the flow of fluids.

---

[3] Throughout this book, the idea of a *point* is interpreted in light of the continuum limit, that is, a point is of some small dimension rather than of zero dimension required by the mathematical definition of a point.

FIGURE 2.2
Pressure at a point is a scalar quantity independent of orientation. Regardless of the orientation of ΔA, application of the defining relationship (Eq. 2.10) results in the same value of the pressure.

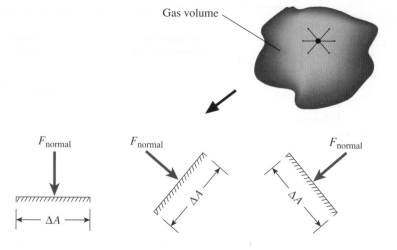

From a microscopic (molecular) point of view, the pressure exerted by a gas on the walls of its container is a measure of the rate at which the momentum of the molecules colliding with the wall is changed.

The SI unit for pressure is a *pascal*, defined by

$$P \; [=] \; \text{Pa} \equiv 1 \; \text{N/m}^2. \tag{2.11a}$$

Since one pascal is typically a small number in engineering applications, multiples of 10^3 and 10^6 are employed that result in the usage of kilopascal, kPa (10^3 Pa), and megapascal, MPa (10^6 Pa). Also commonly used is the **bar**, which is defined as

$$1 \; \text{bar} \equiv 10^5 \; \text{Pa}. \tag{2.11b}$$

Pressure is also frequently expressed in terms of a standard atmosphere:

$$1 \; \text{standard atmosphere (atm)} \equiv 101,325 \; \text{Pa}. \tag{2.11c}$$

As a result of some practical devices measuring pressures relative to the local atmospheric pressure, we distinguish between **gage pressure** and **absolute pressure**. Gage pressure is defined as

$$P_{\text{gage}} \equiv P_{\text{abs}} - P_{\text{atm, abs}}, \tag{2.12}$$

where the absolute pressure P_{abs} is that as defined in Eq. 2.10. In a perfectly evacuated space, the absolute pressure is zero. Figure 2.3 graphically illustrates the relationship between gage and absolute pressures. The term *vacuum* or *vacuum pressure* is also employed in engineering applications (and leads to confusion if one is not careful) and is defined as

$$P_{\text{vacuum}} = P_{\text{atm, abs}} - P_{\text{abs}}. \tag{2.13}$$

This relationship is also illustrated in Fig. 2.3.

In addition to SI units, many other units for pressure are commonly employed. Most of these units originate from the application of a particular measurement method. For example, the use of manometers (as discussed in Chapter 6) results in pressures expressed in inches of water or millimeters of mercury. American (or British) customary units (pounds-force per square inch or *psi*) are frequently appended with a "g" or an "a" to indicate gage or absolute pressures, respectively (i.e., *psig* and *psia)*. Conversion factors for common pressure units are provided at the front of this book.

FIGURE 2.3

Absolute pressure is zero in a perfectly evacuated space; gage pressure is measured relative to the local atmospheric pressure.

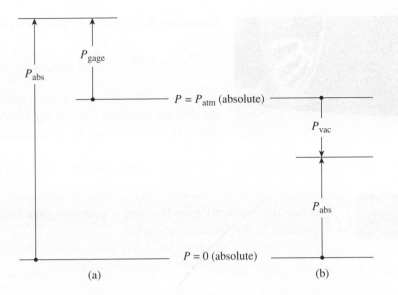

(a)

(b)

Example 2.1

A pressure gage is used to measure the inflation pressure of a tire. The gage reads 35 psig in State College, PA, when the barometric pressure is 28.5 in of mercury. What is the absolute pressure in the tire in kPa and psia?

Solution

Known $P_{\text{tire, g}}, P_{\text{atm, abs}}$

Find $P_{\text{tire, abs}}$

Sketch

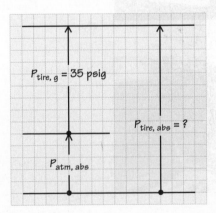

Analysis From the sketch and from Eq. 2.12, we know that

$$P_{\text{tire, abs}} = P_{\text{tire, g}} + P_{\text{atm, abs}}.$$

We need only to deal with the mixed units given to apply this relationship. Applying the conversion factor from the front of this book to express the atmospheric pressure in units of psia yields

$$P_{\text{atm, abs}} = (28.5 \text{ in Hg}) \left[\frac{14.70 \text{ psia}}{29.92 \text{ in Hg}} \right] = 14.0 \text{ psia}.$$

Thus,

$$P_{\text{tire, abs}} = 35 + 14.0 = 49 \text{ psia}.$$

Converting this result to units of kPa yields

$$P_{tire, abs} = 49 \text{ lb}_f/\text{in}^2 \left[\frac{1 \text{ Pa}}{1.4504 \times 10^{-4} \text{ lb}_f/\text{in}^2} \right] \left[\frac{1 \text{ kPa}}{1000 \text{ Pa}} \right] = 337.8 \text{ kPa}.$$

Comments   Note the use of three different units to express pressure: Pa (or kPa), psi (or lb_f/in^2), and in Hg. Other commonly used units are mm Hg and in H_2O. You should be comfortable working with any of these. Note also the usage *psia* and *psig* to denote *absolute* and *gage* pressures, respectively, when working with lb_f/in^2 units.

Self Test 2.1 **A car tire suddenly goes flat and a pressure gage indicates zero psig. Is the absolute pressure in the tire also zero psia?**

(Answer: No. A zero gage reading indicates the absolute pressure in the tire is the atmospheric pressure.)

Example 2.2   SI Engine Application

A vintage automobile has an intake manifold vacuum gage in the dashboard instrument cluster. Cruising at 20.1 m/s (45 mph), the gage reads 14 in Hg vacuum. If the local atmospheric pressure is 99.5 kPa, what is the absolute intake manifold pressure in kPa?

Solution

Given $P_{man, vac}, P_{atm, abs}$

Find $P_{man, abs}$

Sketch

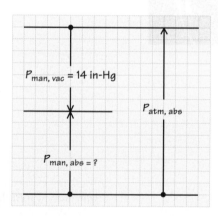

Analysis   From the sketch and a rearrangement of Eq. 2.13, we find the intake manifold absolute pressure to be

$$P_{man, abs} = P_{atm, abs} - P_{man, vac}.$$

Substituting numerical values and converting units yields

$$P_{man, abs} = 99.5 \text{ kPa} - (14 \text{ in Hg}) \left[\frac{1 \text{ atm}}{29.92 \text{ in Hg}} \right] \left[\frac{101.325 \text{ kPa}}{\text{atm}} \right]$$

$$= 99.5 \text{ kPa} - 47.4 \text{ kPa} = 52.1 \text{ kPa}$$

Comments The pressure drops across the throttle plate of an SI engine (see Chapter 11). This lower pressure, in turn, results in a decreased air density and a smaller quantity of air entering the cylinder than would occur without the throttle. The throttle thus controls the power delivered by the engine.

Self Test
2.2

A technician performs a compression test on a vehicle engine and finds that the maximum pressure in one cylinder is 105 psig, while the minimum pressure in another is 85 psig. What is the absolute pressure difference between the two cylinders?

(Answer: 20 psia)

Temperature

Like mass, length, and time, temperature is a fundamental dimension and, as such, eludes a simple and concise definition. Nevertheless, we all have some experiential notion of temperature when we say that some object is hotter than another, that is, that some object has a greater temperature than another. From a macroscopic point of view, we define temperature as that property that is shared by two systems, initially at different states, after they have been placed in thermal contact and allowed to come to thermal equilibrium. Although this definition may not be very satisfying, it is the best we can do from a macroscopic viewpoint. For the special case of an ideal gas, the microscopic (molecular) point of view may be somewhat more satisfying: Here the temperature is directly proportional to the square of the mean molecular speed. Higher temperature means faster moving molecules.

The basis for practical temperature measurement is the zeroth law of thermodynamics.[4] The **zeroth law of thermodynamics** is stated as follows:

> **Two systems that are each in thermal equilibrium with a third system are in thermal equilibrium with each other.**

Alternatively, the zeroth law can be expressed explicitly in terms of temperature:

> **When two systems have equality of temperature with a third system, they in turn have equality of temperature with each other.**

This law forms the basis for thermometry. A thermometer measures the same property, temperature, independent of the nature of the system subject to the measurement. A temperature of 20°C measured for a block of steel means the same thing as 20°C measured for a container of water. Putting the 20°C steel block in the 20°C water results in no temperature change to either.

As a result of the zeroth law, a *practical temperature scale* can be based on a *thermometric* substance. Such a substance expands as its temperature increases; mercury is a thermometric substance. The height of the mercury column in a glass tube can be calibrated against standard *fixed points* of reference. For example, the original Celsius scale (i.e., prior to 1954) defines

---

[4] This law was established after the first and second laws of thermodynamics; however, since it expresses a concept logically preceding the other two, it has been designated the zeroth law.

Table 2.3 Temperature Scales

Temperature Scale	Units*	Relation to Other Scales
Celsius	degree Celsius (°C)	$T\,(°C) = T\,(K) - 273.15$ $T\,(°C) = \dfrac{5}{9}\,[T\,(F) - 32]$
Kelvin	kelvin (K)	$T\,(K) = T\,(°C) + 273.15$ $T\,(K) = \dfrac{5}{9}\,T(R)$
Fahrenheit	degree Fahrenheit (F)	$T\,(F) = T\,(R) - 459.67$ $T\,(F) = \dfrac{9}{5}\,T\,(°C) + 32$
Rankine	degree Rankine (R)	$T\,(R) = T\,(F) + 459.67$ $T\,(R) = \dfrac{9}{5}\,T\,(K)$

* Note that capital letters are used to refer to the units for each scale. The degree symbol (°), however, is used only with the Celsius unit to avoid confusion with the coulomb. Note also that the SI Kelvin scale unit is the kelvin; thus, we say that a temperature, for example, is 100 kelvins (100 K), not 100 degrees Kelvin.

0°C to be the temperature at the **ice point**[5] and 100°C to be the temperature at the **steam point**.[6] The modern Celsius scale assigns a temperature of 0.01°C to the **triple point**[7] of water and the size of a single degree equal to that from the absolute, or Kelvin, temperature scale, as discussed in Chapter 7. With the adoption of the International Temperature Scale of 1990 (ITS-90), the ice point is still 0°C, but the steam point is now 99.974°C. For practical purposes, the original and modern Celsius scales are identical.

Four temperature scales are in common use today: the Celsius scale and its absolute counterpart, the Kelvin scale, and the Fahrenheit scale and its absolute counterpart, the Rankine scale. Both absolute scales start at absolute zero, the lowest temperature possible. The conversions among these scales are shown in Table 2.3.

[5] The **ice point** is the temperature at which an ice and water mixture is in equilibrium with water vapor–saturated air at one atmosphere.
[6] The **steam point** is the temperature at which steam and water are in equilibrium at one atmosphere.
[7] The **triple point** is the temperature at which ice, liquid water, and steam all coexist in equilibrium.

Example 2.3

On a hot day in Boston, a high of 97 degrees Fahrenheit is reported on the nightly news. What is the temperature in units of °C, K, and R?

Solution

Known $T(F)$

Find $T(°C), T(K), T(R)$

Analysis   We apply the temperature-scale conversions provided in Table 2.3 as follows:

$$T(°C) = \frac{5}{9}[T(F) - 32]$$

$$= \frac{5}{9}(97 - 32) = 36.1°C,$$

$$T(R) = T(F) + 459.67$$
$$= 97 + 459.67 = 556.7 \text{ R},$$

$$T(K) = \frac{5}{9}T(R)$$

$$= \frac{5}{9}(556.7) = 309.3 \text{ K}.$$

Comments   Except for the Rankine scale, you are most likely familiar with the conversions in Table 2.3. Note that the size of the temperature unit is identical for the Fahrenheit and Rankine scales. Similarly, the Celsius and Kelvin units are of identical size, and each is 9/5 (or 1.8) times the size of the Fahrenheit or Rankine unit.

Self Test 2.3 ☑ **On the same hot day in Boston, the air conditioning keeps your room at 68 degrees Fahrenheit. Find the temperature difference between the inside and the outside air in (a) R, (b) °C, and (c) K.**

(Answer: (a) 29 R, (b) 16.1°C, (c) 16.1 K)

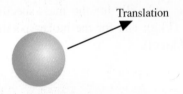

(a) Monatomic species

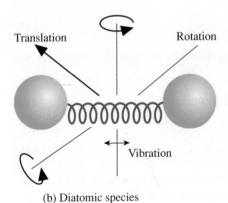

(b) Diatomic species

FIGURE 2.4

(a) The internal energy of a monatomic species consists only of translational (kinetic) energy.
(b) A diatomic species internal energy results from translation together with energy from vibration (potential and kinetic) and rotation (kinetic).

2.2b Properties Related to the First Law and Calorific Equation of State

Internal Energy

In this section, we introduce the thermodynamic property internal energy. Further discussion of internal energy is presented in Chapter 4, which focuses on the many ways that energy is stored and transferred.

Internal energy has its origins with the microscopic nature of matter; specifically, we define **internal energy** as the energy associated with the motions of the microscopic particles (atoms, molecules, electrons, etc.) comprising a macroscopic system. For simple monatomic gases (e.g., helium and argon) internal energy is associated only with the translational kinetic energy of the atoms (Fig. 2.4a). If we assume that a gas can be modeled as a collection of point-mass hard spheres that collide elastically, the translational kinetic energy associated with *n* particles is

$$U_{\text{trans}} = n\frac{1}{2}M_{\text{molec}}\overline{v^2}, \qquad (2.14)$$

where $\overline{v^2}$ is the mean-square molecular speed. Using kinetic theory (see, for example, Ref. [2]), the translational kinetic energy can be related to temperature as

$$U_{\text{trans}} = n\frac{3}{2}k_{\text{B}}T, \qquad (2.15)$$

where k_B is the Boltzmann constant,

$$k_B \equiv 1.3806503 \times 10^{-23} \text{ J/K} \cdot \text{molecule},$$

and T is the absolute temperature in kelvins. [By comparing Eqs. 2.14 and 2.15, we see the previously mentioned microscopic interpretation of temperature, i.e., $T \equiv M_{\text{molec}}\overline{v^2}/(3k_B)$.]

For molecules more complex than single atoms, internal energy is stored in vibrating molecular bonds and rotation of the molecule about two or more axes, in addition to the translational kinetic energy. Figure 2.4b illustrates this model of a diatomic species. In general, the internal energy is expressed

$$U = U_{\text{trans}} + U_{\text{vib}} + U_{\text{rot}}, \tag{2.16}$$

where U_{vib} is the vibrational kinetic and potential energy, and U_{rot} is the rotational kinetic energy. The amount of energy that is stored in each mode varies with temperature and is described by quantum mechanics. One of the fundamental postulates of quantum theory is that energy is quantized; that is, energy storage is modeled by discrete bits rather than continuous functions. The translational kinetic energy states are very close together such that, for practical purposes, quantum states need not be considered and the continuum result, Eq. 2.15, is a useful model. For vibrational and rotational states, however, quantum behavior is important. We will see the effects of this later in our discussion of specific heats.

Another form of internal energy is that associated with chemical bonds and their rearrangements during chemical reaction. Similarly, internal energy is associated with nuclear bonds. We will address the topic of chemical energy storage in a later section of this chapter; nuclear energy storage, however, lies beyond our scope.

The SI unit for internal energy is the joule (J); for the mass-specific internal energy, it is joules per kilogram (J/kg); and for the molar-specific internal energy, it is joules per kilomole (J/kmol).

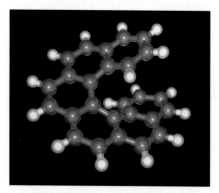

For reacting systems, chemical bonds make an important contribution to the system internal energy.

Enthalpy

Enthalpy is a useful property defined by the following combination of more common properties:

$$H \equiv U + P\mathcal{V}. \tag{2.17}$$

On a mass-specific basis, the enthalpy involves the specific volume or the density, that is,

$$h \equiv u + Pv \tag{2.18a}$$

or

$$h \equiv u + P/\rho. \tag{2.18b}$$

The enthalpy has the same units as internal energy (i.e., J or J/kg). Molar-specific enthalpies are obtained by the application of Eq. 2.3.

The usefulness of enthalpy will become clear during our discussion of the first law of thermodynamics (the principle of energy conservation) in Chapter 5. There we will see that the combination of properties, $u + Pv$, arises naturally in analyzing systems at constant pressure and in analyzing control volumes. In the former, the P–v term results from expansion and/or compression work; for the latter, the P–v term is associated with the work needed to push the fluid into or

> Enthalpy first appears in conservation of energy for systems in Example 5.4.

> Conservation of energy for control volumes (Eq. 5.63) uses enthalpy to replace the combination of flow work (see Chapter 4) and internal energy.

out of the control volume, that is, flow work. Further discussion of internal energy and enthalpy is also given later in the present chapter.

Specific Heats and Specific-Heat Ratio

Here we deal with two intensive properties,

$$c_v \equiv \text{constant-volume specific heat}$$

and

$$c_p \equiv \text{constant-pressure specific heat.}$$

These properties mathematically relate to the specific internal energy and enthalpy, respectively, as follows:

$$c_v \equiv \left(\frac{\partial u}{\partial T} \right)_v \tag{2.19a}$$

and

$$c_p \equiv \left(\frac{\partial h}{\partial T} \right)_p. \tag{2.19b}$$

Similar defining relationships relate molar-specific heats and molar-specific internal energy and enthalpy. Physically, the constant-volume specific heat is the slope of the internal energy-versus-temperature curve for a substance undergoing a process conducted at constant volume. Similarly, the constant-pressure specific heat is the slope of the enthalpy-versus-temperature curve for a substance undergoing a process conducted at constant pressure. These ideas are illustrated in Fig. 2.5. It is important to note that, although the definitions of these properties involve constant-volume and constant-pressure processes, c_v and c_p can be used in the description of *any* process regardless of whether or not the volume (or pressure) is held constant.

For solids and liquids, specific heats generally increase with temperature, essentially uninfluenced by pressure. A notable exception to this is mercury, which exhibits a decreasing constant-pressure specific heat with temperature. Values of specific heats for selected liquids and solids are presented in Appendices G and I.

For both real (nonideal) and ideal gases, the specific heats c_v and c_p are functions of temperature. The specific heats of nonideal gases also possess a pressure dependence. For gases, the temperature dependence of c_v and c_p is a consequence of the internal energy of a molecule consisting of three components—translational, vibrational, and rotational—and the fact that the vibrational and rotational energy storage modes become increasingly active as temperature increases, as described by quantum theory. As discussed previously, Fig. 2.4 schematically illustrates these three energy storage modes by contrasting a monatomic species, whose internal energy consists solely of translational kinetic energy, and a diatomic molecule, which stores energy in a vibrating chemical bond, represented as a spring between the two nuclei, and by rotation about two orthogonal axes, as well as possessing kinetic energy from translation. With these simple models (Fig. 2.4), we expect the specific heats of diatomic molecules to be greater than those of monatomic species, which is indeed true. In general, the more complex the molecule, the greater its molar specific heat. This can be seen clearly in Fig. 2.6, where molar-specific heats for a number of gases are shown as functions of

FIGURE 2.5
The constant-volume specific heat c_v is defined as the slope of u versus T for a constant-volume process (top). Similarly, c_p is the slope of h versus T for a constant-pressure process (bottom). Generally, both c_v and c_p are functions of temperature, as suggested by these graphs.

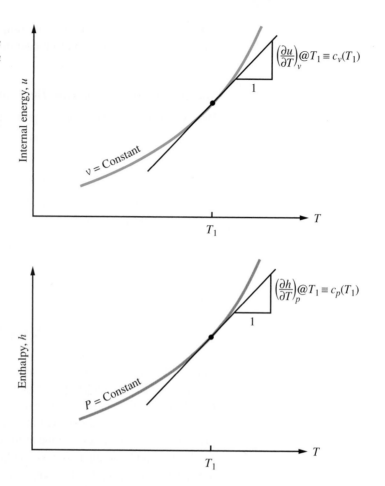

FIGURE 2.6
Molar constant-pressure specific heats as functions of temperature for monatomic (H, N, and O), diatomic (CO, H_2, and O_2), and triatomic (CO_2, H_2O, and NO_2) species. Values are from Appendix B.

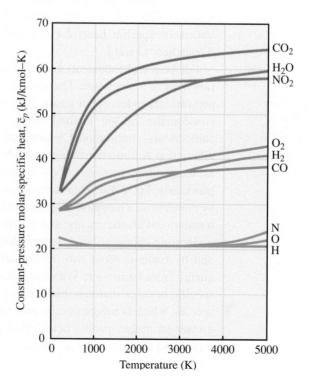

temperature. As a group, the triatomics have the greatest specific heats, followed by the diatomics, and lastly, the monatomics. Note that the triatomic molecules are also more temperature dependent than the diatomics, a consequence of the greater number of vibrational and rotational modes that are available to become activated as temperature is increased. In comparison, the monatomic species have nearly constant specific heats over a wide range of temperatures; in fact, the specific heat of monatomic hydrogen is constant ($\bar{c}_p = 20.786$ kJ/kmol·K) from 200 K to 5000 K.

Constant-pressure molar-specific heats are tabulated as a function of temperature for various ideal-gas species in Tables B.1 to B.12 in Appendix B. Also provided in Appendix B are the curve-fit coefficients, taken from the Chemkin thermodynamic database [3], which were used to generate the tables. These coefficients can be easily used with spreadsheet software to obtain $\bar{c}_p$ values at any temperature within the given temperature range.

Values of c_v and c_p for a number of substances are also available from the National Institute of Standards and Technology (NIST) online database [11] and the NIST12 software provided with this book. We discuss the use of these important resources later in this chapter.

The ratio of specific heats, γ, is another commonly used property[8] and is defined by

$$\gamma \equiv \frac{c_p}{c_v} = \frac{\bar{c}_p}{\bar{c}_v}.$$ (2.20)

[8] The specific-heat ratio is frequently denoted by k or r, as well as gamma (γ), our choice here. We will use k and r to represent the thermal conductivity and radial coordinate, respectively.

Example 2.4

Six liquid hydrogen-fueled engines power the second stage of this Saturn rocket. Courtesy of NASA.

Compare the values of the constant-pressure specific heats for hydrogen (H_2) and carbon monoxide (CO) at 3000 K using the ideal-gas molar-specific values from the tables in Appendix B. How does this comparison change if mass-specific values are used?

Solution

Known H_2 and CO at T

Find $\bar{c}_{p,H_2}$, $\bar{c}_{p,CO}$, c_{p,H_2}, $c_{p,CO}$

Assumption

Ideal-gas behavior

Analysis   To answer the first question requires only a simple table look-up. Molar constant-pressure specific heats found in Table B.3 for H_2 and in Table B.1 for CO are as follows:

$$\bar{c}_{p,H_2}\ (T = 3000\ \text{K}) = 37.112\ \text{kJ/kmol·K},$$

$$\bar{c}_{p,CO}\ (T = 3000\ \text{K}) = 37.213\ \text{kJ/kmol·K}.$$

The difference between these values is 0.101 kJ/kmol·K, or approximately 0.3%.

Using the molecular weights of H_2 and CO found in Tables B.3 and B.1, we can calculate the mass-based constant-pressure specific heats using Eq. 2.4 as follows:

$$c_{p,H_2} = \bar{c}_{p,H_2}/\mathcal{M}_{H_2}$$

$$= \frac{37.112 \text{ kJ/kmol} \cdot \text{K}}{2.016 \text{ kg/kmol}}$$

$$= 18.409 \text{ kJ/kg} \cdot \text{K}$$

and

$$c_{p,CO} = \bar{c}_{p,CO}/\mathcal{M}_{CO}$$

$$= \frac{37.213 \text{ kJ/kmol} \cdot \text{K}}{28.010 \text{ kg/kmol}}$$

$$= 1.329 \text{ kJ/kg} \cdot \text{K}.$$

Comments   We first note that, on a molar basis, the specific heats of H_2 and CO are nearly identical. This result is consistent with Fig. 2.6, where we see that the molar-specific heats are similar for the three diatomic species. On a mass basis, however, the constant-pressure specific heat of H_2 is almost 14 times greater than that of CO, which results from the molecular weight of CO being approximately 14 times that of H_2.

  Self Test 2.4   Calculate the specific heat ratios for (a) H_2, (b) CO, and (c) air using the data from Table E.1 in Appendix E.

(*Answer: (a) 1.402, (b) 1.398, (c) 1.400*)

2.2c Properties Related to the Second Law[9]

Entropy

As we will see in Chapter 7, a thermodynamic property called **entropy** (*S*) originates from the second law of thermodynamics.[10] This property is particularly useful in determining the spontaneous direction of a process and for establishing maximum possible efficiencies, for example.

The property entropy can be interpreted from both macroscopic and microscopic (molecular) points of view. We defer presenting a precise mathematical definition of entropy from the macroscopic viewpoint until Chapter 7; the following verbal definition, however, provides some notion of what this property is all about:

> Chapter 7 revisits entropy and expands upon the discussion here. Equation 7.16 provides a formal macroscopic definition of entropy.

Entropy is a measure of the unavailability of thermal energy to do work in a closed system.

---

[9] For an introductory study of properties, this section may be skipped without any loss of continuity. This section is most useful, however, in conjunction with the study of Chapter 7.

[10] Rudolf Clausius (1822–1888) chose *entropy*, a Greek word meaning transformation, because of its root meaning and because it sounded similar to *energy*, a closely related concept [4].

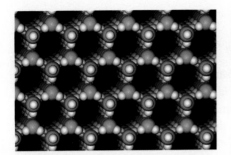

Hexagonal crystal structure of ice. The open structure causes ice to be less dense than liquid water.

Structure of liquid water.

Evaporating water molecules.

From this definition, we might imagine that two identical quantities of energy are not of equal value in producing useful work. Entropy is valuable in quantifying this usefulness of energy.

The following informal definition presents a microscopic (molecular) interpretation of entropy:

> **Entropy is a measure of the microscopic randomness associated with a closed system.**

To help understand this statement, consider the physical differences between water existing as a solid (ice) and as a vapor (steam). In a piece of ice, the individual H_2O molecules are locked in relatively rigid positions, with the individual hydrogen and oxygen atoms vibrating within well-defined domains. In contrast, in steam, the individual molecules are free to move within any containing vessel. Thus, we say that the state of the steam is more disordered than that of the ice and that the steam has a greater entropy per unit mass. It is this idea, in fact, that leads to the **third law of thermodynamics**, which states that all perfect crystals have zero entropy at a temperature of absolute zero. For the case of a perfectly ordered crystal at absolute zero, there is no motion, and there are no imperfections in the lattice; thus, there is no uncertainty about the microscopic state (because there is no disorder or randomness) and the entropy is zero. A more detailed discussion of the microscopic interpretation of entropy is presented in the appendix to this chapter.

The SI units for entropy S, mass-specific entropy s, and molar-specific entropy $\bar{s}$, are J/K, J/kg·K, and J/kmol·K, respectively. Tabulated values of entropies for ideal gases, air, and H_2O are found in Appendices B, C and D, respectively. Entropies for selected substances are also available from the NIST software and online database [11].

Gibbs Free Energy or Gibbs Function

The **Gibbs free energy** or **Gibbs function, G**, is a composite property involving enthalpy and entropy and is defined as

$$G \equiv H - TS, \tag{2.21a}$$

and, per unit mass,

$$g \equiv h - Ts. \tag{2.21b}$$

Molar-specific quantities are obtained by the application of Eq. 2.3. The Gibbs free energy is particularly useful in defining equilibrium conditions for reacting systems at constant pressure and temperature. We will revisit this property later in this chapter in the discussion of ideal-gas mixtures; in Chapter 7 this property is prominent in the discussion of chemical equilibrium.

Helmholtz Free Energy or Helmholtz Function

The **Helmholtz free energy, A**, is also a composite property, defined similarly to the Gibbs free energy, with the internal energy replacing the enthalpy, that is,

$$A \equiv U - TS, \tag{2.22a}$$

LEVEL 3

or, per unit mass,

$$a \equiv u - Ts. \tag{2.22b}$$

Molar-specific quantities relate in the same manner as expressed by Eq. 2.22b. The Helmholtz free energy is useful in defining equilibrium conditions for reacting systems at constant volume and temperature. Although we make no use of the Helmholtz free energy in this book, you should be aware of its existence.

2.3 CONCEPT OF STATE RELATIONSHIPS

2.3a State Principle

An important concept in thermodynamics is the **state principle**:

> **In dealing with a simple compressible substance, the *thermodynamic state* is completely defined by specifying two independent intensive properties.**

The state principle allows us to define **state relationships** among the various thermodynamic properties. Before developing such state relationships, we examine what is mean by *independent properties*.

The concept of independent properties is particularly important in dealing with substances when more than one phase is present. For example, temperature and pressure are not independent properties when water (liquid) and steam (vapor) coexist. As you are well aware, water at one atmosphere boils at a specific temperature (i.e., 100°C). Increasing the pressure results in an increase in the boiling point, which is the principle upon which the pressure cooker is based. One cannot change the pressure and keep the temperature constant: A fixed relationship exists between temperature and pressure; hence, they are not independent. We will examine this concept of independence in greater detail later when we study the properties of substances that exist in multiple phases.

2.3b *P–v–T* Equations of State

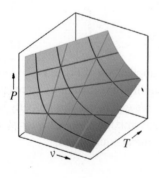

What is generally known as an **equation of state** is the mathematical relationship among the following three intensive thermodynamic properties: pressure P; specific volume v, and temperature T. The state principle allows us to determine any one of the three properties from knowledge of the other two. In its most general and abstract form, we can write the $P–v–T$ equation of state as

$$f_1(P, v, T) = 0. \tag{2.23}$$

In the following sections, we explore the explicit functions relating P, v, and T for various substances, starting with the ideal gas, a concept with which you should already have some familiarity.

2.3c Calorific Equations of State

A second type of state relationship connects energy-related thermodynamic properties to pressure, temperature, and specific volume. The state principle also applies here; thus, for a simple compressible substance, a knowledge of any two intensive properties is sufficient to determine any of the others. The most common **calorific equations of state** relate specific internal energy u to v and T, and, similarly, enthalpy h to P and T, that is,

$$f_2(u, T, v) = 0, \tag{2.24a}$$

or

$$f_3(h, T, P) = 0. \tag{2.24b}$$

These ideas are developed in more detail for various substances in the sections that follow.

2.3d Temperature–Entropy (Gibbs) Relationships

The third and final type of state relationships we consider are those that relate entropy-based properties—that is, properties relating to the second law of thermodynamics—to pressure, specific volume, and temperature. The most common relationships are of the following general form:

$$f_4(s, T, P) = 0, \tag{2.25a}$$

$$f_5(s, T, v) = 0, \tag{2.25b}$$

and

$$f_6(g, T, P) = 0. \tag{2.25c}$$

As with the other state relationships, these, too, are defined concretely in the following sections.

2.4 IDEAL GASES AS PURE SUBSTANCES

In this section, we define all of the useful state relationships for a class of pure substances known as ideal gases. We begin with the definition of an ideal gas.

2.4a Ideal Gas Definition

The following definition of an **ideal gas** is tautological in that it uses a state relationship to define what is meant by an ideal gas:

An *ideal gas* is a gas that obeys the relationship $Pv = RT$.

In this definition P and T are the absolute pressure and absolute temperature, respectively, and R is the **particular gas constant**, a physical constant. The particular gas constant depends on the molecular weight of the gas as follows:

$$R_i \equiv R_u / \mathcal{M}_i \quad [=] \text{ J/kg} \cdot \text{K}, \tag{2.26}$$

where the subscript i denotes the species of interest, and R_u is the **universal gas constant**, defined by

$$R_u \equiv 8314.472\,(15) \quad [=] \text{ J/kmol} \cdot \text{K}. \tag{2.27}$$

This definition of an ideal gas can be made more satisfying by examining what is implied from a molecular, or microscopic, point of view. Kinetic

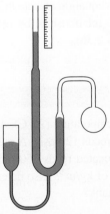

An ideal-gas thermometer consists of a sensing bulb filled with an ideal gas (right), a movable closed reservoir (left), and a liquid column (center). The height of the liquid column is directly proportional to the temperature of the gas in the bulb when the reservoir position is adjusted to maintain a fixed volume for the ideal gas.

theory predicts that $Pv = RT$, first, when the molecules comprising the system are infinitesimally small, hard, round spheres occupying negligible volume and, second, when no forces exist among these molecules except during collisions. Qualitatively, these conditions imply a gas at *low* density. What we mean by low density will be discussed in later sections.

2.4b Ideal-Gas Equation of State

Formally, the P–v–T equation of state for an ideal gas is expressed as

$$Pv = RT. \tag{2.28a}$$

Alternative forms of the ideal-gas equation of state arise in various ways. First, by recognizing that the specific volume is the reciprocal of the density ($v = 1/\rho$), we get

$$P = \rho RT. \tag{2.28b}$$

Second, expanding the definition of specific volume ($v = V/M$) yields

$$PV = MRT. \tag{2.28c}$$

Third, expressing the mass in terms of the number of moles and molecular weight of the particular gas of interest ($M = N\,\mathcal{M}_i$) yields

$$PV = NR_uT. \tag{2.28d}$$

Finally, by employing the molar specific volume $\bar{v} = v\mathcal{M}_i$, we obtain

$$P\bar{v} = R_uT. \tag{2.28e}$$

We summarize these various forms of the ideal-gas equation of state in Table 2.4 and encourage you to become familiar with these relationships by performing the various conversions on your own (see Problem 2.36).

Table 2.4 Various Forms of the Ideal-Gas Equation of State

$Pv = RT$	Eq. 2.28a
$P = \rho RT$	Eq. 2.28b
$PV = MRT$	Eq. 2.28c
$PV = NR_uT$	Eq. 2.28d
$P\bar{v} = R_uT$	Eq. 2.28e

Example 2.5

A compressed-gas cylinder contains N_2 at room temperature (25°C). A gage on the pressure regulator attached to the cylinder reads 120 psig. A mercury barometer in the room in which the cylinder is located reads 750 mm Hg. What is the density of the N_2 in the tank in units of kg/m^3? Also determine the mass of the N_2 contained in the 1.54-ft^3 steel tank?

Solution

Known $T_{N_2}, P_{N_2,g}, P_{atm}, V_{N_2}$

Find ρ_{N_2}, M_{N_2}

Assumption

Ideal-gas behavior

Analysis   To find the density of nitrogen, we apply the ideal-gas equation of state (Eq. 2.28b, Table 2.4). Before doing so we must determine the particular gas constant for N_2 and perform several unit conversions of given information.

From Eqs. 2.26 and 2.27, we find the particular gas constant,

$$R_{N_2} = \frac{R_u}{\mathcal{M}_{N_2}} = \frac{8314.47 \text{ J/kmol} \cdot \text{K}}{28.013 \text{ kg/kmol}}$$

$$= 296.8 \text{ J/kg} \cdot \text{K},$$

where the molecular weight for N_2 is calculated from the atomic weights given in the front of the book (or found directly in Table B.7).

The absolute pressure in the tank is (Eq. 2.12)

$$P_{N_2} = P_{N_2, g} + P_{\text{atm, abs}},$$

where

$$P_{N_2, g} = 120 \frac{\text{lb}_f}{\text{in}^2} \left[\frac{39.370 \text{ in}}{1 \text{ m}} \right]^2 \left[\frac{1 \text{ N}}{0.224809 \text{ lb}_f} \right]$$

$$= 827,367 \text{ Pa (gage)}$$

and

$$P_{\text{atm, abs}} = (750 \text{ mm Hg}) \left[\frac{1 \text{ atm}}{760 \text{ mm Hg}} \right] \left[\frac{101,325 \text{ Pa}}{1 \text{ atm}} \right]$$

$$= 99,992 \text{ Pa}.$$

Thus, the absolute pressure of the N_2 is

$$P_{N_2} = 827,367 \text{ Pa (gage)} + 99,992 \text{ Pa} = 927,359 \text{ Pa},$$

which rounds off to

$$P_{N_2} = 927,000 \text{ Pa}.$$

The absolute temperature of the N_2 is

$$T_{N_2} = 25°C + 273.15 = 298.15 \text{ K}.$$

To obtain the density, we now apply the ideal-gas equation of state (Eq. 2.28b)

$$\rho_{N_2} = \frac{P_{N_2}}{R_{N_2} T_{N_2}}$$

$$= \frac{927,000}{296.8 \, (298.15)}$$

$$= 10.5$$

$$[=] \frac{\text{Pa} \left[\dfrac{1 \text{ N/m}^2}{\text{Pa}} \right]}{\dfrac{\text{J}}{\text{kg} \cdot \text{K}} \left[\dfrac{1 \text{ N} \cdot \text{m}}{\text{J}} \right] \text{K}} = \text{kg/m}^3.$$

Note that we have set aside the units and unit conversions to assure their proper treatment. Unit conversion factors are always enclosed in square brackets. We obtain the mass from the definition of density (Eq. 2.8)

$$\rho_{N_2} \equiv \frac{M_{N_2}}{V_{N_2}},$$

or

$$M_{N_2} = \rho_{N_2} V_{N_2}.$$

The tank volume is

$$V_{N_2} = 1.54 \text{ ft}^3 \left[\frac{1 \text{ m}}{3.2808 \text{ ft}} \right]^3 = 0.0436 \text{ m}^3.$$

Thus,

$$M_{N_2} = 10.5 \frac{\text{kg}}{\text{m}^3} 0.0436 \text{ m}^3 = 0.458 \text{ kg}.$$

Comments   Note that, although the application of the ideal-gas law to find the density is straightforward, unit conversions and calculations of absolute pressures and temperatures make the calculation nontrivial.

Self Test 2.5   **The valve of the tank in Example 2.5 is slowly opened and 0.1 kg of N₂ escapes. Calculate the density of the remaining N₂ and find the final gage pressure in the tank assuming the temperature remains at 25°C.**

(Answer: 8.2 kg/m³, 626.6 kPa)

Example 2.6

It is a cold, sunny day in Merrill, WI. The temperature is -10 F, the barometric pressure is 100 kPa, and the humidity is nil. Estimate the outside air density. Also estimate the molar density, N/V, of the air.

Solution

Known T_{air}, P_{air}

Find ρ_{air}

Assumptions

  i. Air can be treated as a pure substance.
  ii. Air can be treated as an ideal gas.
  iii. Air is dry.

Analysis   With these assumptions, we use the data in Appendix C together with the ideal-gas equation of state (Eq. 2.28b) to find the air density. First, we convert the temperature to SI absolute units:

$$T_{air}(\text{K}) = \frac{5}{9}(-10 + 459.67) = 249.8 \text{ K}.$$

The density is thus

$$\rho_{air} = \frac{P_{air}}{R_{air}T_{air}} = \frac{100,000 \text{ Pa}}{287.0 \text{ J/kg} \cdot \text{K} \,(249.8 \text{ K})} = 1.395 \text{ kg/m}^3,$$

where R_{air}, the particular gas constant for air, was obtained from Table C.1 in Appendix C. The treatment of units in this calculation is the same as detailed in the previous example.

The molar density is the number of moles per unit volume. This quantity is calculated by dividing the mass density (ρ_{air}) by the apparent molecular weight of the air, that is,

$$N_{air}/V_{air} = \rho_{air}/\mathcal{M}_{air} = \frac{1.395 \text{ kg/m}^3}{28.97 \text{ kg/kmol}} = 0.048 \text{ kmol/m}^3.$$

Comments   The primary purpose of this example is to introduce the approximation of treating air, a mixture of gases (see Table C.1 for the composition of dry air), as a simple substance that behaves as an ideal gas. Note the introduction of the apparent molecular weight, $\mathcal{M}_{air} = 28.97$ kg/kmol, and the particular gas constant, $R_{air} = R_u/\mathcal{M}_{air} = 287.0$ J/kg·K. Ideal-gas thermodynamic properties for dry air are also tabulated in Appendix C. In our study of air conditioning (Chapter 12), we investigate the influence of moisture in air.

Self Test 2.6 ✅ **Calculate the mass of the air in an uninsulated, unheated 10 ft × 15 ft × 12 ft garage on this same cold day.**

(Answer: 71.1 kg)

2.4c Processes in *P–v–T* Space

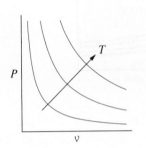

Plotting thermodynamic processes on *P–v* or other thermodynamic property coordinates is very useful in analyzing many thermal systems. In this section, we introduce this topic by illustrating common processes on *P–v* and *T–v* coordinates, restricting our attention to ideal gases. Later in this chapter, we add the complexity of a phase change.

We begin by examining *P–v* coordinates. Rearranging the ideal-gas equation of state (Eq. 2.28a) to a form in which *P* is a function of *v* yields the hyperbolic relationship

$$P = (RT)\frac{1}{v}. \tag{2.29}$$

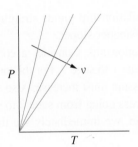

By considering the temperature to be a fixed parameter, Eq. 2.29 can be used to create a family of hyperbolas for various values of *T*, as shown in Fig. 2.7. Increasing temperature moves the isotherms further from the origin.

The usefulness of graphs such as Fig. 2.7 is that one can immediately visualize how properties must vary for a particular thermodynamic process. For example, consider the constant-pressure expansion process shown in Fig. 2.7, where the initial and final states are designated as points 1 and 2, respectively. Knowing the arrangement of constant-temperature lines allows us to see that the temperature must increase in the process 1–2. For the values given, the temperature increases from 300 to 600 K. The important point here, however, is not this quantitative result, but the qualitative information available from plotting processes on *P–v* coordinates. Also shown in Fig. 2.7 is a constant-volume process (assuming that we are dealing with a system of fixed mass). In going from state 2 to state 3 at constant volume, we immediately see that both the temperature and pressure must fall.

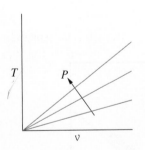

In choosing a pair of thermodynamic coordinates to draw a graph, one usually selects those that allow given constant-property processes to be shown as straight lines. For example, *P–v* coordinates are the natural choice for systems involving either constant-pressure or constant- (specific) volume processes. If, however, one is interested in a constant-temperature process, then *T–v* coordinates may be more useful. In this case, the ideal-gas equation of state can be rearranged to yield

$$T = \left(\frac{P}{R}\right)v. \tag{2.30}$$

FIGURE 2.7

Constant-pressure (1–2) and constant-volume (2–3) processes are shown on P−v coordinates for an ideal gas (N₂). Lines of constant temperature are hyperbolic, following the ideal-gas equation of state, P = (RT)/v.

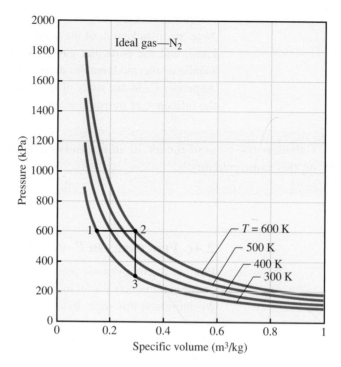

Treating pressure as a fixed parameter, this relationship yields straight lines with slopes of P/R. Higher pressures result in steeper slopes.

Figure 2.8 illustrates the ideal-gas T–v relationship. Consider a constant-temperature compression process going from state 1 to state 2. For this process, we immediately see from the graph that the pressure must increase. Also shown on Fig. 2.8 is a constant- (specific) volume process going from state 2 to state 3 for conditions of decreasing temperature. Again, we immediately see that the pressure falls during this process.

FIGURE 2.8

Constant-temperature (1–2) and constant-volume (2–3) processes are shown on T−v coordinates for an ideal gas (N₂). Lines of constant pressure are straight lines, following the ideal-gas equation of state, T = (P/R)v.

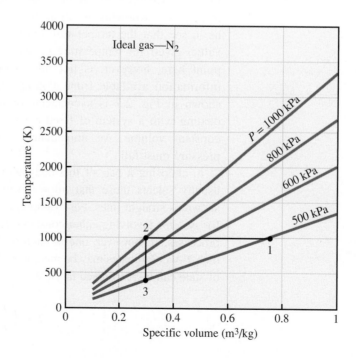

Plotting processes on thermodynamic coordinates develops understanding and aids in problem solving. Whenever possible, we will use such diagrams in examples throughout the book. Also, many homework problems are designed to foster development of your ability to draw and use such plots.

Example 2.7

An ideal gas system undergoes a thermodynamic cycle composed of the following processes:

1–2: constant-pressure expansion,

2–3: constant-temperature expansion,

3–4: constant-volume return to the state-1 temperature, and

4–1: constant-temperature compression.

Sketch these processes (a) on $P–v$ coordinates and (b) on $T–v$ coordinates.

Solution

The sequences of processes are shown on the sketches.

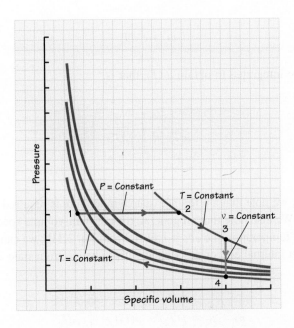
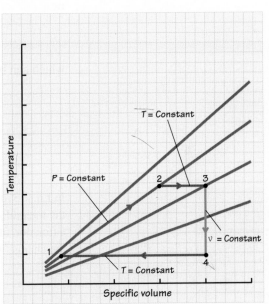

Comments To develop skill in making such plots, the reader should redraw the requested sketches without reference to the solutions given. Note that the sequence of processes 1–2–3–4–1 constitutes a thermodynamic cycle.

Self Test 2.7

✓ **Looking at the sketches of Example 2.7, state whether the pressure P, temperature T, and specific volume v increase, decrease, or remain the same for each of the four processes.**

(Answer: 1–2: P constant, T increases, v increases; 2–3: P decreases, T constant, v increases; 3–4: P decreases, T decreases, v constant; 4–1: P increases, T constant, v decreases)

2.4d Ideal-Gas Calorific Equations of State

From kinetic theory we predict that the internal energy of an ideal gas will be a function of temperature only. That the internal energy is independent of pressure follows from the neglect of any intermolecular forces in the model of an ideal gas. In real gases, molecules do exhibit repulsive and attractive forces that result in a pressure dependence of the internal energy. As in our discussion of the P–v–T equation of state, the ideal gas approximation, however, is quite accurate at sufficiently low densities, and the result that $u = u\,(T$ only$)$ is quite useful.

Ideal-gas specific internal energies can be obtained from experimentally or theoretically determined values of the constant-volume specific heat. Starting with the general definition (Eq. 2.19a),

$$c_v = \left(\frac{\partial u}{\partial T}\right)_v,$$

we recognize that the partial derivative becomes an ordinary derivative when $u = u\,(T$ only$)$; thus

$$c_v = \frac{du}{dT},\tag{2.31a}$$

or

$$du = c_v\,dT,\tag{2.31b}$$

which can be integrated to obtain $u\,(T)$, that is,

$$u(T) = \int_{T_{\text{ref}}}^{T} c_v\,dT.\tag{2.31c}$$

In Eq. 2.31c, we note that a reference-state temperature is required to evaluate the integral. We also note that, in general, the constant-volume specific heat is a function of temperature [i.e., $c_v = c_v(T)$]. From Eq. 2.31b, we can easily find the change in internal energy associated with a change from state 1 to state 2:

$$u_2 - u_1 = u(T_2) - u(T_1) = \int_{T_1}^{T_2} c_v\,dT.\tag{2.31d}$$

If the temperature difference between the two states is not too large, the constant-volume specific heat can be treated as a constant, $c_{v,\text{avg}}$; thus,

$$u_2 - u_1 = c_{v,\text{avg}}(T_2 - T_1).\tag{2.31e}$$

We will return to Eq. 2.31 after discussing the calorific equation of state involving enthalpy.

To obtain the h–T–P calorific equation of state for an ideal gas, we first show that the enthalpy of an ideal gas, like the internal energy, is a function only of the temperature [i.e., $h = h$ (T only)]. We start with the definition of enthalpy (Eq. 2.18),

$$h = u + Pv,$$

and replace the Pv term using the ideal-gas equation of state (Eq. 2.28a). This yields

$$h = u + RT. \tag{2.32}$$

Since $u = u$ (T only), we see from Eq. 2.32 that h, too, is a function only of temperature for an ideal gas. With $h = h$ (T only), the partial derivative becomes an ordinary derivative and so we have

$$c_p \equiv \left(\frac{\partial h}{\partial T} \right)_p = \frac{dh}{dT}. \tag{2.33a}$$

From this, we can write

$$dh = c_p dT, \tag{2.33b}$$

which can be integrated to yield

$$h(T) = \int_{T_{ref}}^{T} c_p dT. \tag{2.33c}$$

The enthalpy difference for a change in state mirrors that for the internal energy, that is,

$$h_2 - h_1 = h(T_2) - h(T_1) = \int_{T_1}^{T_2} c_p dT, \tag{2.33d}$$

and if the constant-pressure specific heat does not vary much between states 1 and 2,

$$h_2 - h_1 = c_{p,\text{avg}}(T_2 - T_1). \tag{2.33e}$$

All of these relationships (Eqs. 2.31–2.33) can be expressed on a molar basis simply by substituting molar-specific properties for mass-specific properties. Note then that R becomes R_u.

Before proceeding, we obtain some useful auxiliary ideal-gas relationships by differentiating Eq. 2.32 with respect to temperature, giving

$$\frac{dh}{dT} = \frac{du}{dT} + R.$$

Recognizing the definitions of the ideal-gas specific heats (Eqs. 2.31 and 2.33a), we then have

$$c_p = c_v + R, \tag{2.34a}$$

or

$$c_p - c_v = R. \tag{2.34b}$$

Since property data sources sometimes only provide values or curve fits for c_p, one can use Eq. 2.34 to obtain values for c_v.

FIGURE 2.9
The area under the c_v-versus-T curve is the internal energy (top), and the area under the c_p-versus-T curve is the enthalpy (bottom). Note that the difference in the areas, $[h(T_1) - h(T_{ref})] - [u(T_1) - u(T_{ref})]$, is $R(T_1 - T_{ref})$.

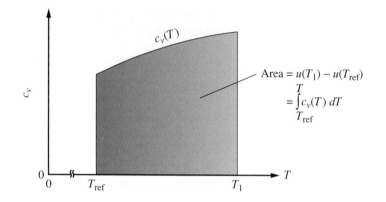

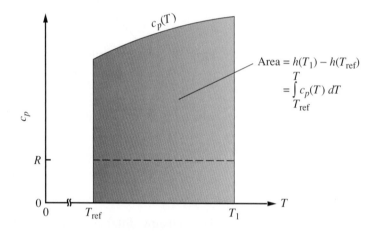

Figure 2.9 provides graphical interpretations of Eqs. 2.31c and 2.33c, our ideal-gas calorific equations of state. Here we see that the area under the c_v-versus-T curve represents the internal energy change for a temperature change from T_{ref} to T_1. A similar interpretation applies to the $c_p(T)$ curve where the area now represents the enthalpy change.

In the same spirit that we graphically illustrated the ideal-gas equation of state using P–v and T–v plots (Figs. 2.7 and 2.8, respectively), we now illustrate the ideal-gas calorific equations of state using u–T and h–T plots. Figure 2.10 illustrates these u–T and h–T relationships for N_2. Because the specific internal energy and specific enthalpy of ideal gases are functions of neither pressure nor specific volume, each relationship is expressed by a single curve.[11] This result (Fig. 2.10) contrasts with the P–v (Fig. 2.7) and T–v (Fig. 2.8) plots generated from the equation of state, where families of curves are required to express these state relationships. Being able to sketch processes on u–T and h–T coordinates, as well as on P–v and T–v coordinates, greatly aids problem solving.

Numerical values for $\bar{h}$ (and $\bar{c}_p$) are available from tables contained in Appendix B for a number of gaseous species, where ideal-gas behavior is assumed. Note the reference temperature of 298.15 K. Curve fits for $\bar{c}_p$ are also provided in Appendix B for these same gases. Although air is a mixture of gases, it can be treated practically as a pure substance (see Example 2.6). Properties of air are provided in Appendix C. Note that the reference temperature used in these air tables is 78.903 K, not 298.15 K. The following examples illustrate the use of some of the information available in the appendices.

---

[11] This result shows that u and T are not independent properties for ideal gases; likewise, h and T are not independent properties. Another property is thus needed to define the state of an ideal gas.

FIGURE 2.10

For ideal gases, the specific internal energy and specific enthalpy are functions of temperature only. The concave-upward curvature in these nearly straight line plots results from the fact that the specific heats (c_v and c_p) for N_2 increase with temperature.

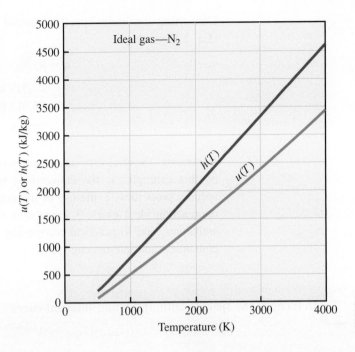

Example 2.8

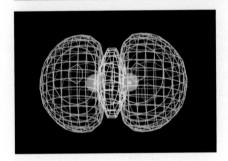

Molecular structure of nitrogen.

Determine the specific internal energy u for N_2 at 2500 K.

Solution

Known N_2, T

Find u

Assumption

Ideal-gas behavior

Analysis   We use Table B.7 to determine u. Before that can be done a few preliminaries are involved. First, we note that molar-specific enthalpies, not mass-specific internal energies, are provided in the table; however, Eq. 2.32 relates u and h for ideal gases and the conversion to a mass basis is straightforward. A second issue is how to interpret the third column in Table B.7, the column containing enthalpy data. The enthalpy values listed under the complex column-three heading, $\bar{h}°(T) - \bar{h}_f°(T_{ref})$, can be interpreted in our present context[12] as simply $\bar{h}(T)$, where $\bar{h}$ is assigned a zero value at 298.15 K (i.e., $T_{ref} = 298.15$ K in Eq. 2.33c). At 2500 K, we see from Table B.7 that

$$\bar{h} = 74{,}305 \text{ kJ/kmol.}$$

We use this value to find the molar-specific internal energy from Eq. 3.32, which has been multiplied through by the molecular weight to yield

$$\bar{u} = \bar{h} - R_u T$$

$$= 74{,}305 \frac{\text{kJ}}{\text{kmol}} - 8.31447 \frac{\text{kJ}}{\text{kmol} \cdot \text{K}} 2500 \text{ K}$$

$$= 74{,}305 \text{ kJ/mol} - 20{,}786 \text{ kJ/mol} = 53{,}519 \text{ kJ/kmol.}$$

---

[12] As we will see later, the column-three heading has an enlarged meaning when dealing with reacting mixtures of ideal gases. For the present, however, this meaning need not concern us.

Converting the molar-specific internal energy to its mass-specific form (Eq. 2.4) yields

$$u = \bar{u}/\mathcal{M}_{N_2}$$

$$= \frac{53,519 \text{ kJ/kmol}}{28.013 \text{ kg/kmol}}$$

$$= 1910.5 \text{ kJ/kg}.$$

Comments   Several very simple, yet very important, concepts are illustrated by this example: 1. the conversions between mass-specific and molar-specific properties, 2. the use of the enthalpy data in Tables B.1–B.12 for nonreacting ideal gases, 3. calculation of ideal-gas internal energies from enthalpies, and 4. practical recognition of the use of reference states for enthalpies (and internal energies).

 Self Test *2.8*   **Determine the specific enthalpy h and internal energy u for air at 1500 K and 1 atm.**

(Answer: 1762.24 kJ/kg, 1331.46 kJ/kg)

Example 2.9

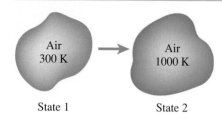

State 1   State 2

Determine the changes in the mass-specific enthalpy and the mass-specific internal energy for air for a process that starts at 300 K and ends at 1000 K. Also show that $\Delta h - \Delta u = R\Delta T$, and compare a numerical evaluation of this with tabulated data.

Solution

Known   air, T_1, T_2

Find $\Delta h \, [= h(T_2) - h(T_1)]$, $\Delta u \, [= u(T_2) - u(T_1)]$, $\Delta h - \Delta u$

Sketch

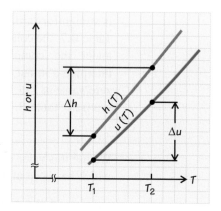

Assumption

Air behaves as a single-component ideal gas.

Analysis   From Table C.2, we obtain the following values for h and u:

T (K)	h (kJ/kg)	u (kJ/kg)
300	426.04	339.93
1000	1172.43	885.22

Using these data, we calculate

$$\Delta h = h(T_2) - h(T_1)$$
$$= h(1000) - h(300)$$
$$= 1{,}172.43 \text{ kJ/kg} - 426.04 \text{ kJ/kg}$$
$$= 746.39 \text{ kJ/kg}$$

and

$$\Delta u = 885.22 \text{ kJ/kg} - 339.93 \text{ kJ/kg}$$
$$= 545.29 \text{ kJ/kg}.$$

Applying Eq. 2.32, we can relate Δh and Δu, that is,

$$h_2 = u_2 + RT_2$$

and

$$h_1 = u_1 + RT_1.$$

Subtracting these yields

$$h_2 - h_1 = u_2 - u_1 + R(T_2 - T_1),$$

or

$$\Delta h - \Delta u = R\Delta T.$$

We evaluate this equation using the particular gas constant for air ($R = R_u/\mathcal{M}_{air}$), giving us

$$\Delta h - \Delta u = \frac{8.31447 \text{ kJ/kmol} \cdot \text{K}}{28.97 \text{ kg/kmol}}(1000 - 300) \text{ K}$$
$$= 200.90 \text{ kJ/kg}.$$

This compares to the value obtained from the Table C.1 data as follows:

$$(\Delta h - \Delta u)_{\text{tables}} = 746.39 \text{ kJ/kg} - 545.29 \text{ kJ/kg}$$
$$= 201.10 \text{ kJ/kg}.$$

This result is within 0.1% of our calculation. Failure to achieve identical values is probably a result of the methods used to generate the tables (i.e., curve fitting).

Comment   This example illustrates once again the treatment of air as a single-component ideal gas and the use of Appendix C for air properties.

> **At this point, the reader has sufficient knowledge of properties to begin a study of Chapters 4 and 5. This is also a good entry point from Chapter 7 following Eq. 7.26.**

Self Test 2.9

 Recalculate the values for the change in mass-specific enthalpy and internal energy for the process in Example 2.9 using Eqs. 2.31e and 2.33e and an appropriate average value of c_v and c_p. Compare your answer with that of Example 2.9.

(Answer: 751.8 kJ/kg, 550.2 kJ/kg)

Josiah Willard Gibbs (1839–1903).

2.4e Ideal-Gas Temperature–Entropy (Gibbs) Relationships

To obtain the desired temperature–entropy relationships (Eqs. 2.25a and 2.25b) for an ideal gas, we apply concepts associated with the first and second laws of thermodynamics. These concepts are developed in Chapters 5 and 7, respectively, and the reader should be familiar with these chapters before proceeding with this section.

We begin by considering a simple compressible system that undergoes an internally reversible process that results in an incremental change in state. For such a process, the first law of thermodynamics is expressed (Eq. 5.5) as

$$\delta Q_{rev} - \delta W_{rev} = dU.$$

The incremental heat interaction δQ_{rev} is related directly to the entropy change through the formal definition of entropy from Chapter 7 (i.e., Eq. 7.16),

$$dS \equiv \left(\frac{\delta Q}{T}\right)_{rev},$$

or

$$\delta Q_{rev} = TdS.$$

For a simple compressible substance, the only reversible work mode is compression and/or expansion, that is,

$$\delta W_{rev} = Pd\mathcal{V}.$$

Substituting these expressions for δQ_{rev} and δW_{rev} into the first-law statement yields

$$TdS - Pd\mathcal{V} = dU.$$

We rearrange this result slightly and write

$$TdS = dU + Pd\mathcal{V}, \tag{2.35a}$$

which can also be expressed on a per-unit-mass basis as

$$Tds = du + Pdv. \tag{2.35b}$$

Equation 2.35 is the first of the so-called Gibbs or T–ds equations. We obtain a second Gibbs equation by employing the definition of enthalpy (Eq. 2.17), that is,

$$U = H - P\mathcal{V},$$

which can be differentiated to yield

$$dU = dH - Pd\mathcal{V} - \mathcal{V}dP.$$

Substituting this expression for dU into Eq. 2.35a and simplifying yields

$$TdS = dH - \mathcal{V}dP, \tag{2.36a}$$

or on a per-unit-mass basis

$$Tds = dh - vdP. \tag{2.36b}$$

Note that Eqs. 2.35 and 2.36 apply to any simple compressible substance, not just an ideal gas; furthermore, these relationships apply to any incremental process, not just an internally reversible one.

We now proceed toward our objective of finding the relationships $f_4(s, T, P) = 0$ and $f_5(s, T, v) = 0$ for an ideal gas by using the ideal-gas equation of state and the ideal-gas calorific equations of state. Starting with Eq. 2.36b, we substitute $v = RT/P$ (Eq. 2.28a) and $dh = c_p dT$ to yield

$$T ds = c_p dT - \frac{RT}{P} dP,$$

which, upon dividing through by T, becomes

$$ds = c_p \frac{dT}{T} - R \frac{dP}{P}. \tag{2.37}$$

Similarly, Eq. 2.35b is transformed using the substitutions $P = RT/v$ and $du = c_v dT$ to yield

$$ds = c_v \frac{dT}{T} + R \frac{dv}{v}. \tag{2.38}$$

To evaluate the entropy change in going from state 1 to state 2 state, we integrate Eqs. 2.37 and 2.38, that is,

$$s_2 - s_1 = \int_1^2 ds = \int_1^2 c_p \frac{dT}{T} - \int_1^2 R \frac{dP}{P}$$

and

$$s_2 - s_1 = \int_1^2 ds = \int_1^2 c_v \frac{dT}{T} + \int_1^2 R \frac{dv}{v}.$$

The second term on the right-hand side of each of these expressions can be easily evaluated since R is a constant, and so

$$s_2 - s_1 = \int_1^2 c_p \frac{dT}{T} - R \ln \frac{P_2}{P_1} \tag{2.39a}$$

and

$$s_2 - s_1 = \int_1^2 c_v \frac{dT}{T} + R \ln \frac{v_2}{v_1}. \tag{2.39b}$$

To evaluate the integrals in Eqs. 2.39a and 2.39b requires knowledge of $c_p(T)$ and $c_v(T)$. Curve-fit expressions for $c_p(T)$ are readily available for many species (e.g., Table B.13); $c_v(T)$ can be determined using the c_p curve fits and the ideal-gas relationship, $c_v(T) = c_p(T) - R$ (Eq. 2.34).

In many engineering applications, an average value of c_p (or c_v) over the temperature range of interest can be used to evaluate $s_2 - s_1$ with reasonable accuracy:

$$s_2 - s_1 = c_{p,\text{avg}} \ln \frac{T_2}{T_1} - R \ln \frac{P_2}{P_1} \tag{2.40a}$$

Dividing by R and rearranging yields

$$\ln\frac{P_2}{P_1} = \frac{c_{p,\text{avg}}}{R}\ln\frac{T_2}{T_1},$$

and removing the logarithm by exponentiation, we obtain

$$\frac{P_2}{P_1} = \left(\frac{T_2}{T_1}\right)^{\frac{c_{p,\text{avg}}}{R}},$$

This equation can be simplified by introducing the specific heat ratio γ and the fact that $c_p - c_v = R$ (Eq. 2.34b), and so

$$\frac{c_{p,\text{avg}}}{R} = \frac{c_{p,\text{avg}}}{c_{p,\text{avg}} - c_{v,\text{avg}}} = \frac{\gamma}{\gamma - 1};$$

thus,

$$\frac{P_2}{P_1} = \left(\frac{T_2}{T_1}\right)^{\frac{\gamma}{\gamma-1}}, \tag{2.41a}$$

or, more generally,

$$T^\gamma P^{1-\gamma} = \text{constant.} \tag{2.41b}$$

Equation 2.40b can be similarly manipulated to yield

$$\frac{v_2}{v_1} = \left(\frac{T_2}{T_1}\right)^{\frac{1}{1-\gamma}}, \tag{2.42a}$$

or

$$Tv^{\gamma-1} = \text{constant.} \tag{2.42b}$$

Applying the ideal-gas equation of state to either Eq. 2.41 or Eq. 2.42, we obtain a third and final ideal-gas isentropic-process relationship:

$$\frac{P_2}{P_1} = \left(\frac{v_2}{v_1}\right)^{-\gamma}, \tag{2.43a}$$

or

$$Pv^\gamma = \text{constant.} \tag{2.43b}$$

This last relationship (Eq. 2.43b) is easy to remember. All of the other isentropic relationships are easily derived from this by applying the ideal-gas equation of state.

The ideal-gas isentropic-process relationships are summarized in Table 2.5 for convenient future reference. One use of these relationships is to model ideal compression and expansion processes in internal combustion engines, air compressors, and gas-turbine engines, for example. Example 2.11 illustrates this use in this chapter; other examples are found throughout the book.

Natural gas compressor station in Wyoming.

2.4g Processes in T–s and P–v Space

With the addition of entropy, we now have four properties (P, v, T, and s) to define states and describe processes. We now investigate T–s and P–v space

Table 2.5 Ideal-Gas Isentropic-Process Relationships

General Form	State 1 to State 2	Equation Reference
$Pv^{\gamma} = \text{constant}$	$\dfrac{P_2}{P_1} = \left(\dfrac{v_2}{v_1}\right)^{-\gamma}$	Eq. 2.43
$Tv^{\gamma-1} = \text{constant}$	$\dfrac{T_2}{T_1} = \left(\dfrac{v_2}{v_1}\right)^{1-\gamma}$	Eq. 2.42
$T^{\gamma}P^{1-\gamma} = \text{constant}$	$\dfrac{P_2}{P_1} = \left(\dfrac{T_2}{T_1}\right)^{\frac{\gamma}{\gamma-1}}$	Eq. 2.41

to see how various fixed-property processes appear on these coordinates. In T–s space, isothermal and isentropic processes are by definition horizontal and vertical lines, respectively. Less obvious are the lines of constant pressure and specific volume shown in Fig. 2.11. Here we see that, through any given state point, lines of constant specific volume have steeper slopes than those of constant pressure. In P–v space, constant-pressure and constant-volume processes are, again by definition, horizontal and vertical lines, whereas constant-entropy and constant-temperature processes follow curved paths as shown in Fig. 2.12. Here we see that lines of constant entropy are steeper (i.e., have greater negative slope) than those of constant temperature. You should become familiar with these characteristics of T–s and P–v diagrams.

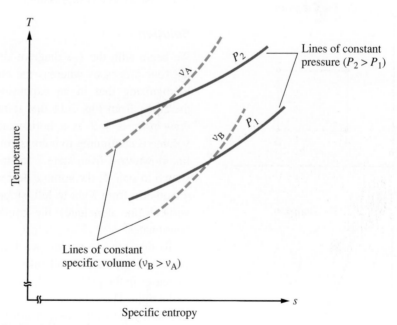

FIGURE 2.11

On a T–s diagram, lines of constant pressure (isobars) and lines of constant specific volume (isochors) both exhibit positive slopes. At a particular state, a constant-volume line has a greater slope than a constant-pressure line.

FIGURE 2.12

On a P–v diagram, lines of constant temperature (isotherms) and lines of constant entropy (isentropes) have negative slopes. At a particular state, an isentrope is steeper (has a greater negative slope) than an isotherm.

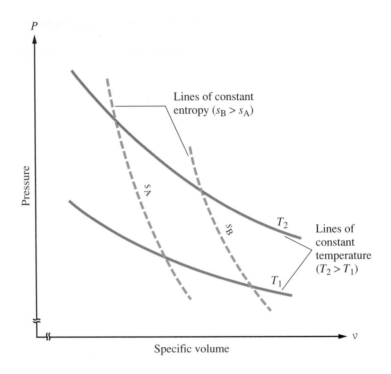

Example 2.11

Sketch the following set of processes on *T–s* and *P–v* diagrams:

1–2: isothermal expansion,

2–3: isentropic expansion,

3–4: isothermal compression, and

4–1: isentropic compression.

Solution

We begin with the *T–s* diagram since the temperature is fixed for two of the four processes whereas the entropy is fixed for the remaining two. Recognizing that in an expansion process v_2 is greater than v_1 and, therefore, from Fig. 2.11 that state 2 must lie to the right of state 1, we draw process 1–2 as a horizontal line from left to right. Because the volume is continuing to increase in process 2–3, our isentrope is a vertical line downward from state 2 to state 3. In the compression process from state 3 to state 4, the volume decreases; thus, we draw the state-3 to state-4 isotherm from right to left, stopping when s_4 is equal to s_1. An upward vertical line concludes the cycle and completes a rectangle on *T–s* coordinates.

To draw the *P–v* plot, we refer to Fig. 2.12, which shows lines of constant temperature and lines of constant entropy. Because the volume increases in the process 1–2, we draw an isotherm directed downward and to the right. The expansion continues from state 2 to state 3, but the process line now follows the steeper isentrope to the lower isotherm. The cycle is completed following this lower isotherm upward and to the left until it intercepts the upper isentrope at the initial state 1.

Comments   This sequence of processes is the famous Carnot cycle, which is discussed at length in Chapter 7.

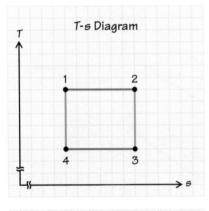

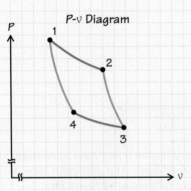

Self Test 2.11

  Referring to Example 2.11, for which process(es) are the equations in Table 2.5 valid? Why or why not?

(Answer: 2–3 and 4–1 only. Equations in Table 2.5 are for isentropic processes only.)

Example 2.12   SI Engine Application

The following processes constitute the *air-standard Otto cycle*. Plot these processes on P–v and T–s coordinates:

1–2:   constant-volume energy addition,

2–3:   isentropic expansion,

3–4:   constant-volume energy removal, and

4–1:   isentropic compression.

Solution

Since there are two constant-volume processes and two constant-entropy processes, we begin by drawing two isentropes on P–v coordinates and two isochors on T–s coordinates. These are shown in the sketches. Because we

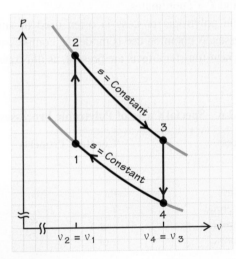

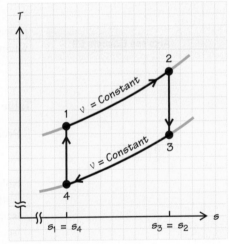

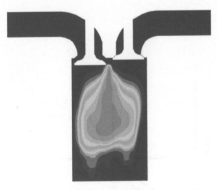

During the intake stroke of a real spark-ignition engine, fresh fuel and air enter the cylinder and mix with the residual gases. Image courtesy of Eugene Kung and Daniel Haworth.

Also see Example 4.2 in Chapter 4.

expect the pressure to increase with energy addition at constant volume, we draw a vertical line upward from state 1 to state 2 on the P–v diagram. Because process 2–3 is an expansion, with state 3 lying below state 2, we thus know that process 1–2 follows the upper constant-volume line on the T–s diagram. Having established the relative locations of states 1, 2, and 3 on each plot, completing the cycle is straightforward.

Comments   The cycle illustrated in this example using air as the working fluid is often used as a starting point for understanding the thermodynamics of spark-ignition (Otto cycle) engines. In the real engine, a combustion process causes the pressure to rise from states 1 to 2, rather than energy addition from the surroundings. Furthermore, the real "cycle" is not closed because the gases exit the cylinder and are replaced by a fresh charge of air-fuel mixture each mechanical cycle. Nevertheless, this air-standard cycle does capture the effect of compression ratio on thermal efficiency and can be used to model other effects as well. Some of these are shown in examples throughout this book.

2.4h Polytropic Processes

In the previous section, we saw that an isentropic process can be described by the equation

$$Pv^{\gamma} = \text{constant.}$$

By replacing γ, the specific heat ratio, by an arbitrary exponent n, we define a generalized process or **polytropic process** as

$$Pv^{n} = \text{constant.} \tag{2.44}$$

For certain values of n, we recover relationships previously developed. These are shown in Table 2.6. For example, when n is unity, a constant-temperature (isothermal) process is described, that is,

$$Pv^{1} = \text{constant} \, (= RT).$$

Figure 2.13 graphically illustrates these special-case polytropic processes on P–v and T–s coordinates.

The polytropic process is often used to simplify and model complex processes. A common use is modeling compression and expansion processes when heat-transfer effects are present, as illustrated in the following example.

Also see Example 5.9. ▶

Table 2.6   Special Cases of Polytropic Processes

Process	Constant Property	Polytropic Exponent (n)
Isobaric	P	0
Isothermal	T	1
Isentropic	s or S	γ
Isochoric	v or V	$\pm\infty$

FIGURE 2.13

P–v and T–s diagrams illustrating polytropic process paths for special cases of constant pressure (n = 0), constant temperature (n = 1), constant entropy (n = γ), and constant volume (n = ±∞).

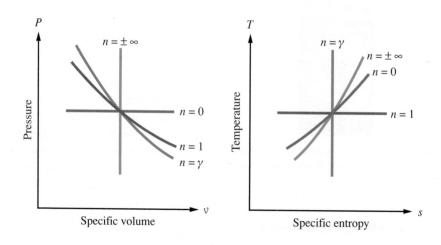

Example 2.13 SI Engine Application

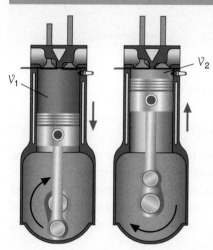

Consider a spark-ignition engine in which the effective compression ratio (*CR*) is 8:1. Compare the pressure at the end of the compression process for an isentropic compression ($n = \gamma = 1.4$) with that for a polytropic compression with $n = 1.3$. The initial pressure in the cylinder is 100 kPa, a wide-open-throttle condition.

Solution

Known P_1, CR, γ, n

Find $P_2(n = \gamma), P_2(n = 1.3)$

Sketch

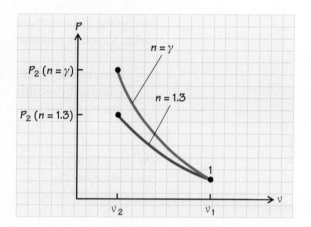

Assumption

The working fluid can be treated as air with $\gamma = 1.4$.

Analysis   We apply the polytropic process relationship, Eq. 2.44, for the two processes, both starting at the same initial state. From Appendix 1A, we know that the compression ratio is defined as V_1/V_2. Since the mass is fixed, the compression ratio also equals the ratio of specific volumes, v_1/v_2. For the isentropic compression,

$$P_2(n = \gamma) = P_1(v_1/v_2)^\gamma$$
$$= 100 \text{ kPa } (8)^{1.4}$$
$$= 1840 \text{ kPa.}$$

For the polytropic compression,

$$P_2(n = 1.3) = P_1(v_1/v_2)^{1.3}$$
$$= 100 \text{ kPa } (8)^{1.3}$$
$$= 1490 \text{ kPa.}$$

Comments   We see that the isentropic compression pressure is approximately 20% higher than the polytropic compression pressure. The ideal compression process is adiabatic and reversible, and, hence, isentropic. In an actual engine, energy is lost from the compressed gases by heat transfer through the cylinder walls, and frictional effects are also present. The combined effects result in a compression pressure lower than the ideal value. The use of a polytropic exponent is a simple way to deal with these effects.

☑ **Given an initial temperature of 400 K, calculate the final temperature for each process in Example 2.13.**

(Answer: 920 K, 745 K)

2.5 NONIDEAL GAS PROPERTIES

2.5a State (*P*–*v*–*T*) Relationships

In this section, we will look at three ways to determine P–v–T state relations for gases that do not necessarily obey the ideal-gas equation of state (Eq. 2.28): 1. the use of tabular data, 2. the use of the van der Waals equation of state, and 3. the application of the concept of generalized compressibility. Before discussing these methods, however, we define the thermodynamic **critical point**, a concept important to all three.

> The *critical point* is a point in P–v–T space defined by the highest possible temperature and the highest possible pressure for which distinct liquid and gas phases can be observed. (See Fig. 2.14.)

At the critical point, we designate the thermodynamic variables with the subscript c, so that

$$\text{critical temperature} \equiv T_c,$$
$$\text{critical pressure} \equiv P_c,$$

and

$$\text{critical specific volume} \equiv v_c.$$

Critical properties for a number of substances are given in Table E.1 in Appendix E. We elaborate on the significance of the critical point later in this chapter in our discussion of substances existing in multiple phases. One immediate use of critical point properties, however, is to establish a rule of thumb for determining when a gas can be considered ideal: A real gas approaches ideal-gas behavior when $P \ll P_c$. In a later section, we will see other conditions where ideal-gas behavior is approached.

FIGURE 2.14
The bold dot indicates the critical point on this P–v–T plot for propane. At temperatures and pressures above their critical values T_c and P_c, respectively, liquid and gas phases are indistinguishable.

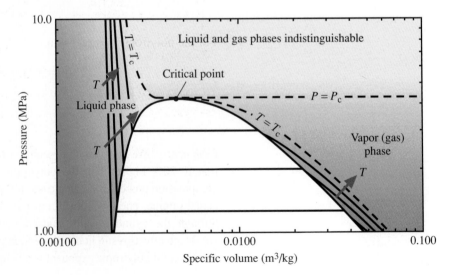

Example 2.14

Determine if N_2 is likely to approximate ideal-gas behavior at 298 K and 1 atm.

Solution

The critical properties for N_2 from Table E.1 are

$$T_c = 126.2 \text{ K},$$
$$P_c = 3.39 \text{ MPa}.$$

Comparing the given properties with the critical properties, we have

$$\frac{T}{T_c} = \frac{298 \text{ K}}{126.2 \text{ K}} = 2.36$$

and

$$\frac{P}{P_c} = \frac{101,325 \text{ Pa}}{3.39 \times 10^6 \text{ Pa}} = 0.030.$$

Clearly, $P \ll P_c$; thus, the ideal-gas equation of state is likely to be a good approximation to the true state relation for N_2 at these conditions. Moreover, T is greater than T_c, which is also in the direction of ideal-gas behavior.

Comment Since P_c is quite high (33 atm) and T_c is quite low (126.2 K), ideal-gas behavior is likely to be good approximation for N_2 over a fairly wide range of conditions covering many applications. However, caution should be exercised. The methods described in the following allow us to estimate *quantitatively* the deviation of real-gas behavior from ideal-gas behavior.

**Self Test
2.14**

Determine if O_2 can be considered an ideal gas at 298 K and atmospheric pressure. Can air at room temperature and atmospheric pressure be considered an ideal gas?

(Answer: $T/T_c = 1.93$ and $P/P_c = 0.020$, so ideal-gas behavior is a good approximation for O_2. Since the main constituents of air (N_2 and O_2) both behave as ideal gases, ideal-gas behavior for air is a good approximation.)

For conditions near the critical point or for applications requiring high accuracy, alternatives to the ideal-gas equation of state are needed. The following subsections present three approaches.

Tabulated Properties

Accurate P–v–T data in tabular or curve-fit form are available for a number of gases of engineering importance. Because of the importance of steam in electric power generation, a large database is available for this fluid, and tables are published in a number of sources (e.g., Refs. [7] and [8]). Particularly useful and convenient sources of thermodynamic properties are available from the NIST [9, 10]. A large portion of the NIST database is available on the CD included with this book (NIST12 v. 5.2) and from the Internet [11]. Table 2.7 lists the fluids for which properties are available from these sources. We use the NIST resources throughout this book, and the reader is encouraged to become familiar with these valuable sources of thermodynamic data. Selected tabular data are also provided in Appendix D for steam.

Table 2.7   Fluid Properties: Fluids Included in NIST12 v. 5.2 Software and NIST Online Database [11]

Fluid	NIST12 V. 5.2	NIST Online
Air	X	
Water*	X	X
Nitrogen	X	X
Hydrogen	X	X
Parahydrogen	X	X
Deuterium		X
Oxygen	X	X
Fluorine		X
Carbon monoxide	X	X
Carbon dioxide	X	X
Methane	X	X
Ethane		X
Ethene		X
Propane	X	X
Propene		X
Butane		X
Isobutane		X
Pentane		X
Hexane		X
Heptane		X
Helium	X	X
Neon		X
Argon	X	X
Krypton		X
Xenon		X
Ammonia	X	X
Nitrogen trifluoride		X
Methane, trichlorofluoro- (R-11)	X	
Methane, dichlorodifluoro- (R-12)	X	
Methane, chlorodifluoro- (R-22)	X	X
Methane, difluoro- (R-32)		X
Ethane, 2,2-dichloro-1,1,1-trifluoro- (R-123)	X	X
Ethane, pentafluoro- (R-125)		X
Ethane, 1,1,1,2-tetrafluoro- (R-134a)	X	X
Ethane, 1,1,1-trifluoro- (R-143a)		X
Ethane, 1,1-difluoro- (R-152a)		X

* See also Appendix D.

Example 2.15

Determine the deviation from ideal-gas behavior associated with the following gases and conditions:

N_2 at 200 K from 1 atm to P_c,

CO_2 at 300 K from 1 to 40 atm, and

H_2O at 600 K from 1 to 40 atm.

Solution

To quantify the deviation from ideal-gas behavior, we define the factor $Z = Pv/RT$. For an ideal gas, Z is unity for any temperature or pressure. We employ the NIST online database [11] to obtain values of specific volume for the conditions specified. These data are then used to calculate Z values. For example, the specific volume of CO_2 at 300 K and 20 atm is 0.025001 m^3/kg; thus,

$$Z = \frac{Pv}{RT} = \frac{20(101{,}325)\,0.025001}{\left(\dfrac{8314.47}{44.011}\right)300} = 0.8939$$

$$[=] \frac{\text{atm}\left(\dfrac{\text{Pa}}{\text{atm}}\right)\left(\dfrac{m^3}{kg}\right)}{\left(\dfrac{J}{kmol \cdot K}\right)\left(\dfrac{kmol}{kg}\right)K} \times \frac{\left[\dfrac{1\ N/m^2}{Pa}\right]}{\left[\dfrac{1\ N \cdot m}{J}\right]} = 1.$$

A few selected results are presented in the following table, and all of the results are plotted in Fig. 2.15.

P(atm)	v (m^3/kg)			Z = Pv/RT		
	N_2	CO_2	H_2O	N_2	CO_2	H_2O
1	0.8786	0.5566	2.7273	0.9998	0.9951	0.9980
10	0.8773	0.5309	0.2676	0.9983	0.9491	0.9792
20	0.4381	0.0250	0.1308	0.9972	0.8939	0.9572
30	0.0292	0.0155	0.0851	0.9965	0.8330	0.9341
40	—	0.0107	0.0621	—	0.7642	0.9097

Comments From the table and from Fig. 2.15, we see that N_2 behaves essentially as an ideal gas over the entire range of pressures from 1 atm to the critical pressure ($P_c = 33.46$ atm). Z values deviate less than 0.4% from the ideal-gas value of unity. Both CO_2 and water vapor, in contrast, exhibit significant departures from ideal-gas behavior, which become larger as the pressure increases. These large departures indicate the need for caution in applying the ideal-gas equation of state, $Pv = RT$.

FIGURE 2.15

Deviations from ideal-gas behavior can be seen by the extent to which Z (= Pv/RT) deviates from unity. At 1 atm, CO_2, N_2, and H_2O all have Z values near unity. At higher pressures, CO_2 and H_2O deviate significantly from ideal-gas behavior, whereas N_2 still behaves essentially as an ideal gas.

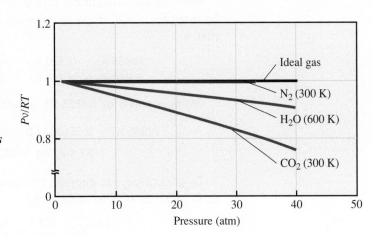

 Self Test 2.15 Using the ideal-gas equation of state, calculate the specific volume for H_2O at 40 atm and 600 K. Compare your answer with the value from the table in Example 2.15 and determine the calculation error.

(Answer: 0.0683 m^3/kg, 10%)

Tutorial 1 | **How to Interpolate**

Relationships among properties are frequently presented in tables. For example, internal energy, enthalpy, and specific volume for steam are often tabulated as functions of temperature at a fixed pressure. Table D.3 illustrates such tables and employs temperature increments of 20 K. If the temperature of interest is one of those tabulated, a simple look-up is all that is needed to retrieve the desired properties. In many cases, however, the temperature of interest will fall somewhere between the tabulated temperatures. To estimate property values for such a situation, you can apply **linear interpolation**. We illustrate this procedure with the following concrete example.

Given: Steam at 835 K and 10 MPa.
Find: Enthalpy using Table D.3P.

From the 10-MPa pressure table we see that the given temperature lies between the tabulated values:

T (K)	h (kJ/kg)
820	3494.1
835	?
840	3543.9

Linear interpolation assumes a straight-line relationship between h and T for the interval $820 \leq T \leq 840$. For the given data, we see that the desired temperature, 835 K, lies three-fourths of the way between 820 and 840 K, that is,

$$\frac{835 - 820}{840 - 820} = \frac{15}{20} = 0.75.$$

With the assumed linear relationship, the unknown h value must also lie three-fourths of the way between the two tabulated enthalpy values, that is,

$$\frac{h(835 \text{ K}) - h(320 \text{ K})}{h(840 \text{ K}) - h(320 \text{ K})} = \frac{h(835 \text{ K}) - 3494.1}{3543.9 - 3494.1} = 0.75.$$

We now solve for h (835 K), obtaining

$$h(835 \text{ K}) = 0.75(3543.9 - 3494.1)\text{kJ/kg} + 3494.1 \text{ kJ/kg}$$

$$= 37.4 \text{ kJ/kg} + 3494.1 \text{ kJ/kg} = 3531.5 \text{ kJ/kg}.$$

To generalize, we express this procedure as follows, denoting $h(T_1)$ as h_1, $h(T_2)$ as h_2, and $h(T_3)$ as h_3:

$$\frac{h(T_3) - h(T_1)}{h(T_2) - h(T_1)} = \frac{h_3 - h_1}{h_2 - h_1} = \frac{T_3 - T_1}{T_2 - T_1},$$

or

$$h_3 = \left(\frac{T_3 - T_1}{T_2 - T_1} \right)(h_2 - h_1) + h_1.$$

For any property pair $Y(X)$, this can be expressed

$$\frac{Y(X_3) - Y(X_1)}{Y(X_2) - Y(X_1)} = \frac{Y_3 - Y_1}{Y_2 - Y_1} = \frac{X_3 - X_1}{X_2 - X_1},$$

or

$$Y_3 = \left(\frac{X_3 - X_1}{X_2 - X_1}\right)(Y_2 - Y_1) + Y_1.$$

Rather than remembering or referring to any equations, knowing the physical interpretation of linear interpolation allows you to create the needed relationships. The following sketch provides a graphic aid for this procedure:

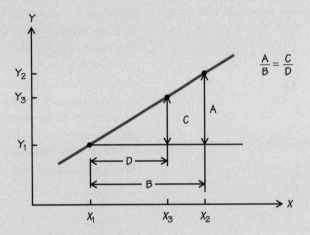

Although the use of computer-based property data may minimize the need to interpolate, some data may only be available in tabular form. Moreover, the ability to interpolate is a generally useful skill and should be mastered.

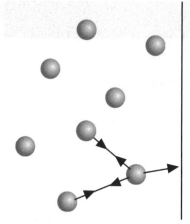

FIGURE 2.16

Intermolecular attractive forces result in a pressure less than would result from an ideal gas in which there are no long-range intermolecular forces.

Other Equations of State

The kinetic theory model of a gas that leads to the ideal-gas equation of state is based on two assumptions that break down when the density of the gas is sufficiently high. The first of these is that the molecules themselves occupy a negligible volume compared to the volume of gas under consideration. Clearly as a gas is compressed, the average spacing between molecules becomes less and the fraction of the macroscopic gas volume occupied by the microscopic molecules increases. The **van der Waals equation of state** accounts for the finite volume of the molecules by subtracting a molecular volume from the macroscopic volume; thus, the molar-specific volume $\bar{v}$ in the state equation is replaced with $\bar{v} - b$, where b is a constant for a particular gas.

The second assumption that is violated at high densities is that the long-range forces between molecules are negligible. Figure 2.16 illustrates how the existence of such forces can affect the equation of state. Since the pressure of a gas is a manifestation of the momentum transferred to the wall by molecular collisions, long-range attractive forces will "pull back" molecules as they approach the wall, thus decreasing the momentum exchange. This effect, in turn, results in a decrease in pressure. In the van der Waals equation of state, this effect of intermolecular attractive forces is accounted for by replacing the pressure P with $P + a/\bar{v}^2$, where a is a constant, again dependent upon the particular molecular species involved. The appearance of the reciprocal of $\bar{v}^2$ is a consequence of the molecular collision frequency with the wall being proportional to the density $(1/\bar{v})$ and the intermolecular attractive force also being proportional to the molar gas density $(1/\bar{v})$.

With these two corrections to the ideal-gas model, we write the van der Waals equation of state as follows:

$$\left(P + \frac{a}{\bar{v}^2}\right)(\bar{v} - b) = R_u T. \tag{2.45}$$

Values for the constants a and b for a number of gases are provided in Appendix E.

Example 2.16

Use the van der Waals equation of state to evaluate $Z \left(= P\bar{v}/R_u T\right)$ for CO_2 at 300 K and 30 atm. Compare this result with the value calculated using the NIST database in Example 2.15.

Solution

Known CO_2, T, P

Find Z

Sketch   See Fig. 2.15.

Assumptions

van der Waals gas

Analysis   We use the van der Waals equation of state (Eq. 2.45) to find the molar-specific volume $\bar{v}$. This value of $\bar{v}$ is then used to calculate Z. We rearrange Eq. 2.45,

$$\left(P + \frac{a}{\bar{v}^2}\right)(\bar{v} - b) = R_u T,$$

to the following cubic form:

$$\bar{v}^3 - \left(\frac{R_u T}{P} + b\right)\bar{v}^2 + \frac{a}{P}\bar{v} - \frac{ab}{P} = 0.$$

Before solving for $\bar{v}$, we calculate the coefficients using values for a and b from Table E.2:

$$a = 3.643 \times 10^5 \, \text{Pa} \cdot (\text{m}^3/\text{kmol})^2,$$
$$b = 0.0427 \, \text{m}^3/\text{kmol}.$$

The coefficients are thus

$$\left(\frac{R_u T}{P} + b\right) = \frac{8314.47 \, (300)}{30 \, (101,325)} + 0.0427$$
$$= 0.863274$$

$$[=] \frac{\dfrac{J}{\text{kmol} \cdot K} K \left[\dfrac{N \cdot m}{1 \, J}\right]}{\text{atm} \left[\dfrac{N/m^2}{\text{atm}}\right]} = \text{m}^3/\text{kmol},$$

$$\frac{a}{P} = \frac{3.643 \times 10^5 \, \text{Pa} \cdot (\text{m}^3/\text{kmol})^2}{30 \, (101,325) \, \text{Pa}}$$
$$= 0.119845 \, (\text{m}^3/\text{kmol})^2,$$

and

$$\frac{ab}{P} = \frac{3.643 \times 10^5 \, \text{Pa} \cdot (\text{m}^3/\text{kmol})^2 \, (0.0427) \, (\text{m}^3/\text{kmol})}{30 \, (101,325) \, \text{Pa}}$$
$$= 0.005117 \, (\text{m}^3/\text{kmol})^3.$$

Substituting these coefficients into the cubic van der Waals equation of state yields

$$\bar{v}^3 - 0.863274 \, \bar{v}^2 + 0.119845 \, \bar{v} - 0.005117 = 0.$$

Many methods exist to find the useful root of this polynomial. Using spreadsheet software to implement the iterative Newton–Raphson method with an initial guess of $\bar{v} = R_u T/P \, (= 0.8206 \, \text{m}^3/\text{kmol})$ results in the following converged value for $\bar{v}$ after four interations:

$$\bar{v} \, (30 \, \text{atm}, \, 300 \, \text{K}) = 0.7032 \, \text{m}^3/\text{kmol}.$$

With this value of $\bar{v}$, we evaluate Z:

$$Z = \frac{P\bar{v}}{R_u T} = \frac{30 \, (101,325) \, 0.7032}{8314.47 \, (300)}$$
$$= 0.8570,$$

which is dimensionless.

This value of Z is only 2.9% higher than the 0.8330 calculated from the NIST database in Example 2.15.

Comments   For this particular example, note that the van der Waals model did an excellent job of predicting the specific volume for conditions far from the ideal-gas regime. Although the procedure used to calculate $\bar{v}$ was straightforward, we still employed the power of a computer to solve the cubic van der Waals equation of state.

Self Test
2.16

✓ **Repeat Example 2.16 for H₂O at 40 atm and 600 K.**

(Answer: Z = 0.930)

Other equations of state have been developed that provide more accuracy than the van der Waals equation. Discussion of these is beyond the scope of this book. For more information, we refer the interested reader to Ref. [1].

Generalized Compressibility

As we will see later, the $T = T_c$ isotherm in P–v space exhibits an inflection point at the critical point. One can relate the constants a and b in the van der Waals equation of state to the properties at the critical state by recognizing that both slope and curvature of the $T = T_c$ isotherm are zero; thus,

$$a = \frac{27}{64}\frac{R^2 T_c^2}{P_c}, \tag{2.46a}$$

$$b = \frac{RT_c}{8P_c}, \tag{2.46b}$$

$$Z = \frac{P_c v_c}{RT_c} = \frac{3}{8}. \tag{2.46c}$$

That a and b depend only on the critical pressure and critical temperature suggests that a generalized state relationship exists when actual pressures and temperatures are normalized by their respective critical values. More explicitly, defining the reduced pressure and temperature as

$$P_R \equiv \frac{P}{P_c} \tag{2.47a}$$

and

$$T_R \equiv \frac{T}{T_c}, \tag{2.47b}$$

respectively, we expect

$$Z = Z(P_R, T_R) \tag{2.47c}$$

to be a single "universal" relationship. This idea is known as the **principle of corresponding states**, and Z is called the **compressibility factor**. Figure 2.17 shows data and the best fit for this relationship, where Z is presented as a function of P_R using T_R as a parameter. For the gases chosen, individual data points are quite close to the curve fits, illustrating the "universal" nature of Eq. 2.47c. Plots such as Fig. 2.17 are known as **generalized compressibility charts**. Figure 2.18 is a working generalized compressibility chart for reduced pressures up to 10 and reduced temperatures up to 15.

Polar molecule (water) and nonpolar molecule (ethane).

In reality, Eq. 2.47c is not truly universal but rather a useful approximation. The best accuracy is obtained when gases are grouped according to shared characteristics. For example, a single plot constructed for polar compounds, such as water and alcohols, provides a tighter "universal" relationship than is obtained when nonpolar compounds are also included. Generalized compressibility charts are useful for substances for which no data are available other than critical properties.

We can also use the generalized compressibility chart to ascertain conditions where real gases deviate from ideal-gas behavior. Using Fig. 2.18 as our guide, we observe the following:

- The largest departures from ideal-gas behavior occur at conditions near the critical point. At the critical point, the ideal-gas equation of state is not at all close to reality ($Z \approx 0.3$).
- Accuracy to within 5% of ideal-gas behavior ($0.95 < Z < 1.05$) is found for the following conditions:
  - at all temperatures, provided $P_R < 0.1$, and
  - when $1.95 < T_R < 2.4$ and $T_R > 15$, both for $P_R < 7.5$.

The following example illustrates the use of generalized compressibility theory.

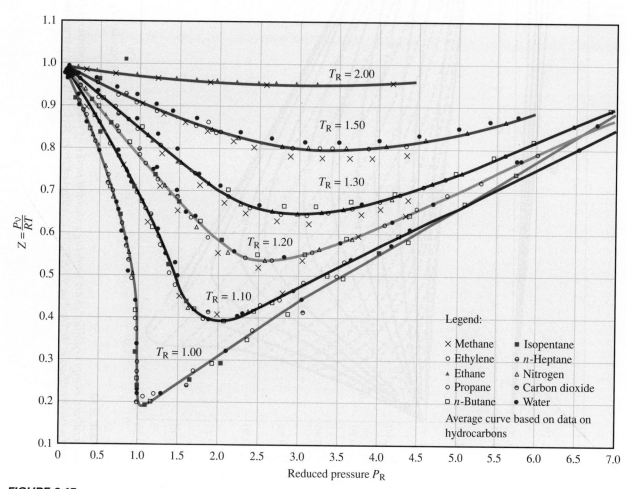

FIGURE 2.17

The compressibility factor Z can be correlated using reduced properties (i.e., P/P_c and T/T_c). Note how the data for various substances collapse when plotted in this manner. **Adapted from Ref. [12] with permission.**

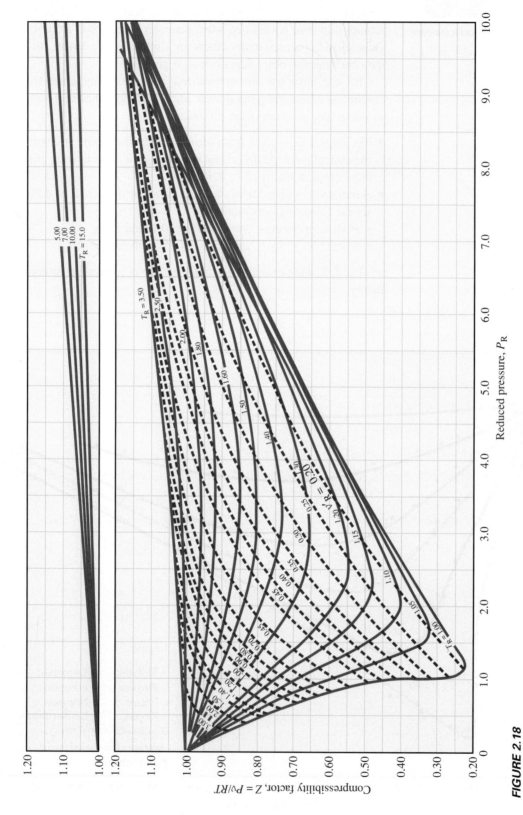

FIGURE 2.18

Generalized compressibility chart showing extended range of reduced temperatures ($1 < T_R < 15$). The pseudo reduced specific volume v'_R shown on the chart is defined as $v/(RT_c/P_c)$. Adapted from **Ref. [13] with permission.**

Example 2.17

Once again, consider CO_2 at 30 atm and 300 K. Use the generalized compressibility chart to determine the density of CO_2 at this condition. Compare this result with that obtained from the ideal-gas equation of state and with that from the NIST tables [11].

Solution

Known CO_2, P, T

Find ρ_{CO_2}

Sketch   See Fig. 2.18.

Analysis   To use the generalized compressibility chart requires values for the critical temperature and pressure, which we retrieve from Appendix E, Table E.1, as follows:

$$T_c = 304.2 \text{ K},$$

$$P_c = 73.9 \times 10^5 \text{ Pa}.$$

The reduced temperature and pressure are thus

$$T_R = \frac{T}{T_c} = \frac{300 \text{ K}}{304.2 \text{ K}} = 0.986,$$

$$P_R = \frac{P}{P_c} = \frac{30 \text{ atm } (101,325 \text{ Pa/atm})}{73.9 \times 10^5 \text{ Pa}} = 0.411.$$

Using these values, we find the compressibility factor from Fig. 2.18 to be

$$Z \approx 0.82.$$

Applying the definition of Z, we obtain the density:

$$Z \equiv \frac{Pv}{RT} = \frac{P}{\rho RT},$$

or

$$\rho = \frac{P}{ZRT} = \frac{P\mathcal{M}_{CO_2}}{ZR_u T}$$

$$= \frac{1}{0.82} \frac{30 \, (101,325) \, 44.01}{8314.47 \, (300)} \text{kg/m}^3$$

$$= \frac{53.63}{0.82} \text{ kg/m}^3 = 65 \text{ kg/m}^3.$$

The reader should verify the units in this calculation. Note that the ideal-gas density is the numerator in the last line of this calculation. Using the results of Example 2.15, we make the requested comparisons:

Method	Z	ρ (kg/m^3)
Fig. 2.18	0.82	65
NIST database	0.833	64.51
Ideal gas	1.000	53.63

Comments Even though reading Z from the chart is an approximate procedure, excellent agreement is found between the density calculated from the chart and that from the NIST database.

Self Test 2.17 ✓ **Repeat Example 2.17 for H$_2$O at 40 atm and 600 K.**

(Answer: $Z \cong 0.95$, $\rho = 15.41 \ kg/m^3$)

2.5b Calorific Relationships

Use the NIST database for values of h, u, c_v, and c_p as functions of T and P for nonideal gases. For generalized corrections to ideal-gas properties, see Refs. [20, 21].

2.5c Second-Law Relationships

Use the NIST database for entropy values as functions of T and P for nonideal gases. For generalized corrections to ideal-gas properties, see Refs. [20, 21].

2.6 PURE SUBSTANCES INVOLVING LIQUID AND VAPOR PHASES

In this section, we investigate the thermodynamic properties of simple, compressible substances that frequently exist in both liquid and vapor states. Because of its engineering importance, we focus on water. When discussing water, we use the common designations, ice, water, and steam to refer to the solid, liquid, and vapor states, respectively, and use the chemical designation H$_2$O when we wish to be general and not designate any particular phase. The terms vapor and gas are used interchangeably in this book.[13]

2.6a State (*P–v–T*) Relationships

Phase Boundaries

We begin our discussion of multiphase properties by conducting a thought experiment. Consider a kilogram of H$_2$O enclosed in a piston–cylinder arrangement as shown in the sketch in Fig. 2.19. The weight on the piston fixes the pressure in the cylinder as we add energy to the system. At the initial state, the H$_2$O is liquid, and remains liquid, as we add energy by heating from A to B. You might envision that a Bunsen burner is used as the energy source, for example. The added energy results in an increase in the temperature of the water, while the water expands slightly, causing a small increase in the specific volume. The temperature and the specific volume increase until state B is reached. At this point, further addition of energy results in a portion of the water becoming vapor or steam, at a fixed temperature. At 1 atm, this phase change occurs at 373.12 K (99.97°C).[14] At

---

[13] Some textbooks distinguish a gas from a vapor using the critical pressure as a criterion. In this usage, gases exist above P_c, whereas vapors exist below P_c.

[14] With the adoption of the International Temperature Scale of 1990 (ITS-90), the normal boiling point of water is not exactly 100°C, but rather 99.974°C.

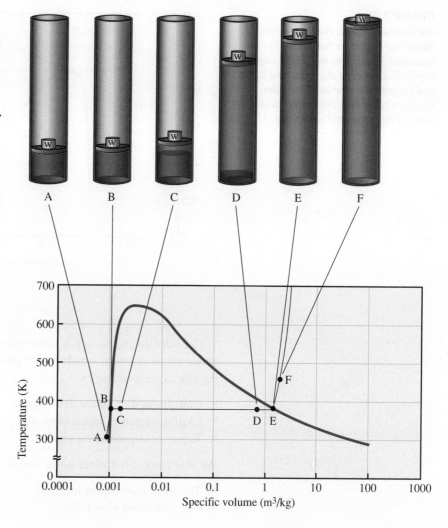

FIGURE 2.19
Heating water at constant pressure follows the path A–B–C–D–E–F on a temperature–specific volume diagram. Note the huge increase in volume in going from the liquid state (B) to the vapor state (E). The volumes shown in the piston–cylinder sketches are not to scale.

state C, we see that most of the H_2O is still in the liquid phase. With continued heating, more and more water is converted to steam. At state D, the transformation is nearly complete, and at state E, all of the liquid has turned to vapor. Adding energy beyond this point results in an increase in both the temperature and specific volume of the steam, creating what is known as superheated steam. The point at F designates a superheated state. Assuming that the heating was conducted quasi-statically, the path A–B–C–D–E–F is a collection of equilibrium states following an isobar (a line of constant pressure) in T–v space. We could repeat the experiment with lesser or greater weight on the piston (i.e., at lower or higher pressures), and map out a phase diagram showing a line where vaporization just begins (states like B) and a line where vaporization is complete (states like E). The bold line in Fig. 2.19 summarizes these thought experiments. Although we considered H_2O, similar behavior is observed for many other fluids, for example, the various fluids used as refrigerants.

Figure 2.20 presents a T–v diagram showing the designations of the various regions and lines associated with the liquid and vapor states of a fluid. Although the numerical values shown apply to H_2O, the general ideas presented in this figure and discussed in the following apply to many liquid–vapor systems.

FIGURE 2.20
On T-v coordinates, the compressed liquid region lies to the left of the saturated liquid line, the liquid–vapor region lies between the saturated liquid and saturated vapor lines, and the vapor or superheat region lies to the right of the saturated vapor line.

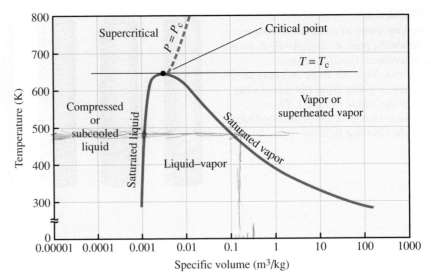

Consider the region in T–v space that lies below the critical temperature but above temperatures at which a solid phase forms. Here we have three distinct regions, corresponding to

- **Compressed** or **subcooled liquid**,
- **Liquid–vapor** or **saturation**, and
- **Vapor** or **superheated vapor**.

We also have two distinct lines, which join at the critical point:

- The **saturated liquid line** and
- The **saturated vapor line**.

The saturated liquid line is the locus of states at which the addition of energy at constant pressure results in the formation of vapor (state B in Fig. 2.19). The temperature and pressure associated with states on the saturated liquid line are called the **saturation temperature** and **saturation pressure**, respectively. To the left of the saturated liquid line is the compressed liquid region, which is also known as the subcooled liquid region. These designations arise from the fact that at any given point in this region, the pressure is higher than the corresponding saturation pressure at the same temperature (thus the designation *compressed*) and, similarly, any point is at a temperature lower than the corresponding saturation temperature at the same pressure (thus the designation subcooled). The saturated liquid line terminates at the critical point with zero slope on both T–v and P–v coordinates.

Immediately to the right of the saturated liquid line are states that consist of a mixture of liquid and vapor. Bounding this liquid–vapor or saturation region on the right is the saturated vapor line. This line is the locus of states that consist of 100% vapor at the saturation pressure and temperature. The saturated vapor line joins the saturated liquid line at the critical point. This junction, the critical point, defines a state in which liquid and vapor properties are indistinguishable.

Figure 2.20 also shows the **supercritical region**. Within this region, the pressure and temperature exceed their respective critical values, and there is no distinction between a liquid and a gas.

FIGURE 2.21

A P–v–T surface for a substance that contracts upon freezing. **After Ref. [22].**

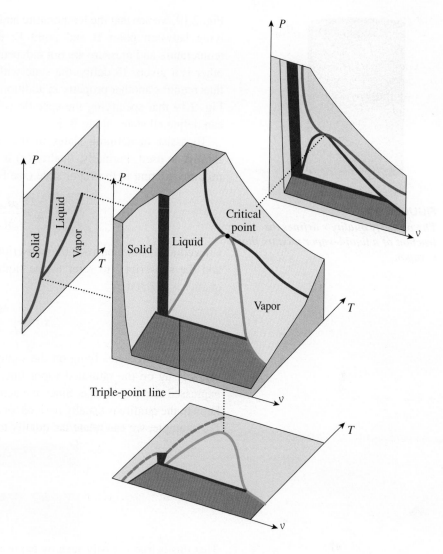

To the right of the saturated vapor line lies the vapor or superheated vapor region. The use of the word *superheated* refers to the idea that within this region the temperature at a particular pressure is greater than the corresponding saturation temperature.

The liquid–vapor or saturation region is frequently referred to as the **vapor dome** because of its shape on a T–log v plot. When dealing with water, the term *steam dome* is sometimes applied. Because two phases coexist in the saturation region, defining a thermodynamic state here is a bit more complex than in the single-phase regions. We examine this issue in the next section.

The three-dimensional P–v–T surface (Fig. 2.21) provides the most general view of a simple substance that exhibits multiple phases. From Fig. 2.21, we see that the T–v, P–v, and P–T plots are simply projections of the three-dimensional surface onto these respective planes.

A New Property–Quality

In our discussion of gases, whether ideal or real, we saw that simultaneously specifying the temperature and pressure defined the thermodynamic state. A knowledge of P and T was sufficient to define all other properties. In the liquid–vapor region, however, this situation no longer holds, although we still require two independent properties to define the thermodynamic state. In

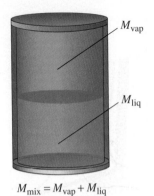

$$M_{\text{mix}} = M_{\text{vap}} + M_{\text{liq}}$$

FIGURE 2.22
The property quality x defines the fraction of a liquid–vapor mixture that is vapor.

Fig. 2.19, we see that the temperature and pressure are both fixed for all states lying between point B and point E; thus, in the two-phase region, the temperature and pressure are not independent properties. If one is known, the other is a given. To define the state within the liquid–vapor mixture region thus requires another property in addition to P or T. For example, we see from Fig. 2.19 that specifying the specific volume together with the temperature can define all states along B–E.

To assist in defining states in the saturation region, a property called **quality** is used. Formally, quality (x) is defined as the mass fraction of the mixture existing in the vapor state (see Fig. 2.22); that is,

$$x \equiv \frac{M_{\text{vapor}}}{M_{\text{mix}}}. \tag{2.48a}$$

Conventional notation uses the subscript g to refer to the vapor (gas) phase and the subscript f to refer to the liquid phase.[15] Using this notation, the quality is defined as

$$x \equiv \frac{M_{\text{g}}}{M_{\text{g}} + M_{\text{f}}}. \tag{2.48b}$$

The quality for states lying on the saturated liquid line is zero, whereas for states lying on the saturated vapor line, the quality is unity, or 100% when expressed as a percentage. Since the actual mass in each phase is not usually known, the quality is usually derived, or related to, other intensive properties. For example, we can relate the quality to the specific volume as follows:

$$v = (1 - x)v_{\text{f}} + xv_{\text{g}}, \tag{2.49a}$$

or

$$x = \frac{v - v_{\text{f}}}{v_{\text{g}} - v_{\text{f}}}. \tag{2.49b}$$

That this is true is easily seen by expressing the intensive properties, v and x, in terms of their defining extensive properties, V and M, that is,

$$\begin{aligned}
v &= \frac{M_{\text{f}}}{M_{\text{f}} + M_{\text{g}}} \frac{V_{\text{f}}}{M_{\text{f}}} + \frac{M_{\text{g}}}{M_{\text{f}} + M_{\text{g}}} \frac{V_{\text{g}}}{M_{\text{g}}} \\
&= \frac{V_{\text{f}}}{M_{\text{f}} + M_{\text{g}}} + \frac{V_{\text{g}}}{M_{\text{f}} + M_{\text{g}}} = \frac{V_{\text{f}} + V_{\text{g}}}{M_{\text{f}} + M_{\text{g}}} \\
&\equiv \frac{V_{\text{mixture}}}{M_{\text{mixture}}}.
\end{aligned}$$

The relationship expressed by Eq. 2.49a can be generalized in that any mass-specific property (e.g., u, h, and s) can be used in place of the specific volume, so that

$$\beta = (1 - x)\beta_{\text{f}} + x\beta_{\text{g}}, \tag{2.49c}$$

or

$$x = \frac{\beta - \beta_{\text{f}}}{\beta_{\text{g}} - \beta_{\text{f}}}, \tag{2.49d}$$

---

[15] The subscripts f and g are actually from the German words *Flussigkeit* (liquid) and *Gaszustand* (gaseous state), respectively.

where β represents any mass-specific property. A physical interpretation of Eq. 2.49d is easy to visualize by referring to Fig. 2.19. Here we interpret the quality at state C as the length of the line B–C divided by the length of the line B–E, where a linear rather than logarithmic scale is used for v. This same interpretation applies to any mass-specific property β plotted in a similar manner.

Commonly employed rearrangements of Eqs. 2.49a and 2.49b are

$$v = v_f + x (v_g - v_f) \qquad (2.50a)$$

and

$$\beta = \beta_f + x (\beta_g - \beta_f). \qquad (2.50b)$$

Defining v_{fg} and β_{fg},

$$v_{fg} \equiv v_g - v_f \qquad (2.50c)$$

and

$$\beta_{fg} \equiv \beta_g - \beta_f, \qquad (2.50d)$$

we rewrite Eqs. 2.50a and 2.50b more compactly as

$$v = v_f + x v_{fg} \qquad (2.50e)$$

and

$$\beta = \beta_f + x \beta_{fg}. \qquad (2.50f)$$

Equations 2.50e and 2.50f are useful when using tables that provide values for v_{fg} and similar properties β_{fg}.

Example 2.18

A rigid tank contains 3 kg of H_2O (liquid and vapor) at a quality of 0.6. The specific volume of the liquid is 0.001041 m^3/kg and the specific volume of the vapor is 1.859 m^3/kg. Determine the mass of the vapor, the mass of the liquid, and the specific volume of the mixture.

Solution

Known M_{mix}, x, v_f, v_g

Find M_g, M_f, v_{mix}

Sketch   See Fig. 2.22.

Assumptions

Equilibrium prevails.

Analysis   We apply the definition of quality (Eq. 2.48b) to find the mass of H_2O in each phase:

$$M_g = x(M_g + M_f) = xM_{mix}$$
$$= 0.6(3 \text{ kg}) = 1.8 \text{ kg}$$

and

$$M_f = M_{mix} - M_g$$
$$= 3 \text{ kg} - 1.8 \text{ kg} = 1.2 \text{ kg}.$$

Using the given quality and given values for the specific volumes of the saturated liquid and saturated vapor, we calculate the mixture specific volume (Eq. 2.49a):

$$v_{mix} = (1 - x)\, v_f + x\, v_g$$

$$= (1 - 0.6)\, 0.001041 \text{ m}^3/\text{kg} + 0.6\, (1.859 \text{ m}^3/\text{kg})$$

$$= 1.1158 \text{ m}^3/\text{kg}.$$

Comments   This example is a straightforward application of definitions. Knowledge of definitions can be quite important in the solution of complex problems.

  Repeat Example 2.18 with $x = 0.8$ and find the volume of the tank.
(Answer: $M_g = 2.4$ kg, $M_f = 0.6$ kg, $v_{mix} = 1.4874$ m³/kg, $V = 4.46$ m³)

Property Tables and Databases

Tabulated properties are available for H_2O and many other fluids of engineering interest (see Table 2.7). Typically, data are presented in four forms:

- Saturation properties for convenient temperature increments,
- Saturation properties for convenient pressure increments,
- Superheated vapor properties at various fixed pressures with convenient temperature increments, and
- Compressed (subcooled) liquid properties at various fixed pressures with convenient temperature increments.

Tables in Appendix D provide data in these forms for H_2O.

Computerized and online databases are making the use of printed tables obsolete. Using such databases also eliminates the need to interpolate, which can be tedious. Throughout this book, we illustrate the use of both tabulated data (Appendix D) and the NIST databases [9, 10]. You should develop proficiency with both sources of properties.

Figures 2.23–2.26 illustrate the use of the NIST online database. Figure 2.23 shows one of the input menus for creating your own tables and graphs. After selecting the fluid of interest and choosing the units desired (not shown), one has the choice of accessing saturation data in temperature or pressure increments (Figs. 2.24 and 2.25, respectively) or accessing data as an isotherm or as an isobar (Fig. 26). Figure 2.24 presents saturation data for temperatures from 300 to 310 K in increments of 10 K; Fig. 2.25 shows similar data but for a range of saturation pressures from 0.1 to 0.2 MPa in increments of 0.1 MPa. Figure 2.26 presents data for the 10-MPa isobar for a 800–900 K temperature range in 100-K increments. Note that the data in Fig. 2.26 all lie within the superheated vapor region as indicated in the final column of the table. Tables generated from the database can also be downloaded. Figures 2.27–2.30, for example, were generated using NIST data with spreadsheet software.

The NIST12 v. 5.2 software packaged with this book provides expanded capability to that available online. Tutorials 2 and 3 describe the use of this powerful software.

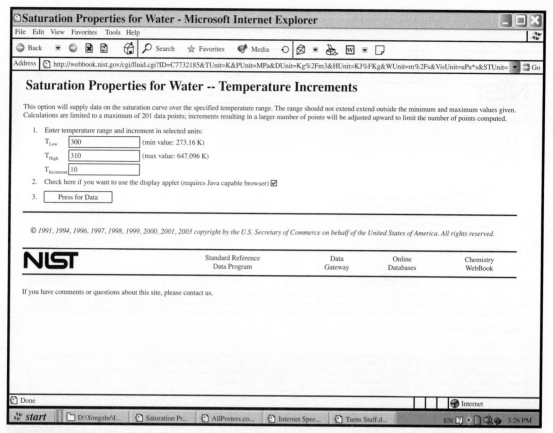

FIGURE 2.23

Graphical user interface for the NIST online thermophysical property database. See Table 2.7 for fluids for which properties are provided. **From Ref. [11].**

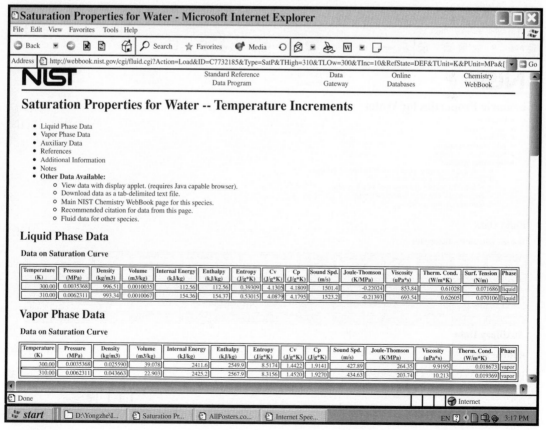

FIGURE 2.24

Saturation properties for water from NIST online database: 300–310 K in 10-K increments. **From Ref. [11].**

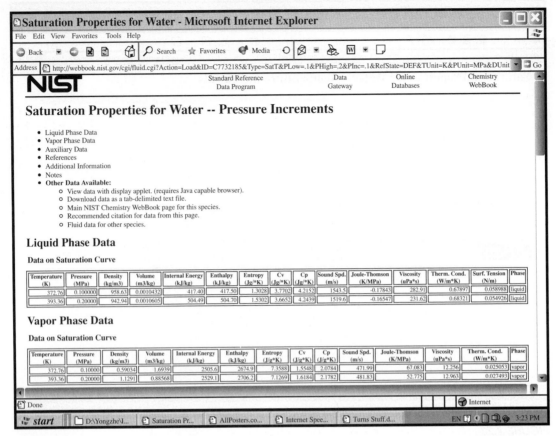

FIGURE 2.25

Saturation properties for water from NIST online database: 0.1–0.2 MPa in 0.1-MPa increments.
From Ref. [11].

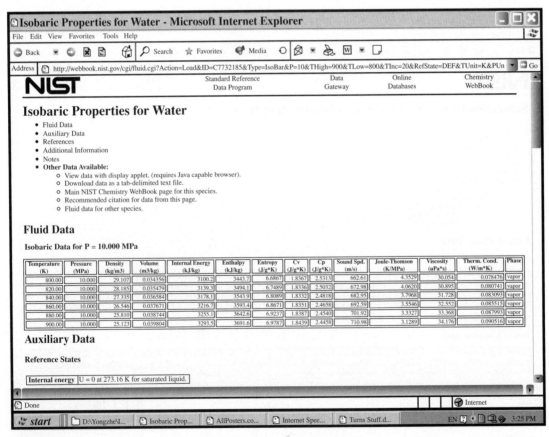

FIGURE 2.26

Properties of water for the 10-MPa isobar from NIST online database: 800–900 K in 20-K
increments. From Ref. [11].

Tutorial 2   How to Use the NIST Software

Packaged with this book is a CD containing the software package NIST12 Version 5.2. This software, an invaluable resource for thermodynamic and transport properties of many pure substances and air, is very easy to use: You can learn to use it in only a few minutes and can be an expert in less than an hour. The purpose of this tutorial is to acquaint you with the basic capabilities of this powerful package and to encourage you to become an expert user.

What fluids are contained in the NIST database?

From the "Substance" menu a selection of seventeen pure fluids is available. A listing is shown in the figure. Air is also available as a pseudo-pure fluid from the substance menu.

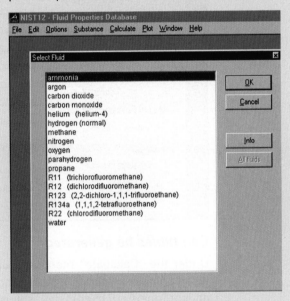

What properties are available?

All of the properties used in this book and many others are provided. "Properties" can be selected from the "Options" menu as illustrated here.

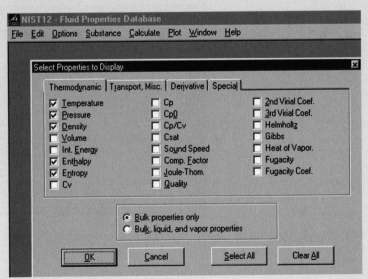

(Continued on next page)

Tutorial 2 **How to Use the NIST Software** *(continued)*

What units can be used?

From the "Options" menu, the user can select SI, U.S. customary, and other units using pull-down selections. Mixed units can be used if desired and either mass- or molar-specific quantities can be selected.

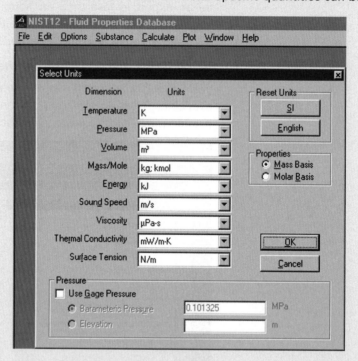

Can tables be generated?

Under the "Calculate" menu, the user has the option to generate two types of tables: saturation tables and isoproperty tables. The creation of an isoproperty (P fixed) table is illustrated here.

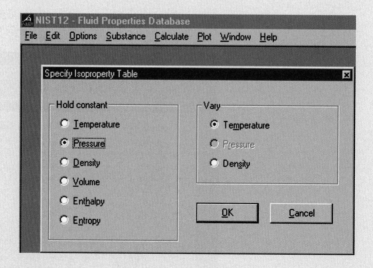

(Continued on next page)

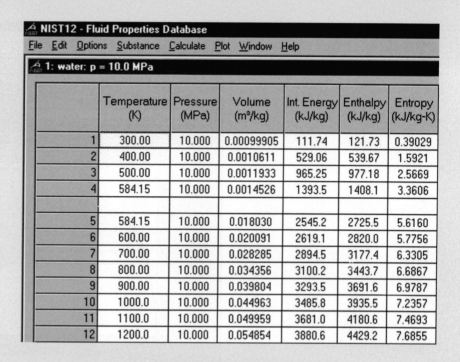

NIST12 - Fluid Properties Database

File   Edit   Options   Substance   Calculate   Plot   Window   Help

1: water: p = 10.0 MPa

	Temperature (K)	Pressure (MPa)	Volume (m³/kg)	Int. Energy (kJ/kg)	Enthalpy (kJ/kg)	Entropy (kJ/kg-K)
1	300.00	10.000	0.00099905	111.74	121.73	0.39029
2	400.00	10.000	0.0010611	529.06	539.67	1.5921
3	500.00	10.000	0.0011933	965.25	977.18	2.5669
4	584.15	10.000	0.0014526	1393.5	1408.1	3.3606
5	584.15	10.000	0.018030	2545.2	2725.5	5.6160
6	600.00	10.000	0.020091	2619.1	2820.0	5.7756
7	700.00	10.000	0.028285	2894.5	3177.4	6.3305
8	800.00	10.000	0.034356	3100.2	3443.7	6.6867
9	900.00	10.000	0.039804	3293.5	3691.6	6.9787
10	1000.0	10.000	0.044963	3485.8	3935.5	7.2357
11	1100.0	10.000	0.049959	3681.0	4180.6	7.4693
12	1200.0	10.000	0.054854	3880.6	4429.2	7.6855

What if I want properties at a single state point?

From the "Calculate" menu the user can select "Specified State Points" or "Saturation Points."

What plots can be generated with the software?

All of the standard thermodynamic coordinates (P–v, T–s, h–s, etc.) and others are available from the "Plot" menu. Plots with iso-lines, as shown in the figure, are easy to create. The option "Other Diagrams" allows the user to select any pair of thermodynamic coordinates. The T–v diagram here illustrates this option.

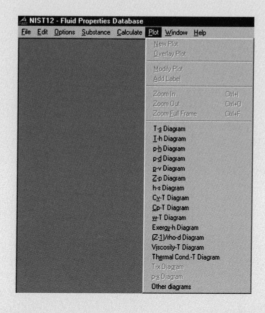

Tutorial 2 | How to Use the NIST Software *(continued)*

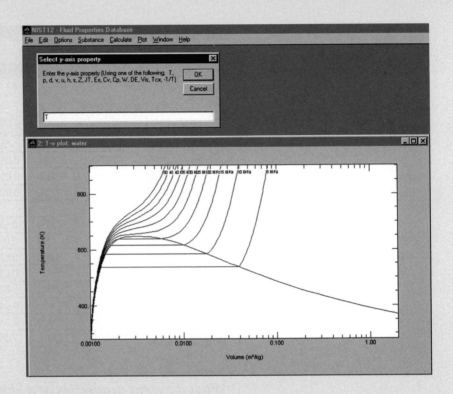

Is the NIST software compatible with other common software?

Tables generated using NIST12 Version 5.2 can be easily pasted into spreadsheets. Similarly, plots can be pasted into word-processing documents.

The reader is encouraged to use this software to solve the homework problems provided in this book.

Tutorial 3 How to Define a Thermodynamic State

Whether employing tabular data or the NIST software, one needs to become proficient at identifying and defining thermodynamic states. What we mean by that is the following: Given a pair of properties (e.g., P_1, T_1), can you determine the region, or region boundary, in or on which the state lies? Specifically, is the state of the substance a compressed liquid, saturated liquid, liquid–vapor mixture, saturated vapor, or superheated vapor? Or does the state lie at the critical point or within the supercritical region? This tutorial offers hints to help you answer such questions efficiently and with confidence.

Given an arbitrary temperature T_1 and pressure P_1, we desire to identify the state and to determine the specific volume v_1. To aid our analysis, we employ the following P–v diagram for H_2O.

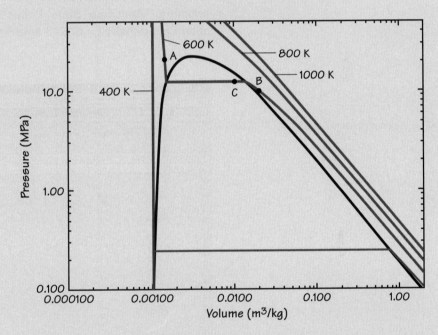

We begin with a few concrete examples using H_2O, with all states having the same temperature of 600 K. The 600-K isotherm is one of the several shown on our P–v plot.

Given: $T = 600$ K, $P = 20$ MPa.
Find: Region, v.

The key to defining the region is determining the relationship of the given properties to those of (1) the critical point and (2) the saturation states. For H_2O, the critical point is 22.064 MPa and 647.1 K. Since the given pressure is less than the critical pressure, we can rule out the supercritical region, which we can easily verify by inspecting the P–v diagram. Next we determine the saturation states corresponding to the given P and T. Using Tables D.1 and D.2, respectively (or the NIST database), we find that

$$P_{sat} (600 \text{ K}) = 12.34 \text{ MPa}, \qquad \text{(Table D.1)},$$
$$T_{sat} (20 \text{ MPa}) = 638.9 \text{ K} \qquad \text{(Table D.2)}.$$

(Continued on next page)

From these values, we see that the given temperature, 600 K, is less than the saturation value, 638.9 K [i.e., T (20 MPa) < T_{sat} (20 MPa)]. The state thus lies in the subcooled region. Alternatively, we see that the given pressure, 20 MPa, is greater than the saturation value, 12.34 MPa [i.e., P (600 K) < P_{sat} (600 K)]; the state thus lies in the compressed-liquid region. The compressed-liquid and subcooled-liquid regions are synonymous. Point A on the P–v plot denotes this state. Now knowing the region, we turn to the compressed-liquid table, Table D.4D, to find the specific volume. No interpolation is needed and we read the value directly:

$$v \text{ (20 MPa, 600 K)} = 0.001481 \text{ m}^3/\text{kg}.$$

A simpler, but less instructive, procedure is to use the NIST software. By selecting "Specified State Points" from the "Calculate" menu, we obtain the density [= 675.11 kg/m³ = $1/v$ = 1/(0.001481 m³/kg)].

		Temperature (K)	Pressure (MPa)	Density (kg/m³)	Volume (m²/kg)	Int. Energy (kJ/kg)	Enthalpy (kJ/kg)
	1	600.00	20.000	675.11	0.0014812	1456.8	1486.4
	2						

Note that finding the region in which the state lies must be a separate step because the "Specified State Point" calculation does not identify the state location. Being able to locate a state relative to the liquid, liquid–vapor mixture, and vapor boundaries is an essential skill.

Given: T = 600 K, P = 10 MPa.
Find: Region, v.

Again, we use Tables D.1 and D.2 to find the corresponding saturation values:

$$
\begin{aligned}
P_{sat} \text{ (600 K)} &= 12.34 \text{ MPa} &&\text{(Table D.1),}\\
T_{sat} \text{ (10 MPa)} &= 584.15 \text{ K} &&\text{(Table D.2).}
\end{aligned}
$$

Because T (10 MPa) > T_{sat} (10 MPa) (i.e., 600 K > 584.15 K), the state lies in the superheated vapor region. We come to the same conclusion by recognizing that P (600 K) < P_{sat} (600 K). Point B on the P–v plot denotes this state. To find the specific volume, we employ the superheated vapor table for 10 MPa, Table D.3P, which directly yields

$$v \text{ (10 MPa, 600 K)} = 0.020091 \text{ m}^3/\text{kg.}$$

The given values were deliberately selected to avoid any interpolation. Finding properties in the superheat region may be complicated by the need to interpolate both temperature and pressure. In such cases, use of the NIST software is recommended to avoid this tedious process.

Given: $T = 600$, $v = 0.01$ m^3/kg.
Find: Region, P.

Since the specific volume is given, we determine the relationship of this value to those of the saturated liquid and saturated vapor at the same temperature to define the region. From Table D.1, we obtain

$$v_{sat\ liq} \text{ (600 K)} = 0.0015399 \text{ m}^3/\text{kg,}$$
$$v_{sat\ vapor} \text{ (600 K)} = 0.014984 \text{ m}^3/\text{kg.}$$

Because the given specific volume falls between these two values, the state must lie in the liquid–vapor mixture region. The pressure must then be the saturation value, P_{sat} (600 K) = 12.34 MPa. If the given specific volume were less than $v_{sat\ liq}$, we would have to conclude that the state lies in the compressed-liquid region. Similarly, if the given specific volume were greater than $v_{sat\ vapor}$, the state would then lie in the superheated-vapor region. Point C denotes this state on the P–v diagram.

Although we determined the regions in which various states lie using T, P, and v, other properties can be employed as well. The key, regardless of the particular properties given, is to find their relationships to the corresponding saturation state boundaries (i.e., the "steam dome").

Example 2.19 Steam Power Plant Application

A turbine expands steam to a pressure of 0.01 MPa and a quality of 0.92. Determine the temperature and the specific volume at this state.

Solution

Known P, x

Find T, v

Sketch

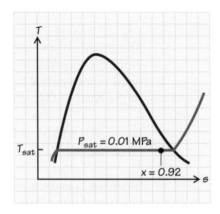

Analysis   The given state lies in the liquid–vapor region on the 0.01-MPa isobar as shown in the sketch. Knowing this, we use Table D.2, *Saturation Properties of Water and Steam—Pressure Increments*, to obtain the following data:

$$T = T_{sat}(P = P_{sat} = 0.01 \text{ MPa}) = 318.96 \text{ K},$$

$$v_f = 0.0010103 \text{ m}^3/\text{kg},$$

$$v_g = 14.670 \text{ m}^3/\text{kg}.$$

Using the given quality and these values for the specific volumes of the saturated liquid and saturated vapor, we calculate the specific volume from Eq. 2.49a as follows:

$$v = (1 - x)\, v_f + x v_g$$
$$= (1 - 0.92)\, 0.0010103 \text{ m}^3/\text{kg} + 0.92\, (14.670 \text{ m}^3/\text{kg})$$
$$= 13.496 \text{ m}^3/\text{kg}.$$

Comments   The keys to this problem are, first, recognizing the region in which the given state lies, and, second, choosing the correct table to evaluate properties. The reader should verify the values obtained from Table D.2 for this example and also verify that the values agree with the NIST online database and NIST12 v. 5.2 software.

Self Test
2.19   ✓   **Calculate the values of the specific internal energy *u* and the specific enthalpy *h* for the conditions given in Example 2.19.**

(*Answer: u = 2257.57 kJ/kg, h = 2392.53 kJ/kg*)

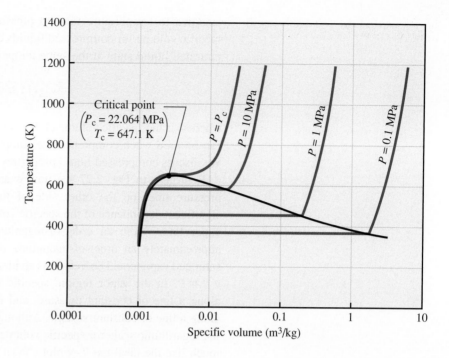

T–v Diagrams

Figure 2.27 presents a T–v diagram for H_2O showing several lines of constant pressure. In the liquid region, lines of constant pressure hug the saturated liquid line quite closely. This fact is illustrated more clearly in the expanded view of the liquid region provided in Fig. 2.28 where the 10-MPa constant-pressure line is shown together with the saturated liquid line. All of the constant-pressure lines for pressures less than 10 MPa, but greater than P_{sat} at any given temperature, are crowded between the two lines shown in Fig. 2.28; thus, we see that specific volume is much more

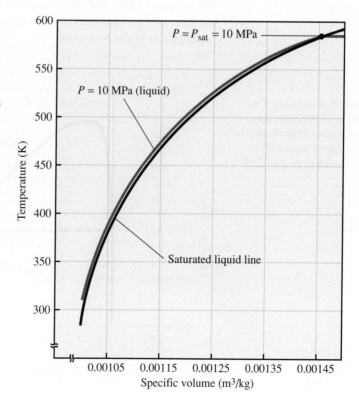

sensitive to temperature than to pressure. As a consequence, values for specific volume for compressed liquids can be approximated as those at the saturated liquid state at the same temperature; that is,

$$v_{liq}(T) \cong v_f(T), \tag{2.51}$$

where v_f is the specific volume of the saturated liquid at T. This rule-of-thumb can be used when comprehensive compressed liquid data are not available. We discuss compressed liquid properties in more detail in a later section.

Returning to Fig. 2.27 we now examine the behavior of the constant-pressure lines on the other side of the steam dome. We first note the significant dependence of the specific volume on pressure along the saturated vapor line, where an order-of-magnitude increase in pressure results in approximately an order-of-magnitude decrease in specific volume. If the saturated vapor could be treated as an ideal gas, this relationship would follow $v^{-1} \propto P$. In the vapor region, specific volume increases with temperature along a line of constant pressure, and pressure increases with temperature along a line of constant volume. Although somewhat obscured by the use of the logarithmic scale for specific volume, the vapor region of Fig. 2.28 looks much like the ideal-gas T–v plot shown earlier (cf. Fig. 2.8). Of course, the real-gas behavior of steam affects the details of such a comparison.

P–v Diagrams

Figure 2.29 presents a P–v diagram for H_2O showing a single isotherm ($T = 373.12$ K or $99.97°C$). The corresponding saturation pressure is 101,325 Pa (1 atm). The saturation condition thus corresponds to the normal boiling point of water. Note that both axes in Fig. 2.29 are logarithmic. Isotherms for temperatures greater than 373.12 K lie above and to the right of the isotherm shown. In the superheated vapor region, the isotherms lie relatively close to the saturated vapor line in the log–log coordinates of Fig. 2.29. Figure 2.30 presents an expanded view of the region $0.01 < P < 0.1$ MPa showing several isotherms in linear P–v space.

FIGURE 2.29

P–v diagram for H₂O showing T = 373.12 K (99.97°C) isotherm. An expanded view of several isotherms in the superheated vapor region (P = 0.01–0.1 MPa) is shown in Fig. 2.30.

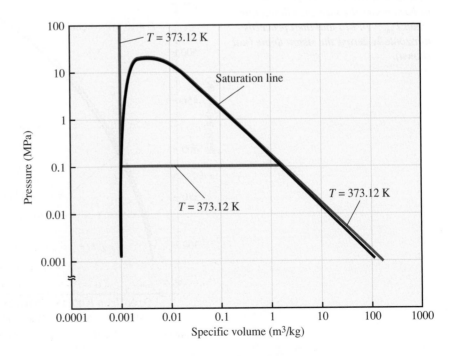

FIGURE 2.30
Isotherms in P–v space for the superheated region of H_2O exhibit a hyperbolic-like behavior (cf. ideal-gas behavior in Fig. 2.7).

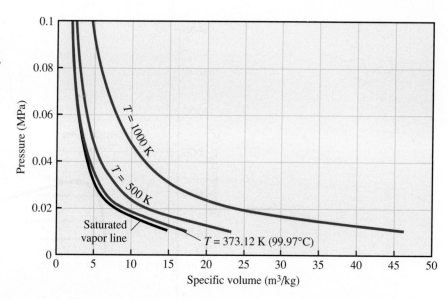

Example 2.20

A 0.5-kg mass of steam at 800 K and 1 MPa is contained in a rigid vessel. The steam is cooled to the saturated vapor state. Plot the process on both $T–v$ and $P–v$ coordinates. Determine the volume of the vessel and the final temperature and pressure of the steam.

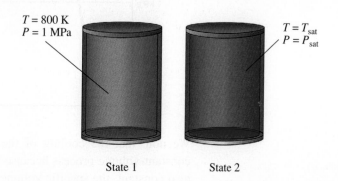

Solution

Known   Steam, M, P_1, T_1, saturated vapor at state 2

Find P_2, T_2, $\mathcal{V}$

Assumptions

i. Rigid tank (given)
ii. Simple compressible substance

Analysis   Before we can create $T–v$ and $P–v$ sketches, we need to determine the region in which the state-1 point lies. Since the state-1 temperature (800 K) is greater than the critical temperature ($T_c = 647.27$ K) and the state-1 pressure (1 MPa) is less than the critical pressure ($P_c = 22.064$ MPa), state 1 must lie in the superheated vapor region. With this information, we now construct the following $T–v$ and $P–v$ plots:

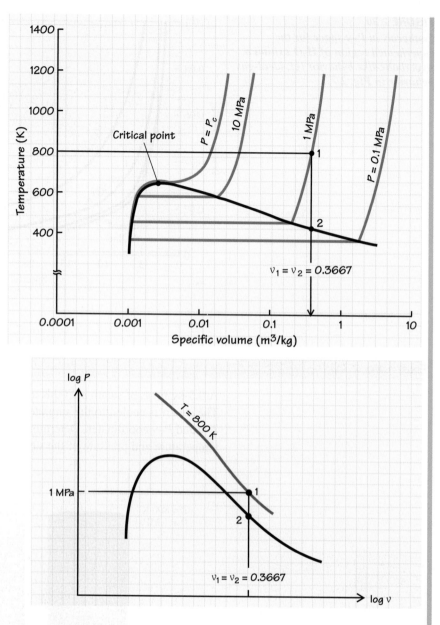

We note that the cooling of the steam from state 1 to state 2 is a constant-volume process because the tank is rigid. Because the mass is also constant, the specific volume at state 1 equals the specific volume at state 2, as shown on the sketches. To find P_2, T_2, and $\mathcal{V}$ requires a numerical value for v_2 ($= v_1$). Using data from Table D.3, we find the specific volume v_1:

T (K)	v (m³/kg)
780	0.35734
800 = T_1	0.36677 = v_1
820	0.37618

The NIST12 v. 5.2 software yields a value identical to the tabulated value. Since state 2 lies on the saturated vapor line, we know that $v_g\,(T_2, P_2) = v_2 = v_1$. From Table D.2, we see that the state-2 specific volume is between those of saturated vapor at 0.5 and 0.6 MPa; thus, we interpolate to find P_2 and T_2 as follows:

v_{sat} (m³/kg)	P_{sat} (MPa)	T_{sat} (K)
0.37481	0.5	424.98
0.36677 = v_2	0.514 = P_2	425.9 = T_2
0.31558	0.6	431.98

Using the NIST database provides a more accurate result:

$$P_2 = 0.511 \text{ MPa,}$$
$$T_2 = 425.8 \text{ K.}$$

To calculate the volume of the rigid vessel containing the steam, we apply the definition of specific volume

$$v \equiv V/M,$$

or

$$V = Mv.$$

Thus,

$$V = (0.5 \text{ kg}) \, 0.36677 \text{ m}^3/\text{kg} = 0.1834 \text{ m}^3.$$

Comments This example illustrates some of the thought processes involved in determining the region in which a state point lies. We also see the importance of recognizing that the process involved was a constant-volume process.

  The system of Example 2.20 is further cooled to a final temperature of 350 K. Determine the final pressure and the quality at this state.

(Answer: $P = P_{sat}$ (350 K) = 41.68 kPa, x = 0.095)

2.6b Calorific and Second-Law Properties

The sources available for calorific properties—specific enthalpies and specific internal energies—and the second-law property—entropy—are the same as for state properties: Appendix D, the NIST online database [11], and NIST12. In the NIST data, both enthalpies and internal energies are provided; however, many sources provide only enthalpies, leaving it to the user to obtain internal energies from the definition

$$u \, (T, P) = h \, (T, P) - Pv.$$

Also frequently tabulated is the **enthalpy of vaporization,** h_{fg}, which is defined as

$$h_{fg} \equiv h_{vap} - h_{liq} = h_g - h_f. \tag{2.52}$$

Physically, this represents the amount of energy required to vaporize a unit mass of liquid at constant pressure. This quantity is also sometimes referred

FIGURE 2.31

*Temperature–entropy (T–s) diagram
for water showing liquid–vapor
saturation lines and the 1-MPa isobar.*

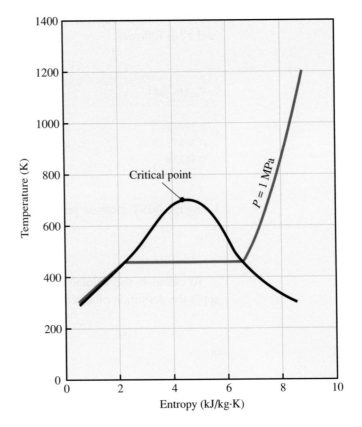

to as the latent heat of vaporization, a term coined during the reign of the caloric theory.[16]

T–s Diagrams

Of equal importance to $T–v$ and $P–v$ diagrams is the temperature–entropy, or $T–s$, diagram. For a reversible process, the integral of $T ds$ provides the energy added as heat to a system; thus, the area under a reversible process line on $T–s$ coordinates represents the energy added or removed by heat interactions. See Chapter 4 for a discussion of heat transfer and see Chapter 7 for a more expansive discussion of $T–s$ diagrams.

Figure 2.31 presents a $T–s$ diagram for water and shows an isobar traversing the compressed liquid region, across the steam dome, and up into the superheat region. Without an expanded scale, the 1-MPa isobar in the compressed liquid region is indistinguishable from the saturated liquid line; however, it does lie above and to the left of the saturated liquid line. Isobars for pressures greater than 1 MPa lie above the isobar shown, and those for lower pressures lie below. Although not shown, isochors (constant-volume lines) in the superheat region have steeper slopes than the isobars (see Fig. 2.11). Note that Fig. 2.31 employs linear scales for both temperature and entropy, rather than the semilog and log–log scales previously used in the analogous $T–v$ (Fig. 2.27) and $P–v$ (Fig. 2.29) diagrams, respectively.

---

[16] See Chapter 4 for a brief history of the caloric theory and how it was superceded by modern constructs of energy.

Example 2.21

Consider an isothermal expansion of steam from an initial state of saturated vapor at 3 MPa to a pressure of 1 MPa. Plot the process on T–s and P–v coordinates and determine the initial and final specific volumes.

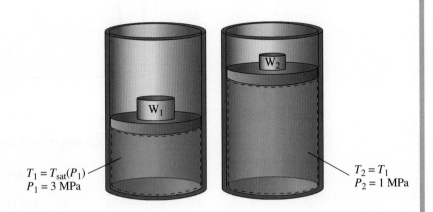

$T_1 = T_{sat}(P_1)$
$P_1 = 3$ MPa

$T_2 = T_1$
$P_2 = 1$ MPa

Solution

Known   Saturated vapor at P_1, T_1 ($= T_2$), P_2

Find v_1 ($= v_g$), v_2

Assumption

Simple compressible substance at equilibrium

Analysis   We begin by drawing the P_1 ($= 3$ MPa) and P_2 ($= 1$ MPa) isobars on a T–s diagram as shown. State 1 is identified on this diagram as a saturated vapor. For the isothermal process, a horizontal line is extended from the state-1 point. The location where this isotherm, $T = T_{sat}$ (3 MPa), crosses the 1-MPa isobar identifies the state-2 point. State 2 is in the superheated vapor region. On P–v coordinates, we draw the same $T = T_{sat}$ (3 MPa) isotherm. Where it crosses the 1-MPa isobar identifies the state-2 point in P–v space.

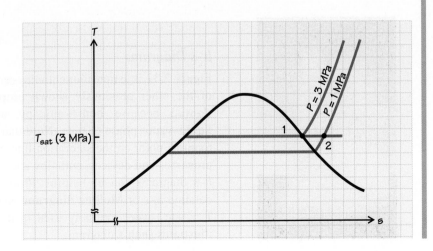

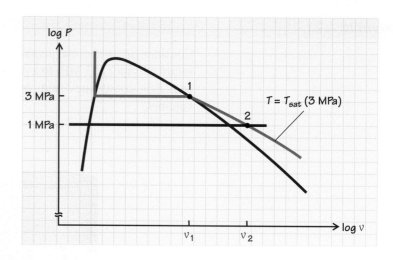

From Table D.2, we obtain the following properties:

$$P_1 = 3 \text{ MPa},$$

$$T_1 = 507.00 \text{ K},$$

$$v_1 = 0.066664 \text{ m}^3/\text{kg}.$$

To obtain the state-2 properties, we can interpolate to find v_2 in the superheated-vapor table for $P = 1$ MPa (see Table D.3). Alternatively, the NIST database can be used directly to determine v_2 by generating isothermal data ($T = 507.00$ K) with pressure increments containing $P = 1$ MPa. The result is

$$v_2 = 0.22434 \text{ m}^3/\text{kg}.$$

Comments   We note the utility of defining constant-property lines on T–s and P–v diagrams. The reader should verify the state-2 property determinations using Table D.3 and the NIST online database and/or software.

Self Test
2.21

 The system of Example 2.21 is allowed to expand isothermally until the final specific volume is 0.32 m³/kg. Find the final pressure.

(Answer: P ≅ 0.70 MPa)

h–s Diagrams

Figure 2.32 illustrates an enthalpy–entropy diagram for water. Unlike all of the other previously shown property diagrams, the steam dome is skewed because both enthalpy and entropy increase during the liquid–vapor phase change. Note that the critical point does not lie at the topmost point on the saturation line in h–s space. In the analysis of many processes and devices, enthalpy and entropy are key properties and hence h–s diagrams can be helpful. Prior to the advent of digital computers, detailed h–s or **Mollier diagrams** showing numerous properties (e.g., P, T, x, etc.) were routinely used in thermodynamic analyses. Figure 2.33 illustrates such a diagram.

FIGURE 2.32
Enthalpy–entropy (h–s) diagram for water showing liquid and vapor saturation lines and the 1-MPa isobar. Also shown is the 453.03-K isotherm.

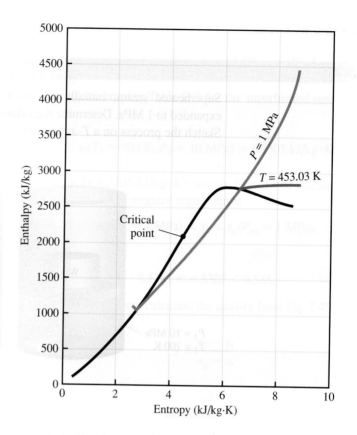

FIGURE 2.33
Mollier (h–s) diagram for water.
Courtesy of The Babcock & Wilcox Company.

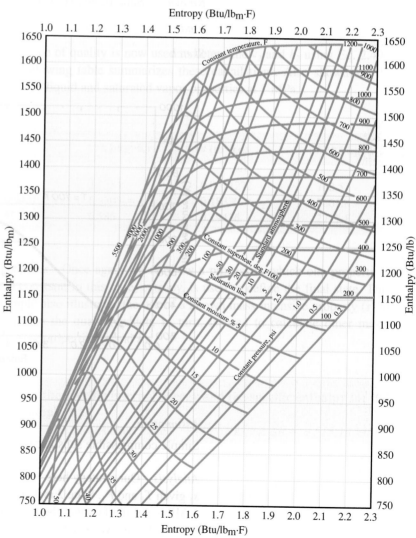

2.7 LIQUID PROPERTY APPROXIMATIONS

Although accurate thermodynamic property data are available for many substances in the compressed liquid region (see Table 2.7), we now discuss some approximations. These approximations for liquid properties are useful to simplify some analyses, and they can be used when detailed compressed liquid data are not available.

For most liquids, the specific volume v and specific internal energy u are nearly independent of pressure; hence, they depend only on temperature, so that,

$$v(T, P) \cong v(T),$$

$$u(T, P) \cong u(T).$$

Since saturation properties are available for many liquids, v and u can be approximated using the corresponding saturation values where the saturation state is evaluated at the temperature of interest, that is,

$$v(T, P) \cong v_f(T_{sat} = T) \tag{2.53a}$$

and

$$u(T, P) \cong u_f(T_{sat} = T). \tag{2.53b}$$

The enthalpy can also be approximated using these relationships, together with the definition $h = u + Pv$; thus,

$$h(T, P) \cong h_f(T_{sat} = T) + (P - P_{sat})v_f(T_{sat} = T). \tag{2.53c}$$

Example 2.23

Determine the error associated with the evaluation of v and h from Eqs. 2.53a–2.53c for water at 10 MPa and 300 K.

Solution

Known   H$_2$O (liquid), P, T

Find v and h (approximations and "exact")

Sketch

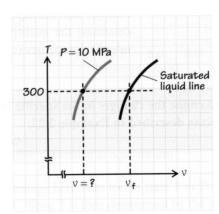

Assumption

Water is approximately incompressible

Analysis   Using the NIST database, we obtain the following values for the compressed liquid at 10 MPa and 300 K:

$$v = 0.00099905 \text{ m}^3/\text{kg},$$

$$h = 121.73 \text{ kJ/kg}.$$

To obtain the approximate specific volume, we employ Eq. 2.53a together with Table D.1, which gives

$$v(T, P) \cong v_f(T_{sat} = T) = v_f(300 \text{ K}) = 0.0010035 \text{ m}^3/\text{kg}.$$

We also employ values from Table D.2 to evaluate Eq. 2.54c for the approximate enthalpy:

$$h(T, P) \cong h_f(T) + [P - P_{sat}(T)]v_f(T)$$

$$= 112.56 \times 10^3 \text{ J/kg} + (10 \times 10^6 \text{ Pa} - 3.5369$$

$$\times 10^3 \text{ Pa})\, 0.0010035 \text{ m}^3/\text{kg}$$

$$= 122.59 \times 10^3 \text{ J/kg or } 122.59 \text{ kJ/kg}.$$

Verification of the units is left to the reader. The errors associated with the approximations are given in the following table:

Property	NIST Value	Approximation	Error
v (m³/kg)	0.00099905	0.0010035	+0.44%
h (kJ/kg)	121.73	122.59	+0.71%

Comments   We see that the approximate values are quite close to the exact values, with errors of less than 1%. We also note that the high pressure (10 MPa) makes a significant contribution to the enthalpy.

Self Test 2.23   ☑ **A beaker contains H_2O at atmospheric pressure and 10°C. Find the specific volume, specific internal energy, and specific enthalpy of the H_2O.**

(Answer: $v = 0.001$ m³/kg, $u = 41.391$ kJ/kg, $h = 41.491$ kJ/kg)

2.8 SOLIDS

We have focused thus far on the thermodynamic properties of gases, liquids, and their mixtures. We now examine solids. Figure 2.34 presents a phase diagram for water showing the solid region, which is designated I; the liquid region, designated II; and the vapor region, III. A key feature of this diagram is the triple point. As you may recall, the triple point is the state at which all three phases (solid, liquid, and vapor) of a substance coexist in equilibrium. For water, the triple point temperature is 0.01°C (273.16 K) and the triple point pressure is 0.00604 atm (0.6117 kPa). The nearly vertical line that originates at the triple point, the **solidification** or **fusion line**, separates the solid region from the liquid region. Another line, the **sublimation line**, separates the solid from the vapor phase and ends at the triple point. A third line, which starts at the triple point and continues up to the critical point, comprises the saturation states that were defined in our discussion of liquid–vapor mixtures. The normal (i.e., 1 atm) freezing point

FIGURE 2.34

Phase diagram for water showing solid (I), liquid (II), and vapor (III) regions. Indicated on the diagram are the critical point ($P_c = 217.8$ atm, $T_c = 647.10$ K), the normal boiling (steam) point ($P_{boil} = 1$ atm, $T_{boil} = 373.12$ K), the normal freezing (ice) point ($P_{freeze} = 1$ atm, $T_{freeze} = 273.15$ K), and the triple point ($P_{triple} = 0.006$ atm, $T_{triple} = 273.16$ K). Note the scale change above 2 atm. **Adapted from Ref. [15] with permission.**

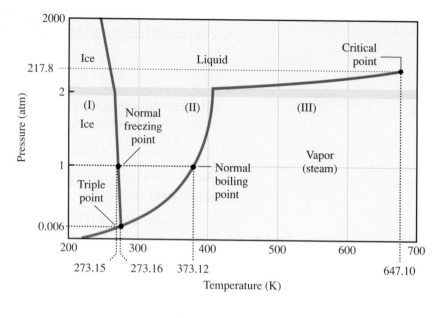

Iodine sublimes from a bluish-black, metallic looking solid to a purple vapor. The vapor pressure of solid iodine at 90°C is 3.57 kPa.

is also indicated on the phase diagram and corresponds to a temperature of 0°C (273.15 K).

Important properties associated with solid–liquid and solid–vapor phase changes are the enthalpy of fusion, h_{fusion}, and the enthalpy of sublimation, h_{sublim}, respectively:

$$h_{fusion} = h_{liq} - h_{solid}, \qquad (2.54a)$$

$$h_{sublim} = h_{vap} - h_{solid}. \qquad (2.54b)$$

Values of these quantities for H_2O at the triple point are 333.4 kJ/kg for h_{fusion} and 2834.3 kJ/kg for h_{sublim}.

In most thermal science applications, the thermodynamic properties of interest for solids are the density (reciprocal specific volume) and specific heats. The dependence of these properties on pressure is very slight over a wide range; in fact, the effect of pressure is so small that solid properties are usually assumed to be functions of temperature alone, that is,

$$\rho = \rho \, (T \text{ only}) \qquad (2.55a)$$

and

$$c_p = c_p \, (T \text{ only}). \qquad (2.55b)$$

Densities and constant-pressure specific heats are tabulated in Appendix I for a number of solids of engineering interest.

To simplify the thermal analysis of solid systems, we frequently assume that the solid is an **incompressible** substance. With this assumption, the density (or specific volume) is a constant, independent of both pressure and temperature; that is,

$$\rho = 1/v = \text{constant.} \qquad (2.56a)$$

Furthermore,

$$c_p = c_v \equiv c = \text{constant.} \qquad (2.56b)$$

When invoking the incompressible substance approximation, properties are usually evaluated at an appropriate average temperature. The proof that $c_p = c_v$ for an incompressible substance is left as an exercise for the reader (Problem 2.118).

Example 2.24

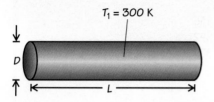

$T_1 = 300$ K

A pure copper rod of diameter $D = 25$ mm and length $L = 150$ mm is initially at a uniform temperature of 300 K. The rod is heated at 1 atm to a uniform temperature of 450 K. Estimate the change in internal energy $U([=] \text{J})$ of the rod in going from the initial to the final state.

Solution

Known   Pure Cu, L, D, T_1, T_2

Find $U_2 - U_1 \equiv \Delta U$

Sketch

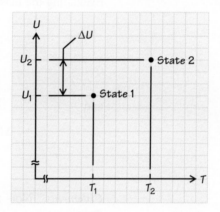

Assumptions

Incompressible solid

Analysis   We develop the required calorific equation of state for copper by starting with the definition of the constant-volume specific heat (Eq. 2.19a),

$$c_v \equiv \left(\frac{\partial u}{\partial T} \right)_v.$$

With the assumption of incompressibility, this becomes an ordinary derivative, which can be separated and integrated as follows:

$$\Delta u = \int_{u_1}^{u_2} du = \int_{T_1}^{T_2} c_v \, dT = c(T_2 - T_1).$$

For a system of mass M, the internal energy change is thus

$$\Delta U = M \Delta u = Mc(T_2 - T_1).$$

We find the mass of the rod as follows:

$$M = \rho \mathbb{V} = \rho \frac{\pi D^2}{4} L.$$

We use Appendix I (Table I.1) to obtain values for c and ρ. Because values of c are given for several temperatures, we interpolate for an average temperature $T[=(T_1 + T_2)/2]$ of 375 K:

T (K)	c_p (J/kg·K)
200	356
375	392
400	397

Because only a single value of the density is given in Table I.1, we use that value:

$$\rho(300 \text{ K}) = 8933 \text{ kg/m}^3 \cong \rho(375 \text{ K}).$$

Evaluating the mass and internal energy change yields

$$M = 8933 \frac{\pi(0.025)^2}{4} 0.150 = 0.6577$$

$$[=] \frac{\text{kg}}{\text{m}^3} (\text{m}^2)\text{m} = \text{kg},$$

and

$$\Delta U = 0.6577 \,(392)(450 - 300) = 38{,}700$$

$$[=] \text{kg} \frac{\text{J}}{\text{kg} \cdot \text{K}} \text{K} = \text{J}.$$

Comments   Note the use of a c_p value based on an average temperature. This approach to find an average c_p is frequently used. That the temperature is uniform at both the start and end of the process also greatly simplifies this problem. (How would you solve the problem if a temperature *distribution* were given at the end of the heating process, rather than a single uniform temperature?)

  Redo Example 2.24 for an iron rod of the same dimensions. Compare your answer with that of Example 2.24 and discuss your results.

Self Test 2.24

(*Answer:* $\Delta U = 41.4$ kJ. Even though the density of iron is lower than that of copper, more energy is required for the same process because iron has a higher c_p value.)

Example 12.15 in Chapter 12 shows that treating the water vapor in moist atmospheric air as an ideal gas is a good approximation.

2.9 IDEAL-GAS MIXTURES

So far we have confined our discussion of properties to pure substances; however, many practical devices involve mixtures of pure substances. Air conditioning and combustion systems are common examples of such. In the former application, water vapor becomes an important component of air, which already is a mixture of several components, whereas combustion systems deal with reactant mixtures of fuel and oxidizer and with product mixtures of various components. In both of these examples, we can treat the various gas streams as ideal-gas mixtures with reasonable accuracy for a wide range of conditions. For air conditioning and humidification systems, this results from the small amounts of water involved in the

Condensation of water vapor from combustion products.

Some specific examples in Chapter 12 include the operation of evaporative coolers (Example 12.17) and household dehumidifiers (Example 12.18).

Conservation of mass for reacting systems uses mole and mass fractions. See Examples 3.14 and 3.15 in Chapter 3.

mixtures. For combustion, the high temperatures typically involved result in mixtures of low density. Applications of the concepts developed in this section are found in Chapters 3, 5, 11, and 12 for combustion systems and in Chapter 12 for air conditioning and humidification systems.

2.9a Specifying Mixture Composition

To characterize the composition of a mixture, we define two important and useful quantities: the constituent mole fractions and mass fractions. Consider a multicomponent mixture of gases composed of N_1 moles of species 1, N_2 moles of species 2, etc. The **mole fraction of species i, X_i,** is defined as the fraction of the total number of moles in the system that are species i:

$$X_i \equiv \frac{N_i}{N_1 + N_2 + \cdots} = \frac{N_i}{N_{\text{tot}}} \qquad (2.57a)$$

Similarly, the **mass fraction of species i, Y_i,** is the fraction of the total mixture mass that is associated with species i:

$$Y_i \equiv \frac{M_i}{M_1 + M_2 + \cdots + M_i + \cdots} = \frac{M_i}{M_{\text{tot}}}. \qquad (2.57b)$$

Note that, by definition, the sum of all the constituent mole (or mass) fractions must be unity; that is,

$$\sum_{i=1}^{J} X_i = 1, \qquad (2.58a)$$

$$\sum_{i=1}^{J} Y_i = 1, \qquad (2.58b)$$

where J is the total number of species in the mixture.

We can readily convert mole fractions and mass fractions from one to another using the molecular weights of the species of interest and the apparent molecular weight of the mixture:

$$Y_i = X_i \mathcal{M}_i / \mathcal{M}_{\text{mix}}, \qquad (2.59a)$$

$$X_i = Y_i \mathcal{M}_{\text{mix}} / \mathcal{M}_i. \qquad (2.59b)$$

The apparent mixture molecular weight, denoted $\mathcal{M}_{\text{mix}}$, is easily calculated from knowledge of either the species mole or mass fractions:

$$\mathcal{M}_{\text{mix}} = \sum_{i=1}^{J} X_i \mathcal{M}_i, \qquad (2.60a)$$

or

$$\mathcal{M}_{\text{mix}} = \frac{1}{\displaystyle\sum_{i=1}^{J} (Y_i / \mathcal{M}_i)}. \qquad (2.60b)$$

2.9b State (P–v–T) Relationships for Mixtures

In our treatment of mixtures, we assume a double ideality: first, that the pure constituent gases obey the ideal-gas equation of state (Eq. 2.8), and, second, that when these pure components mix, an ideal solution results. In an **ideal solution**, the behavior of any one component is uninfluenced by the presence of any other component.

Let us explore the characteristics of an ideal solution considering, first, the thermodynamic property pressure. Consider a fixed volume V containing two or more different species. If the mixture (solution) is ideal, gas molecules of species A are free to roam through the entire volume V, as if no other species were present. The pressure that molecules A exert on the wall is less than the total pressure, however, since other non-A molecules also collide with the wall. We can thus define the **partial pressure** of species A, P_A, by applying the ideal-gas equation of state just to the A molecules:

$$P_A V = N_A R_u T. \qquad (2.61)$$

Since each species behaves independently, similar expressions can be written for each,

$$P_B V = N_B R_u T,$$
$$\vdots \qquad \vdots$$
$$P_i V = N_i R_u T,$$

etc. Summing this set of equations for a mixture containing J species yields

$$(P_A + P_B + \cdots + P_i + \cdots + P_J)V = (N_A + N_B + \cdots + N_i + \cdots + N_J)R_u T. \qquad (2.62)$$

Because the sum of the number of moles of each constituent is the total number of moles in the system, N_{tot}, this equation becomes

$$\sum_{i=1}^{J} P_i V = N_{tot} R_u T, \qquad (2.63)$$

where P_i is the partial pressure of the ith species. Because the mixture as a whole obeys the ideal-gas law,

$$P V = N_{tot} R_u T, \qquad (2.64)$$

the sum of the **partial pressures** must be identical to the total pressure, that is,

$$\sum_{i=1}^{J} P_i = P. \qquad (2.65)$$

This statement is known as *Dalton's law of partial pressures* and is schematically illustrated in Fig. 2.35. Here we see four containers, each having the same volume V. Gas A fills one container, gas B another, and gas C the third. Each gas has the same temperature T. The absolute pressure of the gas in each container is measured to be P_A, P_B, and P_C, respectively. We now conduct the thought experiment in which gases A, B, and C are transferred to the fourth container, again at temperature T. The absolute pressure of the mixture in the fourth container is found to be the sum of the absolute pressures measured when the gases were segregated. That the contribution of a single species to the total pressure in a gas mixture is the same as that of the pure species occupying the same total volume at the same temperature is a defining characteristic of an ideal-gas mixture (i.e., an ideal solution). In nonideal mixtures (solutions), the total pressure is not necessarily equal to the sum of the pure component pressures as in the thought experiment.

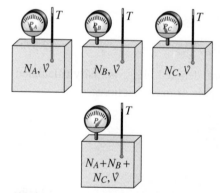

FIGURE 2.35

In an ideal-gas mixture, the sum of the pressures associated with each component isolated in the same volume at the same temperature is identical to the total pressure observed when all species are confined together in the same volume at the same temperature.

In Chapter 12, the partial pressure of water vapor is used to define the relative humidity of moist air. See Eq. 12.62.

The partial pressure can be related to the mixture composition and total pressure by dividing Eq. 2.61 by Eq. 2.64,

$$\frac{P_i \mathcal{V}}{P \mathcal{V}} = \frac{N_i R_u T}{N_{\text{tot}} R_u T},$$

which simplifies to

$$\frac{P_i}{P} = \frac{N_i}{N_{\text{tot}}} \equiv X_i, \qquad (2.66a)$$

or

$$P_i = X_i P. \qquad (2.66b)$$

Ideal-gas mixtures also exhibit the property that when component gases having different volumes but identical pressures and temperatures are brought together, the mixture volume is the sum of the pure component volumes (Fig. 2.36). Mathematically, we express this idea by writing the ideal-gas equation of state for each pure constituent,

$$P \mathcal{V}_A = N_A R_u T,$$
$$P \mathcal{V}_B = N_B R_u T,$$
$$P \mathcal{V}_C = N_C R_u T,$$

and summing to yield

$$P(\mathcal{V}_A + \mathcal{V}_B + \mathcal{V}_C) = (N_A + N_B + N_C) R_u T.$$

Since

$$P \mathcal{V}_{\text{tot}} = N_{\text{tot}} R_u T,$$

we conclude that the individual volumes, or partial volumes, must equal the total volume when combined, that is,

$$\mathcal{V}_{\text{tot}} = \mathcal{V}_A + \mathcal{V}_B + \mathcal{V}_C. \qquad (2.67)$$

This view of an ideal-gas mixture is frequently referred to as *Amagat's model*.

The partial volumes also can be related to the mixture composition by defining a volume fraction,

$$\mathcal{V}_i / \mathcal{V} = N_i / N_{\text{tot}}. \qquad (2.68)$$

Comparing Eqs. 2.66a and 2.68, we see that the three measures of mixture composition—the ratio of the partial pressure to the total pressure, the volume fraction, and the mole fraction—are all equivalent for an ideal-gas mixture:

$$P_i / P = \mathcal{V}_i / \mathcal{V} = N_i / N \quad (= X_i). \qquad (2.69)$$

For insight into the behavior of nonideal mixtures, the reader is referred to Refs. [20, 21].

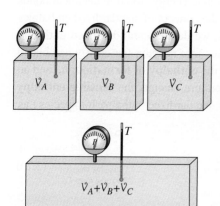

FIGURE 2.36

In an ideal-gas mixture, the sum of the volumes associated with each component isolated at the same temperature and the same pressure is identical to the total volume when all species are confined together at the same temperature and the same pressure.

2.9c Standardized Properties

From our earlier discussion of the calorific equation of state for ideal gases, we see that a reference temperature is required to evaluate the specific internal energy and enthalpy (see Eqs. 2.31c and 2.33c). If one is concerned only with a single pure substance, as opposed to a reacting system where both reactant and product species are present, then Eqs. 2.31c and 2.33c would suffice to describe the internal energy and enthalpy changes for all thermodynamic processes, as only differences in states are of importance. Moreover, any choice of reference temperature would yield the same results. For example, the enthalpy change associated with a change of temperature from T_1 to T_2 is calculated from Eq. 2.33c as

$$h(T_2) - h(T_2) = \int_{T_{ref}}^{T_2} c_p dT - \int_{T_{ref}}^{T_1} c_p dT = \int_{T_1}^{T_2} c_p dT,$$

a straightforward calculation in which the reference temperature drops out. In reacting systems, however, we must include energy stored in chemical bonds in our accounting. To accomplish this, the concept of standardized enthalpies is extremely valuable. For any species, we define a **standardized enthalpy** that is the sum of an enthalpy that takes into account the energy associated with chemical bonds (or lack thereof), the **enthalpy of formation, h_f**, and an enthalpy associated only with a temperature change, the **sensible enthalpy change, Δh_s**. Thus, we write the molar standardized enthalpy for species i as

Table 2.8 Standard Reference State

$$\overline{h}_i(T) \quad = \quad \overline{h}^{\circ}_{f,i}(T_{ref}) \quad + \quad \Delta\overline{h}_{s,i}(T),$$

| Standardized enthalpy at temperature T | Enthalpy of formation at standard reference state (T_{ref}, P°) | Sensible enthalpy change in going from T_{ref} to T |

(2.70)

$T_{ref} \equiv 298.15$ K
$P_{ref} \equiv 101,325$ Pa
For elements in their naturally occurring state (solid, liquid, or gas) at T_{ref} and P_{ref}*
$h_i(T_{ref}) \equiv 0$

where

$$\Delta\overline{h}_{s,i} \equiv \overline{h}_i(T) - \overline{h}^{\circ}_{f,i}(T_{ref}).$$

*For example, the standard-state enthalpies for C(s), N_2(g), and O_2(g) are all zero.

To make practical use of Eq. 2.70, it is necessary to define a **standard reference state** (Table 2.8). We employ a standard-state temperature, $T_{ref} = 25°C$ (298.15 K), and standard-state pressure, $P_{ref} = P^{\circ} = 1$ atm (101,325 Pa), consistent with the JANAF [5], Chemkin [3], and NASA [6] thermodynamic databases. Although not needed to describe enthalpies, the standard-state pressure is required to calculate entropies at pressures other than one atmosphere (see Eq. 2.39a). In addition to defining T_{ref} and P_{ref}, we adopt the convention that enthalpies of formation are zero for the elements in their naturally occurring state at the reference-state temperature and pressure. For example, oxygen exists as diatomic molecules at 25°C and 1 atm; hence,

$$(\overline{h}^{\circ}_{f, O_2})_{298} = 0,$$

where the superscript $^{\circ}$ is used to denote that the value is for the standard-state pressure.[17]

To form oxygen atoms at the standard state requires the breaking of a rather strong chemical bond. The bond dissociation energy for O_2 at 298 K is 498,390 kJ/kmol$_{O_2}$. Breaking this bond creates two O atoms; thus, the enthalpy of formation for atomic oxygen is half the value of the O_2 bond dissociation energy:

$$(\overline{h}^{\circ}_{f,O}) = 249,195 \text{ kJ/kmol}_O.$$

Thus, enthalpies of formation have a clear physical interpretation as the net change in enthalpy associated with breaking the chemical bonds of the

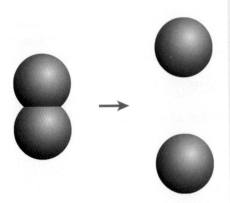

Shared electrons form a covalent bond in molecular oxygen. An energy input is required to break this bond to form two oxygen atoms.

---

[17] The use of this superscript is redundant for ideal-gas enthalpies, which exhibit no temperature dependence; for entropies, however, the superscript is important.

FIGURE 2.37
Graphical interpretation of standardized enthalpy, enthalpy of formation, and sensible enthalpy.

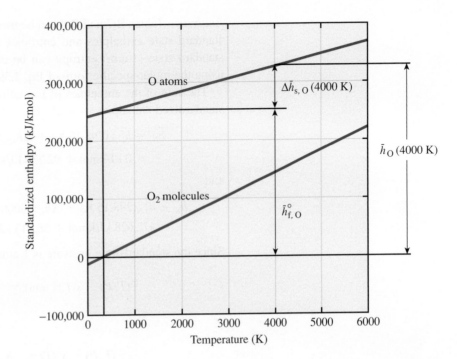

Standardized enthalpies are first used in conservation of energy for combustion systems. See Examples 5.7–5.9 in Chapter 5.

standard-state elements and forming new bonds to create the compounds of interest.

Representing the standardized enthalpy graphically provides a useful way to understand and use this concept. In Fig. 2.37, the standardized enthalpies for atomic oxygen (O) and diatomic oxygen (O_2) are plotted versus temperature starting from absolute zero. At 298.15 K, we see that $\bar{h}_{O_2}$ is zero (by definition of the standard-state reference condition), and the standardized enthalpy of atomic oxygen equals its enthalpy of formation, since the sensible enthalpy at 298.15 K is zero. At the temperature indicated (4000 K), we see the additional sensible enthalpy contribution to the standardized enthalpy. In Appendix B, enthalpies of formation at the reference state are given, and sensible enthalpies are tabulated as a function of temperature for a number of species of importance in combustion. Enthalpies of formation for reference temperatures other than the standard state 298.15 K are also tabulated.

Although the choice of a zero datum for the enthalpy of the elements at the reference state is arbitrary (being merely a pragmatic choice), the third law of thermodynamics is used to set zero values for entropies; thus, the standard-state entropy values are always larger than zero.

Example 2.25

Determine the standardized molar- and mass-specific enthalpies and entropies of N_2 and N at 3000 K and 2.5 atm.

Solution

Known N_2, N, T, P

Find $\bar{h}, h, \bar{s}, s$

Assumptions

Ideal-gas behavior

Analysis Tables B.7 and B.8 can be used to determine the molar-specific standard-state enthalpies and entropies for N_2 and N, respectively. The standard-state (1-atm) entropy can be converted to the value at 2.5 atm using the molar-specific form of Eq. 2.39a.

The sum of the enthalpies of formation and sensible enthalpies are the standardized enthalpies we seek (Eq. 2.70). At 3000 K, these values are

$$\bar{h}_{N_2} = \bar{h}^{\circ}_{f,N_2}(298.15 \text{ K}) + \Delta\bar{h}_{s,N_2}(3000 \text{ K})$$

$$= 0 \text{ kJ/kmol} + 92{,}730 \text{ kJ/kmol} = 92{,}730 \text{ kJ/kmol}$$

and

$$\bar{h}_{N} = \bar{h}^{\circ}_{f,N}(298.15 \text{ K}) + \Delta\bar{h}_{s,N}(3000 \text{ K})$$

$$= 472{,}628 \text{ kJ/kmol} + 56{,}213 \text{ kJ/kmol} = 528{,}841 \text{ kJ/kmol}.$$

Since the standard-state pressure is 1 atm, application of Eq. 2.39a yields

$$\bar{s}(T,P) - \bar{s}(T, 1 \text{ atm}) = -R_u \ln\left(\frac{P \text{ (atm)}}{1 \text{ atm}}\right),$$

or

$$\bar{s}(T,P) - \bar{s}^{\circ}(T) = -R_u \ln\left(\frac{P \text{ (atm)}}{1 \text{ atm}}\right).$$

Thus,

$$\bar{s}(3000 \text{ K}, 2.5 \text{ atm}) = \bar{s}^{\circ}(3000 \text{ K}) - R_u \ln 2.5.$$

Using the values from Tables B.7 and B.8, we evaluate this equation as follows:

$$\bar{s}_{N_2} = 266.810 - 8.31447 \ln 2.5 = 259.192 \text{ kJ/kmol} \cdot \text{K}$$

and

$$\bar{s}_{N} = 201.199 - 8.31447 \ln 2.5 = 193.581 \text{ kJ/kmol} \cdot \text{K}.$$

To compute the mass-specific enthalpies and entropies, we divide the molar-specific values by the molecular weight of the respective species; for example,

$$h_{N_2} = \frac{\bar{h}}{\mathcal{M}_{N_2}} = \frac{92{,}730 \text{ kJ/kmol}}{28.013 \text{ kg/kmol}} = 3310.25 \text{ kJ/kg}.$$

The following table summarizes these calculations:

	$\mathcal{M}$ (kg/kmol)	$\bar{h}$ (kJ/kmol)	h (kJ/kg)	$\bar{s}$ (kJ/kmol·K)	s (kJ/kg·K)
N_2	28.013	92,730	3310.25	259.192	9.2526
N	14.007	528,841	37,755	193.581	13.8203

Comments This example illustrates the use of the tables in Appendix B to calculate standardized properties at temperatures and pressure not equal to the reference-state values. Because we assume ideal-gas behavior, the pressure has no effect on the enthalpy and, furthermore, the entropy is a simple function of pressure (see Eq. 2.39a). The conversion from the reference-state pressure (1 atm) to the pressure at the desired state (2.5 atm) is thus straightforward.

Self Test
2.25
☑ **Determine the standardized mass-specific enthalpies and entropies of O and O_2 at 2000 K and atmospheric pressure.**

(Answer: $h_{O_2} = 1849.09$ kJ/kg, $s_{O_2} = 8.396$ kJ/kg·K, $h_O = 17,806.8$ kJ/kg, $s_O = 12.571$ kJ/kg·K)

2.9d Calorific Relationships for Mixtures

The calorific relationships for ideal-gas mixtures are straightforward mass-fraction or mole-fraction weightings of the pure-species individual specific calorific properties:

$$u_{\text{mix}} = \sum_{i=1}^{J} Y_i u_i, \tag{2.71a}$$

$$h_{\text{mix}} = \sum_{i=1}^{J} Y_i h_i, \tag{2.71b}$$

$$c_{v,\text{mix}} = \sum_{i=1}^{J} Y_i c_{v,i}, \tag{2.71c}$$

$$c_{p,\text{mix}} = \sum_{i=1}^{J} Y_i c_{p,i}, \tag{2.71d}$$

or

$$\bar{u}_{\text{mix}} = \sum_{i=1}^{J} X_i \bar{u}_i, \tag{2.71e}$$

$$\bar{h}_{\text{mix}} = \sum_{i=1}^{J} X_i \bar{h}_i, \tag{2.71f}$$

$$\bar{c}_{v,\text{mix}} = \sum_{i=1}^{J} X_i \bar{c}_{v,i}, \tag{2.71g}$$

$$\bar{c}_{p,\text{mix}} = \sum_{i=1}^{J} X_i \bar{c}_{p,i}, \tag{2.71h}$$

> The analysis of a jet engine combustor in Chapter 12 (Example 12.11) uses concepts developed here: mole and mass fractions and standardized properties of ideal-gas mixtures.

where the subscript i represents the ith species and J is the total number of species in the mixture. Since the specific heats, internal energies, and enthalpies of the constituent ideal-gas species depend only on temperature, the same is true for the mixture calorific properties; for example, $u_{\text{mix}} = u_{\text{mix}}$ (T only), etc. Molar-specific enthalpies for a number of species are tabulated in Appendix B.

2.9e Second-Law Relationships for Mixtures

The mixture entropy also is calculated as a weighted sum of the constituents:

$$s_{\text{mix}}(T, P) = \sum_{i=1}^{J} Y_i s_i(T, P_i), \tag{2.72a}$$

$$\bar{s}_{\text{mix}}(T, P) = \sum_{i=1}^{J} X_i \bar{s}_i(T, P_i). \tag{2.72b}$$

Unlike the ideal-gas calorific relationships, pressure is now required as a second independent variable. Here the pure-species entropies (s_i and $\bar{s}_i$) depend on the species partial pressures, as explicitly indicated in Eq. 2.72. Equation 2.39a can be applied to evaluate the constituent entropies in Eq. 2.72 from standard-state ($P_{\text{ref}} \equiv P^\circ = 1$ atm) values as

$$s_i(T, P_i) = s_i(T, P_{\text{ref}}) - R\ln\frac{P_i}{P_{\text{ref}}}, \tag{2.73a}$$

$$\bar{s}_i(T, P) = \bar{s}_i(T, P_{\text{ref}}) - R_u\ln\frac{P_i}{P_{\text{ref}}}, \tag{2.73b}$$

where $P_i = X_i P$. Ideal-gas, standard-state molar-specific entropies are tabulated in Appendix B for several species.

The Gibbs function is an important second-law property that has many uses in dealing with ideal-gas mixtures. For example, the Gibbs function is used to determine the equilibrium composition of reacting gas mixtures (see Chapter 7). Earlier in the present chapter (see Eq. 2.21) we defined the Gibbs function as

$$G \equiv H - TS,$$

or

$$g \equiv h - Ts, \tag{2.74a}$$

$$\bar{g} \equiv \bar{h} - T\bar{s}. \tag{2.74b}$$

For an ideal-gas mixture, the mass- or molar-specific Gibbs function is a weighted sum of the pure-species mass- or molar-specific Gibbs functions:

$$g_{\text{mix}}(T, P) = \sum_{i=1}^{J} Y_i g_i(T, P_i), \tag{2.75a}$$

$$\bar{g}_{\text{mix}}(T, P) = \sum_{i=1}^{J} X_i \bar{g}_i(T, P_i), \tag{2.75b}$$

where

$$g_i(T, P_i) = h_i(T) - Ts_i(T, P_i) = h_i(T) - T\left[s_i^\circ(T) - R\ln\frac{P_i}{P^\circ}\right] \tag{2.76a}$$

and

$$\bar{g}_i(T, P_i) = \bar{h}_i - T\left[\bar{s}_i^\circ(T) - R_u\ln\frac{P_i}{P^\circ}\right]. \tag{2.76b}$$

The connection to the mixture composition is through the ideal-gas relationship $P_i = X_i P$ (Eq. 2.66b).

2.10 SOME PROPERTIES OF REACTING MIXTURES

2.10a Enthalpy of Combustion

Knowing how to express the enthalpy for mixtures of reactants and mixtures of products allows us to define the enthalpy of reaction, or, when dealing specifically with combustion reactions, the enthalpy of combustion. The definition of the **enthalpy of reaction**, or the **enthalpy of combustion**, ΔH_R, is

$$\Delta H_R(T) = H_{prod}(T) - H_{reac}(T), \qquad (2.77a)$$

where T may be any temperature, although a reference-state value of 298.15 K is frequently used. The enthalpy of combustion is illustrated graphically in Fig. 2.38. Note that the standardized enthalpy of the products lies below that of the reactants. For example, at 25°C and 1 atm, the reactants' enthalpy of a stoichiometric mixture of CH_4 and air is $-74,831$ kJ per kmol of fuel. At the same conditions (25°C, 1 atm), the combustion products have a standardized enthalpy of $-877,236$ kJ for the combustion of 1 kmol of fuel. Thus,

$$\Delta H_R = -877,236 - (-74,831) = -802,405 \text{ kJ (per kmol } CH_4).$$

This value is usually expressed on a per-mass-of-fuel basis:

$$\Delta h_R(kJ/kg_{fuel}) = \Delta H_R/\mathcal{M}_{fuel}, \qquad (2.77b)$$

FIGURE 2.38

The enthalpy of reaction is illustrated using values for the reaction of one mole of methane with a stoichiometric quantity of air. The water in the products is assumed to be in the vapor state.

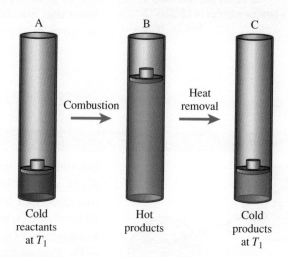

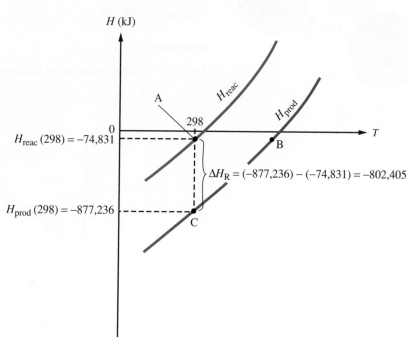

Methane, the largest component of natural gas, has a higher heating value of 55,528 kJ/kg.

or

$$\Delta h_R = (-802,405/16.043) = -50,016 \text{ kJ/kg}_{fuel}.$$

Note that the value of the enthalpy of combustion depends on the temperature chosen for its evaluation. Because the enthalpies of both the reactants and products vary with temperature, the distance between the H_{prod} and H_{reac} lines in Fig. 2.38 is not constant.

2.10b Heating Values

The **heat of combustion, Δh_c** (known also as the **heating value**), is numerically equal to the enthalpy of reaction, but with opposite sign. The **upper** or **higher heating value, HHV**, is the heat of combustion calculated assuming that all of the water in the products has condensed to liquid. In this scenario, the reaction liberates the most amount of energy, hence leading to the designation "upper." The **lower heating value, LHV**, corresponds to the case where none of the water is assumed to condense. For CH_4, the upper heating value is approximately 11% larger than the lower one. Standard-state heating values for a variety of hydrocarbon fuels are given in Appendix H.

> Fuel heating values are used to define the efficiency of gas-turbine engines in Chapter 12. See Eq. 12.38.

Example 2.26

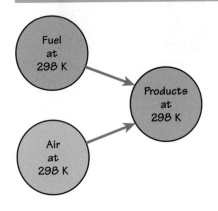

Determine the upper and lower heating values of gaseous n-decane ($C_{10}H_{22}$) for stoichiometric combustion with air at 298.15 K. For this condition, 15.5 kmol of O_2 reacts with each kmol of $C_{10}H_{22}$ to produce 10 kmol of CO_2 and 11 kmol of H_2O. Assume that air can be represented as a mixture of O_2 and N_2 in which there are 3.76 kmol of N_2 for each kmol of O_2. Express the results per unit mass of fuel. The molecular weight of n-decane is 142.284.

Solution

Known T, compositions of reactant and product mixtures, $\mathcal{M}_{C_{10}H_{22}}$

Find Δh_c (upper and lower)

Sketch

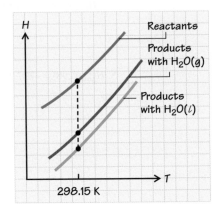

Assumption

Ideal-gas behavior

Conservation of elements is treated in detail in Chapter 3. See Example 3.14.

Analysis   For 1 kmol of $C_{10}H_{22}$, stoichiometric combustion can be expressed from the given information as (see also Eqs. 3.53 and 3.54):

$$C_{10}H_{22}(g) + 15.5\,(O_2 + 3.76\,N_2)$$
$$\rightarrow 10\,CO_2 + 11\,H_2O\,(\ell\ \text{or}\ g) + 15.5\,(3.76)\,N_2.$$

For either the upper or lower heating value,

$$\Delta H_c = -\Delta H_R = H_{\text{reac}} - H_{\text{prod}},$$

where the numerical value of H_{prod} depends on whether the H_2O in the products is liquid (defining the higher heating value) or gaseous (defining the lower heating value). The sensible enthalpy changes for all species involved are zero since we desire ΔH_c at the reference state (298.15 K). Furthermore, the enthalpies of formation of the O_2 and N_2 are also zero at 298.15 K. Recognizing that

$$H_{\text{reac}} = \sum_{\text{reac}} N_i \bar{h}_i \quad \text{and} \quad H_{\text{prod}} = \sum_{\text{prod}} N_i \bar{h}_i,$$

we obtain

$$\Delta H_{c,H_2O(\ell)} = \text{HHV} = (1)\bar{h}^{\circ}_{f,C_{10}H_{22}} - \left[10\bar{h}^{\circ}_{f,CO_2} + 11\bar{h}^{\circ}_{f,H_2O(\ell)}\right].$$

Note that the N_2 contribution as a reactant cancels with the N_2 contribution as a product. Table B.6 (Appendix B) gives the enthalpy of formation for gaseous water; the enthalpy of vaporization, h_{fg}, is obtained from Table D.1 or the NIST database. We thus calculate the enthalpy of formation of the liquid water as follows:

$$\bar{h}^{\circ}_{f,H_2O(\ell)} = \bar{h}^{\circ}_{f,H_2O(g)} - \bar{h}_{fg}$$
$$= -241{,}847\ \text{kJ/mol} - (45{,}876 - 1889)\ \text{kJ/mol}$$
$$= -285{,}834\ \text{kJ/mol}.$$

Using this value together with enthalpies of formation given in Appendices B and H, we obtain the higher heating value:

$$\Delta H_{c,H_2O(\ell)} = (1)(-249{,}659) - [10(-393{,}546) + 11(-285{,}834)]$$
$$= 6{,}829{,}975$$
$$[=]\ \text{kmol}\left(\frac{\text{kJ}}{\text{kmol}}\right) = \text{kJ}.$$

To express this on a per-mass-of-fuel basis, we need only divide by the number of moles of fuel in the combustion reaction and the fuel molecular weight, that is,

$$\Delta h_c = \frac{\Delta H_c}{M_{C_{10}H_{22}}} = \frac{\Delta H_c}{N_{C_{10}H_{22}}\mathcal{M}_{C_{10}H_{22}}}$$
$$= \frac{6{,}829{,}975}{(1)\,142.284} = 48{,}002$$
$$[=]\ \frac{\text{kJ}}{\text{kmol(kg/kmol)}} = \text{kJ/kg}_{\text{fuel}}.$$

For the lower heating value, we repeat these calculations using $\bar{h}^{\circ}_{f,H_2O(g)} = -241,847$ kJ/kmol in place of $\bar{h}^{\circ}_{f,H_2O(\ell)} = -285,834$ kJ/kmol. The result is

$$\Delta H_c = 6,345,986 \text{ kJ}$$

and

$$\Delta h_c = 44,601 \text{ kJ/kg}_{fuel}.$$

Comment We note that the difference between the higher and lower heating values is approximately 7%. What practical implications does this have, say, for a home heating furnace?

Self Test 2.26 ☑ **Determine the enthalpy of reaction per unit mass of fuel for the conditions given in Example 2.26 when the products are at a temperature of 1800 K.**

(Answer: $\Delta h_R = -14,114 \text{ kJ/kg fuel}$)

LEVEL 2

2.11 TRANSPORT PROPERTIES[18]

In addition to thermodynamic properties, we require **transport properties** to characterize many thermal-fluid systems. Of particular interest are the thermal conductivity k and the viscosity μ. These two properties are not defined by thermodynamic analyses of systems in equilibrium, as are all of the properties

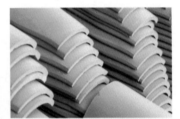

Common materials in order of decreasing thermal conductivity: pure gold, copper alloy, steel, stone block, ceramic tile, and fiberglass insulation.

---

[18] This section is appropriately used in conjunction with the study of conduction heat transfer in Chapter 5 and viscous fluid flow in Chapter 6.

discussed previously, but result from the dynamic, or rate, processes of thermal energy transfer and momentum transfer, respectively. Both thermal conductivity and viscosity can be predicted from models of the behavior of atoms, molecules, and electrons in the solid, liquid, and gaseous states of matter. Other commonly used transport properties are related in simple ways to the thermal conductivity and the viscosity. For example, the thermal diffusivity $\alpha(\equiv k/\rho c_p)$ is introduced in Chapter 5; the kinematic viscosity, also known as the momentum diffusivity, $\nu(\equiv \mu/\rho)$, is introduced in Chapter 6.

2.11a Thermal Conductivity

Physically, **thermal conductivity k** is the proportionality factor relating the rate of energy transfer per unit area by conduction, $\dot{Q}''_{cond}$, to its driving potential, the temperature gradient; that is,

$$\dot{Q}''_{cond} \propto -\nabla T \tag{2.78a}$$

or

$$\dot{Q}''_{cond} = -k\nabla T. \tag{2.78b}$$

The SI units associated with the thermal conductivity are

$$k \, [=] \, \text{W/m·K}.$$

Figure 2.39 illustrates the wide range of thermal conductivity values associated with various substances. From this figure, we see that gases have the smallest magnitudes (say, 0.02–0.13 W/m·K), liquids have intermediate values (say, 0.12–9 W/m·K), and pure metals have the highest values (say, 35–400 W/m·K). Falling between the most conducting pure metal and the least conducting gases are metal alloys and nonmetallic solids. Also indicated in Fig. 2.39 are insulation systems. These systems are neither pure substances nor intimate mixtures of pure substances but, rather, complex nonhomogeneous systems comprised, for example, of fiberglass fibers and air-filled voids, or various layers of foils and/or fabrics. For these systems, the thermal conductivity is an **effective conductivity** since the energy exchange across a layer of such material results from the combined effects of conduction, convection, and radiation.

> Chapter 4 discusses these three modes of heat transfer: conduction, convection, and Radiation.

FIGURE 2.39
Typical ranges of thermal conductivity values for gases, liquids, insulation systems, nonmetallic solids, alloys, and pure metals. **Adapted from Ref. [19] with permission.**

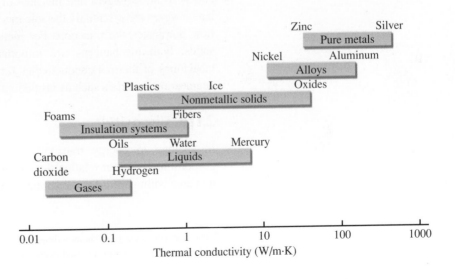

In general, values of thermal conductivity depend on temperature, with weak, or essentially no, dependence on pressure. For gases, a simple kinetic-theory model of energy transfer can be used to explain the dependence of the thermal conductivity on temperature. In this model, molecules in regions of higher temperature have greater kinetic energies than nearby molecules in regions of lower temperature, where the differences in temperature are associated with the temperature gradient ∇T, or dT/dx for a 1-D system. With no net bulk motion, higher energy molecules collide with lower energy molecules, and vice versa. The net effect is an energy transfer from high to low temperature, as expressed by Eq. 2.78. Since the mechanism for energy transfer is the collision of molecules, we expect the energy transfer rate to be proportional to the collision rate. The collision rate, in turn, is proportional to the number of molecules per unit density, N/V, and the mean molecular speed $\bar{v}_{\text{molec}}$. A more complete analysis [1] shows that

$$k \propto (N/V)\bar{v}_{\text{molec}}\ell_{\text{mf}},$$

where ℓ_{mf} is the mean free path, the average distance traveled by a molecule between collisions. Because the mean free path is inversely proportional to the number density N/V, the temperature and pressure dependence inherent in N/V ($\propto T/P$) cancels in our simple model for conductivity. We are left with the conductivity being proportional to the mean molecular speed. The molecular speed, in turn, increases with temperature and with a decrease in the mass and size of the molecules involved. Thus, among the gases, small, light molecules, such as H_2 or He, at high temperatures exhibit the greatest conductivity. Appendix F provides tabulations of k for a range of temperatures for various gases; values for air are given in Appendix C. The NIST software and online database [11] can also be used to obtain values of thermal conductivity for selected gases (see Table 2.7).

The mechanism for energy transfer by conduction in liquids is similar to that of gases, except that the range of interaction is much shorter and the strength and frequency of the molecular collisions greater. Appendix G presents tabulations of k for water and a few other common liquids as functions of temperature. The NIST software and online database [11] also provide values of k for selected liquids (see Table 2.7).

In solids, the mechanism for energy transfer by conduction is the combined effects of lattice waves and the flow of free electrons. In amorphous solids, lattice waves are essentially the sole mechanism, whereas, for metals, electron flow dominates the transport. For metal alloys and nonmetallic crystalline solids, both mechanisms are important. Appendix I provides extensive tabulations of thermal conductivities for pure solids and for commonly used engineering materials such as insulations and building materials.

2.11b Viscosity[19]

In analogy to energy transport by conduction, the **viscosity** μ is the proportionality factor relating the rate at which momentum is transferred per unit area within a fluid to the application of a driving potential (i.e., a velocity

---

[19] The *viscosity* as used here is sometimes referred to as the *absolute viscosity* or the *coefficient of viscosity*. These terms are used interchangeably.

gradient). Alternatively, the viscosity may be thought of as the proportionality factor in the stress–strain relationship associated with a fluid:

$$\text{stress} \propto \text{strain rate}, \tag{2.79a}$$

or

$$\text{stress} = \mu(\text{strain rate}). \tag{2.79b}$$

The SI units of viscosity are $N \cdot s/m^2$. For a more detailed definition of viscosity, the reader is referred to Chapter 6.

Ideas or constructs similar to those discussed here for gases and liquids apply to understanding viscosity except that, now, momentum transport replaces energy transport, and the velocity gradient replaces the temperature gradient. From such arguments, we again conclude that values of viscosity depend on temperature and are weakly sensitive, or insensitive, to pressure. Tabulations of viscosity are presented for air in Appendix C and for various gases and various liquids in Appendices F and G, respectively. Again, values are also available from the NIST software and online database [11] for selected fluids. Note that, for a solid, viscosity has no meaning, and the stress–strain relationship is defined by Young's modulus.

SUMMARY

This chapter introduced the reader to the myriad thermodynamic and thermophysical properties that are used throughout this book. The chapter also showed how these properties relate to one another through equations of state, calorific equations of state, and second-law (or Gibbs) relationships. The concept of an ideal gas was presented, and methods were presented to obtain properties for substances that do not behave as ideal gases. The properties of H_2O in both its liquid and vapor states were emphasized. You should be familiar with the use of both tables and computer-based resources to obtain property data for a wide variety of substances. You should also be proficient at sketching simple processes on thermodynamic coordinates. A more detailed summary of this chapter can be obtained by reviewing the learning objectives presented at the outset. It is recommended that you revisit this chapter many times in the course of your study of later chapters. Appropriate junctures for return are indicated in subsequent chapters.

Chapter 2
Key Concepts & Definitions Checklist[20]

2.1 Key Definitions

❑ Property ➤ *Q2.2*

❑ State ➤ *Q2.2*

❑ Process ➤ *Q2.2*

❑ Pure substance ➤ *Q2.3*

❑ Simple compressible substance ➤ *Q2.3*

❑ Extensive and intensive properties ➤ *Q2.6*

❑ Mass- and molar-specific properties ➤ *Q2.7*

2.2 Frequently Used Thermodynamic Properties

❑ Common thermodynamic properties (list)

❑ Continuum limit ➤ *Q2.5*

❑ Absolute, gage, and vacuum pressures ➤ *2.3, 2.4*

❑ Zeroth law of thermodynamics ➤ *Q2.12*

❑ Internal energy ➤ *2.17*

❑ Enthalpy ➤ *2.18*

❑ Constant-volume and constant-pressure specific heats ➤ *Q2.11*

❑ Specific heat ratio ➤ *Q2.13*

❑ Entropy ➤ *Q2.14*

❑ Gibbs free energy or function ➤ *Q2.15*

2.3 Concept of State Relationships

❑ State principle ➤ *Q2.8*

❑ P–v–T relationships ➤ *Q2.9, Q2.10*

❑ Calorific relationships ➤ *Q2.9, Q2.10*

❑ Second-law relationships ➤ *Q2.9, Q2.10, Q2.16*

2.4 Ideal Gases as Pure Substances

❑ Ideal gas definition ➤ *2.23, 2.36*

❑ Ideal-gas equation of state (Table 2.4) ➤ *2.27, 2.31*

❑ Particular gas constant ➤ *2.36*

❑ P–v and T–v diagrams ➤ *2.35*

❑ u, c_v, T relationships (Eqs. 2.31a–2.31e) ➤ *2.46*

❑ h, c_p, T relationships (Eqs. 2.33a–2.33e) ➤ *2.41, 2.42*

❑ u–T and h–T diagrams ➤ *2.22, 2.46*

❑ T–dS relationships (Eqs. 2.35 and 2.36) ➤ *Q2.21*

❑ Δs relationships (Eqs. 2.39 and 2.40) ➤ *2.54, 2.56*

❑ Isentropic process relationships (Table 2.5) ➤ *2.57, 2.58*

❑ T–s and P–v diagrams (Figs. 2.11 and 2.12) ➤ *2.59*

❑ Polytropic processes ➤ *2.60*

2.5 Nonideal Gas Properties

❑ Critical point ➤ *Q2.24*

❑ Use of tables and NIST databases ➤ *2.73, 2.74*

❑ Van der Waals equation of state ➤ *2.75*

❑ Generalized compressibility ➤ *2.72*

2.6 Pure Substances Involving Liquid and Vapor Phases

❑ Regions and phase boundaries ➤ *Q2.27*

❑ T–v and P–v diagrams ➤ *Q2.28, 2.83*

❑ Quality and liquid–vapor mixture properties ➤ *Q2.25, 2.99*

❑ Use of tables and NIST databases ➤ *2.79, 2.80*

❑ T–s and h–s diagrams ➤ *Q2.29*

2.7 Liquid Property Approximations

❑ Specific volume, internal energy, and enthalpy approximations ➤ *2.116*

2.8 Solids

❑ Fusion and sublimation properties ➤ *Q2.32*

[20] Numbers following arrows refer to Questions (prefaced with a Q) and Problems at the end of the chapter.

2.9 Ideal-Gas Mixtures

❑ Partial pressures and mole and volume fractions ➤ *2.126*

❑ Standardized properties ➤ *Q2.37, 2.132*

❑ Enthalpy of formation ➤ *Q2.38*

❑ Sensible enthalpy change ➤ *Q2.38*

❑ Standard reference state (Table 2.8) ➤ *Q2.39*

❑ Mixture properties (Eqs. 2.71–2.76) ➤ *2.135*

2.10 Some Properties of Reacting Mixtures

❑ Enthalpy of reaction ➤ *2.141, Q2.40*

❑ Heat of combustion and higher and lower heating values ➤ *Q2.41, 2.143*

2.11 Transport Properties ➤ *2.144*

❑ Thermal conductivity

❑ Viscosity

REFERENCES

1. Reid, R. C., Prausnitz, J. M., and Poling, B. E., *The Properties of Gases and Liquids*, 4th ed., McGraw-Hill, New York, 1987.

2. Halliday, D., and Resnick, R., *Physics*, combined 3rd ed., Wiley, New York, 1978.

3. Kee, R. J., Rupley, F. M., and Miller, J. A., "The Chemkin Thermodynamic Data Base," Sandia National Laboratories Report SAND87-8215 B, March 1991.

4. Clausius, R., "The Second Law of Thermodynamics," in *The World of Physics*, Vol. 1 (J. H. Weaver, Ed.), Simon & Schuster, New York, 1987.

5. Stull, D. R., and Prophet, H., "JANAF Thermochemical Tables," 2nd ed., NSRDS-NBS 37, National Bureau of Standards, June 1971. (The 3rd ed. is available from NIST.)

6. Gordon, S., and McBride, B. J., "Computer Program for Calculation of Complex Chemical Equilibrium Compositions, Rocket Performance, Incident and Reflected Shocks, and Chapman-Jouguet Detonations," NASA SP-273, 1976.

7. Keenan, J. H., Keyes, F. G., Hill, P. G., and Moore, J. G., *Steam Tables: Thermodynamic Properties of Water Including Vapor, Liquid & Solid Phases*, Krieger, Melbourne, FL, 1992.

8. Irvine, T. F., Jr., and Hartnett, J. P. (Eds.), *Steam and Air Tables in SI Units*, Hemisphere, Washington, DC, 1976.

9. *NIST Thermodynamic and Transport Properties of Pure Fluids Database: Ver. 5.0*, National Institute of Standards and Technology, Gaithersburg, MD, 2000.

10. *NIST/ASME Steam Properties Database: Ver. 2.2*, National Institute of Standards and Technology, Gaithersburg, MD, 2000.

11. Linstrom, P., and Mallard, W. G. (Eds.), *NIST Chemistry Web Book, Thermophysical Properties of Fluid Systems*, National Institute of Standards and Technology, Gaithersburg, MD, 2000, http://webbook.nist.gov/chemistry/fluid.

12. Su, G.-J., "Modified Law of Corresponding States," *Industrial Engineering Chemistry*, 38:803 (1946).

13. Obert, E. F., *Concepts of Thermodynamics*, McGraw-Hill, New York, 1960.

14. Haar, L., Gallagher, J. S., and Kell, G. S. S., *NBS/NRC Steam Tables*, Hemisphere, New York, 1984.

15. Atkins, P. W., Physical Chemistry, 6th ed., Oxford University Press, Oxford, 1998.

16. Sears, F. W., *Thermodynamics, Kinetic Theory, and Statistical Thermodynamics*, Addision-Wesley, Reading, MA, 1975.

17. Wark, K., *Thermodynamics*, McGraw-Hill, New York, 1966.

18. *Webster's New Twentieth Century Dictionary, Unabridged*, 2nd ed., Simon & Schuster, New York, 1983.

19. Incropera, F. P., and DeWitt, D. P., *Fundamentals of Heat and Mass Transfer*, 4th ed., Wiley, New York, 1996.

20. Moran, M. J., and Shapiro, H. N., *Fundamentals of Engineering Thermodynamics*, 3rd ed., Wiley, New York, 1995.

21. Van Wylen, G. J., Sonntag, R. E., and Borgnakke, C., *Fundamentals of Classical Thermodynamics*, 4th ed., Wiley, New York, 1994.

22. Myers, G. E., *Engineering Thermodynamics*, Prentice Hall, Englewood Cliffs, NJ, 1989.

Some end-of-chapter problems were adapted with permission from the following:

23. Look, D. C., Jr., and Sauer, H. J., Jr., *Engineering Thermodynamics*, PWS, Boston, 1986.

Nomenclature

a	Acceleration vector (m/s^2)	$\mathcal{N}_{\text{AV}}$	Avagodro's number (see Eq. 2.7)
a	Specific Helmholtz free energy (J/kg or van der Waals constant (Pa$\cdot$m^6/kmol^2)	P	Pressure (Pa)
		$\dot{Q}''_{\text{cond}}$	Conduction heat flux (W/m^2)
A	Helmoltz free energy (J) or area (m^2)	R	Particular gas constant (J/kg$\cdot$K)
b	Van der Waals constant (m^3/kmol)	R_{u}	Universal gas constant, 8,314.472 (J/kmol$\cdot$K)
c	Specific heat (J/kg$\cdot$K)		
c_p	Constant-pressure specific heat (J/kg$\cdot$K)	s	Specific entropy (J/kg$\cdot$K)
$\bar{c}_p$	Molar constant-pressure specific heat (J/kmol$\cdot$K)	$\bar{s}$	Molar-specific entropy (J/kmol$\cdot$K)
		S	Entropy (J/K)
c_v	Constant-volume specific heat (J/kg$\cdot$K)	t	Time (s)
$\bar{c}_v$	Molar constant-volume specific heat (J/kmol$\cdot$K)	T	Temperature (K)
		u	Specific internal energy (J/kg)
e	Specific energy (J/kg)	$\bar{u}$	Molar-specific internal energy (J/kmol)
$\bar{e}$	Molar-specific energy (J/kmol)	U	Internal energy (J)
E	Energy (J)	v	Velocity (m/s)
F	Force (N)	$\bar{v}_{\text{molec}}$	Mean molecular speed (m/s)
g	Specific Gibbs function (J/kg)	v	Specific volume (m^3/kg)
$\bar{g}$	Molar-specific Gibbs function (J/kmol)	$\bar{v}$	Molar-specific volume (m^3/kmol)
G	Gibbs function (J)	V	Volume (m^3)
h	Specific enthalpy (J/kg)	x	Quality (dimensionless) or spatial coordinate (m)
$\bar{h}$	Molar-specific enthalpy (J/kmol)		
H	Enthalpy (J)	X	Mole fraction (dimensionless)
HHV	Higher heating value (J/kg$_{\text{fuel}}$)	y	Spatial coordinate (m)
k	Thermal conductivity (W/m$\cdot$K)	Y	Mass fraction (dimensionless)
k_{B}	Boltzmann constant, 1.3806503 $\times$ 10^{-23} (J/K)	z	Spatial coordinate (m)
		Z	Compressibility factor, Pv/RT (dimensionless)
ℓ_{mf}	Mean free path (m)		
LHV	Lower heating value (J/kg$_{\text{fuel}}$)		
M	Mass (kg)		

GREEK

$\mathcal{M}$	Molecular weight (kg/kmol)
m_{u}	Unified atomic mass unit (kg)
n	Number of particles or polytropic exponent (dimensionless)
N	Number of moles (kmol)

β	Arbitrary property
γ	Specific-heat ratio (dimensionless)
Δ	Difference or increment
ΔH_{R}	Enthalpy of reaction (or of combustion) (J)

Δh_{R} Enthalpy of reaction (or of combustion) per mass of fuel (J/kg_{fuel})

Δh_{c} Heat of combustion or heating value (J/kg_{fuel})

μ Viscosity ($N \cdot s/m^2$)

ρ Density (kg/m^3)

SUBSCRIPTS

abs	absolute
atm	atmospheric
avg	average
c	critical
f	fluid (liquid) or formation
fg	difference between saturated vapor and saturated liquid states
g	gas (vapor)
gage	gage
i	species i
liq	liquid

mix	mixture
molec	molecule
prod	products
reac	reactants
ref	reference state
rot	rotational
s	sensible
sat	saturated state
sublim	sublimation
tot	total
trans	translational
vac	vacuum
vap	vapor
vib	vibrational

SUPERSCRIPTS

°	Denotes standard-state pressure ($P° = 1$ atm)

QUESTIONS

2.1   Review the most important equations presented in this chapter (i.e., those with a red background). What physical principles do they express? What restrictions apply?

2.2   Discuss how the following concepts are related: properties, states, and process.

2.3   Distinguish between a pure substance and a simple compressible substance.

2.4   Explain the continuum limit to a classmate.

2.5   Without direct reference to any textbook, write an explanation of the continuum limit.

2.6   Explain the difference between extensive and intensive properties. List five (or more) properties of each type.

2.7   How do the properties u and $\bar{u}$ differ? The properties v and $\bar{v}$?

2.8   Write out the state principle for a simple compressible substance.

2.9   Explain the distinction between an equation of state and a calorific equation of state.

2.10   What are the three types of thermodynamic state relationships? What properties are typically used in each?

2.11   Distinguish between constant-volume and constant-pressure specific heats.

2.12   What is the practical significance of the zeroth law of thermodynamics?

2.13   Define the specific-heat ratio. What Greek symbol is used to denote this ratio?

2.14   Write out a qualitative definition of the thermodynamic property entropy.

2.15   Using symbols, define the Gibbs free energy (or function) in terms of other thermodynamic properties.

2.16   Consider an ideal gas. Indicate which of the following thermodynamic properties are pressure dependent: density, specific volume, molar-specific internal energy, mass-specific enthalpy, mass-specific entropy, constant-volume specific heat, and constant-pressure specific heat.

2.17   Sketch two isotherms on a P–v plot for an ideal gas. Label each where $T_2 > T_1$.

2.18   Sketch two isobars on a T–v plot for an ideal gas. Label each where $P_2 > P_1$.

2.19   What is the microscopic interpretation of internal energy for a monatomic gas?

2.20   What is the microscopic interpretation of internal energy for a gas comprising diatomic molecules?

2.21   Write out the so-called first and second Gibbs (or T–dS) relationships.

2.22   List fluids that you know are used as working fluids in thermal-fluid devices.

2.23   Compare your list of fluids from Question 2.22 with the fluids for which property data are available from the NIST online database.

2.24   What is the physical significance of the critical point? Locate the critical point on a P–v diagram.

2.25   Write out a physical interpretation of the thermodynamic property quality.

2.26   Explain the meaning of quality to a classmate.

2.27   Draw from memory (or familiarity) a T–v diagram for H_2O. Show an isobar that begins in the compressed-liquid region and ends in the superheated-vapor region. Label all lines and regions and indicate the critical point.

2.28   Draw from memory (or familiarity) a P–v diagram for H_2O. Show an isotherm that begins in the compressed-liquid region and ends in the superheated-vapor region. Label all lines and regions and indicate the critical point.

2.29   Draw from memory (or familiarity) a T–s diagram for H_2O. Show an isobar that begins in the compressed-liquid region and ends in the superheated-vapor region. Also draw an isotherm that crosses your isobar at a state well within the superheated-vapor region. Label each.

2.30   Explain the principle of corresponding states and how this principle relates to the use of the generalized compressibility chart.

2.31   List and explain the factors that result in the breakdown of the application of the ideal-gas approximation to real gases.

2.32   Distinguish between the enthalpy of fusion and the enthalpy of sublimation.

2.33   Consider an ideal-gas mixture. Explain the differences and similarities among the following measures of composition: mole fraction, volume fraction, and mass fraction.

2.34   How does the partial pressure of a constituent in an ideal-gas mixture relate to the mole fraction of that constituent?

2.35   Compare and contrast Dalton's and Amagat's views of an ideal-gas mixture.

2.36   Explain in words how the mass-specific and molar-specific enthalpies of an ideal-gas mixture relate to the corresponding properties of the constituent species.

2.37 Explain the concept of standardized properties applied to chemically reacting systems.

2.38 Distinguish between enthalpies of formation and sensible enthalpies.

2.39 List the conditions that define the standard reference state for properties of species involved in reacting systems.

2.40 What is the sign (positive or negative) associated with the enthalpy of reaction for exothermic reactions? For endothermic reactions? Explain.

2.41 Explain what is meant by the heating value of a fuel. What distinguishes the "higher" heating value from the "lower" heating value?

Chapter 2 Problem Subject Areas

2.1–2.16	**State properties: definitions, units, and conversions**
2.17–2.22	**Calorific properties: definitions and units**
2.23–2.39	**Ideal gases: equation of state**
2.40–2.56	**Ideal gases: calorific and second-law state relationships**
2.57–2.71	**Ideal gases: isentropic and polytropic processes**
2.72–2.78	**Real gases: tabulated properties, generalized compressibility, and van der Waals equation of state**
2.79–2.114	**Pure substances with liquid and vapor phases**
2.115–2.116	**Liquid property approximations**
2.117–2.118	**Solids**
2.119–2.125	**Ideal-gas mixtures: specifying composition**
2.126–2.131	**Ideal-gas mixtures: $P–v–T$ relationships**
2.132–2.133	**Ideal-gas mixtures: standardized properties**
2.134–2.140	**Ideal-gas mixtures: enthalpies and entropies**
2.141–2.143	**Ideal-gas mixtures: enthalpy of combustion and heating values**
2.144–2.145	**Transport properties**

PROBLEMS

2.1 The specific volume of water vapor at 150 kPa and 120°C is 1.188 m³/kg. Determine the molar-specific volume and the density of the water vapor.

2.2 Determine the number of molecules in 1 kg of water.

2.3 An air compressor fills a tank to a gage pressure of 100 psi. The barometric pressure is 751 mm Hg. What is the absolute pressure in the tank in kPa?

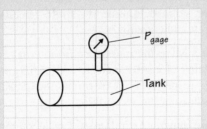

2.4 An instrument used to measure the concentration of the pollutant nitric oxide uses a vacuum pump to create a vacuum of 28.3 in Hg in a reaction chamber. What is the absolute pressure in the chamber if the barometric pressure is 1 standard atmosphere? Express your result in psia, mm Hg, and Pa.

Photograph courtesy of VACUUBRAND GMBH + CO KG, Germany.

2.5 Find the specific volume (in both ft³/lbₘ and m³/kg) of 45 lbₘ of a substance of density 10 kg/m³, where the acceleration of gravity is 30 ft/s².

2.6 Assume that a pressure gage and a barometer read 227.5 kPa and 26.27 in Hg, respectively. Calculate the absolute pressure in psia, psfa (pounds-force per square foot absolute), and atm.

2.7   The pressure in a partially evacuated enclosure is 26.8 in Hg vacuum when the local barometer reads 29.5 in Hg. Determine the absolute pressure in in Hg, psia, and atm.

2.8   A vertical cylinder containing air is fitted with a piston of 68 lb_m and cross-sectional area of 35 in^2. The ambient pressure outside the cylinder is 14.6 psia and the local acceleration due to gravity is 31.1 ft/s^2. What is the air pressure inside the cylinder in psia and in psig?

2.9   The accompanying sketch shows a compartment divided into two sections *a* and *b*. The ambient pressure P_{amb} is 30.0 in Hg (absolute). Gage *C* reads 620,528 Pa and gage *B* reads 275,790.3 Pa. Determine the reading of gage *A* and convert this reading to an absolute value.

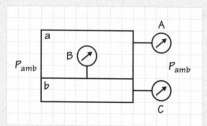

2.10   A cylinder containing a gas is fitted with a piston having a cross-sectional area of 0.029 m^2. Atmospheric pressure is 0.1035 MPa and the acceleration due to gravity is 30.1 ft/s^2. To produce an absolute pressure in the gas of 0.1517 MPa, what mass (kg) of piston is required?

2.11   For safety, cans of whipped cream (with propellant) should not be stored above 120 F. What is this temperature expressed on the Rankine, Celsius, and Kelvin temperature scales?

2.12   Albuterol inhalers, used for the control of asthma, are to be stored and used between 15°C and 30°C. What is the acceptable temperature range on the Fahrenheit scale?

2.13   How fast, on average, do nitrogen molecules travel at room temperature (25°C)? How does this speed compare to the average speed of a modern jet aircraft that travels 2500 miles in 5 hours?

2.14   A thermometer reads 72 F. Specify the temperature in °C, K, and R.

2.15   Convert the following Celsius temperatures to Fahrenheit: (a) −30°C, (b) −10°C, (c) 0°C, (d) 200°C, and (e) 1050°C.

2.16   At what temperature are temperatures expressed in Fahrenheit and Celsius numerically equal?

2.17   At 600 K and 0.10 MPa, the mass-specific internal energy of water vapor is 2852.4 kJ/kg and the specific volume is 2.7635 m^3/kg. Determine the density and mass-specific enthalpy of the water vapor. Also determine the molar-specific internal energy and enthalpy.

2.18   At 0.3 MPa, the mass-specific internal energy and enthalpy of a substance are 3313.6 J/kg and 3719.2 J/kg, respectively. Determine the density of the substance at these conditions.

2.19   A tank having a volume of 2 m^3 contains 6.621 kg of water vapor at 1 MPa. The internal energy of the water vapor is 19,878 J. Determine the density, the mass-specific internal energy, and the mass-specific enthalpy of the water vapor.

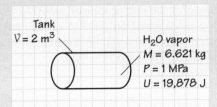

Tank
$V = 2 \, m^3$

H_2O vapor
$M = 6.621 \, kg$
$P = 1 \, MPa$
$U = 19,878 \, J$

2.20 The constant-volume molar-specific heat of nitrogen (N_2) at 1000 K is 24.386 kJ/kmol·K and the specific-heat ratio γ is 1.3411. Determine the constant-pressure molar-specific heat and the constant-pressure mass-specific heat of the N_2.

2.21 For temperatures between 300 and 1000 K and at 1 atm, the molar specific enthalpy of O_2 is expressed by the following polynomial:

$$\bar{h}_{O_2} = R_u(3.697578 \, T + 3.0675985 \times 10^{-4} \, T^2$$
$$-4.19614 \times 10^{-8} \, T^3 + 4.4382025 \times 10^{-12} \, T^4$$
$$-2.27287 \times 10^{-16} \, T^5 - 1233.9301),$$

where $\bar{h}$ is expressed in kJ/kmol and T in kelvins.
Determine the constant-pressure molar-specific heat $\bar{c}_p$ at 500 K and at 1000 K. Compare the magnitudes of the values at the two temperatures and discuss. Also determine c_p, the constant-pressure *mass*-specific heat.

2.22 Plot a graph of the molar specific enthalpy for O_2 given in Problem 2.21 for the temperature range 300–1000 K (i.e., plot $\bar{h}_{O_2}$ versus T). (Spreadsheet software is recommended.) Use your plot to graphically determine $\bar{c}_p$ at 500 K and at 1000 K. Use a pencil and ruler to perform this operation. How do these estimated values for $\bar{c}_p$ compare with your computations in Problem 2.21?

2.23 What is the mass of a cubic meter of air at 25°C and 1 atm?

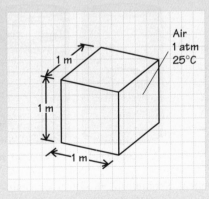

Air
1 atm
25°C

1 m

1 m

1 m

2.24 A. Determine the density of air at Mile High Stadium in Denver, Colorado, on a warm summer evening when the temperature is 78 F. The barometric pressure is 85.1 kPa.

B. Assume to a first approximation that the drag force exerted on a well-hit baseball arcing to the outfield, or beyond, is proportional to the air density. Discuss the implications of this for balls hit at Mile High Stadium in Denver versus balls hit in Yankee Stadium in New York (which is at sea level).

2.25 A compressor pumps air into a tank until a pressure gage reads 120 psi. The temperature of the air is 85 F, and the tank is a 0.3-m-diameter cylinder 0.6 m long. Determine the mass of the air contained in the tank in grams.

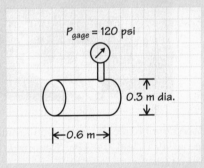

$P_{gage} = 120 \, psi$

0.3 m dia.

0.6 m

2.26 Determine the number of kmols of carbon monoxide contained in a 0.027-m^3 compressed-gas cylinder at 200 psi and 72 F. Assume ideal-gas behavior.

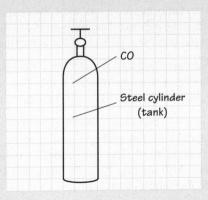

CO

Steel cylinder
(tank)

2.27 A piston-cylinder assembly contains of 9.63 m³ air at 29.4°C. The piston has a cross-sectional area of 0.029 in² and a mass of 160.6 kg. The gravitational acceleration is 9.144 m/s². Determine the mass of air trapped within the cylinder. Atmospheric pressure is 0.10135 MPa.

2.28 Consider the piston–cylinder arrangement shown in the sketch. Determine the absolute pressure of the air (psia) and the mass of air in the cylinder (lb$_m$). The atmospheric pressure is 14.6 psia.

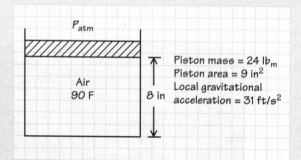

2.29 Determine the mass of air in a room that is 15 m by 15 m by 2.5 m. The temperature and pressure are 25°C and 1 atm, respectively.

2.30 Carbon monoxide is discharged from an exhaust pipe at 49°C and 0.8 kPa. Determine the specific volume (m³/kg) of the CO.

2.31 A cylinder–piston arrangement contains nitrogen at 21°C and 1.379 MPa. The nitrogen is compressed from 98 to 82 cm³ with a final temperature of 27°C. Determine the final pressure (kPa).

2.32 The temperature of an ideal gas remains constant while the pressure changes from 101 to 827 kPa. If the initial volume is 0.08 m³, what is the final volume?

2.33 Nitrogen (3.2 kg) at 348°C is contained in a vessel having a volume of 0.015 m³. Use the ideal-gas equation of state to determine the pressure of the N₂.

2.34 On P–v coordinates, sketch a process in which the product of the pressure and specific volume are constant from state 1 to state 2. Assume an ideal gas. Also assume $P_1 > P_2$ Show this same process on P–T and T–v diagrams.

2.35 Consider the five processes a–b, b–c, c–d, d–a, and a–c as sketched on the P–v coordinates. Show the same processes on P–T and T–v coordinates assuming ideal-gas behavior.

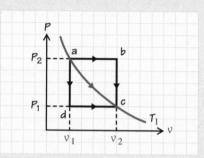

2.36 Starting with Eq. 2.28c, derive all other forms of the ideal-gas law, (i.e., Eqs. 2.28a, 2.28b, 2.28d, and 2.28e).

2.37 Consider three 0.03-m³ tanks filled, respectively, with N₂, Ar, and He. Each tank is filled to a pressure of 400 kPa at room temperature, 298 K. Determine the mass of gas contained in each tank. Also determine the number of moles of gas (kmol) in each tank.

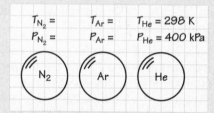

2.38 Nitrogen slowly expands from an initial volume of 0.025 m³ to 0.05 m³ at a constant pressure of 400 kPa. Determine the final temperature if the initial temperature is 500 K. Plot the process on P–v and T–v coordinates using spreadsheet software.

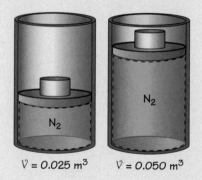

2.39 Repeat Problem 2.38 for a constant-temperature process (500 K). Determine the final pressure for an initial pressure of 400 kPa.

2.40 The constant-pressure specific heat of a gas is 0.24 Btu/lb$_m$·R at room temperature. Determine the specific heat in units of kJ/kg·K.

2.41 Compute the mass-specific enthalpy change associated with N_2 undergoing a change in state from 400 to 800 K. Assume the constant-pressure specific heat is constant for your calculation. Use the arithmetic average of values at 400 and 800 K from Table B.7.

2.42 Compare the specific enthalpy change calculated in Problem 2.41 with the change determined directly from the ideal-gas tables (Table B.7). Also compare these values with that obtained from the NIST software or online database at 1 atm.

2.43 Use Table C.2 to calculate the mass-specific enthalpy change for air undergoing a change of state from 300 to 1000 K. How does this value compare with that estimated using the constant-pressure specific heat at the average temperature, $T_{avg} = (300 + 1000)/2$?

2.44 Determine the mass-specific internal energy for O_2 at 900 K for a reference-state temperature of 298.15 K. Also determine the constant-volume specific heat (mass basis).

2.45 Hydrogen is compressed in a cylinder from 101 kPa and 15°C to 5.5 MPa and 121°C. Determine Δv, Δu, Δh, and Δs for the process.

2.46 As air flows across the cooling coil of an air conditioner at the rate of 3856 kg/hr, its temperature drops from 26°C to 12°C. Determine the internal energy change in kJ/hr. Plot the process on h–T coordinates.

2.47 Air cools an electronics compartment by entering at 60 F and leaving at 105 F. The pressure is essentially constant at 14.7 psia. Determine (a) the change in internal energy of the air as it flows through the compartment and (b) the change in specific volume.

2.48 Nitrogen is compressed adiabatically in a cylinder having an initial volume of 0.25 ft³. The initial pressure and temperature are 20 psia and 100 F, respectively. The final volume is 0.1 ft³ and the final pressure is 100 psia. Assuming ideal-gas behavior and using average specific heats, determine the following:

    A. Final temperature (F)
    B. Mass of nitrogen (lb_m)
    C. Change in enthalpy (Btu)

    D. Change in entropy (Btu/R)
    E. Change in internal energy (Btu)

2.49 In a closed (fixed-mass) system, an ideal gas undergoes a process from 75 psia and 5 ft³ to 25 psia and 9.68 ft³. For this process, $\Delta H = -62$ Btu and $c_{v,avg} = 0.754$ Btu/$lb_m \cdot$R. Determine (a) ΔU, (b) $c_{p,avg}$, and (c) the specific gas constant R.

2.50 In a fixed-mass system, 4 lb_m of air is heated at constant pressure from 30 psia and 40 F to 140 F. Determine the change in internal energy (Btu) for this process assuming $\bar{c}_{v,avg} = 4.96$ Btu/lbmol·R and $k = 1.4$.

2.51 Use Table C.2 to calculate the mass-specific entropy change for air undergoing a change of state from 300 to 1000 K. The initial and final pressures are both 1 atm.

2.52 Calculate the mass-specific enthalpy and entropy changes for air undergoing a change of state from 400 K and 2 atm to 800 K and 7.2 atm.

2.53 Air is compressed in the compressor of a turbojet engine. The air enters the compressor at 270 K and 58 kPa and exits the compressor at 465 K and 350 kPa. Determine the mass-specific enthalpy, internal energy, and entropy changes associated with the compression process.

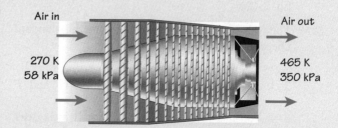

2.54 Consider 0.3 kg of N_2 in a rigid container at 400 K and 5 atm. The N_2 is cooled to 300 K. Determine the specific entropy change of the N_2 associated with this cooling process.

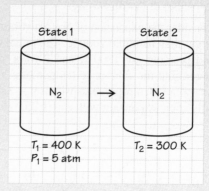

2.55 A fixed mass of air undergoes a state change from 400 K and 1 atm to 500 K and 1.5 atm. Determine the specific entropy change associated with this process.

2.56 A fixed mass of air (0.25 kg) at 500 K is contained in a piston–cylinder assembly having a volume of 0.10 m³. A process occurs to cause a change of state. The final state pressure and temperature are 200 kPa and 400 K, respectively. Determine the entropy change associated with this process. Use both the ideal-gas approximations (Eqs. 2.40a and 2.40b) and the air tables (Table C.2) to calculate ΔS. Note that the data in Table C.2 are for a pressure of 1 atm. Compare the results of the two methods. Also, what is the volume at the final state?

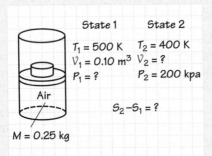

State 1   State 2
$T_1 = 500$ K $T_2 = 400$ K
$V_1 = 0.10$ m³ $V_2 = ?$
$P_1 = ?$ $P_2 = 200$ kpa

Air

$S_2 - S_1 = ?$

$M = 0.25$ kg

2.57 Nitrogen undergoes an isentropic process from an initial state at 425 K and 150 kPa to a final state at 600 K. Determine the density of the N_2 at the final state.

2.58 Starting with the relationship $Pv^\gamma = $ constant, derive Eq. 2.42a.

2.59 The following processes constitute the air-standard Diesel cycle:

1–2: isentropic compression,
2–3: constant-volume energy addition (T and P increase),
3–4: constant-pressure energy addition (v increases),
4–5: isentropic expansion, and
5–1: constant-volume energy rejection (T and P decrease).

Plot these processes on P–v and T–s coordinates. How does this cycle differ from the Otto cycle presented in Example 2.12?

2.60 Consider the expansion of N_2 to a volume ten times larger than its initial volume. The initial temperature and pressure are 1800 K and 2 MPa, respectively. Determine the final-state temperature and pressure for (a) an isentropic expansion ($\gamma = 1.4$) and (b) a polytropic expansion with $n = 1.25$.

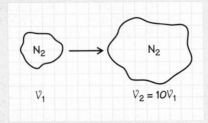

N_2 → N_2

V_1 $V_2 = 10V_1$

2.61 Consider a T–s diagram. Show that the constant specific volume line (v) must have a steeper slope than a line of constant pressure (P).

2.62 Air is drawn into the compressor of a jet engine at 55 kPa and $-23°C$. The air is compressed isentropically to 275 kPa. Determine (a) the temperature after compression (°C), (b) the specific volume before compression (m³/kg), and (c) the change in specific enthalpy for the process (kJ/kg).

2.63 An automobile engine has a compression ratio (V_1/V_2) of 8.0. If the compression is isentropic and the initial temperature and pressure are 30°C and 101 kPa, respectively, determine (a) the temperature and the pressure after compression and (b) the change in enthalpy for the process.

2.64 Air is compressed in a piston–cylinder system having an initial volume of 80 in³. Initial pressure and temperature are 20 psia and 140 F. The final volume is one-eighth of the initial volume at a pressure of 175 psia. Determine the following:

A. Final temperature (F)
B. Mass of air (lb$_m$)
C. Change in internal energy (Btu)
D. Change in enthalpy (Btu)
E. Change in entropy (Btu/lb$_m$ · R)

2.65 An ideal gas expands in a polytropic process ($n = 1.4$) from 850 to 500 kPa. Determine the final volume if the initial volume is 100 m³.

2.66 An ideal gas (3 lb$_m$) in a closed system is compressed such that $\Delta s = 0$ from 14.7 psia and 70 F to 60 psia. For this gas, $c_p = 0.238$ Btu/lbm · F, $c_v = 0.169$ Btu/lb$_m$ · F, and $R = 53.7$ ft · lb$_f$/lb$_m$ · R. Compute (a) the final volume if the initial volume is 40.3 ft³ and (b) the final temperature.

2.67 Air expands from 172 kPa and 60°C to 101 kPa and 5°C. Determine the change in the mass-specific entropy s of the air (kJ/kg · K) assuming constant average specific heats.

2.68 A piston–cylinder assembly contains oxygen initially at 0.965 MPa and 315.5°C. The oxygen then expands such that the entropy s remains constant to a final pressure of 0.1379 MPa. Determine the change in internal energy per kg of oxygen.

2.69 During the compression stroke in an internal combustion engine, air initially at 41°C and 101 kPa is compressed isentropically to 965 kPa. Determine (a) the final temperature (°C), (b) the change in enthalpy (kJ/kg), and (c) the final volume (m³/kg).

2.70 Air is heated from 49°C to 650°C at a constant pressure of 620 kPa. Determine the enthalpy and entropy changes for this process. Ignore the variation in specific heat and use the value at 27°C.

Also determine the percentage error associated with the use of this constant specific heat.

2.71 Air expands through a air turbine from inlet conditions of 690 kPa and 538°C to an exit pressure of 6.9 kPa in an isentropic process. Determine the inlet specific volume, the outlet specific volume, and the change in specific enthalpy.

2.72 Determine the density of methane (CH_4) at 300 K and 40 atm. Compare results obtained by assuming ideal-gas behavior, by using the generalized compressibility chart, and by using the NIST software or online database. Discuss.

2.73 Compare the value of the specific volume of superheated steam at 2 MPa and 500 K found in Appendix D with that calculated assuming ideal-gas behavior. Discuss.

2.74 Consider steam. Plot the 500-K isotherm in P–v space using the NIST online database as your data source. Also plot on the same graph the 500-K isotherm assuming ideal-gas behavior. Start (actually end) the isotherm at the saturated vapor line. Use a sufficiently large range of pressures so that the real isotherm approaches the fictitious ideal-gas isotherm at large specific volumes. Use spreadsheet software for your calculations and plot.

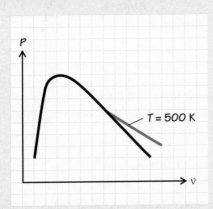

2.75 Use the van der Waals equation of state to determine the density of propane (C_3H_8) at 400 K and 12.75 MPa. How does this value compare with that obtained from the NIST online database?

2.76 Carbon dioxide (CO_2) is heated in a constant-pressure process from 15°C and 101.3 kPa to 86°C Determine, per unit mass, the changes in (a) enthalpy, (b) internal energy, (c) entropy, and (d) volume all in SI units.

2.77 A large tank contains nitrogen at −65°C and 91 MPa. Can you assume that this nitrogen is an ideal gas? What is the specific volume error in this assumption?

2.78 Is steam at 10 MPa and 500°C an ideal gas? Discuss.

2.79 Given the following property data for H_2O designate the region in T–v or P–v space (i.e., compressed liquid, liquid–vapor mixture, superheated vapor, etc.) and find the value(s) of the requested property or properties. Use the tables in Appendix D or the NIST database as necessary. Provide units with your answers.

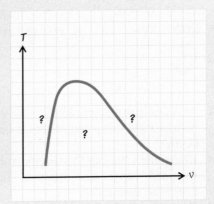

A. $T = 310$ K, $v = 22.903$ m³/kg
   Region = ?
   P = ?

B. $T = 310$ K, $v = 15$ m³/kg
   Region = ?
   h = ?

C. $T = 310$ K, $P = 10$ kPa
   Region = ?
   v = ?

D. $T = 310$ K, $P = 4$ kPa
   Region = ?
   u = ?

E. $T = 647.096$ K, $P = 22.064$ MPa
   Region = ?
   v = ?

F. $T = 800$ K, $P = 25$ MPa
   Region = ?
   ρ = ?

G. $T = 800$ K, $P = 5$ MPa
   Region = ?
   s = ?

H. $T = 743.2$ K, $P = 4.61$ MPa
   Region = ?
   h = ?

2.80 Determine the remaining properties for each of the following states of H_2O:

	A. P =			B. P =	200	psia
	P =	?	psia	P =	200	psia
	T =	200	F	T =	?	F
	v =	?	ft³/lb$_m$	v =	?	ft³/lb$_m$
	h =	?	Btu/lb$_m$	h =	?	Btu/lb$_m$
	u =	?	Btu/lb$_m$	u =	480	Btu/lb$_m$
	s =	1.87	Btu/lb$_m$·R	s =	?	Btu/lb$_m$·R

C. $P =$ 2000 psia
  $T =$ 100 F
  $v =$? ft^3/lb$_m$
  $h =$? Btu/lb$_m$
  $s =$? Btu/lb$_m$·R

D. $P =$ 1 psia
  $T =$ 100 F
  $v =$? ft^3/lb$_m$
  $h =$? Btu/lb$_m$
  $s =$? Btu/lb$_m$·R

E. $P =$? kPa
  $T =$ 95 °C
  $v =$? m^3/kg
  $h =$? kJ/kg
  $s =$ 1.28933 kJ/kg·K

F. $P =$ 1379 kPa
  $T =$? °C
  $v =$? m^3/kg
  $h =$ 1116.5 kJ/kg
  $s =$? kJ/kg·K

G. $P =$ 6.895 kPa
  $T =$ 38 °C
  $v =$? m^3/kg
  $h =$? kJ/kg
  $s =$? kJ/kg·K

H. $P =$ 13,979 kPa
  $T =$ 38 °C
  $v =$? m^3/kg
  $h =$? kJ/kg
  $s =$? kJ/kg·K

$v =$ m^3/kg
$h =$ kJ/kg
$s =$ kJ/kg·K

$v =$ m^3/kg
$h =$ kJ/kg
$s =$ 4.000 kJ/kg·K

G. $P =$ kPa
  $T =$ 500 °C
  $v =$ 0.1161 m^3/kg
  $h =$ kJ/kg
  $s =$ kJ/kg·K

H. $P =$ kPa
  $T =$ 200 °C
  $v =$ m^3/kg
  $h =$ 1500 kJ/kg
  $s =$ kJ/kg·K

2.84 Steam in a boiler has an enthalpy of 2558 kJ/kg and an entropy of 6.530 kJ/kg·K. What is its internal energy in kJ/kg?

2.85 Water at 30 psig is heated from 62 to 115 F. Determine the change in enthalpy in Btu/lb$_m$.

2.81 A. At room temperature (25°C), what pressure (in both kPa and psi) is required to liquefy propane (C_3H_8)?
  B. Determine values of the specific volume of the saturated vapor and saturated liquid. Also determine their ratios.
  C. Determine h_{fg} for these same conditions.

2.86 A hot water heater has 2.0 gal/min entering at 50 F and 40 psig. The water leaves the heater at 160 F and 39 psig. Determine the change in enthalpy in Btu/lb$_m$.

2.87 Water at 6.90 MPa and 95°C enters the steam-generating unit of a power plant and leaves the unit as steam at 6.90 MPa and 850°C. Determine the following properties in SI units:

	Inlet			**Outlet**	
$v =$	?	m^3/kg	$v =$	?	m^3/kg
$h =$	?	kJ/kg	$h =$	?	kJ/kg
$u =$	?	kJ/kg	$u =$	?	kJ/kg
$s =$	?	kJ/kg·K	$s =$	?	kJ/kg·K
Region: ?			Region: ?		

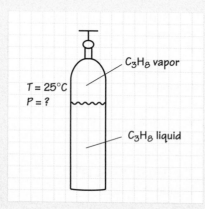

$T = 25°C$
$P = ?$

C_3H_8 vapor

C_3H_8 liquid

2.88 In a proposed automotive steam engine, the steam after expansion would reach a state at which the pressure is 20 psig and the volume occupied per pound mass is 4.8 ft^3. Atmospheric pressure is 15 psia. Determine the following properties of the steam at this state:

$T =$? F
$u =$? Btu/lb$_m$
Region: ?

2.82 Plot the 1-atm isobar for H_2O on T–v coordinates. Start in the compressed-liquid region, continue across the liquid-vapor dome, and end well into the superheated-vapor region. *Hint:* Use a log scale for specific volume.

2.83 For H_2O complete the following:

A. $P =$ 1000 psia
  $T =$ 150 F
  $v =$ ft^3/lb$_m$
  $h =$ Btu/lb$_m$
  $s =$ Btu/lb$_m$·R

B. $P =$ 30 psia
  $T =$ 150 F
  $v =$ ft^3/lb$_m$
  $h =$ Btu/lb$_m$
  $s =$ Btu/lb$_m$·R

C. $P =$ psia
  $T =$ 250 F
  $v =$ ft^3/lb$_m$
  $h =$ Btu/lb$_m$
  $s =$ 1.21 Btu/lb$_m$·R

D. $P =$ 30 psia
  $T =$ F
  $v =$ 1.4 ft^3/lb$_m$
  $h =$ Btu/lb$_m$
  $s =$ Btu/lb$_m$·R

E. $P =$ 200 kPa
  $T =$ 600 °C

F. $P =$ 400 kPa
  $T =$ °C

2.89 A water heater operating under steady-flow conditions delivers 10 liters/min at 75°C and 370 kPa. The input conditions are 10°C and 379 kPa. What are the corresponding changes in internal energy and enthalpy per kilogram of water supplied?

2.90 Water at 3.4 MPa is pumped through pipes embedded in the concrete of a large dam. The water, in picking up the heat of hydration of the curing, increases in temperature from 10°C to 40°C. Determine (a) the change in enthalpy (kJ/kg) and (b) the change in entropy of the water (kJ/kg·K).

2.91 Steam enters the condenser of a modern power plant with a temperature of 32°C and a quality of 0.98 (98% by mass vapor). The condensate (water) leaves at 7 kPa and 27°C. Determine the change in specific volume between inlet and outlet of the condenser in m^3/kg.

2.92 What is the temperature or quality of H_2O in the following states?

A. 20°C, 2000 kJ/kg (u)
B. 2 MPa, 0.1 m^3/kg
C. 140°C, 0.5089 m^3/kg
D. 4 MPa, 25 kg/m^3
E. 2 MPa, 0.111 m^3/kg

2.93 Wet steam exits a turbine at 50 kPa with a quality of 0.83. Determine the following properties of the wet steam: T, v, h, u, and s. Give units.

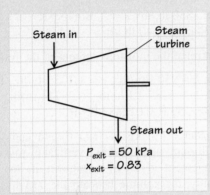

2.94 Wet steam at 375 K has a specific enthalpy of 2600 kJ/kg. Determine the quality of the mixture and the specific entropy s.

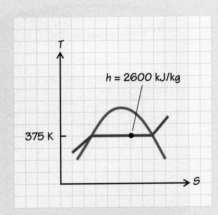

2.95 Consider H_2O at 1 MPa. Plot the following properties as a function of quality x: $T(K)$, h (kJ/kg), s (kJ/kg·K), and v (m^3/kg). Discuss.

2.96 Steam is condensing in the shell of a heat exchanger at 305 K under steady conditions. The volume of the shell is 2.75 m^3. Determine the mass of the liquid in the shell if the specific enthalpy of the mixture is 2500 kJ/kg.

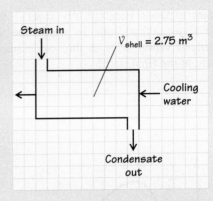

2.97 A 0.6-m^3 tank contains 0.2 kg of H_2O at 350 K. Determine the pressure in the tank and the enthalpy of the H_2O.

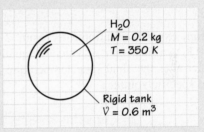

2.98 A 7.57-m^3 rigid tank contains 0.546 kg of H_2O at 37.8°C. The H_2O is then heated to 204.4°C. Determine (a) the initial and final pressures of the H_2O in the tank (MPa) and (b) the change in internal energy (kJ).

2.99 A 2.7-kg mass of H_2O is in a 0.566-m^3 container (liquid and vapor in equilibrium) at 700 kPa. Calculate (a) the volume and mass of liquid and (b) the volume and mass of vapor in the container.

2.100 Consider 0.25 kg of steam contained in a rigid container at 600 K and 4 MPa. The steam is cooled to 300 K. Determine the entropy change of the H_2O associated with this cooling process. Note: Find $S_2 - S_1$, not $s_2 - s_1$.

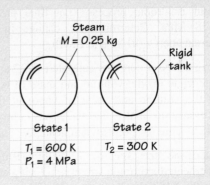

2.101 Steam expands isentropically from 2 MPa and 500 K to a final state in which the steam is saturated vapor. What is the temperature of the steam at the final state?

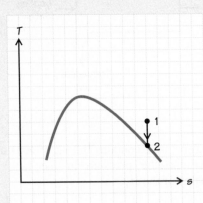

2.102 Steam expands isentropically from 2 MPa and 500 K to a final state in which the quality is 0.90. Determine the final-state temperature, pressure, and specific volume.

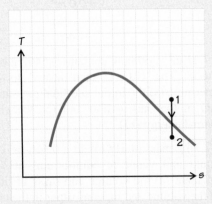

2.103 Steam expands isentropically from 6 MPa and 1000 K to 1 MPa. Determine the temperature at the final state. Also determine the specific enthalpy change for the process, (i.e., $h_2 - h_1$).

2.104 A piston–cylinder assembly contains steam initially at 0.965 MPa and 315.6°C. The steam then expands in an isentropic process to a final pressure of 0.138 MPa. Determine the change in specific internal energy (kJ/kg and Btu/lb$_m$).

2.105 A rigid vessel contains saturated R-22 at 15.6°C. Determine (a) the volume and mass of liquid and (b) the volume and mass of vapor at the point necessary to make the R-22 pass through the critical state (or point) when heated.

2.106 Determine the specific enthalpy (Btu/lb$_m$) of superheated ammonia vapor at 1.3 MPa and 65°C. Use the NIST database.

2.107 Determine the specific entropy of evaporation of steam at standard atmospheric pressure (kJ/kg·K).

2.108 A liquid–vapor mixture of H_2O at 2 MPa is heated in a constant-volume process. The final state is the critical point. Determine the initial quality of the liquid-vapor mixture.

2.109 An 85-m^3 rigid vessel contains 10 kg of water (both liquid and vapor in thermal equilibrium at a pressure of 0.01 MPa). Calculate the volume and mass of both the liquid and vapor.

2.110 Determine the moisture content $(1 - x)$ of the following:

H_2O: 400 kPa, $h = 1700$ kJ/kg
H_2O: 1850 lb$_f$/in^2, $s = 1$ Btu/lb$_m$·R
R-134a: 10 F, 40 lb$_m$/ft^3

2.111 Determine the moisture content of the following:

H_2O: $h = 950$ Btu/lb$_m$, $s = 1.705$ Btu/lb$_m$·F
H_2O: $h = 1187.7$ Btu/lb$_m$, 0.38714 ft^3/lb$_m$
R-134a: 59 F, 0.012883 ft^3/lb$_m$

2.112 Consider 0.136 kg of H_2O (liquid and vapor in equilibrium) contained in a vertical piston–cylinder arrangement at 50°C, as shown in the sketch. Initially, the volume beneath the 113.4-kg piston (area of 11.15 cm^2) is 0.03 m^3. With an atmospheric pressure of 101.325 kPa ($g = 9.14$ m/s^2), the piston rests on the stops. Energy is transferred to this arrangement until there is only saturated vapor inside.

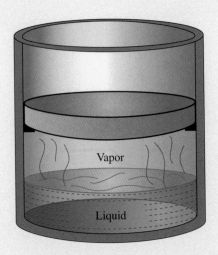

A. Show this process on a T–V diagram.
B. What is the temperature of the H_2O when the piston first rises from the stops?

2.113 H_2O expands isentropically through a steam turbine from inlet conditions of 700 kPa and 550°C to an exit pressure of 7 kPa. Determine the specific volume and specific enthalpy at both the inlet and outlet conditions.

2.114 Fifty kilograms of H_2O liquid and vapor in equilibrium at 300°C occupies a volume of m³. What is the percentage of liquid, that is, the moisture content, $1 - x$?

2.115 Consider water at 400 K and a pressure of 5 atm. Estimate the enthalpy of the water using the *saturated liquid* tables from Appendix D. Compare this result with the value obtained from the NIST database or the compressed liquid tables in Appendix D. Repeat for the density.

2.116 Water exits a water pump at 13.7 MPa and 172°C. Use Eqs. 2.53a–2.53c to estimate the specific volume, internal energy, and enthalpy of the water. Compare with the exact values obtained from the NIST database and calculate the percent error associated with each.

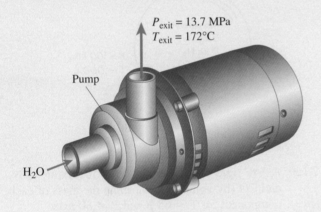

$P_{exit} = 13.7$ MPa
$T_{exit} = 172$°C

Pump

H_2O

2.117 Consider a block of pure aluminum measuring 25 × 300 × 200 mm. Estimate the change in internal energy associated with a temperature change from 600 to 400 K.

2.118 Formally show that $c_p = c_v$ for an incompressible solid.

2.119 A mixture of ideal gases contains 1 kmol of CO_2, 2 kmol of H_2O, 0.1 kmol of O_2, and 7.896 kmol of N_2. Determine the mole fractions and the mass fractions of each constituent. Also determine the apparent molecular weight of the mixture.

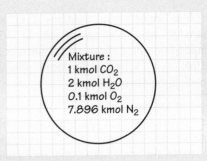

Mixture:
1 kmol CO_2
2 kmol H_2O
0.1 kmol O_2
7.896 kmol N_2

2.120 Determine the apparent molecular weight of synthetic air created by mixing 1 kmol of O_2 with 3.76 kmol of N_2. Also determine the mole and mass fractions of the O_2 in the mixture.

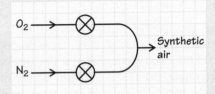

O_2

N_2

Synthetic air

2.121 On a hot (95 F) summer day the relative humidity reaches an uncomfortable 85%. At this condition, the water vapor mole fraction is 0.056. Treating the moist air as an ideal-gas mixture of water vapor and dry air, determine the mass fraction of water

vapor and the apparent molecular weight of the *moist* air. Treat the *dry* air as a simple substance with a molecular weight of 28.97 kg/kmol.

2.122 A $0.5\text{-}m^3$ rigid vessel contains 1 kg of carbon monoxide and 1.5 kg of air at 15°C. The composition of the air on a mass basis is 23.3% O_2 and 76.7% N_2. What are the partial pressures (kPa) of each component?

2.123 A gas mixture of O_2, N_2, and CO_2 contains 5.5, 3, and 1.5 kmol of each species, respectively. Determine volume fractions of each component. Also determine the mass (kg) and molecular weight (kg/kmol) of the mixture.

2.124 For air containing 75.53% N_2, 23.14% O_2, 1.28% Ar, and 0.05% CO_2, by mass, determine the gas constant and its molecular weight. How do these values compare if the mass-based composition is 76.7% N_2 and 23.3% O_2?

2.125 The gravimetric (mass) analysis of a gaseous mixture yields CO_2 = 32%, O_2 = 56.5%, and N_2 = 11.5%. The mixture is at a pressure of 3 psia. Determine (a) the volumetric composition and (b) the partial pressure of each component.

2.126 A 17.3-liter tank contains a mixture of argon, helium, and nitrogen at 298 K. The argon and helium mole fractions are 0.12 and 0.35, respectively. If the partial pressure of the nitrogen is 0.8 atm, determine (a) the total pressure in the tank, (b) the total number of moles (kmol) in the tank, and (c) the mass of the mixture in the tank.

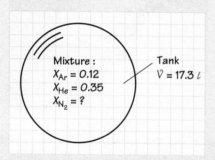

2.127 An instrument for the analysis of trace hydrocarbons in air, or in the products of combustion, uses a flame ionization detector. The flame in this device is fueled by a mixture of 40% (vol.) hydrogen and 60% (vol.) helium. The fuel mixture is contained in

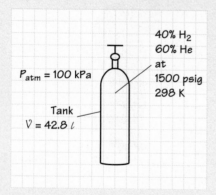

a 42.8-liter tank at 1500 psig and 298 K. The atmospheric pressure is 100 kPa. Determine (a) the partial pressure of each constituent in the tank and (b) the total mass of the mixture. Assume ideal-gas behavior for the mixture.

2.128 A rigid tank contains 5 kg of O_2, 8 kg of N_2, and 10 kg of CO_2 at 100 kPa and 1000 K. Assume that the mixture behaves as an ideal gas. Determine (a) the mass fraction of each component, (b) the mole fraction of each component, (c) the partial pressure of each component, (d) the average molar mass (apparent molecular weight) and the gas constant of the mixture, and (e) the volume of the mixture in m^3.

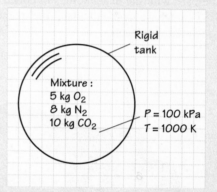

2.129 A $3\text{-}ft^3$ rigid vessel contains a 50–50 mixture of N_2 and CO (by volume). Determine the mass of each component for $T = 65$ F and $P = 30$ psia.

2.130 A $1\text{-}m^3$ tank contains nitrogen at 30°C and 500 kPa. In an isothermal process, CO_2 is forced into this tank until the pressure is 1000 kPa. What is the mass (kg) of each gas present at the end of this process?

2.131 A $0.08\text{-}m^3$ rigid vessel contains a 50–50 (by volume) mixture of nitrogen and carbon monoxide at 21°C and 2.75 MPa. Determine the mass of each component.

2.132 Determine the standardized enthalpies and entropies of the following pure species at 4 atm and 2500 K: H_2, H_2O, and OH.

2.133 Use the curve-fit coefficients for the standardized enthalpy from Table H.2 to verify the enthalpies of formation at 298.15 K in Table H.1 for methane, propane, and hexane.

2.134 Determine the total standardized enthalpy H (kJ) for a fuel–air reactant mixture containing 1 kmol CH_4, 2.5 kmol O_2, and 9.4 kmol N_2 at 500 K and 1 atm. Also determine the mass-specific standardized enthalpy h (kJ/kg) for this mixture.

2.135 A mixture of products of combustion contains the following constituents at 2000 K: 3 kmol of CO_2, 4 kmol of H_2O, and 18.8 kmol of N_2.

Determine the following quantities:

A. The mole fraction of each constituent in the mixture
B. The total standardized enthalpy H of the mixture
C. The mass-specific standardized enthalpy h of the mixture

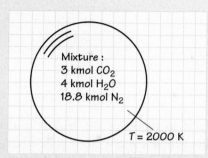

Mixture :
3 kmol CO_2
4 kmol H_2O
18.8 kmol N_2

$T = 2000$ K

2.136 Calculate the change in entropy for mixing 2 kmol of O_2 with 6 kmol of N_2. Both species are initially at 1 atm and 300 K, as is the final mixture.

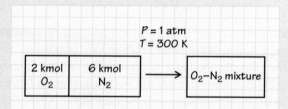

$P = 1$ atm
$T = 300$ K

2 kmol O_2 | 6 kmol N_2 $\longrightarrow$ O_2–N_2 mixture

2.137 A mixture of 15% CO_2, 12% O_2, and 73% N_2, (by volume) expands to a final volume six times greater than its initial volume. The corresponding temperature change is 1000°C to 750°C. Determine the entropy change (kJ/kg·K).

2.138 Determine the change in entropy (kJ/kg·K) of a mixture of 60% N_2 and 40% CO_2 by volume for a reversible adiabatic increase in volume by a factor of 5. The initial temperature is 540°C.

2.139 A partition separating a chamber into two compartments is removed. The first compartment initially contains oxygen at 600 kPa and 100°C; the second compartment initially contains nitrogen at the same pressure and temperature. The oxygen compartment volume is twice that of the one containing nitrogen. The chamber is isolated from the surroundings. Determine the change of entropy associated with the mixing of the O_2 and N_2. *Hint:* Assume that the process is isothermal.

2.140 Consider two compartments of the same chamber separated by a partition. Both compartments contain nitrogen at 600 kPa and 100°C; however, the volume of one compartment is twice that of the other. The partition is removed. Assuming the chamber is isolated from the surroundings, determine the entropy change of the N_2 associated with this process.

2.141 In the stoichiometric combustion of 1 kmol of methane with air (2 kmol of O_2 and 7.52 kmol of N_2). the following combustion products are formed: 1 kmol CO_2, 2 kmol H_2O, and 7.52 kmol N_2. The H_2O is in the vapor state. Determine the following:

A. The enthalpy of reaction (kJ)
B. The apparent molecular weight of the product mixture
C. The enthalpy of reaction per mass of mixture
D. The enthalpy of reaction per mass of fuel

2.142 Repeat Problem 2.141 for the stoichiometric combustion of propane with air where the following reaction occurs:

$$C_3H_8 + 5(O_2 + 3.76N_2) \rightarrow 3CO_2 + 4H_2O + 18.8\,N_2.$$

2.143 Determine the higher and lower heating values for the stoichiometric combustion of methane with air (see Problem 2.141) and for the combustion of propane with air (see Problem 2.142).

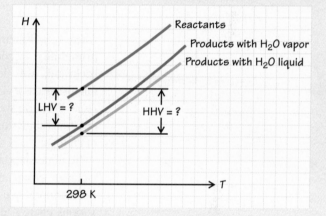

2.144 Use the NIST software or online database to determine the thermal conductivity k and the viscosity μ of H_2O for the following conditions. Express your results in SI units.

A. 1 MPa and 300 K
B. 1 MPa and 400 K
C. 0.1 MPa and 300 K
D. 3 kPa and 300 K

2.145 Use the NIST online database to determine the thermal conductivity k and viscosity μ of N_2 for the following conditions. Express your results in SI units.

A. 1 atm and 300 K
B. 1 atm and 500 K
C. 1 atm and 1000 K
D. 2 atm and 1000 K
E. 5 atm and 1000 K
F. 100 atm and 1000 K

Use your results to discuss the influence of temperature and pressure on the thermal conductivity and viscosity of N_2.

Appendix 2A
Molecular Interpretation of Entropy

The purpose of this appendix is to introduce the reader to a molecular interpretation of entropy without getting bogged down in details. A rigorous development can be found in Ref. [16], for example, and a very readable elementary treatment is provided by Wark [17].

We begin by stating our final result that the entropy S is given by the following expression:

$$S = k_B \ln W_{mp}, \qquad (2A.1)$$

where k_B is Boltzmann's constant and W_{mp} is the most probable **thermodynamic probability**, a concept that requires some elaboration. Continuing to work backward, we define the thermodynamic probability as the number of microstates associated with a given macrostate. We are now at the heart of this issue: What do we mean by a microstate or by a macrostate? Answering these questions allows us to come full circle back to Eq. 2A.1, our statistical definition of entropy.

To understand the concepts of microstates and macrostates, we consider an isolated group of N particles comprising our thermodynamic system. Furthermore, we assume that the individual particles in our system have various energy levels, ε_i, as dictated by quantum mechanics. Although any individual particle may have any particular allowed energy, the system energy must remain fixed and is constrained by

$$\sum N_i \varepsilon_i = U, \qquad (2A.2)$$

where N_i is the number of particles that have the specific energy level ε_i. We continue now with a specific, but hypothetical, example in which our system contains only three particles, A, B, and C, and has a total system energy of six units, or quanta. Furthermore, we assume that the allowed energy levels are equally spaced intervals of one unit, beginning with unity. Thus, any individual particle can have energy of one, two, three, or four quanta. Clearly, energy levels above four are disallowed because, if one or more particles possessed this energy, the overall system constraint of six units would be violated. For example, if particle A possesses five units, the least possible total system energy is seven units, since the least energy particles B and C can possess is one unit $(5 + 1 + 1 = 7)$

We now consider the number of ways the overall system can be configured, assuming particles A, B, and C are distinguishable. Using Table 2A.1 as a guide, we see that there are a total of ten ways that our three particles can be arranged while maintaining the total system energy at six units. Each one of these ten arrangements is identified as a **microstate**. We further note that some

Table 2A.1 Macrostates and Microstates Associated with a System of Three Particles and a Total Energy of Six Units

Macrostate 1

ε_i						
4	—	—	—	—	—	—
3	A	B	C	B	A	C
2	B	A	A	C	C	B
1	C	C	B	A	B	A
$\sum N_i \varepsilon_i =$	6	6	6	6	6	6

$$W = 6$$

Macrostate 2

ε_i			
4	A	B	C
3	—	—	—
2	—	—	—
1	B,C	A,C	B,A
$\sum N_i \varepsilon_i =$	6	6	6

$$W = 3$$

Macrostate 3

ε_i	
4	—
3	—
2	A,B,C
1	—
$\sum N_i \varepsilon_i =$	6

$$W = 1$$

of these microstates are similar; if we remove the restriction that A, B, and C are distinguishable, there are six identical microstates in which one particle possesses three units of energy, another particle possesses two units of energy, and the third particle possesses one unit. The identification of a state purely by enumeration of the number of particles at each energy level without regard to the identification of the individual particle is defined as a **macrostate**. Employing this definition, we see that there are two other macrostates associated with our system: macrostate 2, in which one particle has four units of energy and two particles possess one unit, and macrostate 3, in which each particle has two units of energy. The probability of our system being found in a particular macrostate is related to the number of microstates comprising the macrostate. For our example, there are six internal arrangements that can be identified with macrostate 1, three for macrostate 2, and only one arrangement possible for macrostate 3.

Recall that one of our goals at the outset of this discussion was to understand the definition of thermodynamic probability as the number of microstates associated with a given macrostate. We come to closure on this goal by generalizing and defining the **thermodynamic probability W** for a system containing a total of N particles as follows:

$$W = \frac{N!}{N_1! \, N_2! \ldots N_i! \ldots N_M},$$ (2A.3)

where N_i represents the number of particles having energies associated with the ith energy level. For our example, formal application of Eq. 2A.3 to the three macrostates shown in Table 2A.1 yields (recognizing that $0! = 1$):

$$W_1 = \frac{3!}{1!\ 1!\ 1!\ 0!} = 6,$$

$$W_2 = \frac{3!}{2!\ 0!\ 0!\ 1!} = 3,$$

and

$$W_3 = \frac{3!}{0!\ 3!\ 0!\ 0!} = 1.$$

If we normalize these results by the total number of microstates, a conventional concept of probability gives us:

$$P_1 = \frac{W_1}{W_1 + W_2 + W_3} = 0.6,$$

$$P_2 = \frac{W_2}{W_1 + W_2 + W_3} = 0.3,$$

and

$$P_3 = \frac{W_3}{W_1 + W_2 + W_3} = 0.1.$$

Thus we see that the probability of finding the system in macrostate 1 is 0.6, whereas the probability of finding the system in macrostates 2 and 3 are 0.3 and 0.1, respectively. As the number of particles increases, the probability of the most probable macrostate overwhelms that of all of the other possible macrostates.

We conclude our discussion of entropy with two observations. First, the practical evaluation of the defining relationship for entropy (Eq. 2A.1) requires a knowledge of how particles in a macroscopic system distribute among all of the allowed energy states (i.e., quantum levels) to determine the most probable thermodynamic probability. Various theories provide such distribution functions (e.g., Maxwell–Boltzmann, Bose–Einstein, and Fermi–Dirac statistics). Discussion of these is beyond the scope of this book and the interested reader is referred to Refs. [16, 17], for example. Our second observation is that Boltzmann's definition of entropy (Eq. 2A.1) does indeed say something about "disorder" as being a favored condition.[21] By definition, the most probable macrostate is the one that has the greatest number of microstates. We can view this most probable macrostate as a state of maximum "disorder"—there is no other possible state that has as many different possible arrangements of particles.

---

[21] The idea of disorder is frequently invoked in nontechnical definitions of entropy. For example, one dictionary definition of entropy is a measure of the degree of disorder in a substance or a system [18].

CONSERVATION OF MASS

After studying Chapter 3, you should:

- Have an improved understanding of the meanings of a quasi-static process and local equilibrium.

- Be able to define and calculate volume and mass flow rates for both simple and relatively complex situations.

- Be able to express the conservation of mass principle for thermodynamic **systems** and to apply this principle to analyze practical situations.

- Be able to express the conservation of mass principle for integral **control volumes** for both unsteady and steady flows and to apply this principle to analyze practical situations.

- Have a rudimentary understanding of how the mass conservation principle applies to a differential control volume.

- Be able to apply atom (element) conservation principles to reacting systems or flows.

- Understand the various ways that stoichiometry is expressed for combustion systems and be able to use this information to formulate mass conservation expressions.

Chapter 3 Overview

With the disclaimer that we will consider neither situations involving nuclear transformations nor situations where relativistic effects are important, this chapter presents general and specific statements of mass conservation. We consider, first, thermodynamic systems. The concept of a flow rate and its relationship to the velocity distribution of a flowing fluid is introduced before extending the mass conservation principle to control volumes. In dealing with control volumes, we consider both integral (macroscopic) and differential (microscopic) forms. The principle of mass conservation is extended yet further to include element conservation in chemically reacting systems and flows.

Leonardo da Vinci (1452–1519)

Antoine Lavoisier (1743–1794)

3.1 HISTORICAL CONTEXT

Leonardo da Vinci (1452–1519) was an astute observer of nature. His interest in flowing fluids led him to be the first to express a clear and concise statement of mass conservation (continuity) for incompressible flows. The following statement, and others, reveal his quantitative understanding of this principle [1]:

> **A river of uniform depth will have a more rapid flow at the narrower section than at the wider, to the extent that the greater surpasses the lesser.**

For reacting systems, a correct statement of mass conservation had to wait until 1798, when **Antoine Laurent Lavoisier** (1743–1794) presented his results from careful experiments on closed thermodynamic systems [2]. Lavoisier observed that metals gained mass when heated in the presence of oxygen, whereas, at higher temperatures the metal oxides decomposed back to the original metal and oxygen. These experiments were conducted at a time when the idea of phlogiston was in competition with the newer caloric theory.[1] Phlogiston was thought to be an imponderable (massless) fluid associated with combustion. For example, it was thought that when charcoal burned its phlogiston escaped and combined with the air. Combustion would then be complete, or terminate, when all of the phlogiston had escaped, or when the air was saturated with phlogiston, as would occur for combustion in a closed vessel. Lavoisier's experiments made these ideas untenable. Interestingly, Lavoisier's experiments on mass conservation were performed at the same time (see Appendix A) that Benjamin Thompson (Count Rumford) performed his famous cannon-boring experiments, one result of which was the discovery that heat was not a material substance. We will revisit Thompson's contributions in Chapters 4 and 5.

---

[1] Neither theory has stood the test of time.

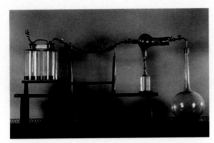

Apparatus from Lavoisier. Image courtesy of Panopticon Lavoisier Institute and Museum of History (Florence).

3.2 MASS CONSERVATION FOR A SYSTEM

We begin by explicitly transforming the generic conservation principles expressed in Eqs. 1.1 and 1.2 to statements of mass conservation. Equation 1.1 thus becomes

$$M_{\text{in}} \quad - \quad M_{\text{out}} \quad + \quad M_{\text{generated}} = \Delta M_{\text{stored}} \equiv M_{\text{sys}}(t_2) - M_{\text{sys}}(t_1). \quad (3.1a)$$

| Quantity of mass crossing boundary and passing into system | Quantity of mass crossing boundary and passing out of system | Quantity of mass generated within system | Quantity of mass stored in system during time interval $\Delta t = t_2 - t_1$ |

By definition, no mass crosses the system boundaries; thus both M_{in} and M_{out} must be zero. Furthermore, if no nuclear transformations occur, no mass is generated (or destroyed) within the system; thus, all terms on the left-hand side of Eq. 3.1a are zero. We formally conclude the obvious:

$$0 = M(t_2) - M(t_1),$$

or

$$M(t_1) = M(t_2) = M = \text{constant}. \quad (3.1b)$$

Furthermore, since the system mass is a constant, its time derivative must be zero, that is,

$$\frac{dM}{dt} = 0. \quad (3.1c)$$

Note that, had we started with Eq. 1.2, the rate form of the generic conservation principles, we would have arrived at Eq. 3.1c directly.

For a system in which the density is uniform (i.e., has the same value at every location), we express Eqs. 3.1b and 3.1c as

$$\rho \mathcal{V} = \text{constant} \ (= M), \quad (3.2a)$$

or

$$\frac{d(\rho \mathcal{V})}{dt} = 0, \quad (3.2b)$$

> Central to our study of mass conservation are the thermodynamic properties M, $\mathcal{V}$, and ρ or v. See Eq. 2.8 and related material in Chapter 2.

where ρ is the mass density and $\mathcal{V}$ is the volume.

Some systems have boundaries that move with time. A familiar example is the gas trapped in the combustion chamber of an internal combustion engine, suggested by the sketch in Fig. 3.1. To assure that we have a thermodynamic system, both intake and exhaust valves must be tightly closed, and no gas can leak past the piston rings. If we assume that the gas in the system has a uniform but time-varying density, Eq. 3.2b can be expanded using the product rule for differentiation to give us

$$\rho \frac{d\mathcal{V}}{dt} + \mathcal{V} \frac{d\rho}{dt} = 0. \quad (3.3)$$

FIGURE 3.1

The gas trapped in the combustion chamber is the thermodynamic system. The motion of the piston creates a moving boundary and a time-varying volume.

To apply Eq. 3.3, say, to a simulation of an engine, $\mathcal{V}$ and $d\mathcal{V}/dt$ can be determined purely from geometric and kinematic relationships (see Appendix 1A). Obtaining the density and density time derivative, however, requires application of additional thermal science concepts: an equation of state (see Chapter 2), conservation of energy (Chapter 5), and expressions for heat-transfer rates (Chapter 4).

The density of the unburned gases can be many times greater than that of the burned gases.

The assumption of a uniform density is an approximation since the temperature of the gas within a real combustion chamber is not uniform. The uniform property approximation is probably most reasonable during the compression and power strokes, but it is far from reality during the combustion event, when a flame propagates through a relatively cool unburned mixture and produces hot products. To generalize to this situation of spatially varying density, Eqs. 3.1b and 3.1c can be rewritten as

$$\iiint_{z \; y \; x} \rho(x, y, z)\, dx\, dy\, dz = \text{constant} \, (= M) \tag{3.4a}$$

and

$$\frac{d}{dt}\left[\iiint_{z \; y \; x} \rho(x, y, z)\, dx\, dy\, dz \right] = 0, \tag{3.4b}$$

where the triple integral explicity shows the integration over the three coordinates associated with the volume. Equations 3.4a and 3.4b are more compactly written by collapsing the triple integral to a single integral over the volume, so that

$$M = \int_{V} \rho\, dV = \text{constant}, \tag{3.4c}$$

or

$$\frac{d}{dt}\left[\int_{V} \rho\, dV \right] = 0. \tag{3.4d}$$

We now illustrate the use of these various relations with a few examples.

Uniform density   Non-uniform density

$M = \rho V$ $M = \int_{V} \rho\, dV$

Example 3.1

Gaseous nitrogen is trapped in a cylinder having a diameter of 75 mm and height of 40 mm. Assume that the temperature (1,250 K) and pressure (500 kPa) are essentially uniform throughout the cylinder. Determine the mass of the N_2.

Solution

Known P, T, D, L

Find M_{N_2}

Sketch

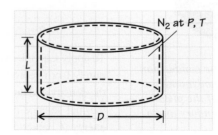

Assumptions

   i. N_2 behaves as an ideal gas.

   ii. Gravity does not affect the pressure.

Analysis The solution to this is quite straightforward. We use the given geometry to calculate the volume of the N_2, then use the ideal-gas equation of state (Eq. 2.28c) to determine the mass as follows:

$$V = (\pi D^2/4)\, L = \pi (0.075 \text{ m})^2 (0.040 \text{ m})/4 = 1.767 \times 10^{-4} \text{ m}^3$$

and

$$PV = MR_{N_2}T.$$

Solving for M and recognizing that the gas constant $R_{N_2} \equiv R_u/\mathcal{M}_{N_2}$ yield

$$M = \frac{PV\mathcal{M}_{N_2}}{R_uT}.$$

Evaluating numerically, we get

$$M = \frac{500 \times 10^3 (1.767 \times 10^{-4})28.013}{8314.47(1250)} = 2.38 \times 10^{-4}$$

$$[=] \frac{\text{Pa}\,(\text{m}^3)\,\text{kg/kmol}}{(\text{J/kmol}\cdot\text{K})\text{K}}\left[\frac{1 \text{ N/m}^2}{\text{Pa}}\right]\left[\frac{1 \text{ J}}{\text{N}\cdot\text{m}}\right] = \text{kg}.$$

Comments Note the importance of the ideal-gas law in solving this nearly trivial problem. An alternative solution to this problem is to apply the operational definition of mass for a system with a uniform density, $M = \rho V$ (Eq. 3.2a), and use the ideal-gas law (Eq. 2.28b) to calculate the density.

  The cylinder in Example 3.1 is compressed to a final height of 15 mm. Determine the final density of the N_2.

(Answer: $\rho = 3.59$ kg/m³)

Example 3.2

Nitrogen gas is contained in a cylinder having a diameter of 75 mm and height of 40 mm. In the absence of gravitational effects, the pressure is uniform at 500 kPa; the N_2, however, is vertically stratified with hotter gas at the top of the cylinder and cooler gas at the bottom. The temperature distribution is linear with height and is described by T ([=] K) $= a + bx$, where $a = 1000$ K, $b = 1.25 \times 10^4$ K/m, and x is the distance in meters measured from the bottom of the cylinder. Determine the mass of N_2 contained in the cylinder.

Solution

Known $P, D, L, T(x)$

Find M_{N_2}

Sketch

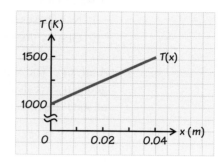

Assumptions

i. Local equilibrium prevails at every location within the cylinder.
ii. N_2 behaves as an ideal gas.
iii. Gravitational effects on pressure are negligible.

Analysis   Since the temperature varies with height through the gas, the density of the N_2 will vary with height. With the assumption of local equilibrium, we can apply the ideal-gas equation of state (Eq. 2.28c) locally, so

$$\rho(x) = \frac{P}{R_{N_2} T(x)}.$$

The mass within the cylinder is obtained by integrating Eq. 3.4c:

$$M = \int_{V} \rho(x)\, dV.$$

The appropriate differential volume dV is a thin disk expressed as

$$dV = \frac{\pi D^2}{4}\, dx.$$

Thus,

$$M = \int_{x=0}^{L} \frac{P}{R_{N_2} T(x)} \frac{\pi D^2}{4}\, dx.$$

Removing all constants from the integral and substituting the given distribution for $T(x)$ yields

$$M = \frac{\pi D^2 P}{4 R_{N_2}} \int_{x=0}^{L} \frac{1}{a + bx}\, dx.$$

Performing the integration and substituting the limits, we obtain

$$M = \frac{\pi D^2 P}{4 R_{N_2}} \left[\frac{1}{b} \ln\left(\frac{a + bL}{a} \right) \right].$$

Substituting numerical values gives us

$$M = \frac{\pi(0.075)^2\,500 \times 10^3}{4(8314.47/28.013)1.25 \times 10^4} \ln\left[\frac{1000 + 1.25 \times 10^4(0.04)}{1000}\right]$$

$$= 5.954 \times 10^{-4} \ln\left(\frac{1500}{1000}\right) = 2.41 \times 10^{-4}$$

$$[=]\frac{\text{m}^2\,(\text{Pa})}{[(\text{J/kmol}\cdot\text{K})/(\text{kg/kmol})]\text{K/m}}\left[\frac{1\ \text{N/m}^2}{\text{Pa}}\right]\left[\frac{1\ \text{J}}{\text{N}\cdot\text{m}}\right] = \text{kg}.$$

Comments   The assumption of local equilibrium is essential to our use of the ideal-gas equation of state, as this equation applies only to equilibrium properties. We also note that the volume-averaged temperature is 1,250 K, the same as the uniform temperature used in Example 3.1; however, this average temperature does not provide the correct mass from direct application of the state equation because the density is inversely proportional to the temperature [i.e., $\rho(\overline{T}) \neq \overline{\rho}$]. Neglecting gravity allows the pressure to be uniform. In Chapter 6, we will see the influence of gravity on the pressure distribution of a column of fluid.

Example 3.3

Consider the same situation described in Example 3.2, but now the N_2 is radially stratified with the hottest gas on the centerline. The temperature distribution is given by $T(r) = a + br^2$, where $a = 1,500$ K and $b = -3.555 \times 10^5$ K/m^2. The radial coordinate r is measured from the centerline of the cylinder. Determine the mass of N_2 contained in the cylinder.

Solution

Known $P, D, L, T(r)$

Find M_{N_2}

Sketch

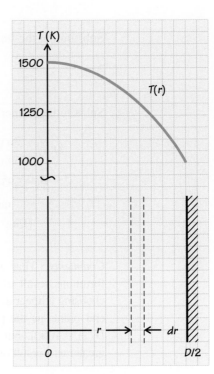

Assumptions

  i. Local equilibrium prevails at every location within the cylinder.
 ii. N_2 behaves as an ideal gas.
iii. Gravitational effects on pressure are negligible.

Analysis   Our solution here is similar to that of Example 3.2, except that now we must integrate over the volume in the radial direction. With our assumptions of local equilibrium, ideal-gas behavior, and uniform pressure, the N_2 density is expressed as

$$\rho(r) = \frac{P}{R_{N_2}T(r)},$$

and the mass is obtained by integrating Eq. 3.4c:

$$M = \int_V \rho(r)\, d\mathcal{V}.$$

As shown in the sketch, the appropriate differential volume is now an annular shell with thickness dr and length L and is expressed as

$$d\mathcal{V} = (2\pi r dr)L.$$

Thus,

$$M = \int_{r=0}^{D/2} \frac{P}{R_{N_2}} \left(\frac{r}{a + br^2} \right) 2\pi L\, dr$$

$$= \frac{2\pi LP}{R_{N_2}} \int_{r=0}^{D/2} \left(\frac{r}{a + br^2} \right) dr.$$

Performing the integration yields

$$M = \frac{2\pi LP}{R_{N_2}} \frac{1}{2b} \ln\left[a + br^2 \right]_{r=0}^{r=D/2}.$$

Substituting the limits and rearranging, we obtain the following:

$$M = \frac{2\pi LP}{R_{N_2}} \frac{1}{2b} \ln\left[1 + \frac{bD^2}{4a} \right].$$

Substituting numerical values yields

$$M = \frac{2\pi (0.04)\,500 \times 10^3}{296.8\,(2)(-3.555 \times 10^5)} \ln\left[1 + \frac{-3.555 \times 10^5 (0.075)^2}{4(1500)} \right]$$

$$= -5.9549 \times 10^{-4} \ln(0.667) = 2.41 \times 10^{-4}$$

$$[=] \frac{m(Pa)}{(J/kg \cdot K)K/m^2} \left[\frac{1\ N/m^2}{Pa} \right]\left[\frac{1\ J}{N \cdot m} \right] = kg.$$

Comments   That we obtain the same mass here as in Example 3.2 is a fortuitous result of our choice of temperature distributions. For example, choosing a linear distribution $T = 1500 - (1.333 \times 10^5)r$ provides the same end points as the parabolic distribution [i.e., $T(r = 0) = 1500$ and $T(r = D/2) = 1000$], but the calculated mass (2.58×10^{-4} kg) is significantly greater as a result of the larger mean density.

Example 3.4 SI Engine Application

Consider a spark-ignition engine. The temperature and pressure are assumed to be uniform within the cylinder with values of 345 K and 184 kPa, respectively, and the properties of the fuel–air mixture can be treated as those of air. Geometric parameters are defined and kinematic relationships for the instantaneous volume $\mathcal{V}(\theta)$ and its time derivative $d\mathcal{V}(\theta)/dt$ are given in Appendix 1A of Chapter 1. For the geometrical parameters

$$B = 70 \text{ mm}, \qquad CR = 8,$$
$$S = 70 \text{ mm}, \qquad \ell/a = 3.5,$$

determine the instantaneous value of $d\rho/dt$ during the compression stroke at a crank angle of $\theta = 270°$ ($3\pi/2$ rad) for a rotational speed of 2,000 rpm.

Solution

Known $P, T, \theta, N, B, S, CR, \ell/a$

Find $d\rho/dt$

Sketch

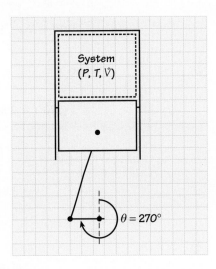

Assumptions

i. Quasi-static equilibrium
ii. Uniform properties
iii. Closed system with no leakage past rings or valves
iv. Air properties approximate charge (ideal gas)

Analysis   Assumptions i, ii, and iii allow us to express conservation of mass for the system shown in the sketch using Eq. 3.3. Solving Eq. 3.3 for $d\rho/dt$ yields

$$\frac{d\rho}{dt} = -\frac{\rho}{\mathcal{V}}\frac{d\mathcal{V}}{dt}.$$

We evaluate the density from the ideal-gas equation of state (Eq. 2.28b); $\mathcal{V}$ and $d\mathcal{V}/dt$ are evaluated from the relationships in Appendix 1A:

$$\rho = \frac{P}{R_{air}T} = \frac{P\mathcal{M}_{air}}{R_uT}$$

$$= \frac{184 \times 10^3(28.97)}{8314.47(345)} = 1.858$$

$$[=]\frac{\text{Pa(kg/kmol)}}{(\text{J/kmol}\cdot\text{K})\text{K}}\left[\frac{1 \text{ N/m}^2}{\text{Pa}}\right]\left[\frac{1 \text{ J}}{\text{N}\cdot\text{m}}\right] = \text{kg/m}^3.$$

To evaluate $\mathcal{V}(\theta)$ from Eq. 1A.4 requires a value for $\mathcal{V}_{TC}$. Using the definition of compression ratio and displacement presented in Appendix 1A, we solve Eq. 1A.3 for $\mathcal{V}_{TC}$ as follows:

$$\mathcal{V}_{TC} = \frac{\mathcal{V}_{disp}}{CR - 1},$$

where

$$\mathcal{V}_{disp} = S\pi B^2/4.$$

Thus,

$$\mathcal{V}_{TC} = \frac{S\pi B^2/4}{CR - 1}$$

$$= \frac{(0.07 \text{ m})\pi(0.07 \text{ m})^2/4}{8 - 1} = 3.848 \times 10^{-5} \text{ m}^3.$$

We can now evaluate $\mathcal{V}(\theta)$ and $d\mathcal{V}(\theta)/dt$ using the expressions from Appendix 1A:

$$\mathcal{V}(\theta) = \mathcal{V}_{TC}\left\{1 + \frac{1}{2}(CR - 1)\left[\frac{\ell}{a} + 1 - \cos\theta - \left(\frac{\ell^2}{a^2} - \sin^2\theta\right)^{1/2}\right]\right\},$$

where

$$\mathcal{V}(270°) = 3.848 \times 10^{-5}\left\{1 + \frac{1}{2}(8 - 1)[3.5 + 1 - \cos 270°\right.$$

$$\left. - (3.5^2 - \sin^2(270°))^{1/2}]\right\}$$

$$= 3.848 \times 10^{-5}\left\{1 + \frac{7}{2}[4.5 - 3.354]\right\} = 1.928 \times 10^{-4} \text{ m}^3,$$

and

$$\frac{d\mathcal{V}(\theta)}{dt} = SN\frac{\pi B^2}{4}\pi\sin\theta\left[1 + \frac{\cos\theta}{\left(\frac{\ell^2}{a^2} - \sin^2\theta\right)^{1/2}}\right].$$

Recognizing that $\cos(270°) = 0$, we see that the term in brackets becomes unity; thus,

$$\frac{d\mathcal{V}}{dt}(270°) = 0.07\left(\frac{2000}{60}\right)\frac{\pi(0.07)^2}{4}\pi\sin(270°)(1) = -0.0282$$

$$[=]\text{m(rev/min)m}^2\left[\frac{1 \text{ min}}{60 \text{ s}}\right] = \text{m}^3/\text{s}.$$

We now evaluate $d\rho/dt$ from Eq. 3.3:

$$\frac{d\rho}{dt} = -\frac{\rho}{\mathcal{V}(270°)}\frac{d\mathcal{V}}{dt}(270°)$$

$$= -\frac{1.858(-0.0282)}{1.928 \times 10^{-4}} \text{ kg/m}^3 \cdot \text{s}$$

$$= +271.8 \text{ kg/m}^3 \cdot \text{s},$$

where the positive sign indicates that the charge is being compressed, as expected.

Comments This example illustrates how the purely mechanical relationships that describe the geometry and kinematics of a reciprocating engine relate to the thermodynamics within the engine cylinder.

3.3 MASS CONSERVATION FOR A CONTROL VOLUME

We now consider conservation of mass applied to a **control volume**. In this section, we develop various mathematical statements of this conservation principle for the most simple and restricted cases, as well as more complex and general situations. Before this can be done, however, we need to define and explore two underlying concepts: the velocity field and flow rates.

3.3a Velocity Field

Velocity Vector

Consider a fluid flowing through a domain defined by a Cartesian coordinate system as shown in Fig. 3.2. At any point[2] in the flow we define the **velocity vector V** to be

$$V \equiv v_x\hat{i} + v_y\hat{j} + v_z\hat{k}, \tag{3.5a}$$

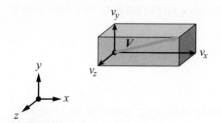

where

$$v_x \equiv \text{velocity component in the } x\text{-direction}, \tag{3.5b}$$

$$v_y \equiv \text{velocity component in the } y\text{-direction}, \tag{3.5c}$$

$$v_z \equiv \text{velocity component in the } z\text{-direction}. \tag{3.5d}$$

FIGURE 3.2
Velocity vector V and components v_x, v_y, and v_z for a Cartesian coordinate system.

To emphasize that the velocity vector and its components depend on location, and may also depend on time, we write

$$V = V(x, y, z, t), \tag{3.6a}$$

with

$$v_x = v_x(x, y, z, t), \tag{3.6b}$$

$$v_y = v_y(x, y, z, t), \tag{3.6c}$$

and

$$v_z = v_z(x, y, z, t). \tag{3.6d}$$

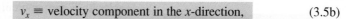

The velocity vector can also be defined in cylindrical and spherical coordinate systems.

The **velocity field** is defined by defining the functional behavior of $V(x, y, z, t)$. Knowing the function $V(x, y, z, t)$ is synonymous with knowing the velocity field. Formulations of conservation of mass, energy, and momentum for a flowing fluid depend on knowing the velocity field. In many of the developments in this book, we are not concerned with the full three-dimensional, time-dependent velocity field. Rather, we frequently deal with steady (no time dependence),

---

[2] As discussed in Chapter 2, a *point* here refers to the continuum limit, not a mathematical point. You may wish to review this distinction before proceeding.

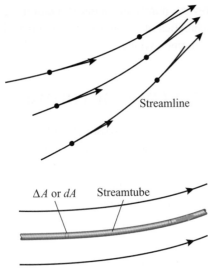

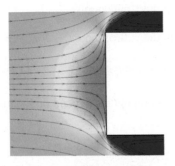

FIGURE 3.3

Velocity vectors are tangent to streamlines at every point in a flow (top). Streamlines define the surface of a streamtube. The cross-sectional area of a streamtube (flow area) is small (ΔA) or infinitessimal (dA).

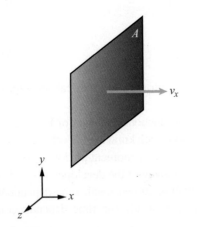

Streamlines for the two-dimensional flow around a slab in a uniform stream.

one- or two-dimensional flows. The formidable looking velocity vector then reduces to $V = v_x(x)$ or $V = v_x(x, y)\hat{i} + v_y(x, y)\hat{j}$, respectively.

Streamlines

We formally define a streamline as follows:

> ***Streamlines* are lines in a flow field that at any instant are tangent to the local velocity vector.**

Figure 3.3 illustrates this definition graphically. Note that, since the velocity vector is tangent to a streamline, no fluid can cross a streamline. Streamlined bodies, such as sleek racing cars and fish, are so designated because the flow follows the outer contours of their bodies.[3] Figure 3.3 also illustrates a **streamtube**. A streamtube is defined by a collection of streamlines around the periphery of a small, or differential, cross-sectional area, ΔA or dA. Since the walls of a streamtube consist of streamlines, no fluid can cross these walls. The amount of fluid flowing through a streamtube, or any contiguous bundle of streamtubes, is fixed.

3.3b Flow Rates

Uniform Velocity

Consider a fluid with a uniform and steady velocity v_x crossing an imaginary (i.e., nonmaterial), planar surface of area A as shown in Fig. 3.4. The fluid velocity is perpendicular to the surface and aligned with the x-direction. In a time interval Δt, a fluid particle will travel a distance $\Delta x = v_x \Delta t$ after crossing the surface. The volume of fluid crossing the surface in this same time interval is simply the product of the flow area A and the distance traveled by the first fluid particle to cross the surface at the start of the time interval, that is,

$$\Delta \mathcal{V} = A \Delta x.$$

Substituting for Δx and rearranging yield

$$\frac{\Delta \mathcal{V}}{\Delta t} = v_x A.$$

The quantity $\Delta \mathcal{V}/\Delta t$ we define as the **volumetric flow rate** $\dot{\mathcal{V}}$. Shrinking the time interval to a very small value, but not going below the limit required to maintain a continuum, allows us to interpret $v_x A$ as the instantaneous volumetric flow rate, removing the restriction that the velocity be steady (i.e., that it does not vary with time). Thus,

$$\dot{\mathcal{V}} = v_x A, \tag{3.7}$$

where we emphasize that v_x is uniform over A and is everywhere perpendicular to the plane of A. This particular condition of uniform velocity is sometimes called **plug** or **slug flow**.

If the fluid density also is uniform over A, the amount of mass crossing the plane in Δt is just

$$\Delta M = \rho \Delta \mathcal{V} = \rho A v_x \Delta t.$$

We thus define the **mass flow rate** $\dot{m}$ to be $\Delta M/\Delta t$ or, instantaneously,

$$\boxed{\dot{m} = \rho v_x A.} \tag{3.8}$$

The SI units for mass flow rate are kg/s.

FIGURE 3.4

Fluid with uniform velocity crosses surface A with the velocity everywhere perpendicular to the plane of A.

---

[3] In Chapter 9, we explore the reasons for the low-drag characteristics of streamlined bodies.

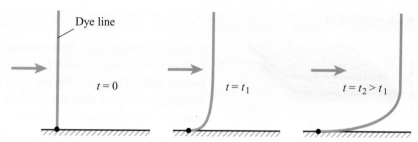

The flow from left to right transports a dye line downstream. The distance traveled by each segment of the line is proportional to the local velocity. The dye line sticks to the wall as the velocity there is zero.

Distributed Velocity

In flows within pipes, tubes, and channels, the velocity distribution is not uniform. In these flows, the fluid sticks to the walls of the flow passage. This results in a velocity distribution with a zero value at the walls and a maximum value some distance from the wall. That the fluid velocity is zero at a surface is termed the **no-slip condition**. For pipe or tube flows, velocity profiles are commonly parabolic (**laminar flow**) or obey a power law (**turbulent flow**). Details and further development of these concepts are presented in Chapter 9 and 10. Presently, our concern is how to evaluate flow rates when the velocity is not uniform over the flow area.

The mass flow rate associated with a differential flow area dA is

$$dm = \rho v_x dA,$$

where v_x is the local velocity at the position of dA. Figure 3.5 illustrates useful differential areas for two-dimensional Cartesian and axisymmetric flows. For the two-dimensional case, dA is a long rectangular strip of width dy, whereas for the axisymmetric geometry, dA is an annulus of width dr. That the area of the differential annulus is $2\pi r dr$ is easy to remember by visualizing the annulus as a strip of length $2\pi r$ (the circumference) with a width dr. With these differential areas, the total mass flow rate can be obtained for the two geometries. For both cases,

$$\dot{m} = \int_A \rho v_x dA. \tag{3.9a}$$

Applying Eq. 3.9a to the Cartesian flow yields

$$\dot{m} = \int_0^Y \rho v_x(y) Z dy, \tag{3.9b}$$

and for the axisymmetric case, we get

$$\dot{m} = \int_0^R \rho v_x(r) 2\pi r dr. \tag{3.9c}$$

Note that Eqs. 3.9a–3.9c allow for a variable density. For constant-density flows, ρ can be removed from the integrand.

In the following examples, we apply Eqs. 3.9b and 3.9c to velocity profiles frequently encountered in real flows.

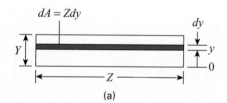

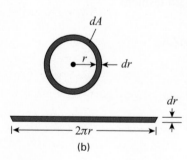

FIGURE 3.5

(a) The differential area associated with a two-dimensional (x, y) flow is the rectangular strip Zdy. For two-dimensional flow, Z ≫ Y is required. (b) The differential area associated with a circular flow area is an annular strip of length 2πr and width dr.

Example 3.5

Consider a steady (i.e., time-invariant), linear velocity distribution for flow between two plates in which the lower plate is stationary and the upper plate moves with a known velocity V_p. The distance between the plates is Y. Determine the mass flow rate per unit flow depth (Z) for motor oil at 300 K with $V_p = 0.8$ m/s and $Y = 0.5$ mm.

Solution

Known   Linear $v_x(y)$, V_p, Y

Find $\dot{m}/Z$

Sketch

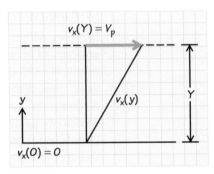

Assumptions

  i. One-dimensional flow

  ii. Constant density

Analysis   Since the velocity profile is linear in the y-direction [i.e., $v_x(y) = ay$] the slope a is simply

$$a = \frac{v_x(Y) - v_x(0)}{Y}$$

$$= \frac{V_p - 0}{Y} = \frac{V_p}{Y}$$

$$= \frac{0.8 \text{ m/s}}{0.0005 \text{ m}} = 1{,}600 \text{ s}^{-1}.$$

We now apply Eq. 3.9b in a straightforward fashion:

$$\dot{m} = \int_0^Y \rho v_x(y) Z dy$$

$$= \frac{\rho V_p Z}{Y} \int_0^Y y dy$$

$$= \frac{\rho V_p Z Y}{2}.$$

With the oil density (884.1 kg/m³) from Appendix G, we numerically evaluate this as

$$\frac{\dot{m}}{Z} = \frac{884.1(0.8)0.0005}{2} = 0.1768$$

$$[=](\text{kg/m}^3)(\text{m/s})\text{m} = \frac{\text{kg/s}}{\text{m}}.$$

Comment This flow, known as Couette flow, has application to fluid-film bearings. If the radius of curvature of the bearing surface is large compared to the fluid-film thickness, then a one-dimensional Cartesian system, as employed in this example, can be used to model the flow.

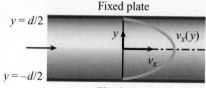

Self Test 3.2 ✓ Consider the solution of Example 3.5 when applied to a 1-m depth ($Z = 1$ m). Does the flow rate of 0.1768 kg/s violate the steady-state assumption?

(Answer: No, it does not. Although the flow is per unit time, it does not depend on time (i.e., is steady state). This flow will always be 0.1768 kg/s no matter when it is observed.)

Example 3.6

Fixed plate

$y = d/2$

$v_x(y)$

v_x

$y = -d/2$

Fixed plate

Consider a steady flow of water at 320 K and 2 atm between two fixed, parallel plates having a separation d of 1 mm. As illustrated in the sketch, the velocity profile $v_x(y)$ is parabolic, with a maximum value of a at the centerline ($y = 0$) and zero values at the surface of each plate ($y = \pm d/2$):

$$v_x(y) = a\left[1 - 4\left(\frac{y}{d}\right)^2\right] \quad [=]\text{m/s}.$$

Determine the mass flow rate of the water per unit flow depth z when a is 0.2 m/s.

Solution

Known $v_x(y)$, a, d, P, T

Find $\dot{m}/z$

Sketch

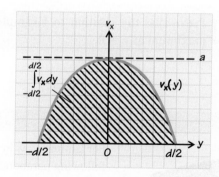

Assumptions

    i. One-dimensional flow

    ii. Constant density

Analysis Recognizing the Cartesian geometry, we apply Eq. 3.9 as follows:

$$\dot{m} = \int_{-d/2}^{d/2} \rho v_x(y)Z dy.$$

Substituting the given velocity distribution, taking advantage of the symmetry of the integral, and removing constants from the integrand, we obtain

$$\dot{m} = 2\rho Za \int_0^{d/2} \left[1 - 4\left(\frac{y}{d}\right)^2 \right] dy.$$

Performing the integration yields

$$\dot{m} = 2\rho Za \left[y - \frac{4y^3}{3d^2} \right]_0^{d/2},$$

and evaluating the limits results in

$$\dot{m} = 2\rho Za \left[\frac{d}{2} - \frac{4d^3}{24d^2} \right] = \frac{2\rho Zad}{3},$$

or

$$\frac{\dot{m}}{Z} = \frac{2\rho ad}{3}.$$

Using a value for the density from the NIST online database and substituting numerical values give

$$\frac{\dot{m}}{Z} = \frac{2(989.5)0.2(0.001)}{3} = 0.132$$

$$[=](kg/m^3)(m/s)m = \frac{kg/s}{m}.$$

Comment   Note the implicit use of the equation of state $\rho = \rho(T, P)$ in our use of the NIST database to obtain a value for the density.

Self Test 3.3   Prove that the solution to Example 3.6 obeys the no-slip law at the wall and has a maximum at $y = 0$.

Example 3.7

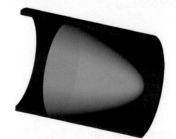

For laminar flow in circular tube or pipe, the velocity distribution obeys the following parabolic form:

$$v_x(r) = a\left(1 - \frac{r^2}{R^2} \right),$$

where r is the radial distance from the tube center and R is the inside radius of the tube or pipe. Determine an algebraic expression for the mass flow rate for a constant-density fluid.

Solution

Known   Parabolic $v_x(r)$, ρ

Find $\dot{m}$

Sketch

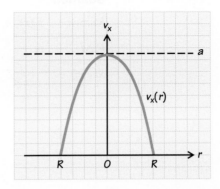

Assumptions

i. Laminar flow
ii. Constant density

Analysis   The straightforward integration of Eq. 3.9c is all that is required here. Starting with

$$\dot{m} = 2\pi\rho \int_0^R v_x(r)r\,dr,$$

we substitute $v_x(r)$ to obtain

$$\dot{m} = 2\pi\rho \int_0^R a\left(1 - \frac{r^2}{R^2}\right)r\,dr.$$

Integrating yields

$$\dot{m} = 2\pi\rho a\left[\frac{r^2}{2} - \frac{r^4}{4R^2}\right]_0^R.$$

Substitution of the limits generates our final result:

$$\dot{m} = \pi\rho a R^2/2.$$

Comment   Rearranging our result to $\dot{m} = \rho(a/2)\pi R^2$, we notice that the flow rate is the product of the density, one-half of the centerline velocity (i.e., $a/2$), and the tube cross-sectional area (πR^2). From this we recognize that $a/2$ must be the area-weighted average velocity, as discussed in the next section.

Example 3.8

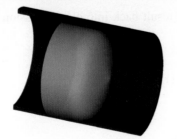

In steady turbulent flows through circular tubes, the following power law [3] approximates the velocity profile:

$$v_x(r) = a\left(1 - \frac{r}{R}\right)^{1/n},$$

where n is an integer ranging between 6 and 10, depending on the flow conditions. Assuming a uniform density, find an expression for the mass flow rate for $n = 7$. Also draw a graph of $v_x(r)$ for this particular power law and compare it with the parabolic distribution from Example 3.7.

Solution

Known $v_x(r)$, n, uniform ρ

Find $\dot{m}$

Sketch

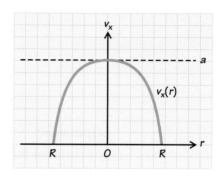

Assumptions

  i. Steady flow

 ii. Uniform density

Analysis   To find an expression for $\dot{m}$, we directly apply Eq. 3.9c:

$$\dot{m} = \int_0^R \rho a \left(1 - \frac{r}{R} \right)^{1/n} 2\pi r \, dr.$$

We can put this in a standard form by defining

$$X = r/R,$$

so

$$dX = dr/R.$$

Substituting these into Eq. 3.9c and removing constants from the integrand yields

$$\dot{m} = 2\rho a \pi R^2 \int_0^1 (1 - X)^{1/n} X \, dX.$$

The solution to this integral, which can be found in standard integral tables (e.g., Ref. [4]), is

$$\int_0^1 (1 - X)^{1/n} X \, dX = \left[\frac{1}{\left(\dfrac{1}{n} + 2 \right)} (1 - X)^{\frac{1}{n} + 2} - \frac{1}{\left(\dfrac{1}{n} + 1 \right)} (1 - X)^{\frac{1}{n} + 1} \right]_0^1.$$

Evaluating the limits and substituting the result back into our expression for $\dot{m}$ yields

$$\dot{m} = \left(\frac{2n^2}{(n + 1)(2n + 1)} \right) \rho a \pi R^2.$$

For our particular case with $n = 7$,

$$\dot{m} = \frac{49}{60} \rho a \pi R^2.$$

The second part of the problem is easily solved using spreadsheet software. Graphical results are as follows:

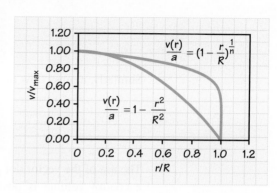

Comments From the graph, we see that the velocity profile for turbulent flow in a tube is much flatter than the parabolic profile for laminar flows. The steep velocity gradient at the tube wall ($r/R = 1$) has implications for frictional effects and pressure losses in tube and pipe flows. These concepts are developed in Chapter 10.

Self Test 3.4 ☑ **Consider the expression for the mass flow rate derived in Example 3.8 and compare it with that of Example 3.7. Which one would have the greatest average velocity?**
(Answer: For the same value of $a = v_{max}$, the turbulent flow has the greater average velocity.)

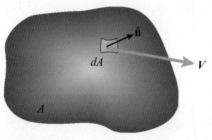

FIGURE 3.6

Arbitrary flow through arbitrary area defined by the variation of velocity vector V and area unit normal vector $\hat{n}$ over the area A.

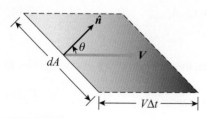

FIGURE 3.7

Two-dimensional representation of flow of unit depth with velocity V through the differential area dA. The volume of fluid passing through dA in a time interval Δt is the volume of the parallelpiped V Δt dA $\cos\theta$.

Generalized Definition

A flow rate can be defined for an arbitrary velocity distribution associated with an arbitrary flow area. Consider the velocity vector V associated with flow through the surface A. Surface A may have any shape. The particular shape is defined by specifying the direction of the unit normal at each location on the surface, as suggested in Fig. 3.6. The volume of fluid passing through the differential surface element in a time interval Δt is equal to the product of the projected area of dA in the direction of V (i.e., $dA\cos\theta$) and the distance traveled by a fluid element in time Δt (i.e., $V\,\Delta t$), so

$$\Delta\mathcal{V} = dA\cos\theta\, V\,\Delta t.$$

The flow rate through dA is then

$$\frac{\Delta\mathcal{V}}{\Delta t} = V\, dA\cos\theta,$$

or

$$\frac{\Delta\mathcal{V}}{\Delta t} = (\boldsymbol{V}\cdot\hat{\boldsymbol{n}})dA.$$

These relationships can be more easily visualized by considering the two-dimensional differential area shown in Fig. 3.7, rather than the three-dimensional situation presented in Fig. 3.6. In Fig. 3.7, we see that the volume of fluid passing through dA in the time Δt is a parallelpiped of unit depth (perpendicular to the page) as indicated by the dashed line. To obtain the total flow rate associated with the surface A, we integrate our relationship for over A, $\Delta \mathcal{V} / \Delta t$, that is,

$$\dot{\mathcal{V}} = \int_A (\boldsymbol{V} \cdot \hat{\boldsymbol{n}}) dA. \tag{3.10}$$

Similarly, a mass flow rate can be generally defined as

$$\dot{m} = \int_A \rho (\boldsymbol{V} \cdot \hat{\boldsymbol{n}}) dA. \tag{3.11}$$

The use of the vector dot product provides a sign convention that flow out of a control volume is positive [$\cos(0) = +1$], whereas flow into a control volume is negative [$\cos(180°) = -1$]. This convention is particularly useful in the application of the conservation of momentum principle in vector form (Chapter 6).

3.3c Average Velocity

In many situations, an area-averaged velocity is employed to characterize the flow. An average velocity is particularly useful in expressing conservation of mass in flows where the density is uniform.

The mathematical definition of the average of a function $f(s)$ over the interval from s_1 to s_2 is

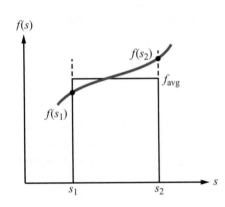

$$f_{avg} \equiv \frac{1}{s_2 - s_1} \int_{s_1}^{s_2} f(s) \, ds. \tag{3.12}$$

For our purposes, f is identified as the velocity distribution and s is the cross-sectional flow area perpendicular to v, $A_{x\text{-sec}}$, that is,

$$v_{avg} \equiv \frac{1}{A_{x\text{-sec}}} \int_A v \, dA. \tag{3.13}$$

By comparing Eq. 3.13 with Eq. 3.7, we see that the volumetric flow rate $\dot{\mathcal{V}}$ is just the average velocity times the flow area:

$$\dot{\mathcal{V}} = \int_A v \, dA = v_{avg} A_{x\text{-sec}}. \tag{3.14}$$

If the density is uniform over the flow area, the mass flow rate also simplifies to

$$\dot{m} = \rho v_{avg} A_{x\text{-sec}}. \tag{3.15}$$

In Examples 3.7 and 3.8, we effectively derived average velocities for the special cases in which the velocity distribution was parabolic or obeyed a power law. These results are summarized in Table 3.1.

Table 3.2 shows some typical average velocities associated with pipe flows in a steam power plant. Note that the typical velocities for steam flows are much greater than those for flows of liquid water.

Table 3.1   Average Velocities for Some Channel and Tube Flows

Geometry	Velocity Distribution	Average Velocity
2-D channel with height d	$v(y) = v_{max}\left[1 - 4\left(\dfrac{y}{d}\right)^2\right]$	$v_{avg} = \dfrac{2}{3}v_{max}$
Circular tube with radius R	$v(r) = v_{max}\left(1 - \dfrac{r^2}{R^2}\right)$	$v_{avg} = \dfrac{1}{2}v_{max}$
Circular tube with radius R	$v(r) = v_{max}\left(1 - \dfrac{r}{R}\right)^{1/n}$	$v_{avg} = \dfrac{2n^2}{(n+1)(2n+1)}v_{max}$
	for $n = 6$	$v_{avg} = \dfrac{72}{91}v_{max} = 0.791v_{max}$
	for $n = 7$	$v_{avg} = \dfrac{49}{60}v_{max} = 0.817v_{max}$
	for $n = 10$	$v_{avg} = \dfrac{200}{231}v_{max} = 0.866v_{max}$

Table 3.2   Typical Average Velocities for Selected Pipe Flows*

Fluid	Application	Velocity (m/s)
Steam	Superheated process steam	45–100
	Auxiliary heat steam	30–75
	Saturated and low-pressure steam	30–50
Water	Centrifugal pump suction lines	0.9–1.5
	Feedwater	2.4–4.6
	General service	1.2–3.1
	Potable water	Up to 2.1

*Adapted from Department of the Army, TM 5-810-15, Central Steam Boiler Plants, August 1995.

Example 3.9   Solar-Heated Buildings Application

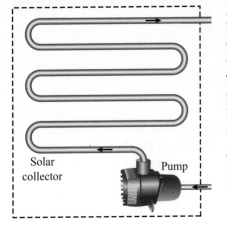

Solar collector   Pump

A solar collector for heating water consists of a 292.6-m length of black EPDM tubing as shown in the sketch. The outside diameter of the tubing is 8.9 mm and the inside diameter is 5.7 mm. Water is pumped through the tubing at a steady volumetric flow rate of 3.9×10^{-5} m³/s. The water temperature is approximately 365 K. The dimensionless parameter that determines whether the flow is laminar or turbulent, and hence the shape of the velocity profile, is the **Reynolds number**.[4] For flow through pipes and tubes, the Reynolds number is defined as $Re = \rho v_{avg}D/\mu$, where ρ and μ are the fluid density and viscosity, respectively, and D is the diameter of the flow passage. For Reynolds numbers less than about 2300, the flow is generally laminar, whereas

[4] The Reynolds number is a very important parameter in fluid mechanics. Origins of the Reynolds number and its physical interpretation are presented in Chapter 8.

for values greater than 2300, the flow is generally turbulent.[5] For Reynolds numbers ranging from about 10^4 to 10^5, the velocity profile exponent n is 7. See Example 3.8.

Determine the average velocity v_{avg} of the water flowing through the collector. Also determine the centerline (maximum) velocity associated with this flow.

Solution

Known $\dot{V}, D, T$

Find v_{avg}, v_{max}

Sketch

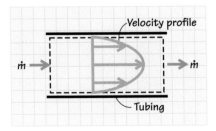

Assumptions

   i. Uniform water properties are evaluated as those of a saturated liquid.

   ii. Tubing bends do not affect velocity profiles.

Analysis The average velocity is calculated from the straightforward application of Eq. 3.14. Solving for v_{avg} yields

$$v_{avg} = \frac{\dot{V}}{A_{x\text{-sec}}},$$

or

$$v_{avg} = \frac{\dot{V}}{\pi D^2/4}.$$

Substituting numerical values gives

$$v_{avg} = \frac{4(3.9 \times 10^{-5})}{\pi(0.0057)^2} = 1.528$$

$$[=]\frac{m^3/s}{m^2} = m/s.$$

To determine v_{max}, we need to determine whether the flow is laminar or turbulent using the Reynolds number criterion. If the flow is laminar, the velocity profile is parabolic and $v_{max} = 2 v_{avg}$. If the flow is turbulent, the velocity profile follows a power law, with v_{max} related to v_{avg} depending on the value of n as shown in Table 3.1. Using saturated water properties

---

[5] Chapter 1 describes Reynolds's classical experiments that show transition from laminar to turbulent flow depends on the value of the Reynolds number.

at 365 K from the NIST database (or Appendix G), we get a Reynolds number of

$$Re = \frac{\rho v_{avg} D}{\mu}$$

$$= \frac{964.1(1.528)0.0057}{0.000308} = 2.7 \times 10^4$$

$$[=] \frac{(kg/m^3)(m/s)m}{(N \cdot s/m^2)} \left[\frac{1 \, N}{kg \cdot m/s^2} \right] = 1 \, (dimensionless).$$

We see that Re is greater than 2,300, thus the flow is turbulent. Moreover, the value of 2.7×10^4 is within the range in which $n = 7$ as given in the problem statement; thus, from Table 3.1 we get

$$v_{max} = \frac{v_{avg}}{0.817}.$$

Substituting the previously calculated value for v_{avg} yields

$$v_{max} = \frac{1.528}{0.817} \, m/s = 1.87 \, m/s.$$

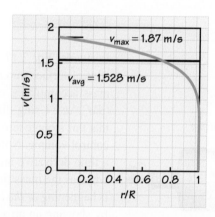

Comments   As you will see in subsequent chapters, many (in fact most) engineering devices operate in the turbulent flow regime, as does this solar collector.

Self Test 3.5   ☑ **Calculate the maximum allowable velocity for the solar collector of Example 3.9 to maintain a laminar flow.**

(Answer: $v_{max} \leq 0.129 \, m/s$)

3.3d General View of Mass Conservation for Control Volumes

With the knowledge of how to express and calculate mass flow rates, we are now able to write explicit mass conservation statements for control volumes. In the following sections, we develop the simplest statements and then add complexity. In all cases, we carefully define the restrictions that apply and state the mass conservation principle with mathematical rigor. Before doing so, however, it is instructive to transform the generic statement of the control-volume conservation principles presented in Chapter 1 to a general statement of mass conservation. Although lacking in definition and rigor, this general statement is very useful for developing an understanding of the concept of mass conservation applied to control volumes. Our subsequent developments will add the necessary rigor.

Choosing the conserved quantity to be mass, we rewrite the generic conservation principle, Eq. 1.2, as follows:

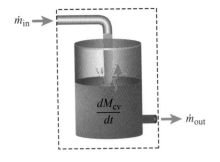

$$\dot{m}_{in} - \dot{m}_{out} + \dot{m}_{generated} = \frac{dM_{cv}}{dt}. \qquad (3.16)$$

| Time rate of mass crossing boundary and passing into control volume, i.e., the mass flow rate in | Time rate of mass crossing boundary and passing out of control volume, i.e., the mass flow rate out | Time rate of mass generated within control volume | Time rate of storage of mass within the control volume |

Since we limit all of our analyses to situations in which no nuclear reactions occur, we can eliminate the generation term; thus, Eq. 3.16 simplifies to

$$\dot{m}_{in} - \dot{m}_{out} = \frac{dM_{cv}}{dt}. \qquad (3.17)$$

In words, Eq. 3.17 states that the net rate at which mass flows across the control surface into a control volume ($\dot{m}_{in} - \dot{m}_{out}$) must equal the rate at which mass accumulates, or is stored, within the boundaries of the control volume (dM_{cv}/dt). Using different terminology to refer to the same physics, we see that this storage term is equivalently the time rate of change of mass within the control volume. Note that the storage rate is thus expressed mathematically as a time derivative and refers to what is happening *within* the control volume, whereas the mass flow rates are not expressed as derivatives and refer to what is happening *at the boundaries* of the control volume. We reinforce these ideas by showing an arrow interior to the control volume (storage) and arrows that stop or start at the control volume boundaries (flow rates) as illustrated in the sketch above.

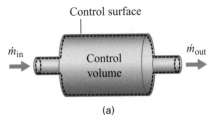

(a)

3.3e Integral Control Volumes

Steady-State, Steady Flow

We start with the simplest case of steady-state, steady flow for a large-scale (i.e., integral) control volume possessing a single inlet and a single outlet stream, as illustrated in Fig. 3.8a. The restriction of **steady state** requires that all thermodynamic properties at all locations within the control volume and on the boundary do not vary with time. The assumption of **steady flow** requires that the velocity at each location where the fluid enters or exits the control volume does not change with time. With these restrictions, conservation of mass is expressed by

$$\dot{m}_{in} = \dot{m}_{out}, \qquad (3.18a)$$

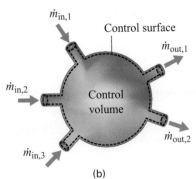

(b)

FIGURE 3.8
The boundary of the control volume is indicated by the dashed line for control volumes with (a) a single inlet and a single outlet and (b) multiple inlets and outlets.

where the mass flow rates relate to the inlet and outlet velocities as previously discussed (e.g., Eq. 3.15).

If more than one stream enters or exits the control volume (Fig. 3.8b), the total mass flow entering must equal the total mass flow exiting, that is,

$$\sum_{j=1}^{N \text{ inlets}} \dot{m}_{\text{in},j} = \sum_{k=1}^{M \text{ outlets}} \dot{m}_{\text{out},k}. \tag{3.18b}$$

The most general expression of steady-state, steady-flow mass conservation is

$$\int_{\text{cs}} \rho(\boldsymbol{V} \cdot \hat{\boldsymbol{n}}) dA = 0. \tag{3.18c}$$

Equations 3.18a and b are easily derived special cases of Eq. 3.18c.

Example 3.10 Solar-Heated Buildings Application

Air at nearly atmospheric pressure (100 kPa) enters a solar collector at 60°C and exits at 89°C, as shown in the sketch. The width of the collector (perpendicular to the page) is 1 m. The air mass flow rate is 0.056 kg/s. Determine the average velocity of the air at the entrance and at the exit of the collector.

Solution

Known $\dot{m}$, T_{in}, T_{out}, P, geometry

Find $v_{\text{avg,in}}$, $v_{\text{avg,out}}$

Sketch

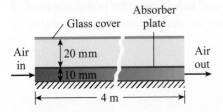

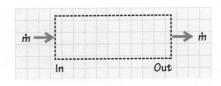

Assumptions

   i. Steady flow
   ii. Negligible pressure drop from inlet to outlet
   iii. Ideal-gas behavior
   iv. Uniform density at inlet and outlet

Analysis   With the assumption of steady flow, conservation of mass for the control volume shown is given by Eq. 3.18a:

$$\dot{m}_{\text{in}} = \dot{m}_{\text{out}} = \dot{m}.$$

Since the air density is assumed to be uniform across the inlet and outlet, we apply Eq. 3.15 as follows:

$$\dot{m} = \rho_{\text{in}} v_{\text{avg,in}} A_{\text{x-sec}} = \rho_{\text{out}} v_{\text{avg,out}} A_{\text{x-sec}},$$

where the cross-sectional area is the product of the air-channel height and the width of the collector, that is,

$$A_{x\text{-sec}} = dW = (0.010 \text{ m})(1.0 \text{ m}) = 0.010 \text{ m}^2.$$

The density is calculated from the ideal-gas equation of state (Eq. 2.28b), taking care to use the absolute temperature:

$$\rho_{in} = \frac{P_{in}}{R_{air} T_{in}} = \frac{100 \times 10^3}{287.0(273 + 60)} = 1.046$$

$$[=] \frac{\text{Pa}}{(\text{J/kg·K})\text{K}} \left[\frac{1 \text{ N/m}^2}{\text{Pa}} \right] \left[\frac{1 \text{ J}}{\text{N·m}} \right] = \text{kg/m}^3.$$

Thus,

$$v_{avg,\, in} = \frac{\dot{m}}{\rho_{in} A_{x\text{-sec}}} = \frac{0.056}{1.046(0.01)} = 5.35$$

$$[=] \frac{\text{kg/s}}{(\text{kg/m}^3)\text{m}^2} = \text{m/s}.$$

Since both $\dot{m}$ and $A_{x\text{-sec}}$ are constants, the average velocity changes only as a result of the reduced density at the outlet; thus,

$$v_{avg,\, out} = v_{avg,\, in} \frac{T_{out}}{T_{in}}$$

$$= 5.35 \frac{(273 + 89)}{(273 + 60)} \text{ m/s} = 5.82 \text{ m/s}.$$

Here we make use of our assumption that $P_{in} \cong P_{out}$.

Comments   A more rigorous calculation might include determination of the pressure drop from the inlet to the outlet. Good design practice, however, seeks to keep this pressure drop small to minimize the power required to push the air through the collector, so our assumption is probably reasonable.

Self Test
3.6

☑ **Consider the system of Example 3.10, which had constant mass flow rate of 0.056 kg/s. Is the volumetric flow rate for this system constant also?**

(Answer: If the mass flow rate of an incompressible flow is constant, then the volumetric flow rate will also be constant because the density is fixed. For a gas, however, even though the mass flow rate is constant, the volumetric flow rate will not necessarily be constant as the density may vary along the flow path.)

Example 3.11

Consider a flow of room-temperature water through a duct of circular cross section and constant taper. The diameters of the duct at the entrance and at the exit are 50 and 80 mm, respectively. The length of the duct is 0.75 m. The average velocity at the entrance is 3 m/s. Determine the mass flow rate through the duct and the average velocity at the exit.

Solution

Known $v_{avg, in}$, D_{in}, D_{out}, T

Find $\dot{m}$, $v_{avg, out}$

Sketch

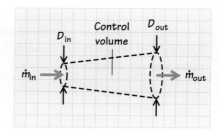

Assumptions

  i. One-dimensional flow
  ii. Steady-state, steady flow
  iii. Incompressible flow (i.e., constant density)
  iv. $\rho \approx \rho(T_{sat} = T)$

Analysis Because the density is constant, we can apply Eq. 3.15 to find the mass flow rate at the inlet:

$$\dot{m} = \rho_{in} v_{avg, in} \frac{\pi D_{in}^2}{4}.$$

From Appendix D, we find the room-temperature (298 K) value for the water density to be 997 kg/m³; thus,

$$\dot{m} = 997(3.0)\pi(.050)^2/4$$
$$\dot{m} = 5.87$$
$$[=] \frac{kg}{m^3} \frac{m}{s} m^2 = kg/s.$$

Steady-state conservation of mass for our integral control volume is expressed by Eq. 3.18a as

$$\dot{m}_{in} = \dot{m}_{out},$$

or

$$\rho v_{avg, in} \frac{\pi D_{in}^2}{4} = \rho v_{avg, out} \frac{\pi D_{out}^2}{4}.$$

Solving for $v_{avg, out}$ yields

$$v_{avg, out} = v_{avg, in} \frac{D_{in}^2}{D_{out}^2}$$

$$= 3.0 \frac{(0.050)^2}{(0.080)^2} \text{ m/s}$$

$$= 1.17 \text{ m/s}.$$

> **Nozzles and diffusers are important components of jet engines. See Example 12.10 and related material in Chapter 12.**

Comment   The tapered duct described in this example is called a **diffuser**. These passive devices are deliberately used to slow the flow and increase the pressure. We will discuss the operation of diffusers and their opposite-taper counterparts, **nozzles**, in Chapters 5 and 11.

Self Test 3.7   ✓   **Consider a household plumbing tee that has one 3/4-in inlet and two 1/2-in outlets. If 0.01 m³/s of water flows steadily into the tee at 10°C, what are the outlet mass flow rates and velocities?**

(Answer: 5.0 kg/s and 39.5 m/s for each outlet).

Photograph courtesy of NASA.

Unsteady Flows

In the preceding analyses time played no role. Our assumption of steady state and steady flow eliminated time as a variable. Although many, many situations can be treated as steady to a good approximation, time plays a key role in a variety of important problems. For example, almost all engineering devices undergo a **transient**, that is, a time-dependent start-up and/or shutdown. An interesting example here is the ignition and start-up of the engines and solid rocket boosters for the Space Shuttle. Timing of events and the transient buildup of thrust are critical to a successful lift-off.

Some of the most challenging engineering problems arise in the control of transients. By their very nature, the thermal-fluid processes associated with reciprocating internal combustion engines never achieve a steady state. Other, less complicated, unsteady flows involve the emptying and filling of tanks and pressure vessels. We now present control-volume mass conservation statements that include time as an explicit independent variable.

For simplicity, we start with a control volume that has a single inlet flow and a single exit flow as illustrated in Fig. 3.9. What makes this different from the situation described in Fig. 3.8 is that now the inlet and exit flow rates are not equal so that the mass within the control volume will be increasing with time, if $\dot{m}_{in}$ exceeds $\dot{m}_{out}$, or decreasing with time, if $\dot{m}_{out}$ exceeds $\dot{m}_{in}$. Our general

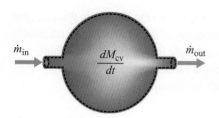

FIGURE 3.9

In an unsteady flow, the mass within a control volume can vary with time. Note that the control-volume boundary is not necessarily fixed and can move with time.

statement of mass conservation (Eq. 3.17) applies without further simplifications:

$$\dot{m}_{in} \quad - \quad \dot{m}_{out} \quad = \quad \frac{dM_{cv}}{dt}. \tag{3.19a}$$

Mass flow into the control volume | Mass flow out of the control volume | Rate of change of mass within the control volume

Because Eqs. 3.9a and 3.15 express instantaneous quantities, they can be used to evaluate the flow rates in Eq. 3.19a.

Equation 3.19a is easily extended to control volumes with multiple inlets and outlets by writing

$$\overset{N\ inlets}{\underset{j=1}{\sum}} \dot{m}_{in,\,j} - \overset{M\ outlets}{\underset{k=1}{\sum}} \dot{m}_{out,\,k} = \frac{dM_{cv}}{dt}. \tag{3.19b}$$

Since dM_{cv}/dt is a new term in our analysis, it is useful to explore its physical meaning in greater detail. For simplicity, we assume that the density is uniform throughout the control volume, that is, ρ takes on the same value irrespective of location, but we allow ρ to be a function of time. With this assumption, the mass within the control volume at any instant is just $M_{cv} = \rho \mathcal{V}$. Applying the product rule to the differentiation of $\rho \mathcal{V}$ gives

$$\frac{dM_{cv}}{dt} = \rho \frac{d\mathcal{V}}{dt} + \mathcal{V} \frac{d\rho}{dt}. \tag{3.20}$$

From Eq. 3.20, we see that the mass within the control volume increases by increasing the volume (i.e., $d\mathcal{V}/dt$ is positive) or increasing the density (i.e., $d\rho/dt$ is positive). Practical examples of either are easy to visualize. Consider the filling of a bathtub (Fig. 3.10a) where the control-volume boundary includes only the water in the tub. In this case, the density of the water is constant but the control volume continually expands as long as the faucet remains open. Just the opposite

(a)

(b)

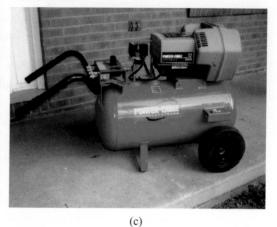

(c)

FIGURE 3.10

Examples of unsteady flows: (a) the filling of a bathtub, (b) the inflation of a toy balloon, and (c) the filling of a compressed air tank.

occurs in the filling of an air-compressor storage tank (Fig. 3.10c). In this case, the control volume remains fixed, whereas the density of the air in the tank continually increases as air is pumped in. A common example in which both the volume and the density vary with time is the inflation of a rubber balloon (Fig. 3.10b).

We illustrate these concepts now with an example.

Example 3.12

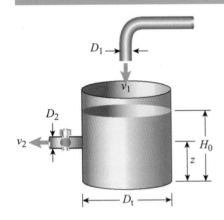

Consider the tank and water supply system as shown in the sketch. The diameter of the supply pipe D_1 is 20 mm, and the average incoming velocity v_1 is 0.595 m/s. A shut-off valve is located at $z = 0.1$ m, and the exit pipe diameter D_2 is 10 mm. The tank diameter D_t is 0.3 m. The water density is 997 kg/m^3.

A. Determine the time to fill the tank to a depth of 1 m ($\equiv H_0$) assuming the tank is initially empty and the shut-off valve is closed. Neglect the volume associated with the short pipe connecting the tank to the shut-off valve.

B. At the instant the water level reaches $H_0 = 1$ m the shut-off valve is opened. The instantaneous average velocity of the outflow depends on the water depth above z, that is, $H(t) - z$, and is given by

$$v_2 = 0.85[g(H(t) - z)]^{1/2},$$

where g is the gravitational acceleration. Determine whether the tank continues to fill or begins to empty immediately after the valve is opened.

C. Determine the steady-state value of the water depth.

D. Determine the time required to achieve steady state after the valve is opened.

Solution

Known $D_1, D_2, D_t, z, H_0, v_1$, expression for v_2

Find t_0, H_{ss}, t_{ss}

Sketch   See the sketches that follow for each part of the problem

Assumptions

  i. Incompressible flow
  ii. Outlet velocity instantaneously adjusts to changes in H

Analysis (Part A)   For this part, we select an expanding control volume that contains all of the water in the tank as shown in the sketch.

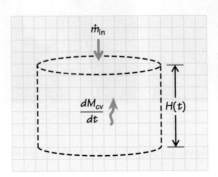

Conservation of mass for the unsteady filling process is expressed by Eq. 3.19a, where $\dot{m}_{\text{out}} = 0$:

$$\dot{m}_{\text{in}} = \dot{m}_1 = \frac{dM_{\text{cv}}}{dt}.$$

The inlet mass flow rate is expressed as (Eq. 3.15)

$$\dot{m}_1 = \rho v_1 A_1,$$

and the instantaneous mass within the control volume is simply the product of the density and the instantaneous volume of water within the tank, that is,

$$M_{\text{cv}} = \rho \mathcal{V}(t) = \rho A_t H(t),$$

where $A_t (= \pi D_t^2/4)$ is the cross-sectional area of the tank. Substituting these expressions for $\dot{m}_1$ and M_{cv} into Eq. 3.19a yields

$$\rho A_t \frac{dH(t)}{dt} = \rho v_1 A_1.$$

This first-order, ordinary differential equation is easily integrated from the initial condition $H(t = 0) = 0$ to $H(t_0) = H_0$ as follows:

$$\int_0^{H_0} dH(t) = \int_0^{t_0} v_1 \frac{A_1}{A_t} dt$$

$$H_0 = v_1 \frac{A_1}{A_t} t_0.$$

Solving for the unknown t_0 yields

$$t_0 = \frac{H_0 A_t}{v_1 A_1}.$$

Since the ratio A_t/A_1 is the ratio D_t^2/D_1^2, we evaluate this as

$$t_0 = \frac{H_0 D_t^2}{v_1 D_1^2}$$

$$= \frac{1.0(0.3)^2}{0.595(0.020)^2} \text{ s}$$

$$= 378 \text{ s or } 6.30 \text{ min.}$$

Analysis (Part B)   To determine whether the water level continues to increase or begins to fall when the valve is opened, we need to know whether the mass in the tank is increasing or decreasing since

$$\frac{dM_{\text{cv}}}{dt} = \rho A_t \frac{dH(t_0)}{dt}.$$

To determine this, we apply the conservation of mass principle,

$$\dot{m}_1 - \dot{m}_2 = \frac{dM_{\text{cv}}}{dt},$$

for the control volume sketched here:

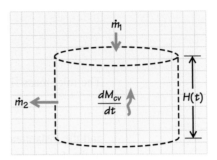

The instantaneous outlet mass flow rate $\dot{m}_2$ is expressed as

$$\dot{m}_2 = \rho v_2 A_2$$

$$= \rho \left(0.85 [g(H(t_0) - z)]^{1/2} \right) \frac{\pi D_2^2}{4}.$$

We evaluate $\dot{m}_2(t_0)$ and $\dot{m}_1(t_0)$ as follows:

$$\dot{m}_2 = 997 \left[0.85[9.81(1.0 - 0.1)]^{1/2} \right] \frac{\pi(0.01)^2}{4} = 0.198$$

$$[=] \frac{kg}{m^3} \left[\left(\frac{m}{s^2} \right) m \right]^{1/2} m^2 = kg/s$$

and

$$\dot{m}_1 = \rho v_1 \pi D_1^2 / 4$$

$$= 997(0.595)\pi(0.020)^2/4 \text{ kg/s}$$

$$= 0.186 \text{ kg/s}.$$

Thus,

$$\frac{dM_{cv}}{dt} = 0.186 - 0.198 \text{ kg/s}$$

$$= -0.012 \text{ kg/s},$$

where the negative sign indicates that the water level must start to fall when the valve is opened.

Analysis (Part C) For steady state, dM_{cv}/dt is zero. The appropriate sketch and mathematical expression for mass conservation are

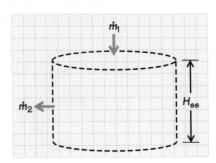

and

$$\dot{m}_1 = \dot{m}_2.$$

The inlet mass flow rate is as previously calculated (0.186 kg/s), whereas $\dot{m}_2$ is expressed in terms of the steady-state water level H_{ss}. Thus,

$$\dot{m}_1 = \dot{m}_2 = \rho\left(0.85\left[g(H_{ss} - z)\right]^{1/2}\right)\frac{\pi D_2^2}{4}.$$

Solving for H_{ss} yields

$$H_{ss} = \frac{1}{g}\left(\frac{4\dot{m}_1}{0.85\rho\pi D_2^2}\right)^2 + z$$

$$= \frac{1}{9.81}\left(\frac{4(0.186)}{0.85(997)\pi(0.010)^2}\right)^2 + 0.1$$

$$= 0.90$$

$$[=] \frac{1}{(m/s^2)}\left(\frac{kg/s}{(kg/m^3)m^2}\right)^2 = m.$$

Analysis (Part D) From the time that the valve is opened until steady state is achieved, conservation of mass is expressed as in part B, except that H is a function of t, rather than being a fixed value, that is,

$$\dot{m}_1 - \rho\left(0.85\left[g(H(t) - z)\right]^{1/2}\right)A_2 = \frac{dM_{cv}}{dt}.$$

From the geometry, we also know that

$$\frac{dM_{cv}}{dt} = \rho A_t \frac{dH(t)}{dt}.$$

Combining these two equations and solving for $dH(t)/dt$ yield

$$\frac{dH(t)}{dt} = \frac{\dot{m}_1}{\rho A_t} - \frac{0.85}{A_t}\left[g(H(t) - z)\right]^{1/2}A_2.$$

Integration of this ordinary differential equation with the limits $H(t = t_0) = H_0$ and $H(t = t_{ss}) = H_{ss}$ enables us to find the desired time interval $\Delta t \equiv t_{ss} - t_0$. Separating the H and t variables yields

$$\frac{dH}{\dfrac{\dot{m}_1}{\rho A_t} - \dfrac{0.85}{A_t}\left[g(H - z)\right]^{1/2}A_2} = dt.$$

This can be expressed more compactly by defining

$$a \equiv \frac{-g(0.85)^2 A_2^2 z}{A_t^2} = \frac{-g(0.85)^2 z D_2^4}{D_t^4},$$

$$b \equiv \frac{g(0.85)^2 A_2^2}{A_t^2} = \frac{g(0.85)^2 D_2^4}{D_t^4},$$

and

$$c \equiv \frac{\dot{m}_1}{\rho A_t}.$$

Thus,

$$\frac{dH}{c - (a + bH)^{1/2}} = dt.$$

Integrating between our limits yields

$$\int_{H_0}^{H_{ss}} \frac{dH}{c - (a + bH)^{1/2}} = \int_{t_0}^{t_{ss}} dt = \Delta t.$$

Using the substitution $u \equiv c - (a + bH)^{1/2}$ makes the evaluation of this integral relatively straightforward. The final result is

$$\Delta t = \frac{-2c}{b} \ln \left[\frac{c - (a + bH_{ss})^{1/2}}{c - (a + bH_0)^{1/2}} \right]$$

$$+ \frac{2}{b} \left[(a + bH_0)^{1/2} - (a + bH_{ss})^{1/2} \right].$$

Substituting $H_0 = 1.0$ m and $H_{ss} = 0.90$ m, along with the numerical values

$$a = \frac{-9.81(0.85)^2 \, 0.1(0.01)^4}{(0.3)^4} = -8.750 \times 10^{-7},$$

$$b = \frac{9.81(0.85)^2 \, (0.01)^4}{(0.3)^4} = 8.750 \times 10^{-6},$$

and

$$c = \frac{0.186(4)}{997(0.3)^2 \pi} = 2.639 \times 10^{-3},$$

yields to two significant digits

$$\Delta t = 2{,}000 \text{ s} \text{ or } 33 \text{ min.}$$

Comments   This example illustrates the use of both steady and unsteady expressions of mass conservation for a control volume. Note that, in all parts of the problem, we chose a control surface that contained only the water in the tank; thus, our control volume expanded or contracted with time. An alternative, but more complex, choice would have been to choose a larger fixed volume containing some air above the water. Our original choice is clearly superior because of its simplicity. We also note that formulation of the unsteady problem generated an ordinary differential equation, a common result for this class of problems.

Self Test
3.8

A constant flow rate of 3 kg/s of water is entering a partially full bathtub measuring 6 ft by 2 ft. The drain plug is removed and at one instant the water level in the tub is decreasing at 0.5 in/min. Determine the exit mass flow rate at this instant.

(Answer: 3.236 kg/s).

Many different control volumes can be selected to analyze this unsteady mid-air refueling process. Photograph courtesy of NASA.

To conclude this section, we present the most general integral form of conservation of mass:

$$-\int_{cs} \rho(\mathbf{V}_{rel} \cdot \hat{\mathbf{n}})dA = \frac{d}{dt}\left[\int_{V} \rho d V\right], \tag{3.21}$$

Net mass flow across the control surface   Rate of increase of mass within the control volume

where $\mathbf{V}_{rel}$ is the local velocity relative to the control surface, that is, the velocity seen by an observer fixed to the control surface. Use of a relative velocity is required when the control surface is moving with respect to a fixed reference frame. A control system boundary may be moving because the entire control volume is in motion, or the control volume is being deformed, or a combination of these. The physical meaning here is the same as our more simple statements (Eqs. 3.19a and 3.19b); however, both the time rate of change of mass within the control volume and the net mass flow into the control volume are expressed in the most general way possible. For many engineering analyses, Eqs. 3.19a and 3.19b are the more useful starting points.

3.3f Differential Control Volumes

Steady-State, Steady Flow

We begin our analysis of differential control volumes with the one-dimensional case. We then extend this simple analysis to three-dimensional flows.

One-Dimensional Analysis Consider a differential control volume having a length dx (Fig. 3.11). For the situation depicted in Fig. 3.11a, the y- and z-coordinates of the flow domain extend indefinitely, and flow properties vary only in the x-direction [e.g., $v_x = v_x(x \text{ only})$]. Since we assume the flow is steady

$$\dot{m}_x = \dot{m}_{x+dx} = \text{constant}, \tag{3.22a}$$

or

$$(\rho v_x A)_x = (\rho v_x A)_{x+dx} = \text{constant}, \tag{3.22b}$$

where A is an arbitrary, but fixed, area perpendicular to the flow. Since A does not vary with x (Fig. 3.11a), we can rewrite Eq. 3.22b as

$$(\rho v_x)_x = (\rho v_x)_{x+dx}. \tag{3.23}$$

The quantity ρv_x, the mass flow rate per unit area, has a special meaning in fluid dynamics and is frequently known as the **mass flux**,

$$\dot{m}'' \equiv \rho v_x. \tag{3.24}$$

The mass flux has units of kg/s·m². For our special one-dimensional flow, then

$$\dot{m}'' = \text{constant}. \tag{3.25}$$

The usual expression of mass conservation for our differential control volume involves taking the spatial derivative, $d(\ )/dx$, of Eqs. 3.24 or 3.25; thus,

$$\frac{d\dot{m}''}{dx} = \frac{d(\rho v_x)}{dx} = 0. \tag{3.26}$$

Expanding this equation yields

$$\rho \frac{dv_x}{dx} + v_x \frac{d\rho}{dx} = 0. \tag{3.27}$$

Differential expressions of conservation of mass, such as Eq. 3.27, are frequently referred to as **continuity equations** or simply **continuity**.

Figure 3.11b depicts another one-dimensional flow. In this case, the flow area A is allowed to vary with x [i.e., $A = A(x)$]. Mass conservation as expressed by Eqs. 3.22a and 3.22b still holds. The mass flux, however, does

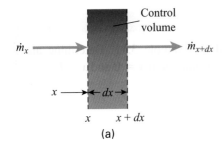

Control volume

$\dot{m}_x \longrightarrow \qquad \longrightarrow \dot{m}_{x+dx}$

$x \longrightarrow |\!\leftarrow dx \rightarrow\!|$

$x \qquad x + dx$

(a)

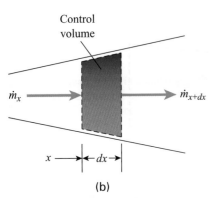

Control volume

$\dot{m}_x \longrightarrow \qquad \longrightarrow \dot{m}_{x+dx}$

$x \longrightarrow |\!\leftarrow dx \rightarrow\!|$

(b)

FIGURE 3.11
Control volumes for (a) one-dimensional unbounded flow and (b) one-dimensional bounded flow.

Examples 11.9–11.12 in Chapter 11 illustrate the use of the 1-D approximation to analyze compressible flows.

Example 12.10 in Chapter 12 relates the compressible-flow nozzle analysis to the performance of a jet engine.

Table 3.3   Steady-State Steady-Flow Expressions of Conservation of Mass for Differential Control Volumes

Coordinate System	Mass Conservation Expression	
Cartesian		
1–D (x)	$\dfrac{d(\rho v_x)}{dx} = 0$	(T3.3a)
2–D (x, y)	$\dfrac{\partial(\rho v_x)}{\partial x} + \dfrac{\partial(\rho v_y)}{\partial y} = 0$	(T3.3b)
3–D (x, y, z)	$\dfrac{\partial(\rho v_x)}{\partial x} + \dfrac{\partial(\rho v_y)}{\partial y} + \dfrac{\partial(\rho v_z)}{\partial z} = 0$	(T3.3c)
Cylindrical		
1–D (r)	$\dfrac{1}{r}\dfrac{d}{dr}(\rho r v_r) = 0$	(T3.3d)
2–D (r, x)	$\dfrac{1}{r}\dfrac{\partial}{\partial r}(\rho r v_r) + \dfrac{\partial}{\partial x}(\rho v_x) = 0$	(T3.3e)
3–D (r, θ, x)	$\dfrac{1}{r}\dfrac{\partial}{\partial r}(\rho r v_r) + \dfrac{1}{r}\dfrac{\partial}{\partial \theta}(\rho v_\theta) + \dfrac{\partial}{\partial x}(\rho v_x) = 0$	(T3.3f)
Spherical		
1–D (r)	$\dfrac{1}{r^2}\dfrac{d}{dr}(\rho r^2 v_r) = 0$	(T3.3g)
3–D (r, θ, ϕ)	$\dfrac{1}{r^2}\dfrac{\partial}{\partial r}(\rho r^2 v_r) + \dfrac{1}{r \sin\theta}\dfrac{\partial}{\partial \theta}(\rho v_\theta \sin\theta) + \dfrac{1}{r \sin\theta}\dfrac{\partial}{\partial \phi}(\rho v_\phi) = 0$	(T3.3h)

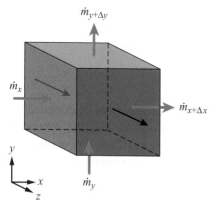

FIGURE 3.12

Three-dimensional control volume in Cartesian coordinates. The control volume has dimensions Δx, Δy, and Δz in the x-, y-, and z-directions, respectively. Mass flow rates through each face are indicated by the labeled arrows. The z-direction flow rates are unlabeled to avoid confusion.

mass for our particular control volume as

$$\underbrace{\sum_{j=1}^{3\text{ inlets}} \dot{m}_{\text{in},j}}_{\substack{\text{Total mass flow} \\ \text{into control} \\ \text{volume}}} - \underbrace{\sum_{k=1}^{3\text{ outlets}} \dot{m}_{\text{out},k}}_{\substack{\text{Total mass flow} \\ \text{out of control} \\ \text{volume}}} = 0. \tag{3.30}$$

In words, Eq. 3.30 states that the rate at which mass enters the control volume must equal the rate at which mass exits the control volume to maintain steady state. Assuming that the control volume is sufficiently small such that v_x, v_y, and v_z represent the average velocities over their respective faces (i.e., $A_{x\text{ face}} \equiv \Delta y \cdot \Delta z$, $A_{y\text{ face}} \equiv \Delta x \cdot \Delta z$, and $A_{z\text{ face}} \equiv \Delta x \cdot \Delta y$), then we can express each of the inlet flow rates (Eq. 3.15) as

$$\dot{m}_x = (\rho v_x \Delta y \Delta z)_x, \tag{3.31a}$$
$$\dot{m}_y = (\rho v_y \Delta x \Delta z)_y, \tag{3.31b}$$
$$\dot{m}_z = (\rho v_z \Delta x \Delta y)_z. \tag{3.31c}$$

The outflows are similarly expressed:

$$\dot{m}_{x+\Delta x} = (\rho v_x \Delta y \Delta z)_{x+\Delta x}, \tag{3.31d}$$
$$\dot{m}_{y+\Delta y} = (\rho v_y \Delta x \Delta z)_{y+\Delta y}, \tag{3.31e}$$
$$\dot{m}_{z+\Delta z} = (\rho v_z \Delta x \Delta y)_{z+\Delta z}. \tag{3.31f}$$

Substituting these expressions into Eq. 3.30, pairing the flows in the x-, y-, and z-directions, and removing Δx, Δy, and Δz from the brackets, recognizing that they do not depend upon spatial location, yield

$$[(\rho v_x)_{x+\Delta x} - (\rho v_x)_x]\Delta y \Delta z + [(\rho v_y)_{y+\Delta y} - (\rho v_y)_y]\Delta x \Delta z$$
$$+ [(\rho v_z)_{z+\Delta z} - (\rho v_z)_z]\Delta x \Delta y = 0, \tag{3.32}$$

Table 3.4   Gradient Operator ∇ in Various Coordinate Systems*

Coordinate System	∇ Operator
Cartesian (x, y, z)	$\nabla(\ ) = \hat{\imath}\,\dfrac{\partial(\ )}{\partial x} + \hat{\jmath}\,\dfrac{\partial(\ )}{\partial y} + \hat{k}\,\dfrac{\partial(\ )}{\partial z}$
Cylindrical (r, θ, x)	$\nabla(\ ) = \hat{\imath}_r\,\dfrac{\partial(\ )}{\partial r} + \hat{\imath}_\theta\,\dfrac{1}{r}\dfrac{\partial(\ )}{\partial \theta} + \hat{\imath}_x\,\dfrac{\partial(\ )}{\partial x}$

Cylindrical
coordinates

| Spherical (r, ϕ, θ) | $\nabla(\ ) = \hat{\imath}_r\,\dfrac{\partial(\ )}{\partial r} + \hat{\imath}_\theta\,\dfrac{1}{r}\dfrac{\partial(\ )}{\partial \theta} + \hat{\imath}_\phi\,\dfrac{1}{r\sin\theta}\dfrac{\partial(\ )}{\partial \phi}$ |

Spherical
coordinates

*The unit vectors in the x-, y-, and z-directions are given by $\hat{\imath}, \hat{\jmath}$, and $\hat{k}$, respectively, for the Cartesian system. For the cylindrical and spherical systems, the subscript of the unit vector $\hat{\imath}$ denotes the coordinate direction.

where we have multiplied through by -1 to change the order of the terms. We next divide Eq. 3.32 by $\Delta x \cdot \Delta y \cdot \Delta z$, cancel terms, and apply the limiting processes $\Delta x \to 0$, $\Delta y \to 0$, and $\Delta z \to 0$ to yield

$$\lim_{\substack{\Delta x \to 0 \\ \Delta y \to 0 \\ \Delta z \to 0}} \left[\frac{(\rho v_x)_{x+\Delta x} - (\rho v_x)_x}{\Delta x} + \frac{(\rho v_y)_{y+\Delta y} - (\rho v_y)_y}{\Delta y} + \frac{(\rho v_z)_{z+\Delta z} - (\rho v_z)_z}{\Delta z} \right] = 0.$$

(3.33)

Recognizing the definition of the partial derivatives in Eq. 3.33, we express our final result as

$$\frac{\partial(\rho v_x)}{\partial x} + \frac{\partial(\rho v_y)}{\partial y} + \frac{\partial(\rho v_z)}{\partial z} = 0.$$

(3.34)

A similar procedure can be followed to derive the other results appearing in Table 3.3 for cylindrical and spherical coordinates.

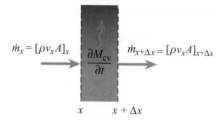

$$\dot{m}_x = [\rho v_x A]_x \qquad \dot{m}_{x+\Delta x} = [\rho v_x A]_{x+\Delta x}$$

$$x \qquad x + \Delta x$$

FIGURE 3.13

Conservation of mass applied to a one-dimensional, planar Cartesian control volume.

Unsteady Flows

To obtain the most general differential expressions for the conservation of mass, we only need to add an unsteady term to the relationships shown in Table 3.3. This unsteady term represents, of course, the time rate of change of mass within the differential control volume. To suggest how this might be done in general, we derive the unsteady differential conservation of mass expression for the one-dimensional Cartesian control volume illustrated in Fig. 3.13. The extension to other coordinate systems, or particular geometries, is then easy to do.

We begin by applying the mass conservation expression of Eq. 3.17 to the control volume illustrated in Fig. 3.13, that is,

$$\dot{m}_{in} \quad - \quad \dot{m}_{out} \quad = \quad \frac{dM_{cv}}{dt}. \qquad (3.17)$$

| Mass flow across control surface into the control volume | Mass flow across control surface out of the control volume | Rate of increase of mass within the control volume |

For our situation, $\dot{m}_{in} \equiv \dot{m}_x$ and $\dot{m}_{out} \equiv \dot{m}_{x+\Delta x}$. The mass within the control volume at any instant is the product of the instantaneous density and the volume itself:

$$M_{cv} = \rho(x, t) A \Delta x, \qquad (3.35)$$

where the volume $A\Delta x$ does not depend on time. Employing the mass flow rate expressions given in Fig. 3.13, we assemble the following mass conservation expression:

$$[\rho(x, t) v_x(x, t)]_x A - [\rho(x, t)v_x(x, t)]_{x+\Delta x} A = A \Delta x \frac{\partial \rho(x, t)}{\partial t}, \qquad (3.36)$$

which states that the time rate of change of mass within our control volume equals the difference between the mass flow in and the mass flow out. We now divide Eq. 3.36 by $A\Delta x$, cancel the As, and take the limit $\Delta x \to 0$:

$$\lim_{\Delta x \to 0} \left\{ -\left[\frac{[\rho v_x]_{x+\Delta x} - [\rho v_x]_x}{\Delta x} \right] \right\} = \frac{\partial \rho}{\partial t}, \qquad (3.37)$$

where we have dropped the x, t functional dependence for clarity and removed a minus sign from the large brackets. Recognizing that the left-hand side of Eq. 3.37 is simply the definition of a (partial) derivative, we express our final result as

$$-\frac{\partial}{\partial x}(\rho v_x) \quad = \quad \frac{\partial \rho}{\partial t}. \qquad (3.38a)$$

| Net mass flow across control surface into control volume, per unit volume | Time rate of change of mass within the control volume, per unit volume |

Here we see that the physical meaning of mass conservation is retained but each term is expressed *per unit volume*. In the usual statement of continuity (mass conservation), the two terms in Eq. 3.38a are rearranged to yield

$$\frac{\partial \rho}{\partial t} + \frac{\partial}{\partial x}(\rho v_x) = 0. \qquad (3.38b)$$

It is a relatively simple matter to show that all of the steady-flow differential expressions of mass conservation (continuity) given in Table 3.3 can be converted to their general unsteady (time-dependent) forms merely by the addition of the term $\partial \rho / \partial t$ to the left-hand side. For example, the unsteady form for a 3-D Cartesian system (Eq. T3.3c) is

$$\frac{\partial \rho}{\partial t} + \frac{\partial}{\partial x}(\rho v_x) + \frac{\partial}{\partial y}(\rho v_y) + \frac{\partial}{\partial z}(\rho v_z) = 0. \tag{3.39}$$

To conclude this section, we write the general vector form of mass conservation

$$\frac{\partial \rho}{\partial t} + \nabla \cdot (\rho \boldsymbol{V}) = 0. \tag{3.40}$$

Again, expanding Eq. 3.40 is a relatively simple matter using the definitions of the ∇ operator for various coordinate systems (see Table 3.4). It is interesting to point out that, for incompressible (i.e., constant-density) flows, the time-dependent term in each expression of mass conservation is zero, and the density cancels in the remaining terms. For example, the incompressible, 3-D Cartesian, unsteady continuity equation is

$$\frac{\partial v_x}{\partial x} + \frac{\partial v_y}{\partial y} + \frac{\partial v_z}{\partial z} = 0, \tag{3.41}$$

Example 2.23 in Chapter 2 ▶ illustrates the weak dependence of density on pressure for H_2O.

which is identical to the corresponding steady-state continuity equation for an incompressible fluid. As this example illustrates, treating a flow as incompressible greatly simplifies an analysis. Most liquid flows are essentially incompressible since densities of liquids vary only slightly with temperature, and even less with pressure.

3.4 TURBULENCE AND TIME-AVERAGED PROPERTIES

In practical devices that involve flowing fluids, turbulent flows are more frequently encountered than are laminar flows. In this section, we examine how turbulence impacts our application of the conservation of mass principle.

3.4a Mean and Fluctuating Quantities

See Figs. 1.30–1.32 and the ▶ corresponding text in Chapter 1.

The basic nature of turbulent flows was introduced in Chapter 1, and this material should be reviewed at this time. Turbulent flow results when instabilities in a flow are not sufficiently damped by viscous action and the

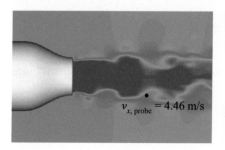
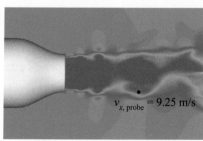
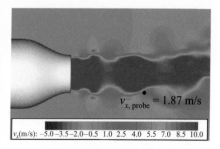

$v_{x, \text{probe}} = 4.46$ m/s $v_{x, \text{probe}} = 9.25$ m/s $v_{x, \text{probe}} = 1.87$ m/s

v_x(m/s): −5.0 −3.5 −2.0 −0.5 1.0 2.5 4.0 5.5 7.0 8.5 10.0

The velocity at a point (bold dot) varies greatly from instant to instant in this turbulent jet simulation.

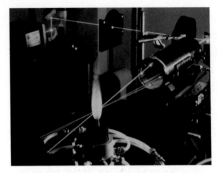

Multiple laser beams are used to measure the instantaneous velocity in a flame.

fluid velocity at each point in the flow exhibits random fluctuations. In the laminar regime, the flow is steady at every point; hence, all of the previously presented steady-flow statements of mass conservation can be applied to laminar flows. In the turbulent regime, the flow is highly unsteady; in particular, the velocity fluctuates greatly. The random unsteadiness associated with various flow properties is the hallmark of a turbulent flow and is illustrated for the axial velocity component in Fig. 3.14. The question we ask now is, How can such complex flows be understood and characterized for engineering purposes?

One particularly useful way to characterize a turbulent flow field is to define **mean** and **fluctuating** quantities. Mean properties are defined by taking a time average of the flow property over a sufficiently large time interval, $\Delta t = t_2 - t_1$; that is,

$$\bar{p} \equiv \frac{1}{\Delta t} \int_{t_1}^{t_2} p(t)\, dt, \qquad (3.42)$$

where p is any flow property (e.g., velocity, temperature, or pressure). The fluctuation $p'(t)$ is the difference between the instantaneous value of the property $p(t)$ and the mean value $\bar{p}$, or

$$p(t) = \bar{p} + p'(t). \qquad (3.43a)$$

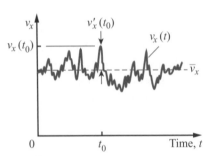

FIGURE 3.14
Velocity component v_x as a function of time at a fixed point in a turbulent flow. The dashed line indicates the mean value $\bar{v}_x$.

Figure 3.14 illustrates the fluctuating component of v_x at a specific time t_0: $v_x(t_0) = \bar{v}_x + v'_x(t_0)$. This manner of expressing variables as a mean and a fluctuating component is termed the **Reynolds decomposition** [5]. In the following section, we apply the Reynolds decomposition to the problem of mathematically expressing mass conservation for a turbulent flow.

3.4b Time-Averaged Mass Conservation

Our overall objective here is to determine how, if at all, turbulence affects our previously derived mass conservation expressions for both integral and differential control volumes. In our analysis, we restrict the flows to be steady in a time-mean sense; that is, the *time-averaged* velocities, or any other averaged flow property, do not change with time. Figure 3.14 illustrates a flow that is steady in a time-average sense.

Integral Relationships

We begin by applying the Reynolds decomposition to our definition of an instantaneous mass flow rate, Eq. 3.9a:

$$\dot{m} \equiv \int_A \rho v_x\, dA,$$

where dA is always perpendicular to v_x. The density and velocity are expressed as

$$\rho = \bar{\rho} + \rho' \qquad (3.43b)$$

and

$$v_x = \bar{v}_x + v'_x, \qquad (3.43c)$$

respectively, where the primed quantities are the turbulent fluctuations associated with each variable. Substituting these expressions into Eq. 3.9a and taking its time average yields

$$\overline{\dot{m}} = \overline{\int_A \rho v_x dA} = \overline{\int_A (\overline{\rho} + \rho')(\overline{v}_x + v_x')dA}. \tag{3.44}$$

Performing the indicated multiplication results in

$$\overline{\dot{m}} = \overline{\int_A \overline{\rho}\,\overline{v}_x dA} + \overline{\int_A \rho'\,\overline{v}_x dA} + \overline{\int_A \overline{\rho}v_x' dA} + \overline{\int_A \rho' v_x' dA}. \tag{3.45a}$$

Because the time average of an integral is equivalent to the integral of the time-averaged integrand, Eq. 3.45a is transformed to

$$\overline{\dot{m}} = \int_A \overline{\overline{\rho}\,\overline{v}_x}dA + \int_A \overline{\rho'\,\overline{v}_x}dA + \int_A \overline{\overline{\rho}v_x'}dA + \int_A \overline{\rho' v_x'}dA. \tag{3.45b}$$

To further simplify Eq. 3.45b, we apply the rules of time averaging. First, we recognize that the time average of the product of two mean quantities is just their product; for example,

$$\overline{\overline{\rho}\,\overline{v}_x} \equiv \overline{\rho}\,\overline{v}_x.$$

Second, we note that the time average of any fluctuating quality is, by definition, zero; for example,

$$\overline{\rho'} \equiv 0.$$

Similarly, the average of the product of a fluctuating quantity and a mean quantity is also zero, so

$$\overline{\rho'\,\overline{v}_x} \equiv 0.$$

With these ideas, Eq. 3.45b simplifies to the following:

$$\overline{\dot{m}} = \int_A \overline{\rho}\,\overline{v}_x dA + \int_A \overline{\rho' v_x'}\,dA. \tag{3.46}$$

Analyzing Eq. 3.46, we see that the first term on the right-hand side is identical to our original definition of mass flow rate (Eq. 3.7) except that now time-mean values of ρ and v_x replace the instantaneous ones. The second term on the right-hand side of Eq. 3.46 represents the effect of turbulence on the mean mass flow rate. In general, the term $\overline{\rho' v_x'}$ is an unknown function and can be very difficult to deal with; however, for flows that can be considered **incompressible** (i.e., $\rho = $ constant $= \overline{\rho}$), there is no fluctuating quantity ρ', and $\overline{\rho' v'}_x$, by definition, is zero. Thus, for incompressible turbulent flows

$$\overline{\dot{m}} = \int_A \rho\,\overline{v}_x dA. \tag{3.47}$$

Flows of liquids are almost always considered to be incompressible because of their insensitivity of their densities to changes in temperature and pressure. Furthermore, low-speed gas flows can be assumed incompressible if the velocities are sufficiently small. A usual criterion for this is that the **Mach number** Ma (i.e., the local flow velocity v divided by the local speed of sound, a) be less than 0.3. Sometimes a more stringent requirement of 0.1 is used. Thus, we define the criterion for assuming incompressible flow to be

$$Ma\left(\equiv \frac{v}{a}\right) \lesssim 0.1\text{–}0.3. \tag{3.48}$$

We conclude that, if the flow can be treated as incompressible, our previously developed expressions for conservation of mass for integral control volumes can be used by simply substituting time-mean quantities for average quantities in the calculation of flow rates.

Differential Relations

We now consider the effects of turbulence on mass conservation as expressed for a differential control volume, again for flows that are steady in a time-mean sense. We begin our analysis by substituting the Reynolds-decomposed variables (Eqs. 3.43b and 3.43c) into Eq. 3.39 and taking the time average, that is,

$$\overline{\frac{\partial(\overline{\rho} + \rho')}{\partial t}} + \overline{\frac{\partial(\overline{\rho} + \rho')(\overline{v}_x + v'_x)}{\partial x}} + \overline{\frac{\partial(\overline{\rho} + \rho')(\overline{v}_y + v'_y)}{\partial y}} + \overline{\frac{\partial(\overline{\rho} + \rho')(\overline{v}_z + v'_z)}{\partial z}} = 0.$$

(3.49)

By the use of the same procedures and rules for averaging as used previously, Eq. 3.49 becomes

$$\frac{\partial\overline{\rho}}{\partial t} + \frac{\partial}{\partial x}(\overline{\rho}\,\overline{v}_x) + \frac{\partial}{\partial y}(\overline{\rho}\,\overline{v}_y) + \frac{\partial}{\partial z}(\overline{\rho}\,\overline{v}_z)$$

$$\boxed{+ \frac{\partial}{\partial x}(\overline{\rho' v'_x}) + \frac{\partial}{\partial y}(\overline{\rho' v'_y}) + \frac{\partial}{\partial z}(\overline{\rho' v'_z})} = 0.$$

(3.50)

Again, we see the appearance of additional terms (the boxed quantities) that arise as a result of the flow being turbulent. Also once again, we have the fortunate result that, for incompressible flows, these new terms disappear; thus, the 3-D continuity equation for a turbulent incompressible flow is expressed as

$$\frac{\partial}{\partial x}(\rho\overline{v}_x) + \frac{\partial}{\partial y}(\rho\overline{v}_y) + \frac{\partial}{\partial z}(\rho\overline{v}_z) = 0.$$

(3.51)

As in our integral analysis, the mean quantities here play the same roles as the instantaneous quantities in the corresponding steady-flow expressions of mass conservation (i.e., Table 3.3). The methods used to deal with the effects of density fluctuations in turbulent compressible flows are well beyond the scope of this book. References [6–8] discuss such methods.

3.5 REACTING SYSTEMS

In many systems of engineering interest, chemical reactions can be important. In particular, combustion reactions liberate thermal energy that is converted into useful work. Three examples of this were highlighted in Chapter 1: fossil-fueled steam power plants, spark-ignition engines, and jet engines. Other examples in which combustion plays an important role are engines of many types (e.g., rocket engines, diesel engines, and gas-turbine engines), space heating with gas- or oil-fired furnaces, a wide variety of industrial heating and processing operations, and metals smelting, to name just a few. This section of the book provides the mass conservation framework to analyze systems and control volumes in which combustion occurs.

A recommended digression at this juncture is to return to Chapter 2 to read or review the section on specifying mixture composition. See specifically Eqs. 2.57–2.60.

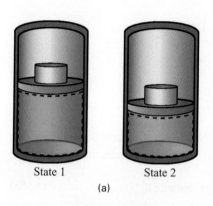

State 1   State 2

(a)

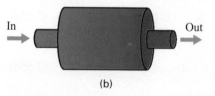

In → Out

(b)

FIGURE 3.15

(a) The mass of any element is conserved in a system in which a chemical reaction takes place in going from state 1 to state 2. (b) For a steady flow reactor, the mass flow rate of any element has the same value at the inlet and outlet.

Flaring of refinery gases. Note the solid carbon (soot) escaping the flare downstream.

3.5a Atom Balances

In dealing with a system, the mass of each chemical element remains constant during any process, assuming, of course, that no nuclear transformations occur. For a process in which the system goes from state 1 to state 2 (see Fig. 3.15a),

$$M_i \text{ (state 1)} = M_i \text{ (state 2)} \qquad \text{for } i = 1, 2, \ldots, I, \qquad (3.52)$$

where M_i is the mass of element i and I is the number of elements involved. For example, consider a pressure vessel filled with a mixture of fuel and air. If we ignore the small amount of CO_2 in the air, all of the carbon is associated with the fuel molecules (e.g., C_xH_y). If the mixture is ignited with a spark, a flame will travel through the pressure vessel, transforming the fuel and air to combustion products. After the passage of the flame, all of the carbon that was initially present in the fuel is now found in the CO_2, CO, and soot, if any, of the combustion products. To keep track of the carbon, and all other elements involved, we write an expression that describes the overall chemical reaction as follows:

$$C_xH_y + aO_2 + bN_2 \rightarrow cCO_2 + dCO + eH_2O + fH_2 + gO_2 + hN_2, \qquad (3.53)$$

where $a, b, c, \ldots$ are the number of moles of each respective species. In Eq. 3.53, we assume for simplicity that the only product species are those shown; however, other minor species, such as OH, O, NO, and H, can exist, depending on the specific conditions. These species could easily be included in Eq. 3.53. Since there are four elements involved in Eq. 3.53, C, H, O, and N, we are able to write four specific element conservation relations:

$$N_{C_xH_y}\left(\frac{N_C}{N_{C_xH_y}}\right) = N_{CO_2}\left(\frac{N_C}{N_{CO_2}}\right) + N_{CO}\left(\frac{N_C}{N_{CO}}\right), \qquad (3.54a)$$

$$N_{C_xH_y}\left(\frac{N_H}{N_{C_xH_y}}\right) = N_{H_2O}\left(\frac{N_H}{N_{H_2O}}\right) + N_{H_2}\left(\frac{N_H}{N_{H_2}}\right), \qquad (3.54b)$$

$$N_{O_2}(\text{Reac})\frac{N_O}{N_{O_2}(\text{Reac})} = N_{CO_2}\left(\frac{N_O}{N_{CO_2}}\right) + N_{CO}\left(\frac{N_O}{N_{CO}}\right)$$
$$+ N_{H_2O}\left(\frac{N_O}{N_{H_2O}}\right) + N_{O_2}\left(\frac{N_O}{N_{O_2}}\right), \qquad (3.54c)$$

and

$$N_{N_2}(\text{Reac})\left(\frac{N_N}{N_{N_2}(\text{Reac})}\right) = N_{N_2}\left(\frac{N_N}{N_{N_2}}\right). \qquad (3.54d)$$

Each of these expressions states that the number of atoms of a particular element in the reactants (i.e., the left-hand sides of Eqs. 3.54a–3.54d) equals the same number of atoms of that element in products (i.e., the right-hand sides of Eqs. 3.54a–3.54d). For example, the term on the left-hand side of Eq. 3.54a is the number of moles of carbon in the reactants, which is obtained by multiplying the number of moles of fuel ($N_{C_xH_y}$) in the reactant mixture by the number of moles of carbon present in each mole of fuel [i.e., $(N_C/N_{C_xH_y})(\equiv x)$]. For the reaction as written in Eq. 3.53, $N_{C_xH_y}$ is unity; thus,

$$N_{C_xH_y}\left(\frac{N_C}{N_{C_xH_y}}\right) = 1 \cdot x.$$

A drag line excavates coal at a mine in British Columbia. The greatest fuel utilization occurs when all of the carbon in a fossil fuel is oxidized to CO_2. Concerns about climate change focus on the role of CO_2 and other greenhouse gases discharged into the atmosphere.

This amount of reactant carbon is equal to the amount of carbon in the products (i.e., the element carbon is conserved). The amount of carbon in the products is then the number of moles of each carbon-containing species multiplied by the respective fraction of carbon in each species. For our example, carbon is present in two product species; thus, the right-hand terms in Eq. 3.54a become

$$N_{CO_2}\left(\frac{N_C}{N_{CO_2}}\right) = c \cdot 1$$

and

$$N_{CO}\left(\frac{N_C}{N_{CO}}\right) = d \cdot 1.$$

We similarly interpret Eqs. 3.54b–3.54d using the combustion reaction expressed in Eq. 3.53. Equations 3.54a–3.54d are then rewritten as follows:

$$x = c + d \quad \text{(C balance)}, \tag{3.55a}$$
$$y = 2e + 2f \quad \text{(H balance)}, \tag{3.55b}$$
$$2a = 2c + d + e + 2g \quad \text{(O balance)}, \tag{3.55c}$$

and

$$2b = 2h \quad \text{(N balance)}. \tag{3.55d}$$

Equations 3.55a–3.55d can be generalized as

$$N_i \text{ (state 1)} = N_i \text{ (state 2)} \qquad \text{for } i = 1, 2, \ldots, I, \tag{3.55e}$$

where N_i is the number of moles of element i and I is the number of elements involved. Note that Eq. 3.55e and Eq. 3.52 are identically equivalent since multiplying the number of moles of element i by its atomic weight yields the mass of element i (i.e., $M_i = N_i \mathcal{M}_i$).

We can easily extend the element conservation principle to control volumes. Consider a steady-state, steady flow through a control volume having a single inlet and a single outlet as shown in Fig. 3.8a. For this situation, we write

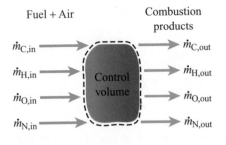

Fuel + Air   Combustion products

$\dot{m}_{C,in}$ $\dot{m}_{C,out}$
$\dot{m}_{H,in}$ Control volume $\dot{m}_{H,out}$
$\dot{m}_{O,in}$ $\dot{m}_{O,out}$
$\dot{m}_{N,in}$ $\dot{m}_{N,out}$

$$\dot{m}_{i,\text{in}} = \dot{m}_{i,\text{out}} \quad \text{for } i = 1, 2, \ldots, I. \tag{3.56}$$

Mass flow of element i Mass flow of element i
into the control volume   out of the control
    volume

Equation 3.56 can also be expressed in terms of the individual element mass fractions $Y_{e,i}$, and the total mass flow rate; that is,

$$(\dot{m} Y_{e,i})_{\text{in}} = (\dot{m} Y_{e,i})_{\text{out}}.$$

Since $\dot{m}_{\text{in}} = \dot{m}_{\text{out}}$ for steady flow (Eq. 3.18a), this equation becomes

$$(Y_{e,i})_{\text{in}} = (Y_{e,i})_{\text{out}} \quad \text{for } i = 1, 2, \ldots, I. \tag{3.57}$$

Note that the element mass fraction $Y_{e,i}$ comprises contributions from every species in the mixture that contains the ith element. Equation 3.57 is the steady-flow control volume equivalent to Eq. 3.52 for a system.

Example 3.14

One mole of propane (C_3H_8) is burned with 47.6 mol of air to produce a mixture of CO_2, H_2O, O_2, and N_2. Assume that the air consists only of O_2 and N_2 and that there are 3.76 kmol of N_2 for each kmol of O_2 (i.e., the air is 21% O_2 and 79% N_2 by volume).

A. Determine the number of moles of each of the product species in the product mixture.
B. Determine the mole fractions of each of the product species in the product mixture.
C. Determine the mass fractions of each of the product species in the product mixture.

Solution

Known $N_{C_3H_8}$, N_{air}

Find N_j, X_j, Y_j for $j = CO_2$, H_2O, O_2, N_2

Sketch

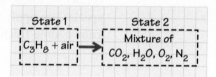

Assumptions

i. The only products are CO_2, H_2O, O_2, and N_2.
ii. No nuclear transformations occur.
iii. Air is 21% O_2 and 79% N_2 (given).

Analysis   Although the problem statement does not indicate whether we are considering a system or a steady-flow reactor (control volume), such knowledge is not required to solve the problem. We begin by writing a reaction equation following Eq. 3.53:

(1) $C_3H_8 + 47.6\,(0.21\,O_2 + 0.79\,N_2) \rightarrow a\,CO_2 + b\,H_2O + c\,O_2 + d\,N_2$.

We now write atom conservation expressions (Eq. 3.54) for C, H, O, and N, respectively:

C: $1\,(3) = a\,(1)$.
H: $1\,(8) = b\,(2)$ or $b = 4$.
O: $47.6 \cdot 0.21\,(2) = a\,(2) + b\,(1) + c\,(2)$;
   substituting a and b and solving for c yields
   $c = 5$.
N: $47.6 \cdot 0.79\,(2) = d\,(2)$
   or $d = 37.6$.

Since the coefficients a, b, c, and d represent the number of moles of CO_2, H_2O, O_2, and N_2, respectively, we have completed part A:

$$N_{CO_2} = 3, \qquad N_{O_2} = 5,$$
$$N_{H_2O} = 4, \qquad N_{N_2} = 37.6.$$

To determine the mole fractions of each of the product species, we apply the definition of a mole fraction (Eq. 2.57a):

$$X_j = \frac{N_j}{N_{tot}}.$$

For our product mixture,

$$N_{tot} = N_{CO_2} + N_{H_2O} + N_{O_2} + N_{N_2}$$
$$= 3 + 4 + 5 + 37.6 \text{ kmol}$$
$$= 49.6 \text{ kmol}$$

Thus,

$$X_{CO_2} = \frac{3}{49.6} = 0.0605 \text{ (dimensionless)},$$

$$X_{H_2O} = \frac{4}{49.6} = 0.0806,$$

$$X_{O_2} = \frac{5}{49.6} = 0.1008,$$

$$X_{N_2} = \frac{37.6}{49.6} = 0.7581.$$

We can check this result by noting that the sum of the mole fractions should equal unity (by definition): $0.0605 + 0.0806 + 0.1008 + 0.7581 = 1.000$, as expected. To find the mass fractions of the product species, we apply Eq. 2.59a, which relates the mass fractions to the now known mole fractions:

$$Y_j = X_j \frac{\mathcal{M}_j}{\mathcal{M}_{mix}},$$

where the apparent molecular weight of the mixture is given by

$$\mathcal{M}_{mix} = \sum_{j=1}^{J} X_j \mathcal{M}_j. \qquad (2.60a)$$

Using values from Appendix B for the molecular weights $\mathcal{M}_j$ for the species in the mixture, we can calculate $\mathcal{M}_{mix}$ as follows:

$$\mathcal{M}_{mix} = 0.0605(44.011) + 0.0806(18.016) + 0.1008(31.999)$$
$$+ 0.7581(28.013) \text{ kg/kmol}$$
$$= 28.577 \text{ kg/kmol}.$$

We now calculate the mass fractions:

$$Y_{CO_2} = X_{CO_2} \frac{\mathcal{M}_{CO_2}}{\mathcal{M}_{mix}} = 0.0605 \left(\frac{44.011}{28.577} \right) = 0.0932,$$

$$Y_{H_2O} = X_{H_2O} \frac{\mathcal{M}_{H_2O}}{\mathcal{M}_{mix}} = 0.0806 \left(\frac{18.016}{28.577} \right) = 0.0508,$$

$$Y_{O_2} = X_{O_2} \frac{\mathcal{M}_{O_2}}{\mathcal{M}_{mix}} = 0.1008 \left(\frac{31.999}{28.577} \right) = 0.1129,$$

$$Y_{N_2} = X_{N_2} \frac{\mathcal{M}_{N_2}}{\mathcal{M}_{mix}} = 0.7581 \left(\frac{28.013}{28.577} \right) = 0.7431,$$

The mass fractions, like the mole fractions, should also sum to unity (i.e., $0.0932 + 0.0508 + 0.1129 + 0.7431 = 1.0000$).

Comments   Although care must be exercised in writing the combustion reaction and the element conservation equations, the procedure is quite straightforward. Note that species with molecular weights larger than the mean mixture molecular weight (CO_2, O_2) have mass fractions larger than their corresponding mole fractions, whereas the converse is true for the lighter species (H_2O, N_2). We note also the large amount of N_2 present in both the reactants and products. Although in this example the N_2 was not involved in the reaction, small quantities do react at high temperatures to form the pollutant NO.

Self Test 3.9 ✓ **Write a balanced expression for the combustion of methane (CH_4) with 4.76a moles of air using the assumptions of Example 3.14.**

(Answer: $CH_4 + a(O_2 + 3.76\,N_2) \rightarrow CO_2 + 2H_2O + (a-2)O_2 + 3.76aN_2$)

Automotive catalytic converter.

3.5b Stoichiometry

The **stoichiometric** quantity of air is just that amount needed to completely burn a given quantity of fuel. If more than a stoichiometric quantity of air is supplied, the mixture is said to be fuel lean, or just **lean**; supplying less than the stoichiometric air results in a fuel-rich, or **rich**, mixture. Spark-ignition engines operate with essentially stoichiometric mixtures when used in automobiles that employ an exhaust catalyst to reduce pollutant (CO, NO, and unburned hydrocarbons) emissions. Fossil-fueled power plants operate with slightly lean mixtures for optimum efficiency.

The stoichiometric air–fuel ratio (by mass) is determined by writing element conservation expressions, as has been discussed here, assuming that the fuel reacts to form an ideal set of products. For a hydrocarbon fuel given by C_xH_y, the stoichiometric reaction is expressed as

$$C_xH_y + a(O_2 + 3.76N_2) \rightarrow xCO_2 + (y/2)H_2O + 3.76aN_2, \quad (3.58)$$

where

$$a = x + y/4. \quad (3.59)$$

For simplicity, we assume that the air consists of 21% O_2 and 79% N_2 (by volume); that is, for each mole of O_2 in air, there are 3.76 moles of N_2. From Eq. 3.58, we see that the stoichiometric air–fuel ratio is given by

$$(A/F)_{stoic} = \left(\frac{M_{air}}{M_{fuel}}\right)_{stoic} = \frac{4.76a}{1}\frac{M_{air}}{M_{fuel}}, \quad (3.60)$$

where M_{air} and M_{fuel} are the molecular weights of the air and fuel, respectively.

The **equivalence ratio** Φ is commonly used to indicate quantitatively whether a fuel–oxidizer mixture is rich, lean, or stoichiometric. The equivalence ratio is defined as

$$\Phi = \frac{(A/F)_{stoic}}{(A/F)} = \frac{(F/A)}{(F/A)_{stoic}}. \quad (3.61a)$$

From this definition, we see that for fuel-rich mixtures, $\Phi > 1$; and for fuel-lean mixtures, $\Phi < 1$. For a stoichiometric mixture, $\Phi = 1$. In many combustion applications, the equivalence ratio is the single most important

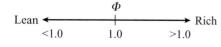

Lean ←————————→ Rich
<1.0   1.0   >1.0
        Φ

QUESTIONS

3.1 Review the most important equations presented in this chapter (i.e., those with red or orange backgrounds). What physical principles do they express? What restrictions apply?

3.2 Express conservation of mass for a thermodynamic system for the following conditions:

A. For a change in state with a uniform density

B. At an instant with a uniform density

C. For a change in state with a nonuniform density

D. At an instant with a nonuniform density

3.3 Using unit direction vectors, write formal expressions for the velocity vector in Cartesian and cylindrical coordinates. Explicitly show the position dependence (e.g., x, y, and z) of the velocity components.

3.4 Define a streamline. How does a streamtube differ from a streamline?

3.5 What is meant by the no-slip condition? What does this condition imply about the radial distribution of the axial velocity in a circular pipe of diameter D.

3.6 Although they share the same units (kg/s), the mass flow rate, $\dot{m}$, and the unsteady term, dM_{cv}/dt, in a general statement of mass conservation are quite different. Explain (discuss) the physical difference between these two quantities.

3.7 Consider the quotation from Leonardo da Vinci given in the historical context section. Rewrite this passage so that it both sounds modern and is precise in an engineering sense.

3.8 Explain the difference between an integral control volume and a differential control volume.

3.9 Write from memory (or familiarity) the following expressions of mass conservation applied to an integral control volume:

A. Steady flow with one inlet and one outlet

B. Steady flow with two inlets and three outlets

C. Unsteady flow with a single inlet and no outlet

D. Unsteady flow with a single inlet and a single outlet

3.10 For flow through a tube, what is the physical interpretation of the average velocity?

3.11 How do typical velocities for steam flowing through pipes compare to those for water?

3.12 Write out the divergence of ρV [i.e., $\nabla \cdot (\rho V)$], in Cartesian coordinates.

3.13 Compare and contrast the velocity at a point in laminar and turbulent flows.

3.14 Describe what is meant by the Reynolds decomposition. How is it useful in describing a turbulent flow?

3.15 What parameter defines whether or not a flow can be treated as incompressible? Define this parameter.

3.16 How does turbulence affect the time-averaged mass flow rate of an incompressible flow? A compressible, that is, nonconstant density, flow?

3.17 What quantities are conserved when a fuel and air burn to form combustion products?

3.18 What does it mean that a fuel burns with air in stoichiometric proportions?

3.19 Distinguish between the terms "rich" and "lean."

3.20 What is the physical meaning of the equivalence ratio Φ? What is implied when Φ is greater than unity? Less than unity?

Chapter 3 Problem Subject Areas

3.1–3.16	**Mass conservation: systems**
3.17–3.31	**Volume and mass flow rates**
3.32–3.52	**Mass conservation: integral control volumes with steady flow**
3.53–3.62	**Mass conservation: integral control volumes with unsteady flows**
3.63–3.65	**Mass flux**
3.66–3.75	**Differential control volumes**
3.76–3.79	**Turbulence and time-averaged properties**
3.80–3.110	**Reacting systems**

PROBLEMS

3.1 A 18-cm-diameter spherical balloon contains air at 1.2 atm and 25°C. Determine the mass of the air in the balloon.

3.2 A $30 \times 30 \times 70$-mm rectangular box containing N_2 at 1 atm is insulated on the four long sides and the ends are maintained at fixed temperatures. Assuming the N_2 is stagnant, this situation results in the following linear temperature distribution through the N_2: $T(x) = 320 + 570x$, where $T [=] K$ and $x [=]$ m. Determine the mass of the N_2 in the box. Neglect any gravitational effects.

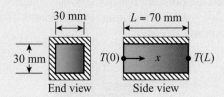

3.3 A vertical, closed, cylindrical tank 0.4 m high with a 0.3-m diameter contains a saturated mixture of water and steam at 400 K. The depth of the liquid is 25 mm. Determine the mass and quality of the mixture contained in the tank.

3.4 During the expansion process at a crank angle $\theta = 400°$, the temperature and pressure in the cylinder of a spark-ignition engine are 2400 K and 1.8 MPa, respectively. (Note: Top center is 360°.) The geometric properties of the engine are the following:

Bore $B = 90$ mm,
Stroke $S = 80$ mm,
Compression ratio $CR = 8.5$, and
Connecting rod length to crank radius ratio $\ell/a = 3.5$.

A. Determine the mass of the combustion products in the cylinder assuming the products' molecular weight is 28.5 kg/kmol.

B. Determine the time rate of change of the products' density, $d\rho/dt$, when the engine is operating at 1600 rpm.

Hint: See Appendix 1A for $V(\theta)$ and $dV(\theta)/dt$ relationships.

3.5 Determine the mass (lb_m) of air in a classroom that is 30 ft wide, 50 ft long, and 15 ft high if the pressure is 14.7 psia and the temperature is 70 F.

3.6 A gas is confined in a 2-ft^3 rigid tank at 140 F and 100 psia. The mass of the gas is 0.5 lb_m.

A. Determine the apparent molecular weight (mass) of this gas in lb_m/lbmole.

B. After a heating process, the gas temperature is 340 F. Determine the pressure (psia).

3.7 Initially, a rigid 150-in^3 tank is filled with saturated liquid water at 14.7 psia. A cover is then placed on the tank, and the water then cools to 70 F. The diameter of the opening underneath the cover is 2.5 in. The atmospheric pressure outside the tank is 14.7 psia.

A. Determine the mass (lb_m) of water initially in the tank.

B. Determine the final volume (in^3) of liquid in the tank.

C. Determine the force (lb_f) that must be applied to raise the cover at the final state. Neglect the mass of the cover.

3.8 With the valves open, the radiator of a steam heating system has a volume of 0.06 m³ and contains dry, saturated vapor (x = 1.0) at 0.14 MPa. When the valves are then closed on the radiator, the pressure drops to 0.07 MPa as a result of heat transfer to the room.

A. Determine the total mass (kg) contained in the radiator.

B. Determine the final mass (kg) and volume (m³) of vapor.

C. Determine the final mass (kg) and volume (m³) of liquid.

3.9 A rigid 1-ft³ tank initially containing water at 200 F and 10 psia cools until it reaches 70 F.

A. Determine the final pressure (psia) in the tank.

B. Determine the final liquid mass (lb_m) and volume (ft³).

C. Determine the final vapor mass (lb_m) and volume (ft³).

3.10 A rigid tank initially contains a mixture of liquid water and water vapor at a pressure of 14.7 psia. Determine the proportions by volume of liquid and vapor necessary to make the mixture pass through the critical point when heated.

3.11 An uninsulated rigid 0.8-m³ tank initially contains steam at 500°C and 0.9 MPa. The tank then cools until the temperature is 40°C.

A. Determine the pressure (MPa) at the final state.

B. Determine the mass (kg) of liquid water at the final state.

3.12 Initially, a rigid 6-m³ tank contains helium (assumed to be an ideal gas) at 300 K and 0.10 MPa. The tank is then heated until the temperature reaches 400 K. A pressure-relief valve at the top of the tank opens when the pressure reaches 0.15 MPa, and helium flows out, preventing the pressure from ever exceeding 0.15 MPa. Determine the final mass (kg) of helium in the tank.

3.13 Air is contained in a vertical cylinder fitted with a frictionless piston and a set of stops as shown in the sketch. The cross-sectional area of the piston is

0.05 m². The air is initially at 400°C and 0.2 MPa. The air is then cooled by heat flow to the surroundings.

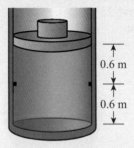

0.6 m

0.6 m

A. Determine the temperature (°C) of the air inside when the piston reaches the stops.

B. Determine the final pressure (MPa) if the cooling is continued until the temperature reaches 20°C.

3.14 A piston–cylinder device contains air at 20°C and 0.14 MPa as shown in the sketch. Initially the volume is 0.3 m³. The piston is then slowly pushed upward until the volume reaches 0.06 m³. Heat transfer with the surroundings maintains the temperature at 20°C during this process. The pressure-relief valve at the top of the cylinder opens when the pressure reaches 0.6 MPa to allow mass to escape, thus preventing the pressure from ever exceeding 0.6 MPa.

A. Determine the final pressure (MPa), temperature (°C), and specific volume (m³/kg) of the air in the cylinder.

B. Determine the amount of mass (kg) that flows through the valve.

3.15 Superheated steam initially at 150°C and 0.3 MPa (state 1) is held in the device shown in the sketch.

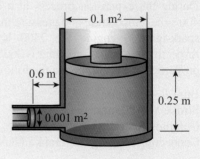

0.1 m²

0.6 m

0.001 m²

0.25 m

At state 1, the plunger is 0.6 m away from the cylinder, and the piston is 0.25 m above the bottom of the cylinder. The plunger moves toward the cylinder until it just reaches the cylinder (i.e., the plunger moves 0.6 m). At this condition (state 2), the steam temperature is 200°C. Determine the distance (m) and direction the piston moves.

3.16 A small pipe with a valve connects two rigid tanks as shown in the sketch. The volume of tank A is 0.6 m³ and it initially contains 0.03 m³ of liquid water and 0.57 m³ of water vapor in equilibrium at 200°C. Tank B is initially completely evacuated. After the valve is opened, the tanks eventually come to the same pressure of 0.7 MPa. During the process enough heat transfer occurs that the final temperature is still 200°C. Determine the volume (m³) of tank B.

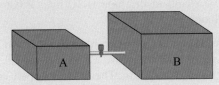

3.17 The flow of water at the outlet of a nozzle has an essentially uniform velocity of 0.7 m/s. The nozzle outlet diameter is 15 mm and the temperature of the water is 305 K. Determine the volume flow rate and the mass flow rate of the water. Give units.

3.18 A. Water at 25°C and 2 atm flows through a tube of 25.4 mm inner diameter at a flow rate of 0.035 kg/s. What is the magnitude of the centerline velocity of the water if the flow is laminar and the velocity distribution is parabolic?

$\dot{m} = 0.035$ kg/s

B. At a flow rate three times greater than that given in part A, the flow becomes turbulent and the velocity distribution is described by the following power law: $v(r) = a(1 - r/R)^{1/6}$, where r is the radial coordinate and R is the tube inner radius. Determine the average velocity of the water and the centerline velocity of the water for this flow.

3.19 The velocity profile of a circular jet of water exiting a nozzle can be approximated as a truncated cone as shown in the graph. The centerline velocity is 1.8 m/s and the nozzle radius R is 28 mm. Determine the mass flow rate associated with the water jet. Assume the jet is at room temperature (25°C).

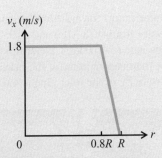

3.20 The steady velocity profile for a fluid flowing in the annular space between a concentric outer tube and cylinder is given by [3]

$$v_x(r) = C\left[1 - \left(\frac{r}{R_2}\right)^2 + \frac{1 - (R_1/R_2)^2}{\ln(R_2/R_1)} \ln\frac{r}{R_2}\right],$$

where R_1 and R_2 are the radii of the inner cylinder and outer tube, respectively, and C is a constant having dimensions of velocity.

A. Derive an expression for the volume and mass flow rates through the annulus.

B. Evaluate your expressions using the following data:

$C = 0.02$ m/s,
$R_1 = 12$ mm,
$R_2 = 16$ mm, and
$\rho = 1000$ kg/m³.

C. Determine the magnitude of the average velocity.

3.21 Consider a flow created in the annular space between a moving central cylinder and a stationary outer tube. The cylinder and the tube are concentric as shown in the sketch. At steady state with the cylinder moving at a velocity V, the velocity distribution in the annulus is given by [3]

$$v_x(r) = V\frac{\ln(r/R_2)}{\ln(R_1/R_2)}.$$

Derive an expression for the volume flow rate and for the average velocity. Numerically evaluate the average velocity given $V = 0.15$ m/s, $R_1 = 9$ mm, and $R_2 = 9.5$ mm.

3.22 A feedwater pump for a 30-MW steam power plant pumps water from 0.78 MPa and 170°C to 13.7 MPa with a flow rate of 40.0 kg/s. Determine the minimum inlet pipe diameter required if the maximum allowed average velocity in the inlet pipe is 4.6 m/s.

Photograph courtesy of Flowserve Corporation.

3.23 An air compressor supplies dry air at 350 psig and 310 K at 0.5 lb_m/s. The high-pressure air flows through a 2-in (nominal), schedule-80, steel pipe having an inside diameter of 1.939 in. Determine the average velocity of the air in the pipe.

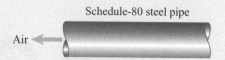

Schedule-80 steel pipe

Air

3.24 Natural gas enters a 150-km-long pipeline (of 0.30-m inner diameter) at 6.5 MPa (gage) and exits at 3.0 MPa (gage). The natural gas flow rate is 24.06 kg/s. The temperature of the natural gas is 17°C and the atmospheric pressure is 100 kPa. Assuming the natural gas to be methane (CH_4), determine the maximum average velocity in the pipeline.

3.25 Liquid water at 212 F and 2 atm flows at a rate of 1 gal/min in a nominal 1-in, schedule-40 pipe (of 1.049-in inside diameter). Determine the average velocity (ft/min) of the water.

3.26 Steam at 600°F and 100 psia flows through a 3-in-diameter pipe at 50 ft/s. Determine the mass flow rate in lb_m/hr.

3.27 Air at 1000 F and 20 psia flows through a 2-ft-inside-diameter pipe at a mass flow rate of 2.32 lb_m/s. Determine the average velocity (ft/s) of the air. Assume ideal-gas behavior.

3.28 Air at 100°C and 0.2 MPa flows at an average velocity of 30 m/s through a pipe whose cross-sectional flow area is 0.2 m². Determine the mass flow rate of the air in kg/s.

3.29 Water with a density of 990 kg/m³ is discharged by a pump at a rate of 3×10^5 cm³/min from a pipe. Determine the mass flow rate in units of kg/min and lb_m/hr.

3.30 Compressed air is introduced into the ballast tank of a submarine and drives the water out of the tank at the rate of 2 m³/s when the submarine is at the depth of 10 m. The pressure at this depth is 1.981×10^5 Pa. What is the flow rate of the ejected water when the same mass flow of compressed air is introduced at the depth of 100 m where the pressure is 10.81×10^5 Pa?

3.31 A steam boiler is fed with water at the rate of 1 kg/s. The boiler supplies steam at atmospheric pressure and 105°C, at the same rate. The average velocity of the steam in the pipe is 10 m/s. What is the pipe diameter?

3.32 Water enters a garden-hose nozzle with an average velocity of 1.8 m/s. The diameter at the nozzle entrance is 15 mm. The nozzle is open such that the effective flow area at the exit is an annulus with outer and inner diameters of 6 mm and 5 mm, respectively. What is the average water velocity at the exit annulus?

3.33 At steady-state, fuel oil is supplied at 0.01 gal/min to an oil furnace for home heating. (Note: 1 gal = 3.785×10^{-3} m³.) The fuel oil density is 870 kg/m³. Combustion air at 65 F and 1 atm enters the furnace with a volumetric flow rate of 15.2 ft³/min. Determine the mass flow rate of combustion product gases

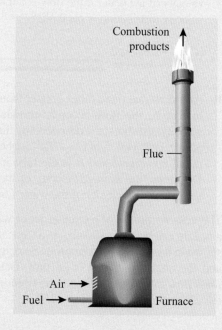

flowing up the chimney flue. Also determine the average velocity of the flue gas if the gas temperature is 225 F and the flue diameter is 9 in. Assume that the flue gases behave as an ideal gas with an apparent molecular weight of 29 kg/kmol. Also assume that the pressure in the flue is essentially 1 atm (absolute).

3.34 Hot water at 120 F (48.9°C) from a water heater is used to supply simultaneously a shower with 2.5 gal/min, an automatic dishwasher with 6.5 gal/min, and a washing machine with 9.4 gal/min. Cold water at 60 F (15.5°C) is supplied to the water heater through a copper tube of 25 mm inner diameter. Determine the average velocity of the incoming water.

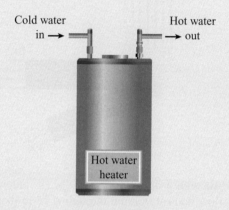

3.35 Synthetic air is made by blending pure O_2 and pure N_2. The synthetic air has a composition of 79% N_2 and 21% O_2 by volume. Determine the mass flow rates of O_2 and N_2 required to produce a 0.2 kg/s flow of synthetic air.

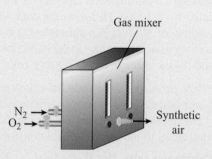

3.36 Air enters a constant-diameter tube with an average velocity of 2 m/s at a temperature of 400 K and a pressure of 0.12 MPa. The air exits the tube at a temperature of 350 K and a pressure of 0.10 MPa. Determine the average velocity of the air at the exit.

3.37 Steam enters a 60-mm-inner-diameter tube with an average velocity of 4 m/s. The specific volume of the entering steam is 0.127 m³/kg. The steam exits the tube with a density of 9.94 kg/m³. Determine the

average velocity of the steam at the tube exit assuming steady flow.

3.38 A saturated liquid–vapor H_2O mixture enters a 25-mm-inner-diameter tube at a temperature of 490 K with a quality of 0.95. The mixture is heated as it travels through the tube and exits at 620 K and 2 MPa. The average velocity of the fluid exiting the tube is 17 m/s. Determine the average velocity of the liquid–vapor mixture at the tube entrance.

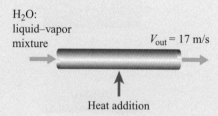

3.39 Consider a steam turbine in which steam enters at 10.45 MPa and 780 K with a flow rate of 38.739 kg/s. As shown in the sketch, a portion of the steam is extracted from the turbine after partial expansion at three different locations. The extracted steam is then led to various heat exchangers. The mass flow rates and the temperatures and pressures at each extraction point are given in the following table:

Extraction Location	$\dot{m}$(kg/s)	P (MPa)	T (K)
1	4.343	3.054	620
2	4.345	0.332	482
3	2.871	0.136	x = 0.949

The remaining wet steam exits the turbine at 11.5 kPa with a quality of 0.88. Determine the

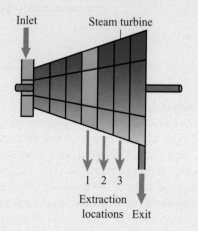

minimum pipe diameters required to restrict the maximum inlet velocity to 50 m/s, the maximum extraction velocities each to 75 m/s, and the turbine outlet velocity to 130 m/s.

3.40 Consider a rectangular sheet-metal duct carrying air at 60 F at 100 kPa. The volume flow rate at the entrance to the duct is 1200 ft³/min. A portion of the air exits through a grille as shown in the sketch. Based on noise considerations, the average velocity in the duct is to be limited to 600 ft/min. The height of the duct is 12 in.

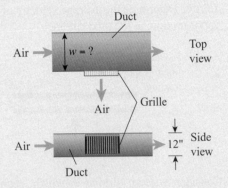

A. Determine the minimum duct width required, subject to the constraint that the duct is only available with dimensions in even increments of 2 in. (e.g., 14, 16, 18 in, etc.).

B. Determine the volumetric flow rate through the grille if the velocity of the air entering the room is 35 ft/min, the maximum value for comfort. The dimensions of the grille are 10 by 15 in.

C. Determine the mass flow rate through the duct downstream of the grille. Assume that all pressure drops in the duct system are sufficiently small that a uniform pressure is a reasonable approximation.

3.41 A system of pipes is shown in the sketch. All pipes have the same diameter of 0.1 m. The average flow velocity in pipe 1 is $v_1 = 10$ m/s to the right. In pipe 2, $v_2 = 6$ m/s to the right. Find v_3. Now set the velocity v_2 to the left. What is v_3 then?

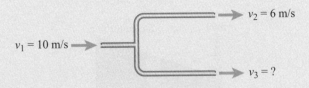

3.42 The average air velocity in the intake duct of an air conditioner is 3 m/s. The intake air temperature is 35°C. The air comes out of the air conditioner at 20°C and flows through a duct, also at 3 m/s. The pressure is atmospheric at 10^5 Pa, everywhere. What is the ratio of the cross-sectional areas of the ducts assuming no condensation of moisture? Discuss

how moisture condensation affects the velocity in the outlet duct?

3.43 A two-dimensional water flow divider is shown in the sketch. The flow in through opening A varies linearly from 3 to 6 m/s and is inclined with an angle $a = 45°$ to the opening itself. The area of the A opening is 0.4 m² per unit depth, that of B is 0.2 m² per unit depth, and that of C is 0.15 m² per unit depth. The flow at B, normal to the opening, varies linearly from 3 to 5 m/s; and the flow at C, normal to the opening, is approximately uniform. Find v_c.

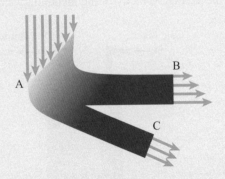

3.44 Air with a density of 0.075 lb$_m$/ft³ enters a steady-flow control volume through a 12-in-diameter duct with an average velocity of 10 ft/s. It leaves with a specific volume of 5.0 ft³/lb$_m$ through a 4-in-diameter duct. Determine (a) the mass flow rate (lb$_m$/hr) and (b) the average outlet velocity (ft/s).

3.45 A gas ($\rho = 1.20$ kg/m³) enters a steady-flow system through a 5-cm-diameter tapering duct with a average velocity of 3.5 m/s. It leaves the duct with a specific volume of 0.31 m³/kg through a 1.6-cm-diameter constriction. Determine (a) the mass flow rate (kg/hr) and (b) the average outlet velocity (m/s).

3.46 While preparing a bath you have both hot- and cold-water faucets on. For the conditions indicated, what is the necessary volume flow rate at the exit for the total mass in the tub to remain constant?

Cold Water Inlet	Hot Water Inlet	Exit
$T_1 = 60$ F	$T_2 = 140$ F	$\rho_3 = 61$ lb$_m$/ft³
$\rho_1 = 62.4$ lb$_m$/ft³	$v_2 = 0.01666$ ft³/lb$_m$	
$\dot{m}_1 = 2$ lb$_m$/s	$\dot{m}_2 = 1.5$ lb$_m$/s	

3.47 For the mixing chamber shown in the sketch, determine the unknown mass flow rate assuming there is no accumulation of fluid in the chamber.

3.48 A hot-water heater has 2.0 gal/min entering at 50 F and 40 psig. The water leaves the heater at 160 F and 39 psig. Determine the flow rate of the exiting water in gal/min.

3.49 Air flows through a 3-in-diameter pipe at a rate of 1 lb_m/s. At section 1, the air has an average velocity of 18 ft/s and a temperature of 100 F. Downstream at section 2, the air reaches a temperature of 240 F and a pressure of 19 psia. Determine the average velocity (ft/s) at section 2.

3.50 Air is heated as it flows through a constant-diameter tube in steady flow. The air enters the tube at 50 psia and 80 F and has an average velocity of 10 ft/s at the entrance. The air leaves at 45 psia and 255 F.

   A. Determine the average velocity of the air (ft/s) at the exit.

   B. If 23 lb_m/min of air is to be heated, what diameter (in) tube is required?

3.51 A variable-diameter duct has a flow cross-sectional area of 0.006 m^2 at state 1 and 0.002 m^2 at state 2. Steam at 400°C and 6 MPa enters at state 1 at a velocity of 30 m/s, flows steadily through the duct, and leaves at 300°C and 3 MPa at state 2. Determine the velocity (m/s) at state 2.

3.52 A steady-flow boiler is essentially a long, heated, variable-area duct with liquid water coming in at one end and steam coming out at the other end. Water enters at 140°C and 20 MPa and leaves at 600°C and 20 MPa. Determine the ratio of the outlet area to the inlet area if the outlet velocity is equal to the inlet velocity.

3.53 The water faucet on a stoppered kitchen sink was accidentally left slightly open. The faucet volumetric flow rate is 0.12 gal/min. If the sink is a rectangular basin 45 cm by 56 cm and 22 cm deep, how long will it take the sink to overflow? How long will it take to flood the 350-ft^2 apartment to a depth of 1 in. if there is no leakage through the floor or baseboards? Formally apply the principles of conservation of mass to the solution of both parts of this problem.

3.54 To fill an initially empty 275-gal tank with fuel oil requires 12 min. If the flow from the filling nozzle is steady, estimate the velocity of the fuel oil exiting the 20-mm-diameter filling nozzle. Formally apply the principle of conservation of mass to solve this problem.

Hose

3.55 Consider a cylindrical tank with a diameter D of 1.0 m and a height H of 1.5 m as shown on the sketch. Water ($\rho = 1000$ kg/m^3) enters the tank at the bottom through a tube with diameter d of 5 mm at a constant average velocity of 4 m/s. Initially (time $t = 0$), the tank is half full as shown; water continues to enter until the tank begins to overflow at time t_{full}. Selecting the water inside the tank at any instant as the control volume of interest, write or develop a formal explicit expression of mass conservation for this control volume. Your result should be in the form of an ordinary differential equation. Solve your differential equation for t_{full} and determine a numerical value using the data given.

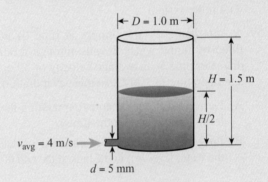

3.56 Consider the stepped cylindrical tank as shown in the sketch. The tank is initially filled with water ($\rho = 997$ kg/m^3) to a height of 0.7 m. A plug in the bottom of the tank is quickly removed and water begins to flow from the 6-mm-diameter hole. The velocity of the exiting water relates to the instantaneous height of the water in the tank, $h(t)$, as $v = 0.98 [2 \ g \ h(t)]^{1/2}$, where g is the gravitational acceleration (9.807 m/s^2). Estimate the time required to drain the tank.

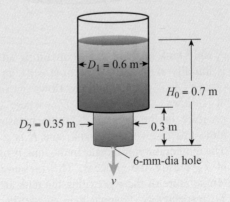

3.57 An automobile is cruising along a level highway steadily at 62 miles/hr. The on-board computer indicates a fuel economy of 26 miles/gal. The engine is operating at an air–fuel ratio of 14.7 (mass basis). The density of the fuel is 860 kg/m³. Assuming the exhaust gases have a molecular weight of 28.5 kg/kmol and a temperature of 580 K, determine their velocity in the 50-mm-diameter exhaust pipe. The exhaust pressure is nominally 100 kPa. Also estimate the change in weight of the automobile and its contents in the time it takes to travel 100 miles.

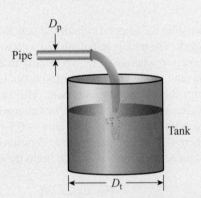

3.58 Mass enters a control volume at a rate of 100 kg/s through pipe 1. Mass leaves the control volume through pipe 2 at 50 kg/s and through pipe 3 at 70 kg/s. Determine the rate of change of mass (kg/s) within the system.

3.59 A 1-m-diameter tank (empty at time 0) is being filled by liquid water flowing through a 0.025-m-diameter pipe at an average velocity of 30 m/s with a temperature of 100°C. Neglect evaporation.

   A. Derive a differential equation relating the water level z in the tank to the time t.

   B. Integrate this differential equation to determine the water level (m) in the tank at the end of 2 min.

3.60 A rigid tank (volume = $\mathcal{V}$) containing an ideal gas is initially at T_1 and P_1. At time zero, an exit pipe (area = A) is opened and gas flows out of the tank at velocity $v = K(P - P_{atm})^{1/2}$, where P is the pressure in the tank, P_{atm} is the pressure of the atmosphere outside the tank, and K is a constant. The temperature of the gas in the tank is maintained at T_1 during the process. The pressure and temperature of the gas exiting the tank are P_{atm} and T_1, respectively. Assume that, inside the tank, the pressure and specific volume do not vary with position.

   A. Derive a differential equation for the tank pressure P as a function of time t.

   B. Determine the time required for the tank pressure to reach P_{atm}.

3.61 Consider the draining of a liquid from a tank. The level of the liquid above the bottom of the tank at any instant of time is z. The cross-sectional area of the tank is A_t. The flow area at the exit is A_e. The exit velocity v is given by $v = Cz^{1/2}$, where C is a constant.

   A. Derive a differential equation for $z(t)$.

   B. Integrate the differential equation to determine the time to drain the tank from level z_1 to a lower level z_2.

3.62 A cylindrical water tank, shown in the sketch, has a cross-sectional area of 5 m². Water flows out of the tank through a pipe of 0.2-m diameter with an average velocity of $v_1 = 6$ m/s. Water flows into the tank through a 0.1-m-diameter pipe with an average velocity of $v_2 = 12$ m/s. Define a control volume and find the rate at which the water level rises in the tank. Use a formal control-volume analysis.

3.63 Water ($\rho = 998$ kg/m³) enters a garden-hose nozzle with an average velocity of 1.8 m/s. The diameter at the nozzle entrance is 15 mm. The nozzle is open such that the effective flow area at the exit is an annulus with outer and inner diameters of 6 and 5 mm, respectively. Determine the water mass flux at the inlet and exit of the nozzle.

3.64 An air compressor supplies dry air at 350 psig and 310 K at 0.5 lb$_m$/s. The high-pressure air flows through a 2-in (nominal), schedule-80, steel pipe having an inside diameter of 1.939 in. Determine the mass flux of the air in the pipe.

3.65 A variable-diameter duct has a flow cross-sectional area of 0.006 m² at state 1 and 0.002 m² at state 2. Steam at 400°C and 6 MPa enters at state 1 at a velocity of 30 m/s, flows steadily through the duct, and leaves at 300°C and 3 MPa at state 2. Determine the mass flux at state 1 and state 2.

3.66 Consider incompressible, one-dimensional flow in a nozzle. The flow area varies linearly with the flow direction x; thus, $dA/dx = -C$, where C is a constant. Determine an expression for the axial velocity v_x as a function of position x. At $x = 0$, the quantities $A(0)$ and $v_x(0)$ are known.

3.67 Determine if the following velocity fields satisfy continuity. Note that K and A are constants.

A. $v_x = Kx,$
  $v_y = -Ky$

B. $v_x = Kx,$
  $v_y = Kx$

C. $v_x = K(x^2 + xy - y^2),$
  $v_y = K(y^2 + x^2)$

D. $v_x = A \ln xy,$
  $v_y = \dfrac{Ay}{x}$

3.68 Determine if the following velocity fields satisfy continuity. Note that K is a constant.

A. $v_x = x^2y + y^2,$
  $v_y = x^2 - y^2x$

E. $v_x = Ky,$
  $v_y = Kx$

B. $v_x = x^3 \sin y,$
  $v_y = 3x^2 \cos y$

F. $v_r = 2r \sin\theta \cos\theta,$
  $v_x = 2r \sin^2\theta$

C. $v_x = x^2 + 2xy,$
  $v_y = y^2 + 2xy$

G. $v_x = \dfrac{x}{x^2 + y^2},$

  $v_y = \dfrac{y}{x^2 + y^2}$

D. $v_x = K,$
  $v_y = 0$

3.69 Consider an ideal, steady, incompressible flow around a long cylinder of radius a in a cross-flow with velocity V_∞, as shown in the sketch.

A. Simplify the continuity equation (Eq. T3.3f) for this two-dimensional geometry.

B. Given the velocity components,

$$v_r = V_\infty \cos\theta \left(1 - \frac{a^2}{r^2} \right)$$

and

$$v_\theta = - V_\infty \sin\theta \left(1 + \frac{a^2}{r^2} \right),$$

show that the continuity equation is satisfied.

3.70 A tornado can be modeled as a steady, incompressible, two-dimensional (r, θ) flow in cylindrical coordinates.

For this model, the velocity components are given by

$$v_x = 0,$$
$$v_r = -C_1/r,$$
$$v_\theta = C_2/r,$$

where C_1 and C_2 are constants. Show that this model satisfies conservation of mass (i.e., the continuity equation).

3.71 Consider a long cylinder in a uniform cross-flow as shown in the sketch. The cylinder now rotates, creating a vortical component to the flow. The velocity components of this idealized steady, incompressible flow given by

$$v_x = 0,$$

$$v_r = V_\infty \cos\theta \left(1 - \frac{a^2}{r^2} \right),$$

$$v_\theta = -V_\infty \sin\theta \left(1 + \frac{a^2}{r^2} \right) + \frac{C}{r},$$

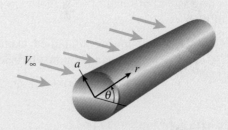

where C is a constant. Show that this model of the flow satisfies conservation of mass (i.e., the continuity equation).

3.72 The x- and y-velocity components of a steady, incompressible, two-dimensional jet are given by

$$v_x = C_1 x^{-1/3}(1 - \tanh^2 Z),$$
$$v_y = C_2 x^{-2/3}[2Z(1 - \tanh^2 Z) - \tanh Z],$$

where

$$Z = C_3 \, yx^{-2/3}.$$

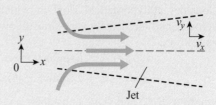

Jet

Show that this description of the jet satisfies conservation of mass. Because the calculus and algebra are somewhat complex, you may want to employ software that performs symbolic algebra (e.g., Mathematica or Maple) to solve the problem or to check your work.

3.73 Determine the divergence of the vector $f = C(x^2 - y^2)\hat{i} - 2Cxy\hat{j}$, where x and y are Cartesian coordinates and C is a constant. Could the vector f possibly represent the velocity vector in an incompressible flow? Explain.

3.74 Determine the divergence of the vector $f = (x^2 - y^2 + x)\hat{i} - (2xy + x)\hat{j}$, where x and y are Cartesian coordinates. Could the vector f possibly represent the velocity vector in an incompressible flow? Explain.

3.75 Determine the divergence of the vector $f = xyz\hat{i} + x^2y^2z\hat{j} + yz^3\hat{k}$, where x, y, and z are Cartesian coordinates. Could the vector f possibly represent the velocity vector in an incompressible flow? Explain.

3.76 Consider the following simplified expression for the axial velocity v_x at a point in a turbulent flow:

$$v_x = 8.1 + 1.7\sin(6t) + 1.1\sin(20t) + 0.65\sin(32t),$$

where v_x [=] m/s and t [=] s. Plot v_x as a function of time.

3.77 Consider the following simplified expression for the axial velocity v_x at a point in a turbulent flow:

$$v_x = 8.1 + 1.7\sin(6t) + 1.1\sin(20t) + 0.65\sin(32t),$$

where v_x [=] m/s and t [=] s.

A. Determine the mean value of v_x (i.e., $\bar{v}_x$).

B. Determine the value of the fluctuating component v'_x at $t = 0.1, 0.5,$ and 1 s.

3.78 In wind-tunnel experiments, the maximum velocity of the flow over the wing of an aircraft is measured to be 550 miles/hr. The temperature of the air in the wind tunnel is 300 K. The speed of sound in an ideal gas is given by $a = (\gamma R T)^{1/2}$, where γ is the specific heat ratio, R is the gas constant, and T is the absolute temperature. Is the flow over the wing of the aircraft compressible or incompressible? Assume ideal-gas behavior for the air.

3.79 Consider a hypothetical flow in which the axial velocity and the density vary sinusoidally at a point as follows:
$$v_x(t) = A + B\sin(\omega t),$$
$$\rho(t) = C + D\sin(\omega t + \phi),$$

where $A, B, C,$ and D are constants; ω is the angular frequency of the fluctuations (rads/s); and ϕ is a phase angle.

A. Develop an expression for the quantity $\overline{\rho' v'_x}$.

B. Evaluate your expression for $\phi = 0, \pi/4, \pi/2,$ and π when $A = 10$ m/s, $B = 1$ m/s, $C = 0.7$ kg/m³, and $D = 0.5$ kg/m³.

3.80 Hexane (C_6H_{14}) burns with air (21% O_2, 79% N_2) in stoichiometric proportions. Write the overall chemical reaction for this situation and determine the mole fractions for the C_6H_{14}, O_2, and N_2 in the reactant mixture. Also determine the mole fraction of each species in the product mixture.

3.81 Air (21% O_2, 79% N_2) and natural gas (CH_4) are supplied to an industrial furnace at an equivalence ratio of 0.95. The air mass flow rate is 4 kg/s. Determine the mass flow rate of the CO_2 exiting the furnace.

3.82 Propane (C_3H_8) and air (21% O_2 and 79% N_2) burn at an air–fuel mass ratio of 20:1. Determine (a) the equivalence ratio Φ, (b) the percent stoichiometric air, and (c) the percent excess air.

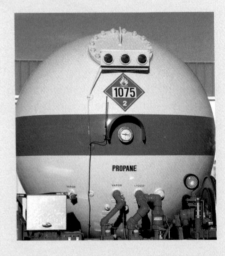

3.83 Derive an expression for the stoichiometric air–fuel ratio (by mass) of an arbitrary hydrocarbon fuel C_xH_y.

3.84 Write the stoichiometric combustion reaction for methanol (CH_3OH) and air (21% O_2, 79% N_2) and determine the stoichiometric air–fuel ratio (by mass).

3.85 In a propane-fueled truck, 2% (by volume) oxygen is measured in the exhaust stream of the running engine. Assuming complete combustion without dissociation, determine the air–fuel ratio (mass) supplied to the engine. Also determine the percent theoretical air, the percent excess air, and the equivalence ratio.

Exhaust with 2% O_2

3.86 For correct operation of the catalytic converter in an automobile, the air-fuel ratio of the engine must be precisely controlled to near the stoichiometric value. This control is achieved using feedback from an O_2 sensor located in the exhaust stream.

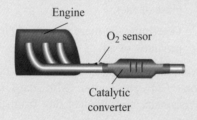

Engine

O_2 sensor

Catalytic converter

A. If the equivalent composition of gasoline is given by C_8H_{15}, determine the value of the stoichiometric air–fuel ratio (by mass). Assume that air is a simple mixture of O_2 and N_2 in molar proportions of 1:3.76.

B. Assuming complete combustion at stoichiometric conditions, determine the mole fractions of each constituent in the exhaust stream.

C. Determine the molar mass (apparent molecular weight) of the exhaust gas mixture.

3.87 Natural gas is burned to produce hot water to heat a clothing store. Assuming that the natural gas can be approximated as methane (CH_4) and that the air is a simple mixture of O_2 and N_2 in molar proportions of 1:3.76, determine the following:

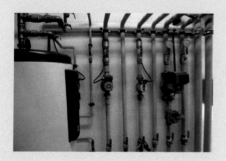

A. The molar air–fuel ratio for stoichiometric combustion

B. The mass air–fuel ratio of the heater when 4% (molar or volume) oxygen, O_2, is present in the flue gases

C. The operating air–fuel ratio (mass) of the heater when 4% (molar or volume) oxygen, O_2, is present in the flue gases

D. The percent excess air, the percent theoretical air, and the equivalence ratio associated with the conditions given in Part C

3.88 The Dassault Falcon aircraft is powered by two TF37 turbofan jet engines. At a cruise condition, each engine consumes fuel at a rate of 0.232 kg/s with a concomitant air consumption of 16.4 kg/s. The fuel blend can be approximated as $C_{12}H_{22}$, and the air composition can be assumed to be 21% O_2 and 79% N_2.

A. Determine the equivalence ratio for the combustion process.

B. Determine the composition of the combustion products exiting the combustor assuming complete combustion with no dissociation. Express your results as mole fractions.

3.89 Set up the necessary combustion equations and determine the mass of air required to burn 1 lb_m of pure carbon to equal masses of CO and CO_2. Assume that the air is 79% N_2 and 21% O_2 by volume.

3.90 Ethane burns with 150% stoichiometric air. Assume the air is 79% N_2 and 21% O_2 by volume. Combustion goes to completion. Determine (a) the air–fuel ratio by mass and (b) the mole fraction (percentage) of each product.

3.91 Rework Problem 3.90 but use propane as the fuel.

3.92 Ethanol (C_2H_5OH) is burned in a space heater at atmospheric pressure. Assume the air is 79% N_2 and 21% O_2 by volume.

A. For combustion with 20% excess air, determine the mass air–fuel ratio and the mass of water formed per mass of fuel.

B. For combustion with 180% stoichiometric air, determine the dry analysis of the exhaust gases in percentage by volume. (Note: For a dry analysis, all of the water is removed from the products.)

3.93 Find the quantities requested in Parts A and B of Problem 3.92 but use 100% stoichiometric air.

3.94 Methane (CH_4) is burned with air (79% N_2 and 21% O_2 by volume) at atmospheric pressure. The **molar** analysis of the flue gas yields $CO_2 = 10.00\%$, $O_2 = 2.41\%$, $CO = 0.52\%$, and $N_2 = 87.07\%$. Balance the combustion equation and determine the **mass** air–fuel ratio, the percentage of stoichiometric air, and the percentage of excess air.

3.95 Determine the air–fuel ratio by mass when a liquid fuel with a composition of 16% hydrogen and 84% carbon by mass is burned with 15% excess air.

3.96 Compute the composition of the flue gases (percentage by volume on a dry basis) resulting from the combustion of C_8H_{18} with 114% stoichiometric air. (Note: For a dry analysis, all of the water is removed from the products.)

3.97 A liquid petroleum fuel having a hydrogen/carbon ratio by weight of 0.169 is burned in a heater with an air–fuel ratio of 17 by mass. Determine the volumetric analysis of the exhaust gas on both wet and dry bases. (Note: For a dry analysis, all of the water is removed from the products.)

3.98 Balance the chemical equations and find the **mass** fuel–oxidizer ratio for the following equations:

A. $C_5H_{12} + a\,O_2 = b\,CO_2 + c\,H_2O$

B. $C_3H_8 + a\,O_2 = b\,CO_2 + c\,H_2O$

3.99 Determine whether the following mixtures are stoichiometric:

A. $C_2H_4 + 4\,O_2$

B. $C_8H_{18} + 12\,O_2$

3.100 Determine whether the following reactions are stoichiometric:

A. $2\,C_3H_8 + 7\,O_2 = 6\,CO + 8\,H_2O$

B. $CH_4 + 2\,O_2 = CO_2 + 2\,H_2O$

3.101 A particular solid fuel (70% carbon, 20% hydrogen, and 10% water by mass) reacts stoichiometrically with air. Assume that the air is 79% N_2 and 21% O_2 by volume. Determine the balanced chemical equation. Your final equation should be for an amount of fuel containing 1 mol of carbon.

3.102 A gas mixture (60% methane, 30% ethane, and 10% nitrogen by volume) undergoes a complete reaction with 120% theoretical air (79% N_2 and 21% O_2 by volume). Determine the composition (mole fractions) of the dry products. (Note: For a dry analysis, as requested here, all of the water is removed from the products.)

3.103 Propane reacts completely with a stoichiometric amount of hydrogen peroxide to form carbon dioxide and water. Determine the mass of water formed per mass of propane.

3.104 One mole of a hydrocarbon fuel (CH_x) is burned with excess air. The volumetric analysis of the dry products (with H_2O removed) yields:

N_2: 83.6% O_2: 5.0%
CO_2: 10.4% CO: 1.0%

A. Determine the approximate composition of the fuel on a mass basis.

B. Determine the percent theoretical air.

C. Determine the equivalence ratio.

3.105 Carbon is burned with exactly the right amount of air (79% N_2 and 21% O_2 by volume) to form carbon dioxide.

A. Determine the air–fuel ratio (by mass).

B. Determine the mass of carbon dioxide per mass of fuel.

C. Determine the mass of carbon dioxide formed per mass of air.

3.106 Assuming complete combustion, determine the total CO_2 production (Mlb_m/day) for Madison, Wisconsin, on a typical day from the following sources:

A. Automotive: Each one of 30,000 cars goes 15 miles at an average 10 miles/gal of octane gasoline, which has a specific gravity (relative to water at 4°C) of 0.703 at 20°C.

B. Human: Each of 200,000 people uses 1 lb_m of glucose ($C_6H_{12}O_6$) per day.

C. Heating systems: Each of 40,000 home-heating systems burns natural gas (assumed to be methane) at a rate of 1000 ft³/day at 70 F and 14.7 psia.

D. Steam power plant: The Madison Gas and Electric power plant produces 180 MW using coal (assumed to be 100% C) at a rate of 1.5 kW·hr/lb_m of coal.

3.107 Determine the equivalence ratio of a mixture of 1 mol of methane and 7 mol of air. Is the mixture lean or rich?

3.108 Liquid hydrogen peroxide is used as the oxidizer to burn 1 kmol of propane in a stoichiometric reaction. Initially, the hydrogen peroxide and propane are each at 40°C and 2 atm. The final state of the exhaust product mixture (carbon dioxide and water) is 800 K and 10 atm. Determine the water (kmol) produced.

3.109 Calculate and compare the stoichiometric air–fuel mass ratios for the following common fuels. Assume that air is a simple mixture of O_2 and N_2 in molar proportions of 1:3.76.

A. Natural gas (assume essentially all methane, CH_4)

B. Liquified petroleum gas (assume essentially all propane, C_3H_8)

C. Typical gasoline blend, $C_{7.9}H_{14.8}$ equivalent

D. Typical light diesel blend, $C_{12.3}H_{22.2}$ equivalent

E. Methanol, CH_3OH

3.110 Coal is a physical mixture of many compounds. However, from an elemental (ultimate) analysis of dry ash-free coal, we can artificially represent the fuel as

$$C_vH_wN_xS_yO_z.$$

With a generating capacity of approximately 2,400 MW, the Navajo Generating Station near Page, Arizona is one of the largest coal-fired power plants west of the Mississippi.

Assuming that air is a simple mixture of O_2 and N_2 in molar proportions of 1:3.76, answer the following:

A. Determine the coeficient a in the general combustion equation

$$C_vH_wN_xS_yO_z + (a/\Phi)(O_2 + 3.76N_2) \rightarrow \text{Products}.$$

Assume that the fuel nitrogen appears as molecular N_2 in the products and that the fuel sulfur is oxidized to SO_2.

B. Determine the mass air–fuel ratio for stoichiometric combustion of Pittsburgh #8 coal, defined as the following equivalent fuel:

$$C_{65}H_{52}NSO_3.$$

C. A 500-MW power plant burns Pittsburgh #8 coal with an equivalence ratio Φ of 0.9. The mass flow rate of the combustion products exiting the combustor/boiler is 383.3 kg/s. Determine the air and fuel mass flow rates supplied to the combustor.

D. Determine the mole fraction (expressed as parts per million) of the SO_2 in the product stream from Part C.

E. Determine the mass flow rate of SO_2 (see Parts C and D) entering the pollution control device (scrubber) for the power plant.

ENERGY AND ENERGY TRANSFER

*After studying Chapter 4,
you should:*

- Understand the differences between bulk (macroscopic) and internal (microscopic) energies possessed by systems and control volumes.

- Understand the formal definitions of both heat and work and be able to state these precisely.

- Be able to identify various forms of work in practical situations and be able to calculate the following knowing their state variables: compression and/or expansion work, viscous stress work, shaft work, electrical work, and flow work.

- Be able to identify the three modes of heat transfer in practical situations.

- Understand and be able to use the concept of heat flux.

- Be able to write and explain Fourier's law of conduction for 1-D Cartesian, 1-D cylindrical, and 1-D spherical systems.

- Be able to explain the physical meaning of the temperature gradient, dT/dx, and, more generally, ∇T.

- Be able to write rate expressions for convection heat transfer (both $\dot{Q}_{conv}$ and $\dot{Q}''_{conv}$) and to define all of the factors therein.

- Understand the following radiation concepts and definitions: blackbody properties, Stefan–Boltzmann law, emissive power, gray-body properties, and emissivity.

- Be able to write and explain the simplified rate law for radiation exchange between two surfaces, recognizing that many restrictions apply.

Chapter 4 Overview

IN THIS CHAPTER, we review the concept of energy and the various ways in which a system or control volume can possess energy at both microscopic (molecular) and macroscopic levels. We also carefully define heat and work, which are boundary interactions and, therefore, not properties of a system or control volume. The chapter concludes with an examination of the rate laws that govern heat transfer. We begin with a brief historical overview of our subject matter.

Benjamin Thompson (1753–1814)

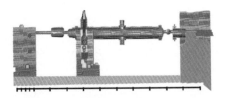

Thompson's canon boring experiments connected the concepts of *heat* and *work*.

4.1 HISTORICAL CONTEXT

With the word *energy* being commonplace, you may find it difficult to believe that the scientific underpinning of this concept as we know it today is just 200 years old. In 1801, **Thomas Young** (1773–1829) (of Young's modulus fame) presented the idea that *the energy of a system is the capacity to do work* [1]. In the early 1800s, various forms of energy had yet to be defined in useful ways. For example, the concept of heat was a muddle of the concepts that we now distinguish as temperature, internal energy, and heat. The discovery that energy is conserved—a premier conservation principle of classical physics—had to wait until **Robert Mayer's** (1814–1878) statement of the theory of conservation of energy in 1842. One of the keys to the recognition of this law was that work could be converted to heat and vice versa. **Benjamin Thompson** (1753–1814), a traitor to the colonists in the Revolutionary War between Great Britain and her American colonies, discovered in 1798, by a series of carefully planned and conducted experiments, that friction produces an inexhaustible supply of heat. Exactly what heat is, however, was not clear in 1798.

The invention and development of the steam engine spurred theoretical development of the thermal sciences at the end of the eighteenth and beginning of the nineteenth centuries. Ironically, the steam engine's very low thermal efficiency (a few percent) may have prevented **Sadi Carnot** (1796–1832), the discoverer of the **second law of thermodynamics**, from also discovering the **first law of thermodynamics** (i.e., the conservation of energy principle). With early steam engines converting such a small percentage of the heat supplied to work, it appeared to Carnot that there was no conversion of heat to work, but merely a heat addition at high temperature and an equal heat rejection at a low temperature. For the interested reader, a timeline of important lives and events in the history of the thermal sciences is presented in Appendix A.

4.2 SYSTEM AND CONTROL-VOLUME ENERGY

A system or a control volume can possess energy, first, as a consequence of its bulk motion or position within a force field, such as gravity, and second, as a consequence of the motion of its constituent molecules and their

Oliver Evan's high-pressure *Columbian* steam engine patented in 1790. *Drawing courtesy of Library of Congress.*

interactions (i.e., the internal energy). We can therefore express the total energy as the sum of these two energy components:

$$E_{\text{sys}} = E_{\text{bulk}} + U_{\text{internal}}. \tag{4.1a}$$

The same relationship holds for a control volume:

$$E_{\text{cv}} = E_{\text{bulk,cv}} + U_{\text{internal, cv}}. \tag{4.1b}$$

We now examine each of the terms on the right-hand side of Eq. 4.1.

4.2a Energy Associated with System or Control Volume as a Whole

A system or control volume can possess energy by virtue of its bulk motion. The most common forms of this bulk energy in engineering applications are kinetic energy *KE* and potential energy *PE*. We can then write

$$E_{\text{bulk}} = (KE)_{\text{bulk}} + (PE)_{\text{bulk}}. \tag{4.2}$$

Assuming a rigid system with no relative motion of subsystem elements, we may easily relate the kinetic and potential energies to the linear and angular velocities of the system and its position as follows:

$$E_{\text{bulk}} = \frac{1}{2}MV^2 + \frac{1}{2}I\omega^2 + Mg(z - z_{\text{ref}}), \tag{4.3}$$

Here we assume that the only contribution to the potential energy is the action of gravity. The first two terms on the right-hand side of Eq. 4.3 are the translational and rotational contributions to the system kinetic energy, where *V* is the magnitude of the bulk translational velocity of the system center of mass, *I* is the system rotational moment of inertia, ω is the system angular velocity, and z_{ref} is some reference elevation. Implicitly, the reference kinetic energies are zero (i.e., $V_{\text{ref}} = 0$ and $\omega_{\text{ref}} = 0$). Equation 4.3 should be familiar to you from your previous study of physics and rigid-body mechanics. If other external fields exist in addition to gravity (e.g., magnetic or electrostatic fields), additional contributions to the potential energy can result depending on the nature of the matter within the system. Discussion of these effects is beyond the scope of this book. Additional information can be found in more advanced texts such as Ref. [2].

The specification of a single translational velocity and a single angular velocity in Eq. 4.3 is possible only for a rigid system. In a control volume, the bulk velocity is likely to vary with position; thus, the kinetic energies are obtained by integrating over the control volume:

$$(KE)_{\text{bulk}} = \frac{1}{2}\int_{M_{\text{cv}}} V^2 dM, \tag{4.4a}$$

or

$$(KE)_{\text{bulk}} = \frac{1}{2}\int_{\text{CV}} \rho V^2 d\mathcal{V}, \tag{4.4b}$$

FIGURE 4.1
Examples of systems and a control volume having macroscopic energies: (a) a baseball thrown with spin imparted, (b) a bicycle rolling down a hill, and (c) a solid-fuel rocket.

Molecular nanotechnology seeks to build engineering devices and components, like this bearing and sleeve, using sequences of chemical reactions. *Drawing courtesy of Zyvex.*

where the differential mass $dM = \rho d\mathcal{V}$. The application of Eq. 4.4 to a rigid translating and rotating system captures the two kinetic energy terms of Eq. 4.3; thus, Eq. 4.4 applies equally well to a control volume or a system.

Figure 4.1 shows several examples that illustrate the energies associated with bulk motions. The spinning baseball in Fig. 4.1a is a system of fixed mass. At any instant in time, the baseball possesses a particular translational velocity V, a particular angular velocity ω, and a particular elevation relative to a datum, $z - z_{\text{ref}}$. Knowing the mass of the ball and its mass moment of inertia, we could easily calculate the instantaneous values of the terms contributing to E_{bulk}.

A similar, but more complex, example is the system defined by the bicycle (Fig. 4.1b), where we have deliberately excluded the rider from our system. In this case, the translational kinetic energy and the potential energy are easy to determine. Since the entire system does not rotate with a single velocity about a common axis, the rotational kinetic energy would have to be determined by subdividing the system to treat each rotating part separately. For example, each wheel makes a separate contribution to the system rotational kinetic energy.

A third example (Fig. 4.1c) is a control volume containing a solid-fuel rocket. We assume that the rocket contains no internal macroscopic moving parts, such as pumps, as would be found in a liquid-fuel rocket. With this assumption, evaluating the control-volume kinetic energy is simplified. Nevertheless, evaluating the integral in Eq. 4.4 remains a daunting task. To do this requires knowledge of the local velocity and density at every point within the complex flow created by the burning of the solid propellant. Evaluating the potential energy is much easier. If the control-volume center of mass does not change as the propellant is consumed, or if we ignore any small change in its location, the potential energy is simply $M_{\text{cv}}g(z - z_{\text{ref}})$.

The three examples of Fig. 4.1 were deliberately chosen to illustrate situations in which macroscopic system energies can be important. In many

In this single-piston microsteam engine, electric current heats and vaporizes water within the cylinder pushing the piston out. Capillary forces then retract the piston once current is removed. *Courtesy Sandia National Laboratories, SUMMiT™ Technologies, www.mems.sandia.gov.*

False-color scanning tunneling microscope (STM) image of the surface of pyrolytic graphite reveals the regular pattern of individual carbon atoms.

> **Rereading the section on internal energy in Chapter 2 is useful at this point.**

thermal science applications, however, these macroscopic system energies can be neglected because they, or their changes, are small compared to heat and work exchanges at boundaries and to changes in internal (molecular) energy during a process. Alternatively, we may select system or control-volume boundaries to exclude some bulk energies. Nevertheless, being able to identify all of the various energies and energy exchanges is crucial in analyzing thermal systems. In our future theoretical developments and examples, we will be careful to point out when we neglect the macroscopic energies. Such practice should help you develop critical thinking in your approach to thermal science problems.

4.2b Energy Associated with Matter at a Microscopic Level

As defined in Chapter 2, **internal energy** is the energy associated with the motion of the microscopic particles (atoms, molecules, electrons, etc.) comprising a system or control volume. Seeing how macroscopic properties relate to the microscopic structure of matter can be intellectually satisfying, although an understanding of microscopic behavior is not necessary to solve most engineering problems in the thermal sciences. (Recall that the subjects of classical thermodynamics and heat transfer predate modern ideas concerning the microscopic nature of matter.) Interestingly, many of the challenges associated with microminiaturization of engineering devices (a hot topic in mechanical engineering at the time of this book's writing) require a detailed understanding of microscopic behavior. See, for example, Refs. [3, 4].

In Chapter 2, we saw that gas molecules possess energy in ways analogous to our macroscopic systems: translational kinetic energy, rotational kinetic energy, and vibrational kinetic and potential energies. As the temperature is increased, not only is more energy associated with some storage modes (e.g., translational kinetic energy) but new states become accessible to the molecule (e.g., vibrational kinetic and potential energies). Figure 2.4 illustrated these degrees of freedom and energy storage modes; their impact on specific heats was shown in Fig. 2.6.

For solids, the internal energy is associated with lattice vibrations that give rise to vibrational kinetic and potential energies. Figure 4.2 illustrates in cartoon fashion the basic structure of a solid being composed of masses (molecular centers) connected by springs (intermolecular forces). The internal energy associated with liquids has its origin in the relatively close range interactions among the molecules making up the liquid. A simplified view is to consider the structure of a liquid lying between the extremes of a disorganized gas and a well-ordered crystalline solid.

4.3 ENERGY TRANSFER ACROSS BOUNDARIES

4.3a Heat

Definition

The recognition that heat is not a property of a system—not something that the system possesses—but, rather, an exchange of energy from one system to another, or to the surroundings, was a breakthrough in thermodynamics. Our

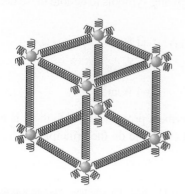

FIGURE 4.2
Energy storage in the vibrating lattice of a solid.

formal definition of heat [2] is the following:

> *Heat* **is energy transferred, without transfer of mass, across the boundary of a system (or across a control surface) because of a temperature difference between the system and the surroundings or a temperature gradient at the boundary.**

Because of the importance of this definition, let us elaborate some of the important implications: First, heat occurs only at the boundary of a system; that is, it is a boundary phenomenon. As a consequence, a system cannot *contain* heat. The addition of heat to a system with all else held constant, however, increases the energy of a system, and, conversely, heat removal from a system decreases the energy of a system. Another important element in this definition of heat is that the "driving force" for this energy exchange is a temperature difference or temperature gradient. Other energy transfers across a system boundary may occur, but only heat is controlled solely by a temperature difference. We can gain some physical insight into this boundary energy exchange by examining the molecular processes involved. The energy exchange proper is carried out by the collision of molecules in which the higher kinetic energy molecules, in general, impart some of their energy to the lower kinetic energy molecules. Because the higher kinetic energy molecules are at a higher temperature than the lower energy molecules, the direction of the energy exchange is from high temperature to low temperature. A more rigorous treatment of this process would involve the subject of irreversible thermodynamics [5–7] and is beyond the scope of this book.

Semantics

In spite of our attempts here to be very precise about the definition of heat, some semantic problems arise out of traditions and nomenclature developed prior to the advent of modern thermodynamic principles. Specifically, in the common term *heat transfer*, the word *transfer* is redundant. The science of heat transfer was developed in the early 1800s [8], prior to our understanding that heat is not possessed by a body and to the development of energy conservation.[1]

Rates of heat transfer and rates per unit area are also of importance and need to be distinguished. We adopt the following symbols to denote these heat interactions:

$$Q = \text{heat (or heat transfer)} \quad [=] \text{ J},$$
$$\dot{Q} = \text{rate of heat transfer} \quad [=] \text{ J/s or W},$$
$$\dot{Q}'' = \text{heat flux} \quad [=] \text{ W/m}^2.$$

We also use the word **adiabatic** to describe a process in which there is no heat transfer. Later in this chapter we will elaborate on the principles of heat transfer.

Heat transfer is important in food preparation and body temperature regulation.

---

[1] When Fourier published his theory of heat transfer (1811–1822), the common wisdom was that *caloric* (or heat) was a material substance, despite the fact that Benjamin Thompson in 1798 had shown that friction produces an inexhaustible supply of heat. Regardless of the fundamental nature of heat, Fourier's mathematical analyses describing temperature distributions in solids are accurate descriptions and stand yet today as the foundation of heat-transfer theory.

4.3b Work

Definition

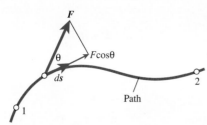

Another fundamental transfer of energy across a system boundary is work. All forms of work, regardless of their origin, are fundamentally expressions of a force acting through a distance,

$$\delta W = \boldsymbol{F} \cdot d\boldsymbol{s}; \qquad (4.5)$$

that is, work is the scalar product of a vector force and the displacement vector $d\boldsymbol{s} = \hat{\boldsymbol{i}}\,dx + \hat{\boldsymbol{j}}\,dy + \hat{\boldsymbol{k}}\,dz$, where $\hat{\boldsymbol{i}}$, $\hat{\boldsymbol{j}}$ and $\hat{\boldsymbol{k}}$ are the unit vectors in a Cartesian coordinate system. We adopt the notation δW to indicate the incremental quantity of work done along the differential path associated with the tangent of $d\boldsymbol{s}$. For a particular process, Eq. 4.5 can be integrated following the process path from position 1 to position 2 to obtain the total work done:

$$_1W_2 \equiv \oint \boldsymbol{F} \cdot d\boldsymbol{s}, \qquad (4.6)$$

where the $\oint$ symbol indicates a path integral. Equations 4.5 and 4.6 should be familiar to you from your previous study of physics.

We adopt the particular notation $_1W_2$ to emphasize that this quantity is the work done in going from point 1 to point 2, which cannot be represented as a difference. In contrast, energy changes associated with a system undergoing a process are expressed as differences. For example, the change in system energy for a process that takes the system from state 1 to state 2 is expressed $\Delta E = E_2 - E_1$. To write a similar expression for work would be nonsense, as there is no such thing as W_1 or W_2.

The force appearing in Eqs. 4.5 and 4.6 may be a purely mechanical one, or it may have other origins, such as the force acting on a charge moving through an electric field or the force on a particle with a magnetic moment moving through a magnetic field. In mechanical engineering applications, work is most commonly associated with mechanical and electrical forces. In this book, we deal only with work arising from these two forces, although we define a mechanical force fairly broadly.

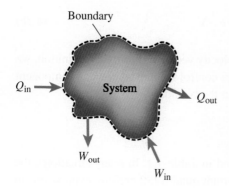

In our study of thermal sciences, it is important to emphasize that work occurs *only* at the boundary of a system or a control volume. Like heat, work is not possessed by a thermodynamic system or a control volume but is just the name of a particular form of energy transfer from a system to the surroundings, or vice versa. For this reason we draw arrows representing work or heat that start or stop at the system or control volume boundary without crossing. In this context, we offer the following formal definition of work:

> *Work* is the transfer of energy across a system or control-volume boundary, exclusive of energy carried across the boundary by a flow, and not the result of a temperature gradient at the boundary or a difference in temperature between the system and the surroundings.

Before presenting examples of work, it is useful to convert Eq. 4.6 to a form expressing the rate at which work is done. The time rate of doing work

Table 4.1   Common Types of Work

Type	Expression for W	Expression for $\dot{W}$ or $\mathcal{P}^*$
Expansion or compression work	$\delta W = Pd\mathcal{V}$ ${}_1W_2 = \displaystyle\int_1^2 Pd\mathcal{V}$	$\dot{W} = P\dfrac{d\mathcal{V}}{dt}$
Viscous work	${}_1W_2 = \displaystyle\int_{t_1}^{t_2} \tau_{\text{visc}}AVdt$	$\dot{W} = \tau_{\text{visc}}AV$
Shaft work	${}_1W_2 = \displaystyle\int_{\Omega_1}^{\Omega_2} \mathcal{T}d\Omega$	$\dot{W} = \mathcal{T}\omega$
Electrical work	${}_1W_2 = \displaystyle\int_{t_1}^{t_2} i\Delta\boldsymbol{V}dt$	$\dot{W} = i\Delta\boldsymbol{V}$
Flow work	${}_1W_2 = \displaystyle\int_{t_1}^{t_2} \dot{m}Pvdt$	$\dot{W} = \dot{m}Pv$

*Rate of work, $\dot{W}$ and power $\mathcal{P}$ are used synonymously throughout this book.

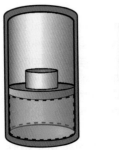

State 1   State 2

FIGURE 4.3
Work is done as a gas expands and pushes back the surroundings in a piston–cylinder assembly.

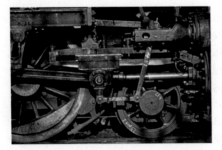

is called **power**, defined as

$$\mathcal{P} \equiv \dot{W} = \lim_{\Delta t \to 0} \frac{\delta W}{\Delta t} = \lim_{\Delta t \to 0} \frac{\boldsymbol{F} \cdot d\boldsymbol{s}}{\Delta t} = \boldsymbol{F} \cdot \frac{d\boldsymbol{s}}{dt}, \tag{4.7a}$$

or

$$\mathcal{P} = \boldsymbol{F} \cdot \boldsymbol{V}, \tag{4.7b}$$

where we recognize that $d\boldsymbol{s}/dt$ is the velocity vector $\boldsymbol{V}$. From this definition, we see that power enters or exits a system or control volume wherever a component of a force is aligned with the velocity at the boundary.

Types

Some common types of work are listed in Table 4.1. In many situations, the power, or rate of working, is the important quantify; therefore, expressions to evaluate the power are also shown.

Expansion (or Compression) Work   In systems or control volumes where a boundary moves, work is performed *by* the system if it expands, whereas work is done *on* the system if the system is compressed. Concomitantly, work is done *on* the surroundings by an expanding system, and work is done *by* the surroundings when the system contracts. As an example of this type of work, consider the expansion of a gas contained in a piston–cylinder assembly as shown in Fig. 4.3. We use this specific example to illustrate application of the fundamental definition of work and some of the subtleties that need to be considered. Examining the entire boundary of the system, we see that work can only be done at the portion of the boundary that is in contact with the piston, as this is the only part of the boundary where there is motion. To have a force act over a distance (Eq. 4.5), the boundary must move. In our examination of the boundary, we also note that the only force present is that due to the pressure of the gas in the system, where we have ignored any possible viscous forces created by friction between the cylinder wall and the moving gas. For our simple geometry, the magnitude of the force exerted by the gas on the piston is given by

$$F = PA,$$

where A is the cross-sectional area of the piston and P is the pressure. The pressure force acts vertically upward in the same direction as the piston motion; thus, the dot product $\mathbf{F} \cdot d\mathbf{s}$ in our definition reduces to the product of the magnitude of the force and the vertical displacement (i.e., $F dx$). The incremental work done is then

$$\delta W = PA dx.$$

For our simple cylindrical geometry, we immediately recognize that $A dx$ is the volume displaced; that is,

$$d\mathcal{V} = A dx.$$

The incremental work then is

$$\delta W = P d\mathcal{V}, \tag{4.8a}$$

and so the total work done in going from state 1 to state 2 is

$$_1 W_2 = \int_1^2 P d\mathcal{V}. \tag{4.8b}$$

We can also express the instantaneous power produced as

$$\mathscr{P} = \dot{W} = P \frac{d\mathcal{V}}{dt}. \tag{4.8c}$$

These expressions for work and power (Eqs. 4.8a–4.8c) represent the single reversible work mode associated with a simple compressible substance.

At this juncture it is important to ask, What assumptions are built into Eqs. 4.8 that might restrict their use? First, as suggested in our development, we assume that the only force acting at the system boundary is that resulting from pressure. Second, we require that the pressure be a meaningful thermodynamic property of the system as a whole. For this to be true, the motion of the piston must be sufficiently slow so that there is enough time for a sufficient number of molecular collisions to cause the pressure to be uniform within the gas volume. A characteristic time to achieve mechanical (pressure) equilibrium is of the order of the height of the volume divided by the speed of sound in the gas. As an example, consider room-temperature air and a volume height of 150 mm (~6 in). For this situation, approximately 0.4 ms are required for the change at the moving boundary to be communicated to the gas

Slow compression

Rapid compression

FIGURE 4.4

(a) A ball thrown at a stationary wall rebounds with the same speed that it strikes the wall. (b) If the wall is moving away from the incoming ball, the rebound velocity is less than the approach velocity.

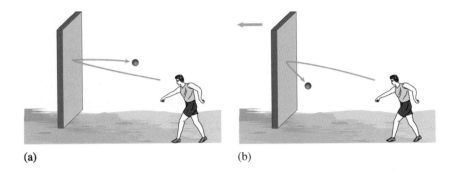

(a)   (b)

> Equilibrium and quasi-equilibrium processes are discussed in Chapter 1. You may find a review of that material useful here.

molecules at the bottom of the cylinder. Thus, our theoretical restriction is that the process must be quasi-static, where the practical meaning of quasi-static is determined by the time scale $t_c \equiv L_c/a$, where L_c is the characteristic length and a is the speed of sound.

Our piston–cylinder example is also useful to illustrate that motion is required to produce work, that is, to transfer energy from the gas molecules to the surroundings in the form of work. Consider throwing a ball against a stationary wall in which the ball rebounds in a perfectly elastic manner as suggested in Fig. 4.4. If, say, you threw the ball at 100 m/s, the ball would rebound back at 100 m/s. The kinetic energy of the ball, $MV^2/2$, is thus the same before and after the collision. The ball experiences no loss of energy. We now allow the wall to move. What then happens to the magnitude of the rebound velocity? In this case, the ball will return with a velocity less than its incoming velocity. For example, when the wall moves at 25 m/s, the rebound velocity is 50 m/s, a value substantially less than the incoming velocity of 100 m/s. The kinetic energy of the ball undergoes a considerable reduction after the collision with the moving wall. Analogous to our ball-throwing example, gas molecules lose energy in their collisions with a receding boundary. If the boundary is advancing, the molecules, of course, gain energy. This example highlights the fundamental idea that work is an energy transfer across a boundary.

Example 4.1

Q_{in}

Consider a piston–cylinder arrangement containing 5.057×10^{-4} kg of dry air. For the following two quasi-static processes, determine the quantity of work performed by or on the air in the cylinder:

A. Constant-pressure heat-addition process

B. Isothermal expansion process

Also completely define the final state (i.e., P_2, V_2, and T_2) and sketch the process on P–V coordinates. For both processes, the following conditions apply:

At initial state	**At final state**
$V_1 = 2.54 \times 10^{-4}$ m³	$V_2 = 5.72 \times 10^{-4}$ m³
$T_1 = 350$ K	

Solution (Part A)

Known   Constant-pressure process, M, V_1, T_1, V_2

Find $_1W_2$, P_2, T_2

Sketch

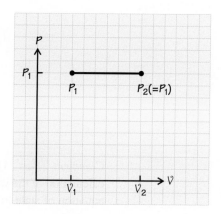

Assumptions

  i. Quasi-static process (given)
  ii. Ideal-gas behavior

Analysis   Because the process is quasi-static, we can apply Eq. 4.8b, which is easily integrated since the pressure is constant, so

$$_1W_2 = \int_1^2 Pd\mathcal{V} = P\int_1^2 d\mathcal{V} = P(\mathcal{V}_2 - \mathcal{V}_1).$$

To evaluate this equation, we only need to find the pressure, as both $\mathcal{V}_1$ and $\mathcal{V}_2$ are given. To find P, we apply the ideal-gas equation of state (Eq. 2.28c) at state 1:

$$P_1\mathcal{V}_1 = MRT_1,$$

or

$$P_1 = \frac{MRT_1}{\mathcal{V}_1},$$

where $R\ (\equiv R_u/\mathcal{M} = 287.0\ \text{J/kg} \cdot \text{K})$ is the gas constant for air (Appendix C). Substituting numerical values, we get

$$P_1 = \frac{5.057 \times 10^{-4}\ (287.0)\ 350}{2.54 \times 10^{-4}} = 200 \times 10^3$$

$$[=] \frac{\text{kg (J/kg} \cdot \text{K)}\ \text{K}}{\text{m}^3}\left[\frac{1\ \text{N} \cdot \text{m}}{\text{J}}\right]\left[\frac{1\ \text{Pa}}{\text{N/m}^2}\right] = \text{Pa}.$$

The work is then

$$_1W_2 = 200 \times 10^3\ (5.72 \times 10^{-4} - 2.54 \times 10^{-4}) = 63.6$$

$$[=] \text{Pa (m}^3)\left[\frac{1\ \text{N/m}^2}{\text{Pa}}\right]\left[\frac{1\ \text{J}}{\text{N} \cdot \text{m}}\right] = \text{J}.$$

We now define the final state. If we know two independent, intensive properties, the state principle tells us that all other properties can be found. Since P is constant, we know $P_2 = P_1 = 200$ kPa, and $\mathcal{V}_2$ and M are given. With this information, we again apply the ideal-gas equation of state

(Eq. 2.28c), this time to find T_2:

$$T_2 = \frac{P_2 \mathcal{V}_2}{MR}$$

$$= \frac{200 \times 10^3 (5.72 \times 10^{-4})}{5.057 \times 10^{-4} (287.0)} = 788.2$$

$$[=] \frac{\text{Pa}(\text{m}^3)}{\text{kg}(\text{J/kg} \cdot \text{K})} \left[\frac{1 \text{ N/m}^2}{\text{Pa}} \right] \left[\frac{1 \text{ J}}{\text{N} \cdot \text{m}} \right] = \text{K}.$$

Comment (Part A) Knowing that the pressure is constant made the calculation of the work quite easy. Note also the importance of the ideal-gas equation of state to determine properties at both state 1 and state 2.

Solution (Part B)

Known   Isothermal process, M, $\mathcal{V}_1$, T_1, $\mathcal{V}_2$

Find $_1W_2$, P_2, T_2.

Assumptions

  i. Quasi-static process
  ii. Ideal-gas behavior

Analysis   We delay drawing a P–$\mathcal{V}$ sketch until the appropriate mathematic relationship between P and $\mathcal{V}$ is determined. We appeal again to the ideal-gas law to do this:

$$P = MRT \left(\frac{1}{\mathcal{V}} \right).$$

Here we recognize that 1. MRT is a constant, since an isothermal process is one carried out at constant temperature, and 2. the P–$\mathcal{V}$ relation is hyperbolic ($P \sim \mathcal{V}^{-1}$). The work can now be found from Eq. 4.8b as

$$_1W_2 = \int_1^2 P d\mathcal{V} = MRT_1 \int_1^2 \frac{d\mathcal{V}}{\mathcal{V}}$$

$$= MRT_1 \left[\ln \mathcal{V} \right]_1^2 = MRT_1 (\ln \mathcal{V}_2 - \ln \mathcal{V}_1) = MRT_1 \ln \frac{\mathcal{V}_2}{\mathcal{V}_1}.$$

Substituting numerical values, we obtain

$$_1W_2 = 5.057 \times 10^{-4} (287.0)(350) \ln \left[\frac{5.72 \times 10^{-4}}{2.54 \times 10^{-4}} \right] = 41.2$$

$$[=] \text{kg}(\text{J/kg} \cdot \text{K})\text{K} = \text{J}.$$

To completely define state 2, we now need P_2. Following the same procedure as in Part A, we apply the ideal-gas equation of state (Eq. 2.28c):

$$P_2 = \frac{MRT_2}{\mathcal{V}_2}.$$

Since $T_2 = T_1 = 350$ K,

$$P_2 = \frac{5.057 \times 10^{-4}(287.0)(350)}{5.72 \times 10^{-4}} = 88.8 \times 10^3$$

$$[=] \frac{\text{kg}(\text{J/kg}\cdot\text{K})\text{K}}{\text{m}^3}\left[\frac{1\,\text{N}\cdot\text{m}}{\text{J}}\right]\left[\frac{1\,\text{Pa}}{\text{N/m}^2}\right] = \text{Pa}.$$

We can now plot this process on P–$\mathcal{V}$ coordinates:

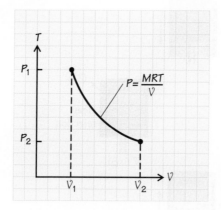

Comment (Part B) Note that the area under this curve is the work. Comparing this graph with that of Part A, we immediately see that less work is performed in the isothermal process, which is consistent with our calculations. Being able to sketch a process on P–$\mathcal{V}$ coordinates is particularly useful in dealing with thermodynamic systems. Such sketches immediately show the work (area under the curve), provided the process is carried out quasi-statically, a requirement for $_1W_2 \equiv \int Pd\mathcal{V}$. Note also that in the solutions to both Parts A and B, we employed only fundamental definitions and the state principle.

Determine whether the work in Example 4.1 is performed on or by the system and on or by the surroundings.

(Answer: Since the numerical values for the work for the system are positive, the work is being performed by the system. A commensurate quantity of work is performed on the surroundings.)

Example 4.2 SI Engine Application

Chapter 7 shows the importance of isentropic processes in defining ideal efficiencies for power producing devices.

The compression and expansion processes associated with spark-ignition engines can be modeled crudely as quasi-static, adiabatic (no heat transfer), isentropic (constant-entropy) processes. For compression and expansion processes described in this way, neither the pressure nor the temperature will remain constant during the process (cf. Example 4.1). Furthermore, we assume that the working fluid is dry air, rather than a mixture of fuel and air (compression) or combustion products (expansion).[2] With these assumptions,

---

[2] These assumptions of adiabatic, isentropic processes with air form the basis for the *air-standard Otto cycle,* aspects of which we will explore at appropriate locations throughout the book.

Table 2.5 in Chapter 2 presents ideal-gas property relationships for isentropic processes. The relationship used here is Eq. 2.43.

the compression/expansion processes obey

$$PV^{\gamma} = \text{constant},$$

where γ ($\equiv c_p/c_v$) is the ratio of specific heats and has a value of 1.4 for air over a wide range of temperatures. Using this model, determine the compression work $_1W_2$, the expansion work $_3W_4$, and the net work associated with the cycle defined by the following processes:

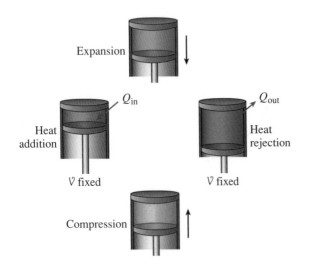

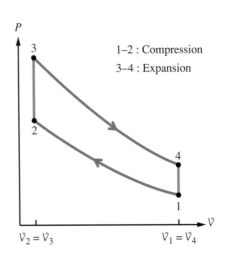

State	1	2	3	4
P (kPa)	100	—	—	—
T (K)	300	—	2800	—
V (m^3)	6.543×10^{-4}	0.8179×10^{-4}	0.8179×10^{-4}	6.543×10^{-4}

Solution

Known   Adiabatic isentropic processes ($PV^{\gamma} = \text{constant}$); selected properties at states 1, 2, 3, and 4; working fluid is air

Find $_1W_2$, $_3W_4$, net work

Sketch

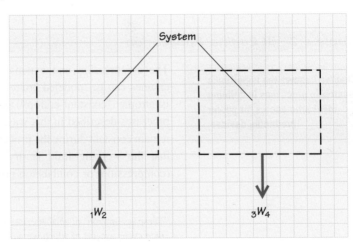

Assumptions

    i. Quasi-static processes
   ii. Ideal-gas behavior

Analysis First we recognize that we are dealing with a thermodynamic system, the air trapped in the cylinder, and not a control volume. This fact, combined with the assumption of quasi-static compression and/or expansion, allows us to use Eq. 4.8b to evaluate the work $_1W_2$ and $_3W_4$, so

$$_1W_2 = \int_1^2 P d\mathcal{V}$$

and

$$_3W_4 = \int_3^4 P d\mathcal{V}.$$

The functional relationship between P and $\mathcal{V}$ is given by $P\mathcal{V}^\gamma = $ constant. Knowing both P and $\mathcal{V}$ at state 1, we express the process from state 1 to state 2 as

$$P\mathcal{V}^\gamma = P_1\mathcal{V}_1^\gamma,$$

or

$$P = \frac{(P_1\mathcal{V}_1^\gamma)}{\mathcal{V}^\gamma}.$$

Substituting this into Eq. 4.8b and integrating yield

$$_1W_2 = \int_1^2 P d\mathcal{V} = (P_1\mathcal{V}_1^\gamma)\int_1^2 \frac{d\mathcal{V}}{\mathcal{V}^\gamma}$$

$$= (P_1\mathcal{V}_1^\gamma)\left[\frac{\mathcal{V}^{1-\gamma}}{1-\gamma}\right]_{\mathcal{V}_1}^{\mathcal{V}_2}$$

$$= \frac{P_1\mathcal{V}_1^\gamma}{1-\gamma}[\mathcal{V}_2^{1-\gamma} - \mathcal{V}_1^{1-\gamma}].$$

Since all of the quantities on the right-hand side are known, we numerically evaluate $_1W_2$ as follows:

$$_1W_2 = \frac{100 \times 10^3 (6.543 \times 10^{-4})^{1.4}}{-0.4}[(0.8179 \times 10^{-4})^{-0.4}$$

$$- (6.543 \times 10^{-4})^{-0.4}]$$

$$= -212.2$$

$$[=]\mathrm{Pa}(\mathrm{m})^3\left[\frac{1\ \mathrm{N/m^2}}{\mathrm{Pa}}\right]\left[\frac{1\ \mathrm{J}}{\mathrm{N\cdot m}}\right] = \mathrm{J}.$$

Note that the sign of $_1W_2$ is negative since work is done *on* the air.

    Our analysis of the expansion process is similar; however, it is complicated by not knowing the pressure at state 3. We need this to evaluate the constant associated with $P\mathcal{V}^\gamma = $ constant $= P_3\mathcal{V}_3^\gamma$. To find P_3 we recognize that, by definition of a system, $M_1 = M_2 = M_3 = M_4 = M$. We thus apply the ideal-gas equation of state (Eq. 2.28) twice, once to find M, using state 1 properties, and a second time to obtain P_3. We express these operations

mathematically as

$$M = \frac{P_1 V_1}{RT_1}$$

and

$$P_3 = M\frac{RT_3}{V_3} = \left(\frac{P_1 V_1}{RT_1}\right)\frac{RT_3}{V_3},$$

which simplifies to

$$P_3 = P_1\left(\frac{V_1}{V_3}\right)\left(\frac{T_3}{T_1}\right).$$

Substituting numerical values, we obtain

$$P_3 = 100 \times 10^3\left(\frac{6.543 \times 10^{-4}}{0.8179 \times 10^{-4}}\right)\left(\frac{2800}{300}\right)\text{Pa} = 7.466 \times 10^6 \text{ Pa}.$$

Following the same procedures as for the compression process, we integrate Eq. 4.8b to obtain

$$_3W_4 = \frac{P_3 V_3^\gamma}{1 - \gamma}[V_4^{1-\gamma} - V_3^{1-\gamma}],$$

which is numerically evaluated as

$$_3W_4 = \frac{7.466 \times 10^6(0.8179 \times 10^{-4})^{1.4}}{-0.4}\left[(6.543 \times 10^{-4})^{-0.4}\right.$$
$$\left. - (0.8179 \times 10^{-4})^{-0.4}\right]\text{J}$$
$$= +862.1 \text{ J}.$$

The plus sign here emphasizes that work is done *by* the air during the expansion process.

Since there is no volume change for both processes 2–3 and 4–1, no work is done in either process; thus, the net work for the cycle 1–2–3–4–1 is

$$W_{\text{net}} = {}_1W_2 + {}_3W_4 = -212.2 + 862.1 \text{ J}$$
$$= +649.9 \text{ J}.$$

Comments   We first note that the net work done is positive, which meets our expectations that engines produce work. It is also important to point out how our crude model differs from the actual processes in a real spark-ignition engine: First, heat transfer is present in the real engine; in particular, there is a substantial heat loss from the hot gases to the cylinder walls during the expansion process. Also affecting the actual net work is the timing of the combustion process, which begins before the piston reaches top center and ends somewhat after the piston begins its descent during the expansion. Both of these factors affect the $P–V$ relationship; nevertheless, if we were to measure P versus V and apply Eq. 4.8b, this would yield a close approximation to the work performed. In fact, experimental $P–V$ data are used in just this way in engine research. Heat losses and a finite combustion time result in the actual work being less than predicted by our crude model.

Pressure transducers mounted within the combustion chamber of an engine provide a record of pressure versus time.

Self Test 4.2

A piston–cylinder device contains 5 kg of saturated liquid water at a pressure of 100 kPa. Heat is added until a saturated vapor state exists. Determine the work performed and whether it is done by or on the system.

(Answer: 846.4 kJ, done by the system)

Pressure and viscous forces are both important in a simple journal bearing where a rotating shaft is supported by a thin film of oil.

Viscous Work   As you will see later in Chapter 6, a moving fluid can exert a shear force at a solid surface or against neighboring fluid elements. Therefore, for control volumes with boundaries that expose these viscous forces, work may or may not be done depending on the velocity at the boundary. The viscous shear force acting on a differential area dA is given by

$$dF_{\text{visc}} = \tau_{\text{visc}} dA,$$

where τ_{visc} is the viscous shear stress.

Since a viscous shear force can only result as a consequence of fluid motion, we consider only the rate of work. If we assume that the viscous shear stress is the same everywhere over the area A, then

$$F_{\text{visc}} = \tau_{\text{visc}} A.$$

Furthermore, if the velocity is also uniform over A, then the rate of doing work is

$$\dot{W}_{\text{visc}} = F_{\text{visc}} \cdot V,$$

or

$$\dot{W}_{\text{visc}} = (\tau_{\text{visc}} \cdot V)A.$$

In general, both τ_{visc} and V vary over a control surface; thus, we need to integrate over this surface to obtain the rate of working, that is,

$$\dot{W}_{\text{visc}} = \int_{\text{CS}} (\tau_{\text{visc}} \cdot V)\,dA. \qquad (4.9)$$

In analyses of macroscopic (integral) control volumes, we may be able to eliminate viscous work from consideration by a judicious choice of boundaries. To illustrate this, we consider the two cases shown in Fig. 4.5. In Fig. 4.5a, we

FIGURE 4.5

Viscous shear forces are associated with moving fluids. At a solid surface (a), although a shear force exists, the velocity is zero. Away from the surface (b), a shear force acts at the control surface opposite to the flow velocity. In (a), no viscous work is done, whereas in (b) work is done by the surroundings on the control volume.

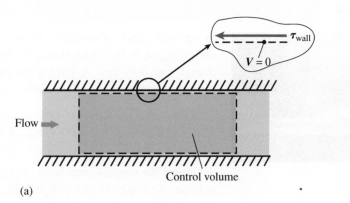

(a)

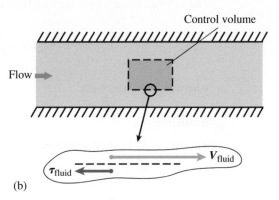

(b)

have chosen the control-volume boundary to be the interface between the fluid and the solid wall. At this boundary, a shear stress exists whose magnitude is proportional to the fluid viscosity and the velocity gradient at the wall in the fluid.[3] The velocity, however, is zero at the wall because of the no-slip condition. Since the velocity V is zero, the dot product $\boldsymbol{\tau}_{\text{visc}} \cdot V$ and, consequently, $\dot{W}_{\text{visc}}$ are zero. No work is done at the wall. This conclusion is important to remember when choosing control-volume boundaries, as a wise choice can avoid dealing with $\dot{W}_{\text{visc}}$.

Figure 4.5b shows a situation in which viscous work is present. Assuming a velocity gradient exists to produce a viscous shear stress, we have all of the conditions necessary for $\dot{W}_{\text{visc}}$ to exist: aligned components of both $\boldsymbol{\tau}_{\text{visc}}$ and V. This is clearly the case for the horizontal portion of the control surface selected for this example. A different situation arises, however, for the portions of the control surface that are perpendicular to the flow, that is, the flow entrance and exit. Here any shear stresses acting over the control surface are now *perpendicular* to the velocity so that $\boldsymbol{\tau}_{\text{visc}} \cdot V = 0$; that is, the shear direction is vertical, whereas the velocity direction is horizontal.[4] Pressure forces here will, however, perform work. We consider this in a subsequent section.

---

[3] Chapter 6 deals with viscous forces in considerable detail. For our purposes here, it is sufficient to know that such forces result whenever a velocity gradient exists in a fluid.

[4] In general, determining all of the viscous forces that act on a control surface is a complex process lying beyond the scope of this book. For most engineering applications, however, ignoring viscous shear forces when the flow is perpendicular to the control surface is reasonable. Furthermore, viscous normal stresses are frequently small and can be neglected.

> The no-slip condition is illustrated in Examples 3.6 and 3.7 and is discussed in the associated text in Chapter 3.

Example 4.3

In a grinding and polishing operation, water at 300 K is supplied at a flow rate of 4.264×10^{-3} kg/s through a long, straight tube having an inside diameter of 6.35 mm. Assuming the flow within the tube is laminar and exhibits a parabolic velocity profile (Table 3.1),

$$v_x(r) = v_{\text{max}} \left[1 - \left(\frac{r}{R} \right)^2 \right],$$

determine the viscous shear work, per meter length, for cylindrical control volumes having radii of $r = R/8, R/2$, and R, where R is the tube radius. The viscous shear stress for this flow is expressed as

$$\tau_{\text{visc}}(r) = \mu \frac{dv_x(r)}{dr},$$

where μ is the viscosity.

Solution

Known   Velocity distribution, shear stress expression, $\dot{m}, T, R (= d/2)$

Find $\dot{W}_{\text{visc}}$ per unit length

Sketch

For r = R/2:

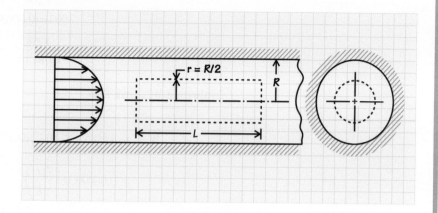

Assumptions

 i. Steady-state, steady flow
 ii. Laminar flow with parabolic velocity profile
 iii. Newtonian fluid (i.e., the viscosity is constant and τ_{visc} is directly proportional to the velocity gradient dv_x/dr)
 iv. Density and viscosity at unknown pressure approximately equal those of a saturated liquid at the given temperature

Analysis   We start with Eq. 4.9, recognizing that both the viscous shear stress and velocity are uniform over the circumferential area of the control volume; thus,

$$\dot{W}_{\text{visc}} = \tau_{\text{visc}}(r)\, v_x(r)\, A(r),$$

where $v_x(r)$ is given and $A(r) = 2\pi r L$. We evaluate the viscous shear stress from the given expression:

$$\tau_{\text{visc}} = \mu \frac{dv_x(r)}{dr} = \mu \frac{d}{dr}\left[v_{\text{max}}\left(1 - \left(\frac{r}{R}\right)^2\right)\right] = -\mu v_{\text{max}} 2r/R^2.$$

> **See Table 3.1 in Chapter 3 for relationships between maximum and average velocities for various internal flows.**

We can relate v_{max} to the given mass flow rate $\dot{m}$ by recognizing that $v_{\text{avg}} = v_{\text{max}}/2$ and by applying the definition of a flow rate (Eq. 3.15):

$$\dot{m} = \rho v_{\text{avg}} A_{\text{x-sec}} = \rho v_{\text{max}} A_{\text{x-sec}}/2,$$

or

$$v_{\text{max}} = \frac{2\dot{m}}{\rho A_{\text{x-sec}}} = \frac{2\dot{m}}{\rho \pi R^2}.$$

Returning to our original expression (definition) for $\dot{W}_{\text{visc}}$ and substituting the detailed expressions for $\tau_{\text{visc}}(r)$, $v_x(r)$, and $A(r)$, we write

$$\dot{W}_{\text{visc}} = 4\pi\mu v_{\text{max}}^2 L \left(\frac{r}{R}\right)^2 \left[1 - \left(\frac{r}{R}\right)^2\right].$$

From the NIST database or Appendix G, we obtain

$$\rho(300\ \text{K}) = 997\ \text{kg/m}^3,$$

$$\mu(300\ \text{K}) = 855 \times 10^{-6}\ \text{N}\cdot\text{s/m}^2.$$

Before numerically evaluating $\dot{W}_{\text{visc}}$, we determine v_{max}:

$$v_{\text{max}} = \frac{2(4.264 \times 10^{-3})}{997\pi(0.00635/2)^2} = 0.270$$

$$[=]\frac{\text{kg/s}}{(\text{kg/m}^3)\text{m}^2} = \text{m/s}.$$

Evaluating $\dot{W}_{\text{visc}}$, we find

$$\dot{W}_{\text{visc}} = 4\pi(855 \times 10^{-6})(0.270)^2 L\left(\frac{r}{R}\right)^2\left[1 - \left(\frac{r}{R}\right)^2\right]$$

$$= 0.000783L\left[\left(\frac{r}{R}\right)^2 - \left(\frac{r}{R}\right)^4\right]$$

$$[=](\text{N}\cdot\text{s/m}^2)(\text{m/s})^2\,\text{m}\left[\frac{1\text{ J}}{\text{N}\cdot\text{m}}\right]\left[\frac{1\text{ W}}{\text{J/s}}\right] = \text{W}.$$

The following table shows our final results for the three values of r/R:

r/R	$\left(\dfrac{r}{R}\right)^2 - \left(\dfrac{r}{R}\right)^4$	$\dot{W}_{\text{visc}}/L$
1/8	0.01538	1.20×10^{-5} W/m
1/2	0.1875	1.47×10^{-4} W/m
1	0	0

Comments We see that this viscous work is quite small. In many practical situations, the viscous work is neglected in the application of the conservation of energy principle because this work is so small compared to other terms.

Shaft Work[5] In many practical devices, power is transmitted across a control surface via a rotating shaft. Figure 4.6 shows gas-turbine and diesel engines, fans, and a propeller-driven aircraft, all of which rely on shaft power for their operation. A control surface that cuts through a shaft exposes a force acting over a distance. From a formal analysis of the forces within the shaft and the application of Eq. 4.7, the work rate or shaft power is expressed as

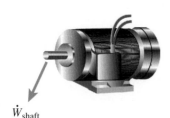

$\dot{W}_{\text{shaft}}$

$$\boxed{\dot{W}_{\text{shaft}} = \mathcal{P}_{\text{shaft}} = \mathcal{T}\omega,} \tag{4.10}$$

where $\mathcal{T}$ is the torque and ω is the angular velocity of the shaft. In many of the applications in this book, $\dot{W}_{\text{shaft}}$ (or $\mathcal{P}_{\text{shaft}}$) will be a given quantity or a quantity derived from, usually, a conservation of energy expression; thus, we seldom refer to either the torque or the angular velocity.

Electrical Work The flow of an electrical current across the boundary of either a system or a control volume results in a flow of energy. This transfer of energy is work or power. That this is true can be seen from a careful application of our definition of work (Eq. 4.5) to the electrical forces and the

---

[5] The rate of work $\dot{W}$, or power $\mathcal{P}$, is frequently implied by the use of the word *work*, as is done here. The context usually makes clear whether the reference is to W or $\dot{W}$ (i.e., $\mathcal{P}$).

FIGURE 4.6

Examples of devices in which shaft work or power is important. Stationary gas-turbine engine (top), diesel engine (left), wind tunnel fans, and propeller-driven aircraft (right). **Drawings and photographs courtesy of General Electric Co., Scania, and NASA, respectively.**

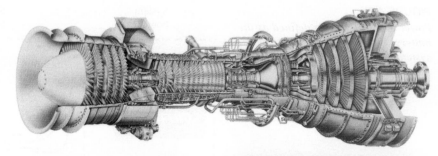

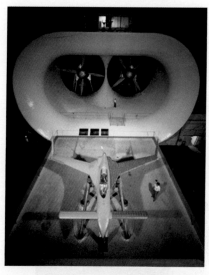

$\dot{W}_{\mathrm{elec}}$

motion of electrons through a conductor. For our purposes, it is sufficient to know how the electrical work and power relate to voltage and current:[6]

$$W_{\mathrm{elec}} = \int_{t_1}^{t_2} i\,\Delta V dt \qquad (4.11\text{a})$$

and

$$\dot{W}_{\mathrm{elec}} = \mathscr{P}_{\mathrm{elec}} = i\,\Delta V. \qquad (4.11\text{b})$$

Note that electrical power flows into the system when ΔV ($= V_{\mathrm{in}} - V_{\mathrm{out}}$) is positive and, conversely, flows out when ΔV ($= V_{\mathrm{in}} - V_{\mathrm{out}}$) is negative.

Figure 4.7 shows examples of electrical work. Choosing a boundary to select just the battery as a thermodynamic system, we see that electrical power is delivered across the boundary from the system to the surroundings, since $V_{\mathrm{out}} > V_{\mathrm{in}}$. In contrast, selecting the light bulb to be a system, we see that electrical power is now delivered in the opposite direction (i.e., from the surroundings to the system). Here V_{in} is greater than V_{out}. If we were to choose a system boundary that enclosed *both* the battery and the light bulb, no work interaction would exist.

As another example, consider the steam power plant, one of our integrating applications. The choice of control volumes in Fig. 4.8 shows electrical work crossing to the surroundings, while high-pressure, high-temperature steam enters the control volume and low-pressure, low-temperature steam exits the control volume.

Electrical power drives a motor; the shaft power from the motor, in turn, drives a pump.

---

[6] Implicit in our discussion of electrical work, current, and voltage is that we are dealing with DC circuits or with AC circuits containing only resistance elements. Treatment of AC circuits, nonresistive loads, and power factors is beyond the scope of this book.

FIGURE 4.7
Power exits the system defined as the battery, whereas power enters the system defined as the light bulb.

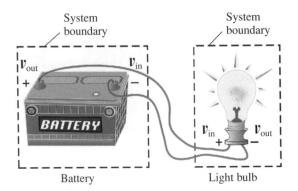

Battery   Light bulb

Chapter 6 provides a detailed discussion of pressure forces. See Fig. 6.1 and Eqs. 6.1 and 6.2.

Flow Work   The work associated with moving a fluid into and out of a control volume is called **flow work**. Pressure forces acting over flow inlets and exits produce this work. To evaluate the flow work, we appeal to our fundamental definition, Eq. 4.7b, which requires the identification of the appropriate forces.

When ascertaining what forces act *on* a control volume, our point of view is always from a position *outside* of the control volume. Forces due to pressure always act perpendicular to a control surface and are directed to the interior of the control volume. As an example, consider the flow in a pipe illustrated in Fig. 4.9. We focus our attention on the inlet and exit areas designated as stations 1 and 2. Assuming a uniform pressure distribution at stations 1 and 2, we can calculate the pressure forces as simply the products of the respective pressures and areas as indicated.

Having identified the forces at the inlet and exit, we now can evaluate the flow work done *on* the fluid at the inlet surface designated 1. Since the velocity over the inlet area A_1 has the same direction as $F_{P,1}$, the dot product in Eq. 4.7b is simply

$$\dot{W}_{\text{flow},1} = P_1 A_1 V_1,$$

where, for the time being, we have assumed that V_1 is uniform over A_1. The simple mathematical manipulation of multiplying and dividing by the density ρ yields

$$\dot{W}_{\text{flow},1} = \rho_1 A_1 V_1 \frac{P_1}{\rho_1}.$$

In this equation, we recognize that $\dot{m}_1 = \rho_1 A_1 V_1$ and that $1/\rho_1$ is just v_1, the specific volume. Thus,

$$\dot{W}_{\text{flow},1} = \dot{m}_1 P_1 v_1, \tag{4.12a}$$

Steam in   Control surface

Electrical generator

Electrical output

Steam turbine

Steam out

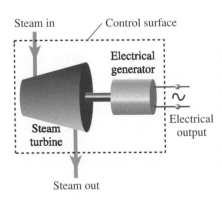

FIGURE 4.8
The control volume shown includes a steam turbine and an electrical generator. Electrical work exits the control volume (left); 500-MW turbine generator (middle); electrical generator windings (right).

FIGURE 4.9
Pressure forces associated with a control volume inside of a pipe with a flowing fluid.

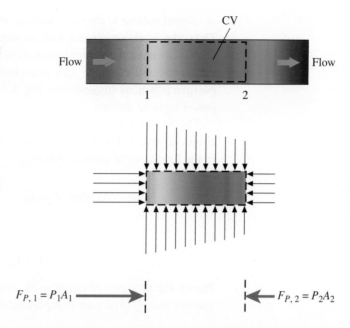

FIGURE 4.9
Pressure forces associated with a control volume inside of a pipe with a flowing fluid.

where we emphasize that this quantity is the rate of work done *on* the control volume by the surroundings. It is a simple matter to show that Eq. 4.12 applies even if the velocity distribution is not uniform; of course, both P_1 and v_1 must be uniform over A_1.

A similar derivation can be applied to obtain the flow work at the control volume exit (i.e., at station 2). In this case, however, the pressure force is directed opposite to the velocity; thus, the dot product $\boldsymbol{F}_{P,2} \cdot \boldsymbol{V}_2 = F_{P,2}\, V_2 \cos(180°) = -F_{P,2}\, V_2$. The rate of flow work performed on the fluid is then

$$\dot{W}_{\text{flow},2} = -\dot{m}_2 P_2 v_2. \tag{4.12b}$$

We end this development by noting that flow work only occurs when a fluid crosses a boundary. If there is no flow, there cannot be any component of the velocity aligned with the pressure force.

Foreshadowing the development of the conservation of energy principle in Chapter 5, we point out that the Pv product in the flow work is frequently grouped with the internal energy of the entering or exiting fluid (i.e., $u + Pv$). You may recall that this particular grouping of thermodynamic properties is termed the enthalpy.

To review enthalpy, see Eq. 2.17 and the associated discussion in Chapter 2.

4.4 SIGN CONVENTIONS AND UNITS

In the next chapter, we will examine the principle of energy conservation and the many ways that this principle can be expressed. Since writing a conservation of energy expression is analogous to maintaining an accountant's ledger, we need to know whether various energy terms are credits or debits to our energy account. In this brief section, we present a consistent set of sign conventions for heat and work interactions.

Heat transfer and its time rate are *positive* when the direction of the energy exchange is *from the surroundings to a system or control volume*. Conversely, heat transfer from a system to the surroundings is a negative quantity. Work and power are defined to be *positive* when they are delivered *from a system or control volume to the surroundings* and negative when the converse is true. Thus, the work associated with an expanding gas is positive, whereas the work associated with compression is negative, as we saw in Example 4.2 for the spark-ignition engine. In a steam turbine, the steam expands and produces positive power, as suggested by Fig. 4.8. In contrast, a water pump requires a power input to operate.

The SI unit associated with energy, heat, and work is the joule, which is abbreviated as J. The joule is derived from the definition of work and relates to the fundamental units as follows:

$$\text{joule} = \text{newton} \times \text{meter} = \frac{\text{kilogram} \cdot \text{meter}}{\text{second}^2} \times \text{meter},$$

or

$$J = kg \cdot m^2/s^2.$$

Power, the time rate of doing work, is expressed in watts (W), as is the heat transfer rate, $\dot{Q}$. The watt is expressed in terms of the fundamental units as

$$\text{watt} = \frac{\text{joule}}{\text{second}},$$

or

$$W = \frac{kg \cdot m^2}{s^3}.$$

In the United States, other units also are used for heat work and power. The British thermal unit (Btu) is frequently used for energy, heat, and work; and horsepower is used for mechanical power. Conversions to SI units are as follows:

$$1 \text{ British thermal unit (Btu)} = 1055.056 \text{ joules (J)},$$
$$1 \text{ horsepower (hp)} = 745.7 \text{ watts (W)}.$$

Other units are used as well. An extensive list of unit conversions is provided on the inside covers of this book. Unfortunately, a wide variety of units are commonly used in commerce and industry. You should be comfortable and proficient in converting units. Some end-of-chapter problems are designed for you to practice these conversions.

Example 4.4 Steam Power Plant Application

Consider the simple steam power plant cycle shown in Fig. 4.10. Using the indicated control volumes, identify all of the energy transfers (i.e., heat and work interactions) for each of the components.

Solution

We start at the water (feedwater) pump and proceed around the loop.

Water Pump We have redrawn the control volume for the pump in Fig. 4.11a. Here we identify a work input from an electric motor (i.e., shaft work in) and flow work in and out. We assume that the casing of the pump will be hotter than the surroundings, so a heat loss is also indicated. We will see in Chapter 5 that this heat loss is negligible compared to the other energy transfers.

Feedwater pump. *Photograph courtesy of Flowserve Corporation.*

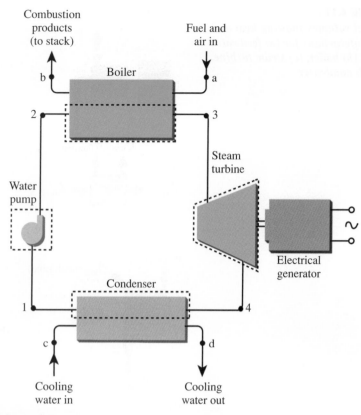

FIGURE 4.10
Rankine cycle schematic for Example 4.4 with individual control volumes for each component.

Steam is generated in these boilers by the combustion of residues from lumber waste and pulp paper (biomass). *Photograph courtesy of NREL.*

Steam turbine rotor.

Boiler In the boiler, water enters the water drum and steam exits the steam drum. Water is converted to steam in a series of tubes connecting the water and steam drums. The outside surfaces of these tubes are exposed to hot products of combustion. Figure 1.2 in Chapter 1 shows a cutaway view of such a boiler. Our control volume (Fig. 4.11b) includes the drums and tubes, along with the water and the steam that they contain, and is represented by a single tube in Fig. 4.11b. The hot combustion products are external to our control volume. Because the tube and drum geometries are significantly different, we have arbitrarily divided the heat transfer from the combustion products to our control volume into three components. Again, flow work exists where the control surface cuts through the flowing fluid. Note that we could have just as easily chosen a control volume that contains only the water and steam. Sometimes an analysis is simplified by either including or excluding the hardware surrounding the working fluid.

Steam Turbine We now consider the steam turbine. Our control surface here (Fig. 4.11c) cuts through the inlet steam line, exposing the flow work at the inlet, and similarly at the outlet. The control surface also cuts through the shaft connecting the turbine to the electrical generator and, thus, shaft work occurs at this location. Since the outer casing of the turbine is likely to be hotter than the surroundings, there will be some heat transfer from the control volume. Although this heat loss is shown in Fig. 4.11c for completeness, the loss is quite small in comparison to all of the other energy transfers and is usually neglected in thermodynamic analyses.

FIGURE 4.11

Control volumes showing heat and work interactions for (a) feedwater pump, (b) boiler, (c) steam turbine, and (d) condenser.

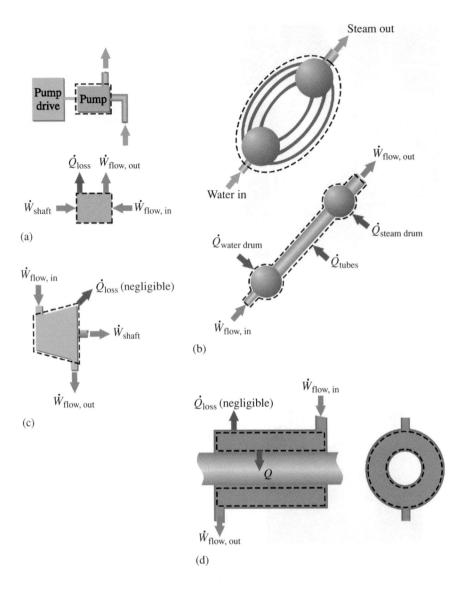

(a)

(b)

(c)

(d)

Condensers and cooling towers at The Geysers power plant in California. *Photograph courtesy of NREL.*

Condenser In the condenser, the entering high-quality steam condenses to all water. Figure 4.11d schematically shows the condenser in which the steam is the shell-side fluid occupying the annular control volume, while cold water flows through the tube. The single tube in this schematic represents all of the tubes in the real condenser (see Fig. 1.5). The primary heat transfer is from the condensing steam to the cold water. Again, there is a small, and usually negligible, heat loss to the surroundings. Flow work again is present as the fluid must be pushed into and out of the control volume.

Comment Note that we have identified the small heat losses that occur in all of these real devices. Although these are usually neglected in applying the conservation of energy principle to these devices, it is important that you become skillful in identifying *all* heat and work interactions. You can always discard a term as you proceed, but if you missed an important term at the beginning of an analysis, there is no later recourse.

It is also important to point out that in conservation of energy analyses the flow work terms identified in all of these devices are conventionally grouped with the rate of internal energy flowing into or out of the control volume as the rate of enthalpy flow (i.e., $\dot{m}u + \dot{W}_{\text{flow}} = \dot{m}u + \dot{m}Pv = \dot{m}h$). We will deal with this at some length in the next chapter.

Self Test 4.3

☑ **Write an expression for the net work and net heat transfer for the steam power plant of Fig. 4.10.**

(Answer: $\dot{W}_{net} = {}_1\dot{W}_2 + {}_3\dot{W}_4 = \dot{W}_{turb} - \dot{W}_{pump}$, $\dot{Q}_{net} = {}_2\dot{Q}_3 + {}_4\dot{Q}_1 = \dot{Q}_{boil} - \dot{Q}_{cond}$. *Note that the directions of work and heat transfer are explicitly defined in Fig. 4.11.)*

Example 4.5 Steam Power Plant Application

For the Rankine cycle illustrated in Fig. 4.10, calculate and compare the net flow work associated with the feedwater pump and the steam turbine for the following conditions:

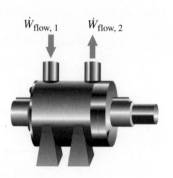
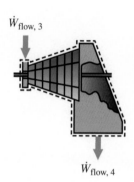

Working fluid (water/steam) flow rate: 2.36 kg/s.

State 1	**State 2**	**State 3**	**State 4**
Saturated liquid	Compressed liquid	Saturated vapor	Wet mixture
$P_1 = 5$ kPa	$P_2 = 1$ MPa	$P_3 = 1$ MPa	$P_4 = 5$ kPa
			$x_4 = 0.9$

These conditions are typical for an oil-fired industrial power plant producing approximately 1000 kW electrical power [9].

Solution

Known $\dot{m}$, physical states, P_1, P_2, P_3, P_4, x_4

Find $\dot{W}_{flow,\ net}$ for pump and turbine

Sketch

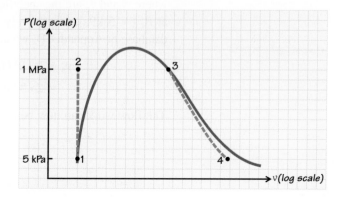

Assumptions

i. The specific volume of the compressed liquid at state 2 is approximately equal to that of the saturated liquid at state 1.

ii. Properties are uniform over the inlet and outlet stations.

Analysis   We calculate the flow work from the straightforward application of Eqs. 4.12a and 4.12b, recognizing that the net work is the sum of the flow work in and flow work out with careful attention being paid to signs. Since all of the required pressures are given, we need only determine the specific volumes. Starting with the feedwater pump, we find the inlet specific volume from the saturated steam tables (NIST database or Appendix D):

$$v_1 = v_f(P_{sat} = 5 \text{ kPa}) = 0.0010053 \text{ m}^3/\text{kg}.$$

The specific volume at state 2 can be approximated as being the same as that at state 1. If we knew another property at state 2, we could use tabulated (or computer-based) data in the compressed liquid region to obtain a precise value. Applying the definition of flow work (Eq. 4.12), we calculate

$$\dot{W}_{flow, 1} = \dot{m}P_1v_1$$
$$= 2.36(5 \times 10^3)0.0010053$$
$$= 11.9$$

$$[=] (\text{kg/s})(\text{N/m}^2)(\text{m}^3/\text{kg}) \left[\frac{1 \text{ J}}{\text{N} \cdot \text{m}} \right] \left[\frac{1 \text{ W}}{\text{J/s}} \right] = \text{W},$$

$$\dot{W}_{flow, 2} = -\dot{m}P_2v_2$$
$$= -2.36(1 \times 10^6)0.0010053 \text{ W}$$
$$= -2372.5 \text{ W},$$

and

$$\dot{W}_{flow, \text{ pump net}} = \dot{W}_{flow, 1} + \dot{W}_{flow, 2}$$
$$= 11.9 - 2372.5 \text{ W}$$
$$= -2360.6 \text{ W}.$$

For the steam turbine, the inlet specific volume is found from the NIST database or Appendix D to be

$$v_3 = v_g(P_{sat} = 1 \text{ MPa}) = 0.19436 \text{ m}^3/\text{kg}.$$

Since the outlet condition lies in the wet region, we use the quality (x_4) to determine v_4, that is,

$$v_4 = (1 - x_4)v_f + x_4v_g,$$

where v_f and v_g are the specific volumes for the saturated liquid and vapor at P_4 (= 5 kPa), respectively. Using values for v_f and v_g from the NIST database or Appendix D, we calculate

$$v_4 = 0.1(0.0010053) + 0.9(28.185) \text{ m}^3/\text{kg}$$
$$= 25.37 \text{ m}^3/\text{kg}.$$

The flow work can now be calculated:

$$\dot{W}_{flow, 3} = \dot{m}P_3v_3$$
$$= 2.36(1 \times 10^6)0.19436 \text{ W}$$
$$= 458,690 \text{ W}$$

and

$$\dot{W}_{\text{flow, 4}} = -\dot{m}P_4 v_4$$
$$= 2.36(5 \times 10^3)25.37 \text{ W}$$
$$= 299{,}370 \text{ W}.$$

Thus, the net turbine flow work is

$$\dot{W}_{\text{flow, turbine net}} = \dot{W}_{\text{flow, 3}} + \dot{W}_{\text{flow, 4}}$$
$$= 458{,}690 - 299{,}370 \text{ W}$$
$$= 159{,}320 \text{ W}.$$

Comment   In comparing the net flow work for the pump and turbine, we note, first, that the magnitude of the turbine flow work is about 70 times that of the pump, and, second, that the signs differ between the pump and turbine flow work. That the turbine flow work is so large results from the specific volume of the steam being much larger than that of the liquid water. In the next chapter, we will return to the sign issue in our first-law (conservation of energy) analysis of pumps and turbines.

Self Test 4.4   ✅ **Calculate the change in enthalpy of the working fluid for the turbine of Example 4.5 by using (a) the definition of enthalpy (i.e., $h = u + Pv$) and (b) tabulated values for h.**
(Answer: (a) −454.32 kJ/kg, (b) −459.32 kJ/kg. Note: The relatively large discrepancy (1%) results from interpolation for properties at 5 kPa. Using NIST data at 5 kPa without interpolation yields no discrepancy.)

Example 4.6   Solar Heating Application

Solar-heated domestic hot water is provided using the system shown schematically in Fig. 4.12 [10]. The solar collector consists of a 292.6-m length of pliable black plastic tubing (EPDM) through which water is continuously pumped. An electric motor drives the pump. The solar-heated water is returned to the solar storage tank. The high-pressure (~400 kPa) domestic water is physically separated from the solar-heated water, allowing the solar circuit to operate close to atmospheric pressure. For the control volume indicated by the dashed line in Fig. 4.12, identify all of the heat and work interactions.

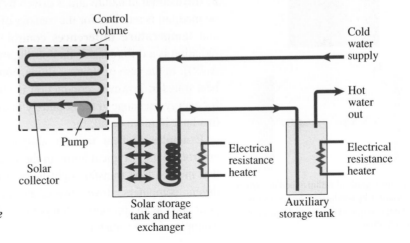

FIGURE 4.12
System for solar heating of domestic hot water [10] considered in Example 4.6.

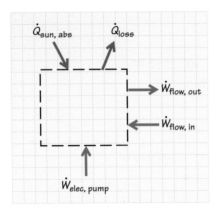

Solution

We begin by redrawing the control volume. We identify a work (power) input associated with the pump, and flow work in and out where the control surface cuts through the tubes. The heat transfer is a bit more complicated, and we will define two components rather than just considering a single net rate. A portion of the radiant energy from the sun is absorbed by the control volume, which we designate as $\dot{Q}_{sun, abs}$. Since the collector will be at a higher temperature than the surroundings (air, ground, etc.), there will be heat transferred to the surroundings, $\dot{Q}_{loss}$. These two heat-transfer components are shown as arrows pointing in the known directions.

Comments The details of the heat-transfer processes depend critically on the specific geometry, the characteristics of the solar radiation, and the properties of the surroundings (e.g., ambient temperature and wind speed). We will investigate some of these factors later in this chapter.

> **Chapter 9 provides ways to calculate heat transfer from the external surface of the tube; Chapter 10 deals with heat-transfer at the inside wall.**

Self Test 4.5

Consider the solar storage tank and heat exchanger in Fig. 4.12. Identify the energy transfers associated with the cold-water coil in the tank.

(Answer: Flow work in, flow work out, and heat transfer in)

4.5 RATE LAWS FOR HEAT TRANSFER

Recall our definition of heat (or heat transfer) as the transfer of energy across a system or control-volume boundary resulting from a difference in temperature or a temperature gradient. In this section, we present the basic relationships that allow us to calculate the heat transfer, knowing either temperature differences or temperature gradients.

There are two physical mechanisms for heat transfer: 1. the transfer of energy resulting from molecular collisions, lattice vibrations, and unbound electron flow, that is, **conduction,** and 2. the net exchange of electromagnetic radiation, that is, **radiation**. Since conduction depends on interactions among neighboring particles (i.e., a local exchange of energy), the process is said to be **diffusional** in nature and is driven by **temperature gradients**. In contrast, no medium is required for the transfer of energy by electromagnetic radiation and **temperature differences** control. Actually, the driving potential for radiation is a difference of the fourth power of the absolute temperature [i.e., $\Delta(T^4)$]. In the case of flowing fluids, **convection** is treated as a third mode of heat transfer; however, conduction is still the only fundamental mechanism for energy exchange at the boundary in a convection problem. We will discuss this later in more detail.

Examples of heat transfer abound in both our natural and synthetic environments. A typical home in the United States is chocked full of devices that involve heat transfer: light bulbs, furnaces, space heaters, toasters, hair dryers, computers, stoves, ovens, air conditioners, heat pumps, etc. Keeping your body at an appropriate temperature is a close-to-home and very practical example of heat transfer.

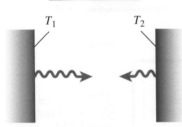

A temperature gradient within a medium drives energy transfer by conduction (top), whereas no medium is required for energy transfer by radiation (bottom).

Fourier's law is a key relationship in our study of conduction. Equation 4.13 is a specific form, whereas Eq. 4.14 expresses the general case.

See Chapter 2 for a brief discussion of thermal conductivity.

4.5a Conduction

The fundamental rate law governing conduction heat transfer in solids and stagnant fluids is **Fourier's law**, which is expressed as

$$\dot{Q}_{cond} = -kA\frac{dT}{dx} \tag{4.13}$$

for a one-dimensional Cartesian system. In Eq. 4.13, k is the **thermal conductivity** of the conducting medium, A is the area perpendicular to the direction of heat transfer (i.e., perpendicular to the x-direction for this one-dimensional case), and dT/dx is the temperature gradient. The negative sign appearing in Eq. 4.13 is the result of the fact that heat transfer has an associated direction. Heat transfer is directed in the positive x-direction when the temperature gradient is negative, as illustrated in Fig. 4.13. That heat transfer has a natural direction from a high-temperature region to a low-temperature region is a consequence of the second law of thermodynamics, which we will state in Chapter 7. In general, thermal conductivity values depend on temperature; however, problems can frequently be simplified by using an appropriate average value. You can find thermal conductivity values for a number of common substances in the appendices to this book. You can also evaluate thermal conductivities for various fluids using the NIST database.

Fourier's law can be more generally expressed as a vector quantity as follows:

$$\dot{Q}''_{cond} = -k\boldsymbol{\nabla}T, \tag{4.14}$$

where we define the **heat flux** vector $\dot{Q}''$ to be the heat transfer rate per unit area:

$$\dot{Q}''_{cond} \equiv \dot{Q}_{cond}/A. \tag{4.15}$$

The area in Eq. 4.14 is perpendicular to the direction of heat flow (cf. Eq. 3.24). The temperature gradient $\boldsymbol{\nabla}T$ for a Cartesian system is simply

$$\boldsymbol{\nabla}T = \hat{\boldsymbol{i}}\frac{dT}{dx} + \hat{\boldsymbol{j}}\frac{dT}{dy} + \hat{\boldsymbol{k}}\frac{dT}{dz},$$

where $\hat{\boldsymbol{i}}, \hat{\boldsymbol{j}}$, and $\hat{\boldsymbol{k}}$ are the directional unit vectors. For the time being, we will restrict our discussion to one-dimensional geometries: 1-D Cartesian, 1-D cylindrical, and 1-D spherical systems. These geometries are illustrated in Fig. 4.14. The corresponding expression of Fourier's law in cylindrical and spherical systems is

$$\dot{Q}_{cond}(r) = -kA(r)\frac{dT}{dr}, \tag{4.16a}$$

where

$$A(r) = 2\pi rL \quad \text{(1-D cylindrical)}, \tag{4.16b}$$

$$A(r) = 4\pi r^2 \quad \text{(1-D spherical)}. \tag{4.16c}$$

In Eq. 4.16b, L is an arbitrary length of the cylindrical domain. Many real-world systems can be approximated as being one dimensional. For

FIGURE 4.13

The temperature gradient dT/dx determines the direction of heat transfer. A negative dT/dx (top) results in a positive heat transfer (i.e., from left to right), whereas a positive dT/dx (bottom) results in a negative heat transfer (i.e., from right to left).

Calculated temperature gradients for a flow over a heated, horizontal plate. The largest gradients (red) are near the leading edge of the plate; thus, the heat flux is greatest near the leading edge (see Eq. 4.17). Flow is from left to right with the vertical coordinate expanded by a factor of 2.

Table 4.2   Typical Values of Convective Heat-Transfer Coefficients [13]

Process	Heat-Transfer Coefficient, h_{conv} (W/m²·K or °C)
Free convection	
Gases	2–25
Liquids	50–1000
Forced convection	
Gases	25–250
Liquids	50–20,000
Convection with phase change	
Boiling or condensation	2500–100,000

where $\bar{h}_{conv}$ is the **heat-transfer coefficient** averaged over the surface area A, T_s is the surface temperature, and T_∞ is the fluid temperature far from the wall in the bulk of the fluid. In this expression, a temperature difference is the driving potential for the heat transfer. A **local heat-transfer coefficient,** $h_{conv, x}$, also can be defined in terms of the local heat flux:

$$\dot{Q}''_{conv} = h_{conv, x}(T_s - T_\infty),\qquad(4.19)$$

(a)

(b)

(c)

FIGURE 4.16

Examples of forced (a, b) and free convection (c–e): (a) A fan bows air over a heating coil in a hair dryer. (b) An automobile radiator relies on a pump to move the coolant through the internal flow passages and motion of the car to force the air through the external passages. (c) A steam radiator transfers energy to the environment by both free convection and radiation. (d) Fins passively cool power transistors in many electronic devices. (e) A free-convection, thermal plume rises from a standing person. The schlieren technique allows density gradients to be made visible. **Photographs (a,c,d) by Paul Ruby, (b) by Sibtosh Pal, and (e) by Gary Settles.**

(d)

(e)

Rising air currents (thermals) are created when the surface of the earth, warmed by the sun, heats the air at ground level (top). Fans and pumps frequently drive forced convection (bottom).

where the value of each quantity appearing in Eq. 4.19 can vary with location on the surface. Some typical values for heat-transfer coefficients are presented in Table 4.2 for a wide range of situations.

The heat-transfer coefficient is not a thermo-physical property of the fluid but, rather, a proportionality coefficient relating the heat flow, $\dot{Q}_{conv}$, to a driving potential difference, $T_s - T_\infty$; thus, Eqs. 4.18 and 4.19 can be considered to be definitions of $\bar{h}_{conv}$ and $h_{conv,x}$ rather than statements of some physical law. As we will see in Chapters 9 and 10, the heat-transfer coefficient depends on the specific geometry, the flow velocity, and several thermo-physical properties of the fluid (e.g., thermal conductivity, viscosity, and density). The wide range of values for h_{conv} shown in Table 4.2 results from the variation of these factors. Note that the temperature units associated with the heat-transfer coefficient are interchangeably K or °C, since the units refer to a temperature difference, rather than a temperature, as can be seen from Eqs. 4.18 and 4.19.

Table 4.2 also categorizes convection heat transfer by general process: free convection, forced convection, and convection with phase change. **Free (or natural) convection** refers to flows that are driven naturally by buoyancy, whereas **forced convection** employs some external agent to drive the flow. The convection heat transfer from an incandescent light bulb is an example of free convection; the cooling of circuit boards in a computer with a fan is an example of forced convection. Figure 4.16 shows other examples of free and forced convection.

Example 4.9

Consider the cooling of an 80-mm by 120-mm computer circuit board that dissipates 6 W as heat from one side. Determine the mean surface temperature of the circuit board when the cooling is by free convection with $\bar{h}_{conv} = 7.25 \ W/m^2 \cdot °C$. Compare this result with that obtained for forced convection ($\bar{h}_{conv} = 57.7 \ W/m^2 \cdot °C$) when a fan is employed. Assume the cooling air temperature in both cases is 25°C.

Solution

Known $\dot{Q}, L, W, T_\infty, \bar{h}_{conv}$ (free convection), $\bar{h}_{conv}$ (forced convection)

Find T_s for both free and forced convection

Sketch

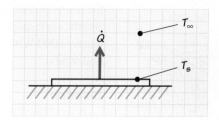

Assumptions

 i. Uniform-temperature board
 ii. Heat transfer from one side only

Analysis   We need only apply the definition of an average heat-transfer coefficient (Eq. 4.18) to find the circuit board temperature, that is,

$$\dot{Q} = \bar{h}_{conv} A(T_s - T_\infty).$$

Solving for T_s yields

$$T_s = \frac{\dot{Q}}{h_{conv} A} + T_\infty.$$

Recognizing that the area $A = LW$, we numerically evaluate this expression for the two conditions. For free convection, we get

$$T_s = \left(\frac{6}{7.25(0.120)0.080} + 25 \right) ^\circ C = 111.2 ^\circ C$$

For forced convection, we get

$$T_s = \left(\frac{6}{57.7(0.120)0.080} + 25 \right) ^\circ C = 35.8 ^\circ C$$

$$[=] \frac{W}{(W/m^2 \cdot {}^\circ C)m^2} = (\Delta) \, K \text{ or } (\Delta) \, ^\circ C.$$

Comments   We see that the circuit-board temperature is much lower with forced convection, 35.8°C, compared to 111.2°C for free convection. In fact, the 111.2°C value most likely exceeds a reliable operating temperature for many circuit components.

Self Test 4.8

Considering the discussion of free versus forced convection, explain why a person would seek relief on a hot day by standing in front of a fan.

(Answer: Since a forced-convection heat-transfer coefficient can be many times larger than that associated with free convection, standing in front of the fan results in a much greater heat-transfer rate from the body to the air.)

Example 4.10   Steam Power Plant Application

Consider a boiler generating steam at a pressure of 1 MPa. The heat flux at the inside surface of a boiler tube (see Fig. 1.2) is approximately 25 kW/m², and the corresponding local heat-transfer coefficient has a value of approximately 6000 W/m²·K. Estimate the temperature of the inside surface of the boiler tube.

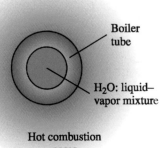

Boiler tube

H₂O: liquid–vapor mixture

Hot combustion gases

Solution

Known $P_{sat}, \dot{Q}'', h_{conv,x}$

Find T_{wall}

Sketch

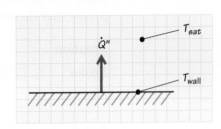

Assumptions

T_{sat} corresponds to T_∞ in Eq. 4.19.

Analysis We solve this problem by the straightforward application of the definition of the local heat-transfer coefficient (Eq. 4.19):

$$\dot{Q}'' = h_{conv,x}(T_{wall} - T_\infty).$$

To apply Eq. 4.19, we assume that the ambient temperature $T_\infty = T_{sat}$ ($P_{sat} = 1$ MPa). From the NIST database or Appendix D, we find T_{sat} (1 MPa) = 179.88°C. Solving Eq. 4.19 for the boiler tube inside surface temperature yields

$$T_{wall} = \frac{\dot{Q}''}{h_{conv,x}} + T_{sat},$$

which we numerically evaluate as follows:

$$T_{wall} = \left(\frac{25{,}000}{6000} + 179.88\right)°C = (4.17 + 179.88)°C \approx 184°C.$$

Note that the units associated with $h_{conv,x}$, W/m²·K, are equivalent, without modification, to units of W/m²·°C, as a temperature *difference* is implied in the units of $h_{conv,x}$ (or $\overline{h}_{conv}$), and ΔT (K) ≡ ΔT (°C).

Comments The results of our calculation show that the tube-wall temperature is quite close to the temperature of the steam. This is a characteristic of most boilers, resulting from the very high values of h_{conv} on the steam side, and serves to keep metal tube temperatures low compared to the high temperatures of the combustion product gases surrounding the tubes. The low tube temperatures preserve the strength of the metal and permit a long service life.

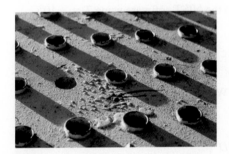

Boiler tube sheet

Example 4.11 *Biological Systems Application*

Golden-crowned kinglets, small North American birds, fluff their feathers to stay warm on cold winter days. Estimate the convective heat-transfer rate from a kinglet in a 20 mile/hr (8.94 m/s) wind, treating the bird as a 60-mm-diameter sphere. The surface temperature of the kinglet's feathers is −7°C and the temperature of the air is −10°C. The barometric pressure is 100 kPa. The following empirical expression relates the average heat-transfer coefficient to the wind speed V:

$$\frac{\overline{h}_{conv}D}{k} = 0.53\left(\frac{\rho VD}{\mu}\right)^{0.5},$$

where D is the bird diameter and k, ρ, and μ are the thermal conductivity, density, and viscosity of the air, respectively.

Solution

Known T_s, T_∞, P, V, expression for h_{conv}

Find $\dot{Q}$

Sketch

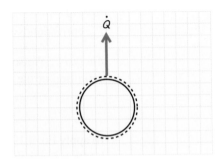

Assumptions

  i. Ideal-gas behavior
 ii. Spherical geometry approximates complex shape (given)

Analysis   To estimate the heat-transfer rate, we apply the basic convective rate law (Eq. 4.18):

$$\dot{Q} = \overline{h}_{conv} A_{surf}(T_s - T_\infty).$$

With the assumption of a spherical shape for the bird, the surface area is πD^2. We calculate the average heat-transfer coefficient using the given correlation. The needed properties are obtained from the NIST software using $T = -10°C$ and $P = 0.1$ MPa:

$$\rho = 1.3245 \text{ kg/m}^3,$$
$$\mu = 16.753 \times 10^{-6} \text{ N·s/m}^2,$$
$$k = 0.02361 \text{ W/m·K}.$$

We now evaluate $\overline{h}_{conv}$:

$$\overline{h}_{conv} = \left(\frac{k}{D}\right)0.53\left(\frac{\rho V D}{\mu}\right)^{0.5}$$

$$= \frac{0.02361(0.53)}{0.06}\left[\frac{1.3245(8.94)0.06}{16.753 \times 10^{-6}}\right]^{0.5}$$

$$= 0.2086(42{,}408)^{0.5} = 42.9$$

$$[=]\frac{W}{m·K}\frac{1}{m}\left(\frac{kg}{m^3}\frac{m}{s}m\left[\frac{m^2}{N·s}\frac{1 N}{kg·m/s^2}\right]\right)^{0.5}$$

$$= \frac{W}{m^2·K}(1).$$

The heat loss rate from the kinglet is thus

$$\dot{Q} = 42.9\pi(0.06)^2[-7 - (-10)]$$

$$= 1.46$$

$$[=]\frac{W}{m^2·K}m^2°C\left[\frac{1 K}{1°C}\right] = W.$$

Comments   An interesting aspect of this example is approximating the small bird as a sphere. Although imprecise, approximations such as these are frequently used to obtain "ballpark" estimates for hard-to-calculate quantities. Notice also the treatment of units where we recognize that a temperature difference expressed in kelvins and °C are equivalent (i.e., $\Delta T = 3 \text{ K} = 3°\text{C}$).

> **These particular dimensionless parameters are known as the Nusselt and Reynolds Numbers, respectively. See Table 8.4 and Eqs. 8.16 and 8.28.**

In our calculation of $\bar{h}_{\text{conv}}$, you may have noticed that the empirical relationship is expressed using two dimensionless parameters: $(h_{\text{conv}} D/k)$ and $(\rho VD/\mu)$. The appearance of these parameters foreshadows our developments in Chapter 8, where you will see how dimensionless parameters and correlations among them are of fundamental engineering importance. Chapter 9 deals explicitly with such correlations for external flows. The wind blowing over the kinglet is a flow of this type.

Example 4.12 SI Engine Application

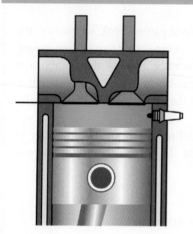

Consider a spark-ignition engine having a disk-shaped combustion chamber as shown in the sketch. The following empirical relationship [14] relates the instantaneous convective heat-transfer coefficient for the combustion chamber to the mean piston speed $\bar{V}_{\text{pist}}$, the engine bore B, and thermophysical properties of the gases in the cylinder, ρ, k, and μ:

$$\frac{\bar{h}_{\text{conv}} B}{k} = 0.5 \left(\frac{\rho \bar{V}_{\text{pist}} B}{\mu} \right)^{0.7}.$$

The mean piston speed is calculated from the engine rotational speed (N) and stroke (S) as

$$\bar{V}_{\text{pist}} = 2SN.$$

The characteristic gas temperature used in Eq. 4.18 with the $\bar{h}_{\text{conv}}$ from our equation here is the bulk-mean temperature defined by

$$\bar{T}_{\text{gas}} = \frac{P\mathcal{V}}{MR}.$$

The gas density ρ, viscosity μ, and thermal conductivity k are all evaluated at $\bar{T}_{\text{gas}}$. The engine has a compression ratio $CR = 9$, and the combustion chamber walls are all at 400 K. Other data include

$N = 2500$ rev/min,	$T_{\text{in}} = 300$ K,
$B = 102$ mm,	$P_{\text{in}} = 100$ kPa,
$S = 88$ mm,	$P_{\text{TC}} = 3800$ kPa.

Estimate the instantaneous heat-transfer rate from the combustion gases to the combustion chamber walls when the piston is at top center.

Solution

Known   Correlation for $\bar{h}_{\text{conv}}$, definitions for $\bar{V}_{\text{pist}}$ and $\bar{T}_{\text{gas}}$, N, B, S, CR, T_{in}, P_{in}, T_{TC}, P_{TC}, T_{wall}

Find $\dot{Q}_{\text{gas–walls}}$

Sketch

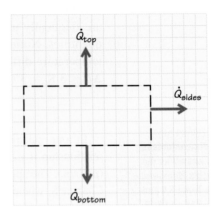

Assumptions

i. The combustion gas properties are the same as air.
ii. The gas is ideal.
iii. The cylinder is filled with a mixture at P_{in} and T_{in} at bottom center.
iv. Transport properties μ and k are independent of pressure.

Analysis   We divide the problem into several pieces. First, we analyze the geometry to obtain values for the volumes at top center and bottom center, V_{TC} and V_{BC}, respectively, and for the surface area of the combustion chamber, A_{surf}. We then apply the ideal-gas law to find $\overline{T}_{gas}$ at top center. Knowing $\overline{T}_{gas}$, we can calculate $\overline{h}_{conv}$ from the given correlation and finally calculate $\dot{Q}_{gas-walls}$ from Eq. 4.18.

Using the various engine geometry definitions from the appendix to Chapter 1, we find the displacement (Eq. 1A.2) and the volume at top center (Eq. 1A.3 rearranged),

$$V_{disp} = S\pi B^2/4$$
$$= 0.088\pi(0.102)^2/4 \text{ m}^3$$
$$= 7.19 \times 10^{-4} \text{ m}^3$$

and

$$V_{TC} = \frac{V_{disp}}{CR - 1}$$
$$= \frac{7.19 \times 10^{-4}}{9 - 1} \text{ m}^3$$
$$= 8.99 \times 10^{-5} \text{ m}^3.$$

Since the combustion chamber is disk shaped, the heat-transfer area consists of a top and bottom circular area ($= \pi B^2/4$) and a circumferential strip equal to πB times the clearance height. The clearance height z is easily found since V_{TC} is the product of the core cross-sectional area and the clearance height; thus,

$$z = \frac{V_{TC}}{\pi B^2/4}$$
$$= \frac{4(8.99 \times 10^{-5})}{\pi(0.102)^2} \text{ m}$$
$$= 0.011 \text{ m or 11 mm.}$$

The total heat-transfer area can now be calculated as

$$A_{\text{surf}} = 2\left(\frac{\pi B^2}{4}\right) + \pi B z$$

$$= \left[\frac{2\pi(0.102)^2}{4} + \pi(0.102)0.011\right] \text{m}^2$$

$$= 0.0199 \text{ m}^2.$$

We now find $\overline{T}_{\text{gas}}$. Applying the ideal-gas equation of state (Eq. 2.28c), we write

$$\overline{T}_{\text{gas}} = \frac{P_{\text{TC}}V_{\text{TC}}}{MR},$$

where R ($= R_u/\mathcal{M}$) is the gas constant for air and has a value of 287.0 J/kg·K (see Appendix C). All quantities in this expression are known except for the mass trapped within the cylinder. This mass, however, can be found from our simplifying assumption that the cylinder is filled with air at the inlet conditions at bottom center:

$$M = \frac{P_{\text{in}}V_{\text{BC}}}{T_{\text{in}}R}.$$

Substituting this expression into our expression for $\overline{T}_{\text{gas}}$ yields

$$\overline{T}_{\text{gas}} = T_{\text{in}}\frac{P_{\text{TC}}V_{\text{TC}}}{P_{\text{in}}V_{\text{BC}}}.$$

Recognizing that $V_{\text{BC}}/V_{\text{TC}}$ is the compression ratio CR, we have

$$\overline{T}_{\text{gas}} = \frac{T_{\text{in}}P_{\text{TC}}}{CR P_{\text{in}}}$$

$$= \frac{300(3800 \times 10^3)}{9(100 \times 10^3)} \text{ K}$$

$$= 1267 \text{ K}.$$

We now evaluate the convective heat-transfer coefficient $\overline{h}_{\text{conv}}$ from the given empirical expression:

$$\frac{\overline{h}_{\text{conv}}B}{k} = 0.5\left(\frac{\rho \overline{V}_{\text{pist}}B}{\mu}\right)^{0.7}.$$

We calculate the mean piston speed from the definition in the problem statement:

$$\overline{V}_{\text{pist}} = 2SN$$

$$= 2(0.088)\frac{2500}{60}$$

$$= 7.33$$

$$[=] \text{m (rev/min)}\left[\frac{1 \text{ min}}{60 \text{ s}}\right] = \text{m/s},$$

where we recognize that *revolution* is dimensionless. Values for the viscosity and thermal conductivity at 1267 K, found by interpolation from Table C.3, are

$$\mu = 50.68 \times 10^{-6} \text{ N·s/m}^2,$$

$$k = 0.0826 \text{ W/m·K}.$$

The density is obtained from the ideal-gas equation of state as

$$\rho = \frac{P_{TC}}{RT_{gas}} = \frac{3800 \times 10^3}{287.0(1267)} \text{ kg/m}^3$$

$$= 10.45 \text{ kg/m}^3.$$

The right-hand side of the empirical correlation is evaluated to give

$$\frac{\overline{h}_{conv}B}{k} = 0.5\left(\frac{10.45(7.33)0.102}{50.68 \times 10^{-6}}\right)^{0.7}$$

$$= 0.5(1.54 \times 10^5)^{0.7} = 2141$$

$$[=]\left(\frac{(\text{kg/m}^3)\,(\text{m/s})\,\text{m}}{\text{N}\cdot\text{s/m}^2}\left[\frac{1\,\text{N}}{\text{kg}\cdot\text{m/s}^2}\right]\right)^{0.7} = 1(\text{dimensionless}).$$

The convective heat-transfer coefficient is thus

$$\overline{h}_{conv} = \frac{k}{B}2141$$

$$= \frac{0.0826}{0.102}2141 = 1734$$

$$[=]\frac{\text{W/m}\cdot\text{K}}{\text{m}} = \text{W/m}^2\cdot\text{K}.$$

Finally, we calculate the instantaneous heat-transfer rate from the hot gases to the cold walls using Eq. 4.18:

$$\dot{Q}_{gas-walls} = \overline{h}_{conv}A_{surf}(\overline{T}_{gas} - T_{wall})$$

$$= 1734\,(0.0199)\,(1267 - 400)$$

$$= 29,900$$

$$[=]\,(\text{W/m}^2\cdot\text{K})(\text{m}^2)\text{K} = \text{W},$$

or

$$29.9\text{ kW.}$$

Comments   The calculated heat-transfer rate is a relatively large value, say, compared to household appliances such as a 1.6-kW electrical space heater. More importantly, the total heat transfer from the hot gases to the coolant in typical automotive engines is a substantial percentage of the chemical energy supplied by the fuel. Heywood [15] cites percentages ranging from 17% to 26%. To calculate the total heat transfer requires that we integrate over the entire mechanical cycle, that is,

$$Q_{total}\,(\text{J}) = \oint \dot{Q}dt$$

$$= \oint \overline{h}_{conv}(t)A_{surf}(t)[T_{gas}(t) - T_{wall}]\,dt,$$

where we recognize that $\overline{h}_{conv}$, A_{surf}, and T_{gas} all vary throughout the cycle.

Notice once again the use of the two dimensionless parameters $(\rho\overline{V}_{pist}B/\mu$ and $\overline{h}_{conv}B/k)$ to correlate geometry, flow conditions, and thermophysical fluid properties.

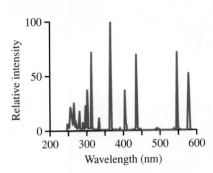

Radiant energy from a mercury vapor lamp is concentrated at several narrow regions of the electromagnetic spectrum.

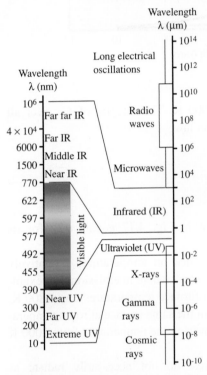

Visible light occupys only a small region of the complete electromagnetic spectrum.

Diffuse emitter

4.5c Radiation

Of the three modes of heat transfer, radiation is the most difficult to describe in general terms. This difficulty arises, in part, because of the wavelength dependence of both the radiant energy and the radiant properties of substances. Also contributing is the action-at-a-distance nature of radiation. Unlike conduction, which depends only on the local material conductivity and local temperature gradient (see Eq. 4.13), radiant energy can be exchanged without any intervening medium and at distances that may be long relative to the dimensions of the device under study. A fortunate example of this is the radiant energy that travels 93 million miles from the sun to the earth through the essential vacuum of outer space. In this chapter, we limit our scope in calculating radiation to a special situation in which many simplifying assumptions have been made to produce a relatively simple result. A more expanded view can be found in Ref. [13]. Several comprehensive textbooks are available for those interested in a detailed treatment [16, 17].

Before presenting a radiation "rate law," we need to define what is meant by a blackbody and a gray surface. You may have encountered the concept of a blackbody in a physics course, although the concept of a gray surface is likely to be new. A **blackbody** has the following properties:

1. It absorbs all incident radiation (hence, the appellation black).

2. It emits the maximum possible radiation at every wavelength for its temperature.

3. It emits radiation in accord with the Stefan–Boltzmann law,

$$E_b = \sigma T^4, \tag{4.20}$$

where E_b is the radiant energy emitted per unit area by the blackbody, or **blackbody emissive power** ($[=] \, W/m^2$), T is the absolute temperature, and σ is the Stefan–Boltzmann constant, given by

$$\sigma = 5.67051 \times 10^{-8} \, W/m^2 \cdot K^4.$$

Furthermore, the intensity of the emitted radiation is independent of direction; thus, a blackbody is said to be a **diffuse emitter**.

To understand the concept of a **gray surface**, we need more information about the characteristics of a blackbody. Planck's distribution law gives the spectral distribution of the radiation emitted by a blackbody as follows:[7]

$$E_{\lambda b} = \frac{2\pi \hbar c_0^2}{\lambda^5 (e^{\hbar c_0/\lambda k_B T} - 1)}, \tag{4.21}$$

where $E_{\lambda b}$ is the spectral blackbody radiant emission per unit area per unit wavelength, or **blackbody spectral emissive power** $[=] \, W/m^2 \cdot \mu m$; λ is the wavelength; $\hbar$ is Planck's constant; c_0 is the vacuum speed of light; and k_B is the Boltzmann constant. Figure 4.17 is a plot of Eq. 4.21 for blackbody

---

[7] Max Planck won the Nobel Prize in 1918 for the discovery of energy quanta. Such quanta were necessary to explain the spectral radiation characteristics of a blackbody.

FIGURE 4.17

Planck's spectral distribution of radiation from a blackbody at temperatures of 1000 and 2000 K.

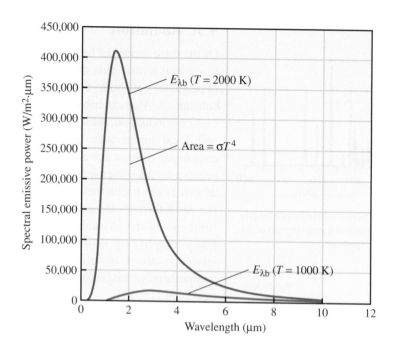

radiation from surfaces at 1000 and 2000 K. The integral of $E_{\lambda b}$ over all wavelengths yields the Stefan–Boltzmann law,

$$\int_0^\infty E_{\lambda b}\, d\lambda = \sigma T^4. \tag{4.22}$$

Graphically, the area under each curve in Fig. 4.17 equals σT^4 for its respective temperature.

Because of the fourth-power dependence on temperature, a relatively small change in temperature results in a large change in emission. For example, the emission from a black surface at furnace temperatures, say, 1800 K, is 1330 times more than the emission from the same surface at room temperature (298 K). The significant effect of lowering the temperature from 2000 K to 1000 K is clearly seen in Fig. 4.17.

Many surfaces of engineering interest do not necessarily radiate as blackbodies. In fact, the emission from real surfaces frequently displays a complex wavelength dependence that greatly complicates a detailed analysis [16, 17]. A common simplifying assumption is that the emission from a surface is a constant fraction of that from a blackbody. Such a surface is termed a **gray** surface. Figure 4.18 illustrates the spectral distribution of radiant emission from black and gray surfaces at 2000 K. The primary property characterizing a gray surface is the **emissivity**. This dimensionless quantity expresses the ratio of the gray surface spectral emissive power (E_λ) to that of a blackbody ($E_{\lambda b}$):

$$\varepsilon \equiv \frac{E_\lambda}{E_{\lambda b}} = \text{constant.} \tag{4.23}$$

Since the emissivity is independent of wavelength, it follows that

$$\int_0^\infty E_\lambda\, d\lambda = \varepsilon \sigma T^4, \tag{4.24}$$

In this iron bar, the highest temperatures occur where the visible radiation is the brightest.

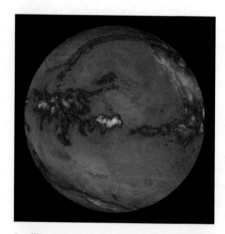

Satellite image of long-wavelength (infrared) radiation from Earth. Radiation from cold clouds is indicated in blue, whereas that from the relatively warm oceans is shown in red. Radiation from the southwestern United States exceeds that from the oceans. *Image courtesy of NASA Goddard.*

FIGURE 4.18
The radiant emission from a gray surface is a fixed fraction of that from a blackbody at every wavelength. The emissivity ε is the ratio of gray surface emission to the blackbody emission.

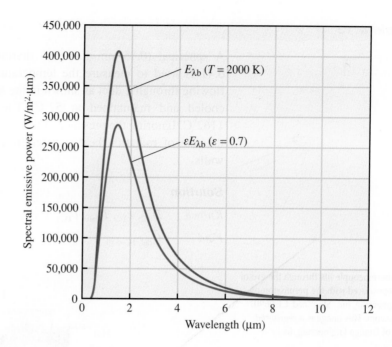

which simply states that the total emissive power of a gray surface is the product of the emissivity and the blackbody emissive power at the same temperature.

We now define a simplified rate law for radiation heat transfer for a solid surface that exchanges radiant energy with its surroundings. The net radiation heat transfer from the surface to the surroundings can be expressed as

$$\dot{Q}_{rad} = \varepsilon_s A_s \sigma \left(T_s^4 - T_{surr}^4\right), \tag{4.25}$$

where ε_s is the emissivity of the surface, A_s is the area of the surface, and T_s and T_{surr} are the absolute temperatures of the surface and surroundings, respectively.

Use of Eq. 4.25 to estimate radiation heat-transfer rates relies on the following conditions being met:

1. The surface and the surroundings are, respectively, isothermal; thus, one single temperature characterizes the surface and another single temperature characterizes the surroundings.

2. All of the radiation that leaves A_s is incident upon the surroundings; that is, there are no intervening surfaces that intercept a portion of the radiation from A_s and there is no radiation leaving any one region of A_s that is incident upon any other region of A_s. Furthermore, any medium between the two surfaces (e.g., air) neither absorbs nor emits any radiation.

3. The surroundings behave as a blackbody *or* the area of the surroundings is much greater than A_s.

4. Surface A_s is a diffuse-gray emitter and reflector.

All of the examples involving radiation in this book are such that these four conditions are met or approximated. To deal with more complex situations, we refer the interested reader to Ref. [13].

Example 4.13

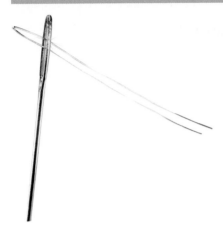

Miniature thermocouple fits through the eye of a needle. Reproduced with the permission of Omega Engineering, Inc., Stamford, CT 06907 www.omega.com. This image is a registered trademark of Omega Engineering, Inc.

A spherical (0.75-mm-diameter) thermocouple with an emissivity of 0.41 is used to measure the temperature of hot combustion products flowing through a duct as shown in the sketch. The walls of the duct are cooled and maintained at 52°C. The thermocouple temperature is 1162°C. Ignoring the presence of the thermocouple lead wires, determine the rate of radiation heat transfer from the thermocouple to the duct walls.

Solution

Known $D_{TC}, \varepsilon_{TC}, T_{wall}, T_{TC}$

Find $\dot{Q}_{rad,\,TC–wall}$

Sketch

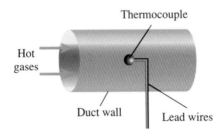

Thermocouple

Hot gases

Duct wall   Lead wires

Assumptions

i. The effects of lead wires are negligible.
ii. All of the restrictions associated with Eq. 4.25 are met.
iii. The combustion product gases neither absorb nor emit any radiation.

Analysis We apply Eq. 4.25, recognizing that T_s is the thermocouple temperature, T_{TC}, and that T_{surr}, the surroundings temperature, is identified with the duct wall temperature, that is, $T_{surr} \equiv T_{wall}$. Since absolute temperatures are required for Eq. 4.25, we convert the given temperatures:

$$T_{wall} = 52 + 273 = 325 \text{ K},$$
$$T_{TC} = 1162 + 273 = 1435 \text{ K}.$$

The radiant heat-transfer rate is calculated from

$$\dot{Q} = \varepsilon_{TC} A_{TC}\, \sigma\, (T_{TC}^4 - T_{wall}^4),$$

where A_{TC} is the surface area of the spherical thermocouple $(= \pi D^2)$. Substituting numerical values, we obtain

$$\dot{Q} = 0.41\, \pi(0.00075)^2\, 5.67 \times 10^{-8}\, [(1435)^4 - (325)^4]$$
$$= 0.174$$
$$[=]\, \text{m}^2(\text{W/m}^2 \cdot \text{K}^4)\text{K}^4 = \text{W}.$$

Comments In our analysis, we assumed that the hot gases themselves do not participate in the exchange of radiation and that the only exchange occurs between the two solid surfaces, the thermocouple and the duct wall. This is an excellent assumption for diatomic molecules such as N_2 and O_2, the primary components of air. Asymmetrical molecules, such as CO, CO_2, and H_2O, which are found in combustion products, can also participate in the radiation process; thus our analysis is a first approximation to the actual radiation exchange associated with the thermocouple. Treatment of participating media (i.e., gas molecules and particulate matter), can found in advanced textbooks [16, 17].

Self Test 4.9

☑ Consider a cast-iron stove with emissivity $\varepsilon = 0.44$ and surface area $A_s = 16$ ft² located in a large room. If the stove surface temperature is 500 F and the room wall temperature is 60 F, determine the radiation heat-transfer rate to the room walls.

(Answer: $\dot{Q}_{rad} = 9380$ Btu/h)

Example 4.14

Consider the same situation described in Example 4.13. For steady state to be achieved, the heat lost by the thermocouple must be balanced by an energy input. This input comes from convection from the hot gases to the thermocouple, so,

$$\dot{Q}_{conv} = \dot{Q}_{rad}.$$

Assuming an average heat-transfer coefficient of 595 W/m²·K, estimate the temperature of the hot flowing gases.

Solution

Known $\dot{Q}_{rad}$ (Ex. 4.13), $\bar{h}_{conv}$, D_{TC}

Find T_{gas}

Sketch

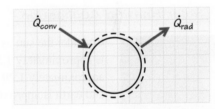

Assumptions

i. The situation is steady state.
ii. The effects of the lead wires are negligible.
iii. The combustion products neither absorb nor emit any radiation.

Analysis   Given that

$$\dot{Q}_{conv} = \dot{Q}_{rad},$$

we substitute the convection rate law (Eq. 4.18) and write

$$\bar{h}_{conv} A_{TC}(T_{gas} - T_{TC}) = \dot{Q}_{rad}.$$

Here we recognize that the gas must be hotter than the thermocouple for the heat transfer to be in the direction shown; thus, the appropriate temperature difference is $T_{gas} - T_{TC}$. Solving for T_{gas} yields

$$T_{gas} = \frac{\dot{Q}_{rad}}{\bar{h}_{conv} A_{TC}} + T_{TC}.$$

Recognizing that the surface area of the thermocouple is πD_{TC}^2, we numerically evaluate this as

$$T_{\text{gas}} = \frac{0.174}{595\,\pi(0.00075)^2} + 1435$$
$$= 1600$$
$$[=]\frac{W}{(W/m^2 \cdot K)\,m^2} = K.$$

Comments This example foreshadows the power of the principle of conservation of energy, from which the statement that $\dot{Q}_{\text{conv}}$ equals $\dot{Q}_{\text{rad}}$ was derived. We also see the inherent difficulty in making temperature measurements in high-temperature environments, as the presence of radiation from the thermocouple to the relatively cold wall results in a substantial error in the hot-gas temperature measurement. In this specific example, the error is 165 K.

Self Test 4.10

☑ **The air temperature of a house is kept at 22°C year-round. Determine the rate of radiation heat transfer from a person ($A_s = 1.2$ m²) to the walls of a room (a) during summer when the walls are at 28°C and (b) during winter when the walls are at 18°C. Discuss.**

(Answer: (a) 66.4 W, (b) 133.6 W. With more heat loss by radiation during the winter than in the summer, a person will feel colder, even though the room air temperature is the same.)

Example 4.15 Solar Heating Application

Consider the sun as a blackbody radiator at 5800 K. The diameter of the sun is approximately 1.39×10^9 m (864,000 miles), the diameter of the earth is approximately 1.29×10^7 m (8020 miles), and the distance between the sun and earth is approximately 1.5×10^{11} m (9.3×10^7 miles).

A. Calculate the total energy emitted by the sun in watts.

B. Calculate the amount of the sun's energy intercepted by the earth.

C. Calculate the solar flux in W/m² at the upper reaches of the earth's atmosphere for the sun's rays striking a surface perpendicularly. Compare this quantity with the maximum solar flux of 720 W/m² recorded during a clear day in Fort Collins, Colorado.

Solution

Known $T_{\text{sun}}, D_{\text{sun}}, D_{\text{earth}}, X_{\text{sun–earth}}$

Find $\dot{Q}_{\text{sun}}, \dot{Q}_{\text{sun–earth}}, \dot{Q}''_{\text{sun @earth}}$

Assumptions

i. The sun emits as a blackbody.

ii. The sun's energy is radiated uniformly in all directions.

iii. The sun is sufficiently far from the earth that the sun's rays are essentially parallel when they strike the earth.

Sketch

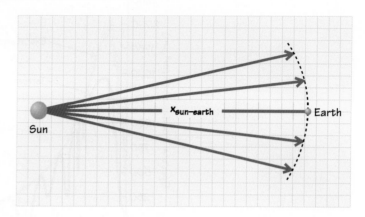

Analysis

A. Given that the sun is a blackbody, the total energy emitted is the product of the sun's emissive power (Eq. 4.20) and surface area, πD^2_{sun}:

$$\dot{Q}_{sun} = E_{b,\,sun} A_{surf,\,sun}$$
$$= \sigma T^4_{sun} \pi D^2_{sun}.$$

Substituting numerical values yields

$$\dot{Q}_{sun} = 5.67 \times 10^{-8}(5800)^4\pi(1.39 \times 10^9)^2 = 3.89 \times 10^{26}$$
$$[=](W/m^2 \cdot K^4)K^4 m^2 = W.$$

B. Referring to our sketch, we see that all of the energy from the sun is distributed over the interior surface of a sphere having a radius equal to the sun–earth distance. The fraction of this energy intercepted by the earth is the ratio of the projected area of the earth, $\pi D^2_{earth}/4$, to the total area, $4\pi x^2_{sun-earth}$. Thus, the energy intercepted by the earth is

$$\dot{Q}_{sun-earth} = \dot{Q}_{sun}\frac{\pi D^2_{earth}/4}{4\pi x^2_{sun-earth}}$$
$$= \frac{\dot{Q}_{sun}}{16}\left(\frac{D_{earth}}{x_{sun-earth}}\right)^2$$
$$= \frac{3.89 \times 10^{26}}{16}\left(\frac{1.29 \times 10^7}{1.5 \times 10^{11}}\right)^2 W$$
$$= 1.80 \times 10^{17} W.$$

C. We now apply the definition of a heat flux (Eq. 4.15) and evaluate this expression using the surface area associated with the sphere having the sun–earth distance as its radius:

$$\dot{Q}''_{sun@earth} = \frac{\dot{Q}_{sun}}{4\pi x^2_{sun-earth}}$$
$$= \frac{3.89 \times 10^{26} W}{4\pi(1.5 \times 10^{11})^2 m^2}$$
$$= 1376 W/m^2.$$

Comments First we note the incredibly large rate at which energy is emitted by the sun. The energy emitted by the sun in just 38 ns is equivalent to the U.S. electricity production for an entire year (cf. Table 1.1)!

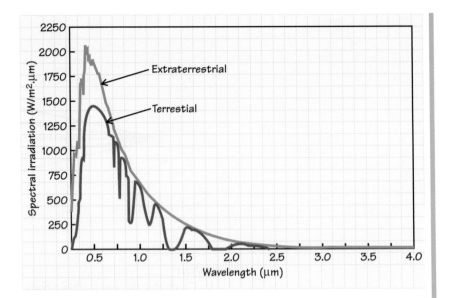

Comparing our calculated solar flux of 1376 W/m² with the measured peak value for a North American location, we see that the measured value (720 W/m²) is substantially less. There are several factors that contribute to this difference. First, the sun's rays at a northern latitude strike the earth's surface obliquely, thus spreading the energy over a greater area. In addition, the sun's radiation is attenuated from absorption and scattering by molecules and particles in the atmosphere. (See sketch.) In the design of solar heating equipment, historical records of measured solar energy at the location of interest are used because of the near impossibility of making theoretical predictions. Local weather conditions also play a major role in determining the average solar flux at a location.

SUMMARY

This chapter began by considering the various ways in which energy is stored in systems and control volumes. We saw the importance of distinguishing between the bulk system energy, which is associated with the system as a whole, and the internal energy, which is associated with the constituent molecules. We also carefully defined heat and work and their time rates. We saw that both of these quantities are energy transfers that occur only at boundaries of systems and control volumes. Neither is possessed by or a property of a system or control volume. Various modes of working were discussed along with the three principal modes of heat transfer: conduction, convection, and radiation. To consolidate your knowledge further, reviewing the learning objectives presented at the beginning of this chapter is recommended.

Chapter 4
Key Concepts & Definitions Checklist[8]

4.2 System and Control-Volume Energy

❑ Bulk versus internal energies ➤ *Q4.4*

❑ Linear and rotational kinetic energies ➤ *4.1*

❑ Potential energy ➤ *4.1*

❑ Molecular description of internal energy ➤ *Q4.5*

4.3 Energy Transfer Across Boundaries

4.3a Heat

❑ Definition of heat (or heat transfer) ➤ *Q4.6*

❑ Driving force for heat transfer ➤ *Q4.7, Q4.8*

❑ Distinctions among heat, heat-transfer rate, and heat flux ➤ *Q4.9, Q4.22*

❑ Identification of heat interactions in real devices and systems ➤ *4.47*

4.3b Work

❑ Definition of work ➤ *Q4.13*

❑ Definition of power ➤ *Q4.14*

❑ Five common types of work (list) ➤ *Q4.15*

❑ P–$d\mathcal{V}$ (compression and expansion) work ➤ *4.8, 4.9*

❑ Viscous work ➤ *4.26*

❑ Shaft work ➤ *4.15*

❑ Electrical work ➤ *4.27*

❑ Flow work ➤ *4.28*

❑ Identification of work interactions in real devices and systems ➤ *4.31*

4.4 Sign Conventions and Units

❑ Units for energy and energy rates ➤ *Q4.16*

4.5 Rate Laws for Heat Transfer

4.5a Conduction

❑ Fourier's law: 1-D Cartesian form ➤ *Q4.19, 4.33*

❑ Fourier's law: 1-D cylindrical and spherical forms ➤ *4.34*

❑ Fourier's law: general vector form ➤ *Q4.18*

❑ Heat flux vector ➤ *Q4.18*

❑ Temperature gradient vector ➤ *Q4.18*

4.5b Convection ➤ *4.38, 4.44*

❑ Convection definition (Eq.4.17) ➤ *4.37*

❑ Average heat-transfer coefficient ➤ *Q4.21*

❑ Local heat-transfer coefficient ➤ *Q4.21*

❑ Forced convection ➤ *Q4.25*

❑ Free convection ➤ *Q4.25*

4.5c Radiation

❑ Blackbody ➤ *Q4.26, 4.46*

❑ Emissive power ➤ *Q4.28*

❑ Gray surface ➤ *Q4.29, Q4.30*

❑ Spectral emissive power ➤ *Q4.29, Q4.30*

❑ Emissivity ➤ *Q4.29, Q4.30*

❑ Net radiation exchange rate (Eq. 4.25) ➤ *Q4.31, 4.47*

[8] Numbers following arrows refer to Questions (prefaced with a Q) and Problems at the end of the chapter.

P	pressure		x	local value
rad	radiation		1	station 1
s	surface		2	station 2
shaft	shaft		∞	ambient or freestream value
surr	surroundings			
sys	system			
visc	viscous			
wall	at the wall			

OTHER

$\nabla( )$	Gradient operator
$(\overline{})$	Average value

QUESTIONS

4.1 Review the most important equations presented in this chapter (i.e., those with a red background). What physical principles do they express? What restrictions apply?

4.2 Name three pioneers in the thermal-fluid sciences and indicate their major contributions. Provide an approximate date for these contributions.

4.3 List and define all of the boldfaced words in this chapter.

4.4 Distinguish between the bulk and internal energies associated with a thermodynamic system. Give a concrete example of a system that possesses both.

4.5 From a molecular point of view, how is energy stored in a volume of gas composed of monatomic species? How is energy stored in diatomic gas molecules? In triatomic gas molecules?

4.6 From memory, write out a formal definition of heat or heat transfer. Compare your definition with that in the text.

4.7 What thermodynamic property "drives" a heat-transfer process?

4.8 Explain the difference between a temperature difference, ΔT, and a temperature gradient, ∇T.

4.9 How does heat (or heat transfer) Q differ from the heat-transfer rate $\dot{Q}$?

4.10 List the units for Q, $\dot{Q}$, and $\dot{Q}''$.

4.11 What distinguishes work from internal energy? What distinguishes heat from internal energy? Discuss.

4.12 What distinguishes work from heat? Discuss.

4.13 From memory, write out a formal definition of work. Compare your definition with that in the text.

4.14 How does work W differ from power $\dot{W}$ (or $\mathscr{P}$)? What units are associated with each?

4.15 List, define, and discuss five types of work. For example, one of these should be shaft work.

4.16 Provide the units associated with the following quantities: E, u, W, $\dot{W}$, Q, and $\dot{Q}''$.

4.17 List the three modes of heat transfer. Compare and contrast these modes.

4.18 Define all of the terms appearing in Fourier's law (Eqs. 4.13 and 4.14). What is the origin of the minus sign?

4.19 Consider the steady-state temperature distribution $T(x)$ in a solid wall as shown in the sketch. Assume the thermal conductivity is uniform throughout the wall.

A. At what location is the heat-transfer rate largest? Smallest? Explain.

B. What is the direction of the heat transfer at location A? At location C? Explain.

C. What special significance, if any, is associated with the plane B? Discuss.

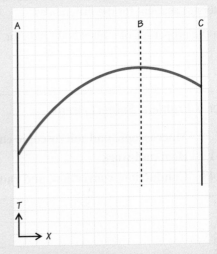

4.20 The temperature distribution $T(x)$ through a solid wall is shown in the sketch. Using the grid given, estimate numerical values for the temperature gradient dT/dx at $x = 0$ and $x = 0.5$ m. Give units. Would the numerical values for dT/dx change if $T(x)$ were given in degrees Celsius rather than kelvins? Discuss.

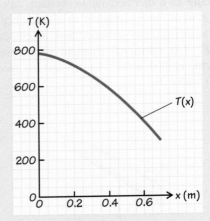

4.21 Using symbols, define the local heat-transfer coefficient $h_{\text{conv}, x}$ and the average heat-transfer coefficient $\bar{h}_{\text{conv}}$. Under what conditions would $h_{\text{conv}, x}$ and $\bar{h}_{\text{conv}}$ have the same numerical values? Discuss.

4.22 Explain the difference between heat-transfer rate $\dot{Q}$ and heat flux $\dot{Q}''$.

4.23 Consider the one-dimensional, steady-state heat conduction through a long hollow cylinder of insulating material (see Fig. 4.14b). The temperature at the inner radius, $T_1 = T(r_1)$, is higher than the temperature at the outer radius, $T_2 = T(r_2)$. If the heat-transfer rate is the same at all radial locations, sketch $\dot{Q}$, $\dot{Q}''$, and T as functions of r. Discuss.

4.24 Repeat Question 4.23 for a planar (Cartesian) geometry. Discuss.

4.25 Compare typical values of heat-transfer coefficients for free convection and forced convection. What factors(s) account for the differences?

4.26 List the properties of a blackbody radiator.

4.27 Radiation from the sun strikes a black surface. Is it possible to determine the fraction of the solar energy that is reflected from the surface? If so, what is the fraction?

4.28 Define the blackbody emissive power. What units are associated with this quantity?

4.29 Discuss the attributes of the radiant energy emitted by a gray surface.

4.30 Explain the concept of emissivity.

4.31 What are the restrictions implied in the use of the expression $\dot{Q}_{rad} = \varepsilon_s A_s \sigma (T_s^4 - T_{surr}^4)$?

Chapter 4 Problem Subject Areas

4.1–4.7	Macroscopic and microscopic energies
4.8–4.31	Heat and work
4.32–4.48	Conduction, convection, and radiation

PROBLEMS

4.1 A 3.084-kg steel projectile is fired from a gun. At a particular point in its trajectory it has a velocity of 300 m/s, an altitude of 700 m, and a temperature of 350 K. The projectile is spinning about its longitudinal axis at an angular speed of 100 rad/s. Treating the projectile as a thermodynamic system, calculate the total system energy relative to a reference state of zero velocity, zero altitude, and a temperature of 298 K. The rotational moment of inertia for the projectile is 9.64×10^{-4} kg·m² and the specific heat is 460.5 J/kg·K.

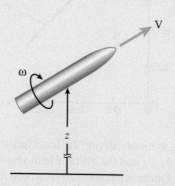

4.2 Determine the kinetic energy (kJ) of a 90-kg object moving at a velocity of 10 m/s on a planet where $g = 7$ m/s².

4.3 Initially, 1 lb_m of water is at $P_1 = 14.7$ psia, $T_1 = 70$ F, $v_1 = 0$ ft/s, and $z_1 = 0$ ft. The water then undergoes a

process that ends at $P_2 = 30$ psia, $T_2 = 700$ F, $v_2 = 100$ ft/s, and $z_2 = 100$ ft. Determine the increases in internal energy, potential energy, and kinetic energy of the water in Btu/lb_m. Compare the increases in internal energy, potential energy, and kinetic energy.

4.4 A 100-kg man wants to lose weight by increasing his energy consumption through exercise. He proposes to do this by climbing a 6-m staircase several times a day. His current food consumption provides him an energy input of 4000 Cal/day. (Note: 1 Cal = unit for measuring the energy released by food when oxidized in the body = 1 kcal.) Determine the vertical distance (m) the man must climb to convert his food intake into potential energy. What is your advice regarding exercise versus a reduced food intake to effect a weight loss?

4.5 Consider a 25-mm-diameter steel ball that has a density of 7870 kg/m³ and a constant-volume specific heat of the form

$$c_v \ (\text{J/kg·K}) = 447 + 0.4233 \ [T(\text{K}) - 300].$$

What velocity would the ball have to achieve such that its bulk kinetic energy equals its internal energy at 500 K? Use an internal energy reference state of 300 K.

4.6 For the temperature range 1000–5000 K, the ratio of the molar constant-pressure specific heat to the universal gas constant is given as follows for CO_2:

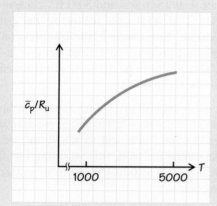

$$\bar{c}_p/R_u = 4.45362 + (3.140168 \times 10^{-3})T$$
$$- (1.278410 \times 10^{-6})T^2 + (2.393996$$
$$\times 10^{-10})T^3 - (1.6690333 \times 10^{-14})T^4,$$

where $T[=]$ K. For 1 kg of CO_2, plot $U(T) - U$ (1000 K) versus T up to 5000 K. Discuss.

4.7 Consider nitrogen gas contained in a piston–cylinder arrangement as shown in the sketch. The cylinder diameter is 150 mm. At the initial state the temperature of the N_2 is 500 K, the pressure is 100 kPa, and the height of the gas column within the cylinder is 200 mm. Heat is removed from the N_2 by a quasi-static, constant-pressure process until the temperature of the N_2 is 300 K. Calculate the work done by the N_2 gas.

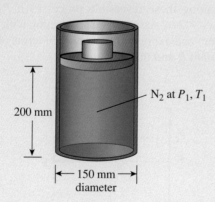

4.8 Saturated steam at 2 MPa is contained in a piston–cylinder arrangement as shown in the sketch. The initial volume is 0.25 m³. Heat is removed from the steam during a quasi-static, constant-pressure process such that 197.6 kJ of work is done by the

surroundings on the steam. Determine the quality of the steam at the final state.

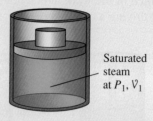

4.9 A polytropic process is one that obeys the relationship $P\mathcal{V}^n = $ constant, where n may take on various values.

A. Define the type of process associated with each of the following values of n assuming the working fluid is an ideal gas (i.e., determine which state variable is constant):

   i. $n = 0$
   ii. $n = 1$
   iii. $n = \gamma \; (\equiv c_p/c_v)$

B. Starting from the same initial state, which process results in the most amount of work performed by the gas when the gas is expanded quasi-statically to a final volume twice as large as the initial volume? Which process results in the least amount of work?

C. Sketch the three processes on P–$\mathcal{V}$ coordinates.

4.10 A 500-lb_m box is towed at a constant velocity of 5 ft/s along a frictional, horizontal surface by a 200-lb_f horizontal force. Consider the box as a system.

A. Draw a force and motion sketch showing directions and magnitudes of all forces (lb_f) acting on the system and the velocity (ft/s) vector.

B. Draw an energy sketch showing magnitudes and directions of all rates of work (hp).

4.11 Initially, 0.05 kg of air is contained in a piston–cylinder device at 200°C and 1.6 MPa. The air then expands at constant temperature to a pressure of 0.4 MPa. Assume the process occurs slowly enough that the acceleration of the piston can be neglected. The ambient pressure is 101.35 kPa.

A. Determine the work (kJ) performed by the air in the cylinder on the piston.

B. Determine the work (kJ) performed by the piston on the ambient environment. Neglect the cross-sectional area of the connecting rod.

C. Determine the work transfer (kJ) from the piston to the connecting rod. Neglect friction between the piston and cylinder. Assume that no heat transfer to or from the piston occurs and that the energy of the piston does not change.

D. Discuss why the assumption of negligible piston acceleration was made.

E. Which of the work interactions might be useful for driving a car? Why?

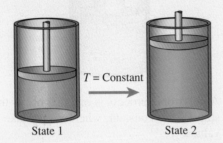

State 1   State 2

4.12 Initially, 0.5 kmol of an ideal gas at 20°C and 2 atm is contained in a piston–cylinder device. The piston is held in place by a pin. The pin is then removed and the gas expands rapidly. After a new equilibrium is attained, the gas is at 20°C and 1 atm. Taking the gas and piston as the system, determine the net work transfer (kJ) to the atmosphere.

4.13 A spherical balloon is inflated from a diameter of 10 in to a diameter of 20 in by forcing air into the balloon with a tire pump. The initial pressure inside the balloon is 20 psia. During the process the pressure is proportional to the balloon diameter.

A. Determine the amount of work transfer (Btu) from the air inside the balloon to the balloon.

B. Determine the amount of work transfer (Btu) from the balloon to the surrounding atmosphere.

4.14 Initially, 3 lb$_m$ of steam is contained in a piston–cylinder device at 500 F with a quality of 0.70. Heat is added at constant pressure until all of the liquid is vaporized. The steam then expands adiabatically at constant temperature to a pressure of 400 psia, behaving essentially as an ideal gas. The process occurs slowly enough that the acceleration of the piston may be neglected.

A. Determine the work transfer (Btu) from the steam in the cylinder to the piston.

B. Determine the work transfer (Btu) between the piston and the ambient environment. Neglect the cross-sectional area of the connecting rod.

C. Determine the work transfer (Btu) from the piston to the connecting rod. Neglect friction between the piston and cylinder. Assume that no heat transfer to or from the piston occurs and that the energy of the piston does not change.

D. Discuss why the assumption of negligible piston acceleration was made.

E. Which of the work transfers might be useful for driving a car? Why?

F. How good is the approximation that the steam can be treated as an ideal gas?

4.15 An automobile drive shaft rotates at 3000 rev/min and delivers 75 kW of power from the engine to the wheels. Determine the torque (N·m and ft·lb$_f$) in the drive shaft.

4.16 A compression process occurs in two steps. From state 1 at 15.0 psia and 3.00 ft^3, compression is to state 2 at 45.0 psia following the path $PV^2 =$ constant. From state 2, the compression occurs at constant pressure to a final volume at state 3 of 1.00 ft^3. Determine the work (Btu) for each step of this two-step process.

4.17 A gas is trapped in a cylinder–piston arrangement as shown in the sketch. The initial pressure and volume are 13,789.5 Pa and 0.02832 m^3, respectively. Determine the work (kJ) assuming that the volume is increased to 0.08496 m^3 in a constant-pressure process.

4.18 For the same conditions as in Problem 4.17, determine the work done if the process is polytropic with $n = 1$ (i.e., $PV =$ constant).

4.19 For the same conditions as in Problem 4.17, determine the work done if the process is polytropic with $n = 1.4$ (i.e., $PV^{1.4} =$ constant).

4.20 Consider the two processes shown in the sketch: ac and abc. Determine the work done by an ideal gas executing these reversible processes if $P_2 = 2P_1$ and $v_2 = 2v_1$? Assume you are dealing with a closed system and express your answer in terms of the gas constant R and the temperature T_1.

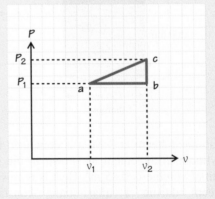

4.21 Determine the work done by an ideal gas in a reversible adiabatic expansion from T_1 to T_2.

4.22 Determine the work done in ft · lb$_f$ by a 2-lb$_m$ steam system as it expands slowly in a cylinder–piston arrangement from the initial conditions of 324 psia and 12.44 ft^3 to the final conditions of 25.256 ft^3 in accordance with the following relations: (a) $P = 20\mathcal{V} + 75.12$, where $\mathcal{V}$ and P are expressed in units of ft^3 and psia, respectively; and (b) $P\mathcal{V}$ = constant.

4.23 A vertical cylinder–piston arrangement contains 0.3 lb$_m$ of H$_2$O (liquid and vapor in equilibrium) at 120 F. Initially, the volume beneath the 250-lb$_m$ piston is 1.054 ft^3. The piston area is 120 in^2. With an atmospheric pressure of 14.7 lb$_f$/in^2, the piston is resting on the stops. The gravitational acceleration is 30.0 ft/s^2. Heat is transferred to the arrangement until only saturated vapor exists inside. (a) Show this process on a P–$\mathcal{V}$ diagram and (b) determine the work done (Btu).

4.24 One cubic meter of an ideal gas expands in an isothermal process from 760 to 350 kPa. Determine the work done by this gas in kJ and Btu.

4.25 Air is compressed reversibly in a cylinder by a piston. The 0.12 lb$_m$ of air in the cylinder is initially at 15 psia and 80 F, and the compression process takes place isothermally to 120 psia. Assuming ideal-gas behavior, determine the work required to compress the air (Btu).

4.26 Consider a 30-mm-diameter shaft rotating concentrically in a hole in a block as shown in the sketch. The clearance between the shaft and the block is 0.2 mm and is filled with motor oil at 310 K. The length of the annular space between the shaft and the block is 60 mm. Since the clearance is quite small compared to the shaft radius, the flow produced by the rotating shaft is a Couette flow (see Example 3.6); that is, a linear velocity distribution exists in the oil between the stationary surface ($v_x = 0$ at the block) to the surface of the rotating shaft ($v_x = R\omega$). The shear stress in the fluid film obeys the relation $\tau = \mu \, dv_x/dy$, where y is the distance from the stationary wall. Estimate the viscous shear work performed on the fluid in the annular space between the shaft and the block for a shaft speed of 1750 rev/min.

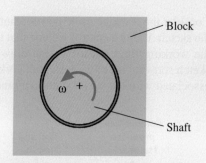

4.27 Consider the electrical circuit shown in the sketch in which a light bulb and a motor are wired in series with a battery. The motor drives a fan. Analyze the following thermodynamic systems for all heat and work interactions. Sketch the system boundaries and use labeled arrows for the various $\dot{Q}$s and $\dot{W}$s. Use subscripts to denote the type of work and make sure that your arrows point in the correct directions.

A. Light bulb alone

B. Light bulb and motor, with the system boundary cutting through the motor shaft

C. Light bulb and motor, but the system boundary is now contiguous with the fan blade rather than cutting through the shaft

D. Battery alone

E. Battery, light bulb, motor, and fan

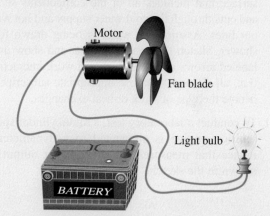

4.28 Steam exits a turbine and enters a 1.435-m-diameter duct. The steam is at 4.5 kPa and has a quality of 0.90. The mean velocity of the steam through the duct is 40 m/s. Determine the flow work required to push the steam into the duct.

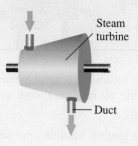

4.29 Consider the Rankine cycle power plant as shown in the sketch. Using a control volume that includes *only* the working fluid (steam/water), draw and label a sketch showing all of the heat and work interactions associated with the cycle control volume.

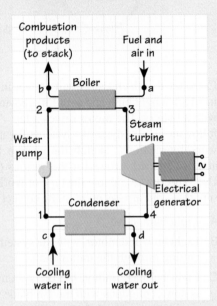

4.30 Consider the solar water heating scheme illustrated in Fig. 4.12. To analyze this scheme, choose a control surface that includes all of the components shown and cuts through the cold water supply and hot water out lines. Assume hot water is being drawn for a shower. Sketch the control volume and show using labeled arrows all of the heat and work interactions (i.e., all $\dot{Q}$s and $\dot{W}$s). Use appropriate subscripts to denote the type of work or heat exchanges.

4.31 To conduct a laboratory test, a multicylinder spark-ignition engine is connected to a dynameter (a device that measures the shaft power output) as shown in the sketch.

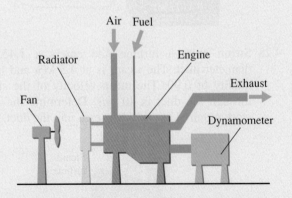

For the following control volumes, draw a sketch indicating all heat and work interactions (i.e., $\dot{Q}$s and $\dot{W}$s):

A. The control volume contains only the engine, and the control surface cuts through the hoses to

and from the radiator, through the air and fuel supply lines, through the exhaust pipe, and through the shaft connecting the engine and dynameter.

B. Same as in Part A but now the radiator is included within the control volume.

4.32 In a high-performance computer, silicon chips are actively cooled by water flowing through the ceramic substrate as shown in the sketch. The chip assembly is inside a case through which air freely circulates.

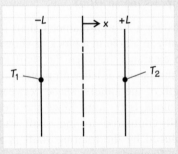

Consider the following thermodynamic systems (or control volumes), sketching each and showing all of the heat interactions with arrows. Label each heat interaction to indicate the mode of heat transfer (i.e., $\dot{Q}_{cond}$, $\dot{Q}_{conv}$, or $\dot{Q}_{rad}$).

A. Silicon chip only

B. Silicon chip and substrate (excluding the water)

C. Ceramic substrate only

D. Ceramic substrate and the flowing water

4.33 The temperature distribution in a planar slab of nuclear generating material is given by

$$T(x) = 50\left(1 - \frac{x^2}{L^2}\right) + \frac{T_2 - T_1}{2}\frac{x}{L} + \frac{T_2 + T_1}{2},$$

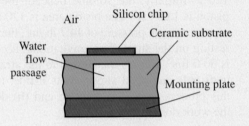

where L is the half-thickness of the slab and T_1 and T_2 are the surface temperatures as shown in the sketch. Note that x is measured from the center plane and $T(x)$ has units of kelvins.

A. Draw a graph of T versus x when $T_1 = 350$ K, $T_2 = 450$ K, and $L = 22$ mm.

B. What is the value of the temperature gradient at $x = -L$? Give units.

C. Calculate the conduction heat flux at $x = -L$ for the conditions in Part A. Give units. The thermal conductivity of the slab is 25 W/m·K. What is the direction of the heat transfer at this location?

D. Determine the location of the adiabatic plane within the slab for the same conditions as Part A [i.e., the location at which $\dot{Q}''(x) = 0$]. Also determine the temperature at this location.

E. At $x = +L$, is the heat transfer from the slab to the surroundings, or from the surroundings to the slab? Discuss.

4.34 A long cylindrical nuclear fuel rod has the following steady-state temperature distribution:

$$T(r) = 500\left(1 - \frac{r^2}{R^2}\right) + T_s,$$

where R is the fuel-rod radius and T_s is the surface temperature. $T(r)$ is expressed in kelvins. For a fuel rod with $R = 25$ mm, $T_s = 800$ K, and thermal conductivity of 40 W/m·K, determine both the heat flux at $r = R$ and the heat-transfer rate per unit length of rod, (i.e., $\dot{Q}/L$ [=] W/m). Plot the temperature distribution.

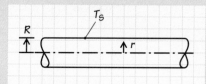

4.35 A 35-mm length of 1-mm-diameter nichrome (80% Ni, 20% Cr) wire is heated to 915 K by the passage of an electrical current through the wire (Joule heating). The wire is in an enclosure, the walls of which are maintained at 300 K. Warm air at 325 K flows slowly through the enclosure. The emissivity of the wire is 0.5 and the convective heat-transfer coefficient has a value of 54.9 W/m²·K. Determine both the heat flux and the heat-transfer rate from the nichrome wire.

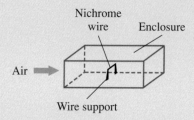

Nichrome wire Enclosure

Air

Wire support

4.36 Consider the same physical situation as in Problem 4.35, except that the electrical current is reduced such that the nichrome wire temperature is now 400 K. This temperature reduction also

results in a reduction of the heat-transfer coefficient associated with forced convection, which now has a value of 36.9 W/m²·K. Compare the fraction of the total heat transfer due to radiation for this situation to the fraction from Problem 4.35. Discuss.

4.37 Air at 300 K ($\equiv T_\infty$) flows over a flat plate maintained at 400 K ($\equiv T_s$). At a particular location on the plate, the temperature distribution in the air measured from the plate surface is given by

$$T(y) = (T_s - T_\infty)\exp[-22{,}000\, y(m)] + T_\infty,$$

where y is the perpendicular distance from the plate, as shown in the sketch.

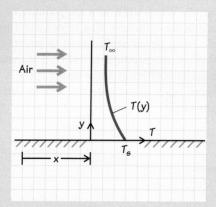

A. Determine the conduction heat flux through the air at the plate surface where the no-slip condition applies (i.e., at $y = 0^+$).

B. The physical mechanism of convection heat transfer starts with conduction in the fluid at the wall, as described in Part A. The details of the flow, however, determine the temperature distribution through the fluid. Use your result from Part A to determine a value for the local heat-transfer coefficient $h_{\text{conv},x}$ at the x location where $T(y)$ is given.

4.38 Determine the instantaneous rate of heat transfer from a 1.5-cm-diameter ball bearing with a surface temperature of 150°C submerged in an oil bath at 75°C if the surface convective heat transfer coefficient is 850 W/m²·°C.

4.39 Air at 480 F flows over a 20-in by 12-in flat surface. The surface is maintained at 75 F and the convective heat-transfer coefficient at the surface is 45 Btu/hr·ft²·F. Find the rate of heat transfer to the surface.

4.40 A convective heat flux of 20 W/m² (free convection) is observed between the walls of a room and the ambient air. What is the heat-transfer coefficient when the air temperature is 32°C and the wall temperature is 35°C?

4.41 A 30-cm-diameter, 5-m-long steam pipe passes through a room where the air temperature is 20°C. If the exposed surface of the pipe is a uniform 40°C, find the rate of heat loss from the pipe to the air if the surface heat-transfer coefficient is 8.5 W/m$^2 \cdot$°C.

4.42 A 10-ft-diameter spherical tank is used to store petroleum products. The products in the tank maintain the exposed surface of the tank at 75 F. Air at 60 F blows over the surface, resulting in a heat-transfer coefficient of 10 Btu/hr $\cdot$ ft$^2 \cdot$ F. Determine the rate of heat transfer from the tank.

4.43 The heat-transfer coefficient for water flowing normal to a cylinder is measured by passing water over a 3-cm-diameter, 0.5-m-long electric resistance heater. When water at a temperature of 25°C flows across the cylinder with a velocity of 1 m/s, 30 W of electrical power are required to maintain the heater surface temperature at 60°C. Estimate the heat-transfer coefficient at the heater surface. What is the significance of the water velocity in determining the answer?

4.44 A cylindrical electric resistance heater has a diameter of 1 cm and a length of 0.25 m. When water at 30°C flows across the heater a heat-transfer coefficient of 15 W/m$^2 \cdot$°C exists at the surface. If the electrical input to the heater is 5 W, what is the surface temperature of the heater?

4.45 A typical value of the heat-transfer coefficient for water boiling on a flat surface is 900 Btu/hr $\cdot$ ft$^2 \cdot$ F. Estimate the heat flux on such a surface when the surface is maintained at 222 F and the water is at 212 F.

4.46 Find the rate of radiant energy emitted by an ideal blackbody having a surface area of 10 m^2 when the surface temperature is maintained at (a) 50°C, (b) 100°C, and (c) 500°C.

4.47 A surface is maintained at 100°C and is enclosed by very large surrounding surfaces at 80°C. What is the net radiant flux (W/m^2) from this surface if its emissivity is (a) 1.0 or (b) 0.8?

4.48 Repeat Problem 4.47 for a surface temperature of 150°C, with all other data remaining unchanged.

CONSERVATION OF ENERGY

After studying Chapter 5, you should:

- Be able to write from memory (or intimate familiarity) conservation of energy (the first law of thermodynamics) applied to a system for an incremental change in state, for a change from state 1 to state 2, and at an instant.

- Understand when and how to apply the first-law statements to practical situations.

- Be able to write from memory (or intimate familiarity) both steady and unsteady forms of energy conservation (first law) for a control volume with a single inlet and a single outlet.

- Be able to explain the physical significance of each term in the various first-law expressions.

- Be able to explain the physical origins of the enthalpy in the first-law analysis of both systems and control volumes.

- Understand the origin and the application of the kinetic energy correction factor in the statement of the first law for control volumes.

- Be able to simplify and apply the integral form of the first law for steady-flow devices. (This objective links to Chapter 11.)

- Be able to derive the differential equations that express energy conservation for conduction heat transfer in planar and cylindrical solids.

- Be able to solve the conduction equation for the temperature distribution given appropriate boundary conditions.

- Be able to solve first-law problems.

- Be able to apply standardized enthalpies to solve first-law problems for reacting systems.

- Be able to recognize differential formulations of conservation of energy and be able to simplify these for common situations.

Chapter 5 Overview

IN THIS CHAPTER, we apply the fundamental principle of energy conservation to thermodynamic systems and control volumes. In our analyses of fixed-mass systems, we also include the problem of heat conduction within the system. In dealing with control volumes, we again follow a hierarchical development by starting with simple, steady-state, steady-flow cases and then adding detail and complexity to arrive at the more general statements of energy conservation. We also present and discuss energy conservation expressions that have been commonly adopted for particular applications. In this way, we hope to show the unity of the many seemingly varied ways that this important principle is expressed in the thermal sciences. Because this chapter relates to many practical applications, the author highly recommends concurrent study of the indicated sections of Chapter 11.

5.1 HISTORICAL CONTEXT

The development of an energy conservation principle depended critically on the evolving idea that work was convertible to heat, and vice versa. **Benjamin Thompson** (1753–1814) (Count Rumford) in 1798 [1], **Julius Robert Mayer** (1814–1878) in 1842 [2], and **James Prescott Joule** (1818–1889) in 1849 [3] published results of their research on the mechanical equivalent of heat and presented numerical values. The issue here was to define the specific number of foot-pounds of work that is equivalent to the heat required to raise the

Robert Mayer (1814-1878)

Hermann Helmholtz (1821-1894)

Rudolf Clausius (1822-1888)

temperature of 1 lb$_m$ of water 1°F. (The accepted value today is 778.16 ft-lb$_f$ and defines the British thermal unit or Btu.) Mayer was the first person to state a formal conservation of energy principle. In 1842, he wrote [2]: "Forces (energies) are therefore indestructible, convertible, and (in contradistinction to matter) imponderable objects… a force (energy) once in existence, cannot be annihilated." (Parenthetical material has been added; in the mid 1800s, the word *force* commonly meant *energy* [4, 5].) **Hermann Helmholtz** (1821–1894), who produced an array of scientific achievements, extended the conservation of energy principle in his 1847 paper, "On the Conservation of Force," to include all known forms of energy: mechanical, thermal, chemical, electrical, and magnetic. (See Ref. [4].) **Rudolf Clausius** (1822–1888) wrote in 1850 one of the most succinct and modern-sounding statement of the energy conservation principle [6]: "The energy of the universe is constant." (Clausius also named the property *entropy* and presented clear statements of the second law of thermodynamics. We will consider these concepts in Chapter 7.)

5.2 ENERGY CONSERVATION FOR A SYSTEM

> At this point, you may find it useful to review the detailed discussion of systems and control volumes in Chapter 1.

We begin our study of the conservation of energy principle by considering a system of fixed mass. We start by explicitly transforming the generic conservation principles from Chapter 1 (Eqs. 1.1 and 1.2) to statements of energy conservation. By defining X to be energy E, Eq. 1.1, which applies to the time interval $t_2 - t_1$, becomes

$$E_{in} - E_{out} + E_{generated} = \Delta E_{stored} \equiv E_{sys}(t_2) - E_{sys}(t_1). \tag{5.1}$$

Quantity of energy crossing boundary and passing into system / Quantity of energy crossing boundary and passing out of system / Quantity of energy generated within system / Change in quantity of energy stored within system during time interval $\Delta t = t_2 - t_1$

Similarly, by defining $\dot{X}$ to be the time rate of energy $\dot{E}$, Eq. 1.2, which applies to an instant, becomes

$$\dot{E}_{in} - \dot{E}_{out} + \dot{E}_{generated} = \dot{E}_{stored}. \tag{5.2}$$

Time rate of energy crossing boundary and passing into the system / Time rate of energy crossing boundary and passing out of system / Time rate of energy generation within system / Time rate of energy storage within system

Figure 5.1 shows a system with superimposed arrows representing the various terms in these equations.

Our task now is to associate each term in these conservation of energy equations with particular forms of energy for various physical situations. We first consider rather general statements of energy conservation in which all forms of energy and their interconversions are allowed. Our only restriction is the exclusion of nuclear transformations, which violate our fixed-mass assumption. With this restriction, no generation term appears. Subsequent to our general treatment, we focus on the conservation of thermal energy in homogeneous solids (or stagnant fluids). This treatment sets the stage for the analysis of conduction heat transfer and the determination of temperature distributions and local heat fluxes within or at the boundaries of systems.

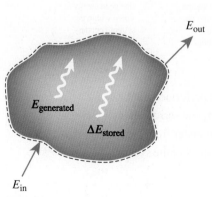

FIGURE 5.1
Energy enters (E_{in}) and exits (E_{out}) across the system boundaries; energy is generated ($E_{generated}$) or stored (ΔE_{stored}) within the system.

Chapter 1 discusses the distinctions between integral and differential systems. See Figs. 1.15–1.20 and related discussion.

5.2a General Integral Forms

Consider the general thermodynamic system illustrated in Fig. 5.2. The system here is a macroscopic (i.e., integral) system. Several forms of energy may be associated with the system. For example, the system will always possess internal energy—which can be further subdivided into thermal and chemical contributions—and the system may also possess bulk kinetic and potential energies. Thus we can write

$$E_{\text{sys}} = U_{\text{sys}} + (KE)_{\text{sys}} + (PE)_{\text{sys}}. \tag{5.3}$$

We cannot emphasize too strongly that the energies expressed in Eq. 5.3 are contained *within* the system boundary. The system possesses these energies. The system may lose or gain energy, however, by energy transfers *across* the system boundary. As we saw in Chapter 4, these transfers occur only as a result of heat and/or work interactions with the surroundings. Thus, the energy of the system can be increased by either heat transfer to the system from the surroundings or by the surroundings performing work on the system. These transfers of energy across the system boundary correspond to E_{in} in Fig. 5.1. Conversely, the system may lose energy by heat transfer from the system to the surroundings or by the system performing work on the surroundings (i.e., E_{out} in Fig. 5.1). With only three quantities involved, writing a mathematical expression of conservation of energy is easy to do and intuitively satisfying. The system energy is analogous to your bank account balance, where heat and work interactions are the credits and debits. Note that $E_{\text{generated}}$ is zero since we exclude nuclear reactions. We now apply these ideas to several situations.

Heat and work only have meaning as energy crossing a boundary.

Heat and work do not exist within a system or control volume.

For an Incremental Change

Consider the transfer of incremental quantities of energy across the system boundary either into the system, δE_{in}, or out of the system, δE_{out}, during the time interval dt. These transfers result in an incremental change in the amount of energy stored within the system, dE_{sys}, as indicated in Fig. 5.2a. Conservation of energy is then expressed as

$$\underset{\substack{\text{Incremental energy}\\\text{crossing the system}\\\text{boundary from the}\\\text{surroundings to the}\\\text{system}}}{\delta E_{\text{in}}} \quad - \quad \underset{\substack{\text{Incremental energy}\\\text{crossing the system}\\\text{boundary from the}\\\text{system to the}\\\text{surroundings}}}{\delta E_{\text{out}}} \quad = \quad \underset{\substack{\text{Incremental}\\\text{change in energy}\\\text{stored within the}\\\text{system boundary}}}{dE_{\text{sys}}}, \tag{5.4a}$$

where

$$\delta E_{\text{in}} \equiv \begin{cases} \delta W_{\text{in}}, \text{ work performed on the} \\ \qquad \text{system by the surroundings,} \\ \delta Q_{\text{in}}, \text{ heat transfer from the} \\ \qquad \text{surroundings to the system,} \end{cases} \tag{5.4b}$$

and

$$\delta E_{\text{out}} \equiv \begin{cases} \delta W_{\text{out}}, \text{ work performed by the} \\ \qquad \text{system on the surroundings,} \\ \delta Q_{\text{out}}, \text{ heat transfer from the} \\ \qquad \text{system to the surroundings.} \end{cases} \tag{5.4c}$$

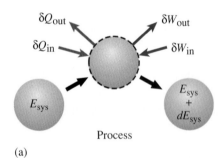

(a)

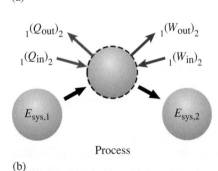

(b)

State 1   State 2

FIGURE 5.2

Heat and work interactions at the system boundaries result in a change in the amount of energy possessed by the system for (a) incremental interactions and for (b) finite interactions.

Substituting the definitions from Eqs. 5.4b and 5.4c back into Eq. 5.4a yields

$$(\delta W_{in} + \delta Q_{in}) - (\delta W_{out} + \delta Q_{out}) = dE_{sys}. \quad (5.4d)$$

Equation 5.4d preserves the *in minus out equals stored* formulation, while introducing heat and work. Traditionally, the heat in and out and work in and out are combined; thus,

$$(\delta Q_{in} - \delta Q_{out}) - (\delta W_{out} - \delta W_{in}) = dE_{sys}, \quad (5.5a)$$

or

$$\delta Q_{in,net} - \delta W_{out,net} = dE_{sys}. \quad (5.5b)$$

> **Note that the arrows representing Q_{in} and W_{in} end precisely at the boundary. Similarly, Q_{out} and W_{out} arrows begin precisely at the boundary. Q and W have no meaning inside the boundary.**

This grouping of terms establishes a sign convention for heat and work, where Q is conventionally treated as a credit (i.e., Q is from the surroundings to the system), whereas work W is conventionally treated as a debit (i.e., W is performed by the system on the surroundings). In some situations we will adopt this convention; however, to avoid any ambiguity or confusion, we employ the subscript *net* to emphasize that both positive and negative heat and work interactions are implied within a single term as done in Eq. 5.5b. Regardless of any sign convention, you should always be prepared to think through any new or ambiguous situation to ensure that heat and work interactions are properly credited or debited in your energy account as expressed by Eq. 5.4d.

We now consider a system that undergoes a finite change in state.

For a Change in State

Conservation of energy is expressed for a process in which the system state changes from state 1 to state 2 by integrating Eq. 5.5a and 5.5b, with the results that

$$[_1(Q_{in})_2 + {_1}(W_{in})_2] - [_1(Q_{out})_2 + {_1}(W_{out})_2] = \Delta E_{sys}, \quad (5.6a)$$

> **Equation 5.6 is a key relationship. Strive to understand the concept it expresses.**

or

$$_1(Q_{in,net})_2 - {_1}(W_{out,net})_2 = \Delta E_{sys}, \quad (5.6b)$$

where

$$\Delta E_{sys} \equiv E_{sys,2} - E_{sys,1}. \quad (5.7)$$

Recall from Chapter 4 that the notation $_1Q_2$ and $_1W_2$ does not represent a change from state 1 to 2 but rather refers to the integration of δQ and δW over particular paths. However, ΔE_{sys} does represent a change in the system energy as defined in Eq. 5.7. Mathematically, Q and W are **path functions**, and δQ and δW are referred to as **inexact differentials**; whereas E is a **state function**, and dE is referred to as an **exact differential**.

Equation 5.6 is an extremely important relationship. You should become very familiar with both its physical interpretation and its application. The following examples should assist you in this endeavor.

Example 5.1

A 450-kg instrument package is dropped in NASA's 132-m drop tower facility. To minimize the drag force acting on the falling package, the air has been evacuated from the tower ($P = 10^{-2}$ torr). The package is dropped from rest. Determine the velocity of the package at the bottom of the tower just before it impacts with the decelerator mechanism. Also determine the kinetic energy of the package at this same instant.

Solution

Known $M, z_2 - z_1, V_1$

Find V_2, KE

Sketch

Photographs courtesy of NASA.

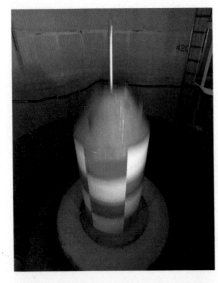

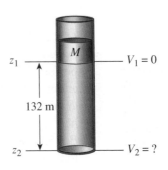

z_1 ———— $V_1 = 0$

132 m

z_2 ———— $V_2 = ?$

Assumptions

i. No frictional or drag forces act on the package.
ii. The process is adiabatic.

Analysis   We begin by identifying the instrument package as a thermodynamic system. This system undergoes a process from the initial state ($V_1 = 0, z_1$) to the final state (V_2, z_2). The appropriate conservation of energy expression is given by Eq. 5.6:

$$_1(Q_{in,net})_2 - {}_1(W_{out,net})_2 = E_{sys,2} - E_{sys,1}.$$

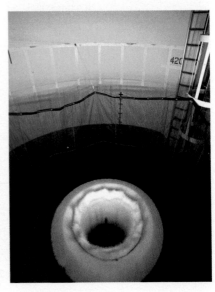

The net heat transfer is zero, as is the net work. That the work is zero follows from the fact that no boundary forces act on the package during its fall:

$$W = \int_1^2 F \cdot ds = 0.$$

Thus,

$$0 - 0 = E_{sys,2} - E_{sys,1},$$

or

$$E_{sys,2} = E_{sys,1}.$$

The system energy is given by Eq. 4.3. Assuming no rotational kinetic energy, we write

$$U_2 + \frac{1}{2}MV_2^2 + Mg(z_2 - z_{ref}) = U_1 + \frac{1}{2}MV_1^2 + Mg(z_1 - z_{ref}).$$

With our assumptions, there is no mechanism to change the system temperature; hence, $U_2 = U_1$. Recognizing that $V_1 = 0$, we can simplify the previous equation to

$$\frac{1}{2}MV_2^2 = Mg(z_1 - z_2).$$

Solving for V_2 yields

$$V_2 = [2g(z_1 - z_2)]^{1/2},$$

which is numerically evaluated as

$$V_2 = [2(9.8067)(132)]^{1/2}$$
$$= 50.88$$
$$[=] \left[\frac{m}{s^2} m \right]^{1/2} = m/s.$$

The kinetic energy is also evaluated as

$$\frac{1}{2}MV_2^2 = Mg(z_1 - z_2)$$
$$= 450(9.8067)(132)$$
$$= 5.825 \times 10^5$$
$$[=] kg\frac{m}{s^2} m \left[\frac{1 \text{ N}}{kg \cdot m/s^2} \right] = N \cdot m = J.$$

Comment Note how thermal energy terms drop out of this problem, leaving only the conversion between system potential and kinetic energies. We note also the substantial impact velocity of the package at the bottom of the tower (50.88 m/s = 113.8 mph). A catch basin of polystrene beads is used to decelerate the package in NASA's 5.2-s drop tower.

Astronaut Edwin Aldrin, Jr., walking on the moon. Apollo XI mission, July 1969.

Self Test 5.1 ✓ A **0.2-kg projectile is launched vertically upward from the surface of the moon** ($g =$ **1.62 m/s²) with an initial velocity of 30 m/s. Determine (a) the initial kinetic energy of the projectile and (b) the maximum altitude it attains.**

(Answer: (a) 90 J, (b) 277.8 m)

Example 5.2

Consider a piston–cylinder assembly containing 0.14 kg of air initially at a pressure of 350 kPa. Heat is added during a quasi-equilibrium, isothermal process in which the volume increases from an initial value of 0.1 m³ to a final value of 0.3 m³. Sketch the process on T–V and P–V coordinates and calculate the amount of heat added during the process.

Solution

Known   Quasi-equilibrium, isothermal process, M, P_1, V_1, V_2

Find $_1Q_2$

Sketch

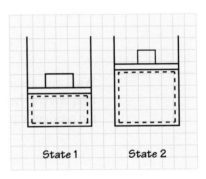

Assumptions

  i. Ideal-gas behavior
  ii. No system kinetic energy
  iii. Negligible change in system potential energy

Analysis   We first identify the air in the cylinder as the thermodynamic system of interest, as indicated by the dashed line in the sketch. Since the process is stated to be isothermal, we know that the temperature remains constant during the process and the T–V plot is a simple horizontal line as shown in the left sketch here. With our assumption of ideal-gas behavior (Eq. 2.28c), we also know the relationship between P and V for a fixed T, that is,

$$P = (MRT)\frac{1}{V}.$$

Since MRT is a constant, the P–V relationship is a simple hyperbola as shown in the sketch on the right.

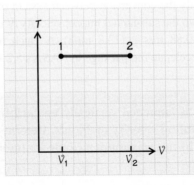

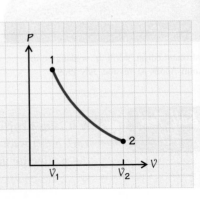

To determine the heat added, we write conservation of energy for the process (Eq. 5.6), that is,

$$_1(Q_{in})_2 - {}_1(W_{out})_2 = U_2 - U_1.$$

For an ideal gas, the internal energy is a function of temperature only; thus, the internal energy change for the process must be zero since the temperature is constant. More formally, we evaluate Eq. 2.31d as

$$U_2 - U_1 = M\Delta u = M \int_{T_1}^{T_2=T_1} c_v dT = 0.$$

With the assumption of a quasi-equilibrium process, we can evaluate the work done by the system from Eq. 4.8b:

$$_1W_2 = \int_1^2 Pd\mathcal{V}.$$

Substituting the previously established relationship between P and $\mathcal{V}$ for this process, the work is evaluated as

$$_1W_2 = \int_1^2 (MRT)\frac{1}{\mathcal{V}}d\mathcal{V}$$

$$= MRT \int_1^2 \frac{d\mathcal{V}}{\mathcal{V}} = MRT \ln\frac{\mathcal{V}_2}{\mathcal{V}_1}.$$

We also know that $P_1\mathcal{V}_1 = MRT$; thus,

$$_1W_2 = P_1\mathcal{V}_1 \ln\frac{\mathcal{V}_2}{\mathcal{V}_1}$$

$$= 350 \times 10^3(0.1)\ln\left(\frac{0.3}{0.1}\right)$$

$$= 38,450$$

$$[=] \frac{N}{m^2}m^3 = N\cdot m = J.$$

We now obtain the heat added from energy conservation:

$$_1(Q_{in})_2 - {}_1(W_{out})_2 = 0,$$

or

$$_1(Q_{in})_2 = {}_1(W_{out})_2$$

$$= 38,450 \text{ J}.$$

Comment   This example illustrates the combined use of a state equation, a calorific equation of state, and conservation of energy. We also note, first, that the process being conducted in a quasi-equilibrium manner is critical to our evaluation of the work, and, second, that this work can be visualized as the area under the curve on our P–$\mathcal{V}$ sketch.

Self Test
5.2

☑ **Repeat Example 5.2 for a constant-pressure process, finding T_1, T_2, ${}_1W_2$, and ${}_1Q_2$.**
    (Answer: 871.1 K, 2613.2 K, 70 kJ, 300.2 kJ using c_v of 0.944 kJ/kg·K)

Example 5.3

Steam is contained in a rigid tank at 500 K and 1 MPa. A cooling coil removes energy from the steam as heat until a final temperature of 400 K is reached. The tank volume is 0.15 m³. Determine the amount of heat removed.

Solution

Known $T_1, P_1, V_1 (=V_2), T_2$

Find $_1Q_2$

Sketch

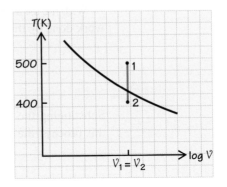

Assumptions

  i. Equilibrium prevails at initial and final states.
  ii. There are no changes in system kinetic or potential energies.

Analysis   We select the steam as the system of interest. The heat removed can then be determined from application of conservation of energy for a constant-volume process, together with a knowledge of the thermodynamic properties at the initial and final states. Since the tank is rigid no expansion or compression work is performed; hence, Eq. 5.6a can be simplified as follows:

$$_1(Q_{in})_2 + {}_1(W_{in})_2 - {}_1(Q_{out})_2 - {}_1(W_{out})_2 = E_{sys,2} - E_{sys,1}$$
$$\Rightarrow 0 + 0 - {}_1(Q_{out})_2 - 0 = U_2 - U_1,$$

or

$$- {}_1(Q_{out})_2 = M(u_2 - u_1).$$

Since two independent thermodynamic properties are known at the initial state, all other properties can be determined. For the given conditions, the steam is superheated; hence, properties can be obtained from the NIST database or Table D.3 as follows:

$$v_1 = 0.22064 \ \text{m}^3/\text{kg},$$
$$u_1 = 2670.6 \ \text{kJ/kg}.$$

The system mass is obtained as

$$M = \frac{V_1}{v_1} = \frac{0.15}{0.22064} = 0.6798$$

$$[=] \frac{\text{m}^3}{\text{m}^3/\text{kg}} = \text{kg}.$$

Because the process occurs at constant volume with a fixed mass, v_2 equals v_1. This knowledge and the given state-2 temperature allow us to determine the state-2 specific internal energy. From the NIST database or Table D.1 we see that, at 400 K,

$$v_f(400 \text{ K}) < v_2 < v_g(400 \text{ K}),$$

so

$$0.0010667 < 0.22064 < 0.73024.$$

State 2 thus lies in the liquid–vapor mixture region, as shown in the preceding sketch.

The useful saturation properties from the NIST database or Table D.1 are as follows:

$$v_f = 0.0010667 \text{ m}^3/\text{kg}, \qquad v_g = 0.73024 \text{ m}^3/\text{kg},$$
$$u_f = 532.69 \text{ kJ/kg}, \qquad u_g = 2536.2 \text{ kJ/kg}.$$

To find u_2, we must first find the quality at state 2, x_2. Applying Eq. 2.49b yields

$$x_2 = \frac{v_2(=v_1) - v_{f,2}}{v_{g,2} - v_{f,2}}$$

$$= \frac{0.22064 - 0.0010667}{0.73024 - 0.0010667} = 0.3011.$$

The state-2 specific internal energy is obtained as (Eq. 2.49c)

$$u_2 = (1 - x_2)u_{f,2} + x_2 u_{g,2}$$

$$= (1 - 0.3011)\,532.69 + (0.3011)\,2536.2 \text{ kJ/kg}$$

$$= 1135.9 \text{ kJ/kg}.$$

Thus, we find the heat removed to be

$$_1(Q_{out})_2 = -M(u_2 - u_1)$$

$$= -0.6798\,(1135.9 - 2670.6)$$

$$= 1043.3$$

$$[=] \text{ kg(kJ/kg)} = \text{kJ}.$$

Comments   Recognizing that a constant-volume process occurs is important, as the work is then simply evaluated (i.e., it is zero). This example emphasizes the use of the property quality in dealing with liquid–vapor mixtures.

Self Test 5.3 A piston–cylinder device originally contains 5 kg of saturated-liquid water at 100 kPa. Determine the heat addition required to bring the fluid to a saturated-vapor state.

(Answer: 11,287 kJ)

Example 5.4

Air is contained in a piston–cylinder arrangement as shown in the sketch. A weight placed on the piston keeps the pressure constant (120 kPa) inside the cylinder as 11,820 J of energy is added to the air by heat transfer. The initial temperature is 300 K, and the initial volume is 0.12 m³. Determine the temperature at the end of the heat-addition process.

Solution

Known $T_1, V_1, \,_1(Q_{in})_2, P_1 = P_2 = $ constant

Find T_2

Sketch

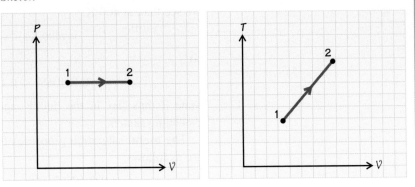

Assumptions

i. Quasi-equilibrium process
ii. Ideal-gas behavior
iii. Negligible change in system potential energy

Analysis We define the air contained in the cylinder as the thermodynamic system of interest and apply conservation of energy (Eq. 5.6b) to the process as follows:

$$_1(Q_{in})_2 - {_1}(W_{out})_2 = E_{sys,2} - E_{sys,1}.$$

Since the pressure is constant, the work out is simply evaluated as

$$_1W_2 = \int_1^2 Pd\mathcal{V} = P(\mathcal{V}_2 - \mathcal{V}_1),$$

or

$$_1W_2 = MP(v_2 - v_1).$$

We also recognize that the only system energy is internal energy u. Upon rearrangement, our conservation of energy expression thus becomes

$$
\begin{aligned}
_1(Q_{in})_2 &= M(u_2 - u_1) + MP(v_2 - v_1) \\
&= M(u_2 + Pv_2 - u_1 - Pv_1) \\
&= M(h_2 - h_1),
\end{aligned}
$$

or

$$\frac{_1(Q_{in})_2}{M} = h_2 - h_1.$$

The mass is evaluated from the ideal-gas equation of state (Eq. 2.28c) as

$$
\begin{aligned}
M &= \frac{P\mathcal{V}_1}{RT_1} \\
&= \frac{120 \times 10^3 (0.12)}{287(300)} = 0.1672
\end{aligned}
$$

$$[=] \frac{\left(\dfrac{N}{m^2}\right)m^3}{\left(\dfrac{J}{kg \cdot K}\right)K}\left[\dfrac{1\,J}{N \cdot m}\right] = kg,$$

where $R = R_{air} = 287$ J/kg·K (Table C.1). To find T_2, we recognize that h_1 and h_2 depend only on temperature and that Table C.2 provides the necessary values of h_1 and h_2; that is, it is a tabular form of the calorific equation of state for air. Using the given heat added, we evaluate the enthalpy difference as

$$h_2 - h_1 = \frac{11{,}820}{0.1672} \text{ J/kg} = 70{,}694 \text{ J/kg}$$

$$= 70.69 \text{ kJ/kg}.$$

From Table C.2, we obtain

$$h_1 = h_1 \, (300 \text{ K}) = 426.04 \text{ kJ/kg}.$$

Thus,

$$h_2 = h_1 + (h_2 - h_1)$$

$$= 426.04 + 70.69 \text{ kJ/kg}$$

$$= 496.73 \text{ kJ/kg}.$$

We see that this value is just slightly greater than the tabulated value for 370 K. Interpolating data from Table C.2 yields

$$T_2 = 370.1 \text{ K}.$$

Comments   We note that the enthalpy appears as a useful thermodynamic property for this constant-pressure process. Note that we could also have solved this problem using an average value of c_p rather than tabular values for the enthalpy. The reader should verify this.

See Eq. 2.33e in Chapter 2. ▶

**Self Test
5.4**   ☑ **A well-insulated (i.e., adiabatic) rigid tank contains 2 kg of air at 300 K and 150 kPa. An electric resistance heater in the tank is turned on and receives 200 kJ of electrical work. Find the final temperature and pressure of the air.**

(Answer: 439.3 K, 219.7 kPa)

At an Instant

We now consider conservation of energy at an instant. The general rate form of energy conservation, Eq. 5.2, can be used directly for this situation by identifying

$$\dot{E}_{\text{in}} \equiv \dot{W}_{\text{in}} + \dot{Q}_{\text{in}}, \tag{5.8a}$$

$$\dot{E}_{\text{out}} \equiv \dot{W}_{\text{out}} + \dot{Q}_{\text{out}}, \tag{5.8b}$$

$$\dot{E}_{\text{generated}} \equiv 0, \tag{5.8c}$$

and

$$\dot{E}_{\text{stored}} \equiv \frac{dE_{\text{sys}}}{dt}, \tag{5.8d}$$

where

$$\dot{Q} \equiv \lim_{\Delta t \to 0} \frac{\delta Q}{\Delta t}, \tag{5.9a}$$

$$\dot{W} \equiv \lim_{\Delta t \to 0} \frac{\delta W}{\Delta t}, \tag{5.9b}$$

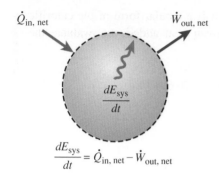

$$\frac{dE_{sys}}{dt} = \dot{Q}_{in,\,net} - \dot{W}_{out,\,net}$$

FIGURE 5.3
The rate at which the energy of a system increases equals the difference between the net rate at which energy enters the system across the system boundary and the net rate at which energy exits the system.

Equation 5.10 is another key conservation of energy expression for systems. ▶

and

$$\frac{dE_{sys}}{dt} = \lim_{\Delta t \to 0} \frac{E_{sys}(t + \Delta t) - E_{sys}(t)}{\Delta t}. \quad (5.9c)$$

Equation 5.2 is thus reassembled as

$$\underbrace{(\dot{W}_{in} + \dot{Q}_{in})}_{\substack{\text{Time rate of}\\ \text{energy crossing}\\ \text{boundary into}\\ \text{system}}} \quad - \quad \underbrace{(\dot{W}_{out} + \dot{Q}_{out})}_{\substack{\text{Time rate of}\\ \text{energy crossing}\\ \text{boundary out of}\\ \text{system}}} \quad = \quad \underbrace{\frac{dE_{sys}}{dt}}_{\substack{\text{Time rate of}\\ \text{change of energy}\\ \text{stored within}\\ \text{system}}}, \quad (5.10a)$$

or alternatively,

$$\underbrace{\dot{Q}_{in,net}}_{\substack{\text{Net heat transfer}\\ \text{rate to the system}}} \quad - \quad \underbrace{\dot{W}_{out,net}}_{\substack{\text{Net rate of work done}\\ \text{by the system, that is,}\\ \text{net power produced}}} \quad = \quad \underbrace{\frac{dE_{sys}}{dt}}_{\substack{\text{Time rate of change}\\ \text{of energy stored}\\ \text{within system}}}. \quad (5.10b)$$

This relationship is shown schematically in Fig. 5.3. Equation 5.10, like Eq. 5.6, is a key relationship, and again, its meaning and application should become firmly fixed in your storehouse of knowledge. The following examples help to illuminate the usefulness of this expression.

Example 5.5

Consider the electrical circuit shown in the sketch in which a light bulb and electric motor are wired in series with a battery. The motor drives a fan. For the following thermodynamic systems, write an appropriate conservation of energy expression:

A. Light bulb alone
B. Battery alone

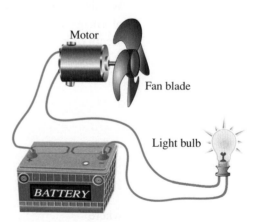

Solution

The nature of this problem suggests that we depart from the standard solution format. We begin by drawing a system boundary around the light bulb and identifying all of the energy flows:

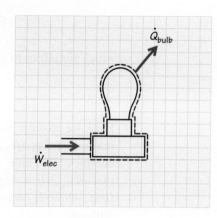

Here we identify a rate of electrical work into the system, $\dot{W}_{\text{elec}}$, and a rate of heat transfer out of the system, $\dot{Q}_{\text{bulb}}$. Assuming steady state, there are no other energy terms to consider. We now apply Eq. 5.10a,

$$(\dot{W}_{\text{in}} + \dot{Q}_{\text{in}}) - (\dot{W}_{\text{out}} + \dot{Q}_{\text{out}}) = \frac{dE_{\text{sys}}}{dt},$$

which simplifies to

$$(\dot{W}_{\text{elec}} + 0) - (0 + \dot{Q}_{\text{bulb}}) = 0,$$

or

$$\dot{W}_{\text{elec}} = \dot{Q}_{\text{bulb}}.$$

From this, we formally conclude that all of the electrical power input is converted to a heat-transfer rate from the light bulb to the surroundings. Note that visible radiation is included in this heat-transfer rate.

We now consider the battery alone as our system of interest. The most obvious flow of energy is the net electrical power out from the battery terminals, $\dot{W}_{\text{elec}}$. Furthermore, if the battery has some small internal resistance, the battery will be warmer than the surroundings, resulting in a small heat transfer to the surroundings, $\dot{Q}_{\text{batt}}$. Since both $\dot{W}_{\text{elec}}$ and $\dot{Q}_{\text{batt}}$ are outflows of energy, we identify the rate of change of the internal energy of the battery, dU_{batt}/dt, as their source. The physical mechanism for this change in internal energy entails the rearrangement of chemical bonds within the battery. We indicate these energy flows on a sketch:

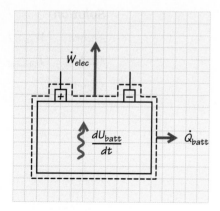

Conservation of energy is again expressed by Eq. 5.10a,

$$(\dot{W}_{in} + \dot{Q}_{in}) - (\dot{W}_{out} + \dot{Q}_{out}) = \frac{dE_{sys}}{dt},$$

which simplifies to yield

$$(0 + 0) - (\dot{W}_{elec} + \dot{Q}_{batt}) = \frac{dU_{batt}}{dt}.$$

Here we assume that the energy of the battery (E_{sys}) comprises only internal energy ($E_{syst} \equiv U_{batt}$); that is, there are no kinetic and potential energies. We note that the formal application of Eq. 10.a indicates that dU_{batt}/dt is negative; that is, the internal energy decreases with time, which is consistent with our experience that batteries "run down" with use.

Self Test 5.5   Consider a 100-W light bulb in a well-insulated 34-m³ room with the air initially at 100 kPa and 25°C. Determine (a) the rate of change of the air temperature and (b) the temperature of the air after 10 h.

(Answer: (a) 0.210 K/min, (b) 151°C)

Example 5.6

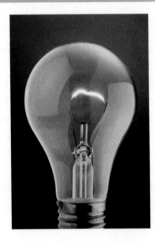

Assume that the filament in a 100-W light bulb can be modeled as a cylinder 1 mm in diameter and 25 mm long as shown in the sketch. Estimate the steady-state operating temperature of the filament assuming that it loses energy by both radiation and convection heat transfer. The emissivity of the wire is 0.25, and the surrounding glass bulb has a temperature of 400 K. The average convective heat-transfer coefficient at the filament surface is approximately 15 W/m²·K, and the ambient gas temperature inside the bulb is approximately 425 K.

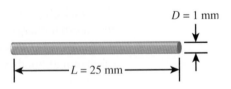

$D = 1$ mm

$L = 25$ mm

Solution

Known $\dot{W}_{elec}, L, D, \varepsilon, h_{conv}, T_{surr}, T_\infty$

Find $T_{filament}$

Sketch

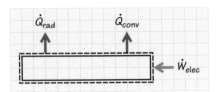

$\dot{Q}_{rad}$ $\dot{Q}_{conv}$

$\dot{W}_{elec}$

Assumptions

    i. Steady state

    ii. Coiled filament can be modeled as a simple cylinder

    iii. Uniform surface temperature

    iv. Negligible heat transfer from ends

    v. Glass bulb absorbs and emits as a blackbody (i.e., the fraction of the transmitted radiation is a small)

Analysis We begin by identifying the filament as our thermodynamic system of interest and indicating all of the energy (rate) terms, as shown in the sketch. We show the electric power input as a single arrow as the net effect of the current flowing in and out of the system (see Eq. 4.11b). Since we are dealing with energy rates (J/s = W), conservation of energy is expressed by Eq. 5.10a:

$$(\dot{W}_{in} + \dot{Q}_{in}) - (\dot{W}_{out} + \dot{Q}_{out}) = \frac{dE_{sys}}{dt}.$$

Since we assume steady-state operation, the energy storage term, dE_{sys}/dt, is zero. Using the previous sketch as a guide, this relationship becomes

$$(\dot{W}_{elec} + 0) - (0 + \dot{Q}_{conv} + \dot{Q}_{rad}) = 0,$$

or

$$\dot{Q}_{conv} + \dot{Q}_{rad} = \dot{W}_{elec}.$$

Since $\dot{W}_{elec}$ is given as 100 W, the right-hand side is fixed. The heat-transfer rates depend on the filament temperature T_f and can be expressed using the "rate laws" presented in Chapter 4 (Eqs. 4.18 and 4.25). These are expressed for the present problem as

$$\dot{Q}_{conv} = h_{conv}\pi DL(T_f - T_\infty)$$

and

$$\dot{Q}_{rad} = \varepsilon_f \pi DL\sigma(T_f^4 - T_{surr}^4),$$

where we identify T_∞ as the gas temperature within the bulb and T_{surr} as the temperature of the bulb itself. Substituting these expressions and dividing through by the surface area, πDL, yield

$$h_{conv}(T_f - T_\infty) + \varepsilon_f \sigma(T_f^4 - T_{surr}^4) = \frac{\dot{W}_{elec}}{\pi DL}.$$

Conceptually, we are done since the only unknown quantity in this relationship is the filament temperature T_f. We substitute numerical values as follows:

$$15(T_f - 425) + 0.25(5.67 \times 10^{-8})(T_f^4 - 400^4) = 100/[\pi(0.001)0.025],$$

where each term has units of W/m². (The reader should verify this.) Performing the indicated arithmetic and rearranging yields the following polynomial:

$$(1.4175 \times 10^{-8})T_f^4 + 15T_f - 1.279978 \times 10^6 = 0.$$

The final task is to find the particular root of this equation that satisfies our problem. (There are multiple roots.) Any number of root-solving methods can be applied. The final result obtained by application of the Newton–Raphson iterative method is

$$T_\mathrm{f} = 3054.6 \ \mathrm{K}.$$

Comment   This result is reasonable, although it is somewhat higher than the 2800 K value frequently used to characterize standard tungsten incandescent lights.

Coal-fired power plant in North Rhine Westphalia, Germany. The white plumes form when water vapor from the cooling towers condenses.

5.2b Reacting Systems

Although a detailed study of reacting systems resides in the realm of chemical engineering, combustion systems are of wide engineering interest. In Chapter 1, we discussed the importance of combustion in the production of electricity in the United States. Table 1.1 shows that approximately 71% of the electricity produced in 2002 had its origin in the combustion of some fuel [7]. The widespread use of combustion to provide useful forms of energy makes understanding the application of the energy conservation principle to reacting systems particularly important.

Our treatment here is quite basic. More information can be found in books dedicated to combustion (e.g., Refs. [8, 9]). Interestingly, our previous statements of energy conservation, Eqs. 5.6 and 5.10, already apply to reacting systems; however, some elaboration is helpful in their application. We now consider two special cases: constant-pressure combustion and constant-volume combustion.

Constant-Pressure Combustion

Consider a piston–cylinder arrangement (Fig. 5.4) that contains a gaseous system consisting of a mixture of fuel and oxidizer at the initial state and of a mixture of combustion products (i.e., CO_2, H_2O, etc.) at the final state. The freely moving piston ensures that the initial and final states are at the same pressure. We simplify Eq. 5.6 by neglecting the system kinetic energy— we assume the system is at rest—and by assuming that the small change in

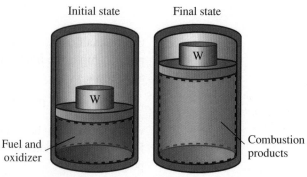

FIGURE 5.4

At the initial state, the piston–cylinder assembly contains an unreacted mixture of fuel and oxidizer. After the combustion process occurs, the cylinder contains the products of combustion.

the system potential energy associated with any change in the system center of mass is insignificant. This is easily shown to be true. With these assumptions, the only contribution to the system energy is the internal energy U; thus, Eq. 5.6 becomes

$$_1(Q_{in,net})_2 - {}_1(W_{out,net})_2 = U_2 - U_1, \tag{5.11}$$

where the subscripts 1 and 2 refer to the unburned and burned mixture states, respectively. We now further assume that the piston is frictionless and that the combustion process occurs sufficiently slowly so that the pressure in the cylinder is uniform and constant. With these assumptions, the work $_1W_2$ is simply evaluated as $P(\mathbb{V}_2 - \mathbb{V}_1)$, and Eq. 5.11 becomes

$$_1(Q_{in,net})_2 - P(\mathbb{V}_2 - \mathbb{V}_1) = U_2 - U_1.$$

The $P\mathbb{V}$ terms are moved to the right-hand side of the equation and we recognize the property enthalpy:

$$_1(Q_{in,net})_2 = U_2 + P\mathbb{V}_2 - U_1 - P\mathbb{V}_1 = H_2 - H_1, \tag{5.12a}$$

or

$$_1(Q_{in,net})_2 = M(h_2 - h_1), \tag{5.12b}$$

where h_1 and h_2 are the specific enthalpies of the reactant and product mixtures, respectively. Since the heat transfer for a combustion process is usually from the hot system to the cooler surroundings, we rewrite Eq. 5.12b and explicitly denote the reactant and product states as

$$Q_{out} = M(h_{reac} - h_{prod}). \tag{5.12c}$$

Recall that standardized enthalpies account for the energy changes resulting from rearrangement of chemical bonds. See Eq. 2.70.

The key to using Eq. 5.12 is the recognition that the enthalpies therein are the standardized enthalpies discussed in Chapter 2. With a knowledge of the initial and final mixture compositions and temperatures, Eq. 5.12c could easily be evaluated to find the heat transfer Q_{out} using tabulated values of enthalpies (Appendix B).

A maximum temperature for a constant-pressure combustion process is defined when the process is adiabatic (i.e., when there is no heat loss from the system). This temperature is referred to as the **constant-pressure adiabatic flame temperature** and is denoted T_{ad}. For this situation, $_1Q_2$ in Eqs. 5.12a and 5.12b is zero. Thus,

$$H_2 - H_1 = 0,$$

or

$$h_2 - h_1 = 0.$$

The subscript 1 refers to reactants at an initial temperature T_{init}, and the subscript 2 refers to products at the adiabatic flame temperature T_{ad}, so we write

$$H_{prod}(T_{ad}) = H_{reac}(T_{init}), \tag{5.13a}$$

or

$$h_{prod}(T_{ad}) = h_{reac}(T_{init}). \tag{5.13b}$$

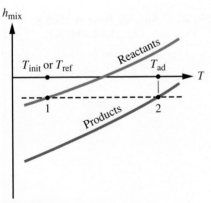

FIGURE 5.5
Standardized enthalpy versus temperature for reactant and product mixtures illustrating the adiabatic flame temperature T_{ad}.

Using h–T plots is invaluable in analyzing combustion problems.

This adiabatic combustion process is shown as a horizontal line on a graph of H (or h) versus T, as illustrated in Fig. 5.5.

Example 5.7

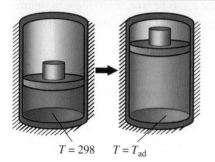

$T = 298$ $T = T_{ad}$

Estimate the constant-pressure adiabatic flame temperature for the combustion of a stoichiometric CH_4–air mixture. The pressure is 1 atm and the initial reactant temperature is 298 K. Assume that the products consist only of undissociated CO_2, H_2O, and N_2 and that the composition of air is 3.76 kmol of N_2 for each kmol of O_2. Neglect any moisture in the air. Furthermore, use constant specific heats evaluated at 1200 K ($\approx [T_{init} + T_{ad}]/2$, where T_{ad} is guessed to be about 2100 K).

Solution

Known CH_4–air, stoichiometric ($\Phi = 1$), T_{init}

Find T_{ad}

Sketch   See Fig. 5.5.

Assumptions

  i. $N_2/O_2 = 3.76$
  ii. Ideal gases
  iii. $c_{p,i} = c_{p,i} (1200 \text{ K})$
  iv. No dissociation of product species

Analysis   For a stoichiometric mixture, all of the carbon and hydrogen in the fuel appear in the CO_2 and H_2O in the products, respectively (see Eqs. 3.58 and 3.59):

$$CH_4 + 2 (O_2 + 3.76 \text{ N}_2) \rightarrow CO_2 + 2H_2O + 2 (3.76) \text{ N}_2;$$

thus, $N_{CO_2} = 1$, $N_{H_2O} = 2$, and $N_{N_2} = 7.52$. We obtain the needed properties from Appendices B and H as follows:

Species	Enthalpy of Formation at 298 K, $\bar{h}^{\circ}_{f,i}$ (kJ/kmol)	Specific Heat at 1200 K, $\bar{c}_{p,i}$ (kJ/kmol·K)
CH_4	$-74{,}831$	—
CO_2	$-393{,}546$	56.21
H_2O	$-241{,}845$	43.87
N_2	0	33.71
O_2	0	—

We apply the first law (Eq. 5.13),

$$H_{prod} = \sum_{prod} N_i \bar{h}_i = H_{reac} = \sum_{reac} N_i \bar{h}_i,$$

where the species enthalpies are evaluated using the assumed (simplified) calorific equation of state:

$$\bar{h}_i(T) = \bar{h}^{\circ}_{f,i}(298) + \bar{c}_{p,i}(T - 298).$$

We now evaluate the reactant and product enthalpies, which are:

$$H_{reac} = (1)(-74{,}831) + 2(0) + 7.52(0) \text{ kJ}$$

$$= -74{,}831 \text{ kJ}$$

$$H_{prod} = \sum N_i[\bar{h}^\circ_{f,i} + \bar{c}_{p,i}(T_{ad} - 298)]$$
$$= (1)[-393,546 + 56.21(T_{ad} - 298)]$$
$$+ (2)[-241,845 + 43.87(T_{ad} - 298)]$$
$$+ (7.52)[0 + 33.71(T_{ad} - 298)].$$

Equating H_{prod} to H_{reac} and solving for T_{ad} yields

$$T_{ad} = 2318 \text{ K}.$$

Comment The value of the adiabatic flame temperature for a stoichiometric mixture given in Table H.1 is 2226 K, which is approximately 100 K lower than our estimate. Our neglect of dissociation and the simplified evaluation of product enthalpies are responsible for this difference.

Self Test
5.6

☑ **Using the assumptions of Example 5.7, determine the heat transferred per kilogram of fuel for the stoichiometric combustion of the CH₄–air mixture if the combustion products are at a temperature of 500 K.**

(Answer: 45,011 kJ/kg$_{CH_4}$. Note use of c_ps at 1200 K is a poor choice.)

Constant-Volume Combustion

We now consider another idealized case: combustion at a fixed volume. In reciprocating internal combustion engines, the combustion process occurs with the piston moving relatively slowly near the top of its stroke. Although the volume is not fixed in this case, the volume change typically is not large. For this reason, constant-volume combustion is frequently used as a primitive model for the combustion event in reciprocating internal combustion engines.

We start our analysis with the same conservation of energy expression presented at the outset of our constant-pressure combustion analysis, Eq. 5.11:

$$_1(Q_{in,net})_2 - {}_1(W_{out,net})_2 = U_2 - U_1.$$

With the volume fixed, there is no compression or expansion work; thus, this expression simplifies to

$$_1(Q_{in,net})_2 = U_2 - U_1. \tag{5.14a}$$

Recognizing again that the heat transfer is most likely from the system to the surroundings and that state 1 is identified with the reactants and state 2 with the products, we rewrite Eq. 5.14a as

$$Q_{out} = U_{reac} - U_{prod}, \tag{5.14b}$$

or

$$Q_{out} = M(u_{reac} - u_{prod}). \tag{5.14c}$$

To evaluate the **constant-volume adiabatic flame temperature**, we set $Q_{out} = 0$; thus,

$$U_{reac}(T_{init}, P_{init}) = U_{prod}(T_{ad}, P_f), \tag{5.15a}$$

where U is the standardized internal energy of the mixture and the subscripts init and f refer to the initial and final states, respectively. Graphically, Eq.

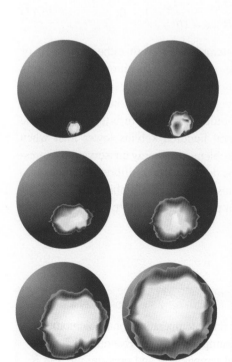

In a spark-ignition engine, a flame initiated at the spark plug propagates across the combustion chamber. Combustion begins before the piston reaches top center and ends after top center. Here the blue regions represent the unburned fuel–air mixture and the red and yellow regions represent the burned gases.

5.15a resembles the sketch used to illustrate the constant-pressure adiabatic flame temperature (Fig. 5.5), except that the internal energy now replaces the enthalpy. Since most compilations or calculations of thermodynamic properties provide values for H (or h) rather than U (or u), we rearrange Eq. 5.15a to the following form:

$$H_{\text{reac}} - H_{\text{prod}} - \mathcal{V}(P_{\text{init}} - P_{\text{f}}) = 0. \tag{5.15b}$$

Applying the ideal-gas law to eliminate the $P\mathcal{V}$ terms,

$$P_{\text{init}}\mathcal{V} = \sum_{\text{reac}} N_j R_u T_{\text{init}} = N_{\text{reac}} R_u T_{\text{init}}$$

$$P_{\text{f}}\mathcal{V} = \sum_{\text{prod}} N_j R_u T_{\text{ad}} = N_{\text{prod}} R_u T_{\text{ad}},$$

we obtain

$$H_{\text{reac}} - H_{\text{prod}} - R_u(N_{\text{reac}} T_{\text{init}} - N_{\text{prod}} T_{\text{ad}}) = 0. \tag{5.16}$$

Equation 5.16 can be expressed on a per-mass-of-mixture basis by dividing by the mass of mixture, M_{mix}, and recognizing that

$$M_{\text{mix}}/N_{\text{reac}} \equiv \mathcal{M}_{\text{reac}}$$

and

$$M_{\text{mix}}/N_{\text{prod}} \equiv \mathcal{M}_{\text{prod}}.$$

We thus obtain

$$h_{\text{reac}} - h_{\text{prod}} - R_u\left(\frac{T_{\text{init}}}{\mathcal{M}_{\text{reac}}} - \frac{T_{\text{ad}}}{\mathcal{M}_{\text{prod}}}\right) = 0. \tag{5.17}$$

To evaluate the adiabatic flame temperature from either Eq. 5.16 or Eq. 5.17 requires knowledge of the mixture composition. Ignoring any species dissociation, we can use simple atom balances as discussed in Chapter 3 (see Eqs. 3.54 and 3.55) to define the composition. At high temperatures, however, dissociation can be quite important, resulting in adiabatic flame temperatures as much as several hundred kelvins lower than when dissociation is ignored. To solve this problem, our energy conservation expressions must be coupled to relationships defining the equilibrium composition of the products. These same caveats also apply to the problem of evaluating constant-pressure adiabatic flame temperatures. For more information, the reader is referred to Ref. [8].

Example 5.8

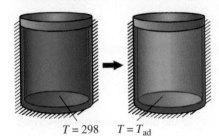

$T = 298 \quad T = T_{\text{ad}}$

Estimate the constant-volume adiabatic flame temperature for a stoichiometric CH_4–air mixture using the same assumptions as in Example 5.7. Also determine the final pressure. The initial temperature and pressure are 298 K and 1 atm, respectively.

Solution

Known CH_4–air, stoichiometric ($\Phi = 1$), T_{init}, P_{init}

Find $T_{\text{ad}}, P_{\text{f}}$

Sketch

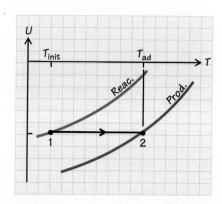

Assumptions

See Example 5.7.

Analysis The same composition and properties used in Example 5.7 apply here. We note, however, that the $\bar{c}_{p,i}$ values should be evaluated at a temperature somewhat greater than 1200 K, since the constant-volume T_{ad} will be higher than the constant-pressure T_{ad}. Nevertheless, we will use the same values as before.

From the first law (Eq. 5.16), we get

$$H_{reac} - H_{prod} - R_u(N_{reac}T_{init} - N_{prod}T_{ad}) = 0,$$

or

$$\sum_{reac} N_j\bar{h}_j - \sum_{prod} N_j\bar{h}_j - R_u(N_{reac}T_{init} - N_{prod}T_{ad}) = 0.$$

Substituting numerical values (see the table in Example 5.7), we have

$$H_{react} = (1)(-74{,}831) + 2(0) + 7.52(0) \text{ kJ}$$
$$= -74{,}831 \text{ kJ},$$

$$H_{prod} = (1)[-393{,}546 + 56.21(T_{ad} - 298)]$$
$$+ (2)[-241{,}845 + 43.87(T_{ad} - 298)]$$
$$+ (7.52)[0 + 33.71(T_{ad} - 298)] \text{ kJ}$$
$$= -877{,}236 + 397.45(T_{ad} - 298) \text{ kJ},$$

and

$$R_u(N_{reac}T_{init} - N_{prod}T_{ad}) = 8.315(10.52)(298 - T_{ad}),$$

where $N_{reac} = N_{prod} = 10.52$ kmol. Reassembling Eq. 5.16 and solving for T_{ad} yields

$$T_{ad} = 2889 \text{ K}.$$

For other fuels, N_{reac} does not equal N_{prod}.

The final pressure is obtained by application of the ideal-gas equation of state (Eq. 2.28a). Since the specific volume is constant,

$$\frac{P_{init}}{T_{init}} = \frac{P_f}{T_{ad}},$$

or

$$P_f = P_{init}\frac{T_{ad}}{T_{init}}.$$

Thus,

$$P_{\mathrm{f}} = (1 \text{ atm}) \frac{2889 \text{ K}}{298 \text{ K}} = 9.69 \text{ atm}.$$

Comments   We note that, for the same initial conditions, the constant-volume adiabatic flame temperature is much higher than that for constant-pressure combustion. We also note the large increase in pressure when the volume is fixed.

 Self Test 5.7   Redo Example 5.8 for a stoichiometric butane (C_4H_{10}) and air mixture.

(Answer: 2987 K, 10 atm)

Example 5.9   SI Engine Application

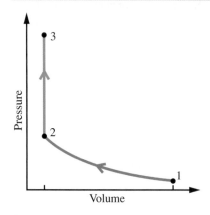

Pressure

Volume

Consider a spark-ignition engine in which the compression and combustion processes have been idealized as a polytropic compression from bottom center (state 1) to top center (state 2) and adiabatic, constant-volume combustion (state 2 to state 3), respectively, as shown in the sketch. Determine the temperature and pressure at states 2 and 3. The engine compression ratio ($CR \equiv V_1/V_2$) is 8, the polytropic exponent is 1.3, and the initial temperature and pressure (state 1) are 298 K and 0.5 atm, respectively. The fuel is isooctane (C_8H_{18}) and the air–fuel ratio is the stoichiometric value. Use the simplified composition for air given in Example 5.7, and assume complete combustion with no dissociation.

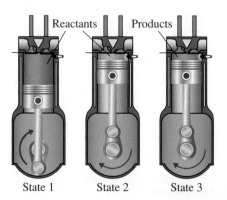

Reactants   Products

State 1   State 2   State 3

Solution

Known   Stoichiometric C_8H_{18}–air mixture, T_1, P_1, V_1, V_2, n

Find P_2, T_2, P_3, T_3

Sketch   See sketch in Example 5.8 for process 2–3.

Assumptions

  i. Ideal-gas behavior
  ii. Polytropic process (1–2)
  iii. Adiabatic, constant-volume process (2–3)

Analysis To determine the properties at state 2, we apply the polytropic process relationship, $PV^n = $ constant (Eq. 2.44), and the ideal-gas equation of state (Eq. 2.28c) as follows:

$$P_2 = P_1 \left(\frac{V_1}{V_2}\right)^n = 0.5(8)^{1.3} \text{ atm}$$

$$= 7.46 \text{ atm or } 756 \text{ kPa}$$

and

$$T_2 = T_1 \left(\frac{P_2}{P_1}\right)\left(\frac{V_2}{V_1}\right) = 298\left(\frac{7.46}{0.5}\right)\left(\frac{1}{8}\right) \text{ K}$$

$$= 556 \text{ K}.$$

For process 2–3, we apply the stoichoimetric combustion equations (Eqs. 3.58 and 3.59) to determine the composition of the reactant and product mixtures:

$$C_xH_y + a(O_2 + 3.76 N_2) \rightarrow xCO_2 + \left(\frac{y}{2}\right)H_2O + 3.76a N_2,$$

where

$$a = x + \frac{y}{4} = 8 + \frac{18}{4} = 12.5,$$

or

$$C_8H_{18} + 12.5(O_2 + 3.76 N_2) \rightarrow 8CO_2 + 9H_2O + 47N_2.$$

To obtain the temperature at state 3, we evaluate the conservation of energy statement (Eq. 5.15a) that

$$U_{reac}(T_2) = U_{prod}(T_3),$$

or, equivalently from a rearrangement of Eq. 5.16,

$$H_{reac}(T_2) = H_{prod}(T_3) + R_u(N_{reac}T_2 - N_{prod}T_3). \tag{A}$$

Expanding the left-hand side of Eq. A yields

$$H_{reac}(T_2) = \sum_{reac} N_j \bar{h}_j(556 \text{ K})$$

$$= N_{C_8H_{18}} \bar{h}_{C_8H_{18}}(556 \text{ K}) + N_{O_2} \bar{h}_{O_2}(556 \text{ K})$$

$$+ N_{N_2} \bar{h}_{N_2}(556 \text{ K}).$$

Evaluating the fuel standardized molar enthalpy (enthalpy of formation plus sensible enthalpy) using the curve-fit coefficients given in Table H.2, and using Tables B.11 and B.7 for the oxygen and nitrogen enthalpies, we obtain

$$H_{reac} = 1(-159,182) + 12.5(0 + 7865) + 47(0 + 7592) \text{ kJ}$$

$$= 295,955 \text{ kJ}.$$

The right-hand side of Eq. A expands as follows:

$$H_{prod}(T_3) + R_u(N_{reac}T_2 - N_{prod}T_3)$$

$$= N_{CO_2} \bar{h}_{CO_2}(T_3) + N_{H_2O} \bar{h}_{H_2O}(T_3) + N_{N_2} \bar{h}_{N_2}(T_3)$$

$$+ R_u(N_{reac}T_2 - N_{prod}T_3).$$

Substituting numerical values for the various numbers of moles, we get

$$H_{prod}(T_3) + R_u(N_{reac}T_2 - N_{prod}T_3)$$
$$= 8\bar{h}_{CO_2}(T_3) + 9\bar{h}_{H_2O}(T_3) + 47\bar{h}_{N_2}(T_3)$$
$$+ R_u(60.5\,T_2 - 64\,T_3),$$

where we have made use of the fact that

$$N_{reac} = 1 + 12.5(1 + 3.76)\text{ kmol} = 60.5\text{ kmol}$$

and

$$N_{prod} = 8 + 9 + 47\text{ kmol} = 64\text{ kmol}.$$

We now solve Eq. A iteratively. Guessing $T_3 = 3000$ K and evaluating the product species enthalpies from Tables B.2 (CO_2), B.6 (H_2O), and B.7 (N_2), we obtain

$$H_{prod}(T_3) + R_u(N_{reac}T_2 - N_{prod}T_3)$$
$$8(-393,546 + 152,891) + 9(-241,845 + 126,563)$$
$$+ 47(0 + 92,730) + 8.3145[60.5(298) - 64(3000)]\text{ kJ}$$
$$= -50,950\text{ kJ}.$$

Comparing the value for the left-hand side of Eq. A ($+295,955$ kJ) with this estimated value for the right-hand side ($-50,950$ kJ), we conclude that our guess for T_3 is too low. Repeating our calculation for $T_3 = 3500$ K yields

$$8(-393,546 + 184,120) + 9(-241,845 + 154,795)$$
$$+ 47(0 + 111,315) + 8.3145[60.5(298) - 64(3500)]\text{ kJ}$$
$$= 1,060,401\text{ kJ}.$$

We thus conclude that the value of T_3 must lie between our two guesses as shown in the following table:

T_3 (K)	$H_{prod} + R_u(N_{reac}T_2 - N_{prod}T_3)$ (kJ)
3000	−50,950
?	295,955
3500	1,060,401

Applying linear interpolation, we estimate that

$$T_3 = 3156\text{ K}.$$

Applying the ideal-gas equation of state for a fixed mass, we obtain the pressure at state 3; that is,

$$M = \frac{P_3 V_3}{RT_3} = \frac{P_2 V_2}{RT_2}.$$

For $V_2 = V_3$, this yields

$$P_3 = P_2 \frac{T_3}{T_2}$$
$$= 7.46\text{ atm }\frac{3156\text{ K}}{556\text{ K}}$$
$$= 42.3\text{ atm or }4290\text{ kPa}.$$

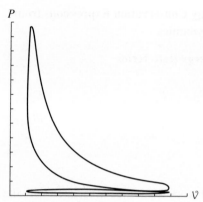

Typical pressure-volume diagram for a 4-stroke-cycle, spark-ignition engine.

Comments   Note that the final temperature and pressure are quite high. Although these values are only estimates, they are typical of real engines. We also note that our estimation of T_3 neglects dissociation of the combustion products, which results in overestimating both T_3 and P_3. Taking dissociation into account yields a final temperature of 2804 K and a pressure of 40.5 atm.

5.2c Special Forms for Conduction Analysis

In this section, we develop the conservation of thermal energy expressions commonly employed in the analysis of conduction heat transfer within solids or stagnant fluids.

Integral (Macroscopic) Systems

Since we are usually interested in heat-transfer *rates*, we begin with the time-rate form of energy conservation, Eq. 5.10, applied, first, to an integral (macroscopic) system. Assuming that the rate of change of the kinetic and potential energies of the system as a whole is negligible, Eq. 5.10 becomes

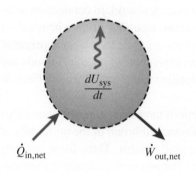

$$\dot{Q}_{\text{in,net}} - \dot{W}_{\text{out,net}} = \frac{dU_{\text{sys}}}{dt}. \tag{5.18}$$

We note that the work can be divided into two contributions: $\dot{W}_{\text{elec}}$, associated with an input of electrical energy, and $\dot{W}_{\text{mech}}$, associated with the free contraction or expansion of the system (i.e., P–$d\mathcal{V}$ work). The mechanical work, however, is usually neglected by assuming that the system is incompressible. With the density constant, $d\mathcal{V}$ is zero. Thus, we write

$$\dot{W}_{\text{out,net}} = -\dot{W}_{\text{elec}},$$

where the negative sign is required since we assume that the electrical work is an input to the system. Substituting this into Eq. 5.18 and rearranging yield

$$\dot{Q}_{\text{in,net}} + \dot{W}_{\text{elec}} = \frac{dU_{\text{sys}}}{dt}. \tag{5.19}$$

Most heat-transfer textbooks use the following conservation of energy statement as a starting point for heat-transfer analyses of systems, which is consistent with the general conservation principle presented in Chapter 1 (Eq. 1.2):

$$\dot{E}_{\text{in}} - \dot{E}_{\text{out}} + \dot{E}_{\text{gen}} = \dot{E}_{\text{stored}}. \tag{5.20}$$

We now wish to reconcile this statement from the domain of heat transfer (Eq. 5.20) with Eq. 5.19 from the domain of thermodynamics. First, the usual meaning of $\dot{E}_{\text{in}}$ and $\dot{E}_{\text{out}}$ is that these represent heat-transfer rates; thus,

$$\dot{E}_{\text{in}} - \dot{E}_{\text{out}} \equiv \dot{Q}_{\text{in,net}} = \dot{Q}_{\text{in}} - \dot{Q}_{\text{out}}. \tag{5.21a}$$

Table 5.1 Comparison of System Energy Conservation Expressions from Heat Transfer and Thermodynamics

Equation	Energy Rate Terms				
5.20	$\dot{E}_{\text{in}} - \dot{E}_{\text{out}}$	$+$	$\dot{E}_{\text{gen}}$	$=$	$\dot{E}_{\text{stored}}$
5.19	$\dot{Q}_{\text{in,net}}$	$+$	$\dot{W}_{\text{elec}}$	$=$	$\dfrac{dU_{\text{sys}}}{dt}$

Furthermore, we recognize the electrical work term to be associated with the thermal energy generation term (actually a transformation of the electrical power to thermal energy); that is,

$$\dot{E}_{\text{gen}} \equiv \dot{W}_{\text{elec}}. \tag{5.21b}$$

Finally, the energy storage term is the usual time rate of change of internal energy:

$$\dot{E}_{\text{stored}} \equiv \frac{dU_{\text{sys}}}{dt}. \tag{5.21c}$$

Table 5.1 summarizes these definitions by aligning the various terms vertically.

In addition to considering the effects of electrical work in conduction heat-transfer analyses, thermal energy generation resulting from chemical or nuclear reactions is sometimes included. In these cases, the physical origin of $\dot{E}_{\text{gen}}$ is a part of dU_{sys}/dt resulting from rearrangements of chemical bonds and nuclear transformations, respectively. A detailed treatment of these effects is beyond the scope of this book.

It is important to point out that no restrictions have been imposed upon either Eq. 5.19 or 5.20 concerning equilibrium or nonequilibrium or the uniformity of any property within our macroscopic (integral) system. Thus, these are very general expressions of energy conservation. The practical difficulty in their use, however, is the evaluation of the system internal energy and the net heat transfer when properties vary throughout the system and over the system boundary, respectively.

For the time being, we can gain some insight into the use and meaning of Eqs. 5.19 and 5.20 by making the assumption that the temperature is uniform throughout the system as the system heats or cools. This is sometimes referred to as the **lumped parameter approximation**. This approximation is valid when the internal resistance to heat transfer is much less than the resistance to heat transfer at the boundary. For a solid immersed in a convective environment, the **Biot number** Bi, a dimensionless parameter, expresses this ratio of internal to external thermal resistances:

$$Bi \equiv \frac{\overline{h}_{\text{conv}} L_{\text{char}}}{k_{\text{s}}}, \tag{5.22}$$

where $\overline{h}_{\text{conv}}$ is the average convective heat-transfer coefficient at the system boundary, k_{s} is the thermal conductivity of the system, and L_{char} is a characteristic length. This characteristic length is defined as the ratio of the volume of the system to the surface area exposed to the convective environment; that is,

$$L_{\text{char}} \equiv \frac{V_{\text{system}}}{A_{\text{conv}}}. \tag{5.23}$$

For example, L_{char} for a sphere of radius R is

$$L_{\text{char}} = \frac{(4/3)\pi R^3}{4\pi R^2} = R/3.$$

The usual engineering criterion to invoke the lumped approximation is that

$$Bi < 0.1. \tag{5.24}$$

The smaller the Biot number is, the more uniform the system temperature must be during a transient process.

To introduce the temperature as an explicit variable in our conservation of energy expression, we recognize that

$$\frac{dU}{dt} \equiv M\frac{du}{dt},$$

and we apply the chain rule and the definition of the specific heat (Eq. 2.19) to yield

$$\frac{du}{dt} \equiv \left(\frac{\partial u}{\partial T}\right)\frac{dT}{dt} = c_v\frac{dT}{dt}.$$

With the assumption of incompressibility, $c_v = c_p = c$ (Eq. 2.56b), our final result becomes

$$\dot{Q}_{\text{in,net}} + \dot{W}_{\text{elec}} = M_{\text{sys}}c\frac{dT}{dt}. \tag{5.25}$$

We emphasize that Eq. 5.25 is only applicable to a macroscopic system if the temperature is uniform (or sufficiently so for an engineering approximation, i.e., $Bi < 0.1$). As we will see in the next section, Eq. 5.25 is a useful starting point in the description of differential systems.

As a practical note, $\dot{W}_{\text{elec}}$ can be related to the electrical current i and the system electrical resistance R_e by

$$\dot{W}_{\text{elec}} = i^2 R_e. \tag{5.26}$$

Equation 5.26 expresses the concept of **Joule heating**, named in honor of its discoverer James Prescott Joule [10].

Example 5.10 Biological Systems Application

Consider a 6-cm-diameter orange growing on a tree. A cold front causes the ambient temperature to drop rapidly from 50 F to 30 F. Estimate the initial rate of temperature change (dT/dt) of the orange if the convective heat-transfer coefficient is approximately 1.5 W/m²·K. Assume the orange is initially at 50 F. The density and specific heat of the orange are approximately 850 kg/m³ and 3770 J/kg·K, respectively [11].

Solution

Known $D, \rho, c, T_i, T_\infty, h_{\text{conv}}$

Find dT/dt

Sketch

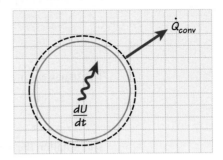

Assumptions

Orange cools with uniform temperature ($Bi < 0.1$)

Analysis We denote the orange as our thermodynamic system and begin by testing our assumption that $Bi < 0.1$. Using an estimated thermal conductivity of 0.5 W/m·K (see Appendix I for foodstuffs) and a characteristic length of $R/3 = D/6 = 0.01$ m for the orange, we obtain a Biot number of

$$Bi = \frac{\overline{h}_{\text{conv}}(D/6)}{k_{\text{orange}}} = \frac{1.5(0.01)}{0.5} = 0.03.$$

Because 0.03 is smaller than 0.1, the lumped approximation applies. We now apply conservation of energy by simplifying Eq. 5.25:

$$-\dot{Q}_{\text{conv}} + 0 = M_{\text{sys}}c\frac{dT_{\text{orange}}}{dt},$$

noting that $\dot{Q}_{\text{conv}}$ has a negative sign since it is a transfer *out* of the system. We use Eq. 4.18 to express the convection heat-transfer rate as

$$Q_{\text{conv}} = h_{\text{conv}}A(T_{\text{orange}} - T_{\infty}),$$

where A is the surface area $(= \pi D^2)$. The mass of the orange is the product of the density and the volume,

$$M_{\text{sys}} = \rho \mathcal{V} = \rho\pi D^3/6.$$

Substituting these expressions into our energy balance and solving for dT_{orange}/dt yield

$$\frac{dT_{\text{orange}}}{dt} = -\frac{h_{\text{conv}}(T_{\text{orange}} - T_{\infty})}{(D/6)(\rho c)_{\text{orange}}}.$$

Converting the given temperatures to SI units, we numerically evaluate this as

$$\frac{dT_{\text{orange}}}{dt} = -\frac{1.5(283.15 - 272.0)}{(0.06/6)850(3770)} = 5.2 \times 10^{-4}$$

$$[=]\frac{(\text{W/m}^2\cdot\text{K})\text{K}}{\text{m}(\text{kg/m}^3)\text{J/kg}\cdot\text{K}}\left[\frac{1\text{ J/s}}{\text{W}}\right] = \text{K/s},$$

or

$$= 1.9 \text{ K/hr}.$$

Comment To determine the temperature–time relationship describing the cooling of the orange requires the solution of the original energy conservation equation, an ordinary differential equation, and not simply its algebraic rearrangement as was done here to obtain the initial cooling rate.

Self Test 5.8

✓ Show that $d(T - T_\infty)/dt = dT/dt$ and then perform the integration to develop an expression for the temperature of the orange in Example 5.10 as a function of time.

(Answer: $(T(t)_{orange} - T_\infty)/(T_{orange,init} - T_\infty) = \exp[-h_{conv}/((D/6)(\rho c)_{orange})t])$

Example 5.11

$t = 0$ $t = t$ $t = \infty$

Consider the tungsten light-bulb filament as modeled in Example 5.6. Estimate the initial rate of temperature rise of the filament at the instant after the current starts to flow. The current is 0.9 A and the voltage drop is 110 V. Assume the filament is initially in equilibrium with its surroundings at 25°C. The density and heat capacity of tungsten are 19,300 kg/m³ and 132 J/kg·K, respectively.

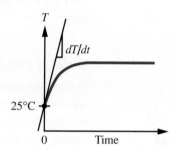

Solution

Known $D, L, i, \mathcal{V}, T_{init}, \rho, c$

Find dT_f/dt

Sketch

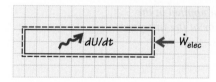

Assumptions

i. Incompressible solid
ii. Filament heats with uniform temperature

Analysis   The filament is chosen as the system of interest. Since the filament is initially at the same temperature as the surroundings, there is no heat transfer. (After the filament heats, however, heat transfer will be significant.) As shown in the sketch, all of the electrical power is used to increase the internal energy of the filament. This is expressed by Eq. 5.25 as

$$\dot{W}_{elec} = Mc\frac{dT_f}{dt},$$

which we can solve for the time rate of temperature change:

$$\frac{dT_f}{dt} = \frac{\dot{W}_{elec}}{Mc},$$

where

$$M = \rho\mathcal{V} = \rho\pi D^2 L/4$$
$$= 19,300\,\pi(0.001)^2\,0.025/4 = 3.789 \times 10^{-4}\,kg.$$

The electrical power is the product of the current flow and voltage drop (Eq. 4.11b), so

$$\dot{W}_{elec} = i\Delta\mathcal{V}$$
$$= 0.9(110) \text{ W} = 99 \text{ W}.$$

Numerically evaluating dT_f/dt yields

$$\frac{dT_f}{dt} = \frac{99}{3.789 \times 10^{-4}(132)} = 1979$$

$$[=] \frac{\text{W}}{\text{kg(J/kg}\cdot\text{K)}}\left[\frac{1 \text{ J/s}}{\text{W}}\right] = \text{K/s}.$$

Comment Consistent with our experience, the filament heats very quickly. The complete heating problem requires the solution of Eq. 5.10 for $T_f(t)$ with the inclusion of both convection and radiation heat-transfer rates.

Surfaces and Interfaces

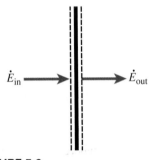

FIGURE 5.6
Energy conservation applied at the surface of a system or at the interface between two systems.

In many analyses, applying the energy conservation principle at the boundary separating the system from the surroundings is useful. Figure 5.6 illustrates a surface energy balance. Since the surface has no thickness (nor mass), it cannot store energy; hence, $\dot{E}_{stored}$ is zero. Similarly, since the surface has no volume, there cannot be any thermal energy generated from Joule heating, or other volumetric conversion effects; hence, $\dot{E}_{gen} = 0$. With these ideas, our system energy conservation expression (Eq. 5.20),

$$\dot{E}_{in} - \dot{E}_{out} + \dot{E}_{gen} = \dot{E}_{stored},$$

reduces to

$$\dot{E}_{in} - \dot{E}_{out} + 0 = 0,$$

or

$$\dot{E}_{in} = \dot{E}_{out}. \tag{5.27}$$

Figure 5.6 schematically illustrates this mathematical statement. Note that here *in* and *out* refer to the surface itself. Whether energy is directed into or out of the system proper depends on which side of the surface the system of interest lies.

The mode of heat transfer switches from convection to conduction at the interface between the boiling water and the potato (left). An aluminized mylar blanket temporarily shielded the damaged Skylab 2 orbital workshop from radiation (center); image courtesy of NASA. Front brake of Indy race car (right); the heat generated by friction at the interface between the pad and the rotor is conducted through both.

FIGURE 5.7

Surface energy balances for a conducting solid on the left side of the boundary and a convective or radiative surroundings (top) or another solid system (bottom) on the right side.

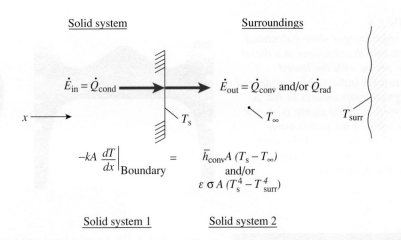

Solid system   Surroundings

$\dot{E}_{in} = \dot{Q}_{cond}$ $\dot{E}_{out} = \dot{Q}_{conv}$ and/or $\dot{Q}_{rad}$

$x \longrightarrow$ T_s T_∞ T_{surr}

$$-kA \left.\frac{dT}{dx}\right|_{Boundary} = \begin{array}{c} \bar{h}_{conv} A\,(T_s - T_\infty) \\ \text{and/or} \\ \varepsilon\,\sigma\,A\,(T_s^4 - T_{surr}^4) \end{array}$$

Solid system 1   Solid system 2

$\dot{E}_{in} = \dot{Q}_{cond,\,1}$ $\dot{E}_{out} = \dot{Q}_{cond,\,2}$

x

$$-k_1 A \left.\frac{dT}{dx}\right|_{sys\,1} = -k_2 A \left.\frac{dT}{dx}\right|_{sys\,2}$$

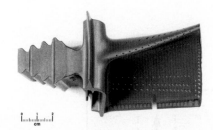

Air flows through holes to cool a jet-engine turbine blade. Photograph courtesy of NASA.

Concentrating photovoltaic system employs small tracking mirrors to focus sunlight on fin-cooled solar cells. Photograph courtesy of DOE/NREL.

Because we are dealing with solid (or stagnant-liquid) systems, conduction heat transfer will always be the heat-transfer mode on the system side of the surface. On the surroundings side, the energy transfer mode may be convection and/or radiation. If the system of interest is in intimate contact with another solid, then conduction will also be the energy transfer mode on the surroundings side. Figure 5.7 illustrates these scenarios where appropriate rate laws express $\dot{E}_{in}$ and $\dot{E}_{out}$.

We emphasize that Eq. 5.27 and the relationships shown in Fig. 5.7 apply to both steady and unsteady conditions, both with and without volumetric thermal energy generation.

Differential (Microscopic) Systems

Governing Equations In this section, we develop conservation of energy expressions that, when coupled with Fourier's law for conduction (Eqs. 4.13 and 4.14), provide a means to determine temperature distributions and local heat fluxes within solid systems. Being able to calculate temperature distributions and heat fluxes is vitally important in many engineering applications. For example, the design of pistons in reciprocating engines involves the problem of keeping the piston cool and controlling thermal expansion to maintain proper mechanical clearances [12], both of which depend on knowledge of the temperature distribution within the piston (Fig. 5.8). The design of cooling systems for electronic components (Fig. 5.9) also depends critically on knowing temperature distributions and heat fluxes within complex systems. A long list of applications could be compiled: compressor and turbine blades, rocket engine combustors and nozzles, solar energy collection systems, automotive and aircraft brake systems, etc.

Within the larger domain of an incompressible system ($0 \leq x \leq L$), consider the planar, one-dimensional system defined by the dashed lines in Fig. 5.10a. The length of this system in the x-direction is Δx, whereas the system is of infinite extent in the y- and z-directions. The area perpendicular to the x-direction

FIGURE 5.8
False color images show calculated temperature distributions in a diesel engine piston with the lowest temperatures indicated in purple and the highest temperatures in red. A cross-section of the piston is shown on the left and the piston top is shown in perspective on the right. **Images courtesy of Rolf Reitz and James Wiedenhoefer.**

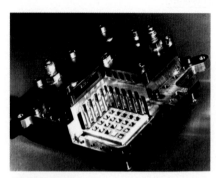

FIGURE 5.9
The Thermal Conduction Modules used in IBM 308X computers contained 133 chips collectively producing 300 W of heat. Aluminum pistons conduct the heat from the chips to a chamber filled with helium. The chamber, in turn, is cooled by a flow of chilled water. **Photograph courtesy of IBM Corporation.**

is designated A. We assume local equilibrium conditions (see Chapter 1) prevail throughout the entire domain; hence, temperature is a meaningful property in our differential system. Arrows at the system boundaries (Fig. 5.10a) represent the heat transfer at x and $x + \Delta x$. Since the system boundary is embedded within the bulk solid, the only possible mode of heat transfer is conduction. Squiggly arrows drawn inside the system boundaries represent thermal energy storage $\dot{E}_{stored}$ and thermal energy generation $\dot{E}_{gen}$, where the latter has its origin with the electrical work done on the system.

We begin our analysis with Eq. 5.20, substituting Fourier's law for the conduction terms and $dU_{sys}/dt = M_{sys}\, c\, \partial T/\partial t$ for the storage term in Eq. 5.21c:

$$\dot{E}_{in} \quad - \quad \dot{E}_{out} \quad + \quad \dot{E}_{gen} \quad = \quad \dot{E}_{stored},$$

or

$$\left[-kA\frac{\partial T}{\partial x} \right]_x - \left[-kA\frac{\partial T}{\partial x} \right]_{x+\Delta x} + \dot{E}_{gen} = \rho A\, \Delta x\, c\frac{\partial T}{\partial t}. \quad (5.28)$$

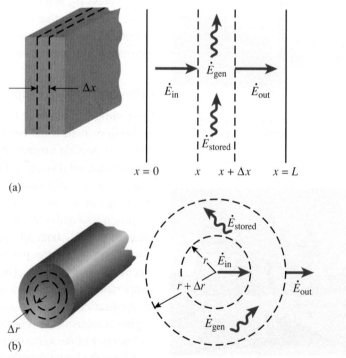

(a)

(b)

FIGURE 5.10
One-dimensional systems for differential analysis of conduction heat transfer in solids for (a) a planar (Cartesian) system and (b) cylindrical or spherical systems.

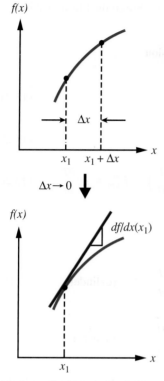

Graphical interpretation of a derivative.

Two features are worth noting. First, we use partial derivatives (e.g., $\partial T/\partial t$) because the temperature is a function of both x and t; second, we have substituted for the system mass, $M_{sys} = \rho \Delta \mathcal{V} = \rho A \Delta x$. Assuming that $\dot{E}_{gen}$ is a given quantity, all of the remaining terms contain a single dependent variable: the temperature T. We now mathematically manipulate Eq. 5.28 to obtain a partial differential equation, the solution of which is the temperature distribution as a function of time, $T(x, t)$. Dividing Eq. 5.28 by $A\Delta x$, rearranging, and applying the limit $\Delta x \to 0$, we obtain

$$\lim_{\Delta x \to 0} \left[\frac{\left[k\dfrac{\partial T}{\partial x} \right]_{x+\Delta x} - \left[k\dfrac{\partial T}{\partial x} \right]_{x}}{\Delta x} + \frac{\dot{E}_{gen}}{A\Delta x} = \rho c \frac{\partial T}{\partial t} \right].$$

In this equation, we recognize the definition of a derivative for the bracketed quantity $k\partial T/\partial x$; that is,

$$\lim_{\Delta x \to 0} \frac{\left[k\dfrac{\partial T}{\partial x} \right]_{x+\Delta x} - \left[k\dfrac{\partial T}{\partial x} \right]_{x}}{\Delta x} \equiv \frac{\partial \left[k\dfrac{\partial T}{\partial x} \right]}{\partial x}.$$

The second term we define to be the **thermal energy generation rate per unit volume**, given by

$$\lim_{\Delta x \to 0} \frac{\dot{E}_{gen}}{A\Delta x} \equiv \dot{E}_{gen}''', \tag{5.29a}$$

or, more generally,

$$\lim_{\Delta \mathcal{V} \to 0} \frac{\dot{E}_{gen}}{\Delta \mathcal{V}} \equiv \dot{E}'''. \tag{5.29b}$$

Reassembling, we write

$$\frac{\partial}{\partial x}\left(k\frac{\partial T}{\partial x} \right) + \dot{E}_{gen}''' = \rho c \frac{\partial T}{\partial t}. \tag{5.30}$$

In many situations, the thermal conductivity k can be treated as a constant by using an appropriate mean value. With the assumption of constant conductivity, Eq. 5.30 further simplifies to

$$\frac{\partial^2 T}{\partial x^2} + \frac{\dot{E}_{gen}'''}{k} = \frac{\rho c}{k} \frac{\partial T}{\partial t}. \tag{5.31}$$

Equation 5.31 is known as the **heat-conduction equation** applied to a one-dimensional (Cartesian) system. Appearing in Eq. 5.31 is the combination of thermophysical properties $\rho c/k$. The reciprocal of this combination,

A review of the Chapter 2 discussion of thermal conductivity is useful at this point.

$$\alpha \equiv \left[\frac{\rho c}{k} \right]^{-1} = \frac{k}{\rho c}, \tag{5.32}$$

is known as the **thermal diffusivity**, as indicated previously in Chapter 2.

Similar one-dimensional analyses can be performed for cylindrical and spherical systems (Fig. 5.10b) where r is the only position variable. These analyses are left for the reader (see Problems 5.63 and 5.64). The results of

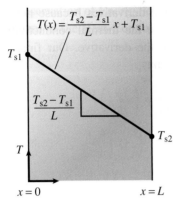

FIGURE 5.11
The steady-state temperature distribution through a plane wall is linear in the absence of internal thermal energy generation, provided the thermal conductivity is constant.

We apply these boundary conditions to Eq. 5.36. At the left face ($x = 0$),

$$T(0) = T_{s1} = C_1(0) + C_2.$$

Thus,

$$C_2 = T_{s1}. \tag{5.38a}$$

At the right face ($x = L$),

$$T(L) = T_{s2} = C_1 L + C_2,$$

or

$$T_{s2} = C_1 L + T_{s1}.$$

Thus,

$$C_1 = \frac{T_{s2} - T_{s1}}{L}. \tag{5.38b}$$

Substituting these expressions for C_1 and C_2 back into the general solution (Eq. 5.36), we obtain our final result:

$$T(x) = \frac{T_{s2} - T_{s1}}{L}x + T_{s1}. \tag{5.39}$$

From Eq. 5.39, we see that the steady-state temperature distribution in a plane wall is linear, subject, of course, to the assumptions listed earlier. Figure 5.11 illustrates this result.

The heat-transfer rate through the wall, $\dot{Q}_{cond}(x)$, can be obtained by applying Fourier's law (Eq. 4.13) to this distribution, that is,

> **This is a key relationship from Chapter 4.** ▶

$$\dot{Q}_{cond}(x) = -kA\frac{dT(x)}{dx}.$$

From Eq. 5.39, we see that the temperature gradient in the wall is

$$\frac{dT(x)}{dx} = \frac{T_{s2} - T_{s1}}{L}. \tag{5.40}$$

Thus, $\dot{Q}_{cond}(x)$ is a constant and is given by

$$\dot{Q}_{cond}(x) = \dot{Q}_{cond} = -kA\left(\frac{T_{s2} - T_{s1}}{L}\right),$$

or

$$\dot{Q}_{cond} = \frac{kA}{L}(T_{s1} - T_{s2}). \tag{5.41}$$

The heat flux ($\dot{Q}_{cond}/A$) is also constant through the wall since the area A does not vary with x-location; thus,

$$\dot{Q}''_{cond}(x) = \dot{Q}''_{cond} = \frac{k}{L}(T_{s1} - T_{s2}). \tag{5.42}$$

Equations 5.41 and 5.42 are useful for calculating heat-transfer rates when the surface temperatures are known or are easily determined. We illustrate this utility with the following example.

Example 5.12

Consider one-dimensional, steady conduction through a 6-mm-thick plane slab of glass ($k = 1.4$ W/m·K). A heat flux of 150 W/m² enters at the left face of the slab, which is maintained at 35°C. Determine the temperature at the right face. Repeat the computation for a wood slab of the same thickness ($k = 0.11$ W/m·K) and compare the results.

Solution

Known $L, k, T_{s1}, \dot{Q}''$

Find T_{s2}

Sketch

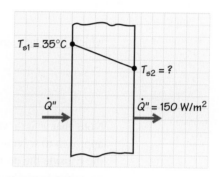

$T_{s1} = 35°C$

$T_{s2} = ?$

$\dot{Q}''$

$\dot{Q}'' = 150$ W/m²

Assumptions

  i. One-dimensional geometry
  ii. Steady state
  iii. Constant thermal conductivity

Analysis   With these assumptions, a linear temperature distribution results. To determine the temperature of the right face, we rearrange Eq. 5.42 as

$$T_{s2} = T_{s1} - \frac{L\dot{Q}''}{k}.$$

For the glass,

$$T_{s2} = 35 - \frac{0.006(150)}{1.4} = 34.4$$

$$[=]\frac{\text{m (W/m}^2)}{\text{W/(m}^2\cdot\text{K)}}\left[\frac{1°C}{K}\right] = °C,$$

and for the wood,

$$T_{s2} = \left[35 - \frac{0.006(150)}{0.11}\right]°C = 26.8°C.$$

Comments   As a result of the much smaller thermal conductivity of the wood, the temperature difference across the wood slab ($\Delta T = 8.2°C$) is many times greater than that across the glass ($\Delta T = 0.6°C$). As can be seen from Eq. 5.42, the temperature difference is inversely proportional to the thermal conductivity.

Self Test
5.9

   A conducting solid wall is in contact with a fluid as in Fig. 5.7 (top). A thermocouple measures a temperature T_L at a distance L within the wall. Neglecting radiation, develop an expression for temperature T_s at the solid–fluid interface.

(Answer: $T_s = [(\bar{h}_{conv}L/k)T_\infty + T_L]/[(\bar{h}_{conv}L/k) + 1])$

We now consider cylindrical geometry (Fig. 5.10b). This geometry is useful for representing the wires, pipes, tubes, and other objects frequently encountered in engineering applications. Using the same assumptions as before, we eliminate the thermal energy generation and storage terms from Eq. 5.33, that is,

$$\frac{1}{r}\frac{\partial}{\partial r}\left(kr\frac{\partial T}{\partial r}\right) + \dot{E}'''_{\text{gen}} = \rho c \frac{\partial T}{\partial t}.$$

$$0 \qquad\qquad 0$$

No generation   Steady state

Thus,

$$\frac{1}{r}\frac{d}{dr}\left(kr\frac{dT}{dr}\right) = 0,$$

which is further simplified with the constant-property assumption to yield

$$\frac{d}{dr}\left(r\frac{dT}{dr}\right) = 0. \qquad (5.43)$$

Note that r is a variable and cannot be removed from the derivative. The approach to solving Eq. 5.43 is similar to that for the planar geometry; although the r inside the derivative makes the solution a bit more interesting. As before, we separate and integrate twice. The first separation and integration yields

$$\int d\left(r\frac{dT}{dr}\right) = \int 0\, dr,$$

or

$$r\frac{dT}{dr} = C_1.$$

The second separation and integration yields

$$\int dT = \int C_1 \frac{dr}{r},$$

or

$$T(r) = C_1 \ln r + C_2, \qquad (5.44)$$

where C_1 and C_2 are constants of integration whose values are determined from the two boundary conditions.

We now consider the particular situation of a hollow cylinder with specified surface temperatures (see Fig. 5.12) to provide the following two boundary conditions for the solution of Eq. 5.44:

$$T(r_1) = T_{s1} \qquad (5.45a)$$

and

$$T(r_2) = T_{s2}. \qquad (5.45b)$$

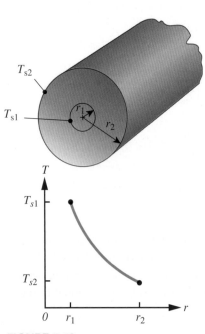

FIGURE 5.12
The steady-state temperature distribution through a hollow cylinder is logarithmic when the thermal conductivity is constant and there is no thermal energy generation ($\dot{E}'''_{\text{gen}} = 0$).

We use these boundary conditions to eliminate C_1 and C_2 as follows:

$$T(r_1) = T_{s1} = C_1 \ln r_1 + C_2$$

and

$$T(r_2) = T_{s2} = C_1 \ln r_2 + C_2.$$

Simultaneous solution of these two equations yields

$$C_1 = \frac{T_{s2} - T_{s1}}{\ln r_2 - \ln r_1} = \frac{T_{s2} - T_{s1}}{\ln r_2/r_1}$$

and

$$C_2 = T_{s2} - C_1 \ln r_2.$$

Substituting these expressions back into the general solution (Eq. 5.44) yields our final result for the temperature distribution:

$$T(r) = \frac{T_{s2} - T_{s1}}{\ln(r_2/r_1)} \ln(r/r_2) + T_{s2}. \tag{5.46}$$

Figure 5.12 illustrates this logarithmic distribution for the case where the temperature at the inner radius is greater than that at the outer radius.

We now use this temperature distribution to obtain an expression for the heat-transfer rate. For the cylindrical geometry, Fourier's law (Eq. 4.16a) is expressed as

$$\dot{Q}_{cond}(r) = -k2\pi rL \frac{dT(r)}{dr}, \tag{5.47}$$

where L is the length of the cylinder (where $L \gg r_2$ for the 1-D approximation to hold) and $2\pi rL$ is the area perpendicular to r at any location within the cylinder. We differentiate Eq. 5.46 to obtain the following expression for the local temperature gradient:

$$\frac{dT(r)}{dr} = \frac{T_{s2} - T_{s1}}{\ln(r_2/r_1)} \frac{1}{r}. \tag{5.48}$$

Combining Eqs. 5.47 and 5.48 yields

$$\dot{Q}_{cond}(r) = \dot{Q}_{cond} = \frac{2\pi Lk}{\ln(r_2/r_1)}(T_{s1} - T_{s2}). \tag{5.49}$$

As expected, the steady-state conduction rate through the cylinder is a constant (i.e., it is independent of radial location). The heat flux, however, does depend on radial position since the area is directly proportional to r, that is, $A(r) = 2\pi rL$; thus,

$$\dot{Q}''_{cond}(r) = \frac{\dot{Q}_{cond}}{A(r)} = \frac{k}{\ln(r_2/r_1)}(T_{s1} - T_{s2})\frac{1}{r}. \tag{5.50}$$

The analysis of a hollow sphere follows that of the hollow cylinder and the following relationships result:

$$T(r) = \left[\frac{1 - (r_1/r)}{1 - (r_1/r_2)}\right](T_{s2} - T_{s1}) + T_{s1}, \tag{5.51}$$

$$\dot{Q}_{cond}(r) = \dot{Q}_{cond} = \frac{4\pi k}{(1/r_1 - 1/r_2)}(T_{s1} - T_{s2}), \tag{5.52}$$

and

$$\dot{Q}''_{cond}(r) = \frac{\dot{Q}_{cond}}{4\pi r^2} = \frac{k}{r^2[(1/r_1) - (1/r_2)]}(T_{s1} - T_{s2}). \tag{5.53}$$

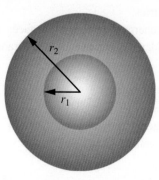

The inner and outer radii, r_1 and r_2, respectively, define constant-temperature surfaces for a 1-D conduction analysis of a hollow sphere.

Table 5.3 summarizes the results for the three 1-D geometries analyzed.

Table 5.3   Summary of 1-D, Steady-State Conduction Analyses with No Heat Generation

Geometry	Governing Equation[*]	Temperature Distribution (T)	Heat-Transfer Rate ($\dot{Q}$)	Heat Flux ($\dot{Q}''$)
Plane wall	$\dfrac{d}{dx}\left(\dfrac{dT}{dx}\right) = 0$	$T(x) = \dfrac{T_{s2} - T_{s1}}{L}x + T_{s1}$	$\dfrac{kA}{L}(T_{s1} - T_{s2})$	$\dfrac{k}{L}(T_{s1} - T_{s2})$
Hollow cylinder	$\dfrac{d}{dr}\left(\dfrac{1}{r}\dfrac{dT}{dr}\right) = 0$	$T(r) = \dfrac{T_{s2} - T_{s1}}{\ln(r_2/r_1)}\ln(r/r_2) + T_{s2}$	$\dfrac{2\pi Lk}{\ln(r_2/r_1)}(T_{s1} - T_{s2})$	$\dfrac{k(T_{s1} - T_{s2})}{\ln(r_2/r_1)}\dfrac{1}{r}$
Hollow sphere	$\dfrac{d}{dr}\left(\dfrac{1}{r^2}\dfrac{dT}{dr}\right) = 0$	$T(r) = \left[\dfrac{1 - (r_1/r)}{1 - (r_1/r_2)}\right](T_{s2} - T_{s1}) + T_{s1}$	$\dfrac{4\pi k}{(1/r_1 - 1/r_2)}(T_{s1} - T_{s2})$	$\dfrac{k(T_{s1} - T_{s2})}{r^2[(1/r_1) - (1/r_2)]}$

[*] A constant thermal conductivity is assumed.

Table 5.4   Thermal Resistances for 1-D Geometries

Configuration	Thermal Resistances (R_{th})
Plane wall	$\dfrac{L}{kA}$
Hollow cylinder	$\dfrac{\ln(r_2/r_1)}{2\pi Lk}$
Hollow sphere	$\dfrac{(1/r_1 - 1/r_2)}{4\pi k}$
Convection at surface	$\dfrac{1}{\overline{h}_{conv}A_{surf}}$

> **Equation 5.54 is a key relationship and greatly facilitates problem solving.**

Electric Circuit Analogy

The mathematical form of the various heat-transfer rate expressions suggests an electric circuit analogy for heat transfer. For example, consider the rearrangement of Eq. 5.41 expressing the conduction heat-transfer rate through a plane wall:

$$\dot{Q}_{\substack{plane \\ wall}} = \frac{kA}{L}(T_{s1} - T_{s2}) = \frac{T_{s1} - T_{s2}}{(L/kA)}.$$

Here, the heat transfer rate is analogous to an electric current (i), the numerator on the right-hand side is analogous to a voltage difference (ΔV), and the denominator is analogous to an electrical resistance (R_e). Ohm's law for current flow is

$$i = \frac{\Delta V}{R_e},$$

and the heat-transfer analog is thus

$$\dot{Q} = \frac{\Delta T}{R_{th}}, \tag{5.54}$$

where R_{th} is defined to be the thermal resistance. For the plane wall, the thermal resistance is given by

$$R_{th,\,plane\,wall} = L/(kA). \tag{5.55}$$

Figure 5.13 schematically illustrates this analogy. Table 5.4 summarizes the thermal resistances for our three 1-D geometries.

The circuit analogy can be extended to energy balances at surfaces and interfaces. In any of the three geometries considered, the convective heat-transfer rate is expressed (Eq. 4.18) as

> **This is an important relationship from Chapter 4.**

$$\dot{Q}_{conv} = \overline{h}_{conv}A(T_s - T_\infty),$$

where T_s is the temperature of the surface exposed to a fluid at T_∞. Rearranging this equation to the Ohm's law analog yields

$$\dot{Q}_{conv} = \frac{T_s - T_\infty}{\left(\dfrac{1}{\overline{h}_{conv}A}\right)}. \tag{5.56}$$

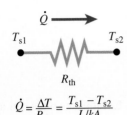

$$\dot{Q} = \frac{\Delta T}{R_{th}} = \frac{T_{s1} - T_{s2}}{L/kA}$$

FIGURE 5.13
Electric circuit analog for steady conduction heat transfer through a plane wall.

Thus, the thermal resistance associated with convection is

$$R_{th,conv} \equiv \frac{1}{\bar{h}_{conv}A}. \qquad (5.57)$$

Although a linearized form of the "rate law" for radiation heat transfer can be used to define a radiation thermal resistance similar to the one derived here, we will not consider this form here and refer the interested reader to other textbooks (e.g., Ref. [13]).

The circuit analog is particularly useful in dealing with systems where surface temperatures are unknown but information is provided about the convective environment(s) at the surface(s). Similarly, composite systems are easily analyzed using the circuit analogy. These two cases are illustrated in Fig. 5.14. For a series arrangement of thermal resistances, the heat-transfer

FIGURE 5.14
Electric circuit analogs for (a) conduction through a plane wall with convection at surfaces, (b) conduction through a 1-D composite wall, and (c) a 1-D approximation to conduction through a 2-D composite wall.

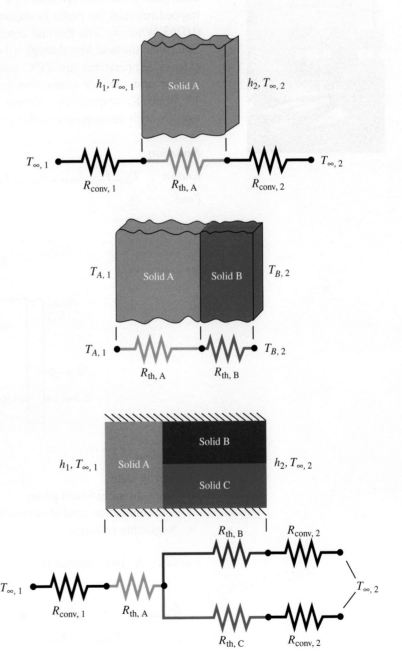

Using the Ohm's law analogs (Eqs. 5.54 and 5.58) is equivalent to writing an overall energy balance.

rate is expressed as

$$\dot{Q} = \frac{\Delta T_{OA}}{\sum R_{th}},$$

(5.58)

where ΔT_{OA} is the overall temperature difference. Composite solids can also be modeled as series–parallel combinations of thermal resistances. For this situation, the sum of the thermal resistances appearing in Eq. 5.58 represents the series–parallel combination. These ideas are illustrated in the following examples.

Example 5.13 Solar Heating Application

Double-pane windows are installed in a solar-heated house. The 3-mm-thick glass panes are separated by a 10-mm air space. Assume that the air trapped between the panes is stagnant and has a thermal conductivity of 0.0263 W/m^2·K. The thermal conductivity of the glass is 1.4 W/m·K. Determine the heat loss through a 1-m by 1-m window if the interior and exterior temperatures are 21°C and 0°C, respectively, and the average interior and exterior convective heat-transfer coefficients are 2.6 and 15 W/m^2·K, respectively. Repeat the calculations for a single-pane window with the same convective conditions and compare results.

Solution

Known $T_{\infty,i}$, $T_{\infty,o}$, $\bar{h}_{conv,i}$, $\bar{h}_{conv,o}$, k_{glass}, k_{air}, geometry (see sketch)

Find $\dot{Q}$

Sketch

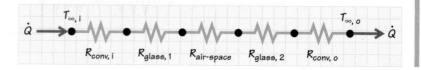

Assumptions

i. Steady state
ii. Stagnant air between panes
iii. 1-D conduction (end effects negligible)
iv. Negligible radiation

Analysis   A series thermal circuit can be used to represent this situation:

We calculate numerical values for the five thermal resistances, recognizing that the area A is 1 m^2, as follows:

$$R_{\text{conv,i}} = \frac{1}{\bar{h}_i A} = \frac{1}{2.6(1)} = 0.385 \text{ K(or °C)/W},$$

$$R_{\text{glass,1}} = R_{\text{glass,2}} = \frac{L_g}{k_g A} = \frac{0.003}{1.4(1)} = 0.002 \text{ K(or °C)/W},$$

$$R_{\text{air space}} = \frac{L_{a\text{-}s}}{k_{\text{air}} A} = \frac{0.010}{0.0263(1)} = 0.380 \text{ K(or °C)/W},$$

$$R_{\text{conv,o}} = \frac{1}{\bar{h}_{\text{conv,o}} A} = \frac{1}{15(1)} = 0.067 \text{ K(or °C)/W}.$$

Using the thermal analog of Ohm's law (Eq. 5.58), we calculate the heat-transfer rate through the double-pane window as

$$\dot{Q}_{\text{2-pane}} = \frac{\Delta T}{\sum R_{\text{th}}} = \frac{21 - 0}{0.385 + 0.002 + 0.380 + 0.002 + 0.067}$$

$$= 25.1$$

$$[=] \frac{°\text{C}}{\text{K(or °C)/W}} = \text{W}.$$

We calculate the heat loss through the single-pane window in a similar manner, but deleting the thermal resistances of the air space and the second glass pane, so

$$\dot{Q}_{\text{1-pane}} = \frac{21 - 0}{0.385 + 0.002 + 0.067} \text{ W}$$

$$= 46.3 \text{ W}.$$

The ratio of the single- to double-pane window heat-transfer rates is

$$\frac{\dot{Q}_{\text{1-pane}}}{\dot{Q}_{\text{2-pane}}} = \frac{46.3}{25.1} = 1.8.$$

Comments   We see that the heat loss through the double-paned window is almost half that of the single pane. Note the importance of the thermal resistance of the air gap. In many solar-heated houses, triple-pane windows are used, with the addition of a second air space resulting in even less heat loss. We also note that free convection is likely to play a role in the heat transfer in the air space; thus, our calculations underestimate the actual heat loss.

Self Test
5.10

Consider heat transfer through a composite wall as shown in Fig. 5.14c. Develop an expression for the total thermal resistance R_{th} between the two fluid temperatures $T_{\infty,1}$ and $T_{\infty,2}$

(Answer: $R_{\text{th}} = R_{\text{conv,1}} + R_{\text{th,A}} + R_{\text{BT}}R_{\text{CT}}/(R_{\text{BT}} + R_{\text{CT}})$, where $R_{\text{BT}} = R_{\text{th,B}} + R_{\text{conv,2}}$ and $R_{\text{CT}} = R_{\text{th,C}} + R_{\text{conv,2}}$)

Example 5.14 Steam Power Plant Application

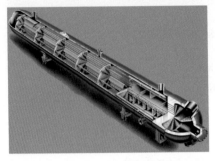

High pressure feedwater heater. Image courtesy of Yuba Heat Transfer, a Division of Connell LP.

Steam flows from a turbine to a feedwater heater through a 5-m-long pipe with an outer diameter of 150 mm. The pipe is insulated with a 50-mm-thick layer of insulation (85% magnesia, $k_{ins} = 0.055$ W/m·K). Determine the heat loss rate from the pipe if the external surface temperature of the pipe is 410 K, the ambient temperature is 300 K, and the external convective heat-transfer coefficient has a value of 1.2 W/m²·K. Also determine the temperature of the outside surface of the insulation.

Solution

Known $L, r_i, r_o, k_{ins}, T_i, T_\infty, \overline{h}_{conv,o}$

Find $\dot{Q}, T_o$

Sketch   See Fig. 5.12.

Assumptions

   i. One-dimensional steady state
   ii. Negligible radiation
   iii. Negligible thermal contact resistance at the pipe–insulation interface

Analysis   The thermal circuit for the insulation and its external convective environment is as follows:

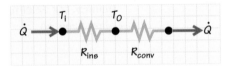

The thermal resistance of the pipe insulation is determined as

$$R_{ins} = \frac{\ln(r_o/r_i)}{2\pi k_{ins} L}$$

$$= \frac{\ln\left(\dfrac{75 + 50}{75}\right)}{2\pi(0.055)5} = 0.2956$$

$$[=]\frac{1}{(\text{W/m}\cdot\text{K})\text{m}} = \text{K(or °C)/W}.$$

The external convective resistance is

$$R_{conv} = \frac{1}{\overline{h}_{conv,o} 2\pi r_o L}$$

$$= \frac{1}{(1.2)2\pi(0.075 + 0.050)5} = 0.2122$$

$$[=]\frac{1}{(\text{W/m}^2\cdot\text{K})\text{m(m)}} = \text{K(or °C)/W}.$$

The heat loss rate is now found from Eq. 5.58:

$$\dot{Q} = \frac{\Delta T}{\sum R_{th}} = \frac{T_i - T_\infty}{R_{ins} + R_{conv}}$$

$$= \frac{410 - 300}{0.2956 + 0.2122} = 217$$

$$[=]\frac{\text{K}}{\text{K/W}} = \text{W}.$$

The insulation surface temperature can now be determined by applying Eq. 5.58 once again, but now only for the convective part of the circuit, that is,

$$\dot{Q} = \frac{\Delta T}{\sum R_{\text{th}}} = \frac{T_{\text{o}} - T_{\infty}}{R_{\text{conv}}}.$$

Solving for T_{o} yields

$$T_{\text{o}} = T_{\infty} + \dot{Q}R_{\text{conv}}$$
$$= 300 + 217(0.2122) = 346$$
$$[=] \text{ K} + \text{W}(\text{K/W}) = \text{K}.$$

> **Using Kirchhoff's current law analog that the currents in equal the currents out is equivalent to writing a surface energy balance.**

Comment Note how the thermal analog guided the solution by conveniently supplying the necessary energy balances: In finding $\dot{Q}$, it provided an energy balance for the insulation as a thermodynamic system, and in finding T_{o}, it provided a surface energy balance.

5.3 ENERGY CONSERVATION FOR CONTROL VOLUMES

Our treatment here parallels the development of the mass conservation principle for control volumes from Chapter 3. We start by developing steady-state expressions of energy conservation for integral control volumes, followed by development of the general unsteady case. We then repeat these developments for differential control volumes.

5.3a Integral Control Volumes with Steady Flow

Recall from Chapter 3 the idea that for steady-state and steady flow, properties do not change with time within the control volume or within the inlet and outlet streams. Furthermore, for the single-inlet, single-exit, control volume shown in Fig. 5.15, the mass flow rate in must equal the mass flow rate out (Eq. 3.18a); that is,

> **Review Chapter 3 for details of the concept of steady flow.**

$$\dot{m}_{\text{in}} = \dot{m}_{\text{out}} = \dot{m}.$$

For simplicity, we assume that the fluid properties are uniform where the flow crosses the control surface, or that appropriately corrected average values characterize the inlet and outlet streams (see Eqs. 3.12 and 3.13, for

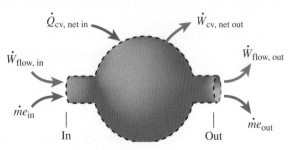

FIGURE 5.15
Control volume with one inlet and one outlet for steady-state, steady-flow analysis of energy conservation.

example). In most engineering applications of steady-state and steady flow, energy rates are of importance (e.g., $\dot{W}$ and $\dot{Q}$ expressed in joules per second or watts).

We begin our analysis by identifying all of the energy interactions around the control surface: The inlet stream carries energy into the control volume ($\dot{m}e_{in}$), and the exit stream carries energy out of the control volume ($\dot{m}e_{out}$). Flow work is performed by the surroundings to push the fluid into the control volume ($\dot{W}_{flow,in}$), whereas the control volume performs work to push the exiting fluid from the control volume ($\dot{W}_{flow,out}$). In addition, the control volume may perform work that crosses the control surface at locations other than the fluid inlet and exit ($\dot{W}_{cv,net\,out}$), for example, shaft work. Finally, a net rate of heat transfer into the control volume ($\dot{Q}_{cv,net\,in}$) will exist if there are temperature gradients and/or differences at the control surface.

Having identified all of the energy interactions at the control surface—heat, work, and mass flow (there can be no others)—we now apply the following broad conservation of energy principle:

In steady state, the net rate at which energy crosses the control surface must be zero; that is, the rate at which energy enters the control volume must equal the rate at which energy exits.

This principle is mathematically expressed as

$$\dot{E}_{CS,in} = \dot{E}_{CS,out}, \tag{5.59}$$

where $\dot{E}_{CS}$ is the energy transfer across the control surface. Using this as the starting point, and making some rearrangements and substitutions, we arrive at a standard form of energy conservation for steady-state, steady flow. Substituting the particular energy flows yields

$$\dot{Q}_{cv,in} + \dot{W}_{cv,in} + \dot{W}_{flow,in} + \dot{m}e_{in} = \dot{Q}_{cv,out} + \dot{W}_{cv,out} + \dot{W}_{flow,out} + \dot{m}e_{out}. \tag{5.60}$$

Rearranging, we write

$$(\dot{Q}_{cv,in} - \dot{Q}_{cv,out}) + (\dot{W}_{cv,in} - \dot{W}_{cv,out}) = \dot{m}(e_{out} - e_{in}) + \dot{W}_{flow,out} - \dot{W}_{flow,in}. \tag{5.61}$$

The specific energy (energy per unit mass) of the flowing streams can be expanded to show explicitly the specific internal, specific kinetic, and specific potential energies:

$$e = u + (ke) + (pe). \tag{5.62}$$

Furthermore, the flow work, as derived in Chapter 4 (Eq. 4.12), is expressed as

$$\dot{W}_{flow} = \dot{m}Pv.$$

Substituting Eqs. 5.62 and 4.12 into Eq. 5.61 yields

$$(\dot{Q}_{cv,in} - \dot{Q}_{cv,out}) + (\dot{W}_{cv,in} - \dot{W}_{cv,out})$$
$$= \dot{m}[(u + Pv)_{out} - (u + Pv)_{in} + (ke)_{out} - (ke)_{in} + (pe)_{out} - (pe)_{in}].$$

In this expression, we recognize the specific enthalpy $h\,(= u + Pv)$. With this, and the substitution of the definitions of the specific kinetic and potential energies [i.e., $ke \equiv V^2/2$ and $pe \equiv g(z - z_{ref})$], we write our final steady-state,

This is the most basic steady-state, steady-flow statement of energy conservation.

steady-flow conservation of energy expression:

$$(\dot{Q}_{cv,in} - \dot{Q}_{cv,out}) + (\dot{W}_{cv,in} - \dot{W}_{cv,out})$$
$$= \dot{m}[(h_{out} - h_{in}) + \tfrac{1}{2}(V_{out}^2 - V_{in}^2) + g(z_{out} - z_{in})]. \tag{5.63a}$$

Equation 5.63a is one of those key relationships whose meaning and applications should be deeply integrated in your knowledge base. A slightly more compact form that may be easier to recall is

> **Equations 5.63a and 5.63b are key relationships.**

$$\dot{Q}_{cv,net\ in} - \dot{W}_{cv,net\ out} = \dot{m}[\Delta h + \Delta(ke) + \Delta(pe)], \tag{5.63b}$$

where $\Delta \equiv (\ )_{out} - (\ )_{in}$ and

$$\dot{Q}_{cv,net\ in} \equiv \dot{Q}_{cv,in} - \dot{Q}_{cv,out} \tag{5.63c}$$

and

$$\dot{W}_{cv,net\ out} \equiv \dot{W}_{cv,out} - \dot{W}_{cv,in}. \tag{5.63d}$$

The reader should work to become quite familiar with Eqs. 5.63a–5.63d as these are powerful tools and can be used to solve many practical problems.

Note that Eq. 5.63 applies to reacting flows as well as to nonreacting flows. The only special requirement for reacting flows is the use of standardized enthalpies. We illustrate this later with an example.

The steady-flow conservation of energy expressions represented in Eqs. 5.63a and 5.63b can be modified to allow for nonuniform velocity profiles at the inlet and outlet stations by correcting the kinetic energy terms. When the velocity is not uniform, $\dot{m}V_{avg}^2/2$ is only an approximation to the flow of kinetic energy. The true kinetic energy flow over an inlet or outlet area A is given by the following, which is also used to define a **kinetic-energy correction factor α**:

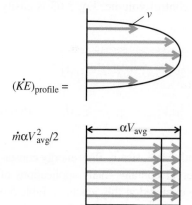

$(\dot{KE})_{profile} =$

$\dot{m}\alpha V_{avg}^2/2$

$$\int_A \frac{1}{2}v^2\rho vdA \equiv \dot{m}\alpha V_{avg}^2/2. \tag{5.64}$$

Thus, Eq. 5.63a can be rewritten as

$$\dot{Q}_{cv,net\ in} - \dot{W}_{cv,net\ out}$$
$$= \dot{m}\left[(h_{out} - h_{in}) + \frac{1}{2}(\alpha_{out}V_{avg,out}^2 - \alpha_{in}V_{avg,in}^2) + g(z_{out} - z_{in})\right]. \tag{5.63e}$$

> **See Example 3.8 in Chapter 3.**

Values for the correction factor α depend on the shape of the velocity profile. As discussed in Chapter 3, velocity profiles for turbulent flow can be expressed as a power law for circular-cross-section inlets and outlets:

$$v(r) = a\left(1 - \frac{r}{R}\right)^{1/n}.$$

> **See Example 3.7 in Chapter 3.**

Table 5.5 shows values of α computed using this power law. We see that the correction is relatively small (<10%) for turbulent flows and frequently can be neglected. For laminar flows (parabolic velocity profile), the correction factor is 2.0 and, hence, the correction cannot be neglected.

Table 5.5   Kinetic Energy Correction Factors for Power-Law Velocity Distributions: $v(r) = a[1 - (r/R)]^{1/n}$

Exponent (n)	Correction Factor (α)
n	$\dfrac{(n+1)^3(2n+1)^3}{4n^4(2n+3)(n+3)}$
6	1.077
7	1.058
10	1.031

Table 5.6   Important Devices Amenable to Steady-Flow Analysis[*]

Nozzle	Fan
Diffuser	Turbine
Throttle	Heat exchanger
Pump	Furnace or boiler
Compressor	Combustor

[*] Simplified analyses of each of these devices are presented in Chapter 11.

If multiple streams enter and/or exit a control volume, Eq. 5.63 is easily extended to treat such a case by writing

$$\dot{Q}_{cv,net\,in} - \dot{W}_{cv,net\,out} = \sum_{k=1}^{M\;outlets} \dot{m}_{out,k}\left[h_k + \tfrac{1}{2}\alpha_k V_{avg,k}^2 + g(z_k - z_{ref})\right]$$
$$- \sum_{j=1}^{N\;inlets} \dot{m}_{in,j}\left[h_j + \tfrac{1}{2}\alpha_j V_{avg,j}^2 + g(z_j - z_{ref})\right]. \qquad (5.65)$$

> In spite of its apparent complexity, Eq. 5.65 simply states $\dot{E}_{CS,in} = \dot{E}_{CS,out}$.

A few examples of the application of steady-state, steady-flow energy conservation are presented next; however, Chapter 11 contains many applications of engineering interest that can and should be explored at this juncture. Table 5.6 lists important devices that are amenable to a simple steady-flow analysis. Combinations of these devices are used to create complex systems such as power plants, propulsion systems, and systems for heating and cooling, the subject of Chapter 12.

Example 5.15   Jet Engine Application

In a jet engine, the incoming air is compressed before it enters the combustor. Located between the compressor outlet and the combustor inlet is a diffuser. The purpose of the diffuser is to slow the high-velocity air stream exiting the compressor to velocities sufficiently low to allow combustion to take place within the combustor. To accomplish this velocity change, the flow area of the passive diffuser increases in the flow direction as shown in the drawing. For a particular engine, the velocity and temperature at the diffuser entrance are 30 m/s and 450 K, respectively. The velocity at the diffuser exit is 4 m/s. Assuming adiabatic operation, estimate the temperature at the diffuser exit.

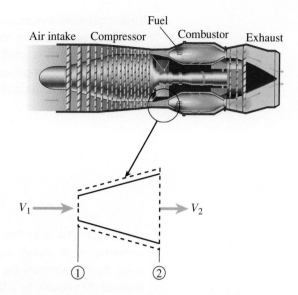

Air intake Compressor Fuel Combustor Exhaust

$V_1 \longrightarrow$ $\longrightarrow V_2$

①   ②

Solution

Known V_1, V_2, T_1

Find T_2

Sketch   See control volume in sketch above.

Assumptions

  i. Steady state
 ii. $\dot{Q}_{cv} = 0$ (adiabatic)
iii. $\dot{W}_{cv} = 0$
 iv. $\Delta pe = 0$
  v. Uniform velocity profiles ($\alpha_1 = \alpha_2 = 1$)

Analysis   Since the diffuser is a passive device (with no moving parts and no moving boundaries) no work is done. We also neglect the change in potential energy since elevation changes are likely to be small, say, 5–10 cm. With these assumptions and designating the inlet as state 1 and the outlet as station 2, we simplify the basic steady-flow, steady-state conservation of energy expression (Eq. 5.63) as follows:

$$\dot{Q}_{cv} - \dot{W}_{cv} = \dot{m}\left(h_2 - h_1 + \frac{V_2^2 - V_1^2}{2} + g(z_2 - z_1) \right),$$

or

$$0 - 0 = \dot{m}\left(h_2 - h_1 + \frac{V_2^2 - V_1^2}{2} + 0 \right).$$

Dividing by $\dot{m}$ and rearranging, we obtain

$$h_2 - h_1 = \frac{V_1^2 - V_2^2}{2}.$$

Treating air as an ideal gas, we know that finding $h_2 - h_1$ is tantamount to finding the temperature difference since $h = h(T$ only). Substituting numerical values yields

$$h_2 - h_1 = \frac{(30)^2 - (4)^2}{2} = 442$$

$$[=] \frac{m^2}{s^2}\left[\frac{1\ J}{N \cdot m} \right]\left[\frac{1\ N}{kg \cdot m/s^2} \right] = J/kg.$$

To determine the temperature change, we could employ the calorific equation of state represented by the air property table (Table C.2). Because the enthalpy difference is small ($\Delta h = 0.442$ kJ/kg), however, a simpler approach is to use the constant-pressure specific heat at 450 K from Table C.3 and the approximate relationship (Eq. 2.33e) $\Delta h = c_{p,\,\text{avg}}\Delta T$; thus,

$$T_2 = T_1 + \frac{h_2 + h_1}{c_p}$$

$$= 450 + \frac{0.442}{1.021} = 450.4$$

$$[=]\frac{\text{kJ/kg}}{\text{kJ/kg}\cdot\text{K}} = \text{K}.$$

Comments We note first the procedure used to apply the control-volume conservation of energy equation: listing reasonable assumptions and then using these to simplify Eq. 5.63. We also note that, even though the diffuser reduced the velocity by a factor of 7.5, the temperature rise was quite small. This calls into question our original assumption of adiabaticity, since heat-transfer effects might easily produce a temperature change of the same order as was found in our adiabatic analysis. The diffuser in most jet engines is an annulus surrounding the rotating components in the engine core.

Self Test 5.11   Water flowing at 38.7 kg/s enters a pump as a saturated liquid at 320 K and exits at 325 K and 10 MPa. Neglecting any heat losses and potential and kinetic energy changes, determine the work performed by the pump.

(Answer: $\dot{W} = -1140$ kW, where the negative sign indicates work is done on the system)

Example 5.16 Steam Power Plant Application

A steam turbine is used to generate electricity at a pulp and paper manufacturing plant. Superheated steam (10 MPa and 780 K) enters the turbine with a flow rate of 38.7 kg/s. The steam exits the turbine at 320 K with a quality of 0.88. The turbine operation is nearly adiabatic, and changes in both kinetic and potential energies of the entering and exiting steam are negligible. Estimate the shaft power delivered by the turbine to the generator.

Solution

Known $\dot{m}$, T_1, P_1, T_2, x_2

Find $\dot{W}_{\text{turb}}$

Sketch

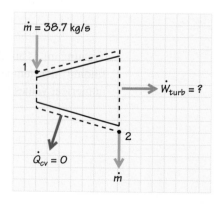

Assumptions

i. Steady-state steady flow
ii. $\dot{Q}_{cv} = 0$ (adiabatic)
iii. $\Delta ke = \Delta pe = 0$

Analysis   We begin with the steady-state, steady-flow conservation of energy expression for a control volume having a single inlet and a single outlet (Eq. 5.63):

$$\dot{Q}_{cv,\text{net in}} - \dot{W}_{cv,\text{net out}} = \dot{m}\left(h_2 - h_1 + \frac{V_2^2 - V_1^2}{2} + g(z_2 - z_1) \right),$$

which simplifies to the following by applying the assumptions:

$$0 - \dot{W}_{\text{turb}} = \dot{m}(h_2 - h_1 + 0 + 0),$$

or

$$\dot{W}_{\text{turb}} = \dot{m}(h_1 - h_2).$$

The problem is now reduced to finding the values of h_1 and h_2. To find the inlet enthalpy, h_1, we use the NIST database or Table D.3. For steam at 10 MPa and $T = 780$ K, we find

$$h_1 = 3392.4 \text{ kJ/kg}.$$

Given the outlet state ($T_2 = 320$ K, $x_2 = 0.88$), we employ data from the NIST database (or Table D.1) with Eq. 2.49c, which expresses mass-specific properties in the liquid–vapor region, to find h_2:

$$h_2 = (1 - x_2)h_{f,2} + x_2 h_{g,2}$$
$$= (1 - 0.88)\,196.17 + 0.88\,(2585.7) \text{ kJ/kg}$$
$$= 2299.0 \text{ kJ/kg}.$$

The turbine power is thus

$$\dot{W}_{\text{turb}} = 38.7(3392.4 - 2299.0)$$
$$= 42,315$$
$$[=]\frac{\text{kg}}{\text{s}}\frac{\text{kJ}}{\text{kg}} = \frac{\text{kJ}}{\text{s}} \text{ or kW}.$$

Comments   We can now justify our original assumptions that kinetic and potential energies of the steam (or their changes) are negligible. The National Electrical Manufacturers Association (NEMA) has set standards for maximum inlet and outlet velocities for steam turbine applications of 175 ft/s (53.3 m/s) and 250 ft/s (76.2 m/s), respectively. The kinetic energy rate associated with the 53.3-m/s velocity is $\dot{m}V^2/2 = 38.7$ $(53.3)^2/2 = 55.0 \times 10^3$ W or 55.0 kW. This energy rate is 0.13% of the power delivered by the turbine. The kinetic energy difference, that is, $|\dot{m}(V_2^2 - V_1^2)/2|$, associated with these standard maximum velocities is equally small, amounting to about 0.14% of the power delivered by the turbine. Similar estimates can be made to show that potential energy differences are also much less than $\dot{W}_{\text{turb}}$. This exercise is left to the reader.

Example 5.17 Steam Power Plant Application

Steam condenser. Image courtesy of Yuba Heat Transfer, a Division of Connell LP.

Consider the same turbine and flow condition described in Example 5.16. After exiting this turbine, steam is condensed in the shell side of a condenser as shown in the sketch. The state of the steam entering the condenser (station 1) is essentially the same as the turbine exit state defined in Example 5.16 ($x = 0.88$, $T = 320$ K). The condensate exits the condenser as saturated liquid at 320 K (station 2). Cooling water is pumped through the tubes of the condenser at a rate of 2119.4 kg/s and a pressure of 0.3 MPa. The water enters the condenser at 305 K (station 3). Determine the outlet temperature of the cooling water (station 4).

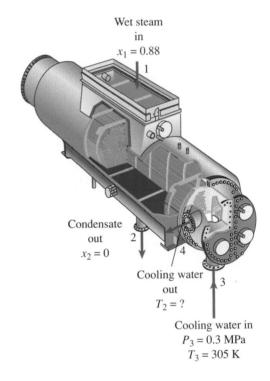

Wet steam
in
$x_1 = 0.88$

1

Condensate
out
$x_2 = 0$

2

4

Cooling water
out
$T_2 = ?$

3

Cooling water in
$P_3 = 0.3$ MPa
$T_3 = 305$ K

Solution

Known $\dot{m}_1$, $\dot{m}_3$, T_1, T_2, x_1, x_2, P_3, T_3

Find T_4

Sketch

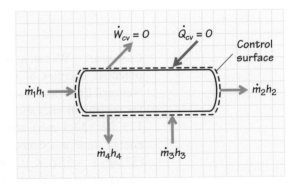

$\dot{W}_{cv} = 0$ $\dot{Q}_{cv} = 0$ Control
surface

$\dot{m}_1 h_1$ $\dot{m}_2 h_2$

$\dot{m}_4 h_4$ $\dot{m}_3 h_3$

Assumptions

  i. Steady-state, steady flow

  ii. $\dot{Q}_{cv} = 0$ (adiabatic operation)

iii. Negligible pressure drops ($P_2 = P_1$ and $P_4 = P_3$)
iv. Negligible kinetic and potential energies

Analysis We apply conservation of mass and conservation of energy to the condenser for the control volume shown in the sketch. Since the condensing steam and the cooling water streams remain separated as they pass through the condenser, mass conservation requires

$$\dot{m}_2 = \dot{m}_1 \equiv \dot{m}_{\text{hot}}$$

and

$$\dot{m}_4 = \dot{m}_3 \equiv \dot{m}_{\text{cold}},$$

where we denote the two flow rates, respectively, as $\dot{m}_{\text{hot}}$ and $\dot{m}_{\text{cold}}$. For a control volume with two inlets and two outlets, energy conservation is expressed by Eq. 5.65, which we simplify as follows by applying the assumptions listed, so that

$$\dot{Q}_{\text{cv,net in}} - \dot{W}_{\text{cv,net out}} = \sum \dot{m}_{\text{out},k}\left(h_k + \frac{1}{2}\alpha_k V_k^2 + g(z_k - z_{\text{ref}})\right)$$
$$- \sum \dot{m}_{\text{in},j}\left(h_j + \frac{1}{2}\alpha_j V_j^2 + g(z_j - z_{\text{ref}})\right)$$

becomes

$$0 + 0 = \dot{m}_2(h_2 + 0 + 0) + \dot{m}_4(h_4 + 0 + 0) - \dot{m}_1(h_1 + 0 + 0)$$
$$- \dot{m}_3(h_3 + 0 + 0),$$

or

$$0 = \dot{m}_{\text{hot}}(h_2 - h_1) + \dot{m}_{\text{cold}}(h_4 - h_3).$$

We solve this for h_4 to get

$$h_4 = h_3 + \frac{\dot{m}_{\text{hot}}}{\dot{m}_{\text{cold}}}(h_1 - h_2).$$

From Example 5.16, we have

$$h_1 = 2299.0 \text{ kJ/kg}$$

and

$$h_2 = h_f(320 \text{ K}) = 196.17 \text{ kJ/kg}.$$

We use the NIST online database to determine the enthalpy of the entering cold water (compressed liquid):

$$h_3(0.3 \text{ MPa}, 305 \text{ K}) = 133.74 \text{ kJ/kg}.$$

Using these values to determine h_4 yields

$$h_4 = 133.74 + \frac{38.7}{2119.4}(2299 - 196.17) \text{ kJ/kg} = 172.14 \text{ kJ/kg}.$$

We use the NIST database once again now to find

$$T_4(0.3 \text{ MPa}, 172.14 \text{ kJ/kg}) = 314.2 \text{ K}.$$

The cooling water thus experiences a temperature increase of $\Delta T = 9.2$ ($= 314.2 - 305$) K.

Comments Note the large flow rate used for the cooling water (2119.4 kg/s) to condense a much smaller amount of steam (38.7 kg/s). Maintaining a small ΔT for the cooling water mitigates thermal pollution for an open system that dischargers its water into a river or other natural body of water.

You are now prepared to begin a study of Chapter 11.

✓ Consider the steady flow of hot and cold water through a mixing faucet. Neglecting any heat losses and potential and kinetic energy changes, find the final temperature of the flow when 0.03 kg/s of water at 52°C mixes with 0.045 kg/s of water at 10°C.

(*Answer: $T \cong 300$ K (27°C)*)

5.3b Road Map for Study

Chapter 11 applies the steady-flow form of energy conservation to a number of important engineering devices. Depending on your study objectives,[1] you might want to explore portions of Chapter 11 to see how the first law of thermodynamics applies to system components other than the ones treated in the examples here. Chapter 11 revisits these and also discusses pumps, compressors, throttles, heat exchangers, furnaces, and other commonly encountered components.

5.3c Special Form for Flows with Friction

In many engineering applications, particularly flows through pipes, tubes, and ducts, the effects of fluid friction at the fluid–solid wall interface are manifest as a loss of pressure. Because of this, the steady-flow energy equation for integral control volumes is frequently cast in a form in which the pressure appears explicitly. To obtain one such form, we assume the following:

- The flow is steady and the control volume has a single inlet and exit.
- The fluid is incompressible. This is a very good approximation for liquids; however, gases also can be treated as incompressible if density changes are relatively small, say, less than 10%.
- All properties, except velocity, are uniform over the inlet and exit surfaces.

> **Note that $v = 1/\rho$. See Eq. 2.9.**

We begin our analysis with Eq. 5.63a; explicitly expressing the enthalpy $h = u + P/\rho$ and taking the nonuniform velocity distribution into account, we have

$$\dot{Q}_{\text{cv,net in}} - \dot{W}_{\text{cv,net out}} = \dot{m}\left[(u_{\text{out}} + P_{\text{out}}/\rho) - (u_{\text{in}} + P_{\text{in}}/\rho) \right.$$
$$\left. + \frac{1}{2}(\alpha_{\text{out}} V^2_{\text{avg,out}} - \alpha_{\text{in}} V^2_{\text{avg,in}}) + g(z_{\text{out}} - z_{\text{in}}) \right].$$

Dividing this expression by the mass flow rate $\dot{m}$ and rearranging yield

$$\frac{P_{\text{in}}}{\rho} + \frac{1}{2}\alpha_{\text{in}} V^2_{\text{avg,in}} + gz_{\text{in}} = \frac{P_{\text{out}}}{\rho} + \frac{1}{2}\alpha_{\text{out}} V^2_{\text{avg,out}} + gz_{\text{out}} + \frac{\dot{W}_{\text{cv,net out}}}{\dot{m}}$$
$$+ \left[u_{\text{out}} - u_{\text{in}} - \frac{\dot{Q}_{\text{cv,net in}}}{\dot{m}} \right]. \tag{5.66}$$

> **Providing a more rigorous meaning for the head loss must wait until Chapter 6, where we show its origins in a manipulation of the conservation of momentum principle.**

As we will see later in Chapter 6, the final term in Eq. 5.66 can be identified with the **head loss**, defined as

$$h_{\text{L}} \equiv \left[\frac{\dot{Q}_{\text{cv, net out}}}{\dot{m}} + u_{\text{out}} - u_{\text{in}} \right] \bigg/ g, \tag{5.67}$$

---

[1] For a single survey course, the present discussion of steady-flow devices probably suffices; however, for students studying the core thermal-fluid sciences in a sequence of courses, a brief visit to Chapter 11 is a useful and logical option.

where we have changed the sign on $\dot{Q}_{cv}$ to treat this as energy out of the control volume. Applying this definition to Eq. 5.66 yields

$$\frac{P_{in}}{\rho} + \frac{1}{2}\alpha_{in}V_{avg,in}^2 + gz_{in} = \frac{P_{out}}{\rho} + \frac{1}{2}\alpha_{out}V_{avg,out}^2 + gz_{out} + \frac{\dot{W}_{cv,net\ out}}{\dot{m}} + gh_L.$$

(5.68)

Equation 5.68 is frequently the starting point for analyzing pipe flows. Conservation of momentum and other considerations developed in Chapters 6 and 8 allow an independent means to calculate the head loss resulting from friction. As an aside, we foreshadow these developments: For a flow through a horizontal pipe ($\Delta z = 0$) with the same inlet and outlet velocity profiles ($\Delta[\alpha V_{avg}^2] = 0$), Eq. 5.68 reduces to the simple result

| **Chapter 10 treats pipe flows in detail.** |

$$P_{in} - P_{out} = \rho gh_L,$$

a key expression used in the design of piping systems.

Example 5.18

As shown in the sketch, water is pumped from one reservoir to another at an elevation 30 m above the first. The 34.3-mm-diameter interconnecting pipe is 200-m long and the flow rate is 0.921 kg/s. If the head loss associated with both the pipe and the pump is 9.7 m, determine the power supplied to the pump.

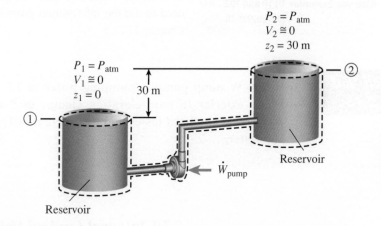

Solution

Known $\Delta z, \dot{m}, D, L$

Find $\dot{W}_{pump}$

Sketch   See control volume above.

Assumptions

   i. Steady-state steady flow
  ii. Large reservoirs
 iii. $P_1 = P_2 = P_{atm}$
  iv. $V_1 = V_2 \approx 0$
   v. $\dot{Q}_{cv} = 0$ (adiabatic)
  vi. Incompressible flow

Analysis The key to solving this problem is the choice of a control volume that includes both reservoirs. The "inlet" (station 1) is an imaginary surface just below the physical surface of the water in the reservoir. This choice allows the pressure to be essentially atmospheric at this location. Furthermore, it allows for flow into the control volume; however, the velocity is very small since the surface area of the reservoir is very large. As a result, the entering kinetic energy rate is negligible. The same arguments hold for the "exit" at the upper reservoir (i.e., at station 2). We apply these assumptions to Eq. 5.68 as follows:

$$\cancel{\frac{P_1}{\rho}} + \frac{1}{2}\cancel{V_1^2} + gz_1 = \cancel{\frac{P_2}{\rho}} + \frac{1}{2}\cancel{V_2^2} + gz_2 - \frac{\dot{W}_{\text{pump}}}{\dot{m}} + gh_{\text{L}}.$$

$\sim 0 \qquad\qquad \sim 0$

Cancel

Thus,

$$\dot{W}_{\text{pump}} = \dot{m}g(z_2 - z_1 + h_{\text{L}})$$
$$= 0.921(9.806)(30 - 0 + 9.7)$$
$$= 358.5$$
$$[=]\frac{\text{kg}}{\text{s}}\frac{\text{m}}{\text{s}^2}\,\text{m}\left[\frac{1\,\text{J}}{\text{N}\cdot\text{m}}\right]\left[\frac{1\,\text{N}}{\text{kg}\cdot\text{m/s}^2}\right] = \text{J/s or W}.$$

 See Example 11.18 in Chapter 11. Also see Examples 10.18 and 10.9 in Chapter 10.

Comments Note that since the head loss was a given quantity, we had no need to use the information given about the pipe. We revisit this example in Chapter 11.

Self Test
5.13

☑ **A 200-W sump pump is pumping water at 1.5 kg/s from a flooded basement to the house exterior (a total elevation change of 9 m) through a 40-mm-diameter pipe. Neglecting any head losses and assuming $\alpha = 1$, determine the average exit velocity of the water.**

(Answer: 21.0 m/s)

5.3d Integral Control Volumes with Unsteady Flow

Although many engineering applications involve steady-flow processes, unsteady (i.e., time-dependent) flows can be important in some situations. Transient start-ups and shutdowns fall in this category, as do emptying or filling of tanks and pressure vessels. At this point, it is a relatively simple matter to extend our previous analyses to handle unsteady flows. Figure 5.16 shows the addition of the term needed to deal with the unsteady case: dE_{cv}/dt, indicated with the squiggly arrow inside the control volume (cf. Fig. 5.15). With the addition of this single term to the previously identified energy interactions at the control surface, we state conservation of energy as follows:

The net rate at which energy from the surroundings crosses the control surface into the control volume must equal the time rate of increase of energy within the control volume.

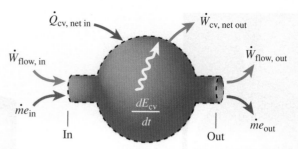

FIGURE 5.16
Control volume with one inlet and one outlet for unsteady analysis of energy conservation.

Symbolically, this is expressed most simply as

Equation 5.69 expresses a key concept in a very compact way.

$$\dot{E}_{CS,in} - \dot{E}_{CS,out} = \frac{dE_{cv}}{dt}, \tag{5.69a}$$

or

$$\dot{E}_{CS,net\ in} = \frac{dE_{cv}}{dt}. \tag{5.69b}$$

From this point, our analysis follows directly from the previous development for the steady-flow case (i.e., Eqs. 5.60–5.65). Therefore, we need only state the final results, first, for a control volume with a single inlet and outlet:

$$\dot{Q}_{cv,net\ in} - \dot{W}_{cv,net\ out} + \dot{m}_{in}\left[h_{in} + \frac{1}{2}V_{in}^2 + g(z_{in} - z_{ref}) \right]$$
$$- \dot{m}\left[h_{out} + \frac{1}{2}V_{out}^2 + g(z_{out} - z_{ref}) \right] = \frac{dE_{cv}}{dt}, \tag{5.70}$$

and, second, for control volumes with multiple inlets and/or outlets:

$$\dot{Q}_{cv,net\ in} - \dot{W}_{cv,net\ out} + \sum_{j=1}^{N\ inlets} \dot{m}_{in,j}\left[h_j + \frac{1}{2}V_j^2 + g(z_j - z_{ref}) \right]$$
$$- \sum_{k=1}^{M\ outlets} \dot{m}_{out,k}\left[h_k + \frac{1}{2}V_k^2 + g(z_k - z_{ref}) \right] = \frac{dE_{cv}}{dt}. \tag{5.71}$$

Although not shown, kinetic-energy correction factors can easily be added to these relationships, if needed. It is also important to emphasize that Eqs. 5.70 and 5.71 are *instantaneous* expressions of energy conservation. The following example illustrates the application of these expressions.

Example 5.19

Specialty coffee beverages are made by adding steamed milk to espresso coffee. In the preparation of the steamed milk (see sketch), saturated water vapor at 100 kPa jets into a pitcher containing 0.17 kg of skim milk, initially at 278 K (~40 F). If the steam flow rate is 4 g/min, estimate how long it takes to heat the milk to 322 K (~120 F). Assume that the process effectively occurs at constant pressure (100 kPa) and that the steam jet provides good stirring action.

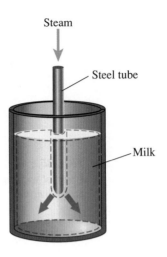

Solution

Known $M_1, T_1, T_2, P_1 = P_2 = P_{in}, \dot{m}_{in}$

Find $\Delta t(= t_2 - t_1)$

Sketch

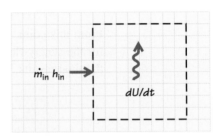

Assumptions:

  i. Steam enters with constant properties and constant flow rate.
 ii. The process occurs at constant pressure (i.e., pressure variations within the milk can be neglected).
iii. The kinetic energy of the entering steam is negligible (i.e., $V_{in}^2/2 \approx 0$).
 iv. Properties are uniform within the control volume from the jet stirring action.
  v. $\dot{W}_{cv} \approx 0$ [i.e., boundary work ($Pd\mathcal{V}/dt$) is negligibly small].
 vi. $\dot{Q}_{cv} \approx 0$ (i.e., the process is adiabatic).
vii. Kinetic and potential energies of the control volume are negligible.
viii. The properties of the milk are approximated by those of water.
 ix. There is no evaporation into the surroundings.

Analysis   To solve this problem requires the application of the unsteady forms of both conservation of energy and conservation of mass. We begin with conservation of energy, Eq. 5.70:

$$\dot{Q}_{cv,in} - \dot{W}_{cv,out} + \dot{m}_{in}\left[h_{in} + V_{in}^2/2 + g(z_{in} - z_{ref})\right]$$
$$- \dot{m}_{out}\left[h_{out} + V_{out}^2/2 + g(z_{out} - z_{ref})\right] = \frac{dE_{cv}}{dt}.$$

Applying our assumptions and noting that $\dot{m}_{out} = 0$, this simplifies to

$$0 - 0 + \dot{m}_{in}h_{in} - 0 = \frac{dU_{cv}}{dt},$$

or

$$\frac{dU_{cv}}{dt} = \dot{m}_{in} h_{in}.$$

Since both $\dot{m}_{in}$ and h_{in} are constants, we can easily integrate this expression and, as a result, introduce the unknown time difference as follows:

$$\int_{U_1}^{U_2} dU_{cv} = \int_{t_1}^{t_2} \dot{m}_{in} h_{in} \, dt,$$

and

$$U_{cv,2} - U_{cv,1} = \dot{m}_{in} h_{in} (t_2 - t_1).$$

Since we assume the properties within the control volume are uniform, $U_{cv} = Mu$; thus,

$$M_2 u_2 - M_1 u_1 = \dot{m}_{in} h_{in} \Delta t,$$

where $\Delta t \equiv t_2 - t_1$.

We now apply mass conservation (Eq. 3.19a) to find M_2:

$$\frac{dM_{cv}}{dt} = \dot{m}_{in} - \dot{m}_{out},$$

and

$$\int_{M_1}^{M_2} dM_{cv} = \int_{t_1}^{t_2} \dot{m}_{in} \, dt.$$

Performing the integration and solving for M_2 yields

$$M_2 = M_1 + \dot{m}_{in} \Delta t.$$

Substituting this expression for M_2 into our energy conservation expression and solving for Δt results in

$$\Delta t = \frac{M_1 (u_2 - u_1)}{\dot{m}_{in} (h_{in} - u_2)}.$$

Values for u_1, u_2, h_{in} are easily obtained from the NIST database:

$$u_1 \, (100 \text{ kPa}, 278 \text{ K}) = 20.388 \text{ kJ/kg},$$

$$u_2 \, (100 \text{ kPa}, 322 \text{ K}) = 204.51 \text{ kJ/kg},$$

$$h_{in} = h_g (100 \text{ kPa}) = 2674.9 \text{ kJ/kg}.$$

Before substituting numerical values we express the given flow rate in SI units:

$$\dot{m}_{in} = 4 \, (\text{g/min}) \left[\frac{1 \text{ kg}}{1000 \text{ g}} \right] \left[\frac{1 \text{ min}}{60 \text{ s}} \right] = 6.67 \times 10^{-5} \text{ kg/s}.$$

Our final result is thus

$$\Delta t = \frac{0.17(204.51 - 20.388)}{6.67 \times 10^{-5}(2674.9 - 204.51)} = 190$$

$$[=] \frac{\text{kg}}{\text{kg/s}} \frac{\text{kJ/kg}}{\text{kJ/kg}} = \text{s}.$$

To continue a study of integral control volumes, you can skip the next section and proceed to Chapter 6.

Comments The final result (slightly more than 3 min) seems quite reasonable. Note the many simplifying assumptions used to solve this problem. The application of reasonable assumptions to complex problems is an important part of the art of engineering.

5.3e Differential Control Volumes with Steady Flow

Earlier in this chapter, we applied the energy conservation principle to differential fixed-mass systems. In this section, we now consider differential control volumes.

One-Dimensional Flow

We begin with a simple one-dimensional analysis of a planar differential control volume. Choosing this simple control volume allows us to show all of the steps of our analysis without getting bogged down in the complexity of a multidimensional treatment.

Consider a planar differential control volume that has a length Δz and extends infinitely in the x- and y-directions as shown in Fig. 5.17. A vertical orientation has been selected so that the potential energy associated with the inlet and outlet streams explicitly appears. Our goal in this analysis is to derive a differential equation expressing energy conservation. With the application of appropriate boundary conditions, this equation can be solved to provide the temperature distribution within a macroscopic control volume.[2] For simplicity, we employ the following assumptions:

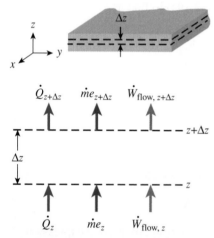

FIGURE 5.17
One-dimensional, planar, differential control volume showing energy transfers across the control surfaces.

- The flow is steady. At any fixed location, no property of the flow changes with time.
- The flow is one dimensional and planar. All properties are uniform on any x–y plane.
- Species diffusion is negligible. (This is a reasonable assumption provided there are no strong species concentration gradients as exist in flames.)
- There are no work interactions other than flow work. In particular, we assume that work associated with viscous stresses (see Eq. 4.9) is negligible (i.e., we neglect viscous dissipation). (This too is a reasonable assumption, except in high-speed flows, or other flows where there are very large velocity gradients.)

With these assumptions, the only energy transfers are those shown in Fig. 5.17: conduction heat transfer $\dot{Q}$, energy entering or exiting because of the flow, $\dot{m}e$, and the flow work $\dot{W}_{\text{flow}}$. In steady state, the rate at which energy enters the control volume equals the rate at which energy exits (Eq. 5.59), so

$$\dot{E}_{\text{CS,in}} = \dot{E}_{\text{CS,out}}.$$

Thus,

$$\dot{Q}_z + \dot{m}e_z + \dot{W}_{\text{flow},z} = \dot{Q}_{z+\Delta z} + \dot{m}e_{z+\Delta z} + \dot{W}_{\text{flow},z+\Delta z}. \qquad (5.72)$$

---

[2] The 1-D, planar system is not particularly useful in practice; however, it is a very useful learning tool. Furthermore, our 1-D, planar analysis is easily extended to two dimensions and axisymmetric geometries that, indeed, are quite useful. Examples of such applications include external boundary layers (Chapter 9) and pipe and tube flows (Chapter 10).

In Eq. 5.72, we have implicitly applied conservation of mass (Eq. 3.18a) in that

$$\dot{m}_{\text{CS,in}} = \dot{m}_{\text{CS,out}} \equiv \dot{m}.$$

We now substitute into Eq. 5.72 the terms comprising the specific energy e (Eq. 5.62),

$$e \equiv u + \frac{1}{2}v^2 + g(z - z_{\text{ref}}),$$

and the definition of the flow work $\dot{W}_{\text{flow}}$ (Eq. 4.12),

$$\dot{W}_{\text{flow}} = \dot{m}\,Pv.$$

After rearranging, this yields

$$\dot{Q}_z - \dot{Q}_{z+\Delta z} = \dot{m}(u_{z+\Delta z} - u_z) + \dot{m}[(Pv)_{z+\Delta z} - (Pv)_z]$$

$$+ \frac{1}{2}\dot{m}(v_{z+\Delta z}^2 - v_z^2) + \dot{m}g(z_{z+\Delta z} - z_z). \tag{5.73}$$

In this equation, we immediately recognize that the internal energy and flow work terms can be combined using the property enthalpy ($h \equiv u + Pv$). Furthermore, we can apply Fourier's law to express the 1-D heat conduction rate $\dot{Q}$ ($\equiv -kA\,dT/dz$). With these two substitutions and careful attention to signs, Eq. 5.73 becomes

$$\left(kA\frac{dT}{dz}\right)_{z+\Delta z} - \left(kA\frac{dT}{dz}\right)_z = \dot{m}(h_{z+\Delta z} - h_z) + \frac{1}{2}\dot{m}(v_{z+\Delta z}^2 - v_z^2)$$

$$+ \dot{m}g(z_{z+\Delta z} - z_z). \tag{5.74}$$

We now divide Eq. 5.74 by $A\,\Delta z$, where A is an arbitrary area in the x–y plane, and take the limit $\Delta z \to 0$. The left-hand side of Eq. 5.74 then becomes

$$\lim_{\Delta z \to 0} \frac{\left(kA\dfrac{dT}{dz}\right)_{z+\Delta z} - \left(kA\dfrac{dT}{dz}\right)_z}{A\,\Delta z} = \frac{d}{dz}\left(k\frac{dT}{dz}\right), \tag{5.75a}$$

where we have recognized the definition of a derivative. Similarly, the three terms on the right-hand side become

$$\lim_{\Delta z \to 0} \frac{\dot{m}(h_{z+\Delta z} - h_z)}{A\,\Delta z} = \dot{m}''\frac{dh}{dz}, \tag{5.75b}$$

$$\lim_{\Delta z \to 0} \frac{\dot{m}(v_{z+\Delta z}^2 - v_z^2)}{2\,A\,\Delta z} = \dot{m}''v_z\frac{dv_z}{dz}, \tag{5.75c}$$

and

$$\lim_{\Delta z \to 0} \frac{\dot{m}g(z_{z+\Delta z} - z_z)}{A\,\Delta z} = \dot{m}''g, \tag{5.75d}$$

where we have applied the definition of the mass flux $\dot{m}'' = \dot{m}/A$ (Eq. 3.24). Assembling these terms, we arrive at the following ordinary differential equation expressing energy conservation:

$$\frac{d}{dz}\left(k\frac{dT}{dz}\right) = \dot{m}''\left(\frac{dh}{dz} + v\frac{dv}{dz} + g\right). \tag{5.76}$$

Although Eq. 5.76 is fundamentally sound, it is usually converted to the **thermal energy equation** by further manipulation. Specifically, momentum conservation is used to eliminate the "mechanical" energy terms $v\,dv/dz$ and g. As you are most likely already to have studied the appropriate sections of Chapter 6, we apply the following 1-D z-momentum conservation expression (see Eq. 6.63) to simplify Eq. 7.76:

$$v\frac{dv}{dz} = -\frac{1}{\rho}\frac{dP}{dz} - g.$$

Performing this substitution yields

$$\frac{d}{dz}\left(k\frac{dT}{dz}\right) = \dot{m}''\left(\frac{dh}{dz} - \frac{1}{\rho}\frac{dP}{dz}\right). \tag{5.77}$$

Equation 5.77 is quite general in that it applies to any flowing fluid, subject, of course, to our starting assumptions. Note that the velocity and gravity terms are no longer involved. Moreover, by assuming a particular equation of state for the fluid (see Chapter 2), the enthalpy and pressure can be expressed in terms of the temperature. Recall that our original objective is to obtain a differential equation in which the temperature is the dependent variable. The following special cases achieve that objective.

Fluid at Constant Pressure For a fluid at constant pressure, dP/dz is obviously zero. Furthermore, from our study of properties in Chapter 2, we recall that for a pure substance $h = h(T, P)$; thus,

$$dh = \left(\frac{\partial h}{\partial T}\right)_P dT + \left(\frac{\partial h}{\partial P}\right)_T dP. \tag{5.78}$$

Since the pressure is constant, $dP = 0$, and Eq. 5.78 simplifies to

$$dh = \left(\frac{\partial h}{\partial T}\right)_P dT \equiv c_p dT. \tag{5.79}$$

Substitution of Eq. 5.79 into Eq. 5.77 yields our final result:

$$\frac{d}{dz}\left(k\frac{dT}{dz}\right) = \dot{m}'' c_p \frac{dT}{dz}, \tag{5.80}$$

where the mass flux $\dot{m}''$ is treated as a given quantity for any particular situation.

Ideal Gas For an ideal gas, $dh = c_p dT$, independent of any changes in pressure. With this substitution, Eq. 5.77 becomes

$$\frac{d}{dz}\left(k\frac{dT}{dz}\right) = \dot{m}''\left(c_p\frac{dT}{dz} - \frac{P}{RT}\frac{dP}{dz}\right), \tag{5.81}$$

where we have also applied the ideal-gas equation of state $P = \rho RT$ (Eq. 2.28b).

Incompressible Fluid For an incompressible fluid, a result identical to that for the constant-pressure case is obtained:

$$\frac{d}{dz}\left(k\frac{dT}{dz}\right) = \dot{m}'' c_p \frac{dT}{dz}. \tag{5.82}$$

Since our particular derivation does not lend itself to seeing how this particular result comes about, the interested reader is referred to Ref. [14].

Example 5.20

FIGURE 5.18
Worker cleaning heat shield used on a Mercury space capsule. The fiberglass shield is held for cleaning in a metal frame. **Photograph courtesy of NASA.**

An ablating heat shield protected the space capsules used in the first U.S. manned-flight missions (see Fig. 5.18) from the intense frictional heating during reentry. Ablation here refers to the vaporization and/or decomposition of a material exposed to a high heat flux. As a large fraction of the imposed heat flux gasifies the ablating material, very little energy penetrates beyond the ablating surface, thereby protecting the underlying material from dangerous heating.

Considering the ablation as a steady, one-dimensional process, determine and plot the temperature distribution as a function of distance from an ablating surface maintained at an "ablation temperature" of 1900 K. The temperature deep in the bulk material is 250 K. Assume that the ablation front moves at a velocity of 0.048 mm/s and that the thermal diffusivity α ($\equiv k/\rho c_p$) of the ablating material is 1.345×10^{-6} m²/s.

Solution

Known V_{ab}, T_s, T_{bulk}, α, ρ

Find $T(z)$

Sketch

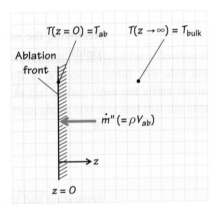

Assumptions:

  i. Steady
  ii. One dimensional
  iii. Constant properties
  iv. Constant pressure

Analysis   As indicated in the sketch, we model the process as a flow of material from right to left (i.e., in the negative z-direction). At the origin

The Faith 7 capsule used in Mercury 9 (left) and the Galileo Jupiter probe (right) employed ablating heat shields. Photographs courtesy of NASA.

See Appendix 9A in Chapter 9 for derivation of the boundary-layer conservation equations.

Constant Pressure or Incompressible Fluid Although for different reasons, flow at constant pressure or for an incompressible fluid results in the simplification of Eq. 5.84a to the following:

$$\rho c_p\left(v_x\frac{\partial T}{\partial x} + v_y\frac{\partial T}{\partial y}\right) = \frac{\partial}{\partial x}\left(k\frac{\partial T}{\partial x}\right) + \frac{\partial}{\partial y}\left(k\frac{\partial T}{\partial y}\right) + \Phi_{\text{visc}}. \quad (5.85)$$

Equation 5.85 is the usual starting point for further simplification in the study of boundary layers.

5.3f Differential Control Volumes with Unsteady Flow

For the sake of completeness, Table 5.7 presents relatively general expressions of thermal energy conservation for three-dimensional Cartesian, cylindrical, and spherical coordinates. Appearing in each is the unsteady term $\partial T/\partial t$, which arises from the inclusion of the time rate of change of the energy within the control volume, dE_{cv}/dt, in the original analysis. In spite of their formidable appearance, these equations originate from the straightforward idea that the net rate at which energy enters the control volume from the surroundings equals the time rate of change of the control volume energy (i.e., $\dot{E}_{\text{CS,in}} - \dot{E}_{\text{CS,out}} = dE_{\text{cv}}/dt$).

Because viscous dissipation is only important in flows in which large velocity gradients exist, the corresponding mathematical expressions are given separately in Table 5.8 so as not to clutter Table 5.7. Physically, the viscous dissipation originates with the work done by or against the viscous stresses, which in turn is irreversibly converted to internal energy. The equations presented in Table 5.7 are frequently simplified for specific applications; a prime example is heat transfer in tubes and pipes.

The filling of a spark-ignition engine cylinder is a complex unsteady, three-dimensional flow. Image courtesy of Daniel C. Haworth.

Table 5.7 Expressions of Conservation of Thermal Energy for Differential Control Volumes for Incompressible Fluids (ρ = constant) or for Conditions of Constant Pressure

Coordinate System	Energy Conservation Expression	
Cartesian (x, y, z, t)	$\rho c_p\left(\dfrac{\partial T}{\partial t} + v_x\dfrac{\partial T}{\partial x} + v_y\dfrac{\partial T}{\partial y} + v_z\dfrac{\partial T}{\partial z}\right)$	
	$\quad = \dfrac{\partial}{\partial x}\left(k\dfrac{\partial T}{\partial x}\right) + \dfrac{\partial}{\partial y}\left(k\dfrac{\partial T}{\partial y}\right) + \dfrac{\partial}{\partial z}\left(k\dfrac{\partial T}{\partial z}\right) + \Phi_{\text{visc}}$	(T5.7a)
Cylindrical (r, θ, x, t)	$\rho c_p\left(\dfrac{\partial T}{\partial t} + v_r\dfrac{\partial T}{\partial r} + \dfrac{v_\theta}{r}\dfrac{\partial T}{\partial \theta} + v_x\dfrac{\partial T}{\partial x}\right)$	
	$\quad = \dfrac{1}{r}\dfrac{\partial}{\partial r}\left(rk\dfrac{\partial T}{\partial r}\right) + \dfrac{1}{r^2}\dfrac{\partial}{\partial \theta}\left(k\dfrac{\partial T}{\partial \theta}\right) + \dfrac{\partial}{\partial x}\left(k\dfrac{\partial T}{\partial x}\right) + \Phi_{\text{visc}}$	(T5.7b)
Spherical (r, θ, ϕ, t)	$\rho c_p\left(\dfrac{\partial T}{\partial t} + v_r\dfrac{\partial T}{\partial r} + \dfrac{v_\theta}{r}\dfrac{\partial T}{\partial \theta} + \dfrac{v_\phi}{r\sin\theta}\dfrac{\partial T}{\partial \phi}\right)$	
	$\quad = \dfrac{1}{r^2}\dfrac{\partial}{\partial r}\left(r^2 k\dfrac{\partial T}{\partial r}\right) + \dfrac{1}{r^2\sin\theta}\dfrac{\partial}{\partial \theta}\left(\sin\theta k\dfrac{\partial T}{\partial \theta}\right) + \dfrac{1}{r^2\sin^2\theta}\dfrac{\partial}{\partial \phi}\left(k\dfrac{\partial T}{\partial \phi}\right) + \Phi_{\text{visc}}$	(T5.7c)

Table 5.8   Expressions for Viscous Dissipation for Incompressible Newtonian* Fluids

Coordinate System	Viscous Dissipation (Φ_{visc})
Cartesian (x, y, z, t)	$2\mu\left[\left(\dfrac{\partial v_x}{\partial x}\right)^2 + \left(\dfrac{\partial v_y}{\partial y}\right)^2 + \left(\dfrac{\partial v_z}{\partial z}\right)^2\right] + \mu\left[\left(\dfrac{\partial v_x}{\partial y} + \dfrac{\partial v_y}{\partial x}\right)^2\right.$ $\left. + \left(\dfrac{\partial v_x}{\partial z} + \dfrac{\partial v_z}{\partial x}\right)^2 + \left(\dfrac{\partial v_y}{\partial z} + \dfrac{\partial v_z}{\partial y}\right)^2\right]$ (T5.8a)
Cylindrical (r, θ, x, t)	$2\mu\left\{\left(\dfrac{\partial v_r}{\partial r}\right)^2 + \left[\dfrac{1}{r}\left(\dfrac{\partial v_\theta}{\partial \theta} + v_r\right)\right]^2 + \left(\dfrac{\partial v_x}{\partial x}\right)^2\right\} + \mu\left\{\left(\dfrac{\partial v_\theta}{\partial x} + \dfrac{1}{r}\dfrac{\partial v_x}{\partial \theta}\right)^2\right.$ $\left. + \left(\dfrac{\partial v_x}{\partial r} + \dfrac{\partial v_r}{\partial x}\right)^2 + \left[\dfrac{1}{r}\dfrac{\partial v_r}{\partial \theta} + r\dfrac{\partial}{\partial r}\left(\dfrac{v_\theta}{r}\right)\right]^2\right\}$ (T5.8b)
Spherical (r, θ, ϕ, t)	$2\mu\left[\left(\dfrac{\partial v_r}{\partial r}\right)^2 + \left(\dfrac{1}{r}\dfrac{\partial v_\theta}{\partial \theta} + \dfrac{v_r}{r}\right)^2 + \left(\dfrac{1}{r\sin\theta}\dfrac{\partial v_\phi}{\partial \phi} + \dfrac{v_r}{r} + \dfrac{v_\theta\cot\theta}{r}\right)^2\right]$ $+ \mu\left\{\left[r\dfrac{\partial}{\partial r}\left(\dfrac{v_\theta}{r}\right) + \dfrac{1}{r}\dfrac{\partial v_r}{\partial \theta}\right]^2 + \left[\dfrac{1}{r\sin\theta}\dfrac{\partial v_r}{\partial \phi} + r\dfrac{\partial}{\partial r}\left(\dfrac{v_\phi}{r}\right)\right]^2\right.$ $\left. + \left[\dfrac{\sin\theta}{r}\dfrac{\partial}{\partial \theta}\left(\dfrac{v_\phi}{\sin\theta}\right) + \dfrac{1}{r\sin\theta}\dfrac{\partial v_\theta}{\partial \phi}\right]^2\right\}$ (T5.8c)

* In a Newtonian fluid, the viscous stresses are directly proportional to strain rates. The viscosity μ is the proportionality constant in the stress–strain relations (see Chapter 6).

For a more detailed discussion of energy conservation applied to differential control volumes, the reader is referred to Ref. [14], where many different expressions are developed. For reacting flows in which mass diffusion is important, Ref. [8] is recommended as a starting point.

SUMMARY

Although many different expressions for conservation of energy were presented in this chapter, it should be clear that they all express a single principle. You should be quite familiar with the meanings of the simpler expressions for systems, in particular, Eqs. 5.5, 5.6, and 5.10, and for control volumes, Eqs. 5.59, 5.63, and 5.70. You should also be able to select from the many expressions the most appropriate one for any particular problem or analysis. Concurrent study of Chapter 11 aids in your development of this skill. We also applied the conservation of energy principle to 1-D conduction in solids and stagnant fluids. You should understand the origin of the temperature distributions for 1-D systems and be proficient at calculations employing the electrical analogy for conduction heat transfer.

Chapter 5
Key Concepts & Definitions Checklist[3]

5.2 Energy Conservation for a System

☐ Energy conservation for a time interval (Eq. 5.1) ➤ *Q5.2A,B*

☐ Energy conservation at an instant—rate form (Eq 5.2) ➤ *Q5.2C, 5.9*

5.2a General Integral Forms

☐ Various system energies ➤ *Q5.3*

☐ First law for incremental change in state ➤ *Q5.4*

☐ First law for finite change in state ➤ *Q5.5, 5.8*

☐ Creating system boundaries with dashed lines and representing energy transfers or changes with arrows

☐ Systems undergoing constant-*P*, constant-ν, or other simple processes, for ideal-gases or two-phase substances ➤ *5.1, 5.2*

☐ Application of the first law to systems involving ideal gases ➤ *5.3, 5.6*

☐ Application of the first law to systems involving two-phase substances ➤ *5.7, 5.46*

5.2b Reacting Systems

☐ Standardized enthalpies (Chapter 2) ➤ *Q2.37, 2.133*

☐ Constant-pressure adiabatic flame temperature ➤ *5.54*

☐ *H–T* diagrams for constant-*P* combustion ➤ *Q5.10, 5.53*

☐ Heating values (HHV and LHV) (Chapter 2) ➤ *Q2.41, 2.144*

☐ Constant-volume adiabatic flame temperature ➤ *5.55, 5.57*

☐ *U–T* diagrams for constant-ν combustion ➤ *Q5.11, Q5.12*

5.2c Special Forms for Conduction Analysis

☐ Electrical power and thermal energy generation rate ➤ *Q5.13*

☐ Relationship of energy storage rate to rate of temperature rise ➤ *5.11*

☐ Lumped approximation ➤ *Q5.14, 5.12*

☐ Biot number and characteristic length ➤ *5.13*

☐ Joule heating (Eq. 5.26) ➤ *Q5.13*

☐ Fourier's law (Chapter 4) ➤ *Q4.18, Q4.19, 4.33*

☐ Surface and interface energy balances ➤ *Q5.15*

☐ 1-D heat conduction equations ➤ *Q5.16, 5.63*

☐ Thermal diffusivity ➤ *Q5.17*

☐ Steady, 1-D temperature distributions for plane wall, hollow cylinders, and hollow spheres ➤ *5.65, 5.69*

☐ Electric circuit analogy ➤ *5.86, 5.87, 5.88*

5.3 Energy Conservation for Control Volumes

5.3a Integral Control Volumes with Steady Flow

☐ Creating control-volume boundaries with dashed lines and representing energy transfers or changes with arrows

☐ Standard simplifications of the first law to analyze steady-flow devices ➤ *5.122, 5.123, 5.124, 5.138, 5.141*

☐ Control volumes with multiple inlets and outlets ➤ *5.126, 5.166*

5.3c Special Form for Flows with Friction

☐ Steady-flow energy conservation with head loss (Eq. 5.68) ➤ *5.127*

5.3d Integral Control Volumes with Unsteady Flow

☐ Basic first-law statements (Eqs. 5.69–5.71) ➤ *5.187*

☐ Applications involving fixed-property inlet streams ➤ *5.182, 5.188*

5.3e and 5.3f Differential Control Volumes

☐ 1-D formulations ➤ *5.193, 5.194*

☐ 2- and 3-D formulations ➤ *5.196*

☐ Most general forms (Tables 5.7 and 5.8) ➤ *5.196*

---

[3] Numbers following arrows refer to Questions (prefaced with a Q) and Problems at the end of the chapter.

REFERENCES

1. Thompson, B. (Count Rumford), "An Inquiry Concerning the Source of Heat Which Is Excited by Friction," in *The Complete Works of Count Rumford*, American Academy of Arts and Sciences, Boston, 1870, pp. 471–491. (Original publication 1798.)

2. Mayer, J. R., "The Forces of Inorganic Nature," in *The Correlation and Conservation of Forces*, E. L. Youmans, (Ed.), Appleton, New York, pp. 251–258, 1865. (Original publication 1842.)

3. Joule, J. P., "On the Mechanical Equivalent of Heat," *Philosophical Transactions of the Royal Society*, 140: 61–82 (1850).

4. Helmholtz, H., "Interaction of Natural Forces," in *The Correlation and Conservation of Forces* (E. L. Youmans, Ed.), Appleton, New York, 1865, pp. 211–247. (Original publication 1854.)

5. Mayer, J. R., "The Mechanical Equivalent of Heat," in *The Correlation and Conservation of Forces* (E. L. Youmans, Ed.), Appleton, New York, 1865, pp. 316–355. (Original publication 1851.)

6. Clausius, R., "The Second Law of Thermodynamics," in *A Source Book of Physics* (W. F. Magie, Ed.), Harvard University Press, Cambridge, MA, 1963, pp. 228–236. (Original publication 1850.)

7. Energy Information Agency, U.S. Department of Energy, "Annual Energy Review 2002," http://www.eia.doe.gov/emeu/aer/contents.html, posted Oct. 24, 2003.

8. Turns, S. R., *An Introduction to Combustion: Concepts and Applications*, 2nd ed., McGraw-Hill, New York, 2000.

9. Warnatz, J., Maas, U., and Dibble, R. W., *Combustion*, Springer-Verlag, Berlin, 1996.

10. Joule, J. P., "On the Production of Heat by Voltaic Electricity," *Philosophical Transactions of the Royal Society of London*, IV:280–282 (1840).

11. Johnson, A. T., *Biological Process Engineering: An Analogical Approach to Fluid Flow, Heat Transfer, and Mass Transfer Applied to Biological Systems*, Wiley, New York, 1998.

12. Li, C.-H., "Piston Thermal Deformation and Friction Considerations," SAE Paper 820086, 1982.

13. Incropera, F. P., and DeWitt, D. P., *Fundamentals of Heat and Mass Transfer*, 5th ed., McGraw-Hill, New York, 2002.

14. Bird, R. B., Stewart, W. E., and Lightfoot, E. N., *Transport Phenomena*, Wiley, New York, 1960.

Some end-of-chapter problems were adapted with permission from the following:

15. Chapman, A. J., *Fundamentals of Heat Transfer*, Macmillan, New York, 1987.

16. Look, D. C., Jr., and Sauer, H. J., Jr., *Engineering Thermodynamics*, PWS, Boston, 1986.

17. Myers, G. E., *Engineering Thermodynamics*, Prentice Hall, Englewood Cliffs, NJ, 1989.

Nomenclature

A	Area (m²)	P	Pressure (Pa)
Bi	Biot number (dimensionless)	pe	Specific potential energy (J/kg)
c	Specific heat (J/kg·K)	PE	Potential energy (J)
c_p	Constant-pressure specific heat (J/kg·K)	Q	Heat (J)
$\overline{c}_p$	Molar constant-pressure specific heat (J/kmol·K)	$\dot{Q}$	Heat transfer rate (W)
c_v	Constant-volume specific heat (J/kg·K)	$\dot{Q}''$	Heat flux (W/m²)
		r	Radial coordinate (m)
$\overline{c}_v$	Molar constant-volume specific heat (J/kmol·K)	R	Radius (m) or particular gas constant (J/kg·K)
e	Specific energy (J/kg)	R_e	Electrical resistance (ohm)
E	Energy (J)	R_{th}	Thermal resistance (K/W)
$\dot{E}$	Energy rate (W)	R_u	Universal gas constant, 8314.472 (J/kmol·K)
$\dot{E}'''$	Energy rate per unit volume (W/m³)	t	Time (s)
g	Gravitational acceleration (m/s²)	T	Temperature (K)
h	Specific enthalpy (J/kg)	u	Specific internal energy (J/kg)
$\overline{h}$	Molar-specific enthalpy (J/kmol)	$\overline{u}$	Molar-specific internal energy (J/kmol)
$\overline{h}_{conv}$	Average convective heat-transfer coefficient (W/m²·K or °C)	U	Internal energy (J)
h_L	Head loss (m)	v, V	velocity (m/s)
H	Enthalpy (J)	v_r, v_θ, v_x	Cylindrical coordinate velocity components (m/s)
i	Electric current (A)	v_r, v_θ, v_ϕ	Spherical coordinate velocity components (m/s)
k	Thermal conductivity (W/m·K or °C)	v_x, v_y, v_z	Cartesian coordinate velocity components (m/s)
ke	Specific kinetic energy (J/kg)		
KE	Kinetic energy (J)	v	Specific volume (m³/kg)
L	Length (m)	$\mathbb{V}$	Volume (m³)
$\dot{m}$	Mass flow rate (kg/s)	$\mathbf{v}$	Voltage (V)
$\dot{m}''$	Mass flux (kg/s·m²)	W	Work (J)
M	Mass (kg)	$\dot{W}$	Rate of work or power (W)
$\mathcal{M}$	Molecular weight (kg/kmol)	x	Spatial coordinate (m)
n	Exponent in power-law velocity distribution (dimensionless)	y	Spatial coordinate (m)
		z	Spatial coordinate (m)
N	Number of moles (kmol)		

GREEK

α	Thermal diffusivity (m^2/s) or kinetic-energy correction factor (dimensionless)
γ	Specific-heat ratio (dimensionless)
δ	Small increment
Δ	Difference or change
θ	Angular coordinate in cylindrical and spherical systems
ρ	Density (kg/m^3)
ϕ	Angular coordinate in spherical system
Φ_{visc}	Viscous dissipation (W/m^3)

SUBSCRIPTS

ad	adiabatic
avg	average
char	characteristic
cond	conduction
conv	convection
CS	control surface
cv	control volume
elec	electrical
f	formation
flow	associated with the flow
gen	generated within system or control volume

i	species i
in	into system or control volume
init	initial
j	index for inlets
k	index for outlets
mech	mechanical
mix	mixture
OA	overall
out	out of system or control volume
prod	products
reac	reactants
ref	reference value or state
s	sensible
stored	stored within system or control volume
surr	surroundings for radiation exchange
sys	system
s1	surface 1
s2	surface 2
1	inner
2	outer
∞	ambient

SUPERSCRIPTS

$\circ$	Standard-state pressure ($P^\circ = 1$ atm)

QUESTIONS

5.1 Review the most important equations presented in this chapter (i.e., those with a red background). What physical principles do they express? What restrictions apply?

5.2 Without reference to the text, write symbolic expressions of conservation of energy (the first law of thermodynamics) for a system for the following conditions:

  A. For a change from state 1 to state 2

  B. For an incremental change (use δ and d as appropriate)

  C. At an instant

  Below each term, write its meaning in words.

5.3 Distinguish between bulk system energy and internal energy. Illustrate your discussion with a practical example.

5.4 Write out the first law of thermodynamics for an incremental change in state. Include heat, work, and energy terms in your expression.

5.5 Write out the first law of thermodynamics for a finite change in state. Include heat, work, and energy terms in your expression.

5.6 Without reference to the text, write a symbolic expression of conservation of energy (the first law of thermodynamics) for a control volume having a single inlet and a single outlet. Assume steady state. Below each term, write its meaning in words.

5.7 Repeat Question 5.6, eliminating the assumption of steady state.

5.8 Explain the origin of enthalpy in the expression of conservation of energy for a control volume.

5.9 Show how the first law reduces to $_1Q_2 = \Delta H$ for a system comprising a compressible substance undergoing a constant-pressure process. What assumptions are required?

5.10 Sketch an adiabatic, constant-pressure combustion process on H–T coordinates. Show lines representing $H_{\text{reac}}(T)$ and $H_{\text{prod}}(T)$.

5.11 Sketch an adiabatic, constant-volume combustion process on U–T coordinates. Show lines representing $U_{\text{reac}}(T)$ and $U_{\text{prod}}(T)$.

5.12 Using the sketches created in Questions 5.10 and 5.11, show how decreasing the temperature of the initial reactants results in a decreased temperature for the products.

5.13 Explain the relationship between electrical work, or power, and Joule heating.

5.14 Explain the meaning of the lumped parameter approximation.

5.15 How does an energy balance at a surface, or interface, differ from an energy balance for a thermodynamic system or a control volume?

5.16 Explain the origin or physical significance of each term in the one-dimensional Cartesian conduction equation (Eq. 5.30).

5.17 Define the thermal diffusivity α. Under what condition(s) does this thermophysical property appear in the one-dimensional heat conduction equation (cf. Eqs. 5.30 and 5.31)?

5.18 Sketch the steady-state temperature distribution through a composite plane wall as shown in the sketch. The two materials are in direct contact at the interface. The highest temperature occurs at the right face of material B.

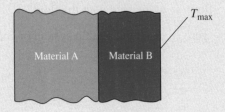

Chapter 5 Problem Subject Areas

5.1–5.7	**System processes and the first law of thermodynamics**
5.8–5.52	**Energy conservation applied to systems**
5.53–5.62	**Reacting systems (fixed mass)**
5.63–5.120	**One-dimensional conduction**
5.121–5.168	**Steady-flow integral control volumes**
5.169–5.181	**Steady-flow integral control volumes with chemical reactions**
5.182–5.188	**Unsteady integral control volumes**
5.189–5.192	**General principles of energy conservation**
5.193–5.196	**Differential control volumes**

PROBLEMS

5.1 Consider a piston–cylinder assembly containing 0.1 kg of dry air initially at 300 K and 200 kPa (state 1). Energy is added to the air (heat transfer) at constant pressure until the final temperature is 450 K at state 2. Plot the process on P–v and T–v coordinates and determine the following quantities: $_1W_2$, ΔU, ΔH, and $_1Q_2$. Give units.

5.2 Consider a piston–cylinder assembly containing 0.1 kg of saturated steam at 300 kPa (state 1). Heat is added to the steam at constant pressure until a temperature of 600 K is reached (state 2). Plot the process on P–v and T–v coordinates and determine the following quantities: $_1W_2$, ΔU, ΔH, and $_1Q_2$. Give units.

5.3 Consider a piston–cylinder assembly containing 0.12 kg of nitrogen at 300 K and 1 atm (state 1). The nitrogen is heated at constant pressure to 700 K (state 2). Plot the process on P–v and T–v coordinates. Assuming a constant value of the constant-pressure specific heat of 1.067 kJ/kg·K for the N_2, determine the following quantities: $_1W_2$, ΔU, ΔH, and $_1Q_2$. Give units.

5.4 Repeat Problem 5.3 but use tabulated values of $\overline{h}$ (Table B.7), rather than the given constant specific heat, to find the desired quantities. Give units.

5.5 Consider a sealed rigid tank containing 0.15 kg of dry air at 300 K and 100 kPa (state 1). The air is heated to 600 K (state 2). Plot the process on P–v and T–v coordinates and determine the following quantities: $_1W_2$, ΔU, ΔH, and $_1Q_2$. Give units.

5.6 Consider a sealed rigid tank containing 0.18 kg of N_2 at 300 K and 1 atm (state 1). The N_2 is heated to a temperature of 700 K (state 2). Plot the process on P–v and T–v coordinates and determine the following quantities: $_1W_2$, ΔU, ΔH, and $_1Q_2$. Compare these results with the results obtained in Problem 5.3 or 5.4. Discuss.

5.7 Superheated steam is contained in a sealed rigid tank at 600 K and 0.3 MPa (state 1). Energy is removed from the steam by heat transfer until a temperature of 350 K (state 2) is reached. Plot the process on P–v and T–v coordinates. Include the steam dome on your plots. Determine the following quantities for this process: $_1W_2/M$, Δu, Δh, and $_1Q_2/M$.

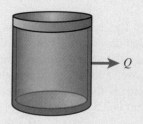

5.8 Air is contained in a piston–cylinder assembly at 2 MPa and 400 K (state 1). The 0.15-m-diameter piston is locked in place by stops at the 0.2-m position, and a 2-kg steel block sits on top of the piston, as shown in the sketch. The mass of the piston is 0.1 kg. The stops are removed and the air rapidly expands until the piston comes to rest at the upper position ($x = 0.4$ m), while the block continues upward as shown in the sketch on the right. The temperature is 310 K (state 2). Assume that the process is so rapid that it can be considered adiabatic and that the constant-pressure specific heat is constant ($c_p = 1.013$ kJ/kg·K). The atmospheric pressure is 100 kPa.

A. Considering the air, piston, cylinder, and the block combined to be the system, determine the work done by the system on the surroundings.

B. Determine the velocity of the steel block immediately after the piston comes to rest.

C. Neglecting drag, determine the maximum height achieved by the block measured from the bottom of the cylinder.

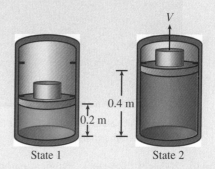

State 1   State 2

5.9 An electric current of 0.25 A flows through a 5-ohm electrical resistor placed in an evacuated chamber. The resistor is a 30-mm-long cylinder with a diameter of 5 mm. Estimate the steady-state surface temperature of the resistor if it has an emissivity of unity (i.e., is black) and the temperature of the chamber walls is 300 K. Neglect any conduction heat transfer through the lead wires.

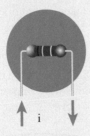

5.10 Consider the same situation as in Problem 5.9 except that now air at 300 K fills the chamber. Does the steady-state temperature of the resistor increase or decrease? Explain. Estimate the temperature of the resistor if the average convective heat-transfer coefficient is 2.5 W/m²·K.

5.11 Pulverized coal particles at 300 K are injected into hot furnace gases at 1500 K. Estimate the initial heating rate of a 70-μm-diameter particle. Express your result in kelvins per second. The convective heat-transfer coefficient is approximately 1500 W/m²·K. The specific heat and density of the coal are 1.3 kJ/kg·K and 1650 kg/m³, respectively.

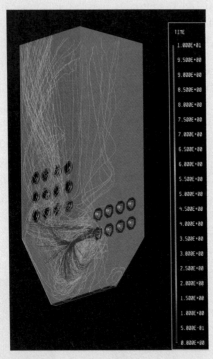

5.12 Small (200-μm-diameter), spherical carbon particles are injected into a hot stream of air. The carbon particles are initially at 300 K, and the air is at 1200 K.

A. Assuming that the lumped parameter approximation holds (i.e., the carbon particles are of essentially uniform temperature at each instant) derive an expression of the carbon-particle temperature history [i.e., $T_p(t)$]. Neglect radiation and assume that the convective heat-transfer coefficient is constant.

B. Estimate the time for a carbon particle to reach 550 K given the following additional data:

$$c_{p,C} = 0.7 \text{ kJ/kg}\cdot\text{K},$$
$$\rho_C = 1950 \text{ kg/m}^3, \text{ and}$$
$$\bar{h}_{\text{conv}} = 600 \text{ W/m}^2\cdot\text{K}.$$

5.13 A very long, 5-cm-diameter copper cylinder is heated to 30°C and then plunged into a liquid bath at 0°C. After 3 min, the cylinder center temperature is 4°C. Assuming the Biot number is small, determine the average heat-transfer coefficient for cooling in the liquid bath. Was the assumption of a small Biot number justified?

5.14 A fine copper wire (0.07-cm diameter) is initially heated to 175°C. The wire is suddenly exposed to an environment at 40°C with a heat-transfer coefficient of 60 W/m²·°C.

A. Calculate the Biot number to establish whether the internal resistance can be neglected for the cooling process.

B. Determine (a) the wire temperature and (b) the heat loss per unit length (W/m) 10 s after the wire is exposed to the convective environment.

C. Repeat Parts A and B for an aluminum wire.

5.15 A 1-in-thick stainless-steel plate is heated to some initial temperature and then exposed to convective cooling by air at 100 F with a heat-transfer coefficient $h_{conv} = 2$ Btu/hr·ft²·F. Find the initial temperature of the plate if its temperature after 10 min is 1000 F. Justify the use of the lumped-capacitance approximation.

5.16 A 3-cm-diameter pure-aluminum sphere is heated to an initial temperature of 175°C and then immersed in a fluid at 25°C. After 42 s, the center of the sphere is measured to be 100°C. Estimate the convective heat-transfer coefficient for the cooling process.

5.17 A nearly spherical copper–constantan thermocouple is used to measure the temperature of a flowing gas stream. The properties of the thermocouple junction are $\rho = 8920$ kg/m³, $c_p = 384$ J/kg·K, and $k = 23$ W/m·K. The convective heat-transfer coefficient for the junction in the gas stream is 500 W/m²·°C. The thermocouple and the gas stream initially have the same temperature. A step increase then occurs in the gas stream temperature. It is desired that, after 3 s, the temperature increase indicated by the thermocouple be 95% of the step increase in gas temperature. Calculate the diameter of the junction required to meet this condition.

5.18 An electric circuit consists of a 12-V storage battery, a 10-ohm resistor, and an electric motor connected in parallel. If the motor produces 100 mW of power, estimate the rate of energy depletion from the battery.

Motor

10-ohm Resistor

BATTERY

5.19 During the compression stroke in an automobile engine, air initially at 14.5 psia and 90 F is compressed reversibly according to $PV^{1.48} = $ constant. The engine compression ratio $(= V_1/V_2)$ is 7.5. Determine (a) the work and (b) the heat transfer, each in Btu/lb$_m$.

5.20 Consider 5 kg of air initially at 101.3 kPa and 38°C. Heat is transferred to the air until the temperature reaches 260°C. Determine the change of internal energy, the change in enthalpy, the heat transfer, and the work done for (a) a constant-volume process and (b) a constant-pressure process. Use SI units.

5.21 While trapped in a cylinder, 2.27 kg of air is compressed isothermally (with a water jacket being used around the cylinder to maintain constant temperature) from initial conditions of 101 kPa and 16°C to a final pressure of 793 kPa. For this

process, determine (a) the work required and (b) the heat removed, both in units of kJ.

5.22 A closed system rejects 25 kJ of energy in a heat interaction while experiencing a volume change of 0.1 m³ (0.15 to 0.05 m³). Assuming a reversible, constant-pressure process at 350 kPa, determine the change in internal energy.

5.23 An inventor claims to have a closed system that operates continuously and produces the following energy effects during the cycle: net $Q = 3$ Btu and net $W = 2430$ ft·lb$_f$. Support or refute this claim.

5.24 A tank contains a fluid that is stirred by a paddle wheel. The work input to the paddle wheel is 4309 kJ. The heat transferred from the tank is 1371 kJ. Considering the tank and the fluid as a closed system, determine the change in the internal energy (kJ) of the system.

5.25 Steam (6 lb$_m$) at 200 psia with an 80% quality comprises a thermodynamic system. The steam is heated in a frictionless process until the temperature is 500 F.

A. Calculate the heat transfer in Btu if the process occurs at constant pressure.

B. What is the heat transfer if the process is carried out at constant volume?

5.26 Consider 0.5 lb$_m$ of steam initially at 20 psia and 228 F ($u_1 = 1081$ Btu/lb$_m$ and $v_1 = 20.09$ ft³/lb$_m$) in a closed, rigid container. This steam is heated to 440 F ($u_2 = 1158.9$ Btu/lb$_m$). Determine the amount of heat added to the system in Btu.

5.27 Consider a thermodynamic system executing the two reversible processes shown in the sketch (*ab* and *adb*). What is the heat transferred to an ideal gas if $P_2 = 2P_1$ and $v_2 = 2v_1$? Express your answer in terms of the gas constant R and T_1.

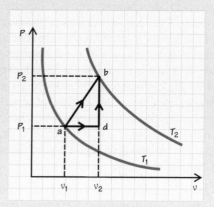

5.28 Consider a thermodynamic system consisting 3 lb$_m$ of an ideal gas. The gas is compressed frictionlessly and adiabatically from 14.7 psia and 70 F to 60 psia. For this gas, $c_p = 0.238$ Btu/lb$_m$·F, $c_v = 0.169$ Btu/lb$_m$·F, and $R = 53.7$ ft·lb$_f$/lb$_m$·R. Compute (a) the initial volume, (b) the final volume, (c) the final temperature, and (d) the work.

5.29 The heat-transfer rate to the surroundings from a person at rest is about 100 W. Suppose the ventilation system fails in an auditorium containing 2000 people. Assume there is no heat transfer through the auditorium walls.

A. Determine the increase in internal energy of the air in the auditorium during the first 20 min after the ventilation system fails. Express your result in MJ.

B. Considering the auditorium and all of the people as a system, determine the increase in internal energy (MJ) of the system. How do you explain the fact that the temperature of the air increases?

5.30 A car battery is charged by applying a current of 40 A at 12 V for 30 min. During the charging process, there is a heat transfer of 200 Btu from the battery to the surroundings. Determine the increase in internal energy (Btu) of the battery.

5.31 A balloon at sea level contains 2 kg of helium at 30°C and 1 atm. The balloon then rises to 1500 m above sea level. At this height, the helium temperature is 6°C. Determine the change in internal energy (kJ) of the helium.

5.32 Front-wheel-drive cars do not distribute the work involved in stopping the car equally among all wheels. The front-wheel brakes dissipate about 60% of the energy transfer involved in braking, while the rear wheels take care of the remaining 40%. If you are in such a car, traveling at 55 miles/hr, and if you slow the car down to 20 miles/hr by applying the brakes, determine the amount of kinetic energy (J and Btu) dissipated in the front brakes. Neglect rolling resistance and aerodynamic drag. Assume that the car and driver have a total mass of 1087 kg (2400 lb$_m$).

5.33 One kilogram of air in a rigid tank receives 10 kJ of energy as heat transfer. Determine the increase in internal energy of the air in kJ/kg.

5.34 Initially, 1 kg of water (liquid and/or vapor) at 0.1 MPa is contained in a rigid 1.5-m³ tank. The tank is then heated to 300°C. Determine the final pressure (MPa) and the heat transfer (kJ).

5.35 Initially, 1 lb$_m$ of air at 14.7 psia and 800 F is contained in a rigid tank. The tank is not insulated and will cool down because of heat transfer to the atmosphere ($T_{atm} = 70$ F, $P_{atm} = 14.7$ psia) until it reaches thermal equilibrium. Determine the final pressure (psia) and the heat transfer (Btu) for the process.

5.36 A 1-kg piece of iron, initially at 700°C, is quenched by dropping it into an insulated tank containing 2 kg of liquid water. The initial temperature of the water is 20°C. Determine the final temperature of the iron.

5.37 A steam radiator has a volume of 5 ft³ and initially contains dry, saturated vapor at 30 psia. The valves are then closed on the radiator and, as a result of heat transfer to the room, the temperature drops to 100 F. Determine the heat transfer (Btu) from the radiator to the room.

5.38 An 80-ft³ steam boiler initially contains 60 ft³ of liquid water and 20 ft³ of water vapor in equilibrium at 14.7 psia. The boiler is fired up and the liquid and vapor in the boiler are heated. Somehow the valves on the inlet and discharge of the boiler are both left closed. The relief valve lifts when the pressure reaches 800 psia. Determine the heat transfer (MBtu) to the boiler before the relief valve lifts.

5.39 The temperature of 150 liters of liquid water, initially at 10°C, is increased to 60°C by a 2500-W electric heater.

A. Determine the total energy (MJ and kBtu) required.

B. Determine the length of time (hr) required.

C. If electricity can be purchased for 6 cents per kW·hr, determine the cost (cents).

5.40 A closed, rigid steel tank has an inner volume of 1 ft³ and has a mass of 50 lb$_m$ when empty. Initially, the tank contains 1.0 lb$_m$ of water (liquid plus vapor), and the tank and the water are both at 70 F. The tank containing the water is then placed on a 100-lb$_m$ slab of steel, which is at another temperature. Assume the tank–water–slab system comes to equilibrium

Tank

Slab

with no heat transfer to the surroundings. The final temperature of the tank–water–slab system is 440 F. The specific heat of the steel in both the tank and the slab is 0.1 Btu/lb$_m$·R. Determine the initial slab temperature (F).

5.41 An insulated container, filled with 10 kg of liquid water at 20°C, is fitted with a stirrer. The stirrer is made to turn by lowering a 25-kg object outside the container a distance of 10 m using a frictionless pulley system. The local acceleration of gravity is 9.7 m/s^2. Assume that all work done by the object is transferred to the water and that the water is incompressible.

A. Determine the work transfer (kJ) to the water.

B. Determine the increase in internal energy (kJ) of the water.

C. Determine the final temperature (°C) of the water.

D. Determine the heat transfer (kJ) from the water required to return the water to its initial temperature.

5.42 A cylinder fitted with a freely floating piston initially contains 1 lb$_m$ of water at 1200 psia and a quality of 0.25. The water is then heated to a temperature of 900 F. Determine the work transfer (Btu) and the heat transfer (Btu).

5.43 Two pounds-mass of water (liquid and/or vapor) at 20 psia is contained in a piston–cylinder device. The initial volume is 25 ft^3. The water is then heated until its temperature reaches 800 F. The piston is free to move up or down unless it reaches the stops. If the piston is up against the stops the cylinder volume is 50 ft^3.

A. Determine the initial internal energy (Btu) of the water.

B. Determine the final internal energy (Btu) of the water.

C. Determine the work transfer (Btu) from the water.

D. Determine the heat transfer (Btu) to the water.

5.44 A frictionless 3000-N piston maintains a gas at constant pressure in a cylinder. The cross-sectional area of the piston is 52 cm^2 and the atmospheric pressure acting on this area is 0.1 MPa. A process occurs in which a paddle wheel transfers 6800 N·m of work to the gas, 10 kJ of heat is transferred from the gas, and the internal energy of the gas decreases by 1 kJ. Determine the distance (cm) moved by the piston.

5.45 Initially, 1 lb$_m$ of water at 300 F with a quality of 0.60 is contained in a piston–cylinder device. The piston is then slowly moved until the water is entirely changed into liquid. Heat transfer maintains the temperature at 300 F. Determine the following:

A. The increase in volume (ft^3) of the cylinder

B. The work transfer (Btu)

C. The heat transfer (Btu)

5.46 Initially, saturated liquid water at 200°C is contained in a piston–cylinder device. The water then expands isothermally until its volume is 100 times larger than its initial volume. Determine the following:

A. The increase in energy (kJ/kg) of the water

B. The work transfer (kJ/kg) and direction into or out of the water

C. The heat transfer (kJ/kg) and direction into or out of the water

5.47 A piston–cylinder device has an initial volume of 0.003 m^3 and contains dry, saturated water vapor at 200°C. The piston then moves out until the volume reaches 0.015 m^3. The final pressure is 0.25 MPa. The process occurs rapidly enough that it may be assumed to be adiabatic. Determine the following:

A. The work transfer (kJ) from the steam

B. The work transfer (kJ) to the atmosphere

C. The work transfer (kJ) from the piston to the connecting rod

5.48 Initially, 1 lb$_m$ of steam at 100 psia with a quality of 0.90 is held in an adiabatic cylinder fitted with a freely floating piston and a paddle wheel. The paddle wheel is then turned on for 1 min. During this time interval, 200 Btu of work is transferred to the steam from the paddle wheel. Determine the following:

A. The final pressure (psia)

B. The final temperature (F)

C. The final volume (ft^3)

D. The final energy (Btu) of the steam

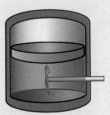

5.49 Air, initially at 400 F and 500 psia, expands isothermally in a piston–cylinder device until its volume is 100 times larger than its initial volume. Determine the following:

A. The increase in internal energy (Btu/lb$_m$) of the air

B. The work transfer (Btu/lb$_m$) and direction into or out of the air

C. The heat transfer (Btu/lb$_m$) and direction into or out of the air

5.50 Initially, 0.5 kg of dry, saturated steam at 115°C is contained inside a spherical elastic balloon whose internal pressure is proportional to its diameter. Heat is then transferred to the steam until the steam pressure reaches 0.2 MPa. Determine the final temperature (°C) and the heat transfer in kJ.

5.51 A rigid cylinder fitted with a freely floating piston is divided into two parts (A and B) by a rigid metal partition. Initially, part A contains 2 lb$_m$ of water (liquid and/or vapor) at 250 psia and part B contains 1 lb$_m$ of dry, saturated water vapor at 90 psia. The piston and the sides of A and B are perfectly insulated. The partition is a good heat conductor and allows enough heat transfer to keep the temperature of part A always equal to the temperature of part B. The bottom of part B is then heated until the pressure of part A equals the pressure of part B. Determine the heat transfer (Btu) into the bottom of part B.

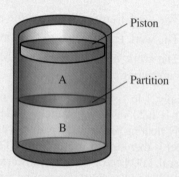

Piston

A

Partition

B

5.52 A 10-ft-diameter open-top tank is initially empty. The tank is filled to a height of 50 ft with water at 60 F pumped from the surface of a large lake. The lake surface and the tank bottom are at the same elevation. Using a system (fixed-mass) analysis, determine the pump work (Btu) required to adiabatically fill the tank.

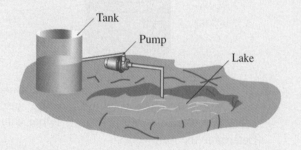

Tank

Pump

Lake

5.53 Consider the combustion of a fuel at constant pressure. Reproduce the coordinate system in the sketch and indicate and appropriately label the following items:

A. H_{reac}

B. H_{prod}

C. Heating value (HV)

D. The initial adiabatic flame temperature (T_{ad}) for the reactants at 298 K.

5.54 A piston–cylinder arrangement initially contains 0.002 kmol of H$_2$ and 0.01 of O$_2$ at 298 K and 1 atm. The mixture is ignited and burns adiabatically at constant pressure. Determine the final temperature assuming the products contain only H$_2$O and the excess reactant. Also determine the work done during the process. Sketch the process on H–T and P–V coordinates.

5.55 A rigid spherical pressure vessel initially contains 0.002 kmol of H$_2$ and 0.01 kmol of O$_2$ at 298 K and 1 atm. The mixture is ignited and burns adiabatically. Determine the final temperature and pressure assuming the products contain only H$_2$O and the excess reactant. Sketch the process on U–T and P–V coordinates.

5.56 Hydrogen burns with 200% stoichiometric air in a piston–cylinder arrangement. The process is carried out adiabatically and at constant pressure. The initial temperature and pressure are 298 K and 1 atm, respectively. Determine the final temperature and the work performed per mass of mixture. Assume the air is a simple mixture of 21% O$_2$ and 79% N$_2$ by volume. Also sketch the process on H–T and P–V coordinates.

5.57 Determine the adiabatic, constant-volume flame temperature for a stoichiometric mixture of propane (C$_3$H$_8$) and air. The reactants are at 1 atm and 298 K. Assume the air to be a simple mixture of 21% O$_2$ and 79% N$_2$. Furthermore, assume complete combustion with no dissociation. Sketch the process on H–T and P–V coordinates.

5.58 A hydrocarbon fuel is burned with air. Assume the air to be a simple mixture of 21% O$_2$ and 79% N$_2$. On a dry basis (all of the water vapor has been removed from the products), a volumetric analysis of the products of combustion yields the following:

CO$_2$—7.8%,

CO—1.1%,

O$_2$—8.3%,

N$_2$—82.8%.

Determine the following:

A. Composition of the fuel on a mass basis

B. Percent theoretical air

C. Air–fuel ratio by mass

5.59 Determine the heat transfer in the constant-volume combustion of 1 kg of carbon as indicated in the following reaction:

$$C + 1.5O_2 \rightarrow CO_2 + 0.5O_2.$$

The reactants are at 38°C and the products are at 205°C.

5.60 A mixture of 5 g of ethane and a stoichiometric amount of oxygen are contained in a piston–cylinder device at 25°C and 1 atm. A spark ignites the mixture and a complete reaction occurs at a pressure of 1 atm. The products are cooled to 25°C by the end of the reaction. Determine the following:

A. The initial volume (cm³) of the reactants

B. The masses of liquid water and water vapor in grams

C. The work transfer in kJ

D. The increase in internal energy in the cylinder in kJ

E. The heat transfer (MJ)

5.61 Initially, 10 cm³ of dry air and a stoichiometric amount of gunpowder (carbon) are contained behind a 20-g bullet in a long barrel at 25°C and 1 atm. The gunpowder is then ignited and completely burned. As the pressure builds up behind the bullet, the bullet accelerates forward. Assume the process occurs so fast that there is no time for any heat transfer to the barrel or to the bullet to occur. The temperature of the products at the time when the bullet leaves the barrel is estimated to be 700 K and the pressure has returned to 1 atm. Determine (a) the initial gunpowder mass (g) and (b) the bullet velocity (m/s) leaving the barrel.

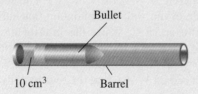

Bullet

10 cm³   Barrel

5.62 A stoichiometric mixture of gaseous octane and oxygen reacts completely at 25°C in a nonflow, constant-volume process. The initial pressure is 1 atm. Assume all the water in the products is liquid. Determine the heat transfer in MJ per kg of octane.

5.63 Consider a long, cylindrical, homogeneous solid. Derive the governing partial differential equation whose solution describes the time-dependent temperature distribution in the cylinder. Assume one-dimensional behavior (i.e., the radial coordinate r is the only spatial variable). Assume constant properties and allow for internal thermal energy generation. Use $\dot{E}'''_{gen}$ ([=] W/m³) to express the energy generation rate per unit volume.

5.64 Consider a spherical homogeneous solid. Derive the governing partial differential equation whose solution describes the time-dependent temperature distribution in the sphere. Assume one-dimensional behavior (i.e., the radial coordinate r is the only spatial variable). Assume constant properties and allow for internal thermal energy generation. Use $\dot{E}'''_{gen}$ ([=] W/m³) to express the energy generation rate per unit volume.

5.65 Find the conduction loss through a 50-cm by 100-cm, 0.6-cm-thick plate glass window when the inside and outside surface temperatures are 20°C and 0°C, respectively.

5.66 Consider a 3-cm-thick slab of insulation applied to the walls of a furnace. The thermal conductivity of the insulation is 0.25 W/m·°C. Determine the temperature difference across the slab for a conduction heat flux of 4000 W/m².

5.67 The heat conduction rate through a 10-in-thick slab of 1.0% carbon steel is 650 Btu/hr·ft². If the cooler surface is maintained at 200 F, what is the temperature of the other surface?

5.68 A 3-m by 2-m, 15-cm-thick furnace wall is composed of fire-clay brick having a thermal conductivity of 1.65 W/m·°C. Measurements of the surface temperatures show that the inside surface temperature is 1120°C and the outside surface temperature is 875°C. Find the rate of heat loss through the wall.

5.69 The conduction heat-transfer rate through a 3-cm-thick layer of insulating material with a face area of 15 m² is 5 kW. The insulation has a thermal conductivity of 0.2 W/m·°C. If the temperature of the hot side of the insulation is 420°C, what is the temperature of the cooler side?

5.70 A 1-ft-thick concrete wall ($k = 0.6$ Btu/hr·ft·F) has a surface area of 300 ft². One surface of the wall is maintained at 75 F and the other at 5 F. What is the heat loss through the wall?

5.71 The thermal conductivity of a material can be measured by heating one surface of a slab of the material with an electric heater, while cooling the other surface with a water-cooled plate. With this apparatus, the thermal conductivity can be calculated from measurements of the two surface temperatures, the slab thickness, and the heat flux imposed by the electric heater. Determine the apparent thermal conductivity of a 5-cm-thick slab of glass wool when one surface is maintained at 30°C and the other at 80°C for an imposed heat flux of 50 W/m².

5.72 A 3-in-thick slab of concrete is tested in the manner described in Problem 5.71. A heat flux of 50 Btu/hr·ft² is measured when one surface is maintained at 80°F and the other at 60°F. What is the thermal conductivity of the concrete?

5.73 The walls of a freezer compartment are to be insulated with hair felt ($k = 0.026$ Btu/hr·ft·F). Determine the thickness of the insulation required to limit the heat flow into the freezer to a flux of 5 Btu/hr·ft² when the inner and outer insulation surface temperatures are 15 F and 75 F, respectively.

5.74 The inside surface of an 8-in-thick, common brick wall is maintained at 100 F. The outer surface of the

wall loses heat to the ambient air by convection. The heat-transfer coefficient at the outer surface is 5 Btu/hr·ft²·F and the air temperature is 40 F. Find (a) the heat flux through the wall and (b) the temperature of the exposed surface.

5.75 One surface of a 12-cm-thick slab of material ($k = 2.5$ W/m·°C) is maintained at 100°C. The other surface is exposed to a fluid at 70°C. The heat-transfer coefficient at this surface is 150 W/m²·°C. What is the temperature of the surface exposed to the fluid? What is the heat flux through the slab?

5.76 The dividing surface in a simple heat exchanger is a 0.6-cm-thick, AISI-304 stainless-steel plate. Hot combustion gases at 425°C flow over one side of the plate and produce a convective heat-transfer coefficient of 48.8 W/m²·°C at that surface. Air at 60°C flows over the other side of the plate, producing a convective heat transfer coefficient of 36.9 W/m²·°C at that surface. Find the heat-transfer rate between the two fluids per unit surface area.

5.77 It is desired to limit the heat loss through a boiler furnace wall to 2000 W/m². The wall is composed of fire brick having a thermal conductivity of 1.15 W/m·°C. If the inner surface temperature is 1100°C and the outer surface 370°C, what brick thickness should be used?

5.78 The addition of a 5-cm-thick insulation ($k = 0.09$ W/m·°C) to the outer surface of the wall in Problem 5.77 reduces the exposed surface temperature to 180°C. Determine the rate of heat loss through the insulated wall per unit area.

5.79 The wall of a pressure vessel is so large that it may be treated as planar. The 0.75-in-thick wall is made of 1%-carbon steel. Determine the heat flux through the wall when the inside temperature is maintained at 350 F and the outside temperature at 300 F.

5.80 The basement wall of a building is to be made of concrete ($k = 0.95$ W/m·°C). The soil on the outside of the wall will maintain that surface temperature at 15°C, while the interior environment in the basement will maintain the inner surface at 26°C. If the heat flux through the wall is not to exceed 25 W/m², find the required thickness of the wall.

5.81 To avoid possible burn injury to workers, as well as to reduce heat losses, insulation is applied to the pressure-vessel wall described in Problem 5.79. A 0.5-in-thick layer of insulation ($k = 2.0$ Btu/hr·ft·F) applied to the outer surface of the vessel wall results in an exposed insulation temperature of 140 F. Assuming that the inside wall surface remains at 350 F, determine the heat flux.

5.82 A wall is composed of a 12-cm-thick layer of material A and a 20-cm-thick layer of material B. The temperature of the outer surface of layer A is known to be 260°C when the outer temperature at layer B is at 30°C. A layer of 2.5-cm-thick insulation ($k = 0.09$ W/m·°C) is added to the outer surface of layer B. Under these conditions, it is observed that the surface temperature of layer A rises to 310°C and the junction between layer B and the insulation (formerly the outer surface of layer B) becomes 220°C. The insulation surface is measured to be 25°C. What are the rates of heat flow, per square meter of wall area, before and after the insulation is added?

5.83 The composite wall of a furnace consists of three layers of different materials. The inner layer ($k_{12} = 20$ W/m·°C) has a thickness $\Delta x_{12} = 30$ cm. The middle layer has a thickness $\Delta x_{23} = 15$ cm; however, the conductivity of this layer is unknown. The outer layer ($k_{34} = 50$ W/m·°C) has a thickness $\Delta x_{34} = 15$ cm. Under steady operating conditions, the inside temperature is 600°C, the outside surface temperature is 20°C, and the heat flux through the wall is measured to be 5000 W/m². What is the thermal conductivity of the middle layer?

5.84 Consider a 12-cm-thick plane wall with its inner surface temperature fixed at 0°C. When its outer surface is maintained at 100°C, the heat flux through the wall is measured to be 16,000 W/m². When the outer surface is raised to 200°C, the heat flux increases to 40,000 W/m². Is the thermal conductivity independent of temperature? Explain.

5.85 A 1.25-cm-thick steel ($k = 43$ W/m·°C) plate is exposed on one side to steam at 650°C. The convective heat-transfer coefficient at this surface is 570 W/m²·°C. As a safety precaution, it is desired to insulate the outer surface so that the exposed surface of the outer insulation does not exceed 38°C. To minimize cost, an expensive high-temperature insulation ($k = 0.26$ W/m·°C) is applied to the steel surface. A second layer of a less-expensive insulation ($k = 0.09$ W/m·°C) is then placed on top of the first layer of insulation. The maximum allowable temperature of the less-expensive insulation is 315°C. The heat-transfer coefficient at the outermost surface is 11.3 W/m²·°C, and the ambient air temperature is 30°C. Find the thickness of each layer of insulation.

5.86 A stone wall (Salem limestone) in an old house can be modeled as a homogeneous plane wall 225 mm thick. Estimate the heat loss through the wall per unit area if the convective heat-transfer coefficients in the inside and outside are 10 and 65 W/m²·K, respectively, and the inside and outside air temperatures are 20°C and 0°C, respectively. The

thermal conductivity of the stone is 2.15 W/m·K. Neglect radiation effects.

5.87 Determine the inside and outside surface temperatures of the stone wall from Problem 5.86.

5.88 A composite wall consists of a 13-mm-thick inner plaster board, a 192-mm-thick fiberglass blanket insulation, and a 20-mm-thick high-density particle board exterior panel. The thermal conductivity of the insulation is 0.035 W/m·K. Estimate the heat flux through the wall for the same convective conditions given in Problem 5.86.

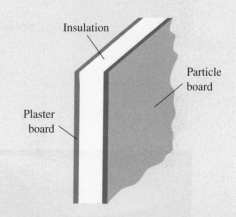

5.89 Determine the interior wall temperature for the situation described in Problem 5.88. How does this temperature compare with that estimated for the stone wall in Problem 5.87?

5.90 Plot the temperature distribution through the composite wall described in Problem 5.88. Discuss.

5.91 Consider steady conduction through a plane wall. The temperature of the left-hand face ($x = 0$) is greater than the temperature at the right-hand face ($x = L$). Sketch the temperature distribution for the following situations:

A. The thermal conductivity of the wall is constant.

B. The thermal conductivity increases with increasing temperature.

C. The thermal conductivity decreases with increasing temperature.

Discuss the logic involved in creating your plots.

5.92 A 2.5-cm-thick layer of insulation ($k = 1.5$ W/m·°C) is applied to a plane surface so that the inner surface of the insulation is maintained at 150°C. The insulation outer surface is exposed to ambient air at 35°C to which it loses heat by convection. Find the value of the convective heat-transfer coefficient that will ensure that the surface temperature will not exceed 45°C. What is the heat flux through the insulation?

5.93 Joule heating in a thin, electrically conducting foil is used to maintain the external surface of a 5-mm-thick plate ($k = 6.5$ W/m·K) hot enough to prevent moisture condensation (see sketch). Determine the power supplied to the foil per unit surface area to maintain the plate surface temperature at 310 K. The convective conditions at this surface are $\bar{h}_{conv} = 25$ W/m²·K and $T_\infty = 25$°C. The foil is insulated on its left surface. Also determine the temperature of the foil.

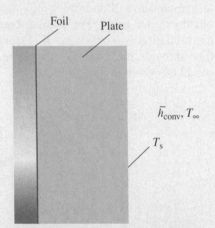

Foil   Plate

$\bar{h}_{conv}, T_\infty$

T_s

5.94 A long, cylindrical, ceramic tube with a 50-mm inside diameter and a 110-mm outside diameter is used as a

Photograph courtesy of CLASIC CZ Ltd.

laboratory furnace. Estimate the exterior surface temperature of the tube if the inner wall temperature is 900 K and, exterior to the tube, the convective heat-transfer coefficient and ambient temperature are 30 W/m²·K and 300 K, respectively. The thermal conductivity of the tube is 0.3 W/m·K.

5.95 Condensing steam flows through a steel pipe (of 100 mm inside diameter with a 4-mm wall thickness) such that the interior pipe wall temperature is fixed at 450 K. Compare the heat loss per unit length of pipe for the case of an uninsulated pipe and a pipe insulated with a 50-mm-thick layer of insulation. The thermal conductivities of the steel pipe and insulation are 16 and 0.06 W/m·K, respectively. The ambient temperature is 25°C. The convective heat-transfer coefficient for the bare pipe is 8 W/m²·K; that for the insulated pipe is 5 W/m²·K.

5.96 Figure 5.12 illustrates the temperature distribution through a hollow cylinder when the temperature at the inner radius is hotter than that at the outer radius. Sketch the temperature distribution $T(r)$, for the reversed situation where the higher temperature is on the outside. Discuss the reasoning used to create your sketch.

5.97 A nominal 5-in-diameter, wrought-iron schedule-40 pipe (inside diameter = 5.047 in, outside diameter = 5.563 in) is covered with a 5-cm-thick layer of 85% magnesia insulation. When steam flows through the pipe, the temperature of the inner surface of the pipe is 425°C and the temperature of the outer insulation surface is 40°C. Determine the heat loss rate from a 30-m length of this pipe.

5.98 Consider a nominal 8-in-diameter, steel, schedule-80 steam pipe (inside diameter = 7.625 in, outside diameter = 8.625 in) covered first with a 7.5-cm layer of 85% magnesia insulation and then a 2.5-cm layer of air-cell insulation ($k = 0.064$ W/m·°C). Find the heat loss, per unit length of pipe, if the inside pipe surface temperature is 530°C and the outside insulation surface temperature is 90°C. What error results from neglecting the thermal resistance of the pipe wall?

5.99 Find the temperature at the junction between the two insulations in Problem 5.98.

5.100 A nominal 4-in-diameter, wrought-iron, schedule-40 pipe (inside diameter = 4.026 in, outside diameter = 4.500 in) is covered with a 2-in-thick layer of 85% magnesia insulation. If the pipe carries steam so that the inner surface is at 700 F and the outer insulation surface is 150 F, what is the heat loss rate from a 100-ft length of this pipe?

5.101 A 5-cm-outside-diameter steel pipe is covered with a 0.6-cm layer of one insulation ($k = 0.17$ W/m·°C) followed by a 2.5-cm layer of a second insulation ($k = 0.048$ W/m·°C). The pipe surface temperature is 315°C, and the outer insulation surface temperature is 40°C. Calculate the temperature at the junction of the two insulating materials.

5.102 Hot water flows through a nominal 1-in-diameter, schedule-40, wrought-iron pipe (inside diameter = 1.049 in, outside diameter = 1.315 in). The inside surface temperature of the pipe is 250 F. A ½-in-thick layer of magnesia insulation surrounds the pipe. The outer surface temperature of this insulation is 100 F. Determine (a) the heat loss per unit length of pipe and (b) the temperature at the pipe–insulation interface.

5.103 A hollow cylinder of inside and outside radii r_1 and r_2, respectively, is heated such that its inner and outer surfaces are at uniform temperatures T_1 and T_2. The thermal conductivity of the cylinder material varies with temperature as follows:

$$k = k_0(1 + bT).$$

Derive an expression for the rate of heat flow through the cylinder. At what mean temperature should one evaluate k to use a constant-thermal-conductivity model (i.e., Eq. 5.49)?

5.104 A nominal 4-in-diameter, schedule-40, wrought-iron pipe (inside diameter = 4.026 in, outside diameter = 4.500 in) is covered with 3.8-cm-thick layer of 85% magnesia insulation. The pipe carries superheated steam at a temperature of 400°C, and the outer insulation surface is exposed to air at 25°C. The inside and outside heat-transfer coefficients are 1400 and 10 W/m²·°C, respectively. Find (a) the rate of heat loss per unit of pipe length, (b) the temperature of the inside pipe surface, (c) the temperature of the outside pipe surface, and (d) the temperature of the outside insulation surface.

5.105 The insulated pipe configuration described in Problem 5.98 carries steam at 650°C with an inside convective heat-transfer coefficient of 2800 W/m²·°C. The outer insulation surface is exposed to air at 50°C with a convective heat-

transfer coefficient of 10 W/m²·°C. Determine the heat loss per unit length.

5.106 Calculate the heat loss per unit of length of a nominal 4-in-diameter, schedule-40, steel pipe (inside diameter = 4.026 in, outside diameter = 4.500 in) covered with a 1.25-cm-thick layer of insulation ($k = 0.09$ W/m·°C) if the inside pipe surface temperature is 200°C and the outside air temperature is 20°C. Assume an outside convective heat-transfer coefficient of 170 W/m²·°C.

5.107 A long pipe (inside diameter = 7.5 cm, outside diameter = 10 cm, $k = 35$ W/m·°C) is covered with a 1.25-cm layer of insulation ($k = 5$ W/m·°C). The temperature of a fluid flowing inside the pipe is 20°C, and the temperature of the inner pipe wall is 45°C. The temperature of the fluid surrounding the outer surface of the insulation is 150°C, and the convective heat-transfer coefficient at that surface is 85 W/m²·°C. Find (a) the heat-transfer rate between the two fluids per meter of pipe length and (b) the temperature of the outer surface of the insulation.

5.108 The sketch shows a pipe covered with two layers of insulation. The pipe has an inside radius (r_2) of 3 in, an outside radius (r_3) of 4 in, and a thermal conductivity (k_{23}) of 8 Btu/hr·ft·F. The first layer of insulation is 3 in thick with an unknown thermal conductivity. The second layer of insulation is 4 in thick and has a thermal conductivity (k_{45}) of 0.4 Btu/hr·ft·F. The pipe carries a fluid of known temperature $T_1 = 1000$ F, and the inner pipe temperature is measured to be 900 F. The heat-transfer coefficient at the inner pipe surface is $h_{conv,12} = 5.5$ Btu/hr·ft²·F. The temperature of the outermost insulation surface is $T_5 = 100$ F, and the heat-transfer coefficient at that surface is $h_{conv,56} = 10.0$ Btu/hr·ft²·F. The ambient fluid temperature, T_6, is unknown. Find (a) the heat-transfer rate from the pipe per unit length and (b) the thermal conductivity of the inner insulation, k_{34}.

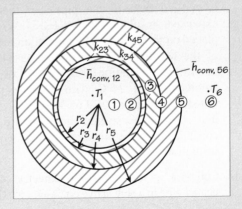

5.109 A refrigerant at $-40°C$ flows inside a copper pipe (inside diameter = 3 cm, outside diameter = 3.4 cm) with a convective heat-transfer coefficient of 110 W/m$^2 \cdot$°C at the inner surface. To reduce the heat flow into the pipe to 20 W/m, insulation (k = 0.04 W/m$\cdot$°C) is added to the outside of the pipe. For an ambient air temperature of 25°C and a heat-transfer coefficient of 18 W/m$^2 \cdot$°C at the insulation surface, determine the required thickness of the insulation.

5.110 Steam at 250°C flows through a carbon-steel pipe (inside diameter = 6.0 cm, outside diameter = 7.5 cm) covered with a 2.5-cm layer of insulation (k = 0.05 W/m$\cdot$°C). The convective heat-transfer coefficient at the inner pipe surface is 400 W/m$^2 \cdot$°C. The outside insulation surface experiences both convective and radiative heat losses. The ambient air temperature is 20°C, the convective heat-transfer coefficient is 20 W/m$^2 \cdot$°C, and the outer surface has an emissivity of 0.8. The surrounding room walls are also at 20°C. Find the surface temperature of the insulation and the rate of heat loss from the pipe per unit length.

5.111 A bare 2.5-cm-diameter pipe with a surface temperature of 175°C loses heat by convection to the ambient air. The air temperature is 30°C and the convective heat-transfer coefficient is 5.6 W/m$^2 \cdot$°C. It is desired to reduce the heat loss by 50% by the addition of insulation (k = 0.17 W/m$\cdot$°C). Assuming that the pipe surface temperature and the exposed surface heat-transfer coefficient remain unchanged as insulation is added, find the required insulation thickness.

5.112 Repeat Problem 5.111 for a 10-cm-diameter pipe, with all other data unchanged.

5.113 A nominal 12-in-diameter, schedule-40, steel pipe (inside diameter = 303.2 mm, outside diameter = 323.9 mm) has an outside surface temperature of 480°C. As a safety precaution, it is desired to insulate the pipe such that the exposed insulation surface does not exceed 40°C. Two insulating materials are to be used. First, a high-temperature insulation (k = 0.26 W/m$\cdot$°C) is applied next to the pipe and then a less-expensive insulation (k = 0.073 W/m$\cdot$°C) is placed on the outside. The maximum temperature of the less-expensive insulation is 315°C. The outermost insulation surface is exposed to ambient air at 24°C with a convective heat-transfer coefficient of 9.65 W/m$^2 \cdot$°C. Assuming that the pipe surface temperature remains constant, find the thickness of the two types of insulation.

5.114 Consider a long, 0.040-in-diameter, electrically heated wire in a convective environment. The surface temperature of the wire is 400 F and the surrounding ambient air temperature is 60 F. The convective heat-transfer coefficient is 5 Btu/hr$\cdot$ft$^2 \cdot$F.

A. Ignoring radiation, find the heat loss from the wire per unit length (Btu/hr$\cdot$ft).

B. The wire is covered with a 0.125-in-thick layer of insulation (k = 0.025 Btu/hr$\cdot$ft$\cdot$F). Determine the heat loss if the wire surface temperature, the air temperature, and the heat transfer coefficient remain unchanged.

5.115 Consider a hollow, stainless-steel (AISI 304) sphere (inside diameter = 50 cm, outside diameter = 60 cm) in a convective environment. The ambient fluid temperature is 15°C and temperature of the outer surface of the sphere is 50°C. The convective heat-transfer coefficient is 140 W/m$^2 \cdot$°C. Determine the temperature (a) at the inside surface of the sphere and (b) at a point midway through the sphere wall.

5.116 A 10-cm-diameter sphere is electrically heated to maintain its surface temperature at 225°C. The surface is exposed to a convective environment with an ambient fluid temperature of 25°C and a heat-transfer coefficient of 50 W/m$^2 \cdot$°C.

A. Neglecting radiation, find the heat loss (W) from the sphere's surface.

B. Insulation (k = 0.08 W/m $\cdot$°C) is added to the sphere and the electrical current is adjusted to maintain the same sphere surface temperature. What insulation thickness is required to reduce the heat loss to 20% of its original value, assuming the heat-transfer coefficient is unchanged?

5.117 Consider a 75-W light bulb modeled as a 7-cm-diameter sphere. Assume only 10% of the power input to the bulb is converted to light, with the remaining energy absorbed by the glass envelope. Estimate the surface temperature of the glass envelope if the only heat loss from the bulb surface is convection to the ambient air at 20°C. The convective heat-transfer coefficient is 8 W/m$^2 \cdot$°C. Repeat your calculation taking radiation into account. The glass emissivity is 0.9 and the radiant surroundings are at 20°C.

5.118 A 15-cm-thick furnace wall is made of brick with a thermal conductivity of 1.3 W/m$\cdot$°C. The outside surface of the wall loses heat by convection to the ambient air at 25°C with a convective heat-transfer coefficient of 25 W/m$^2 \cdot$°C. The surface also loses heat by radiation to a very large room with walls at 25°C. The emissivity and temperature of the outside surface are 0.8 and 100°C, respectively. Determine the temperature of the inside surface of the brick wall.

5.119 A 30-m² exterior surface of the wall of a house is exposed to a convective environment. The ambient air temperature is 40°C and the convective heat-transfer coefficient is 50 W/m²·°C. The surface simultaneously exchanges heat by radiation to very large surroundings also at 40°C. The surface emissivity is 0.75. Determine the combined convective and radiative heat loss (W) when the exterior surface temperature is 80°C.

5.120 Repeat Problem 5.75 with the side of the wall exposed to the convecting fluid also exchanging radiant heat with surroundings at 50°C. The surface emissivity is 0.8.

5.121 Air enters a nozzle at 300 K with a velocity of 10 m/s. Determine the temperature at the nozzle exit where the velocity is 250 m/s. Work this problem, first, assuming a constant specific heat c_p of 1.005 kJ/kg·K. Repeat using the air tables in Appendix C.

5.122 Low-velocity steam (with negligible kinetic energy) enters an adiabatic nozzle at 300°C and 3 MPa. The steam leaves the nozzle at 2 MPa with a velocity of 400 m/s. The mass flow rate is 0.4 kg/s. Determine the quality or temperature (°C) of the steam leaving the nozzle and the exit area of the nozzle in mm².

5.123 Estimate the power required to compress a steady flow of air ($\dot{m}$ = 0.02 kg/s) from 100 kPa and 300 K to 598 kPa at 500 K. Neglect changes in kinetic and potential energies of the air stream.

5.125 Consider a steam turbine in which steam enters at 10.45 MPa and 780 K with a flow rate of 38.739 kg/s. A portion of the steam is extracted from the turbine after partial expansion at three different locations as shown in the sketch. The extracted steam is then led to various heat exchangers. The mass flow rates and the temperatures and pressures at each extraction point are given in the following table:

Extraction Location	$\dot{m}$ (kg/s)	P (MPa)	T (K)
1	4.343	3.054	620
2	4.345	0.332	482
3	2.871	0.136	$x = 0.949$

The remaining wet steam exits the turbine at 11.5 kPa with a quality of 0.88. Estimate the power produced by the turbine assuming adiabatic operation and negligible kinetic and potential energies for all streams.

5.124 Superheated steam (8 MPa, 900 K) enters a turbine with a flow rate of 0.16 kg/s. The steam exits at 15 kPa. Determine the power produced by the turbine if the expansion process is isentropic. Neglect all heat losses and changes in kinetic and potential energies.

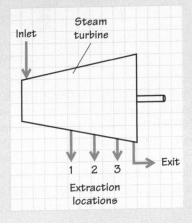

5.126 Wet steam ($P = 12.8$ kPa, $x = 0.90$) enters a heat exchanger with a flow rate of 1.2 kg/s. Cold water at 12.8 kPa and 295 K also enters the exchanger with a flow rate of 30 kg/s. The steam and the water mix adiabatically at constant pressure within the heat exchange. A single stream exits as shown in the sketch. Determine the flow rate and the temperature of the hot water exiting the heat exchanger.

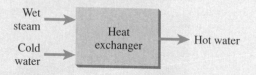

5.127 As shown in the sketch, water flows at 2.56 m³/s from a large reservoir through a penstock (pipe) to a hydro turbine. The water exits the turbine through a 0.5-m-diameter pipe. The reservoir surface is 30 m above the location where the water exits. Assuming no head losses, determine the ideal power produced by the hydro turbine. Repeat your calculations if the total head losses are 1 m.

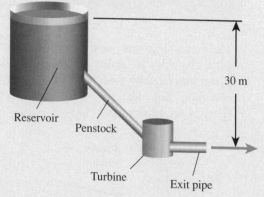

5.128 Mass flows through a control volume at 1 lb$_m$/s. The enthalpy, velocity, and elevation at entrance are 100 Btu/lb$_m$, 100 ft/s, and 300 ft, respectively. At the exit, these quantities are 99 Btu/lb$_m$, 1 ft/s, and −10 ft. Heat is transferred to the system at 5 Btu/s. How much work is done by this system (a) per pound of fluid and (b) per minute? (c) What is the rate of working (power) in kilowatts?

5.129 Water at 500 psia and 260 F enters the steam-generating unit of a power plant and leaves the unit as steam at 500 psia and 1500 F. The water mass flow rate is 30,000 lb$_m$/hr. Determine the capacity of the steam-generating unit in Btu/hr.

5.130 Electrical heating elements heat a 5-gal/min flow of water at 30 psig from 62 F to 164 F. Determine (a) the wattage required and (b) the current in amperes if a single-phase, 220-V circuit is used.

5.131 Air at the rate of 18 kg/s is drawn into the compressor of a jet engine at 55 kPa and −23°C and is compressed reversibly and adiabatically (i.e., isentropically) to 276 kPa. Determine the required power input to the compressor expressed in horsepower.

5.132 In a coal-fired power plant, 1,500,000 lb$_m$/hr of steam enters a turbine at 1000 F and 500 psia. The steam expands isentropically to 15 psia. Determine the ideal power produced by the turbine in kilowatts.

5.133 After flowing from the combustion chamber and expanding through the turbine of a jet engine, combustion products at low velocity and 1600 F enter the jet nozzle. Determine the maximum velocity (ft/s) that can be obtained at the

nozzle exit if the products discharge at 700 F. Approximate the thermodynamic properties of the combustion products as those of air.

5.134 Air, entering at 16°C, is used to cool an electronic compartment. The maximum allowable air temperature is 38°C. If the equipment in the compartment dissipates 3600 W of energy to the air, determine the necessary air flow rates in (a) kg/hr and (b) m³/min at inlet conditions.

5.135 What is the minimum power motor (hp) that would be necessary to operate a pump that handles 85 gal/min of city water while increasing the water pressure from 15 to 90 psia?

5.136 When the pressure in a steam line reaches 690 kPa, the safety valve opens and releases steam to the atmosphere in a constant-enthalpy process across the valve. The temperature of the escaping steam (after the valve) was measured at 700°C. Determine the temperature in °C and the specific volume in m³/kg of the steam in the line. Also identify the state region (e.g., superheated vapor, etc.) of the steam in the line.

5.137 A pump is used to remove water after a flood. To estimate the work done by this pump, model the process as frictionless and steady flow. Assume the water enters the pump at 0.0689 MPa and leaves at 3.516 MPa with an average density of 995 kg/m³. Estimate the work of the pump (kJ/kg) assuming no changes in kinetic or potential energies.

5.138 Considering a pump to be a frictionless, steady-flow device, estimate the work input in kilojoules per kilogram of water entering the pump. The water enters at 0.01 MPa and 40°C and leaves at 0.35 MPa. Neglect kinetic and potential energy changes.

5.139 During the operation of a steam power plant, the steam flow rate is 500,000 lb$_m$/hr with turbine inlet conditions of 500 psia and 1000 F and turbine exhaust (condenser inlet) conditions of 1.0 psia and 90% quality. Determine (a) turbine output in

kW and (b) the condenser heat-rejection rate in Btu/hr.

5.140 A control volume is described by the following information:

	Input	Output
Velocity	36.58 m/s	12.19 m/s
Elevation	30.48 m	54.86 m
Enthalpy	2791.2 kJ/kg	2795.9 kJ/kg
Mass rate	0.756 kg/s	0.756 kg/s

If the net work rate out is 4.101 kW, what is the heat-transfer rate?

5.141 Steam is throttled from a saturated liquid at 212 F to a temperature of 50 F. What is the quality of the steam after passing through this expansion valve? Assume $\Delta PE = \Delta KE = Q = W = 0$.

5.142 Steam at 100 psia and 400 F enters a rigid, insulated nozzle with a velocity of 200 ft/s. The steam leaves at a pressure of 20 psia. Assuming that the enthalpy at the entrance is 1227.6 Btu/lb$_m$ and at the exit it is 1148.4 Btu/lb$_m$, determine the exit velocity.

5.143 During the operation of a steam power plant, steam enters the turbine with a flow rate of 230,000 kg/hr at 3.5 MPa and 550°C. The turbine exhaust (condenser inlet) condition is 0.01 MPa with 85% quality. Determine (a) the turbine output in kW and (b) the condenser heat-rejection rate in kJ/hr.

5.144 Steam at 0.7 MPa and 205°C enters a rigid, insulated nozzle with a velocity of 60 m/s. The steam leaves at a pressure of 0.14 MPa. Assuming that the enthalpy at the entrance is 0.793 kJ/kg and that at the exit it is 0.742 kJ/kg, determine the exit velocity.

5.145 The heat from students, from lights, through the walls, and so forth to the air moving through a classroom is 22,156 kJ/hr. Air is supplied to the room from the air conditioner at 12.8°C. The air leaves the room at 25.5°C. Determine the following:

A. Air mass flow rate (kg/hr)

B. Air volumetric flow rate (m³/hr) at inlet conditions

C. Duct diameter (m) for an air velocity of 183 m/min

5.146 Saturated steam at 0.276 MPa flows through a 5.08-cm-inside-diameter pipe at the rate of 7818 kg/hr. Determine the specific kinetic energy of the steam in kJ/kg.

5.147 The mass flow rate of air into a nozzle is 100 kg/s. The discharge pressure and temperature are 0.1 MPa and 270°C, respectively, and the inlet conditions are 1.4 MPa and 800°C. Determine the outlet diameter of the nozzle in meters.

5.148 Superheated steam at 500°C and 3 MPa enters a steady-state turbine with a mass flow rate of 500 kg/s. The steam leaves the turbine at 250°C and 0.6 MPa. The turbine is not well insulated and, thus, there is a heat-transfer rate of 20 MW to the atmosphere. Determine the power delivered by the turbine in MW.

5.149 Air enters an adiabatic, steady-flow gas turbine at 700 F and 80 psia and leaves at 350 F and 14.7 psia. The mass flow rate of the air is 100 lb_m/s. Treat air as an ideal gas with constant specific heats.

   A. Neglecting potential energy and kinetic energy at the inlet and outlet, determine the power (Btu/s) produced by the turbine.

   B. If the air enters the turbine through a pipe with a flow area of 0.75 ft^2 and exits through a 4.00-ft^2 pipe, determine the increase in kinetic energy (Btu/s) of the air.

5.150 Air at 400°C and 0.4 MPa is steadily supplied to an adiabatic gas turbine. The air leaves the turbine at 200°C and 0.10135 MPa. Neglecting changes in potential energy and kinetic energy, determine the following:

   A. The ratio of the exit flow area to the inlet flow area required for an exit velocity equal to the inlet velocity

   B. The specific work (kJ/kg), assuming constant specific heats

5.151 Steam enters an adiabatic turbine at 1000 F and 900 psia and leaves at 800 F and 400 psia. The mass flow rate is 1000 lb_m/hr. Potential and kinetic energies can be neglected. Determine the following:

   A. The power delivered (kBtu/hr)

   B. The ratio of the outlet flow area to inlet flow area to keep the exit velocity equal to the inlet velocity

5.152 Air enters a steady-flow, adiabatic compressor at 15°C and 0.1 MPa at 2 m^3/s. The air leaves at 150°C and 0.4 MPa. Neglecting changes in potential energy and kinetic energy, determine the power (kW) required to drive the compressor.

5.153 Steam flowing at 1000 lb_m/hr is compressed from 400 F and 80 psia to 900 F at 200 psia in an adiabatic, steady-flow process. Determine the input power required in horsepower. Neglect changes in potential energy and kinetic energy.

5.154 Dry, saturated steam from a turbine enters a condenser at 4 kPa and exits as saturated liquid, also at 4 kPa. In the condenser, energy is removed from the steam by heat transfer to a stream of lake water. The lake water enters at 5°C and is then returned to the lake at 10°C (the maximum allowed by local regulations). Determine the mass of lake water required per mass of steam condensed.

5.155 Saturated liquid water at 50 psia enters a steady-flow boiler. Inside the boiler, the water is heated at constant pressure to 600 F. Potential and kinetic energies are negligible.

   A. Determine the heat input in Btu/lb_m.

   B. If the exit area is 2 ft^2 and the average exit velocity is 100 ft/s, determine the mass flow rate in lb_m/hr.

5.156 Air undergoes a steady-flow, constant-pressure heating process at 150 kPa from 20°C to 150°C. Determine the increase in specific internal energy of the air in kJ/kg.

5.157 Steam at 600 F and 200 psia enters a well-insulated, steady-state device through a standard 3-in pipeline (inside diameter = 3.068 in) at 10 ft/s. The exhaust from the device flows through a standard 10-in pipeline (inside diameter = 10.02 in) at 200 F and 5 psia. Determine the horsepower output of the device.

5.158 As a fluid flows steadily past a turbine blade with friction present, the fluid velocity drops from 400 m/s to 100 m/s while the fluid enthalpy increases 25 kJ/kg. Assuming the process is adiabatic, determine the specific work transfer (kJ/kg) from the fluid to the blade.

5.159 An ideal gas passes steadily through a device that increases the gas velocity from 5 to 300 m/s without transfers of heat or work.

   A. Determine the increase in specific enthalpy (kJ/kg) of the gas.

   B. If the specific heat of the gas is a linear function of temperature given by c_p [kJ/kg·K] = 1.00 + 0.01T [K], determine the increase in temperature of the gas if it enters the device at 20°C.

5.160 A number of years ago, Frank Lloyd Wright designed a one-mile-high building. Suppose that in such a building steam for the heating system enters a pipe at ground level as dry, saturated vapor at 30 psia. On the top floor of the building, the pressure in the pipe is 10 psia. The heat transfer from the steam as it flows up the pipe is 50 Btu per lb_m of steam. Taking the one-mile-high pipe as a control volume (open system), determine the quality of the steam at the top of the pipe.

5.161 A gas expands through an adiabatic nozzle. During the expansion there is a decrease in specific enthalpy of 50 Btu/lb_m from entrance to exit.

A. If the initial velocity of the gas entering the nozzle is nearly zero, determine the exit velocity in ft/s.

B. If the initial velocity of the gas entering the nozzle is 100 ft/s, determine the exit velocity in ft/s.

5.162 Consider a waterfall having a drop of 84.7 m.

A. Determine the specific potential energy (J/kg and Btu/lb_m) of the water at the top of the falls with respect to the base of the falls.

B. Assuming no energy is exchanged with the surroundings, determine the velocity (ft/s) of the water just before it reaches the bottom.

C. What happens to the kinetic energy of the water after it reaches the bottom?

5.163 Water (assumed to be incompressible) is pumped at a constant rate of 50 lb_m/min through a pipeline that has an internal diameter of 2 in. The pipe discharges through a nozzle that has a diameter of 1 in at the exit and is at an elevation of 100 ft above the inlet to the pump. At the inlet to the pump the water is at 70 F and 20 psia. At the exit of the nozzle, the water is at 70 F and 14.7 psia.

Neglecting head losses, determine the horsepower that must be supplied to the pump.

5.164 An amusement park at the bottom of Niagara Falls wants to install a water turbine to produce 100 kW. Water (assumed to be incompressible) would enter the pipeline leading to the turbine at 20°C and 0.10135 MPa at the top of the falls, 51 m above the turbine exit, with a velocity of 3 m/s. The water should leave the turbine at 20°C and 0.10135 MPa. Assume the pipeline and the turbine are both adiabatic.

A. Determine the mass flow rate of the water in kg/min.

B. Determine the diameter of the pipeline.

5.165 Water at 200 F and 30 psia flows into a rigid, insulated tank through pipe 1 at a rate of 100 lb_m/s. Steam at 400 F and 30 psia flows into the same tank through pipe 2 at a rate of 200 lb_m/s. The two flows mix together within the tank and leave through pipe 3 at 30 psia. This is a steady-flow process with negligible potential and kinetic energies. Determine the temperature (and quality, if applicable) of the flow leaving through pipe 3.

5.166 Superheated steam at 400°C and 1.6 MPa flows steadily into a control volume at a rate of 0.2 kg/s. A second stream (dry, saturated water vapor at 1.6 MPa) enters the control volume at 0.1 kg/s. The control volume is adiabatic and the pressure at the only exit stream is 1.6 MPa. Determine the mass flow rate (kg/s) and temperature (°C) of the exit flow.

5.167 Water at 50 F and 500 psia enters a steady-flow control volume at a rate of 2 lb_m/s. Wet steam at 14.7 psia with a quality of 0.25 also enters the control volume but at a rate of 1 lb_m/s. Steam at 700 F and 600 psia leaves the control volume. The work rate into the control volume is 2500 Btu/s. Determine the rate of heat transfer (Btu/s). Is the heat flow into or out of the control volume?

5.168 In certain situations when only superheated steam is available, a need for saturated steam may arise. This need can be met in an adiabatic desuperheater in which liquid water is sprayed into the superheated steam in such amounts that dry, saturated steam leaves the desuperheater. The following data apply to such a steadily operating desuperheater. Superheated steam at 300°C and 3 MPa enters the desuperheater at 0.25 kg/s. Liquid water enters the desuperheater at 40°C and 5 MPa. Dry, saturated vapor leaves at 3 MPa. Determine the mass flow rate (kg/s) of liquid water.

5.169 Determine the lower and higher heating values (kJ/kg) of butane at 298 K.

5.170 Consider a stoichiometric reaction involving liquid octane and oxygen in a steady-flow reactor. Reactants enter at 25°C and 1 atm, and products exit at the same conditions. Assuming liquid water in the products, determine the heat transfer from the reactor in MJ per kg of fuel.

5.171 For a steady-flow, stoichiometric reaction of gaseous octane with oxygen, each reactant enters at 25°C and 1 atm and the products leave at 25°C and 1 atm. Determine the heat transfer (MJ per kg fuel).

5.172 Ethane flows into a combustion chamber at 1.5 kg/ min along with 0% excess air. Fuel, oxidizer, and products are all at 25°C. The reaction is complete and the water is liquid. Determine the heat-transfer rate from the combustion chamber (MW).

5.173 A 50–50 blend (by volume) of methane and propane is burned with 100% excess air in a steady-flow combustion chamber. The fuel mixture and the air each enter the combustion chamber at 298 K and 1 atm. Assuming the reaction is complete and exhaust products leave the combustion chamber at 1000 K and 1 atm, determine the heat transfer per mass of fuel burned (MJ/kg).

5.174 Propane is completely burned in a steady-flow process with 100% excess air. The propane and the air each enter the control volume at 25°C and 1 atm, and the combustion products leave at 600 K and 1 atm. Determine the heat transfer (MJ/kg$_{C_3H_8}$) to the surroundings.

5.175 A stoichiometric mixture of carbon monoxide and air, initially at 25°C and 1 atm, reacts completely in an adiabatic, constant-pressure, steady-flow process. Determine the exit temperature (K) of the products.

5.176 Determine the adiabatic flame temperature (K) for a mixture of methane and 200% theoretical air that reacts completely in a steady-flow process at 1 atm. The methane and air enter the reaction at 298 K.

5.177 A cutting torch burns a steady flow of acetylene gas at 25°C and 1 atm with stoichiometric air at 25°C and 1 atm. The products are at 1 atm. Assume the reaction is complete and adiabatic. Determine the exit temperature (K).

5.178 Hydrogen gas and 100% excess air each enter a steady-flow combustion chamber at 25°C and 1 atm. A complete reaction occurs with a heat loss of 40 MJ/kmol of fuel. Determine the temperature (K) of the product gas.

5.179 Anticipating the coming of the "hydrogen economy," an enthusiastic mechanical engineer invents a steam generator as shown in the sketch. Pure hydrogen, H_2, enters the well-insulated combustion chamber at 298 K with a flow rate of 0.3 kg/s. The oxidizer is pure oxygen, O_2, and enters in stoichiometric proportions (i.e., $\Phi = 1$). A steam generator coil is located in a well-insulated tailpipe of the apparatus. Saturated liquid water enters the steam-generator coil at 10 MPa. Steam (saturated vapor at 10 MPa) exits the coil. The products of combustion (H_2O) exit the tailpipe at 2000 K. Neglect any dissociation in either the combustion chamber or the tailpipe. Determine the following (including units):

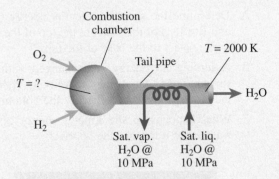

A. The mass flow rate of the O_2

B. The maximum temperature in the combustion chamber assuming all of the H_2 and O_2 react to form H_2O at constant pressure

C. The heat removed from the product stream in the tailpipe section

D. The mass flow rate of the steam through the steam-generator coil

5.180 Consider the theoretical combustion of methane in a steady-flow process at 1 atm. Determine the heat transfer per kmol and per kg of fuel from the combustion chamber for the following cases:

A. The products and the reactants are at the same temperature of 15.6°C.

B. The air and the methane enter at 5°C and 60°C, respectively, and the products exit at 449°C.

5.181 Determine the mass air–fuel ratio if gaseous propane (C_3H_8) is burned with air at 1 atm in a steady-flow reactor for the following conditions: The propane and air enter the reactor at 25°C, and the combustion products must exit at less than 1425°C.

5.182 A well-insulated, 80-gal (0.3028 m³), electric hot-water heater initially contains water at 120 F. Hot water is withdrawn from the tank at 3 gal/min while cold water at 50 F enters to keep the tank full. The cold water mixes completely with the water in the tank at each instant. Because of a power outage, there is no energy input to heat the incoming cold water. Estimate the time required for the temperature of the outgoing water to reach 80 F. Assume constant water properties with $\rho = 994$ kg/m^3 and $c_p = c_v = 4149$ J/kg·K.

5.183 A tank having a volume of 200 ft^3 contains saturated water vapor at 20 psia. A line is attached to the tank in which vapor flows at 100 psia and 400 F. Steam from this line enters the tank until the pressure is 100 psia. Calculate the mass of steam that enters the tank assuming that the process is adiabatic and that the heat capacity of the tank is negligible.

5.184 A 216-in^3 tank contains saturated water vapor at 0.143 MPa. A line is attached to the tank in which vapor flows at 0.7 MPa and 200°C. Steam from this line enters the tank until the pressure is 0.7 MPa. Calculate the mass of steam that enters the tank assuming that the process is adiabatic and that the heat capacity of the tank is negligible.

5.185 A rigid tank contains water (liquid and vapor) at 600 F. Liquid is withdrawn from the bottom at a slow rate. The cross-sectional area of the tank is 50 in^2 and the liquid level drops 6 in. During this time heat transfer maintains the temperature at 600 F. Neglecting changes in potential energy, determine the heat transfer (Btu).

5.186 Tank A has a volume of 0.15 m^3 and contains dry, saturated water vapor at 200°C. A frictionless piston initially rests at the bottom of cylinder B. When the valve is opened, water vapor flows slowly into cylinder B. The piston mass is such that a pressure of 0.3 MPa in cylinder B is required to raise the piston. The process ends when the pressure in tank A has fallen to 0.3 MPa. During this process, heat transfer with the surroundings keeps the temperature of the water on each side always at 200°C. Determine the heat transfer (kJ).

5.187 Consider the process of filling a scuba diving tank with air. The tank is initially at T_1 and P_1. A large reservoir can supply air at $T_r > T_1$ and $P_r > P_1$. The scuba tank is connected to the reservoir and quickly filled to P_r. Assume the filling process is adiabatic. The tank temperature returns to T_1 as a result of heat transfer with the surroundings. Assuming air is an ideal gas with constant specific heats (c_v and c_p), derive an expression for the final pressure in the tank in terms of known quantities.

5.188 Initially, 0.1 kg of water at 40°C and 0.2 MPa is contained in an insulated piston–cylinder device. During a time interval of 35 s, 0.1 kg of steam is added to the contents of the cylinder through the valve from a source at 250°C and 0.5 MPa. Determine the final volume (m^3) of the cylinder contents.

5.189 A refrigerator in a kitchen consists of a beer compartment and a cooling system. To keep the beer cold, the cooling system transfers energy from the beer compartment to the cooling system (a heat-transfer process) at a rate of 3000 W. To do this, the cooling system is plugged into an electrical outlet from which it receives electrical energy at the rate of 700 W. Since the kitchen, cooling system, and beer compartment are operating steadily, the energy contained anywhere in the kitchen or refrigerator does not change with time. Define an appropriate system and use an energy balance to determine the following:

A. The heat-transfer rate in watts between the kitchen air and the other rooms

B. The heat-transfer rate in watts between the cooling system and the kitchen air

C. The heat-transfer rate in watts between the beer compartment and the kitchen air

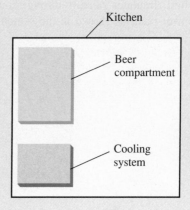

5.190 Fluid circulates steadily through four devices in a power plant as shown in the sketch. Mass flow rates and enthalpies per unit mass are tabulated for some of the states. Heat- and work-interaction

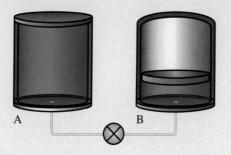

A   B

rates are tabulated for some of the devices. Complete the following tables:

State	$\dot{m}$ (kg/s)	h (J/kg)
1		15
2		13
3	25	9
4		
5	5	

Device	$\dot{Q}$ (W)	$\dot{W}$ (W)
A	150	0
B	30	
C		0
D	0	5

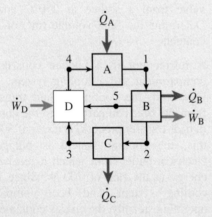

5.191 Steam enters a turbine at a rate of 200,000 lb$_m$/hr with an enthalpy of 1306.6 Btu/lb$_m$. Steam leaves the turbine at two exits. The mass flow rate at the first exit is 50,000 lb$_m$/hr with an enthalpy of 1181.9 Btu/lb$_m$. At the second exit, the enthalpy is 1013.9 Btu/lb$_m$. The only other energy transfer is work. There is no storage of mass or energy within the turbine. Determine the power (kW) delivered by the turbine.

5.192 Fluid circulates steadily through four devices in a power plant as shown in the sketch. All transfers of heat and work are indicated. The mass flow rates at states 1 and 5 are 100 lb$_m$/s and 20 lb$_m$/s, respectively. The fluid enthalpies per unit mass at states 1 through 5 are 4, 10, 5, 1, and 7 Btu/lb$_m$, respectively. Determine the power delivered by device B in Btu/s.

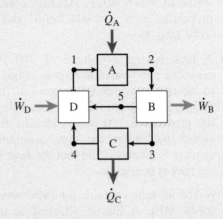

5.193 Starting with the differential system described in Fig. 5.17, derive Eq. 5.76.

5.194 Consider an ablating heat shield similar to that described in Example 5.20. The ablation temperature is now 2000 K, with the bulk material temperature remaining at 250 K far from the ablation front. The ablation front velocity is 0.05 mm/s and the thermal diffusivity of the shield is 1.4×10^{-6} m^2/s. Determine the temperature 25 mm from the ablation front.

5.195 Simplify Eqs. T5.7a and T5.8a for a two-dimensional (x, y) Cartesian coordinate system. Assume the flow is steady.

5.196 Simplify Eqs. T5.7b and T5.8b for an axisymmetric (r, x) steady flow in a pipe.

See Chapter 11 for additional problems dealing with steady-flow devices.

CONSERVATION OF MOMENTUM

After studying Chapter 6, you should:

- Understand the difference between body and surface forces.

- Describe what forces act on a fluid element when at rest and in motion.

- Have a firm grasp of the sign conventions associated with pressure forces, viscous stresses, and momentum flows.

- Be able to state in words and write symbolically the conservation of momentum principles for a rigid system.

- Understand how pressure varies within a fluid when at rest (or moving with constant velocity) and when accelerating at a constant rate as a rigid body.

- Be able to solve complex manometry problems and be able to calculate the forces and moments created by fluid pressure on submerged planar surfaces.

- Understand the concept of a momentum flow and be able to calculate the components of momentum flows given a velocity distribution.

- Be able to state in words and express symbolically both steady-flow and unsteady conservation of momentum principles for a control volume.

- Be able to apply the conservation of momentum principle to problems involving integral control volumes with steady flow.

- Understand the origins of and the physical interpretation of the various terms comprising the Navier–Stokes equation, the Euler equation, the mechanical energy equation, and the Bernoulli equation.

- Given the governing differential equations for mass, energy, and momentum, be able to recognize which conservation principle is expressed by each.

- Be able to express in words the meaning of the total derivative and its components, the local and convective accelerations.

- Understand the physical origins of the head loss.

- Be able to apply the Bernoulli equation to appropriate steady flows.

- Understand the origins of the turbulent or Reynolds stresses in turbulent flows.

Chapter 6 Overview

This chapter focuses on the conservation of momentum principle, applying it to control volumes for fluids at rest and in motion. We introduce the concept of a momentum flow and relate it to the various forces that can act on a fluid element: pressure forces, viscous shear forces, and the gravitational body force. We also introduce the mechanical energy and Bernoulli equations. The physical interpretation of these equations are emphasized, along with the restrictions associated with them. This chapter provides the foundation for understanding and estimating head losses and pressure drops associated with flowing fluids. Many of the concepts presented in this chapter play key roles in the design and analysis of propulsion systems and power generation systems, among others. The performance of rockets and turbojet engines, for example, relate critically to the conservation of momentum principle. We revisit and build upon many of the ideas presented here in later chapters.

6.1 HISTORICAL CONTEXT

An early topic in this chapter, hydrostatics—the study of fluids at rest, has its origins with the Greek mathematician **Archimedes** (287–212 BCE). Everyone is familiar with the tale of Archimedes running naked through the streets of Syracuse shouting "Eureka!" after his bath-time discovery that the buoyant force acting on a floating body equals the weight of the volume of water displaced by the body. The next breakthrough in understanding fluid statics did not occur until late in the Renaissance, in 1586 [1]. **Simon Stevin** (1548–1620), a Flemish mathematician and scientist, resolved the "hydrostatic paradox" by showing how the force exerted by a fluid on the base of a container depends only on the height of the fluid within and the area of the base, independent of the shape of the container. Another development related to fluid statics is **Daniel Bernoulli's** (1700–1782) measurement of pressure by observing the height to which a column of fluid rises when connected to a small hole in the wall of a conduit containing a flowing fluid. After Bernoulli's discovery (ca. 1738), physicians of the day measured blood pressure by tapping directly into arteries.

    Isaac Newton (1642–1727) is famous for his laws of motion, which include the principle of momentum-impulse, or momentum conservation. His theory put an end to other views of motion, such as that of **Descartes** (1596–1650), who hypothesized that the planets were pulled by vortices containing a fixed "quantity of movement" [1]. Newton also contributed to the understanding of viscous forces, and his name is now associated with fluids that exhibit a linear relationship between stress and strain rate. The modern form of the famous Bernoulli equation, which relates pressure, velocity, and elevation in a flowing fluid, is the work of **Leonhard Euler** (1707–1783), a

Archimedes (287–212 BCE)

Leonhard Euler (1707–1783)

brilliant co-worker of Daniel Bernoulli. Bernoulli's original equation, like that of **Gottfried Wilhelm von Leibnitz** (1646–1716), contained only kinetic and potential energy terms [1].

A progression of workers has led to our current understanding of the conservation of momentum principle applied to viscous fluids. Extensions of Euler's original formulation to viscous fluids include the contributions of **Louis Marie Henri Navier** (1785–1836) in 1822, those of **Augustin Louis de Cauchy** (1789–1857), and **Simeon Denis Poisson** (1781–1840), and that of **George Gabriel Stokes** (1819–1903) in 1845 [2]. In honor of the contributions of Navier and Stokes, their names are associated with the modern partial differential equation that expresses the conservation of momentum principle applied to the flow of a viscous fluid. Although the work of **Jean-Claude Barré de Saint-Venant** (1799–1886) is not generally recognized, Rouse and Ince [2] indicate that Saint-Venant developed a more general formulation that could also account for the effects of turbulence. The publication of his work in 1843 predates Stokes's publication by two years.

The modern control-volume method of analysis, an approach unique to engineering, and employed throughout this book, has its origins with **Ludwig Prandtl** (1875–1953) and its subsequent development and codification in the textbooks of **Jerome Hunsaker** and **Brandon Rightmire** (1947) and **Ascher Shapiro** (1953), the latter three engineering professors at the Massachusetts Institute of Technology [3].

Ludwig Prandtl (1875–1953)

6.2 FORCES

The conservation of momentum principle applied to a control volume relates the following three quantities: the forces acting on this control volume, the flows of momentum in and out of this control volume, and the time rate of change of momentum within this control volume. In this section, we focus on the first of these three, examining the most important forces acting on fluids. To begin, we distinguish between those forces that act at the surface of a control volume and those that act on the control volume as a whole.

6.2a Surface Forces

Two types of forces act at the surfaces of systems or control volumes.[1] In all cases, a pressure force will exist, whereas viscous forces will exist only when there is relative motion at the surface of interest. We explore each of these forces in the following.

Pressure Forces

In Chapter 2, we discussed the thermodynamic property pressure and saw that it is a scalar quantity. Pressure, however, gives rise to a pressure force, a vector quantity acting in a particular direction. In Chapter 4, we introduced the idea of a pressure force in our discussion of flow work. In defining a pressure force, we adopt the following convention:

See Fig. 4.9 in Chapter 4. ▷

Pressure forces always act in a direction toward the interior of the system or control volume.

[1] In situations where a free surface exists, surface tension forces can be important. Treatment of such forces is beyond the scope of this book.

FIGURE 6.1

(a) The unit normal vector n̂ is directed outward from the system boundary or control surface. (b) The pressure force dF_P acts inward opposite n̂.

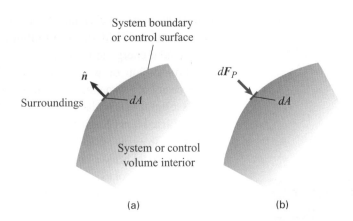

(a)

(b)

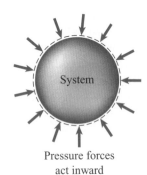

Pressure forces
act inward

The no-slip condition applies to all fluids regardless of their viscosities.

To mathematically express this convention, we introduce the idea of an area vector, the product of an outward-pointing unit vector and the scalar surface area. For a differential area dA, this is expressed as

$$dA = \hat{n}dA. \tag{6.1}$$

Figure 6.1a shows this unit vector, which is always perpendicular to the surface. The differential pressure force acting on dA toward the interior is thus the product of the pressure and the area dA, as shown in Fig. 6.1b. Mathematically,

$$dF_P = -PdA = -\hat{n}PdA, \tag{6.2}$$

where the minus sign indicates that the pressure force is directed inward in a direction opposite to the surface normal $\hat{n}$. In many of our developments, we will be interested in the total force acting over a finite planar area, such as at the cross-sectional area associated with flow in a pipe or duct, as previously illustrated in Fig. 4.9. Referring to Fig. 4.9, we see that if the pressure is uniform over the cross section, the total pressure force at station 1 is easily obtained by integrating Eq. 6.2 over the area A_1 and is given by

$$F_{P,1} = P_1 A_1 \quad \text{(directed to the right)},$$

and the total pressure force at station 2 is

$$F_{P,2} = P_2 A_2 \quad \text{(directed to the left)}.$$

In addition to dealing with flowing fluids, Eq. 6.2 is the starting point in determining the static forces exerted on submerged surfaces by fluids contained in tanks or behind dams. We visit this topic later in the chapter.

Viscous Forces

A fluid flowing over a solid surface exerts a shear force on that surface; reciprocally, we can also say that the surface exerts a shear force on the fluid. In Chapter 3, we introduced the idea of the no-slip condition at a solid–fluid interface. Recall that this condition implies that the fluid immediately at a solid surface has zero velocity; that is, it *sticks* to the wall. This stickiness gives rise to a velocity gradient, which, in turn, generates the shear stress. Consider the simplified situation of a two-dimensional (x, y) flow, as shown in Fig. 6.2a. The general flow direction is from left to right (i.e., in the positive x-direction). The x-direction velocity distribution at a fixed x location varies from zero at the wall (no slip) to a fixed value V_∞ at some distance from the wall. The lengths of the arrows in Fig. 6.2a are proportional to the velocity at the particular distance y from the surface. This velocity distribution is replotted in Fig. 6.2b so that v_x appears explicitly as a function of distance from the surface y [i.e., $v_x(y)$]. With fluids of the type known as **Newtonian** (such as air and water), for this particular 2-D flow the shear stress at the surface is given by

$$\tau_{yx} = \mu \left[\frac{\partial v_x}{\partial y} \right]_{y=0}, \tag{6.3}$$

where μ is the fluid viscosity. This relationship is sometimes referred to as Newton's law of viscosity. The shear stress τ_{yx}, as indicated in Fig. 6.3, is the viscous shear stress acting on the lower y face of the control volume with length dx, depth W, and height y. This shear stress acts in the negative x-direction since the wall exerts a retarding force directed against the fluid motion. From the definition of a shear stress (dF/dA), the force acting on the area Wdx is

$$dF_{x,\text{visc}} = \tau_{yx} W dx. \tag{6.4}$$

Combining Eq. 6.4 and 6.3 yields

$$dF_{x,\text{visc}} = \mu W \left[\frac{\partial v_x}{\partial y} \right]_{y=0} dx. \tag{6.5}$$

The viscous force described here is the origin of the **drag force,** or **drag,** associated with flows over surfaces aligned with the flow. For example, consider a plate of length L. The drag force exerted by the fluid on the plate is the integral of $dF_{x,\text{visc}}$ over the area of the plate exposed to the fluid; that is,

$$F_{\text{drag}} = \int_A dF_{x,\text{visc}}, \tag{6.6a}$$

or

$$F_{\text{drag}} = \mu W \int_{x=0}^{L} \left[\frac{\partial v_x}{\partial y} \right]_{y=0} dx. \tag{6.6b}$$

Note that a knowledge of the velocity distribution as a function of both x and y is required to evaluate the integral in Eq. 6.6b. In fact, the evaluation of the drag force is one of the practical motivations for developing the conservation of momentum principle for control volumes. Solving the so-called momentum equation provides $v_x(x, y)$ and, hence, its partial derivative, $\partial v_x(x, y)/\partial y$.

So far we have considered the viscous force acting on a single face of a control volume for a very specific flow. A general, 3-D, Cartesian, differential control volume is shown in Fig. 6.4a, where the various viscous stresses acting on three of the six faces are indicated. Note that viscous normal stresses, τ_{ii}, act on each face in addition to the shear stresses, τ_{ij}. The following matrix compactly expresses all of these:

$$\tau_{\text{visc},ij} \equiv \begin{bmatrix} \tau_{xx} & \tau_{xy} & \tau_{xz} \\ \tau_{yx} & \tau_{yy} & \tau_{yz} \\ \tau_{zx} & \tau_{zy} & \tau_{zz} \end{bmatrix}, \tag{6.7}$$

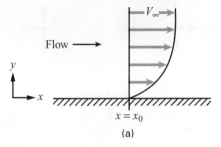

> The no-slip condition is a key concept in fluid mechanics:
> $V_{\text{fluid,wall}} = 0$.

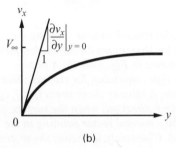

FIGURE 6.2
*(a) As a result of the no-slip condition at the wall, a velocity distribution exists in a fluid flowing over a solid surface.
(b) The x-direction velocity v_x is plotted here as a function of distance from the surface y. Also indicated is the slope of the profile at the wall $[dv_x/dy]_{y=0}$.*

> In Chapter 9, we explore how both the pressure forces and viscous forces contribute to the total drag.

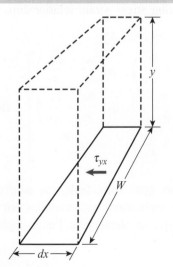

FIGURE 6.3
Control volume for the 2-D flow illustrated in Fig. 6.2. The shear stress τ_{yx} acts on the bottom control surface of area Wdx.

Viscous forces combine with pressure forces to produce the total drag force (see Chapter 9).

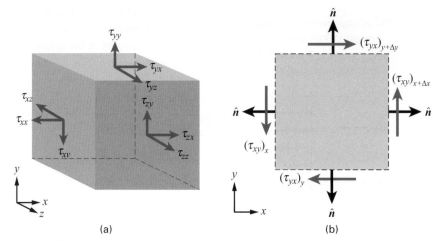

(a)   (b)

FIGURE 6.4

(a) Three-dimensional, Cartesian, differential control volume showing the viscous stresses acting on three faces. Similar stresses act on the hidden faces. All stresses follow the sign convection presented in Fig. 6.4b. (b) A two-dimensional control volume illustrates the sign convention for viscous shear stresses. Consider the τ_{xy} pair as an example. A viscous shear stress ($[\tau_{xy}]_{x+\Delta x}$) points in the positive coordinate direction (+y-direction) when the surface on which it acts (x face at x+Δx) has its surface normal vector pointing in the positive coordinate direction (+x-direction). Conversely, a viscous shear stress ($[\tau_{xy}]_x$) points in the negative coordinate direction (−y-direction) when the surface on which it acts (x face at x) has its surface normal pointing in the negative coordinate direction (−x-direction).

where the second subscript indicates the direction in which the stress acts, and the first subscript indicates the face upon which the stress acts. For example, τ_{yx} is the viscous stress in the x-direction acting on the y face (i.e., the face of the control volume that is perpendicular to the y axis). Figure 6.4b defines the sign convention associated with the viscous stresses.

All of these stresses relate to various velocity gradients, or more precisely, to the fluid strain rates. Jean-Claude Barré de Saint-Venant and George G. Stokes were the first to derive the following stress–strain relationships for incompressible,[2] Newtonian[3] fluids [2]:

Fluids containing suspended solids typically exhibit non-Newtonian behavior. The stress-strain rate relationships such fluids are more complex than that given by Eq. 6.8.

$$\tau_{\text{visc},ij} \equiv \begin{bmatrix} 2\mu\dfrac{\partial v_x}{\partial x} & \mu\left(\dfrac{\partial v_x}{\partial y} + \dfrac{\partial v_y}{\partial x}\right) & \mu\left(\dfrac{\partial v_z}{\partial x} + \dfrac{\partial v_x}{\partial z}\right) \\ \mu\left(\dfrac{\partial v_x}{\partial y} + \dfrac{\partial v_y}{\partial x}\right) & 2\mu\dfrac{\partial v_y}{\partial y} & \mu\left(\dfrac{\partial v_y}{\partial z} + \dfrac{\partial v_z}{\partial y}\right) \\ \mu\left(\dfrac{\partial v_z}{\partial x} + \dfrac{\partial v_x}{\partial z}\right) & \mu\left(\dfrac{\partial v_y}{\partial z} + \dfrac{\partial v_z}{\partial y}\right) & 2\mu\dfrac{\partial v_z}{\partial z} \end{bmatrix}. \quad (6.8)$$

Although this matrix of stresses appears quite formidable, the matrix is symmetric (i.e., $\tau_{ij} = \tau_{ji}$), and many of the applications and examples that we deal with require only a small subset of the stresses shown here. For example in

---

[2] If compressibility is significant, additional terms that make small contributions need to be added. See Refs. [4, 5].

[3] A Newtonian fluid is one in which the fluid stress depends linearly on strain rates (i.e., velocity gradients). Air and water are common Newtonian fluids; catsup is a familiar example of a non-Newtonian fluid.

the flow introduced at the beginning of this section (Figs. 6.2 and 6.3), only one stress is important,

$$\tau_{yx} = \tau_{xy} = \mu\left(\frac{\partial v_x}{\partial y} + \frac{\partial v_y}{\partial x}\right). \tag{6.9a}$$

This expression can be simplified further since the second term on the right-hand side is much smaller than the first term; thus,

$$\tau_{yx} \approx \mu\frac{\partial v_x}{\partial y}. \tag{6.9b}$$

That this is the case is not obvious at this juncture; however, in our discussion of **boundary-layer flows** in Chapter 9, the justifications of these approximations are presented in some detail.

With Eq. 6.8, we have all the information needed to determine the viscous forces acting on any control volume for a large class of fluids (i.e., incompressible, Newtonian fluids). We emphasize that these viscous forces (stresses), like pressure forces, are those acting *on* the control volume *by* the surroundings.

6.2b Body Forces

We will consider only one body force: the gravitational force, or weight. Other body forces, however, can result from electrical fields or magnetic fields in fluids containing charged particles or particles with magnetic moments, respectively. Consideration of these forces is beyond the scope of this book.

The gravitational body force is easy to deal with. For a differential control volume (Fig. 6.5a), this force is the product of the acceleration of gravity **g** and the mass contained within the control volume, dM:

$$d\boldsymbol{F}_{\text{grav}} = \boldsymbol{g}dM, \tag{6.10a}$$

or

$$d\boldsymbol{F}_{\text{grav}} = \boldsymbol{g}\rho dxdydz. \tag{6.10b}$$

For an arbitrary orientation, the gravitational acceleration vector is expressed as

$$\boldsymbol{g} = \hat{\boldsymbol{i}}g_x + \hat{\boldsymbol{j}}g_y + \hat{\boldsymbol{k}}g_z. \tag{6.10c}$$

Usually **g** is aligned with one of the primary axes. For example, if we align **g** with the y axis, then $g_x = g_z = 0$, and $g_y = -9.81 \text{ m/s}^2$, where the minus sign indicates that the gravity force acts downward in the negative y-direction.

For an integral control volume, the weight, or gravitational body force, once again, is the product of the gravitational acceleration and the mass of fluid within the control volume, assuming that **g** does not vary within the control volume, so

$$\boldsymbol{F}_{\text{grav}} = \boldsymbol{g}M, \tag{6.11}$$

where the mass is obtained by integrating the density over the control volume, as discussed in Chapter 3, that is,

$$M = \int dM = \int_{\text{CV}} \rho d\mathcal{V}. \tag{3.4c}$$

A spring scale measures the gravitational body force.

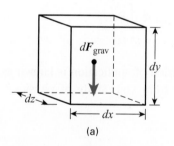

(a)

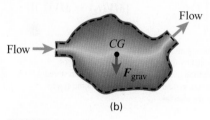

(b)

FIGURE 6.5
(a) Differential control volume showing the gravitational body force. (b) Integral control volume showing a gravitational body force acting at the center of mass, or center of gravity (CG).

The usual convention when sketching a control volume is to represent the gravity force as an arrow with its point of action at the center of mass (i.e., the center of gravity), as illustrated in Fig. 6.5.

6.2c Other Forces

As you will see later, a conveniently chosen integral control volume can cut through a pipe wall or some other solid structure. In such situations, additional mechanical forces then act on the control volume that are the result of neither fluid pressure, viscous stresses, or body forces. Another example of such forces is the normal upward force exerted by a floor on a tank. Our point here is that you should carefully examine any control surface for the existence of such external forces, as they may be important in expressing conservation of momentum.

6.3 MOMENTUM CONSERVATION FOR RIGID SYSTEMS

To develop the conservation of momentum principle for a rigid system,[4] we begin with the generic conservation principle given by Eq. 1.1 from Chapter 1. This is consistent with our developments of mass conservation in Chapter 3 and energy conservation in Chapter 5. We note, however, that momentum is a vector quantity, unlike mass and energy. For a time interval, the generic conservation principle is expressed as

$$X_{in} - X_{out} + X_{generated} = \Delta X_{stored} \equiv X(t_2) - X(t_1), \tag{6.12}$$

where X now represents momentum, MV. For a thermodynamic system, no mass crosses the boundaries; thus, $(MV)_{in}$ and $(MV)_{out}$ must both be zero. The application of a force to a system over a period of time generates momentum; thus, we define

$$X_{generated} \equiv (MV)_{generated} \equiv \int_{t_1}^{t_2} F \, dt. \tag{6.13}$$

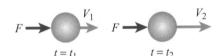

The integral of the force over the time period of application is known as the **impulse,** which may be familiar to you from your study of physics.[5] Re-assembling Eq. 6.12 thus yields

$$
\begin{array}{ccccccc}
0 & + & 0 & + & \displaystyle\int_{t_1}^{t_2} F \, dt & = & [MV(t_2) - MV(t_1)].
\end{array}
$$

Momentum crossing system boundary and passing into system	Momentum crossing system boundary and passing out of system	Momentum generated by the application of a force to the system during the interval $\Delta t = t_2 - t_1$	Change in the amount of momentum stored within the system

---

[4] By rigid we mean no relative motion between different parts of the system. For nonrigid systems (i.e., flowing fluids), we employ a control-volume approach, which is developed in a subsequent section.

[5] What we designate as the conservation of momentum principle is sometimes referred to as the momentum-impulse principle.

Taking the time derivative of this expression, and explicitly noting that several forces may simultaneously act on the system, generate the following familiar form of Newton's second law:

$$\sum \boldsymbol{F}_{\text{sys}} \quad = \quad \frac{d(M\boldsymbol{V})_{\text{sys}}}{dt} \quad = \quad M\boldsymbol{a}, \qquad (6.14)$$

| Sum of vector forces acting on the system | Time rate of change of system momentum | Product of system mass and system acceleration |

where $\boldsymbol{a}$ is the acceleration of the system.

We now apply this principle to the differential fluid system shown in Fig. 6.5a:

$$\sum d\boldsymbol{F}_{\text{sys}} = \frac{d[(dM)\boldsymbol{V}]_{\text{sys}}}{dt}. \qquad (6.15a)$$

The right-hand side of Eq. 6.15a can be expanded by recognizing that, by definition, the system mass ($dM_{\text{sys}} = \rho d\mathcal{V} = \rho dxdydz$) does not change with time. Thus,

$$\sum d\boldsymbol{F}_{\text{sys}} = \rho dxdydz \frac{d\boldsymbol{V}}{dt}, \qquad (6.15b)$$

or

$$\sum d\boldsymbol{F}_{\text{sys}} = \rho dxdydz\,\boldsymbol{a}, \qquad (6.15c)$$

where $\boldsymbol{a}$ is the acceleration of the fluid system.

In the sections that follow, we consider two physical situations that allow the straightforward application of Eq. 6.15 to problems of engineering interest: fluid statics and rigid-body motion of a fluid.

6.3a Fluid Statics

Our objective in this section is to determine the pressure distribution within a fluid at rest.[6] Consider a fluid element at an arbitrary location within a rectangular tank of fluid, as shown in Fig. 6.6. The tank and the fluid are at rest, so $\boldsymbol{V}$ is zero; thus, momentum conservation becomes

$$\sum d\boldsymbol{F}_{\text{sys}} = 0. \qquad (6.16)$$

> Local equilibrium prevails so pressure is a meaningful property at a point.

We need now only identify the forces acting on the system. Since the fluid does not move, there can be no *relative* motion between any fluid elements. As a result, no viscous forces act, as these require the existence of velocity gradients (see Eq. 6.8). We thus conclude that the only forces acting are pressure forces and the body force due to gravity. As done in previous analyzes, we start with a fluid element having dimensions Δx, Δy, and Δz and then apply a limiting process to create the desired differential relations. Figure 6.7 shows the pressure forces acting on six faces of the fluid element. Each pressure force acts inward and is the product of the pressure at the face and the area of the face. For example, the pressure force in the positive x-direction is given by

$$dF_{P,x\,\text{face}} = P_x A_{x\,\text{face}} = P_x \Delta y \Delta z.$$

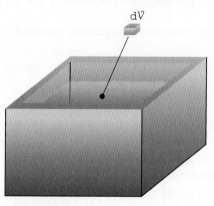

FIGURE 6.6
Tank containing a fluid at rest.

---

[6] Up to this point, we have always assumed that the pressure was uniform throughout the region of interest or, equivalently, that the pressure variations caused by gravity were negligible. The result of the present analysis allows us to assess these assumptions quantitatively.

FIGURE 6.7

A fluid element having dimensions Δx, Δy, and Δz is at rest ($V = 0$). Pressure forces act inward on all six faces and each is the product of the local pressure and the area of the face.

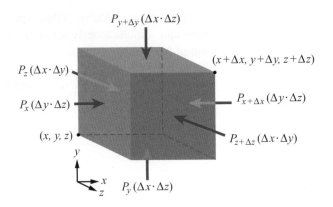

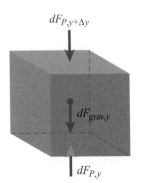

Because momentum conservation (Eqs. 6.15 and 6.16) is a vector relationship, we must consider each component of Eq. 6.16 separately. The most interesting is the component in the y-direction, assuming that $\boldsymbol{g}$ is aligned with the y axis. The y-component of Eq. 6.16 is expressed as

$$\sum dF_{P,y\text{-dir}} + dF_{\text{grav},y\text{-dir}} = 0, \tag{6.17a}$$

or

$$P_y \Delta x \Delta z - P_{y+\Delta y} \Delta x \Delta z + \rho g_y \Delta x \Delta y \Delta z = 0. \tag{6.17b}$$

Note that the gravitation force does act downward since g_y is defined to be negative ($g_y = -9.81 \text{ m/s}^2$). Rearranging Eq. 6.17b and dividing through by $\Delta x \Delta y \Delta z$ yield

$$\frac{-(P_{y+\Delta y} - P_y)}{\Delta y} + \rho g_y = 0. \tag{6.17c}$$

Allowing our fluid element to shrink to the continuum limit, or mathematically taking the limits Δx, Δy, and $\Delta z \rightarrow 0$, results in the first term of Eq. 6.17c becoming the partial derivative $\partial P / \partial y$, or

$$-\frac{\partial P}{\partial y} + \rho g_y = 0. \tag{6.18}$$

Recognizing that $g_x = g_z = 0$, the x- and z-component equations are simply

$$\frac{\partial P}{\partial x} = 0 \tag{6.19}$$

and

$$\frac{\partial P}{\partial z} = 0, \tag{6.20}$$

respectively. The physical significance of Eqs. 6.18–6.20 is that the pressure within the macroscopic system (the fluid in tank in Fig. 6.6) depends only on depth (y-direction) and is independent of x and z. Because P is independent of x and z (Eqs. 6.19 and 6.20), the partial derivative in Eq. 6.18 can be replaced with an ordinary derivative:

$$-\frac{dP}{dy} + \rho g_y = 0. \tag{6.21}$$

FIGURE 6.8
For an incompressible liquid contained in a tube, the pressure difference $P_{bottom} - P_{top}$ equals the height of the liquid column Y multiplied by ρg.

If the variation of the density with height (y) is known and g_y is assumed constant, we can easily integrate Eq. 6.21 to obtain the pressure variation with height, $P(y)$.

Consider the case of an incompressible fluid (ρ = constant). We separate the variables in Eq. 6.21 and integrate as follows:

$$\int_{P_0}^{P(y)} dP = \int_{y_0}^{y} \rho g_y d\hat{y}, \tag{6.22a}$$

or

$$P(y) - P_0 = \rho g_y (y - y_0). \tag{6.22b}$$

To avoid confusion with the sign associated with g_y, we define the scalar g as

$$g \equiv -g_y = +9.80665 \text{ m/s}^2, \tag{6.23}$$

for earth-standard gravity. Making this substitution and taking P_0 to the right-hand side of Eq. 6.22b yield

$$P(y) = P_0 - \rho g (y - y_0). \tag{6.24}$$

Application of this relationship is facilitated by considering a column of fluid inside a tube, as shown in Fig. 6.8. If we ignore any extraneous forces due to surface tension, Eq. 6.24 expresses the pressure within the fluid column at any height. Designating the pressure at the free surface P_{top}, the pressure at the bottom of the tube is given by

$$P_{bottom} = P_{top} + \rho g Y, \tag{6.25a}$$

> **Use of Eqs. 6.25a and 6.25b helps to avoid sign errors.**

where Y is the height of the column. That the pressure at the bottom of the tube is greater than at the top is consistent with our intuition or experience. For any two points within a single, incompressible fluid, Eq. 6.25a generalizes to

$$P_{below} = P_{above} + \rho g \Delta y, \tag{6.25b}$$

where Δy ($= Y$) is the vertical distance between the two points. As used here, the quantity Δy is always positive. Equations 6.24, 6.25a, and 6.25b are equivalent; however, you may find Eqs. 6.25a and 6.25b easier or more convenient to use.

In our development thus far, we have constrained the gravitational acceleration to be aligned with the y-direction. For an arbitrary orientation of the gravity vector, the following general result describes the pressure distribution in a system:

$$-\left(\hat{i} \frac{\partial P}{\partial x} + \hat{j} \frac{\partial P}{\partial y} + \hat{k} \frac{\partial P}{\partial z} \right) + \rho \left(\hat{i} g_x + \hat{j} g_y + \hat{k} g_z \right) = 0, \tag{6.26a}$$

or, in vector notation,

$$\underbrace{-\nabla P}_{\substack{\text{Net pressure forces} \\ \text{per unit volume}}} + \underbrace{\rho \mathbf{g}}_{\substack{\text{Gravitational body} \\ \text{force per unit volume}}} = 0. \tag{6.26b}$$

In the application of Eq. 6.26, three separate component equations are usually employed.

Example 6.1

Set in 2001, the world record for women's unassisted saltwater diving is 70 m. Estimate the pressure at this depth assuming a saltwater density of 1025 kg/m³ and a sea surface pressure of 1 atm.

Solution

Known $Y \, (= \Delta y)$, ρ_{sw}, P_{above}

Find P_{below}

Sketch   See Fig. 6.8.

Assumptions
The density is constant.

Analysis   The pressure at the diving depth is obtained by the direct application of Eq. 6.25b:

$$P_{below} = P_{above} + \rho_{sw}\,g\,\Delta y$$
$$= 101{,}325 + 1025\,(9.807)70$$
$$= 101{,}325 + 703{,}652 = 804{,}977$$

$$[=]\, \frac{kg}{m^3}\, \frac{m}{s^2}\,(m)\left[\frac{1\,N}{kg \cdot m/s^2}\right] = \frac{N}{m^2}\ or\ Pa.$$

Rounding, we obtain $P_{below} = 805.0$ kPa.

Comment   To obtain a physical appreciation for this pressure, we convert to units of atmospheres: $P_{below} = 805/101.3 = 7.95$ atm. Thus, a diver is exposed to a pressure nearly 8 times that at the surface.

**Self Test
6.1**

☑ **A large water storage tank is open to atmospheric pressure (100 kPa) at the top. An absolute pressure gage at the bottom of the tank reads 175 kPa. Determine the height of the water in the tank.**

(Answer: 7.65 m)

Example 6.2

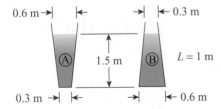

0.6 m→| |←   →| |← 0.3 m

Ⓐ   1.5 m   Ⓑ $L = 1$ m

0.3 m →| |←   →| |← 0.6 m

Consider the two tanks shown in cross section in the sketch. The length of the tanks in the direction perpendicular to the page is 1 m. Each tank is filled to a depth of 1.5 m with water ($\rho = 997$ kg/m³) and open to the atmosphere at the top. The atmospheric pressure is 100 kPa. Determine (a) the pressure forces exerted on the inside bottom surface of each tank, (b) the gage-pressure forces acting on the same surfaces, and (c) the weight of the water in each tank.

Solution

Known   Tank geometry and dimensions, Y, ρ, P_{atm}

Find $F_{P,\text{bottom}}, F_{P(\text{gage}),\text{bottom}}, W_{\text{H}_2\text{O}}$

Sketch

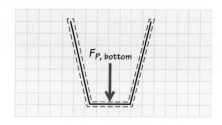

Assumptions

Constant density

Analysis The pressure at the bottom of each tank can be calculated from Eq. 6.25a. Since the only geometrical parameter involved is the depth Y, the pressures will be identical at the bottom of each tank:

$$P_{\text{bottom}} = P_{\text{top}} + \rho g Y$$
$$= 100{,}000 + 997(9.807)1.5$$
$$= 100{,}000 + 14{,}666 = 114{,}666$$
$$[=] \frac{\text{kg}}{\text{m}^3} \frac{\text{m}}{\text{s}^2} \, \text{m} \left[\frac{1\,\text{N}}{\text{kg} \cdot \text{m/s}^2} \right] = \frac{\text{N}}{\text{m}^2} \text{ or Pa.}$$

The pressure force acting over the horizontal, rectangular bottom is simply the product of the pressure and the area of the tank base (see Eq. 6.2); thus, for tank A,

$$F_{P,\text{bottom},A} = P_{\text{bottom}} A_{\text{bottom},A}$$
$$= 114{,}666(0.3)1.0$$
$$= 34{,}400$$
$$[=] \frac{\text{N}}{\text{m}^2} (\text{m})\text{m} = \text{N},$$

and for tank B,

$$F_{P,\text{bottom},B} = P_{\text{bottom}} A_{\text{bottom},B}$$
$$= 114{,}666(0.6)1.0 \, \text{N}$$
$$= 68{,}800 \, \text{N}.$$

The gage pressure at the bottom of each tank is (Eq. 2.12)

$$P_{\text{bottom,gage}} = P_{\text{bottom}} - P_{\text{atm}}$$
$$= 114{,}666 - 100{,}000 \, \text{Pa}$$
$$= 14{,}666 \, \text{Pa},$$

and the forces exerted by the water (exclusive of the atmosphere) are

$$F_{P(\text{gage}),\text{bottom},A} = P_{\text{bottom,gage}} A_{\text{bottom},A}$$
$$= 14{,}666(0.3)(1.0) \, \text{N}$$
$$= 4400 \, \text{N}$$

and

$$F_{P(\text{gage}),\text{bottom},B} = P_{\text{bottom,gage}} A_{\text{bottom},B}$$
$$= 14{,}666(0.6)1.0\,\text{N}$$
$$= 8800\,\text{N}$$

The mass of the water in each tank is

$$M = \rho\mathcal{V} = \rho A_{\text{x-sec}} L,$$

and the weight is given by

$$W_{\text{H}_2\text{O}} = Mg = \rho A_{\text{x-sec}} Lg.$$

Each tank has the same cross-sectional area; thus,

$$W_{\text{H}_2\text{O}} = 997\left[\left(\frac{0.6 + 0.3}{2}\right)1.5\right]1.0(9.807)$$
$$= 6600$$
$$[=]\frac{\text{kg}}{\text{m}^3}\,\text{m}\,(\text{m})(\text{m})\frac{\text{m}}{\text{s}^2}\left[\frac{1\,\text{N}}{\text{kg}\cdot\text{m/s}^2}\right] = \text{N}.$$

Comment We note that the weight of the water falls between the two gage-pressure forces; that is, 4400 N < 6600 N < 9900 N. What is responsible for these values being different? By answering this, you solve the "hydrostatic paradox." *Hint*: Draw a detailed free-body diagram for the water or the tank.

Self Test 6.2 A test tube contains mercury at 300 K. Determine the gage pressure at the bottom of the test tube if the depth of the mercury is 10 cm when the tube is held vertically.

(Answer: 13.27 kPa)

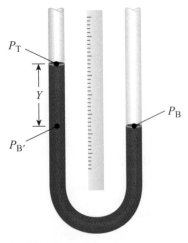

FIGURE 6.9

A U-tube manometer uses the difference in height between its two legs to measure a pressure difference. For a manometer fluid of known density ρ_m, $P_B - P_T = \rho_m gY$. A scale between the legs is used to determine Y.

Two applications of this theory of fluid statics are frequently encountered: the measurement of pressure differences using manometers and the calculation of fluid pressure forces exerted on submerged surfaces. We treat these two applications next.

Manometry

Figure 6.9 illustrates a simple U-tube manometer. Manometers are used to determine gage pressures when one leg of the manometer is connected to a location of unknown pressure and the other leg is open to the atmosphere. Manometers also can be used to determine pressure differences (differential pressure) when the legs are connected to locations having different pressures. The manometer fluid is usually chosen to have a density substantially greater than the fluid to which it is connected. Mercury, water, and oils are common manometer fluids. The use of a manometer to measure pressure differences relies on principles of fluid statics (i.e., the degenerate expression of momentum conservation given by Eq. 6.24 or 6.25).

Consider the manometer shown in Fig. 6.9. We will use Eq. 6.24 or 6.25 to determine the pressure difference $P_B - P_T$. First, recognize that the pressure in the manometer fluid at a depth Y below the point designated P_T has the same

Manometer board set up in the supersonic wind tunnel at NASA's Lewis (now Glenn) Research Center in 1949. Photograph courtesy of NASA.

value as the pressure in the right leg designated P_B [i.e., $P_{B'}$ (left leg) = P_B (right leg)]. This is true, however, only if a single fluid connects the points in the two legs. Knowing $P_{B'} = P_B$, we write simply

$$P_B - P_T = \rho_m g Y, \tag{6.27}$$

where ρ_m is the manometer fluid density. Furthermore, if the left leg is open to the atmosphere, then we have

$$P_T \cong P_{atm},$$

noting that the atmospheric pressure varies with height as well but that such variation is generally small because the density of air is relatively low (i.e., $\rho_{air} \ll \rho_m$). Thus,

$$P_B - P_{atm} \equiv P_{B,gage} = \rho_m g Y.$$

The following example illustrates the use of a manometer.

Example 6.3

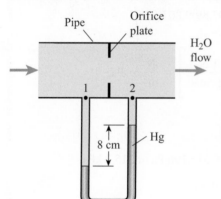

As shown in the sketch, a mercury manometer measures the pressure drop across an orifice plate in a flow of water. The gage pressure at location 1 is 325 kPa and the atmospheric pressure is 100 kPa. The water and the manometer are at 72 F (295.3 K). Determine both the absolute and gage pressures at location 2.

Solution

Known $P_{1,g}, Y, T, P_{atm}$

Find $P_{2,g}, P_2$

Sketch

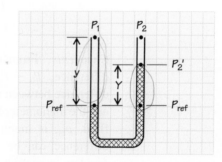

Assumptions

i. Constant ρ_{H_2O} and ρ_{Hg}

ii. $\rho_{Hg}(T, P) \approx \rho_{Hg}(T_{sat})$

Analysis   As indicated in the sketch, the left column and the right column of the manometer have a common pressure, P_{ref}. The sketch also defines an intermediate pressure, $P_{2'}$, in the right leg of the manometer. We begin by equating P_{ref} (left) and P_{ref} (right) and applying Eq. 6.25 to each leg of the manometer:

$$P_{ref} \text{ (left leg)} = P_{ref} \text{ (right leg)},$$

or

$$P_1 + \rho_{H_2O}gy = P_{2'} + \rho_{Hg}gY.$$

The intermediate pressure $P_{2'}$ can be related to the unknown pressure P_2 by

$$P_{2'} = \rho_{H_2O}g(y - Y),$$

where $Y - y$ is the height of the water column above the $P_{2'}$ location; thus,

$$P_{ref} \text{ (right leg)} = P_2 + \rho_{H_2O}g(y - Y) + \rho_{Hg}gY.$$

Combining these equations and solving for $P_1 - P_2$ yields

$$P_1 - P_2 = (\rho_{Hg} - \rho_{H_2O})gY.$$

From the NIST database we find

$$\rho_{H_2O}(295.3\ K, 425\ kPa) = 997.89\ kg/m^3,$$

and from Table G.2,

$$\rho_{Hg}(295.3\ K, P_{sat}) = 13{,}540\ kg/m^3.$$

Thus,

$$P_1 - P_2 = (13{,}540 - 997.89)9.807(0.08)$$
$$= 9840$$

$$[=]\ \frac{kg\,m}{m^3\,s^2}(m)\left[\frac{1\ N}{kg\cdot m/s^2}\right] = \frac{N}{m^2} = Pa,$$

and

$$P_{2,g} = P_{1,g} - (P_1 - P_2)$$
$$= 325{,}000 - 9804\ Pa = 315{,}196\ Pa \text{ or } 315.2\ kPa.$$

The absolute pressure at location 2 is then

$$P_2 = P_{2,g} + P_{atm}$$
$$= 315.2 + 100\ kPa = 415.2\ kPa.$$

Comments   From our analysis, we conclude that the desired pressure difference, $P_1 - P_2$, is proportional to the product of the density difference between the manometer fluid and the system fluid, $\rho_{Hg} - \rho_{H_2O}$, and the manometer fluid column height Y. If the density of the manometer fluid, ρ_m, is much greater than that of the other fluid, the pressure difference can be simply calculated as $\Delta P = \rho_m gY$. For the present problem, assuming $\rho_{Hg} \gg \rho_{H_2O}$ results in an error of approximately 8% $(= [\rho_{Hg}/(\rho_{Hg} - \rho_{H_2O})]\cdot 100\%)$. This is not likely to be acceptable for most measurements. If the fluid of interest were air rather than water, the error would then be only 0.037%.

Self Test 6.3 ✓   **A large tank contains an immiscible mixture of oil and water at 300 K. A scale indicates that the combined fluid depth is 5 m and a pressure gage at the bottom of the tank reads 47 kPa gage. Find the heights of the water and oil.**

(Answer: $Y_{H_2O} = 3.3\ m, Y_{Oil} = 1.7\ m$)

Forces on Submerged Surfaces

In the design of dams and other submerged structures, knowing the forces exerted by the surrounding fluid on the structure is vitally important. In this section, we apply the theory just developed to planar submerged surfaces. For treatments of curved surfaces, we refer the reader to standard fluid mechanics textbooks (e.g., Refs. [4, 5]).

We begin by considering a simple situation and then generalizing. Consider a tank filled to the top with a liquid of density ρ, as shown in Fig. 6.10a. The end wall on the left is vertical, whereas the end wall on the right is sloping at an angle θ as shown. We wish to determine the net, or resultant, pressure forces acting on these end walls together with their directions and points of application. Figure 6.10b illustrates the distributed pressure force and its single-force equivalent for the left wall.

To calculate the pressure force acting on the left wall, we begin by applying the formal definition of a differential pressure force (Eq. 6.2) and integrating over the surface; thus,

$$F_P = \int_A -\hat{n}P\,dA, \qquad (6.28)$$

where the pressure varies with distance along the surface (ζ) as

$$P(\zeta) = P_{\text{ref}} + \rho g Y(\zeta)$$

and the depth below the surface, Y, depends on the ζ coordinate [i.e., $Y = Y(\zeta)$] The differential area dA is given by

$$dA = W\,d\zeta,$$

where W is the width of the surface perpendicular to the page (Fig. 6.10a). Because the left-hand wall is vertical, the coordinate along the surface ζ is identical to the depth coordinate Y (i.e., $\zeta \equiv Y$). With these substitutions, we have

$$F_P = -\int_0^{L_1} (P_{\text{ref}} + \rho g\zeta)W\,d\zeta,$$

which we integrate to yield

$$F_P = -WL_1\left(P_{\text{ref}} + \rho g\frac{L_1}{2}\right). \qquad (6.29)$$

The minus sign here indicates that the force is directed in the negative x-direction since $\hat{n} \equiv \hat{i}$. Analyzing Eq. 6.29, we see that the pressure force is the product of the area over which the pressure acts and the pressure at the depth of the midpoint of the surface. The validity of this result is easily seen without recourse to our formal derivation by recognizing that the mean pressure associated with the linear distribution (Fig. 6.10b) is the pressure at the midpoint. Performing a similar analysis on the right wall $(A_2 = WL_2)$, we write

$$F_{P,\perp} = \int_0^{L_2} P(\zeta)W\,d\zeta, \qquad (6.30)$$

Glen Canyon Dam, Arizona.

A huge tank containing more than 2-million gallons of molasses collapsed in the "Great Molasses Flood of 1919" in Boston resulting in the destruction of an elevated train structure and the death of 21 persons.

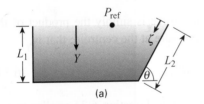

(a)

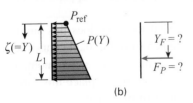

(b)

FIGURE 6.10

(a) Fluid-filled tank with constant width W perpendicular to the page. (b) Pressure distribution acting on left wall and the equivalent single force acting at Y_F.

To find the y coordinate of the center of pressure, we apply Eq. 6.40:

$$y_{CP} = \frac{-\rho g I_{zz} \sin \theta}{P_{CG} A}.$$

The area moment of inertia, I_{zz}, is given by (see Fig. 6.11)

$$I_{zz} = \frac{L^3(a^2 + 4ab + b^2)}{36(a + b)}$$

$$= \frac{(1.2)^3[(0.5)^2 + 4(0.5)1 + (1)^2]}{36(0.5 + 1)} \, \mathrm{m}^4 = 0.104 \, \mathrm{m}^4.$$

Thus,

$$y_{CP} = \frac{-997(9.807)0.104 \sin 30°}{114{,}800}$$

$$= -0.00443$$

$$[=] \frac{\dfrac{\mathrm{kg}}{\mathrm{m}^3}\dfrac{\mathrm{m}}{\mathrm{s}^2}\mathrm{m}^4}{\mathrm{N}\left[\dfrac{1\,\mathrm{kg \cdot m/s^2}}{\mathrm{N}}\right]} = \mathrm{m}.$$

The minus sign here indicates that y_{CP} is located below the centroid.

Comments It is important to point out that the pressure force is not the only force acting on this plug. For static equilibrium, reaction forces act at the edges or, if present, hinge joints to balance the pressure force. Example 6.5 illustrates this bigger picture.

Nuclear-powered submarine USS Salt Lake City. Photograph courtesy U.S. Navy.

Self Test
6.4

✓   A submarine has a flat 1.5-m high and 1.0-m wide hatch on its side, which is hinged at the top. The top of the hatch is 5 m below the seawater surface ($\rho_{sw} = 1025$ kg/m³) and the pressure inside the submarine is atmospheric. Determine the minimum force that would be required to open the hatch.

(Answer: 45 kN)

Example 6.5

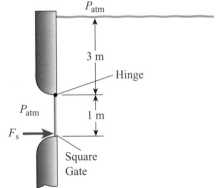

A 1-m-square steel gate is located in a vertical wall of a large water tank as shown in the sketch. The 15-mm-thick gate ($\rho_{gate} = 7850$ kg/m³) is hinged at the top and restrained at the bottom by the force of a stop, F_s. Atmospheric pressure acts above the water ($\rho_{H_2O} = 997$ kg/m³) and behind the gate as indicated. Determine the magnitude of F_s. Also find values for the reaction forces at the hinge, $F_{x,hinge}$ and $F_{y,hinge}$.

Solution

Known ρ_{H_2O}, ρ_{gate}, L, b, t (thickness), Y

Find F_s, $F_{x,hinge}$, $F_{y,hinge}$

Sketch

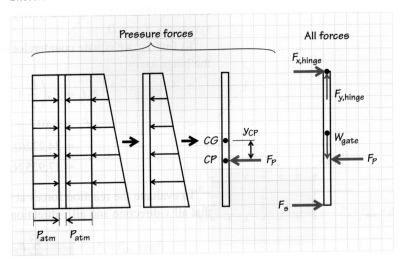

Assumptions

   i. Constant water density

  ii. Frictionless hinge

 iii. Frictionless seal at bottom

Analysis In the sketch, we see that atmospheric pressure acts equally on both sides of the gate; thus, the net effect of pressure results only from the water. Applying Eq. 6.34 with $P_{CG} = P_{CG,\text{gage}}$ yields

$$F_P \equiv F_{P,\text{net}} = \rho_{H_2O} g Y_{CG} A_{\text{gate}}.$$

Since the centroid of the plate is at $L/2$,

$$Y_{CG} = Y + L/2$$
$$= 3 + 1/2\,\text{m} = 3.5\,\text{m}$$

and

$$F_P = 997(9.807)3.5(1)^2$$
$$= 34{,}222$$
$$[=]\frac{\text{kg}}{\text{m}^3}\left(\frac{\text{m}}{\text{s}^2}\right)\text{m}\,(\text{m}^2)\left[\frac{1\,\text{N}}{\text{kg}\cdot\text{m/s}^2}\right] = \text{N}.$$

This net (gage) pressure force acts at the center of pressure defined by Eqs. 6.40 and 6.41. The moments of inertia (Fig. 6.11) are

$$I_{zz} = \frac{bL^3}{12} = \frac{1(1)^3}{12}\,\text{m}^4 = 0.083\,\text{m}^4$$

and

$$I_{zy} = 0.$$

Thus,

$$y_{CP} = \frac{-\rho_{H_2O} g I_{zz} \sin\theta}{P_{CG} A}$$
$$= \frac{-997(9.807)0.0833\sin 90°}{34{,}222}\,\text{m}$$
$$= -0.0238\,\text{m},$$

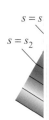

where $V(= [v_x^2 + v_y^2]^{1/2})$ is the magnitude of **V**. Because ρ, v_x, and V are all uniform,

$$x\text{-momentum flow} = \hat{i}\rho v_x V \frac{\pi D^2}{4}.$$

Similarly, the y-component is

$$y\text{-momentum flow} = \hat{j}\rho v_y V \frac{\pi D^2}{4}.$$

In both components, we recognize that $\rho V \pi D^2/4$ is $\dot{m}$, the mass flow rate through the area A, and from the geometry, that

$$v_x = V\cos\theta,$$
$$v_y = V\sin\theta.$$

Our final result is, thus,

$$x\text{-momentum flow} = \dot{m}v_x = \dot{m}V\cos\theta,$$
$$y\text{-momentum flow} = \dot{m}v_y = \dot{m}V\sin\theta,$$

with numerical values

$$x\text{-direction: } 2.05\cos 20°\,\text{N} = 1.93\,\text{N},$$
$$y\text{-direction: } 2.05\sin 20°\,\text{N} = 0.70\,\text{N}.$$

Comments From this example, we see that the momentum flux vector is simply $\dot{m}\mathbf{V}$ or $\dot{m}(\hat{i}v_x + \hat{j}v_y + \hat{k}v_z)$ when the velocity and density are uniform over the region of interest.

(g – ·)

FIGURI

The pre
and is (
perpen(

The e
can b

Exan

cc

z
x

Self Test
6.5

A 5-cm-diameter horizontal jet of water with velocity of 25 m/s strikes a curved deflector, which diverts the stream 90° upward. Determine the horizontal and vertical momentum flow through a control volume surrounding the deflector.

(Answer: x-momentum flow = −1227.2 N, y-momentum flow = 1227.2 N)

Example 6.9

As shown in Fig. 6.16c, water enters a long tube with a uniform velocity profile. At the tube exit, the velocity profile is parabolic and is expressed as

$$v_x(r) = v_{x,0}\left[1 - \left(\frac{r}{R}\right)^2\right],$$

where $v_{x,0}$ is the centerline velocity and R is the tube radius. The flow rate through the tube is 8.3×10^{-3} kg/s and the water density is 997 kg/m^3. The tube diameter is 25 mm. Determine the momentum flow at the inlet and outlet of the tube.

Solution

Known $v_x(r)$, $\dot{m}$, ρ, D

Find   Inlet and exit momentum flows

Sketch See Fig. 6.16c.

Assumptions
  i. Constant ρ
  ii. Steady state

Analysis At the inlet, the flow is essentially the same as in Example 6.8, except now the velocity vector and control surface normal vector are in opposite directions; thus,

$$\boldsymbol{V} \cdot \hat{\boldsymbol{n}} = v_{\text{inlet}} \cos 180° = -v_{\text{inlet}}.$$

The inlet momentum flow is then

$$\int_A \boldsymbol{V}\rho(\boldsymbol{V} \cdot \hat{\boldsymbol{n}})dA = \hat{\boldsymbol{i}}v_{\text{inlet}}(-\rho v_{\text{inlet}}A) = -\hat{\boldsymbol{i}}\dot{m}v_{\text{inlet}}.$$

Since the inlet velocity is not given, we apply the definition of mass flow rate (Eq. 3.15) to obtain this quantity:

$$v_{\text{inlet}} = v_{\text{avg}} = \frac{\dot{m}}{\rho A_{\text{x-sec}}},$$

where we recognize that the uniform inlet velocity equals the average velocity. The value of v_{inlet} is

$$v_{\text{inlet}} = \frac{8.3 \times 10^{-3}}{997\pi(0.025)^2/4} = 0.017 \text{ m/s},$$

and the inlet momentum flow is

$$\int_A \boldsymbol{V}\rho(\boldsymbol{V} \cdot \hat{\boldsymbol{n}})dA = -\hat{\boldsymbol{i}}(8.3 \times 10^{-3})0.017 = -\hat{\boldsymbol{i}}1.41 \times 10^{-4} \text{ N}.$$

At the outlet, the situation is more complex because the velocity is no longer uniform. Recognizing that $\boldsymbol{V} = \hat{\boldsymbol{i}}v_x(r)$ and that $\boldsymbol{V} \cdot \hat{\boldsymbol{n}} = +v_x(r)$, we apply the definition of momentum flow (Eq. 6.50) as follows:

$$\int_A \boldsymbol{V}\rho(\boldsymbol{V} \cdot \hat{\boldsymbol{n}})dA = \int_0^R \hat{\boldsymbol{i}}v_x(r)\rho v_x(r)2\pi r dr,$$

where $dA = 2\pi r dr$ for the axisymmetric geometry. Substituting the given expression for $v_x(r)$ and removing constant quantities from the integrand yield

$$\int_A \boldsymbol{V}\rho(\boldsymbol{V} \cdot \hat{\boldsymbol{n}})dA = \hat{\boldsymbol{i}}2\pi\rho v_{x,0}^2 \int_0^R \left[1 - \left(\frac{r}{R}\right)^2\right]^2 r dr$$

$$= \hat{\boldsymbol{i}}2\pi\rho v_{x,0}^2 \int_0^R \left[1 - 2\left(\frac{r}{R}\right)^2 + \left(\frac{r}{R}\right)^4\right] r dr$$

$$= \hat{\boldsymbol{i}}2\pi\rho v_{x,0}^2 \left[\frac{r^2}{2} - \frac{2}{R^2}\frac{r^4}{4} + \frac{1}{R^4}\frac{r^6}{6}\right]_0^R$$

$$= \hat{\boldsymbol{i}}2\pi\rho v_{x,0}^2 \left(\frac{R^2}{6}\right).$$

The only remaining task is to relate the centerline velocity $v_{x,0}$ to known quantities. We do this by applying mass conservation to the cylindrical control volume (Fig. 6.16c). With the assumption of steady state, Eq. 3.18a applies:

$$\dot{m}_{inlet} = \dot{m}_{outlet}$$

or

$$\rho v_{inlet} A = \rho v_{avg,out} A.$$

Thus, we find that

$$v_{avg,out} = v_{inlet}.$$

In Chapter 3, we determined that the centerline velocity $v_{x,0}$ is twice the average velocity for a parabolic velocity distribution. Thus,

$$v_{x,0} = 2v_{avg,out} = 2v_{inlet},$$

and the outlet momentum flux is easily evaluated as

$$\int V\rho(V \cdot \hat{n})dA = \hat{i}2\pi\rho(2v_{avg,out})^2 R^2/6$$

$$= \hat{i}\frac{4}{3}\rho v_{avg,out}^2 \pi R^2$$

$$= \hat{i}\frac{4}{3}\dot{m}v_{avg,out},$$

recognizing that $\rho v_{avg,out}\pi R^2 = \dot{m}$. Numerically evaluating the exit momentum flow yields

$$\hat{i}\frac{4}{3}\dot{m}v_{avg,out} = \hat{i}\frac{4}{3}8.3 \times 10^{-3}(0.017)$$

$$= \hat{i}\frac{4}{3}(1.41 \times 10^{-4})\,N$$

$$= \hat{i}1.88 \times 10^{-4}\,N.$$

The reader should verify the units here.

Comments   We see that the magnitude of the outlet momentum flux exceeds that of the inlet by the factor 4/3. This result requires that the contribution of the higher velocity in the central portion of the tube to the momentum flow more than compensates for the less-than-average velocity toward the tube walls. This observation also suggests that a correction factor (β) for nonuniform velocity distributions can be used to relate the actual momentum flow to that based on the average velocity, that is,

$$\int V\rho(V \cdot \hat{n})dA \equiv \beta\dot{m}V_{avg}.$$

See Table 5.5 in Chapter 5.

This is similar to the kinetic energy correction factors defined in Chapter 5. We explore momentum flow correction factors later in this chapter.

Self Test 6.6   Redo Self Test 6.5 if the curved deflector turns the jet a full 180°, (i.e., opposite to the direction from which it came).

(*Answer: x-momentum flow = −2454.4 N, y-momentum flow = 0 N*)

6.5 LINEAR MOMENTUM CONSERVATION FOR CONTROL VOLUMES

In this section, we restrict our analysis to nonaccelerating control volumes. This is equivalent to the requirement that we work with an inertial coordinate system, that is, a coordinate system that is fixed or moves at a constant velocity with respect to a stationary coordinate system. This restriction prevents us from analyzing a whole class of problems such as accelerating rockets; however, many interesting and challenging problems can still be treated with our restricted analysis. For the interested reader, the appendix to this chapter presents conservation of momentum expressions for noninertial (accelerating) coordinate systems.

6.5a Simplified General View

With the knowledge of how to express and calculate momentum flows, we are now able to write explicit momentum conservation statements for control volumes. In the following subsections, we develop the simplest statements and then add complexity. In all cases, we carefully define the restrictions that apply and state the momentum conservation principle with mathematical rigor. Before doing so, however, it is instructive to transform the generic statement of the control-volume conservation principles presented in Chapter 1 to a general statement of momentum conservation. Although lacking in detailed definition and rigor, this general statement is very useful for developing an understanding of the concept of momentum conservation applied to control volumes. Our subsequent developments will add the necessary rigor.

The transformation of the generic conservation principle to the specific case of momentum conservation parallels the development of the mass and energy conservation principles presented in Chapters 3 and 5, respectively. Applying the conservation of momentum principle to a control volume, however, is inherently more complex than applying either the conservation of mass or energy principles because forces and momentum flows are vectors. We begin with Eq. 1.2, the rate form of the generic conservation principle:

$$\dot{X}_{\text{in}} - \dot{X}_{\text{out}} + \dot{X}_{\text{generated}} = \dot{X}_{\text{stored}}.$$

Identifying the conserved quantity as momentum, we recognize that $\dot{X}_{\text{in}}$ and $\dot{X}_{\text{out}}$ are the respective momentum flows in and out of the control volume (see Fig. 6.17), that is,

$$\dot{X}_{\text{in}} \equiv (\dot{m}V)_{\text{in}}, \tag{6.51a}$$

$$\dot{X}_{\text{out}} \equiv (\dot{m}V)_{\text{out}}. \tag{6.51b}$$

Forces acting on the control volume generate momentum on a time-rate basis; thus,

$$\dot{X}_{\text{generated}} \equiv \sum \boldsymbol{F}_{\text{cv}}. \tag{6.51c}$$

Depending on the direction of application, a force can also act as a sink for momentum. The time rate of change of momentum within the control volume is simply

$$\dot{X}_{\text{stored}} \equiv \frac{d(MV)_{\text{cv}}}{dt}. \tag{6.51d}$$

To maintain steady, level flight, the drag force is balanced by a net horizontal momentum flow from the aircraft's engines.

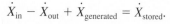See Eq. 3.19 for mass conservation and Eq. 5.69 for energy conservation.

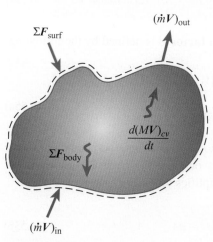

FIGURE 6.17

Control volume schematically showing momentum flowing in, momentum flowing out, and surface and body forces that act as sources (or sinks) of momentum on a time-rate basis. The rate at which momentum is stored in the control volume is represented with a squiggly arrow inside the control volume.

Using these definitions (Eqs. 6.51a–6.51d), we write the conservation of momentum principle for a control volume as

$$(\dot{m}\boldsymbol{V})_{\text{in}} \quad - \quad (\dot{m}\boldsymbol{V})_{\text{out}} \quad + \quad \sum \boldsymbol{F}_{\text{cv}} \quad = \quad \frac{d(M\boldsymbol{V})_{\text{cv}}}{dt}. \qquad (6.52)$$

| Rate of momentum flowing into control volume | Rate of momentum flowing out of control volume | Rate at which momentum is generated by forces acting on control volume | Time rate of change of momentum within control volume |

6.5b Integral Control Volumes with Steady Flow

We begin by considering steady flow for an integral control volume having a single inlet and a single outlet. For this simplified situation, conservation of momentum can be stated as follows:

In steady state, the vector momentum flow in minus the vector momentum flow out plus the vector sum of all forces acting on the control volume must equal zero.

Assuming that the velocity is uniform over the inlet and the outlet, this is expressed symbolically as

$$\dot{m}\boldsymbol{V}_{\text{in}} \quad - \quad \dot{m}\boldsymbol{V}_{\text{out}} \quad + \quad \sum \boldsymbol{F}_{\text{cv}} \quad = \quad 0 \quad (\textit{uniform } \boldsymbol{V}) \qquad (6.53)$$

| Vector momentum flow into control volume | Vector momentum flow out of control volume | Vector sum of forces acting on control volume |

$\dot{m}V_{\text{in}}$ → → $\dot{m}V_{\text{out}}$

The horizontal reaction force exerted by the test stand on this jet engine equals the difference between the momentum flow out and the momentum flow in.

If the velocity is not uniformly distributed at the inlet or outlet, then

$$\dot{m}\beta_{\text{in}}\boldsymbol{V}_{\text{avg,in}} - \dot{m}\beta_{\text{out}}\boldsymbol{V}_{\text{avg,out}} + \sum \boldsymbol{F}_{\text{cv}} = 0 \quad (\textit{distributed } \boldsymbol{V}), \qquad (6.54)$$

where the **momentum flow correction factor** β is defined by the following:

$$\beta \equiv \frac{\displaystyle\int_A \boldsymbol{V}\rho|(\boldsymbol{V} \cdot \hat{\boldsymbol{n}})|\,dA}{\dot{m}V_{\text{avg}}}. \qquad (6.55a)$$

Here we assume that the velocity vector is everywhere perpendicular to the control surface where the flow enters or exits the control volume. If the density is constant, this expression simplifies to

$$\beta = \frac{1}{A}\int_A \frac{V^2}{V_{\text{avg}}^2}\,dA, \qquad (6.55b)$$

where V is the magnitude of the local velocity, and V_{avg} is the magnitude of the average velocity over A (see Eq. 3.13). Values for β associated with common velocity distributions for circular flow areas are given in Table 6.1.

The forces acting on the control volume are the surface forces resulting from pressure or viscous stress, the gravitational body force, and other forces exposed where a control surface cuts through a solid, as discussed at the beginning of this chapter.

Table 6.1 Linear Momentum Correction Factors for Parabolic and Power-Law* Velocity Distributions over Circular Flow Areas

Velocity Distribution	Correction Factor (β)
Parabolic[†]	1.333
Power law	$\dfrac{(n + 1)^2(2n + 1)^2}{2n^2(2n + 2)(n + 2)}$
$n = 6$	1.027
$n = 7$	1.020
$n = 10$	1.011

* The power-law distribution is given by $v(r) = a(1 - r/R)^{1/n}$.
[†] See Example 6.9.

Equations 6.53 and 6.54 can be made more general by considering multiple inlets and outlets. For this situation, momentum conservation is expressed as

$$\sum_{j=1}^{\text{N outlets}} \dot{m}_{\text{in}, j}\beta_{\text{in}, j}\boldsymbol{V}_{\text{avg,in}, j} - \sum_{k=1}^{\text{M outlets}} \dot{m}_{\text{out}, k}\beta_{\text{out}, k}\boldsymbol{V}_{\text{avg,out}, k} + \sum \boldsymbol{F}_{\text{cv}} = 0. \qquad (6.56)$$

All of the previous expressions of momentum conservation are specific cases of the following general mathematical representation:

$$-\int_{\text{CS}} \boldsymbol{V}\rho(\boldsymbol{V}_{\text{rel}} \cdot \hat{\boldsymbol{n}})dA + \int_{\text{CS}} d\boldsymbol{F}_P + \int_{\text{CS}} d\boldsymbol{F}_{\text{visc}} + \int_{\text{CS}} d\boldsymbol{F}_{\text{other}} + \int_{\text{CV}} d\boldsymbol{F}_{\text{grav}} = 0, \qquad (6.57)$$

where $\int_{\text{CS}}$ indicates integration over the entire control surface. Note that the minus sign in the first term causes momentum flows *in* to be positive and momentum flows *out* to be negative. The appearance of the relative velocity $\boldsymbol{V}_{\text{rel}}$ handles the situation where the control volume moves with a constant velocity with respect to a fixed observer. If the control volume is fixed, $\boldsymbol{V}_{\text{rel}} \equiv \boldsymbol{V}$. This use of a relative velocity is already embodied in the simpler expressions (Eqs. 6.53 and 6.54) where the mass flow rate across the surface of interest, $\dot{m}$, takes this into account.

> **Appendix 6A presents the extension of momentum conservation to noninertial (accelerating) control volumes.**

Example 6.10

> **Converging–diverging nozzles are treated in detail in Chapter 11— see Examples 11.10 and 11. 11.**

The Space Shuttle orbiter has three main engines located at the rear of the vehicle. These engines burn hydrogen with oxygen at high temperatures and pressures. The products of combustion, primarily steam and excess hydrogen, expand in a converging–diverging nozzle to produce a high-velocity jet at the nozzle exit. Consider a single Space Shuttle main engine (SSME) secured in a test stand and firing vertically downward as shown in Fig. 6.18. During a test at full power, the hydrogen and oxygen flow rates are 72.7 kg/s and 440.9 kg/s, respectively. The exit diameter of the nozzle is 2.3 m, and the temperature and pressure of the exhaust gases at the exit plane are 1166 K and 0.1915 atm, respectively. The apparent molecular weight of the exhaust gases is 13.78 kg/kmol. The engine weighs 31,140 N.

FIGURE 6.18
The Space Shuttle main engine secured in a test stand burns hydrogen with oxygen to create a high-velocity gas flow at the nozzle exit. **Photograph courtesy of NASA.**

Determine the thrust exerted by the single SSME on the test stand for an ambient pressure of 1 atm.

Solution

Known $\dot{m}_{H_2}$, $\dot{m}_{O_2}$, P_{atm}, P_e, T_e, $\mathcal{M}_e$, W

Find F_t

Sketch

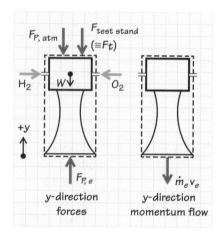

Assumptions

i. Steady state
ii. Uniform velocity at exit plane
iii. No y-direction forces or momentum flows associated with incoming H_2 and O_2.

Analysis   We select a cylindrical control volume that surrounds the entire engine. The control surface cuts through the engine test stand mount to expose the desired thrust force. The surface also cuts through the H_2 and O_2 supply lines, which we assume are oriented in a radial direction as shown in the sketch. Atmospheric pressure acts over the entire control surface except at the exit plane where $P_e < P_{atm}$. All y-direction forces are shown in the sketch. Note that the area associated with the pressure at the top of the control volume is identical to that of the nozzle exit. Since the H_2 and O_2 enter radially, there are no y-directed forces or momentum flows associated with these streams. The only y-direction momentum flow is at the exit plane directed downward.

Using the sketch as a guide, we apply the conservation of momentum principle (Eq. 6.53) as follows:

$$0 - [\dot{m}_e(-v_e)] + F_{P,e} - F_{P,atm} - F_t - W = 0.$$

Note the minus sign attached to v_e indicating that v_e is directed in the negative y-direction. From mass conservation, the exit mass flow rate is

$$\dot{m}_e = \dot{m}_{H_2} + \dot{m}_{O_2}$$
$$= 72.7 + 440.9 \text{ kg/s} = 513.6 \text{ kg/s}.$$

With the assumption of a uniform exit velocity,

$$\dot{m}_e = \rho_e v_e A_e.$$

Combining this with the ideal-gas equation of state to find ρ_e, we obtain the exit velocity:

$$v_e = \frac{\dot{m}_e}{\left(\dfrac{P_e \mathcal{M}_e}{R_u T_e}\right)\dfrac{\pi D_e^2}{4}}$$

$$= \frac{513.6}{\left(\dfrac{0.1915(101,325)13.78}{8314.5(1166)}\right)\dfrac{\pi(2.3)^2}{4}}$$

$$= \frac{513.6 \text{ kg/s}}{0.0276 \text{ kg/m}^3 \, 4.15 \text{ m}^2}$$

$$= 4480 \text{ m/s}.$$

The reader should verify the units in the calculation of ρ_e. The exit momentum flow is thus

$$\dot{m}_e v_e = 513.6 \,(4480) = 2,300,000$$

$$[=]\frac{\text{kg}}{\text{s}}\frac{\text{m}}{\text{s}}\left[\frac{1 \text{ N}}{\text{kg} \cdot \text{m/s}^2}\right] = \text{N}.$$

The pressure forces are evaluated as

$$F_{P,e} - F_{P,\text{atm}} = (P_e - P_{\text{atm}})\pi D_e^2/4$$

$$= (0.1915 - 1)101,325 \, \pi\,(2.3)^2/4$$

$$= -81,921(4.15) = -340,000$$

$$[=]\frac{\text{N}}{\text{m}^2}\text{m}^2 = \text{N}.$$

Solving our momentum conservation expression for the unknown test-stand force (i.e., the thrust) and evaluating yield

$$F_t = \dot{m}_e v_e + (F_{P,e} - F_{P,\text{atm}}) - W$$

$$= 2,300,000 - 340,000 - 31,140 \text{ N}$$

$$= 1,930,000 \text{ N}.$$

Comments First, we note that this force is very large ($\sim$434,000 lb$_f$). In comparison, automobiles weigh a few thousand pounds. Other big numbers are associated with this engine; for example, the turbopump that supplies the H_2 to the combustion chamber is rated at 75,000 hp. Clearly, huge rates of energy transfer are commonplace in the Space Shuttle propulsion system. Second, we note the importance of the choice of control volume and the simplifying assumptions invoked. Choosing a good control volume and sketching all of the forces and momentum flows are key to solving problems such as this.

Self Test
6.7

✓ **Using the momentum flows determined in Example 6.9, calculate the force required to hold the long tube illustrated in Fig. 6.16c. Neglect gravitational effects and ignore any pressure drop between the inlet and the exit.**

(Answer: $F_x = 4.7 \times 10^{-5}$ N, $F_y = 0$ N)

In many applications, a fluid stream exits a control volume as a **free jet,** unconstrained by any solid boundaries. Examples of free jets are a flow from a pipe into the atmosphere and the jet of air one creates to blow out a candle. For jets of incompressible fluids, or jets with low Mach numbers ($Ma < 0.3$, say), the pressure at the exit plane where the jet enters the ambient fluid is essentially uniform and equal to the pressure in the quiescent ambient fluid (i.e., $P_{exit} \approx P_{amb}$).

In the following example, we take advantage of this **jet-exit boundary condition** to evaluate the pressure force where a jet exits a control volume. The previous example involved a high-speed compressible flow. In that case, the jet exit plane and the atmospheric pressures were not equal. Chapter 11 provides further insight into compressible flows.

Example 6.11

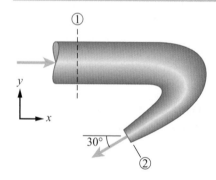

Water ($\rho = 997$ kg/m³) exits into the atmosphere (100 kPa) from a backward-curving nozzle as shown in the sketch. At station 1, the water velocity is 2 m/s, the absolute pressure is 257 kPa, and the inside diameter of the pipe is 75 mm. The nozzle exit diameter is 25 mm. The weight of the pipe–nozzle assembly and the water it contains between stations 1 and 2 is 54 N. Determine the equivalent reaction forces, $F_{w,x}$ and $F_{w,y}$, in the pipe wall at station 1.

Solution

Known ρ_w, P_{atm}, P_1, V_1, D_1, D_2, W_{tot}

Find $F_{w,x}$, $F_{w,y}$

Sketch

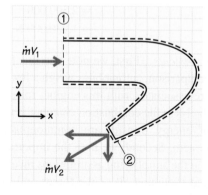

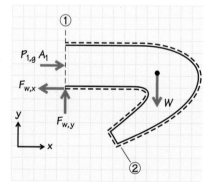

Assumptions

  i. Steady state
 ii. Uniform entrance and exit velocities
iii. $P_2 = P_{atm}$
 iv. $F_{w,x}$ acts to the left, $F_{w,y}$ acts upward

Analysis   We choose a control volume that cuts through the pipe wall and the entering water at station 1, is then contiguous with the outer surface of the pipe and nozzle, and finally cuts through the exiting jet. This choice results in the only net pressure force being the product of the gage pressure and cross-sectional area at station 1 because atmospheric pressure acts uniformly over the entire control volume. The forces acting on the control volume are shown in the sketch. The directions of $F_{w,x}$ and $F_{w,y}$ are chosen arbitrarily. If our analysis shows either to be negative, they must then act opposite to the direction chosen. The momentum flows are illustrated in the top sketch. Note that $\dot{m}V_2$ can be resolved into x- and y-components as shown. With all forces and all momentum flows identified, we can apply linear momentum conservation for steady flow with uniform entrance and exit velocities expressed by Eq. 6.53:

$$\dot{m}V_1 - \dot{m}V_2 + \sum F_{cv} = 0.$$

Resolving this into x- and y-components yields

$$\dot{m}v_{x,1} - \dot{m}v_{x,2} - F_{w,x} + P_{1g}A_1 = 0 \quad (x\text{-direction})$$

and

$$\dot{m}v_{y,1} - \dot{m}v_{y,2} + F_{w,y} - W_{tot} = 0 \quad (y\text{-direction}).$$

We note from the sketch that both $v_{x,2}$ and $v_{y,2}$ are negative quantities.

To solve these equations requires values for the velocity components. Using mass conservation, we first find the magnitude of V_2 as follows:

$$\dot{m} = \rho_w |V_1| A_1 = \rho_w |V_2| A_2.$$

Thus,

$$V_2 = V_1 \frac{A_1}{A_2} = V_1 \frac{\pi D_1^2/4}{\pi D_2^2/4} = V_1 \frac{D_1^2}{D_2^2}$$

$$= 2\left(\frac{0.075}{0.025}\right)^2 \text{m/s} = 18 \text{ m/s}.$$

From the geometry, the x- and y-velocity components are

$$v_{x,1} = 2 \text{ m/s},$$
$$v_{y,1} = 0,$$
$$v_{x,2} = -V_2\cos 30° = -18(0.866) \text{ m/s} = -15.588 \text{ m/s},$$
$$v_{y,2} = -V_2\sin 30° = -18(0.500) \text{ m/s} = 9.000 \text{ m/s}.$$

The mass flow rate is

$$\dot{m} = \rho_w V_1 \frac{\pi D_1^2}{4} = 997(2)\frac{\pi(0.075)^2}{4} = 8.809$$

$$[=]\frac{\text{kg}}{\text{m}^3}\frac{\text{m}}{\text{s}}\text{m}^2 = \text{kg/s}.$$

Rearranging the x-component expression of momentum conservation and substituting numerical values yield

$$F_{w,x} = \dot{m}(v_{x,1} - v_{x,2}) + (P_1 - P_{atm})A_1$$

$$= 8.809[2 - (-15.588)] + (257 \times 10^3 - 100 \times 10^3)\frac{\pi(0.075)^2}{4}$$

$$= 154.9 + 693.6 = 848.5$$

$$[=]\frac{\text{kg}}{\text{s}}\frac{\text{m}}{\text{s}}\left[\frac{1 \text{ N}}{\text{kg}\cdot\text{m/s}^2}\right] = \text{N}.$$

and

$$F_{w,y} = W - \dot{m}(v_{y,1} - v_{y,2})$$
$$= 54.0 - 8.809[0 - (-9.000)]\ N$$
$$= -25.3\ N.$$

The minus sign here indicates that $F_{w,y}$ must act *downward* rather than upward as originally assumed.

Comments   We note how the choice of control volumes made dealing with the pressure forces easy, as the atmospheric pressure components canceled. Note also the importance of keeping track of the vector nature of momentum conservation and the use of positive and negative signs with the scalar velocity components.

The momentum within a control volume defined by the exterior surfaces of the Space Shuttle changes as a result of two factors: the mass decreases as the propellants are consumed and expelled, and the velocity increases as the system accelerates.

6.5c Integral Control Volumes with Unsteady Flow

If the linear momentum associated with the control volume as a whole varies with time, an unsteady term must be included in our statement of momentum conservation. Here we are concerned with the control-volume momentum itself, distinct from any momentum flow across the control surface. The control-volume momentum is most generally expressed as

$$\begin{array}{c}\text{control-volume}\\\text{momentum}\end{array} \equiv \int_{CV} \mathbf{V} dM = \int_{CV} \mathbf{V} \rho d\mathcal{V}. \qquad (6.58)$$

The integrand of Eq. 6.58 is $\mathbf{V}dM$ ($= \mathbf{V}\rho d\mathcal{V}$), the momentum of a fluid element with mass dM. Allowing the velocity and/or the density of the fluid to vary from point to point within the control volume requires integration over the control volume to evaluate the total control-volume momentum. If both the velocity and density are the same at every point within the control volume, but still allowed to vary with time, the control-volume momentum is simply $M_{cv}\mathbf{V}_{cv}$.

With this understanding of the control-volume momentum, we state the general conservation of momentum principle for a nonaccelerating control volume:

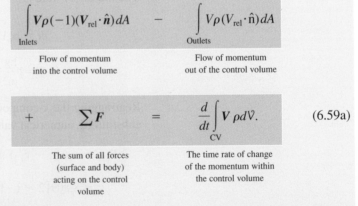

$$\underbrace{\int_{\text{Inlets}} \mathbf{V}\rho(-1)(\mathbf{V}_{\text{rel}} \cdot \hat{\mathbf{n}})\,dA}_{\substack{\text{Flow of momentum}\\\text{into the control volume}}} \quad - \quad \underbrace{\int_{\text{Outlets}} \mathbf{V}\rho(\mathbf{V}_{\text{rel}} \cdot \hat{\mathbf{n}})\,dA}_{\substack{\text{Flow of momentum}\\\text{out of the control volume}}}$$

$$+ \quad \underbrace{\sum \mathbf{F}}_{\substack{\text{The sum of all forces}\\\text{(surface and body)}\\\text{acting on the control}\\\text{volume}}} \quad = \quad \underbrace{\frac{d}{dt}\int_{CV} \mathbf{V}\,\rho d\mathcal{V}.}_{\substack{\text{The time rate of change}\\\text{of the momentum within}\\\text{the control volume}}} \qquad (6.59a)$$

This particular presentation of momentum conservation preserves the physical interpretation that the momentum within the control volume can be changed in three ways: 1. by having a force act on the control volume, 2. by an inflow of momentum, and 3. by an outflow of momentum. Note that the negative sign within the inlet momentum flow integral is required to make this a positive quantity since the entering fluid velocity V_{rel} is directed opposite to the surface normal $\hat{n}$ (i.e., $V_{rel} \cdot \hat{n} < 0$). To provide a statement of conservation of momentum in a form consistent with our steady-flow analysis (Eq. 6.57), we combine the entering and exiting momentum flows in a single term and rearrange Eq. 6.59a to yield

$$-\int_{CS} V\rho(V_{rel} \cdot \hat{n})\,dA \;+\; \sum F \;=\; \frac{d}{dt}\int_{CV} V\rho\,dV. \qquad (6.59b)$$

| Net flow of momentum into the control volume | Sum of all forces (surface and body) acting on the control volume | Time rate of change of the momentum within the control volume |

The force term and the momentum flow term in Eqs. 6.59a and 6.59b are handled in the same manner as discussed in the previous section, the only difference being that these terms are now instantaneous expressions that can vary with time.

Recall that our development of Eq. 6.59 assumes a nonaccelerating coordinate system. Appendix 6A shows how to deal with noninertial coordinate systems.

The following example illustrates the application of Eq. 6.59.

Example 6.12

A water truck empties its load of water ($\rho_{H_2O} = 996$ kg/m³) through a 50-mm-diameter pipe at the rear as shown in the sketch. The truck is traveling forward at 10 miles/hr (4.47 m/s), and the velocity of the water jet with respect to the moving truck is 5 m/s rearward. Air enters the tank through a vent at the top of the truck, replacing the water that exits from the tank. The total drag force F_D opposing the motion of the truck is 95.9 N. Estimate the traction force required to keep the truck moving at a constant speed.

Solution

Known ρ_{H_2O}, D, $v_{x,truck}$, $V_{x,rel}$, F_D

Find F_{tr}

Sketch

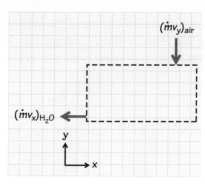
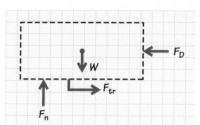

Assumptions

  i. The water-jet velocity is uniform.

  ii. F_D accounts for all horizontal forces acting on control volume other than F_{tr}.

  iii. $\rho_{air} \ll \rho_{H_2O}$ so that $\dot{m}_{air} \ll \dot{m}_{H_2O}$.

  iv. Momentum flows associated with the engine air and exhaust products are negligible as is the change in truck mass from fuel burning.

Analysis We choose a control volume that cuts between the tires and the road to expose the unknown traction force F_{tr}. With assumption ii, the only forces acting in the horizontal direction are F_{tr} and F_D. Two momentum flows are shown in the sketch: one associated with the incoming air (vertical) and one associated with the outgoing water (horizontal). We note that the momentum of the control volume (truck plus tank and its contents) is changing, so the problem is unsteady; thus, linear momentum conservation is expressed by Eq. 6.59b. The x-component of this equation can be written as

$$0 - (\dot{m}_{H_2O}\, v_{x,H_2O})_{out} + F_{tr} - F_D = \frac{d(M_{cv} v_{x,cv})}{dt},$$

where the zero indicates that there is no flow of x-momentum into the control volume. The velocities here are all with respect to a fixed observer, so that

$$v_{x,cv} = v_{x,truck} = 4.47 \text{ m/s}$$

and

$$v_{x,H_2O,out} = v_{x,H_2O,rel} + v_{x,truck}$$
$$= -5 + 4.47 \text{ m/s} = -0.53 \text{ m/s}.$$

The flow rate $\dot{m}_{H_2O}$, however, is based on the velocity at which the water crosses the control surface, $v_{x,H_2O,rel}$; thus,

$$\dot{m}_{H_2O} = \rho_{H_2O} v_{x,H_2O,rel} \pi D^2/4$$
$$= 996(5)\pi(0.05)^2/4 = 9.78$$
$$[=]\frac{\text{kg}}{\text{m}^3}\frac{\text{m}}{\text{s}}\text{m}^2 = \text{kg/s}.$$

We focus now on the time rate of change of the momentum within the control volume. Since $v_{x,cv}$ is constant,

$$\frac{d(M_{cv} v_{x,cv})}{dt} = v_{x,cv}\frac{dM_{cv}}{dt}.$$

The time rate of change of the mass with the control volume can be related to the water outflow through Eq. 3.19a, an unsteady expression of the conservation of mass principle. For our control volume,

$$\dot{m}_{air,in} - \dot{m}_{H_2O,out} = \frac{dM_{cv}}{dt}.$$

Neglecting the air mass flow rate, we have

$$\frac{dM_{cv}}{dt} = -\dot{m}_{H_2O,out}.$$

Reassembling the momentum conservation expression yields

$$-\dot{m}_{H_2O,out}v_{x,H_2O,out} + F_{tr} - F_D = v_{x,cv}\frac{dM_{cv}}{dt} = v_{x,cv}(-\dot{m}_{H_2O,out}).$$

Solving for F_{tr} and substituting numerical values, we obtain

$$\begin{aligned}
F_{tr} &= F_D + \dot{m}_{H_2O,out}v_{x,H_2O,out} + v_{x,cv}(-\dot{m}_{H_2O,out}) \\
&= F_D + \dot{m}_{H_2O,out}(v_{x,H_2O,out} - v_{x,cv}) \\
&= 95.9 + 9.78(-0.53 - 4.47) \\
&= 47.0 \\
&[=]\frac{kg}{s}\frac{m}{s}\left[\frac{1\,N}{kg \cdot m/s^2}\right] = N.
\end{aligned}$$

Comments In this example, we see the importance of interpreting velocities in the reference frames of a fixed observer and for an observer traveling with the control volume. We also see that keeping track of signs ($\pm$) in both mass and momentum conservation is essential to obtaining the proper solution. To further reinforce the concepts presented in this example, the reader is encouraged to solve this problem using the inertial reference frame of the moving control volume. (See Problem 6.87.)

6.5d Differential Control Volumes

See Chapters 3 and 5.

In our previous applications of mass and energy conservation to differential control volumes, we began with a simple one-dimensional steady flow and then added complexity. Since you should now be quite familiar with the general process used to derive such differential equations, we start with the three-dimensional (Cartesian), unsteady case rather than building up to it. We begin by applying the general conservation of momentum relation (Eq. 6.59) to the small control volume shown in Fig. 6.19. The pressure forces acting on this control volume are as shown previously in Fig. 6.7; the viscous forces are obtained by multiplying the viscous stresses shown in Fig. 6.4a by the area upon which they act. For simplicity, let us consider only the forces that act in the x-direction, all of which are shown in Fig. 6.20 except for the x-component of the body force. Summing these x-direction forces yields the following:

$$\sum F_{x\text{-dir}} = [P_x - P_{x+\Delta x}]\Delta y\Delta z + [(\tau_{xx})_{x+\Delta x} - (\tau_{xx})_x]\Delta y\Delta z$$

Net pressure force in   Net viscous normal force in
the x-direction   x-direction

$$+ [(\tau_{yx})_{y+\Delta y} - (\tau_{yx})_y]\Delta x\Delta z + [(\tau_{zx})_{z+\Delta z} - (\tau_{zx})_z]\Delta x\Delta y \quad (6.60)$$

Net viscous shear forces in the x-direction

$$+ \quad \rho g_x \Delta x\Delta y\Delta z.$$

x-component of
gravitational body force

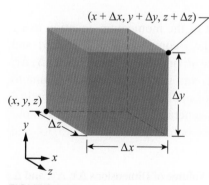

FIGURE 6.19
We apply momentum conservation to a stationary control volume of dimensions Δx, Δy, and Δz. Viscous and pressure forces act on all six faces, and momentum flows through all six faces. Gravity is the single body force acting on the control volume.

Similar expressions can be written for the y- and z-directions, a task presented as a homework exercise.

FIGURE 6.20
Pressure and viscous forces acting in the x-direction on a fluid element of dimensions Δx, Δy, and Δz.

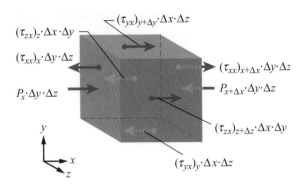

We now investigate the momentum flows associated with our small control volume. At each of the six faces, the momentum flow is generically expressed as $(\dot{m}V)_{i\,\text{face}}$, where $V = \hat{i}v_x + \hat{j}v_y + \hat{k}v_z$. To illustrate the momentum flow through a specific face, consider the x face located at x, for example. The flow rate through this face is $(\rho v_x)_x \Delta y \Delta z$ and the momentum flow is $(V\rho v_x)_x \Delta y \Delta z$. Table 6.2 presents the specific expressions for the momentum flows through all six faces. Faces designated as inlets lie in planes at x, y, and z, whereas faces designated as outlets lie in planes at $x + \Delta x$, $y + \Delta y$, and $z + \Delta z$. Using Table 6.2 as a guide, the net momentum flow in (inflow minus outflow) is given by

$$\begin{aligned}
\begin{matrix}\textbf{net momentum}\\ \textbf{flow in}\end{matrix} = &-[(V\rho v_x)_{x+\Delta x} - (V\rho v_x)_x]\Delta y \Delta z \\
&- [(V\rho v_y)_{y+\Delta y} - (V\rho v_y)_y]\Delta x \Delta z - [(V\rho v_z)_{z+\Delta z} - (V\rho v_z)_z]\Delta x \Delta y.
\end{aligned} \tag{6.61}$$

The final contribution to our conservation of momentum expression (see Eq. 6.59b) is the time rate of change of the momentum within the control volume, which is given simply by

$$\begin{matrix}\textbf{rate of change of}\\ \textbf{momentum within the}\\ \textbf{control volume}\end{matrix} = \frac{\partial}{\partial t}(V\rho)\Delta x \Delta y \Delta z. \tag{6.62}$$

We now assemble the x-component of the momentum equation using Eq. 6.60 for the x forces and taking the x-components of Eqs. 6.61 and 6.62. Before doing this, however, we divide every term by $\Delta x \Delta y \Delta z$, cancel terms involving Δx, Δy and Δz, and shrink the control volume to the continuum limit. This shrinking, or application of the mathematical limits $\Delta x \to 0$, $\Delta y \to 0$, and $\Delta z \to 0$, generates partial derivatives in our results, as we have seen several times before (cf. Eq. 3.33, 5.30, and

Table 6.2 Momentum Flows for Control Volume of Dimensions Δx, Δy, and Δz

Face	Area	Inlet Momentum Flow	Outlet Momentum Flow
x	$\Delta y \Delta z$	$(V\rho v_x)_x \Delta y \Delta z$	$(V\rho v_x)_{x+\Delta x}\Delta y \Delta z$
y	$\Delta x \Delta z$	$(V\rho v_y)_y \Delta x \Delta z$	$(V\rho v_y)_{y+\Delta y}\Delta x \Delta z$
z	$\Delta x \Delta y$	$(V\rho v_z)_z \Delta x \Delta y$	$(V\rho v_z)_{z+\Delta z}\Delta x \Delta y$

5.83). Performing these operations yields for the x-component (per unit volume)

$$-\frac{\partial P}{\partial x} + \frac{\partial \tau_{xx}}{\partial x} + \frac{\partial \tau_{yx}}{\partial x} + \frac{\partial \tau_{zx}}{\partial x} + \rho g_x$$

Sum of all forces in x-direction per unit volume

$$= \quad \frac{\partial}{\partial t}(\rho v_x) \quad + \quad \frac{\partial}{\partial x}(\rho v_x^2) + \frac{\partial}{\partial y}(\rho v_x v_y) + \frac{\partial}{\partial z}(\rho v_x v_z). \qquad \text{(6.63a)}$$

Rate of change of x-direction momentum within the control volume per unit volume

Net x-direction momentum flow out of control volume per unit volume

Note that we have retained the forces on the left-hand side of the equation but have moved the momentum flows to the right-hand side, as is conventionally done in differential analyses. We also designate the momentum flows as a net outflow with a sign change. Similar expressions for the y- and z-components are easily derived following the same procedures. For the y-component (per unit volume), we get

$$-\frac{\partial P}{\partial y} + \frac{\partial \tau_{xy}}{\partial y} + \frac{\partial \tau_{yy}}{\partial y} + \frac{\partial \tau_{zy}}{\partial y} + \rho g_y$$

Sum of all forces acting in y-direction per unit volume

$$= \quad \frac{\partial}{\partial t}(\rho v_y) \quad + \quad \frac{\partial}{\partial x}(\rho v_y v_x) + \frac{\partial}{\partial y}(\rho v_y^2) + \frac{\partial}{\partial z}(\rho v_y v_z), \qquad \text{(6.63b)}$$

Rate of change of y-direction momentum within the control volume per unit volume

Net y-direction momentum flow out of control volume per unit volume

and for the z-component (per unit volume), we get

$$-\frac{\partial P}{\partial z} + \frac{\partial \tau_{xz}}{\partial z} + \frac{\partial \tau_{yz}}{\partial z} + \frac{\partial \tau_{zz}}{\partial z} + \rho g_z$$

Sum of all forces acting in z-direction per unit volume

$$= \quad \frac{\partial}{\partial t}(\rho v_z) \quad + \quad \frac{\partial}{\partial x}(\rho v_z v_x) + \frac{\partial}{\partial y}(\rho v_z v_y) + \frac{\partial}{\partial z}(\rho v_z^2). \qquad \text{(6.63c)}$$

Rate of change of z-direction momentum within the control volume per unit time

Net z-direction momentum flow out of control volume per unit volume

Note that Eqs. 6.63a–6.63c, by and large, preserve the physics of momentum conservation as expressed in Eqs. 6.59a or 6.59b. The physical meaning of each term is retained, only expressed on a per unit volume basis. These particular differential expressions of momentum conservation are frequently referred to as the **conservative forms.**

Total (or Material) Derivative

We now mathematically manipulate the right-hand sides of Eqs. 6.63a–6.63c to provide a frequently employed (nonconservative) form of momentum conservation. Combining the three component equations yields the following vector equation:

$$\sum \hat{f} = \frac{\partial}{\partial t}(V\rho) + \frac{\partial}{\partial x}(V\rho v_x) + \frac{\partial}{\partial y}(V\rho v_y) + \frac{\partial}{\partial z}(V\rho v_z), \qquad \text{(6.64)}$$

① ② ③ ④

where $\hat{f}$ represents the various forces per unit volume. We apply the product rule for differentiation, that is,

$$\frac{\partial(ab)}{\partial c} = a\frac{\partial b}{\partial c} + b\frac{\partial a}{\partial c},$$

to each numbered term as follows:

① $\quad V\dfrac{\partial \rho}{\partial t} \qquad\qquad + \qquad \rho\dfrac{\partial V}{\partial t},$

② $\quad V\dfrac{\partial}{\partial x}(\rho v_x) \qquad + \qquad \rho v_x\dfrac{\partial V}{\partial x},$

③ $\quad V\dfrac{\partial}{\partial y}(\rho v_y) \qquad + \qquad \rho v_y\dfrac{\partial V}{\partial y},$

④ $\quad V\dfrac{\partial}{\partial z}(\rho v_z) \qquad + \qquad \rho v_z\dfrac{\partial V}{\partial z}.$

The boxed terms can be combined as

$$V\left[\frac{\partial \rho}{\partial t} + \frac{\partial}{\partial x}(\rho v_x) + \frac{\partial}{\partial y}(\rho v_y) + \frac{\partial}{\partial z}(\rho v_z)\right].$$

We now note that the bracketed expression is the left-hand side of the general continuity equation (Eq. 3.39):

$$\frac{\partial \rho}{\partial t} + \frac{\partial}{\partial x}(\rho v_x) + \frac{\partial}{\partial y}(\rho v_y) + \frac{\partial}{\partial z}(\rho v_z) = 0.$$

Thus, the sum of the boxed terms is identically zero, and Eq. 6.64 becomes

$$\sum \hat{f} = \rho\left[\frac{\partial V}{\partial t} + v_x\frac{\partial V}{\partial x} + v_y\frac{\partial V}{\partial y} + v_z\frac{\partial V}{\partial z}\right]. \tag{6.65}$$

A special operator, $D(\ )/Dt$, is defined as

$$\frac{D(\ )}{Dt} \equiv \frac{\partial(\ )}{\partial t} + v_x\frac{\partial(\ )}{\partial x} + v_y\frac{\partial(\ )}{\partial y} + v_z\frac{\partial(\ )}{\partial z} \tag{6.66}$$

and is variously called the **total derivative,** the **material derivative,** or the **substantial derivative.** Using this total derivative, a new interpretation can be given to Eq. 6.65:

$$\sum \hat{f} = \rho\frac{DV}{Dt}. \tag{6.67}$$

This relationship is equivalent to the system expression of momentum conservation (Eq. 6.15) for a fixed-mass fluid element ($dM = \rho\,dx\,dy\,dz$) tracked through the flow. The acceleration associated with the fluid element is $a \equiv DV/Dt$. This identification of a fluid element is why $D(\ )/Dt$ is referred to as the material or substantial derivative.

Convective Acceleration

We digress to examine this DV/Dt term further. Thinking of DV/Dt as a total acceleration, we break it into two components, a local acceleration and a

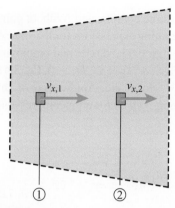

FIGURE 6.21
In moving from location 1 to location 2, the fluid element experiences a convective deceleration, $v_x dv_x/dx < 0.$

> Review Examples 3.11 and 3.13 to add perspective to the present discussion.

convective acceleration, as follows:

$$
\underset{\substack{\text{Total}\\\text{acceleration}}}{a \equiv \frac{DV}{Dt}} = \underset{\substack{\text{Local}\\\text{acceleration}}}{\frac{\partial V}{\partial t}} + \underset{\text{Convective acceleration}}{v_x \frac{\partial V}{\partial x} + v_y \frac{\partial V}{\partial y} + v_z \frac{\partial V}{\partial z}}
\tag{6.68}
$$

The local acceleration results from a time dependence. If the velocity at a point (x_0, y_0, z_0) in a flow increases with time, then the local acceleration is positive; if the velocity decreases with time, it is negative. If the flow is steady, $\partial V/\partial t$ is zero, and there is no local acceleration. The convective acceleration, however, does not depend on time; rather, it is determined by the nature of the flow field.

To illustrate this, we return to Examples 3.11 and 3.13 from Chapter 3. Figure 6.21 shows the control volume associated with the 1-D flow through the diffuser from these examples. Also shown is a fixed-mass fluid element at location 1, at an initial time, and the same fluid element at location 2, some time later. Because the cross-sectional area of the diffuser increases in the flow direction, conservation of mass requires that the velocity decrease in the flow direction for an incompressible flow. Thus, the fluid element in moving from location 1 to location 2 experiences a deceleration, that is, the convective acceleration (or deceleration in this case), $v_x \, \partial v_x/\partial x$. Conversely, if the flow area contracts downstream, the convective acceleration is positive—the fluid element speeds up as it moves downstream.

The Navier–Stokes Equation

We now return to our momentum conservation component equations (Eqs. 6.63a–6.63c) and focus on the three forces (pressure, viscous, and gravity) that act on the differential fluid element. Our objectives here are to express the force terms compactly using vector notation and to incorporate the relationship between the viscous stresses and the velocity gradients (Eq. 6.8) into our final result.

We easily obtain a vector expression for the pressure force *per unit volume*. By inspection of the x-, y-, and z-components of the pressure force (Eqs. 6.63a–6.63c), we see that

$$
\hat{f}_P = -\left(\hat{i} \frac{\partial P}{\partial x} + \hat{j} \frac{\partial P}{\partial y} + \hat{k} \frac{\partial P}{\partial z} \right).
\tag{6.69a}
$$

We can express Eq. 6.69a more compactly using the gradient (or del) vector operator[8] as follows:

$$
\hat{f}_P = -\nabla P.
\tag{6.69b}
$$

We also obtain by inspection of Eqs. 6.63a–6.63c the vector form of the gravitational body force *per unit volume*:

$$
\hat{f}_{\text{grav}} = \rho(\hat{i}g_x + \hat{j}g_y + \hat{k}g_z) = \rho g.
\tag{6.70}
$$

---

[8] The gradient (or del) operator is defined in Table 3.4 for various coordinate systems.

Dealing with the viscous forces, however, is more complicated. Without going into the details, the procedure involves substituting expressions for the stresses from Eq. 6.8 into the component equations (Eq. 6.63a–6.63c) and employing the incompressible form of continuity ($\partial v_x/\partial x + \partial v_y/\partial y + \partial v_z/\partial z = 0$, Eq. 3.41) to eliminate cross-derivative terms like $\partial^2(\ )/\partial x\partial y$, etc. Application of this procedure yields the net viscous force *per unit volume* for an incompressible, Newtonian fluid as follows:

$$\hat{f}_{\text{visc}} = \hat{i}\mu\left[\frac{\partial^2 v_x}{\partial x^2} + \frac{\partial^2 v_x}{\partial y^2} + \frac{\partial^2 v_x}{\partial z^2}\right] + \hat{j}\mu\left[\frac{\partial^2 v_y}{\partial x^2} + \frac{\partial^2 v_y}{\partial y^2} + \frac{\partial^2 v_z}{\partial z^2}\right]$$
$$+ \hat{k}\mu\left[\frac{\partial^2 v_z}{\partial x^2} + \frac{\partial^2 v_z}{\partial y^2} + \frac{\partial^2 v_z}{\partial z^2}\right]. \tag{6.71a}$$

We can also express Eq. 6.71a more compactly by employing the **Laplacian** (or **del-squared**) **operator.** For a Cartesian coordinate system, this scalar operator is defined as

$$\nabla^2(\ ) \equiv \frac{\partial^2(\ )}{\partial x^2} + \frac{\partial^2(\ )}{\partial y^2} + \frac{\partial^2(\ )}{\partial z^2}.$$

Thus, Eq. 6.71a becomes

$$\hat{f}_{\text{visc}} = \mu\nabla^2(\boldsymbol{V}). \tag{6.71b}$$

Note that Eq. 6.71b expresses a vector relationship with $\nabla^2(\ )$, a scalar operator, operating on the velocity vector $\boldsymbol{V}$.

We reassemble the momentum conservation equation using the vector expressions for the various forces (Eqs. 6.69, 6.70, and 6.71) and a single term combining the time rate of change of control volume momentum and the momentum flows across the control surface (Eq. 6.67 or 6.68):

$$-\nabla P + \mu\nabla^2\boldsymbol{V} + \rho g = \rho\frac{D\boldsymbol{V}}{Dt}, \tag{6.72a}$$

or

$$-\nabla P + \mu\nabla^2\boldsymbol{V} + \rho g = \rho\left(\frac{\partial\boldsymbol{V}}{\partial t} + v_x\frac{\partial\boldsymbol{V}}{\partial x} + v_y\frac{\partial\boldsymbol{V}}{\partial y} + v_z\frac{\partial\boldsymbol{V}}{\partial z}\right). \tag{6.72b}$$

Equation 6.72 is the celebrated **Navier–Stokes equation.** At this point, you should have a good grasp of the physical interpretation of each term in this equation and the ability to break the equation down into its *x*-, *y*-, and *z*-components.

Tables 6.3 and 6.4 present the components of the Navier–Stokes equation for Cartesian and cylindrical coordinate systems. The cylindrical system equations are particularly important as their solutions provide the details for flows through circular tubes and pipes. We employ these equations later in this chapter and at other locations throughout the book.

At this juncture, we digress to view the developments in this chapter in a larger context. With the development of the Navier–Stokes equation (Eq. 6.72), we add the application of the momentum conservation principle to

Chapter 10 investigates pipe flows in detail

Table 6.3 **Conservation of Momentum for Differential Control Volumes: Component Equations in Cartesian (x, y, z) Coordinates for Incompressible Newtonian Fluids with Constant Viscosity**

x-component
$$-\frac{\partial P}{\partial x} + \mu\left(\frac{\partial^2 v_x}{\partial x^2} + \frac{\partial^2 v_x}{\partial y^2} + \frac{\partial^2 v_x}{\partial z^2}\right) + \rho g_x = \rho\left(\frac{\partial v_x}{\partial t} + v_x\frac{\partial v_x}{\partial x} + v_y\frac{\partial v_x}{\partial y} + v_z\frac{\partial v_x}{\partial z}\right) \quad \text{(T6.3a)}$$

y-component
$$-\frac{\partial P}{\partial y} + \mu\left(\frac{\partial^2 v_y}{\partial x^2} + \frac{\partial^2 v_y}{\partial y^2} + \frac{\partial^2 v_y}{\partial z^2}\right) + \rho g_y = \rho\left(\frac{\partial v_y}{\partial t} + v_x\frac{\partial v_y}{\partial x} + v_y\frac{\partial v_y}{\partial y} + v_z\frac{\partial v_y}{\partial z}\right) \quad \text{(T6.3b)}$$

z-component
$$-\frac{\partial P}{\partial z} + \mu\left(\frac{\partial^2 v_z}{\partial x^2} + \frac{\partial^2 v_z}{\partial y^2} + \frac{\partial^2 v_z}{\partial z^2}\right) + \rho g_z = \rho\left(\frac{\partial v_z}{\partial t} + v_x\frac{\partial v_z}{\partial x} + v_y\frac{\partial v_z}{\partial y} + v_z\frac{\partial v_z}{\partial z}\right) \quad \text{(T6.3c)}$$

Table 6.4 **Conservation of Momentum for Differential Control Volumes: Component Equations in Cylindrical Coordinates (r, θ, x) for Incompressible Newtonian Fluids with Constant Viscosity**

r-component
$$-\frac{\partial P}{\partial r} + \mu\left[\frac{\partial}{\partial r}\left(\frac{1}{r}\frac{\partial}{\partial r}(rv_r)\right) + \frac{1}{r^2}\frac{\partial^2 v_r}{\partial \theta^2} - \frac{2}{r^2}\frac{\partial v_\theta}{\partial \theta} + \frac{\partial^2 v_r}{\partial x^2}\right] + \rho g_r = \rho\left(\frac{\partial v_r}{\partial t} + v_r\frac{\partial v_r}{\partial r} + \frac{v_\theta}{r}\frac{\partial v_r}{\partial \theta} - \frac{v_\theta^2}{r} + v_x\frac{\partial v_r}{\partial x}\right) \quad \text{(T6.4a)}$$

θ-component
$$-\frac{1}{r}\frac{\partial P}{\partial \theta} + \mu\left[\frac{\partial}{\partial r}\left(\frac{1}{r}\frac{\partial}{\partial r}(rv_\theta)\right) + \frac{1}{r^2}\frac{\partial^2 v_\theta}{\partial \theta^2} + \frac{2}{r^2}\frac{\partial v_r}{\partial \theta} + \frac{\partial^2 v_\theta}{\partial x^2}\right] + \rho g_\theta = \rho\left(\frac{\partial v_\theta}{\partial t} + v_r\frac{\partial v_\theta}{\partial r} + \frac{v_\theta}{r}\frac{\partial v_\theta}{\partial \theta} + \frac{v_r v_\theta}{r} + v_x\frac{\partial v_\theta}{\partial x}\right) \quad \text{(T6.4b)}$$

x-component
$$-\frac{\partial P}{\partial x} + \mu\left[\frac{1}{r}\frac{\partial}{\partial r}\left(r\frac{\partial v_x}{\partial r}\right) + \frac{1}{r^2}\frac{\partial^2 v_x}{\partial \theta^2} + \frac{\partial^2 v_x}{\partial x^2}\right] + \rho g_x = \rho\left(\frac{\partial v_x}{\partial t} + v_r\frac{\partial v_x}{\partial r} + \frac{v_\theta}{r}\frac{\partial v_x}{\partial \theta} + v_x\frac{\partial v_x}{\partial x}\right) \quad \text{(T6.4c)}$$

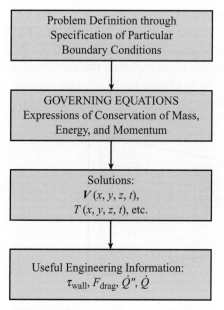

FIGURE 6.22

How the fundamental conservation principles are used to solve engineering problems.

our differential description of flows. This complements the applications of mass conservation and energy conservation from Chapters 3 and 5, respectively. Table 3.3 and Eq. 3.40 for mass conservation, Tables 5.7 and 5.8 for energy conservation, and now Tables 6.3 and 6.4 together provide a firm mathematical basis for describing the details of a flow. Figure 6.22 outlines how this is done. Any particular problem is defined by describing the boundary conditions for the flow for the specific geometry of interest; for example, the no-slip condition is a boundary condition applied at solid surfaces. With the boundary conditions specified, the governing equations are solved to provide the detailed velocity and temperature fields. Knowing the velocity field allows the determination of viscous stresses at surfaces, which, in turn, allows the calculation of drag forces or pressure drops. Similarly, knowledge of the temperature distribution allows the wall heat flux to be determined from Fourier's law (Eq. 4.14). Integration of the heat flux over a surface yields the overall heat-transfer rates. Being able to calculate drag forces, pressure drops, and heat-transfer rates is of great importance in engineering design. We must, however, add the caveat that exact solutions to these equations (Tables 3.3, 5.7, and 5.8) are limited to a few rather simple geometries for laminar flows. Later in this chapter, we touch on the problems introduced by turbulence; in Chapter 8, we show how empirical methods guided by theory are used to provide engineering solutions for complex flows. We also note that the governing equations (Tables 3.3, 5.7, and 5.8, etc.) form the basis for computational fluid dynamics, a far-ranging field in which numerical solutions are generated using powerful computers.

Tutorial 4 | **What Is CFD?**

Computational fluid dynamics (CFD) uses the power of computers to provide numerical solutions to the basic conservation equations (mass, energy, and momentum) that describe flows of engineering interest. CFD is a rapidly expanding subject area within the thermal-fluid sciences and has gained acceptance as a design tool in many fields. Many of the colorful illustrations presented in this book were created using CFD tools. This tutorial provides a very brief overview of this subject.

What is the basic approach of CFD?

Finding the *solution* to a thermal-fluids problem usually means finding the velocity, pressure, density, and temperature fields for a flow in a particular domain. Knowing these fields allows one to determine forces and heat fluxes, quantities of engineering interest. An example here is the flow of hot gases through a rocket engine nozzle. In such a case knowledge of the forces and heat fluxes at the walls of the nozzle are critically important to its design. The first steps in creating a numerical solution are 1. define a discrete domain (rather than working with a continuum) and 2. replace the continuum-based partial differential equations for mass, energy, and momentum conservation with a set of algebraic equations. Subsequent steps include 3. incorporating the known boundary conditions and 4. solving the equations generated in steps 2 and 3 using a computer.

- **Step 1: Creating a grid**
  The sketch shows a grid used to perform a *finite-difference* CFD solution for the flow through a 2-D diffuser. The bold dots at the intersection of the grid lines define nodal points, here designated 1, 2, 3, etc. The velocity components v_x and v_y and the temperature T at each node are designated v_{x1}, v_{x2}, v_{x3}, ...; v_{y1}, v_{y2}, v_{y3}, ...; and T_1, T_2, T_3, ..., respectively. Mathematically, we have replaced the continuous functions $v_x(x,y)$, $v_y(x,y)$, and $T(x,y)$ with a large set of discrete values. Other variables are treated similarly.

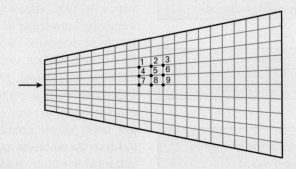

Typical CFD applications employ many thousands, or a few million, nodal points. Creating a grid that provides both an accurate and efficient solution is a nontrivial task. Special software is frequently employed to do this.

(continued on next page)

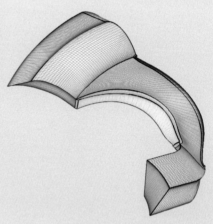

A very complex mesh is used to compute the flow through a centrifugal compressor. A surface mesh (above) overlays the solid components of the pump. The grid for the flow passage through the compressor is broken into many segments. A single segment is shown here (right).

- **Step 2: *Replacing the governing equations with nodal equations***
  Consider the *x*-component of the conservation of momentum equation (Eq. T6.3a) simplified for a steady, incompressible, 2-D flow:

$$-\frac{1}{\rho}\frac{\partial P}{\partial x} \quad + \quad \frac{\mu}{\rho}\left(\frac{\partial^2 v_x}{\partial x^2} \quad + \quad \frac{\partial^2 v_x}{\partial y^2}\right) \quad = \quad v_x\frac{\partial v_x}{\partial x} \quad + \quad v_y\frac{\partial v_x}{\partial y}$$

  Ⓐ   Ⓑ   Ⓒ   Ⓓ   Ⓔ

To illustrate how to discretize this equation, we identify a portion of a much larger grid:

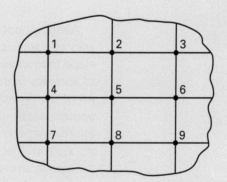

We now focus on term *D*, which we approximate at node 5 as follows:

$$\left(v_x\frac{\partial v_x}{\partial x}\right)_{node\,5} \approx v_{x,5}\frac{v_{x,6} - v_{x,5}}{\Delta x},$$

where Δx is the horizontal distance between nodes 5 and 6 (i.e., $x_6 - x_5$). This *finite-difference* approximation simply replaces the partial

Tutorial 4 **What Is CFD?** *(continued)*

derivative [slope of v_x (x,y) in the x plane] with a straight line as illustrated in the following sketch:

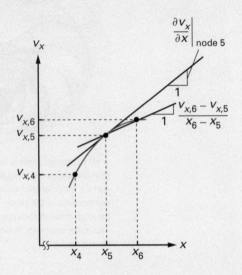

Alternatively, we can approximate this derivative by drawing a straight line between nodes 4 and 5:

$$\left(\frac{\partial v_x}{\partial x}\right)_{\text{node 5}} \approx \frac{v_{x,5} - v_{x,4}}{\Delta x}.$$

This approach is referred to as *backward-difference* approximation, whereas our former approximation is called a *forward difference*. A *central-difference* approximation draws a straight line between nodes 4 and 6:

$$\left(\frac{\partial v_x}{\partial x}\right)_{\text{node 5}} \approx \frac{v_{x,6} - v_{x,4}}{2\Delta x}.$$

Although not shown on the previous sketch, this approximation is easy to visualize; we see that it is a more accurate estimate of the true slope than either the forward or backward difference. Considerations of *accuracy* and *stability* dictate the choice of approximation methods. All of the terms (A–E) in the x-momentum equation can be similarly approximated using finite differences. The result of all of these approximations is an algebraic equation involving unknown values of v_x, v_y, and P at node 5 and the surrounding nodes. Repeating this process at every node in the domain creates a large set of algebraic equations. Applying the same finite-difference approximations to the y-component of momentum conservation, and any other conservation equations and auxiliary relationships required, creates additional sets of equations. Although the details differ in their formulation, *finite-volume* and *finite-element* methods are also used in CFD. Discussion of these techniques is beyond our scope.

To solve the equations created in step 2 requires the application of boundary conditions.

(continued on next page)

- ### Step 3: Incorporating boundary conditions
  At the boundaries of a domain, values for the various variables (e.g., v_x, v_y, T, etc.) may be known. For example, the no-slip condition requires that the velocity be zero at a solid surface. Similarly, the temperature may be a constant known value along a solid boundary. In such cases, these known values can be assigned to nodal points on the boundary of the domain. Other boundary conditions can be more complicated; for example, the heat flux at the boundary may be specified rather than the temperature. Again, finite-difference approximations can be employed to incorporate the boundary information into the problem formulation. For the specified heat flux, Fourier's law (Eq. 4.14) can be discetized. For a horizontal surface as shown in the sketch,

$$\dot{Q}''_{\text{specified}} \approx -k\frac{T_{24} - T_{27}}{\Delta y}.$$

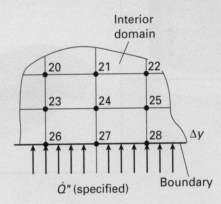

Identifying and approximating boundary conditions are frequently challenging aspects of CFD.

- ### Step 4: Solving the nodal equations
  For a domain containing N nodes, the preceding steps define a set of algebraic equations involving, for example, the unknowns

$$v_{x,1}, \ v_{x,2}, \ ..., \ v_{x,N}$$
$$v_{y,1}, \ v_{y,2}, \ ..., \ v_{y,N}$$
$$P_1, \ P_2, \ ..., \ P_N$$
$$T_1, \ T_2, \ ..., \ T_N$$

etc. The total number of equations required must equal the product of the number of variables considered and the number of nodes, so

$$M \text{ (number of equations)} = J \text{ (number of variables)} \cdot$$
$$N \text{ (number of nodes)}.$$

For a complicated problem, M can be several million! Various numerical methods are used to solve such large equation sets. CFD problems frequently push the capacity and speed limits of the biggest and fastest computers.

Tutorial 4 | **What Is CFD?** *(continued)*

How is CFD useful?

CFD can provide approximate solutions to problems involving complex geometries and unsteady flows. Gas-turbine and internal combustion engine manufacturers routinely use CFD methods to guide their designs. The images that follow illustrate the application of CFD to an internal combustion engine. CFD also critically impacts the design of aircraft. The scope for CFD is huge.

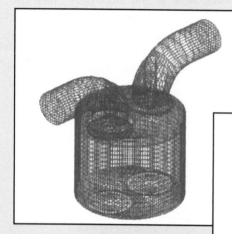

As the piston descends, fresh fuel–air mixture(red) enters the engine cylinder and mixes with the residual gases from the previous cycle (blue). A complex grid (left) is used for this CFD simulation of the intake stroke. Images courtesy of Daniel Haworth and Eugene Kung.

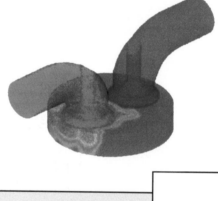

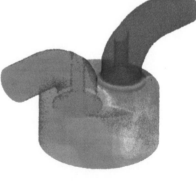

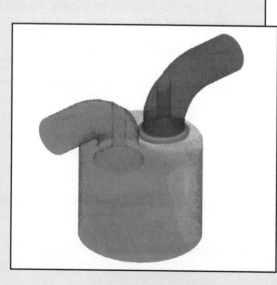

Even though CFD is widely used, engineers must exercise caution in accepting CFD results as accurate representations of reality. Numerical solutions do not always agree with experimental observations; validation of CFD simulations is thus an integral part of the application of these powerful tools.

The following examples illustrate how the differential momentum conservation equations can be simplified and used to provide detailed information about a flow.

Example 6.13

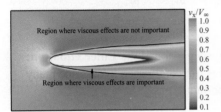

In certain flows, or regions of flows, the viscous forces can be quite small compared to the other fluid forces and momentum flows. Examples include flows outside the region close to solid surfaces and flows at high Reynolds numbers. For these flows, the **inviscid approximation** is employed in which the viscous terms cancel from the conservation of momentum (Navier–Stokes) equation. For a two-dimensional (x, y) flow, write out the component equations for momentum conservation using this approximation.

Solution[9]

We start by deleting the viscous force term in the general expression of momentum conservation (Eq. 6.72a):

$$-\nabla P + \mu \nabla^2 V + \rho g = \rho \frac{DV}{Dt},$$

$$0$$

or

$$-\nabla P + \rho g = \rho \frac{DV}{Dt}.$$

Explicitly indicating the vectors yields

$$-\left[\hat{i}\frac{\partial P}{\partial x} + \hat{j}\frac{\partial P}{\partial y} + \hat{k}\frac{\partial P}{\partial z} \right] + \rho\left[\hat{i}g_x + \hat{j}g_y + \hat{k}g_z \right] = \rho \frac{D(\hat{i}v_x + \hat{j}v_y + \hat{k}v_z)}{Dt}.$$

For a 2-D flow, $\partial P/\partial z = 0$, $g_z = 0$, and $v_z = 0$; thus,

$$-\left[\hat{i}\frac{\partial P}{\partial x} + \hat{j}\frac{\partial P}{\partial y} \right] + \rho\left[\hat{i}g_x + \hat{j}g_y \right] = \rho \frac{D(\hat{i}v_x + \hat{j}v_y)}{Dt}.$$

The x-direction component is then

$$-\frac{\partial P}{\partial x} + \rho g_x = \rho \frac{Dv_x}{Dt},$$

or

$$-\frac{\partial P}{\partial x} + \rho g_x = \rho \left[\frac{\partial v_x}{\partial t} + v_x \frac{\partial v_x}{\partial x} + v_y \frac{\partial v_x}{\partial y} \right],$$

where the $D(\ )/Dt$ operator has been expanded for the 2-D case. The y-direction component, which follows in a similar fashion, is

$$-\frac{\partial P}{\partial y} + \rho g_y = \rho \left[\frac{\partial v_y}{\partial t} + v_x \frac{\partial v_y}{\partial x} + v_y \frac{\partial v_y}{\partial y} \right].$$

[9] Because of the nature of this problem, we abandon the usual solution format.

Further simplification can be obtained if we align the gravity vector with the y-direction ($g_x = 0$ and $g_y = -g$); thus,

$$-\frac{\partial P}{\partial x} = \rho\left[\frac{\partial v_x}{\partial t} + v_x\frac{\partial v_x}{\partial x} + v_y\frac{\partial v_x}{\partial y}\right]$$

and

$$-\frac{\partial P}{\partial y} - \rho g = \rho\left[\frac{\partial v_y}{\partial t} + v_x\frac{\partial v_y}{\partial x} + v_y\frac{\partial v_y}{\partial y}\right].$$

Comments   The first step in our analysis generated the **Euler equation:** $-\nabla P + \rho g = \rho DV/Dt$. This equation is frequently used to describe many flows where viscous effects are unimportant.

Self Test 6.8

 Consider an incompressible, inviscid, steady, two-dimensional flow where $v_x = Ax$ and $v_y = -Ay$. Verify that these velocity components satisfy general continuity (Eq. 3.39) and write out the component equations of momentum conservation.

(Answer: $-\partial P/\partial x = \rho A^2 x$, $-\partial P/\partial y - \rho g = \rho A^2 y$)

Example 6.14

Simplify the momentum conservation component equations given in Table 6.3 for a steady, two-dimensional (x, y) flow. Assume gravity acts in the negative y-direction.

Solution

For steady flow, the $\partial(\ )/\partial t$ terms are zero in each component equation; furthermore, the z-direction velocity component, v_z, and all z derivatives are zero in a 2-D flow; that is,

$$\frac{\partial(\ )}{\partial t} = 0,$$

$$v_z = 0,$$

and

$$\frac{\partial(\ )}{\partial z} = 0.$$

Also,

$$g_x = g_z = 0.$$

With these simplifications, the x-component of momentum conservation (Eq. T6.3a),

$$-\frac{\partial P}{\partial x} + \mu\left(\frac{\partial^2 v_x}{\partial x^2} + \frac{\partial^2 v_x}{\partial y^2} + \frac{\partial^2 v_x}{\partial z^2}\right) + \rho g_x$$

$$= \rho\left(\frac{\partial v_x}{\partial t} + v_x\frac{\partial v_x}{\partial x} + v_y\frac{\partial v_x}{\partial y} + v_z\frac{\partial v_x}{\partial z}\right),$$

becomes

$$-\frac{\partial P}{\partial x} + \mu\left(\frac{\partial^2 v_x}{\partial x^2} + \frac{\partial^2 v_x}{\partial y^2} + 0\right) + \rho(0)$$
$$= \rho\left(0 + v_x\frac{\partial v_x}{\partial x} + v_y\frac{\partial v_x}{\partial y} + 0(0)\right),$$

or

$$-\frac{\partial P}{\partial x} + \mu\left(\frac{\partial^2 v_x}{\partial x^2} + \frac{\partial^2 v_x}{\partial y^2}\right) = \rho\left(v_x\frac{\partial v_x}{\partial x} + v_y\frac{\partial v_x}{\partial y}\right).$$

Similarly, the y-component (Eq. T6.3b),

$$-\frac{\partial P}{\partial y} + \mu\left(\frac{\partial^2 v_y}{\partial x^2} + \frac{\partial^2 v_y}{\partial y^2} + \frac{\partial^2 v_y}{\partial z^2}\right) + \rho g_y$$
$$= \rho\left(\frac{\partial v_y}{\partial t} + v_x\frac{\partial v_y}{\partial x} + v_y\frac{\partial v_y}{\partial y} + v_z\frac{\partial v_y}{\partial z}\right),$$

becomes

$$-\frac{\partial P}{\partial y} + \mu\left(\frac{\partial^2 v_y}{\partial x^2} + \frac{\partial^2 v_y}{\partial y^2} + 0\right) + \rho g_y$$
$$= \rho\left(0 + v_x\frac{\partial v_y}{\partial x} + v_y\frac{\partial v_y}{\partial y} + 0(0)\right),$$

or

$$-\frac{\partial P}{\partial y} + \mu\left(\frac{\partial^2 v_y}{\partial x^2} + \frac{\partial^2 v_y}{\partial y^2}\right) + \rho g_y = \rho\left(v_x\frac{\partial v_y}{\partial x} + v_y\frac{\partial v_y}{\partial y}\right).$$

Since every term in the z-component equation (Eq. T6.3c) is zero, this equation is irrelevant to the 2-D problem.

Comment This example illustrates how the somewhat formidable momentum conservation equations can be simplified for specific flows. The next example shows how the relationships generated here are simplified even further for a special case.

Self Test 6.9   **Consider an incompressible, steady, two-dimensional flow where $v_x = Ax^2y$ and $v_y = -Axy^2$. Verify that these velocities satisfy general continuity and write out the component equations of momentum conservation.**

(Answer: $-\partial P/\partial x + 2\mu Ay = \rho A^2x^3y^2$, $-\partial P/\partial y - 2\mu Ax + \rho g_y = \rho A^2x^2y^3$)

Example 6.15

Consider a steady, 2-D flow between parallel plates separated a distance $2H$ as shown in the sketch. The following assumptions also apply:

- The flow is fully developed; that is, the axial velocity profile does not change with distance downstream; mathematically, $v_x(x, y) = v_x(y$ only).
- The fluid is incompressible and Newtonian.
- The axial pressure gradient $\partial P/\partial x$ is a constant.

- Gravity acts in the y-direction.
- The no-slip boundary condition applies at the plate surfaces; that is, $v_x(\text{walls}) = 0$.
- The walls are impermeable; that is, $v_y(\text{walls}) = 0$.

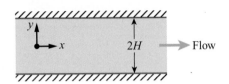

A. Simplify the continuity equation (Eq. 3.41) to find v_y.
B. Using the given assumptions, simplify further the 2-D momentum equations generated in Example 6.14.

Solution

We begin by simplifying the continuity (mass conservation) equation (Eq. 3.41):

$$\frac{\partial v_x}{\partial x} + \frac{\partial v_y}{\partial y} + \frac{\partial v_z}{\partial z} = 0.$$

With both $v_z = 0$ and $\partial(\ )/\partial z = 0$ for a 2-D flow, this becomes

$$\frac{\partial v_x}{\partial x} + \frac{\partial v_y}{\partial y} = 0.$$

The assumption that the flow is fully developed [i.e., $v_x = v_x$ (y only)], results in $\partial v_x/\partial x$ being zero; thus,

$$\frac{\partial v_y}{\partial y} = 0.$$

Integrating this yields

$$v_y = f(x).$$

That the walls are impermeable means that v_y is zero at $y = \pm H$ for all x; therefore, $f(x)$ must be zero, and we conclude that

$$v_y(x, y) = 0.$$

We use this result to simplify further to the 2-D x- and y-momentum equations from Example 6.14. The x-momentum component is

$$-\frac{\partial P}{\partial x} + \mu\left(\frac{\partial^2 v_x}{\partial x^2} + \frac{\partial^2 v_x}{\partial y^2}\right) = \rho\left(v_x\frac{\partial v_x}{\partial x} + v_y\frac{\partial v_x}{\partial y}\right),$$

which now becomes

$$-\frac{\partial P}{\partial x} + \mu\left(0 + \frac{\partial^2 v_x}{\partial y^2}\right) = \rho\left(v_x(0) + 0\frac{\partial v_x}{\partial y}\right).$$

Here we use the fact that $\partial v_x/\partial x = 0$ to eliminate two terms (the first two zeros) and that $v_y = 0$ to eliminate the last term. Recognizing that the partial derivative becomes an ordinary derivative because $v_x = v_x(y$ only), we write our final result as

$$\frac{d^2 v_x}{dy^2} = \frac{1}{\mu}\frac{\partial P}{\partial x}.$$

We can also simplify the *y*-momentum equation considerably:

$$-\frac{\partial P}{\partial y} + \mu\left(\frac{\partial^2 v_y}{\partial x^2} + \frac{\partial^2 v_y}{\partial y^2}\right) + \rho g_y = \rho\left(v_x\frac{\partial v_y}{\partial x} + v_y\frac{\partial v_y}{\partial y}\right)$$

becomes

$$-\frac{\partial P}{\partial y} + \mu(0 + 0) + \rho g_y = \rho(v_x(0) + 0(0)),$$

or

$$\frac{\partial P}{\partial y} = \rho g_y = -\rho g.$$

This significant simplification results from $v_y = 0$.

Comments   We note the key use of the continuity equation to deduce that v_y is zero for this problem. That $v_y = 0$ results, first, in a simple, second-order, ordinary differential equation describing $v_x(y)$ and, second, in a simple relationship describing the pressure variation with depth *y*. Note the similarity of this latter relationship, $\partial P/\partial y = \rho g_y$, with the hydrostatic description of pressure (cf. Eqs. 6.18 and 6.21).

Example 6.16

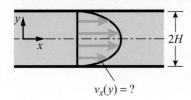

$v_x(y) = ?$

Use the results from Example 6.15 to find the velocity distribution for fully developed, steady flow between two parallel plates separated a distance 2*H*.

Solution

Known   Governing equation for $v_x(y)$ from Example 6.15

Find $v_x(y)$

Sketch

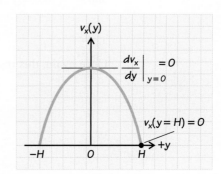

Assumptions

See Examples 6.14 and 6.15.

Analysis   Our specific objective is to solve the following second-order, ordinary differential equation derived in Example 6.15:

$$\frac{d^2 v_x}{dy^2} = \frac{1}{\mu}\frac{\partial P}{\partial x}(= \text{constant}).$$

Two boundary conditions are required. We can use either the no-slip conditions at $y = \pm H$,

$$v_x(y = +H) = 0,$$
$$v_x(y = -H) = 0.$$

or make use of symmetry and a single no-slip condition,

$$\left.\frac{dv_x}{dy}\right|_{y=0} = 0,$$
$$v_x(y = H) = 0.$$

These boundary conditions are illustrated in the sketch. Our second-order, ordinary differential equation can be rewritten as

$$\frac{d\left(\dfrac{dv_x}{dy}\right)}{dy} = \frac{1}{\mu}\frac{\partial P}{\partial x}.$$

Recognizing that $(1/\mu)\partial P/\partial x$ is a constant, we separate and integrate as follows:

$$\int d(dv_x/dy) = \frac{1}{\mu}\frac{\partial P}{\partial x}\int dy$$

and

$$dv_x/dy = \frac{1}{\mu}\frac{\partial P}{\partial x}y + C_1.$$

We immediately solve for the integration constant C_1 by applying the symmetry boundary condition:

$$\left.\frac{dv_x}{dy}\right|_{y=0} = 0 = \frac{1}{\mu}\frac{\partial P}{\partial x}(0) + C_1.$$

Thus,

$$C_1 = 0$$

and

$$\frac{dv_x}{dy} = \frac{1}{\mu}\frac{\partial P}{\partial x}y.$$

Once again, this result can be separated and integrated, and so

$$\int dv_x = \frac{1}{\mu}\frac{\partial P}{\partial x}\int y\,dy,$$

or

$$v_x = \frac{1}{\mu}\frac{\partial P}{\partial x}\frac{y^2}{2} + C_2.$$

We apply the no-slip condition to evaluate the second constant of integration, C_2:

$$v_x(y = H) = 0,$$

or

$$0 = \frac{1}{\mu}\frac{\partial P}{\partial x}\frac{H^2}{2} + C_2.$$

Thus,

$$C_2 = -\frac{1}{\mu} \frac{\partial P}{\partial x} \frac{H^2}{2}.$$

The velocity distribution across the channel is then

$$v_x(y) = \frac{1}{2\mu} \frac{\partial P}{\partial x}(y^2 - H^2),$$

or

$$v_x(y) = -\frac{H^2}{2\mu} \frac{\partial P}{\partial x}\left[1 - \left(\frac{y}{H}\right)^2\right].$$

We see that this result describes a parabolic distribution suggested in the sketch.

Comments   Note how the sequence of Examples 6.14–6.16 have taken us from the formidable, and somewhat intimidating, Navier–Stokes (momentum conservation) equation to a relatively simple practical result, $v_x(y)$. We can use this result to obtain other practical information. For example, the velocity distribution can be used to relate the maximum velocity to the fluid viscosity, channel height, and the pressure gradient, that is,

$$v_{max} = v_x(y = 0) = -\frac{H^2}{2\mu} \frac{\partial P}{\partial x}.$$

Applying Eq. 6.9b, we can also determine the fluid shear stress at any y location:

$$\tau(y) = \mu \frac{dv_x}{dy}$$

$$= \mu \frac{d}{dy}\left[\frac{1}{2\mu} \frac{\partial P}{\partial x}(y^2 - H^2)\right]$$

$$= \frac{\partial P}{\partial x} y.$$

See Eq. 10.39 in Chapter 10. ▶

Thus, the shear stress is linear across the channel with a zero value on the centerline and a maximum value at the walls. In Chapter 10, we derive the fully developed velocity profile for laminar flow through a circular tube in a manner very similar to this example.

6.6 MECHANICAL ENERGY EQUATION[10]

As an adjunct to the momentum conservation principle, we now consider the **mechanical energy** equation. It is important at the outset to emphasize that this equation is not identical to the conservation of energy principle (the first law of thermodynamics) discussed in Chapter 5 but, rather, is based on

---

[10] Mechanical energy here refers to kinetic and potential energies and all forms of work, including flow work.

The mechanical energy equation is used to determine pressure losses through pipes, elbows, and valves.

a mathematical manipulation of the momentum equation. Specifically, the heat transfer rate $\dot{Q}_{cv}$ and internal energy u, which are essential in any first-law statement of energy conservation, do not appear in the mechanical energy equation. The usefulness of the mechanical energy equation is that it explicitly accounts for the irreversible conversion of mechanical energy to thermal energy through a "friction loss" term. In this light, mechanical energy is not actually conserved since some mechanical energy is destroyed never to return through the action of internal friction (i.e., mechanical energy is converted to thermal energy). We now explore the origins of the mechanical energy equation.

Although ultimately desiring an expression that applies to an integral control volume, we start with the differential control volume form of momentum conservation as expressed in Eq. 6.72. The first step in a rather lengthy analysis, which we highlight rather than go through in detail, is to take the scalar (dot) product of the velocity vector V and Eq. 6.72a. This gives us

$$V \cdot \left(-\nabla P + \mu \nabla^2 V + \rho g = \rho \frac{DV}{Dt} \right). \tag{6.73}$$

Note that after performing the implied multiplications, each term in Eq. 6.73 has units of W/m^3 (i.e., energy rate or power per unit volume). Note also that the term on the right-hand side can be expressed as

$$V \cdot \left(\rho \frac{DV}{Dt} \right) = \rho \frac{D\left(\frac{1}{2} V^2 \right)}{Dt}. \tag{6.74}$$

Thus, we see how the kinetic energy, one of the mechanical energy terms, appears from this operation.

Consider now an integral control volume such as illustrated in Fig. 6.5b. Equation 6.73 applies at every point within this control volume; thus, we can integrate this expression over the entire control volume, that is,

$$\int_{CV} V \cdot (-\nabla P + \mu \nabla^2 V + \rho g) d\mathcal{V} = \int_{CV} V \cdot \left(\rho \frac{DV}{Dt} \right) d\mathcal{V}.$$

Performing this integration, and other operations, result in

$$\frac{d}{dt}(KE_{cv} + PE_{cv}) = \dot{m} \left[\frac{P_{in} - P_{out}}{\rho} + \frac{1}{2}(\alpha_{in} V^2_{avg,in} - \alpha_{out} V^2_{avg,out}) \right.$$

$$\left. + g(z_{in} - z_{out}) \right] - \dot{W}_{cv,net\ out} - \dot{E}_{visc}, \tag{6.75}$$

where we assume a single inlet and outlet to the control volume, assume isothermal conditions, and define $\dot{E}_{visc}$ to be the friction loss or irreversible conversion rate of mechanical energy to thermal energy. This equation is similar in form to the conservation of energy (first law) expression given by Eq. 5.70.

For steady-state situations, the time-dependent term in Eq. 6.75 is zero. After eliminating this term, dividing through by the mass flow rate $\dot{m}$, and rearranging, we obtain the following mechanical energy equation:

$$\frac{P_\text{in}}{\rho} + \frac{1}{2}\alpha_\text{in}V^2_\text{avg,in} + gz_\text{in} = \frac{P_\text{out}}{\rho} + \frac{1}{2}\alpha_\text{out}V^2_\text{avg,out} + gz_\text{out} + \frac{\dot{W}_\text{cv,net out}}{\dot{m}} + gh_\text{L},$$

(6.76)

where h_L, called the **head loss**, is defined by

$$h_\text{L} \equiv \frac{\dot{E}_\text{visc}}{\dot{m}g}.$$

(6.77)

It is quite interesting to compare this momentum conservation–based equation with the corresponding steady-flow, incompressible energy conservation equation (first law) developed in Chapter 5 (i.e., Eq. 5.66), which we rewrite here as

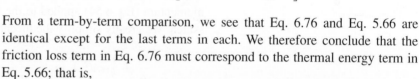

$$\frac{P_\text{in}}{\rho} + \frac{1}{2}\alpha_\text{in}V^2_\text{avg,out} + gz_\text{in} = \frac{P_\text{out}}{\rho} + \frac{1}{2}\alpha_\text{out}V^2_\text{avg,in} + gz_\text{out} + \frac{\dot{W}_\text{cv,net out}}{\dot{m}}$$
$$+ \left[u_\text{out} - u_\text{in} + \frac{\dot{Q}_\text{cv,net out}}{\dot{m}} \right].$$

(5.66)

From a term-by-term comparison, we see that Eq. 6.76 and Eq. 5.66 are identical except for the last terms in each. We therefore conclude that the friction loss term in Eq. 6.76 must correspond to the thermal energy term in Eq. 5.66; that is,

$$\frac{\dot{E}_\text{visc}}{\dot{m}} = u_\text{out} - u_\text{in} + \frac{\dot{Q}_\text{cv,net out}}{\dot{m}}.$$

(6.78)

Head losses can be quite large in long pipelines.

This statement closes the loop on the idea that the irreversible loss of mechanical energy must appear as thermal energy. From Eq. 6.78, we see that the action of the internal fluid friction requires that there be a net heat transfer out of the control volume to maintain the isothermal conditions ($u_\text{out} = u_\text{in}$) assumed in the derivation of Eq. 6.76. However, in many real flows, even under adiabatic conditions, the temperature increase resulting from frictional effects is small; thus, Eq. 6.76 is still sufficiently accurate for engineering purposes when the flow is not exactly isothermal.

The following example illustrates how head loss can be calculated for fully developed laminar flow through a tube. Further treatment of losses in tubes, pipes, and piping systems is found in Chapters 8 and 10.

Example 6.17

Water ($\rho = 996.6$ kg/m^3) flows through a long, horizontal, 1-mm-diameter capillary tube. Sufficiently far from the tube entrance, the flow becomes fully developed (i.e., the velocity profile is fixed), with $v_x = v_x$ (r only) and $v_r = 0$. In this region, the wall shear stress is 11.7 Pa. Determine the head loss between two stations located 0.8 m apart.

Solution

Known ρ, D, L, τ_w, horizontal ($\Delta z = 0$), fully developed

Find h_L

Sketch

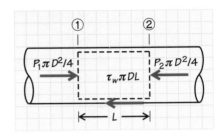

Assumptions

i. Steady flow
ii. Incompressible fluid
iii. Isothermal
iv. No $\dot{W}_{cv}$

Analysis   To solve this problem, we apply the mechanical energy equation (Eq. 6.76) to relate the head loss h_L, to the pressure drop $P_1 - P_2$. In turn, we relate the pressure drop to the wall shear stress through conservation of momentum (Eq. 6.54) applied to the integral control volume shown in the sketch. We greatly simplify the mechanical energy equation,

$$\frac{P_1}{\rho} + \frac{1}{2}\alpha_1 V_1^2 + gz_1 = \frac{P_2}{\rho} + \frac{1}{2}\alpha_2 V_2^2 + gz_2 + \frac{\dot{W}_{net\,out}}{\dot{m}} + gh_L,$$

by noting that $\alpha_1 = \alpha_2$ (fully developed flow), $V_1 = V_2$ ($\dot{m}$ = constant = $\rho V_1 A = \rho V_2 A$), and $z_1 = z_2$ (horizontal tube); therefore, both the kinetic and potential energy terms cancel. Furthermore, $\dot{W}_{net\,out} = 0$; thus,

$$h_L = \frac{P_1 - P_2}{\rho g}.$$

For the control volume in the sketch, we express conservation of momentum (Eq. 6.54) as

$$\beta_1 \dot{m} V_1 - \beta_2 \dot{m} V_2 + (P_1 - P_2)\pi D^2/4 - \tau_w \pi D L = 0.$$

Because the flow is fully developed, $\beta_1 = \beta_2$ and the inlet and outlet momentum flows cancel ($\beta_1 \dot{m} V_1 = \beta_2 \dot{m} V_2$); thus,

$$P_1 - P_2 = \frac{\tau_w \pi D L}{\pi D^2/4} = \frac{4L}{D}\tau_w.$$

Combining this with the result from the mechanical energy equation, we have

$$h_L = \frac{P_1 - P_2}{\rho g} = \frac{4L\tau_w}{\rho g D}$$

$$= \frac{4(0.8)11.7}{996.6(9.807)0.001} = 3.83$$

$$[=]\frac{m(N/m^2)}{(kg/m^3)(m/s^2)m}\left[\frac{1\,kg\cdot m/s^2}{N}\right] = m.$$

Comment We note the complementary nature of the mechanical energy equation and the conservation of momentum principle in this problem. Although the wall shear stress was a given quantity in this example, you will see in Chapter 10 how the shear stress can be determined for a fully developed flow knowing only the flow rate (or average velocity) and the fluid viscosity.

Self Test 6.10 ✓ **Develop an expression for the pumping power required to overcome the head loss in the tube of Example 6.17.**

(Answer: $\dot{W}_{\text{pump}} = 4\dot{m}L\tau_{\text{w}}/\rho D$)

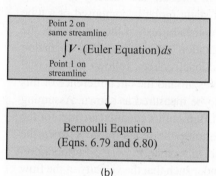

$$\int_{\text{CV}} V \cdot (\text{Navier–Stokes Equation})\,d\mathcal{V}$$

Mechanical Energy Equation
(Eqns. 6.75 and 6.76)

(a)

Point 2 on same streamline

$$\int V \cdot (\text{Euler Equation})\,ds$$

Point 1 on streamline

Bernoulli Equation
(Eqns. 6.79 and 6.80)

(b)

FIGURE 6.23

(a) The mechanical energy equation (Eqs. 6.75 and 6.76) is obtained by integrating over the control volume the scalar product of the velocity vector and the Navier–Stokes equation (Eq. 6.72). (b) The Bernoulli equation (Eqs. 6.79 and 6.80) is obtained by integrating along a streamline from point 1 to point 2 the scalar product of the velocity vector and the Euler equation.

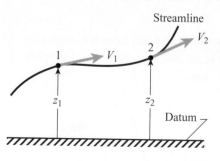

6.7 THE BERNOULLI EQUATION

In Example 6.13, we defined the inviscid approximation. This approximation was used to derive the Euler equation, a differential statement of momentum conservation. To obtain the famous Bernoulli equation, we now employ the Euler equation in the same spirit that the Navier–Stokes equation was employed to obtain the mechanical energy equation (see Fig. 6.23). We first form the scalar product of the velocity vector V and the Euler equation. We then integrate this product *along a streamline* from location 1 to location 2. Hence we have

$$\int_{\text{Location 1}}^{\text{Location 2}} V \cdot (-\nabla P + \rho g = DV/Dt)\,ds,$$

where ds is the incremental distance along the streamline. The result of this operation is the **unsteady** form of the **Bernoulli equation:**

$$\int_1^2 \frac{\partial V}{\partial t}\,ds + \int_1^2 \frac{dP}{\rho} + \frac{1}{2}(V_2^2 - V_1^2) + g(z_2 - z_1) = 0, \qquad (6.79)$$

where z is the elevation above a horizontal reference plane (i.e., a plane perpendicular to the gravity vector). The corresponding unsteady form of the mechanical energy equation is Eq. 6.75.

The steady-flow form of Eq. 6.79 is frequently used in engineering applications. In addition to setting the unsteady term of Eq. 6.79 to zero, incompressible (constant density) flow is usually assumed. The second term then becomes $(P_2 - P_1)/\rho$. With these modifications, we arrive at the following **steady-flow** form of the **Bernoulli equation:**

$$\frac{P_1}{\rho} + \frac{1}{2}V_1^2 + gz_1 = \frac{P_2}{\rho} + \frac{1}{2}V_2^2 + gz_2 = \text{constant}. \qquad (6.80)$$

This particular form is most frequently implied when reference is made to *the* Bernoulli equation. Because of the importance of this equation and the fact that it is frequently misused, we carefully reiterate the assumptions embedded in Eq. 6.80:

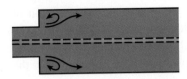
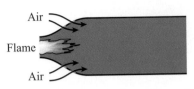

The blue regions (left and center) meet the restrictions for the application of the Bernoulli equation, whereas the pink regions (left, center, and right) do not.

- Points 1 and 2 lie on the same streamline.[11]
- The flow is steady.
- The flow is incompressible (ρ = constant)
- There are neither frictional effects nor heat or work interactions.

Note that the second and third restrictions can be removed by returning to Eq. 6.79.

[11] For the special case of irrotational flow (i.e., flows where $\nabla \times V = 0$), the points do not have to lie on a streamline. The interested reader is referred to Refs. [4, 5].

Example 6.18

Pitot tube with flow direction sensors (balsa flags). Photograph courtesy of NASA Langley Research Center.

Pitot-static probes are frequently used to measure the velocity of flowing fluid. As shown in the sketch, the pitot-static probe is composed of a tube within a tube. The flow stagnates (i.e., comes to rest) at the entrance of the central (or pitot) tube along the "stagnation" streamline. The stagnation streamline splits and, if we ignore frictional effects, follows the external contour of the outer (or static) tube. Holes around the circumference of this outer tube allow the "static" pressure to be measured as shown. Assuming that the conditions for the application of Bernoulli's equation are met for the streamline joining points 1 and 2, determine the relationship between the velocity V and the measured pressured difference $P_1 - P_2$. Furthermore, assume that the probe is sufficiently slender such that the velocity of the flow passing by the static pressure holes is the same as the upstream velocity, V. Also determine the pressure difference expressed in inches of H_2O associated with air flowing at 100 knots (51.48 m/s) at 100 kPa and 300 K.

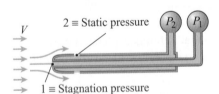

Solution

Known $P_1 - P_2$

Find V–ΔP relationship, ΔP

Assumptions

  i. Steady flow
 ii. Incompressible fluid
iii. Frictionless flow
 iv. Flow along a streamline connecting points 1 and 2
  v. Negligible effects of gravity

Analysis   We apply the Bernoulli equation (Eq. 6.80) from the stagnation point (station 1) to the circumferential static holes (station 2) and simplify by noting that $V_1 = 0$ and $z_1 - z_2 = 0$; thus,

$$\frac{P_1}{\rho} + \frac{1}{2}V_1^2 + gz_1 = \frac{P_2}{\rho} + \frac{1}{2}V_2^2 + gz_2$$

becomes

$$\frac{P_1}{\rho} = \frac{P_2}{\rho} + \frac{1}{2}V_2^2.$$

Solving for V_2 yields

$$V_2 = \left[\frac{2(P_1 - P_2)}{\rho}\right]^{1/2}.$$

Recognizing that V_2 is essentially identical to the upstream velocity V, we conclude

$$V = \left[\frac{2(P_1 - P_2)}{\rho}\right]^{1/2}.$$

To calculate the pressure difference associated with air flowing at 100 knots, we write

$$P_1 - P_2 = \frac{1}{2}\rho V_2^2.$$

Evaluating the density using the ideal-gas law, we get a pressure difference of

$$P_1 - P_2 = 0.5\left(\frac{P}{RT}\right)V_2^2$$

$$= 0.5\left(\frac{100{,}000}{287(300)}\right)(51.48)^2 \text{ Pa} = 1539 \text{ Pa}$$

$$= 1539 \text{ Pa}\left[\frac{4.015 \times 10^{-3} \text{ in H}_2\text{O}}{\text{Pa}}\right]$$

$$= 6.18 \text{ in H}_2\text{O}.$$

Comments   We see that the velocity is proportional to the square root of the pressure difference. Practical applications of pitot-static probes often employ a differential pressure transducer to measure $P_1 - P_2$. A logic circuit performs the square-root operation to provide a direct readout of velocity with proper calibration. Pitot-static probes are often used in wind-tunnel studies and are standard equipment in aircraft to measure airspeed. Typical pitot probes used in aircraft are about 25 cm long with a 1-cm diameter. Static holes are frequently in the walls of the aircraft rather than integrated with the pitot tube. Blockage of either the stagnation or static holes by insects or ice can result in erroneous indications of airspeed, thereby creating an unsafe situation. Aircraft pitot tubes are heated to minimize icing problems.

6.8 TURBULENCE REVISITED

See the discussion surrounding Eqs. 3.47 and 3.51 in Chapter 3.

We introduced the topic of turbulent flow in Chapter 1 and in our study of mass conservation in Chapter 3. You may find it useful to review the appropriate sections of these chapters at this juncture. Now that we have completed our presentation of the three key conservation principles of the thermal-fluid sciences—mass, energy, and momentum conservation—we can investigate how these principles can be applied to analyze turbulent flows. The basic conservation equations (Eq. 3.40 and Tables 5.7, 6.3, and 6.4) describe the fluid motions and energy transport within turbulent as well as laminar flows, provided, of course, that the unsteady terms are retained. For example, we can write the mass conservation and the x-component of momentum

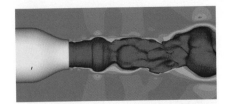

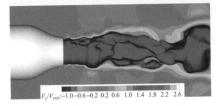

v_x/V_{exit}: −1.0 −0.6 −0.2 0.2 0.6 1.0 1.4 1.8 2.2 2.6

Numerical simulation of a turbulent jet shows velocity field contours viewed from two different circumferential positions.

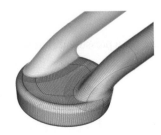

Computational mesh for artificial heart simulation.

conservation equations for unsteady, isothermal flow of an incompressible, Newtonian fluid as follows:

$$\frac{\partial v_x}{\partial x} + \frac{\partial v_y}{\partial y} + \frac{\partial v_z}{\partial z} = 0 \quad \text{(mass conservation)}, \tag{6.81}$$

$$\frac{\partial}{\partial t}(\rho v_x) + \frac{\partial}{\partial x}(\rho v_x v_x) + \frac{\partial}{\partial y}(\rho v_y v_x) + \frac{\partial}{\partial z}(\rho v_z v_x)$$

$$= \mu\left(\frac{\partial^2 v_x}{\partial x^2} + \frac{\partial^2 v_x}{\partial y^2} + \frac{\partial^2 v_x}{\partial z^2}\right) - \frac{\partial P}{\partial x} + \rho g_x \tag{6.82}$$

(x-direction momentum conservation)[12].

Conceptually, these two equations, together with the y- and z-components of momentum conservation, could be solved for the functions $v_x(t)$, $v_y(t)$, $v_z(t)$ and $P(t)$ at discrete points in the flow using numerical techniques; however, the number of grid points to resolve the details of the flow would be enormous. For example, Ref. [7] cites that 10^9 grid points would be necessary to deal with a simple plane-mixing layer of two adjacent fluids. At present, this overwhelms even the best available computers. It should be pointed out, however, that full solutions are being accomplished, provided the Reynolds number is not too large [8].

6.8a Integral Equations

Before we look at how turbulent flows can be analyzed using a differential control volume approach, we comment briefly on how turbulence affects the usage of the integral control volume relations developed earlier in this chapter (e.g., Eqs. 6.56, 6.57, 6.59, and 6.76). In most engineering applications, we employ these relations in a straightforward manner by substituting time-mean properties for the instantaneous ones. For example, $\overline{V}_{avg}$ replaces V_{avg} and $\overline{P}$ replaces P. In general, the effects of turbulence on these integral relations are thought to be small [9]. In addition, empiricism, which uses data from actual turbulent flows, is brought into the calculation of the head loss term in Eq. 6.76 (see Chapter 8). The situation for the differential forms of momentum conservation (Tables 6.3 and 6.4), however, is much more complex.

6.8b Differential Equations

Reynolds Averaging and Turbulent Stresses

Rather than relying entirely on empirical methods, techniques have been developed that allow turbulent flows to be analyzed to obtain useful information and to allow predictions. One fruitful method of analyzing turbulent flows is to write out the partial differential equations that embody the basic conservation principles—mass, momentum, and energy, perform a Reynolds decomposition (see Eq. 3.43), and then average the equations over time. The resulting governing equations are called the **Reynolds-averaged** equations. The averaging process has two major consequences: First, it eliminates the fine details of the flow; for example, the complex, time-dependent velocity at a point, as illustrated in Fig. 3.14, could not be predicted using time-averaged equations of motion. The second consequence

---

[12] To retain the appearance of momentum flows (per unit volume), the conservative form of this equation (Eq. 6.63a) is employed here in combination with Eq. 6.71a for the viscous stresses.

of averaging is the appearance of new terms in the time-averaged governing equations that have no counterpart in the original time-dependent equations (see Eq. 3.50). Finding a way to calculate or approximate these new terms is frequently called the **closure problem** of turbulence.

To develop some feel for the time-averaged approach, we develop the time-averaged x-momentum equation for a two-dimensional, incompressible, boundary layer flow over a flat plate.[13] For this flow, with x in the flow direction and y perpendicular to the plate, Eq. 6.82 reduces to

$$\frac{\partial}{\partial t}\rho v_x + \frac{\partial}{\partial x}\rho v_x v_x + \frac{\partial}{\partial y}\rho v_y v_x = \mu \frac{\partial^2 v_x}{\partial y^2}. \tag{6.83}$$

The first step in the process is to substitute for each fluctuating quantity its representation as a mean and a fluctuating component, that is, the Reynolds decomposition $v_x = \bar{v}_x + v'_x$ and $v_y = \bar{v}_y + v'_y$. We then time-average Eq. 6.83, term by term, using the overbar to signify a time-averaged quantity. Thus, the first term can be averaged as

$$\overline{\frac{\partial}{\partial t}\rho(\bar{v}_x + v'_x)} = \overline{\frac{\partial}{\partial t}\rho\bar{v}_x} + \overline{\frac{\partial}{\partial t}\rho v'_x} = 0, \tag{6.84}$$

where each term on the right-hand side of Eq. 6.84 is identically zero; the first because we assume the flow is steady in an overall sense (i.e., $\bar{v}_x$ is a constant), and the second because the time derivative of a random function with a zero mean is also a random function with a zero mean.

The third term in Eq. 6.83 is the most interesting because it yields the dominant additional momentum flux arising from the velocity fluctuations:

$$\overline{\frac{\partial}{\partial y}\rho(\bar{v}_y + v'_y)(\bar{v}_x + v'_x)} = \overline{\frac{\partial}{\partial y}\rho(\bar{v}_x\bar{v}_y + \bar{v}_x v'_y + v'_x v'_y + \bar{v}_y v'_x)}. \tag{6.85}$$

Splitting out each term under the large overbar on the right-hand side and evaluating their individual averages yields

$$\overline{\frac{\partial}{\partial y}\rho\bar{v}_x\bar{v}_y} = \frac{\partial}{\partial y}\rho\bar{v}_x\bar{v}_y, \tag{6.86a}$$

$$\overline{\frac{\partial}{\partial y}\rho\bar{v}_x v'_y} = 0, \tag{6.86b}$$

$$\overline{\frac{\partial}{\partial y}\rho v'_x v'_y} = \frac{\partial}{\partial y}\rho\overline{v'_x v'_y}, \tag{6.86c}$$

and

$$\overline{\frac{\partial}{\partial y}\rho\bar{v}_y v'_x} = 0. \tag{6.86d}$$

Similar operations can be performed on the second and fourth terms of Eq. 6.83 and are left as an exercise for the reader. Reassembling the time-

[13] In Chapters 9, we develop the concept of a boundary layer and present various solutions for the drag and heat transfer for flow over a flat plate.

averaged Eq. 6.83 yields

$$\frac{\partial}{\partial x}\rho\bar{v}_x\bar{v}_x + \frac{\partial}{\partial y}\rho\bar{v}_y\bar{v}_x + \boxed{\frac{\partial}{\partial x}\rho\overline{v'_x v'_x}} + \boxed{\frac{\partial}{\partial y}\rho\overline{v'_x v'_y}} = \mu\frac{\partial^2 \bar{v}_x}{\partial y^2}. \tag{6.87}$$

The quantities in the dashed boxes are new terms arising from the turbulent nature of the flow. In a laminar flow v'_x and v'_y are zero; thus, Eq. 6.87 becomes identical with Eq. 6.83 for a steady flow.

It is common practice to define

$$\tau_{xx}^{\text{turb}} \equiv -\rho\overline{v'_x v'_x} \tag{6.88a}$$

and

$$\tau_{yx}^{\text{turb}} \equiv -\rho\overline{v'_x v'_y}, \tag{6.88b}$$

which represent additional momentum fluxes resulting from the turbulent fluctuations. These terms have several designations and are referred to as components of the **turbulent momentum flux** (momentum flow per unit area), **turbulent stresses**, or **Reynolds stresses.** In a general formulation of the Reynolds-averaged momentum equation, a total of nine Reynolds stresses appear, analogous to the nine components of the laminar viscous shear stresses (i.e., $\tau_{ij}^{\text{turb}} = -\rho\overline{v'_i v'_j}$, where i and j represent the coordinate directions; cf. Eq. 6.7).

Rearranging Eq. 6.87 and applying continuity to the evaluation of the derivatives of the velocity products (cf. Eq. 6.65), yield our final result for a two-dimensional, turbulent, boundary-layer flow:

$$\rho\left(\bar{v}_x\frac{\partial \bar{v}_x}{\partial x} + \bar{v}_y\frac{\partial \bar{v}_x}{\partial y}\right) = \mu\frac{\partial^2 \bar{v}_x}{\partial y^2} - \frac{\partial}{\partial y}\rho\overline{v'_x v'_y}, \tag{6.89}$$

where we assume that $\partial\tau_{xx}^{\text{turb}}/\partial x$ can be neglected.

The Closure Problem

The Reynolds-averaging process introduces new unknowns into the equations of motion, namely, the turbulent stresses, $-\rho\overline{v'_i v'_j}$. Finding a way to evaluate these stresses and solving for any other additional unknowns that may be introduced in the process is termed the **closure problem** of turbulence. Presently, there are many techniques that are used to "close" the system of governing equations, ranging from the very straightforward to the quite sophisticated [10–14]. The simplest of these, Prandtl's mixing-length hypothesis, is discussed in Chapter 10.

SUMMARY

This chapter has focused on the conservation of momentum principle applied to fluids at rest and to fluids in motion. You should have a firm grasp of how pressure varies with depth in a static fluid or in an nonflowing fluid under the influence of a constant linear acceleration. You should be able to calculate the forces acting on submerged surfaces. To help you identify the most important ideas, we recommend a review of the learning objectives presented at the beginning of the chapter.

In addition to pressure forces and the gravitational body force, we examined the viscous forces generated by shearing action of one fluid layer against a neighboring layer and defined a Newtonian fluid. The concept of a momentum flow was defined and used to state the basic conservation of momentum principle for a flowing fluid, that is, that the time rate of change of momentum within a control volume is equal to the net rate at which momentum flows into the control volume (in minus out) plus the sum of all the forces acting on the control volume. We saw how the vector nature of this conservation principle complicates its application. You should be able to apply this principle to steady-flow integral control volumes.

The partial differential equation expressing conservation of momentum for a differential control volume was developed for incompressible, Newtonian fluids (the Navier–Stokes equation), and we saw how the inviscid approximation simplifies this equation to produce the Euler equation. You should be familiar with these two equations and understand their physical meanings. Furthermore, you should be comfortable with their vector forms and be able to expand them into their component equations (e.g., x-, y-, and z-components).

We saw how the mechanical energy equation and its companion, the Bernoulli equation, were derived from the conservation of momentum principle. We also saw the physical origin of the head loss in the mechanical energy equation as the irreversible conversion of mechanical energy to thermal energy. The chapter concluded with a look at how turbulence impacts the application of the conservation of momentum principle.

Chapter 6
Key Concepts & Definitions Checklist[14]

6.2 Forces

- ☐ Surface and body forces ➤ *Q6.2, Q6.3*
- ☐ Pressure forces: Definition and sign convention ➤ *Q6.9*
- ☐ Viscous forces ➤ *Q6.16, 6.68*
- ☐ No-slip condition ➤ *Q6.4, 6.69*
- ☐ Newtonian fluid ➤ *Q6.3*
- ☐ Drag ➤ *Q6.5*
- ☐ Gravitational acceleration vector ➤ *Q6.6*

6.3 Momentum Conservation for Rigid Systems

- ☐ Newton's second law ➤ *Q6.8*
- ☐ Pressure distributions in static fluid systems ➤ *6.3, 6.6*
- ☐ Manometry ➤ *6.7, 6.15*
- ☐ Forces on submerged surfaces ➤ *6.32, 6.37*
- ☐ Centroid and center of pressure ➤ *6.36*
- ☐ Buoyancy ➤ *6.48, 6.50*
- ☐ The G (or $g - a$) vector ➤ *6.53*
- ☐ Pressure distributions in nonflowing fluid systems ➤ *6.55*

6.4 Momentum Flows

- ☐ Definition and vector nature ➤ *Q6.11*
- ☐ Momentum flow diagrams
- ☐ Momentum flows from velocity distributions ➤ *6.75*

6.5a–c Linear Momentum Conservation for Integral Control Volumes

- ☐ Simplified general formulation (Eq. 6.52) ➤ *Q6.15*
- ☐ Steady-flow formulation (Eqs. 6.54 and 6.56) ➤ *Q6.13*
- ☐ Momentum flow correction factor ➤ *Q6.14*
- ☐ Steady-flow applications ➤ *6.82, 6.83*

- ☐ Unsteady formulation ➤ *Q6.30*
- ☐ Unsteady applications ➤ *6.87*

6.5d Linear Momentum Conservation for Differential Control Volumes

- ☐ Forces acting on a small (Δx, Δy, Δz) control volume ➤ *Q6.7, Q6.16*
- ☐ Momentum flows entering and exiting a small (Δx, Δy, Δz) control volume ➤ *Q6.21*
- ☐ Total (or material or substantial) derivative ➤ *Q6.25, Q6.27*
- ☐ Convective acceleration ➤ *6.124*
- ☐ Physical interpretation of terms in the Navier–Stokes equation (Eq. 6.72) ➤ *Q6.20*

6.6 Mechanical Energy Equation

- ☐ Incompressible, steady formulation (Eq. 6.76) ➤ *Q6.28*
- ☐ Head loss ➤ *6.100, 6.101*

6.7 The Bernoulli Equation

- ☐ Incompressible, steady formulation (Eq. 6.80) ➤ *Q6.23*
- ☐ Restrictions for use ➤ *Q6.24*
- ☐ Incompressible, steady-flow applications ➤ *6.109, 6.112*

6.8 Turbulence Revisited

- ☐ Effects of turbulence on integral control-volume relationships ➤ *Q6.32*
- ☐ Reynolds averaging (differential relationships) ➤ *6.135*
- ☐ Turbulent or Reynolds stresses ➤ *Q6.31*
- ☐ Meaning of closure ➤ *Q6.33*

Tutorial 4: What is CFD?

- ☐ Basic idea of what CFD is and can do

---

[14] Numbers following arrows below refer to end-of-chapter Problems (e.g., 6.7) and Questions (e.g., Q6.2).

REFERENCES

1. Rouse, H., "Highlights in the History of Hydraulics," *Books at Iowa,* 38:3–8,10–16 (1983).

2. Rouse, H., and Ince, S., *History of Hydraulics,* Iowa Institute of Hydraulics Research, State University of Iowa, 1957.

3. Vincenti, W. G., *What Engineers Know and How They Know It: Analytical Studies from Aeronautical History,* Johns Hopkins University Press, Baltimore, 1990.

4. White, F. M., *Fluid Mechanics,* 5th ed., McGraw-Hill, New York, 2003.

5. Fox, R. W., and McDonald, A. T., *Introduction to Fluid Mechanics,* 3rd ed., Wiley, New York, 1985.

6. Zwillinger, D. (Ed.), *CRC Standard Mathematical Tables and Formulae,* 31st ed., CRC Press, Boca Raton, FL, 2002.

7. Lumley, J. E. (Ed.), *Whither Turbulence? Turbulence at the Crossroads*, Lecture Notes in Physics, Vol. 357, Springer-Verlag, New York, 1990.

8. Moin, P., "Towards Large Eddy and Direct Simulation of Complex Turbulent Flows," *Computer Methods in Applied Mechanics and Engineering,* 87:329–334 (1991).

9. Bird, R. B., Stewart, W. E., and Lightfoot, E. N., *Transport Phenomena,* Wiley, New York, 1960, pp. 210–212.

10. Turns, S. R., *An Introduction to Combustion: Concepts and Applications,* 2nd ed., McGraw-Hill, New York, 2000.

11. Patankar, S. V., and Spalding, D. B., *Heat and Mass Transfer in Boundary Layers,* 2nd ed., International Textbook, London, 1970.

12. Launder, B. E., and Spalding, D. B., *Lectures in Mathematical Models of Turbulence,* Academic Press, New York, 1972.

13. Schetz, J. A., *Injection and Mixing in Turbulent Flow,* Progress in Astronautics and Aeronautics, Vol. 68, American Institute of Aeronautics and Astronautics, New York, 1980.

14. Wilcox, D. C., *Turbulence Modeling for CFD,* DCW Industries, La Cañada, CA, 1993.

Some end-of-chapter problems were adapted with permission from the following:

15. Look, D. C., Jr., and Sauer, H. J., Jr., *Engineering Thermodynamics,* PWS, Boston, 1986.

16. Myers, G. E., *Engineering Thermodynamics,* Prentice Hall, Englewood Cliffs, NJ, 1989.

17. Pnueli, D., and Gutfinger, C., *Fluid Mechanics,* Cambridge University Press, Cambridge, 1992.

18. Shepherd, D. G., *Elements of Fluid Mechanics,* Harcourt, Brace & World, New York, 1965.

Nomenclature

a	Acceleration (m/s^2)
$\boldsymbol{a}$	Acceleration vector (m/s^2)
A	Area (m^2)
$\boldsymbol{A}$	Area vector (m^2)
D	Diameter (m)
$\dot{E}$	Energy rate (W)
$\hat{\boldsymbol{f}}$	Vector force per unit volume (N/m^3)
F	Force (N)
$\boldsymbol{F}$	Force vector (N)
g	Gravitational acceleration (m/s^2)
$\boldsymbol{g}$	Gravitational acceleration vector (m/s^2)
$\boldsymbol{G}$	$\boldsymbol{g} - \boldsymbol{a}$ vector (m/s^2)
h_L	Head loss (m)
I_{zz}, I_{zy}	Area moments of inertia (see Fig. 6.11) (m^4)
$\hat{\boldsymbol{i}}, \hat{\boldsymbol{j}}, \hat{\boldsymbol{k}}$	Cartesian coordinate unit vectors (dimensionless)
KE	Kinetic energy (J)
L	Length (m)
$\dot{m}$	Mass flow rate (kg/s)
M	Mass (kg)
n	Exponent in power-law velocity distribution (dimensionless)
$\hat{\boldsymbol{n}}$	Unit normal vector (dimensionless)
P	Pressure (Pa)
PE	Potential energy (J)
$\dot{Q}$	Heat transfer rate (W)
$\dot{Q}''$	Heat flux (W/m^2)
r	Radial coordinate (m)
R	Tube radius (m)
t	Time (s)
T	Temperature (K)
u	Specific internal energy (J/kg)
v, V	Velocity magnitude (m/s)
$\boldsymbol{V}$	Velocity vector (m/s)
v_r, v_θ, v_x	Cylindrical coordinate velocity components (m/s)
v_r, v_θ, v_ϕ	Spherical coordinate velocity components (m/s)
v_x, v_y, v_z	Cartesian coordinate velocity components (m/s)
V_∞	Freestream velocity (m/s^2)
$\mathcal{V}$	Volume (m^3)
W	Width (m)
$\dot{W}$	Rate of work or power (W)
X	Arbitrary conserved quantity (various units)
x, y, z	Cartesian spatial coordinates (m)
Y	Fluid column height (m)

GREEK

α	Kinetic energy correction factor (dimensionless)
β	Momentum flow correction factor (dimensionless)
Δ	Difference or change
ζ	Spatial coordinate along slanted surface (m)
μ	Viscosity ($N \cdot s/m^2$)
ν	Kinematic viscosity, μ/ρ (m^2/s)
θ	Angular coordinate in cylindrical and spherical systems or as defined in Fig. 6.10 or 6.14
ρ	Density (kg/m^3)
τ	Shear stress (N/m^2)
ϕ	Angular coordinate in spherical system

SUBSCRIPTS

atm	atmospheric
avg	average
B	bottom
CG	centroid of area
CS	control surface
cv	control volume
grav	gravitational
in	into system or control volume
out	out of system or control volume
P	pressure
rel	relative
ref	reference value or state
sys	system
T	top
visc	viscous
x-sec	cross-sectional

1	station 1
2	station 2

SUPERSCRIPTS

turb	turbulent

OTHER NOTATION

$\overline{(\ )}$	Time-average quantity
$(\ )'$	Fluctuating quantity
$\nabla(\ )$	Gradient (or del) operator, $\hat{i}\partial(\ )/\partial x + \hat{j}\partial(\ )/\partial y + \hat{k}\partial(\ )/\partial z$ in Cartesian coordinates
$\nabla^2(\ )$	Laplacian (or del-squared) operator, $\partial^2(\ )/\partial x^2 + \partial^2(\ )/\partial y^2 + \partial^2(\ )/\partial z^2$ in Cartesian coordinates
$D(\ )/Dt$	Total (or material or substantial) derivative
$\perp$	Perpendicular

QUESTIONS

6.1 Review the most important equations presented in this chapter (i.e., those with a reddish background). What physical principles do they express? What restrictions apply?

6.2 List the forces acting on a fluid element within a fluid at rest.

6.3 Define a Newtonian fluid? What common fluids exhibit Newtonian behavior?

6.4 Define the no-slip condition. Of what practical consequence is this condition?

6.5 What forces contribute to the drag of a body moving through a fluid? In what direction do these forces act?

6.6 Write out the gravitational acceleration vector for a Cartesian (x, y, z) coordinate system with the x axis aligned in the vertical downward direction. Include all three components explicitly. Repeat for the x axis aligned vertically upward.

6.7 List the forces acting on a fluid element within a fluid flowing through a channel or tube.

6.8 Write out Newton's second law for a rigid system with a fixed mass. Be sure to explicitly use the quantity momentum in your expression.

6.9 Provide a precise physical interpretation of the quantity $-\nabla P$.

6.10 Define streamline and streamtube.

6.11 Describe in words the concept of a momentum flow.

6.12 How do conservation equations derived for integral control volumes differ from those derived for differential control volumes?

6.13 Express both in words and symbolically the conservation of momentum principle for steady flow through a control volume having a single inlet and a single outlet.

6.14 Write out the definition of the momentum flow correction factor. What is the physical meaning of this factor?

6.15 Describe how you would modify your statements (words and symbols) in Question 6.13 for the case of unsteady flow.

6.16 The net viscous forces per unit volume can be expressed as $f_{vs} = \mu \nabla^2 V$. Write out the x-component of f_{vs}, (i.e., $f_{vs,x\text{-dir}}$). Use a Cartesian coordinate system.

6.17 Using words only, define the following quantities. Make sure the physical meaning is clear. Be precise.

A. $\displaystyle\int_{CV} V \rho \, d\mathcal{V}$

B. $\displaystyle\int_{CS} \rho V (V_{rel} \cdot \hat{n}) \, dA$

6.18 Symbolically express the following three operators in Cartesian (x, y, z) coordinates. Also indicate whether the operator is a vector or scalar.

A. $\nabla(\ )$

B. $\nabla^2(\ )$

C. $V \cdot \nabla(\ )$

6.19 Symbolically express the application of the operators in Question 6.18 to the quantities indicated below. Again use Cartesian (x, y, z) coordinates. Also indicate whether the resulting quantity is a vector or a scalar.

A. ∇P

B. $\nabla^2 V$

C. $V \cdot \nabla(V)$

6.20 Write out the algebraic expression of the Navier–Stokes equation. (Use the textbook as necessary.) Without reference to the text, describe the physical origins of each term.

6.21 Write out algebraic expressions for the momentum flows entering or exiting the y faces of a small control volume of dimensions Δx, Δy, and Δz.

6.22 Describe the physical differences between the Navier–Stokes equation and the Euler equation.

6.23 Describe the differences between the mechanical energy equation and the Bernoulli equation (steady-flow forms).

6.24 List the restrictions associated with the Bernoulli equation.

6.25 Explain the meanings of the total derivative, the local acceleration, and the convective acceleration.

6.26 Symbolically express the total derivative, $D(\ )/Dt$, using Cartesian spatial coordinates (x, y, z) and velocity components (v_x, v_y, v_z).

6.27 Apply the definition of the total derivative, $D(\ )/Dt$, to the Cartesian velocity vector, $V = \hat{i}v_x + \hat{j}v_y + \hat{k}v_z$, and determine the z-component of the acceleration vector in a three-dimensional unsteady flow that is, find a_z in $a = \hat{i}a_x + \hat{j}a_y + \hat{k}a_z$.

6.28 Compare and contrast the steady-state, steady-flow conservation of energy equation (Eq. 5.66) with the mechanical energy equation (Eq. 6.76).

6.29 Which of the following are correct expressions of either steady or unsteady conservation of mass?

i. $\sum_{\text{Inlets}} (\dot{m}V)_i = \sum_{\text{Outlets}} (\dot{m}V)_i$

ii. $\sum_{\text{Outlets}} \dot{m}_i - \sum_{\text{Inlets}} \dot{m}_i = \dfrac{dM_{cv}}{dt}$

iii. $\sum_{\text{Inlets}} \dot{m}_i - \sum_{\text{Outlets}} \dot{m}_i = \dfrac{dM_{cv}}{dt}$

iv. $\partial(\rho v_x)/\partial x + \partial(\rho v_y)/\partial y + \partial(\rho v_z)/\partial z + \partial\rho/\partial t = 0$

v. $\partial\rho/\partial t + \boldsymbol{\nabla} \cdot (\rho V) = 0$

6.30 Which of the following are correct expressions of either steady or unsteady conservation of linear momentum?

i. $\sum_{\text{Inlets}} (\dot{m}V)_i - \sum_{\text{Outlets}} (\dot{m}V)_i + \sum_{\text{CV}} F = 0$

ii. $\sum_{\text{Inlets}} (\dot{m}V)_i - \sum_{\text{Outlets}} (\dot{m}V)_i + \sum_{\text{CV}} F = 0$

iii. $-\displaystyle\int_{\text{CS}} V\rho(V \cdot \hat{n})\,dA + \sum_{\text{CV}} F = \dfrac{d}{dt}\int_{\text{CV}} V\rho\,d\mathcal{V}$

iv. $-\displaystyle\int_{\text{CS}} \rho(V \cdot \hat{n})\,dA = \dfrac{d}{dt}\int_{\text{CV}} \rho\,d\mathcal{V}$

v. $\partial\rho/\partial t + \boldsymbol{\nabla} \cdot (\rho V) = 0$

6.31 Explain the origins of the turbulent (Reynolds) stresses in turbulent flows.

6.32 How does turbulence affect the integral forms of the conservation principles?

6.33 Explain the meaning of *closure* as it relates to turbulent flow.

PROBLEMS

6.1 Use the NIST database to determine the density ρ and the viscosity μ for the following fluids at the conditions given. Also calculate the kinematic viscosity ν and identify the state (liquid or vapor). Use a table to present your results. Label the table headings and indicate units.

A. H_2O at 20°C and 1 atm

B. H_2O at 20°C and 10 MPa

C. H_2O at 200°C and 10 MPa

D. H_2O at 300 K and 3 kPa

E. N_2 at 300 K and 1 atm

F. N_2 at 1000 K and 1 atm

G. N_2 at 300 K and 10 atm

Analyze your results and comment on the effects of temperature and pressure on viscosity.

6.2 Water is frequently assumed to be incompressible. Estimate $\Delta\rho/\rho$ for a 100-m high column of water. The pressure at the top of the column is 101,325 Pa and the water temperature is 25°C. Discuss.

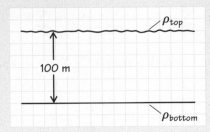

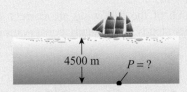

4500 m $P = ?$

6.3 At a latitude of 35° south, the density of seawater varies linearly with depth from 1025 kg/m³ at the surface to 1028 kg/m³ at a depth of 1000 m. The density is constant at 1028 kg/m³ from 1000 m to 4500 m depth. Sketch the density–depth relationship and determine the mass in a 3-m-diameter column of seawater (a) 500 m deep and (b) 4500 m deep. Also determine the weight of the water associated with your results from (a) and (b). Use SI units.

6.4 The maximum depth of the lake at Stone Valley Recreation Area is 65 ft. Determine the pressure at this depth assuming an atmospheric pressure of 100 kPa and a water density of 996 kg/m³.

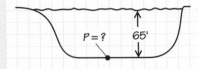

6.5 The water depth in the diving area at a community swimming pool is 12 ft. Determine the pressure at the bottom of the pool in the diving area. Assume the pressure at the water surface is 1 atm. Express your result in units of kPa, atm, bar, and psia.

6.6 Consider the density–depth profile of the ocean described in Problem 6.3. Estimate the absolute pressure at a depth of 4500 m for a surface pressure of 1 atm. Express your result in MPa, atm, and psia.

6.7 Write an expression for the specific volume of an incompressible fluid in a tall cylindrical tank at a distance h below the surface where the pressure is twice the ambient pressure P_{amb}. Assume g is constant. Numerically evaluate your expression for $P_{amb} = 15$ lb$_f$/in² and $h = 5$ m.

6.8 A person with a barometer is driving up a mountain in Colorado. In the foothills of the mountain, the barometer reading is 75 cm Hg absolute. Several hours later the barometer reads 70 cm Hg absolute. Assuming the average density of the atmospheric air is 1.2 kg/m³, estimate the altitude change experienced on this trip.

6.9 The water level in a sealed tank is 26 m above the ground. The pressure in the air space above the water is 0.250 MPa gage. The average density of the water is 1000 kg/m³. Determine the pressure of the water at ground level.

6.10 A water manometer is used to measure the pressure rise across a fan. The manometer reading is 1.1 in of H$_2$O and the density of the water in the manometer is 62.1 lb$_m$/ft³. Determine the pressure difference in psi.

6.11 A U-tube contains water and a fluid of unknown density as shown in the sketch. Both legs of the U-tube are open to the atmosphere. If the density of the water is 997 kg/m³, determine the density of the unknown fluid.

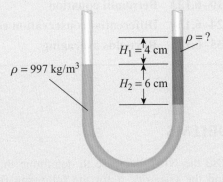

$\rho = ?$

$H_1 = 4$ cm

$\rho = 997$ kg/m³

$H_2 = 6$ cm

6.12 Air flows through a nozzle at high velocities such that compressible flow effects are important. The air exits the nozzle into the atmosphere (1 atm). The inlet pressure is 1.67 times the outlet pressure. What column height of water would be required to measure the inlet-to-exit pressure difference using a manometer? What would the column height be if the manometer fluid were mercury?

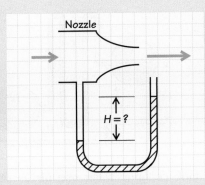

6.13 A simple U-tube manometer is used to measure the pressure drop across a throttling device as shown in the sketch. The fluid passing through the throttle is air, and the manometer fluid is mercury. The density of mercury is 13,550 kg/m³. Determine the pressure drop across the throttle in kPa.

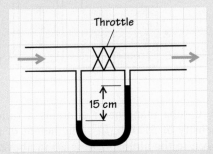

6.14 A simple U-tube manometer is used to measure the pressure drop across a throttling device as shown in the sketch for Problem 6.13. Water flows through the throttle ($\rho_{H_2O} = 997$ kg/m³), and the manometer fluid is mercury ($\rho_{Hg} = 13,550$ kg/m³). Determine the pressure drop across the throttle. Do *not* assume $\rho_{H_2O} \ll \rho_{Hg}$. Compare your result to that from Problem 6.13.

6.15 An inclined manometer containing red oil ($\rho_{oil} = 840$ kg/m³) is used to measure the pressure drop across a capillary tube through which air is flowing. Determine the pressure drop across the capillary tube in kPa and psi.

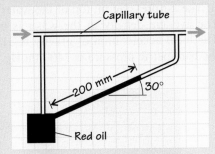

6.16 Determine the pressure at the center of bulb A for the situation described by the sketch. The density of mercury is 13,550 kg/m³, the density of the water is 997 kg/m³, and the atmospheric pressure is 100 kPa.

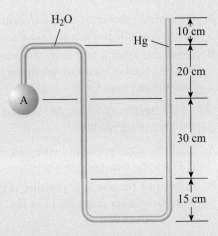

6.17 A piston–cylinder assembly is attached to a tube containing water and mercury as shown in the sketch. The diameter of the piston exposed to the water is 16 cm. The density of mercury is 13,550 kg/m³, the density of the water is 997 kg/m³, and the atmospheric pressure is 100 kPa. What is the weight of the piston?

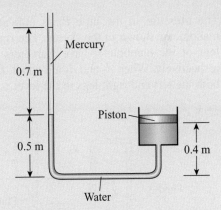

6.18 A U-shaped tube connects two cylindrical reservoirs as shown in the sketch. The tube contains water ($\rho_W = 1000$ kg/m³) from point A to point B and mercury ($\rho_{Hg} = 13,600$ kg/m³) from point B to point C. Weight W_1 (= 300 N) is supported in the left reservoir, and weight W_2 is supported in the right reservoir. There is no fluid leakage around the weights, and there are no frictional forces acting on the weights. The atmosphere pressure P_{atm} is 100 kPa.

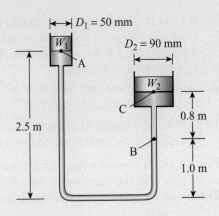

A. Determine the pressure difference between points A and C in Pa (i.e., find $P_A - P_C$).

B. Draw and label a free-body diagram of weight W_1. Use your diagram to determine the absolute pressure at point A.

6.19 Oil with a density of 830 kg/m³ is used in the draft gage shown in the sketch.

A. Find the gage pressure expressed in inches of H_2O for the situation shown.

B. Determine the absolute pressure (kPa and psia) if the barometer reads 29.32 in Hg.

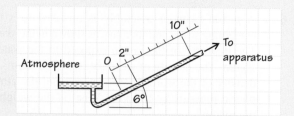

6.20 The pressure in the pipe P_p is indicated by the manometer shown in the sketch. The right and left legs of the manometer have diameters d_1 and d_2, respectively. When d_1 and d_2 are equal, fluid B fills both the left and right legs to the level z_0.

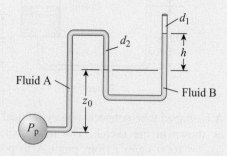

A. Show that the gage pressure in the pipe, P_p, is given by

$$P_p = \rho_A g z_0 + gh\left[\rho_B + (\rho_B - \rho_A)\left(\frac{d_1}{d_2}\right)^2\right].$$

B. If fluid B is mercury ($\rho_B = 13{,}600$ kg/m³) and fluid A is oil ($\rho_A = 830$ kg/m³), find P_p for $h = 4$ ft and $z_0 = 3$ ft for a manometer having both legs of the same diameter.

C. What would be the practical advantage of making d_2 much greater than d_1?

D. Derive an expression relating P_p to h and z_0 directly from the hydrostatic equation for $d_1 = d_2$.

6.21 Pipes A and B contain water. The pressure in pipe A is 40 psig and the pressure in pipe B is 20 psig. Determine the manometer reading, h, in inches.

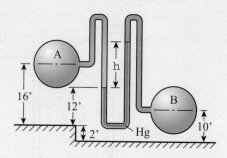

6.22 A tank contains water and air as shown in the sketch. The densities of the water and mercury are 1000 kg/m³ and 13,600 kg/m³, respectively. Find the reading of pressure gage A in both kPa and psi.

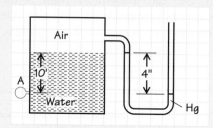

6.23 A U-tube manometer contains a fluid having a density of 51.0 lbm/ft³. The height of the column open to the atmosphere is 20 in above the height of the other column. The barometric pressure is 745 mm Hg.

A. Determine the gage pressure in units of lbf/in² and atmospheres.

B. Determine the absolute pressure in the same units.

C. Determine the height difference in inches when a manometer containing mercury measures this same pressure difference.

6.24 A standpipe is a vertical container filled with water to provide pressure and reserve water for the city water system. The water level in an enclosed standpipe for a particular city water system is 75 ft. The air pressure on top of the water is 20 psig. The average density of the water is 62.4 lbm/ft³. The local acceleration of gravity is 31.5 ft/s² and the barometric pressure is 740 mm Hg.

A. Determine the reading (psig) of a pressure gage located at ground level and connected to the bottom of the standpipe.

B. Determine the absolute pressure (psia) at the bottom of the standpipe.

6.25 You are an engineer who has just been told by a lab technician that the gage pressure of a certain container of helium gas is 8 ft Hg. This sounds too low to you. When you investigate the technician's work you find that the pressure was measured with a large U-tube mercury manometer and that the technician did not notice that some water had been spilled into the end of the manometer that is open to the atmosphere. You measure the height of the water column (which is possible since water does not mix with mercury) and find that it is 27.2 in. You also check a barometer and find that the local atmospheric pressure is 29.0 in Hg.

A. Determine the true gage pressure of the helium (ft H_2O).

B. Determine the absolute pressure of the helium (in Hg).

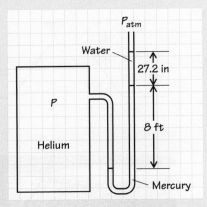

6.26 What is the pressure of the air in reservoir A shown in the sketch? The atmospheric pressure is 100 kPa. Assume the water and mercury densities are 1000 and 13,600 kg/m³, respectively.

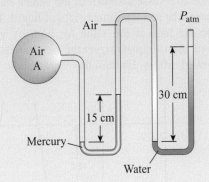

6.27 Consider the inclined water manometer shown in the sketch. When the pressure tap is open to the atmosphere, the water reaches mark B on the

inclined leg. A pressure P is applied and the water level rises a distance L of 100 mm through the inclined tube. The water density is 1000 kg/m³.

A. If $d = 2$ mm, $D = 100$ mm, and $P_0 = 1$ atm, determine the pressure P.

B. What error is introduced by not considering the water level drop in the wide left leg?

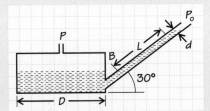

6.28 The pressure difference between points 1 and 2 in a pipe is measured using a manometer shown in the sketch, where $H = 3$ m and $h = 0.2$ m. The pressure at point 2 is 1.5 bars (absolute). The pumped fluid is water ($\rho = 1000$ kg/m³). Determine the pressure at point 1? What is the direction of the flow through the pump?

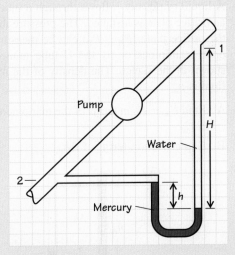

6.29 A kerosene line is fitted with a differential manometer as shown in the sketch. Find the absolute pressure at points A and B, given $P_{atm} = 100$ kPa, $\rho_{water} = 1000$ kg/m³, $\rho_{kerosene} = 900$ kg/m³, and $\rho_{mercury} = 13,600$ kg/m³.

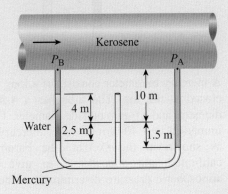

6.30 A mercury barometer gave a true reading of H_1 mm Hg. A quantity of fluid B was then injected into the bottom of the vertical barometer tube. Fluid B rose to the top of the mercury, and the situation became that as shown in the sketch. Determine the vapor pressure of fluid B given the following information: $H_1 = 760$ mm, $H = 660$ mm, $h = 100$ mm, $\rho_{Hg} = 13,600$ kg/m^3, and $\rho_B = 850$ kg/m^3.

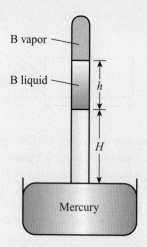

6.31 A liquid of density ρ flows through a tube in which an orifice plate is installed. The pressure drop across the orifice plate ($P_A - P_B$) can be measured by the open standpipes or by the U-tube manometer, as shown in the sketch. The density of the manometer fluid is ρ_m.

A. Find $P_A - P_B$ in terms of the manometer reading Δh_m.

B. Express Δh in terms of Δh_m.

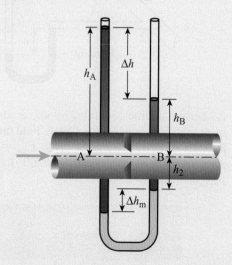

6.32 A mercury barometer consists of a long glass tube closed at the top. The tube has a 4-mm inside diameter and a 6-mm outside diameter. This tube is immersed in a 40-mm-diameter mercury reservoir as shown in the sketch. The barometer was calibrated and was found to give an exact atmospheric pressure determination when reading 760 mm Hg. Now it reads 750 mm. What is the current atmospheric pressure?

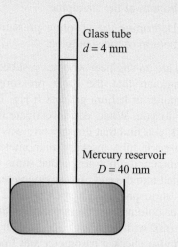

6.33 A 50-m-long concrete dam is used to retain water ($\rho = 997$ kg/m^3) in a reservoir as shown in the sketch. Determine the pressure force exerted by the water on the dam. Also determine the location at which a single resultant force acts to create the same moment on the dam as the distributed pressure force.

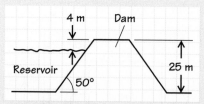

6.34 The two vessels shown in the sketch have the same base area and the same height of liquid. The pressure on the base is the same for both by virtue of the hydrostatic equation, yet the weight of liquid in each vessel differs. Explain this "hydrostatic paradox."

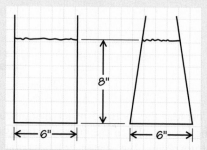

6.35 The containers in the sketch contain water (ρ_{H_2O} = 996 kg/m³) each filled to a depth of 2.5 m. The pressure at the free surface is 100 kPa. For each of the three containers, calculate the following quantities: (a) the pressure in the water at the bottom of the container, (b) the pressure force exerted by the water on the bottom of the container, and (c) the weight of the water. Discuss.

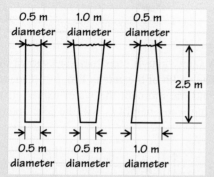

6.36 Consider the tank shown in the sketch. The width of the tank (perpendicular to the page) is 20 m and it is filled to the top with water (ρ_{H_2O} = 1000 kg/m³). The atmospheric pressure is 100 kPa and the gravitational acceleration is 9.81 m/s². Determine the following quantities:

A. The absolute pressure at the bottom of the tank

B. The gage pressure at the bottom of the tank

C. The pressure force exerted on the bottom of the tank (B–C) based on the absolute pressure

D. The pressure force exerted on the left-hand vertical wall (C–D) of the tank based on the absolute pressure

E. The pressure force exerted on the right-hand slanted wall (A–B) of the tank based on the absolute pressure

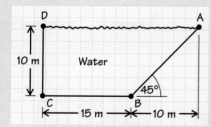

6.37 Flood waters from hurricane Agnes (1972) submerged a basement window in a home as shown in the sketch. Determine (a) the total force exerted on the 2-ft-high, 3-ft-wide glass pane (assuming atmospheric pressure acts on the inside), (b) the location of the center of pressure, and (c) the net moment created by the water about the window centroid.

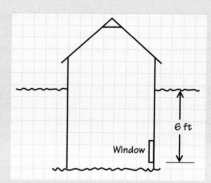

6.38 A circular gate valve is used in a storm sewer system as shown in the sketch. To keep the gate valve closed, a force is applied at the bottom of the valve at $r = D/2$, where r is the radial location from the centroid and D is the valve diameter. (For example, there could be a stop at this location to prevent rotation.) Determine the magnitude of this force and the magnitude of the horizontal force exerted on the hinge pins. Assume that atmospheric pressure acts on the outside (nonwetted side) of the valve. Also assume a storm water density of 996 kg/m³.

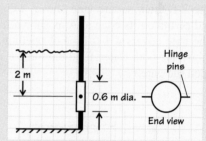

6.39 Consider a situation similar to that described in Example 6.4. The trapezoidal plug is hinged at the top and a stop holds the plug in place at the bottom. An access tunnel leads to the plug; thus, atmospheric pressure acts on the backside of the plug. Find (a) the force exerted by the bottom stop and (b) the reaction forces in the hinge.

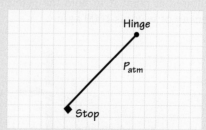

6.40 A rectangular tank is 12-ft long, 6-ft wide, and 8-ft deep. If the tank is filled with water (ρ = 996 kg/m³), what are the magnitude and the location of the resultant force on the 6-ft by 8-ft wall?

6.41 Consider a 5-ft-diameter, horizontal pipe with a vertical flat plate secured over one end. Determine the internal pressure forces acting on the flat plate when the pipe is filled with water to a depth of 3 ft and the remaining space is filled with air at 1 atm pressure.

6.42 The 3-ft by 4-ft flat gate shown in the sketch weighs 200 lb_f. The gate pivots 1 ft from one edge. Find the force that the gate exerts upon its supports and the moment that must be applied to hold it in the position shown. Neglect the thickness of the gate.

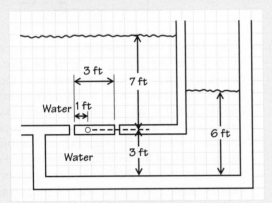

6.43 The 2-ft-thick square gate shown weighs 3000 lb_f and pivots on its edge at A as shown in the sketch. The gate covers a 4-ft by 4-ft opening. The fluids shown are both water. Find the force F necessary to open the gate.

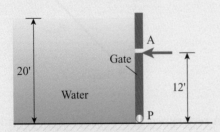

6.44 A rectangular gate is 20-ft wide and pivots along the axis at P as shown in the sketch. Find the force required at point A to hold the gate in place. The density of water is 1000 kg/m³.

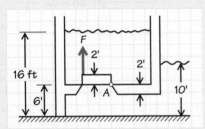

6.45 Consider the reservoir as shown in the sketch. The left-hand wall slopes with angle α, and a trapezoidal gate is located in the right-hand wall. Key dimensions are the following: $h = 1$ m, $b = 0.75$ m, $B = 1.5$ m, and $H = 6$ m.

A. Determine the resultant force acting on the gate and its point of application.

B. Determine the wall angle α to make the line of action of the resultant force R on the left-hand wall pass through point C.

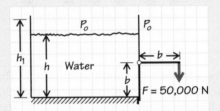

6.46 A safety gate is constructed as shown in the sketch. The width of the gate is 1 m, $b = 1$ m, and $h = 4$ m. Find the moment the water exerts on the gate. At what water height h_1 will the gate open?

6.47 A chamber designed to test underwater windows is shown in the sketch. As indicated, the windows can be set in the following three orientations: horizontal (A), vertical (B), and at an angle α to the vertical (C). The windows themselves are plane and have the shapes shown designated a–d. Find the total force exerted on the windows, its direction, and its point of application in the three positions. Solve the problem for each window configuration.

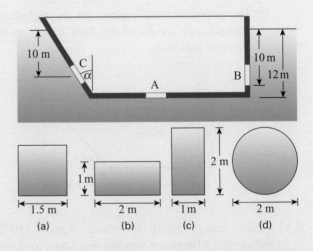

6.48 Two reservoirs, A and B, are filled with water and connected by a pipe, as shown in the sketch. Find the resultant force and the point of application of this force on the wall with the pipe (i.e., the vertical rectangular surface with a circular hole). Use the following dimensions: $L = 4$ m, $H = 3$ m, $D = 1$ m, and $b = 3$ m.

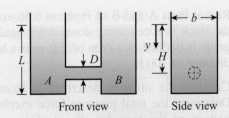

Front view   Side view

6.49 Two bowling balls, one 8-lb$_m$ ball (3.63 kg) and one 16-lb$_m$ ball (7.26 kg), are dropped into the sea from the deck of a yacht. Both balls have a diameter of 8.55 in, and the density of the sea water is 1027 kg/m^3. What is the final fate of each (i.e., do the balls sink or float)? Determine the buoyant force (N) exerted on each of the balls after any initial transient behavior has decayed.

6.50 A 30-cm-diameter, helium-filled, toy balloon is tethered by a string as shown in the sketch and is inflated to 127 kPa. The atmospheric pressure is 100 kPa. The temperature of the balloon and the surroundings is 300 K. The balloon when empty has a mass of 4 g, and the mass of the string is 0.75 g. Determine the tension (N) in the string at the lower attachment point.

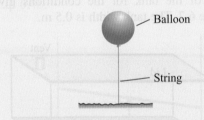

6.51 A nickel coin is glued to the bottom of a cylindrical cork of the same diameter (21.2 mm) as shown in the sketch. The mass and approximate thickness of the nickel are 5 g, and 1.8 mm, respectively. The density of the cork is 200 kg/m^3. The cork and nickel assembly float in water ($\rho = 996$ kg/m^3) with the nickel downward. Determine the length of the cork such that one-third of the cork is exposed to the atmosphere.

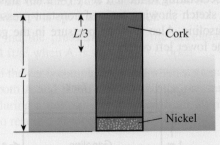

6.52 An 80-mm-wide, 180-mm-long steel plate is glued to the bottom of a 60-mm-thick rectangular balsa-wood block having the same width and length. The densities of the balsa wood and steel are 140 kg/m^3 and 7850 kg/m^3, respectively. Determine the thickness of the steel plate required for the assembly to be neutrally buoyant (i.e., it neither sinks nor floats) in water having a density of 996 kg/m^3.

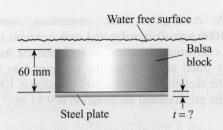

6.53 A hydrometer uses the principle of buoyancy to measure specific gravities of liquids. The specific gravity S is defined as the ratio of the density of the fluid of interest to the density of water (i.e., $S \equiv \rho_{fluid}/\rho_{water}$). The hydrometer consists of a constant-area stem with a bulb weighted so that it is always in stable equilibrium. Different lengths of stem are exposed in liquids of different densities. If a hydrometer is calibrated for a liquid of a specific gravity $S = 1$ with a mark at the point shown, show that the distance h when the hydrometer is placed in another liquid is given by

$$h = \frac{V}{A}\frac{S-1}{S},$$

where V is the volume submerged for a liquid with $S = 1$, and A is the cross-sectional area of the stem.

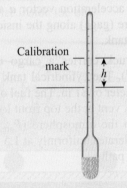

6.54 Consider a tank at rest half filled with water as shown in the sketch. The tank is accelerated horizontally to the left (negative x-direction) at a constant acceleration of 9.81 m/s^2. Gravity acts downward in the z-direction ($g = 9.81$ m/s^2). Assume a steady-state condition is achieved. On a sketch, (a) carefully draw and label the $\mathbf{g} - \mathbf{a}$ vector, (b) draw and label the free surface, and (c) indicate with a bold dot the location of the maximum pressure. Calculate the pressure from (c) for $H = 0.9$ m with $\rho_{H_2O} = 997$ kg/m^3.

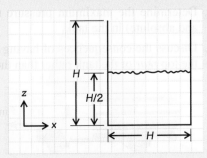

The device is running at steady-state (i.e., time-invariant) conditions. Determine the net force (N) exerted on the fluid. In which direction does the net force act?

6.82 A rocket is mounted on a test stand so that the mass flow leaves the rocket in a downward direction. At a particular instant, the mass flow rate leaving the rocket is 50 lb_m/s and the flow velocity is 500 ft/s. The mass of the rocket and its contents at this instant is 1000 lb_m. Gravitational acceleration is 31.0 ft/s² at this location. Determine the force (lb_f) that the test stand must exert on the rocket to hold the rocket in place.

6.83 A turbofan jet engine is connected to a test stand as shown schematically in the sketch. Air enters the engine at station 1, and a portion of this air is accelerated by a fan and exhausts at station 2. The remaining air is used to burn the fuel that enters as shown. The fuel flow rate is 3.7 kg/s. The hot products of combustion and excess air exit at station 3. (The hot products can be treated as air.) Determine the x-direction reaction force (the thrust) exerted by the test-stand support on the engine assembly. Known properties at stations 1, 2, and 3 are as follows:

	Station 1	Station 2	Station 3
P (kPa)	87.4	101.3 ($= P_{atm}$)	101.3 ($= P_{atm}$)
T (K)	290	403	1000
Flow area (m²)	6.33	3.21	1.31
Mass flow (kg/s)	1157	1000.6	

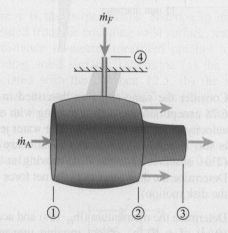

6.84 A jet of water with a velocity of 6 m/s and a mass flow rate of 0.2 kg/s strikes a stationary curved vane. The vane redirects the water jet as shown in the sketch. Values for the v_x and v_y velocity components of the redirected jet are given in the sketch. Determine the horizontal and vertical components of the force exerted on the support

bracket at its base. Neglect the weight of the water, vane, and bracket and any other effects of gravity. Indicate explicitly whether the x-component acts to the left or the right and whether the y-component acts upward or downward. Explicitly show all units and any necessary conversions to yield a final result in newtons.

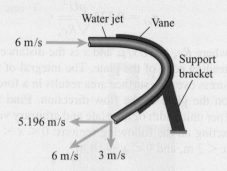

6.85 Water at 25°C flows through a pipe and exits to the atmosphere through a nozzle pointed upward at 30° from the horizontal (station 3). A portion of the water also exits to the atmosphere through a short section of pipe pointed downward at 45° from the horizontal (station 2). This assembly is contained in the x–y plane as shown in the sketch. The pressure and flow rate at station 1 are 40 MPa and 157 kg/s, respectively. The weight of the water inside the assembly (station 1 to stations 2 and 3) is 278 N. The diameters of the circular flow areas at stations 1, 2, and 3 are 100, 40, and 25 mm, respectively. The average velocity at station 2 equals the average velocity at station 3. Determine the net force components in the x- and y-directions that act on the interior walls of the pipes and nozzle. These components will include both the frictional (viscous shear) and the net pressure forces at the walls. Use a control volume defined by the interior space of the assembly between stations 1, 2, and 3. Be sure to list explicitly any assumptions you make. (Note: The pressure where the jet exits the nozzle into the atmosphere is atmospheric pressure.)

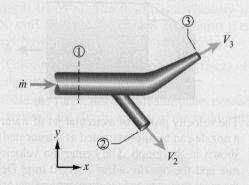

6.86 Bolted flanges hold a U-bend to pipes as shown in the sketch. The diameter of the large pipe is

100 mm and the diameter of the small pipe is 60 mm. Water flows through the bend at a flow rate of 50 kg/s. The weight of the bend, the two lower flanges, and the water contained in the bend total 450 N. The pressure in the water in the plane between the inlet flanges is 400 kPa and the pressure in the water in the plane between the exit flanges is 200 kPa. Determine the net force exerted on the bolts over and above any initial tension force used to close the joint. On average, does this net force act in tension or compression in the bolts?

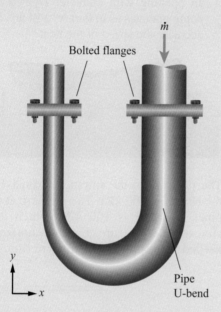

Bolted flanges

$\dot{m}$

Pipe
U-bend

y

x

6.87 Redo Example 6.12 using the moving control volume as the inertial reference frame.

6.88 Consider a water truck similar to that discussed in Example 6.12, but with the following differences: Water exits at the rear through a rake of twenty 11-mm-diameter nozzles angled at 45° toward the ground. The tank is pressurized to provide a total flow rate of 28.3 kg/s. As in the example, the truck is traveling forward at 10 miles/hr (4.47 m/s) and the total drag force opposing the motion of the truck is 95.9 N. Estimate the traction force required to keep the truck moving at a constant speed.

45°

6.89 Water at 60 psig enters a nozzle (station 1 in the sketch) with a flow rate of 300 gal/min and a velocity of 10 ft/s. The water exits the nozzle (station 2), flowing to the atmosphere (P_{atm} = 14.7 psia) with a velocity of 90 ft/s. Find the force required to hold the nozzle in position.

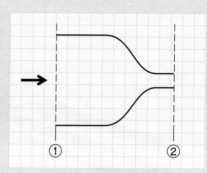

① ②

6.90 Water flows through a 90° reducing elbow in a horizontal plane as shown in the sketch. At the inlet, the velocity is 2.5 m/s, the pressure is 68.9 kPa gage, and the pipe diameter is 152.4 mm. At the outlet, the pipe diameter is 76.2 mm. Assuming ideal flow, find the force necessary to hold the elbow in place. The elbow is part of a continuous piping system.

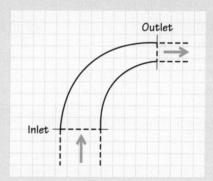

Outlet

Inlet

6.91 A water container has a hole in its side as shown in the sketch. The water flows through this hole with the velocity v_x of 5 m/s. The effective area of the hole is 0.1 m². The pressure around the container is atmospheric. Determine the horizontal force acting on the container.

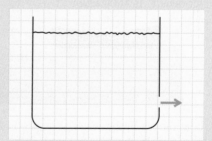

6.92 A flexible fire hose ends with a nozzle as shown in the sketch. The flow areas at stations F and E are 0.01 m² and 0.0025 m², respectively. The average velocity and the absolute pressure at F are 8 m/s and 578 kPa, respectively. The absolute pressure at the exit, E, is atmospheric and has a value of 98 kPa. The hose is connected to the nozzle by a flange at

G. What are the magnitude and direction of the forces acting on the flange? Neglect the weight of the nozzle and the water it contains.

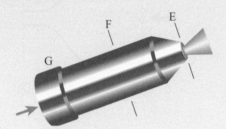

6.93 A ramjet airplane flies horizontally with a velocity of 600 m/s. This is also the same velocity that air enters its engine. Fuel is burned at the rate of 0.12 kg/kg$_{air}$. The combustion gases leave the engine exhaust nozzle at 700 m/s, relative to the airplane. The flow area of the air intake is 0.25 m^2 and that of the nozzle exit is 0.75 m^2. The density of the incoming air is 1.2 kg/m^3. Find the thrust of the engine and the power delivered.

6.94 A floating anchor is a device used in lifeboats to keep the nose of the boat against the waves. The anchor is made of heavy cloth in the shape of a cone with holes at both ends as shown in the sketch. The anchor is tied to the rear of the boat and dragged underwater. The boat in turn is dragged by the waves and the wind. In this scenario, water is forced into the anchor at its larger opening (1-m diameter) and comes out at the narrow end (0.4-m diameter). Experiments show that the pressure at the exit, the narrow end, is only slightly below the hydrostatic pressure for this depth and that the pressure at the wide end is above hydrostatic by about 0.4 ρV^2, where V is the speed of the anchor relative to the water. Furthermore, one can assume that the water exits the anchor at a relative speed essentially equal to V. Estimate the resistance force of this anchor for relative speeds of 1, 3, 10, 15, and 20 m/s. The water density is 1000 kg/m^3.

6.95 A boat moves in the water at a speed of 10 m/s. A pump propels the boat by taking in water at the front (station 1) through a 0.25-m-diameter suction pipe and discharging the water at the rear (station 2) through a 0.2-m-diameter pipe. The inlet and outlet depths are $H_1 = 2$ m and $H_2 = 2.5$ m, respectively. Assume the pressures P_1 and P_2 are the hydrostatic pressures for the given depths. The relative velocity of the fluid at station 1 is that of the boat (i.e., $V_{1,rel}$ is 10 m/s).

A. Find the total driving force applied to the boat, that is, the force transferred through the bolts that connect the pump to the boat, or, equivalently, the total drag force exerted on the boat.

B. Someone suggests connecting a short *converging* cone at point 2 such that the water exits through an opening with diameter $D_3 = 0.15$ m. The pump operates as in Part A. Will this modification increase or decrease the speed of the boat?

C. Someone suggests connecting a short *diverging* cone at point 2 such that the water exits through an opening with diameter $D_4 = 0.25$ m. The pump operates as in Part A. Will this increase or decrease the speed of the boat?

6.96 The pressure at the entrance to pipe 1 shown in the sketch is 200 kPa. The pressure at the exit of pipes 2 and 3 is 100 kPa, which is also the atmospheric pressure. Find the horizontal force needed to keep the pipe structure from moving.

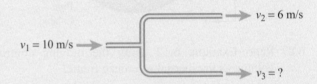

6.97 Consider the water tank and plumbing shown in the sketch. When either valve A or valve B is opened, the flow in the corresponding pipe is 1 m^3/s with a mean velocity of 10 m/s. When both valves are closed, the tank transfers a force of 100,000 N to the floor. Find the magnitude of the force transferred to the floor a short time after (a) valve A is opened with valve B closed, (b) valve B is opened with valve A closed, and (c) both valves are opened.

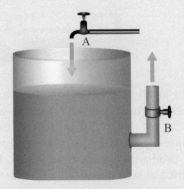

6.98 Consider the pipe bend shown in the sketch. The mean diameter at the entrance flange, point A, is 0.5 m, and that at the outlet, point B, is 0.25 m. The mean velocity of the water at point A is 5.0 m/s. The pressure in the water at point A is 187.5 kPa gage. The atmospheric pressure, which is also the pressure at point B, is 100 kPa absolute. The mass of the pipe section between points A and B is estimated as 10% of that of the body of water inside the pipe. For bend angles of $\alpha = \pi/6$ and $\alpha = \pi/4$, find the mass of water in the bend section that makes the force in the bolts of the flange at A just horizontal (i.e., such that there is no vertical component of the force).

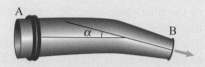

6.99 On yachts and ferries without air conditioning, large scoops on the deck are used to ventilate compartments below. Assuming the maximum wind velocities the ship is expected to encounter are not greater than 100 m/s, find the upward force for which a 1-m-diameter scoop must be designed.

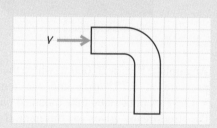

6.100 A light airplane is used to spray insecticide on cotton fields. The spray nozzles are directed toward the rear of the airplane with the spray exiting the nozzles at a speed of 20 m/s relative to the airplane. The total flow rate of the insecticide is 100 kg/s.

A. Draw the velocity vectors of the spray (a) relative to the airplane and (b) as seen by an observer on the ground.

B. Find the thrust added to the airplane by the spray.

C. Find the velocity and the thrust when the nozzles are directed downward.

6.101 Consider a 65-m, straight run of 50-mm-inside-diameter pipe through which water flows at 6.72 kg/s at 300 K. The pipe is horizontal. Determine the head loss in meters if the pressure at the inlet is 49 psig and the pressure at the outlet is 16 psig.

6.102 A 1-hp electric motor drives a water pump. The pump volumetric flow rate is 2×10^{-3} m³/s at 300 K. Determine the head loss associated with the pump if the inlet pressure is 90 kPa and the outlet pressure is 300 kPa. Also determine the pump efficiency. Neglect kinetic and potential energy differences between the inlet and outlet streams.

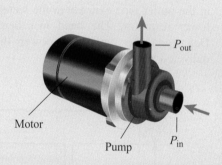

6.103 Water at 20°C is flowing through a 10-cm-diameter pipe at 1.5 m/s. At one point, the pressure is 135 kPa gage. The friction head loss between this point and a second point 10 m below the first is 3.5 m. Find the pressure at the second point.

6.104 A tank contains water filled to a depth of 1.25 m. A siphon is placed in the tank shown in the sketch. The inlet end of the siphon is 1 m below the surface. The friction head loss associated with the siphon is equivalent to 10 kPa. How far below the surface must the outlet end of the siphon be placed to produce a velocity of 0.75 m/s?

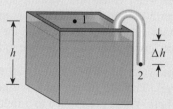

6.105 A fluid flows in a pipe with a sudden increase in its cross section, as shown in the sketch. Assume that the pressure in the wider section right after the jump, P_2, equals P_1. Using a control volume where stations 1 and 3 are the inlets and outlets, respectively, apply conservation of mass and momentum to find expressions for the average velocity at station 3, the pressure at station 3, and the head loss associated with the sudden expansion. Express your results in terms of the diameters d and D, the average inlet velocity v_1, and the pressure at station 1, P_1.

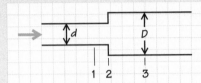

6.106 Water flows from an upper water reservoir through a turbine to a lower reservoir, as shown in the sketch. The pressures at the reservoir surfaces are atmospheric. Friction losses are negligible, except at the outlet pipe exit where the water enters the lower reservoir. The head loss here is $v_2^2/2g$. Determine the maximum power that can be delivered by the turbine.

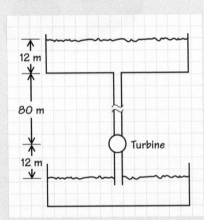

6.107 The pressure in a 0.150-m-diameter water main is 400 kPa. The mean velocity of the water in the main is 10 m/s. The main is 1 m below ground level. An industrial washing machine located 3 m above ground level has a peak water demand of 0.020 m³/s. The water enters the washing machine at atmospheric pressure (100 kPa). Assuming all friction losses are negligible, estimate the diameter of the pipe leading from the main line to the machine.

6.108 The water level in a tank is 20 m above ground level. The top of the tank is open to the atmosphere. Water is supplied to a field at ground level through a 0.012-m-diameter pipe.

   A. Neglecting friction, find the maximum flow rate that can be supplied.

   B. To increase supply rate, the pipe is changed to another having a diameter of 0.019 m. Find the new flow rate.

6.109 Water flows through a main at 10 m/s and enters a booster pump at 300 kPa. The water is then distributed into houses located 100 m above the main line. When a tap is opened in a house, the water comes out at the speed of 20 m/s. The atmospheric pressure is 100 kPa. Friction head losses between the water main and the houses are estimated to be five times the velocity head $(v^2/2g)$ at the houses. The booster pump overall efficiency is 0.8. Find the power needed to run the booster pump per cubic meter of water.

6.110 A pitot-static tube is used to measure the velocity profile across the test section of a wind tunnel. A water manometer is used to measure the difference between the stagnation and static pressures of the pitot-static tube. At a particular location in the tunnel, the manometer reading is 90 mm H₂O. The temperature of the air in the tunnel is 305 K and the pressure is 102 kPa. What is the air velocity in the tunnel at this measurement location?

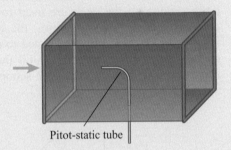

Pitot-static tube

6.111 An airplane flies at 20,000 ft, where the atmospheric pressure is 6.753 psia and the temperature is −12.3 F. A pitot-static tube mounted on the plane in the freestream gives a differential manometer reading equivalent to 8.5 in H₂O. Estimate the speed of the plane, assuming incompressible flow.

6.112 Consider a 1-m-diameter cylindrical tank. The tank is filled with water to a depth of 0.8 m above a 3-mm-diameter hole in the sidewall of the tank. The entrance to the hole is well rounded. Assuming frictionless flow, estimate the velocity and volume flow rate of the water escaping from the hole.

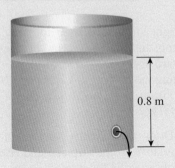

6.113 Water exits vertically downward from a faucet with a flow rate $\dot{V}$. The diameter of the water jet at the faucet exit is D_1. Determine the diameter of the water jet D_2 at a vertical distance L from the faucet.

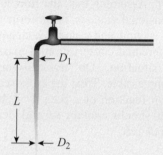

6.114 An insurance-underwriter Class A pump for fire-truck duty should deliver a maximum flow of 1500 gal/min with a 150 psig stagnation pressure at the pump exit. Assuming that no losses occur between the pump and the delivery nozzle at the end of the hose, find (a) the height to which the pump can deliver water and (b) the diameter of the nozzle for the maximum flow.

6.115 Water flows through the system shown in the sketch. The manometer fluid is mercury. Assume ideal (frictionless) flow throughout.

A. Determine the depth of the water in the tank, h, necessary to provide a flow rate of 3 gal/min.

B. What is the manometer reading Δh (in Hg) for this condition?

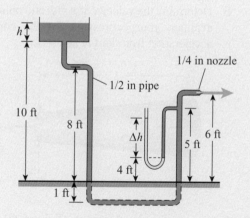

6.116 Oil with a density of 900 kg/m³ flows steadily from the constant-level reservoir through the constant-area pipe to the atmosphere (see sketch). Assuming ideal (frictionless) flow, find the manometer reading Δh (in Hg). The oil is separated from the mercury by a column of water. The water density is 1000 kg/m³.

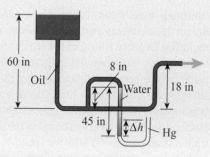

6.117 Water is siphoned through a 75-mm-diameter pipe as shown in the sketch.

A. Assuming ideal (frictionless) flow, determine the volume flow rate $\dot{V}$ (m³/s) and the pressure at point B.

B. Find the pressure at B if the friction head loss between E and B is 3 ft and that between B and A is 4 ft.

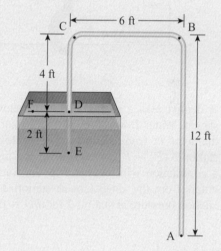

6.118 A venturi tube consists of a converging pipe section followed by a diverging section as shown in the sketch. This device can be used to measure the flow rate of a fluid by measuring the pressure difference between station 1 and station 2. Assuming conditions are met for applying the Bernoulli equation, derive expressions for the velocity at station 1 and the mass flow rate. Express your result in terms of the pressure difference $P_1 - P_2$, the fluid density ρ, and the two diameters d_1 and d_2.

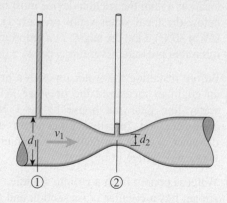

6.119   Railroads used steam locomotives into the 1950s. Steam locomotives operated without condensers, requiring frequent filling of the water tenders. Some locomotives incorporated a system that allowed water to be taken on without stopping the train. After slowing the train from a high speed, a water scoop was lowered from the tender into an open water tank laid beside the railway, as shown in the sketch. For a train moving to the left with the speed V_1 and a water tank that permits a 50-m run with the scoop lowered, estimate (a) the velocity of the water relative to the scoop at station 2 and (b) the volume of water taken into the tender. The width of the scoop is b. Express your results in terms of V_1, H, b, and δ_1. How do your results simplify if you assume δ_1 and δ_2 are both much less than the height H?

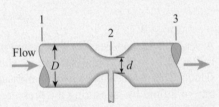

6.120   A dentist uses a suction device as shown in the sketch. Water flows through the primary tube of diameter D from station 1 ($P_1 = 150$ kPa) to station 3 ($P_3 = P_{atm} = 100$ kPa), passing through a contraction with diameter d. The suction flow enters from the small tube at station 2. Find the suction pressure at station 2 for $d/D = 0.8$.

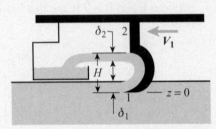

6.121   Consider the dentist's suction device described in Problem 6.120. A 5-m-high water tower maintains a constant pressure in the main water line feeding the device. The atmospheric pressure is 100 kPa ($= P_3$). To ensure smooth operation, the pressure at any location within the suction device must never drop below the local water vapor pressure (i.e., 4.246 kPa at 30°C). Find the largest D/d ratio such that the minimum pressure never drops below 5.0 kPa.

6.122   Before installing the water pipes in a new house, an engineer measured the pressure in the main water line near the house to be 1 MPa. The engineer then installed pipes with an inside diameter of 2 cm. Estimate the maximum possible water supply rate in m³/s.

6.123   Water is poured from a jar into a bottle. The water stream has a circular cross section and enters the bottle with a 0.02-m diameter. The flow rate is 0.0005 m³/s. Estimate the height of the jar above the bottle.

6.124   A fluid flows steadily through a duct of circular cross section and constant taper as shown in the sketch. Assuming that, the flow is essentially one dimensional (i.e., v_x is uniform across the duct cross section, and v_r and v_θ both equal zero), derive a differential conservation of mass expression for this situation. Do not assume the flow is incompressible. Treat the cross-sectional area as a known function of x [i.e., $A = A(x)$.] *Hint:* Your result should contain several derivatives of the form $d(\ )/dx$.

6.125   Consider the same physical situation as described in Problem 6.124 but with the duct tapering to a smaller diameter in the flow direction. Water at 25°C (an incompressible fluid) is flowing steadily through the duct. The diameter of the duct at the entrance is $D_1 = 100$ mm and at the exit $D_2 = 80$ mm. The length of the duct is $L = 0.75$ m. Since the taper of the duct is constant, the diameter varies with axial distance as $D(x) = D_1 + [(D_2 - D_1)/L]x$. The average velocity at the entrance is 3 m/s.

A.   Write an algebraic expression for the convective acceleration at any x location in the duct.

B.   Determine the velocity at $x = L$ and numerically evaluate your expression for the convective acceleration from Part A at $x = L$.

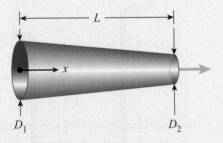

6.126   Following the development of Eq. 6.60, draw a sketch and write out similar expressions for the forces acting in the y- and z-directions.

6.127   Consider the differential form of the conservation of momentum equation, that is, the Navier–Stokes equation:

$$-\nabla P + \mu \nabla^2 V + \rho g = \rho \frac{DV}{Dt}.$$

A. For a Cartesian (x, y, z) coordinate system, rewrite this equation expanding the various operators: $\nabla(\ )$, $\nabla(\ )^2$ and $D(\ )/Dt$. Use the velocity components, v_x, v_y, and v_z, and the unit vectors, $\hat{i}$, $\hat{j}$, and $\hat{k}$, as necessary.

B. Write out the x-component equation from your expression in Part A.

C. Simplify your x-component equation from Part B for a steady, two-dimensional (x, y) flow. Assume that gravity acts in the y-direction.

6.128 Consider flow through a circular pipe. Cylindrical coordinates (Table 6.4) are the most useful to describe this flow. Simplify these equations for a steady, fully developed, flow through a circular tube. The term "fully developed" means that the velocity profile does not vary with axial distance x, (i.e., v_x is a function of the radial coordinate r only). Assume that there is no swirl component to the velocity (i.e., $v_\theta = 0$). *Hint:* Use Eq. 3.29 (mass conservation) to say something useful about the radial velocity v_r.

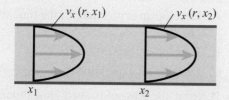

6.129 Consider a 2-D flow between two parallel plates of length L separated a distance $2d$ as shown in the sketch. The following assumptions apply:

 i. The flow is steady.

 ii. The flow is fully developed with $v_x(x, y) = v_x$ (y only).

 iii. The fluid is incompressible and Newtonian.

 iv. The axial pressure gradient $\partial P/\partial x$ is a constant.

 v. Gravity acts in the y-direction.

 vi. The no-slip boundary condition applies at the plate surfaces.

 vii. The plates are impermeable [i.e., $v_y(x, \pm d) = 0$].

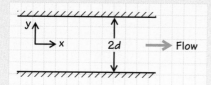

A. Simplify the continuity equation and use it to obtain $v_y(x, y)$. See Eq. 3.41. *Hint:* Use the impermeable wall boundary condition. Your final result will be simple (trivial?).

B. Using the assumptions and your result from Part A, simplify the x-component and y-component

of the Navier–Stokes (momentum conservation) equation.

C. Solve your x-component equation for the axial velocity distribution $v_x(y)$. The plate spacing d and the axial pressure gradient $\partial P/\partial x$ should be parameters in your solution.

D. Solve your y-component equation for the vertical pressure distribution $P(y)$.

6.130 Write the Euler equations of motion (see Example 6.13) for a two-dimensional flow with the following special conditions: The body force is in the x-direction only, the pressure force in the y-direction is constant, and the velocity in the y-direction varies in the y-direction but not in the x-direction.

6.131 A laminar layer of glycerin slides down on a semi-infinite vertical wall as shown in the sketch. The density and viscosity of the glycerin are 1260 kg/m^3 and 0.80 N·s/m^2, respectively. Calculate the shear stress on the wall and the volume flow rate of the glycerin. Assume the thickness of the glycerin layer is constant (i.e., δ = constant). How are your results affected if the wall leans at an angle α?

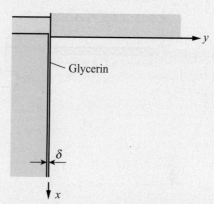

6.132 Consider fully developed, laminar flow through a tube. The assumptions describing this flow are as follows:

 i. The flow is steady.

 ii. The flow is fully developed with no swirl [i.e., $v_x(r, x, \theta) = v_x$ (r only)].

 iii. The fluid is incompressible and Newtonian.

 iv. The axial pressure gradient, $\partial P/\partial x$, is a constant.

 v. Gravity acts perpendicular to the x-direction (i.e., the pipe is horizontal).

 vi. The pipe walls are impermeable [i.e., $v_r(R, x) = 0$].

Simplifying the x-momentum equation results in

$$\frac{1}{r}\frac{d}{dr}\left(r\frac{dv_x}{dr}\right) = \frac{1}{\mu}\frac{\partial P}{\partial x}.$$

A. List the boundary conditions needed to solve this ordinary differential equation.

B. Solve this ordinary differential equation for the velocity distribution $v_x(r)$.

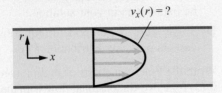

6.133 Viscosity of printing ink is measured by first coating a metal plate with the ink and then passing the plate through a gap between two parallel stationary plates as shown in the sketch. The velocity of the plate is measured by timing the descent of the weight between two marks on a meterstick. Determine the viscosity of the printing ink if the plate velocity is 1 cm/s, the length and width of the stationary plates are 20 cm and 10 cm, respectively, the gaps on each side of the moving plate are 0.5 mm, and the weight is 5 N.

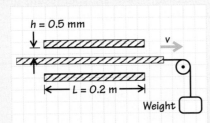

6.134 A fluid flows through the gap between two plates ($2h = 10$ mm), as shown in the sketch. The pressure drop in the fluid between the plates is given by

$$-\frac{\Delta P}{\Delta L} = 20 \text{ Pa/m} = 20 \text{ N/m}^3.$$

Find the maximum velocity in the flow and the shear stress on the plates for (a) water ($\mu = 0.001$ N·s/m^2) and (b) glycerin ($\mu = 1.5$ N·s/m^2).

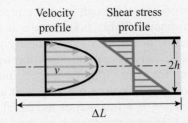

6.135 The axial velocity at a point in a flow is given by

$$v_x(t) = A\sin(\omega_1 t + \Phi_1) + B\sin(\omega_2 t + \Phi_2) + C.$$

A. Develop an expression for the time-mean velocity $\bar{v}_x$.

B. Develop an expression for v_x'.

C. Develop an expression for $v_{x,\text{rms}}'$.

6.136 Perform the Reynolds-averaging process on the second and fourth terms of Eq. 6.83. Compare your results with Eq. 6.87.

Appendix 6A
Linear Momentum Conservation for Control Volumes in Noninertial Coordinate Systems

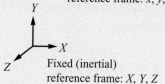

$V = V_{cv} + V_r$

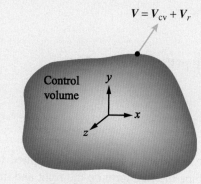

Accelerating (noninertial)
reference frame: x, y, z

Fixed (inertial)
reference frame: X, Y, Z

FIGURE 6A.1
The velocity at any location within the control volume (or on its surface) can be represented as the sum of the velocity of the x, y, z coordinate system with respect to the fixed system (X, Y, Z), V_{cv}, and the local velocity within the x, y, z coordinate system, V_r.

In Fig. 6A.1, we define a fixed (inertial) reference frame with coordinates X, Y, and Z and an accelerating (noninertial) reference frame with coordinates x, y, and z. The acceleration is rectilinear (i.e., nonrotational). For an observer riding with the accelerating control volume, the x, y, z system is the local frame of reference. The velocity relative to the fixed reference frame, V, at any point within the control volume or on the control surface can be expressed as

$$V = V_{cv} + V_r, \tag{6A.1}$$

where

$V_{cv} \equiv$ velocity of the accelerating coordinate system with respect to the fixed coordinate system,
$V_r \equiv$ local velocity within the accelerating coordinate system.

Before proceeding, we define one more velocity:

$V_{r,\mathrm{rel}} \equiv$ Velocity within the x, y, z reference frame relative to the control surface boundary.

For a noninertial reference frame, Eq. 6.59b expresses the general conservation of momentum principle applied to a control volume:

$$-\int_{\mathrm{CS}} V\rho(V_{\mathrm{rel}} \cdot \hat{n})\,dA + \sum F = \frac{d}{dt}\int_{\mathrm{CV}} \rho V\,d\mathcal{V}. \tag{6.59b}$$

To obtain an expression of momentum conservation for a noninertial control volume, we substitute Eq. 6A.1 into Eq. 6.59b as follows:

$$-\int_{\mathrm{CS}} (V_{cv} + V_r)\,\rho(V_{r,\mathrm{rel}} \cdot \hat{n})\,dA + \sum F = \frac{d}{dt}\int_{\mathrm{CV}} \rho(V_{cv} + V_r)\,d\mathcal{V}. \tag{6A.2}$$

To create the desired final result, we expand the two integral terms. Starting with the first term on the left, we write

$$-\int_{\mathrm{CS}} (V_{cv} + V_r)\rho(V_{r,\mathrm{rel}} \cdot \hat{n})\,dA = -V_{cv}\int_{\mathrm{CS}} \rho(V_{r,\mathrm{rel}} \cdot \hat{n})\,dA - \int_{\mathrm{CS}} V_r\rho(V_{r,\mathrm{rel}} \cdot \hat{n})\,dA.$$

The right-hand side of Eq. 6A.2 can be expanded to yield

$$\frac{d}{dt}\int_{\mathrm{CV}} \rho V_{cv}\,d\mathcal{V} + \frac{d}{dt}\int_{\mathrm{CV}} \rho V_r\,d\mathcal{V}.$$

The first term here can be expressed as

$$\frac{d}{dt}\left[\boldsymbol{V}_{\text{cv}}\int_{\text{CV}}\rho\mathcal{V}\right] = \boldsymbol{V}_{\text{cv}}\frac{d}{dt}\int_{\text{CV}}\rho d\mathcal{V} + \int_{\text{CV}}\rho d\mathcal{V}\frac{d\boldsymbol{V}_{\text{cv}}}{dt},$$

or, after performing the indicated integrations,

$$\boldsymbol{V}_{\text{cv}}\frac{dM_{\text{cv}}}{dt} + M_{\text{cv}}\frac{d\boldsymbol{V}_{\text{cv}}}{dt},$$

or

$$\boldsymbol{V}_{\text{cv}}\frac{dM_{\text{cv}}}{dt} + M_{\text{cv}}\boldsymbol{a}_{\text{cv}}$$

where $\boldsymbol{a}_{\text{cv}}$ is the acceleration of the control volume seen in the fixed reference frame.

We now reassemble Eq. 6A.2:

$$-\boldsymbol{V}_{\text{cv}}\int_{\text{CS}}\rho(\boldsymbol{V}_{r,\,\text{rel}}\cdot\hat{\boldsymbol{n}})dA - \int_{\text{CS}}\boldsymbol{V}_r\rho(\boldsymbol{V}_{r,\,\text{rel}}\cdot\hat{\boldsymbol{n}})dA + \sum\boldsymbol{F}$$

$$= \frac{d}{dt}\int_{\text{CV}}\rho\boldsymbol{V}_r d\mathcal{V} + \boldsymbol{V}_{\text{cv}}\frac{dM_{\text{cv}}}{dt} + M_{\text{cv}}\boldsymbol{a}_{\text{cv}}.$$

The final step is to recognize that conservation of mass can be applied to simplify this equation; that is, we combine the first term on the right and the first term on the left as

$$\boldsymbol{V}_{\text{cv}}\left[\frac{dM_{\text{cv}}}{dt} + \int_{\text{CS}}\rho(\boldsymbol{V}_{r,\text{rel}}\cdot\hat{\boldsymbol{n}})dA\right]$$

and recognize that the bracketed term is zero (see Eq. 3.21). Thus,

$$-\int_{\text{CS}}\boldsymbol{V}_r\rho(\boldsymbol{V}_{r,\text{rel}}\cdot\hat{\boldsymbol{n}})dA + \sum\boldsymbol{F} = \frac{d}{dt}\int_{\text{CV}}\rho\boldsymbol{V}_r d\mathcal{V} + M_{\text{cv}}\boldsymbol{a}_{\text{cv}}. \qquad (6A.3)$$

The only important restriction associated with this result is that the acceleration be rectilinear and not rotational.

SECOND LAW OF THERMODYNAMICS AND SOME OF ITS CONSEQUENCES

*After studying Chapter 7,
you should:*

- *Be able to state in words the Kelvin–Planck statement of the second law of thermodynamics and at least one other meaningful statement of the second law.*

- *Be able to state two or more ways in which the second law is useful in engineering applications.*

- *Be able to explain the difference between a reversible and an irreversible process and illustrate it with a concrete example.*

- *List four or more irreversible processes.*

- *Understand the concept of a heat engine and its reversed-cycle counterparts, the heat pump and the refrigerator.*

- *Be able to draw the processes comprising the Carnot cycle on pressure–volume and temperature–entropy coordinates.*

- *Be able to calculate the thermal efficiencies of reversible heat engines.*

- *Be able to state in words and write out symbolically the macroscopic (Clausius) definition of entropy.*

- *Be proficient in illustrating ideal and real processes on $P–v$, $T–s$, and $h–s$ coordinates for the following common devices: turbines, compressors, pumps, heat exchangers, throttles, and nozzles.*

- *Be able to explain the increase of entropy principle and show how it is an indicator of spontaneous change.*

- *Be able to calculate equilibrium constants from tabulations of Gibbs function of formation data.*

- *Be able to calculate the equilibrium composition of systems involving a single equilibrium reaction.*

- *Be able to calculate the equilibrium composition of systems involving two or more equilibrium reactions.*

- *Enjoy the beauty of the second law of thermodynamics.*

Chapter 7 Overview

The first law of thermodynamics can be mathematically expressed in a variety of ways. All of these expressions, however, are easily viewed as rearrangements of the statement that energy can neither be created nor destroyed, but only converted from one form to another. In contrast, there is no single universally agreed upon statement of the second law of thermodynamics. Kline [1] indicates that many seemingly different statements have been accepted as the second law, all of which, however, can be shown to be equivalent after careful and sometimes subtle application of logic. This multiplicity of apparently disparate statements can lead to confusion in understanding the second law. In this chapter, we examine several statements of the second law and discuss the consequences of each.

To understand these various second law statements and their consequences, we define and develop many concepts in this chapter. These include various devices that execute thermodynamic cycles (heat engines, heat pumps, and refrigerators), the distinctions between reversible and irreversible processes, the thermodynamic temperature scale, and second-law–based maximum theoretical efficiencies for cycles and individual components. We define entropy and explore its usefulness. The chapter concludes by examining the principles of chemical equilibrium and phase equilibrium as extensions of the second law.

7.1 HISTORICAL CONTEXT

Sadi Carnot (1796–1832)

Sadi Carnot (1796–1832), a French military engineer during the rise and fall of Napoleon, worked to understand the relationship between the heat supplied to an engine and the work it produced. Steam engines during Carnot's time converted only about six percent of the energy in the fuel to useful work [2]. In 1824, Carnot published his now celebrated paper in which he set forth the ideas that a reversible cycle is the most efficient of any cycle and that the efficiency of the conversion of heat to work depends only on the temperatures at which an engine receives and rejects heat. Although Carnot worked under the misconception of the caloric theory, his ideas concerning the second law have stood the test of time as fundamental principles. Today we still use the *Carnot efficiency* as a limiting case for practical devices.

Rudolf Clausius (1822–1888) recognized the importance of Carnot's work, and building upon this, he presented a clear statement of the second law, a statement that we will explore in some detail. Although others made significant contributions to the development of the second law, Clausius is frequently regarded as its discoverer because of his naming of the property *entropy* [2]. Because the word *energy* is central to the first law of thermodynamics, he chose a like-sounding word, *entropy*, to designate the property that is central to the second law. Other key figures in the development of the second law are

Rudolf Clausius (1822–1888)

Ludwig Boltzmann (1844–1906)

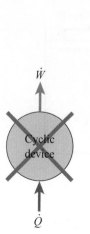

William Thomson (**Lord Kelvin**) (1824–1907), who used second-law concepts to define an absolute thermodynamic temperature scale, and **Josiah Willard Gibbs** (1839–1903), a Yale professor who rigorously generated and extended thermodynamic concepts to reacting systems. **Ludwig Boltzmann** (1844–1906), one of the originators of statistical mechanics, derived a physical meaning of entropy for gases from the behavior of assemblies of molecules. His famous equation $S = k \ln W$, which states that the entropy of a system is a measure of the probability of its state, is engraved on his tombstone. **Max Planck** (1858–1947), in addition to winning the Nobel Prize for his quantum theory of energy, advanced the understanding of the second law and entropy by approaching the subject from a macroscopic point of view, that is, from a point of view that does not take into account the detailed behavior of the atoms and molecules comprising a system. **Joseph Keenan** (1900–1977) is often credited as a modern codifier of second-law concepts for engineers [3] and with introducing irreversibility as a thermodynamic property [1].

7.2 USEFULNESS OF THE SECOND LAW

To provide a firm focus for the remainder of the chapter, we set out the following two ways in which the second law is particularly useful to engineers:

- The second law establishes the theoretical limits of performance of cycles, engines, and other energy conversion devices—over and above those imposed by energy conservation—and provides means to quantitatively compare real devices with these theoretical ideals.
- The second law determines the direction for any spontaneous change and, furthermore, can be used to determine the equilibrium state of any system.

The second law is useful in many other ways as well; for example, the second law provides a means to define a thermodynamic temperature scale independent of the properties of any substance. Although we discuss this and other uses, a major objective of this chapter is for the reader to develop an appreciation of the two particular uses just set out.

7.3 ONE FUNDAMENTAL STATEMENT OF THE SECOND LAW

As indicated in the chapter overview, the second law of thermodynamics can be stated in many ways. We choose the following statement to be a useful starting point for engineering purposes:

Statement IA: Although all work can be converted completely to heat, heat cannot be completely and *continuously* converted into work.

This statement paraphrases the often quoted and more precise **Kelvin–Planck statement** of the second law [4]:

Statement IB: It is impossible to construct a cyclically operating device for which the sole effect is exchange of heat with a single reservoir and the creation of an equivalent amount of work.

The immediate and obvious consequence of statement IA is that, in effect, not all forms of energy are equally valuable. Here we see that work is inherently a more valuable form of energy than is heat because we can always convert all work to heat, but not vice versa. The first law of thermodynamics, the conservation of energy principle, puts no limits on the interconversion, only that energy cannot appear or disappear, that is, our energy accounting ledger must always balance.

We can gain some insight into this qualitative difference between heat and work by returning to our discussion of these two energy transfer modes in Chapter 4. There we saw that, in a gas, heat transfer by conduction is a result of the random collisions of molecules and requires no macroscopic organization of these molecules to cause the energy exchange that we call heat transfer; however, for work to be extracted from a volume of gas requires a macroscopic organization superimposed on the random molecular motion. This macroscopic organization is the macroscopic flow velocity resulting from a moving boundary. As illustrated in Fig. 4.4, the system boundary must be moving for energy to be extracted from the molecules that collide with it. No such organized motion is required for heat transfer to take place. That energy has quality as well as quantity has important implications for engineering design.[1]

A second consequence of statements IA and IB is that they place no restriction on the exchange of heat for work in a *process*; that is, all heat can be converted to work in a process. They do, however, place a severe restriction on devices that execute a thermodynamic cycle, that is, a series of processes that returns the working fluid to its initial state. This restriction to cycles is explicit in the Kelvin–Planck statement (statement IB) and implicit in statement IA by the use of the word *continuously*.

> **Chapter 4 presents precise definitions of heat and work.**

> **See Chapter 1 to review the definitions of thermodynamic processes and cycles.**

---

[1] Quality as used here should not be confused with the quality (x) used to describe a liquid–vapor mixture.

Example 7.1

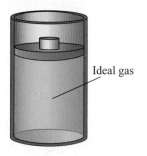

Ideal gas

Consider a piston–cylinder arrangement filled with an ideal gas. Show that for a quasi-static, isothermal, heat-addition process all of the heat added is converted to work.

Solution

Sketch

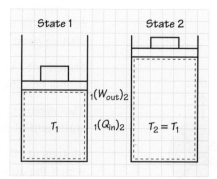

Assumption

Changes in kinetic and potential energies are negligible.

Analysis   We define the gas in the cylinder to be the thermodynamic system of interest. Conservation of energy relates the heat added and the work performed by the gas (Eq. 5.6) as follows:

$$_1(Q_{in,net})_2 - _1(W_{out,net})_2 = \Delta E_{syst}.$$

With negligible changes in kinetic and potential energies, the change in the system energy for the process is expressed as

$$\Delta E_{syst} = U_2 - U_1 = m(u_2 - u_1).$$

For an ideal gas, the internal energy for a process is expressed by Eq. 2.31d:

$$u_2 - u_1 = \int_{T_1}^{T_2} c_v \, dT.$$

As the given process is isothermal (i.e., $T_1 = T_2$), we conclude that

$$u_2 - u_1 = 0.$$

Energy conservation is thus

$$_1(Q_{in})_2 - _1(W_{out})_2 = 0,$$

or

$$_1(W_{out})_2 = _1(Q_{in})_2.$$

We thus conclude that, for this particular process, all of the heat added is converted to work.

Comment   The purpose of this example is to illustrate that for a thermodynamic *process* it is possible to convert all heat added to work. Later in this chapter, we see that, for a *continuously operating device*, the second law restricts how much of the heat added can be converted to work. The important distinction here is between a thermodynamic *process* and a thermodynamic *cycle*. We discuss this later.

Self Test 7.1

☑ **Consider a piston–cylinder device and a rigid tank, both containing saturated-liquid water at room temperature and pressure. Heat is added to both devices until a saturated-vapor state exists. Is heat converted entirely to work in both cases?**

(Answer: For the piston–cylinder device boundary work is performed, but the internal energy of the system has also increased because $\Delta U = m(u_g - u_f) \neq 0$; for the rigid tank, no work is performed and all of the heat transfer results in an increase in the internal energy of the water.)

To explore the deeper meaning and significance of the second law given by statements IA and IB, we need to define formally what is meant by heat reservoirs, heat engines, thermal efficiency, and reversibility.

7.3a Reservoirs

A **heat reservoir** is a source of heat energy that is sufficiently large such that the extraction of any desired amount of energy as heat does not change the temperature of the reservoir. For most practical purposes, the earth's atmosphere,

Oceans, large lakes, and the earth's atmosphere serve as thermal reservoirs.

oceans, lakes, and rivers can be considered heat, or thermal, reservoirs. Large amounts of heat can be added or removed from these without any significant change in their temperatures. For example, an air conditioning unit exchanges heat with the atmosphere without affecting the temperature of the atmosphere.[2] Frequently a boiling or condensation phase change has the practical effect of acting as a constant-temperature reservoir.

7.3b Heat Engines

The Kelvin–Planck statement explicitly mentions a *cyclically operating device*, which can be construed as a **heat engine**. Many consequences of the various statements of the second law involve heat engines. Figure 7.1 illustrates this concept. Here we see that energy is transported as heat from a high-temperature reservoir to the engine. The engine converts a portion of this heat energy to work and rejects the remainder to the low-temperature reservoir. Because the engine operates in a thermodynamic cycle—with the working

FIGURE 7.1

A heat engine (a) receives energy from a high-temperature reservoir, some of which is converted to work, while the remainder is rejected to the low-temperature reservoir. Reversing the cycle converts the heat engine into a heat pump or refrigerator (b). Work is used to transfer energy from the low-temperature to the high-temperature reservoir.

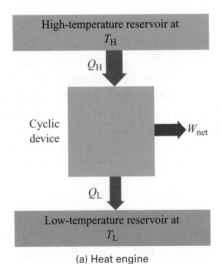
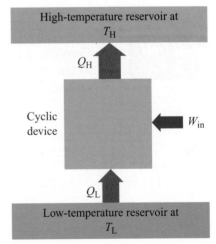

(a) Heat engine

(b) Reversed heat engine (heat pump or refrigerator)

[2] At intermediate scales, however, thermal pollution and the heat island effects can come into play such that local regions of the atmosphere are affected. For example, the huge energy released at night from roads, parking lots, and buildings affects nighttime temperatures in Phoenix, Arizona.

fluid always returning to its initial state—the device is capable of continual operation. The first law of thermodynamics, the energy conservation principle, relates the heat supplied and rejected during the cycle to the net work produced. For any incremental portion of the cycle, Eq. 5.5 applies:

$$\delta Q - \delta W = dU. \tag{7.1}$$

We integrate over the cycle to yield

$$\oint \delta Q - \oint \delta W = \Delta U = 0, \tag{7.2}$$

where the internal energy change is zero since the initial and final states are identical. The path integrals of the heat and work are just $Q_H - Q_L$ and W_{net}, respectively; thus,

$$(Q_H - Q_L) - W_{net} = 0, \tag{7.3a}$$

or

$$W_{net} = Q_H - Q_L. \tag{7.3b}$$

> **Generation of electrical power from steam is an important application of the thermal-fluid sciences (see Chapter 1).**

A practical example of a heat engine is the closed-loop portion of a Rankine-cycle steam power plant. Figure 7.2 shows the heat addition to the steam in the boiler. Not all of the heat is added at a single temperature, however, as the water from the feedwater pump is first heated to the saturation

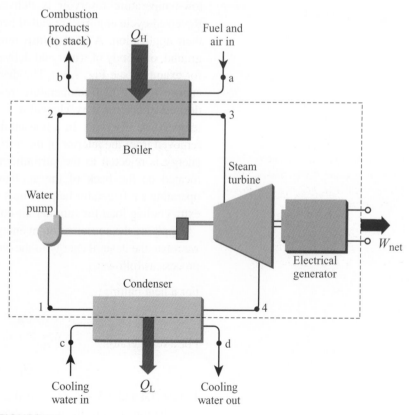

FIGURE 7.2

Steam power plant (Rankine cycle) schematically represented as a heat engine. The dashed line contains the components corresponding to the cyclic device of Fig. 7.1a.

FIGURE 7.3
(a) Geothermal heat pump for space and water heating. **Photograph courtesy of U.S. Department of Energy and Craig Miller Productions.** *(b) Home refrigerator.*

temperature, and the steam may be superheated. Energy is rejected as heat at a fixed temperature in the condenser at the low-pressure, saturation condition. The net work crossing the boundary is that from the turbine-driven electrical generator. No other work crosses the boundary. Note that the pump is driven by the steam turbine *within* the system (Fig. 7.2) so that we do not consider this work interaction in W_{net}.

A heat engine can be reversed by providing an input of work from the surroundings, as shown in Fig. 7.1b. In this reversed engine, energy from the low-temperature reservoir is delivered to the high-temperature reservoir. Reversed-cycle engines are called heat pumps, or refrigerators, depending on their application. A **heat pump** removes energy from the atmosphere, the ground, or a body of water and delivers energy to provide residential heating, for example (see Fig. 7.3a). The desired effect for a heat pump is to supply energy to the high-temperature reservoir (e.g., a building interior). The desired effect for a **refrigerator** is the removal of energy from the low-temperature reservoir. In a household refrigerator (see Fig. 7.3b), energy is removed from the interior of the refrigerator where food items are stored, and energy is rejected to the surroundings, typically through a heat exchanger located on the back of the appliance. As a result of this heat rejection, operating a refrigerator helps to heat your home in the winter and provides an extra cooling load for summer air conditioning.

Applying the conservation of energy principle (Eq. 7.2) to a reversed cycle, we relate the desired energy to the work input and the secondary heat transfer process as follows:

For a heat pump,

$$Q_H = W_{in} + Q_L. \tag{7.4}$$

For a refrigerator,

$$Q_L = Q_H - W_{in}. \tag{7.5}$$

In Eqs. 7.4 and 7.5, Q_H, Q_L, and W_{in} are all numerically positive in agreement with the direction of the arrows shown in Fig. 7.1b.

Our current purpose is to use these devices to develop an understanding of the second law. A detailed and practical analysis of heat pumps and refrigerators is presented in Chapter 12.

See Fig. 12.32 and Examples 12.12–12.14 in Chapter 12. ▷

7.3c Thermal Efficiency and Coefficients of Performance

We define a **thermal (or first-law) efficiency** for heat engines as the ratio of the net useful work produced to the heat energy supplied:

$$\eta_{th} \equiv \frac{\text{Useful work produced}}{\text{Energy supplied}}. \tag{7.6}$$

Using Eq. 7.3, this is expressed as

$$\eta_{th} = \frac{W_{net}}{Q_H} = \frac{Q_H - Q_L}{Q_H}, \tag{7.7a}$$

or

$$\eta_{th} = 1 - \frac{Q_L}{Q_H}. \tag{7.7b}$$

Equations 7.6 and 7.7 are quite general and are not restricted to devices that receive or reject heat at fixed temperatures. All that is required is for Q_H, Q_L, and W_{net} to be determined by a proper integration over the cycle, that is,

$$Q_H \equiv \oint_{\text{Cycle}} \delta Q_H,$$

where the temperature at which heat is added can vary throughout the cycle. Thus, we could apply Eq. 7.7 to calculate the thermal efficiency of a real Rankine-cycle steam power plant were we able to measure any two of the quantities Q_H, Q_L, or W_{net}.

The **coefficient of performance,** or *COP,* defines a measure of performance for a reversed cycle. The following definition applies to both the heat pump and refrigerator:[3]

$$COP \equiv \beta \equiv \frac{\text{Desired energy}}{\text{Energy that costs}}. \tag{7.8}$$

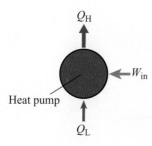

For the heat pump, the desired energy is that delivered to the high-temperature reservoir, and the energy that costs is the work input; thus, the coefficient of performance for a heat pump is given by

$$\beta_{\text{heat pump}} = \frac{Q_H}{W_{in}}. \tag{7.9}$$

For a refrigerator, the desired energy is the energy removed from the cold space, and the energy that costs, once again, is the work input; thus,

$$\beta_{\text{refrig}} = \frac{Q_L}{W_{in}}. \tag{7.10}$$

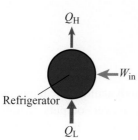

Unlike the thermal efficiency of a heat engine (Eq. 7.7), the coefficient of performance exceeds unity. Typical values for practical devices range, say, from 2 to 5.

---

[3] Note that the right-hand side of Eq. 7.8 also defines the thermal efficiency for a heat engine (cf. Eq. 7.6).

Example 7.2

In a Stirling engine, heat is supplied to the working fluid, typically helium, from an external combustor. (See schematic diagram.) A particular Stirling engine continuously produces 5 kW of shaft power with a thermal efficiency of 0.24. Determine the rate at which heat is added and the rate at which heat is rejected by this engine.

A Stirling engine consists of displacer pistons, power pistons, and a complex drive mechanism. The displacer piston(s) shuttle the working gas (shown in orange) back and forth between the heated space above the piston and the cooled space below the piston through a regenerator. Schematic diagram courtesy of Kockums AB, Sweden.

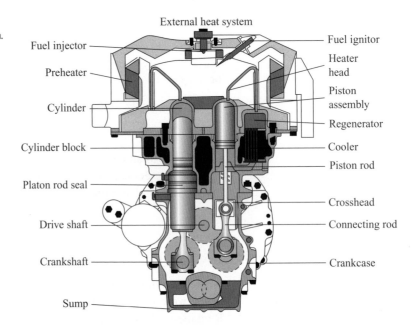

Solution

Known $\dot{W}_{net}$, η_{th}

Find $\dot{Q}_H$, $\dot{Q}_L$

Sketch

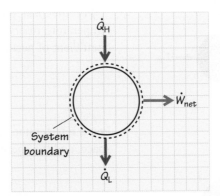

Analysis   Actual heat engines execute a thermodynamic cycle many times over to produce an essentially continuous power output; therefore, relationships involving the heat and work interactions, W_{net}, Q_H, and Q_L, apply equally well when these quantities are expressed on a rate basis (i.e., $\dot{W}_{net}$, $\dot{Q}_H$, and $\dot{Q}_L$). To find the heat addition rate, we apply the definition of thermal efficiency for a heat engine (Eq. 7.7a),

$$\eta_{th} = \dot{W}_{net}/\dot{Q}_H,$$

or

$$\dot{Q}_H = \dot{W}_{net}/\eta_{th}$$
$$= \frac{5 \text{ kW}}{0.24} = 20.8 \text{ kW}.$$

We obtain the heat-rejection rate from conservation of energy (Eq. 7.3b):

$$\dot{Q}_L = \dot{Q}_H - \dot{W}_{net}$$
$$= 20.8 - 5 \text{ kW} = 15.8 \text{ kW}.$$

Comments   Note the relatively low value of the thermal efficiency for this real heat engine; only 24% of the supplied energy is converted to useful work.

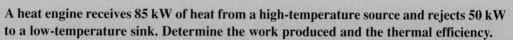

Self Test 7.2   ✓   **A heat engine receives 85 kW of heat from a high-temperature source and rejects 50 kW to a low-temperature sink. Determine the work produced and the thermal efficiency.**
(Answer: $\dot{W}_{net} = 35$ kW, $\eta_{th} = 41.2\%$)

Example 7.3

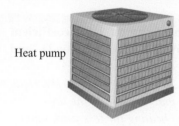

Heat pump

A heat pump operates with a coefficient of performance of 3 and requires a power input of 3.5 kW.

A. Determine the rate of heat removal from the low-temperature reservoir.
B. The heat pump is reconfigured to operate as a refrigerator (i.e., an air conditioner) with the same power input and the same heat-removal rate as in Part A. Determine the coefficient of performance of this refrigerator.

Solution

Known $\beta_{heat\ pump}$, $\dot{W}_{in}$

Find $\dot{Q}_L$, β_{refrig}

Sketch

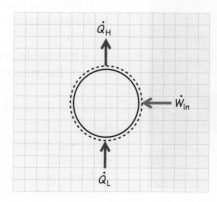

Analysis   We use the definition of the coefficient of performance for a heat pump (Eq. 7.9) and a cycle energy balance (Eq. 7.3b) to find $\dot{Q}_L$ (Part A):

$$\beta_{heat\ pump} = \dot{Q}_H/\dot{W}_{in},$$

so

$$\dot{Q}_H = \beta_{\text{heat pump}} \dot{W}_{in}$$
$$= 3(3.5 \text{ kW}) = 10.5 \text{ kW}$$

and

$$\dot{Q}_L = \dot{Q}_H - \dot{W}_{in}$$
$$= 10.5 - 3.5 \text{ kW} = 7 \text{ kW}.$$

To find the coefficient of performance for the device reconfigured as a refrigerator, we apply Eq. 7.10:

$$\beta_{\text{refrig}} = \dot{Q}_L / \dot{W}_{in}$$
$$= \frac{7 \text{ kW}}{3.5 \text{ kW}} = 2.0.$$

Comment   Manipulating the definitions of the coefficient of performance for the heat pump and the refrigerator shows that $\beta_{\text{heat pump}}$ always exceeds β_{refrig} by one unit, provided the energy quantities are all the same, that is,

$$\beta_{\text{heat pump}} = \frac{Q_H}{W_{in}} = \frac{Q_L + W_{in}}{W_{in}} = \frac{Q_L}{W_{in}} + 1,$$

so

$$\beta_{\text{heat pump}} = \beta_{\text{refrig}} + 1.$$

Self Test
7.3

☑ **Heat pump A has a coefficient of performance of 2.5 and heat pump B has a coefficient of performance of 3.2. If both heat pumps provide 85 kW of heat, determine $\dot{Q}_L$ and the work required for each. Which one is more efficient?**

(Answer: $\dot{W}_A = 34$ kW, $\dot{Q}_{L,A} = 51$ kW, $\dot{W}_B = 26.6$ kW, $\dot{Q}_{L,B} = 58.4$ kW. Heat pump B is more efficient because it extracts more heat from the low-temperature reservoir and provides the same heating for less work input.)

7.3d Reversibility

Many of the consequences of the second law require an understanding of a **reversible process,** and its counterpart, an **irreversible process.** We define a reversible process as follows:

> **A reversible process is one such that the system *and* all parts of the surroundings can be restored to their initial states.**

From this definition, we see that the effects of a reversible process can be undone such that there is no evidence of the process ever having occurred. Reversing a reversible process leaves no trace in *either* the system *or* the surroundings.

A true reversible process is bit a piece of fiction since all real processes have some aspect that prevents them from being reversible. Table 7.1 lists common irreversible processes, several of which are illustrated in Figs. 7.4–7.8.

Friction is a source of irreversibility that is present in essentially all mechanical processes (i.e., processes involving forces and motion). Friction is most obviously present in dry sliding, as you can easily demonstrate by

FIGURE 7.4

The thermal heating produced by friction in the hand-driven drill is sufficient to ignite easily inflammable tinder.

FIGURE 7.5

Electrical current flowing through a soldering iron tip results in a heating of the tip (i.e., Joule heating). This illustrates the irreversible conversion of work (the flow of electricity; see Chapter 4) to heat. The shadowgraph technique makes visible the free convection currents set up by the hot soldering iron. **Photograph courtesy of Gary Settles.**

Table 7.1   Common Irreversible Processes

Any process involving friction (Fig. 7.4)
Heat transfer across a finite temperature difference
Unrestrained expansion of a gas to a low pressure
Joule ($i^2 R_{elec}$) heating (Fig. 7.5)
Plastic deformation of a solid
Mixing of two different substances (Fig. 7.6)
Turbulent flow (Chapter 1, Fig. 1.32)
Shock waves (Fig. 7.7)
Spontaneous chemical reaction (Fig. 7.8)
Magnetic hysteresis
Freezing of a subcooled liquid
Condensation of a supersaturated vapor

firmly pressing your hands together and then sliding one over the other. The organized macroscopic energy resulting from the relative motion of your hands ($\dot{W} = F_{fric}V$) is converted to the randomized thermal motion of molecules, which you sense as a heating up of your hands. There is no way for you to move your hands back to their original position that can recover the thermal energy generated by the friction. A "fire" drill (Fig. 7.4) utilizes the irreversible effects of friction for a useful purpose.

Friction also occurs within fluids. We saw this in Chapter 6, in our investigation of the effects of fluid viscosity. At a system boundary, the effect of viscosity is manifest as work (shear work); however, viscous forces acting internal to the system or control-volume boundary manifest themselves as viscous dissipation in the conservation of energy equation and as the head loss in the mechanical energy equation. As a practical example, consider the steady flow of a fluid through a pipe. For this situation, the mechanical energy equation (Eq. 6.76) reduces to

$$\frac{P_{in} - P_{out}}{\rho} = h_L,$$

> In Chapter 8, we define special dimensionless parameters to deal with the effects of friction: The *friction coefficient* is used in Chapter 9 to calculate drag forces and the *friction factor* is used in Chapter 10 to calculate pressure drops.

where h_L is the head loss. In an ideal frictionless flow (a reversible process), the head loss would be zero, and no work would be required to keep the fluid flowing once it was accelerated to a steady-state condition, as there would be no pressure drop through the pipe. We explore the connection between the second law and head loss in greater detail later in this chapter.

By examining the other irreversible processes shown in Table 7.1, it is easy to see that reversing any of the processes would leave a significant history in the surroundings, the absence of such a history being the requirement for a process to be reversible. For example, consider a metal rod plastically deformed in a tensile test machine. Clearly, a plastically-deformed rod will not spontaneously return to its original shape. It is also unlikely that any mechanical process could restore the rod to its original state short of melting and refabrication. If one were to reconstruct the rod, a permanent change most certainly would occur in the surroundings.

One purpose of the preceding discussion is to show that a reversible process is an *ideal* process and that all *real* processes involve some degree of irreversibility. We will see in the next section how the reversible process is a standard to which real processes can be compared and how such comparisons affect engineering choices.

FIGURE 7.6

A dye jet mixing with a reservoir fluid is an irreversible process, as are all mixing processes that involve two or more different substances.

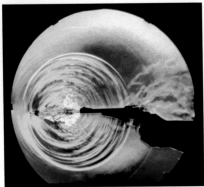

FIGURE 7.7
The schlieren optical technique makes visible the shock waves created by a bullet. A shock wave is a very thin region in a gas over which the gas properties change dramatically. Hearing a sonic boom is the result of the remnants of a shock wave passing by your ear. **Photograph courtesy of Gary Settles.**

Table 7.2 Consequences of the Kelvin–Planck Statement of the Second Law

No.	Consequence
1	Energy transfer by work is more valuable than energy transfer by heat.
2	Any and all heat engines must reject a portion of the heat energy supplied; thus, their thermal efficiency can never be 100%.
3	The energy contained in the earth's heat reservoirs (atmosphere, oceans, rivers, ground, etc.) cannot be utilized to produce a continuous supply of work.
4	For any engine working between two reservoirs having the same high and same low temperatures, a reversible engine will have the greatest thermal efficiency.
5	All reversible heat engines have the same thermal efficiency when operating between the same two reservoirs.
6	A thermodynamic temperature scale can be defined that is independent of the thermometric properties of any substance; the attainment of negative absolute temperatures is impossible; and it is impossible for a finite system to attain a zero value on the absolute scale.

7.4 CONSEQUENCES OF THE KELVIN–PLANCK STATEMENT

At this point, you may be asking how all of this relates to the second law of thermodynamics. Table 7.2 presents some answers to this question.

The first entry in Table 7.2, that work is inherently more valuable than heat, has already been discussed; the remaining entries, however, need some elaboration. The second entry in Table 7.2, that the thermal efficiency of a heat engine can never be 100%, follows from the Kelvin–Planck statement (IB) of the second law stating that it is impossible to convert heat from a *single* reservoir continuously to work. Implicit is the requirement for a *second* reservoir and the rejection of some heat. The idea here is that it is impossible to close the cycle (i.e., bring the working fluid back to its original state) without rejecting some heat in the course of performing some work. There is no possible path that does not include a heat rejection. Theoretical maximum thermal efficiencies fall far short of 100%.

The third entry in Table 7.2 is a way of restating the Kelvin–Planck statement that makes the importance of the second law obvious. If one could construct a device that converted low-grade heat from the earth's atmosphere, oceans, etc. to work, we would enjoy an essentially inexhaustible supply of useful energy with no cost other than that of the conversion device. Such a device is called a **perpetual-motion machine**[4] of the *second kind* and violates the second law of thermodynamics. (A perpetual-motion machine of the *first kind* violates the first law of thermodynamics, that is, the conservation of energy principle.)

The fourth and fifth consequences shown in Table 7.2 provide standards for thermal efficiency to which all real heat engines can be compared. Recall that our Rankine-cycle steam power plant is such a real heat engine. As we will show shortly, the thermal efficiency of a reversible engine is solely determined by the temperatures of the hot and cold reservoirs. Carnot was the first to discover this, and his name is associated with the efficiency of a reversible engine.

FIGURE 7.8
Combustion is a highly irreversible process. The products of combustion cannot be converted back to their original reactant state without a large expenditure of work.

---

[4] For an interesting history of the search for a perpetual-motion machine, the reader is referred to Ref. [7]. Figure 7.9 whimsically presents an idea of perpetual motion.

FIGURE 7.9

M. C. Escher's art creates the illusion of a perpetual motion machine. **M. C. Escher's** *Waterfall* **© 2003 Cordon Art B. V., Baarn, Holland.**

Before exploring this efficiency in greater detail, we show how the fourth and fifth consequences (Table 7.2) do indeed follow from the Kelvin–Planck statement of the second law. Figure 7.10 illustrates the logic applied. In Fig. 7.10a, an irreversible engine and a reversible engine operate between the two reservoirs at T_H and T_L. We postulate that the irreversible engine has a greater thermal efficiency than the reversible engine; that is, for the same heat addition, this engine produces more work than the reversible engine. These conditions are expressed mathematically as

$$Q_{H,irrev} = Q_{H,rev},$$
$$W_{irrev} > W_{rev},$$

and it follows from the application of the first law to the two cycles that

$$Q_{L,irrev} < Q_{L,rev}.$$

We now choose to operate the reversible engine in reverse Fig. 7.10b, that is, like a heat pump, in that work is supplied (W_{rev}), heat is taken from the low-temperature reservoir ($Q_{L,rev}$), and heat added to the high-temperature reservoir ($Q_{H,rev}$). That the cycle is reversed in no way violates either the first or second laws of thermodynamics.

Since the heat added to the high-temperature reservoir by the reversible engine (heat pump) has the same magnitude as the heat taken from the high-temperature reservoir by the irreversible engine, we can replace this heat exchange by a direct path not involving a reservoir, as shown in Fig. 7.10c. We also use some of the work produced by the irreversible engine to drive the heat pump, as indicated by the arrow labeled W_{rev} in Fig. 7.10c. Since the magnitude of the work produced by the irreversible engine exceeds that required to drive the reversible heat pump, net work $W_{excess} = W_{irrev} - W_{rev}$ is produced by the combined cycles. To complete our analysis, we make use of the fact that $|Q_{L,irrev}| < |Q_{L,rev}|$ to replace these two arrows with a single arrow directed *from* the low-temperature reservoir as shown in Fig. 7.10d. What have we now created? We clearly have a device that directly violates the Kelvin–Planck

FIGURE 7.10

Logical sequence illustrating that no engine can have a thermal efficiency greater than that of a reversible engine without violating the Kelvin–Planck statement of the second law.

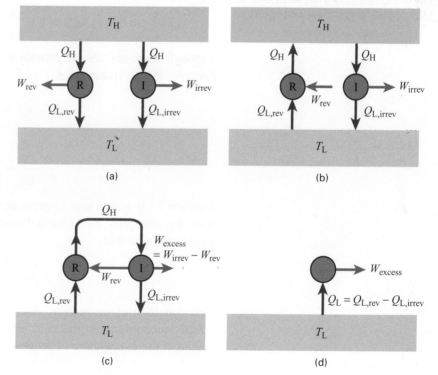

statement, a device that produces work continuously from a single heat reservoir. Since such a device is impossible to construct, our original postulate that the thermal efficiency of the irreversible engine exceeds that of the reversible engine must be false. We therefore conclude that the fourth consequence listed in Table 7.2 must be true. Similar arguments can be mustered to show that the fifth consequence is true as well.

Having just shown that a heat engine operating in a reversible cycle is the most efficient of any engine, we seek to quantify this maximum efficiency. Such quantification is invaluable to engineering analyses as it provides hard and fast standards to which real devices can be compared. Imagine the wasted effort of an engineer seeking to improve the thermal efficiency of a power plant to 60% when the second law indicates that 55% is the theoretical maximum achievable. To achieve our goal of quantifying the thermal efficiency of a reversible cycle, we must first define an absolute thermodynamic temperature scale (see item 6 in Table 7.2).

7.4a Kelvin's Absolute Temperature Scale

William Thomson (1824–1907), Baron Kelvin of Largs, Scottish mathematician and physicist.

We have just shown the validity of the statement that, for any engine working between two reservoirs having the same high temperatures and the same low temperatures, a reversible engine will have the greatest thermal efficiency. William Thomson, Lord Kelvin, used this statement to define a temperature scale that is independent of any substance or particular measuring instrument, that is, an **absolute temperature scale**[5] [8]. We now explore how this was done by mathematically expressing this statement as

$$\eta_{rev} = f(T_H, T_L),\tag{7.11}$$

where f indicates an arbitrary function of the two variables T_H and T_L. Equation 7.11 implies that the *only* factors affecting the thermal efficiency of a reversible cycle are the temperatures of the two reservoirs. We combine this second-law conclusion, Eq. 7.11, with the first-law definition of thermal efficiency, Eq. 7.7b, as follows:

$$\eta_{rev} = 1 - \frac{Q_L}{Q_H} = f(T_H, T_L).\tag{7.12}$$

Although there are several choices that can be made for the function f, Kelvin's choice [9] was to set

$$f(T_H, T_L) \equiv 1 - \frac{T_L}{T_H},\tag{7.13a}$$

or

$$\frac{Q_L}{Q_H} \equiv \frac{T_L}{T_H}.\tag{7.13b}$$

Equation 7.13b is then used to create an absolute thermodynamic temperature scale by arbitrarily assigning a numerical value to one of the reservoir temperatures, so that

$$T = T_{fixed}\left(\frac{Q_T}{Q_{T_{fixed}}}\right)_{rev}.\tag{7.14a}$$

This definition of temperature states that a thermodynamic temperature is directly proportional to the ratio of the heat received by a reversible heat engine at the temperature of interest to the heat rejected at a known fixed

---

[5] An absolute temperature scale is also referred to as a **thermodynamic temperature scale**.

FIGURE 7.11

Kelvin's absolute thermodynamic temperature scale is independent of the properties of any substance and depends only on the ratio of the heat transferred from a high-temperature reservoir to the low-temperature reservoir in a reversible heat engine. The slope of the Kelvin scale is set by choosing the triple point of water as a reference condition and assigning to it a temperature of 273.16 K.

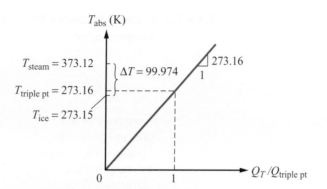

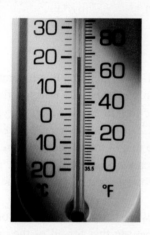

The International Temperature Scale of 1990 (ITS-90) defines a practical temperature scale (see Chapter 1).

temperature. The fixed point for the Kelvin scale adopted in 1954 by the Conférence Générale des Poids et Mesures (CGPM) is the triple point of water and is assigned the thermodynamic temperature of 273.16 K, that is,

$$T_{\text{fixed}} \equiv 273.16 \text{ K}.$$

Equation 7.14a then becomes

$$T\,(\text{K}) = 273.16\left(\frac{Q_{T\,(\text{K})}}{Q_{273.16}}\right)_{\text{rev}}. \tag{7.14b}$$

Figure 7.11 graphically illustrates this absolute thermodynamic temperature scale.

Since one cannot in reality measure the ratio of heat received and rejected in a reversible heat engine, Eq. 7.14b and Fig. 7.11 are theoretical constructs. Practical temperature scales are defined to allow accurate approximations to this thermodynamic ideal using laboratory instruments.

7.4b The Carnot Efficiency

With the establishment of a thermodynamic temperature scale and its relation to a practical means of measurement, we have a way to actually quantify the ideal (maximum) thermal efficiency of a heat engine operating between two heat reservoirs:

$$\eta_{\text{rev}} = 1 - \frac{T_{\text{L}}}{T_{\text{H}}}, \tag{7.15a}$$

where T_{L} and T_{H} are absolute temperatures. The maximum thermal efficiency defined by Eq. 7.15a is also named the **Carnot efficiency** in honor of Carnot's contributions. From Eq. 7.15a, we see that increasing the temperature of the heat source increases the ideal efficiency, as does decreasing the temperature of the heat sink. To approach an efficiency of unity (100%) requires either that the temperature of the high-temperature reservoir approach infinity ($T_{\text{H}} \rightarrow \infty$) or that the low-temperature reservoir approach absolute zero ($T_{\text{L}} \rightarrow 0$).

For any real device, maximum temperatures are usually dictated by materials considerations. For example, the strength of metal parts decreases as temperature increases, resulting in metallurgical limits. For highly stressed boiler tubes and steam turbine blades [10], this metallurgical limit is approximately 900 K, whereas for modern gas-turbine engines, maximum turbine blade temperatures for nickel-based alloys are approximately 1100 K [11]. Cooling the blades allows working fluid temperatures to be significantly higher than the metallurgical limits. Advanced gas-turbine systems permit the use of turbine inlet temperatures above 1775 K. Use of ceramic materials provides the hope of extending metallurgical limits several hundred kelvins [12]. If we consider the sun as an extreme high-temperature source, energy is

Table 7.3 Carnot Efficiencies for Various High-Temperature Reservoir Temperatures*

	$T_H(K)$	$\eta_{rev} = 1 - \dfrac{T_L}{T_H}$
Temperature of sun	~5800	0.949
Advanced gas-turbine inlet gas temperatures	~1775	0.835
Metallurgical limit (Hastelloy X, etc.)	~1100	0.734
Metallurgical limit (steels)	~900	0.674

* The low-temperature reservoir is fixed at 293 K (68°F).

Turbine rotor blade failure from excessive temperatures. Photograph courtesy of Rolls-Royce (*The Jet Engine*).

See Chapter 12 for analyses of power plants, jet engines, and other systems.

available at approximately 5800 K (see Ex. 4.15). The ambient temperatures of the atmosphere, or the body of water, into which heat is rejected fix the practical minimum temperatures for heat rejection. In North America, typical ambient temperatures range, say, from about 279 K (40 F) to 300 K (80 F). To create a reservoir at a temperature less than that of the ambient requires the use of refrigeration, which, in turn, requires a work input. This work input more than cancels any improvement in efficiency gained by lowering the sink temperature of the original cyclic device.

Table 7.3 presents the Carnot efficiencies associated with some of the practical temperatures just discussed. Here we see that for all of the cycles constrained by materials considerations ($T < 1775$ K), thermal efficiencies are well below unity, with approximately 20–30% of the supplied heat rejected to the cold reservoir. It should be kept in mind that real cyclic devices (steam power plants, for example) do not operate between two fixed-temperature reservoirs; nevertheless, the theoretical maximum thermal efficiencies shown are useful approximations for real devices receiving energy on the average at such temperatures. Older steam power plants operate with actual thermal efficiencies on the order of 40%, whereas modern dual-cycle power plants that employ both steam turbines and gas turbines can achieve actual efficiencies of approximately 60% [13].

Equation 7.13b can also be use to define ideal (Carnot) coefficients of performance for heat pumps and refrigerators as follows:

$$\beta_{rev,\,heat\,pump} = \frac{T_H}{T_H - T_L}, \tag{7.15b}$$

$$\beta_{rev,\,refrig} = \frac{T_L}{T_H - T_L}. \tag{7.15c}$$

Example 7.4

A reversible heat engine rejects heat at 25°C and has a thermal efficiency of 50%. At what temperature is heat added to the engine cycle?

Solution

Known T_L, η_{rev}

Find T_H

Analysis   We apply the definition of thermal efficiency (Eq. 7.15a) for a reversible heat engine to find the temperature of the high-temperature reservoir as follows:

$$\eta_{\text{rev}} = 1 - \frac{T_{\text{L}}}{T_{\text{H}}},$$

which is rearranged to yield

$$T_{\text{H}} = \frac{T_{\text{L}}}{1 - \eta_{\text{rev}}}$$

$$= \frac{(25 + 273.15)}{1 - 0.5} \, \text{K} = 596 \, \text{K or } 323°\text{C}.$$

Comment   Note that *absolute* temperatures define the thermal efficiency.

Self Test
7.4

✓ A refrigerator operates between two reservoirs, one at 275 K and the other at 310 K. Determine the maximum coefficient of performance for the refrigerator. If $\dot{Q}_{\text{L}} = 10$ kW for this maximum *COP*, determine the input work required.

(Answer: $\beta_{\text{refrig}} = 7.86$, $\dot{W}_{\text{in}} = 1.27\,kW$)

7.4c Some Reversible Cycles

In this section, we discuss two ideal cycles that are of particular practical and historical interest: the Carnot cycle and the Stirling cycle. In both of these cycles, a cyclic device exchanges heat with two constant-temperature reservoirs through a series of reversible processes, thus meeting the requirements of our previous analysis. Exploring these ideal cycles provides insight into the conversion of heat to work, in general, and also provides a framework for understanding practical energy-conversion devices.

Carnot Cycle

In his 1824 publication, Carnot defined a cycle that now bears his name. This Carnot cycle is illustrated in Fig. 7.12 for a single-phase working fluid.

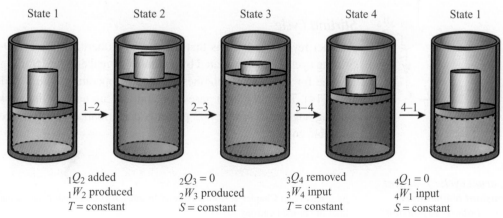

FIGURE 7.12

This sequence of states illustrates the Carnot cycle applied to a fixed mass of gas. The size of the weights placed on top of the piston indicates the relative pressure levels of the gas within the cylinder.

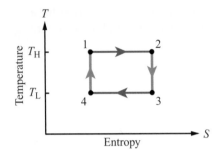

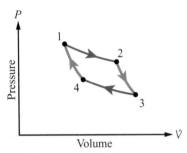

FIGURE 7.13
The four reversible processes of the Carnot cycle are presented using (top) temperature–entropy (T–S) and (bottom) pressure–volume (P–V) thermodynamic coordinates.

To review plotting processes on thermodynamic coordinates, see Examples 2.11 and 2.12 in Chapter 2; also see Fig. 2.13.

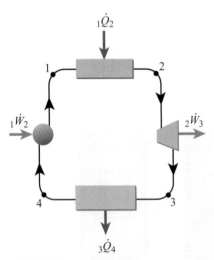

FIGURE 7.14
Steady-flow Carnot cycle: 1–2, heat addition at constant temperature; 2–3, reversible adiabatic expansion; 3–4, heat rejection at constant temperature; and 4–1, reversible adiabatic compression.

Here the thermodynamic system under consideration is the gas within the piston–cylinder assembly. The reversible processes that comprise the Carnot cycle are the following:

State Change	Process
1–2	Reversible heat addition at constant temperature
2–3	Reversible adiabatic expansion
3–4	Reversible heat rejection at constant temperature
4–1	Reversible adiabatic compression

These four processes are conveniently illustrated using temperature–entropy (T–S) and pressure–volume (P–V) coordinates as shown in Fig. 7.13. Although we have yet to discuss fully the thermodynamic property entropy, we see from Fig. 7.13 that the reversible, adiabatic expansion and compression processes appear as constant-entropy (isentropic) processes; thus, the Carnot cycle is described by a rectangle on T–S coordinates. On the P–V coordinates, the isothermal and isentropic processes follow curved paths, with isotherm slopes that are not as steep as those of the isentropes.[6] Since all of the processes involved are reversible, the work associated with each process can be calculated as

$$_aW_b = \int_a^b P\, d\mathcal{V}.$$

The area enclosed in the P–V diagram, therefore, represents the net work produced by the cycle. We will see later in this chapter that the area enclosed in the T–S diagram is the net heat addition for the cycle, and, thus, by application of the first law to the cycle, also equals the net work produced.[7]

The Carnot cycle can also be applied to a sequence of steady-flow processes as illustrated in Fig. 7.14. This steady-flow cycle is a key building block in our analysis of the steam power plant (Rankine cycle).

Stirling Cycle

Another reversible cycle is that derived by Robert Stirling (1790–1878). The Stirling cycle is approximated by a number of real engines. Stirling engines fill a niche for quiet, small engines capable of operating with a wide variety of fuels [14]. Typically, light gases (H_2 and He) are used as the working fluid in Stirling engines. Example 7.2 presents performance data for a commercially available engine.

---

[6] As discussed in Chapter 2, isotherms refer to lines of constant temperature and isentropes to lines of constant entropy.

[7] The emblem of the Mechanical Engineering Honor Society, Pi Tau Sigma, uses the outline of the Carnot cycle in P–V coordinates in honor of Sadi Carnot's contribution to the field of mechanical engineering.

Mirrors focus sunlight onto a thermal receiver of a Stirling engine. The engine drives a 25 kW electrical generator. Photograph courtesy of Bill Timmerman.

The ideal Stirling cycle consists of the following processes:

State Change	Process
1–2	Reversible heat addition at constant temperature
2–3	Reversible heat rejection at constant volume
3–4	Reversible heat rejection at constant temperature
4–1	Reversible heat addition at constant volume

These processes are shown on P–$\mathcal{V}$ and T–S coordinates in Fig. 7.15.

7.5 ALTERNATIVE STATEMENTS OF THE SECOND LAW

As discussed at the outset of this chapter, the second law can be expressed in a number of ways, any of which expressions can be used (ultimately) to generate the others. Table 7.4 presents four classes of second-law statements. We have already considered the class I statements in some detail.

The original statements of Clausius (II) are frequently invoked in discussions of the second law. The following provides the spontaneous direction for heat flow:

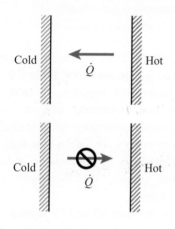

Heat flows spontaneously from high to low temperature, but not conversely.

The first law provides no guideline for the direction of spontaneous processes, only stipulating that energy must be conserved. One of the particularly useful aspects of the second law is that it *does* provide clear guidelines for what processes will occur naturally. Clausius' statement clearly asserts that one will never observe a cup of coffee spontaneously getting hotter at the expense of energy drawn from the colder atmosphere (Fig. 7.16). Here the use of the word **spontaneous** refers to the idea that no work is applied. Clearly, a heat pump could be used to heat the coffee at the expense of the atmosphere. Clausius' original statements can be deduced from the Kelvin–Planck statements (I). These deductions are left as exercises for the reader.

Both the class III and IV statements focus on the ways that the second law is useful in determining the spontaneous direction for real processes and the

FIGURE 7.15

The ideal Stirling cycle consists of four reversible processes shown here using (a) T–S coordinates and (b) P–$\mathcal{V}$ coordinates.

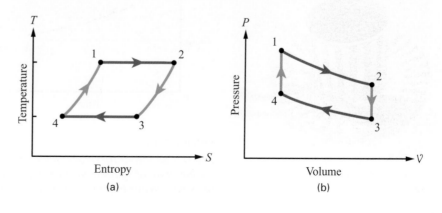

(a)   (b)

Table 7.4 Various Statements of the Second Law of Thermodynamics

Designation	Statement
IA: Paraphrase of Kelvin–Planck statement	Although all work can be converted completely to heat, heat cannot be completely and continuously converted into work.
IB: Kelvin–Planck statements [4]	It is impossible to construct a cyclically operating device for which the sole effect is exchange of heat with a single reservoir and the creation of an equivalent amount of work.
II: Clausius (original) statements [5]	It is impossible to operate a cyclic device in such a manner that the sole effect external to the device is the transfer of heat from one energy reservoir to another at a higher temperature. Heat flows spontaneously from high to low temperature, but not conversely.
III: Clausius entropy statement [6]	The entropy of a system and the environment with which it is in contact increases, or, in the limit of a reversible process, remains constant [1]. The entropy of the universe tends toward a maximum [6].
IV A, B, C: Equilibrium statements	A. Equilibrium occurs when the entropy is a maximum for a simple system at constant internal energy and volume. B. Equilibrium occurs when the Gibbs free energy is a minimum for a simple system at constant pressure and temperature. C. Equilibrium occurs when the Helmholtz free energy is a minimum for a simple system at constant temperature and volume.

conditions required for equilibrium. This usefulness is one of the two emphasized at the beginning of this chapter. As we see from Table 7.4, an understanding of the class III and IV statements requires, in turn, an understanding of entropy and the Gibbs and Helmholtz free energies. Engineering analyses use these thermodynamic properties in quantitative ways. For example, we use the Gibbs free energy, or Gibbs function, to calculate the detailed composition of the products of combustion at high temperatures where chemical dissociation occurs. We investigate this particular application in detail later in this chapter.

Before we can continue our discussion of the class III and IV second-law statements, we need to define and explore entropy and related thermodynamic properties.

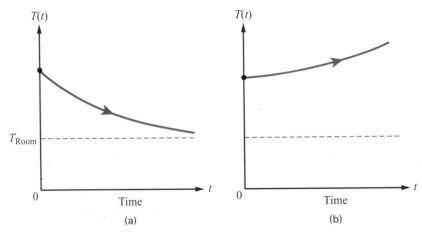

FIGURE 7.16
The second law of thermodynamics provides unambiguous criteria for the direction of spontaneous change. (a) Coffee cools spontaneously. (b) Spontaneous heating of coffee in cooler surroundings is impossible.

7.6 ENTROPY REVISITED

The thermodynamic property **entropy** was introduced in Chapter 2 without the context provided by the second law. Entropy can be defined most generally from a macroscopic viewpoint, and in more restrictive ways from a microscopic (molecular) viewpoint. We will focus on the macroscopic definition in this chapter.

7.6a Definition[8]

The following definition of entropy, S, applies to a macroscopic thermodynamic system:

$$dS \equiv \left(\frac{\delta Q}{T}\right)_{rev}. \tag{7.16a}$$

To properly implement this definition, we enforce the convention that heat *into* the system is positive and heat *out* is negative. Integrating Eq. 7.16a for any process involving a change from state 1 to state 2 yields

$$\Delta S(\equiv S_2 - S_1) \equiv \int_1^2 \left(\frac{\delta Q}{T}\right)_{rev}. \tag{7.16b}$$

Equation 7.16b is curious. To employ it to calculate a change in entropy, ΔS, a *reversible* path must first be defined between states 1 and 2, and then the heat transferred divided by the temperature at which the heat exchange takes place must be integrated over this path. This definition of entropy also is quite abstract and begs for a physical interpretation. Unfortunately, classical thermodynamics offers none. Definitions, however, are free to stand on their own as part of a logical framework. Fortunately, microscopic (statistical) thermodynamics does lend some physical insight into the meaning of entropy for some systems. (See Chapter 2.)

> **Appendix 2A in Chapter 2 provides a microscopic interpretation of entropy for a gas.**

That entropy, as defined in Eq. 7.16a, is a thermodynamic property is not obvious. To prove that entropy is a thermodynamic property requires that we demonstrate that

$$\oint_{cycle} \left(\frac{\delta Q}{T}\right)_{rev} = 0 \tag{7.17}$$

is true. For *any* quantity to be a thermodynamic property requires that, upon executing a thermodynamic cycle, the value of the quantity returns to its initial value, that is,

$$\oint_{cycle} d(\text{Property}) \equiv 0. \tag{7.18}$$

It is in this sense that energy is a thermodynamic property, since from the first law,

$$\oint_{cycle} (\delta Q - \delta W) = 0,$$

---

[8] This definition from classical thermodynamics is usually attributed to Clausius [6].

or

$$\oint_{\text{cycle}} dE = 0.$$

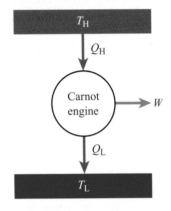

To begin our proof that Eq. 7.17 is true, we consider a Carnot heat engine operating between a high-temperature reservoir at T_H and a low-temperature reservoir at T_L. We choose a Carnot engine because the heat transferred at T_H and T_L are reversible processes, as are all the processes in the engine's execution of a cycle. For the Carnot cycle, the integral of $(\delta Q/T)_{\text{rev}}$ has two components, one for the heat exchange at T_H and one for the heat exchange at T_L:

$$\oint_{\substack{\text{Carnot} \\ \text{cycle}}} \left(\frac{\delta Q}{T}\right)_{\text{rev}} = \frac{Q_H}{T_H} - \frac{Q_L}{T_L}, \tag{7.19a}$$

where the minus sign indicates that Q_L is rejected during the cycle. We also know from the definition of thermodynamic temperature, Eq. 7.13b, that

$$\frac{Q_H}{T_H} = \frac{Q_L}{T_L}. \tag{7.19b}$$

Substituting this result into Eq. 7.19a yields

$$\oint_{\substack{\text{Carnot} \\ \text{cycle}}} \left(\frac{\delta Q}{T}\right)_{\text{rev}} = \frac{Q_L}{T_L} - \frac{Q_L}{T_L} = 0, \tag{7.19c}$$

which provides the desired proof. To generalize to *any* reversible cycle, we note that such cycles can always be modeled as an appropriate assembly of Carnot engines.

7.6b Connecting Entropy to the Second Law

Having established that entropy is indeed a thermodynamic property, we now wish to link this property with the second law, for its connection with the second law is what makes entropy such a useful property. To accomplish this connection, we first state the **Clausius inequality** [3] and then prove that it is true:

> **Whenever a system executes a complete cyclic process, the integral of $\delta Q/T$ around the cycle is less than zero, or in the limit is equal to zero; that is,**

$$\oint_{\text{cycle}} \frac{\delta Q}{T} \leq 0. \tag{7.20}$$

The truth of the Clausius inequality has its origins in the Kelvin–Planck statement of the second law. We now show this connection between Eq. 7.20 and the second law.

Consider an arbitrary thermodynamic system undergoing a thermodynamic cycle that includes both heat and work interactions. We constrain all of the heat interactions of this system to be from a reversible heat engine that receives heat

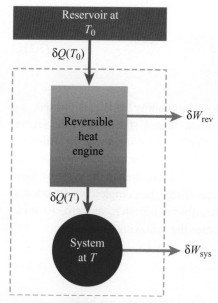

FIGURE 7.17

A thermodynamic system exchanges heat through the agency of a reversible heat engine. The reversible engine operates in a reversed cycle when heat is removed from the system. The dashed line incorporating both the reversible engine and the system results in a violation of the Kelvin–Planck statement of the second law when the total work $\delta W_{rev} + \delta W_{sys}$ integrated over a cycle is positive.

from a reservoir at T_0, as illustrated in Fig. 7.17.[9] We define these heat and work interactions as follows:

$\delta Q(T_0) \equiv$ heat received from reservoir during one or more complete cycles of the reversible engine,

$\delta Q(T) \equiv$ heat rejected in one or more complete cycles of the reversible engine consistent with $\delta Q(T_0)$,

$\delta W_{rev} \equiv$ work performed in one or more complete cycles of the reversible engine consistent with $Q(T_0)$, and

$\delta W_{sys} \equiv$ increment of work performed by the system as it executes a cycle.

Note, first, that $\delta Q(T)$, $\delta Q(T_0)$, and δW_{rev} are *cyclic* quantities with respect to the reversible engine, but they are *incremental* quantities with respect to the system. Note also that, although these quantities are positive, consistent with the arrows in the sketch (Fig. 7.13), they may assume negative values if the reversible engine were to be reversed for any portion of the system cycle.

To begin our proof, we first focus on the reversible engine(s). The first law (Eq. 5.4d) applied to one or more engine cycles is

$$\delta W_{rev} - \delta Q(T_0) + \delta Q(T) = 0. \tag{7.21a}$$

From our discussion of an absolute temperature scale, we relate $\delta Q(T_0)$ and $\delta Q(T)$ to T_0 and T as follows (see Eq. 7.13b):

$$\frac{T}{T_0} = \frac{\delta Q(T)}{\delta Q(T_0)}. \tag{7.21b}$$

Solving Eq. 7.21b for $\delta Q(T_0)$ and substituting this result into Eq. 7.21a yield, upon rearrangement,

$$\delta W_{rev} = \left(\frac{T_0}{T} - 1 \right) \delta Q(T). \tag{7.21c}$$

We next find the total net work of the combined system and reversible engine(s) for one complete cycle of the *system* by summing their contributions:

$$W_{net} = \oint_{\substack{system \\ cycle}} \delta W_{rev} + \oint_{\substack{system \\ cycle}} \delta W_{sys}. \tag{7.21d}$$

The second term on the right-hand side of Eq. 7.21d, the system work over the system cycle, relates to $\delta Q(T)$ through the application of the first law to the system alone, that is,

$$\oint_{\substack{system \\ cycle}} \delta W_{sys} = \oint_{\substack{system \\ cycle}} \delta Q(T). \tag{7.21e}$$

We now substitute Eqs. 7.21e and 7.21c into Eq. 7.21d to yield

$$W_{net} = \oint_{\substack{system \\ cycle}} \left[\left(\frac{T_0}{T} - 1 \right) \delta Q(T) \right] + \oint_{\substack{system \\ cycle}} \delta Q(T), \tag{7.21f}$$

[9] As the system temperature changes in the course of its cycle, additional reversible engines may be added to this picture to ensure that reversible heat exchange always occurs between any engine and the system.

which simplifies to

$$W_{net} = T_0 \oint_{\substack{\text{system} \\ \text{cycle}}} \frac{\delta Q(T)}{T}. \tag{7.21g}$$

What happens if we allow W_{net} to be a positive quantity? If W_{net} is positive, we have created a device that violates the Kelvin–Planck statement of the second law that *it is impossible to construct a cyclically operating device for which the sole effect is the exchange of heat with a single reservoir and the creation of an equivalent amount of work*. If W_{net} is negative (i.e., an input to the system), there is no violation of the second law. Work can always be completely converted to heat. Since the work in Eq. 7.21g can only be negative, or zero, to avoid violating the second law, Eq. 7.21g becomes the following inequality:

$$T_0 \oint \frac{\delta Q(T)}{T} \leq 0, \tag{7.21h}$$

or, recognizing that T_0 must always be greater than zero, we have

$$\oint \frac{\delta Q(T)}{T} \leq 0. \tag{7.21i}$$

With this result, we have proven the validity of the Clausius inequality, Eq. 7.20. It is extremely important to note that the second law is imbedded in the Clausius inequality; thus, the inequality could stand on its own as yet another statement of the second law, although that is not usually done.

Another form of the second law, however, is created by combining the Clausius inequality with the definition of entropy. Consider a system that undergoes a cycle between two states, 1 and 2. In one case, the cycle is accomplished by two reversible process, $(1-2)_{rev}$ and $(2-1)_{rev}$; in a second case, the process 1–2 follows the same path as in the first case, and hence is reversible, but the return process 2–1 is irreversible. These two cases are illustrated in Fig. 7.18. For case I, we apply the Clausius inequality as follows:

$$\oint \frac{\delta Q}{T} \leq 0,$$

or

$$\int_1^2 \left(\frac{\delta Q}{T} \right)_{rev} + \int_2^1 \left(\frac{\delta Q}{T} \right)_{irrev} \leq 0. \tag{7.22a}$$

For case II, we apply the definition of entropy (Eq. 7.16) for each process, the sum of which must be zero when the system executes a cycle (Eq. 7.17); that is,

$$\int_1^2 \left(\frac{\delta Q}{T} \right)_{rev} + \int_2^1 \left(\frac{\delta Q}{T} \right)_{rev} = 0. \tag{7.22b}$$

Subtracting Eq. 7.22b from Eq. 7.22a yields

$$\int_2^1 \left(\frac{\delta Q}{T} \right)_{irrev} - \int_2^1 \left(\frac{\delta Q}{T} \right)_{rev} \leq 0,$$

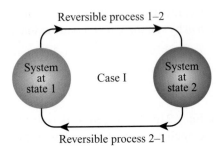

Reversible process 1–2

System at state 1 — Case I — System at state 2

Reversible process 2–1

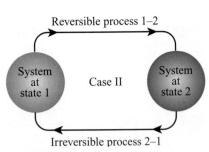

Reversible process 1–2

System at state 1 — Case II — System at state 2

Irreversible process 2–1

FIGURE 7.18
Schematic of a system undergoing a reversible cycle between states 1 and 2 (case I) and the same system undergoing an irreversible cycle between states 1 and 2 (case II).

or

$$\int_2^1 \left(\frac{\delta Q}{T}\right)_{\text{irrev}} - \int_2^1 dS_{\text{sys}} \leq 0. \tag{7.22c}$$

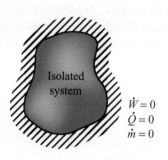

Here we have explicitly substituted the definition of entropy (Eq. 7.16) in the second integral. Applying algebraic rules for the manipulation of inequalities and differentiating the integrals result in

$$dS_{\text{sys}} \geq \left(\frac{\delta Q}{T}\right)_{\text{irrev}}. \tag{7.23}$$

Our final step in creating a second-law statement is to apply Eq. 7.23 to an isolated system. For an isolated system, no boundary interactions can occur; hence, δQ is zero for any process. We thus conclude that

$$dS_{\substack{\text{isolated}\\\text{sys}}} \geq 0. \tag{7.24a}$$

> **Equations 7.24a and 7.24b are alternative ways to state the second law.**

This is the celebrated **increase in entropy principle** and corresponds identically with the third statement of the second law (the Clausius entropy statement) presented in Table 7.4. To emphasize that Eq. 7.24 applies to an isolated system, we rewrite this expression to show explicitly the entropy changes associated with a system and its surroundings contained within the isolated system:

$$dS_{\substack{\text{isolated}\\\text{sys}}} \equiv dS_{\text{sys}} + dS_{\text{surr}} \geq 0. \tag{7.24b}$$

We cannot emphasize too strongly the idea that these entropy-based second-law statements (Eqs. 7.24a and 7.24b) rest firmly on the Kelvin–Planck statement of the second law, as a review of their derivation will show.

Most engineering applications involve systems that are not isolated from their surroundings. Table 7.5 summarizes the ways that the entropy of such systems can change. Here we see that there are only two possible ways for the entropy to remain unchanged after a process: 1. if the process is both adiabatic and reversible and 2. if the process entails heat removal in an amount that balances exactly the entropy increase generated by irreversibilities. These are the first and last entries in Table 7.5. Note that the increase-of-entropy principle does *not* require that the entropy change for any process to be positive. Heat removal, in the absence of any irreversibilities, always results in an entropy decrease.

Table 7.5 Entropy Changes for a System*

Process	Entropy Change
Adiabatic and reversible	$\Delta S = 0$
Adiabatic and irreversible	$\Delta S > 0$
Reversible or irreversible heat addition	$\Delta S > 0$
Reversible heat removal	$\Delta S < 0$
Irreversible heat removal	$\Delta S \gtrless 0$

* Results shown here follow from the application of Eqs. 7.16b and 7.23.

Example 7.5

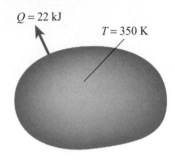

$Q = 22$ kJ

$T = 350$ K

During an irreversible process, 22 kJ of energy is removed as heat from a thermodynamic system at a constant temperature of 350 K. Determine whether the entropy change of the system is positive, negative, or zero. Also determine whether this entropy change is less than, greater than, or equal to the entropy change for the same heat removal for a reversible process occurring at the same temperature.

Solution

Known T, Q

Find ΔS

Sketch

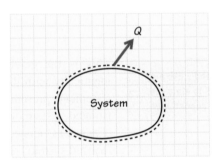

Q

System

Analysis   Because the temperature is constant, we can integrate the inequality (Eq. 7.23) that relates the entropy change to the heat transferred and the temperature, so

$$\int_1^2 dS_{sys} \geq \int_1^2 \left(\frac{\delta Q}{T} \right)_{irrev},$$

or

$$\Delta S_{sys} \geq \frac{Q}{T}.$$

Since energy is *removed* from the system as heat, Q is negative; thus,

$$\Delta S_{sys} \geq \frac{-22{,}000 \text{ J}}{350 \text{ K}} = -63 \text{ J/K}.$$

> **Inequalities containing negative quantities can be confusing. Exercise care in their interpretation.**

Because reversible heat removal causes the entropy of the system to decrease, whereas irreversibilities cause an increase, we cannot determine whether the actual entropy change for the system is positive, negative, or zero. The best we can say is $\Delta S_{sys} \geq -63$ J/K. For a reversible process, $\Delta S_{sys,rev} = -63$ J/K; thus, the entropy change for the irreversible process must be greater than this. For example, the change might be less negative, say -50 J/K. The entropy change could also be zero, or positive, depending on the magnitude of the irreversibilities. If we assume that the irreversibility is external to the system proper, we can then treat the process as **internally reversible**. In this case, the entropy change of the system is -63 J/K.

Comment   Note the importance of the sign convention that heat added to a system is positive and heat removed is negative.

Self Test 7.5

Consider a piston–cylinder device containing 2 kg of saturated-liquid water at 300 K. Heat is added from a surrounding reservoir at 800 K until the water becomes saturated vapor. Determine the entropy change of the system, the entropy change of the surroundings, and the total entropy change. Assume internal reversibility.

(Answer: $\Delta S_{sys} = 16.25\,kJ/K$, $\Delta S_{surr} = -6.09\,kJ/K$, $\Delta S_{total} = 10.16\,kJ/K$)

7.6c Entropy Balances

Using entropy "balances," we can add insight to the ideas expressed in Table 7.5.

Systems Undergoing a Change in State

We develop an entropy balance for a system adopting the generic conservation principle introduced in Chapter 1 (Eq. 1.1) that

$$X_{in} - X_{out} + X_{generated} = \Delta X_{stored},$$

where X is the conserved quantity of interest. For a process taking a system from state 1 to state 2, we write

$$\underbrace{\int_1^2 \frac{\delta Q}{T}}_{\substack{\text{Net transfer of} \\ \text{entropy into the} \\ \text{system by heat} \\ \text{interactions}}} + \underbrace{\int_1^2 \delta \mathscr{S}_{gen}}_{\substack{\text{Entropy generated} \\ \text{by irreversibilities} \\ \text{within the system}}} = \underbrace{S_2 - S_1.}_{\substack{\text{Change of} \\ \text{system} \\ \text{entropy}}} \qquad (7.25a)$$

The first term in this entropy balance corresponds to $X_{in} - X_{out}$ and can be positive or negative depending on the direction of the heat interaction, δQ. The second term is always positive because irreversibilities always generate an increase in entropy; for a reversible process, this term is zero. In the same sense that heat is path dependent and not a property of the system, the entropy generation is also a path function and not a system property; that is,

$$\int_1^2 \delta Q \equiv {}_1Q_2,$$

the magnitude of which is path dependent, and

$$\int_1^2 \delta \mathscr{S}_{gen} \equiv {}_1\mathscr{S}_{gen\,2},$$

the magnitude of which is path dependent. We use the symbol $\mathscr{S}$, to emphasize that $\mathscr{S}_{gen}$ is not a property and is not identical to the entropy S, which is a property.

Control Volumes with a Single Inlet and Outlet

For a control volume, we adopt the rate form of the generic conservation principle (Eq. 1.2):

$$\dot{X}_{in} - \dot{X}_{out} + \dot{X}_{gen} = dX_{cv}/dt.$$

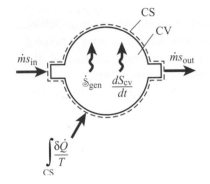

For our entropy balance, this becomes

$$\int_{\substack{\text{control} \\ \text{surface}}} \frac{\delta \dot{Q}}{T} + \dot{m}s_{\text{in}} - \dot{m}s_{\text{out}} + \dot{\mathcal{S}}_{\text{gen}} = \frac{dS_{\text{cv}}}{dt}. \qquad (7.25\text{b})$$

Here we see that entropy enters (or leaves) the control volume by a heat interaction, the first term in Eq. 7.25b, or is carried in ($\dot{m}s_{\text{in}}$) or out ($\dot{m}s_{\text{out}}$) by the flow. The term $\dot{\mathcal{S}}_{\text{gen}}$ represents the rate of entropy production within the control volume by irreversibilities. A common source of $\dot{\mathcal{S}}_{\text{gen}}$ in real systems is fluid friction, either at the walls of a device or within the moving fluid itself. The right-hand term is the rate at which the entropy within the control volume increases. For steady state, this term is zero. The thermodynamic properties appearing in this entropy balance are the specific entropies s_{in} and s_{out}, the total entropy within the control volume, S_{cv}, and the temperature T; $\delta \dot{Q}$ and $\dot{\mathcal{S}}_{\text{gen}}$ again are process-dependent quantities and are not thermodynamic properties.

7.6d Criterion for Spontaneous Change

We now explore how the entropy-based second-law statements (Table 7.4) relate to the spontaneity of any real process. To begin, we define **spontaneity** or a **spontaneous process** as follows:

A process is spontaneous if, in the absence of any interactions with the surroundings (i.e., heat or work), a system undergoes a change in state.

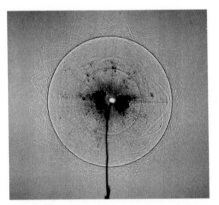

Explosions of all sorts are highly irreversible processes. This instantaneous image shows an explosion of one gram of triacetone triperoxide. Image courtesy of Gary Settles.

What is meant by a spontaneous process can be made clear by considering the simple examples shown in Fig. 7.19: the mixing of two dissimilar gases, the free expansion of a gas into an evacuated space, and the transfer of heat from a hot body to a cold body. Based on your experience, all three of these processes have a preferred, or spontaneous, direction. You expect that, after the partition separating the N_2 and O_2 gases is removed, the gases will mix, and given enough time, the composition in the combined space will become uniform. However, it is inconceivable that starting with an N_2–O_2 mixture the N_2 molecules will gather in the left half of the enclosure while the O_2 molecules congregate in the right. The same is true for the other two processes. The evacuated space will spontaneously fill after the removal of the partition, but not the reverse. A hot and cold body in contact will spontaneously exchange energy until they achieve a common uniform temperature, but two

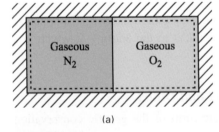

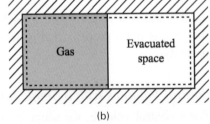

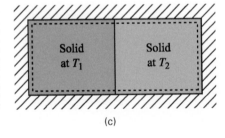

(a)　　　　　(b)　　　　　(c)

FIGURE 7.19

Three examples of spontaneous processes: (a) the mixing of two dissimilar gases, (b) the free expansion of a gas into an evacuated space, and (c) the transfer of heat from a hot body to a cold body.

bodies initially in contact at the same temperature will not spontaneously change temperatures. In all three examples, the first law (conservation of energy) does not preclude the reverse processes from occurring; the second law, however, does. Moreover, a quantitative evaluation of spontaneity is possible by considering the second-law statements involving entropy. This is the beauty of the property entropy.

In Fig. 7.19, each of the three example systems is isolated from its surroundings as suggested by the cross-hatching at the boundaries. For such isolated systems, the second-law statement expressed by Eq. 7.24a,

$$\Delta S_{\substack{\text{isolated} \\ \text{sys}}} \geq 0,$$

applies. If we were able to quantify the entropy of the initial and final states—something that is possible to do through the relationships presented in Chapter 2—we would find in each case that

$$S_{\text{final}} > S_{\text{initial}}. \tag{7.26}$$

Exploring the appropriate sections of Chapter 2 is recommended at this juncture to provide the quantitative link to our present discussion.

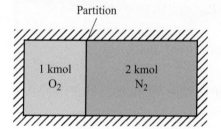

To calculate ΔS for ideal gases, use Eqs. 2.39 and 2.40. For many other substances, use the NIST database.

Example 7.6

Consider 1 kmol of O_2 and 2 kmol of N_2 both at 298 K and 1 atm separated by a partition. The partition is removed and the O_2 and N_2 mix. Determine the entropy change associated with this mixing process.

Solution

Known $N_{O_2}, N_{N_2}, T_1, P_1$

Find $S_{\text{final}} - S_{\text{init}}$

Sketch

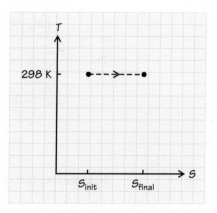

Assumptions

i. Ideal-gas behavior
ii. Adiabatic process

Analysis   For the mixing of ideal gases with no heat or work interactions, conservation of energy tells us that the initial and final temperatures are equal. To calculate the entropy change, we therefore only need to be concerned with the changing partial pressures. The entropy change is expressed as follows:

$$\Delta S = S_{final} - S_{init}$$
$$= [N_{O_2}\bar{s}_{O_2}(T, P_{O_2}) + N_{N_2}\bar{s}_{N_2}(T, P_{N_2})]$$
$$- [N_{O_2}\bar{s}_{O_2}(T, P_{init}) + N_{N_2}\bar{s}_{N_2}(T, P_{init})].$$

For an ideal gas, the entropy of an individual species is given by Eq. 2.73b, that is,

$$\bar{s}_i(T, P_i) = \bar{s}_i(T, P_{ref}) - R_u \ln \frac{P_i}{P_{ref}}.$$

At the initial unmixed state, the pressures of both the N_2 and O_2 are the given value (i.e., $P_{O_2} = P_{N_2} = 1$ atm). At the final mixed state, the total pressure is still 1 atm, whereas the partial pressures of each constituent are given by (Eq. 2.66b)

$$P_{O_2} = X_{O_2}P_{tot} = \frac{N_{O_2}}{N_{mix}}P_{tot}$$

$$= \left(\frac{1}{1 + 2}\right)1 \text{ atm} = 0.333 \text{ atm}$$

and

$$P_{N_2} = X_{N_2}P_{tot} = \frac{N_{N_2}}{N_{mix}}P_{tot}$$

$$= \left(\frac{2}{1 + 2}\right)1 \text{ atm} = 0.667 \text{ atm}.$$

Explicitly substituting the ideal-gas expressions for $\bar{s}_i$ into our expanded relationship for ΔS gives

$$\Delta S = \left\{N_{O_2}\left[\bar{s}_{O_2}(T, P_{ref}) - R_u \ln \frac{P_{O_2}}{P_{ref}}\right] + N_{N_2}\left[\bar{s}_{N_2}(T, P_{ref}) - R_u \ln \frac{P_{N_2}}{P_{ref}}\right]\right\}$$
$$- \left\{N_{O_2}\left[\bar{s}_{O_2}(T, P_{ref}) - R_u \ln \frac{P_{init}}{P_{ref}}\right] + N_{N_2}\left[\bar{s}_{N_2}(T, P_{ref}) - R_u \ln \frac{P_{init}}{P_{ref}}\right]\right\}.$$

The $\bar{s}_i$ terms all cancel in this expression to yield

$$\Delta S = N_{O_2}\left[-R_u \ln \frac{P_{O_2}}{P_{ref}} - \left(-R_u \ln \frac{P_{init}}{P_{ref}}\right)\right]$$
$$+ N_{N_2}\left[-R_u \ln \frac{P_{N_2}}{P_{ref}} - \left(-R_u \ln \frac{P_{init}}{P_{ref}}\right)\right].$$

Recognizing that $\ln (a/b) = \ln a - \ln b$, we reduce this to

$$\Delta S = R_u\left[N_{O_2} \ln \frac{P_{init}}{P_{O_2}} + N_{N_2} \ln \frac{P_{init}}{P_{N_2}}\right].$$

Substituting numerical values, we obtain our final result:

$$\Delta S = 8.314\left[(1)\ln \frac{1 \text{ atm}}{0.333 \text{ atm}} + (2)\ln \frac{1 \text{ atm}}{0.666 \text{ atm}}\right] = 15.88$$

$$[=] \frac{kJ}{kmol \cdot K}(kmol) = kJ/K.$$

Comment What if we were to perform the same mixing process, but using a single constituent, say, O_2? In this scenario, the partial pressure of the O_2 at the end of the mixing is the same as at the initial state (i.e., $P_{init} = 1$ atm $= P_{final} = P_{O_2,final}$). With no change in pressure, ΔS is then zero. We can also view this process as a reversible one since we could put the partition back in place after the mixing takes place and retrieve the initial state, treating the O_2 molecules as indistinguishable.

Self Test 7.6

 Consider two solid metals (copper and iron) separated by a partition in a container, as depicted in Fig. 7.19c. The copper is originally at 400 K and the iron is at 300 K. Determine the total change in entropy when the partition is removed, the two metals are brought into contact, and the temperature is allowed to equilibrate.

(Answer: $\Delta S_{total} = 19$ J/K)

7.6e Isentropic Efficiency

In addition to entropy's usefulness in predicting the direction of spontaneous change and establishing equilibrium conditions (statements IV A–C of Table 7.4), we also use entropy to quantify the performance of practical devices. Our discussions of entropy thus reinforce the two ways that the second law is particularly useful to engineers as outlined at the outset of this chapter. In an earlier section, we saw that the second law establishes theoretical limits to thermodynamic *cycles* (Carnot efficiency); in an analogous manner, the second law also establishes theoretical limits to *processes*. For example, consider the expansion of steam through a turbine. The maximum work will be obtained from this process when it occurs adiabatically—with no loss of potential to do work through heat transfer—and reversibly—with no loss of potential to do work through friction or other irreversibilities. For all of the steady-flow devices we wish to consider (i.e., turbines, compressors, pumps, and nozzles; see Chapter 11), the reversible adiabatic, or isentropic, process is the standard to which real processes can be compared.

The isentropic efficiency for a turbine is thus defined as

$$\eta_{isen,t} \equiv \frac{\dot{W}_{act}}{\dot{W}_{isen}}. \tag{7.27}$$

In Eq. 7.27, both the actual power $\dot{W}_{act}$ and the isentropic power $\dot{W}_{isen}$ are evaluated using the same inlet states and for the same pressures at the exit states.

Example 7.7   Steam Power Plant Application

Example 5.16 in Chapter 5 provides useful background for this example. ▷

Consider a steam turbine in which the steam enters as superheated vapor at 800 K and 6 MPa and exits at 0.1 MPa. The flow rate of the steam is 15 kg/s, and the isentropic efficiency of the turbine is 90%. Determine the outlet state of the steam and the power produced by the turbine.

Solution

Known $T_1, P_1, P_2, \eta_{isen,t}$

Find   State-2 properties, $\dot{W}_{act}$

A 660-MW steam turbine. Photograph courtesy of General Electric Company.

Sketch

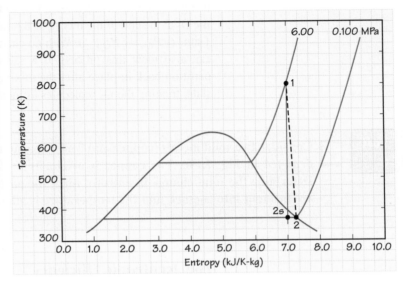

Assumptions

i. Steady state

ii. Negligible Δke and Δpe

Analysis   Using the NIST database, we find that $s_1 = 6.9635$ kJ/kg·K for the given inlet temperature and pressure. For an isentropic expansion from state 1 to state 2, we see that s_{2s} falls between s_{2f} and s_{2g} at 0.1 MPa (i.e., $1.3028 < 6.9635 < 7.3588$ kJ/kg·K). The actual state-2 entropy will lie to the right of s_{2s} somewhere on the 0.1-MPa isobar. To determine the actual state 2, we apply the definition of isentropic efficiency (Eq. 7.27):

$$\eta_{isen,t} = \frac{\dot{W}_{act}}{\dot{W}_{isen}}.$$

Here both the actual power $\dot{W}_{act}$ and the isentropic power $\dot{W}_{isen}$ are evaluated using the same inlet states and with the same exit-state pressure. With our assumptions, the turbine power is expressed as $\dot{W}_t = \dot{m}(h_{in} - h_{out})$; thus,

See Example 5.16 and Table 11.1.

$$\eta_{isen,t} = \frac{\dot{m}(h_1 - h_2)}{\dot{m}(h_1 - h_{2s})} = \frac{h_1 - h_2}{h_1 - h_{2s}}.$$

The enthalpy at state 1 is found from the NIST database and is listed in the table that follows this discussion. To find h_{2s}, we use Eq. 2.49d to find the quality x_{2s} using known entropies:

$$x_{2s} = \frac{s_{2s} - s_{2f}}{s_{2g} - s_{2f}} = \frac{6.9635 - 1.3028}{7.3588 - 1.3028} = 0.9347,$$

and

$$h_{2s} = (1 - x_{2s})h_{2f} + x_{2s}h_{2g}$$
$$= 0.0653(417.5) + (0.9347)2674.9 \text{ kJ/kg}$$
$$= 2527.5 \text{ kJ/kg},$$

where s_{2f}, s_{2g}, h_{2f}, and h_{2g} are found from the NIST database for $P_2 = P_{sat} = 0.1$ MPa. Using the given value of $\eta_{isen,t}$ ($= 0.90$), we calculate h_2 as follows:

$$0.90 = \frac{3486.7 - h_2}{3486.7 - 2527.5},$$

or

$$h_2 = 2623.4 \text{ kJ/kg}.$$

To completely define the properties at state 2, we again apply Eq. 2.49d, now using the known h_2, to find the quality x_2. This quality is used in turn to find s_2 (Eq. 2.49c):

$$x_2 = \frac{h_2 - h_{2f}}{h_{2g} - h_{2f}} = \frac{2623.4 - 417.5}{2674.9 - 417.5} = 0.9772$$

and

$$s_2 = (1 - x_2)s_{2f} + x_2 s_{2f}$$

$$= (0.0228)1.3028 + 0.9772(7.3588) \text{ kJ/kg} \cdot \text{K}$$

$$= 7.2207 \text{ kJ/kg} \cdot \text{K}.$$

The following table presents all of the properties:

Property	State 1	State 2s	State 2
T (K)	800	372.76	372.76
P (MPa)	6	0.10	0.10
s (kJ/kg·K)	6.9635	6.9635	7.2207
h (kJ/kg)	3486.7	2527.5	2623.4
x	—	0.9347	0.9772

The actual turbine power is

$$\dot{W}_{act} = \dot{m}(h_1 - h_2)$$

$$= 15(3486.7 - 2623.4)$$

$$= 12,949$$

$$[=] \frac{\text{kg}}{\text{s}} \frac{\text{kJ}}{\text{kg}} = \frac{\text{kJ}}{\text{s}} = \text{kW}.$$

Comment   We note that the final state still lies within the liquid–vapor dome on our *T–s* diagram to the right of the 2s state, as expected for an irreversible, nearly adiabatic expansion. Note also the importance of the thermodynamic property quality in the solution of this problem.

 Self Test 7.7

✓ **Steam enters a turbine at 10 MPa and 800 K and exits at a quality of 0.91 at 100 kPa. Determine the isentropic efficiency of the turbine.**

(*Answer:* $\eta_{\text{isent,t}} = 95.4\%$)

For devices that consume power (e.g., pumps and compressors) one desires to minimize the power input; thus, the isentropic efficiency is now defined as

$$\eta_{\text{isen,p or c}} \equiv \frac{\dot{W}_{\text{in,isen}}}{\dot{W}_{\text{in,act}}}. \qquad (7.28)$$

Again, the same inlet conditions and identical outlet pressures apply to the evaluation of both the isentropic and actual power.

Example 7.8 Steam Power Plant Application

Feedwater pump. Photograph courtesy of Flowserve Corporation.

A feedwater pump operates with a flow rate of 464 kg/s. The inlet pressure is 689 kPa and the outlet pressure is 26 MPa. The water enters the pump at 422 K. Assuming an isentropic efficiency of 85%, estimate the power required to drive the pump.

Solution

Known $\dot{m}, P_1, P_2, T_1, \eta_{isen,p}$

Find $\dot{W}_{in,act}$

Sketch

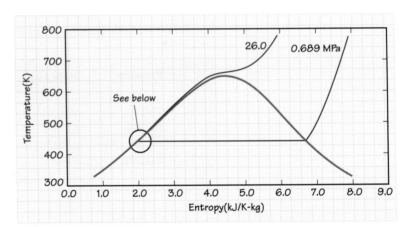

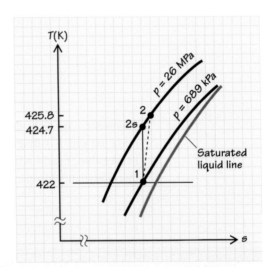

Assumptions

i. Adiabatic process
ii. Negligible changes in kinetic and potential energies

Analysis   To find the pump power, we apply the definition of isentropic efficiency (Eq. 7.28) and the simplified first-law analysis of the pump (Eq. 5.63b), that is,

$$\eta_{isen,p} = \frac{\dot{W}_{in,isen}}{\dot{W}_{in,act}},$$

where

Also see Table 11.1.

$$\dot{W}_{in,isen} = \dot{m}(h_{2s} - h_1).$$

We use the NIST database to find the properties at state 1. Recognizing that $s_{2s} = s_1$, we are able to define all of the properties at state 2. These are shown in the following table:

Property	State 1	State 2s
P (MPa)	0.689	26
T (K)	422	424.67
s (kJ/kg·K)	1.8298	1.8298
h (kJ/kg)	627.36	654.74

With the inlet and exit properties now known, we evaluate $\dot{W}_{in,act}$ as follows:

$$\dot{W}_{in,act} = \frac{\dot{m}(h_{2s} - h_1)}{\eta_{isen,p}}$$

$$= \frac{464(654.74 - 627.36)}{0.85} = 14{,}950$$

$$[=] \frac{kg}{s} \frac{kJ}{kg} = \frac{kJ}{s} = kW.$$

Comments The irreversibilities in the pump, such as friction, convert some of the input power to thermal energy. To find the additional temperature rise associated with the irreversibilities, we first calculate the actual state-2 enthalpy:

$$h_2 = \frac{\dot{W}_{in,act}}{\dot{m}} + h_1$$

$$= 14{,}950/464 + 627.36 \text{ kJ/kg} = 659.58 \text{ kJ/kg}.$$

With h_2 and P_2 (= 26 MPa) defining the state, we find that $T_2 = 425.8$ K using the NIST database. This value is about 1.1 K greater than T_{2s} and is shown in the sketch.

Self Test 7.8 ✓ **Air enters a compressor at 100 kPa and 300 K and exits at 600 kPa and 550 K. Assuming constant specific heats at 300 K and neglecting potential and kinetic energy changes, determine the isentropic efficiency of the compressor.**

(Answer: $\eta_{isen,c} = 80\%$)

Isentropic efficiencies are also illustrated in Examples 11.15–11.19.

For a nozzle, the outlet kinetic energy is the quantity maximized; thus, the isentropic efficiency for this device is

$$\eta_{isen,n} \equiv \frac{KE_{act}}{KE_{isen}}. \qquad (7.29)$$

Table 7.6 Typical Isentropic Efficiencies for Turbines, Compressors, Pumps, and Nozzles

Device	Isentropic Efficiency (%)
Turbine	70–90
Compressor	75–85
Pump	75–85
Nozzle	>95

Table 7.6 shows typical ranges of isentropic efficiency for the steady-flow devices discussed here.

7.6f Entropy Production, Head Loss, and Isentropic Efficiency

Friction losses are important in long pipe runs.

In Chapters 5 and 6, we introduced the concept of *head loss* in the mechanical energy equation to account for the effects of fluid friction and other irreversibilities. We also showed how the head loss relates to the first law of thermodynamics. We now relate the effects of *irreversibilities* to *entropy production* and *isentropic efficiency*, exploring the connections among these three concepts using two illustrations: 1. steady, isothermal flow through a horizontal pipe and 2. the steady-flow, adiabatic operation of a pump. For both illustrations we assume an incompressible fluid.

Figure 7.20a shows the control volume for our pipe flow analysis. We assume the flow is both steady and isothermal; furthermore, the velocity profiles at the entrance and exit stations are identical. Because friction retards the flow, the pressure at the inlet must be greater than the that at the outlet to maintain the flow through the pipe. We first apply the second law of thermodynamics by writing an entropy balance (Eq. 7.25b) for this control volume:

$$\frac{\dot{Q}}{T} + \dot{S}_{gen} + \dot{m}(s_1 - s_2) = \frac{dS_{CV}}{dt}.$$

For an incompressible fluid, the entropy is independent of pressure and depends only on the temperature. Since the inlet and exit temperatures are equal (isothermal flow), s_2 equals s_1, causing the third term to be zero. Furthermore, the term on the right-hand side is zero because the flow is steady. We thus conclude that

$$\frac{\dot{Q}}{T} + \dot{S}_{gen} = 0.$$

Because the fluid within the control volume is heated by friction, heat must be transferred out of the control volume to maintain a constant temperature. Thus, $\dot{Q}$ is negative (i.e., $\dot{Q}_{out} = -\dot{Q}$). Our entropy balance is then

$$\dot{Q}_{out} = T\dot{S}_{gen}.$$

Energy conservation expressed by Eq. 5.68 simplifies with our assumptions ($V_1 = V_2$, $z_1 = z_2$, and $\dot{W}_{net} = 0$) to

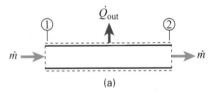

(a)

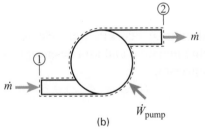

(b)

FIGURE 7.20

Control volumes for (a) isothermal, steady flow through a pipe with friction and (b) adiabatic, steady flow through a pump.

$$\frac{P_1 - P_2}{\rho g} = h_L,$$

where the head loss is defined as (Eq. 5.67)

$$h_L = \frac{\dot{Q}_{out}}{\dot{m}g} + \frac{u_2 - u_1}{g}.$$

With the isothermal assumption, $u_2 = u_1$; thus,

$$\frac{P_1 - P_2}{\rho g} = h_L = \frac{\dot{Q}_{out}}{\dot{m}g}.$$

Substituting our second-law result for $\dot{Q}_{out}$ yields

$$\frac{P_1 - P_2}{\rho g} = \frac{T\dot{S}_{gen}}{\dot{m}g}, \qquad (7.30a)$$

or

$$P_1 - P_2 = \left(\frac{\rho T}{\dot{m}}\right)\dot{S}_{gen}. \qquad (7.30b)$$

Equation 7.30b tells us that the pressure drop in the pipe is directly proportional to the rate at which irreversibilities generate entropy. This result makes our desired connection between the first and second laws of thermodynamics.

We next consider a pump that operates steadily with an incompressible fluid (see Fig. 7.20b). We furthermore assume adiabatic operation, equal inlet and outlet velocities ($V_1 = V_2$), and negligible elevation change ($z_1 = z_2$). With these assumptions, the entropy balance for the pump simplifies to

$$0 + \dot{S}_{gen} + \dot{m}(s_1 - s_2) = 0,$$

or

$$s_2 - s_1 = \frac{\dot{S}_{gen}}{\dot{m}}.$$

Circulation pumps for geothermal heat pump system for district heating. Photograph courtesy of Geo-Heat Center, Oregon Institute of Technology.

Because the pump power is directed into the control volume, conservation of energy (Eq. 5.68) becomes

$$\frac{P_1 - P_2}{\rho g} = \frac{-\dot{W}_{pump}}{\dot{m}g} + h_L.$$

For the adiabatic control volume, the head loss is expressed as

$$h_L = \frac{\dot{Q}_{out}}{\dot{m}g} + \frac{u_2 - u_1}{g}$$

$$= 0 + \frac{u_2 - u_1}{g} = \frac{u_2 - u_1}{g}.$$

To relate $u_2 - u_1$ to $s_2 - s_1$, we employ the general property relationship (Eq. 2.35b)

$$Tds = du + Pdv.$$

For an incompressible fluid, dv is zero for any process; thus,

$$T ds = du.$$

We can also relate the specific internal energy to the specific heat as

$$du = c\, dT,$$

See Equation 2.56b in Chapter 2. ▶

recognizing that $c_p = c_v = c$ for an incompressible substance. Combining these two relationships and treating the specific heat as a constant, we integrate from the initial to the final state to yield

$$s_2 - s_1 = c \ln \frac{T_2}{T_1}.$$

Manipulating and rearranging this expression to relate $s_2 - s_1$ to $T_2 - T_1$ give

$$T_2 = T_1 \exp\left(\frac{s_2 - s_1}{c}\right),$$

or

$$T_2 - T_1 = T_1\left[\exp\left(\frac{s_2 - s_1}{c}\right) - 1\right].$$

As $\Delta u = c\Delta T$, the internal energy change relates to the entropy change as follows:

$$u_2 - u_1 = cT_1\left[\exp\left(\frac{s_2 - s_1}{c}\right) - 1\right].$$

Substituting the result from the second-law analysis that $s_2 - s_1 = \dot{S}_{gen}/\dot{m}$ yields

$$u_2 - u_1 = cT_1\left[\exp\left(\frac{\dot{S}_{gen}}{\dot{m}c}\right) - 1\right].$$

Returning to the first-law expression, we have for the pump power

$$\dot{W}_{pump} = \frac{\dot{m}(P_2 - P_1)}{\rho} + \dot{m}cT_1\left[\exp\left(\frac{\dot{S}_{gen}}{\dot{m}c}\right) - 1\right].$$

Since the entropy generation rate by irreversibilities is always positive, the last term causes the actual pump work to be greater than the ideal (isentropic) pump work,

$$\dot{W}_{isen} = \dot{m}(P_2 - P_1)/\rho.$$

The isentropic efficiency for the pump is thus

$$\eta_{isen,p} = \frac{\dot{W}_{isen}}{\dot{W}_{pump}}$$

$$= \frac{\dot{m}(P_2 - P_1)/\rho}{\dot{m}(P_2 - P_1)/\rho + \dot{m}cT_1\left[\exp\left(\frac{\dot{S}_{gen}}{\dot{m}c}\right) - 1\right]}. \tag{7.31}$$

From inspection of Eq. 7.31, we see that the isentropic efficiency decreases as the entropy generation rate $\dot{S}_{gen}$ increases, as expected. In the limit that $\dot{S}_{gen} \to 0$, we recover the result that the isentropic efficiency is unity (or 100%), also as expected.

This is a good exit or entry point for this Chapter. ▶

u, v fixed

T, P fixed

T, v fixed

See Equations 2.21 and 2.22 in Chapter 2. ▶

These two analyses show from slightly different perspectives how the principles of energy conservation and the second law of thermodynamics combine to describe the behavior of real systems or devices.

7.7 THE SECOND LAW AND EQUILIBRIUM

As indicated in Table 7.4, the second law can be cast as a series of statements defining conditions for equilibrium. These statements are mathematically expressed, respectively, in terms of entropy, the Gibbs free energy (G), and the Helmholtz free energy (A) as follows:

$$(dS)_{u,v} \geq 0, \tag{7.32a}$$

$$(dG)_{T,P} \leq 0, \tag{7.32b}$$

and

$$(dA)_{T,v} \leq 0. \tag{7.32c}$$

The subscripts refer to the properties that are held fixed during the establishment of equilibrium. The Gibbs and Helmholtz free energies are introduced in Chapter 2, and their definitions are repeated here for clarity. The Gibbs free energy and its allied partial molar quantity, the chemical potential, are essential to our discussion of equilibrium. The **Gibbs free energy**, or **Gibbs function**, is defined in terms of other thermodynamic properties as

$$G \equiv H - TS, \tag{2.21}$$

and the **Helmoltz free energy** is defined as

$$A \equiv U - TS. \tag{2.22}$$

Equation 7.32a and its companions, Eqs. 7.32b and 7.32c, are very general statements of equilibrium and apply to pure substances, solutions, and reacting and nonreacting mixtures. The only restriction is their application to a simple thermodynamic system, that is, a system unaffected by gravity, magnetic forces, electrical forces, and surface forces other than normal forces (see Chapter 2). In the following sections, we focus on the application of Eq. 7.32b to chemical equilibrium, with particular attention given to combustion applications and to liquid–vapor phase equilibrium.

Before proceeding, we want to clarify the connections between the verbal statement that *equilibrium occurs when the entropy is a maximum for a simple system at constant internal energy and volume* (Table 7.4 statement IV A), and the mathematical statements in Eq. 7.32a and Eq. 7.24. In Eq. 7.32a, the subscripts u and v denote that any changes to the system under consideration occur at both constant internal energy and volume. To maintain the internal energy constant requires either that any heat and work interactions cancel ($\delta Q = \delta W$) or that both are zero ($\delta Q = \delta W = 0$). Since the volume is constant, no work is possible (i.e., $\delta W = Pdv = 0$); we thus conclude that neither heat nor work is allowed. The absence of these interactions is our definition of an isolated system. Earlier we stated the second law in terms of an isolated system; thus, Eqs. 7.24 and 7.32a are equivalent. From the perspective of equilibrium,

Hydrogen is produced by the steam reforming of natural gas. A key step in the process involves the equilibrium reaction $CO + H_2O \rightleftharpoons CO_2 + H_2$. Photograph courtesy of Linde AG.

A shifting chemical equilibrium is a useful characterization of the major products of combustion (H_2O and CO_2) in a diesel engine. Photograph courtesy of Scania.

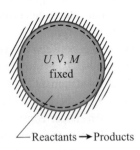

$U, \mathcal{V}, M$ fixed

Reactants → Products

Eq. 7.32a indicates that an isolated system will proceed to an equilibrium state only by processes that involve an increase of entropy; thus, we conclude that the equilibrium state must be a state of maximum entropy. Similar arguments can be applied to the equilibrium criteria given by Eqs. 7.32b and 7.32c.

7.7a Chemical Equilibrium

Chemical equilibrium is important in any process in which chemical reactions take place. Criteria for chemical equilibrium provide no information about the rate of reaction but, rather, determine the final state given a set of initial reactants and the final temperature and pressure. In reacting systems, specifying the composition (i.e., the values of the various species mole fractions) is essential to specifying the thermodynamic state. Many thermal-fluid applications require an understanding of the principles of chemical equilibrium. Combustion chemistry is of particular importance to the many devices that burn a fuel to produce work or heat. Although the theoretical development in this section is reasonably general, our applications focus on those involving combustion.

The most useful treatment of chemical equilibrium considers conditions of constant temperature and pressure. For these conditions, the Gibbs function is the natural second-law system property used to describe equilibrium (Eq. 7.32b). Before we take that approach, we investigate a system of fixed internal energy and volume to reinforce our discussion of the second law as embodied in the increase of entropy principle.

Conditions of Fixed Internal Energy and Volume

Consider a fixed-volume, adiabatic reaction vessel in which a fixed mass of reactants forms products. As the reactions proceed, both the temperature and pressure rise until a final equilibrium condition is reached. We apply the second-law statement (Eq. 7.32a) in combination with the first law to determine this final state (temperature, pressure, and composition). It is the addition of the composition variable that makes the present analysis different from those for nonreacting systems. Rather than work in general terms, we will illustrate this problem by considering a single specific reaction. Carbon monoxide burns in oxygen to form carbon dioxide via the reaction

$$CO + \frac{1}{2}O_2 \rightarrow CO_2. \tag{7.33}$$

If the final temperature is high enough, the CO_2 will dissociate. Assuming the products to consist only of CO_2, CO, and O_2, we can write

$$\left[CO + \tfrac{1}{2}O_2\right]_{\substack{\text{cold} \\ \text{reactants}}} \rightarrow \left[(1 - \alpha)CO_2 + \alpha CO + \frac{\alpha}{2}O_2\right]_{\substack{\text{hot} \\ \text{products}}}, \tag{7.34}$$

where α is the fraction of the CO_2 dissociated. We can calculate the adiabatic flame temperature as a function of the dissociation fraction α using Eq. 5.15a. For example, with $\alpha = 1$, no heat is released and the mixture temperature, pressure, and composition remain unchanged; whereas with $\alpha = 0$, the maximum amount of heat release occurs and the temperature and pressure would be the highest possible allowed by the first law. This variation in temperature with α is plotted in Fig. 7.21.

FIGURE 7.21

We determine the equilibrium composition of a chemically reacting, isolated, fixed-mass system by locating the condition of maximum entropy. **Adapted from Ref. [22] with permission.**

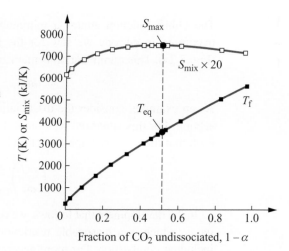

See Equations 2.39a and 2.73 in Chapter 2.

What constraints are imposed by the second law on this thought experiment where we vary α? The entropy of the product mixture can be calculated by summing the product species entropies as follows:

$$S_{mix}(T_f, P) = \sum_{i=1}^{3} N_i \bar{s}_i (T_f, P_i) = (1 - \alpha)\bar{s}_{CO_2} + \alpha\bar{s}_{CO} + \frac{\alpha}{2}\bar{s}_{O_2}, \quad (7.35)$$

where N_i is the number of moles of species i in the mixture. As discussed in Chapter 2, the individual species entropies can be expressed as

$$\bar{s}_i = \bar{s}_i^\circ (T_{ref}) + \int_{T_{ref}}^{T_f} \bar{c}_{p,i} \frac{dT}{T} - R_u \ln \frac{P_i}{P^\circ}, \quad (7.36)$$

where ideal-gas behavior is assumed and P_i is the partial pressure of the ith species. Plotting the mixture entropy (Eq. 7.35) as a function of the dissociation fraction (Fig. 7.21), we see that a maximum value is reached at some intermediate value of α. For the reaction chosen, $CO + \frac{1}{2}O_2 \rightarrow CO_2$, the maximum entropy occurs near $1 - \alpha = 0.5$.

The statement that $(dS)_{u,v} \geq 0$, or more precisely that

$$\left(\frac{dS}{d\alpha}\right)_{u,v} = 0, \quad (7.37)$$

causes us to choose the state of maximum entropy as the equilibrium state, as indicated in Fig. 7.21. The idea that spontaneous change occurs in the direction of increasing entropy requires that the composition of the system shift toward the point of maximum entropy when approaching from either side, since dS is positive. Once the maximum entropy is reached, no further change in composition is allowed, since this would require the system entropy to decrease, in violation of the second law (Eq. 7.32a).

From this example, we see that a maximization of entropy principle was applied in our plotting S versus α and then choosing α so as to maximize S to fix the equilibrium composition.

Conditions of Fixed Temperature and Pressure

For conditions of fixed temperature and pressure, we apply the second law as expressed by Eq. 7.32b, $(dG)_{T,P} \leq 0$, to determine the equilibrium composition.

The Gibbs function attains a minimum in equilibrium, in contrast to the maximum in entropy we saw for the fixed-energy and fixed-volume case (Fig. 7.21). This minimization is mathematically expressed as

$$(dG)_{T,P} = 0. \tag{7.38}$$

As an example, consider the dissociation of CO_2 discussed in the previous section. For this situation, we obtain the value of the dissociated fraction α such that the mixture Gibbs function is a minimum by the application of

$$\left(\frac{dG}{d\alpha} \right)_{T,P} = 0.$$

In the development that follows, we consider only reacting mixtures of ideal gases, which is a reasonable restriction for many engineering applications, including combustion. For more general treatments of chemical equilibrium, we refer the reader to other textbooks [15, 16].

We begin by considering the following general equilibrium reaction among the species A, B, C, etc.:

$$a\text{A} + b\text{B} + \cdots \Leftrightarrow e\text{E} + f\text{F} + \ldots. \tag{7.39}$$

The Gibbs function for the mixture containing the species A, B, C, etc. can be expressed as

$$G_{\text{mix}} = \sum N_i \mu_{i,T}, \tag{7.40}$$

where N_i is the number of moles of the ith species and $\mu_{i,T}$ is the chemical potential of the ith species at the temperature of interest. For a mixture of *ideal* gases, the **chemical potential** for the ith species is simply the Gibbs function per mole of i; that is,

$$\mu_{i,T} = \bar{g}_{i,T} = \bar{g}^{\circ}_{i,T} + R_{\text{u}} T \ln(P_i/P^{\circ}), \tag{7.41}$$

where $\bar{g}^{\circ}_{i,T}$ is the Gibbs function of the *pure* species at the standard-state pressure (i.e., $P_i = P^{\circ}$) and P_i is the partial pressure. The standard-state pressure P°, by convention taken to be 1 atm, appears in the denominator of the logarithm term. In dealing with reacting systems, a **Gibbs function of formation**, $\Delta \bar{g}^{\circ}_{\text{f},i}$, is frequently employed:

$$\Delta \bar{g}^{\circ}_{\text{f},i}(T) \equiv \bar{g}^{\circ}_i(T) - \sum_{j \text{ elements}} \nu'_j \bar{g}^{\circ}_j(T), \tag{7.42}$$

where the ν'_j are the stoichiometric coefficients of the elements required to form one mole of the compound of interest. For example, the coefficients are $\nu'_{O_2} = 1/2$ and $\nu'_C = 1$ for forming a mole of CO from O_2 and C, respectively. As with enthalpies, the Gibbs functions of formation of the naturally occurring elements are assigned values of zero at the reference state. Appendix B provides tabulations of Gibbs functions of formation over a range of temperatures for selected species. Having tabulations of $\Delta \bar{g}^{\circ}_{\text{f},i}(T)$ as a function of temperature is quite useful. In later calculations, we will need to evaluate differences in $\bar{g}^{\circ}_{i,T}$ among various species at the same temperature. These differences can be easily obtained by using the Gibbs function of formation at the temperature of interest. In addition to those in Appendix B, tabulations for over 1000 species can be found in the JANAF tables [17].

Substituting Eq. 7.41 into Eq. 7.40 yields

$$G_{\text{mix}} = \sum N_i \bar{g}_{i,T} = \sum N_i [\bar{g}^{\circ}_{i,T} + R_{\text{u}} T \ln(P_i/P^{\circ})]. \tag{7.43}$$

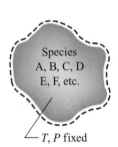

Species
A, B, C, D
E, F, etc.

T, P fixed

We now apply the equilibrium criterion $(dG)_{T,P} = 0$ (Eq. 7.38) to Eq. 7.43; thus,

$$\sum dN_i[\bar{g}^{\circ}_{i,T} + R_u T \ln(P_i/P^{\circ})] + \sum N_i d[\bar{g}^{\circ}_{i,T} + R_u T \ln(P_i/P^{\circ})] = 0. \quad (7.44)$$

The second term in Eq. 7.44 can be shown to be zero by recognizing that $d(\ln P_i) = dP_i/P_i$ and that $\Sigma dP_i = 0$, since all changes in the partial pressures must sum to zero because the total pressure is constant. Thus,

$$dG_{\text{mix}} = 0 = \sum dN_i[\bar{g}^{\circ}_{i,T} + R_u T \ln(P_i/P^{\circ})]. \quad (7.45)$$

From the chemical system described by Eq. 7.39, the change in the number of moles of each species is directly proportional to its stoichiometric coefficient,[10] so we have

$$dN_A = -\kappa a, \quad dN_B = -\kappa b, \ldots$$
$$dN_E = +\kappa e, \quad dN_F = +\kappa f, \ldots \quad (7.46)$$

Substituting Eq. 7.46 into Eq. 7.49 and canceling the proportionality constant κ, we obtain

$$-a[\bar{g}^{\circ}_{A,T} + R_u T \ln(P_A/P^{\circ})] - b[\bar{g}^{\circ}_{B,T} + R_u T \ln(P_B/P^{\circ})] - \ldots$$
$$+ e[\bar{g}^{\circ}_{E,T} + R_u T \ln(P_E/P^{\circ})] + f[\bar{g}^{\circ}_{F,T} + R_u T \ln(P_F/P^{\circ})] + \ldots = 0. \quad (7.47)$$

Equation 7.47 can be rearranged and the log terms grouped together to yield

$$-(e\bar{g}^{\circ}_{E,T} + f\bar{g}^{\circ}_{F,T} + \ldots -a\bar{g}^{\circ}_{A,T} - b\bar{g}^{\circ}_{B,T} - \ldots)$$
$$= R_u T \ln \frac{(P_E/P^{\circ})^e \cdot (P_F/P^{\circ})^f \cdots}{(P_A/P^{\circ})^a \cdot (P_B/P^{\circ})^b \cdots}. \quad (7.48)$$

The term in parentheses on the left-hand-side of Eq. 7.48 is called the **standard-state Gibbs function change** ΔG°_T, defined by

$$\Delta G^{\circ}_T \equiv (e\bar{g}^{\circ}_{E,T} + f\bar{g}^{\circ}_{F,T} + \ldots - a\bar{g}^{\circ}_{A,T} - b\bar{g}^{\circ}_{B,T} - \ldots), \quad (7.49a)$$

or, alternatively,

$$\Delta G^{\circ}_T \equiv (e\Delta\bar{g}^{\circ}_{f,E} + f\Delta\bar{g}^{\circ}_{f,F} + \ldots -a\Delta\bar{g}^{\circ}_{f,A} - b\Delta\bar{g}^{\circ}_{f,B} - \ldots)_T. \quad (7.49b)$$

The argument of the natural logarithm is defined as the **equilibrium constant K_p** for the reaction expressed in Eq. 7.39; that is,

$$K_p \equiv \frac{(P_E/P^{\circ})^e \cdot (P_F/P^{\circ})^f \cdots}{(P_A/P^{\circ})^a \cdot (P_B/P^{\circ})^b \cdots}. \quad (7.50)$$

With these definitions, Eq. 7.48, our statement of chemical equilibrium at constant temperature and pressure, is given by

$$\Delta G^{\circ}_T = -R_u T \ln K_p, \quad (7.51a)$$

---

[10] We can show that this is true by considering the following simple specific equilibrium reaction: $1\ O_2 \Leftrightarrow 2\ O$. Assume for sake of argument that there are 10 moles of O_2 in our reacting mixture and that the only other species is the O atom. We now force a change of the number of O_2 moles from 10 to 7 (i.e., $dN_{O_2} = -3$). Since the total number of atoms must be conserved, the number of moles of O atoms must increase by a factor of 6 (i.e., $dN_O = +6$). Every time one mole of O_2 disappears, two moles of O atoms are generated. This result is consistent with the requirement of Eq. 7.46 that $-dN_A/a = +dN_E/e = \kappa$ (i.e., $-dN_{O_2}/1 = +dN_O/2$ or $[-(-3)/1 = +6/2]$).

or

$$K_p = \exp(-\Delta G^\circ_T / R_u T). \tag{7.51b}$$

From the definition of K_p (Eq. 7.50) and its relation to ΔG°_T (Eq. 7.51), we can obtain a qualitative indication of whether a particular reaction favors products (goes strongly to completion) or reactants (undergoes very little reaction) at equilibrium. If ΔG°_T is positive, reactants will be favored since $\ln K_p$ is negative, which requires that K_p itself be less than unity. Similarly, if ΔG°_T is negative, the reaction tends to favor products. We obtain physical insight into this behavior by appealing to the definition of ΔG in terms of the enthalpy and entropy changes associated with the reaction. From Eq. 2.21, we can write

$$\Delta G^\circ_T = \Delta H^\circ - T \Delta S^\circ,$$

which can be substituted into Eq. 7.51b to yield

$$K_p = e^{-\Delta H^\circ / R_u T} \cdot e^{\Delta S^\circ / R_u}.$$

For K_p to be greater than unity, which favors products, the enthalpy change for the reaction, ΔH°, should be negative; that is, the reaction is exothermic and the system energy is lowered. Also, positive changes in entropy, which indicate greater number of molecular microstates, lead to values of $K_p > 1$.

In solving problems involving chemical equilibrium, **element conservation** is usually required in conjunction with the second-law concepts just discussed. The following examples illustrate the use of these combined principles.

> **Element conservation is a subset of the general principle of mass conservation. See Chapter 3.**

Example 7.9

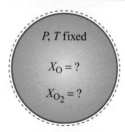

P, T fixed

$X_O = ?$

$X_{O_2} = ?$

At high temperatures, oxygen molecules dissociate to produce O atoms via the reaction

$$O_2 \Leftrightarrow O + O.$$

Determine the equilibrium partial pressures and mole fractions of O and O_2 at 1 atm and 3000 K. Repeat for a pressure of 0.1 atm.

Solution

Known   Equilibrium, P, T

Find $P_O, P_{O_2}, X_O, X_{O_2}$

Assumptions

Ideal-gas behavior

Analysis   To solve this problem, we perform the following sequence of calculations: First, we calculate the equilibrium constant K_p defined in Eq. 7.51b using thermodynamic property data from Appendix B; second, we use this K_p value to calculate the partial pressures P_O and P_{O_2}, using Eq. 7.50; and last, we relate these partial pressures to their corresponding mole fractions (Eq. 2.66b).

Using Tables B.11 and B.12 we evaluate the standard-state Gibbs function change (Eq. 7.49b) for the dissociation of O_2 at 3000 K as

$$\Delta G_T^\circ = 2\Delta \bar{g}_{f,O}^\circ - 1\Delta \bar{g}_{f,O_2}^\circ$$
$$= 2(54{,}554) - 1(0) \text{ kJ/kmol}$$
$$= 109{,}108 \text{ kJ/kmol}.$$

From Eq. 7.51b, we now evaluate K_p:

$$K_p = \exp\left[-\Delta G_T^\circ/(R_u T)\right]$$
$$= \exp\left[\frac{-109{,}108}{8.314(3000)}\right] = 0.0126$$

Note that K_p must be dimensionless since the argument of the exponential cannot have dimensions. The partial pressures relate to K_p (Eq. 7.50) as follows:

$$K_p = \frac{(P_O/P^\circ)^2}{(P_{O_2}/P^\circ)^1} = \frac{P_O^2}{P_{O_2}}\left(\frac{1}{P^\circ}\right).$$

Recognizing that the partial pressures sum to the total pressure (Eq. 2.65), we have

$$P_{tot} = P = P_O + P_{O_2},$$

or

$$P_{O_2} = P - P_O.$$

Substituting this expression for P_{O_2} into our K_p definition (Eq. 7.50) yields

$$K_p = \frac{P_O^2}{P - P_O}\left(\frac{1}{P^\circ}\right),$$

or

$$P_O^2 + K_p P^\circ P_O - K_p P^\circ(P) = 0.$$

For a total pressure P of 1 atm and a reference pressure of 1 atm, this becomes

$$P_O^2 + 0.0126P_O - 0.0126 = 0.$$

We find the useful root of this quadiatic equation to be

$$P_O = 0.1061 \text{ atm.}$$

Thus,

$$P_{O_2} = P - P_O = 1 - 0.1061 \text{ atm}$$
$$= 0.8939 \text{ atm.}$$

The corresponding mole fractions are

$$X_O = P_O/P = 0.1061/1 = 0.1061$$

and

$$X_{O_2} = P_{O_2}/P = 0.8939/1 = 0.8939.$$

We now reevaluate

$$P_O^2 + K_p P^\circ P_O - K_p P^\circ(P)$$

with $P = 0.1$ and obtain

$$P_O^2 + 0.0126 P_O - 0.0126(0.1) = 0.$$

We find the useful root to be

$$P_O = 0.02975 \text{ atm.}$$

Thus,

$$P_{O_2} = 0.1 - 0.02975 \text{ atm} = 0.07025 \text{ atm}$$

and

$$X_O = \frac{0.02975}{0.1} = 0.2975,$$

$$X_{O_2} = \frac{0.07025}{0.1} = 0.7025.$$

Comment Note how dissociation is enhanced at the lower pressure. At 0.1 atm, the O-atom mole fraction is nearly three times its value at 1 atm.

Self Test 7.9

Consider the reverse reaction to that presented in Example 7.9 (i.e., $O + O \Leftrightarrow O_2$). Evaluate K_p through Eq. 7.51b and develop an expression for the partial pressure as in Eq. 7.50. Compare these results to those from Example 7.9 and comment on them.
(Answer: $K_{p,\text{rev}} = 79.398 = (P_{O_2}/P_O^2)(1/P^\circ)$. Note that the equilibrium constant for this reverse reaction is the reciprocal of the forward reaction; for this reverse reaction, $K_{p,\text{rev}} = 1/K_{p,\text{forward}}$. Additionally, the large value of $K_{p,\text{rev}}$ indicates that reaction strongly favors the products, in this case, O_2.)

Example 7.10

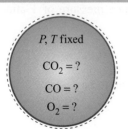

P, T fixed

$CO_2 = ?$

$CO = ?$

$O_2 = ?$

Consider the dissociation of CO_2 as a function of temperature and pressure via the reaction

$$CO_2 \Leftrightarrow CO + \tfrac{1}{2}O_2.$$

Find the composition of the mixture (i.e., the mole fractions of CO_2, CO, and O_2) that results from subjecting originally pure CO_2 to various temperatures ($T = 1500, 2000, 2500,$ and 3000 K) and pressures ($P = 0.1,$ 1, 10, and 100 atm).

Solution

Known Initially all CO_2, P, T

Find X_i

Assumptions

Ideal-gas behavior

Analysis To find the three unknown mole fractions, X_{CO_2}, X_{CO}, and X_{O_2}, we will need three equations. The first equation will be an equilibrium expression, Eq. 7.51. The other two equations will come from element conservation expressions that state that the total amounts of C and O are constant, regardless of how they are distributed among the three species, since the original mixture was pure CO_2.

To implement Eq. 7.51, we recognize the $a = 1$, $b = 1$, and $c = 1/2$, since

$$(1)CO_2 \Leftrightarrow (1)CO + (\tfrac{1}{2})O_2.$$

We use this to evaluate the standard-state Gibbs function change. For example at $T = 2500$ K,

$$\Delta G_T^\circ = [(\tfrac{1}{2})\Delta \bar{g}_{f,O_2}^\circ + (1)\Delta \bar{g}_{f,CO}^\circ - (1)\Delta \bar{g}_{f,CO_2}^\circ]_{T=2500}$$

$$= (\tfrac{1}{2})0 + (1)(-327{,}245) - (-396{,}152) \text{ kJ/kmol}$$

$$= 68{,}907 \text{ kJ/kmol}.$$

These values are taken from Tables B.11, B.1, and B.2.

From the definition of K_p, we have

$$K_p = \frac{(P_{CO}/P^\circ)^1 (P_{O_2}/P^\circ)^{0.5}}{(P_{CO_2}/P^\circ)^1}.$$

We rewrite K_p in terms of the mole fractions by recognizing that $P_i = X_i P$. Thus,

$$K_p = \frac{X_{CO} X_{O_2}^{0.5}}{X_{CO_2}} \cdot (P/P^\circ)^{0.5}.$$

Substituting this expression into Eq. 7.51b, we have

$$\frac{X_{CO} X_{O_2}^{0.5} (P/P^\circ)^{0.5}}{X_{CO_2}} = \exp\left[\frac{-\Delta G_T^\circ}{R_u T}\right]$$

$$= \exp\left[\frac{-68{,}907}{(8.3145)(2500)}\right] \qquad (I)$$

$$= 0.03635.$$

We create a second equation to express conservation of elements:

$$\frac{\text{number of carbon atoms}}{\text{number of oxygen atoms}} = \frac{1}{2} = \frac{X_{CO} + X_{CO_2}}{X_{CO} + 2X_{CO_2} + 2X_{O_2}}.$$

We can make the problem more general by defining the C/O ratio to be a parameter Z that can take on different values depending on the initial composition of the mixture:

$$Z = \frac{X_{CO} + X_{CO_2}}{X_{CO} + 2X_{CO_2} + 2X_{O_2}},$$

or

$$(Z - 1)X_{CO} + (2Z - 1)X_{CO_2} + 2ZX_{O_2} = 0. \qquad (II)$$

To obtain a third and final equation, we require that all of the mole fractions sum to unity:

$$\sum_i X_i = 1,$$

or

$$X_{CO} + X_{CO_2} + X_{O_2} = 1. \qquad (III)$$

Table 7.7 Equilibrium Compositions at Various Temperatures and Pressures for $CO_2 \Leftrightarrow CO + \frac{1}{2}O_2$

	P = 0.1 atm	P = 1 atm	P = 10 atm	P = 100 atm
T = 1500 K, $\Delta G_T^\circ = 1.5268 \times 10^8$ J/kmol				
X_{CO}	7.755×10^{-4}	3.601×10^{-4}	1.672×10^{-4}	7.76×10^{-5}
X_{CO_2}	0.9988	0.9994	0.9997	0.9999
X_{O_2}	3.877×10^{-4}	1.801×10^{-4}	8.357×10^{-5}	3.88×10^{-5}
T = 2000 K, $\Delta G_T^\circ = 1.10462 \times 10^8$ J/kmol				
X_{CO}	0.0315	0.0149	6.96×10^{-3}	3.243×10^{-3}
X_{CO_2}	0.9527	0.9777	0.9895	0.9951
X_{O_2}	0.0158	0.0074	3.48×10^{-3}	1.622×10^{-3}
T = 2500 K, $\Delta G_T^\circ = 6.8907 \times 10^7$ J/kmol				
X_{CO}	0.2260	0.1210	0.0602	0.0289
X_{CO_2}	0.6610	0.8185	0.9096	0.9566
X_{O_2}	0.1130	0.0605	0.0301	0.0145
T = 3000 K, $\Delta G_T^\circ = 2.7878 \times 10^7$ J/kmol				
X_{CO}	0.5038	0.3581	0.2144	0.1138
X_{CO_2}	0.2443	0.4629	0.6783	0.8293
X_{O_2}	0.2519	0.1790	0.1072	0.0569

Simultaneous solution of Eqs. I, II, and III for selected values of P, T, and Z yields values for the mole fractions X_{CO}, X_{CO_2}, and X_{O_2}. By using Eqs. II and III to eliminate X_{CO_2} and X_{O_2}, Eq. I becomes

$$X_{CO}(1 - 2Z + ZX_{CO})^{0.5}(P/P^\circ)^{0.5}$$
$$- [2Z - (1 + Z)X_{CO}]\exp(-\Delta G_T^\circ/R_uT) = 0.$$

This expression is easily solved for X_{CO} by applying Newton–Raphson iteration, which can be implemented simply using spreadsheet software. The other unknowns, X_{CO} and X_{O_2}, are then recovered using Eqs. II and III.

Results are shown in Table 7.7 for four levels each of temperature and pressure. Figure 7.22 shows the CO mole fractions over the range of parameters investigated.

Comments Two general observations concerning these results can be made: First, at any fixed temperature, increasing the pressure suppresses the dissociation of CO_2 into CO and O_2; second, increasing the temperature at a fixed pressure promotes the dissociation. Both of these trends are consistent with the **principle of Le Châtelier,** which states that any system initially in a state of equilibrium when subjected to a change (e.g., increasing pressure or temperature) will shift in composition in such a way as to minimize the change. For an increase in pressure, this translates to the equilibrium shifting in the direction to produce fewer moles. For the $CO_2 \Leftrightarrow CO + \frac{1}{2}O_2$ reaction, this means a shift to the left, to the CO_2 side. For equimolar reactions, pressure has no effect. When the temperature is increased, the composition shifts in the endothermic direction. Since energy is absorbed when CO_2 breaks down into CO and O_2, increasing the temperature produces a shift to the right, to the $CO + \frac{1}{2}O_2$ side.

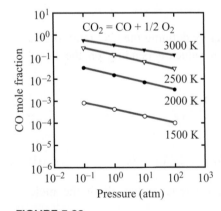

FIGURE 7.22

Increasing the temperature and/or decreasing the pressure promotes the dissociation of CO_2 to CO and O_2. See Example 7.10. Adapted from Ref. [22] with permission.

Multiple Equilibrium Reactions

The preceding discussion focused on simple situations involving a single equilibrium reaction; however, in most combustion systems, many species and several simultaneous equilibrium reactions are important. We can easily extend the previous example to include additional reactions. For example, the reaction $O_2 \Leftrightarrow 2O$ is likely to be important at the temperatures considered. Including this reaction introduces only one additional unknown, X_O. We add an additional equilibrium equation to account for the O_2 dissociation as follows:

$$(X_O^2/X_{O_2})P/P^\circ = \exp(-\Delta G_T^{\circ'}/R_u T),$$

where $\Delta G_T^{\circ'}$ is the appropriate standard-state Gibbs function change for the $O_2 \Leftrightarrow 2O$ reaction. We also modify the element-conservation expression (Eq. II) to account for the additional O-containing species,

$$\frac{\text{number of C atoms}}{\text{number of O atoms}} = \frac{X_{CO} + X_{CO_2}}{X_{CO} + 2X_{CO_2} + 2X_{O_2} + X_O},$$

and Eq. III becomes

$$X_{CO} + X_{CO_2} + X_{O_2} + X_O = 1.$$

We now have a new set of four equations: the three equations here plus Eq. I from Example 7.10. The solution of this particular problem is left as an exercise for the reader (see Problem 7.121).

An example of the approach described here applied to the C, H, N, O and Ar system is the computer code developed by Olikara and Borman [18]. This code solves for twelve species, involving seven equilibrium reactions and five atom-conservation relations, one each for C, O, H, N, and Ar. This code was developed specifically for internal combustion engine simulations and is readily imbedded as a subroutine in simulation codes. Other frequently encountered software packages capable of solving complex equilibria include the NASA chemical equilibrium code [19] and STANJAN [20]. Another easy-to-use code, GASEQ, is available as freeware [21] from the Internet.

7.7b Phase Equilibrium

We now consider conditions for equilibrium between two phases of a pure substance. A familiar example of this is the equilibrium between the liquid and vapor phases of H_2O. The questions we wish to address are the following: For a given temperature, what pressure must exist at equilibrium? Or, alternatively, for a given pressure, what temperature must prevail for equilibrium? Like the chemical equilibrium problem, phase equilibrium considers a fixed-mass system at constant temperature and pressure. As a result, the appropriate mathematical statement of the second law of thermodynamics controlling equilibrium is Eq. 7.38,

$$(dG_{sys})_{T,P} = 0,$$

where G_{sys} is the Gibbs function for the two-phase system.

For a liquid–vapor system, the Gibbs function can be expressed as

$$G_{sys} = N_f \bar{g}_f + N_g \bar{g}_g, \tag{7.52}$$

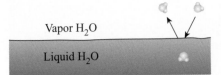

Vapor H_2O

Liquid H_2O

where the subscripts f and g refer to the liquid and vapor phases, respectively. Applying the condition of Eq. 7.38 yields

$$dG_{sys} = dN_f \bar{g}_f + dN_g \bar{g}_g = 0.$$

Because any evaporating liquid appears as vapor and, conversely, any condensing vapor appears as liquid, the change in the number of moles of one phase is the negative of the change in the other phases, and so

$$dN_f = -dN_g.$$

The condition for phase equilibrium then becomes

$$dG_{sys} = 0 = dN_f(\bar{g}_f - \bar{g}_g),$$

or

$$\bar{g}_f = \bar{g}_g. \tag{7.53a}$$

Our conclusion is that, for liquid–vapor equilibrium, the molar Gibbs function of the liquid phase must equal the molar Gibbs function of the vapor phase. Since the molecular weight of any substance is the same in both phases, we can also express Eq. 7.53a on a mass basis,

$$g_f = g_g. \tag{7.53b}$$

The ideas developed here apply not only to liquid–vapor equilibrium but to any pair of coexisting phases (i.e., vapor–solid and liquid–solid); thus, the general expression of phase equilibrium is

$$\boxed{\bar{g}_{phase\ 1} = \bar{g}_{phase\ 2},} \tag{7.54a}$$

or

$$\boxed{g_{phase\ 1} = g_{phase\ 2}.} \tag{7.54b}$$

The following example illustrates the criterion that $g_f = g_g$ applies to the equilibrium between the liquid and vapor phases of H_2O.

Example 7.11

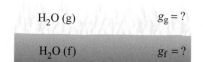

H₂O (g) $g_g = ?$

H₂O (f) $g_f = ?$

Use the saturation properties for H_2O at 100°C from the NIST database to show that the Gibbs function criterion (Eq. 7.53b) is met.

Solution

From the NIST online database we obtain the following property data at 100°C:

Phase	h (kJ/kg)	s (kJ/kg·K)
Liquid	419.17	1.3072
Vapor	2675.6	7.3541

The Gibbs function is defined by Eq. 2.21:

$$g \equiv h - Ts,$$

where T is the absolute temperature. We use this definition and the property data to calculate the Gibbs function for each phase as follows:

$$g_f = h_f - Ts_f$$
$$= 419.17 - (100 + 273.15)\,1.3072 \text{ kJ/kg}$$
$$= -68.612 \text{ kJ/kg}$$

and

$$g_g = h_g - Ts_g$$
$$= 2675.6 - (100 + 273.15)\,7.3541 \text{ kJ/kg}$$
$$= -68.582 \text{ kJ/kg}.$$

Comment Although these two values (-68.612 and -68.582) are not identical, they differ by only 0.044% and, most likely, are within the accuracy of the curve fits used to generate the NIST property data.

SUMMARY

In this chapter, we introduced the second law of thermodynamics and indicated its usefulness in establishing theoretical performance limits for real devices: the Carnot efficiency for cyclically operating devices, such as the closed-loop steam power plant, and the isentropic efficiency for noncyclic devices, such as turbines, compressors, pumps, and nozzles. Achieving an understanding of these performance limits required a fairly large number of concepts and definitions. Among these, you should be familiar with thermal reservoirs; heat engines and their reversed-cycle counterparts, heat pumps and refrigerators; the distinctions between reversible and irreversible processes; and Kelvin's absolute temperature scale. At the chapter outset, we also indicated the usefulness of the second law in establishing the direction of spontaneous change and defining thermodynamic equilibrium. To explore this usefulness, we defined and examined the thermodynamic property entropy and its companion, the Gibbs free energy, or Gibbs function. We also used entropy to calculate isentropic efficiencies. We examined how the second law relates to chemical (reacting) and liquid–vapor (nonreacting) equilibria. From this analysis, the equilibrium constant was defined and used to determine the detailed composition of a system at a fixed temperature and pressure.

An organizing theme of this chapter was the idea that the second law can be expressed in a number of ways—all of which are ultimately equivalent. We started with the Kelvin–Planck statement, with which you should be quite familiar, and ended with the various equilibrium statements that involve entropy and its companion properties. From this presentation, the reader should now be able to recognize the various forms of the second law and be comfortable with the idea that no single statement is universally recognized as *the* second law of thermodynamics.

Chapter 7
Key Concepts & Definitions Checklist*

7.2 Usefulness of the Second Law
❑ Performance limits
❑ Direction for spontaneous change

7.3 One Fundamental Statement of the Second Law
❑ Kelvin–Planck statement ➤ *Q7.2*
❑ Heat reservoir ➤ *Q7.5*
❑ Heat engine ➤ *Q7.6*
❑ Heat pump ➤ *Q7.7*
❑ Refrigerator ➤ *Q7.7*
❑ Thermal (first-law) efficiency ➤ *Q7.8, Q7.12*
❑ Coefficients of performance ➤ *Q7.8, Q7.13*
❑ Reversible and irreversible processes ➤ *Q7.9*

7.4 Consequences of the Kelvin–Planck Statement
❑ Six consequences (Table 7.2) ➤ *Q7.10*
❑ Absolute temperature scale ➤ *Q7.19*
❑ Carnot efficiency ➤ *7.35, 7.37*
❑ Carnot cycle (gas-phase working fluid) ➤ *7.12*
❑ Carnot cycle (two-phase working fluid) ➤ *7.13*
❑ Stirling cycle ➤ *Q7.15, 7.1*

7.5 Alternative Statements of the Second Law
❑ Original Clausius statements ➤ *Q7.16*
❑ Clausius entropy statement ➤ *Q7.16*
❑ Equilibrium statements ➤ *Q7.24, Q7.25*

7.6 Entropy Revisited
❑ Formal definition of entropy (Eq. 7.16) ➤ *Q7.17*
❑ Increase in entropy principle ➤ *Q7.18*
❑ Entropy changes for a system (Table 7.5) ➤ *Q7.20, Q7.21*
❑ Entropy balance (Eq. 7.25) ➤ *7.53, 7.55*
❑ Spontaneous process ➤ *Q7.19, 7.56*
❑ Isentropic efficiency ➤ *7.74, 7.75*

7.7 The Second Law and Equilibrium
❑ Gibbs free energy (Gibbs function) ➤ *Q7.25*
❑ Chemical equilibrium ➤ *Q7.25*
❑ Standard-state Gibbs function change ➤ *Q7.26*
❑ Equilibrium constant ➤ *Q7.27, 7.25*
❑ Equilibrium composition: single equilibrium reaction ➤ *7.110, 7.112*
❑ Equilibrium composition: multiple equilibrium reactions ➤ *7.121*
❑ Phase equilibrium ➤ *Q7.28, 7.126*

---

* Numbers following arrows refer to Questions (prefaced with a Q) and Problems at the end of the chapter.

REFERENCES

1. Kline, S. J., *The Low-Down on Entropy and Interpretive Thermodynamics*, DCW Industries, La Cañada, CA, 1999.

2. Weaver, J. H. (Ed.), *World of Physics*, Simon & Schuster, New York, 1987, p. 734.

3. Keenan, J. H., *Thermodynamics*, Wiley, New York, 1957.

4. Planck, M., *Treatise on Thermodynamics*, translated by Alexander Ogg, 3rd ed., Dover, New York, 1945.

5. Clausius, R. J. E., "Ueber die bewegende Kraft der Wärme," *Annalen der Physik und Chemie*, 79:368 (1850), translated excerpts in *A Source Book of Physics*, W. F. Magie, McGraw-Hill, New York, 1935, pp. 228–233.

6. Clausius, R. J. E., "Ueber verschiedene für die Anwendung bequeme Formen der Hauptgleichungen der mechanischen Wärmetheorie," *Annalen der Physik und Chemie*, 125:353 (1865); translated excerpts in *A Source Book of Physics*, W. F. Magie, McGraw-Hill, New York, 1935, pp. 234–236.

7. Ord-Hume, A. W. J. G., *Perpetual Motion: The History of an Obsession*, St. Martin's Press, New York, 1977.

8. Thomson, W. (Lord Kelvin), "On an Absolute Thermometric Scale Founded on Carnot's Theory of the Motive Power of Heat, and Calculated from Regnault's Observations," *Cambridge Philosophical Society Proceedings*, June 5, 1848.

9. Thomson, W. (Lord Kelvin), "On the Dynamical Theory of Heat, with Numerical Results Deduced from Mr. Joule's Equivalent of a Thermal Unit, and M. Regnault's Observations on Steam," *Transactions of the Royal Society of Edinburgh*, March, 1851.

10. Goodall, P. M., *The Efficient Use of Steam*, IPC Science and Technology Press, Surrey, England, 1980.

11. Lefebvre, A. H., *Gas Turbine Combustion*, Taylor & Francis, Bristol, PA, 1983.

12. Ohhashi, I., and Arakawa, S., "Development of 300 kW Class Ceramic Gas Turbine (CGT 303)," *Journal of Engineering for Turbines and Power—Transactions of the ASME*, 117:777–782 (1995).

13. Chase, D. L., and Kehoe, P. T., "GE Combined-Cycle Product Line and Performance," GE Power Systems, GER-3574G, Oct., 2000.

14. Podesser, E., "Electricity Production in Rural Villages with a Biomass Stirling Engine," *Renewable Energy*, 16:1049–1052 (1999).

15. Atkins, P. W., *Physical Chemistry*, 6th ed., Oxford University Press, Oxford, 1999.

16. DeNevers, N., *Physical and Chemical Equilibrium for Chemical Engineers*, Wiley, New York, 2002.

17. Stull, D. R., and Prophet, H., *JANAF Thermochemical Tables*, 2nd ed., NSRDS-NBS 37, National Bureau of Standards, June 1971.

18. Olikara, C., and Borman, G. L., "A Computer Program for Calculating Properties of Equilibrium Combustion Products with Some Applications to I. C. Engines," SAE Paper 750468, 1975.

19. Gordon, S., and McBride, B. J., "Computer Program for Calculation of Complex Chemical Equilibrium Compositions, Rocket Performance, Incident and Reflected Shocks, and Chapman-Jouguet Detonations," NASA SP-273, 1976.

20. Reynolds, W. C., "The Element Potential Method for Chemical Equilibrium Analysis: Implementation in the Interactive Program STANJAN," Department of Mechanical Engineering, Stanford University, January 1986.

21. Morley, C., "Gaseq, A Chemical Equilibrium Program for Windows," available for download at http://www.c.morley.ukgateway.net/.

22. Turns, S. R., *An Introduction to Combustion*, 2nd ed., McGraw-Hill, New York, 2000.

Some end-of-chapter problems were adapted with permission from the following:

23. Look, D. C., Jr., and Sauer, H. J., Jr., *Engineering Thermodynamics*, PWS, Boston, 1986.

24. Myers, G. E., *Engineering Thermodynamics*, Prentice Hall, Englewood Cliffs, NJ, 1989.

Nomenclature

A Helmholtz free energy (J)

b Van der Waals constant (m^3/kmol)

c Specific heat (J/kg·K)

c_p Constant-pressure specific heat (J/kg·K)

c_v Constant-volume specific heat (J/kg·K)

COP Coefficient of performance

E Energy (J)

F Force (N)

g Specific Gibbs function (J/kg)

G Gibbs function (J)

$\bar{g}$ Molar-specific Gibbs function (J/kmol)

h Specific enthalpy (J/kg)

h_L Head loss (m)

H Enthalpy (J)

i Electrical current (A)

K_p Equilibrium constant (dimensionless)

KE Kinetic energy (J)

$\dot{m}$ Mass flow rate (kg/s)

$\mathcal{M}$ Molecular weight (kg/kmol)

N Number of moles (kmol)

P Pressure (Pa)

P_i Partial pressure (Pa)

Q Heat interaction (J)

$\dot{Q}$ Heat transfer rate (W)

R_{elec} Electrical resistance (ohm)

R_u Universal gas constant, 8314.472 (J/kmol·K)

s Specific entropy (J/kg·K)

$\bar{s}$ Molar-specific entropy (J/kmol·K)

S Entropy (J/K)

$\mathcal{S}_{gen}$ Entropy generated by irreversibilities (J/K)

$\dot{\mathcal{S}}_{gen}$ Rate of entropy generation by irreversibilities (J/K·s)

t Time (s)

T Temperature (K)

u Specific internal energy (J/kg)

V Velocity (m/s)

v Specific volume (m^3/kg)

$\mathcal{V}$ Volume (m^3)

W Work (J)

$\dot{W}$ Power (W)

x Quality (dimensionless)

X Mole fraction (dimensionless)

GREEK

α Fraction dissociated

β Coefficient of performance

δ Incremental quantity

Δ Difference

ΔG_T° Standard-state Gibbs function change (J/kmol)

η_{th} Thermal efficiency

κ Proportionality constant

μ Chemical potential (J/kmol)

ν_j Stoichiometric coefficient of species j

ρ Density (kg/m^3)

SUBSCRIPTS

act   actual

c   compressor

f   liquid, or formation, or final

fg	difference between saturated vapor and saturated liquid states		P	pressure
			p	pump
fric	friction		ref	reference state
g	gas (vapor)		rev	reversible
H	high-temperature reservoir		sat	saturated state
i	initial		sys	system
i	species i		t	turbine
irrev	irreversible			
isen	isentropic			
L	low-temperature reservoir			
mix	mixture			

SUPERSCRIPTS

°	Standard-state (e.g., pressure $P^o = 1$ atm)
′	Product

QUESTIONS

7.1 Review the most important equations presented in this chapter i.e., those with a reddish background. What physical principles do they express? What restrictions apply?

7.2 State in words the Kelvin–Planck statement of the second law of thermodynamics.

7.3 Write a paragraph explaining the second law of thermodynamics for your friends not majoring in engineering- or science-related fields.

7.4 List two or more ways that the second law is useful to engineers.

7.5 Define a heat reservoir.

7.6 Explain the concept of a heat engine. Of what significance is this concept to the study of thermodynamics?

7.7 Explain the thermodynamic concept of a heat pump. Also explain the term refrigerator.

7.8 Write in words the definition of thermal efficiency and coefficient of performance for cyclic devices.

7.9 Explain the difference between a reversible and an irreversible process. List three or more examples of irreversible processes.

7.10 List as many consequences of the Kelvin–Planck statement of the second law as you can remember. Compare your list with Table 7.2.

7.11 Explain the difference between a perpetual-motion machine of the first kind and a perpetual-motion machine of the second kind.

7.12 Consider a reversible heat engine operating between two heat reservoirs. How does the thermal efficiency of the engine relate to the temperatures of the reservoirs?

7.13 Consider a reversible heat pump operating between two heat reservoirs. How does the heat pump coefficient of performance relate to the temperatures of the reservoirs? How does your answer change if the device is operated as a refrigerator rather than a heat pump?

7.14 List the processes comprising the Carnot cycle.

7.15 List the processes comprising the Stirling cycle.

7.16 State in words two statements of the second law of thermodynamics other than the Kelvin–Planck statement.

7.17 Write symbolically and explain in words the macroscopic, or Clausius, definition of entropy.

7.18 Explain the increase of entropy principle. Be precise.

7.19 Explain how the increase of entropy principle is an indicator of spontaneous change. Illustrate with an example.

7.20 List the kinds of processes that produce an increase in entropy of a system.

7.21 List the ways that a process can be isentropic, that is, list the combined types of processes or conditions that are required for the overall process to be isentropic.

7.22 Explain how the head loss for isothermal flow through a pipe relates to the entropy production rate by irreversibilities (friction).

7.23 Explain how the isentropic efficiency of a pump relates to the production of entropy by irreversibilities. Assume that the fluid is incompressible and that the pump operates adiabatically.

7.24 Explain how the second law of thermodynamics governs the equilibrium composition of a reacting system at fixed internal energy and volume.

7.25 Explain how the second law of thermodynamics governs the equilibrium composition of a reacting system at fixed temperature and pressure.

7.26 Define the standard-state Gibbs function change ΔG_T°. What is the relationship of ΔG_T° to the equilibrium constant K_p?

7.27 Consider an arbitrary chemical equilibrium. What is the physical significance of $K_p \ll 1$? $K_p \gg 1$?

7.28 Consider the following equilibrium reaction at 1 atm and 2500 K:

$$CO_2 \leftrightarrow CO + \frac{1}{2} O_2.$$

A. Which is the exothermic direction for this reaction? To the left or to the right?

B. If the temperature is increased to 4000 K, but the pressure is unchanged, will the equilibrium shift to the left or to the right?

C. If the pressure is increased to 5.0 atm, but the temperature is unchanged, will the equilibrium shift to the left or to the right?

7.29 Explain how the second law of thermodynamics governs the equilibrium between the phases of a simple substance.

7.30 Consider a thermodynamic system consisting of liquid H_2O and H_2O vapor. Which of the following conditions are necessary for thermodynamic equilibrium to prevail in the system? Note that there may be more than one correct answer.

A. $T_{liq} = T_{vap}$

B. $\bar{g}_{liq} > \bar{g}_{vap}$

C. $\bar{g}_{liq} = \bar{g}_{vap}$

D. $\bar{g}_{liq} < \bar{g}_{vap}$

Chapter 7 Problem Subject Areas

7.1–7.9   Heat engines, heat pumps, and refrigerators

7.10–7.52   Carnot cycle, ideal efficiencies, and *COPs*

7.53–7.73   Entropy and entropy balances

7.74–7.107   Reversible and irreversible processes; isentropic efficiency

7.108–7.125   Chemical equilibrium

7.126–7.127   Phase equilibrium

PROBLEMS

7.1 Each cycle, the working fluid of a Stirling engine receives 2.67 kJ of energy in the heat interaction with its combustor. The engine operates with a thermal efficiency of 0.20 and delivers 4 kW of shaft power. Determine (a) the number of cycles executed per minute and (b) the energy rejected as heat to the low-temperature reservoir each cycle.

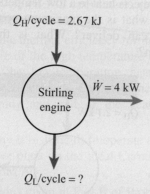

Q_H/cycle = 2.67 kJ

Stirling engine $\dot{W} = 4$ kW

Q_L/cycle = ?

7.2 Energy is transferred to the working fluid (H_2O) in the boiler of a Rankine-cycle power plant at a rate of 131 MW. Energy is rejected in the condenser at a rate of 97 MW. Determine the net power produced and the thermal efficiency of this power plant.

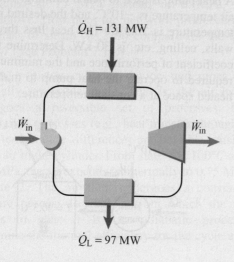

$\dot{Q}_H = 131$ MW

$\dot{W}_{in}$ $\dot{W}_{in}$

$\dot{Q}_L = 97$ MW

7.3 Using atmospheric air as the heat source, a heat pump delivers 43,000 Btu/hr to heat a home. The coefficient of performance is 3.2. Determine (a) the electrical power required to operate the heat pump and (b) the rate at which energy is removed from the outside air.

7.4 A residential heat pump uses the ground as an energy source. At a particular operating condition, the heat pump delivers energy at a rate of 13.37 kW to the heated interior while simultaneously removing energy from the ground at a rate of 10.05 kW. Determine (a) the electrical power required to operate the heat pump and (b) the coefficient of performance.

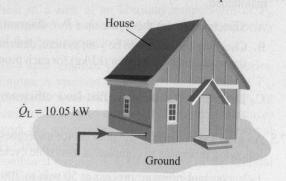

House

$\dot{Q}_L = 10.05$ kW

Ground

7.5 A refrigerator operates with a coefficient of performance of 4.2 and requires an electrical input of 700 W. Determine (a) the rate of heat transfer from the cold space of the refrigerator and (b) the rate at

7.101 A pump delivers 11,000 lb_m/hr of water from an elevation 30 ft below the pump to an elevation 50 ft above the pump through various diameter pipes. At the lower elevation, the water is at 60 F and 10 psia and the flow area is 0.0233 ft². At the higher elevation, the water is at 80 psia and the flow area is 0.0060 ft². The pump and pipes are insulated. Determine the minimum pump power (Btu/hr) required.

7.102 Air at 80 F and 14.7 psia enters an adiabatic compressor at 125 lb_m/hr and leaves at 60 psia. The compressor isentropic efficiency is 80%. Determine the rate of power required to drive the compressor (hp).

7.103 Dry, saturated water vapor at 40°C steadily enters a centrifugal compressor with a mass flow rate of 150 kg/hr. The vapor leaves at 200°C and 0.04 MPa. During the process, heat is transferred from the vapor at the rate of 1 kW. Determine the following:

A. The power (kW) required to drive the compressor

B. The minimum power (kW) required for an adiabatic compression from the same initial state to the same final pressure

C. The compressor isentropic efficiency

7.104 Two steady flows of steam enter a rigid, adiabatic black box at 300°C. One of these flows is at $P_1 =$ 2 MPa and has a flow rate of 100 kg/hr. The other flow is at $P_2 = 0.5$ MPa and has a flow rate of 50 kg/hr. The box is to produce useful work and a single flow stream at 0.2 MPa is to leave it. Determine the maximum useful power (kW) that could be produced by this box.

7.105 A well-insulated, 3-m³ tank contains carbon monoxide at 500 K and 0.5 MPa. A valve is opened and gas is slowly bled out of the tank until the tank pressure is reduced to 0.2 MPa. Determine the temperature (K) of the gas remaining in the tank.

7.106 A well-insulated, 1-m³ tank contains carbon dioxide at 500 K and 0.4 MPa. A valve is opened and the gas is slowly bled out of the tank until the pressure in the tank is reduced to 0.1 MPa. Determine the temperature (K) of the gas remaining in the tank.

7.107 A 0.5-m³ tank contains air at 300 K and 0.75 MPa. A valve is suddenly opened and air rushes out until the tank pressure drops to 0.15 MPa. The air in the tank at the end of the process may be assumed to have undergone a reversible, adiabatic process. Determine the final temperature (K) and mass (kg) in the tank.

7.108 Using the property data in Appendix B, reproduce Fig. 7.21. Note that the initial temperature and pressure are 298 K and 1 atm, respectively. Spreadsheet software is recommended to facilitate your calculations. Plot the mixture temperature, entropy, and pressure as functions of the fraction of CO_2 dissociated.

7.109 Consider the adiabatic, constant-pressure combustion of 1 kmol of CO with 0.5 kmol of O_2 to form CO_2 (see Eq. 7.33). The reactants are at 298 K and 1 atm. Create a plot of the Gibbs function of the product mixture, G_{mix}, as of function of the fraction of CO_2 dissociated, α_{CO_2}. (See Eq. 7.34.) Use your plot to estimate the equilibrium composition and temperature of the products.

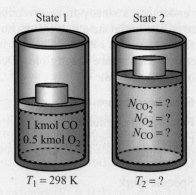

State 1   State 2

1 kmol CO
0.5 kmol O_2

$N_{CO_2} = ?$
$N_{O_2} = ?$
$N_{CO} = ?$

$T_1 = 298$ K $T_2 = ?$

7.110 Consider the equilibrium reaction

$$N_2 \rightleftarrows N + N.$$

Determine the equilibrium partial pressures and mole fractions of N and N_2 at 1 atm and 3000 K. Repeat for a pressure of 0.1 atm. Compare and contrast your results with those for $O_2 \rightleftarrows O + O$ presented in Example 7.9. Discuss.

7.111 Consider the dissociation of oxygen molecules,

$$O_2 \rightleftarrows O + O.$$

Determine the equilibrium mole function of O atoms as a function of pressure for a fixed temperature of 2500 K. Use pressures of 0.01, 0.1, 1.0, and 10 atm. Plot your results using logarithmic scales.

7.112 Consider the isolated equilibrium reaction

$$\frac{1}{2}N_2 + \frac{1}{2}O_2 \rightleftarrows NO.$$

Determine the equilibrium mole fraction of NO at 1 atm and 4000 K. Assume that there are equal proportions of N and O atoms in the mixture.

7.113 Consider a mixture of molecular nitrogen (N_2) and atomic nitrogen (N) in equilibrium at 3 atm at an unknown temperature. Determine the value of the equilibrium constant K_p for the reaction $N_2 \leftrightarrow 2N$ if the partial pressure of the N_2 is 2.995 atm. Does K_p have units, or is it dimensionless?

7.114 Carbon monoxide and oxygen (O_2) exist in equilibrium with carbon dioxide. Determine the equilibrium composition (mole fractions) at 3200 K and 1 atm of an initial mixture of 2 mol of carbon monoxide and 2 mol of oxygen.

7.115 Carbon monoxide and oxygen (O_2) react to form carbon dioxide. Determine the equilibrium composition (mole fractions) at 3200 K and 1 atm of an initial mixture of 2 mol of carbon monoxide, 1 mol of oxygen, and 3.774 mol of nitrogen. *Hint:* Treat the nitrogen as an inert species.

7.116 Carbon monoxide and oxygen (O_2) react to form carbon dioxide. Determine the equilibrium composition (mole fractions) of a mixture at 298 K and 1 atm initially containing 2 mol of carbon monoxide and 1 mol of oxygen.

7.117 Carbon monoxide and water react to form carbon dioxide and hydrogen (H_2), the so-called water–gas shift reaction. Determine the equilibrium composition (mole fractions) of a mixture at 1100 K and 1 atm initially containing 1 mol of carbon monoxide and 1 mol of water.

7.118 Carbon monoxide and water react to form carbon dioxide and hydrogen (H_2), the so-called water–gas shift reaction. Determine the equilibrium composition (mole fractions) of a mixture at 1100 K and 1 atm initially containing 1 mol of carbon monoxide and 2 mol of water.

7.119 Hydrogen and oxygen react to form water. Determine the equilibrium composition (mole fractions) at 4000 K and 1 atm of an initial mixture of 1 mol of hydrogen (H_2) and 1 mol of oxygen (O_2).

7.120 Water at high temperature dissociates into hydrogen (H_2) and oxygen (O_2). A mixture of 1 mol of water vapor and 10 mol of nitrogen (N_2) is placed in a piston–cylinder device at an initial state of 298 K and 1 atm. The mixture is then heated with an electric heater at constant pressure to 4000 K. Assume the nitrogen does not react. Determine the mixture composition (mole fractions) and the percent dissociation at the final state.

7.121 Add the $O_2 \rightleftarrows O + O$ equilibrium reaction to the problem discussed in Example 7.10. Determine the equilibrium mole fractions of the product species (CO_2, CO, O_2, and O) at 0.1 atm and 2500 K.

7.122 Consider the reaction

$$2H_2 + O_2 \rightleftarrows 2H_2O.$$

Write an expression for the equilibrium constant K_p for the reaction as indicated using appropriate partial pressures. Also write an expression for the equilibrium constant when the reaction is written from right to left.

7.123 Using appropriate partial pressures, write expressions for the equilibrium constant K_p for each of the two following reactions:

$$2H_2 + O_2 \leftrightarrow 2H_2O$$

and

$$H_2 + \tfrac{1}{2}O_2 \leftrightarrow H_2O.$$

How do the two K_ps' relate?

7.124 For the reaction

$$CO + \tfrac{1}{2}O_2 \leftrightarrow CO_2,$$

determine the equilibrium partial pressures of each species at 2222 K for a total pressure of 1 atm.

7.125 In Problem 7.124, what changes in the partial pressures occur if the total pressure is increased to 5 atm?

7.126 Using data for H_2O from the NIST database, verify that the condition for phase equilibrium (Eq. 7.53) is met. Use temperatures of 300 and 600 K.

7.127 Consider the liquid–vapor equilibrium of H_2O at 20°C in which N_2 is added to the gas phase to obtain a total pressure of 1 atm. Assuming that the N_2 is both inert and insoluable in the liquid H_2O, how does the N_2 affect the equilibrium pressure (or partial pressure) of the H_2O vapor? To answer this, compute the partial pressure of the H_2O vapor in the $H_2O(g)$–N_2 mixture. The total pressure is fixed at 1 atm.

H_2O Vapor + N_2 $P = 1$ atm

H_2O Liquid $T = 20°C$

SIMILITUDE AND DIMENSIONLESS PARAMETERS

After studying Chapter 8, you should:

- Understand the role of dimensional analysis as a bridge between incomplete theory and the need for information for engineering design.

- Understand the concepts of geometric similarity and scale models.

- Understand the concept of similitude and its requirement of dynamic similarity.

- Be able to apply similitude to scale-model testing and the design of experiments.

- Be able to identify characteristic scales or values in thermal-fluid science problems.

- Be able to state two ways that the controlling dimensionless parameters can be determined for a particular problem.

- Be able to transform the dimensional governing equations for a boundary-layer flow to their dimensionless forms.

- Be able to recognize the symbolic representation of the following dimensionless parameters: Reynolds number, Peclet number, Prandtl number, friction coefficient, and Nusselt number.

- Be able to relate the various dimensionless parameters to a ratio of physical "forces" or effects, or provide other appropriate physical interpretations.

- Be able to apply the Buckingham pi theorem to determine the controlling dimensionless parameters for a novel situation.

Chapter 8 Overview

Engineers must frequently design devices for which a complete mathematical description based on the conservation principles of Chapters 3, 5, and 6 is lacking. A common example of this is the design of devices involving turbulent flows. There are many approximate theories for turbulent flows, and modern computer-based solutions can provide much information; nevertheless, these are incomplete. This chapter provides an overview of the experimental methods used to bridge the gap created by the lack of rigorous theoretical descriptions and the need to design a device to perform a function in some optimal manner. These experimental methods include the concept of parametric testing, the application of laws of similitude, and the use of dimensionless parameters. We describe these ideas in both general and specific ways. This chapter focuses on how similitude and dimensionless parameters can be used to solve many engineering problems. Chapters 9 and 10 rely heavily on the ideas presented here.

8.1 HISTORICAL CONTEXT

The first prominent example of systematic experimentation using scale models is attributed to **John Smeaton** (1724–1792) [1], founder of the British Society of Civil Engineers. From parametric tests on scale models, Smeaton sought to provide information for the design of full-sized waterwheels [2]. While measuring the power output of a model waterwheel, Smeaton independently varied both the speed and flow rate of the water and the rotational speed of the wheel. (How to make the most of this basic methodology is a major topic of this chapter.) Laws for estimating ship hull

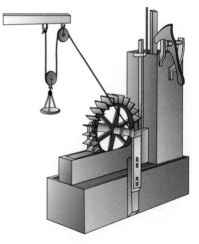

John Smeaton (1724–1792) and model water wheel from his 1759 publication [2].

Osborne Reynolds (1842–1912). Photograph of portrait by John Collier in 1904, courtesy of Emeritus Professor J. D. Jackson of the University of Manchester.

drag from model tests were first proposed by **William Froude** (1810–1871) in the late 1860s [3]. **Aímée Vaschy** (1857–1899), **Lord Rayleigh** (1842–1919), **Dimitri Riabouchinsky** (1882–1962), and **Edgar Buckingham** (1867–1940) all made significant contributions to engineering applications of dimensional analysis and similitude [4–7]. The Buckingham *pi theorem,* also known as the Vaschy–Buckingham theorem, is routinely presented in modern fluids textbooks—as is also done here—as a standard method for determining what dimensionless parameters are important in any particular problem. **Jean Baptiste Joseph Fourier** (1768–1830), whose work in conduction heat transfer is discussed in Chapter 4, was the first to formulate a theory of dimensional analysis, as set forth in his works spanning 1807–1822 [1]. **Osborne Reynolds** (1842–1912), another previously discussed engineering pioneer, was the first to use dimensionless parameters to analyze experimental results [8]. In 1919, **Moritz Weber** (1871–1951) assigned the names Reynolds number and Froude number, respectively, to a dimensionless parameter that Reynolds showed to be so important to fluid mechanics and to a dimensionless parameter that Froude never employed [3]. (Weber also has been honored with a parameter named after him.) In a delightful review [1], Vincenti traces the modern development and application of dimensional analysis in aeronautical engineering to the research of **William F. Durand** (1859–1917) and **Everett P. Lesley**. **Percy Bridgman** (1882–1961), a physicist, published a classic monograph on dimensional analysis in 1922 [9]. The application of dimensionless parameters and the laws of similitude have become standard tools in modern engineering practice.

8.2 THE LIMITS OF THEORY

In Chapters 3, 5, and 6, detailed mathematical descriptions of the flow of fluids were presented (see Tables 5.7, 6.3, and 6.4). Even though these equations offer quite precise descriptions of reality, they cannot be used exclusively in the design of complex practical devices. This state of affairs arises from the complexity of real devices and, in particular, the lack of knowledge of the necessary boundary conditions needed to solve these equations. For example, consider using theory to predict the detailed flow of gases into and out of the piston ring grooves of an internal combustion engine. Flow areas will depend on the exact position of the rings, which tilt in various directions as the piston goes up and down. The motion of the piston rings, in turn, depends on their interaction with the cylinder wall and the lubricating oil film on this wall. The gas flows also depend on a precise knowledge of the pressure within the cylinder during the cycle. The cylinder pressure depends on the details of the intake process, the combustion event, and the exhaust process. This linking of one complex process to another poses a great challenge in developing predictive mathematical models. Complicating all of these factors even further is the fact that many of the flows in the chain of events are turbulent. In spite of the difficulties involved, engineers have developed numerous mathematical tools to aid in engineering design and will continue to do so in the future. In the cases where these models fall short of the needed precision for prediction, or are nonexistent, the design process must rely on experimental results. How best to conduct experiments and how to employ their results are the subjects of this chapter.

> **In Chapter 6, we discussed the intrinsic difficulty of dealing with turbulent flows and the so-called closure problem.**

An example of parametric testing is illustrated by this sequence of images. Here a jet blows with increasing strength through a slot at the rounded trailing edge of a wing to increase the lift force. The velocity field is presented in false color with the lowest velocities shown in blue and the highest velocities shown in red. Jet-to-free-stream velocity ratios are (top to bottom) 0, 3.25, 5.3, and 6.3.

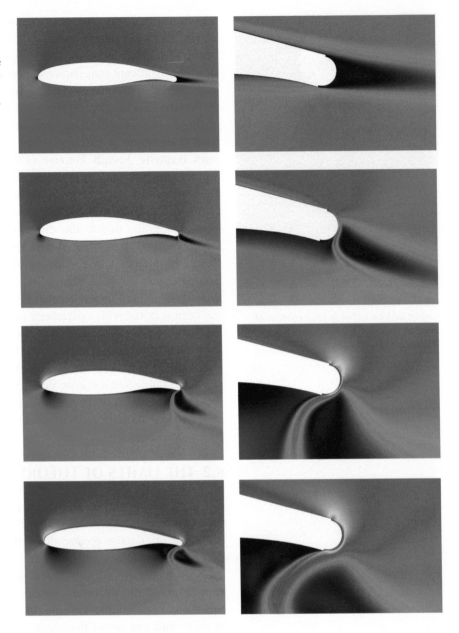

8.3 PARAMETRIC TESTING

The idea of parametric testing is not limited to experiments; it also applies to the exercise of theoretical models. The basic procedure is to fix all but one of the parameters (variables) that are under the control of the experimenter. This single parameter is then varied over a range of values to determine its influence on a particular output variable or set of output variables. The experiment is then repeated after changing the level of one of the fixed parameters. This process continues until all of the important parameters are varied over appropriate ranges.

We now illustrate the use of parametric testing with a specific example from the field of heat transfer. The overall context of our example is the design of shell-and-tube heat exchangers (Fig. 8.1). In Chapter 4, an empirical rate law for convective heat transfer was defined as

$$\dot{Q}'' = h_{\mathrm{conv},x}(T_s - T_\infty), \tag{4.19}$$

FIGURE 8.1

A simple shell-and-tube heat exchanger. The function of a heat exchanger is to heat (or cool) one fluid at the expense of another. Here the entering hot fluid cools as it passes through the shell of the heat exchanger, while the entering cold fluid becomes hotter on its journey through the tubes.

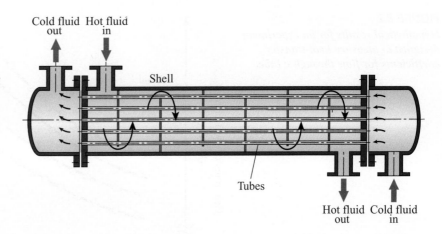

where $\dot{Q}''$ is the heat flux (W/m^2) associated with a fluid flowing over a solid surface, $h_{\mathrm{conv},x}$ is the local convective heat-transfer coefficient (W/m$^2 \cdot$ K), and T_s and T_∞ are the surface and fluid temperatures, respectively. For the flow of a fluid across the outside surface of a tube, Eq. 4.19 applies directly. For a fluid flowing within the tube, a mean fluid temperature averaged over the tube cross-sectional area, T_m, replaces T_∞. The details of this need not concern us presently; that there are two temperatures involved is what is important. Thus, for the internal flow, the local convective heat-transfer coefficient is defined by

$$h_{\mathrm{conv},x} = \frac{\dot{Q}''}{T_s - T_m}. \qquad (8.1)$$

Let us now assume that you have been assigned the task of experimentally determining values of $h_{\mathrm{conv},x}$ to be used to design a family of heat exchangers. It is desirable to have data for a variety of fluids, say, water, oil, ethylene glycol (engine coolant), and air, as a minimum. (It is conceivable that a potential customer may wish to use some other fluid.) Heat exchangers will be sold in a variety of sizes, so tube diameter will be a key parameter. Furthermore, customers will use the heat exchangers under various conditions, so flow rate, or mean velocity, will be an important parameter. Similarly, the surface temperature T_s and mean fluid temperature T_m will need to be taken into account in some way. For the sake of argument, we assume that these are the only important variables in your experiments: fluid type, tube inside diameter D, average fluid velocity V_{avg}, tube wall temperature T_s, and mean fluid temperature T_m.

To perform the experiments, a rig is available in which a heat-flux gauge measures $\dot{Q}''$ and other instrumentation measures T_s and T_m. The rig has been designed so that measurements of $h_{\mathrm{conv},x}$ are independent of the axial location along the tube. For any fluid, tube size, and set of conditions (V_{avg}, T_s, and T_m), you can use the data from your rig to calculate $h_{\mathrm{conv},x}$ from Eq. 8.1.

A typical treatment of data so obtained is to create a plot of $h_{\mathrm{conv},x}$ versus V_{avg} using, say, D as a parameter. All other parameters remain fixed. Figure 8.2 illustrates such a plot. If we assume that 10 data points are required to generate a smooth curve of $h_{\mathrm{conv},x}$ versus V_{avg} and that 10 tube sizes are required to satisfy the expected range of customer requirements, then 100 (10 × 10) separate experimental runs are required to create a single plot. Recall now that this plot is for a single fluid and for single values of T_s and T_m. To deal with

FIGURE 8.2
Hypothetical results for an experiment designed to measure heat-transfer coefficients for flow through a tube.

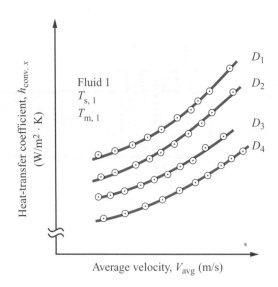

a number of fluids and to cover a reasonable range of both T_s and T_m now presents a Herculean task! If only one fluid and one set of temperatures were of interest, generating 100 data points would be very doable. Expanding this to 4 fluids and 10 levels each for T_s and T_m requires that 40,000 (= $10 \times 10 \times 4 \times 10 \times 10$) experiments be conducted. To say that such a task would be very time consuming and expensive is understatement indeed.

Fortunately, the principles of dimensional analysis allow this task to be accomplished with the same degree of accuracy by conducting only 100 experiments. We return to this specific problem after discussing how the use of scale models in experiments also requires dimensional analysis.

8.4 SIMILITUDE

The engineering design of sophisticated devices is a complicated process, often involving many iterations to ensure that the final device performs as desired, is safe in its operation, and can be manufactured economically. For many devices, particularly those of large scale, construction and testing of full-size prototypes as a part of the design process is not practical. In such cases, small-scale, geometrically similar models are used to provide the information needed to improve and evaluate the design. Wind-tunnel testing of models is

A wide variety of models are used in engineering design. Models here include a helicopter, a dam spillway, and an automobile. Photographs courtesy of NASA, the U.S. Army Corps of Engineers, and Adam Opel A. G., respectively.

FIGURE 8.3
Wind-tunnel testing is an essential part of aircraft design. A 0.36-scale model of the Space Shuttle orbiter is tested in NASA Ames 40 × 80 ft wind tunnel. **Photograph courtesy of NASA.**

> **We can tell whether a flow will be laminar or turbulent based on the Reynolds number. See Figure 1.32.**

routine in the design of aircraft (see Fig. 8.3). In fact, wind-tunnel testing has always been an integral part of the aircraft design process as attested by the Wright Brothers use of this technique in creating their *Flyer*. The use of model testing also has a long history in the design of ship hulls. Hydroelectric dams and their associated stilling basins, spillways, penstocks,[1] gates, and hydroturbines are all designed using scale-model testing. Models are used to understand wind loads on buildings and the interactions created when many buildings are clustered. Every field of engineering offers many examples.

How can such model tests be planned and executed so that the results are somehow useful to the full-scale device? Indeed, how can model results be applied to a full-size prototype? The answers to both questions are contained in the following statement:

> **If all appropriate dimensionless variables have been defined, *similitude* occurs between *geometrically similar* models and prototypes when the same values of these dimensionless variables are attained for both.**

This statement embodies the concepts of geometric similarity and dynamic similarity. **Geometric similarity** implies that the model is a *scale* model of the prototype, that is, that any linear dimension of the model is a constant fraction of the corresponding linear dimension of the prototype. For example, the use of quarter-scale (1:4) models is common in the aerodynamic design of automobiles. You are likely familiar with toy plastic models of aircraft, automobiles, and ships. In these toy models, *most* of the dimensions are true scalings from the full-size devices; however, many small features, such as rivet detail, may be grossly out of scale. The failure to achieve total geometric similarity in such details can be quite important in engineering design.

Dynamic similarity, obviously, relates to motion. In thermal-fluid applications, this motion is usually associated with a fluid moving around or through a device. In some applications, achieving complete dynamic similarity is not possible. Regardless, the idea that the various dimensionless variables that contain a velocity (or a time) must be equal is a statement of dynamic similarity. Even though we have yet to formally define any of these dimensionless variables (although we have, however, mentioned the Reynolds number in previous chapters) this statement of similarity, or scaling, can best be understood by appealing to a concrete example.

Figure 8.4 illustrates the way in which the lift force in a full-scale prototype can be predicted from measurements of the lift force on a much smaller scale model. Here three dimensionless parameters are important:

the lift coefficient, $c_L \equiv \dfrac{F_L/L^2}{\rho V^2/2}$,

the Reynolds number, $Re \equiv \dfrac{\rho V L}{\mu}$, and

the angle of attack, α.

In these parameters, F_L is the lift force (N), L is a characteristic length[2] (m), V is the velocity of the air stream relative to the aircraft (m/s), and ρ and

---

[1] A penstock is the pipe that supplies the water stored behind the dam to the turbine.
[2] Here one might logically choose the mean wing chord (width) as the characteristic length. We discuss characteristic dimensions in a subsequent section.

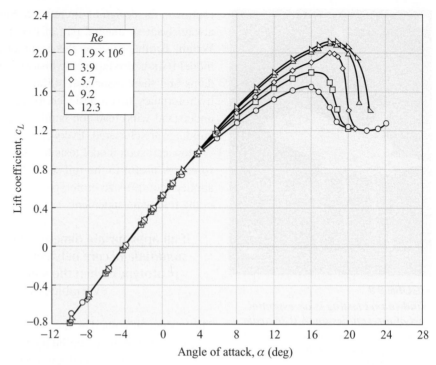

FIGURE 8.4

The dimensionless lift force, expressed by the lift coefficient c_L is a function of the angle of attack of the wing, α, and the Reynolds number Re. Adapted from Ref. [10] with permission.

This sequence of images shows how the pressure field varies around an airfoil as the angle of attack increases. The highest pressures (or pressure coefficients) are indicated by red, while the lowest pressures (sub-atmospheric) are indicated by blue. These results, computed for a Reynolds number of 1×10^7, agree well with the experimental results presented in Fig. 8.4. Note the stall condition in the last panel (α = 20°).

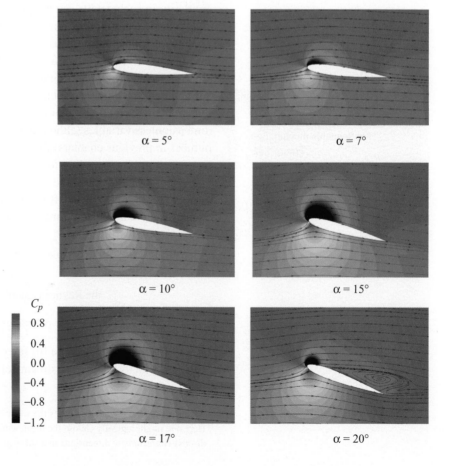

μ are the density and viscosity of the air, respectively. Data from wind-tunnel tests of a particular scale model are collected and plotted to show the lift coefficient as a function of the angle of attack, using the Reynolds number as a parameter, as illustrated in Fig. 8.4. We can use these data to obtain the lift force for the full-scale prototype by equating the dimensionless parameters of the model and prototype; that is, we set

$$\frac{F_{L,m}/L_m^2}{\rho_m V_m^2/2} = \frac{F_{L,p}/L_p^2}{\rho_p V_p^2/2}, \tag{8.2a}$$

when

$$\frac{\rho_m V_m L_m}{\mu_m} = \frac{\rho_p V_p L_p}{\mu_p} \tag{8.2b}$$

and

$$\alpha_m = \alpha_p. \tag{8.2c}$$

Rearranging Eq. 8.2a, we explicitly express the dimensional lift force on the prototype as

$$F_{L,p} = F_{L,m} \frac{L_p^2 \rho_p}{L_m^2 \rho_m} \frac{V_p^2}{V_m^2}. \tag{8.2d}$$

From Eq. 8.2d, we see that the lift force on the prototype scales with the square of the geometric scale, that is, $(L_p/L_m)^2$. Thus, if the model is 1/20 the size of the prototype, the lift force measured in the wind tunnel should be multiplied by $400V_p^2/V_m^2$ to obtain the corresponding lift force on the full-scale aircraft, provided that the air densities are the same (Eq. 8.2d) and that the Reynolds numbers (Eq. 8.2b) and angles of attack (Eq. 8.2c) are the same. To achieve identical Reynolds numbers, however, is very difficult in aircraft testing since both V and L are quite large for the prototype compared to practically achievable values in the wind tunnel. (See Problem 8.8.)

From this discussion, the need for and use of dimensionless parameters should be quite clear. We now focus directly on these.

8.5 DIMENSIONLESS PARAMETERS

8.5a Origins

The question we seek to answer in this section is the following: How does one determine what dimensionless parameters are important in any particular problem? We present two methods to identify the important parameters. The first of these is the application of the theory of dimensional analysis [6, 7, 9]. Appendix 8A presents this particular approach in detail. The second approach is to create dimensionless forms of the governing conservation equations and their appropriate boundary conditions. The dimensionless parameters then appear as parameters in these dimensionless governing equations. The remainder of the chapter deals with this second approach and its application to certain problems in fluid mechanics and heat transfer. In spite of our focus on the use of dimensionless governing equations, a much more general approach invoking the Buckingham (or Vaschy) pi theorem (Appendix 8A) is frequently used to deal with complex problems where theory is incomplete or nonexistent. Examples of the application of this theory are also contained in Appendix 8A.

8.5b Dimensionless Governing Equations

To illustrate how parameters can originate from a manipulation of the basic conservation equations, we choose to work with simplified versions of momentum conservation (Tables 6.3 and 6.4) and energy conservation (Table 5.7). Using simplified equations allows us to focus on the method without getting bogged down in a large number of terms. By choosing the simplified set of equations known as the boundary-layer equations, we get the added benefit of dealing with a very important class of flows (i.e., boundary-layer flows). The sections of Chapter 9 devoted to boundary-layer flows extend the developments presented in this chapter.

Dimensional Forms of the Boundary-Layer Equations

The simplest boundary-layer flow is that associated with the flow over a sharp-edged flat plate in a stream having a uniform velocity and a uniform temperature. Figure 8.5 illustrates this situation. Although understanding the details of this flow is not necessary for our present purposes, such understanding is quite useful in its own right because of the overwhelming importance of the boundary-layer concept; thus, we digress a bit to explore the flat-plate boundary layers.

Figure 8.5 (top) illustrates the development of the hydrodynamic boundary layer. Upstream of the plate ($x < 0$), the velocity field is uniform with a magnitude V_∞ and directed parallel to the surface of the plate. Upon encountering the plate, the fluid immediately adjacent to the plate comes to rest (no-slip condition), creating a shearing action on the fluid layers above. The effect of this shearing is to create a thin layer of fluid in which the magnitude of the velocity is zero at the surface ($y = 0$) and rapidly approaches the free-stream value, V_∞, with the perpendicular distance from the plate surface, y. The thickness of this layer, δ_H, grows with distance downstream, x, yet remains thin compared to this distance, (i.e., $\delta_H/x \ll 1$). The existence of this thin layer adjacent to a solid surface was the insight of the famous fluid mechanician Ludwig Prandtl (1875–1953). That the effects of viscosity are confined to a thin layer provides a tremendous simplification to a wealth of flow problems, among them the calculation of lift and drag from airfoil surfaces.

FIGURE 8.5

(Top) The development of a hydrodynamic boundary layer of thickness δ_H resulting from fluid sticking to the surface and subsequent viscous shearing of the layers above. The boundary-layer thickness grows with distance downstream. (Bottom) The development of the thermal boundary layer of thickness δ_T on a heated flat plate. Typical velocity and temperature profiles are shown at the plate leading edge and at two positions downstream.

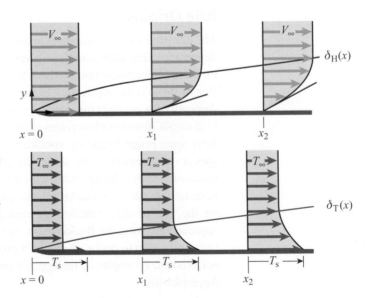

The false color velocity field (v_x/V_∞) shows the thin boundary layers that develop on the long sides of a wedge in a uniform flow. Note the recirculation zone, a non-boundary layer flow, at the rear of the wedge.

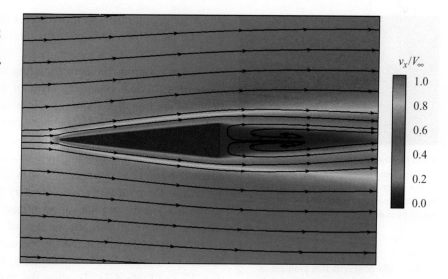

Within a flat-plate boundary layer, the velocity directed in the mean flow direction is much greater than that directed perpendicular to the plate, that is,

$$v_x \gg v_y. \tag{8.3a}$$

Furthermore, since the boundary layer is thin, gradients in the y-direction are much greater than those in the x-direction, that is,

$$\frac{\partial(\ )}{\partial y} \gg \frac{\partial(\ )}{\partial x}. \tag{8.3b}$$

As a consequence of these two conditions, the incompressible axial momentum equation (Eq. T6.3a) can be greatly simplified. Details of this simplification are presented in the appendix to Chapter 9. For steady flow in the absence of a pressure gradient and negligible effects of gravity, Eq. T6.3a becomes

$$v_x \frac{\partial v_x}{\partial x} + v_y \frac{\partial v_x}{\partial y} = \frac{\mu}{\rho} \frac{\partial^2 v_x}{\partial y^2}. \tag{8.4}$$

To fully describe this incompressible, two-dimensional flow, we must add the conservation of mass, or continuity, equation (see Table 3.3). This is given as Eq. T3.3b, which simplifies to

$$\frac{\partial v_x}{\partial x} + \frac{\partial v_y}{\partial y} = 0. \tag{8.5}$$

Subject to the application of the appropriate boundary conditions, Eqs. 8.4 and 8.5 can be solved to provide the two velocity functions

$$v_x = v_x(x, y) \tag{8.6a}$$

and

$$v_y = v_y(x, y). \tag{8.6b}$$

Knowledge of these velocity distributions then allows one to determine the shear stress at the wall using Eq. 6.3,

$$\tau_w = \mu \frac{\partial v_x}{\partial y}\bigg|_{y=0}. \tag{8.7}$$

A thin thermal boundary layer develops over the long sides of a heated wedge as indicated by this image of the temperature field. The highest temperature is denoted by red and the lowest by violet.

We deal with this problem in Chapter 9.

The engineering utility of this result is that the shear stress can be integrated over the plate surface to obtain the total drag force exerted by the fluid on the plate.

We now examine the corresponding thermal problem, that is, the development of a thermal boundary layer of thickness δ_T. Consider the situation depicted in Fig. 8.5 (bottom). Here we see a flat plate maintained at a temperature T_s. For the temperature profiles sketched, the plate temperature T_s is greater than the freestream temperature T_∞; thus, energy is transferred from the plate to the flowing fluid in a heat interaction. The thickness of the thermal boundary layer grows with distance downstream, similar to the growth of the hydrodynamic boundary layer. Also, as the viscous effects in the hydrodynamic problem are confined to a thin layer, so too are the thermal effects. Assuming that the effects of temperature variation on the fluid density are small, we can use the incompressible energy equation to describe the temperature field. From Chapter 5, the appropriate governing equation is Eq. T5.7a (Table 5.7). Treating the flow as steady and applying the same simplifying assumptions we just used (i.e., Eqs. 8.3a and 8.3b), we can express energy conservation within the boundary layer as

$$v_x \frac{\partial T}{\partial x} + v_y \frac{\partial T}{\partial y} = \frac{k}{\rho c_p} \frac{\partial^2 T}{\partial y^2}, \tag{8.8}$$

where viscous dissipation has been neglected and the thermophysical properties have been treated as constants.

Solving Eq. 8.8 yields the temperature distribution,

$$T = T(x, y). \tag{8.9}$$

Knowing the temperature distribution allows us to compute the heat flux at the plate surface by applying Fourier's law to the stagnant fluid immediately adjacent to the surface ($y = 0$). This procedure is analogous to using the velocity distribution to find the shear stress (Eq. 8.7). Equating this theoretical heat flux (Eq. 4.17) to the empirical definition of the convective heat flux (Eq. 4.19) introduces the convective heat-transfer coefficient into the problem solution, that is,

$$\dot{Q}''_{\text{conv}} = -k \frac{\partial T}{\partial y}\bigg|_{y=0} \equiv h_{\text{conv},x}(T_s - T_\infty). \tag{8.10}$$

Because of their importance and our future need to refer to them conveniently, the **boundary-layer governing equations** are summarized in Table 8.1.

Characteristic Scales or Values

To proceed in our development of the dimensionless forms of the boundary-layer equations, we must first consider the concept of characteristic scales or values. In most thermal-fluid problems, characteristic lengths, velocities, times, and/or temperatures can be identified. Often knowledge of these values, or scales, is all that is required to obtain "ballpark" values for quantities of engineering interest (e.g., shear stresses or heat-transfer coefficients). To illustrate this idea, let us return to the twin problems of friction and heat transfer for flow over a flat plate.

What characteristic velocities are associated with this flow? In the absence of a pressure gradient, the free-stream velocity V_∞ is a fixed quantity;

Table 8.1 Boundary-Layer Equations for Flow over a Flat Plate*

Conservation Principle	Equation	
Mass	$$\frac{\partial v_x}{\partial x} + \frac{\partial v_y}{\partial y} = 0$$	(8.5)
Axial momentum	$$v_x \frac{\partial v_x}{\partial x} + v_y \frac{\partial v_x}{\partial y} = \frac{\mu}{\rho} \frac{\partial^2 v_x}{\partial y^2}$$	(8.4)
Energy	$$v_x \frac{\partial T}{\partial x} + v_y \frac{\partial T}{\partial y} = \frac{k}{\rho c_p} \frac{\partial^2 T}{\partial y^2}$$	(8.8)

*The fluid is incompressible with constant properties.

furthermore, V_∞ is the maximum velocity in the flow field since the boundary layer always retards the flow and never results in any speeding up. We can also identify the minimum velocity, the value of zero resulting from the no-slip condition at the surface. With these two values, V_∞ and zero, we define a characteristic velocity difference: $\Delta v \equiv V_\infty - 0 = V_\infty$.

What length scales are important to this problem? In the streamwise, or x-, direction, we can identify the distance from the leading edge of the plate to the location of interest as a characteristic length. This is simply the x coordinate. For a plate of finite length L, we might choose L as a characteristic length. In the transverse, or y-, direction, the important length scales are the boundary-layer thicknesses: δ_H for the hydrodynamic boundary layer and δ_T for the thermal boundary layer. Why is the plate thickness not an important characteristic transverse length? To achieve a pure boundary-layer flow, the plate thickness is deliberately made thin (i.e., the thickness is unimportant by definition). Certainly if the plate were thick, the plate thickness would be an important characteristic length scale.

To show the utility of defining these characteristic scales, we estimate a value of the wall shear stress at a particular location by approximating Eq. 8.7 as follows:

$$\tau_w = \mu \frac{\partial v_x}{\partial y}\bigg|_{y=0}$$

$$\approx \mu \frac{V_\infty - 0}{\delta_H} = \frac{\mu V_\infty}{\delta_H}. \tag{8.11}$$

To provide a numerical estimate, consider a flow of air at 300 K with a freestream velocity of 3 m/s; thus,

$$\tau_w \approx \frac{18.46 \times 10^{-6}(\text{N} \cdot \text{s/m}^2)\, 3(\text{m/s})}{\delta_H(\text{m})}$$

$$= \frac{5.54 \times 10^{-5}}{\delta_H} \,[=]\, \text{N/m}^2 \text{ or Pa.}$$

Arbitrarily picking an x location where the boundary-layer thickness is 2 mm, we obtain the estimate that

$$\tau_w \approx \frac{5.54 \times 10^{-5}}{0.002} \,\text{N/m}^2 = 0.028 \,\text{N/m}^2 \text{ or Pa.}$$

This value is only 40% lower than the value of 0.046 Pa calculated from a detailed analysis (Chapter 9), which is a pretty good estimate considering that the only information employed was the characteristic velocity V_∞, the characteristic thickness δ, and a knowledge of the fluid viscosity. Equation 8.11 also shows how τ_w "scales" with V_∞ and δ.

We can perform a similar analysis of the thermal problem. This introduces another variable, T, and, thus, requires that we identify one or more characteristic temperatures. From the sketch in Fig. 8.5, we identify the freestream temperature T_∞ and the plate surface temperature, T_s. The characteristic temperature difference is, thus, $\Delta T = T_s - T_\infty$. We now estimate the local heat flux by applying Fourier's Law (see Eq. 8.10) as follows:

$$\dot{Q}''_{\text{conv}} = -k \left.\frac{\partial T}{\partial y}\right|_{y=0}$$

$$\approx -k\frac{T_\infty - T_s}{\delta_{\text{T}}}.$$

Going a step further, we can approximate the local heat-transfer coefficient (Eq. 8.10) using our characteristic temperature difference and transverse length scale:

$$
\begin{aligned}
h_{\text{conv},x} &= \frac{\dot{Q}''_{\text{conv}}}{T_s - T_\infty} \\
&\approx \frac{-k(T_\infty - T_s)}{(T_s - T_\infty)\delta_{\text{T}}} = \frac{k}{\delta_{\text{T}}}.
\end{aligned}
\tag{8.12}
$$

From this approximate relationship, we see that the local heat-transfer coefficient becomes smaller as the boundary-layer thickness grows in the streamwise direction. Again, we obtain an interesting and useful result simply by identifying the characteristic scales of the problem.

More sophisticated arguments [11] can be applied to ascertain how δ_{H} and δ_{T} scale with the streamwise length scales x or L. For example,

$$\delta_{\text{H}} \propto x\left(\frac{\mu}{\rho V_\infty x}\right)^{1/2}$$

or

$$\delta_{\text{H}} \propto L\left(\frac{\mu}{\rho V_\infty L}\right)^{1/2}.$$

Actual solution of the boundary-layer conservation equations provides the proportionality constants associated with these scaling relationships. These solutions are presented in Chapter 9.

Making the Equations Dimensionless

We apply the following procedure to transform the dimensional boundary-layer equations shown in Table 8.1 to their corresponding dimensionless forms:

- Identify the dimensional variables appearing in the equations.
- Choose characteristic values associated with each variable.
- Form dimensionless variables by dividing each dimensional variable by its characteristic value.
- Express each dimensional variable in terms of its dimensionless counterpart.

- Substitute the dimensional variables as expressed in terms of their dimensionless counterparts into the equations.
- Rearrange the resulting equation to form dimensionless coefficients.

The dimensionless coefficients in the final step are the dimensionless parameters that we seek. The procedure outlined here is quite general and can be used to identify dimensionless parameters from any set of conservation equations. We now proceed with our particular problem.

Reynolds, Peclet, and Prandtl Numbers

Treating the thermophysical properties as constants, we see that the boundary-layer equations (Table 8.1) contain five variables: two independent spatial-coordinate variables, x and y; and three dependent variables, $v_x(x, y)$, $v_y(x, y)$ and $T(x, y)$. We list these dimensional variables and units in the first two columns of Table 8.2. Note that the temperature variable is given as a temperature difference, $T - T_s$, rather than simply as T. As will be seen later (Chapter 9), use of temperature difference results in our dimensionless temperature ranging conveniently between zero and unity.

Applying the second step, we choose useful *characteristic values* for each of the variables. For a flat plate in a uniform flow the obvious choice for a characteristic velocity is the freestream velocity V_∞. Similarly, a good choice for the characteristic temperature difference is $T_\infty - T_s$. We also need a characteristic length. Our choice here depends on how we might want to use the final result. For a flat plate, the characteristic length is usually chosen as the x value itself, or the total length of the plate, L. For the time being, we will simply denote the characteristic length as L_c. These three characteristic values are also listed with the dimensional variables in Table 8.2.

With these characteristic values, we define the dimensionless variables. The dimensionless spatial variables x^* and y^* are obtained by dividing x and y by L_c (see column 4 of Table 8.2), the dimensionless velocities v_x^* and v_y^* by dividing v_x and v_y by V_∞, and the dimensionless temperature difference by dividing $T - T_s$ by $T_\infty - T_s$. Note that only *one* characteristic length (L_c) and only *one* characteristic velocity (V_∞) are involved.

Table 8.2 **Definition of Dimensionless Variables for Flat-Plate Boundary-Layer Problem**

Dimensional Variable	Units	Characteristic Value	Dimensionless Variable	Dimensional Variable in Terms of Dimensionless Variable
x	m	L_c	$x^* \equiv \dfrac{x}{L_c}$	$x = L_c x^*$
y	m	L_c	$y^* \equiv \dfrac{y}{L_c}$	$y = L_c y^*$
v_x	m/s	V_∞	$v_x^* \equiv \dfrac{v_x}{V_\infty}$	$v_x = V_\infty v_x^*$
v_y	m/s	V_∞	$v_y^* \equiv \dfrac{v_y}{V_\infty}$	$v_y = V_\infty v_y^*$
$T - T_s$	K or °C	$T_\infty - T_s$	$T^* \equiv \dfrac{T - T_s}{T_\infty - T_s}$	$T = (T_\infty - T_s)\, T^* + T_s$

From the definition of each dimensionless variable, we solve for the dimensional variable, giving us

$$x = L_c x^*,$$ (8.13a)

$$y = L_c y^*,$$ (8.13b)

$$v_x = V_\infty v_x^*,$$ (8.13c)

$$v_y = V_\infty v_y^*,$$ (8.13d)

$$T = (T_\infty - T_s)T^* + T_s.$$ (8.13e)

These, in turn, we substitute back into our original dimensional boundary-layer equations (Table 8.1). Starting with mass conservation, Eq. 8.5, we write

$$\frac{\partial(V_\infty v_x^*)}{\partial(L_c x^*)} + \frac{\partial(V_\infty v_y^*)}{\partial(L_c y^*)} = 0.$$

Since V_∞ and L_c are constants, they can be removed from the derivatives, so

$$\frac{V_\infty}{L_c}\frac{\partial v_x^*}{\partial x^*} + \frac{V_\infty}{L_c}\frac{\partial v_y^*}{\partial y^*} = 0.$$

Furthermore, V_∞ and L_c cancel, leaving us with the result that

$$\frac{\partial v_x^*}{\partial x^*} + \frac{\partial v_y^*}{\partial y^*} = 0.$$ (8.14)

This is the dimensionless continuity (conservation of mass) equation. Note that our mathematical manipulation of the original dimensional equation introduced no new physics—we merely played with definitions. Note also that no dimensionless parameters appear in our final dimensionless continuity equation; however, that will not be the case for the dimensionless momentum and energy equations, which we now derive.

Consider, first, the second derivative that appears on the right-hand side of the momentum conservation equation, Eq. 8.4. Before substituting any of the variables expressed by Eqs. 8.13a–8.13d, we rewrite the second derivative as

$$\frac{\partial^2 v_x}{\partial y^2} = \frac{\partial\left(\dfrac{\partial v_x}{\partial y}\right)}{\partial y}.$$

We now use Eqs. 8.13a–8.13d to eliminate the dimensional variables in this second derivative, as follows:

$$\frac{\partial^2 v_x}{\partial y^2} = \frac{\partial\left(\dfrac{\partial(V_\infty v_x^*)}{\partial(L_c y^*)}\right)}{\partial(L_c y^*)} = \frac{V_\infty}{L_c^2}\frac{\partial\left(\dfrac{\partial v_x^*}{\partial y^*}\right)}{\partial y^*} = \frac{V_\infty}{L_c^2}\frac{\partial^2 v_x^*}{\partial y^{*2}}.$$

Using this result and similar substitutions for the remaining terms in Eq. 8.4, we rewrite the complete momentum equation as

$$(V_\infty v_x^*)\frac{V_\infty}{L_c}\frac{\partial v_x^*}{\partial x^*} + (V_\infty v_y^*)\frac{V_\infty}{L_c}\frac{\partial v_x^*}{\partial y^*} = \frac{\mu}{\rho}\frac{V_\infty}{L_c^2}\frac{\partial^2 v_x^*}{\partial y^{*2}}.$$

The appearance of the velocity field depends critically on the value of the Reynolds number for this flow over a thick plate with a curved trailing edge. Note the large difference in the boundary layers and wakes for flows at $Re = 100$ (top) and $Re = 10 \times 10^5$.

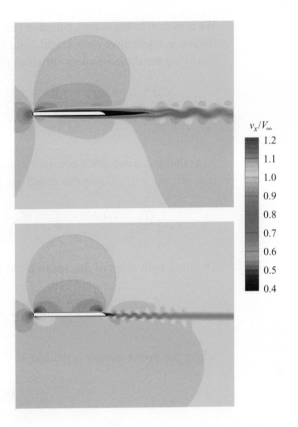

This can be simplified by dividing through by V_∞^2/L_c to yield

$$v_x^* \frac{\partial v_x^*}{\partial x^*} + v_y^* \frac{\partial v_x^*}{\partial y^*} = \left[\frac{\mu}{\rho V_\infty L_c} \right] \frac{\partial^2 v_x^*}{\partial y^{*2}}.$$

The bracketed term is one of the dimensionless parameters we seek. By convention, this term is known by its inverse, the **Reynolds number:**

$$Re_{L_c} \equiv \frac{\rho V_\infty L_c}{\mu}, \tag{8.15}$$

where the subscript L_c associated with Re emphasizes this particular choice of characteristic length.[3] The Reynolds number is frequently interpreted as expressing the ratio of fluid inertia to viscous forces:

$$Re \equiv \frac{\textit{fluid inertia}}{\textit{viscous forces}}. \tag{8.16}$$

Using the definition of the Reynolds number (Eq. 8.15), our dimensionless momentum conservation equation for the flat-plate boundary layer becomes

$$v_x^* \frac{\partial v_x^*}{\partial x^*} + v_y^* \frac{\partial v_x^*}{\partial y^*} = \frac{1}{Re_{L_c}} \frac{\partial^2 v_x^*}{\partial y^{*2}}. \tag{8.17}$$

---

[3] Since the Reynolds number is important in many, many problems involving the flow of fluids, the characteristic length is unique to the problem under consideration. For example, the characteristic length for flow inside a tube is the diameter D; thus, the appropriate Reynolds number is denoted Re_D ($\equiv \rho V_{\text{avg}} D / \mu$).

We now turn to the energy equation (Eq. 8.8, Table 8.1). As before, we expand the second derivative on the right-hand side and then substitute the expressions involving the dimensionless variables (Eqs. 8.13a–8.13e) as follows:

$$\frac{\partial^2 T}{\partial y^2} = \frac{\partial \left(\frac{\partial T}{\partial y} \right)}{\partial y} = \frac{\partial \left[\frac{\partial[(T_\infty - T_s)T^* + T_s]}{\partial(L_c y^*)} \right]}{\partial(L_c y^*)} = \frac{(T_\infty - T_s)}{L_c^2} \frac{\partial \left(\frac{\partial T^*}{\partial y^*} \right)}{\partial y^*} = \frac{(T_\infty - T_s)}{L_c^2} \frac{\partial^2 T^*}{\partial y^{*2}}.$$

Note that the stand-alone constant T_s disappears in the differentiation process (i.e., $\partial T_s / \partial y = 0$). Using this result, we write the complete energy equation as

$$V_\infty v_x^* \frac{(T_\infty - T_s)}{L_c} \frac{\partial T^*}{\partial x^*} + V_\infty v_y^* \frac{(T_\infty - T_s)}{L_c} \frac{\partial T^*}{\partial y^*} = \frac{k}{\rho c_p} \frac{(T_\infty - T_s)}{L_c^2} \frac{\partial^2 T^*}{\partial y^{*2}}.$$

Dividing both sides of this result by $V_\infty (T_\infty - T_s)/L_c$ yields

$$v_x^* \frac{\partial T^*}{\partial x^*} + v_y^* \frac{\partial T^*}{\partial y^*} = \frac{k}{\rho c_p V_\infty L_c} \frac{\partial^2 T^*}{\partial y^{*2}},$$

where the Peclet number is defined as

$$Pe \equiv \frac{\rho c_p V_\infty L_c}{k}. \tag{8.18}$$

With this definition, the final dimensionless energy equation is

$$v_x^* \frac{\partial T^*}{\partial x^*} + v_y^* \frac{\partial T^*}{\partial y^*} = \frac{1}{Pe} \frac{\partial^2 T^*}{\partial y^{*2}}. \tag{8.19}$$

As a result of creating the dimensionless form of the energy equation, we discover a second important dimensionless parameter, the **Peclet number.** The Peclet number can be physically interpreted as the ratio of the advection of energy (energy carried by the flow) to conduction:

$$Pe \equiv \frac{advection}{conduction}. \tag{8.20}$$

In introductory treatments of heat transfer, the Prandtl number is more commonly used than the Peclet number. We can easily show the relationship between these two important dimensionless parameters. Multiplying and dividing the Peclet number by μ create the product of the Reynolds number and the Prandtl number, that is

$$Pe = \frac{\rho c_p V_\infty L_c}{k} \left(\frac{\mu}{\mu} \right) = \left(\frac{\rho V_\infty L_c}{\mu} \right) \left(\frac{\mu c_p}{k} \right),$$

or

$$Pe = Re_L Pr, \tag{8.21}$$

where the **Prandtl number** is defined by

$$Pr \equiv \frac{\mu c_p}{k}. \tag{8.22}$$

Table 8.3   Dimensionless Boundary-Layer Equations for Flow over a Flat Plate

Origin	Dimensionless Equation	
Conservation of mass	$$\dfrac{\partial v_x^*}{\partial x^*} + \dfrac{\partial v_y^*}{\partial y^*} = 0$$	(8.14)
Conservation of x momentum	$$v_x^*\dfrac{\partial v_x^*}{\partial x^*} + v_y^*\dfrac{\partial v_x^*}{\partial y^*} = \dfrac{1}{Re_{L_c}}\dfrac{\partial^2 v_x^*}{\partial y^{*2}}$$	(8.17)
(Surface boundary condition)	$$\left.\dfrac{\partial v_x^*}{\partial y^*}\right|_{y^*=0} = \dfrac{1}{2}c_{\mathrm{f},L_c}Re_{L_c}$$	(8.26)
Conservation of energy	$$v_x^*\dfrac{\partial T^*}{\partial x^*} + v_y^*\dfrac{\partial T^*}{\partial y^*} = \dfrac{1}{Re_{L_c}Pr}\dfrac{\partial^2 T^*}{\partial y^{*2}}$$	(8.19)
(Surface boundary condition)	$$\left.\dfrac{\partial T^*}{\partial y^*}\right|_{y^*=0} = Nu_{L_c}$$	(8.27)

A useful physical interpretation can be given to the Prandtl number by introducing the thermal diffusivity as follows:

$$Pr = \frac{\mu c_p}{k}\left(\frac{\rho}{\rho}\right) = \frac{\mu}{\rho}\frac{\rho c_p}{k}.$$

From this we see that the Prandtl number is the ratio of the kinematic viscosity ($v = \mu/\rho$), which can be interpreted as the momentum diffusivity, to the thermal diffusivity ($\alpha = k/\rho c_p$):

$$Pr = \frac{v}{\alpha} \equiv \frac{momentum\ diffusivity}{thermal\ diffusivity}. \tag{8.23}$$

The dimensionless governing equations that we have just derived are summarized in Table 8.3. An important result from their derivation is the appearance of two dimensionless parameters: the Reynolds number and the Prandtl number. Numerical values for these parameters alone control the solutions to the dimensionless equations, that is, $v_x^*(x^*, y^*)$, $v_y^*(x^*, y^*)$, and $T^*(x^*, y^*)$. Note that the effects of the thermophysical properties are now contained in the dimensionless parameters. Recall the example presented at the beginning of this chapter in which we were seeking an answer of how best to generate experimental results that would be useful for a variety of fluids. We now have a partial answer to that question in that we need not be concerned with the fluids individually but rather with how changing from one fluid to another changes the values of the Reynolds and Prandtl numbers. To fully answer the question, however, requires that we explore the boundary conditions expressed by Eq. 8.7, for the wall shear stress, and Eq. 8.10, for the wall heat flux.

Friction Coefficient and Nusselt Number

We apply the same dimensionless variable transformations given by Eqs. 8.13a–8.13e to Eqs. 8.7 and 8.10. Introducing the dimensionless variables into the defining relationship for the shear stress (Eq. 8.7) yields

$$\tau_{\mathrm{w}} = \mu\left.\frac{\partial v_x}{\partial y}\right|_{y=0} = \mu\left.\frac{\partial (V_\infty v_x^*)}{\partial (L_c y^*)}\right|_{y^*=0} = \frac{\mu V_\infty}{L_c}\left.\frac{\partial v_x^*}{\partial y^*}\right|_{y^*=0}.$$

Principles of similarity allowed wind-tunnel drag tests of large-scale models to be used in the design of the *USS Albacore*. The *Albacore* was designed to maximize underwater speed and, in the 1950s, was the world's fastest submarine. Photograph courtesy of NASA.

Solving for the dimensionless velocity gradient at the wall yields

$$\left.\frac{\partial v^*_x}{\partial y^*}\right|_{y^*=0} = \frac{\tau_w L_c}{\mu V_\infty}. \tag{8.24}$$

The right-hand side of Eq. 8.24 is another dimensionless parameter. Usually this parameter is not used as it stands but is manipulated to create another dimensionless parameter, the friction coefficient. To form the traditional parameter, we multiply and divide the right-hand side of Eq. 8.24 by $2\rho V_\infty^2$ to get

$$\frac{\tau_w L_c}{\mu V_\infty}\left(\frac{2\rho V_\infty^2}{2\rho V_\infty^2}\right) = \frac{1}{2}\left(\frac{\tau_w}{\frac{1}{2}\rho V_\infty^2}\right)\left(\frac{\rho V_\infty L_c}{\mu}\right).$$

In this expression, we recognize the appearance of the Reynolds number, $\rho V_\infty L_c/\mu$, and define the **friction coefficient**[4] as follows:

$$c_{f,L_c} \equiv \frac{\tau_w}{\frac{1}{2}\rho V_\infty^2}. \tag{8.25}$$

The dimensionless velocity gradient of Eq. 8.24 thus becomes

$$\left.\frac{\partial v^*_x}{\partial y^*}\right|_{y^*=0} = \frac{1}{2}c_{f,L_c}Re_{L_c}. \tag{8.26}$$

We now consider the surface heat flux. Starting with the defining relationship for the surface heat flux (Eq. 8.10), we introduce the dimensionless variables as follows:

$$-k\left.\frac{\partial T}{\partial y}\right|_{y=0} = h_{\text{conv},x}(T_s - T_\infty),$$

becomes

$$-k\left.\frac{\partial[(T_\infty - T_s)T^* + T_s]}{\partial(L_c y^*)}\right|_{y^*=0} = h_{\text{conv},x}(T_s - T_\infty),$$

or

$$-\frac{k(T_\infty - T_s)}{L_c}\left.\frac{\partial T^*}{\partial y^*}\right|_{y^*=0} = h_{\text{conv},x}(T_s - T_\infty).$$

Solving for the dimensionless temperature gradient at the wall yields

$$\left.\frac{\partial T^*}{\partial y^*}\right|_{y^*=0} = \frac{h_{\text{conv},x}L_c}{k}. \tag{8.27}$$

The dimensionless parameter appearing on the right-hand side, known as the **Nusselt number,** is defined as

$$Nu_{L_c} \equiv \frac{h_x L_c}{k}. \tag{8.28}$$

---

[4] This parameter is also known as the skin friction coefficient. In aeronautical jargon, the *skin* is the outer surface of an aircraft.

FIGURE 8.6

For a smooth, flat plate, the friction coefficient depends only on the Reynolds number for a given characteristic length (top). The characteristic length is the distance from the leading edge to the location where the shear stress is desired. For heat transfer, the problem is complicated by the introduction of a second dimensionless parameter, the Prandtl number (bottom). Here the Nusselt number is a function of both the Reynolds number and the Prandtl number.

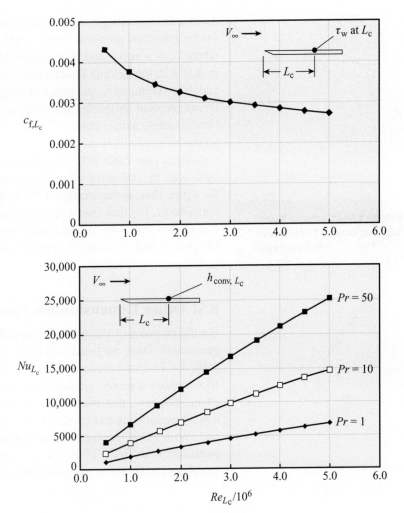

In this definition, the thermal conductivity is that of the *fluid* in contact with the solid surface, not that of the surface itself.

From the dimensionless conservation equations and the dimensionless boundary conditions (see the summary in Table 8.3), we conclude that only four parameters (Re, Pr, c_{f,L_c}, and Nu_{L_c}) are required to describe the friction and heat transfer associated with a smooth,[5] flat plate in a parallel flow, either analytically or experimentally. More explicitly, this conclusion suggests that one should organize results from calculations or experiments so that c_{f,L_c} can be plotted as a function of Re_{L_c} and so that Nu_{L_c} can be plotted as a function of Re_{L_c} using Pr as a parameter. This idea is illustrated in Fig. 8.6.

Although we have dealt with flow and heat transfer for a flat plate rather than with the flow and heat transfer inside a tube, we can still use our results to answer the questions set forth at the outset of this chapter of how best to conduct experiments to determine heat-transfer coefficients for flow through a tube. The bottom plot in Fig. 8.6 provides the answer: One should vary the flow rates, tube sizes, and fluids so that a sufficiently wide range of Reynolds numbers and Prandtl numbers is covered. Assuming again that 10 data points are sufficient to create a smooth curve on a plot of Nu versus Re, only 100 data points would have to be gathered if 10 values of Pr were selected. Of course,

---

[5] To deal with a rough plate requires the introduction of another dimensionless parameter. See the fourth entry in Table 8.4.

the characteristic length, velocity, and temperature difference need to be defined for the tube-flow problem. These would be D, V_{avg}, and $T_m - T_s$, respectively, where T_m is a mean fluid temperature. Missing so far in our discussion is how to deal with the fact that fluid temperatures and tube wall temperatures are also variables. Fortunately, the effect of temperature is contained only in how the thermophysical properties are affected by temperature. Our choice of the dimensionless temperature, $T^* = (T - T_s)/(T_\infty - T_s)$, prevents T_s or T_∞ from appearing explicitly in the dimensionless governing equations or their boundary conditions (see Table 8.3). The effect of temperature on properties is usually dealt with by evaluating all properties at a so-called film temperature, which, for a pipe flow, is the average of the mean fluid temperature and the tube wall temperature. For flow over a flat plate, the film temperature is the average of the free-stream and plate surface temperatures. We will deal with this in detail in Chapters 9 and 10 where we present methods to obtain numerical values for heat-transfer coefficients.

> **Chapter 9 treats external flows, whereas Chapter 10 deals with internal flows.**

8.5c Other Dimensionless Parameters

For the purposes of this book, we consider a limited number of dimensionless parameters. These are listed with physical interpretations, where appropriate, in Table 8.4. Five of these parameters have been introduced in this chapter, and the Biot number appeared previously in Chapter 5 (see Eq. 5.22). You should be aware, however, that many more dimensionless parameters are defined and used in engineering and science. For example, the *CRC Handbook of Chemistry and Physics* [12] lists the definition of over 200 dimensionless parameters. Of particular note from Table 8.4 are the **friction factor** f and the **relative roughness** $\varepsilon/\mathbf{D}$. These two parameters have their origins in the application of

Table 8.4 Key Dimensionless Parameters Used in Fluid Mechanics and Heat Transfer

Name	Symbol	Physical Interpretation	Application
Biot number	Bi	Ratio of conduction to convective thermal resistances	Unsteady conduction
Drag coefficient	c_D	Dimensionless drag force	External flows
Friction coefficient	c_f	Dimensionless wall shear stress or dimensionless velocity gradient at surface	External flows
Relative roughness	ε/L_c	Ratio of average height of surface asperity to characteristic length	Internal and external flows
Friction factor	f	Dimensionless head loss, $h_L / [(V_{avg}^2/2g)(L/D)]$	Internal flows
Fourier number	Fo	Dimensionless time	Unsteady conduction
Grashof number	Gr	Ratio of buoyancy force to viscous force*	Free convection
Dimensionless length	L/D	Ratio of duct length to diameter	Internal flows
Nusselt number	Nu	Dimensionless temperature gradient at surface	Internal and external forced and free convection
Peclet number	Pe	Ratio of advection of energy to conduction	Internal and external forced convection
Prandtl number	Pr	Ratio of momentum diffusivity to thermal diffusivity	Forced and free convection
Rayleigh number	Ra	Product of Grashof and Prandtl numbers	Free convection
Reynolds number	Re	Ratio of fluid inertia to viscous force	All flows

* Chapter 9 provides a more complex interpretation of this parameter.

Chapters 9 and 10 capitalize on the ideas presented in this chapter. ➤

the Buckingham (Vaschy) pi theorem (Appendix 8A) to turbulent pipe flows. We will employ friction factors and relative roughnesses in Chapter 10 to determine the pressure drop for internal flows, such as through tubes, pipes, and ducts. The relative roughness is also important in determining the drag associated with flows over external surfaces, the previously discussed flat-plate flow being an example. For such flows, the **drag coefficient** c_D plays a role similar to the friction factor for internal flow. You may already be familiar with the use of drag coefficients to characterize the "slipperiness" of automobiles. The following example illustrates how drag-coefficient correlations can be extracted from experimental data. We develop these ideas in Chapter 9.

Example 8.1   Biological Systems Application

Reference [14] provides drag force data for a number of swimming mammals expressed using dimensionless parameters. The following graph shows the data for blue whales and beavers, where the drag coefficient is plotted as a function of Reynolds number.

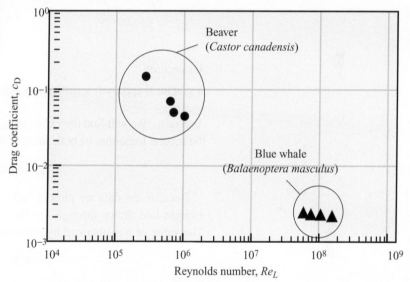

The drag coefficient and Reynolds number are defined, respectively, as

$$c_D \equiv \frac{F_D/A_{\text{wetted}}}{\frac{1}{2}\rho V^2}$$

and

$$Re_L \equiv \rho \frac{VL}{\mu},$$

where F_D is the drag force, A_{wetted} and L are the wetted surface area and length of the mammal, respectively, V is the swimming speed, and ρ and μ are the density and viscosity, respectively, of the swimming medium (water).

Use the graphical data for blue whales to obtain a correlation $c_D = f(Re_L)$ of the form

$$c_D = C Re_L^n,$$

where C and n are constants.

Solution

Known   Log–log plot with c_D–Re_L data

Find C, n

Sketch

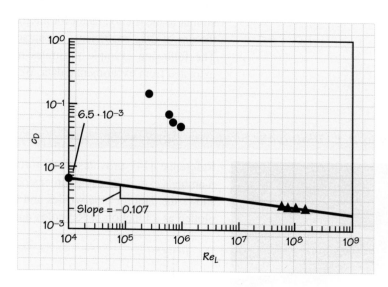

Assumption

A graphical approach is sufficiently accurate (since raw data are unavailable).

Analysis   We will find the exponent n graphically. First, we note that taking the natural logarithm of both sides of the expression $c_D = C\, Re_L^n$ yields

$$\ln c_D = \ln C + n \ln Re_L.$$

Because the data are plotted on log–log coordinates, n is the slope of a straight line drawn through the blue-whale data, as shown in the sketch. The value of n is obtained by measuring the run and rise with a ruler:

$$n = -0.107.$$

To obtain the coefficient C, we evaluate $c_D = C\, Re_L^{-0.107}$ at any convenient location on the straight line. For example, at $Re_L = 1 \times 10^4$ the drag coefficient c_D is 6.5×10^{-3}; thus

$$6.5 \times 10^{-3} = C(1 \times 10^4)^{-0.107},$$

or

$$C = 0.0174.$$

Our final result is thus

$$c_D = 0.0174\, Re_L^{-0.107}.$$

Comments   Power-law correlations are quite common in fluid mechanical and heat-transfer applications. This example illustrates how experimental data are transformed into such working mathematical expressions. We also note how experimental uncertainty is embedded in our final result: First, each data point has some unknown uncertainty in both the c_D and Re_L values. Second, we merely "eyeballed" a straight line through the data. Another person might draw a line with a slightly different slope. Also, more rigorous methods could be used if we had the raw data (c_D and Re_L values). For example, we could employ a least-squares regression analysis to obtain a value for n as well as some statistical measure of the goodness of fit. As we will see in Chapters 9 and 10, many engineering correlations have validities within some range of uncertainly (e.g., ±15%).

Self Test 8.1

☑ **Estimate the power-law exponent for drag on beavers (*Castor canadensis*).**

(Answer: $n \approx -1.1$)

Returning to Table 8.4, we see that free convection heat transfer introduces two additional important dimensionless parameters, the **Grashof number** and the **Rayleigh number.** Like the Reynolds number, the Grashof (or Rayleigh) number appears naturally in the dimensionless momentum conservation equation for the buoyancy-driven boundary layer. This equation differs from Eq. 8.4 by the inclusion of both pressure force and gravity terms, as we will see in Chapter 9.

SUMMARY

Many engineering problems are so large, or so complex, that they yield to no single solution technique. Dimensional analysis helps engineers apply a full arsenal of tools in intelligent ways. No matter what the tool—theoretical analyses, numerical simulations, or experimentation—dimensional analysis can add insight and generality. Using experimentation to bridge the gap between theory and practical design creates the problem of deciding how best to conduct these experiments and to interpret their results. This problem, too, is solved in part by dimensional analysis. We saw that dimensional analysis provides the basis for the laws of similitude, which are routinely applied in scale-model testing. You should be familiar with the concepts of geometrical and dynamic similarity and how to use these concepts to design and interpret results from scale-model testing. You should also be familiar with the process of forming dimensionless governing equations and boundary conditions to identify the controlling dimensionless parameters. Using this approach, we identified several key dimensionless parameters associated with flowing fluids: the Reynolds number, the Peclet number, the Prandtl number, the friction coefficient, and the Nusselt number. You should be able to recognize these and provide a physical interpretation for each. Other important dimensionless parameters, which are used in later chapters in this book, were identified. For those desiring greater understanding of dimensional analysis, the Buckingham pi theorem was presented and applied to the problems of drag and convective heat transfer in the appendix to this chapter. This method has greater generality for finding dimensionless parameters than does the method of nondimensionalizing the governing equations. A review of the learning objectives presented at the beginning of this chapter is useful at this juncture.

Chapter 8
Key Concepts & Definitions Checklist[6]

8.2 The Limits of Theory
- [ ] Complexity ➤ *Q8.2*
- [ ] Need for experiments ➤ *Q8.2*

8.3 Parametric Testing
- [ ] Output variable ➤ *Q8.3*
- [ ] Parameter ➤ *Q8.3*

8.4 Similitude
- [ ] Geometric similarity ➤ *Q8.4*
- [ ] Dynamic similarity ➤ *Q8.5, Q8.10*
- [ ] Similitude ➤ *8.8, 8.10*

8.5 Dimensionless Parameters
- [ ] Buckingham (or Vaschy) pi theorem ➤ *8.27, 8.30*
- [ ] Dimensionless governing equations ➤ *8.2, 8.5*
- [ ] Characteristic scales or values ➤ *Q8.6, Q8.7*
- [ ] Dimensionless boundary conditions ➤ *Q8.8, Q8.9*
- [ ] Key dimensionless parameters (Table 8.4) ➤ *Q8.11–Q8.15, 8.7*

---

[6] Numbers following arrows refer to Questions (prefaced with a Q) and Problems at the end of the chapter.

REFERENCES

1. Vincenti, W. G., *What Engineers Know and How They Know It: Analytical Studies from Aeronautical History,* Johns Hopkins University Press, Baltimore, 1990.

2. Smeaton, J., "An Experimental Inquiry Concerning the Natural Powers of Water and Wind to Turn Mills, and Other Machines, Depending on a Circular Motion," *Philosophical Transactions of the Royal Society,* 51(1): 100–174 (1759).

3. Rouse, H., and Ince, S., *History of Hydraulics,* Iowa Institute of Hydraulic Research, State University of Iowa, Iowa City, IA, 1957, pp. 182–187.

4. Vaschy, A., "Sur les lois de similitude en physique, *Annales Télégraphiques,* 19:25–28 (1892).

5. Rayleigh, J. W. S., "The Principle of Similitude," *Nature, London,* 95:66, 591, 644 (1915).

6. Riabouchinsky, D., "Méthod des variables de dimension zéro et son application en aérodynamique, *L'Aerophile,* Sept. 1, 1911, pp. 407–408.

7. Buckingham, E., "On Physically Similar Systems: Illustration on the Use of Dimensional Equations, *Physical Review,* 4:345–376 (1914).

8. Reynolds, O., "An Experimental Investigation of the Circumstance Which Determine Whether the Motion of Water Shall Be Direct or Sinuous and of the Law of Resistance in Parallel Channels," *Philosophical Transactions of the Royal Society of London,* 174:935–982 (1883).

9. Bridgman, P. W., *Dimensional Analysis,* Yale University Press, New Haven, 1922.

10. McCormick, B. W., *Aerodynamics, Aeronautics, and Flight Mechanics,* Wiley, New York, 1975.

11. Mills, A. F., *Heat and Mass Transfer,* Irwin, Chicago, 1995.

12. Weast, R. C. (Ed.), *Handbook of Chemistry and Physics,* 56th ed., CRC Press, Cleveland, OH, 1976.

13. Fox, R. W., and McDonald, A. T., *Introduction to Fluid Mechanics,* Wiley, New York, 1985, p. 302.

14. Fish, F. E., Chapter 3, "Aquatic Locomotion," in *Mammalian Energetics* (T. E. Tamasi and T. H. Horton, Eds.) Cornell University Press, Ithaca, NY, 1992.

Some end-of-chapter problems were adapted with permission from the following:

15. Chapman, A. J., *Fundamentals of Heat Transfer,* Macmillan, New York, 1987.

16. Pnueli, D., and Gutfinger, C., *Fluid Mechanics,* Cambridge University Press, Cambridge, England, 1992.

17. Shepherd, D. G., *Elements of Fluid Mechanics,* Harcourt, Brace & World, New York, 1965.

Nomenclature

c_p Constant-pressure specific heat (J/kg · K)

C_p Pressure coefficient (dimensionless)

c_D Drag coefficient (dimensionless)

c_f Friction coefficient (dimensionless)

c_L Lift coefficient (dimensionless)

F Force (N)

$h_{\text{conv},x}$ Local heat-transfer coefficient (W/m^2· K or °C)

k Thermal conductivity (W/m · K or °C)

L Length

Nu Nusselt number (dimensionless)

Pe Peclet number (dimensionless)

Pr Prandtl number (dimensionless)

$\dot{Q}''$ Heat flux (W/m^2)

Re Reynolds number (dimensionless)

T Temperature (K)

T_m Bulk mean temperature (K)

v_x, v_y Cartesian velocity components (m/s)

V velocity (m/s)

x Cartesian coordinate in axial direction (m)

y Cartesian coordinate in transverse direction (m)

ε Roughness height (m)

μ Viscosity (N · s/m^2)

ν Kinematic viscosity (m^2/s)

Π Pi group or dimensionless parameter

τ Shear stress (N/m^2 or Pa)

ρ Density (kg/m^3)

SUBSCRIPTS

avg   average

c   characteristic value

D   drag

H   hydrodynamic

m   model

p   prototype or full scale

s   surface

T   thermal

w   wall

∞ free stream

GREEK

α Angle of attack (rad) or thermal diffusivity (m^2/s)

δ Boundary-layer thickness (m)

OTHER

$\overline{(\ )}$ Average quantity

$(\ )^*$ Dimensionless quantity

QUESTIONS

8.1 Review the most important equations presented in this chapter (i.e., those with a reddish background). What physical principles do they express? What restrictions apply?

8.2 Explain how complex devices push or exceed the limits of theoretical description. Provide an example and discuss.

8.3 Explain the concept of parametric testing. Give several examples. Indicate the output variables and parameters.

8.4 Explain the concept of geometric similarity. Provide some examples in which geometric similarity would be difficult to achieve for a length scale ratio of, say, 100:1.

8.5 Define dynamic similarity. What dimensionless parameter is frequently used to achieve dynamic similarity?

8.6 Explain how the wall shear stress for a uniform flow over a flat plate scales with distance from the leading edge of the plate.

8.7 Consider a heated flat plate with a constant surface temperature T_s in a uniform parallel flow with a temperature T_∞. Explain how the convective heat flux scales with distance from the leading edge of the plate.

8.8 What dimensionless parameter originates from the specification of the no-slip boundary condition?

8.9 What dimensionless parameter originates from the definition of the convective heat flux at the surface of a heated flat plate?

8.10 Describe two methods of obtaining the dimensionless parameters that are important to any engineering problem.

8.11 Define the Reynolds number in terms of geometrical and fluid properties. Also provide a physical interpretation of the Reynolds number.

8.12 List two or more dimensionless parameters that relate in some way to the shear stresses and/or drag forces associated with external flows. Also provide a physical interpretation of these parameters.

8.13 List two or more dimensionless parameters that relate in some way to the pressure drop or head loss associated with internal flows. Also provide a physical interpretation of these parameters.

8.14 List two or more dimensionless parameters that relate in some way to the heat transfer from surfaces exposed to an external flow. Also provide a physical interpretation of these parameters.

8.15 List two or more dimensionless parameters associated with free convection heat transfer. Also provide a physical interpretation of these parameters.

8.16 List two or more dimensionless parameters that relate in some way to forced convection heat transfer associated with flow through pipes, tubes, and ducts.

8.17 Write out the dimensions associated with force in terms of mass (M), length (L), time (T), and temperature (θ).

Chapter 8 Problem Subject Areas

8.1–8.5	**Dimensionless governing equations**
8.6–8.26	**Similitude**
8.27–8.38	**Buckingham (Vaschy) pi theorem**

PROBLEMS

8.1 Starting with conservation of mass for a boundary layer on a flat plate (Eq. 8.5), use the characteristic values in Table 8.2 to transform this equation to its dimensionless equivalent (Eq. 8.14).

8.2 Starting with conservation of axial momentum for a boundary layer on a flat plate (Eq. 8.4), use the characteristic values in Table 8.2 to transform this equation to its dimensionless equivalent (Eq. 8.17).

8.3 Starting with conservation of energy for a boundary layer on a flat plate (Eq. 8.8), use the characteristic values in Table 8.2 to transform this equation to its dimensionless equivalent (Eq. 8.19).

8.4 Consider a steady, two-dimensional flow between two parallel plates of length L separated by a distance $2d$ as shown in the sketch. For fully developed flow (a concept discussed in Chapter 10), the x- and y-momentum equations are given as follows:

$$-\frac{\partial P}{\partial x} + \mu \frac{\partial^2 v_x}{\partial y^2} = 0,$$

$$\frac{\partial P}{\partial y} = -\rho g.$$

Create the nondimensional forms of these equations using V_{avg} as the characteristic velocity, d as the characteristic length in the y-direction, L as the characteristic length in the x-direction, and P_{atm} as the characteristic pressure.

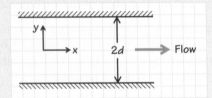

8.5 An important step in the preparation of frozen chickens is the quick chilling process. Immediately after the chickens are cleaned, they pass through a blast tunnel where very cold air is blown on them. Larger chickens stay in the tunnel longer than smaller ones. The differential equation governing the cooling process is given by

$$\frac{\partial T}{\partial t} = \alpha \nabla^2 T,$$

where T is temperature, t is time, and α is the thermal diffusivity ($[=] m^2/s$). Transform this equation to its dimensionless counterpart using a characteristic time t_c, a characteristic length L_c, and a characteristic temperature T_c.

8.6 Use the result from Problem 8.5 to find the similarity rules for chilling chickens. Assume all chickens are geometrically similar. If a chicken of mass M_1 takes a time t_1 to chill, how long does it take to chill a chicken that is twice as large (i.e., $M_2 = 2M_1$)?

8.7 Calculate the indicated dimensionless parameters for the following specified fluids and conditions:

A. Find the Reynolds number Re_x for $V_\infty = 20$ m/s, $x = 0.6$ m, and air at 25°C and 1 atm.

B. Find the Nusselt number Nu_L for $h = 10$ W/m²·°C, $L = 1$ m, and CO_2 at 50°C and 1 atm.

C. Find the Nusselt number Nu_x for $h = 110$ Btu/hr·ft²·F, $x = 3$ ft, and steam at 100 psia, and 400 F.

D. Find the Prandtl number Pr for $\mu = 2.04 \times 10^{-5}$ N·s/m², $c_p = 0.958$ kJ/kg·°C, and $k = 0.02631$ W/m·°C.

8.8 The drag coefficient associated with a particular aircraft wing section is measured in a wind tunnel using a 1/4-scale model. The model is tested at 300 K and 100 kPa. The full-size aircraft flies at 130 miles/hr (58.1 m/s) at an altitude of 6000 m where the temperature and pressure are nominally 250 K and 47 kPa, respectively. What wind-tunnel speed is required for the Reynolds number of the model to match that of the full-size aircraft?

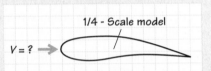

8.9 Consider the wind-tunnel testing described in Problem 8.8. In a particular test with a velocity of 60 miles/hr (26.8 m/s) and 300 K and 1 atm, the drag force F_D on a model wing is measured to be 2.15 N. The test wing has a span S of 2 m and a chord C of 0.6 m. For this situation, the drag coefficient is defined as

$$c_D = \frac{F_D/CS}{\frac{1}{2}\rho V^2}.$$

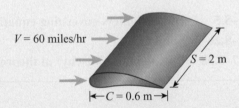

A. Determine the drag coefficient and Reynolds number based on chord length for this test.

B. Assuming the drag coefficient is independent of Reynolds number, determine the drag force on the full-size aircraft having a chord of 2.4 m and a span of 16.8 m. The flight conditions are the same as given in Problem 8.8.

8.10 A 1/4-scale model of a Saturn SL2 sedan is tested in a wind tunnel. The nominal temperature and pressure in the tunnel are 300 K and 1 atm, respectively. The length L of the full-size vehicle is 4.52 m. The frontal area A_f of the full-size vehicle is 1.914 m². (Note: The frontal area is the projected area of the vehicle when viewed from the front.)

A. What would the flow velocity in the wind tunnel have to be to achieve dynamic similarity with a full-size Saturn SL2 traveling at 55 miles/hr? Express your result in m/s and in miles/hr.

B. Experiments show that the drag coefficient for the Saturn SL2 is essentially independent of the Reynolds number for $Re_L > 2.5 \times 10^6$, where the characteristic length L is defined as the length of the vehicle. The drag coefficient for an automobile is defined as

$$c_D = \frac{F_D/A_f}{\frac{1}{2}\rho V^2}.$$

a. The drag force measured for the model is 283.3 N for tests conducted at a tunnel air velocity of 50 m/s. Does this test condition lie in the regime $Re_L > 2.5 \times 10^6$?

b. What is the frontal area of the model?

c. Estimate the drag force in newtons experienced by a full-size vehicle traveling at 55 miles/hr.

d. Determine the power required to overcome this drag force (Part c). Express your result in both kilowatts and horsepower.

8.11 Using the drag coefficient–Reynolds number data for beavers (*Castor canadensis*) shown in Example 8.1 and tabulated here, obtain a correlation of the form $c_D = CRe_L^n$ (i.e., find C and n).

Re_L	c_D
3.6×10^5	0.12
7×10^5	0.058
8×10^5	0.046
1×10^6	0.040

8.12 An aircraft with a representative dimension of 10 ft (the width of the wing section) is to have a design speed of 450 miles/hr at 30,000 ft. A laboratory wind tunnel has a working section in which the air

is at 60 psia and 40 F. Find the correct size of the model for similar flow conditions to hold in the wind tunnel and at the operational altitude.

8.13 When a body moves through air with a velocity of 100 ft/s at ground level ($T = 288$ K and $P = 101.3$ kPa), a drag force of 1000 lb$_f$ is measured. What prediction can be made about the drag force at an altitude of 6500 m ($T = 246$ K and $P = 44$ kPa)?

8.14 Consider the convective heat transfer case for a cross-flow of air past a horizontal cylinder. Two experiments are performed in which the air temperature is $T_\infty = 100°C$, the cylinder surface temperature is $T_s = 40°C$, and the cylinder diameter is $D = 0.3$ m. The data from the two experiments provide the following average heat-transfer coefficients between the surface and the air:

$$h = \begin{cases} 45 \text{ W/m}^2 \cdot °C \text{ when } V_\infty = 20 \text{ m/s,} \\ 35 \text{ W/m}^2 \cdot °C \text{ when } V_\infty = 15 \text{ m/s.} \end{cases}$$

From experience, the Nusselt number is known to be of the form $Nu_D = CRe_D^m Pr^n$, in which C, m, and n are empirical constants. Estimate the surface heat-transfer coefficients that would be expected if, for identical specified temperatures, a cylinder with diameter $D = 0.6$ m is placed in an air stream with a free-stream velocity V_∞ of (a) 15 m/s and (b) 30 m/s.

8.15 A body of some shape has a characteristic length of $L_c = 0.5$ m. When placed in an air stream at $T_\infty = 100°C$ with $V_\infty = 25$ m/s, the heat flux from the surface is found to be 2000 W/m² when the surface temperature is 140°C. If a geometrically similar body five times as large is placed in an air stream with $V_\infty = 5$ m/s, find the resultant surface heat-transfer coefficient. Assume the temperatures are the same as in the original case.

8.16 Water with a bulk mean temperature of 100 F flows in a circular tube, the surface of which is maintained at 80 F. Two experiments are performed when the tube diameter is $D = 2$ in. These two experiments are performed at different average flow velocities. The following surface heat-transfer coefficients and V_{avg} are measured:

$$h = \begin{cases} 570 \text{ Btu/hr} \cdot \text{ft}^2 \cdot \text{F when } V_{avg} = 3 \text{ ft/s,} \\ 410 \text{ Btu/hr} \cdot \text{ft}^2 \cdot \text{F when } V_{avg} = 2 \text{ ft/s.} \end{cases}$$

From experience, the Nusselt number is known to be of the form $Nu_D = CRe_D^m Pr^n$, in which C, m, and n are empirical constants. Estimate the surface heat-transfer coefficients that would be expected for water flowing in a 4-in-diameter tube if the average flow velocity V_{avg} is (a) 2 ft/s and (b) 4 ft/s. Assume the same temperatures are employed as in the original experiments.

8.17 Water at 90 F flows over the outside of a circular tube normal to the tube axis. The surface temperature is maintained at 40 F. Two experiments are performed for a 2-in-diameter tube at different free-stream velocities of the water. The following average heat transfer coefficients are measured:

$$h = \begin{cases} 940 \text{ Btu/hr} \cdot \text{ft}^2 \cdot \text{F when } V_\infty = 5 \text{ ft/s,} \\ 1420 \text{ Btu/hr} \cdot \text{ft}^2 \cdot \text{F when } V_\infty = 10 \text{ ft/s.} \end{cases}$$

From experience, the functional form for the Nusselt number is known to be $Nu_D = CRe_D^m Pr^n$, in which C, m and n are constants. Estimate the average heat-transfer coefficients that would be expected if, for identical temperatures, a 1.5-in-diameter tube is placed in a water stream at (a) 10 ft/s and (b) 20 ft/s.

8.18 Steam at a pressure of 4000 kPa and a bulk mean temperature of 500°C flows in a 19-cm-diameter pipe. The surface temperature of the pipe is maintained at 450°C. Two experiments are performed at different flow velocities, and the following surface heat fluxes are measured:

$$V_{avg} = \begin{cases} 3.6 \text{ m/s with } \dot{Q}'' = 9400 \text{ W/m}^2, \\ 5.0 \text{ m/s with } \dot{Q}'' = 12{,}200 \text{ W/m}^2. \end{cases}$$

From experience, the functional form for the Nusselt number is known to be $Nu_D = CRe_D^m Pr^n$, where C, m, and n are constants. Estimate the surface heat flux, for identical temperatures, when (a) $V_{avg} = 6.0$ m/s with $D = 12$ cm and (b) $V_{avg} = 2.5$ m/s with $D = 25$ cm.

8.19 Wind-tunnel tests with a 1/5-scale model are to be used to estimate the air resistance (drag force) of a full-size train traveling at 3, 10 and 25 m/s. At what air speeds should the wind-tunnel tests be conducted? Assume the air to be at 298 K and 1 atm for both the full-size and model trains. Also determine the relationship between the forces measured on the model and those that will act on the full-size train.

8.20 A 1/4-scale model automobile is used to estimate the drag forces on a full-size automobile for vehicle speeds between 50 and 100 km/hr. Separate sequences of tests are to be conducted in (a) a wind tunnel and (b) a water tunnel. Find the running speeds in each tunnel and the interpretation of the measured resistance forces. The tunnels operate nominally at 1 atm and 298 K.

8.21 Complete similarity between a ship model and a full-size ship requires the same Reynolds number ($\equiv LV/\nu$) and the same Froude number ($\equiv gL/V^2$). Typical Reynolds and Froude numbers for a full-size ship are 1.1×10^8 and 33, respectively. Assuming a model 1/100 the size of the ship, can you design an experiment having full similarity? If not, why?

8.22 A submerged body is designed to move in oil ($\rho_{oil} = 880$ kg/m^3, $\mu_{oil} = 0.0082$ N·s/m^2) at a speed of 2 m/s. Water-tunnel tests using a 1/8-scale model are used to estimate the drag force on the full-sized submerged body.

A. At what speed must the model be run to achieve similarity with the full-sized body? The water density and viscosity are 998 kg/m^3 and 0.001 N·s/m^2, respectively.

B. What is the drag force on the submerged body if the force measured on the model at the test speed from Part A is 300 N?

8.23 An airplane wing is tested in a wind tunnel, as shown in the sketch. The purpose of the test is to obtain drag forces and lift forces acting on the wing at various flight speeds. Drag forces are those acting on the wing in the direction of the velocity of the oncoming air; lift forces act in the direction perpendicular to that velocity. These forces are measured on the model. How do the measured forces relate to the forces that act on the full-size wing?

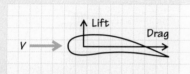

8.24 A 2-m-diameter weather balloon is to be used in air at 20°C. To find the drag force on the balloon, an experiment was conducted in which a 0.02-m-diameter sphere was held in water moving with a velocity of 10 m/s. A drag force of 6.5 N was measured for the sphere at this condition. Determine (a) the speed of the wind past the balloon that corresponds to the experiment and (b) the drag force on the balloon at this wind velocity.

8.25 Consider a 1/10-scale model glider. The model is connected to a car by a fine wire and towed at a speed of 140 km/hr. The tension force in the wire is 1000 N. Find (a) the speed of the full-size glider corresponding to that of the model and (b) the power expended by an airplane pulling the glider at that speed. Assume that the drag of the wire and that of the towing cable are negligible compared to the drag of the model and full-size glider, respectively.

8.26 A 3-m-long model is used to test the design of a new ship hull. The length of the full-size hull is 100 m. The model is tested in a tow tank, which can accommodate towing speeds of up to 2 m/s. To achieve complete similarity requires equality of both Froude ($\equiv gL/V^2$) and Reynolds (VL/ν) numbers for the model and full-size hulls. However, it is impractical to provide equal Reynolds numbers; thus, tests are conducted only at equal Froude numbers. A correction is later made for the different Reynolds numbers. For a model speed of 2 m/s, determine the corresponding full-size ship speed. What are the ship and the model Reynolds numbers at these speeds?

8.27 Apply the procedure that is part of the Buckingham (Vaschy) pi theorem to find the exponents a, b, and c for the following dimensional parameter (Π-group):

$$(\rho)^a(V)^b(L)^c(F_{\mathrm{D}})^1,$$

where ρ is the density, V is the velocity, L is length, and F_{D} is the drag force. Write out the resulting dimensionless parameter (i.e., the Π group).

8.28 A sphere is submerged in a liquid and released. Depending on the densities of the sphere and the liquid, the sphere either rises (and ultimately floats) or sinks. Find the dimensionless parameters (Π groups) associated with this problem.

8.29 Consider the drag force F_{D} associated with a circular cylinder of diameter D and length L exposed to a flow with velocity V. The velocity is directed perpendicular to the cylinder length. It is desired to determine all of the dimensionless parameters associated with this problem. Use the Buckingham (Vaschy) pi theorem to find these dimensionless parameters.

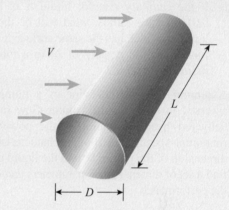

8.30 The following dimensional variables are important in the heat transfer from a heated sphere of diameter D in a cross-flow: D, V_∞, ρ, μ, c_p, k, and $\overline{h}_{\mathrm{conv}}$. Use the Buckingham (Vaschy) pi theorem to find all of the dimensionless parameters associated with this problem.

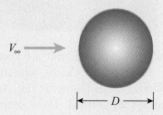

8.31 The thrust T of a propeller is assumed to depend on the tip diameter D, the fluid density ρ, the fluid viscosity μ, the rotational speed (rpm) N, and the forward speed V, relative to the distant fluid. Use the Buckingham (Vaschy) pi theorem to find all of the dimensionless parameters associated with this problem. Assess the physical significance of the parameters that you obtain.

Hint: Important dimensional parameters include drop diameter, drop density, drop velocity, drop viscosity, surface tension, gravitational acceleration, and air density and viscosity.

8.32 The height h to which a liquid rises in a small-bore tube owing to surface-tension forces (the capillary rise) is a function of the specific weight of the liquid $\gamma(= \rho g$, i.e., the product of the liquid density and the gravitational acceleration), the radius of the tube, r, and the surface tension of the liquid, σ.

A. Use the Buckingham (Vaschy) pi theorem to find all of the dimensionless parameters associated with this problem.

B. If the capillary rise for liquid A is 1 in in a 0.010-in tube radius, what will be the rise for liquid B, which has the same surface tension but four times the density of A, in a tube with a radius of 0.005 in?

8.33 Assuming that the behavior of a pump can be assessed in terms of the volume flow rate $\dot{V}$, the delivered head of liquid, h, the power W, the rotational speed N in rpm, a characteristic linear dimension of the pump, D, and the liquid density ρ, find a set of dimensionless parameters that expresses the performance.

8.34 Specially equipped aircraft are used to help extinguish forest fires. Such aircraft release large quantities of water above the fire. As the water falls, large drops are formed. These drops, in turn, break into smaller and smaller drops. Once a drop reaches a certain size, it does not break any more. The details of this phenomenon are to be investigated experimentally. Using the Buckingham (Vaschy) pi theorem, find the similarity parameters (i.e., the dimensionless parameters) associated with this phenomenon. Before reading the hint, what *dimensional* parameters did you think are important?

8.35 To set posts in sand, holes are drilled by pushing a water hose, with the water running, into the sand as shown in the sketch. Find the similarity parameters (i.e., the dimensionless parameters) associated with this hole-drilling technique. Before reading the hint, what *dimensional* parameters did you think are important? *Hint:* Important dimensional parameters include hose diameter, water density, water velocity, sand particle diameter, sand shear stress, gravitational acceleration, and hole depth.

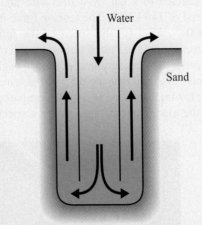

8.36 Consider a ship fuel tank with a free surface above the fuel. Rolling of the ship generates a complex motion of the fuel within the tank. The moving fuel, in turn, generates forces that act on the tank. These forces are measured on a scale model. Important dimensional parameters are tank width, fuel depth, gravitational acceleration, fuel density, fuel viscosity, roll frequency, and roll amplitude (m). Determine the dimensionless parameters associated with this problem. How do the forces measured for the model relate to the full-size ship forces?

8.37 The size of raindrops depends on the following drop parameters: diameter, speed, density, trajectory length, and surface tension. Also important is the gravitational acceleration. Determine the dimensionless parameters associated with this problem.

8.38 In some oil fields, crude oil may be mixed with pebbles, gravel, and sand at the wellhead. To separate the oil from the solids, oil flows through settling tanks. Experimental investigations of these tanks utilize water instead of oil so that the settling process may be viewed. Using water also eliminates the continuous cleaning of instrumentation required when using crude oil. Use the Buckingham (Vaschy) pi theorem to find the similarity parameters (i.e., the dimensionless parameters) associated with this problem.

Table 8A.1   Primary Dimensions for Common Thermal-Fluid Quantities

Quantity	Symbol	SI Units	Primary Dimensions MLTθ	FLTθ
Acceleration	a	m/s^2	LT^{-2}	LT^{-2}
Angle	θ, ϕ	rad	None	None
Angular velocity	$\dot{\omega}$	rad/s	T^{-1}	T^{-1}
Area	A	m^2	L^2	L^2
Convective heat-transfer coefficient	h	W/m$^2\cdot$K	$MT^{-3}\theta^{-1}$	$FL^{-1}T^{-1}\theta^{-1}$
Density	ρ	kg/m^3	ML^{-3}	$FL^{-4}T^2$
Energy	E or U	J	ML^2T^{-2}	FL
Enthalpy	H	J	ML^2T^{-2}	FL
Force	F	N	MLT^{-2}	F
Gravitational acceleration	g	m/s^2	LT^{-2}	LT^{-2}
Heat	Q	J	ML^2T^{-2}	FL
Heat flux	$\dot{Q}''$	W/m^2	MT^{-3}	$FL^{-1}T^{-1}$
Heat transfer rate	$\dot{Q}$	W	ML^2T^{-2}	FLT^{-1}
Kinematic viscosity	ν	m^2/s	L^2T^{-1}	L^2T^{-1}
Length	L	m	L	L
Mass	M	kg	M	$FL^{-1}T^2$
Mass flow rate	$\dot{m}$	kg/s	MT^{-1}	$FL^{-1}T$
Moment	M	N$\cdot$m	ML^2T^{-2}	FL
Power	$\dot{W}$	J	ML^2T^{-3}	FLT^{-1}
Pressure	P	Pa	$ML^{-1}T^{-2}$	FL^{-2}
Shear stress	τ	Pa	$ML^{-1}T^{-2}$	FL^{-2}
Speed of sound	a	m/s	LT^{-1}	LT^{-1}
Specific heat	c_p, c_v	J/kg$\cdot$K	$L^2T^{-2}\theta^{-1}$	$L^2T^{-2}\theta^{-1}$
Surface tension	σ	N/m	MT^{-2}	FL^{-1}
Temperature or temperature difference	$T, \Delta T$	K	θ	θ
Thermal conductivity	k	W/m$\cdot$K	$MLT^{-3}\theta^{-1}$	$FT^{-1}\theta^{-1}$
Thermal diffusivity	α	m^2/s	L^2T^{-1}	L^2T^{-1}
Torque	$\mathcal{T}$	N$\cdot$m	ML^2T^{-2}	FL
Velocity	V	m/s	LT^{-1}	LT^{-1}
Viscosity	μ	N$\cdot$s/m^2	$ML^{-1}T^{-1}$	$FL^{-2}T$
Volume	$\mathcal{V}$	m^3	L^3	L^3
Volume flow rate	$\dot{\mathcal{V}}$	m^3/s	L^3T^{-1}	L^3T^{-1}
Volumetric coefficient of expansion	β	K^{-1}	θ^{-1}	θ^{-1}
Work	W	J	ML^2T^{-2}	FL

choose a parameter that you wish to consider as a dependent parameter. For example, the convective heat-transfer coefficient would be considered a dependent parameter in the problem presented at the outset of this chapter.

Step 5: Define the m dimensionless groups by forming a product of each $q_{l,\text{base}}$ raised to an unknown power and one of the remaining dimensional variables in turn, that is,

$$\Pi_1 = [q_{1,\text{base}}^{a_1} q_{2,\text{base}}^{b_1} \cdots q_{m,\text{base}}^{c_1}] q_1^1,$$

$$\Pi_2 = [q_{1,\text{base}}^{a_2} q_{2,\text{base}}^{b_2} \cdots q_{m,\text{base}}^{c_2}] q_2^1,$$

$$\vdots \qquad \vdots$$

$$\Pi_m = [q_{1,\text{base}}^{a_m} q_{2,\text{base}}^{b_m} \cdots q_{m,\text{base}}^{c_m}] q_m^1.$$

Step 6: Solve for the exponents appearing in each Π group by requiring the power product to be dimensionless. (Example 8A.1 illustrates the details of carrying out this step.)

Step 7: Write out each Π group using the exponents determined in step 6. Verify that each Π group is dimensionless.

We now illustrate this procedure with a couple of examples.

Example 8A.1

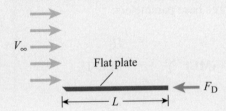

V_∞

Flat plate

F_D

L

Apply the Buckingham pi theorem to determine the dimensionless parameters associated with the problem of measuring the drag force on flat plates in water-tunnel tests. Assume all plates are of unit width and that the flow is parallel to the plate surface (i.e., the angle of attack is zero).

Solution

We apply the procedure just outlined step by step.

Step 1: We identify parameters affecting the drag force F_D on a flat plate as follows:

$$F_D = f(V_\infty, \rho, \mu, \varepsilon, L).$$

The roughness height ε is included because, as we will see later (Chapter 9), this parameter is important in turbulent flows. Since all plates are of unit width, the only important plate dimension is the length L. A total of six parameters are involved (i.e., $n = 6$).

Step 2: Using Table 8A.1, we identify the primary dimensions associated with each of the six parameters:

F_D	V_∞	ρ	μ	ε	L
MLT^{-2}	LT^{-1}	ML^{-3}	$ML^{-1}T^{-1}$	L	L

From these, we see that three primary dimensions, **M, L,** and **T,** are involved (i.e., $k = 3$).

Step 3: The minimum number of dimensionless parameters (Π groups) involved is $m = n - k = 6 - 3 = 3$.

Step 4: We choose three base parameters: $V_\infty, \rho,$ and L. Note that the time dimension (**T**) appears in V_∞ but not in either of the other two base parameters; thus, it is impossible to create a dimensionless parameter from this set of base parameters.

Steps 5 and 6: We now determine the three dimensionless parameters using the remaining parameters ($\mu, F_D,$ and ε), in turn, with the three base parameters. By using the viscosity μ, the first dimensionless parameter is formed:

$$V_\infty^a \qquad \rho^b \qquad L^c \quad \mu^1 \qquad\qquad = \quad \Pi_1$$

$$(LT^{-1})^a \quad (ML^{-3})^b \quad (L)^c \quad (ML^{-1}T^{-1})^1 \quad = \quad M^0T^0L^0.$$

Since the exponents of each primary dimension must sum to zero to assure that Π_1 is dimensionless, we write

M: $b + 1 = 0,$

L: $a - 3b + c - 1 = 0,$ and

T: $-a - 1 = 0.$

Solving this simple set of three equations yields $a = b = c = -1$; thus,

$$\Pi_1 = V_\infty^{-1} \rho^{-1} L^{-1} \mu^1 = \frac{\mu}{\rho V_\infty L}.$$

The second dimensionless parameter is found by following the same procedure. We now use F_D with the three base parameters:

$$V_\infty^a \qquad \rho^b \qquad L^c \qquad F_D^1 \qquad = \qquad \Pi_2$$
$$(\mathbf{LT^{-1}})^a \quad (\mathbf{ML^{-3}})^b \quad (\mathbf{L})^c \quad (\mathbf{MLT^{-2}})^1 \quad = \quad \mathbf{M^0L^0T^0},$$

and we have

M: $b + 1 = 0,$

L: $a - 3b + c + 1 = 0,$ and

T: $-a - 2 = 0.$

Solving these yields $a = -2$, $b = -1$, and $c = -2$; thus,

$$\Pi_2 = V_\infty^{-2} \rho^{-1} L^{-2} F_D^1 = \frac{F_D}{\rho V_\infty^2 L^2}.$$

The third and final dimensionless group involves the roughness height. Proceeding as before, we have

$$V_\infty^a \qquad \rho^b \qquad L^c \qquad \varepsilon^1 \qquad = \qquad \Pi_3$$
$$(\mathbf{LT^{-1}})^a \quad (\mathbf{ML^{-3}})^b \quad (\mathbf{L})^c \quad (\mathbf{L})^1 \quad = \quad \mathbf{M^0L^0T^0}$$

with

M: $b = 0,$

L: $a - 3b + c + 1 = 0,$ and

T: $-a = 0.$

Since $a = b = 0$, $c = -1$; thus,

$$\Pi_3 = V_\infty^0 \rho^0 L^{-1} \varepsilon^1 = \frac{\varepsilon}{L}.$$

Step 7: We have already identified the three Π groups in step 6. Using SI units, we now verify that each is dimensionless:

$$\Pi_1 = \frac{\mu}{\rho V_\infty L} \; [=] \; \frac{(\text{kg/m} \cdot \text{s})}{(\text{kg/m}^3)(\text{m/s})(\text{m})} = 1,$$

$$\Pi_2 = \frac{F_D}{\rho V_\infty^2 L^2} \; [=] \; \frac{(\text{kg} \cdot \text{m/s}^2)}{(\text{kg/m}^3)(\text{m/s})^2(\text{m})^2} = 1,$$

$$\Pi_3 = \frac{\varepsilon}{L} \; [=] \; \frac{\text{m}}{\text{m}} = 1.$$

Each Π group is dimensionless, as required.

Comments The significance of our results is that the dimensionless drag force can be expressed as a function of two, and only two, dimensionless parameters, the Reynolds number (the reciprocal of Π_1) and the relative roughness; that is,

$$\frac{F_D/L^2}{\rho V_\infty^2} = f\left(\frac{\rho V_\infty L}{\mu}, \frac{\varepsilon}{L}\right).$$

With the introduction of a factor of 1/2 in the denominator, our dimensionless drag force, $(F_D/L^2)/\rho V_\infty^2$, is converted to the traditional **drag coefficient** for a flat plate, c_D, and so

$$c_D \equiv \frac{F_D/L^2}{\frac{1}{2}\rho V_\infty^2}.$$

A plot of c_D versus Re_L with ε/L as a parameter is shown in Fig. 9.19 in Chapter 9.

Example 8A.2

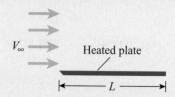

Apply the Buckingham pi theorem to determine the dimensionless parameters associated with the problem of convective heat transfer from a heated flat plate of length L in a uniform flow. Assume that the average convective heat-transfer coefficient $\bar{h}_{conv}$ is an important dependent variable.

Solution

We proceed as in the previous example.

Step 1: Having some previous knowledge of the important parameters involved in this problem, we assume that

$$\bar{h}_{conv} = f(V_\infty, \rho, \mu, k, c_p, \varepsilon, L).$$

Including $\bar{h}_{conv}$, we have identified $n = 8$ dimensional parameters.

Step 2: The primary dimensions associated with the eight parameters are (Table 8A.1) the following:

$\bar{h}_{conv}$	V_∞	ρ	μ	k	c_p	ε	L
$MT^{-3}\theta^{-1}$	LT^{-1}	ML^{-3}	$ML^{-1}T^{-1}$	$MLT^{-3}\theta^{-1}$	$L^2T^{-2}\theta^{-1}$	L	L

From this table we see that $k = 4$ dimensions are involved: **M, L, T,** and **θ**.

Step 3: The minimum number of dimensionless parameters involved is $m = n - k = 8 - 4 = 4$.

Step 4: We choose the same base parameters as before, V_∞, ρ, and L, and add to these the thermal conductivity k. Note that our choice of a fourth parameter was restricted to those variables containing the temperature dimension **θ** (i.e., $\bar{h}_{conv}$, k, and c_p). Any other choice would allow a dimensionless parameter to be formed directly from the base parameters. (We chose k so that our dependent group would contain both $\bar{h}_{conv}$ and k to form the Nusselt number. Choosing c_p would create a different, but altogether acceptable, dimensionless parameter.)

Steps 5 and 6: We determine the four dimensionless parameters using the remaining parameters, $\overline{h}_{conv}$ μ, c_p, and ε, one by one, with the four base parameters. Starting with $\overline{h}_{conv}$, we form the first dimensionless group:

$$V_\infty^a \qquad \rho^b \qquad L^c \qquad k^d \qquad\qquad \overline{h}_{conv}^1 \qquad = \quad \Pi_1$$

$$(LT^{-1})^a \quad (ML^{-3})^b \quad (L)^c \quad (MLT^{-3}\theta^{-1})^d \quad (MT^{-3}\theta^{-1}) \quad = \quad M^0L^0T^0\theta^0.$$

Summing the exponents of each primary dimension on each side of the equation yields

M: $b + d + 1 = 0$,

L: $a - 3b + c + d = 0$,

T: $-a - 3d - 3 = 0$, and

θ: $-d - 1 = 0$.

Solving for a, b, c, and d yields $a = b = 0$, $c = 1$, and $d = -1$; thus,

$$\Pi_1 = V_\infty^0 \rho^0 L^1 k^{-1} \overline{h}_{conv}^1 = \frac{\overline{h}_{conv} L}{k}.$$

The remaining Π groups are found similarly. From

$$V_\infty^a \qquad \rho^b \qquad L^c \qquad k^d \qquad\qquad \mu^1 \qquad\qquad \equiv \quad \Pi_2$$

$$(LT^{-1})^a \quad (ML^{-3})^b \quad (L)^c \quad (MLT^{-3}\theta^{-1})^d \quad (ML^{-1}T^{-1})^1 \quad = \quad M^0L^0T^0\theta^0,$$

We get

M: $b + d + 1 = 0$,

L: $a - 3b + c + d - 1 = 0$,

T: $-a - 3d - 1 = 0$, and

θ: $-d = 0$.

Thus, $a = b = c = -1$ and $d = 0$, and therefore

$$\Pi_2 = V_\infty^{-1} \rho^{-1} L^{-1} k^0 \mu^1 = \frac{\mu}{\rho V_\infty L}.$$

The third Π group is formed as follows:

$$V_\infty^a \qquad \rho^b \qquad L^c \qquad k^d \qquad\qquad c_p^1 \qquad\qquad \equiv \quad \Pi_3$$

$$(LT^{-1})^a \quad (ML^{-3})^b \quad (L)^c \quad (MLT^{-3}\theta^{-1})^d \quad (L^2T^{-2}\theta^{-1}) \quad = \quad M^0L^0T^0\theta^0,$$

and so

M: $b + d = 0$,

L: $a - 3b + c + d + 2 = 0$,

T: $-a - 3d - 2 = 0$, and

θ: $-d - 1 = 0$.

Thus, $a = b = c = 1$ and $d = -1$ and therefore

$$\Pi_3 = V_\infty^1 \rho^1 L^1 k^{-1} c_p^1 = \frac{\rho V_\infty L c_p}{k}.$$

The fourth Π group is formed as follows:

$$V_\infty^a \qquad \rho^b \qquad L^c \qquad k^d \qquad\qquad \varepsilon^1 \qquad\qquad \equiv \quad \Pi_4$$

$$(LT^{-1})^a \quad (ML^{-3})^b \quad (L)^c \quad (MLT^{-3}\theta^{-1})^d \quad (L)^1 \quad = \quad M^0L^0T^0\theta^0,$$

which gives

M: $b + d = 0$,

L: $a - 3b + c + d + 1 = 0$,

T: $-a - 3d = 0$, and

θ: $-d = 0$.

Thus, $a = b = d = 0$ and $c = -1$, and therefore

$$\Pi_4 = V_\infty^0 \rho^0 L^{-1} k^0 \varepsilon^1 = \frac{\varepsilon}{L}.$$

Step 7: We now verify that each parameter is dimensionless:

$$\Pi_1 = \frac{\overline{h}_{conv} L}{k} \; [=] \; \frac{(W/m^2 \cdot K)m}{W/m \cdot K} = 1,$$

$$\Pi_2 = \frac{\mu}{\rho V_\infty L} \; [=] \; \frac{(kg/m \cdot s)}{(kg/m^3)(m/s)(m)} = 1,$$

$$\Pi_3 = \frac{\rho V_\infty L c_p}{k} \; [=] \; \frac{(kg/m^3)(m/s)(m)(J/kg \cdot K)}{(W/m \cdot K)} = 1,$$

$$\Pi_4 = \frac{\varepsilon}{L} \; [=] \; \frac{(m)}{(m)} = 1.$$

Each Π group is dimensionless, as required.

Comments Comparing the dimensionless parameters found here with those found from making the boundary-layer equations dimensionless, we see that Π_1 is the Nusselt number Nu_L, Π_2 is the reciprocal of the Reynolds number Re_L, and Π_3 is the Peclet number Pe_L ($= Re_L Pr$). We would expect to form these particular groups, assuming that our original choice of the dimensional variables includes all of the important ones. From our analysis, we conclude that the original functional relationship, $\overline{h}_{conv} = f(V_\infty, \rho, \mu, k, c_p, \varepsilon, L)$, can be reduced to

$$\frac{\overline{h}_{conv} L}{k} = f\left(\frac{\rho V_\infty L}{\mu}, \frac{\rho V_\infty L c_p}{k}, \frac{\varepsilon}{L} \right),$$

or

$$Nu_L = f(Re_L, Pe_L, \varepsilon/L),$$

where

$$Pe_L = Re_L Pr.$$

We also note that the relative roughness ε/L could have been identified at the outset to simplify the analysis since the only way to make ε dimensionless is to divide by L.

Self Test 8A.1

The drag force F_D experienced by a smooth sphere is thought to depend on the diameter D, the fluid velocity V, the fluid density ρ, and the fluid viscosity μ. Determine the dimensionless parameters associated with these dimensional variables, that is, $F_D = f(D, V, \rho, \mu)$.

(Answer: $F_D/\rho V^2 D^2 = f(\mu/\rho V D)$*)*

BEYOND THE FUNDAMENTALS

EXTERNAL FLOWS: FRICTION, DRAG, AND HEAT TRANSFER

After studying Chapter 9, you should:

- Be able to define the following concepts: boundary layer, bluff body, separation, wake, friction drag, and form drag.

- Be able to explain the differences between forced and free convection.

- Understand what is meant by the transition from a laminar to a turbulent boundary layer, the criteria for its occurrence, and its consequences for friction and heat transfer.

- Be able to recognize and name the conservation equations that apply to laminar boundary-layer flows for both forced and free convection.

- Be able to explain the difference between the friction coefficient and the drag coefficient and the difference between the local heat-transfer coefficient and the average heat-transfer coefficient.

- Be able to sketch the velocity profiles through a boundary layer as it develops in a forced flow over a flat plate and, likewise, for a boundary layer developing in a free convection flow on a vertical flat plate.

- Be able to sketch the temperature profiles through a laminar thermal boundary layer as it develops in a forced flow over a flat plate for conditions of heating or cooling and, likewise, for a thermal boundary layer developing in a free convection flow on a vertical flat plate.

- Be able to explain how and why the Prandtl number affects the relative thickness of the hydrodynamic and thermal boundary layers.

- Be able to calculate the local shear stress and the total drag associated with a flat plate in a uniform flow for laminar, mixed, and turbulent conditions.

- Be able to calculate local and average heat-transfer coefficients for a flat plate for both forced flow and free convection (vertical plate).

- Be able to calculate drag forces and heat-transfer coefficients for spheres, cylinders, and several other selected shapes subjected to a uniform flow.

- Be able to calculate average heat-transfer coefficients for free convection from vertical and horizontal cylinders, inclined and horizontal planar surfaces, and spheres.

Chapter 9 Overview

In this chapter, we describe various external flows and present ways to calculate the friction, drag, and heat transfer associated with such flows. Geometries considered include flat plates, cylinders, spheres, and other two- and three-dimensional shapes. For the simplest of these flows, boundary layers control the friction, drag, and heat transfer at the fluid–solid surface interface. Early in the chapter, we explore these boundary layers in some detail where an agent such as a fan drives the flow (i.e., forced flows). Later in the chapter, we apply a similar analysis to buoyancy-driven flows where density differences give rise to the fluid motion. More complicated flows, however, are not necessarily dominated by boundary layers and are less amenable to analysis. For these, we present empirical correlations to calculate useful engineering quantities. These correlations, as well as results from theoretical analyses, utilize dimensionless parameters. Throughout the chapter, we stress the physical significance of these dimensionless parameters. The chapter also emphasizes how the fundamental conservation principles of mass, momentum, and energy underlie the phenomena of interest.

9.1 HISTORICAL CONTEXT

In 1752, **Jean le Rond d'Alembert** (1717–1783) published a theoretical paper in which he proved that an immersed body should experience no drag force, a result that he recognized was contrary to experiment. D'Alembert wrote [1]:

> *Thus, I do not see, I admit, how one can satisfactorily explain by theory the resistance of fluids. On the contrary, it seems to me that the theory, in all rigor, gives in many cases zero resistance; a singular paradox which I leave to future Geometers for elucidation.*

Helping to resolve this paradox were the studies of **Pierre Louis Georges Du Buat** (1734–1809), a French military engineer [1, 2]. In the late 1700s, Du Buat made detailed pressure measurements over the forward- and rearward-facing surfaces of immersed bodies. These measurements showed that suction exists at the rear of a body that contributes to the drag of a body as much as the positive pressure on the forward portion. (D'Alembert's theory predicted a symmetrical pressure distribution and, hence, no drag force.) The full elucidation of d'Alembert's paradox had to await the arrival of **Ludwig Prandtl** (1875–1953), an extraordinary "Geometer." Prandtl's discovery of the boundary layer [3] was a monumental breakthrough in the understanding of flows over solid surfaces. As a result, he is frequently referred to as the

Jean d'Alembert (1717–1783)

Theodor von Kármán (1881–1963)

father of modern fluid mechanics. Prandlt's idea that the effects of viscosity are limited to a very thin layer near the surface of a body, and that phenomena within the boundary layer can lead to separation (i.e., a condition in which the flow no longer follows the physical surface of the body), introduced the controlling phenomena that were not envisioned in d'Alembert's theory. Many of the developments presented in this chapter have their origins with Prandtl and the work of his students, particularly, **Paul Richard Heinrich Blasius** (1883–1970). Our historical overview of external flow, however, cannot be complete without recognizing the contributions of **Theodor von Kármán** (1881–1963), a faculty colleague of Prandtl at the University of Göttingen in Germany. Von Kármán's contributions are numerous, and his name is associated with the alternating votices shed from cylinders and spheres in a cross-flow: the *von Kármán vortex street*. To do justice to the history associated with this chapter is impossible in this brief summary; we, therefore, refer the interested reader to the excellent survey of Rouse and Ince [1].

9.2 BASIC EXTERNAL FLOW PATTERNS

As the name suggests, an **external flow** is a flow over the outside surface of an object. Common examples of external flows are the flows around automobiles, trucks, and aircraft as they speed along. Similarly, sports balls of all types (baseballs, footballs, soccer balls, etc.) are subjected to external flows when in motion. Rivers flowing around bridge piers, the wind blowing around buildings and over wires and cables, and the fan-driven flow over circuit boards in computers are all examples of external flows. The wind-induced vibration and subsequent collapse of the Tacoma Narrows Bridge in 1940 is a particularly dramatic example of the importance of external flows. Buoyant forces arising naturally from heating or cooling can also generate external flows. For example, an incandescent light bulb creates a rather strong flow with a plume rising above the bulb. People, similarly, create buoyancy-driven external flows. Some illustrations of these free-convection flows have been presented previously in Figs. 4.16 and 7.5.

Depending on the geometry, external flows can be very simple or quite complex. The simpler flows, such as those over thin, flat plates and streamlined bodies, can be treated as boundary-layer flows. Flows that are required to make sharp turns, however, exhibit a variety of behaviors. In the sections that follow, we revisit the boundary layer and introduce so-called bluff-body flows.

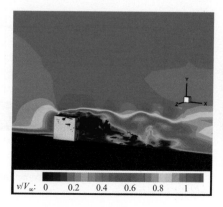

See Fig. 4.16 for a visualization of the thermal plume from a person. Figure 7.5 shows the plume from a hot soldering iron.

Calculated velocity field for wind blowing around an isolated building (left) and streamlines for wind flow through a building complex (right).

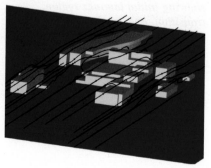

See Fig. 8.5. ▶

FIGURE 9.1

Air bubbles in water make visible this flow of water over a flat plate at zero incidence [4a]. Flow is from left to right with a Reynolds number Re_L ($= \rho V_\infty L/\mu$) of 10,000. The laminar boundary layer thickness is only a few percent of the length of this plate. **Courtesy of H. Werle, Le Tunnel Hydrodynamique au Service de la Recherche Aerospatiale, Pub. No. 156, 1974, ONERA.**

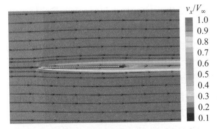

This numerical simulation shows the boundary layers developing on the top and bottom surfaces of a flat plate for the same Reynolds number (10^4) as the flow visualization in Fig. 9.1.

9.2a Boundary Layer Concept Revisited

We introduced the concept of a boundary layer in Chapter 8. In the present chapter, we explore boundary-layer flows in detail. Our desire to understand the physics of such flows is grounded in a need to be able to calculate drag and heat transfer for engineering purposes.

We begin our discussion with a definition:

> A *boundary-layer* flow is a flow in which the effects of viscosity and/or thermal effects are concentrated in a thin region near a surface of interest.

The key word in this definition is *thin*. Figure 9.1 shows a visualization of a boundary layer developing over a flat plate in a parallel, uniform, external flow. The bright lines denote the upper and lower surfaces of the plate, and the leading and trailing edges of the plate are beveled. The thin, dark region immediately above the top surface is the boundary layer. It is within this region that the velocity changes from zero at the plate surface (no-slip condition) to the free-steam value V_∞. Figure 8.5 illustrates these properties of the velocity profile.

The velocity profile itself is used to define the **hydrodynamic boundary-layer**[1] **thickness** δ_H. The hydrodynamic boundary-layer thickness is the perpendicular distance y from the surface of interest to the point in the flow at which the velocity $v_x(x, y)$ is 99% of the free-stream velocity V_∞. Figure 9.2 illustrates the growth of the hydrodynamic boundary-layer thickness with downstream distance x. The boundary-layer thickness is greatly exaggerated in this sketch (cf. Fig. 9.1). Also illustrated in this figure is the transition from a laminar to a turbulent boundary layer. The value of the Reynolds number $Re_{x_{crit}}$ determines where the transition occurs. The characteristic length in this Reynolds number is x_{crit}, the distance from the plate leading edge to the transition location. We return to this later. Upstream of the transition point, the flow is steady and laminar. Downstream of the transition point, the flow is unsteady and turbulent. Figure 9.3 shows a portion of a turbulent boundary layer. Even though the boundary layer is turbulent, it still remains thin in comparison to its length. As suggested by the sketch in Fig. 9.2, the turbulent boundary-layer thickness grows more rapidly with downstream distance than does the laminar thickness. One objective of this chapter is to quantify such relationships.

FIGURE 9.2

Sketch of hydrodynamic boundary-layer development on a flat plate showing initial laminar region, transition, and continued growth as a turbulent layer. The characteristics of a thermal boundary layer are similar to this hydrodynamic layer.

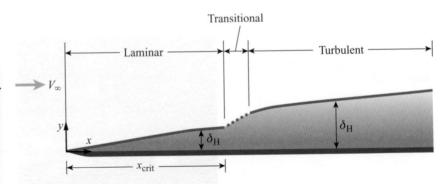

[1] The hydrodynamic boundary layer is sometimes referred to as the *velocity boundary layer*, or just the *boundary layer*.

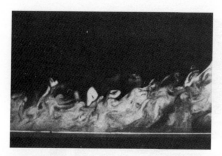

FIGURE 9.3

A section of a turbulent boundary layer growing on a wind-tunnel wall. Light reflecting from small oil droplets shows the flow pattern. Note the chaotic appearance of this flow and the frequent penetration of the outer flow into the regions close to the wall as evidenced by the dark streaks. Copyright© 1982 from Flow Visualization II **(M. R. Head, ed.), Wolfgang Merzkirch. Reproduced by permission of Routledge/Taylor & Francis Group, LLC.**

If the surface on which the hydrodynamic boundary layer grows is hotter or cooler than the free-stream fluid, a thermal boundary layer will also develop. Note that both the hydrodynamic and thermal boundary-layer thickness are functions of the distance from the plate leading edge, x, that is,

$$\delta_H = \delta_H(x),$$
$$\delta_T = \delta_T(x).$$

Furthermore, both of these layers are thin, which can be expressed mathematically as

$$\frac{\delta_H(x)}{x} \ll 1$$

and

$$\frac{\delta_T(x)}{x} \ll 1.$$

This thinness, a key characteristic of boundary layers, enables a great simplification in analyzing real flows by treating the situation as two distinct flow regimes: one close to the surface, where frictional and thermal effects are important, and a second regime outside the boundary layer, where the effects of friction and heat transfer are unimportant. Since Prandtl's discovery of the boundary layer, tremendous strides have been made in understanding real flows and using this understanding in engineering design. We will continue our discussion of boundary layers after we look at other types of external flows.

9.2b Bluff Bodies, Separation, and Wakes

If you have ever walked on a windy day along a city street occupied by tall buildings, you have probably experienced some of the flows created by the wind striking or going around these large structures. Some street locations might be quite peaceful, whereas at others, you are hurried along or stymied by the force of the wind. These external flows are generally not boundary-layer flows.

Objects that are not thin in the direction of flow often greatly disturb the free-stream flow. (Note that a flat plate at zero incidence is a thin object.) Examples of such **bluff bodies** are cars, trucks, buildings, bridge piers, and the palm of your hand held vertically out the window of a moving car. The flow passing around a bluff body cannot follow the sharp contours and thus **separates** from the surface. Figure 9.4 shows separation of the boundary layer from the upper surface of an airfoil at an angle of attack. Separation causes a loss of lift on an aircraft wing and, thus, is to be avoided in this application. Modern aircraft have complex, variable-geometry, wing structures to prevent separation while simultaneously creating high lift forces. Another illustration of separation is shown in Fig. 9.5. Here we see the boundary layer separating from a cylinder. Depending on the specific geometry, a separated flow may, in turn, create a large disturbance downstream of the bluff body called a **wake.** You are likely familiar with the wakes produced by ships, and you have probably been buffeted by the wake of fast-moving trucks and cars as they pass by. By inclining a flat plate at some angle to the incoming flow, the flow separates and a wake is formed, as illustrated in

FIGURE 9.4

At a 7° angle of attack the boundary layer separates from the upper surface of this airfoil, whereas the boundary layer remains attached to the lower surface. Low velocities (top panel) are evident over much of the upper surface. The corresponding pressure distribution (bottom panel) is expressed using the pressure coefficient C_p. A C_p value of zero corresponds to the ambient pressure, with positive pressures indicated by $C_p > 0$ and negative pressures (vacuum) by $C_p < 0$. From the pressure distribution, we see that very little lift (net upward-directed pressure force) is generated at this condition.

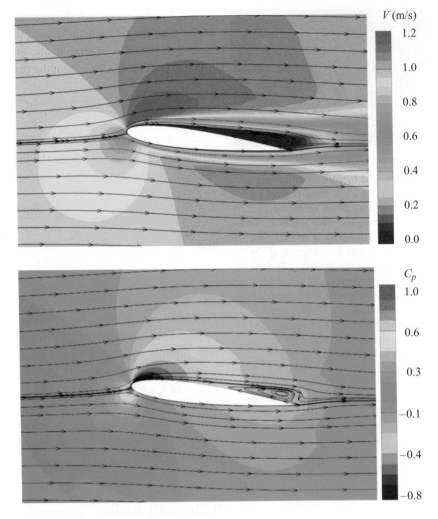

FIGURE 9.5

The boundary layer around this cylinder separates and two counter-rotating vortices form behind the cylinder. The Reynolds number Re_D ($= \rho V_\infty D/\mu$) is 25.

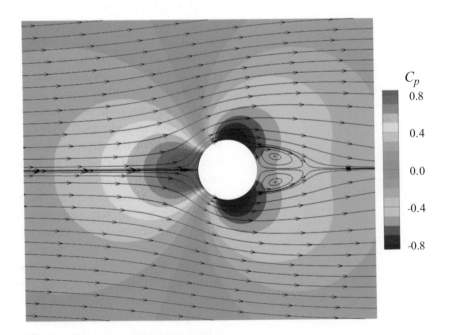

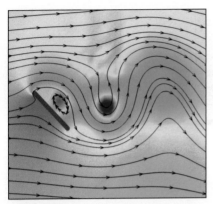

FIGURE 9.6
Streamlines from a numerical (CFD) simulation show a complex wake from a flat plate inclined at 45° from the oncoming flow.

FIGURE 9.7
Full-size truck and trailer are tested in the NASA Ames 80 by 120 ft wind tunnel. **Photograph courtesy of NASA.**

Fig. 9.6. Note the huge disturbance downstream, completely dissimilar to the flow over a plate at zero incidence where boundary-layer flows dominate (cf. Fig. 9.1).

The detailed behaviors of separated flows depend strongly on geometry, and often small changes in geometry can effect large changes in the flow. Automobile manufactures know that great attention must be paid to geometrical details (side-view mirrors, windshield angles, roof shape, etc.) to minimize the drag resulting from separated flows. Wind-tunnel testing of full-size automobiles is now routine in the automobile industry (see Fig. 9.7). As the various flow visualizations suggest (Figs. 9.4–9.6), applying the fundamental conservation principles to calculate the drag and heat transfer associated with bluff bodies is considerably more difficult than dealing with a pure boundary-layer flow. As a consequence, empiricism plays a much greater role in establishing data for engineering design.

In the following sections, we present theoretical and empirical results that can be used to make useful engineering predictions of the drag and heat-transfer associated with flat-plate and selected bluff-body flows.

9.3 FORCED LAMINAR FLOW—FLAT PLATE

Laminar flat-plate boundary-layer flows are sufficiently simple that they can be analyzed accurately using theoretical methods alone, without resorting to empiricism. To make such a bold statement, however, is circular since it relies on the fact that the results of such theoretical analyses have been verified by experiment and shown to be quite accurate.

9.3a Problem Statement

Our objective is to apply the fundamental mass, momentum, and energy conservation principles to a laminar boundary-layer flow to obtain useful mathematical relationships to calculate the following:

- hydrodynamic boundary-layer thickness $\delta_H(x)$,
- thermal boundary-layer thickness $\delta_T(x)$,
- local wall shear stress $\tau_w(x)$,
- total drag force F_D,
- local heat flux $\dot{Q}''_x$, and
- total heat-transfer rate $\dot{Q}$.

These quantities are useful in engineering design.

What information about the flow is required to calculate these quantities? Fortunately, the answer is quite simple. We need only determine the velocity profile through the boundary layer, $v_x(x, y)$, and the temperature profile $T(x, y)$ through the boundary layer. The sketches in Figs. 9.8 and 9.9 help us to see why this is so. In Fig. 9.8, we see the development of the velocity profile through the boundary layer. At the plate leading edge, the profile is uniform, with a discontinuity where the velocity jumps from zero at the surface ($y = 0$) to the free-stream value V_∞ at $y = 0^+$. As can be seen in the middle panel of Fig. 9.8, the theoretical velocity gradient at the wall, $[\partial v_x/\partial y]_{y=0}$, is infinite at the plate leading edge ($x = 0^+$). Since the fluid sticks to the plate surface, a shearing action is created whereby fluid layers are slowed below the free-stream velocity V_∞. This slowing diminishes with the perpendicular distance from the plate until, far from the plate, the local velocity equals the free-stream

FIGURE 9.8

A hydrodynamic boundary layer forms over a flat plate as a result of the no-slip condition at the surface. The boundary-layer thickness $\delta_H(x)$ is shown greatly exaggerated at the top. Also shown are the velocity distributions at the plate leading edge ($x = 0$) and at two positions downstream (x_1, x_2). These are replotted in the form v_x versus y in the middle. The boundary-layer thickness is defined as the y location where $v_x = 0.99\,V_\infty$. The wall shear stress can be obtained from the velocity gradient at the wall (bottom) (i.e., $\tau_w = \mu\,[\partial v_x/\partial y]_{y=0}$).

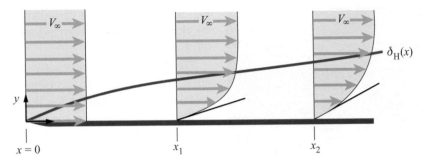

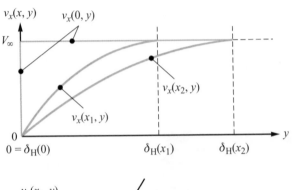

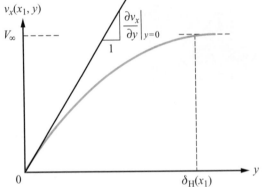

velocity. We formally define the boundary-layer thickness δ_H as the y location where the local velocity is 99% of the free-stream value, that is,

$$\frac{v_x(x, \delta_H)}{V_\infty} = 0.99. \tag{9.1}$$

From this definition, we see that knowing the velocity field $v_x(x, y)$ allows us to determine the boundary-layer thickness $\delta_H(x)$.

In addition to providing the boundary-layer thickness, the velocity field also determines the wall shear stress. From $v_x(x, y)$, we can calculate the velocity gradient at the wall (see bottom panel of Fig. 9.8). This, in turn, allows computation of the local wall shear stress (Eq. 6.3), that is,

$$\tau_w = \mu \frac{\partial v_x(x, y)}{\partial y}\bigg|_{y=0}. \tag{9.2}$$

To add physical perspective, recall our simple scaling of this equation in Chapter 8 (Eq. 8.11):

$$\tau_w \approx \frac{\mu V_\infty}{\delta_H}.$$

FIGURE 9.9
A thermal boundary layer develops
over a flat plate when the plate's
surface temperature T_s differs from
the free-stream temperature T_∞. The
growing thermal boundary layer
thickness $\delta_T(x)$ is shown at the top
along with temperature distributions
superimposed at three x locations for
a heated plate ($T_s > T_\infty$). The same
information is contained in the center
panel where T is shown as a function
of distance from the plate surface, y.
The thermal boundary-layer
thickness is defined as the y location
where $T_s - T = 0.99(T_s - T_\infty)$. The
temperature gradient at the wall
determines the local heat flux
(bottom) (i.e., $\dot{Q}_s'' = -k_f[\partial T/\partial y]_{y=0}$).

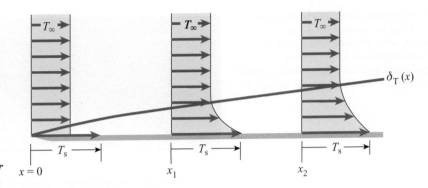

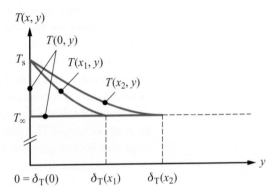

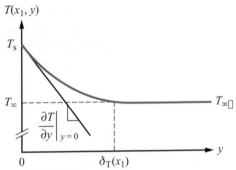

From this, we see that as the boundary layer grows the local shear stress falls. Frequently, the total drag force is of more interest than the local shear stress. We obtain the total force exerted by the fluid on the plate, F_D, by integrating the wall shear stress over the plate surface area. For a plate of length L and width W, the drag force exerted on one side of the plate is given by

$$F_D = \int_A \tau_w \, dA = \int_0^L \tau_w W \, dx. \qquad (9.3)$$

Substituting Eq. 9.2 into Eq. 9.3 yields

$$F_D = \mu W \int_0^L \left. \frac{\partial v_x}{\partial y} \right|_{y=0} dx. \qquad (9.4)$$

To keep the plate stationary, a force of equal magnitude in the opposite direction must counteract this drag force. (See Fig. 9.10.)

In principle, the friction and drag problem is simple: Find $v_x(x, y)$ and apply Eqs. 9.1, 9.2, and 9.4 to determine all of the quantities of engineering interest. Later we will see ways to determine $v_x(x, y)$.

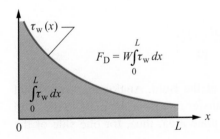

FIGURE 9.10
The force needed to hold a plate
stationary in a moving flow is equal
to the drag force generated by the
fluid shear stress at the surface, but
directed in the opposite (upstream)
direction. The differential drag force
dF_D is shown at various locations on
the top surface of the plate.

The temperature in a thermal boundary layer on a flat plate goes from the surface temperature at the plate surface to the free-stream temperature T_∞. The dimensionless temperature distribution goes from zero (red) at the surface to unity (blue) in the free stream.

$$\frac{T-T_s}{T_\infty-T_s}$$

1.0
0.8
0.5
0.2
0.0

The thermal (heat-transfer) problem is very similar to the hydrodynamic (friction and drag) problem. Figure 9.9 illustrates the development of a thermal boundary layer for the situation in which the plate surface temperature T_s is greater than the free-stream temperature T_∞. Heat is thus transferred from the plate to the fluid. At the plate leading edge, we see again a uniform profile with a discontinuity at the surface. As the total energy transferred to the fluid increases with downstream distance (x), the thermal boundary layer thickens. As a result of this thickening, the absolute value of the temperature gradient at the wall decreases. This can be seen by applying the graphical definition of the slope (see bottom panel of Fig. 9.9) to the three temperature profiles shown in the central panel. Thus, even without any detailed knowledge of the temperature profiles, we expect the local heat flux to decrease with distance from the plate leading edge. Later we quantify this effect.

The thermal boundary-layer thickness δ_T is defined as the perpendicular distance from the plate surface at which the temperature difference $T(x, y) - T_s$ is 99% of the maximum possible temperature difference, $T_\infty - T_s$. More formally, we write

$$\frac{T(x, \delta_T) - T_s}{T_\infty - T_s} = 0.99. \tag{9.5}$$

Like the hydrodynamic problem, if we knew $T(x, y)$, we could determine $\delta_T(x)$. This same knowledge allows us to determine the local surface heat flux by applying Fourier's law (Eq. 4.17):

> **This equation expresses a surface energy balance at the plate–fluid interface.**

$$\dot{Q}_s''(x) = -k_f \frac{\partial T(x, y)}{\partial y}\bigg|_{y=0}, \tag{9.6}$$

where k_f is the thermal conductivity of the fluid. Analogous to the total drag force, we obtain the total heat-transfer rate from the plate to the fluid by integrating the heat flux over the plate area; thus, for one side of the plate,

$$\dot{Q} = \int_A \dot{Q}_s'' dA = \int_0^L \dot{Q}_s'' W dx, \tag{9.7a}$$

or

$$\dot{Q} = -k_f W \int_0^L \frac{\partial T(x, y)}{\partial y}\bigg|_{y=0} dx. \tag{9.7b}$$

We can use Eqs. 9.6 and 9.7 to calculate local and average heat-transfer coefficients, $h_{\mathrm{conv},x}$ and $\bar{h}_{\mathrm{conv}}$, respectively, by applying the definitions of these quantities (Eqs. 4.19 and 4.18):

See Fig. 4.15 and the associated remarks in Chapter 4.

$$h_{\mathrm{conv},x} \equiv Q''_s/(T_s - T_\infty) \tag{9.8}$$

and

$$\overline{h}_{\mathrm{conv}} \equiv \frac{\dot{Q}}{WL(T_s - T_\infty)}. \tag{9.9}$$

Again, to add physical perspective, recall our simple scaling of the local heat-transfer coefficient in Chapter 8 (Eq. 8.12):

$$h_{\mathrm{conv},x} \approx \frac{k_f}{\delta_T}.$$

Here we see that the heat-transfer coefficient falls as the boundary layer grows. This behavior follows the same pattern as the wall shear stress.

Before looking at ways to obtain $v_x(x, y)$ and $T(x, y)$, we cast the problem in dimensionless terms, putting into practice what we learned from Chapter 8. In Chapter 8, we saw from forming the dimensionless governing equations and the dimensionless surface boundary condition that the important dimensionless parameters describing friction and drag are the friction coefficient c_{f,L_c} and the Reynolds number, Re_{L_c}. More explicitly, we derived the dimensionless velocity gradient at the wall related to c_{f,L_c} and Re_{L_c} as follows (see also Table 8.3):

$$\left.\frac{\partial v_x^*}{\partial y^*}\right|_{y^*=0} = \frac{c_{f,L_c} Re_{L_c}}{2}. \tag{9.10}$$

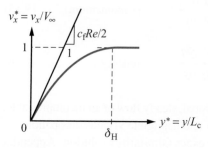

Figure 9.11 illustrates this relationship graphically where we see the dimensionless velocity $v_x^*(= v_x/V_\infty)$ as a function of the dimensionless distance from the wall, $y^* (= y/L_c)$. This plot of dimensionless variables is equivalent to the plot of dimensional variables shown in the bottom panel of Fig. 9.8. A similar situation exists for the heat-transfer problem. In this case, the dimensionless temperature gradient at the wall is defined to be the Nusselt number (see also Table 8.3):

$$\left.\frac{\partial T^*}{\partial y^*}\right|_{y^*=0} = Nu. \tag{9.11}$$

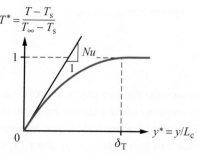

Figure 9.11 also illustrates this relationship. In the bottom panel, we see the dimensionless temperature $T^* [= (T - T_s)/(T_\infty - T_s)]$ as a function of y^*. Our purpose in presenting these relationships (Eqs. 9.10 and 9.11 and Fig. 9.11) is to show that solving the dimensionless problem [i.e., finding $v_x^*(x^*, y^*)$ and $T^*(x^*, y^*)$] is not only equivalent to solving the problem cast in ordinary dimensional variables but is also more general. This generality is manifest in the fact that only one parameter, Re, controls the value of c_f, and that only two parameters, Re and Pr, control the value of Nu; that is,

$$c_{f,L_c} = f_1(Re_{L_c}) \tag{9.12}$$

and

$$Nu_{L_c} = f_2(Re_{L_c}, Pr). \tag{9.13}$$

FIGURE 9.11

Dimensionless velocity (top) and temperature (bottom) profiles for hydrodynamic and thermal boundary layers. The dimensionless velocity gradient at the wall is given by $c_f Re/2$; the dimensionless temperature gradient at the wall defines the Nusselt number Nu.

The use of dimensionless variables takes into account all of the physical properties (ρ, k, c_p, and μ), the geometric and dynamic parameters (L_c and V_∞), and the desired engineering outputs (τ_w and h_{L_c}).

9.3b Solving the Problem

Two methods have traditionally been applied to determine velocity and temperature fields within a laminar boundary layer on a flat plate. The first of these is known as the **similarity solution**. This method provides an *exact* solution to the governing boundary-layer equations. Blausius was the first to apply the similarity solution to the friction and drag problem, publishing his results in 1908 [5]; E. Pohlhausen [6] extended this method to the heat-transfer problem in 1921. The second method provides *approximate* solutions to the hydrodynamic and thermal problems by applying integral control volume analyses. Von Kármán first conceived this **integral method** and presented results for the friction problem in 1921 [7]. In the same year, K. Pohlhausen [8] executed von Kármán's suggestion that the integral method could also be used to solve the heat-transfer problem.

Both the exact and approximate approaches start with the equations governing the conservation of mass, momentum, and energy in a boundary layer. We introduced these equations in Chapter 8, and their derivation from the more general—and much more complex—governing equations is presented in Appendix 9A. The boundary-layer equations are as follows:

$$\frac{\partial v_x}{\partial x} + \frac{\partial v_y}{\partial y} = 0 \qquad \text{(mass)}, \qquad (9.14)$$

$$v_x \frac{\partial v_x}{\partial x} + v_y \frac{\partial v_x}{\partial y} = \frac{\mu}{\rho} \frac{\partial^2 v_x}{\partial y^2} \qquad \text{(x momentum)}, \qquad (9.15)$$

$$v_x \frac{\partial T}{\partial x} + v_y \frac{\partial T}{\partial y} = \frac{k}{\rho c_p} \frac{\partial^2 T}{\partial y^2} \qquad \text{(energy)}. \qquad (9.16)$$

> **Table 8.3 in Chapter 8 shows the dimensionless forms of these equations.**

These equations apply to the two-dimensional, steady flow of an incompressible, Newtonian fluid with constant properties and negligible viscous dissipation.

Next we present an overview of the exact (similarity) solution; Appendix 9B provides a detailed presentation of the approximate solution.

9.3c Exact (Similarity) Solution
Friction and Drag

In the derivation of Eqs. 9.14–9.16, we assume the thermo-physical properties (k, ρ, μ, and c_p) to be constants. This assumption uncouples the momentum and energy equations (i.e., knowledge of the temperature field is not required to solve the momentum equation). Thus we can first find $v_x(x, y)$ and $v_y(x, y)$ from the momentum and continuity equations, and then apply these results to solve the energy equation for $T(x, y)$.

The heart of the similarity solution is the assumption that the velocity profile is geometrically similar at each x location when normalized by the local boundary-layer thickness $\delta_H(x)$. Figure 9.12 illustrates this idea. Here we see velocity profiles at two x locations. Although these two profiles are not identical when v_x is plotted as a function of y (Fig. 9.12 top panel), if y is divided by the local thickness δ_H and v_x divided by V_∞, and these normalized

FIGURE 9.12
The similarity assumption requires that the shape of the dimensionless velocity profile v_x/V_∞ be identical at any x location when the y coordinate is scaled by the local boundary-layer thickness $\delta_H(x)$. The two profiles in the top panel are transformed to a single profile when $v_x(x, y)$ is divided by V_∞ and y is divided by $\delta_H(x)$.

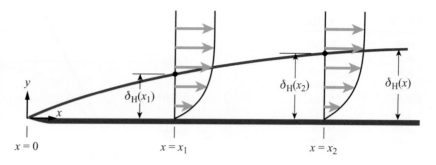

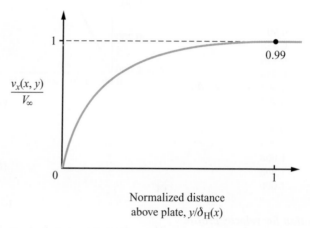

variables plotted, a single curve results (Fig. 9.12 lower panel). Mathematically, this is stated as

$$\frac{v_x}{V_\infty} = f\left(\frac{y}{\delta_H}\right). \qquad (9.17)$$

With this scaling, the partial differential momentum equation (Eq. 9.15) is reduced to a third-order, ordinary differential equation using a transformation of variables that combines x and y to form a new independent variable [i.e., $y(\rho V_\infty/\mu x)^{1/2}$]. The details of the application of this transformation and the solution of the ordinary differential equation are beyond the scope of this book. For more information, we refer the interested reader to Refs. [9–12].

Before presenting the similarity solution for v_x/V_∞, we note that the so-called similarity variable, $y(\rho V_\infty/\mu x)^{1/2}$, can be manipulated to introduce a **local Reynolds number,** defined by

$$Re_x \equiv \rho V_\infty x/\mu. \qquad (9.18)$$

Multiplying and dividing the similarity variable by x yield

$$y(\rho V_\infty/\mu x)^{1/2}\frac{x}{x} = \left(\frac{y}{x}\right)\left(\frac{\rho V_\infty x}{\mu}\right)^{1/2} \equiv \frac{y}{x}Re_x^{1/2}.$$

Figure 9.13 presents the similarity solution showing the functional relationship between v_x/V_∞ and $(y/x)Re_x^{1/2}$ in both tabular and graphical forms. These results agree well with experimental measurements [11]. The velocity profile in Fig. 9.13 can be used to determine how the hydrodynamic boundary-layer thickness grows with distance from the leading edge of the plate. We see that v_x/V_∞ is approximately 0.99 when $(y/x)Re_x^{1/2}$ is 5.0. Since $y \equiv \delta_H$ when $v_x/V_\infty = 0.99$, we conclude that

$(y/x)\,Re_x^{1/2}$	v_x/V_∞
0.00	0.000
0.40	0.133
0.80	0.265
1.20	0.394
1.40	0.456
1.60	0.517
2.00	0.630
2.40	0.729
2.80	0.812
3.00	0.846
3.20	0.876
3.40	0.902
4.00	0.956
4.40	0.976
4.80	0.988
5.00	0.992
5.20	0.994
5.40	0.996
5.60	0.997
6.00	0.999
6.40	1.000
6.80	1.000
7.00	1.000

 $\leftarrow y \cong \delta_H$ (at the 5.00 row)

FIGURE 9.13

Similarity solution for velocity profile in a boundary layer over a flat plate. The dimensionless velocity v_x/V_∞ is approximately 0.99 when $(y/x)\,Re_x^{1/2}$ takes on a value of 5.0, thus defining the boundary-layer thickness δ_H [i.e., $(\delta_H/x)\,Re_x^{1/2} = 5.0$]. The slope at the wall is 0.332. Tabular values are after Howarth [12].

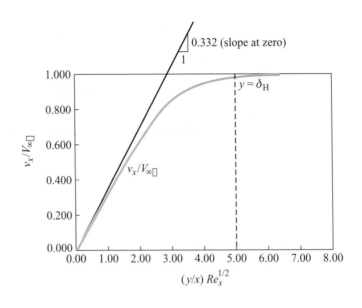

$$(\delta_H/x)\,Re_x^{1/2} = 5.0, \tag{9.19a}$$

or

$$(\delta_H/x) = 5.0\,Re_x^{-1/2}. \tag{9.19b}$$

The x dependence of δ_H from Eq. 9.19 is

$$\delta_H = 5\left(\frac{\mu}{\rho V_\infty}\right)^{1/2} x^{1/2},$$

or

$$\delta_H \propto x^{1/2}. \tag{9.20}$$

Thus, we see that the hydrodynamic boundary-layer thickness grows as the square root of x.[2]

[2] Scaling arguments alone can generate the same result without actually solving the momentum equation for δ_H. See Refs. [11,16].

Example 9.1

Consider a 150-mm-long, thin, flat plate oriented parallel to a uniform flow. Calculate the hydrodynamic boundary-layer thickness at the trailing edge of the flat plate for a flow of air and for a flow of water. The free-stream velocity for both flows is 2 m/s, and the temperature and pressure are 300 K and 1 atm, respectively.

Solution

Known   Fluid, L, V_∞, T, P

Find δ_H

Sketch

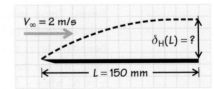

Assumptions

Laminar flow

Analysis   We can use Eq. 9.19b to find the boundary-layer thickness by recognizing that $x = L$. To evaluate Re_L requires values for the density and viscosity, which we obtain from the ideal-gas law and Appendix C for air, and from the NIST database for water:

	ρ (kg/m³)	μ (N·s/m²)	$\nu = \mu/\rho$ (m²/s)
Air	1.177	18.58×10^{-6}	15.79×10^{-6}
Water	996.6	8.5×10^{-4}	0.85×10^{-6}

Thus,

$$Re_L = \frac{\rho V_\infty L}{\mu},$$

$$Re_{L,air} = \frac{1.177\,(2)\,0.150}{18.58 \times 10^{-6}} = 19{,}000,$$

$$Re_{L,water} = \frac{996.6\,(2)\,0.150}{8.5 \times 10^{-4}} = 351{,}700$$

$$[=] \frac{(kg/m^3)(m/s)(m)}{N \cdot s/m^2}\left[\frac{1\,N}{kg \cdot m/s^2}\right] = 1$$

and

$$(\delta_H/L) = 5\,Re_L^{-1/2},$$

$$(\delta_{H,air}/L) = 5\,(19{,}000)^{-0.5} = 0.036,$$

$$(\delta_{H,water}/L) = 5\,(351{,}700)^{-0.5} = 0.0084.$$

The dimensional boundary-layer thicknesses are thus

$$\delta_{H,air} = 0.36\,(150\,mm) = 5.4\,mm,$$

$$\delta_{H,water} = 0.0084\,(150\,mm) = 1.3\,mm.$$

Comments   We first note that the boundary-layer thicknesses are small percentages of the plate length (i.e., less than 4% for air and about 0.8% for water). We also note the influence of the fluid properties. Although the absolute viscosity (μ) of the water is about 46 times greater than that of the air, the much greater density of water results in a much smaller kinematic viscosity (0.85×10^{-6} versus 15.89×10^{-6} m²/s). The smaller kinematic viscosity, in turn, results in a much higher Reynolds number and a thinner boundary layer for the water flow. Lastly we note that the original assumption of laminar flow is justified since Re_L in both cases is less than the critical value for transition to turbulent flow given later in this chapter (i.e., $Re_{crit} = 0.5 \times 10^6$).

Self Test
9.1

✓ Redo Example 9.1 for engine oil. What velocity would be required to have the boundary-layer thickness δ_H for all the fluids equal to 2 mm at a length of 0.15 m?

(Answer: $\delta_{H,oil} = 32.1\,mm$, $V_{air} = 14.7\,m/s$, $V_{water} = 0.80\,m/s$, $V_{oil} = 515.4\,m/s$)

We can also use the information contained in Fig. 9.13 to determine the wall shear stress and the friction coefficient. Finding these is one of our objectives. From Fig. 9.13, we see that the slope of the function v_x/V_∞ at the plate surface is

$$\frac{\partial[v_x/V_\infty]}{\partial[(y/x)Re_x^{1/2}]}\bigg|_{y=0} = 0.332.$$

For a fixed value of x, the velocity gradient at the wall is

$$\frac{\partial v_x}{\partial y}\bigg|_{y=0} = 0.332\,(V_\infty/x)Re_x^{1/2}.$$

The wall shear stress (Eq. 9.2) is then expressed as

$$\tau_w = \mu\frac{\partial v_x}{\partial y}\bigg|_{y=0} = 0.332\,(\mu V_\infty/x)Re_x^{1/2}. \tag{9.21}$$

We now apply the definition of the friction coefficient (Eq. 8.25) to obtain another desired result:

$$c_f \equiv \frac{\tau_w}{\frac{1}{2}\rho V_\infty^2} = \frac{0.332(\mu V_\infty/x)Re_x^{1/2}}{\frac{1}{2}\rho V_\infty^2},$$

or

$$c_f = 0.664\,Re_x^{-1/2}. \tag{9.22}$$

We use these results to calculate the drag force exerted by the fluid on one side of the plate (see Fig. 9.10). Substituting the definition of the local Reynolds number (Eq. 9.18) into Eq. 9.21 yields the following expression for the variation of the wall shear stress with distance from the plate leading edge:

$$\tau_w(x) = 0.332\mu^{1/2}\rho^{1/2}V_\infty^{3/2}x^{-1/2}.$$

We obtain the drag force by integrating this shear stress over the plate area (Eq. 9.3):

$$F_D = \int_A \tau_w dA = \int_0^L \tau_w W dx.$$

$$= 0.332\,\rho^{1/2}\mu^{1/2}V_\infty^{3/2}W\int_0^L x^{-1/2}dx,$$

where W is the plate width and all of the constants have been removed from the integral. Performing the integration and multiplying and dividing by $L\rho V_\infty^2/2$ yield

$$F_D = 1.328\,Re_L^{-1/2}(LW)\frac{1}{2}\rho V_\infty^2, \tag{9.23}$$

where

$$Re_L \equiv \frac{\rho V_\infty L}{\mu}. \tag{9.24}$$

By convention, the drag coefficient c_D for a flat plate is defined as

$$c_D \equiv \frac{F_D/(LW)}{\frac{1}{2}\rho V_\infty^2}. \tag{9.25}$$

Substituting this drag force into this definition yields

$$c_D = 1.328 \, Re_L^{-1/2}. \tag{9.26}$$

Comparing this result to Eq. 9.22, we note that the drag coefficient is twice the value of the friction coefficient at $x = L$; that is,

$$c_D = 2c_{f,L}. \tag{9.27}$$

Physically, there is nothing special about this result, but it is easy to remember and sometimes convenient to employ.

Example 9.2

We revisit the situation described in Example 9.1. Considering the drag on both sides of the plate, determine the force required to hold the plate stationary (a) in a flow of air and (b) in a flow of water. The plate width is 70 mm.

Solution

Known   Fluid, L, W, V_∞, T, P

Find F_D (two sides)

Sketch

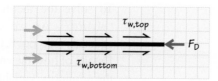

Assumptions

i. Laminar flow
ii. Sharp leading edge with no separation

Analysis   To reinforce the use of dimensionless parameters, we first calculate the drag coefficient (Eq. 9.26) and then apply its definition (Eq. 9.25) to obtain the dimensional quantity of interest, the drag force on the plate. Using the properties and Re_L values for air from Example 9.1, we have

$$c_D = 1.328 \, Re_L^{-0.5}$$
$$= 1.328 \, (19{,}000)^{-0.5} = 0.0096$$

and rearranging Eq. 9.25 yields

$$F_{D,\text{one side}} = c_D \, (LW) \left(\frac{1}{2} \rho V_\infty^2 \right)$$

$$= 0.0096 \, (0.150) \, 0.070 \, (0.5) \, 1.177 \, (2)^2$$

$$= 2.37 \times 10^{-4}$$

$$[=] \, \text{m(m)} \frac{\text{kg}}{\text{m}^3} \left(\frac{\text{m}}{\text{s}} \right)^2 \left[\frac{1 \, \text{N}}{\text{kg} \cdot \text{m/s}^2} \right] = \text{N}.$$

The same drag force is exerted on both sides; thus

$$F_{\text{tot}} = 2 \, F_{D,\text{one side}} = 2(2.37 \times 10^{-4}) \, \text{N}$$
$$= 4.74 \times 10^{-4} \, \text{N}.$$

Repeating the calculations using the properties of water yields

$$Re_L = 351{,}700,$$
$$c_D = 0.00224,$$
$$F_{D,\text{one side}} = 469 \times 10^{-4}\ \text{N},$$

and

$$F_{\text{tot}} = 937 \times 10^{-4}\ \text{N}.$$

Comment   The drag force exerted by the water is nearly 200 times greater than that of the air. The higher density of the water is the primary factor responsible for this result, as the drag force is directly proportional to the density (Eq. 9.25).

  Calculate the drag for the plate in Example 9.2 when the fluid is engine oil. Compare your result to that for water in Example 9.2.
(Answer: $F_{\text{total}} = 2.1$ N. This value is 22 times larger than that for water because of the much larger value for absolute viscosity.)

Having solved the friction and drag problem, we now turn to the heat-transfer problem.

Heat Transfer

To solve the thermal problem, we employ the same similarity variable, $(y/x)Re_x^{1/2}$. As we expect from our dimensional analysis of the energy equation (see Eq. 8.19), the Prandtl number Pr is a parameter in the solution. Details of how the energy equation (Eq. 9.16) is transformed and solved using the similarity variables can be found in Ref. [13]. The similarity solution is shown in Fig. 9.14. Here we see the dimensionless temperature, $T^* = (T(x, y) - T_s)/(T_\infty - T_s)$, as a function of the similarity variable, $(y/x)Re_x^{1/2}$, for various Prandtl numbers. For a fixed value of x, each curve can be interpreted as the dimensionless temperature distribution expressed as a function of the perpendicular distance from the wall. At the surface ($y = 0$), T equals T_s, so T^* is zero; far from the surface, T approaches T_∞, so T^* equals unity. The y value at which $T^* = 0.99$ determines the thermal boundary-layer thickness. For a unity Prandtl number, the function $T^* = f[(y/x)Re_x^{1/2}]$ is identical to that for the hydrodynamic solution, $v_x/V = f[(y/x)Re_x^{1/2}]$. Although the $Pr = 1$ curve in Fig. 9.14 does not extend beyond $(y/x)Re_x^{1/2} = 4$, T^* is approximately

FIGURE 9.14

Similarity methods provide the temperature distribution in a laminar boundary layer over a plate maintained at a temperature T_s. The temperature distribution and its slope at the wall depend on the Prandtl number Pr of the fluid.

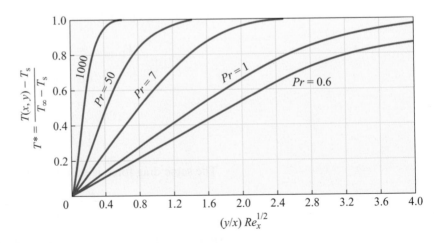

Temperature distributions in thermal boundary layers for Prandtl numbers of 10, 1, 0.1 and 0.01 (top to bottom) at a fixed Reynolds number of 500.

0.99 when $(y/x)Re_x^{1/2}$ is 5.0. That the two solutions are identical for $Pr = 1$ follows directly from the theory. Comparing the dimensionless momentum and energy equations (see Table 8.3), we see that when $Pr = 1$, v_x^* and T^* play the same mathematical roles in these two equations. This correspondence is the basis for the Reynolds analogy discussed later.

Using the results of Fig. 9.14, we can quantitatively define the thermal boundary-layer thickness $\delta_T(x)$. We have already indicated that

$$\frac{\delta_T(x)}{x} = 5.0 Re_x^{-1/2} \quad \text{for} \quad Pr = 1. \tag{9.28a}$$

The effect of the Prandtl number can be included in this relationship by noting that the numerical constant in Eq. 9.28a scales approximately with $Pr^{-1/3}$, that is,

$$\frac{\delta_T(x)}{x} = 5.0 Re_x^{-1/2} Pr^{-1/3}. \tag{9.28b}$$

How do the thermal and hydrodynamic boundary-layer thicknesses compare? For Prandtl numbers greater than unity, the thermal boundary layer is thinner than the hydrodynamic boundary layer (see Fig. 9.15). Physically this results from the fact that the diffusivity for thermal energy (α) is less that the diffusivity for momentum ($\nu = \mu/\rho$) when $Pr\,(= \nu/\alpha) > 1$. Typical Prandtl numbers for several fluids are given in Table 9.1, and detailed tabulations are provided in Appendices F and G. Prandtl numbers can depend on temperature, depending upon the fluid, but exhibit little or no pressure dependence. Note the strong temperature dependence for the engine oil shown in Table 9.1.

We also see from Fig. 9.14 that the magnitude of the slope at the wall of T^* versus $(y/x)Re_x^{1/2}$ increases with Prandtl number; that is, the slope becomes increasingly negative. Since the local heat flux (and h_x) is directly proportional to this slope, the Prandtl number effect is taken into account by correlating the slope with Pr. The result of such correlation [9] is that, for $Pr \gtrsim 0.6$,

$$|\text{slope}| \propto Pr^{1/3}.$$

To obtain a relationship expressing Nu as a function of $Re_x^{1/2}$ (and Pr), we follow a procedure analogous to that of the hydrodynamic problem. Combining Fourier's law with the definition of the local heat-transfer coefficient yields

$$-k_f \left.\frac{\partial T}{\partial y}\right|_{y=0} = h_{\text{conv},x}(T_s - T_\infty).$$

We use Fig. 9.14 to determine the slope $\partial T/\partial y|_{y=0}$ for the unity Prandtl number curve and apply the Prandtl number dependence just given. These substitutions yield

$$Nu_x = 0.332\,Re_x^{1/2} Pr^{1/3} \quad \text{for} \quad Pr \geq 0.6, \tag{9.29}$$

FIGURE 9.15

For unity Prandtl number, the thermal and hydrodynamic boundary layers grow at the same rate [i.e., $\delta_T(x) = \delta_H(x)$]. For large-Prandtl-number fluids ($Pr > 1$), the thermal boundary layer δ_T grows at a slower rate than the hydrodynamic boundary layer δ_H. For small-Prandtl-number fluids ($Pr < 1$), the converse is true.

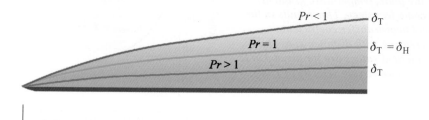

Table 9.1   **Prandtl Numbers for Various Fluids***

Substance	Prandtl Number	
	At 300 K	**At 400 K**
Air	0.707	0.690
Hydrogen	0.701	0.695
Saturated steam	0.857	1.033
Mercury (liquid)	0.0248	0.0163
Saturated water	5.83	1.47
Engine oil	6400	152

*Data from Ref. [9]. Pressure is 1 atm for the gases and the saturation value for the liquids.

where Nu_x is the **local Nusselt number** defined by

$$Nu_x \equiv \frac{h_{\text{conv},x}\, x}{k_{\text{f}}}. \tag{9.30}$$

Providing the details to arrive at Eq. 9.29 is an exercise for the reader (see Problem 9.30).

Although we have been constructing relationships among dimensionless parameters for generality, it is instructive to see how the local convective heat-transfer coefficient varies with distance from the plate leading edge. Combining Eqs. 9.29 and 9.30, we see that $h_{\text{conv},x} \propto x^{-1/2}$. Figure 9.16 shows $h_{\text{conv},x}$ versus x for air at 350 K. In this figure, we see that $h_{\text{conv},x}$ falls rapidly from very high values at the leading edge, and then decreases less rapidly as the boundary layer continues to grow.

As we integrated the local shear stress (or friction coefficient) over the surface of the plate to obtain the total drag force (or drag coefficient), we can integrate the local heat flux over the plate area to obtain the total heat-transfer rate. We then use this rate to define the average Nusselt number. We start by expressing the heat-transfer rate for a differential area of the plate, $W dx$, as

$$d\dot{Q} = \dot{Q}''(x)W dx.$$

The local heat flux relates to the local heat-transfer coefficient as

$$\dot{Q}''(x) = h_{\text{conv},x}(T_{\text{s}} - T_{\infty}).$$

FIGURE 9.16

As the thermal boundary layer grows with distance from the leading edge of the plate, temperature gradients become less steep. This results in the local heat-transfer coefficient decreasing with distance x, as shown. Numerical values are for air at 350 K with a free-stream velocity of 1 m/s.

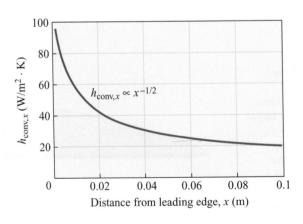

The local heat-transfer coefficient $h_{\text{conv},x}$ is obtained by combining Eqs. 9.29 and 9.30 to yield

$$h_{\text{conv},x} = \frac{k_{\text{f}}}{x} 0.322 Re_x^{1/2} Pr^{1/3},$$

or, with the x dependence explicitly shown,

$$h_{\text{conv},x} = 0.332 k_{\text{f}} \rho^{1/2} \mu^{-1/2} V_\infty^{1/2} Pr^{1/3} x^{-1/2}.$$

The total heat-transfer rate is thus

$$\dot{Q} = \int_0^L \dot{Q}'' W dx = \int_0^L h_{\text{conv},x}(T_{\text{s}} - T_\infty) W dx,$$

or

$$\dot{Q} = \int_0^L 0.322 k_{\text{f}} \rho^{1/2} \mu^{-1/2} V_\infty^{1/2} Pr^{1/3}(T_{\text{s}} - T_\infty) W x^{-1/2} dx.$$

Removing the constants from the integrand, performing the integration, and multiplying and dividing by L yield

$$\dot{Q} = \left\{ 0.664 \frac{k_{\text{f}}}{L} Re_L^{1/2} Pr^{1/3} \right\} WL(T_{\text{s}} - T_\infty).$$

We identify the bracketed quantity as the average heat-transfer coefficient $\bar{h}_{\text{conv},L}$, that is,

$$\bar{h}_{\text{conv}, L} = 0.664 \frac{k_{\text{f}}}{L} Re_L^{1/2} Pr^{1/3}.$$

Using $\bar{h}_{\text{conv},L}$, we define the average Nusselt number, as

$$\overline{Nu_L} \equiv \frac{\bar{h}_{\text{conv},L} L}{k_{\text{f}}} = 0.664 Re_L^{1/2} Pr^{1/3} \quad \text{for} \quad Pr \gtrsim 0.6. \tag{9.31}$$

Comparing Eq. 9.31 with Eq. 9.29, we see that the average heat-transfer coefficient from the plate leading edge to an arbitrary location x is always twice the value of the local heat-transfer coefficient at x, that is,

$$\bar{h}_{\text{conv},x} = 2 h_{\text{conv},x}. \tag{9.32}$$

Note that this relationship between $\bar{h}_{\text{conv},x}$ and $h_{\text{conv},x}$ is identical to that between c_{D} and $c_{\text{f},L}$ for the friction problem (Eq. 9.27).

At this juncture, we have achieved all of our objectives set out at the beginning of this section. We are now armed with the tools needed to find $\delta_{\text{H}}(x)$, $\delta_{\text{T}}(x)$, τ_{w} or $c_{\text{f},x}$, $\dot{Q}''(x)$ or $h_{\text{conv},x}$, and $\dot{Q}$ or $\bar{h}_{\text{conv},L}$. These results are summarized in Table 9.2.

Before discussing the equivalent turbulent flow problem, we introduce the Reynolds analogy, the uniform heat-flux boundary condition, and procedures to apply the results here to real situations.

Reynolds Analogy

As already pointed out, the dimensionless temperature distribution through a laminar boundary layer, $T^* = (T - T_{\text{s}})/(T_\infty - T_{\text{s}})$, is identical to the dimensionless

9.3e Calculation Method

To calculate any of the quantities of engineering interest (τ_w, F_D, $h_{conv,x}$, and $\overline{h}_{conv}$) from the various relationships shown in Table 9.2 requires values for the thermo-physical properties (ρ, μ, k, and c_p). In the solution to the boundary-layer equations, we assumed that the properties were constants, rather than functions of temperature, as they are in reality. This raises the following question: At what temperature should these properties be evaluated? When T_s differs from T_∞, the temperature varies through the boundary layer from T_s at the plate surface to T_∞ in the free stream, suggesting that some mean value be used. Experimental results agree well with theory when the arithmetic mean is used. This mean, called the **film temperature**, is defined as

$$T_f \equiv \frac{T_s + T_\infty}{2}. \tag{9.38}$$

Later in this chapter, we will see other ways that the temperature dependence of properties is taken into account for other external flows. Use of the film temperature, however, is one of the easier and more common ways.

After establishing the property values, a second important step is to verify that all of the parameters (usually just Re and Pr) are within the valid range of the expression. For example, the following restrictions apply to the similarity solution to the laminar, flat-plate problem:

$$Re_x < Re_{crit},$$

and

$$0.6 \lesssim Pr \lesssim 50,$$

where Re_{crit} is the critical Reynolds number at which transition to turbulent flow begins. We consider the phenomenon of transition after the following example.

Example 9.3

Consider a 70-mm-wide, 150-mm-long, flat plate. Air at 300 K and 1 atm flows over the plate at a velocity of 2 m/s. The upper surface of the plate is maintained at a temperature of 320 K by electrical resistance heaters, whereas the lower surface is insulated. Determine the heat flux at the trailing edge of the plate and the total heat-transfer rate. Repeat for a flow of water at the same conditions. Compare the results obtained for the two fluids and discuss the origins of the differences.

Solution

Known Fluid, T_s, T_∞, P, V_∞, L, W

Find $\dot{Q}''(x = L)$, $\dot{Q}_{0-L}$

Sketch

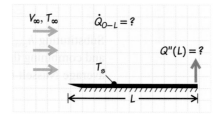

Assumptions

  i. Laminar flow

  ii. No separation at plate leading edge

Analysis   We begin by determining the required thermo-physical properties evaluated at the film temperature (Eq. 9.38);

$$T_f = (T_s + T_\infty)/2$$
$$= (320 + 300)/2\,\mathrm{K} = 310\ \mathrm{K}.$$

We collect the desired properties for air and water in the table that follows. Properties for air, other than density, are obtained by interpolation in Table C.3; properties for H_2O are obtained from the NIST online database. The air density is obtained by the application of the ideal-gas equation of state (Eq. 2.28b).

	ρ (kg/m³)	μ (N·s/m²)	k (W/m·K)	c_p (J/kg·K)	$Pr = \mu c_p/k$
Air	1.1398	19.059×10^{-6}	0.0269	1007.0	0.713
Water	993.38	6.935×10^{-4}	0.6261	4179.2	4.629

To evaluate the heat flux at the trailing edge ($x = L$), we apply Eq. 9.8,

$$\dot{Q}''(L) = h_{\mathrm{conv},x}(L)(T_s - T_\infty),$$

where Eqs. 9.29 and 9.30 are used to obtain $h_x(L)$ (i.e., h_L). To apply these, we need to evaluate Re_x at $x = L$:

$$Re_L = \frac{\rho V_\infty L}{\mu} = \frac{1.1398(2)(0.15)}{19.059 \times 10^{-6}} \qquad \text{(Eq. 9.18)}$$
$$= 17{,}940 \text{ for air.}$$

We note that since Re_L is less than Re_{crit} ($\approx 500{,}000$) the flow is laminar over the entire plate. To find $h_{\mathrm{conv},x}(L)$, we evaluate

$$Nu_x = 0.332 Re_x^{1/2} Pr^{1/3}$$

for $x = L$ to get

$$Nu_{x=L} = 0.332(17{,}940)^{0.5}(0.713)^{1/3}$$
$$= 39.7 \text{ (dimensionless)}$$

and apply the definition of the local Nusselt number (Eq. 9.30):

$$h_{\mathrm{conv},x} = k\,Nu_x/x,$$

or

$$h_{\mathrm{conv},L} = k\,Nu_L/L$$
$$= 0.0269\,(39.7)/0.15 = 7.12 \text{ for air}$$
$$[=]\ (\mathrm{W/m \cdot K})/m = \mathrm{W/m^2 \cdot K}.$$

The trailing-edge heat-flux is thus

$$\dot{Q}''(L) = 7.12(320 - 300) = 142$$
$$[=]\ (\mathrm{W/m^2 \cdot K})K = \mathrm{W/m^2}.$$

The total heat-transfer rate is found from Eq. 9.9,

$$\dot{Q}_{0-L} = \overline{h}_{conv,L} WL(T_s - T_\infty),$$

where the average heat-transfer coefficient for the plate, $\overline{h}_{conv,L}$, is recognized to be twice the local value at $x = L$ (Eq. 9.32) or is evaluated from Eq. 9.31. Thus,

$$\dot{Q}_{0-L} = 14.24(0.07)0.15(320 - 300)$$
$$= 3.0$$
$$[=](W/m^2 \cdot K)m(m)K = W.$$

All of these relationships apply for the water flow. Results for both fluids are summarized in the following table where the last line is the ratio of the values for water to those for air.

	Re_L	Nu_L	$h_{conv,L}$ (W/m²·K)	$\dot{Q}''(L)$(W/m²)	$\dot{Q}_{0-L}$(W)
Air	17,940	39.7	7.12	142	3.0
Water	429,700	363	1500	30,300	636
Ratio (water:air)	24.0	9.1	210	213	210

Comment We see from this table that, for the same flow velocity and temperature difference, the heat flux and heat-transfer rate for the water flow are more than 200 times that associated with the air flow. What factors are responsible for this? We first see that the Nusselt number for the air flow is about 10 times that for the water flow. This is a result of (a) the smaller kinematic viscosity of the water, which gives a larger Reynolds number (by a factor of 24), and (b) a larger Prandtl number. Since both Re and Pr are raised to fractional powers, the combined effect is lessened. The largest single factor causing the heat transfer for the water to be so much greater is differences in the thermal conductivities. From Eq. 9.30, we see that the heat-transfer coefficient is proportional to the product of the Nusselt number and the thermal conductivity; and from our property data table, we see that k_{H_2O} (0.6261 W/m·K) is more than 20 times k_{air} (0.0269 W/m·K). Understanding how fluid properties relate to heat-transfer coefficients is helpful in developing engineering intuition.

Self Test 9.3

Redo Example 9.3 for engine oil. Compare this result to that of the air and water and discuss.

(Answer: $Nu_L = 161.1$, $h_{conv,L} = 155.7 \, W/m^2 \cdot K$, $\dot{Q}''(L) = 3114 \, W/m^2$, $\dot{Q}_{0-L} = 65.4 \, W$. Even though oil has a huge Prandtl number, the large viscosity yields a low Reynolds number, which, in turn, results in a moderate Nusselt number with the overall heat transfer falling between that of air and water.)

9.4 FORCED TURBULENT FLOW—FLAT PLATE

9.4a Transition Criteria

At the beginning of this chapter, we indicated that a flat-plate boundary layer initially develops as a laminar flow, and at some location downstream, it undergoes a transition to turbulent flow (Fig. 9.2). To eliminate the need to deal with any details of the transition, we assume that transition from laminar to turbulent conditions occurs abruptly at a critical Reynolds number Re_{crit}. Values of the critical Reynolds number depend on the turbulence levels in the free stream, the roughness of the surface, any vibrations present, and other factors. The specific effect of free-stream turbulence is shown in Ref. [15]. Well-controlled laboratory experiments show that

$$3.2 \times 10^5 \leq Re_{crit} \leq 10^6. \tag{9.39}$$

For practical situations where conditions are less controlled, Re_{crit} may be as low as 50,000 [16]. Common practice assumes the easily remembered value for the critical Reynolds of one-half million (5×10^5).

Sometimes we desire to know the distance from the plate leading edge to the transition location, x_{crit}. Using the definition of Re_{crit}, x_{crit} can be estimated as

$$x_{crit} = \frac{\nu}{V_\infty} Re_{crit}, \tag{9.40a}$$

or

$$x_{crit} = 5 \times 10^5 \, \nu/V_\infty, \tag{9.40b}$$

where we have assumed the half-million value for Re_{crit} and $\nu \, (= \mu/\rho)$ is the kinematic viscosity.

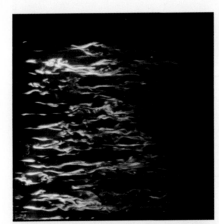

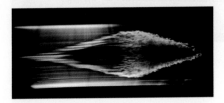

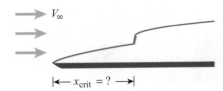

Turbulent spots (bottom) develop in the transition from laminar to turbulent flow near the wall in boundary-layer flows. In fully-developed turbulent boundary layers, the near-wall region contains low-speed streaks (top), streamwise elongated regions of slower-than-average flow. Flow is from left to right for both images. Photographs courtesy of M. Gad-el-Hak.

Example 9.4

Consider a flat plate in a uniform flow. Estimate the distance from the leading edge of the plate to the transition location, x_{crit}, for flows of the following fluids at atmospheric pressure: air, water, motor oil, and glycerin. The free-stream velocity is 2.5 m/s and the fluid film temperature is 300 K.

Solution

Known   Fluid, V_∞, T_f, P

Find x_{crit}

Sketch   See Fig. 9.17.

Assumptions

  i. No separation at the leading edge
 ii. $Re_{crit} = 5 \times 10^5$

Analysis   With our assumptions, we use Eq. 9.40b directly to find x_{crit} for each fluid. We need only find values for ρ and μ, or ν, to evaluate

$$x_{crit} = \frac{5 \times 10^5 \mu}{\rho V_\infty} = \frac{5 \times 10^5 \nu}{V_\infty}.$$

FIGURE 9.17

In modeling the drag on a smooth flat plate, we assume a laminar boundary layer grows initially, followed by an abrupt transition to turbulent flow at x_{crit}. At the transition location, the wall shear increases discontinuously and then decreases continuously toward the end of the plate ($x = L$).

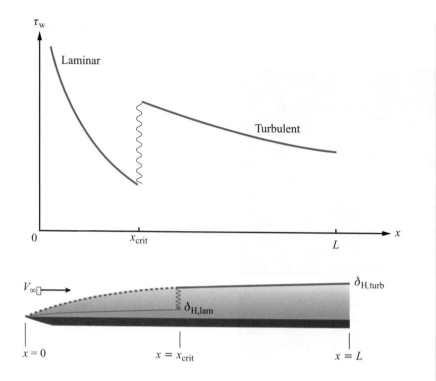

The following table lists property values and their sources together with a numerical evaluation of this expression:

Fluid	Source	ρ (kg/m³)	μ (N·s/m²)	ν (m²/s)	x_{crit}(m)
Air	Table C.3	1.177	18.58×10^{-6}	1.579×10^{-5}	3.2
Water	NIST	996.6	8.538×10^{-4}	8.568×10^{-7}	0.17
Oil	Table G.2	884.1	0.486	5.5×10^{-4}	110
Glycerin	Table G.2	1259.9	0.799	6.34×10^{-4}	127

Comments These four fluids exhibit a huge range of transition lengths: 0.17 m to 127 m! Water has the smallest value because of its relatively low viscosity (compared to the other two liquids) and relatively large density (compared to air).

Self Test 9.4

✓ For the conditions of Example 9.4, recalculate the distance to the transition location for oil at 400 K. What is the major cause for the change in location?

(Answer: $x_{crit} = 2.12$ m. This dramatic decrease in x_{crit} is almost entirely due to the large decrease in viscosity with increased temperature.)

9.4b Velocity Profiles and Boundary-Layer Development

A key feature of turbulent flows is their enhanced transport of momentum and energy compared to laminar flows. This enhanced transport is a consequence of the relatively large scale motion associated with the turbulent eddies rapidly carrying momentum, or energy, over large distances. The effect of turbulence on the time-mean velocity profile is to steepen velocity gradients

FIGURE 9.18
*The velocity profile for a turbulent boundary layer is much steeper near the wall than that of a laminar boundary layer. The y coordinate for both profiles is scaled by the **laminar boundary-layer thickness** $\delta_{H,lam}$ to show that the turbulent boundary-layer thickness is approximately 3.8 times that of the laminar boundary layer. The local Reynolds number for both is 5×10^5.*

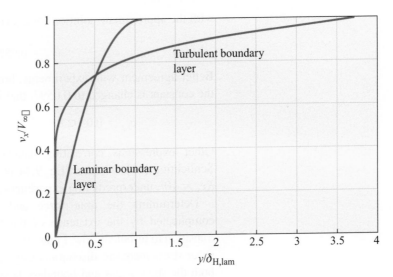

at the surface and to cause the boundary layer to extend farther from the surface in comparison to the laminar boundary layer. Figure 9.18 illustrates these two effects where turbulent and laminar velocity profiles are compared at the same local Reynolds number, $Re_x = 5 \times 10^5$. For this condition, the turbulent boundary layer is almost four times as thick as the laminar one. In spite of this increased thickness, the velocities in the rear-wall region ($y/\delta_{H,lam} < 0.5$) are substantially greater for the turbulent boundary layer. As we will see later, the greater momentum flow near the surface has the important consequence of delaying separation of turbulent boundary layers in comparison to their laminar counterparts.

The turbulent velocity profile can be approximated by the following power law [11]:

$$\frac{v_x(x, y)}{V_\infty} = \left(\frac{y}{\delta_H(x)} \right)^{1/7}. \tag{9.41}$$

With the scaling of y with the local boundary-layer thickness $\delta_H(x)$, the dimensionless profiles are once again *similar*. (In Chapter 10, we discuss other velocity profiles that are more general than Eq. 9.41.) To obtain an expression for the boundary-layer thickness $\delta_H(x)$, an integral momentum analysis (Appendix 9B) is performed using Eq. 9.41 in conjunction with the following experimental result of Blasius [5] for the wall shear stress:

$$\tau_w = 0.0225\rho V_\infty^2 \left(\frac{\nu}{V_\infty \delta_H} \right)^{1/4}. \tag{9.42}$$

The result of such an analysis is

$$\frac{\delta_H(x)}{x} = 0.37 Re_x^{-1/5}. \tag{9.43}$$

To arrive at Eq. 9.43, we assume that the turbulent boundary layer grows starting at the leading edge of the plate, ignoring any laminar or transitional regions. Thus, Eq. 9.43 provides approximate values for $\delta_H(x)$ beyond $x = x_{crit}$.

9.4c Friction and Drag

To determine an expression for the friction coefficient, we combine the experimental measurements of Blasius [5] for the wall shear stress, Eq. 9.42,

> **See Fig. 9.13 and Table 9.2 for laminar boundary-layer profiles.**

with the approximate solution for $\delta_H(x)$, Eq. 9.43, to yield

$$c_{f,x} = 0.0577\, Re_x^{-1/5}.$$

Better agreement with experiments, however, is obtained when the value of the constant is changed to 0.0592, that is, with

$$c_{f,x} = 0.0592\, Re_x^{-1/5} \quad \text{for} \quad 5 \times 10^5 \leq Re_x \leq 10^7. \tag{9.44}$$

Other expressions correlating data beyond $Re_x = 10^7$ are presented in Schlichting [11]; however, Eq. 9.44 is a reasonable approximation to about $Re_x = 10^8$, underpredicting experimental results, say, by 10%–15%.

Determining the drag force and, hence, the drag coefficient c_D is complicated by the existence of the laminar boundary layer prior to the transition to turbulent flow. Figure 9.17 illustrates this situation where the wall shear stress increases discontinuously at the transition location. The jumps in both the shear stress and boundary-layer thickness at x_{crit} are consistent with the assumption that the turbulent boundary layer behaves as if it began at the leading edge. The velocity profiles of Fig. 9.18 explicitly show this increase in boundary-layer thickness for $Re_{crit} = 5 \times 10^5$. To find the total drag force thus requires that we break the integration of the shear stress over the plate into two parts: a laminar portion and a turbulent portion. Thus we write

$$F_D = \int_A \tau_w(x)\, dA$$

$$= \int_0^{x_{crit}} \tau_{w,lam} W\, dx + \int_{x_{crit}}^L \tau_{w,turb} W\, dx.$$

We use the definition of $c_{f,x}$ to express τ_w as

$$\tau_w = c_{f,x} \frac{1}{2} \rho V_\infty^2,$$

where $c_{f,x}$ is obtained from Eq. 9.22 for the laminar case and from Eq. 9.44 for the turbulent case. Performing the integration and applying the definition of the drag coefficient (Eq. 9.25),

$$c_D \equiv \frac{F_D/LW}{\frac{1}{2}\rho V_\infty^2},$$

yields

$$c_D = 0.074 Re_L^{-1/5} - 1742 Re_L^{-1}, \tag{9.45}$$

where $Re_{crit} = 5 \times 10^5$ is used to determine x_{crit}. If the boundary layer is *tripped* and forced to be turbulent from the plate leading edge, we can use Eq. 9.45 by ignoring the second term on the right-hand side.

See Table 9.2 for a summary of boundary-layer relationships for turbulent flow over smooth, flat plates.

The final topic of this section is how to deal with plates that are rough. As discussed in Chapter 8, the roughness height ε influences both friction and heat transfer in turbulent flows, enhancing both over corresponding values for smooth plates.

In a classic experiment, Nikuradse [17] glued various-sized sand grains in tubes to establish the effect of wall roughness on the resistance to flow.

Table 9.3   Apparent Roughness Heights of Commercial Pipes and Tubes

Material (New)	Roughness Height, ε (mm)
Glass	0.0003 or smooth
Drawn tubing	0.0015
Commercial steel or wrought iron	0.046
Asphalted cast iron	0.12
Galvanized iron	0.15
Cast iron	0.26
Wood stave	0.18–0.9
Concrete	0.3–3.0
Riveted steel	0.9–9.0

Moody [18] used Nikuradse's data to establish apparent roughness heights for various materials, which are shown in Table 9.3. Prandtl and Schlichting [11,19] extended the results of Nikuradse's pipe-flow experiments to flat plates. Figure 9.19 summarizes their work. Here we see the drag coefficient plotted as a function of Reynolds number with the inverse relative roughness, L/ε, used as a parameter. At any given Reynolds number, the effect of roughness is to increase the drag coefficient; note that as the roughness height increases, the parameter L/ε decreases. The physical reason that roughness increases drag is that, if the roughness elements are large enough to penetrate

FIGURE 9.19

The drag coefficient for a flat plate shown as a function of Reynolds number Re_L for laminar, transitional, and turbulent flow. Also shown is the effect of plate roughness ε. **Reprinted from Ref. [20] with permission.**

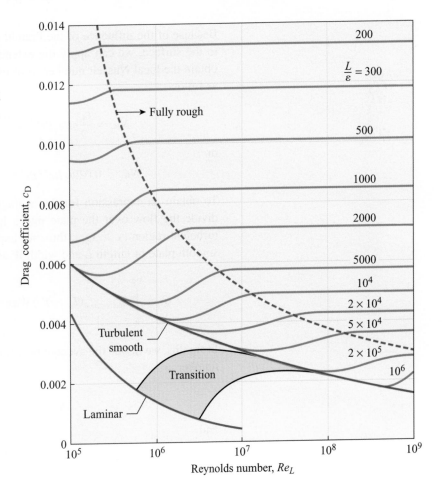

through the **laminar sublayer** immediately adjacent to the wall, an additional resistance results from form drag. (Form drag is discussed later in this chapter.) As long as the roughness elements are fully contained within the laminar sublayer, they create no additional resistance. Since the thickness of the laminar sublayer decreases with the Reynolds number, departures of c_D from the smooth curve occur at increasingly larger Reynolds numbers as roughness decreases. We also see from Fig. 9.19 a *fully rough* region where the drag coefficient is independent of Reynolds number. Figure 9.19 in conjunction with Table 9.3 suffice to calculate the drag force on rough plates for the purposes of this book. More sophisticated methods, however, are provided in Refs. [15,16,19, 20].

9.4d Heat Transfer

Unlike the hydrodynamic problem, where a single universal velocity profile results close to the surface, the thermal problem is complicated by the dependence of the temperature profile on the molecular Prandtl number. The thickness of the boundary layers, however, is controlled by the turbulent motion rather than molecular properties (e.g., ν and α). Thus, except for the case of liquid metals ($Pr \ll 1$), the hydrodynamic and thermal boundary-layer thicknesses are roughly equal for turbulent flow, (i.e., $\delta_H \approx \delta_T$). Therefore, we can use Eq. 9.43 to estimate both the hydrodynamic and thermal boundary-layer thicknesses, so

$$\frac{\delta_T}{x} = 0.37 Re_x^{-1/5}. \tag{9.46}$$

Because of the influence of the Prandtl number in the viscous layer adjacent to the surface, we can apply the extended Reynolds analogy (Eq. 9.33b) to obtain the local Nusselt number from the local friction coefficient (Eq. 9.44) as follows:

$$Nu_x = \frac{c_{f,x}}{2} Re_x Pr^{1/3} = \frac{0.0592 \, Re_x^{-1/5}}{2} Re_x Pr^{1/3},$$

or

$$Nu_x = 0.0296 \, Re_x^{4/5} Pr^{1/3} \quad \text{for} \quad 0.6 \lesssim Pr \lesssim 60. \tag{9.47}$$

To obtain an expression for the average heat-transfer coefficient, we again divide the flow over the plate into a laminar region ($0 < x < x_{crit}$) and a turbulent region ($x > x_{crit}$); thus, we express the total heat-transfer rate for a smooth plate of length L and width W as

$$\dot{Q} = \int_0^{x_{crit}} h_{conv,x,lam}(T_s - T_\infty) \, W dx + \int_{x_{crit}}^{L} h_{conv,x,turb}(T_s - T_\infty) \, W dx.$$

From this, we define the average heat-transfer coefficient and average Nusselt number as

$$\overline{h}_{conv} = \frac{\dot{Q}}{LW(T_s - T_\infty)}$$

and

$$\overline{Nu_L} = \frac{\overline{h}_{conv} L}{k_f},$$

respectively. Using Eqs. 9.29 and 9.47 for the local heat-transfer coefficients, $h_{conv,x,lam}$ and $h_{conv,x,turb}$, respectively, in the integration yields

$$\overline{Nu_L} = (0.037\, Re_L^{4/5} - 871)Pr^{1/3} \quad \text{for} \quad 0.6 \lesssim Pr \lesssim 60, \qquad (9.48)$$

where a critical Reynolds number of 5×10^5 is assumed. If the boundary layer is tripped to be turbulent from the leading edge by roughness or other means, or if $L \gg x_{crit}$, then the constant 871 in Eq. 9.48 can be neglected.

Because the Reynolds analogy does not apply for rough plates (since there is no mechanism for heat transfer analogous to the form drag created by the roughness elements), Fig. 9.19 cannot be used for the heat-transfer problem. For the interested reader, Mills [16] gives a comprehensive set of relationships for heat transfer from rough plates with roughness elements of various shapes.

9.4e Uniform Surface Heat Flux

As discussed for laminar flow, we need to consider a uniform heat flux at the surface boundary condition in addition to that of a fixed temperature. For the uniform heat flux condition, the local Nusselt number is about 4% larger than that for fixed temperature and is given by [9]

$$Nu_x = 0.0308\, Re_x^{4/5} Pr^{1/3} \quad \text{for} \quad 0.6 \lesssim Pr \lesssim 60. \qquad (9.49)$$

The definitions of the local and average heat-transfer coefficients, Eqs. 9.35 and 9.36, can be employed to calculate the local surface temperature T_s and the mean temperature difference $\overline{T_s - T_\infty}$. For the latter, correlations for $\overline{Nu_L}$ for the constant surface temperature condition are sufficiently accurate to determine $\overline{h}_{conv,L}$ [9].

Example 9.5

An ice floe—a large, flat sheet of free-floating ice—is exposed to a warm ($-5°C$), strong (35 miles/hr) wind on the Bering Sea. The temperature of the surface of the ice floe exposed to the air is $-30°C$, and the atmospheric pressure is 100 kPa. The ice flow is roughly rectangular with a length of 20 m and a width of 15 m. Estimate the drag force exerted by the wind on the ice floe. Also estimate the total heat-transfer rate from the air to the ice floe.

Solution

Known $V_\infty, T_\infty, T_s, P, L, W$

Find $F_D, \dot{Q}$

Sketch

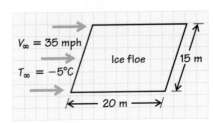

Assumptions

i. The floe can be modeled as a flat plate.

ii. The drag force exerted by the wind on the edges of the floe can be ignored.

iii. The velocity of the ice floe is small compared to the wind velocity such that $V_{rel} \approx V_\infty$.

Analysis With our assumptions, we can calculate F_D *and* $\dot{Q}_{0-L}$ using drag and heat-transfer relationships developed for a flat plate. We first calculate the Reynolds number to ascertain whether the flow regime is laminar, predominantly turbulent, or mixed laminar and turbulent. Properties are evaluated at the film temperature, which is

$$T_f = \frac{T_\infty + T_s}{2} = \frac{-5 + (-30)}{2} = -17.5°C \text{ or } 255.5 \text{ K.}$$

From the ideal-gas law (Eq. 2.28b), we calculate the density to be

$$\rho = \frac{P}{RT} = \frac{100,000}{287(255.5)} \text{ kg/m}^3 = 1.364 \text{ kg/m}^3.$$

Interpolating in Table C.3, we obtain the following air properties:

$$\mu(255.5 \text{ K}) = 16.36 \times 10^{-6} \text{ N·s/m}^2,$$
$$k(255.5 \text{ K}) = 0.023 \text{ W/m·K,}$$
$$Pr(255.5 \text{ K}) = 0.714.$$

The kinematic viscosity is

$$\nu = \frac{\mu}{\rho} = \frac{16.36 \times 10^{-6}}{1.364} \text{ m}^2/\text{s} = 1.20 \times 10^{-5} \text{ m}^2/\text{s.}$$

The Reynolds number based on the length of the floe is

$$Re_L = \frac{\rho V_{rel} L}{\mu} \cong \frac{\rho V_\infty L}{\mu}$$

$$= \frac{1.364\,(15.6)20}{16.36 \times 10^{-6}} = 2.60 \times 10^7$$

$$[=] \frac{(\text{kg/m}^3)\,(\text{m/s})\text{m}}{(\text{N·s/m}^2)} \left[\frac{1 \text{ N}}{\text{kg·m/s}^2} \right] = 1,$$

where the wind speed of 35 miles/hr was converted to SI units (15.6 m/s). We see that the boundary layer above the flow is predominantly turbulent since

$$Re_L \gg Re_{crit}(\sim 5 \times 10^5),$$

or

$$x_{crit} = \frac{\nu}{V_\infty} Re_{crit} = \frac{1.20 \times 10^{-5}}{15.6} 5 \times 10^5 \text{ m} = 0.38 \text{ m,}$$

and

$$\frac{x_{crit}}{L} = \frac{0.38}{20} = 0.019.$$

Therefore, the appropriate relationship expressing the drag coefficient is Eq. 9.45:

$$c_D = 0.074 \, Re_L^{-1/5} - 1742 \, Re_L^{-1}$$

$$= 0.074\,(2.60 \times 10^7)^{-0.2} - 1742\,(2.60 \times 10^7)^{-1}$$
$$= 0.00236.$$

The drag force is calculated from the definition of the drag coefficient as follows:

$$F_D = c_D \frac{1}{2} \rho V_\infty^2 LW$$

$$= 0.00236(0.5)1.364(15.6)^2\,20(15)$$

$$= 117.5$$

$$[=]\frac{\text{kg}}{\text{m}^3}\left(\frac{\text{m}}{\text{s}}\right)^2 \text{m(m)}\left[\frac{1\,\text{N}}{\text{kg}\cdot\text{m/s}^2}\right] = \text{N}.$$

To estimate the heat-transfer rate, we use Eq. 9.48 to evaluate the average heat-transfer coefficient:

$$\overline{Nu_L} \equiv \frac{\overline{h}_{conv} L}{k} = (0.037\,Re_L^{4/5} - 871)Pr^{1/3}$$

$$= [0.037(2.60 \times 10^7)^{0.8} - 871](0.714)^{1/3}$$

$$= 27{,}498$$

and

$$\overline{h}_{conv} = \frac{k}{L}\overline{Nu_L} = \frac{0.023}{20}27{,}498 = 31.6$$

$$[=]\frac{\text{W/m}\cdot\text{K}}{\text{m}} = \text{W/m}^2\cdot\text{K or W/m}^2\cdot{}^\circ\text{C}.$$

The total heat-transfer rate from the air to the ice is then given by (Eq. 9.9)

$$\dot{Q}_{0-L} = \overline{h}_{conv}(LW)(T_\infty - T_s)$$

$$= 31.6(20)15[-5 - (-30)]\,\text{W}$$

$$= 237{,}000\,\text{W or }237\,\text{kW}.$$

Comments   We first note that engineering judgment was used to model the ice floe as a flat plate, hence leading to the approximate nature of our final results. In reality, the floe will not be rectangular and the surface will not be perfectly smooth. We also note that it is likely that the wind will also exert an additional drag force on the floe where the wind strikes the vertical front surface (i.e., a bluff-body component to the total drag). Last, we note that the submerged portion of the ice flow experiences drag forces associated with water flow and that the combined effects of the wind and water define the dynamics ($F = Ma$) of the ice floe.

Example 9.6

A circuit board in an electronic control system has a large number of embedded chip resistors such that the top surface experiences an essentially uniform heat flux of 850 W/m². The lower surface is insulated. Air flows over the top surface at 4 m/s with a free-stream temperature of 300 K. The circuit board is 300 mm long. Determine the location and value of the maximum top surface temperature of the board with and without a boundary-layer trip wire. A "trip" at the leading edge of the plate results in a turbulent boundary layer starting at the leading edge (see sketch).

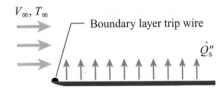

V_∞, T_∞

Boundary layer trip wire

$\dot{Q}''_s$

Solution

Known $Q''_s, V_\infty, T_\infty, L$

Find $T_{s,max}, x_{T_{s,max}}$

Assumptions

   i. The trip wire results in a turbulent boundary layer.

   ii. $P \simeq P_{atm} \simeq 100$ kPa.

Analysis   We begin with the definition of a local heat transfer coefficient (Eq. 9.8),

$$\dot{Q}''_s = h_{conv,x}(T_s(x) - T_\infty).$$

From this definition, we see that the maximum surface temperature will occur at the location where the local heat-transfer coefficient $h_{conv,x}$ is a minimum since the product of $(T_s(x) - T_\infty)$ and $h_{conv,x}$ must equal the constant value of $\dot{Q}''_x$. From Fig. 9.16, we observe that $h_{conv,x}$ falls with distance from the leading edge of the plate for a laminar boundary layer. The same general behavior also results for a turbulent boundary that starts from the leading edge. We thus conclude that the location of the maximum surface temperature will be at the trailing edge of the plate $(x = L)$. To evaluate $T_s(L)$, we need to determine a value for $h_{conv,x}$ at $x = L$ (i.e., $h_{conv,L}$). Since the surface temperature is unknown, we evaluate the thermophysical properties at T_∞, recognizing that an iteration may be required using properties evaluated at the film temperature $T_f = (T_s(L) + T_\infty)/2$. Using properties from Table C.3, we have

$$Re_L = \frac{V_\infty L}{\nu} = \frac{4(0.3)}{1.579 \times 10^{-5}} = 76,000$$

$$[=]\frac{m/s(m)}{m^2/s} = 1.$$

Since $Re_L < Re_{crit}$, we expect that in the absence of the trip wire the boundary layer will remain laminar over the entire length of the plate. For this case, Eq. 9.34 can be used to evaluate the local heat-transfer coefficient:

$$Nu_L = 0.453 \, Re_L^{1/2} Pr^{1/3}$$

$$= 0.453(76,000)^{0.5}(0.713)^{1/3}$$

$$= 111.6$$

and

$$h_{conv,L} = \frac{k}{L}Nu_L$$

$$= \frac{0.0262}{0.3}(111.6) = 9.74$$

$$[=]\frac{W/m \cdot K}{m} = W/m^2 \cdot K,$$

where air properties are obtained from Table C.3. Using this value of $h_{conv,L}$, we obtain the maximum temperature for the untripped plate from the definition of the local heat-transfer coefficient (Eq. 9.8 or 9.35a):

$$T_s(L) = \frac{\dot{Q}''_s}{h_{conv,L}} + T_\infty$$

$$= \frac{850}{9.74} + 300 = 387.3 \text{ K}$$

$$[=] \frac{\text{W/m}^2}{\text{W/m}^2 \cdot \text{K}} = \text{K}.$$

The calculation can now be repeated using properties evaluated at the film temperature,

$$T_f = (387.3 + 300)/2 = 344 \text{ K}.$$

This is left as an exercise for the reader.

For the tripped boundary layer, the appropriate heat-transfer correlation is Eq. 9.49:

$$Nu_L = 0.0308 \, Re_L^{0.8} Pr^{1/3}$$
$$= 0.0308(76,000)^{0.8}(0.713)^{1/3}$$
$$= 220.9,$$

and

$$h_{\text{conv},L} = \frac{k}{L} Nu_L$$

$$= \frac{0.0262}{0.3}(220.9)\,\text{W/m}^2 \cdot \text{K} = 19.3 \text{ W/m}^2 \cdot \text{K}.$$

For this case, the maximum surface temperature is

$$T_s(L) = \frac{\dot{Q}''_s}{h_{\text{conv},L}} + T_\infty \qquad h_{conv} = \frac{Q}{T_s - T_\infty}$$

$$= \frac{850}{19.3} + 300 \text{ K} = 344.0 \text{ K}.$$

Again, iteration is required to obtain a more accurate result.

Comments Because of the higher heat-transfer coefficient associated with the turbulent boundary layer, the maximum surface temperature is significantly reduced over the laminar case (i.e., 344 K versus 387 K).

Self Test 9.5 A 10-m/s wind blows along a 8-m-long by 4-m-high wall on the side of a house. The local air pressure and temperature are 101 kPa and 5°C, respectively. Assuming the wall is maintained at a constant temperature of 15°C, determine the drag force on the wall and the rate of heat loss.

(Answer: $F_D = 5.99$ N, $\dot{Q}_{0-L} = 7.54$ kW)

9.5 FORCED FLOW—OTHER GEOMETRIES

We now consider flows over bluff bodies, or other configurations, where boundary layers may separate. As mentioned at the beginning of this chapter, complex flow patterns can develop downstream of a separation point. As a consequence, empirical relationships are usually employed to calculate the drag and heat transfer for these flows. After introducing the concept of form drag, we then present these relationships.

9.5a Friction and Form Drag

In our discussion of momentum conservation in Chapter 6, we saw that two types of forces act on a surface: viscous forces and pressure forces. The components of these two forces in the direction of the flow are the drag force. The drag resulting from the unbalanced pressure forces is termed **form drag.** In the previous sections, we considered a flat plate aligned with the flow. For this situation, viscous shearing of the fluid at the plate surface creates all of the drag force. For a very thin plate, essentially no pressure force acts in the positive or negative flow directions. For a plate whose surface is perpendicular to the oncoming flow, all of the drag results from unbalanced pressure forces, with no contribution from fluid friction. Figure 9.20 shows the flow pattern and pressure distribution associated with a disk facing a uniform flow. Here the integrated pressure force on the surface facing the flow is substantially greater than that on the downstream surface. To keep the plate stationary, an external force pointed upstream—the drag force—is required.

For the general case, we obtain the net pressure and viscous forces acting on an arbitrary body by an integration over the surface area A:

$$\boldsymbol{F}_{\text{net}} = \int_A -P\hat{\boldsymbol{n}}\,dA + \int_A \boldsymbol{\tau}_{\text{w}}\,dA. \tag{9.50a}$$

For a flow directed in the x-direction, the drag force is then the x-component of this net force,

$$F_{\text{D}}\,(x\text{-}direction) \equiv F_{\text{net},\,x\text{-component}}. \tag{9.50b}$$

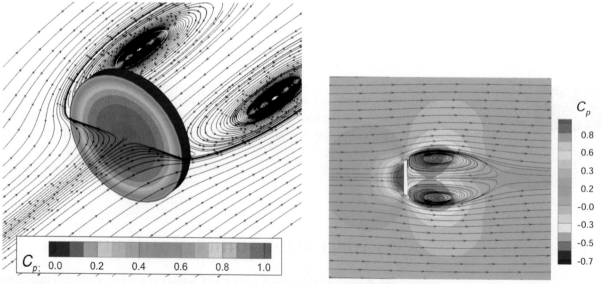

FIGURE 9.20

The $(Re = 10^5)$ flow around a disk held perpendicular to the oncoming flow separates at the edge, forming a complex flow pattern in the wake. As a result, a much lower pressure is exerted on the rear of the plate, creating a drag force. The pressure distribution is shown in color in dimensionless form where $C_p \equiv (P - P_\infty)/[(1/2)\rho V_\infty^2]$. The side view shows a strong positive pressure on the upstream face and somewhat less than ambient pressure (P_∞) at the downstream face.

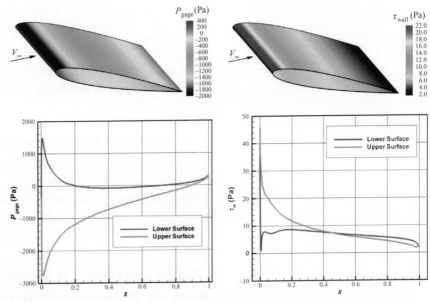

CFD results for $Re = 10^6$ show that pressure forces dominate the lift and drag for an airfoil (NACA 0012) at a 5° angle of attack. Here wall shear stresses (right) are about two orders of magnitude less than surface pressures (left).

Similarly, we define a **lift force** as the component of the net force directed perpendicular to the flow direction. Flying creatures and flying devices of all types depend on this lift force for flight. In spite of the interesting applications of lift, it is beyond the scope of this book.

If our theories were sufficiently complete, calculating the drag for any body would simply be a matter of computing the pressure and viscous shear-stress distributions over the external surface and then applying Eqs. 9.50. Although tremendous progress is being made in this regard, particularly with developments in CFD, we still rely heavily upon experimental data to determine drag and lift forces.

For the flows we wish to consider, dimensional analysis shows that two dimensionless groups are needed to deal with the drag problem for smooth surfaces: the drag coefficient and the Reynolds number. We therefore expect correlations of the type

$$c_D \equiv f(Re_{L_c}). \qquad (9.51a)$$

The drag coefficient is usually defined as

$$c_D \equiv \frac{F_D/A_p}{\frac{1}{2}\rho V_\infty^2}, \qquad (9.51b)$$

where A_p is the **projected (or frontal) area.** This area is the silhouetted area seen by an upstream observer looking downstream at the body of interest. For example, the projected area of a sphere with radius R is the circular area πR^2; whereas for a cylinder whose axis is perpendicular to the flow, the projected area is the rectangular area $2RL$, where L is the length of the cylinder. For complex shapes, such as automobiles, the frontal area can be obtained from a head-on photograph, for example. In the actual design, of course, more sophisticated methods would be employed.

The Reynolds number in Eq. 9.51a depends on a characteristic length L_c, that is,

The projected, or frontal, area of an automobile depends on the viewing angle. Compared to a condition with no wind, a crosswind increases the drag force exerted on a cruising automobile because of a higher velocity, a larger projected area, and an increased drag coefficient.

$$Re_{L_c} \equiv \frac{\rho V_\infty L_c}{\mu}, \tag{9.52}$$

This length is defined for each configuration and care must be exercised to use the correct length in solving problems.

9.5b Empirical Correlations for Drag and Heat Transfer

In this section, we present empirical correlations to calculate drag forces and convective heat-transfer rates for a selected number of two- and three-dimensional bodies. Much more information is available in the literature than is presented here. For readers desiring a more in-depth treatment, we cite textbooks that present comprehensive listings of drag and heat-transfer correlations.

Cylinders and Other 2-D Shapes

At very low Reynolds number (**creeping flow,** $Re_D \le 1$), essentially all of the drag force results from fluid friction, whereas at very high Reynolds numbers ($Re_D \le 5 \times 10^6$), form drag dominates. In between these extremes, friction and form drag both contribute to the total in varying degrees. As we see in Fig. 9.21, the drag coefficient for a cylinder, c_D, decreases with Reynolds number up to $Re_D \approx 2000$, after which c_D increases slightly and then remains essentially constant at a value of about 1.2 for $10^4 \le Re_D \le 10^5$. In the range $10^5 < Re_D < 10^6$, some very interesting behavior occurs. Here we see a dramatic drop in the drag coefficient from a value of about 1.2 to a value of about 0.3. This rapid drop in c_D is termed the **drag crisis.**

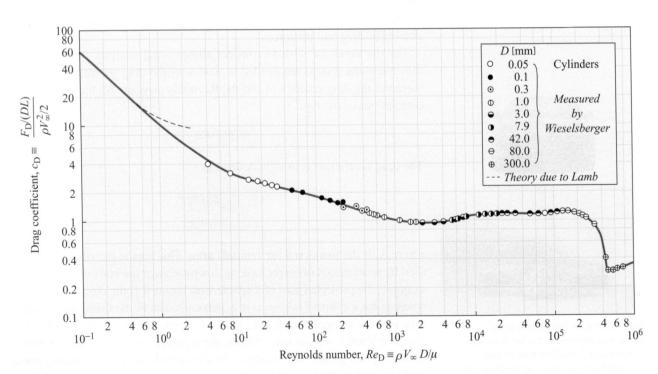

FIGURE 9.21

The drag coefficient for a circular cylinder in a cross flow shows a complicated behavior as a function of Reynolds number Re_D. Note the so-called drag crisis in the range $10^5 < Re_D < 10^6$. **Reprinted from Ref. [11] with permission.**

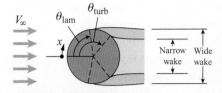

FIGURE 9.22
The laminar boundary layer separates before the equator of a cylinder ($\theta_{lam} = 80°$) and produces a wide wake. Because of its greater momentum flow near the surface, the turbulent boundary layer separates downstream of the equator ($\theta_{turb} \approx 140°$, say). The resulting wake is substantially narrower than that associated with the laminar boundary layer.

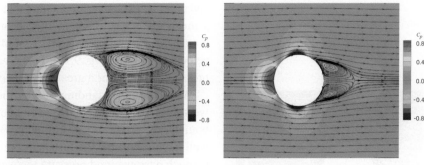

FIGURE 9.23
This CFD simulation of flow around a sphere shows the large difference in the size of the wake regions associated with the separation of a laminar boundary layer (left) and of a turbulent boundary layer (right). The color contours show the dimensionless pressure distribution expressed as $C_p \equiv (P - P_\infty)/[(1/2)\rho V_\infty^2]$. Flow patterns are similar for cylinders.

What causes the drag crisis? To explain this phenomenon we examine the behavior of the boundary layer that develops over the forward portion of the cylinder. For conditions before the drag crisis ($Re_D \lesssim 2 \times 10^5$), this boundary layer is laminar and separates at approximately 80° (see Fig. 9.22), independent of Re_D. For conditions beyond the drag crisis ($Re_D \gtrsim 5 \times 10^5$), the boundary layer is turbulent and separates further around the cylinder at approximately 140° (see Fig. 9.22.) We see this difference in the location of the separation point in the flow simulations of Fig. 9.23. The cause of the separation for either the laminar or turbulent boundary layer is that the pressure increases in the flow direction along the circumference beyond 90° (i.e., $dP/dx > 0$). Figure 9.24 shows the effects of this **adverse pressure gradient** on the boundary-layer-velocity profile. The **separation point** is formally defined by the criterion that the velocity gradient at the surface vanishes:

$$\left.\frac{\partial v_x}{\partial y}\right|_{y=0} = 0 \quad \text{at separation,} \tag{9.53}$$

where v_x is the velocity component parallel to the surface and y is the perpendicular distance from the surface. This separation criterion is general and applies to other flows in addition to the present one. Beyond the separation point, the pressure force overcomes the fluid inertia near the surface, causing a **reverse flow,** that is, a secondary flow directed opposite to that of the mean flow direction.

We now ask, why does the turbulent boundary layer separate later than the laminar one? That the separation is farther rearward for the turbulent boundary

FIGURE 9.24
Velocity profiles illustrating attached flow, the separation point, and separated flow with a reverse flow region. The height of the velocity distributions is greatly exaggerated as the boundary layer is quite thin. The reverse flow regions, however, can be quite large (cf. Fig. 9.23 left).

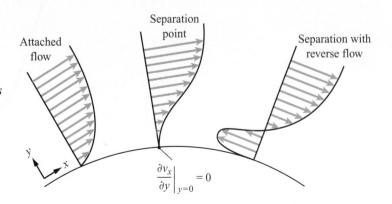

layer can be explained by the fact that the turbulent boundary layer processes a much greater momentum flow near the surface compared to the laminar boundary layer (see the comparison of laminar and turbulent boundary-layer-velocity profiles in Fig. 9.18). This greater momentum flow resists the adverse pressure gradient for a greater distance before separation occurs.

With this understanding of why the flow separates at different locations, we return to the original question of why the drag is reduced when the boundary layer becomes turbulent. Because the separation point is moved well beyond 90° for the turbulent boundary layer, the resulting wake region is much narrower and pressure recovery over the rear of the cylinder is more complete. As a consequence, the form drag is greatly reduced. Figure 9.25 shows the nondimensional pressure distribution around a sphere, first, for conditions preceding the drag crisis where a laminar boundary layer separates, and after, where a turbulent boundary layer separates. (Although these pressure distributions are for a sphere, similar distributions result for flow over a circular cylinder.) Integrating these pressure distributions in the manner prescribed by Eqs. 9.50a and 9.50b provides a good estimate of the total drag force. Also shown in Fig. 9.25 is the pressure distribution for an inviscid or ideal flow around the sphere. Neglecting the boundary layer and

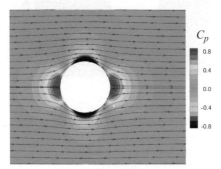

The pressure distribution is symmetric for an inviscid (ideal) flow around a sphere.

FIGURE 9.25

The dimensionless pressure distribution around a sphere shows a substantial pressure recovery over the rear surface before the turbulent boundary layer separates. With the early separation of the laminar boundary layer, essentially no pressure recovery is possible. **Adapted from Ref. [21] with permission.**

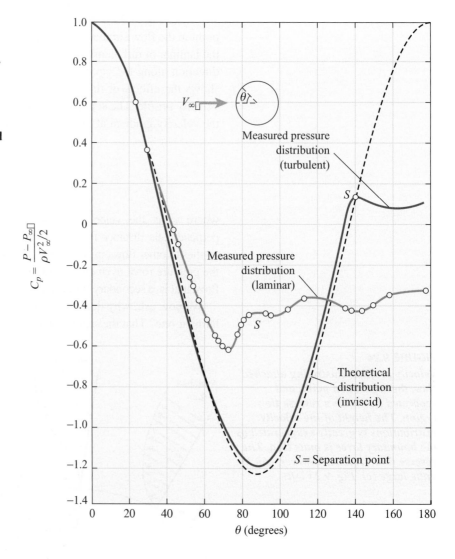

the separation phenomena gives a symmetric pressure distribution and, hence, no net pressure force or drag. As discussed at the beginning of this chapter, this unrealistic situation (*D'Alembert's paradox*) caused experimentalists to dismiss theoretical studies as useless in the early days of fluid mechanics [1].

As might be expected, local heat-transfer coefficients (and Nusselt numbers) show complex patterns as functions of both position (θ) and Reynolds number (Re_D), as shown in Fig. 9.26. In this figure, we see the initial decline in Nu_θ as the laminar boundary layer grows over the forward

FIGURE 9.26

Distribution of local Nusselt number Nu_θ around the circumference of a circular cylinder in a cross-flow for various Reynolds numbers Re_D. **Adapted from Ref. [22] with permission.**

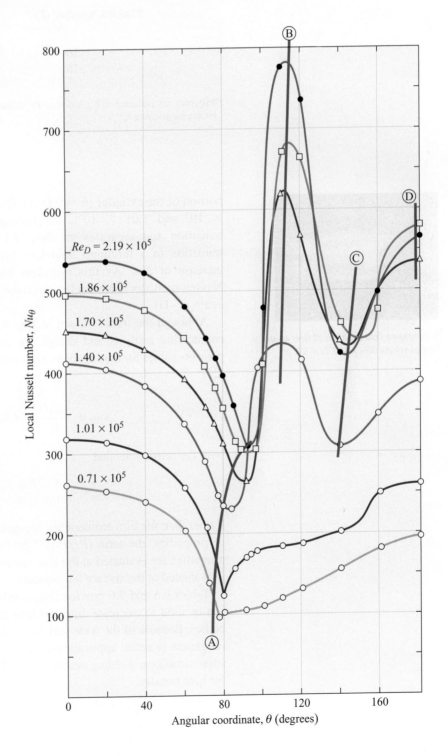

Table 9.4 **Constants for Nusselt Number Correlation for Circular Cylinder in a Cross-flow [23]:** $CRe_D^m Pr^n(Pr/Pr_s)^{1/4}$ **(Eq. 9.54)**

Reynolds Number (Re_D)*	C	m
0.4–4	0.989	0.330
4–40	0.911	0.385
40–4000	0.683	0.466
4000–40,000	0.193	0.618
40,000–400,000	0.027	0.805

Prandtl Number (Pr)*	n
≤10	0.37
>10	0.36

*Properties are evaluated at T_∞; however, Pr_s is the value of the Prandtl number evaluated at the surface temperature T_s.

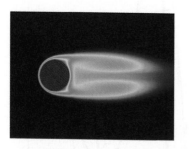

Temperature field for a forced flow around a heated sphere ($Re_D = 100$, $Pr = 5$)

portion of the cylinder ($\theta = 0$ to A). For the two smallest Re_D values (0.71×10^5 and 1.01×10^5), the laminar boundary layer grows without transition and separates at about 80°; whereas for the higher values, transition to a turbulent boundary layer occurs (A–B), resulting in an increase of Nu_θ. As this turbulent boundary layer grows (B–C), local Nusselt numbers again decline until the flow separates at C and start to rise again (C–D).

Although the trends shown in Fig. 9.26 are interesting, more useful for engineering analysis and design are average Nusselt numbers. Table 9.4 presents values for the constants used with the following empirical correlation [23]:

$$Nu_D \equiv \frac{\overline{h}_{conv}D}{k_f} = CRe_D^m Pr^n\left(\frac{Pr}{Pr_s}\right)^{1/4}, \tag{9.54}$$

with the restrictions that

$$1 < Re_D < 10^6,$$
$$0.7 < Pr < 500.$$

Rather than the film temperature accounting for the temperature dependence of properties, the term $(Pr/Pr_s)^{1/4}$ performs this function. In Eq. 9.54, all properties are evaluated at the free-stream temperature T_∞, except Pr_s, which is evaluated at the surface temperature.

Tables 9.5 and 9.6 provide drag coefficients and various expressions that can be used to calculate drag and heat transfer for various two-dimensional bodies. Because of the wide variability in free-stream turbulence and surface roughness in actual applications, these relationships should be considered as approximations, yielding uncertainties of 10%–15%, or perhaps even greater for heat transfer.

Table 9.5 Summary of Drag and Heat-Transfer Relationships for Various Two-Dimensional Bodies in a Cross-flow

Geometry	c_D^*	Comment	Reference	Nu_D	Comment	Reference
Circular cylinder	Fig. 9.21	Full Re_D range	[11]	$CRe_D^m Pr^n \left(\dfrac{Pr}{Pr_s}\right)^{1/4}$	See Eq. 9.54. C, m, and n depend on Re_D (Table 9.4)	[23]
Square cylinders	2.1	$Re_D \geq 10^4$	[25, 26]	$0.102 Re_D^{0.675} Pr^{1/3}$	Gas flows, $5 \times 10^3 \leq Re_D \leq 10^5$	[24]
	1.6	$Re_D \geq 10^4$	[25, 26]	$0.246 Re_D^{0.588} Pr^{1/3}$	Gas flows, $5 \times 10^3 \leq Re_D \leq 10^5$	[24]
Hexagonal cylinders	1.0	$Re_D \geq 10^4$	[25, 26]	$0.160 Re_D^{0.638} Pr^{1/3}$ $0.0385 Re_D^{0.782} Pr^{1/3}$	Gas flows, $5 \times 10^3 \leq Re_D \leq 1.95 \times 10^4$ Gas flows, $1.95 \times 10^4 \leq Re_D \leq 10^5$	[24] [24]
	0.7	$Re_D \geq 10^4$	[25, 26]	$0.153 Re_D^{0.638} Pr^{1/3}$	Gas flows, $5 \times 10^3 \leq Re_D \leq 10^5$	[24]

(continued)

Table 9.5 (continued)

Geometry	c_D*	Comment	Reference	Nu_D	Comment	Reference
Vertical plate	2.0	$Re_D \gtrsim 10^4$	—	$0.228 Re_D^{0.731} Pr^{1/3}$	Gas flows, $4 \times 10^3 \lesssim Re_D \lesssim 1.5 \times 10^4$	[24]
Half tubes	1.2	$Re_D \gtrsim 10^4$	[25, 26]	—	—	—
	2.3	$Re_D \gtrsim 10^4$	[25, 26]	—	—	—
Half cylinders	1.2	$Re_D \gtrsim 10^4$	[25, 26]	—	—	—
	1.7	$Re_D \gtrsim 10^4$	[25, 26]	—	—	—
Equilateral triangles	1.6	$Re_D \gtrsim 10^4$	[25, 26]	—	—	—
	2.0	$Re_D \gtrsim 10^4$	[25, 26]	—	—	—

*Drag coefficient c_D is based on projected (frontal) area A_p.

Table 9.6 Drag Coefficients* for Elliptical Cylinders [26]

Aspect Ratio		c_D	Reynolds Number (Re_D)
2:1 → ⬭		0.6	4×10^4
		0.46	10^5
4:1 → ⬬		0.32	2.5×10^4–10^5
8:1 → ▭		0.29	2.5×10^4
		0.20	2×10^5

*Drag coefficients are based on frontal area A_p. The characteristic length for the Reynolds number is the length in the flow direction.

Example 9.7

Hurricane Fran (1996) struck North Carolina as a category-3 storm with sustained winds of 115 miles/hr. Photograph courtesy of NASA.

Among hurricanes, those designated as category 1 have the least damage potential and are characterized by wind speeds ranging from 75 to 95 miles/hr. Category 5 hurricanes, at the other end of the scale, are devastating and have wind speeds in excess of 155 miles/hr. Estimate the drag force exerted on a 6-m-long flagpole by winds of 75, 95, and 155 miles/hr. The corresponding barometric pressures are 0.99, 0.98, and 0.92 atm. Assume an air temperature of 290 K for all cases. The pole diameter is 80.3 mm.

Solution

Known $V_\infty, T_\infty, P_\infty, L, D$

Find F_D

Sketch

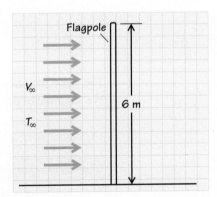

Assumptions
  i. Uniform wind speed
 ii. End effects negligible
iii. Dry air

Analysis   The drag force can be calculated from Eq. 9.51,

$$F_D = \frac{1}{2} \rho V_\infty^2 c_D A_p,$$

where the projected area A_p is the product of the flagpole length L and width $W \, (= D)$. The drag coefficient c_D is obtained from Fig. 9.21 at the appropriate Reynolds number

$$Re_D = \frac{\rho V_\infty D}{\mu}.$$

To evaluate Re_D, we start by converting the wind speed to SI units:

$$V_\infty = 75 \text{ miles/hr} \left[\frac{1 \text{ m/s}}{2.237 \text{ miles/hr}} \right] = 33.5 \text{ m/s}.$$

The viscosity, found from Table C.3, is $\mu(290 \text{ K}) = 18.09 \times 10^{-6} \text{ N} \cdot \text{s/m}^2$. The air density is calculated from the ideal-gas equation of state:

$$\rho = \frac{P}{RT}$$

$$= \frac{0.99(101{,}325)}{287(290)} = 1.205$$

$$[=] \frac{\text{atm}\left(\dfrac{\text{Pa}}{\text{atm}}\right)}{\left(\dfrac{\text{J}}{\text{kg} \cdot \text{K}}\right)\text{K}} \left[\frac{1 \text{ N/m}^2}{\text{Pa}} \right]\left[\frac{1 \text{ J}}{\text{N} \cdot \text{m}} \right] = \text{kg/m}^3.$$

Thus,

$$Re_D = \frac{1.205(33.5)0.0803}{18.09 \times 10^{-6}} = 1.79 \times 10^5$$

$$[=] \frac{(\text{kg/m}^3)(\text{m/s})\text{m}}{(\text{N} \cdot \text{s/m}^2)} \left[\frac{1 \text{ N}}{\text{kg} \cdot \text{m/s}^2} \right] = 1.$$

From Fig. 9.21, $c_D \approx 1.2$; thus the drag force is evaluated as

$$F_D = \frac{1}{2}\rho V_\infty^2 c_D L D$$

$$= 0.5(1.205)(33.5)^2\, 1.2(6)0.0803$$

$$= 391$$

$$[=] \frac{\text{kg}}{\text{m}^3}\left(\frac{\text{m}}{\text{s}}\right)^2 \text{m(m)} \left[\frac{1 \text{ N}}{\text{kg} \cdot \text{m/s}^2} \right] = \text{N}.$$

The same procedure is used to evaluate F_D for the other two wind speeds. We summarize the results of these computations in the following table:

V_∞		P	ρ	Re_D	c_D	F_D	
(miles/hr)	(m/s)	(atm)	(kg/m³)			(N)	(lb$_f$)
75	33.5	0.99	1.205	1.79×10^5	~1.2	391	88
95	42.5	0.98	1.193	2.25×10^5	~1.0	519	117
155	69.3	0.92	1.120	3.45×10^5	~0.95	1231	277

Comments We first note that the values for the Reynolds number [(1.79×10^5)–(3.45×10^5)] are in the range where the drag crisis is just beginning, hence, the drag coefficients fall with increasing wind speed. We also observe that the ratio of the drag force for the 155- and 75-miles/hr wind speeds is approximately 3.1, whereas the speed ratio is about 2.1. If the density and drag coefficients were constant, then the drag forces would scale as $F_D \sim V_\infty^2$. The table also shows drag forces in pounds force for those who appreciate these units.

Self Test 9.6 ☑ **Determine the drag force exerted on your legs as you stand in a 0.5-m–deep river with water running at 4 m/s at 300 K. Model a leg as a 10-cm-diameter cylinder.**

(Answer: $F_D = 239.2$ N or about 27 lb_f per leg)

Example 9.8

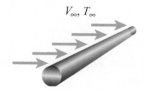

V_∞, T_∞

An electrically heated nichrome wire is to be used as an ignition source in a combustion experiment. The wire has a length of 30 mm, a diameter of 0.5 mm, an electrical resistance of 0.229 ohm, and an emissivity of 0.25. The wire is placed in a combustion chamber such that the flow is perpendicular to the length of the wire. Prior to introducing the fuel–air mixture, air flows through the combustor at 5 m/s at 300 K and 100 kPa. Determine the electrical current that must be supplied to the wire to maintain its surface temperature at 980 K in the air flow.

Solution

Known $L, D, R_{elec}, \varepsilon, V_\infty, T_\infty, P, T_s$

Find i

Sketch

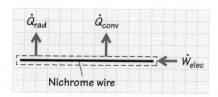

Nichrome wire

Assumptions

  i. Steady-state conditions apply.

  ii. Convection correlations for an infinite cylinder apply since $L/D > 1$.

 iii. Conduction losses from the ends of the wire are negligible.

 iv. For radiation, $T_{surr} = T_\infty$.

Analysis   Identifying the wire as a thermodynamic system, we express steady-state conservation of energy as

$$\dot{E}_{in} = \dot{E}_{out},$$

or

$$\dot{W}_{elec} = \dot{Q}_{conv} + \dot{Q}_{rad}.$$

The radiation heat-transfer rate can be calculated directly using Eq. 4.25 by recognizing that the surface area $A_s = \pi DL$;

$$\dot{Q}_{rad} = \varepsilon \pi DL \sigma (T_s^4 - T_{surr}^4)$$

$$= 0.25\pi\, 0.0005(0.030)5.67 \times 10^{-8}[(980)^4 - (300)^4]$$

$$= 0.61$$

$$[=]\, m(m)\left(\frac{W}{m^2 \cdot K^4}\right)K^4 = W.$$

Similarly, we evaluate the convective heat-transfer rate from Eq. 4.18,

$$\dot{Q}_{conv} = \bar{h}_{conv} \pi DL (T_s - T_\infty),$$

where the average heat-transfer coefficient $\overline{h}_{conv}$ is obtained from Eq. 9.54 in conjunction with Table 9.4. To evaluate $\overline{h}_{conv}$, we require the following air properties (see Table C.3):

$$\rho \text{ (300 K, 100 kPa)} = 1.1614 \text{ kg/m}^3,$$

$$\mu \text{ (300 K)} = 18.58 \times 10^{-6} \text{ N} \cdot \text{s/m}^2,$$

$$k \text{ (300 K)} = 0.0262 \text{ W/m} \cdot \text{K},$$

$$Pr \text{ (300 K)} = 0.713,$$

$$Pr_s \text{ (980 K)} = 0.722.$$

To choose the appropriate values for C and m from Table 9.4 to use with Eq. 9.54 requires a value of the Reynolds number:

$$Re_D = \frac{\rho V_\infty D}{\mu} = \frac{1.1614(5)(0.0005)}{18.58 \times 10^{-6}} = 156.3$$

$$[=] \frac{(\text{kg/m}^3)(\text{m/s})\text{m}}{\text{N} \cdot \text{s/m}^2} \left[\frac{1 \text{ N}}{\text{kg} \cdot \text{m/s}^2} \right] = 1.$$

From Table 9.4, we thus obtain

$$C = 0.683,$$

$$m = 0.466,$$

$$n = 0.37.$$

The Nusselt number is evaluated using these values in Eq. 9.54:

$$Nu_D = 0.683 Re_D^{0.466} Pr^{0.37} (Pr/Pr_s)^{0.25}$$

$$= 0.683(156.3)^{0.466}(0.713)^{0.37}(0.713/0.722)^{0.25}$$

$$= 6.32$$

The average heat-transfer coefficient is then

$$\overline{h}_{conv} = \frac{k}{D} Nu_D = \frac{0.0262}{0.0005}(6.32)$$

$$= 331$$

$$[=] \frac{\text{W/m} \cdot \text{K}}{\text{m}} = \text{W/m}^2 \cdot \text{K}.$$

The convective heat-transfer rate is

$$\dot{Q}_{conv} = 331\pi(0.0005)0.030(980 - 300) \text{ W}$$

$$= 10.6 \text{ W},$$

and the total heat-transfer rate from the wire is

$$\dot{Q}_{tot} = \dot{Q}_{conv} + \dot{Q}_{rad} = 10.6 + 0.61 \text{W} = 11.2 \text{ W}.$$

From conservation of energy and the definition of electrical power (Eq. 5.26), we write

$$\dot{W}_{elec} = i^2 R_{elec} = \dot{Q}_{tot}.$$

The electrical current i is thus

$$i = (\dot{Q}_{tot}/R_{elec})^{1/2}$$

$$= (11.2/0.229)^{1/2} \text{A} = 7.0 \text{ A}.$$

Comments We note here the use of the term $(Pr/Pr_s)^{0.25}$ to account for the temperature dependence of the air properties. For the conditions of this problem, this factor was unimportant [i.e., $(0.713/0.722)^{0.25} = 0.9958$]. We also observe that the radiation heat transfer here is significantly smaller than the convective heat transfer and contributes about 5% to the total heat-transfer rate. Neglecting the radiation results in a calculated current of 6.8 A.

Spheres and Other 3-D Shapes

The physical phenomena controlling the drag and heat transfer associated with a sphere are quite similar to those of the circular cylinder. Comparing Figs. 9.27 and 9.21, we see that the shape of the c_D versus Re_D relationships is very similar for spheres (Fig. 9.27) and cylinders (Fig. 9.21). Furthermore, the drag crisis occurs within the same range of Reynolds numbers. An interesting practical application related to the drag crisis of spheres is the deliberate roughening of golf balls with dimples. The dimples promote the laminar-to-turbulent boundary-layer transition at lower Reynolds numbers, moving the drag crisis to lower values of Re_D. This results in reduced drag at typical golf-ball flight speeds (see Example 9.9).

A comparison of Figs. 9.27 and 9.21 also shows that the drag coefficient for a sphere is substantially less than that for a cylinder (~0.45 versus ~1.2) in the range of Reynolds numbers (10^4–10^5) where the drag coefficient is essentially constant. In general, three-dimensional shapes have lower drag coefficients than their two-dimensional counterparts, (e.g., a square plate versus a long plate), as can be seen by comparing the appropriate corresponding values shown in Tables 9.5 and 9.7.

Average heat-transfer coefficients for heated (or cooled) spheres can be calculated from the following empirical correlation [27].

$$\overline{Nu_D} \equiv \frac{\overline{h}_{conv}D}{k_f} = 2 + \left(0.4Re_D^{1/2} + 0.06Re_D^{2/3}\right) Pr^{0.4}\left(\frac{\mu}{\mu_s}\right)^{1/4}, \quad (9.55)$$

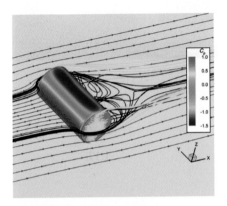

Streamlines in a horizontal plane for forced flow ($Re_D = 10^6$) around a stubby cylinder. Color contours define the dimensionless pressure distribution.

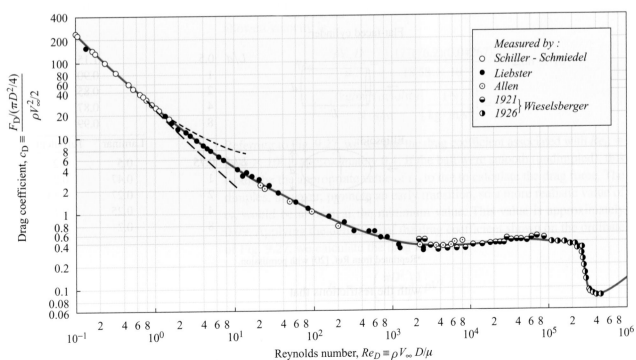

FIGURE 9.27
Drag coefficient for a sphere in a uniform flow. **Reprinted from Ref. [11] with permission.**

Using the two values of c_D, we obtain the following drag force values:

	c_D	F_D (N)	F_D (lb$_f$)
Undimpled	0.4	1.96	0.44
Dimpled	0.09	0.44	0.090

Comments Early transition to a turbulent boundary layer caused by the dimples reduces the drag force by a factor of 4.5. We also note that the magnitude of the drag force of the dimpled ball is approximately the same as its weight. Drag coefficients for actual golf balls are higher than those estimated from smooth-sphere data as done in this example, specifically, 0.25 versus 0.09 [28].

Self Test 9.7

 A parachutist descends at a constant rate when his or her weight equals the drag force. Calculate the descent rate for a 150-lb$_m$ parachutist with a 3-m-diameter parachute when the air temperature is 300 K.

(Answer: V = 9.9 m/s.)

Example 9.10 Biological Systems Application

Photograph courtesy of Phillip Colla, OceanLight.com.

With an average length of about 24 m and a weight of 100,000 kg, blue whales (*Balaenoptera masculus*) are the largest animals ever known. To propel itself through the water at a steady speed, a whale must generate a propulsive thrust equal to the drag force. Hydrodynamic studies provide the following dimensionless relationship between the total drag force F_D and the swimming speed V of blue whales:

$$c_D = 0.0174 Re_L^{-0.107},$$

where

$$c_D \equiv \frac{F_D / A_{\text{wetted}}}{\frac{1}{2}\rho V^2}$$

and

$$Re_L \equiv \frac{VL}{\nu}.$$

Consider a 22-m-long blue whale swimming at its top speed of 13 m/s, a speed that can be maintained for short bursts. The wetted surface area of the whale, A_{wetted}, is 186 m². Estimate the drag force the whale must overcome to swim at this speed. Also determine the instantaneous power required to maintain this speed. The density and kinematic viscosity of the water are 1025 kg/m³ and 8.6×10^{-7} m²/s, respectively.

Solution

Known $c_D = f(Re_L)$, L, V, A_{wetted}, ρ, ν

Find F_D, $\dot{W}_D$

Assumptions

   i. Flow is steady.

   ii. Swimming motion generates constant thrust.

Analysis We sequentially determine the Reynolds number, use this in the correlation to find the drag coefficient, and then use the definition of c_D to retrieve the desired drag force, as follows:

$$Re_L = \frac{VL}{\nu} = \frac{13\,(22)}{8.6 \times 10^{-7}} = 3.3 \times 10^8$$

$$[=]\frac{(m/s)m}{m^2/s} = 1 \text{ (dimensionless)},$$

$$c_D = 0.0174\,Re_L^{-0.107} = 0.0174(3.3 \times 10^8)^{-0.107}$$
$$= 0.00213 \text{ (dimensionless)},$$

$$F_D = \frac{1}{2}\rho V^2 c_D A_{\text{wetted}}$$
$$= 0.5(1025)(13)^2\,0.00213(186)$$
$$= 34{,}300$$
$$[=]\frac{kg}{m^3}\left(\frac{m}{s}\right)^2 m^2 \left[\frac{1\ N}{kg \cdot m/s^2}\right] = N.$$

The power required to overcome this drag force is the product of the drag force and the velocity (see Eq. 4.7):

$$\dot{W}_D = F_D V$$
$$= 34{,}300(13) = 446{,}000$$
$$[=] N(m/s) = J/s = W.$$

Comments The 34,300-N propulsive thrust of the whale seems like a large number. To provide a sense of scale, we note that this value is nearly equal to the maximum thrust generated by a General Electric CF34-3A-1 turbofan engine (39,000 N). Two of these engines power the Canadair regional jet aircraft. We also note the use of the wetted surface area in the whale drag coefficient correlation, rather than the frontal area employed in many other correlations.

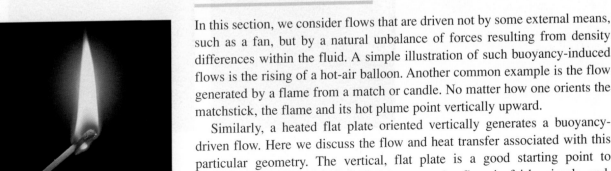

9.6 FREE CONVECTION

In this section, we consider flows that are driven not by some external means, such as a fan, but by a natural unbalance of forces resulting from density differences within the fluid. A simple illustration of such buoyancy-induced flows is the rising of a hot-air balloon. Another common example is the flow generated by a flame from a match or candle. No matter how one orients the matchstick, the flame and its hot plume point vertically upward.

   Similarly, a heated flat plate oriented vertically generates a buoyancy-driven flow. Here we discuss the flow and heat transfer associated with this particular geometry. The vertical, flat plate is a good starting point to understand free convection, first, because the flow is fairly simple and, second, because it has much in common with the previous discussion of

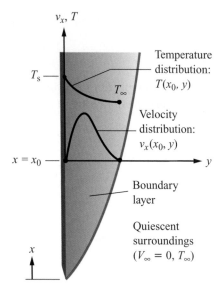

FIGURE 9.28

In the boundary layer associated with a heated, vertical flat plate, the velocity distribution exhibits a maximum, whereas the temperature distribution monotonically decreases. The origins for both distributions are located at a distance x_0 from the plate leading edge.

forced flows over flat plates. Although the flow and thermal aspects are intimately coupled and equally important in free convection (i.e., the mass, momentum, and energy conservation principles are all needed to describe the free-convection problem), our focus will be heat transfer since friction and drag are usually not important.

9.6a Vertical, Flat Plate
Physical Description

Consider a flat plate oriented vertically. The plate is heated to a temperature T_s above that of the **quiescent**[3] surroundings, T_∞, as shown in Fig. 9.28. The x-direction is vertically upward, and the y coordinate is the perpendicular distance from the plate surface. At the plate surface ($y = 0$), the fluid temperature is also at T_s; and because of the no-slip condition, the fluid velocity is zero. Beyond the surface, the hot fluid rises, creating an upward flow. The fluid motion is concentrated in a relatively thin region close to the plate. Thus, the flow forms a boundary layer. Figure 9.28 shows the velocity distribution through this boundary layer at a location x_0 from the plate leading edge. In this figure, we see that the vertical velocity component v_x possesses a maximum within the boundary layer, with zero values at the plate surface and at the outer edge of the boundary layer. The temperature distribution, however, decreases continually from T_s at the plate surface to T_∞ at the outer edge of the boundary layer. The relative position of the velocity maximum within the thermal boundary layer depends on the fluid Prandtl number. As the Prandtl number increases, the velocity maximum moves outward

FIGURE 9.29

Computed temperature (left) and velocity (right) fields for a vertical heated plate. The y-coordinate has been stretched by a factor of 5 to show the details within the boundary layers. (Pr = 0.72).

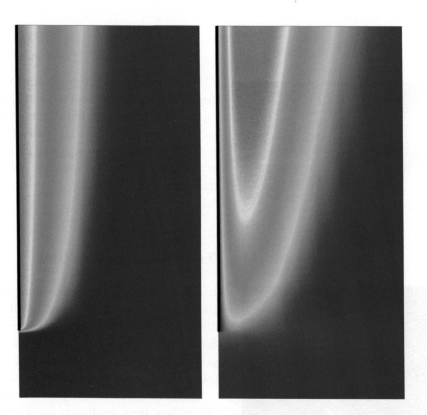

[3]The use of the word quiescent indicates that the velocity of the fluid in the surroundings is zero (i.e., $V_\infty = 0$). This is a clear distinction from the forced convection problem where a nonzero V_∞ controls the flow.

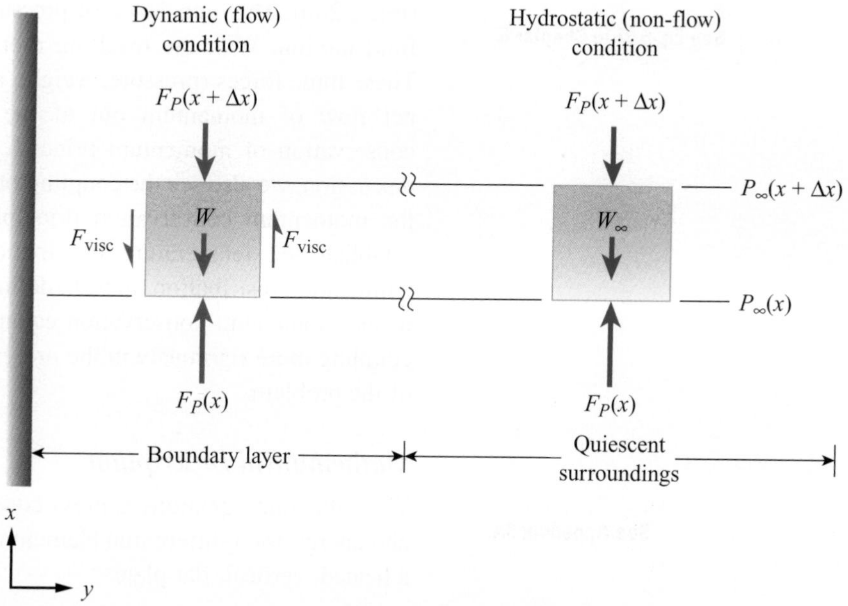

FIGURE 9.30

The hydrostatic pressure distribution in the surroundings is impressed through the boundary layer up to the wall. Since the fluid elements within the boundary layer have a lower density than those in the surroundings, a net upward buoyant force is created (net pressure force minus weight).

relative to the temperature profile. As in the forced-convection problem, the solution to the free-convection problem is the determination of these velocity and temperature distributions. Knowing these, the wall shear stress and surface heat flux can be determined. Figure 9.29 shows computed velocity and temperature distributions within the boundary layer of a heated, vertical plate.

The driving of the flow from a buoyant force that arises from the density variations within the boundary layer is the key feature that distinguishes the free-convection problem from the forced-convection problem. Figure 9.30 illustrates the origin of this force. Here we see two control volumes: one within the boundary layer and the second in the quiescent surroundings. Both control volumes have the same dimensions, Δx by Δy, with a unit width perpendicular to the page. The lower surfaces of each volume are at the same elevation x, as are their upper surfaces at $x + \Delta x$. Outside the boundary layer, the fluid within the control volume is in a state of hydrostatic equilibrium (i.e., there is no motion) and the weight of the fluid within the control volume ($W_\infty = \rho_\infty g(1)\Delta x \Delta y$) is exactly balanced by the difference in hydrostatic pressure forces. As we saw in Chapter 6 (cf. Eq. 6.24), this can be expressed mathematically as

> The principles of fluid statics describe conditions outside of the boundary layer.

> See Example 6.1 in Chapter 6.

$$P_\infty(x) = P_\infty(x + \Delta x) + \rho_\infty g \, \Delta x.$$

A characteristic of a boundary-layer flow is that the pressure in the surroundings at any given elevation is impressed through the boundary layer. This means that the pressures at the top and bottom of the control volume within the boundary layer are identical to the corresponding pressures acting on the control volume in the quiescent surroundings. Mathematically, this implies that P is independent of the y coordinate and is a function of x only. With the pressure forces within the boundary layer established by the hydrostatic condition outside the boundary layer, the pressure forces no longer balance the weight of the fluid within the control volume. Within the boundary layer, the weight is given by

$$W = \rho g(1)\Delta x \Delta y,$$

where ρ is the local density. Since the temperature of the fluid within the boundary layer is greater than that in the surroundings, its density is less. For an ideal gas, the density is inversely proportional to the absolute temperature

See Eq. 6.57 in Chapter 6.

(Eq. 2.28b). This unbalance of pressure and gravitational forces creates the fluid motion. With the resulting motion, viscous shear forces are created. These three forces (pressure, weight, and viscous shear) must then equal the net flow of momentum out of the control volume as required by the conservation of momentum principle (see Eqs. 6.53 and 6.57). From this discussion, we also see the coupling of the energy conservation principle with the momentum conservation principle: Energy conservation is needed to establish the temperature distribution through the boundary layer; the temperature distribution, in turn, defines the density (body force) distribution in the momentum conservation equation. The following section shows this coupling more rigorously in the presentation of the mathematical description of the problem.

Mathematical Description

See Appendix 9A.

The following equations express conservation of mass, vertical momentum, and energy for a differential element ($dx \cdot dy \cdot 1$) within the boundary layer of a heated, vertical, flat plate:

$$\frac{\partial v_x}{\partial x} + \frac{\partial v_y}{\partial y} = 0 \qquad \text{(mass),} \qquad (9.56)$$

$$v_x\frac{\partial v_x}{\partial x} + v_y\frac{\partial v_x}{\partial y} = -\frac{1}{\rho}\frac{\partial P}{\partial x} - g + \nu\frac{\partial^2 v_x}{\partial y^2} \qquad \text{(x momentum),} \qquad (9.57)$$

$$v_x\frac{\partial T}{\partial x} + v_y\frac{\partial T}{\partial y} = \alpha\frac{\partial^2 T}{\partial y^2} \qquad \text{(energy).} \qquad (9.58)$$

Comparing this set of equations with those for forced convection over a flat plate (Eqs. 9.14–9.16), we see that the only difference is the addition of the buoyancy term,

$$-\frac{1}{\rho}\frac{\partial P}{\partial x} - g,$$

to the momentum equation. To arrive at Eqs. 9.56–9.58, we assume that the flow is incompressible, with the exception that the density can vary in the buoyancy term. This assumption is known as the **Boussinesq approximation**.

We now link the x-momentum equation (Eq. 9.57) directly to the energy equation (Eq. 9.58) by relating the density to the temperature. First, however, we eliminate the pressure-gradient term through the hydrostatic variation of pressure discussed earlier:

$$\frac{\partial P}{\partial x} = -\rho_\infty g. \qquad (9.59)$$

Substituting Eq. 9.59 into Eq. 9.57 and rearranging yield

$$v_x\frac{\partial v_x}{\partial x} + v_y\frac{\partial v_x}{\partial y} = \frac{g}{\rho}(\rho_\infty - \rho) + \nu\frac{\partial^2 v_x}{\partial y^2}. \qquad (9.60)$$

Our link to temperature is through the **volumetric thermal expansion coefficient** β, a thermodynamic property defined as follows:

$$\beta \equiv -\frac{1}{\rho}\left(\frac{\partial \rho}{\partial T}\right)_P. \qquad (9.61)$$

We approximate Eq. 9.61 as

$$\beta \approx -\frac{1}{\rho}\left(\frac{\rho_\infty - \rho}{T_\infty - T}\right).$$

Substitution of this approximation into Eq. 9.60 produces the desired final form of the x-momentum equation:

$$v_x\frac{\partial v_x}{\partial x} + v_y\frac{\partial v_x}{\partial y} = g\beta(T - T_\infty) + \nu\frac{\partial^2 v_x}{\partial y^2}. \tag{9.62}$$

From Eq. 9.62, we clearly see that a knowledge of the temperature distribution $T(x, y)$ is required to solve the momentum (and continuity) equations for the velocity distributions $v_x(x, y)$ and $v_y(x, y)$; in other words, the energy and momentum equations are *coupled.*

Consistent with the formulations of the boundary-layer equations, the expansion coefficient β is treated as a constant. For an ideal gas, the equation of state (Eq. 2.28b) can be differentiated to yield

$$\beta = -\frac{1}{\rho}\left(\frac{\partial \rho}{\partial T}\right)_P = -\frac{1}{\rho}\left[\frac{\partial(P/RT)}{\partial T}\right]_p = \frac{1}{T}. \tag{9.63}$$

For fluids other than ideal gases, values of β are tabulated in Appendix G.

Like the forced-convection problem, a similarity solution exists for the solution of the free-convection problem described by Eqs. 9.56, 9.58, and 9.62. For the interested reader, the detailed solution can be found in Ref. [9], for example. Before summarizing these results, we investigate the dimensionless forms of our final equation set to see what dimensionless parameters control the solution.

Dimensionless Parameters

Our immediate objective is to rewrite the conservation equations in dimensionless form. We use the same approach as in Chapter 8, with the exception that there is no obvious characteristic velocity V_c; however, V_c will be defined in the course of our nondimensionalizing procedure. With this caveat, we define the following dimensionless variables:

$$x^* \equiv x/L, \qquad v_x^* \equiv v_x/V_c,$$

$$y^* \equiv y/L, \qquad v_y^* \equiv v_y/V_c$$

and

$$T^* = \frac{T - T_\infty}{T_s - T_\infty},$$

where L is the plate length and $T_s - T_\infty$ is the characteristic temperature difference.

We begin by substituting the dimensional variables expressed in terms of their dimensionless counterparts (e.g., $x = x^*L$, $v_x = v_x^*V_c$, etc.) back into the momentum equation (Eq. 9.62). This operation yields

$$\frac{V_c^2}{L}v_x^*\frac{\partial v_x^*}{\partial x^*} + \frac{V_c^2}{L}v_y^*\frac{\partial v_x^*}{\partial y^*} = g\beta(T_s - T_\infty)T^* + \frac{\nu V_c}{L^2}\frac{\partial^2 v_x^*}{\partial y^{*2}}.$$

We now divide through by V_c^2/L to get

$$v_x^* \frac{\partial v_x^*}{\partial x^*} + v_y^* \frac{\partial v_x^*}{\partial y^*} = \frac{g\beta(T_s - T_\infty)L}{V_c^2} T^* + \frac{\nu}{V_c L} \frac{\partial^2 v_x^*}{\partial y^{*2}}. \tag{9.64}$$

At this juncture, we define V_c by setting the coefficient of the final term equal to unity, that is,

$$\frac{\nu}{V_c L} \equiv 1.$$

This forces the characteristic velocity to be

$$V_c = \nu/L. \tag{9.65}$$

We complete our nondimensionalizing procedure by substituting Eq. 9.65 back into Eq. 9.64, which gives us

$$v_x^* \frac{\partial v_x^*}{\partial x^*} + v_y^* \frac{\partial v_x^*}{\partial y^*} = \frac{g\beta(T_s - T_\infty)L^3}{\nu^2} T^* + \frac{\partial^2 v_x^*}{\partial y^{*2}}. \tag{9.66}$$

From inspection of Eq. 9.66, we identify a new dimensionless parameter, the **Grashof number**, defined as

$$Gr \equiv \frac{g\beta(T_s - T_\infty)L^3}{\nu^2}, \tag{9.67a}$$

or

$$Gr \equiv \frac{g\beta\Delta T L^3}{\nu^2}, \tag{9.67b}$$

where ΔT is the absolute difference between the surface and surroundings temperatures. A typical physical interpretation of the Grashof number is that it represents the ratio of buoyant forces to viscous forces, that is,

$$Gr \equiv \frac{\text{buoyant force}}{\text{viscous force}}. \tag{9.68a}$$

More precisely, the Grashof number is the product of two force ratios, as follows:

$$Gr \equiv \left(\frac{\text{buoyant force}}{\text{viscous force}} \right) \left(\frac{\text{fluid inertia}}{\text{viscous force}} \right). \tag{9.68b}$$

Leaving the details as an exercise for the reader, we transform the continuity and energy equations, respectively,

$$\frac{\partial v_x^*}{\partial x^*} + \frac{\partial v_y^*}{\partial y^*} = 0 \tag{9.69}$$

and

$$v_x^* \frac{\partial T^*}{\partial x^*} + v_y^* \frac{\partial T^*}{\partial y^*} = \frac{1}{Pr} \frac{\partial^2 T^*}{\partial y^{*2}}. \tag{9.70}$$

Note the appearance, once again, of the Prandtl number Pr ($\equiv \nu/\alpha$). The boundary condition for the free-convection heat-transfer problem is identical to that for the forced-convection problem; thus, the Nusselt number, Nu ($\equiv h_{conv}L/k_f$) is the third, and final, dimensionless parameter. With this knowledge, we expect heat-transfer correlations to be of the following forms:

$$Nu_x = f_1(Gr_x, Pr) \tag{9.71a}$$

Table 9.8 Heat-Transfer Relationships for Free Convection: Vertical, Flat Plate

Relationship*	Equation	Restrictions	Comments	References
$Nu_x = 0.707\,Gr_x^{1/4}\dfrac{0.75Pr^{1/2}}{(0.609 + 1.221Pr^{1/2} + 1.238Pr)^{1/4}}$	9.74	Laminar, local	Similarity solution	[9, 29, 30]
$\overline{Nu_L} = 1.333Nu_L$	9.75	Laminar, average	Use Eq. 9.74 for Nu_L	—
$\overline{Nu_L} = 0.68 + \dfrac{0.670Ra_L^{1/4}}{\left[1 + (0.492/Pr)^{9/16}\right]^{4/9}}$	9.76	Laminar ($0 < Ra_L < 10^9$), average	Empirical correlation	[31]
$\overline{Nu_L} = \left[0.825 + \dfrac{0.387\,Ra_L^{1/6}}{\left[1 + (0.492/Pr)^{9/16}\right]^{8/27}}\right]^2$	9.78	Turbulent (or laminar with decreased accuracy), average	Empirical correlation	[31]

*Properties in all relationships are evaluated at the film temperature $T_f = (T_s + T_\infty)/2$ (Eq. 9.38).

and

$$\overline{Nu}_L = f_2(Gr_L, Pr). \tag{9.71b}$$

It is common practice, however, to introduce an additional parameter, the Rayleigh number Ra. This parameter is the product of the Grashof and Prandtl numbers:

$$Ra \equiv GrPr = \frac{g\beta\Delta T L^3}{\nu\alpha}. \tag{9.72}$$

Solutions and Correlations

> See the discussion of Eq. 9.18 to review the similarity concept.

A similarity solution exists for the solution of the free-convection problem (Eqs. 9.66, 9.69, and 9.70). The procedure to obtain this solution is analogous to the forced-convection problem except that now the similarity variable $(y/x)Ra_x^{1/4}$ replaces $(y/x)Re_x^{1/2}$ [41]. The similarity solution yields the following approximate expression for the hydrodynamic boundary-layer thickness [41]:

$$\delta_H/x = 5(Ra_x)^{-1/4}. \tag{9.73}$$

The thermal boundary-layer thickness is always thinner than the hydrodynamic boundary-layer thickness for Pr greater than unity. For more details on the complex scaling of free-convection boundary layers, we refer the reader to Ref. [41].

Heat-transfer coefficients can be obtained by applying the solutions and correlations presented in Table 9.8 (Eqs. 9.74–9.76). By using the absolute value of the temperature difference (i.e., $\Delta T \equiv |T_s - T_\infty|$) to evaluate the Grashof number, these relationships apply for either heating or cooling. For cooling, the x coordinate is measured downward from the leading edge of the upward-facing plate (i.e., the situation described by inverting Fig. 9.28).

Turbulence

When the boundary layer grows to a certain thickness, instabilities develop and the flow undergoes a transition from laminar to turbulent conditions. This is similar to the forced-convection problem, except now, the local Grashof (or Rayleigh) number defines the transition location rather than the local Reynolds number. The critical Rayleigh number is typically of the order of 10^9, that is,

$$Ra_{x,crit} = Gr_{x,crit} Pr \approx 10^9. \tag{9.77}$$

Average heat-transfer coefficients for turbulent flow can be obtained from Eq. 9.78 given in Table 9.8. This correlation applies over the full range of Rayleigh numbers and thus can be used even when laminar flow exists over a large portion of the plate.

A smoke plume initially rises in a laminar fashion and then becomes unstable.

Example 9.11

During steady operation of a household oven, the poorly insulated oven door attains a nearly uniform external surface temperature of 317 K. The vertical length of the door is 0.6 m. The ambient conditions in the kitchen are 294 K and 0.1 MPa.

A. Estimate the hydrodynamic boundary-layer thickness at the top of the door.
B. What would the vertical length of the door have to be for laminar-to-turbulent transition to occur at $x = L$?

C. For a forced flow, what velocity would be required for transition at $x = L$ as determined in Part B?

Solution

Known T_s, T_∞, L

Find δ_H, L for $Ra_{crit} = Ra_L$, V_∞ for Re_{crit} (forced) $= Re_L$

Sketch

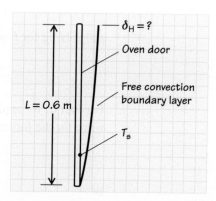

Assumption

The oven door can be treated as a vertical flat plate (i.e., the effect of the blunt leading edge is negligible).

Analysis   To evaluate the Rayleigh and/or Grashof numbers, we use properties of air evaluated at the film temperature

$$T_f = (T_s + T_\infty)/2$$
$$= (317 + 294)/2 = 305.5 \text{ K}.$$

Using values from the NIST software, we obtain the following:

$$\rho(305.5 \text{ K}) = 1.140 \text{ kg/m}^3,$$
$$\mu = 18.85 \times 10^{-6} \text{ N·s/m}^2,$$
$$\nu = 1.65 \times 10^{-5} \text{ m}^2/\text{s},$$
$$\alpha = 2.32 \times 10^{-5} \text{ m}^2/\text{s},$$
$$Pr = 0.713.$$

Assuming a laminar boundary layer, we employ Eq. 9.73 to evaluate the boundary-layer thickness at $x = L$:

$$\delta_H/x = 5Ra_x^{-0.25},$$

where (Eq. 9.72)

$$Ra_x = Ra_L = \frac{g\beta\Delta TL^3}{\nu\alpha}$$

$$= \frac{9.807\left(\dfrac{1}{305.5}\right)(317 - 294)(0.6)^3}{(1.65 \times 10^{-5})2.32 \times 10^{-5}} = 4.17 \times 10^8$$

$$[=]\frac{(\text{m/s}^2)(1/\text{K})\text{K}(\text{m}^3)}{(\text{m}^2/\text{s})\,\text{m}^2/\text{s}} = 1.$$

Thus,

$$\delta_H = L5Ra_L^{-0.25}$$
$$= 0.6(5)(4.17 \times 10^8)^{-0.25} \text{ m} = 0.02 \text{ m}.$$

Since the Raleigh number is less than Ra_{crit} ($\approx 10^9$), we justify our original assumption of a laminar boundary layer.

To answer Part B, we equate Ra_L to Ra_{crit},

$$Ra_L = \frac{g\beta\Delta T}{\nu\alpha}L^3 = Ra_{crit} = 1 \times 10^9$$

$$= \frac{9.807\left(\dfrac{1}{305.5}\right)(317 - 294)}{1.65 \times 10^{-5}(2.32 \times 10^{-5})}L^3 = 1.93 \times 10^9 L^3,$$

and solving for L we obtain

$$L = \left[\frac{1 \times 10^9}{1.93 \times 10^9}\right]^{1/3} = 0.80 \text{ m}.$$

For Part C, we apply the forced-flow criterion for transition (Eq. 9.40) and solve for the forced-convection velocity V_∞:

$$Re_{crit} \approx 5 \times 10^5 = \frac{\rho V_\infty L}{\mu},$$

or

$$V_\infty = Re_{crit}\left(\frac{\mu}{\rho L}\right) = Re_{crit}\left(\frac{\nu}{L}\right)$$

$$= 5 \times 10^5 \frac{1.65 \times 10^{-5}(\text{m}^2/\text{s})}{0.80 \text{ m}} = 10.3 \text{ m/s}.$$

Comments   From Part A, we see that the boundary-layer thickness is relatively thin, being only about 3% (0.02 m/0.60 m) of the vertical length. Part B shows that the door would have to be a third longer (0.80 m/0.60 m) for transition to a turbulent boundary layer. The result from Part C is interesting in that a substantial velocity would be required to have transition at $L = 0.8$ m for a forced flow.

Self Test 9.8

Consider a single-pane, 5-ft-wide by 3-ft-high glass window. The exterior surface of the glass is at −4.5°C and the exterior air temperature is −10°C. Estimate the hydrodynamic boundary-layer thickness at the top of the window and verify that the flow is in the laminar regime.

(Answer: $\delta_H = 29$ mm)

Example 9.12

Determine the convective heat loss from the oven door described in Example 9.11. The width of the door is 0.76 m.

Solution

Known W, L, Ra_L, k, Pr (see results from Ex. 9.11)

Find $\dot{Q}_{conv}$

Assumptions

Heat-transfer correlations for a vertical, flat plate apply to the oven door.

Analysis   Three different free-convection heat-transfer correlations can be used for this situation: Eqs. 9.75, 9.76, and, with lesser accuracy, Eq.

9.78. We will use all three and compare the results. Using Eq. 9.75 (and 9.74), we obtain the average Nusselt number as follows:

$$\overline{Nu}_L = 1.333\, Nu_L,$$

where

$$Nu_L = 0.707\, Gr_x^{0.25}\, \frac{0.75\, Pr^{0.5}}{[0.609 + 1.221\, Pr^{0.5} + 1.238\, Pr]^{0.25}}.$$

Using the values $Ra_L = 4.17 \times 10^8$ and $Pr = 0.713$ from Example 9.11, we calculate the Grashof number and use this to determine the Nusselt number:

$$Gr_L = Ra_L/Pr = 4.17 \times 10^8/0.713 = 5.85 \times 10^8$$

and so

$$Nu_L = 0.707\, (5.85 \times 10^8)^{0.25}\, 0.50 = 55.0.$$

The average Nusselt number is

$$\overline{Nu}_L = 1.333(55.0) = 73.3.$$

From the definition of $\overline{Nu}_L$ and k from the NIST database, we have

$$\overline{h}_{\text{conv},L} = \overline{Nu}_L\, \frac{k}{L} = 73.3\, \frac{0.0266\ \text{W/m} \cdot \text{K}}{0.60\ \text{m}}$$

$$= 3.25\ \text{W/m}^2 \cdot \text{K}.$$

The convective heat-transfer rate is thus

$$\dot{Q}_{\text{conv}} = \overline{h}_{\text{conv},L} LW(T_s - T_\infty)$$

$$= 3.25(0.6)(0.76)(317 - 294)$$

$$= 34.1$$

$$[=]\ (\text{W/m}^2 \cdot \text{K})\text{m}(\text{m})\text{K} = \text{W}.$$

Similarly, we use the Rayleigh number $Ra_L = 4.17 \times 10^8$ calculated in Example 9.11 to evaluate Eqs. 9.76 and 9.78. The results of these calculations are summarized and compared with the previous results in the following table:

	$\overline{Nu}_L$	$\overline{h}_L(\text{W/m}^2 \cdot \text{K})$	$\dot{Q}(\text{W})$	Ratio*
Eq. 9.75	73.3	3.25	34.1	1.00
Eq. 9.76	74.2	3.30	34.6	1.01
Eq. 9.78	93.9	4.18	43.8	1.28

*Relative to results from Eq. 9.75.

Comments We see that the two correlations developed specifically for laminar flow (Eqs. 9.75 and 9.76) produce nearly identical results. Equation 9.78 predicts the heat-transfer rate to be 28% greater. Some difference is expected based on the note provided in Table 9.8 indicating decreased accuracy for laminar flow in the application of Eq. 9.78.

Because the average heat-transfer coefficients are relatively small (cf. Table 4.2), we expect radiation heat transfer to be important. To see if this is so, we calculate $\dot{Q}_{\text{rad}}$ from Eq. 4.25 as follows:

$$\dot{Q}_{\text{rad}} = \varepsilon LW\sigma(T_s^4 - T_{\text{surr}}^4)$$

$$= \varepsilon 0.6(0.76)(5.67 \times 10^{-8})(317^4 - 294^4) \text{ W}$$
$$= \varepsilon 67.9 \text{ W}.$$

Using the emissivity of cleaned stainless steel as an estimate for that of the oven door ($\varepsilon = 0.22$, Table J.1), we get a radiation heat-transfer rate that is approximately 40% of the convection rate, confirming our suspicion.

Self Test 9.9 ☑ **Determine the overall heat-transfer rate from the window of Self Test 9.8.**

(Answer: $\dot{Q}_{0-L} = 14.81$ W)

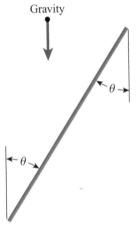

9.6b Other Geometries

Empirical heat-transfer relationships are summarized in Tables 9.9 and 9.10 for the following geometries:

- vertical cylinders,
- inclined plates,
- horizontal surfaces,
- horizontal cylinders, and
- spheres.

In the following examples, we consider each of these configurations.

FIGURE 9.31
Definition of θ, the angle from the vertical, used with Eq. 9.79 in Table 9.9.

Example 9.13 Biological Systems Application

Modeling a standing person in a swimming suit as a 1.8-m-long, 0.3-m-diameter cylinder, estimate the body heat loss rate for a skin temperature of 33°C and an ambient air temperature of 25°C. Consider (a) a quiescent environment, and (b) an environment in which the wind is blowing at a steady 10 miles/hr (4.47 m/s).

Solution

Known $L, D, T_{\text{skin}}, T_\infty, V_\infty$

Find $\dot{Q}(V_\infty = 0), \dot{Q}(V_\infty = 4.47 \text{ m/s})$

Sketch

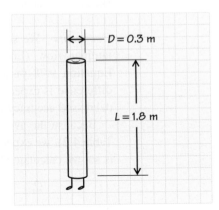

Table 9.9 Heat-Transfer Relationships for Free Convection: Vertical Cylinders, Inclined Plates, Horizontal Surfaces, Horizontal Cylinders, and Spheres

Geometry	Relationship	Equation	Restrictions	Comments	References
Vertical cylinders	See Table 9.8	—	$D/L \gtrsim \dfrac{35}{Gr_L^{1/4}}$	Same as vertical flat plate provided D/L is sufficiently large	[32]
Inclined plates	$\left[0.825 + \dfrac{0.387(Ra_L\cos\theta)^{1/6}}{[1+(0.492/Pr)^{9/16}]^{8/27}}\right]^2$	9.79	Hot surface down or cold surface up, $0 < \theta < 60°$ (see Fig. 9.31)	Does *not* apply to hot surface up and cold surface down‡	[33]
Horizontal surfaces* Upper surface hot or lower surface cold	$\overline{Nu}_{L_c} = 0.54\,Ra_{L_c}^{1/4}$	9.80	$10^4 \lesssim Ra_{L_c} \lesssim 10^7$	Convective flow is relatively unimpeded by the surface	[9, 34, 35]
	$\overline{Nu}_{L_c} = 0.15\,Ra_{L_c}^{1/3}$	9.81	$10^7 \lesssim Ra_{L_c} \lesssim 10^{11}$		
Lower surface hot or upper surface cold	$\overline{Nu}_{L_c} = 0.27\,Ra_{L_c}^{1/4}$	9.82	$10^5 \lesssim Ra_{L_c} \lesssim 10^{10}$	Convective flow is impeded by the surface	[9, 34, 35]
Horizontal cylinders†	$\overline{Nu}_D = CRa_D^{\,n}$	9.83	See Table 9.10 for values of C and n for various values of Ra_D	—	[36]
	$\overline{Nu}_D = \left[0.60 + \dfrac{0.387Ra_D^{1/6}}{[1+(0.559/Pr)^{9/16}]^{8/27}}\right]^2$	9.84	$10^9 < Ra_D < 10^{12}$	—	[37]
Sphere†	$\overline{Nu}_D = 2 + \dfrac{0.589Ra_D^{1/4}}{[1+(0.469/Pr)^{9/16}]^{4/9}}$	9.85	$Ra_D \lesssim 10^{11}$ and $Pr \geq 0.7$	—	[38]

*The characteristic length L_c is defined as the surface area A_{surf} divided by the perimeter $\mathcal{P}$ (i.e., $L_c \equiv A_{surf}/\mathcal{P}$).
†The characteristic length in both the Nusselt and Rayleigh numbers is the diameter D.
‡For these situations, see Refs. [39, 40].

Table 9.10   Free Convection on a Horizontal Cylinder: $\overline{Nu}_D = CRa_D^n$ (Eq. 9.83, Table 9.9)

Coefficient (C)	Exponent (n)	Rayleigh Number Range
0.675	0.058	10^{-10}–10^{-2}
1.02	0.148	10^{-2}–10^{2}
0.850	0.188	10^{2}–10^{4}
0.480	0.250	10^{4}–10^{7}
0.125	0.333	10^{7}–10^{12}

Assumptions

   i. The complex body geometry can be represented as a cylinder.
  ii. Heat loss occurs only from cylinder sides.
 iii. For forced convection (Part B), end effects can be neglected and, thus, correlations for an infinite cylinder apply.
  iv. $P \approx 100$ kPa.

Analysis   To calculate the heat loss for Part a, we employ the correlation indicated in Table 9.9 for free convection from a vertical cylinder. For Part b, Eq. 9.54 applies for forced convection.

The film temperature is

$$T_f = (T_{\text{skin}} + T_\infty)/2$$
$$= (33 + 25)/2°C = 29°C \text{ or } 302 \text{ K}.$$

Air properties at the film temperature (NIST12 or Table C.3) are

$$\rho = 1.154 \text{ kg/m}^3,$$
$$\mu = 18.68 \times 10^{-6} \text{ N} \cdot \text{s/m}^2,$$
$$\nu = 16.19 \times 10^{-6} \text{ m}^2\text{/s},$$
$$k = 0.02636 \text{ W/m} \cdot \text{K},$$
$$\alpha = 22.7 \times 10^{-6} \text{ m}^2\text{/s},$$
$$Pr = 0.713.$$

Table 9.9 indicates that, if

$$D/L > \frac{35}{Gr_L^{0.25}},$$

then a vertical flat-plate correlation can be applied to the side walls of a cylinder. We test this condition as follows:

$$Gr_L = \frac{g\beta \Delta T L^3}{\nu^2}$$

$$= \frac{9.807\left(\dfrac{1}{302}\right)(306 - 298)(1.8)^3}{(16.19 \times 10^{-6})^2}$$

$$= 5.78 \times 10^9$$

$$[=] \frac{(\text{m/s}^2)(1/\text{K})(\text{K})\text{m}^3}{(\text{m}^2\text{/s})^2} = 1,$$

where we employ the ideal-gas relationship that $\beta = 1/T_f$. Given $D/L = 0.3/1.8 = 0.167$, we test the inequality

$$0.167 > \frac{35}{(5.78 \times 10^9)^{0.25}} = 0.127,$$

which we see is indeed satisfied. Before choosing a flat-plate correlation, we determine whether the flow is laminar or turbulent by calculating a value for the Rayleigh number:

$$Ra_L = Gr_L \, Pr = 5.78 \times 10^9 \, (0.713) = 4.12 \times 10^9.$$

Since $Ra_L > 10^9$, we employ Eq. 9.78 (Table 9.8) to obtain the average Nusselt number:

$$\overline{Nu}_L = \left[0.825 + \frac{0.387 \, Ra_L^{1/6}}{\left[1 + \left(\frac{0.492}{Pr} \right)^{9/16} \right]^{8/27}} \right]^2$$

$$= \left[0.825 + \frac{0.387(4.12 \times 10^9)^{0.167}}{1.1925} \right]^2$$

$$= 193.6.$$

The average heat-transfer coefficient is thus

$$\overline{h}_{\text{conv},L} = \overline{Nu}_L \frac{k}{L} = 193.6 \frac{0.02636(\text{W/m} \cdot \text{K})}{1.8(\text{m})},$$

$$= 2.8 \text{ W/m}^2 \cdot \text{K},$$

and the heat-transfer rate is

$$\dot{Q} = \overline{h}_{\text{conv}} \pi D L (T_{\text{skin}} - T_\infty)$$

$$= 2.80 \, \pi \, 0.3 (1.8)(306 - 298)$$

$$= 38$$

$$[=] (\text{W/m}^2 \cdot \text{K}) \text{m}(\text{m})\text{K} = \text{W}.$$

For Part b, we use Eq. 9.54 to determine $\overline{h}_{\text{conv}}$. Since the temperature difference is small (8°C), we will use the properties previously evaluated at the film temperature rather than reevaluating them at T_∞ as required for Eq. 9.54. To determine C and m, we calculate the Reynolds number based on the cylinder diameter, which is

$$Re_D = \frac{\rho V_\infty D}{\mu} = \frac{1.154(4.47)0.3}{18.68 \times 10^{-6}}$$

$$= 82,800.$$

Thus, $C = 0.027$ and $m = 0.805$ (Table 9.4) in Eq. 9.54, and so

$$\overline{Nu}_D = 0.027 \, Re_D^{0.805} Pr^{0.37} \left(\frac{Pr}{Pr_s} \right)^{0.25}.$$

Treating $Pr/Pr_s \approx 1$, we get

$$\overline{Nu}_D = 0.027(82,800)^{0.805}(0.713)^{0.37}$$

$$= 216.8$$

and

$$\overline{h}_{\text{conv}} = \overline{Nu}_D \frac{k}{D} = 216.8 \frac{0.02636}{0.3} \text{ W/m}^2 \cdot \text{K} = 19.0 \text{ W/m}^2 \cdot \text{K}.$$

The heat loss rate from the body when the wind is blowing is then

$$\dot{Q} = \bar{h}_{\text{conv}} \pi DL (T_{\text{skin}} - T_\infty)$$
$$= 19.0\pi(0.3)1.8(306 - 298)\ \text{W}$$
$$= 258\ \text{W}.$$

Comments Consistent with our experience, the body loses much more energy in a windy environment than in a quiescent environment ($\dot{Q}_{\text{forced}}/\dot{Q}_{\text{free}} \approx 7$). Our estimate of the total heat loss in the quiescent environment, however, neglects radiation and thus underestimates the actual loss. We also note the change in the characteristic length when going from free convection ($L_{\text{char}} = L$) to forced convection ($L_{\text{char}} = D$). The boundary layer for the forced flow grows in the circumferential direction rather than along the length of the cylinder as is the case for free convection.

Example 9.14 (Biological systems application)

A cold frame is a small greenhouse-like structure used to germinate seeds and start plants before ambient temperatures become consistently warm. On a calm and sunny spring day, the top surface of a horizontal cover glass on a cold frame has a steady temperature of 75 F when the ambient air temperature is 55 F. The cover is 0.6 m by 0.8 m. Estimate the convective heat loss from the top surface of the glass cover.

Solution

Known T_∞, T_s, L, W

Find $\dot{Q}_{\text{conv}}$

Assumption

  i. Quiescent ambient conditions so that free convection is the only convection mode
  ii. $P \approx 100$ kPa

Analysis   From Table 9.9, we see that either Eq. 9.80 or 9.81 applies to this situation depending on the value of the Rayleigh number Ra_{L_c}. To calculate this Raleigh number requires that the characteristic length L_c be determined. From the first footnote in Table 9.9, we see that L_c is defined as

$$L_c = \text{area/perimeter}$$
$$= \frac{LW}{2L + 2W}$$
$$= \frac{0.8(0.6)}{2(0.8 + 0.6)}\ \text{m} = 0.171\ \text{m}.$$

We also require properties at the film temperature:

$$T_\infty = 55\ \text{F} = 285.9\ \text{K},$$
$$T_s = 75\ \text{F} = 297.0\ \text{K},$$
$$T_f = 0.5\,(T_\infty + T_s) = 291.5\ \text{K}$$

From the NIST database (or Table C.3), we obtain

$$\rho = 1.195\ \text{kg/m}^3,$$
$$\mu = 18.17 \times 10^{-6}\ \text{N} \cdot \text{s/m}^2,$$

$$\nu = 15.2 \times 10^{-6} \ \text{m}^2/\text{s},$$
$$k = 0.0256 \ \text{W/m·K},$$
$$\alpha = 21.3 \times 10^{-6} \ \text{m}^2/\text{s},$$
$$Pr = 0.713.$$

From the definition of Ra_{L_c} (Eq. 9.72) and the relationship that $\beta = 1/T_f$ for an ideal gas (Eq. 9.63), we calculate

$$
\begin{aligned}
Ra_{L_c} &= \frac{g\beta\Delta T L_c^3}{\nu\alpha} \\
&= \frac{9.807(1/291.5)(297.0 - 285.9)(0.171)^3}{15.2 \times 10^{-6}(21.3 \times 10^{-6})} \\
&= 5.77 \times 10^6.
\end{aligned}
$$

Since this value falls in the range 10^4–10^7, we employ Eq. 9.80 to calculate the Nusselt number (see Table 9.9),

$$
\begin{aligned}
\overline{Nu}_{L_c} &= 0.54 Ra_{L_c}^{0.25} \\
&= 0.54 \, (5.77 \times 10^6)^{0.25} = 26.5,
\end{aligned}
$$

and so

$$
\begin{aligned}
\overline{h}_{\text{conv}} &= \overline{Nu}_{L_c} \frac{k}{L_c} \\
&= 26.5 \frac{0.0256(\text{W/m·K})}{0.171(\text{m})} = 3.97 \ \text{W/m}^2\text{·K}.
\end{aligned}
$$

Using this value of the heat-transfer coefficient, we calculate the convective heat-transfer rate as

$$
\begin{aligned}
\dot{Q}_{\text{conv}} &= \overline{h}_{\text{conv}} A(T_s - T_\infty) \\
&= 3.97(0.8)(0.6)(297.0 - 285.9) \\
&= 21.1 \\
&[=] (\text{W/m}^2\text{·K})\,\text{m(m)K} = \text{W}.
\end{aligned}
$$

Comments   Because there is no preferred direction associated with the buoyancy-induced flows on a horizontal surface, the characteristic dimension is a combination of the length and width [i.e., $L_c \equiv LW/(2L + 2W)$]. We also note that, as in our previous examples, radiation heat transfer may also contribute to the total heat-transfer rate since the free-convection rate is rather small.

Example 9.15

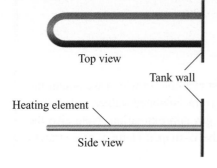

Top view

Tank wall

Heating element

Side view

An immersion heater heats oil contained in a large tank. A 1-m length of tubing bent into a U-shape forms the heating element as shown in the sketch. The 12.5-mm-diameter element is located horizontally in the nominally quiescent oil. A feedback control system maintains the surface temperature of the element at 380 K. Determine the heat-transfer rate to the oil when the oil temperature is 320 K.

Solution

Known   U-tube, D, L, T_s, T_∞

Find $\dot{Q}$

Assumptions

i. The U-tube can be modeled as a straight, 1-m-long, horizontal cylinder.
ii. End effects at the wall are negligible.
iii. Oil properties are the same as those of saturated motor oil (Table G.3).

Analysis Free convection dominates the heat transfer since the oil is nominally stagnant. The appropriate heat-transfer correlation is that for a horizontal cylinder (Eq. 9.83, Table 9.9). The film temperature is

$$T_f = (T_s + T_\infty)/2$$
$$= (380 + 320)/2 \text{ K} = 350 \text{ K}.$$

From Table G.3, we obtain the properties appearing in Eq. 9.83. For $T_f = 350$ K, these are

$$\nu = 41.7 \times 10^{-6} \text{ m}^2/\text{s},$$
$$k = 0.138 \text{ W/m} \cdot \text{K},$$
$$\alpha = 0.763 \times 10^{-7} \text{ m}^2/\text{s},$$
$$\beta = 0.70 \times 10^{-3} \text{ K}^{-1}.$$

Using the diameter as the characteristic length, we get a Rayleigh number of

$$Ra_D = \frac{g\beta\Delta T D^3}{\nu\alpha}$$
$$= \frac{9.807\,(0.7 \times 10^{-3})(380 - 320)(0.0125)^3}{(41.7 \times 10^{-6})(0.763 \times 10^{-7})}$$
$$= 2.53 \times 10^5$$
$$[=]\frac{(\text{m/s}^2)(1/\text{K})\text{K}\,(\text{m}^3)}{(\text{m}^2/\text{s})(\text{m}^2/\text{s})} = 1.$$

From Table 9.10, we select $C = 0.480$ and $n = 0.250$ as the appropriate constants for our value of Ra_D; thus,

$$\overline{Nu}_D = 0.480 Ra_D^{0.250}$$
$$= 0.480\,(2.53 \times 10^5)^{0.25} = 10.8,$$

and so

$$\overline{h}_{\text{conv}} = \overline{Nu}_D \frac{k}{D}$$
$$= 10.8 \frac{0.138\,(\text{W/m} \cdot \text{K})}{0.0125(\text{m})} = 119 \text{ W/m}^2 \cdot \text{K}.$$

The convective heat-transfer rate is then

$$\dot{Q} = \overline{h}_{\text{conv}} A (T_s - T_\infty) = \overline{h}_{\text{conv}}(\pi DL)(T_s - T_\infty)$$
$$= 119\,\pi\,0.0125\,(1)(380 - 320)$$
$$= 280$$
$$[=] (\text{W/m}^2 \cdot \text{K})\text{m(m)K} = \text{W}.$$

Comments We note that this is the first example in which we obtain the volumetric coefficient of expansion β from tabulated values rather than being evaluated as T_f^{-1} using the ideal-gas approximation. Note also the engineering judgment applied to treat the U-shaped element as a straight cylinder.

Example 9.16

In steady operation, the glass envelope of a 150-W incandescent light bulb reaches 300°C. The diameter of the spherical portion of the bulb is 73 mm. Estimate the convective heat loss from the bulb envelope for quiescent ambient conditions at 20°C and 100 kPa.

Solution

Known   Bulb geometry, T_s, T_∞, $\dot{W}_{elec}$

Find $\dot{Q}_{conv}$

Assumptions

i. The complex geometry of the glass envelope is modeled as a simple sphere with a diameter of 73 mm.
ii. The surface temperature of glass is uniform.

Analysis Because the ambient air is quiescent, we calculate the convection heat transfer using free convection correlations. Treating the glass envelope as a sphere allows us to use Eq. 9.85 (Table 9.9). Proceeding as in previous examples, we determine the film temperature, use the NIST software (or Table C.3) to obtain the required properties, use the correlation (Eq. 9.85) to find the average Nusselt number, apply the definition of the Nusselt number to determine $\bar{h}_{conv}$, and finally calculate $\dot{Q}$. These steps follow. First,

$$T_f = (T_s + T_\infty)/2$$
$$= (300 + 20)/2 = 160°C = 433 \text{ K.}$$

The properties are

$$\nu = 30.48 \times 10^{-6} \text{ m}^2/\text{s,}$$
$$k = 0.0350 \text{ W/m} \cdot \text{K,}$$
$$\alpha = 42.7 \times 10^{-6} \text{ m}^2/\text{s,}$$
$$Pr = 0.713,$$
$$\beta = 1/T_f = (1/433) \text{ K}^{-1},$$

and so the Rayleigh number is

$$Ra_D = \frac{g\beta\Delta T D^3}{\nu\alpha}$$

$$= \frac{9.807\left(\dfrac{1}{433}\right)(300 - 20)(0.073)^3}{30.48 \times 10^{-6}(42.7 \times 10^{-6})}$$

$$= 1.90 \times 10^6$$

$$[=]\frac{(\text{m/s}^2)(1/\text{K})(\text{K})\text{m}^3}{(\text{m}^2/\text{s})(\text{m}^2/\text{s})} = 1.$$

Since $Ra_D < 10^{11}$, Eq. 9.85 can be used as follows:

$$\overline{Nu}_D = 2 + \frac{0.589 Ra_D^{0.25}}{\left[1 + \left(\dfrac{0.469}{Pr}\right)^{9/16}\right]^{4/9}}$$

$$= 2 + \frac{0.589(1.90 \times 10^6)^{0.25}}{\left[1 + \left(\dfrac{0.469}{0.713}\right)^{9/16}\right]^{4/9}} = 18.9.$$

The heat-transfer coefficient and heat-transfer rate are thus

$$\overline{h}_{\text{conv}} = \overline{Nu}_D \frac{k}{D}$$

$$= 18.9 \frac{0.0350(\text{W/m} \cdot \text{K})}{0.073(\text{m})} = 9.06 \text{ W/m}^2 \cdot \text{K}$$

and

$$\dot{Q}_{\text{conv}} = \overline{h}_{\text{conv}} A (T_s - T_\infty) = \overline{h}_{\text{conv}} \pi D^2 (T_s - T_\infty)$$

$$= 9.25\pi(0.073)^2 (300 - 20)$$

$$= 43.4$$

$$[=](\text{W/m}^2 \cdot \text{K}) \text{m}^2 {}^\circ\text{C} \left[\frac{\text{K}}{1^\circ\text{C}}\right] = \text{W},$$

respectively.

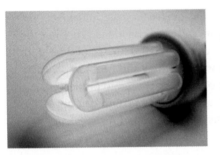

Fluorescent light bulbs convert a much larger fraction of their input electrical energy to useful light than do incandescent bulbs.

Comments We see that approximately 29% of the electrical power input is dissipated to thermal energy and convected away. The remaining input is transferred as thermal radiation emitted from the bulb surface, radiation that passes through the glass envelope, and conduction through the base. The transmitted radiation that falls in the visible region of the electromagnetic spectrum (0.4–0.7 mm) is the useful light output. Most household light bulbs convert a very small amount of the input electrical energy to visible light, with efficiencies of only a few percent.

SUMMARY

This chapter dealt with flows over external surfaces. We saw how wall shear stresses, drag forces, and local and average heat-transfer coefficients can be calculated for a wide array of surface geometries. From the relationships presented, you should be able to make useful engineering calculations for design or analysis. One focus of this chapter was the flow over a flat plate. For both forced and free convection, this simple geometry allowed us to see how the fundamental conservation principles (mass, momentum, and energy) can provide the details of laminar boundary-layer flows. You should have a firm grasp of these details. We also saw how dimensional analysis (Chapter 8) helps to organize theoretical results and empirical correlations. Key dimensionless parameters encountered were the friction and drag coefficients along with the Reynolds, Nusselt, Prandtl, Grashof, and Rayleigh numbers. You should understand the meaning and practical use of all of these. The reader should now review the specific learning objectives presented at the beginning of this chapter.

Chapter 9
Key Concepts & Definitions Checklist[4]

9.2 Basic External Flow Patterns

- ❑ External flow ➤ *Q9.2, Q9.13, Q9.14*
- ❑ Boundary-layer flow ➤ *Q9.3*
- ❑ Bluff bodies ➤ *Q9.4*
- ❑ Separation ➤ *Q9.5*
- ❑ Wakes ➤ *Q9.6*

9.3 and 9.4 Forced Flow—Flat Plate

- ❑ Hydrodynamic boundary-layer thickness (Eq. 9.1) ➤ *Q9.7, 9.2, 9.6*
- ❑ Velocity distribution ➤ *Q9.8, 9.29*
- ❑ Thermal boundary-layer thickness (Eq. 9.5) ➤ *Q9.17, Q9.18*
- ❑ Temperature distribution ➤ *9.29*
- ❑ Local wall shear stress definition (Eq. 9.2) and distribution ➤ *Q9.16*
- ❑ Total drag force ➤ *9.7, 9.8, 9.11*
- ❑ Local heat flux definition (Eq. 9.6) and distribution ➤ *Q9.17, Q9.19, 9.28*
- ❑ Total heat-transfer rate ➤ *Q9.20*
- ❑ Friction and drag coefficients ➤ *Q9.21, 9.9, 9.11*
- ❑ Local and average heat-transfer coefficients ➤ *Q9.23, 9.34*
- ❑ Local and average Nusselt numbers ➤ *Q9.24*
- ❑ Reynolds numbers: Re_x and Re_L ➤ *Q9.11, Q9.25*
- ❑ Prandtl number ➤ *Q9.26, Q9.27*

- ❑ Reynolds analogy (Eq. 9.33) ➤ *Q9.15*
- ❑ Film temperature ➤ *Q9.28*
- ❑ Transition and critical Reynolds number ➤ *9.1, 9.2*
- ❑ Boundary-layer velocity and temperature profiles: laminar versus turbulent ➤ *Q9.29, Q9.30*
- ❑ Use of correlations (Table 9.2) ➤ *9.38, 9.47*

9.5 Forced Flow—Other Geometries

- ❑ Form drag ➤ *Q9.31*
- ❑ Adverse and favorable pressure gradients ➤ *Q9.32*
- ❑ Drag crisis ➤ *Q9.33*
- ❑ Streamlining ➤ *Q9.35*
- ❑ Circular cylinders ➤ *9.48, 9.50*
- ❑ Other 2-D shapes ➤ *9.53, 9.55*
- ❑ Spheres ➤ *9.66, 9.75*
- ❑ Other 3-D shapes ➤ *9.70, 9.72*

9.6 Free Convection

- ❑ Vertical heated plate temperature profiles ➤ *Q9.36*
- ❑ Vertical heated plate velocity profiles ➤ *Q9.37*
- ❑ Grashof number ➤ *Q9.40*
- ❑ Rayleigh number ➤ *Q9.40*
- ❑ Use of correlations (Tables 9.8 and 9.9) ➤ *9.79, 9.83, 9.111*

[4] Numbers following arrows below refer to end-of-chapter Problems (e.g., 9.2) and Questions (e.g., Q9.7).

REFERENCES

1. Rouse, H., and Ince, S., *History of Hydraulics*, Iowa Institute of Hydraulics Research, State University of Iowa, Iowa City, IA, 1957.

2. Du Buat, P.L.G., *Principes d'hydraulique, vérifiés par un grand nombre d'expériences faites par ordre du gouvernement*, 2nd Ed., Paris, 1786.

3. Prandtl, L., "Über Flüssigkeitsbewegung bei sehr kleiner Reibung," *Verhandlungen des III. Internationalen Mathematiker Kongresses (Heidelberg, 1904)*, Leipzig, 1905.

4. Van Dyke, M., *An Album of Fluid Motion*, Parabolic Press, Stanford, CA, 1982.
    a. Ibid., p. 22. ONERA photograph, Werlé, 1974.
    b. Ibid., p. 96. R.E. Falco, in *Head*, 1982.

5. Blasius, H., "Grenzschichten in Flüssigkeiten mit kleiner Reibung," *Zeitshrift für Math. und Physik*, 56:1–37 (1908). English translation: "Fluid Motion with Very Small Friction," NACA Technical Memorandum 1256, 1950.

6. Pohlhausen, E., "Der Wärmeaustausch zwischen festen Körpern und Flüssigkeiten mit kleiner Reibung und kleiner Wärmeleitung," *Zeitshrift für Angewandte Mathematik und Mechanik*, 1:115 (1921).

7. von Kármán, T., "Über laminare und turbulente Reibung," *Zeitshrift für Angewandte Mathematik und Mechanik*, 1:233–252 (1921). English translation: "On Laminar and Turbulent Friction," NACA Technical Memorandum 1092, 1946.

8. Pohlhausen, K., "Zur näherungsweisen Integration der Differentialgleichung der laminaren Reibungsschicht," *Zeitshrift für Angewandte Mathematik und Mechanik*, 1:252–268 (1921).

9. Incropera, F. P., and DeWitt, D. P., *Fundamentals of Heat and Mass Transfer*, 4th ed., Wiley, New York, 1996.

10. Fox, R. W., and McDonald, A. T., *Introduction to Fluid Mechanics*, 3rd ed., Wiley, New York, 1985.

11. Schlichting, H., *Boundary-Layer Theory*, 6th ed., McGraw-Hill, New York, 1968.

12. Howarth, L., "On the Solution of the Laminar Boundary-Layer Equations," *Proceedings of the Royal Society of London*, A164:547–579 (1938).

13. Burmeister, L. C., *Convective Heat Transfer*, 2nd ed., Wiley, New York, 1993, pp. 172–177.

14. Reynolds, O., "On the Extent and Action of the Heating Surface for Steam Boilers," *Proceedings of the Literary and Philosophical Society of Manchester*, 14:7–12 (1874). See also *Papers on Mechanical and Physical Subjects by Osborne Reynolds*, Cambridge University Press, Cambridge, 1900.

15. Shames, I. H., *Mechanics of Fluids*, 3rd ed., McGraw-Hill, New York, 1992.

16. Mills, A. F., *Heat and Mass Transfer*, Irwin, Chicago, 1995.

17. Nikuradse, J., "Strömungsgesetze in rauhen Rohren," *VDI-Forschungsheft*, 361, 1933. English translation: "Laws of Flow in Rough Pipes," NACA Technical Memorandum 1292, 1950.

18. Moody, L. F., "Friction Factors for Pipe Flow," *Transactions of the ASME*, 66: 671–684 (1944).

19. Prandtl, L., and Schlichting, H., "Das Widerstandsgesetz rauher Platten," Werft, Reederei, Hafen 1–4 (1934).

20. White, F. M., *Fluid Mechanics*, 2nd ed., McGraw-Hill, New York, 1986.

21. Fage, A., "Experiments on a Sphere at Critical Reynolds Numbers," Aeronautical Research Council (Great Britain), Reports and Memoranda, No. 1766, 1937.

22. Giedt, W. H., "Investigation of Variation of Point Unit Heat Transfer Coefficient around a Cylinder Normal to an Air Stream," *Transactions of the ASME*, 71:375–381 (1949).

23. Zhukauskas, A., "Heat Transfer from Tubes in Cross Flow," in *Advances in Heat Transfer* (J. P. Hartnett and T. F. Irvine, Jr., Eds.), Vol. 8, Academic Press, New York, 1972.

24. Jacob, M., *Heat Transfer*, Vol. 1, Wiley, New York, 1949.

25. Hoerner, S. F., *Fluid Dynamic Drag*, 2nd ed., published by the author, Midland Park, NJ, 1965.

26. Lindsey, W. F., "Drag of Cylinders of Simple Shapes," NACA Report 619, 1938.

27. Whitaker, S., "Forced Convection Heat Transfer Correlations for Flow in Pipes, Past Flat Plates, Single Cylinders, Single Spheres, and for Flow in Packed Beds and Tube Bundles," *AIChE Journal*, 18:361–371 (1972).

28. Moin, P., and Kim, J., "Tackling Turbulence with Supercomputers," *Scientific American*, Jan., 1997.

29. Ostrach, S., "An Analysis of Laminar Free Convection Flow and Heat Transfer about a Flat Plate Parallel to the Direction of the Generating Body Force," NACA Report 1111, 1953.

30. LeFevre, E. J., "Laminar Free Convection from a Vertical Plane Surface," *Proceedings of the Ninth International Congress of Applied Mechanics*, Brussels, Vol. 4, 168, 1956.

31. Churchill, S. W., and Chu, H. H. S., "Correlating Equations for Laminar and Turbulent Free Convection from a Vertical Plate," *International Journal of Heat and Mass Transfer*, 18:1323–1329 (1975).

32. Sparrow, E. M., and Gregg, J. L., "Laminar Free Convection Heat Transfer from the Outer Surface of a Vertical Circular Cylinder," *Transactions of the ASME*, 78:1823 (1956).

33. Rich, B. R., "An Investigation of Heat Transfer from an Inclined Flat Plate in Free Convection," *Transactions of the ASME*, 75:489 (1953).

34. Goldstein, R. J., Sparrow, E. M., and Jones, D. C., "Natural Convection Mass Transfer Adjacent to Horizontal Plates," *International Journal of Heat and Mass Transfer*, 16:1025 (1973).

35. Lloyd, J. R., and Moran, W. R., "Natural Convection Adjacent to Horizontal Surfaces of Various Planforms," ASME Paper 74-WA/HT-66, 1974.

36. Morgan, V. T., "The Overall Convective Heat Transfer from Smooth Circular Cylinders," in *Advances in Heat*

Transfer (T. F. Irvine and J. P. Hartnett, Eds.), Vol. 11, Academic Press, New York, 1975, pp. 199–264.

37. Churchill, S. W., and Chu, H. H. S., "Correlating Equation for Laminar and Turbulent Free Convection from a Horizontal Cylinder," *International Journal of Heat and Mass Transfer*, 18:1049–1053 (1975).

38. Churchill, S. W., "Free Convection around Immersed Bodies," in *Hemisphere Heat Exchanger Design Handbook* (G. F. Hewitt, Ed.), Hemisphere, Washington, DC (1990).

39. Vliet, G. C., "Natural Convection Local Heat Transfer on Constant-Heat-Flux Inclined Surfaces," *Transactions of the ASME*, 91C:511 (1969).

40. Fugii, T., and Imura, H. "Natural Convection Heat Transfer from a Plate with Arbitrary Inclination,"

International Journal of Heat and Mass Transfer, 15:755–767 (1972).

41. Bejan, A., *Convection Heat Transfer*, Wiley, New York, 1984.

Some end-of-chapter problems were adapted with permission from the following:

42. Chapman, A. J., *Fundamentals of Heat Transfer*, Macmillan, New York, 1987.

43. Pnueli, D., and Gutfinger, C., *Fluid Mechanics*, Cambridge University Press, Cambridge, England, 1992.

44. Shepherd, D. G., *Elements of Fluid Mechanics*, Harcourt, Brace & World, New York, 1965.

Nomenclature

A	Area (m²)	R	Specific gas constant (J/kg·K)
A_p	Projected area (m²)	R_{elec}	Electrical resistance (ohm)
c_D	Drag coefficient (dimensionless)	Ra	Rayleigh number (dimensionless)
c_f	Friction coefficient (dimensionless)	Re	Reynolds number (dimensionless)
c_p	Constant-pressure specific heat (J/kg·K)	t	Time (s)
		T	Temperature (K)
C_p	Pressure coefficient or pressure recovery coefficient (Fig. 9.25) (dimensionless)	T_f	Film temperature (K)
		T_∞	Free-stream temperature (K)
D	Diameter (m)	v_x, v_y, v_z	Cartesian velocity components (m/s)
$\dot{E}$	Energy rate (W)	V_∞	Free-stream velocity (m/s)
F	Force (N)	W	Width (m) or weight (N)
F_D	Drag force (N)	$\dot{W}$	Rate of work or power (W)
g	Gravitational acceleration (m/s²)	x, y, z	Cartesian spatial coordinates (m)
Gr	Grashof number		
F	Force (N)		

GREEK

$h_{conv,x}$	Local convective heat-transfer coefficient (W/m²·K or W/m²·°C)	α	Thermal diffusivity (m²/s) or kinetic energy correction factor (dimensionless)
$\bar{h}_{conv}$	Average convective heat-transfer coefficient (W/m²·K or W/m²·°C)	β	Volumetric thermal expansion coefficient (K⁻¹)
i	Electric current (A)	δ_H	Hydrodynamic boundary-layer thickness (m)
k	Thermal conductivity (W/m·K or W/m·°C)	δ_T	Thermal boundary-layer thickness (m)
L	Length (m)	Δ	Difference or change
M	Mass (kg)	θ	Angular position from forward stagnation point (rad)
$\dot{m}$	Mass flowrate (kg/s)		
$\hat{n}$	Unit normal vector (dimensionless)	ε	Roughness height (m) or emissivity (dimensionless)
Nu	Nusselt number (dimensionless)	μ	Viscosity (N·s/m²)
P	Pressure (Pa)	ν	Kinematic viscosity, μ/ρ (m²/s)
$\mathscr{P}$	Perimeter (m)	ρ	Density (kg/m³)
Pr	Prandtl number (dimensionless)	σ	Stefan–Boltzmann constant (W/m²·K⁴)
$\dot{Q}$	Heat transfer rate (W)	τ	Shear stress (N/m²)
$\dot{Q}''$	Heat flux (W/m²)	Φ_{visc}	Viscous dissipation function

SUBSCRIPTS

avg	average
c	characteristic
conv	convection
crit	critical
elec	electrical
f	fluid
fric	friction
in	into system or control volume
lam	laminar
out	out of system or control volume
P	pressure
s	surface

surr	surroundings for radiation exchange
tot	total
turb	turbulent
visc	viscous
w	wall
∞	ambient or freestream

SUPERSCRIPTS

*	Dimensionless quantity

OTHER NOTATION

$(^{-})$	Average quantity

QUESTIONS

9.1 Review the most important equations presented in this chapter (i.e., those with a reddish background). What physical principles do they express? What restrictions apply?

9.2 Define an external flow. List several examples of external flows.

9.3 Define a boundary-layer flow. Give an example of a boundary-layer flow.

9.4 Define a *bluff body*. List several examples of bluff bodies.

9.5 Define separation. List two or three examples of separated flow.

9.6 Discuss the relationship between a separated flow and a wake flow. List several examples of wake flows.

9.7 Sketch a uniform flow over a thin flat plate at zero incidence. Sketch the hydrodynamic boundary-layer thickness as a function of the distance from the leading edge. Explain why this behavior occurs.

9.8 Consider a laminar boundary layer on a flat plate. Sketch the velocity profile through the boundary layer for $x = 0$, $x = x_1$, and $x = x_2$, where $x_2 > x_1$. Use v_x as the ordinate and y as the abscissa. Put all three profiles on a single graph. What can you say about the velocity gradient at the wall, $[\partial v_x/\partial y]_{y=0}$, at these three locations?

9.9 How does the sketch from Question 9.7 relate to the graph from Question 9.8?

9.10 Explain what is meant by *transition*. What factors distinguish laminar and turbulent boundary layers?

9.11 Explain the physical significance of the Reynolds number.

9.12 Explain the physical significance of the Prandtl number.

9.13 Consider a flat plate held at a positive angle of incidence as shown in the sketch. Describe and contrast the flow patterns expected over the top surface and the bottom surface of the plate.

9.14 Consider the truck and trailer pictured in Fig. 9.7. Sketch and discuss the flow patterns you expect over the top surface of the cab and trailer. Also describe the flow at the rear of the trailer.

9.15 Consider a boundary-layer flow for a fluid having a Prandtl number of unity. Explain the Reynolds analogy for this situation. What is the physical basis of this analogy? That is, why does it work?

9.16 Sketch the wall shear stress distribution $\tau_w(x)$ associated with a flow over a flat plate at zero incidence. Assume the plate is long enough such that transition from a laminar to turbulent boundary layer occurs approximately half way between the leading and trailing edges.

9.17 Sketch the temperature profiles through the thermal boundary layer associated with a flow over a heated $(T_s > T_\infty)$ flat plate at

(a) the plate leading edge $(x = 0)$,

(b) some arbitrary distance from the leading edge, x_1, and

(c) a distance x_2, where $x_2 > x_1$.

Put all three curves on one graph of $T(y)$ versus y. Label T_s and T_∞. What can you say about the relative magnitudes of the temperature gradient at the wall, $[\partial T/\partial y]_{y=0}$, at these three locations?

9.18 What connections can you make between Questions 9.8 and 9.17?

9.19 Sketch the wall heat flux distribution $\dot{Q}''(x)$ associated with a flow over a flat plate at zero incidence. The plate is maintained at a temperature T_s and the flow has a temperature T_∞. Assume the plate is long enough such that the flow is always laminar and that $T_s > T_\infty$.

9.20 How does the local heat flux $\dot{Q}''(x)$ relate to the total heat-transfer rate $\dot{Q}_{0-L}$ from a flat plate of length L and width W?

9.21 Distinguish between the friction coefficient c_f and the drag coefficient c_D for a flat plate in a uniform flow.

9.22 For a flat plate in a uniform flow, is the friction coefficient c_f at all x locations numerically greater than drag coefficient c_D? Explain why or why not. Assume a laminar boundary layer.

9.23 Sketch the local heat-transfer coefficient for a laminar flat-plate flow as a function of distance from the leading edge $(x = 0)$ to the end of the plate $(x = L)$. Show on your graph the average heat-transfer coefficient for the entire plate.

9.24 How do the local and average heat-transfer coefficients relate to local and average Nusselt numbers, respectively?

9.25 Distinguish between Re_x and Re_L. In what situations would you use each?

9.26 Give a physical interpretation of the Prandtl number. List several fluids with Prandtl numbers close to unity, much greater than unity, and less than unity.

9.27 Sketch both the hydrodynamic and thermal boundary-layer thicknesses as functions of distance from the leading edge of a flat plate for a high-Prandtl-number fluid. Repeat for a low-Prandtl-number fluid.

9.28 Define the film temperature T_f. What purpose is served by evaluating thermo-physical properties at the film temperature in heat transfer and friction and drag calculations?

9.29 Consider a laminar boundary layer over a flat plate. Sketch the nondimensional velocity profile $v_x(y)/V_\infty$ as a function of the distance from the plate surface, y. On the same coordinates, plot the velocity profile for a turbulent boundary layer. Discuss the similarities and differences.

9.30 Repeat Question 9.29 for the temperature distribution $T(y)$ through a flat-plate boundary layer.

9.31 Distinguish between skin friction drag and form drag. Give examples of flows dominated by one or the other.

9.32 Define adverse and favorable pressure gradients in a flow. How do they relate to the phenomenon of separation?

9.33 What is the drag crisis? Explain in detail.

9.34 How are golf-ball dimples related to the drag crisis?

9.35 Explain how streamlining an object reduces drag when streamlining usually involves adding more surface area exposed to fluid friction.

9.36 Sketch the temperature profile through the thermal boundary layer associated with a vertically oriented flat plate. The plate surface temperature is *greater* than that of the quiescent surroundings (i.e., $T_s > T_\infty$). Use a single set of $T(y)$ versus y coordinates and draw graphs for two axial locations, x_1 and x_2, where $x_2 > x_1$. Label each curve.

9.37 Repeat Question 9.36, adding velocity distributions for the two locations; that is, plot both $v_x(y, x_1)$ and $v_x(y, x_2)$ versus y on a single graph. Label.

9.38 Sketch the temperature and velocity profiles through the thermal boundary layer associated with a vertically oriented flat plate. The plate surface temperature is *less* than that of the quiescent surroundings (i.e., $T_s < T_\infty$). Plot $T(y)$ and $v_x(y)$ for an arbitrary location from the leading edge. Where is the leading edge relative to that for Question 9.36?

9.39 Show that the units in each term of the free-convection x-momentum (Eq. 9.57) and energy (Eq. 9.58) equations are consistent. Express all quantities in terms of the fundamental dimensions of mass, length, time, and temperature.

9.40 What dimensionless parameters are important to the problem of free convection from a flat plate? What is the physical significance of each of these parameters?

Chapter 9 Problem Subject Areas

9.1–9.25	**Flat-plate friction and drag**
9.26–9.37	**Flat-plate heat transfer**
9.38–9.47	**Flat-plate combined drag and heat transfer**
9.48–9.64	**2-D shapes: drag and heat transfer**
9.65–9.77	**3-D shapes: drag and heat transfer**
9.78–9.113	**Free convection**

PROBLEMS

9.1 Consider a thin, flat plate that is 0.4 m long and 0.2 m wide. Determine the Reynolds number based on plate length for 2-m/s flows of the following fluids: air, water, motor oil, and glycerin. Assume a temperature of 300 K. Indicate whether the flow at the trailing edge is laminar or turbulent. See Appendices C and G for thermo-physical property data.

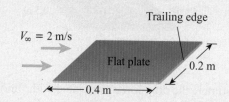

9.2 Consider a flow over a 10-cm-long flat plate. Determine the maximum free-stream velocity V_∞, such that the boundary layer is laminar over the entire length of the plate for flows of all of the following: air, water, and oil. Assume a critical Reynolds number of 5×10^5. The temperature and pressure are 300 K and 1 atm, respectively. Also determine the boundary-layer thickness at $x = L$ for each fluid.

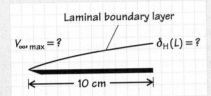

9.3 A fluid at 1 atm and 40°C flows over a 30-cm-long, flat plate. Find the maximum free-stream velocity of the fluid for which laminar flow at the trailing edge is probable for the following fluids: (a) air, (b) ammonia, (c) water, and (d) ethylene glycol.

9.4 Repeat Problem 9.3 for an ambient pressure of 5 atm. All other conditions remain the same.

9.5 A fluid at 60°C flows over a sharp-edged flat plate with a free-stream velocity of 50 m/s. Determine whether the flow is likely to be laminar or turbulent at a location 50 cm from the leading edge for the following fluids: (a) saturated water, (b) engine oil, and (c) air at 1 atm.

9.6 Air at atmospheric pressure and 5°C flows over a flat plate with a free-stream velocity of 45 m/s. Plot the laminar, hydrodynamic boundary-layer thickness as a function of the distance from the leading edge up to the distance corresponding to a local-length Reynolds number of 5×10^5.

9.7 Consider a uniform flow of air at 300 K and 1 atm over a flat plate at zero incidence. The free-stream velocity is 5 m/s. Plot the wall shear stress on the top surface of the plate, $\tau_w(x)$, from $x_1 = 0.01$ m to $x_2 = 0.03$ m. In what way can this graph be interpreted to obtain the drag force associated with the region $x_1 \leq x \leq x_2$? Determine the drag force in newtons associated with this region. The plate width is 0.5 m.

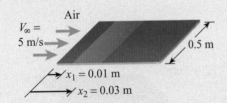

9.8 Estimate the drag force associated with fluid friction on the hull of a model sailboat traveling at 0.8 m/s. Model the hull as two flat plates, each 0.5 m long

and 0.15 m wide. Consider the drag only on the outside surface of each plate. Assume a water temperature of 300 K.

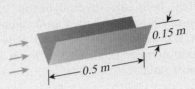

9.9 Determine the shear stress in the fluid at the wall of a model boat twice as long as that in Problem 9.8 at midship ($L/2 = 0.5$ m) and at the stern ($L = 1$ m). Use the same velocity and fluid properties as in Problem 9.8. Which is larger and why?

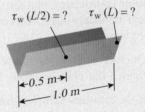

9.10 Estimate the frictional drag on the hull of a supertanker moving at 8 m/s. Model the hull as a 300-m-long by 60-m-wide flat plate. How much larger is the drag force if the hull is encrusted with barnacles? Assume a roughness height ε of 12 mm is associated with the encrustation. (Note: Reynolds numbers here are larger than the largest values used to create drag correlations. Feel free to extrapolate.)

9.11 Consider a uniform flow over a flat plate of width b and length L. The Reynolds number at the rear edge of the plate is 2×10^5. Determine the fraction of the total drag force exerted on the plate that is associated with the rearward half of the plate; that is, find

$$F_{D,L/2-L}/F_{D,0-L}.$$

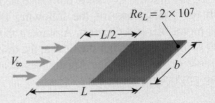

9.12 Repeat Problem 9.11 for the case where $Re_L = 2 \times 10^7$. Assume that the boundary-layer flow is effectively turbulent over the entire plate at this condition. Compare your results with those of Problem 9.11 and discuss.

9.13 Estimate the ratio of the laminar boundary-layer thickness at a given distance from the leading edge of a flat plate for air flowing at 100 ft/s, 70 F, and 1 atm to that for water flowing at 5 ft/s and 70 F.

9.14 A thin, 2-ft-long by 1-ft-wide plate is secured in a uniform stream of water moving at a velocity of 2 ft/s and having a temperature of 50 F. The angle of incidence is zero.

   A. Will the boundary layer be laminar over the entire length of the plate?

   B. What are the minimum and maximum boundary-layer thicknesses, and where are they found?

   C. Find the force component parallel to the direction of water flow necessary to hold the plate in position.

   D. Plot the shear stress at the plate surface as a function of x/L.

9.15 Consider a thin, flat plate at zero incidence in a uniform flow. For a plate of given length L, find the ratio of the total drag force in water to that in air when the plate Reynolds numbers are the same in both fluids. Assume temperatures of 288 K for water and 298 K for air at atmospheric pressure.

9.16 A thin, rectangular plate is towed through water at 295 K at a velocity of 6.25 m/s. Assume that the boundary layer is turbulent from the leading edge.

   A. What is the thickness of the boundary layer 1 m from the leading edge?

   B. At $x = 1$ m, what is the axial velocity v_x at the point where $y = \delta_H/2$?

9.17 A very thin, smooth, flat plate is set parallel to an air stream having a velocity of 62.5 m/s, a pressure of 104.8 kPa, and a temperature of 198 K. The plate has a dimension of 1.6 m in the flow direction and is 3.75 m wide (span).

   A. Assuming that a laminar boundary layer exists over the entire plate, calculate the thickness of the boundary layer at the trailing edge and the total drag on the plate.

   B. Repeat Part A assuming that the boundary layer is wholly turbulent.

9.18 A thin, 0.2-m by 0.5-m, metal plate is held in a parallel flow of water. The water velocity is 4 m/s and the water density and kinematic viscosity are 1000 kg/m^3 and 1×10^{-6} m^2/s, respectively. Find the drag force exerted on the plate (a) for the flow

parallel to the shorter (0.2-m) side and (b) for the flow parallel to the longer (0.5-m) side.

9.19 Repeat Problem 9.18 for a 2-m by 5-m plate (i.e., a plate ten times as large).

9.20 A 1-m by 0.5-m, flat plate is held at zero incidence in a stream of air flowing at 1 m/s. The plate can be oriented with the air flow parallel to the long side or parallel to the short side. What is the minimum possible drag force on the plate?

9.21 A 0.1-m by 0.05-m, flat plate is inserted into a flow such that a boundary layer develops along the longer side. The shear force on the plate is 0.1 N. Estimate the force expected to act on a 0.2-m by 0.1-m plate in the same flow.

9.22 A fluid flows between two wide, 1-m-long, parallel flat plates separated a distance of 0.5 m as shown in the sketch. The two plates are attached to a structure (not shown). The velocity of the oncoming stream is 0.25 m/s, and the fluid density and viscosity are 1000 kg/m^3 and 1×10^{-3} N-s/m^2, respectively. Find the force (per unit width) transferred from the plates to the structure.

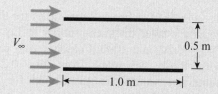

9.23 Consider the aircraft shown in the sketch. An air scoop takes air from the free-steam outside of the boundary layer that grows along the fuselage. For 1-atm air at 300 K and a flight speed of 1000 km/hr, estimate the relationship between the dimensions a and b to ensure that the incoming air is from the free-stream. *Hint:* Treat the fuselage as a flat plate.

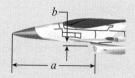

9.24 Use the von Kármán integral analysis (Appendix 9B) to derive expressions for δ_H/x and $c_{f,x}$ for flow over a flat plate. Assume the velocity profile through the boundary layer is linear.

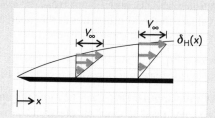

9.25 Assuming a velocity distribution in the boundary layer of the form $V = V_\infty \sin(\pi y/2\delta)$, apply the von Kármán integral analysis (Appendix 9B) to find the shear stress $\tau_w(x)$ on a flat plate.

9.26 A 0.3-m-long and 0.15-m-wide circuit board in a computer is cooled with a fan. Estimate the heat transfer from the circuit board if the board is maintained at 325 K, and the air flows with a velocity of 5 m/s at a temperature of 300 K. Calculate the total heat-transfer rate based on the heat loss from both sides of the board.

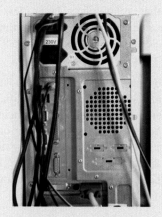

9.27 Use Fig. 9.14 to obtain expressions for the thermal boundary-layer thickness similar to Eq. 9.28a for $Pr = 7$, 50, and 1000. Compare your expressions with those generated from Eq. 9.28b when numerical values are substituted for the three Prandtl numbers.

9.28 Use the information presented in Fig. 9.14 together with the following conditions listed to create a plot of $T(y)$ versus y for the thermal-boundary-layer flat plate:

$$T_s = 80°C, \quad Pr = 7,$$
$$T_\infty = 30°C, \quad v = 8.6 \times 10^{-7}\ \text{m}^2/\text{s},$$
$$V_\infty = 1.7\ \text{m/s} \quad k = 0.6\ \text{W/m·°C},$$
$$x = 0.10\ \text{m}.$$

Use your plot to determine the heat flux, $\dot{Q}''(x)$, at the plate surface. Compare your result with that calculated using $h_{\text{conv},x}$ from Eq. 9.29.

9.29 Saturated water at 80°C flows over a flat plate maintained at 40°C. The free-stream velocity is 1.5 m/s. Using the results of the analytical solution given in Figs. 9.13 and 9.14, plot the velocity and temperature profiles in the boundary layer at stations 2.5, 5.0, and 7.5 cm from the leading edge. Evaluate the fluid properties at the mean of the free-stream and surface temperatures.

9.30 Derive Eq. 9.29 starting with an energy balance at the surface and applying the results of Fig. 9.14.

9.31 Four rectangular fins, each 12 mm high and 100 mm long, are used to cool an electrical power supply.

Air at 300 K and 1 atm is blown by a fan along the length of the fins with a velocity of 4 m/s. The temperature of the fins is 310 K. Determine the average convective heat-transfer coefficient at the surface of the fins. Also, estimate the total heat-transfer rate from the fins.

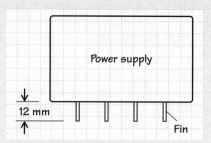

9.32 Imbedded electrical resistance heaters heat a 0.3-m-wide, 0.7-m-long, horizontal glass plate. The electrical current supplied to the heaters is 3 A at 12 V. A fan blows air at 300 K and 1 atm over the top surface of the plate along its length, while the lower surface is insulated. The air velocity is 6 m/s. Determine the following quantities:

A. The location and the value of the maximum surface temperature of the plate

B. The average temperature difference between the plate surface and the air, $\overline{T_s - T_\infty}$

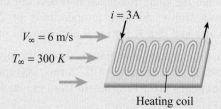

Heating coil

9.33 A blacktop tennis court is located on the top of a concrete slab elevated somewhat from the surrounding ground. The dimensions of the court are 18.3 m by 36.6 m. On a bright, sunny day, the surface of the blacktop is approximately 310 K when the wind is blowing at 15 miles/hr (6.7 m/s) along the length of the court. The air temperature is 302 K. Estimate the average convective heat-transfer coefficient and the total convection heat-transfer rate from the blacktop surface. Assume that the boundary layer is "tripped" by the edge of concrete slab.

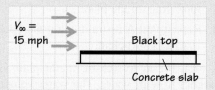

9.34 Atmospheric air at 15°C flows over a 24-cm-long, flat plate maintained at 45°C. The free-stream velocity is 7.5 m/s. Using air properties evaluated at the film temperature, find (a) the local heat transfer coefficient at 6, 12, 18, and 24 cm from the leading edge and (b) the average heat-transfer coefficient for the entire surface.

9.35 Saturated water at 80°C flows over a 7.5-cm-long, flat plate maintained at 40°C. The free-stream velocity is 1.5 m/s. Find (a) the local heat-transfer coefficient at 2.5, 5.0, and 7.5 cm from the plate leading edge and (b) the average heat-transfer coefficient for the entire plate.

9.36 Air at 1 atm and 50°C flows over a 0.2-m-long and 0.1-m-wide, flat plate. The plate is heated to 100°C. The plate Reynolds number Re_L is 40,000.

   A. What is the heat-transfer rate from one side of the plate?

   B. If the air velocity is doubled and the air pressure increased to 10 atm, what is the heat transfer rate?

9.37 An isothermal flat plate has the dimensions L by $2L$. Based on a transition Reynolds number of 5×10^5, determine the conditions of a fluid stream flowing past the surface for which the total heat-transfer rate is the same when either the L side or $2L$ side is the leading edge.

9.38 Glycerin at a free-stream temperature of 40°C flows with a velocity of 3 m/s over a flat plate maintained at 10°C.

   A. Find the local boundary-layer thickness, the local skin friction coefficient, and the local heat-transfer coefficient at the following locations from the leading edge of the plate: (a) 0.1 m, (b) 0.2 m, and (c) 0.3 m.

   B. If the plate is 0.3 m long and 0.3 m wide, find the heat-transfer rate and the drag force associated with one side.

9.39 Air flows over a flat plate with $V_\infty = 80$ ft/s and $T_\infty = 70$ F. The pressure is 1 atm. The surface temperature of the 6-in-long, 12-in-wide plate is maintained at 175 F.

   A. Determine the local boundary-layer thickness, the local heat-transfer coefficient, and the local skin friction coefficient at the following locations from the leading edge of the plate: (a) 2 in, (b) 4 in, and (c) 6 in.

   B. Calculate the heat-transfer rate and the drag force associated with one side.

9.40 Glycerin at 40°C flows at 4 m/s over a 0.3-m-long by 0.5-m-wide, thin, flat plate maintained at 10°C. Calculate the heat-transfer rate and the drag force associated with one side.

9.41 Water at 90°C flows with a velocity of 1 m/s over a 0.15-m-long, flat plate maintained at 50°C. Calculate the local heat-transfer coefficient and the local skin friction coefficient at the following locations from the plate leading edge: 0.01, 0.05, 0.1, and 0.15 m. Also calculate the average heat-transfer coefficient and the drag coefficient for the plate.

9.42 Atmospheric-pressure air at 20°C flows with a velocity of 25 m/s over a 15-cm-long, thin, flat plate maintained at 80°C. Calculate the average heat-transfer coefficient and the drag coefficient for the plate.

9.43 A 1.2-m by 0.5-m, thin, flat plate is towed parallel to its long side at a velocity of 2 m/s through engine oil. The oil temperature is 80°C and the plate is electrically heated to maintain its surface at 100°C. Estimate the power required (a) to tow the plate and (b) to heat the plate.

9.44 Atmospheric air at 0 F flows over a flat surface maintained at 100 F. The surface is 2 ft long and 1 ft wide. Determine the air velocity required to achieve a heat-transfer rate of 500 Btu/hr from one side of the surface?

9.45 Atmospheric air flows at 25 m/s over a 50-cm-long, flat plate. The air temperature is 15°C and the plate surface temperature is 75°C.

   A. Determine the local heat-transfer coefficient and local skin friction coefficient at locations 12.5, 25, 37.5, and 50 cm from the plate leading edge.

   B. Determine the average heat-transfer coefficient and drag coefficient for one side of the surface.

9.46 A 3-in by 18-in, flat plate is maintained at 190 F. The plate is held in a 90-ft/s and 50-F air stream at 1 atm. Find the heat-transfer rate and the drag force associated with one side of the plate if the leading edge is (a) the 18-in side and (b) the 3-in side.

9.47 Nitrogen at atmospheric pressure and 50°C flows at 20 m/s over a flat plate maintained at 0°C. The plate is 0.8 m long and 0.2 m wide. Find (a) the heat-transfer rate to one side of the plate, (b) the drag force exerted on one side of the plate, and (c) the fraction of the heat transferred and drag force in the laminar portion of the boundary layer.

9.48 The Delaware Memorial Bridge connects southern New Jersey with the Wilmington–New Castle area of Delaware. The twin suspension spans of this bridge each accommodate four lanes of traffic. Each span employs two 0.508-m-diameter, 1250-m-long suspension cables. Estimate the drag force exerted on a pair of these cables by a strong breeze (25 miles/hr = 11.2 m/s) and by a storm wind

(50 miles/hr = 22.4 m/s). Assume the wind blows perpendicular to the cables.

Photograph courtesy of the Delaware River and Bay Authority.

9.49 Consider a telephone cable pole line in which a 50-mm-diameter cable is secured to wooden poles at 120-m intervals. The poles have a diameter of 0.30 m and are 9 m tall. The cable is connected at a distance of 2.4 m from the top of the pole. Estimate the bending moment at the base of the pole when the wind speed is 30 miles/hr (= 13.4 m/s) and the wind is directed perpendicular to the pole line.

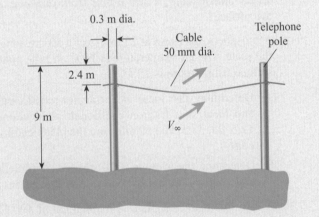

9.50 The elevated portions of the Trans-Alaska Oil Pipeline are constructed from 1.22-m-o.d. steel pipe with a 95-mm-thick layer of composite fiberglass insulation. A thin metal sheet is bonded to the exterior surface of the fiberglass for protection. Estimate the heat-transfer rate per unit length of pipe for an ambient air temperature of −40°C, a wind velocity of 35 miles/hr (15.6 m/s), and a pipeline surface temperature of −20°C.

9.51 A hot-wire anemometer is a device that is used to measure the velocity of flowing fluids. Such a device is employed to measure the air speed in a wind tunnel. The air flow is directed perpendicular to the 5-μm-diameter, 1.25-mm-long wire as shown in the sketch. A current of 24 mA is supplied to the wire, and the electrical resistance of the wire is 6 ohms. The wire temperature is 345 K. The air temperature and pressure are 300 K and 1 atm, respectively. Estimate the tunnel air speed.

9.52 Aircraft at the beginning of the twentieth century were typically biplanes. The upper and lower wings of these planes were held in place by a complex system of cables. Estimate the power required to overcome the drag associated with the cables. Assume a combined length of 100 m of 6-mm-diameter cable and a plane flying at 75 miles/hr (33.5 m/s). What fraction of the total engine power of 90 hp does this represent? How much could this drag be reduced if the cables could be streamlined by adding ellipsoidal fairings?

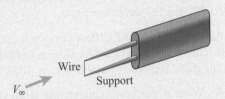

Photograph courtesy of Joe Nonneman.

9.53 Certain small spiders disperse over great distances by "ballooning." These spiders are carried aloft by vertical air currents acting on a long silk strand spun by the spider. Estimate the vertical air velocity needed to support a spider and its strand for a spider diameter of 2.67 mm, a spider mass of 10 mg, and a strand length and diameter of 2 m and 3 μm, respectively. Assume the spider can be treated as a sphere, and assume a density of 1000 kg/m³ for the strand. Treat the spider and strand as a single rigid unit ignoring any moments. The air temperature and pressure are 300 K and 1 atm, respectively.

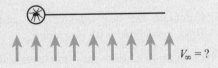

9.54 Consider a fifty-story office building. The 520-ft tall building has a square design with 100-ft-long sides. Ignoring end effects, estimate the drag force exerted on the building for wind velocities of 10, 30, and 100 miles/hr. Assume the wind is aligned with the face of the building and strikes the left side of the building head on. How do your results change if the wind is aligned at 45° to the building?

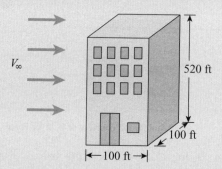

9.55 Consider the building described in Problem 9.54. Determine the average heat-transfer coefficient associated with the convective heat transfer created by the wind. Use the same wind conditions given in Problem 9.54. Evaluate the thermo-physical properties of air at 275 K.

9.56 Consider the heat transfer from a cylinder in a cross-flow. Use data from Fig. 9.26 to obtain a correlation for the local Nusselt number at the stagnation point ($\theta = 0°$) as a function of the Reynolds number in the following form:

$$Nu_\theta(\theta = 0°) = C\,Re_D^n.$$

In other words, find the "best-fit" values for C and n. Spreadsheet or other software can be used to accomplish this task.

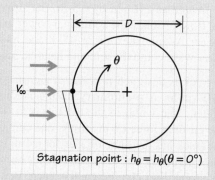

9.57 The leading edge of the wing of a small airplane is electrically heated to prevent ice from forming. Estimate the heat flux required to maintain the front portion of the leading edge (the stagnation point) at a temperature of 5°C for a plane flying at 150 miles/hr (67 m/s). The ambient temperature and pressure are −5°C and 85 kPa, respectively. Model the wing leading edge as a long cylinder ($D =$

0.10 m) in a cross-flow. *Hint:* Use the results from Problem 9.56.

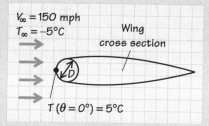

9.58 A hot-wire anemometer is used to measure the velocity of a fluid by determining the heat loss from a fine wire, heated to a known temperature, placed in the fluid stream. A particular hot-wire probe uses a 0.25-mm-diameter wire, placed normal to an air stream at 1 atm and 15°C. When the wire is electrically heated so that its surface temperature is 100°C, the heat loss from the wire is measured to be 40 W per meter of wire length. Estimate the velocity of the air stream.

9.59 Air at 190 F and 1 atm flows past a 1/16-in-diameter, heated wire. For an air velocity is 20 ft/s, find the heat loss per unit length of wire (Btu/hr·ft) if the wire surface temperature is 300 F.

9.60 Air at 80 F flows with a velocity of 90 ft/s normal to a 1-in-diameter cylinder. The cylinder surface temperature is 250 F. Find the heat transfer rate (Btu/hr·ft) and drag force (lb$_f$/ft) on the cylinder per unit length when the air pressure is (a) 1 atm, (b) 2 atm, and (c) 4 atm.

9.61 The surface temperature of a 5-cm-diameter tube is maintained at 100°C in a cross-flow. The oncoming fluid (20°C and 1 atm) flows with a velocity of 10 m/s. Determine the heat-transfer rate (W/m) and drag force per unit length (N/m) for the following fluids: (a) air, (b) water, and (c) ethylene glycol.

9.62 Hot combustion gases, having properties close to those of nitrogen, flow at a pressure of 175 kPa and a temperature of 60°C normal to a 5-cm-diameter cylinder at a velocity of 15 m/s. Determine the average convective heat-transfer coefficient when the pipe surface temperature is 120°C. What is the drag force per unit length (N/m) on the pipe?

9.63 Repeat Problem 9.62 for the flow of the same gases at the same conditions past a square pipe with 5-cm sides. The pipe is oriented with one side normal to the flow, and the pipe surface temperature is 120°C.

9.64 A fluid at 40°C flows with a velocity of 10 m/s normal to a 5-mm-diameter cylinder heated to 120°C. Calculate the heat-transfer coefficient at the cylinder surface and the drag force per unit length for the following fluids: (a) helium at 1 atm, (b) air at 1 atm, and (c) water.

9.65 A 10-mm-diameter glass ball ($\rho = 2225$ kg/m^3) is dropped into a tall cylinder containing glycerin at 300 K. Determine the steady-state velocity (i.e., the terminal velocity) of the glass ball. Be sure to include the buoyant force in your analysis.

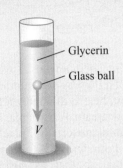

Glycerin

Glass ball

V

9.66 Consider the glass ball in Problem 9.65 now dropped from the top of a tall building. Estimate the terminal velocity of the glass ball when the air temperature and pressure are 25°C and 99 kPa, respectively. Compare this result with that of Problem 9.65 and discuss.

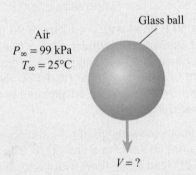

Glass ball

Air
$P_\infty = 99$ kPa
$T_\infty = 25°C$

$V = ?$

9.67 A 4-mm-diameter spherical particle is dropped into a tank of water. The density of the particle is 3000 kg/m^3 and the density of the water is 1000 kg/m^3. Assuming the kinematic viscosity of water is 1×10^{-6} m^2/s, find the terminal velocity of the particle.

9.68 An automobile with a projected area of 20 ft^2 has a drag coefficient of 0.3. What power (hp) is required to overcome air resistance at 70 miles/hr in still air at 60 F?

9.69 Find the ratio of the velocities of a 5-mm-diameter spherical drop of water falling at constant velocity through air and a 5-mm-diameter spherical bubble rising at constant velocity through water. Assume air at 10°C and 29.92 in Hg and water at the same temperature.

9.70 The drag coefficient for a low-porosity parachute is 1.2 for Reynolds numbers greater than 10,000. The drag coefficient is based on the frontal area. What diameter parachute is required to allow an electronics package weighing 50 N to have a soft landing? A soft landing is defined to have an impact velocity of 4 m/s or less.

Photograph courtesy of U. S. Air Force.

9.71 A 16-lb$_m$ (7.26-kg) bowling ball is dropped from the deck of a ship into the ocean. The diameter of the ball is 8.55 in, and the density of the seawater is 1027 kg/m^3. At steady state, has the boundary layer associated with the ball experienced a transition from laminar to turbulent flow? Support your conclusion with calculations. Does your conclusion change if the ball is falling through air? Again, support your result with computations. Note that the buoyant force is important.

9.72 Consider the blue whale described in Example 9.10. Calculate the drag force and power associated with a blue whale swimming at 5 m/s. Repeat your calculations for 10 m/s. Compare the results for the two speeds and for the maximum speed given in Example 9.10.

9.73 A car travels at 80 km/hr. To cool off, the driver puts her arm out of the window with her hand held vertically upright. Assuming the arm and hand can be treated as a 0.1-m by 0.5-m, flat strip normal to the flow, estimate the additional power expended by the car. Assume the air is at 300 K and 1 atm. In what other ways could the arm and hand be modeled?

9.74 A truck is designed to travel at a cruising speed of 110 km/hr. A preliminary design is shown in the sketch. The width of the truck is 4 m. To estimate the air drag force on the truck, one can assume that the truck has the same drag as for a 4-m by 4-m, flat plate perpendicular to the flow. It has been suggested to streamline the geometry such that the drag coefficient becomes 0.6. Estimate the power saved at cruising speeds by streamlining.

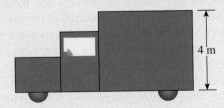

4 m

9.75 To manufacture lead shot, uniform, molten lead droplets fall through atmospheric air and solidify, forming nearly spherical (2.29-mm-diameter) particles. Estimate the convective heat-transfer coefficient for a single solid lead sphere after the terminal velocity has been reached during its free fall. The temperature of the air is 300 K. Assume the sphere temperature is 400 K.

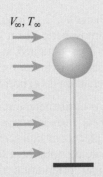

Photograph courtesy of Axel Hennig, www.australienbilder.de.

9.76 You are asked to assist in the design of an experiment. A thin-walled, aluminum, 60-mm-diameter sphere is to be placed in a wind tunnel and its surface temperature is to be controlled by an electrical resistance heater placed inside the sphere. Your job is to estimate the power requirement of the electrical heater if the maximum flow velocity in the tunnel is 20 m/s and the maximum desired surface temperature is 30°C. Assume the air in the tunnel has a temperature of 25°C and the pressure is 1 atm.

9.77 Estimate the drag force exerted on the sphere in Problem 9.76 and determine the bending moment applied to the 20-cm-long rod holding the sphere in place. Neglect the drag associated with the rod.

9.78 Transform the dimensional mass and energy equations for a free convection boundary layer to their dimensionless forms; that is, transform Eqs. 9.56 and 9.58 to Eqs. 9.69 and 9.70.

9.79 Estimate the convective heat transfer from the air in a room at 22.5°C to the three exterior walls at 17°C.

The room is 3.6 m wide, 6 m long, and 2.4 m high. One of the 6-m-long walls is an interior wall and should not be included in your calculations. Neglect end effects at the floor and ceiling.

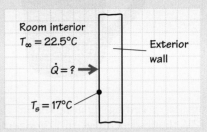

9.80 Estimate the convective heat loss per unit length from a sheet-metal, hot-air heating duct to the ambient air. The width of the duct is 0.6 m and the height is 0.3 m. The duct temperature is 85 F (303 K) and the ambient air temperature is 68 F (293 K). Compare this result to the radiant heat-transfer rate per unit length from the duct to surroundings at 68 F (293 K).

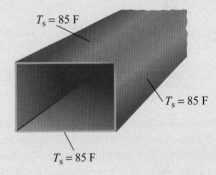

9.81 Thin fins (90 mm long and 30 mm high) are to be used to cool a laser power supply. For a fin temperature of 305 K and an ambient air temperature of 293 K, estimate the number of fins to dissipate 125 W. Assume the fins are sufficiently spaced such that the boundary layers do not interact.

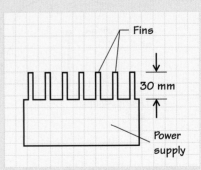

9.82 Can the convective heat transfer from the exterior surface of a coffee mug (100-mm-diameter by 100-mm-long cylinder) be modeled as a flat plate? The surface temperature of the mug is 305 K and the ambient air temperature is 295 K. If so, what is the convective heat transfer rate?

9.83 Consider a ceiling lighting fixture that contains 34-W fluorescent tubes in a horizontal position. The fluorescent tubes are 1.22 m long with a 36-mm diameter. A student using a fine-wire type-K thermocouple measures the surface temperature of a tube to be 34°C when the ambient air temperature is 20°C. Estimate the convective and radiant heat losses from a single tube. Assume the glass envelope radiates as a blackbody ($\varepsilon = 1$). Compare the estimated total heat loss rate from the bulb to the electrical power input (34 W).

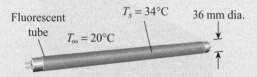

9.84 On a cold winter day, an engineering student uses a 1.5-mm-diameter thermocouple to measure the air temperature in a dormitory room. The thermocouple readout indicates a temperature of 68 F. The average temperature of the walls of the room is 40 F. Estimate an upper bound error in the air temperature measurement. The emissivity of the thermocouple is approximately 0.2.

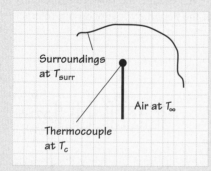

9.85 A 20-cm-long, 10-cm-wide, flat plate is oriented vertically in still atmospheric air. The air temperature is 40°C, and the plate surface is maintained at 90°C. Find the average heat-transfer coefficient for the plate surface.

9.86 Repeat Problem 9.85 using helium as the ambient fluid.

9.87 The plate of Problem 9.85 is now placed vertically in 40°C water. Determine the average convective heat-transfer coefficient.

9.88 For the plate of Problem 9.85, find the local convective heat-transfer coefficient and the boundary-layer thickness at a location (a) 10 cm from the leading edge and (b) 20 cm from the leading edge.

9.89 A 0.3-m-high, 0.2-m-wide plate is maintained at 150°C and placed vertically in still air at 40°C. Find the total heat loss from one side of the plate if the air pressure is (a) 0.1 atm, (b) 1 atm, (c) 5 atm, and (d) 10 atm.

9.90 Estimate the heat loss rate from one side of a 0.6-m-high, 0.3-m-wide, vertical plate maintained at 120°C in still nitrogen at $-15°C$ and 70 atm.

9.91 A 40-cm by 40-cm steel plate is heated to 90°C in a furnace. The plate is removed from the furnace and hung vertically in atmospheric air at 30°C. Estimate the initial cooling rate of the plate.

9.92 Repeat Problem 9.91 with the plate placed in 30°C water rather than air.

9.93 An electrically heated, 20-in-square plate hangs vertically in a still fluid. The fluid pressure and temperature are 1 atm and 70 F, respectively. Estimate the required plate surface temperature if the plate is to dissipate (both sides combined) 990 Btu/hr of heat when the fluid is (a) air and (b) water.

9.94 An engine oil bath at 60 F (289 K) is heated using a thin, 8-in-square, electrically heated, vertical plate immersed in the oil. What is the maximum power input allowed if the plate temperature is not to exceed 250 F (394 K)?

9.95 A 75-mm-diameter, 2-m-long cylinder is placed in quiescent atmospheric air at 40°C. The surface of the cylinder is maintained at 90°C. Find the total heat loss from the cylinder by free convection if the cylinder orientation is (a) horizontal and (b) vertical.

9.96 A 6-ft-long, 6-in-diameter circular cylinder is heated to maintain its surface temperature at 450 F. The cylinder is placed in quiescent atmospheric air at 75 F. Estimate the total heat loss from the cylinder if the cylinder orientation is (a) horizontal and (b) vertical.

9.97 A 10-cm-diameter circular electrical heater plate is located horizontally in a tank containing lubricating (engine) oil at 20°C. Calculate the required electrical power input to the plate to maintain the upper and lower surfaces of the plate at 200°C.

9.98 Repeat Problem 9.97 if the plate is placed in atmospheric air at 20°C.

9.99 A heated circular grill, positioned horizontally, has a diameter of 25 cm and a surface temperature of

130°C. The grill surface loses heat by free convection to atmospheric air at 25°C and simultaneously experiences radiant exchange with a large room, also at 25°C. Assuming the surface emissivity is 0.75, estimate the total heat loss from the top of the grill.

9.100 Find the average heat-transfer coefficient for free convection for a horizontal, 14-cm-diameter cylinder with a surface temperature of 90°C in atmospheric air at 30°C.

9.101 A horizontal cylinder with a surface temperature of 200°C is placed in atmospheric air at 40°C. Calculate the free convective heat-transfer coefficient for cylinder diameters of (a) 0.025 cm, (b) 0.25 cm, (c) 2.5 cm, and (d) 25 cm.

9.102 A nominal 12-in-diameter, schedule-40 steam pipe (with an outside diameter of 12.75 in) passes horizontally through a room with the air at 100 F. The pipe surface temperature is 600 F. Determine the heat loss by free convection per unit length of pipe (Btu/hr·ft)?

9.103 Consider a horizontal, 10-cm-diameter pipe with a surface temperature of 40°C. Find the heat-transfer coefficient for free convection when the pipe is surrounded by (a) methane at −20°C and 350 kPa and (b) water at 90°C.

9.104 Cold air at 50 F flows through a long, 8-in-diameter, air-conditioning duct. The horizontal duct is not insulated and is exposed to ambient air at 90 F. Estimate the heat-transfer rate to the duct per unit length.

9.105 Compare the free-convection heat-transfer rates (W) from a horizontal, 5-cm-diameter, 1.5-m-long cylinder maintained at 80°C when exposed to the following still environments at 20°C: (a) atmospheric pressure air, (b) air at 2 atm pressure, (c) carbon dioxide at 1 atm, and (d) water.

9.106 A horizontal, uninsulated, steam pipe passes through a large room where the ambient still air and the walls are at 25°C. The nominal 5-in pipe (with an outside diameter of 141.3 mm) has a surface temperature of 125°C and a surface emissivity of 0.8. Estimate the total heat loss rate per unit length (W/m) considering both free convection and radiation.

9.107 A fine (0.025-mm-diameter) wire is heated electrically in a horizontal position while exposed to quiescent helium at 3 atm and 10°C. Determine the electrical power per unit length (W/m) that must be supplied to maintain the wire surface temperature at 240°C.

9.108 A 40-W fluorescent light tube is 3.8 cm in diameter and 1.2 m long. Sixty percent of the electrical energy supplied to the tube is converted into light. The tube wall absorbs the remainder. Neglecting radiation, estimate the operating surface temperatures of the tube when the tube is exposed to atmospheric air at 20°C in both horizontal and vertical positions. For the vertical position, assume the tube can be modeled as a flat plate.

9.109 A 550-W, electrical-resistance heater is designed to operate in a quiescent water bath at 20°C. The heater is a horizontal, 30-cm-long, 1-cm-diameter rod.

A. Estimate the surface temperature of the heater when operating in water as designed.

B. Estimate the temperature of the heater if the heater is inadvertently turned on while in atmospheric air at 20°C.

9.110 A horizontal, 4.75-mm-diameter, electrical wire dissipates 13 W of power per unit length to the surrounding atmospheric air at 25°C. Estimate the surface temperature of the wire.

9.111 A 1.5-cm-diameter sphere is electrically heated to a surface temperature of 100°C while suspended in still atmospheric air at 20°C. Calculate the power required to maintain the surface at 100°C.

9.112 A 5-cm-diameter sphere is heated by 640 W of electrical power. The sphere is submerged in a quiescent body of water at 30°C. Estimate the surface temperature of the sphere.

9.113 A 2.5-cm-diameter sphere has an internal heat source that maintains its surface at 90°C. Equilibrium is maintained by free convection from the sphere to a surrounding fluid at 20°C. Find the heat-transfer rate (W), when the fluid is (a) atmospheric air, (b) water, and (c) ethylene glycol.

Appendix 9A
Boundary-Layer Equations

In this appendix, we derive the governing equations for boundary-layer flows by simplifying the most general expressions of mass, momentum, and energy conservation presented in Chapters 3, 6, and 5, respectively.

We begin by explicitly listing our assumptions:

1. The flow is incompressible.
2. The fluid is Newtonian.
3. The thermo-physical properties (ρ, μ, k, and c_p) are constant.
4. The flow is steady state [$\partial(\ )/\partial t = 0$].
5. The flow is two dimensional [$v_z = 0$ and $\partial(\ )/\partial z = 0$].
6. The boundary layer is thin [$\partial(\ )/\partial y >> \partial(\ )/\partial x$ and $v_x >> v_y$].
7. There is negligible viscous dissipation ($\Phi_{\text{visc}} = 0$).

9A.1 MASS CONSERVATION (CONTINUITY)

Equation 3.39 is the most general expression of mass conservation. We write this equation and cross out terms, indicating the specific assumption that causes any particular term to be zero, as follows:

$$\underset{0}{\cancel{\frac{\partial \rho}{\partial t}}}^{①} + \frac{\partial}{\partial x}(\rho v_x) + \frac{\partial}{\partial y}(\rho v_y) + \underset{0}{\cancel{\frac{\partial}{\partial z}(\rho v_z)}}^{⑤} = 0. \tag{9A.1}$$

Since the fluid is incompressible (assumption 1), the density can be removed from the derivatives, and we can write

$$\rho \frac{\partial v_x}{\partial x} + \rho \frac{\partial v_y}{\partial y} = 0.$$

Dividing through by ρ, we arrive at our final result:

$$\frac{\partial v_x}{\partial x} + \frac{\partial v_y}{\partial y} = 0. \tag{9A.2}$$

9A.2 MOMENTUM CONSERVATION

Our starting point for momentum conservation is the Navier–Stokes equation, components of which are presented in Table 6.3. Assumptions 1, 2, and 3 were employed in the derivation of the component equations (Eqs. T6.3a–T6.3c). We write out the x-component equation (Eq. T6.3a) and simplify, again, indicating the assumptions that cause terms to be zero:

$$-\frac{\partial P}{\partial x} + \mu\left(\frac{\partial^2 v_x}{\partial x^2} + \frac{\partial^2 v_x}{\partial y^2} + \frac{\partial^2 v_x}{\partial z^2}\right) + \rho g_x$$

$$= \rho\left(\frac{\partial v_x}{\partial t} + v_x\frac{\partial v_x}{\partial x} + v_y\frac{\partial v_x}{\partial y} + v_z\frac{\partial v_x}{\partial z}\right). \tag{9A.3}$$

Note that we remove $\partial^2 v_x/\partial x^2$ from the viscous force terms because it is small compared to $\partial^2 v_x/\partial y^2$, not because it is identically zero. This follows from assumption 6 that requires $\partial(\ )/\partial x \ll \partial(\ )/\partial y$. The simplified Eq. 9A.3 is thus

$$-\frac{\partial P}{\partial x} + \mu\frac{\partial^2 v_x}{\partial y^2} + \rho g_x = \rho\left(v_x\frac{\partial v_x}{\partial x} + v_y\frac{\partial v_x}{\partial y}\right). \tag{9A.4}$$

This equation is the starting point for the free-convection problem where both the pressure-force term ($\partial P/\partial x$) and body-force term (ρg_x) are important.

For the flat-plate, forced-convection problem, however, we neglect the effect of gravity and set the pressure gradient to zero. To see why $\partial P/\partial x$ is zero for this problem, we analyze the free-stream flow outside of the boundary layer. In the free stream, v_x is constant and v_y and v_z are identically zero. For these conditions, Eq. 9A.3 simplifies to the trivial result that

$$-\frac{\partial P}{\partial x} = 0. \tag{9A.5}$$

You should verify that this is so. For forced convection over a flat plate, Eq. 9.A4 becomes

$$v_x\frac{\partial v_x}{\partial x} + v_y\frac{\partial v_x}{\partial y} = \frac{\mu}{\rho}\frac{\partial^2 v_x}{\partial y^2}. \tag{9A.6}$$

Although we have not explicitly dealt with the y-component of momentum for the boundary layer, applying the assumptions set out at the beginning of this appendix yields the conclusion that the pressure does not depend on the y coordinate [i.e., $P = P(x$ only)]. This is important for both the free-convection and forced-convection problems; for the former, it allows the hydrostatic pressure variation in the surroundings to be imposed through the boundary layer up to the surface, whereas for the latter, it results in the pressure being uniform within the boundary layer, as well as in the free stream.

9A.3 ENERGY CONSERVATION

Proceeding as before, we now simplify the general energy equation (Eq. T5.7a) as follows:

$$\rho c_p \left(\overset{\textcircled{4}}{\frac{\partial T}{\partial t}} + v_x \frac{\partial T}{\partial x} + v_y \frac{\partial T}{\partial y} + \overset{\textcircled{5}}{v_z \frac{\partial T}{\partial z}} \right) = k \left(\overset{\textcircled{6}}{\frac{\partial^2 T}{\partial x^2}} + \frac{\partial^2 T}{\partial y^2} + \overset{\textcircled{5}}{\frac{\partial^2 T}{\partial z^2}} \right) + \overset{\textcircled{7}}{\Phi_{\text{visc.}}} \quad (9A.7)$$

$$0 \qquad\qquad 0 \qquad\qquad 0 \qquad\qquad 0 \qquad\qquad 0$$

With the simplifications shown, Eq. 9A.7 is written

$$v_x \frac{\partial T}{\partial x} + v_y \frac{\partial T}{\partial y} = \frac{k}{\rho c_p} \frac{\partial^2 T}{\partial y^2}, \qquad (9A.8)$$

where we recognize the grouping of thermo-physical properties as the thermal diffusivity $\alpha \ (\equiv k/\rho c_p)$.

Summarizing, we see that Eqs. 9A.2, 9A.6, and 9A.8 are the boundary-layer equations for the problem of forced convection over a flat plate, and Eqs. 9A.2, 9A.4, and 9A.8 apply to the free-convection problem, the only difference being the retention of the pressure-gradient and body-force terms in the momentum equation.

Appendix 9B
Boundary-Layer Integral Analysis

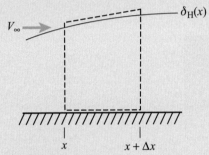

FIGURE 9B.1
Control volume through boundary layer on a flat plate. The height of the control volume is $\delta_H(x)$ and the width is Δx. Unit depth is assumed.

In this appendix, we present von Kármán's [7] elegant approximate solution to the problem of finding the boundary-layer thickness $\delta_H(x)$ and the friction coefficient $c_{f,x}$ for forced flow over a flat plate. The motivation for this presentation is to illustrate the power of the integral control volume analyses discussed earlier for mass conservation in Chapter 3 and for momentum conservation in Chapter 6.

Consider the control volume sketched in Fig. 9B.1 where we have chosen a slice through a hydrodynamic boundary layer. For unit depth (perpendicular to the page), the upstream control surface at x is a plane of area $\delta_H(x) \cdot 1$, and similarly the downstream control surface at $x + \Delta x$ is a plane of area $\delta_H(x + \Delta x) \cdot 1$. The lower control surface is contiguous with the plate surface and has an area $\Delta x \cdot 1$. The top boundary of the control volume follows the contour of the boundary-layer thickness.

Before applying the conservation principles, we assume a shape for the velocity profile within the boundary layer. We choose the following parabolic distribution:

$$v_x(x,y) = V_\infty \left(\frac{2y}{\delta_H(x)} - \frac{y^2}{\delta_H^2(x)} \right). \tag{9B.1}$$

Note that this profile has the desired characteristics that $v_x = 0$ at the surface ($y = 0$) and approaches (actually equals) V_∞ at $y = \delta_H(x)$. The beauty of the integral approach is that the final result does not depend strongly on the assumed form of the velocity profile; for example, even a simple straight-line approximation yields results for δ_H/x and $c_{f,x}$ that are within 35% and 13%, respectively, of the exact solutions. (See the end-of-chapter problems.)

Our first problem, given the assumed form of the velocity profile, is to find the function $\delta_H(x)$ that will simultaneously satisfy conservation of mass and momentum applied to the control volume shown in Fig. 9B.1. We begin by applying mass conservation. To facilitate our analysis, the control volume is redrawn in Fig. 9B.2a, where arrows indicate the mass flow in at x, the mass flow out at $x + \Delta x$, and the mass flow through the top of the control volume. For steady flow, the mass flow entering the control volume must equal the total mass flow exiting (Eq. 3.18b); thus,

$$\dot{m}_x = \dot{m}_{x+\Delta x} + \dot{m}_{\text{top}}. \tag{9B.2}$$

Rearranging Eq. 9B.2 to solve for $\dot{m}_{\text{top}}$, and expressing the other two flowrates in terms of their velocity distributions (Eq. 3.9), yield

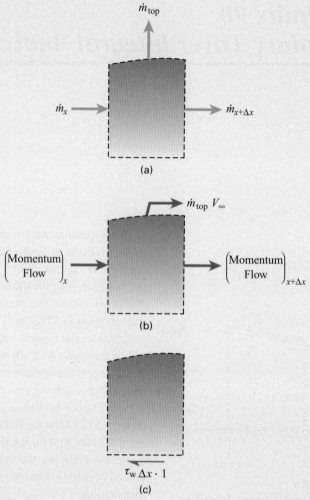

FIGURE 9B.2

(a) Mass flows through control surfaces, (b) x-direction momentum flows through control surfaces, and (c) x-direction forces acting on control surfaces. Note that, since the pressure is uniform over all surfaces, the net pressure force is zero and hence is not shown.

$$
\dot{m}_{\text{top}} = \left[\int_0^{\delta_{\text{H}}(x)} \rho v_x(x, y)(1) \, dy \right]_x - \left[\int_0^{\delta_{\text{H}}(x)} \rho v_x(x, y)(1) \, dy \right]_{x+\Delta x}. \qquad (9\text{B}.3)
$$

We will use this result in the following application of momentum conservation.

Figure 9B.2b shows the three momentum flows directed in the x-direction for our control volume: one entering across the planar surface at x, one exiting at $x + \Delta x$, and the third at the upper surface. Our expression for this third momentum flow, $\dot{m}_{\text{top}} V_{\infty}$, requires some explanation. At the top surface there exists some y-component of the velocity vector (it is this component that produces the mass flowrate $\dot{m}_{\text{top}}$); however, this y-component (v_y) is quite small. As a consequence, the x-component of the velocity at the top surface is approximately equal to the free-stream velocity V_{∞}; thus we can approximate the x-direction momentum flux at the top surface as

$$
\dot{m}_{\text{top}} v_x(x, \delta_{\text{H}}(x)) \approx \dot{m}_{\text{top}} V_{\infty}.
$$

Figure 9B.2c shows the single x-direction force acting on the control volume. Although pressure forces act inward at all of the control surfaces, the pressure is uniform around the control surface; hence, there is no net pressure force. The only x-directed force is the viscous shear force, $F_{\text{visc}} = \tau_w \Delta x \cdot 1$, at the lower boundary where the fluid sticks to the surface. At $y = \delta_H$, the velocity gradient is zero; hence no viscous shear force exists here. Furthermore, we ignore any normal viscous stresses at x and $x + \Delta x$ since $\partial v_x / \partial x$ is small.

With the momentum flows and forces defined (Figs. 9B.2b and 9B.2c), we apply the fundamental steady-flow statement of momentum conservation that the sum of all forces acting in the x-direction must equal the difference between the x-direction momentum flow out of the control volume and the x-direction momentum flow into the control volume (see Eqs. 6.53 and 6.54), that is,

$$-\tau_w \Delta x \cdot 1 = \left(\begin{array}{c} momentum \\ flow \end{array} \right)_{x+\Delta x} + \dot{m}_{\text{top}} V_\infty - \left(\begin{array}{c} momentum \\ flow \end{array} \right)_x. \quad (9B.4)$$

The wall shear stress τ_w is related to our assumed velocity distribution through the definition of a Newtonian fluid (Eq. 6.3):

$$\tau_w = \mu \left. \frac{\partial v_x}{\partial y} \right|_{y=0} = \mu V_\infty \left[\frac{2}{\delta_H} - \frac{2y}{\delta_H^2} \right]_{y=0} = \frac{2\mu V_\infty}{\delta_H}. \quad (9B.5)$$

See Example 6.8 in Chapter 6.

The x-direction momentum flows at x and $x + \Delta x$ are evaluated from the general definition of a momentum flow (Eq. 6.50), that is,

$$\left(\begin{array}{c} x-momentum \\ flow \end{array} \right) \equiv \int_0^{\delta_H(x)} v_x (\rho v_x)(1)\, dy. \quad (9B.6)$$

We now simultaneously substitute several expressions into the basic momentum conservation expression, Eq. 9B.4: Eq. 9B.5 for τ_w, the appropriate forms of Eq. 9B.6 for the momentum flows at x and $x + \Delta x$, and Eq. 9B.3 for $\dot{m}_{\text{top}}$. These substitutions yield

$$-\frac{2\mu V_\infty}{\delta_H(x)} \Delta x = \left[\int_0^{\delta_H(x)} \rho v_x^2 dy \right]_{x+\Delta x} - \left[\int_0^{\delta_H(x)} \rho v_x^2 dy \right]_x$$
$$+ V_\infty \left\{ \left[\int_0^{\delta_H(x)} \rho v_x dy \right]_x - \left[\int_0^{\delta_H(x)} \rho v_x dy \right]_{x+\Delta x} \right\}.$$

We form an ordinary differential equation from this expression by dividing through the Δx, taking the limit $\Delta x \to 0$, and recognizing the definition of a derivative. Performing these operations yields

$$-\frac{2\mu V_\infty}{\delta_H(x)} = \frac{d}{dx} \left[\int_0^{\delta_H(x)} \rho v_x^2 dy \right] - V_\infty \frac{d}{dx} \left[\int_0^{\delta_H(x)} \rho v_x dy \right]. \quad (9B.7)$$

We now substitute the assumed velocity distribution (Eq. 9B.1) into Eq 9B.7 using $\delta \equiv \delta_H(x)$ to simplify the notation to get

$$-\frac{2\mu V_\infty}{\delta} = \rho V_\infty^2 \frac{d}{dx} \left[\int_0^{\delta} \left[\frac{4y^2}{\delta^2} - \frac{4y^3}{\delta^3} + \frac{y^4}{\delta^4} \right] dy \right] - \rho V_\infty^2 \frac{d}{dx} \left[\int_0^{\delta} \left[\frac{2y}{\delta} - \frac{y^2}{\delta^2} \right] dy \right].$$

Performing the integrations and evaluating the limits result in

$$-\frac{2\mu V_\infty}{\delta} = -\frac{2}{15}\rho V_\infty^2 \frac{d\delta}{dx}. \tag{9B.8}$$

Simplifying Eq. 9B.8 and separating the variables δ and x yield

$$\delta d\delta = \frac{15\mu}{\rho V_\infty} dx.$$

We now integrate this expression from $x = 0$ to an arbitrary position x to obtain our desired expression for δ [$= \delta_H(x)$]:

$$\int_0^{\delta_H(x)} \delta d\delta = \frac{15\mu}{\rho V_\infty}\int_0^x dx,$$

or

$$\frac{\delta_H^2(x)}{2} = \frac{15\mu}{\rho V_\infty}x.$$

Solving for δ_H yields

$$\delta_H(x) = \left(\frac{30\mu x}{\rho V_\infty}\right)^{1/2}, \tag{9B.9a}$$

or

$$\frac{\delta_H(x)}{x} = \sqrt{30}Re_x^{-1/2} = 5.48Re_x^{-1/2}, \tag{9B.9b}$$

where

$$Re_x \equiv \frac{\rho V_\infty x}{\mu}.$$

Comparing this result with that from the exact (similarity) solution, Eq. 9.19b, we see that the forms of the two expressions are identical, but with a slightly ($\sim 10\%$) higher coefficient (i.e., 5.48 versus 5.0).

Our second objective is to obtain an expression for the friction coefficient $c_{f,x}$. This is easily done, first, by substituting the expression for $\delta_H(x)$ from Eq. 9B.9a into the defining relationship for τ_w, Eq. 9B.5, that is,

$$\tau_w = \frac{2\mu V_\infty}{\left(\dfrac{30\mu x}{\rho V_\infty}\right)^{1/2}},$$

and, second, by substituting the definition of $c_{f,x}$ (Eq. 8.25),

$$c_{f,x} \equiv \frac{\tau_w}{\frac{1}{2}\rho V_\infty^2}.$$

Performing this substitution yields

$$c_{f,x} = \frac{4\sqrt{30}}{30}Re_x^{-1/2},$$

or

$$c_{f,x} = 0.730Re_x^{-1/2}. \tag{9B.10}$$

Again, compared to the exact solution (Eq. 9.22),

$$c_{f,x} = 0.664Re_x^{-1/2},$$

we see that the approximate result (Eq. 9B.10) is again about 10% too high.

A similar analysis can be performed for the heat-transfer problem to obtain an approximate expression for $Nu_x(x)$, although we will not do this here. The basic approach, however, applies energy conservation to the control volume of Fig. 9B.1 and utilizes an assumed temperature profile to obtain $\delta_T(x)$, similar to our use of an assumed velocity profile to obtain $\delta_H(x)$.

INTERNAL FLOWS: FRICTION, PRESSURE DROP, AND HEAT TRANSFER

After studying Chapter 10, you should:

- Be able to define the following concepts: hydrodynamically and thermally fully developed flow, critical Reynolds number, Darcy (Moody) friction factor, bulk mean temperature, viscous sublayer, hydrodynamic and thermal entrance lengths, minor losses, and hydraulic diameter.

- Understand the similarities and differences between internal and external flows.

- Be able to write and apply integral mass, momentum, mechanical energy, and thermal energy conservation relationships for isothermal and nonisothermal flows.

- Be able to apply theory or empirical correlations to determine friction factors and convective heat-transfer

coefficients for both laminar and turbulent flows in both smooth and rough tubes and pipes.

- Be able to calculate pressure drops and overall heat-transfer rates for both laminar and turbulent flows in tubes and pipes.

- Be able to sketch the axial velocity distribution as a function of radial location for both laminar and turbulent flows, illustrating their essential differences.

- For fully developed internal flows, be able to sketch the pressure and wall shear stress as functions of axial distance.

- For fully developed internal flows, be able to sketch the bulk-mean temperature and the wall temperature for

conditions of uniform heat flux and for conditions of constant wall temperature.

- Be able to sketch the radial profiles of the axial velocity in the hydrodynamic entry region, and be able to sketch the radial profiles of the bulk mean temperature in the thermal entry region.

- Understand the behavior of $P(x)$, $\tau_w(x)$, and $h_{conv,x}$ in the hydrodynamic and thermal entry regions.

- Be able to calculate friction factors, convective heat-transfer coefficients, pressure drops, and overall heat-transfer rates for ducts of noncircular cross section.

- Be able to calculate pressure drops for piping systems with minor losses

Chapter 10 Overview

In this chapter, we deal with fluids flowing through tubes, pipes, and other ducts having constant cross-sectional areas. Our focus here is to understand the physical phenomena responsible for frictional resistance (head loss) and heat transfer for such flows and to be able to calculate these quantities for flows of engineering interest. This chapter provides a good mix of theory and empiricism. We first apply momentum and energy conservation to integral control volumes, relating important engineering quantities to the details of the flow. In turn, we obtain these details from the application of the same conservation principles to differential control volumes. For laminar flows, we obtain wall shear stress, head loss, and heat-transfer coefficients directly from theoretical results. For turbulent flows, however, theoretically guided empirical correlations are presented to calculate these same quantities. We also present in this chapter the concept of hydrodynamic and thermal flow development. Flow development results in enhanced frictional losses and greater heat-transfer rates in the entrance region of a tube, or pipe, compared to those in the fully developed region downstream. Throughout this chapter, you will see that dimensionless parameters play an important role in understanding and describing internal flows, just as they did in the previous chapter on external flows.

10.1 HISTORICAL CONTEXT

Because of the practical importance of moving water from one location to another, the history of internal flows is both long and rich. For centuries, engineers and scientists sought to understand and use the principles controlling the flow of fluids in pipes, tubes, and channels. Long before a theoretical understanding of flow resistance was developed, many empirical formulas relating the mean velocity—a measure of flow—to elevation changes—the driving force—were proposed and used in design. The earliest of these formulas is that of **Antoine Chézy** (1718–1798), who considered both open-channel and pipe flows and reported his results circa 1775 [1]. A decade later, **Pierre Louis Georges Du Buat** (1734–1809), unaware of Chézy's work, developed yet more complex empirical expressions based on a series of 200 experiments. His complicated equations relate the flow velocity to the slope $[\Delta z/L]$ and the mean radius of the pipe or channel [1]. This experimental approach continued until the middle of the nineteenth century. Although largely empirical, the formula proposed in 1839 by the German engineer **Gotthilf Heinrich Ludwig Hagen** (1797–1884) is of a form that is predicted by modern theory of laminar flow: For a fixed pressure drop and tube length, the flow rate is proportional to the fourth power of the tube radius [1,2].

G. H. L. Hagen (1797–1884)

Jean Poiseuille (1799–1869)

Lewis F. Moody (1880–1953). Photograph courtesy of Princeton University Library.

The French physician **Jean Louis Poiseuille** (1799–1869) obtained similar results from detailed investigations of laminar flows in 1841 [3]. Adding the correct theoretical understanding to the laminar-flow resistance formulas of Hagen and Poiseuille came nearly two decades later (1858–1860) in the independent derivations of **Franz Neumann** (1798–1885) and **Edward Hagenbach** (1833–1910), German and Swiss physicists, respectively [1].

Other important developments in understanding internal flows relate to the effects of wall roughness and turbulence. **Henry Philibert Gaspard Darcy** (1803–1858), a French engineer, studied various diameter pipes with various degrees of surface roughness. In 1857, he showed conclusively the importance of wall roughness in the resistance to flow [1]. Although Hagen many years before observed the phenomenon of transition from laminar to turbulent flow, Osborne Reynolds in his famous 1883 paper [4] first showed that this transition occurs at a fixed value of the dimensionless parameter that now bears his name, the Reynolds number. The more complete and modern understanding of turbulent internal flows comes from the Prandtl and von Kármán school and the application of boundary-layer concepts. Although many persons played a part in these later developments, we cite the 1933 work of **Johann Nikuradse** (1894–1979), a student of Prandtl. As we discuss later in this chapter, Nikuradse conducted a meticulous series of experiments in which he glued sand grains, sieved to particular size classes, to the inside of tubes [5]. These experiments provided useful empirical data and extended the theoretical understanding of the influence of wall roughness. **Lewis Ferry Moody** (1880–1953) provides full closure to the engineering problem of pipe resistance in his 1944 paper [6]. In Moody's paper, the work of the many who have gone before is presented in a few figures, one of which we will use later in this chapter.

Again, this space is too small to present a complete and detailed history of all of the important discoveries and advances in the understanding and engineering of internal flows. For more historical information on the fluid mechanical aspects of internal flows, we once again refer the reader to Rouse and Ince [1]. For heat-transfer aspects, not at all touched on here, see Ref. [7] along with the specific citations presented later in this chapter.

10.2 THE BIG PICTURE—INTEGRAL ANALYSES

In this section, we frame the basic problem of internal flows using integral analyses of the fundamental conservation principles (mass, momentum, and energy). These analyses reveal what detailed information is subsequently required to solve problems of engineering interest. The problems we attack are, first, how to calculate the pressure drop resulting from frictional resistance in an internal flow, and, second, how to calculate heat-transfer rates and/or temperature distributions along the length of an internal flow.

10.2a Assumptions

For the analyses that follow, we invoke the following assumptions:

1. The flow is steady.

2. The fluid is incompressible.

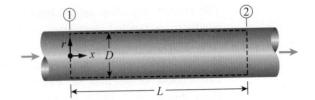

3. Entrance effects are negligible (i.e., the flow is fully developed).[1]

4. The flow is through a straight duct with constant circular cross section and impermeable walls.

5. Uniform pressure exists over the flow cross-sectional area.

10.2b Mass Conservation

Although expressing mass conservation is nearly trivial for the problem considered here, we formally present this basic principle to emphasize its underlying importance. Consider the cylindrical control volume shown in Fig. 10.1. Mass enters at station 1 and exits at station 2 through the circular cross-sectional area $A_{x\text{-sec}} = \pi D^2/4$. No mass enters or exits through the cylindrical wall area $A_{wall} = \pi DL$, since the control surface is contiguous with the impermeable duct wall.

Because the flow is steady (assumption 1), our integral statement of mass conservation is simply (Eq. 3.18a)

$$\dot{m}_{in} = \dot{m}_{out}, \tag{10.1a}$$

or

$$\dot{m}_1 = \dot{m}_2. \tag{10.1b}$$

Equation 10.1b can be restated using the velocity distributions at stations 1 and 2 as

$$\left[\rho \int_{A_{x\text{-sec}}} v_x dA\right]_1 = \left[\rho \int_{A_{x\text{-sec}}} v_x dA\right]_2. \tag{10.2}$$

In general, the velocity distribution depends on both the radial and axial coordinates [i.e., $v_x = v_x(r, x)$]; however, our assumption of fully developed flow implies that the velocity distribution depends only on the radial coordinate [i.e., $v_x = v_x(r)$].

We can also express mass conservation in terms of average velocities:

$$[\rho v_{avg} A_{x\text{-sec}}]_1 = [\rho v_{avg} A_{x\text{-sec}}]_2, \tag{10.3a}$$

or

$$\dot{m} = \rho v_{avg} \pi R^2 = \text{constant}, \tag{10.3b}$$

where R is the tube radius ($= D/2$).

---

[1] For momentum conservation, this assumption implies that the velocity distribution does not change in the flow direction. For energy conservation, the implication is that the *dimensionless* temperature distribution does change in the flow direction. We discuss these details in a later section.

When we explore the details of the flow later in this chapter, momentum conservation will be applied to develop an independent relationship for v_{avg} involving the pressure gradient and gravitational body force. Thus, through Eq. 10.3b, the mass flow rate will be related to these quantities. With that information, one can then answer questions such as the following: What is the mass flow rate through a tube of given dimensions generated by a particular drop in elevation? Given the pressures at the inlet and outlet of a particular horizontal tube, what is the mass flow rate?

10.2c Momentum Conservation

Application of the conservation of momentum principle provides the desired link between the pressure and frictional forces. This is of engineering interest in that friction irreversibly converts some useful mechanical energy of the flow to, generally useless, thermal energy. Subject to constraints, one typically attempts to minimize these losses in the design of practical devices and systems.

Pressure Drop and Wall Shear Stress

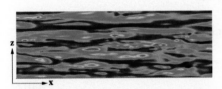

Direct numerical simulation shows skin friction for fully-developed turbulent channel flow. Image from Ref. [29] used with permission.

> See Eq. 6.2 and Fig. 6.1 in Chapter 6.

> See the discussion of Eqs. 6.49 and 6.50 in Chapter 6 to review momentum flow.

We consider the same control volume as in Fig. 10.1; however, now the duct is inclined at an arbitrary angle from the horizontal to explicitly include the gravitational body force in our analysis. This inclined control volume is sketched in Fig. 10.2 along with all of the forces acting on it. Among these forces are the pressure forces at stations 1 and 2, both acting inward by definition. We assume the pressure forces acting over the cylindrical sides are symmetrical and thus do not introduce any net force; therefore, no arrows are shown on the sketch for the side-wall pressure forces. The viscous force acts at the side wall of the cylindrical control surface opposing the flow. Again, because the flow is fully developed, the wall shear stress does not depend on x and hence is a constant; thus, the viscous force is simply the product of the wall shear stress τ_w and the side-wall area A_{wall} ($= \pi D L$), as shown. The body force is just the weight of the fluid within the control volume, W ($= Mg = \rho g A_{x\text{-sec}} L$), and acts vertically downward. The component $W \sin \theta$ acts in the negative x-direction, opposing the flow. No other forces act on the control volume.

The momentum flow entering the control volume in the x-direction is expressed as

$$x\text{-momentum flow at } 1 \equiv \int_{A_{x\text{-sec}}} \rho v_x^2 dA_{x\text{-sec}}. \tag{10.4a}$$

FIGURE 10.2

Forces acting on this cylindrical control volume in the x-direction include the pressure forces at stations 1 and 2, $F_{P,1}$ and $F_{P,2}$; the viscous frictional force, F_{visc}; and the component of the fluid weight, Wsinθ.

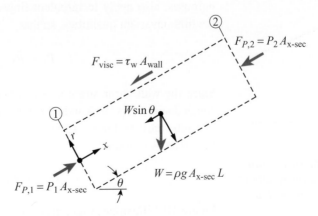

Similarly, the momentum flow exiting in the x-direction is

$$x\text{-momentum flow at } 2 \equiv \int_{A_{\text{x-sec}}} \rho v_x^2 dA_{\text{x-sec}}. \tag{10.4b}$$

With all the x-direction forces and momentum flows defined, we are ready to apply the momentum conservation principle expressed by the following (cf. Eq. 6.53):

$$\left(\begin{array}{c}\text{momentum}\\\text{flow in}\end{array}\right)_{x\text{-dir}} - \left(\begin{array}{c}\text{momentum}\\\text{flow out}\end{array}\right)_{x\text{-dir}} + \sum F_{x\text{-dir}} = 0. \tag{10.5}$$

Substituting the various forces and momentum flows and rearranging yield

$$P_1 A_{\text{x-sec}} - P_2 A_{\text{x-sec}} - \tau_w A_{\text{wall}} - \rho g \sin\theta A_{\text{x-sec}} L$$
$$= \int_{A_{\text{x-sec}}} \rho v_x^2 dA_{\text{x-sec}} - \int_{A_{\text{x-sec}}} \rho v_x^2 dA_{\text{x-sec}}. \tag{10.6}$$

We immediately notice that the right-hand side of Eq. 10.6 is zero since the two momentum-flow integrals are identical. With this recognition, we rearrange Eq. 10.6 to isolate $P_1 - P_2$ as follows:

$$(P_1 - P_2) - \rho g L \sin\theta = \tau_w \frac{A_{\text{wall}}}{A_{\text{x-sec}}}.$$

Relating the two areas to the cylindrical geometry,

$$A_{\text{x-sec}} = \pi D^2/4$$

and

$$A_{\text{wall}} = \pi D L,$$

yields the desired relationship between the pressure drop $(P_1 - P_2)$ and the viscous shear stress τ_w:

$$(P_1 - P_2) - \rho g L \sin\theta = 4\tau_w \frac{L}{D}. \tag{10.7a}$$

For a horizontal tube, $\sin\theta$ is zero; thus,

$$P_1 - P_2 = 4\tau_w \frac{L}{D}. \tag{10.7b}$$

Equations 10.7a and 10.7b apply exactly for laminar flows and, for practical purposes, also apply to turbulent flows where time-mean quantities replace the time-invariant quantities, so that

$$\overline{P}_1 - \overline{P}_2 = 4\overline{\tau}_w \frac{L}{D}. \tag{10.8}$$

Since the wall shear stress is a constant (i.e., it does not depend on x), we conclude from Eqs. 10.7 and 10.8 that the pressure distribution along the length of a tube must be linear. Allowing L to be a variable (i.e., $L \equiv x$), we rewrite Eq. 10.8a as follows:

$$P(x) = P_1 - 4\frac{\tau_w}{D}x. \tag{10.9}$$

Figure 10.3 illustrates this linear pressure distribution.

Pipeline pumping requirements depend on frictional resistance and elevation changes.

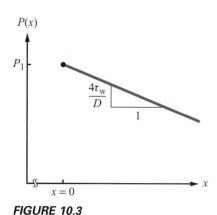

FIGURE 10.3

In the region where the flow is fully developed, the pressure distribution is linear in the flow direction with a slope $-4\,\tau_w/D$.

At this juncture, we have now transferred the problem of calculating the pressure drop to the problem of calculating the wall shear stress τ_w. To determine the wall shear stress, however, requires knowledge of the details of the flow. Specifically, for a Newtonian fluid, knowledge of the velocity distribution allows us to evaluate the shear stress using

We discussed the stress–velocity gradient relationships in Chapter 6—see the discussion of Eq. 6.8.

$$\tau_w = -\mu \frac{\partial v_x}{\partial r}\bigg|_{r=R}, \tag{10.10}$$

where R is the tube radius. In later sections, we investigate the details needed to determine wall shear stresses for both laminar and turbulent flows.

Friction Factors

Analogous to the dimensionless friction and drag coefficients defined for external flows, friction factors are defined for internal flows. The **Darcy (or Moody) friction factor** is a dimensionless wall shear stress defined by

$$f_D \equiv \frac{4\tau_{\text{wall}}}{\frac{1}{2}\rho v_{\text{avg}}^2}. \tag{10.11}$$

Sometimes employed is the **Fanning friction factor** f_F, which is one-fourth the Darcy friction factor, that is,

$$f_F = f_D/4. \tag{10.12}$$

In this book, we employ only the Darcy friction factor f_D, as this is more commonly used by mechanical engineers. Exercise care in using friction-factor data from other sources to avoid being off by a factor of 4.

10.2d Mechanical Energy Conservation

In Chapter 6, we derived the so-called mechanical energy equation from the conservation of momentum principle. Applying the mechanical energy equation to the control volume of Fig. 10.1 (or Fig. 10.2) for an incompressible, nominally isothermal flow yields the following:

See Eq. 6.76 in Chapter 6.

$$P_1 - P_2 + \frac{1}{2}\rho(\alpha_1 v_{\text{avg},1}^2 - \alpha_2 v_{\text{avg},2}^2) + \rho g(z_1 - z_2) = \rho g h_L, \tag{10.13}$$

where α_1 and α_2 are the kinetic energy correction factors (see Table 5.5), h_L is the head loss, and z is the elevation above some reference plane. Since we are dealing with a fully developed flow, mass conservation requires that $\alpha_1 = \alpha_2$ and $v_{\text{avg},1} = v_{\text{avg},2}$ (see Eq. 10.3); thus, the kinetic energy terms in Eq. 10.13 cancel. Furthermore, we note that the elevation change $z_1 - z_2$ relates to the length of the control volume as

$$z_2 - z_1 = L\sin\theta.$$

With the cancellation of the kinetic energy terms and this substitution, Eq. 10.13 becomes

$$P_1 - P_2 - \rho g L \sin\theta = \rho g h_L. \tag{10.14}$$

Comparing Eq. 10.14 with the previously derived momentum conservation relationship (Eq. 10.7a), we see that $\rho g h_L$ must equal $4\tau_w L/D$. The head loss, therefore, relates to the wall shear stress as follows:

$$h_L = 4\frac{\tau_w L}{\rho g D}.$$ (10.15)

From Eq. 10.15, we see that, again, the problem reduces to determining the wall shear stress.

The head loss can also be related to the Darcy friction factor f_D by combining Eqs. 10.11 and 10.15 to yield

$$h_L = f_D \frac{L}{D}\frac{v_{avg}^2}{2g}.$$ (10.16)

Equation 10.16 is known as the Darcy–Weisbach equation and was first formulated by the German professor Julius Weisbach in 1850. For a horizontal pipe, combining Eqs. 10.15 and 10.14 yields the useful relationship

See Examples 10.6 and 10.8 for applications of this key relationship.

$$P_1 - P_2 = f_D \frac{L}{D}\frac{1}{2}\rho v_{avg}^2.$$ (10.17)

10.2e Energy Conservation

To apply energy conservation, we distinguish between a nominally isothermal flow where temperature variations, although not zero, are quite small, and nonisothermal flows where temperature variations in both the radial and axial directions are important. The former case applies to flows in which the only heating of the fluid is a result of frictional effects, whereas the latter applies to flows where the fluid is deliberately heated or cooled by applying a heat flux to the tube wall or fixing the tube wall temperature above or below that of the entering fluid.

Nominally Isothermal Flow

The conditions applied to the derivation of Eq. 5.66 also apply here; thus, we can immediately write conservation of energy for the control volume of Fig. 10.1 (or Fig. 10.2 to consider gravity effects) as follows:

$$P_1 - P_2 + \rho g(z_1 - z_2) = \rho\left[u_2 - u_1 + \frac{\dot{Q}_{cv,net\,out}}{\dot{m}}\right].$$ (10.18)

Equation 10.18 results from canceling the kinetic energy terms and deleting the work term, $\dot{W}_{cv,net\,out}$, of Eq. 5.66. Since the flow is nominally isothermal, the conservation of energy principle is not particularly useful to our analysis. We present it here for completeness and to show how the head loss from the mechanical energy equation (obtained by momentum conservation) relates to the thermal energy terms of energy conservation. Comparing Eqs. 10.13 and 10.18, we see that

$$gh_L = u_2 - u_1 + \frac{\dot{Q}_{cv,net\,out}}{\dot{m}}.$$ (10.19)

Also see the discussion surrounding Eq. 6.76 in Chapter 6.

Once again, we have canceled the kinetic energy terms in Eq. 10.13, as was done to arrive at Eq. 10.18 from Eq. 5.66. From Eq. 10.19, we see that if the flow is adiabatic ($\dot{Q}_{cv,net\,out} \equiv 0$), the head loss is directly proportional to the

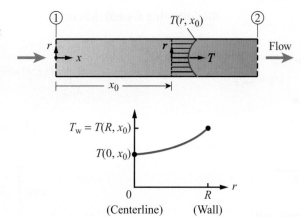

FIGURE 10.4

A fluid is heated as it flows through a tube. Superimposed on the sketch of the control volume (top) is the temperature distribution at the axial location x_0. This distribution $T(r, x_0)$ is redrawn in the more conventional T–r coordinate system (bottom).

increase in the internal energy of the fluid. Furthermore, for an ideal gas (and many liquids), the internal energy change is proportional to the temperature change (i.e., $u_2 - u_1 \propto T_2 - T_1$). Adding momentum conservation (Eq. 10.15) to these observations, we conclude that

$$T_2 - T_1 \propto \tau_w.$$

That is, the fluid friction at the wall produces a small temperature rise in a fluid flowing through a tube, a conclusion consistent with our intuition.

Nonisothermal Flow

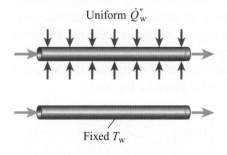

Uniform $\dot{Q}''_w$

Fixed T_w

In this subsection, we apply integral energy conservation analyses, first, to flow through a tube in which a uniform heat flux is applied through the tube wall, and, second, to a flow through a tube in which the wall temperature is maintained at a fixed value above (or below) the temperature of the entering fluid. In both cases, our objective is to provide mathematical relationships that describe the following:

- bulk mean temperature as a function of axial position x,
- tube wall temperature as a function of axial position x,
- local heat flux as a function of axial position x, and
- total heat transferred to the fluid over the distance $x = 0$ to $x = L$.

Bulk Mean Temperature Our analyses are greatly simplified by defining a single temperature that characterizes the thermal state of the fluid at any axial location, that is, the bulk mean temperature alluded to in the first item in our list.

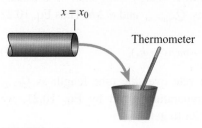

FIGURE 10.5

A thought experiment in which the bulk mean temperature $T_m(x_0)$ is measured. The fluid passing the axial location x_0 is well mixed in a cup. The temperature measured in the cup is the bulk mean temperature.

Consider a fluid that is being heated as it flows through a tube, as suggested in Fig. 10.4. At any axial location, the fluid temperature is greatest at the wall and is a minimum at the centerline. The specific form of this temperature distribution is determined by details of the flow as described by the differential forms of mass, momentum, and energy conservation equations. For the time being, however, we are not concerned with these details but need only to recognize that a temperature distribution $T(r, x)$ exists. To define the **bulk mean temperature**, imagine that we cut the tube at an arbitrary location x_0 and the fluid flows into a cup and mixes adiabatically (Fig. 10.5). The temperature measured by a thermometer in the cup is the bulk mean temperature at x_0, that is, $T_m(x_0)$. Mathematically, the bulk mean temperature is the energy-averaged temperature of the fluid, as we see from

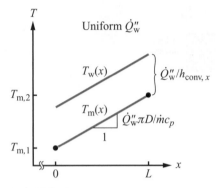

FIGURE 10.6

For the condition of uniform wall heat flux, the bulk mean temperature varies linearly with downstream distance x. For fully developed flow, the corresponding wall temperature distribution differs only by a constant.

Heating tape applied to a valve. Photograph courtesy of BriskHeat Products.

the following formal definition:

$$\dot{m}c_{p,\text{avg}}T_m \equiv \int_{A_{x\text{-sec}}} c_pT(x, r)\,d\dot{m} = \int_0^R \rho v_x c_p T(x, r)\,2\pi r dr. \tag{10.20}$$

If we assume constant thermo-physical properties (ρ, c_p), this equation becomes

$$T_m(x) \equiv \frac{2\pi\rho \int_0^R v_x(x, r)T(x, r)r\,dr}{\dot{m}}. \tag{10.21}$$

We see that this definition involves both the temperature and velocity distributions. Fortunately, we have no need in this book to apply this formal definition. What is important for our purposes is the conceptual definition and the idea that a single temperature, $T_m(x)$, represents the thermal state of the fluid at location x.

Uniform Heat Flux   We once again consider the control volume in Fig. 10.1. The entering fluid is heated as it passes through the tube from $x = 0$ to $x = L$ by the application of a uniform heat flux Q''_w over the circumferential area A_w ($= \pi DL$), as illustrated in Figure 10.6. This uniform heat-flux boundary condition can be approximated in practice by using some form of Joule (i^2R_{elec}) heating. For example, an electrical heating tape can be spirally wound around the tube and insulated on the outside surface. Alternatively, the tube itself can be heated by the passage of an electric current as a result of a voltage difference between the ends.

The mathematical expression of conservation of energy for this situation has already been derived in Chapter 5 and is given by Eq. 5.63e:

$$\dot{Q}_{cv,\text{net in}} - \dot{W}_{cv,\text{net out}} = \dot{m}\left[(h_2 - h_1) + \frac{1}{2}(\alpha_2 v_{\text{avg},2}^2 - \alpha_1 v_{\text{avg},1}^2) + g(z_2 - z_1)\right], \tag{10.22}$$

where the enthalpies h_1 and h_2 are appropriately averaged over the inlet and exit flow areas. We simplify this relationship by noting that no work is performed by (or on) the control volume and that the outlet and inlet kinetic energy terms cancel; furthermore, we neglect the potential energy term since it is usually quite small compared to the thermal terms, $\dot{Q}_{cv,\text{net in}}$ and $\dot{m}\Delta h$. Thus Eq. 10.22 becomes

$$\dot{Q}_{cv,\text{net in}} \equiv \dot{Q}_{0-L} = \dot{m}(h_2 - h_1), \tag{10.23}$$

where we define the total heat-transfer rate over the tube length as $\dot{Q}_{0-L}$. Using the definition of bulk mean temperature given by Eq. 10.21, we eliminate the mean enthalpies in Eq. 10.23 to get

$$\dot{Q}_{0-L} = \dot{m}c_p(T_{m,2} - T_{m,1}), \tag{10.24}$$

where $T_{m,1}$ and $T_{m,2}$ are the bulk mean temperatures at the inlet and outlet, respectively. Equation 10.24 applies to the general problem of fluid heating or cooling in a tube; for the uniform heat flux condition, the net heat transfer from the surroundings to the control volume is simply the product of the uniform heat flux and the surface area over which it acts, that is,

$$\dot{Q}_{0-L} = \dot{Q}''_{w}A_{w} = \dot{Q}''_{w}\pi DL. \tag{10.25}$$

Substituting this relationship into Eq. 10.24 and solving for $T_{m,2}$ yield our final result:

$$T_{m,2} = T_{m,1} + \frac{\dot{Q}''_{w}\pi DL}{\dot{m}c_{p}}. \tag{10.26}$$

Equation 10.26 can be transformed into a relationship expressing the axial variation of the bulk mean temperature $T_{m}(x)$ simply by replacing L by x, which gives

$$T_{m}(x) = T_{m,1} + \frac{\dot{Q}''_{w}\pi D}{\dot{m}c_{p}}x. \tag{10.27}$$

As would be expected by the application of a uniform heat flux to the flow of a constant-property fluid, the bulk mean temperature increases linearly with distance as shown in Fig. 10.6.

Note that in the derivation of Eqs. 10.26 and 10.27 we have implicitly neglected the effect of friction on the enthalpy change. This is justified for the same reason that led us to treat the friction-dominated problem (cf. Eq. 10.19) as isothermal: The conversion of mechanical energy through friction results in only a small increase in the fluid temperature in most engineering applications. For example, head losses in typical pipe-flow problems range from a few centimeters to, say, ten meters. For a flow of water with a 10-m head loss, the corresponding temperature increase is only approximately 0.023 K. Such a small increase clearly justifies the assumption of isothermal flow. For nonisothermal flow, we can similarly justify neglecting elevation changes. A 100-m change in elevation corresponds to a temperature change of 0.23 K for a flow of water.

With Eqs. 10.27 and 10.25, together with the fact that the local heat flux is specified, we have in hand three of the four relationships we seek. Finding the wall-temperature distribution $T_{w}(x)$ is the only remaining task. We accomplish this by defining a local convective heat-transfer coefficient $h_{conv,x}$ in terms of the difference between the wall and bulk mean temperatures:

Equation 10.28 is a key definition.

$$\dot{Q}''_{x} \equiv h_{conv,x}[T_{w}(x) - T_{m}(x)]. \tag{10.28}$$

We apply this definition to obtain the wall temperature distribution

$$T_{w}(x) = T_{m}(x) + \frac{\dot{Q}''_{w}}{h_{conv,x}}. \tag{10.29}$$

For a fully developed flow, the local heat-transfer coefficient is a constant; thus, the wall-temperature distribution differs from the bulk mean temperature distribution only by the constant $\dot{Q}''_{w}/h_{conv,x}$. The two distributions are thus parallel as illustrated in Fig. 10.6.

Correlations of the form *Nu* = *f(Re, Pr)* are used to determine $h_{conv,x}$ (see Table 10.5).

To determine a value for $h_{conv,x}$ requires a detailed analysis of the flow. In later sections, we discuss ways to calculate convective heat-transfer coefficients for both laminar and turbulent flows.

Constant Wall Temperature The condition of constant wall temperature is frequently approximated in practice when a fluid flowing over the outside

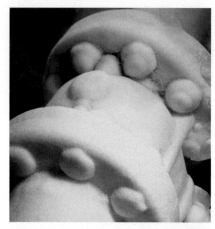

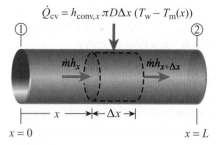

surface of a tube or pipe undergoes a phase change. An example of this is the condensation of steam on the outside surface of a tube through which cold water flows, a typical configuration for condensers used in steam power plants (see Figs. 1.5 and 11.41).

In the following analyses, we derive useful relationships to calculate the total heat transfer rate $\dot{Q}_{0-L}$, the bulk mean temperature distribution $T_m(x)$, and the local wall heat flux $\dot{Q}''_w(x)$.

To determine the total heat-transfer rate $\dot{Q}_{0-L}$, the basic analysis applied to the uniform heat flux situation also applies to the case of constant tube wall temperature; thus, Eq. 10.24 describes both situations, and so

$$\dot{Q}_{0-L} = \dot{m}c_p(T_{m,2} - T_{m,1}). \tag{10.24}$$

Since the bulk mean inlet temperature $T_{m,1}$ is usually known, we have transferred the problem of finding $\dot{Q}_{0-L}$ to determining the bulk mean temperature at $x = L$ (i.e., $T_{m,2}$).

To determine the bulk mean temperature distribution requires that we consider a control volume of length Δx, rather than the original control volume in Fig. 10.1. This new control volume is shown in Fig. 10.7, with arrows indicating all of the energy flows. The enthalpies shown, again, are properly averaged values over the cross section. Conservation of energy is then expressed as

$$\dot{Q}_{cv} = \dot{m}(h_{x+\Delta x} - h_x). \tag{10.30}$$

To arrive at this relationship, we have used the same assumptions previously employed to simplify the general integral control volume energy conservation expression (Eq. 5.63e). For a fluid with constant properties, we can express the mean enthalpy in terms of the bulk mean temperature as follows:

$$h(x) = c_p[T_m(x) - T_{ref}],$$

and the convective heat transfer is expressed (cf. definition in Eq. 10.28) as

$$\dot{Q}_{cv} = h_{conv,x}\pi D\Delta x[T_w - T_m(x)].$$

Substituting these expressions into Eq. 10.30 yields

$$h_{conv,x}\pi D\Delta x[T_w - T_m(x)] = \dot{m}c_p[T_m(x + \Delta x) - T_m(x)].$$

Dividing by Δx, rearranging, and taking the limit $\Delta x \to 0$ results in

$$\frac{h_{conv,x}\pi D}{\dot{m}c_p}[T_w - T_m(x)] = \lim_{\Delta x \to 0}\frac{T_m(x + \Delta x) - T_m(x)}{\Delta x}.$$

$$\dot{Q}_{cv} = h_{conv,x}\pi D\Delta x\,(T_w - T_m(x))$$

① ②

$\dot{m}h_x$ $\dot{m}h_{x+\Delta x}$

x Δx

$x = 0$ $x = L$

FIGURE 10.7
Heat transferred by convection, $\dot{Q}_{cv}$, enters through the sides of a cylindrical control volume Δx in length. Energy flows into ($\dot{m}h_x$) and out of ($\dot{m}h_{x+\Delta x}$) the faces of the control volumes at x and $x + \Delta x$. Conduction heat transfer in the x-direction is neglected.

Recognizing the right-hand side as the definition of a derivative, we obtain the following first-order, ordinary differential equation:

$$\frac{h_{conv,x}\pi D}{\dot{m}c_p}[T_w - T_m(x)] = \frac{dT_m(x)}{dx}. \tag{10.31}$$

We solve Eq. 10.31 for the temperature distribution $T_m(x)$ by separating the variables x and T_m and integrating from $x = 0$ to an arbitrary location x as follows:

$$\int_{T_m(0)}^{T_m(x)}\frac{dT_m(x)}{[T_w - T_m(x)]} = \frac{h_{conv,x}\pi D}{\dot{m}c_p}\int_0^x dx. \tag{10.32}$$

Table 10.1 Summary of Integral Relationships for Nonisothermal Flow in a Tube*

Quantity	Uniform Heat Flux Relationship	Eq.	Constant Wall Temperature Relationship	Eq.
$\dot{Q}_{0-L}$	$\dot{Q}''_w \pi D L$	10.25	$\dot{m}c_p(T_{m,2} - T_{m,1})$	10.24
$\dot{Q}''(x)$	$\dot{Q}''_w$ (given)	—	$h_{conv,x}[T_w - T_m(x)]$	10.28
$T_m(x)$	$T_{m,1} + \dfrac{\dot{Q}''_w \pi D}{\dot{m}c_p}x$	10.27	$\dfrac{T_m(x) - T_w}{T_m(0) - T_w} = \exp\left[\dfrac{-h_{conv,x}\pi D}{\dot{m}c_p}x\right]$	10.33
$T_w(x)$	$T_m(x) + \dfrac{\dot{Q}''_w}{h_{conv,x}}$	10.29	T_w (given)	—

* Although these relationships were derived from an analysis of hydrodynamically and thermally fully developed flow, they apply equally well to developing flows provided that, in Eqs. 10.28 and 10.29, the local heat-transfer coefficient $h_{conv,x}(x)$ is employed and that, in Eq. 10.33, the average heat-transfer coefficient $\bar{h}_{conv,x}(x)$ is used.

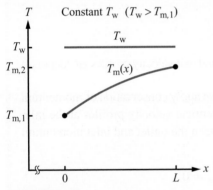

FIGURE 10.8

For the condition of constant wall temperature, the bulk mean temperature approaches the wall temperature in an exponential fashion described by Eq. 10.33b. The sketch illustrates the case where $T_w > T_{m,1}$. A decreasing exponential curve describes the cooling problem when $T_w < T_{m,1}$.

> **How would you redraw Fig. 10.8 for $T_w < T_m(0)$?**

Since the flow is fully developed, we note that the local convective heat-transfer coefficient is a constant and can be removed from the integrand.[2] Performing the indicated integrations and evaluating limits yield

$$\ln\frac{T_m(x) - T_w}{T_m(0) - T_w} = \frac{-h_{conv,x}\pi D}{\dot{m}c_p}x.$$

Removing the natural log by exponentiation provides our final result:

$$\frac{T_m(x) - T_w}{T_m(0) - T_w} = \exp\left[\frac{-h_{conv,x}\pi D}{\dot{m}c_p}x\right], \tag{10.33a}$$

or

$$T_m(x) = T_w + (T_m(0) - T_w)\exp\left[\frac{-h_{conv,x}\pi D}{\dot{m}c_p}x\right]. \tag{10.33b}$$

Figure 10.8 shows a sketch of this distribution when $T_w > T_m(0)$ ($= T_{m,1}$). Equation 10.33 is equally valid for $T_w < T_m(0)$.

To obtain our final desired quantity, $\dot{Q}''_w(x)$, we need only employ the definition of the heat-transfer coefficient for an internal flow (Eq. 10.28), that is,

$$\dot{Q}''_w(x) = h_{conv,x}[T_w - T_m(x)], \tag{10.28}$$

where the local bulk mean temperature $T_m(x)$ is calculated from Eq. 10.33. As for the uniform heat flux case, the value of $h_{conv,x}$ depends on the details of the flow.

The key results from this section describing nonisothermal flows are summarized in Table 10.1. After the following examples, we investigate the details of internal flows. These details provide the wall shear stresses and heat-transfer coefficients appearing in our integral analyses.

---

[2] In a later section, we include entrance effects, one of which is the fact that $h_{conv,x}$ varies with x in the entry region.

Example 10.1

The pressure drop ($P_1 - P_2$) associated with flow of water through a 100-m-long horizontal run of 10-mm-inside-diameter pipe is 23.4 kPa. Determine the frictional force retarding the flow and the average shear stress in the fluid at the wall.

Solution

Known $L, D, P_1 - P_2$

Find $F_{\text{fric}}, \overline{\tau}_w$

Sketch

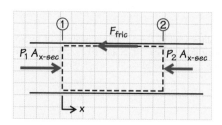

Assumptions

   i. Steady-state, steady flow
  ii. Fully developed flow (i.e., the inlet and exit velocity profiles are identical)

Analysis   To find the frictional force, we apply conservation of momentum (Eq. 10.5), where our assumption of identical velocity profiles at the inlet and outlet results in the difference between the outlet and inlet momentum flows being zero, that is,

$$\sum F_{x\text{-dir}} = 0.$$

The various forces acting on the control volume are shown in the sketch. Here we identify pressure forces acting at the inlet (positive *x*-direction) and at the outlet (negative *x*-direction) and the frictional force retarding the flow (negative *x*-direction); thus,

$$P_1 A_{\text{x-sec}} - P_2 A_{\text{x-sec}} - F_{\text{fric}} = 0,$$

or

$$
\begin{aligned}
F_{\text{fric}} &= (P_1 - P_2)A_{\text{x-sec}} \\
&= (P_1 - P_2)\pi D^2/4 \\
&= 23.4 \times 10^3 \pi (0.01)^2/4 = 1.84 \\
&[=]\text{Pa}\left[\frac{1\,\text{N/m}^2}{\text{Pa}}\right]\text{m}^2 = \text{N}.
\end{aligned}
$$

The average wall shear stress is simply this frictional force divided by the area over which it acts, $A_{\text{wall}} = \pi DL$; thus,

$$
\begin{aligned}
\overline{\tau}_w &= F_{\text{fric}}/(\pi DL) \\
&= \frac{1.84}{\pi(0.01)100}\,\text{N/m}^2 = 0.59\,\text{N/m}^2.
\end{aligned}
$$

Comments   Notice how the limited amount of information given forced us to assume fully developed flow. If we had a value for the flow rate, we could estimate the validity of this assumption. We consider how this might be done in a later section.

For a look ahead, see Eqs. 10.90 and 10.91.

Self Test
10.1

☑ Using the average wall shear stress calculated in Example 10.1, determine the head loss associated with this flow. Assume a water temperature of 300 K.

(Answer: $h_L = 2.4$ m)

Example 10.2

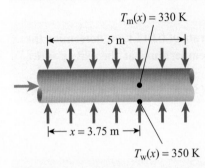

$T_m(x) = 330$ K

5 m

$x = 3.75$ m

$T_w(x) = 350$ K

Water is heated as it flows through a 5-m-long tube. The tube is wrapped with an electrical heating tape, and a layer of insulation covers the tape. This arrangement results in a uniform heat flux of 2878 W/m² at the inside wall of the tube (of inside diameter of 20 mm). At an axial location $x = 3.75$ m, the bulk mean temperature of the water is 330 K, and the tube wall temperature is 350 K. Determine the local heat-transfer coefficient at the 3.75-m location and the total heat-transfer rate to the water over the entire length of the tube.

Solution

Known $\dot{Q}''_w = $ constant, L, D, x

Find $h_{conv,x}(x), \dot{Q}_{0-L}$

Sketch

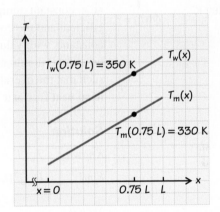

$T_w(0.75\ L) = 350$ K

$T_w(x)$

$T_m(x)$

$T_m(0.75\ L) = 330$ K

$x = 0$ $0.75\ L$ L

Assumptions

   i. Steady flow

  ii. Uniform heat flux

Analysis   To find the local heat-transfer coefficient $h_{conv,x}$, we apply its definition for internal flow (Eq. 10.28):

$$\dot{Q}''_x \equiv h_{conv,x}[T_w(x) - T_m(x)],$$

or

$$h_{conv,x} = \frac{\dot{Q}''_x}{T_w(x) - T_m(x)}$$

$$= \frac{2878}{350 - 330} = 143.9$$

$$[=]\frac{\text{W/m}^2}{\text{K}} = \text{W/m}^2 \cdot \text{K}.$$

Since the heat flux is uniform over the wall area (i.e., $\dot{Q}'' = \dot{Q}''_x$), the total heat-transfer rate is simply the product of this heat flux and the wall area (see Eq. 10.25):

$$\dot{Q}_{0-L} = \dot{Q}'' A_{wall} = \dot{Q}'' \pi DL$$
$$= 2878 \pi (0.02)(5) = 904$$
$$[=](W/m^2)m(m) = W.$$

Comments   This example illustrates the use of key definitions associated with heat transfer in internal flows.

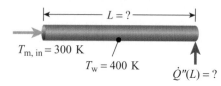

Self Test 10.2   ✓ **For the flow described in Example 10.2 with a mass flow rate of 0.011 kg/s, determine the bulk mean temperature of the fluid and the pipe wall temperature at the exit ($x = L$). *Hint*: Combine Eq. 10.24 and Eq. 10.25.**

(Answer: $T_m (x = 5\ m) = 343.9\ K$, $T_w (x = 5\ m) = 354.9\ K$)

Example 10.3

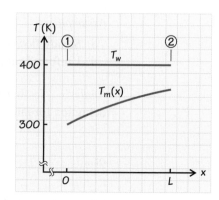

Consider a flow of air through a 2-mm-inside-diameter copper tube. At the location $x = 0$, the flow is fully developed and the air has an average velocity of 12 m/s and a temperature of 300 K. The tube wall is maintained at a constant temperature of 400 K. The convective heat-transfer co-efficient and the constant-pressure specific heat of the air can be treated as constants with values of 532 W/m²·K and 1008 J/kg·K, respectively. The pressure is nominally 100 kPa. Determine the length L of the tube segment, such that the overall heat-transfer rate $\dot{Q}_{0-L}$ is 3.309 W. Also determine the local heat flux at $x = L$.

Solution

Known $T_w, T_m(0), D, v_{ave,in}, h_{conv}, c_p, \dot{Q}_{0-L}$

Find $L, \dot{Q}''(L)$

Sketch

Assumptions

  i. Fully developed flow
  ii. Constant properties

Analysis   We find the required length in two steps: First, we relate the given heat-transfer rate to the outlet temperature through overall energy conservation (Eq. 10.24); second, we relate the unknown length L to the outlet temperature through Eq. 10.33, the integrated form of local energy conservation. The mass flow rate appears in both of these equations. Using the density of air calculated from the ideal-gas law, we get a mass flow rate of

$$\dot{m} = (\rho v_{\text{avg}} A)_{\text{in}}$$

$$= 1.1614(12)\pi(0.002)^2/4$$

$$= 4.38 \times 10^{-5}$$

$$[=](\text{kg/m}^3)(\text{m/s})\text{m}^2 = \text{kg/s}.$$

Solving Eq. 10.24 for the bulk mean temperature at $x = L$ yields

$$T_{\text{m}}(L) = \frac{\dot{Q}_{0-L}}{\dot{m}c_p} + T_{\text{in}}$$

$$= \frac{3.309}{4.38 \times 10^{-5}(1008)} + 300$$

$$= 375$$

$$[=]\frac{\text{W}}{(\text{kg/s})(\text{J/kg·K})}\left[\frac{1\,\text{J/s}}{\text{W}}\right] = \text{K}.$$

With this value of $T_{\text{m}}(L)$, and recognizing that $h_{\text{conv},L}$ is equal to the given constant value of the heat-transfer coefficient, we can apply Eq. 10.33 to find L as follows:

$$\frac{T_{\text{m}}(L) - T_{\text{w}}}{T_{\text{m}}(0) - T_{\text{w}}} = \exp\left[\frac{-h_{\text{conv},L}\pi DL}{\dot{m}c_p}\right],$$

or

$$L = -\frac{\dot{m}c_p}{h_{\text{conv},L}\pi D}\ln\left[\frac{T_{\text{m}}(L) - T_{\text{w}}}{T_{\text{m}}(0) - T_{\text{w}}}\right]$$

$$= -\frac{4.39 \times 10^{-5}(1008)}{532\pi\,0.002}\ln\left[\frac{375 - 400}{300 - 400}\right]$$

$$= 0.0183$$

$$[=]\frac{(\text{kg/s})(\text{J/kg·K})}{(\text{W/m}^2\text{·K})\text{m}}\left[\frac{1\,\text{W}}{\text{J/s}}\right] = \text{m},$$

or

$$L = 18.3\,\text{mm}.$$

To find the heat flux at $x = L$, we apply the definition of the local heat-transfer coefficient (Eq. 10.28):

$$\dot{Q}''(L) = h_{\text{conv},L}[T_{\text{w}} - T_{\text{m}}(L)]$$

$$= 532(400 - 375) = 13,300$$

$$[=](\text{W/m}^2\text{·K})\text{K} = \text{W/m}^2.$$

Comments This example makes use of key definitions and illustrates application of energy conservation for the case of internal flow with a constant wall temperature.

Self Test 10.3

Water flows at 0.3 m/s through a coiled 7-mm-inside-diameter tube immersed in an ice-water bath ($T = 0°C$). If the water enters at 25°C and the tube is 3 m long, determine the water exit temperature. Assume a fully developed flow with $h_{\text{conv},x} = 310\ \text{W/m}^2\text{·K}$.

(Answer: 16.4°C)

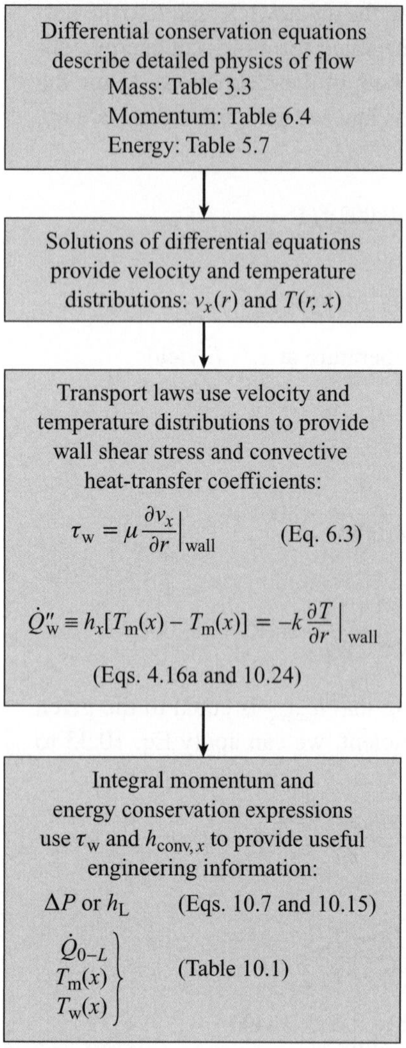

Differential conservation equations
describe detailed physics of flow
Mass: Table 3.3
Momentum: Table 6.4
Energy: Table 5.7

↓

Solutions of differential equations
provide velocity and temperature
distributions: $v_x(r)$ and $T(r, x)$

↓

Transport laws use velocity and
temperature distributions to provide
wall shear stress and convective
heat-transfer coefficients:

$$\tau_w = \mu \frac{\partial v_x}{\partial r}\Big|_{wall} \qquad \text{(Eq. 6.3)}$$

$$\dot{Q}''_w \equiv h_x[T_m(x) - T_m(x)] = -k\frac{\partial T}{\partial r}\Big|_{wall}$$

(Eqs. 4.16a and 10.24)

↓

Integral momentum and
energy conservation expressions
use τ_w and $h_{conv,x}$ to provide useful
engineering information:

ΔP or h_L (Eqs. 10.7 and 10.15)

$\left.\begin{array}{l}\dot{Q}_{0-L} \\ T_m(x) \\ T_w(x)\end{array}\right\}$ (Table 10.1)

FIGURE 10.9

Schematic diagram illustrating how fundamental conservation principles (differential and integral) combine to provide working relationships for engineering applications.

▶ Note that this value is much less than $Re_{x,crit}$ for external flows. See Eq. 9.39.

10.3 DETAILS—FULLY DEVELOPED LAMINAR FLOWS

In this section, we analyze the details of the laminar flow through a straight tube of circular cross section to provide the information needed to complete the integral analyses of the preceding sections. Specifically, first, we seek the wall shear stress τ_w to complete our momentum analysis; second, we seek the convective heat-transfer coefficients to complete our thermal analyses for conditions of uniform heat flux and constant wall temperature. Figure 10.9 illustrates the chain of concepts employed to produce useful engineering relationships used to calculate pressure drops, overall heat-transfer rates, local heat fluxes, and bulk mean and wall-temperature distributions.

10.3a Laminar Flow Criterion

In a manner analogous to external flows (Chapter 9), the Reynolds number determines whether the flow is laminar or turbulent. For an internal flow, the Reynolds number is defined as

$$Re_D \equiv \frac{\rho v_{avg} D}{\mu} = \frac{v_{avg} D}{\nu}, \qquad (10.34a)$$

where D is the tube diameter and v_{avg} is the average velocity as defined by Eq. 3.13. Note that the characteristic length L_c is now the tube diameter, not the length in the flow direction. The Reynolds number is also conveniently expressed using the mass flow rate:

$$Re_D \equiv \frac{4\dot{m}}{\pi\mu D}. \qquad (10.34b)$$

The value of the critical Reynolds number establishing whether the flow is laminar or turbulent is

$$Re_{D,\,crit} \approx 2300. \qquad (10.35)$$

Laminar flows exist in the regime $Re_D < Re_{D,crit}$, and turbulent flows in the regime $Re_D > Re_{D,crit}$, although transitional conditions frequently occur for $2300 \leq Re_D \leq 10{,}000$. In the transitional regime, the flow can be intermittently laminar or turbulent, creating flow pulsations. Good engineering design practice avoids transitional flows because of this complication.

10.3b Differential Conservation Equations

To simplify and solve the general relationships expressing mass, momentum, and energy conservation, the following assumptions are invoked:

1. The flow is steady state.

2. The fluid is incompressible.

3. All fluid properties are constant.

4. The geometry is two dimensional and cylindrical, which implies

  a. $v_\theta = 0$,

  b. $\partial(\ )/\partial\theta = 0$.

5. The tube wall is impermeable [i.e., $v_r(R, x) = 0$].

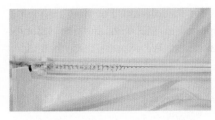

Dye stream shows flow transition from laminar to turbulent. Flow is from right to left. (cf. Fig. 1.32.) Photograph courtesy of Flowmetrics.

6. The flow is fully developed, which implies

a. $\partial v_x / \partial x = 0$,

b. $\dfrac{\partial}{\partial x}\left[\dfrac{T_w(x) - T(r, x)}{T_w(x) - T_m(x)} \right] = 0$.

7. Axial conduction is small compared to radial conduction.

8. Viscous dissipation is negligible.

The detailed application of these assumptions to simplify the general equations is presented in Appendix 10A. Of these assumptions, the sixth one is likely to be vague, so we digress here to elaborate, noting that the concept of developing flows is treated in detail later in this chapter. Assumption 6a applies to the development of the velocity profile and is a formal mathematical statement that the velocity profile $v_x(r, x)$ does not change with x [i.e., $v_x(r, x) = v_x(x$ only)]. That the velocity profile is constant means that the flow is *hydrodynamically fully developed*. For the flow to be *thermally fully developed* requires not that the temperature profile $T(r, x)$ be constant with x but, rather, that the *dimensionless temperature distribution* be constant. This idea is mathematically stated in assumption 6b.

With the assumption of a fully developed flow (assumption 6a), the continuity (mass conservation) equation provides the result that v_r is zero not only at the wall but throughout the flow. The simplified axial momentum and energy equations are as follows:

See Appendix 10A.

Momentum conservation:

$$
\underbrace{-\frac{dP}{dx}}_{\substack{\text{Net pressure} \\ \text{force in } x\text{-direction} \\ \text{per unit volume}}} + \underbrace{\mu \frac{1}{r} \frac{d}{dr}\left(r \frac{dv_x}{dr} \right)}_{\substack{\text{Net viscous force in} \\ x\text{-direction per unit} \\ \text{volume}}} + \underbrace{\rho g_x}_{\substack{\text{Gravitational body} \\ \text{force in } x\text{-direction} \\ \text{per unit volume}}} = 0. \quad (10.36)
$$

Energy conservation:

$$
\underbrace{\rho c_p v_x \frac{\partial T}{\partial x}}_{\substack{\text{Net rate of energy per unit} \\ \text{volume carried by bulk flow} \\ \text{(advection)}}} = \underbrace{k \frac{1}{r} \frac{\partial}{\partial r}\left(r \frac{\partial T}{\partial r} \right).}_{\substack{\text{Net rate of conduction heat} \\ \text{transfer in radial direction per} \\ \text{unit volume}}} \quad (10.37)
$$

Note that, with the assumption of fully developed flow, the axial momentum equation (Eq. 10.36) expresses a simple balance of forces with no momentum flows involved (i.e., there is no acceleration of the fluid as it flows through the tube). Likewise, the energy equation (Eq. 10.37) affords a simple interpretation: Any increase in the energy carried by the flow is a result of radial heat conduction.

See Fig. 5.19 and related developments in Chapter 5.

10.3c Hydrodynamic Problem Solution

Velocity Distribution

From our integral analysis, we see that the pressure gradient dP/dx is a constant (see Fig. 10.3); thus, Eq. 10.36 is a second-order, *ordinary* differential equation (ODE) rather than a *partial* differential equation. A second-order

ODE requires two boundary conditions for its solution. For flow through a circular tube, the no-slip condition applies at the wall and the velocity distribution is symmetric about the centerline. We express these boundary conditions mathematically as

$$v_x(r = R) = 0 \quad \text{(no slip)}, \tag{10.38a}$$

$$\left.\frac{dv_x}{dr}\right|_{r=0} = 0 \quad \text{(symmetry)}. \tag{10.38b}$$

Equation 10.36 is easily solved for the function $v_x(r)$ by separating and integrating twice. We start by moving all constants to the right-hand side:

$$\frac{1}{r}\frac{d}{dr}\left(r\frac{dv_x}{dr}\right) = \frac{1}{\mu}\left(\frac{dP}{dx} - \rho g_x\right) \equiv A,$$

where, for simplicity, we designate the right-hand-side constant as A. Separating and integrating yields

$$\int d\left(r\frac{dv_x}{dr}\right) = A\int r\,dr,$$

or

$$r\frac{dv_x}{dr} = A\frac{r^2}{2} + C_1.$$

Separating and integrating a second time yields

$$\int dv_x = \int\left(A\frac{r}{2} + \frac{C_1}{r}\right)dr,$$

or

$$v_x = \frac{A}{4}r^2 + C_1\ln r + C_2.$$

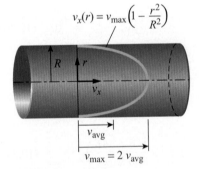

$$v_x(r) = v_{\max}\left(1 - \frac{r^2}{R^2}\right)$$

$v_{\max} = 2\,v_{\text{avg}}$

FIGURE 10.10
Parabolic velocity distribution for laminar flow in a tube of circular cross section with radius R.

Using the two boundary conditions (Eqs. 10.38a and 10.38b), we find that

$$C_1 = 0$$

and

$$C_2 = -AR^2/4.$$

With these substitutions, the velocity distribution is given by

$$v_x(r) = \frac{1}{4}A(r^2 - R^2),$$

or

$$v_x = -\frac{R^2}{4\mu}\left(\frac{dP}{dx} - \rho g_x\right)\left(1 - \frac{r^2}{R^2}\right). \tag{10.39}$$

By symmetry, the velocity is a maximum on the centerline ($r = 0$); thus,

$$v_{\max} = -\frac{R^2}{4\mu}\left(\frac{dP}{dx} - \rho g_x\right) \tag{10.40}$$

and

$$v_x(r) = v_{\max}\left(1 - \frac{r^2}{R^2}\right). \tag{10.41}$$

Figure 10.10 illustrates this parabolic velocity distribution.

> **The reader should review Example 3.7 and related material at this time.**

> **This is the parabolic distribution introduced in Chapter 3.**

Average Velocity

We now link our integral mass conservation analysis (Eq. 10.3b) to the detailed description of the velocity distribution (Eqs. 10.39 and 10.41) through the definition of an average velocity (Eq. 3.13);

$$
v_{avg} \equiv \frac{1}{\pi R^2} \int_0^R v_x(r) 2\pi r \, dr = \frac{1}{2} v_{max}.
$$

(10.42)

The verification of this statement is left as an exercise for the reader. Substituting Eq. 10.40 for v_{max} yields

$$
v_{avg} = -\frac{R^2}{8\mu} \left(\frac{dP}{dx} - \rho g_x \right).
$$

(10.43)

Using this result, we relate the mass flow rate to the tube dimensions, pressure gradient, body force, and fluid properties as follows:

$$
\dot{m} = \rho v_{avg} \pi R^2 = -\frac{\pi R^4}{8\nu} \left(\frac{dP}{dx} - \rho g_x \right),
$$

(10.44)

where $\nu \, (= \mu/\rho)$ is the kinematic viscosity. It is important to point out that the use of Eq. 10.44 to calculate flow rates requires that dP/dx be obtained in the fully developed portion of any tube flow. Later in this chapter, we present a criterion to determine if fully developed conditions exist.

Wall Shear Stress

Analogous to Newton's law of viscosity in a Cartesian coordinate system (Eq. 6.9b),

$$
\tau_{yx} = \mu \frac{\partial v_x}{\partial y},
$$

the x-directed shear stress in a cylindrical coordinate system is given by

$$
\tau_{rx} = -\mu \frac{\partial v_x}{\partial r}.
$$

(10.45)

Using the parabolic velocity distribution, Eq. 10.41, we obtain the following expression for the wall shear stress:

$$
\tau_w = -\mu \frac{\partial v_x}{\partial r} \bigg|_{r=R} = -\mu v_{max} \left[-\frac{2r}{R^2} \right]_{r=R} = \frac{4\mu v_{avg}}{R},
$$

(10.46)

recognizing that $v_{max} = 2 v_{avg}$. We can also relate the wall shear stress to the mass flow rate since $v_{avg} = \dot{m}/(\rho \pi R^2)$; thus,

$$
\tau_w = \frac{4\mu \dot{m}}{\rho \pi R^3}.
$$

(10.47)

Pressure Drop

With Eqs. 10.46 and 10.47, we have our link to the integral momentum relationships discussed earlier, that is, Eqs. 10.7 and 10.9. Substituting

Eq. 10.47 into Eq. 10.7a yields

$$(P_1 - P_2) - \rho g L \sin\theta = \frac{8\mu L \dot{m}}{\rho \pi R^4},$$ (10.48a)

or for a horizontal tube,

$$(P_1 - P_2) = \frac{8\mu L \dot{m}}{\rho \pi R^4}.$$ (10.48b)

Equations 10.48a and 10.48b apply provided the flow is laminar, that is, if

$$Re_D = \frac{\rho v_{avg} D}{\mu} = \frac{4\dot{m}}{\pi \mu D} < 2300.$$

Of course, Eqs. 10.48a and 10.48b can be rearranged such that the flow rate $\dot{m}$ is isolated and expressed in terms of all of the other quantities, and so

$$\dot{m} = \frac{\rho \pi R^4 (P_1 - P_2)}{8\mu L}$$ (10.49a)

See Eqs. 3.14 and 3.15 in Chapter 3. ▶

for a horizontal tube. Since $\dot{m} = \rho \dot{V}$, the volumetric flow rate can be expressed as

$$\dot{V} = \frac{\pi R^4 (P_1 - P_2)}{8\mu L}.$$ (10.49b)

This relationship is frequently called the Hagen–Poiseuille law in honor of the German engineer G. H. L. Hagen [2] and the French scientist J. L. Poiseuille [3], who experimentally related the flow rate to the pressure drop and the tube length and diameter in the manner of Eq. 10.49b.

Since the pressure gradient dP/dx is a constant in the fully developed region, Eqs. 10.48a and 10.48b can be used for its evaluation, and thus

$$\frac{dP}{dx} = \frac{P_2 - P_1}{L} = -\frac{8\mu \dot{m}}{\rho \pi R^4} - \rho g \sin\theta.$$ (10.50)

Example 10.4

Photograph courtesy of Friedrich and Dimmock, Inc.

Small-bore, glass capillary tubes can be used to construct inexpensive flowmeters. Consider the flow of air through a 1-m-long, 2-mm-inside-diameter tube. The air enters at 1 atm and 300 K. The pressure drop from entrance to exit is 585 Pa. Determine the mass flow rate of the air through the tube. Also determine the flow rate for a tube with a 1-mm bore for the same conditions.

Solution

Known $D, P_1, P_1 - P_2, T$

Find $\dot{m}$

Sketch

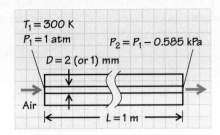

Assumptions

   i. Laminar flow
  ii. Horizontal tube
 iii. Negligible entrance effects
 iv. Constant properties

Analysis   Equation 10.49a was derived subject to the assumptions listed here. We employ this equation to find the mass flow rate and then test to see whether our assumption of laminar flow is valid. With the density calculated from the ideal-gas law and the viscosity from Table C.3, we have

$$\dot{m} = \frac{\rho \pi R^4 (P_1 - P_2)}{8\mu L}$$

$$= \frac{1.1767\pi(0.002/2)^4(585)}{8(18.58 \times 10^{-6})(1)}$$

$$= 1.454 \times 10^{-5}$$

$$[=]\frac{(\text{kg/m}^3)\text{m}^4(\text{N/m}^2)}{(\text{N} \cdot \text{s/m}^2)\text{m}} = \text{kg/s}.$$

Since the flow rate is proportional to the fourth power of the tube radius, we find for the 1-mm-inside-diameter tube

$$\dot{m}(1\ \text{mm}) = \dot{m}(2\ \text{mm})\left(\frac{0.001}{0.002}\right)^4$$

$$= 1.454 \times 10^{-5}(0.5)^4\ \text{kg/s}$$

$$= 9.09 \times 10^{-7}\ \text{kg/s}.$$

We now test our assumption of laminar flow by ascertaining whether $Re_D < 2300$ (see Eq. 10.35). For the 2-mm-inside-diameter tube,

$$Re_D = \frac{4\dot{m}}{\pi \mu D}$$

$$= \frac{4(1.454 \times 10^{-5})}{\pi(18.58 \times 10^{-6})0.002} = 498.$$

Since $498 < 2300$, the flow is laminar and our original assumption is justified. This is equally true for the 1-mm-inside-diameter tube where $Re_D = 62$.

Comments   Note the very strong dependence of the flow rate on the tube diameter (i.e., $\dot{m} \propto D^4$).

Head Loss

We can also relate the head loss appearing in the mechanical energy equation to the average velocity by substituting Eq. 10.46 into Eq. 10.15, which gives

$$h_L = \frac{8\mu L}{\rho g R^2} v_{avg},$$ (10.51)

and, similarly, in terms of the mass flow rate, we have

$$h_L = \frac{8\mu L}{\pi g \rho^2 R^4}\, \dot{m}.$$ (10.52)

Friction Factor

The Darcy friction factor is defined (Eq. 10.11) as

$$f_D \equiv \frac{4\tau_w}{\frac{1}{2}\rho v_{avg}^2}.$$

Substituting Eq. 10.46 for the wall shear stress into this definition yields

$$f_D = \frac{4\left(\dfrac{4\mu v_{avg}}{R}\right)}{\frac{1}{2}\rho v_{avg}^2} = \frac{32\mu}{\rho v_{avg} R},$$

or, recognizing that $R = D/2$, we obtain

$$f_D = \frac{64\mu}{\rho v_{avg} D} = 64/Re_D.$$ (10.53)

Thus, for laminar flow, a simple inverse relationship exists between the friction factor and the Reynolds number Re_D.

Example 10.5

Using the conditions given in Example 10.4, determine the head loss (a) from its defining relationship, Eq. 10.14, and (b) from its relationship to the mass flow rate, Eq. 10.52. Also determine the friction factor.

Solution

Known $D, P_1, P_1 - P_2, T, \dot{m}$ (Example 10.4), Re_D (Example 10.4)

Find h_L, f_D

Assumptions

See Example 10.4.

Analysis   For a horizontal table, Eq. 10.14 simplifies to

$$P_1 - P_2 = \rho g h_L,$$

or

$$h_L = \frac{P_1 - P_2}{\rho g}.$$

Since the pressure drop is a given quantity, the head loss is easily evaluated:

$$h_L = \frac{585}{1.1764(9.807)} = 50.7$$

$$[=] \frac{N/m^2}{(kg/m^3)(m/s^2)} \left[\frac{kg \cdot m/s^2}{1\,N} \right] = m.$$

Alternatively, the head loss can be calculated from the mass flow rate (Eq. 10.52) as follows:

$$h_L = \frac{8\mu L \dot{m}}{\pi g \rho^2 R^4}$$

$$= \frac{8(18.58 \times 10^{-6})(1)1.454 \times 10^{-5}}{\pi(9.807)(1.1767)^2(0.001)^4} = 50.7$$

$$[=] \frac{(N \cdot s/m^2) m (kg/s)}{(m/s^2)(kg/m^3)^2 m^4} \left[\frac{kg \cdot m/s^2}{1\,N} \right] = m.$$

Considering significant digits and round off, these results are identical. The friction factor is straightforwardly calculated from Eq. 10.53:

$$f_D = 64/Re_D$$

$$= 64/498 = 0.1285 \text{ (dimensionless)}.$$

We repeat the same calculations for the 1-mm-bore tube and summarize the results as follows:

D (mm)	$\dot{m}$ (kg/s)	h_L (m)	Re_D	f_D
2	1.454×10^{-5}	50.7	498	0.1285
1	9.09×10^{-7}	50.7	62	1.032

Comments   Note that the head loss can be calculated in different ways depending on the known information. We also note the large value of the friction factor for the tube with the smaller bore. Why is the head loss independent of the diameter?

Self Test 10.5 ✓ **The tube of Example 10.4 is tilted 45° from horizontal with the exit above the inlet. Determine the head loss in this position and recalculate the pressure drop through the tube for the 1-mm-bore tube.**

(Answer: $h_L = 50.7$ m, $\Delta P = 593.2$ Pa)

Example 10.6

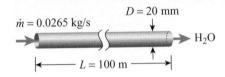

$\dot{m} = 0.0265$ kg/s

$D = 20$ mm

H_2O

$L = 100$ m

Determine the pressure drop and head loss associated with a 0.0269-kg/s flow of water through a smooth, 100-m-long, horizontal tube. The inside diameter of the tube is 20 mm. The temperature of the water is 300 K, and the flow is fully developed, (i.e., there are no entrance effects). Evaluate properties at 1 atm.

Solution

Known $\dot{m}, D, L, T$

Find $\Delta P, h_L$

Assumptions

  i. Steady, fully developed flow
  ii. Constant properties

Analysis   We first determine whether the flow is laminar or turbulent. To evaluate the Reynolds number Re_D, we calculate the average velocity from $\dot{m} = \rho v_{avg} A_{x\text{-sec}}$. Solving for v_{avg} gives

$$v_{avg} = \frac{\dot{m}}{\rho \pi D^2 / 4}$$

$$= \frac{0.0269}{997\pi(0.02)^2/4} = 0.0859$$

$$[=]\frac{\text{kg/s}}{(\text{kg/m}^3)\text{m}^2} = \text{m/s},$$

where the density of water is obtained from the NIST online database, as is the viscosity used in the following step. Thus,

$$Re_D = \frac{\rho v_{avg} D}{\mu}$$

$$= \frac{997(0.0859)(0.02)}{854 \times 10^{-6}} = 2006$$

$$[=]\frac{(\text{kg/m}^3)(\text{m/s})\text{m}}{\text{N}\cdot\text{s/m}^2}\left[\frac{1\ \text{N}}{\text{kg}\cdot\text{m/s}^2}\right] = 1.$$

Since the value of the Reynolds number is less than the critical value for transition to turbulent flow (i.e., 2006 < 2300), the flow is laminar. At this point, we have several (equivalent) options to calculate the pressure drop. Rather than using the same procedure employed in Example 10.4, we choose to calculate the friction factor (Eq. 10.53) and apply Eq. 10.17 to determine ΔP, thus,

$$f_D = \frac{64}{Re_D} = \frac{64}{2006} = 0.032$$

and

$$\Delta P = P_1 - P_2 = f_D\left(\frac{L}{D}\right)\frac{1}{2}\rho v_{avg}^2$$

$$= 0.032(100/0.02)0.5(997)(0.0859)^2$$

$$= 588.5$$

$$[=](\text{m/m})(\text{kg/m}^3)(\text{m/s})^2\left[\frac{1\ \text{N}}{\text{kg}\cdot\text{m/s}^2}\right] = \text{N/m}^2 \text{ or Pa.}$$

We determine the head loss from Eq. 10.14 as

$$h_L = \frac{P_1 - P_2}{\rho g}$$

$$= \frac{588.5}{997(9.807)} = 0.060$$

$$[=] \frac{(N/m^2)}{(kg/m^3)(m/s^2)} \left[\frac{kg \cdot m/s^2}{1\,N} \right] = m.$$

Comments   Although pressure drops for laminar flows can be calculated directly using dimensional variables, the approach taken here illustrates the use of dimensionless parameters. In considering turbulent flows, the dimensionless-parameter approach is almost always used.

Self Test
10.6 ☑   **Oil at 273 K flows through a smooth, 7.5-cm-diameter pipe at 2 kg/s. Calculate the head loss and pressure drop through a 10-m-long section of pipe.**

(Answer: $h_L = 12.5$ m, $\Delta P = 110.2$ kPa)

10.3d Thermal Problem Solution

Temperature Distributions

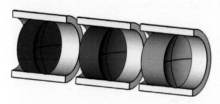

Temperature distributions for laminar flow through a tube with a uniform surface heat flux for $Re_D = 500$ and $Pr = 5.83$. Each distribution is 2.5 m downstream from the preceding.

The shape of the temperature distribution $T(r, x)$ in a laminar flow depends on the thermal boundary condition. We consider two such conditions: a uniform wall heat flux $\dot{Q}_w''$ and a constant wall temperature $T_w(x)$. For the constant-$\dot{Q}_w''$ case, solving the governing equation for the temperature distribution $T(r, x)$ is conceptually no more difficult than solving for the velocity distribution. The steps involved are outlined in the following. For the constant-T_w case, however, finding $T(r, x)$ involves an iterative solution and is beyond the scope of this book. For the interested reader, a detailed solution is presented in Ref. [8].

Our practical interest in obtaining the temperature distributions is to determine expressions (or values) for the heat-transfer coefficients that appear in the various integral relationships summarized in Table 10.1. Using $T(r, x)$ and $v_x(r)$ to evaluate $T_m(x)$, we can determine the heat-transfer coefficient for the constant-$\dot{Q}_w''$ case by applying its definition for an internal flow (Eq. 10.28),

$$h_{conv,x} \equiv \frac{\dot{Q}_w''}{T_w(x) - T_m(x)}. \tag{10.54}$$

For the constant-T_w case, the heat-transfer coefficient is similarly obtained, except that Fourier's law is needed to evaluate $\dot{Q}_w''(x)$, that is,

$$h_{conv,x} = \frac{-\left(-k \dfrac{dT}{dr}\right)_{r=R}}{T_w - T_m(x)}. \tag{10.55}$$

We now derive the temperature distribution for the uniform heat flux case and then present results for the heat-transfer coefficients for both uniform heat flux and constant wall temperature conditions. Solving Eq. 10.37, which expresses conservation of energy, provides the desired temperature distribution. To solve Eq. 10.37, we begin with two substitutions: First, we substitute Eq. 10.41 for v_x; second, we substitute $dT_m(x)/dx$ for $\partial T/\partial x$. This second substitution follows from the assumption that the temperature distribution is fully developed, that is,

$$\frac{\partial}{\partial x}\left[\frac{T_w(x) - T(r,x)}{T_w(x) - T_m(x)}\right] = 0.$$

From Eq. 10.28, the denominator in the brackets is

$$T_w(x) - T_m(x) = \dot{Q}''_w/h_{conv,x},$$

where $h_{conv,x}$ is a constant for a fully developed flow. Substituting this result into the previous equation and performing the indicated differentiation show that indeed

$$\frac{\partial T(r, x)}{\partial x} = \frac{dT_m(x)}{dx}.$$

With this substitution and the substitution for v_x, Eq. 10.37 becomes

$$\frac{1}{r}\frac{d}{dr}\left(r\frac{dT}{dr}\right) = \frac{2v_{avg}}{\alpha}\left(\frac{dT_m}{dx}\right)\left(1 - \frac{r^2}{R^2}\right),$$

where $\alpha\ (= k/\rho c_p)$ is the thermal diffusivity. For simplicity, we define the constant B

$$B \equiv \frac{2v_{avg}}{\alpha}\frac{dT_m}{dx},$$

thus,

$$\frac{1}{r}\frac{d}{dr}\left(r\frac{dT}{dr}\right) = B\left(1 - \frac{r^2}{R^2}\right).$$

Separating and integrating yields

$$\int d\left(r\frac{dT}{dr}\right) = \int B\left(r - \frac{r^3}{R^2}dr\right)$$

and

$$r\frac{dT}{dr} = \frac{Br^2}{2} - \frac{Br^4}{4R^2} + C_1.$$

We once again separate and integrate to get

$$\int dT = \int\left(\frac{Br}{2} - \frac{Br^3}{4R^2} + \frac{C_1}{r}\right)dr$$

and

$$T = \frac{Br^2}{4} - \frac{Br^4}{16R^2} + C_1\ln r + C_2.$$

The boundary conditions needed to evaluate the constants C_1 and C_2 are

$$T(r = R) = T_w \tag{10.56a}$$

and, by symmetry,

$$\left.\frac{dT}{dr}\right|_{r=0} = 0. \tag{10.56b}$$

From these we see that

$$C_1 = 0$$

and

$$C_2 = T_w - \left[\frac{BR^2}{4} - \frac{BR^2}{16}\right].$$

Thus, the final temperature distribution is given by

$$T(r, x) = \frac{-BR^2}{4}\left[\frac{3}{4} - \frac{r^2}{R^2} + \frac{1}{4}\frac{r^4}{R^4}\right] + T_w(x),$$

or

$$T(r, x) = \frac{-v_{avg}R^2}{2\alpha}\frac{dT_m}{dx}\left(\frac{3}{4} - \frac{r^2}{R^2} + \frac{r^4}{4R^4}\right) + T_w(x). \qquad (10.57)$$

Note that

$$\frac{dT_m(x)}{dx} = \frac{\dot{Q}_w''\pi D}{\dot{m}c_p}, \qquad (10.58)$$

which is easily obtained by differentiating Eq. 10.27, the axial distribution of the bulk mean temperature derived from our integral analysis of this case.

Heat-Transfer Coefficient: Uniform Heat Flux

Equation 10.21 defines the bulk mean temperature $T_m(x)$.

We use the temperature distribution given by Eq. 10.57 to evaluate the bulk mean temperature $T_m(x)$ and substitute this result into Eq. 10.54. This yields the simple result that

$$h_{conv,x} = \frac{48}{11}\frac{k}{D} = 4.36\frac{k}{D}. \qquad (10.59)$$

Using this result, we define a Nusselt number for the uniform heat flux case, that is,

$$Nu_D \equiv \frac{h_{conv,x}D}{k} = 4.36. \qquad (10.60)$$

Now knowing $h_{conv,x}$, we have closure to the problem of laminar flow through a tube with a uniform heat flux imposed on the tube wall (see Table 10.1).

Heat-Transfer Coefficient: Fixed Wall Temperature

For the case of a fixed wall temperature, the solution of Eq. 10.37 and the application of the result to Eq. 10.55 yield

$$h_{conv,x} = 3.66\frac{k}{D}, \qquad (10.61)$$

or

$$Nu_D \equiv \frac{h_{conv,x}D}{k} = 3.66. \qquad (10.62)$$

In the practical use of Eq. 10.62 (and Eq. 10.60, as well), the fluid thermal conductivity is evaluated at the local bulk mean temperature $T_m(x)$.

Example 10.7

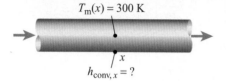

$T_m(x) = 300\text{ K}$

x

$h_{conv,\,x} = ?$

Consider laminar flows of air, water, and oil, respectively, through a 10-mm-diameter tube. Estimate the local heat-transfer coefficient for conditions of (a) uniform heat flux and (b) constant wall temperature. Assume the bulk mean temperature is 300 K for both situations. Use properties at 100 kPa for air and saturated liquid conditions for the water and the oil. Repeat for a bulk mean temperature of 400 K.

Solution

Known   Laminar flow, fluid, T_m

Find $h_{conv,\,x}\,(\dot{Q}'' = \text{constant})$, $h_{conv,\,x}\,(T_w = \text{constant})$

Analysis   For a laminar internal flow, the Nusselt number for the case of a uniform heat flux is $Nu_D = 4.36$ (Eq. 10.60), and for a constant wall temperature, $Nu_D = 3.66$ (Eq. 10.62). To evaluate $h_{conv,\,x}$, we need only apply the definition of the Nusselt number, that is,

$$Nu_D \equiv \frac{h_{conv,\,x}D}{k}.$$

The only thermo-physical property values required are the thermal conductivities, which we obtain from Table C.3 (air), the NIST database (water), and Table G.2 (oil). A sample calculation for air is shown followed by a summary of all of the results.

For air,

$$k_{air}\,(300\text{ K}) = 26.2 \times 10^{-3}\text{ W/m}\cdot\text{K},$$

$$Nu_D(\dot{Q}'' = \text{constant}) = 4.36,$$

$$h_{conv,\,x} = 4.36\frac{k_{air}}{D}$$

$$= \frac{4.36(26.3 \times 10^{-3}\text{ W/m}\cdot\text{K})}{0.01\text{ m}}$$

$$= 11.4\text{ W/m}^2\cdot\text{K}.$$

Fluid	T_m (K)	k (W/m·K)	$h_{conv,\,x}\,(\dot{Q}'' = \text{constant})$ (W/m²·K)	$h_{conv,\,x}\,(T_w = \text{constant})$ (W/m²·K)
Air	300	26.2×10^{-3}	11.4	9.6
Air	400	32.9×10^{-3}	14.3	12.1
Water	300	610×10^{-3}	266	223
Water	400	684×10^{-3}	298	250
Oil	300	145×10^{-3}	63.2	53.1
Oil	400	134×10^{-3}	58.4	49.0

Comments   In Chapter 4, Table 4.2 presents typical values of convective heat-transfer coefficients. Comparing the values listed here with those in Table 4.2, we see that our air values are lower than the low end of the range of the Table 4.2 values, whereas those for water and oil are near the low end of the range. We might anticipate this result since most flows of engineering interest are not laminar. We also note from the summary table that the temperature increase of 100 K has a substantial effect on the magnitudes of h_{conv}, altering them by +27.8%, +12.0%, and −7.6%, respectively, for the three fluids.

Self Test 10.7 ✓ A fully developed flow of water flows at 0.011 kg/s through a 20-mm-diameter, 3-m-long tube heated by an electrical heating tape. Determine the power require to heat the water from 300 to 350 K. Also determine the heat-transfer coefficient and find the tube wall temperature at the inlet and exit.

(Answer: $\dot{Q} = 2.3\ kW$, $h_{conv} = 140.8\ W/m^2$, $T_{w,1} = 386.7\ K$, $T_{w,2} = 436.7\ K$)

Axial velocity: near–wall region   **End view**

Calculated instantaneous axial velocity distribution for turbulent pipe flow. Image courtesy of Parviz Moin.

10.4 DETAILS–FULLY DEVELOPED TURBULENT FLOWS

In the preceding discussion of laminar flow, we started with the governing conservation equations, applied appropriate boundary conditions, and then solved the equations for the velocity and temperature distributions. In turn, we used these distributions to obtain useful relationships for v_{avg}, τ_w, ΔP, f, T_m, T_w, and $h_{conv,x}$. It would be intellectually satisfying to follow this same approach for turbulent flows, using the time-averaged, rather than instantaneous, conservation equations as our starting point. Unfortunately, this cannot be done easily. Although we outline such an approach for momentum conservation, essentially all of the useful engineering expressions we present are based on experimental measurements. For example, theoretical frameworks can be used to determine the form of velocity profiles, but final results depend on empirical constants determined from experimentally measured velocity profiles. These measured profiles form the basis of much of the "theory" of calculating wall shear stress, pressure drops, and friction factors for turbulent pipe flow. For nonisothermal flows, empiricism plays an even greater role. In this case, experimental measurements tend to be more global and are used to develop correlations of the type $Nu_D = f(Re_D, Pr)$, with no need to determine temperature distributions.

10.4a Velocity Distributions and Wall Friction

A Theoretical Framework

We invoke the same basic assumptions employed for the laminar flow analysis, that is, we deal with a fully developed, time-mean steady flow of an incompressible fluid with constant properties in a two-dimensional cylindrical geometry. In Chapter 6, we developed the time-averaged (i.e., Reynolds-averaged) momentum equation appropriate for a boundary-layer flow in Cartesian (x, y) coordinates. Performing the same operations of velocity decomposition (i.e., $v_x = \bar{v}_x + v'_x$, etc.) and time averaging to the axial momentum equation for our tube flow problem yield

See Eq. 6.89 in Chapter 6. ▶

$$-\frac{d\bar{P}}{dx} + \mu\frac{1}{r}\frac{d}{dr}\left(r\frac{d\bar{v}_x}{\partial r}\right) - \frac{1}{r}\frac{d}{dr}\left(r\rho\overline{v'_r v'_x}\right) + \rho g_x = 0. \quad (10.63)$$

| Net pressure force in x-direction per unit volume | Net viscous force in x-direction per unit volume | Net momentum flow in x-direction per unit volume resulting from turbulent velocity fluctuations | Gravitational body force in x-direction per unit volume |

We see that Eq. 10.63 is identical to the corresponding laminar flow equation (Eq. 10.36) except for the additional term in which the fluctuations of the axial and radial velocities appear, that is, the term containing the turbulent stress,

$$\tau_{rx}^{turb} \equiv -\rho\overline{v'_r v'_x}. \quad (10.64)$$

LEVEL 2

See discussion of Reynolds averaging and closure surrounding Eqs. 6.83–6.89 in Chapter 6.

This turbulent stress arises from the transport of axial momentum by the turbulent fluctuations. As mentioned in Chapter 6, finding a way to evaluate this turbulent stress is termed the **closure problem.** In the following, we investigate the simplest of closure schemes: the introduction of an eddy viscosity and the application of Prandtl's mixing-length hypothesis.

Eddy Viscosity   We begin by exploring the concept of eddy viscosity. The eddy viscosity is a fictional construct arising from our treatment of the turbulent momentum fluxes as turbulent stresses. For example, we can rewrite Eq. 10.63 explicitly in terms of the laminar (Newtonian) and turbulent (Reynolds) stresses as follows:

$$\frac{d\overline{P}}{dx} - \rho g_x = \frac{1}{r}\frac{d}{dr}[r(\tau_{\text{lam}} + \tau_{\text{turb}})]. \tag{10.65a}$$

Furthermore, we can define the stresses to be proportional to the mean velocity gradient, that is,

$$\tau_{\text{lam}} = \mu\frac{d\overline{v}_x}{dr} \tag{10.65b}$$

and

$$\tau_{\text{turb}} = \rho\varepsilon\frac{d\overline{v}_x}{dr}, \tag{10.65c}$$

where μ is the molecular viscosity and ε is the kinematic **eddy viscosity** ($\varepsilon = \mu_{\text{turb}}/\rho$, where μ_{turb} is the apparent turbulent viscosity). The relationship expressed by Eq. 10.65b is just the familiar expression derived for a Newtonian fluid for the flow of interest; the relationship expressed by Eq. 10.65c, however, is the definition of the so-called eddy viscosity. The eddy viscosity as defined here was first proposed by Boussinesq [9] in 1877. An **effective viscosity** μ_{eff} is defined as

$$\mu_{\text{eff}} \equiv \mu + \mu_{\text{turb}} = \mu + \rho\varepsilon, \tag{10.66}$$

so that

$$\tau_{\text{tot}} = (\mu + \rho\varepsilon)\frac{d\overline{v}_x}{dr}. \tag{10.67}$$

For turbulent flows far from a wall, $\rho\varepsilon \gg \mu$, so $\mu_{\text{eff}} \cong \rho\varepsilon$; however, near a wall, both μ *and* $\rho\varepsilon$ contribute to the total stress in the fluid.

Note that the introduction of the eddy viscosity per se does not achieve closure; the problem is now transformed into how to determine a value or the functional form of ε. Unlike the molecular viscosity μ, which is a thermo-physical property of the fluid itself, the eddy viscosity ε depends on the flow itself. One would not necessarily expect the same value of ε for two distinctly different flows; for example, a free jet and a swirling confined flow with recirculation are not likely to have the same values for ε. Moreover, since ε depends on the local flow properties, it will take on different values at different locations within the flow. Therefore, one must be careful not to take too literally the analogy between laminar and turbulent flows implied by Eqs. 10.65b and 10.65c. Also, in some flows, the turbulent stresses are not proportional to the mean velocity gradients, as required by Eq. 10.65c [10].

Mixing-Length Hypothesis   The simplest closure scheme is to assume the eddy viscosity is a constant throughout the flow field; unfortunately, experimental evidence shows this not to be a generally useful assumption, as expected. More sophisticated hypotheses, therefore, need to be applied. One of the most useful, and yet simple, hypotheses is that proposed by Prandtl [11]. Prandtl's hypothesis, by analogy with the kinetic theory of gases, states that the eddy viscosity is proportional to the product of the fluid density, a length scale called the **mixing length**, and a characteristic turbulent velocity, that is,

$$\mu_{\text{turb}} = \rho \varepsilon = \rho \ell_{\text{m}} v_{\text{turb}}. \tag{10.68}$$

Furthermore, Prandtl [11] assumed that the turbulent velocity v_{turb} is proportional to the product of the mixing length ℓ_{m} and the magnitude of the mean velocity gradient, $|\partial \bar{v}_x / \partial r|$. Thus,

$$\mu_{\text{turb}} = \rho \varepsilon = \rho \ell_{\text{m}}^2 \left| \frac{d\bar{v}_x}{dr} \right|. \tag{10.69}$$

Note, however, that we have yet to obtain closure because we have introduced yet another unknown! All that has been done so far is to replace the unknown velocity correlation $\overline{v_r' v_x'}$ with an expression involving an unknown eddy viscosity, and, in turn, to relate that eddy viscosity to an unknown mixing length. Our next step finally attains closure by specifying the mixing length. Since the mixing length depends on the flow itself, different specifications generally are required for each kind of flow. Reference [10] provides mixing-length functions for a wide variety of flows. We focus, however, only on the mixing length required to solve the wall-flow problem. For either internal or external flows near a wall, a cross-stream dependence is inherent in the mixing length and the flow is conveniently divided into three zones: the **viscous sublayer** adjacent to the wall, an **overlap layer,** and a fully turbulent **outer layer** far from the wall. These divisions are shown in Fig. 10.11. For flow adjacent to a wall, the mixing lengths for these three regions are given by the following equations [10]:

$$\ell_{\text{m}} = 0.41y \left[1 - \exp\left(-\frac{y\sqrt{\rho \tau_{\text{w}}}}{26\mu} \right) \right] \quad \text{(laminar sublayer),} \tag{10.70a}$$

$$\ell_{\text{m}} = 0.41y \quad \text{for} \quad y \leq 0.2195 \delta_{\text{H}} \quad \text{(buffer layer),} \tag{10.70b}$$

and

$$\ell_{\text{m}} = 0.09 \delta_{\text{H}} \quad \text{(fully turbulent region).} \tag{10.70c}$$

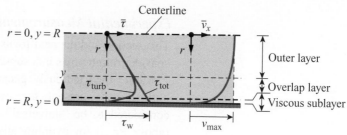

FIGURE 10.11

The maximum shear stress occurs at the wall within the viscous sublayer. At the wall, the contribution of the turbulent stress to the total is zero; within the overlap layer, both viscous and turbulent stresses are important; and in the outer layer, the total stress is dominated by the turbulent contribution. Note the linear distribution of the stress, with a zero value on the tube centerline.

In these expressions, τ_w is the local wall shear stress, and δ_H is the local boundary-layer thickness, defined as the y location at which the velocity equals 99% of the free-stream value. Note that in Eq. 10.70a, proposed by van Driest [12], the mixing length vanishes with approach to the wall; hence, $\mu_{eff} = \mu$ at $y = 0$. For sufficiently large y values, Eq. 10.70a degenerates to Eq. 10.70b.

Although presented in terms of the external flow boundary layer, these equations can be recast and also used for the internal flow problem. For flow through a tube, $y \equiv R - r$, and it follows that the boundary-layer thickness δ_H must equal the tube radius since the free-stream velocity V_∞ corresponds to v_{max} at the tube centerline (i.e., at $r = 0$ or $y = R - 0 = R$, as shown in Fig. 10.11). The mixing-length distribution given by Nikuradse [5],

$$\ell_m/R = 0.14 = 0.08 \, (r/R)^2 - 0.06 \, (r/R)^4, \tag{10.71}$$

where R is the pipe radius, can be used for turbulent flow in a tube or a pipe. Note that this equation also predicts a zero mixing length at the wall ($r = R$). With the closure problem now solved, (i.e., the specification of mixing lengths), we are now in a position to solve the momentum equation to yield the velocity distribution for turbulent pipe flow.

For fully developed tube flow, $d\overline{P}/dx$ is a constant; thus, our momentum conservation equation (Eq. 10.65a) can be separated and integrated as follows:

$$\int d[r(\tau_{lam} + \tau_{turb})] = \int \left[\frac{d\overline{P}}{dx} - \rho g_x \right] r \, dr$$

$$r(\tau_{lam} + \tau_{turb}) = \left[\frac{d\overline{P}}{dx} - \rho g_x \right] \frac{r^2}{2} + C_1,$$

or

$$\overline{\tau}_{tot} = \tau_{lam} + \tau_{turb} = \frac{1}{2} \left[\frac{d\overline{P}}{dx} - \rho g_x \right] r. \tag{10.72}$$

Here C_1 is zero because both the laminar and turbulent stresses vanish on the centerline ($r = 0$) as both are proportional to the mean velocity gradient $d\overline{v}_x/dr$, which, in turn, is zero on the centerline owing to the symmetry of the velocity profile. This linear distribution of the total stress, Eq. 10.72, is sketched in Fig. 10.11. Also indicated here is the contribution of the turbulent stress to the total. Successively substituting Eqs. 10.67, 10.69, and 10.71 into Eq. 10.72 results in a first-order ODE, which can be solved for the function $\overline{v}_x(r)$.

Experimental Measurements

This section and the next focus on hydraulically smooth tubes and pipes. We consider rough pipes in a subsequent section.

Rather than tediously generate the theoretical solution for $\overline{v}_x(r)$ as just outlined, we present the experimental velocity distribution, which is considered to be "universal" in that it applies for all Reynolds numbers (above Re_{crit}) for hydraulically smooth tubes and pipes. The experimental database used in its formulation is the same database used to generate the tube-flow mixing-length distribution given previously (i.e., Eq. 10.71); hence, any theoretical result for $\overline{v}_x(r)$ obtained from the solution of Eq. 10.72 is entirely consistent with the experimental results given in the following.

The velocity distribution is broken into two regions: 1. a near-wall region containing the viscous sublayer and 2. everything beyond. For smooth tubes and pipes, the velocity in these two regions is given by the following dimensionless expressions:

$$\frac{\bar{v}_x(y)}{v_*} = \frac{y v_*}{\nu} \quad \text{near the wall} \tag{10.73}$$

and

$$\frac{\bar{v}_x(y)}{v_*} = 2.5 \ln \frac{y v_*}{\nu} + 5.5 \quad \text{for} \quad \frac{y v_*}{\nu} \gtrsim 20. \tag{10.74}$$

To interpret these expressions, we first recognize that $y \, (= R - r)$ is the perpendicular distance from the tube wall. Second, we identify v_* as the **friction velocity**, defined as

$$v_* \equiv (\tau_w / \rho)^{1/2}. \tag{10.75}$$

Using the results from our integral analyses (Eq. 10.7b), we see that the friction velocity directly couples the velocity distribution with the pressure drop driving the flow. Rearranging Eq. 10.7b shows this explicitly:

$$\frac{\bar{P}_1 - \bar{P}_2}{L} = \frac{2}{R} \tau_w.$$

Here we see that the larger the imposed pressure drop, the larger the wall shear stress, and, hence, the larger the friction velocity (Eq. 10.75). In turn, $\bar{v}_x(y)$ increases with friction velocity, as can be seen by rearranging Eqs. 10.73 and 10.74 as

$$\bar{v}_x(y) = \frac{y}{\nu} v_*^2$$

and

$$\bar{v}_x(y) = v_* \left(2.5 \ln \frac{y v_*}{\nu} + 5.5 \right).$$

Figure 10.12 shows the velocity distributions given by Eqs. 10.73 and 10.74 along with experimental data.

At this point, we ask the following question: For a given flow rate, what does the velocity distribution look like when plotted more familiarly as $\bar{v}_x$ versus r, or, more generally, as $\bar{v}_x / v_{avg}$ versus r? We address this question in the following section.

Average Velocity and Friction Factor

To answer the question just posed, we need to eliminate the friction velocity v_* from Eqs. 10.73 and 10.74 and introduce the average velocity v_{avg}.[3] As a natural consequence of these operations, the Darcy friction factor f_D is introduced; thus, in addition to answering our original question, we generate

---

[3] To avoid confusion, note that we deal with two kinds of average velocities in this chapter: the time-average velocity defined by Eq. 3.42 and denoted with an overbar as $\bar{v}_x$ and the area average of this velocity denoted $v_{avg} [= \dot{m}/(\rho A_{x\text{-sec}})]$.

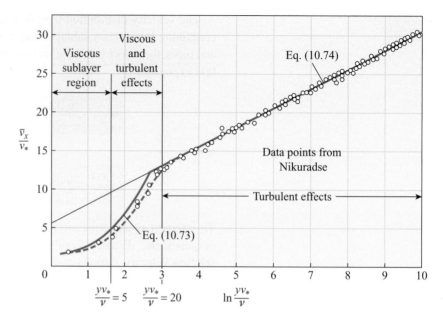

a useful relationship for the friction factor as a bonus. To begin, we note that
the near-wall region is very thin, with only a very small fraction of the total
flow passing through this region. Therefore, we use Eq. 10.74, the logarithmic
law, to represent the entire region from the wall to the tube centerline. With
this, we determine the average velocity as follows:

$$v_{avg} \equiv \frac{1}{\pi R^2} \int_0^R \bar{v}_x(r)\, 2\pi r\, dr$$

$$= \frac{1}{\pi R^2} \int_0^R v_* \left[A \ln \frac{(R-r)v_*}{\nu} + B \right] 2\pi r\, dr, \tag{10.76}$$

where, for greater generality, we define the two numerical constants
appearing in Eq. 10.74 as $A\ (= 2.5)$ and $B\ (= 5.5)$. Performing the indicated
integration provides the following relationship between v_{avg} and v_*:

$$v_{avg} = v_* \left[A \ln \frac{R v_*}{\nu} + B - \frac{3}{2} A \right]. \tag{10.77}$$

We next apply the definitions of the friction velocity (Eq. 10.75),

$$v_* = (\tau_w / \rho)^{1/2},$$

and the Darcy friction factor (Eq. 10.11),

$$f_D = \frac{8\tau_w}{\rho v_{avg}^2}.$$

We now have three equations (Eqs. 10.77, 10.75, and 10.11) involving four
variables: v_{avg}, v_*, τ_w, and f_D. If we treat v_{avg} as a given (i.e., an independent
variable), we then have three equations involving three unknowns. Treating
v_{avg} as a known quantity is equivalent to specifying the mass flow rate since
$\dot{m} = \rho v_{avg} \pi R^2$.

Our next step is to eliminate τ_w and v_* from the set of equations to yield a single relationship between f_D and v_{avg}. Solving Eq. 10.11 for τ_w yields

$$\tau_w = \frac{1}{8}\rho v_{avg}^2 f_D,$$

which we substitute into Eq. 10.75 as follows:

$$v_* = \left(\frac{1}{8}v_{avg}^2 f_D\right)^{1/2} = v_{avg}\left(\frac{f_D}{8}\right)^{1/2}. \tag{10.78}$$

In turn, we substitute this expression for the friction velocity into Eq. 10.77, with the result

$$v_{avg} = v_{avg}\left(\frac{f_D}{8}\right)^{1/2}\left\{A\ln\left[\frac{1}{2}\frac{v_{avg}2R}{\nu}\left(\frac{f_D}{8}\right)^{1/2}\right] + B - \frac{3}{2}A\right\}.$$

In this equation, we recognize $Re_D\ (= v_{avg}\ 2R/\nu)$ and simplify to yield the following transcendental equation for the friction factor:

$$\frac{1}{f_D^{1/2}} = \frac{A}{2\sqrt{2}}\ln\left(Re_D f_D^{1/2}\right) - \frac{A}{2\sqrt{2}}\left(\ln 4\sqrt{2} + \frac{3}{2}\right) + \frac{B}{2\sqrt{2}}. \tag{10.79}$$

Substituting numerical values ($A = 2.5, B = 5.5$) we arrive at our final result:

$$\frac{1}{f_D^{1/2}} = 0.884\ln\left(Re_D f_D^{1/2}\right) - 0.91. \tag{10.80a}$$

Substituting $\ln 10\ \log_{10}(\ )$ for $\ln(\ )$ generates the traditional expression

$$\frac{1}{f_D^{1/2}} = 2.035\log_{10}(Re_D f_D^{1/2}) - 0.91. \tag{10.80b}$$

The accepted form of Eq. 10.80 has slightly different constants from fitting to friction data and is given by

$$\frac{1}{f_D^{1/2}} = 0.87\ln(Re_D f_D^{1/2}) - 0.8 \tag{10.81a}$$

or

$$\frac{1}{f_D^{1/2}} = 2.0\log_{10}(Re_D f_D^{1/2}) - 0.8. \tag{10.81b}$$

Equation 10.81 is referred to as Prandtl's *universal law of friction* for smooth pipes. Note how the transcendental form of Eq. 10.81 complicates its solution. For a limited range of Reynolds numbers, we can use the simpler Blasius [14] expression

$$f_D = 0.316 Re_D^{-1/4} \quad \text{for} \quad 4000 < Re_D < 10^5, \tag{10.82}$$

or the more general approximation [15]

$$f_D = (0.79\ln Re_D - 1.64)^{-2}. \tag{10.83}$$

No iteration is required to obtain f_D from either Eq. 10.82 or 10.83.

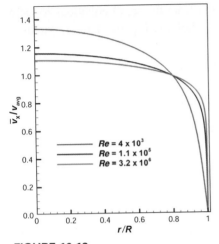

FIGURE 10.13

As the Reynolds number increases, velocity profiles for turbulent flow become increasingly flat in the center region and increasingly steep near the wall. Results shown are for a smooth pipe.

We now return to our original objective of generating an explicit relationship for $\bar{v}_x(r)$. To achieve this objective, we simply substitute Eq. 10.78 into the original distribution (Eq. 10.74),

$$\frac{\bar{v}_x(r)}{v_*} = 2.5\ln\frac{(R-r)v_*}{\nu} + 5.5,$$

and simplify. This results in

$$\frac{\bar{v}_x(r)}{v_{\text{avg}}} = \frac{f_{\text{D}}^{1/2}}{2\sqrt{2}}\left\{2.5\ln\left[\frac{Re_D f_{\text{D}}^{1/2}}{4\sqrt{2}}\left(1-\frac{r}{R}\right)\right] + 5.5\right\}. \tag{10.84}$$

Since the friction factor is a function only of Re_D the shape of the velocity distribution $\bar{v}_x(r)/v_{\text{avg}}$ depends on Re_D. To illustrate this, we plot Eq. 10.84 for three values of Re_D: 4×10^3, 1.1×10^5, and 3.2×10^6. Corresponding friction factors from Eq. 10.81 are shown in Table 10.2, and Fig. 10.13 presents the plots. The general form of each distribution is a somewhat uniform, broad distribution in the central region of the pipe with a steep falloff at the wall—a falloff that is much steeper than for laminar flow. The effect of increasing Reynolds number is to flatten the profiles in the central region and to increase the velocity gradient at the wall.

Power-law expressions are sometimes used to represent turbulent velocity profiles in tube flow, for example,

$$\frac{\bar{v}_x(r)}{v_{\text{avg}}} = \frac{v_{\text{max}}}{v_{\text{avg}}}(1-r/R)^{1/n}. \tag{10.85}$$

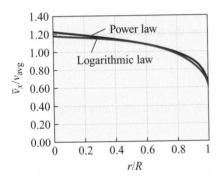

FIGURE 10.14

Power-law and universal velocity profiles for flow in a smooth pipe at $Re_D = 1.1 \times 10^5$.

In our discussion of mass conservation in Chapter 3, we used such expressions to find ratios of average to maximum velocities (see Table 3.1). In Table 10.2, we compare these power-law results with those generated using the universal profile described by Eqs. 10.74 and 10.84. At the higher Reynolds numbers, the power-law expressions overpredict the ratios of maximum (centerline) to average velocity by only 3.8% ($n = 7$) and 1.4% ($n = 10$). In Fig. 10.14, we compare the velocity profiles for $Re_D = 1.1 \times 10^5$. Here we see that the power-law velocity is somewhat less than the logarithmic-law velocity from the centerline out to about $r/R \approx 0.6$, with roles reversing until very close to the wall ($r/R = 0.992$).

Table 10.2 Ratio of Maximum Velocity to Average Velocity for Turbulent Flow in a Smooth Pipe

Re_D	f_D	Universal Profile* $v_{\text{max}}/v_{\text{avg}}$	n	Power Law† $v_{\text{max}}/v_{\text{avg}}$
4×10^3	0.04	1.264	6	1.264
1.1×10^5	0.0176	1.179	7	1.224
3.2×10^6	0.0096	1.138	10	1.155

* Equation 10.74.
† See Table 3.1.

Example 10.8

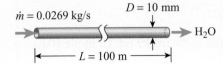

$\dot{m} = 0.0269$ kg/s

$D = 10$ mm

$L = 100$ m

H_2O

Determine the pressure drop and head loss associated with a 0.0269-kg/s flow of water through a smooth, 100-m-long, horizontal tube. The inside diameter of the tube is 10 mm. The temperature of the water is 300 K, and the flow is fully developed (i.e., there are no entrance effects). Evaluate properties at 1 atm. Note that this problem statement is identical to that of Example 10.6, except now the inside diameter of the tube is 10 mm rather than 20 mm.

Solution

Known $\dot{m}, D, L, T$

Find $\Delta P, h_L$

Assumptions

Steady, fully developed flow

Analysis Since the mass flow rate is the same as in Example 10.6, reducing the diameter will result in the average velocity increasing by a factor of 4 because the velocity is inversely proportional to the square of the diameter, and so

$$v_{avg,new} = v_{avg,old}\left(\frac{D_{old}}{D_{new}}\right)^2$$

$$= 0.0859 \text{ m/s}\left(\frac{20 \text{ mm}}{10 \text{ mm}}\right)^2$$

$$= 0.3436 \text{ m/s}.$$

The new Reynolds number is twice the value in Example 10.6:

$$Re_{D,new} = \frac{\rho v_{avg,new} D_{new}}{\mu} = \frac{4\dot{m}}{\pi \mu D_{new}}$$

$$= Re_{D,old}\frac{D_{old}}{D_{new}}$$

$$= 2006(20/10) = 4012.$$

Since the value of the Reynolds number is now greater than the critical value (4012 > 2300), the flow is turbulent. For turbulent flow, we can choose from several equally valid relationships to calculate the friction factor (i.e., Eq. 10.81, 10.82, or 10.83). For convenience in evaluation, we choose Eq. 10.82, the simple Blasius expression:

$$f_D = 0.316 \, Re_D^{-1/4}$$

$$= 0.316(4012)^{-0.25} = 0.0397 \text{ (dimensionless)}.$$

Thus, the pressure drop is (Eq. 10.17)

$$\Delta P = f_D\left(\frac{L}{D}\right)\frac{1}{2}\rho v_{avg}^2$$

$$= 0.0397\left(\frac{100}{0.01}\right)0.5(997)(0.3436)^2 \text{ Pa}$$

$$= 23,400 \text{ Pa or } 23.4 \text{ kPa},$$

and the head loss is

$$h_L = \frac{\Delta P}{\rho g} = \frac{23,400}{997(9.807)}\text{m} = 2.39\text{ m}.$$

The units evaluation for these results is left to the reader (cf. Example 10.6).

Comments We note, first, that the Reynolds number depends inversely on the tube diameter for a *fixed mass flow rate* [i.e., $Re_D = 4\dot{m}/(\pi\mu D)$]. Second, we note the large increase in the pressure drop associated with the diameter change: 23,400 Pa compared to 589 Pa. If the flow had remained laminar when the diameter was reduced by a factor of 2, how would the pressure drops then compare?

Self Test
10.8

  Recalculate the friction factor for Example 10.8 using (a) Eq. 10.81 and (b) Eq. 10.83. Also determine the head losses associated with these friction factors.

(*Answer: (a)* $f_D = 0.0399$, $h_L = 2.40$ m; (b) $f_D = 0.0414$, $h_L = 2.49$ m)

Rough Tubes and Pipes

In our previous discussion of the friction and drag associated with an external flow over a flat plate (Chapter 9), we saw that surface roughness plays an important role. Since the physical processes at the fluid–wall interface are the same for external and internal (tube and pipe) flows, we expect surface roughness to be important for internal flows. This, indeed, is true. At the beginning of the present chapter, we discussed the contributions of Darcy, Nikuradse, and Moody to our modern understanding of flow through rough pipes. We now present the specific results of their efforts.

Consideration of flow over rough surfaces introduces an additional parameter, the roughness height ε, or its dimensionless counterpart, the relative roughness ε/D. Nikuradse [5] performed a careful study in which he measured friction factors for flow through tubes whose surfaces were artificially roughened by sand grains of nominal uniform size. The results of these classical experiments are shown in Fig. 10.15, where the friction factor f_D is plotted as a function of the pipe Reynolds number Re_D, using the relative roughness ε/D as a parameter. From Fig. 10.15, we see, first, that wall roughness is unimportant for laminar flow ($Re_D \lesssim 2300$) since the friction factor follows the theoretical result that $f_D = 64/Re_D$, uninfluenced by sand grain size. For turbulent flow, however, we observe three regimes related to wall roughness: a *smooth regime* where the roughness height is fully contained within the viscous sublayer and, like laminar flow, has no effect on the friction factor; a *smooth–rough transitional regime* where both Reynolds number and relative roughness influence the friction factor; and a *fully rough regime* in which the roughness elements protrude well into the turbulent portion of the boundary layer, causing form (or pressure) drag to be the dominant factor in establishing the wall resistance. In this regime, the friction coefficient is independent of the Reynolds number and depends solely on the relative roughness. Table 10.3 lists the criteria associated with these three regimes expressed as a roughness Reynolds number. The roughness Reynolds number uses the friction velocity and roughness height as the characteristic velocity and length, respectively (i.e., $\varepsilon v_*/\nu$).

See Historical Context. ▷

See Chapter 8. ▷

FIGURE 10.15
Results of Nikuradse's study [5] in which uniformly sized sand grains were used to determine the effect of wall roughness on pipe friction.

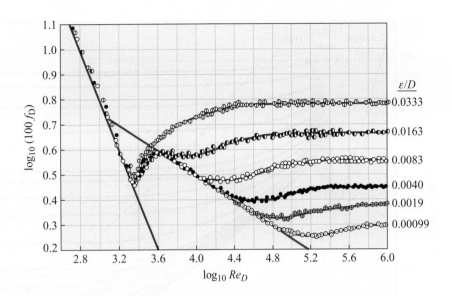

FIGURE 10.16
In the smooth–rough transitional regime, the constant B in the velocity distribution function depends on the roughness Reynolds number $v_* \varepsilon / \nu$ as shown here. Adapted from [13] with permission.

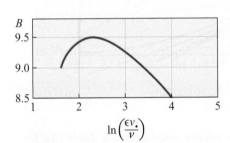

Also shown in Table 10.3 are expressions for the velocity profiles associated with each regime. For the rough–smooth transitional regime, the functional form of the velocity profile is the same as that for the smooth regime except that the constant B varies with $\varepsilon v_*/\nu$, as given in Fig. 10.16. In the fully rough regime, we see that the velocity profile no longer depends on viscosity, as we expect.

L.F. Moody in the abstract to a paper published in 1944 [6] stated modestly that he did

> *not claim to offer anything particularly new or original, his aim merely being to embody the now accepted conclusions in convenient form for engineering use.*

In this paper, Moody presented a figure that has become the standard resource for determining friction factors for flow through rough pipes. This figure is now known as the *Moody chart* (Fig. 10.17). Like Fig. 10.15, the Moody chart correlates friction factors with Reynolds number and relative roughness, but now the relative roughness is that associated with real pipe materials rather than artificially roughened tubes. Table 10.4 presents roughness heights of

Table 10.3 Summary of Relationships for Turbulent Flow in Rough Pipes

Roughness Regime	Criterion*	Velocity Profile $\overline{v}_x(r)/v_*$	B	Friction Factor (f_D)
Smooth	$\dfrac{v_* \varepsilon}{\nu} < 5$	$2.5\ln \dfrac{(R-r)v_*}{\nu} + B$	5.0	Eq. 10.81, 10.82, or 10.83
Smooth–rough transition	$5 < \dfrac{v_* \varepsilon}{\nu} < 70$	$2.5\ln \dfrac{(R-r)v_*}{\nu} + B$	See Fig. 10.16	Use Moody chart (Fig. 10.17) or Eq. 10.86
Fully rough	$\dfrac{v_* \varepsilon}{\nu} > 70$	$2.5\ln \dfrac{R-r}{\varepsilon} + 8.5$	—	Use Moody chart (Fig. 10.17) or Eq. 10.86

* The practical criteria for roughness regimes is built into the Moody chart, where knowledge of the Reynolds number Re_D and relative roughness ε/D determine the regime.

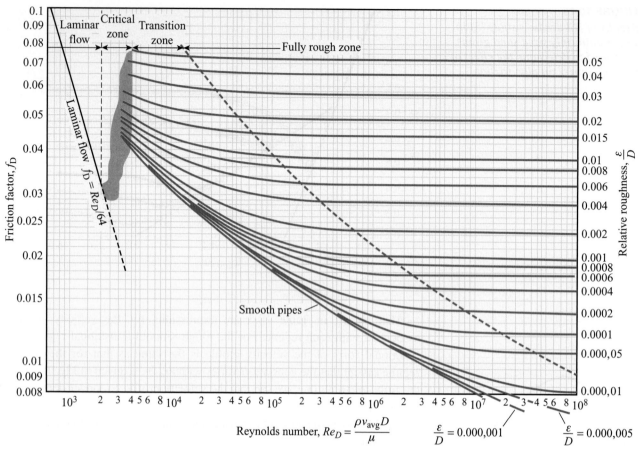

FIGURE 10.17

The Moody chart correlates the friction factor with the Reynolds number and the relative roughness. See Table 10.4 for roughness heights associated with various pipe materials. **Adapted from [6] with permission.**

Table 10.4   Apparent Roughness Heights for Commercial Pipes and Tubes*

Material	Condition	Roughness Height, ε(mm)	Uncertainty (%)
Steel	Sheet metal, new	0.05	±60
	Stainless, new	0.002	±50
	Commercial, new	0.046	±30
	Riveted	3.0	±70
	Rusted	2.0	±50
Iron	Cast, new	0.26	±50
	Wrought, new	0.046	±20
	Galvanized, new	0.15	±40
	Asphalted cast	0.12	±50
Brass	Drawn, new	0.002	±50
Plastic	Drawn tubing	0.0015	±60
Glass	—	Smooth	
Concrete	Smoothed	0.04	±60
	Rough	2.0	±50
Rubber	Smoothed	0.01	±60
Wood	Stave	0.5	±40

* Reproduced from White, F. M., *Fluid Mechanics,* 4th ed., McGraw-Hill, New York, 1999, with permission.

Head losses for a wide range of pipe diameters and surface roughness can be determined using the Moody chart (or Eq. 10.86).

various pipe materials for use with the Moody chart (Fig. 10.17). Although we will rely heavily on the Moody chart to estimate friction factors for turbulent pipe flows, the following closed-form mathematical expression [13] offers an alternative:

$$f_D = \frac{0.25}{\left\{\log_{10}\left[\dfrac{\varepsilon}{3.7\,D} + \dfrac{5.74}{Re_D^{0.9}}\right]\right\}^2}, \tag{10.86}$$

which is valid for

$$5 \times 10^3 \le Re_D \le 10^8 \text{ and } 10^{-6} \le \varepsilon/D \le 10^{-2}.$$

For smooth tubes, Eqs. 10.81–10.83 can be employed.

Example 10.9

Estimate the pressure drop and head loss for the same conditions as in Example 10.8, but for flow through a galvanized iron pipe of the same inside diameter (10 mm) as the smooth tube.

Solution

Known Re_D, v_{avg}, L, D (see Example 10.8)

Find ΔP, h_L

Assumptions

Steady, fully developed flow

Analysis   Since the flow rate and pipe diameter are the same as in Example 10.8, the Reynolds number is unchanged, (i.e., $Re_D = 4012$). Since the pipe is galvanized iron, the friction factor is now a function of Re_D and the surface roughness:

$$f_D = f_D(Re_D, \varepsilon/D).$$

The roughness height from Table 10.4 is $\varepsilon = 0.15$ mm; thus,

$$\varepsilon/D = 0.15 \text{ mm}/10 \text{ mm} = 0.015.$$

From the Moody diagram (Fig. 10.17), we estimate the friction factor to be

$$f_D \cong 0.053.$$

The friction factor and the pressure drop in the smooth pipe (Example 10.8) are 0.0397 and 23.4 kPa, respectively. Since ΔP is directly proportional to f_D, with all other factors being constant, we have

$$\Delta P_{\text{rough}} = \frac{f_{\text{D,rough}}}{f_{\text{D,smooth}}} \Delta P_{\text{smooth}}$$

$$= \frac{0.053}{0.0397} 23.4 \text{ kPa} = 31.2 \text{ kPa}.$$

Comments We first note that the roughness of the galvanized pipe results in an estimated pressure drop that is 33.5% greater than that for the smooth tube—a significant increase. We also note that the Re_D value of 4012 falls in the shaded zone of the Moody chart, indicating that the flow may be unsteady, chugging between laminar and turbulent conditions. This region, therefore, should be avoided in design.

Use the Moody diagram to calculate the friction factors for 5-mm-diameter brass and plastic pipes for (a) $Re_D = 2 \times 10^4$ and (b) $Re_D = 3 \times 10^7$. Also calculate the friction factors for (c) $Re_D = 3 \times 10^7$ using Eq. 10.86.

(Answer: (a) $f_{\text{D,brass}} \cong 0.0265, f_{\text{D,plastic}} \cong 0.0265$; (b) $f_{\text{D,brass}} \cong 0.0160, f_{\text{D,plastic}} \cong 0.0147$; (c) $f_{\text{D,brass}} = 0.0159, f_{\text{D,plastic}} = 0.0150$)

10.4b Heat-Transfer Relationships

See Table 10.1.

In this section, we seek to provide the details needed to complete our integral analyses of energy conservation for nonisothermal internal flows. Specifically, we seek ways to calculate convective heat-transfer coefficients appropriate for use in fully developed turbulent flows for the condition of uniform wall heat flux ($\dot{Q}''_w$ = constant) or the condition of uniform wall temperature (T_w = constant) In our study of laminar flows, we were able to deal with the same problem by applying theory, that is, by solving the differential equation that expresses energy conservation (see Eq. 10.37). For turbulent flows, however, we will sidestep the corresponding theoretical approach and simply present experimental results cast in their most general (and most useful) dimensionless forms. We do this because experimental data abound, whereas theoretical results are sparse.

The relationships we present are for hydrodynamically and thermally fully developed flows. The practical implication of this restriction is treated in more detail later; but, generally, this implies that the local heat-transfer coefficients $h_{\text{conv},x}$ so obtained apply at locations greater than about ten diameters downstream of the pipe entrance.

Before presenting any correlations, we emphasize that the characteristic length employed in both Nusselt and Reynolds numbers is the diameter, *not* the axial distance x, or pipe length L, even though we seek *local* heat-transfer coefficients. The use of D rather than x results from the physics of the boundary-layer development in internal flows. In a pipe or tube flow, the boundary layer grows from the wall axisymmetrically, merging in the center, and thus never exceeds the pipe radius, unlike external flows, where the boundary layer continually grows with downstream distance unimpeded by a boundary layer growing from an opposite wall (see Fig. 10.18). Thus,

FIGURE 10.18
In an external flow (a), the boundary layer grows continually with downstream distance. For an internal flow (b), the boundary layers from the walls merge along the centerline at the end of the developing region, resulting in a fixed boundary-layer thickness in the fully developed region.

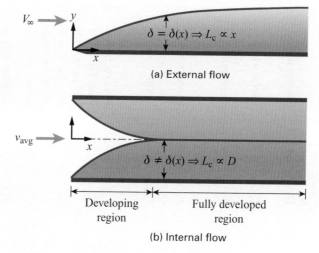

(a) External flow

(b) Internal flow

heat-transfer correlations for fully developed internal flows use the tube diameter as the characteristic length and are generally expressed in the following form:

$$Nu_D = f(Re_D, Pr), \tag{10.87}$$

where

> **Note that the diameter *D* is the characteristic length for pipe and tube flows.**

$$Nu_D \equiv \frac{h_{conv,x} D}{k_f} \tag{10.88}$$

and

$$Re_D \equiv \frac{\rho v_{avg} D}{\mu}. \tag{10.89}$$

Rather than discuss each correlation, we present some of the most common and useful correlations in Table 10.5. In this table, in addition to providing relationships in the form of Eq. 10.87, we cite appropriate references, list all of the restrictions, and indicate how the thermo-physical properties are to be evaluated. Note that separate correlations are provided for liquid metals ($Pr \ll 1$). Note also that the prescribed approach for rough pipes is a first approximation because the analogy between momentum transfer and heat transfer breaks down for rough pipes; the mechanisms for friction and energy transfer are no longer the same. If accurate results for rough pipes are required, the reader is referred to Ref. [17]. This reference is generally useful for those desiring more detailed information on heat transfer in internal flows. Also recommended, but at an introductory level, is the excellent text by Incropera and DeWitt [15].

We now illustrate the use of the relationships in Table 10.5 with some examples.

Table 10.5 Heat-Transfer Correlations for Fully Developed, Turbulent Flow in Tubes and Pipes

Correlation	Eq.	Reference	Constant Wall Condition	Applications and Restrictions	Condition for Property Evaluation	Comments
$Nu_D = 0.023 Re_D^{0.8} Pr^n$ $n = 0.4$ $T_w > T_m$ $n = 0.3$ $T_w < T_m$	T10.5a	Dittus & Boelter [18]	$\dot{Q}_w''$ or T_w	Smooth walls $0.7 \lesssim Pr \lesssim 160$ $Re_D \gtrsim 10{,}000$ $x/D \gtrsim 10$	$T_m(x)$	Useful when $|T_m - T_w|$ is "moderate"
$Nu_D = 0.027 Re_D^{0.8} Pr^{1/3} \left(\dfrac{\mu}{\mu_w} \right)^{0.14}$	T10.5b	Sieder & Tate [19]	$\dot{Q}_w''$ or T_w	Smooth walls $0.7 \lesssim Pr \lesssim 16{,}700$ $Re_D \gtrsim 10{,}000$ $x/D \gtrsim 10$	$T_m(x)$ for all μ_w except at $T_w(x)$	Useful when $|T_m - T_w|$ is "large"
$Nu_D = \dfrac{(f/8)(Re_D - 1000)Pr}{1 + 12.7(f/8)^{0.5}(Pr^{2/3} - 1)}$	T10.5c	Gnielinski [20]	$\dot{Q}_w''$ or T_w	Smooth or rough walls $0.5 < Pr < 2000$ $3000 < Re_D < 5 \times 10^6$	$T_m(x)$	Friction factors from Eq. 10.83 for smooth tubes, Fig. 10.17 or Eq. 10.86 for rough tubes
$Nu_D = 4.82 + 0.0185(Re_D Pr)^{0.827}$	T10.5d	Skupinski et al. [21]	$\dot{Q}_w''$	Liquid metals Smooth walls $3.6 \times 10^3 < Re_D < 9.05 \times 10^5$ $100 < Re_D Pr < 10^4$	$T_m(x)$	Applies only to liquid metals
$Nu_D = 5.0 + 0.025(Re_D Pr)^{0.8}$	T10.5e	Seban & Shimazaki [22]	T_w	Liquid metals Smooth walls $Re_D Pr > 100$	$T_m(x)$	Applies only to liquid metals

Example 10.10

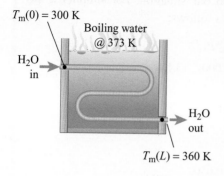

$T_m(0) = 300$ K

Boiling water @ 373 K

H_2O in

H_2O out

$T_m(L) = 360$ K

It is desired to heat a 0.0576-kg/s flow of water from 300 to 360 K in a simple heat exchanger as shown in the sketch. The water to be heated flows through a 12.5-mm-inside-diameter smooth tube, which is immersed in a tank of boiling water at 373 K. Determine the length L of tubing required and the total heat-transfer rate $\dot{Q}_{0-L}$.

Solution

Known H_2O, $T_m(0)$, $T_m(L)$, T_w, $\dot{m}$, D

Find L, $\dot{Q}_{0-L}$

Sketch

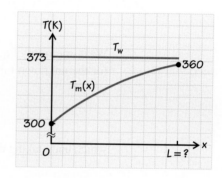

Assumptions

 i. The flow is steady state.
 ii. Entrance effects are negligible.
 iii. The effects of tube bends are negligible.
 iv. The wall temperature equals the temperature of the boiling water.
 v. Water properties are not strongly dependent on temperature.

Analysis   The bulk mean temperature as a function of distance through the tube is given by Eq. 10.33 and is illustrated on the sketch. We will employ this equation to obtain L ($= x$) when $T_m(L) = 360$ K. To use this equation requires a value for the mean convective heat-transfer coefficient $\bar{h}_{conv,L}$. We employ the following properties for saturated water at the average bulk mean temperature $[\bar{T}_m = (300 + 360)/2$ K $= 330$ K] obtained from the NIST database:

$$\rho = 984 \text{ kg/m}^3,$$
$$c_p = 4184 \text{ J/kg} \cdot \text{K},$$
$$\mu = 489 \times 10^{-6} \text{ N} \cdot \text{s/m}^2,$$
$$k = 650 \times 10^{-3} \text{ W/m} \cdot \text{K},$$
$$Pr = 3.15 \text{ (dimensionless)}.$$

To choose a heat-transfer correlation, we first determine whether the flow is laminar or turbulent by calculating the Reynolds number using Eq. 10.34b:

$$Re_D = \frac{4\dot{m}}{\pi \mu D}$$

$$= \frac{4(0.0576)}{\pi(489 \times 10^{-6})0.0125} = 12,000$$

$$[=] \frac{\text{kg/s}}{(\text{N} \cdot \text{s/m}^2)\text{m}} \left[\frac{1 \text{ N}}{\text{kg} \cdot \text{m/s}^2} \right] = 1.$$

Since $12{,}000 > 2300$, the flow is turbulent. From Table 10.5, we see that both Eq. T10.5a and T10.5b apply to our situation. For simplicity, we choose Eq. T10.5a to evaluate $\overline{h}_{\text{conv},L}$ as follows:

$$\overline{Nu}_D = 0.023\, Re_D^{0.8} Pr^{0.4}$$
$$= 0.023(12{,}000)^{0.8}(3.15)^{0.4}$$
$$= 66.74,$$

and from the definition of Nu_D, we get

$$\overline{h}_{\text{conv},L} = \overline{Nu}_D \frac{k}{D}$$

$$= 66.74\, \frac{0.650}{0.0125} = 3470$$

$$[=]\frac{\text{W/m}\cdot\text{K}}{\text{m}} = \text{W/m}^2\cdot\text{K}.$$

We now employ Eq. 10.33, which expresses conservation of energy, to find for the unknown tube length L; that is,

$$\frac{T_m(L) - T_w}{T_m(0) - T_w} = \exp\left[\frac{-\overline{h}_{\text{conv},L}\,\pi D}{\dot{m}c_p}L\right],$$

which upon taking the natural logarithm of both sides and rearranging yields

$$L = \frac{-\dot{m}c_p}{\overline{h}_{\text{conv},L}\,\pi D}\ln\left[\frac{T_m(L) - T_w}{T_m(0) - T_w}\right].$$

Substituting numerical values, we obtain

$$L = \frac{-0.0576\,(4184)}{3470\,\pi\,0.0125}\ln\left[\frac{360 - 373}{300 - 373}\right]$$

$$= 3.05$$

$$[=]\frac{(\text{kg/s})\,(\text{J/kg}\cdot\text{K})}{(\text{W/m}^2\cdot\text{K})\,(\text{m})}\left[\frac{1\,\text{W}}{\text{J/s}}\right] = \text{m}.$$

To determine the total heat-transfer rate, we use an overall energy balance (Eq. 10.24):

$$\dot{Q}_{0-L} = \dot{m}c_p[T_m(L) - T_m(0)]$$

$$= 0.0576(4184)(360 - 300)$$

$$= 14{,}460$$

$$[=](\text{kg/s})\,(\text{J/kg}\cdot\text{K})(\text{K}) = \text{J/s} = \text{W}.$$

Use Table 4.2 as a validity check whenever you calculate heat-transfer coefficients.

Comments Our justification for assuming a constant wall temperature is that the heat-transfer coefficient on the boiling-water side of the tube is likely to be quite large. Table 4.2 (Chapter 4) indicates values as large as $100{,}000$ W/m$^2\cdot$K for conditions of boiling or condensation. With such a large heat-transfer coefficient, only a small temperature difference is required to support the necessary heat flow (i.e., $\dot{Q}_x'' = h_{\text{conv},x}\Delta T$).

This example illustrates how heat-transfer correlations (specifically, Eq. T10.5a) can be employed to obtain an *average* value of a convective heat-transfer coefficient $\overline{h}_{\text{conv}}$ when an *average* value of bulk mean temperature ($\overline{T}_m$) is used to evaluate the fluid properties. To obtain a *local* value for a heat-transfer coefficient $h_{\text{conv},x}$, the *local* bulk mean temperature $T_m(x)$ is employed. This approach to calculating average heat-transfer coefficients applies only if thermal entrance effects are negligible.

Self Test 10.10 ☑ Redo Example 10.10 for a tube with a roughness height of **0.125 mm** using an appropriate correlation for the Nusselt number from Table 10.5.

(Answer: L = 2.35 m)

Example 10.11

Consider the simple heat exchanger designed in Example 10.10. Determine the outlet temperature of the water, $T_m(L)$, and the total heat-transfer rate $\dot{Q}_{0-L}$, if the device is now operated with a flow rate three times the original design flow rate (i.e., $3 \times 0.0576 = 0.1728$ kg/s).

Solution

Known $L, D, T_m(0), T_w, \dot{m}$

Find $T_m(L), \dot{Q}_{0-L}$

Sketch

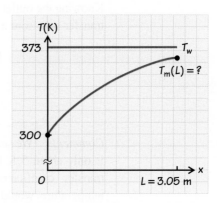

Assumptions

  i. The flow is steady state.
 ii. Entrance effects are negligible.
iii. The effects of tube bends are negligible.
 iv. The wall temperature equals the temperature of the boiling water.
  v. Water properties are not strongly dependent on temperature.
 vi. Use the same properties as in Example 10.10 recognizing that iteration may be required.

Analysis   We will apply Eq. 10.33 to obtain $T_m(L)$; however, we must first calculate $\bar{h}_{conv,L}$. Since $Re_D = 4\dot{m}/(\pi\mu D)$, the Reynolds number is now three times that of Example 10.10; that is,

$$Re_D = 3(12,000) = 36,000$$

To find the Nusselt number and average heat-transfer coefficient, we employ the Dittus–Boelter correlation (T10.5a) as follows:

$$Nu_D = 0.023 Re_D^{0.8} Pr^{0.4}$$
$$= 0.023(36,000)^{0.8}(3.15)^{0.4}$$
$$= 160.7,$$

and

$$\overline{h}_{\text{conv}, L} = \frac{k}{D} \overline{Nu}_D$$

$$= \frac{0.650 \text{ W/m} \cdot \text{K}}{0.0125 \text{ m}} 160.7 = 8356 \text{ W/m}^2 \cdot \text{K}.$$

We note that this value is about 2.4 times greater than the design-point value. Applying Eq. 10.33 yields

$$\frac{T_{\text{m}}(L) - T_{\text{w}}}{T_{\text{m}}(0) - T_{\text{w}}} = \exp\left[\frac{-\overline{h}_{\text{conv}, L} \pi D L}{\dot{m} c_p}\right]$$

$$= \exp\left[\frac{-8356 \,\pi\,(0.0125)\,3.05}{0.1728\,(4184)}\right]$$

$$= \exp(-1.3843) = 0.2505,$$

where the term in brackets can be shown by the reader to be dimensionless. Solving this equation for the unknown outlet temperature yields

$$T_{\text{m}}(L) = [T_{\text{m}}(0) - T_{\text{w}}]0.2505 + T_{\text{w}}$$

$$= (300 - 373)0.2505 + 373 \text{ K}$$

$$= 355 \text{ K}.$$

Knowing the outlet bulk mean temperature, we can easily find the total heat-transfer rate from a simple overall energy balance (Eq. 10.24) as follows:

$$\dot{Q}_{0-L} = \dot{m} c_p [T_{\text{m}}(L) - T_{\text{m}}(0)]$$

$$= 0.1728\,(4184)(355 - 300)$$

$$= 39{,}800$$

$$[=]\,(\text{kg/s})\,(\text{J/kg} \cdot \text{K})\,\text{K} = \text{J/s or W}.$$

Comments We see that with the increased flow rate the outlet temperature is only about 5 K lower than that for the design condition. Had the heat-transfer coefficient not increased with the greater flow, the outlet temperature would have been much lower (~332 K) as a result of the diminished time for heat transfer to occur. Actually, the larger Re_D resulted in the heat-transfer coefficient being about 2.4 times greater, as previously noted. We also note that our new outlet temperature is sufficiently close to the Example 10.10 value to justify no further iteration with properties at a new average bulk-mean temperature.

Example 10.12

Repeat the design exercise of Example 10.10 taking into account the relatively large difference between the bulk mean and wall temperatures by using the Sieder and Tate correlation (Eq. T10.5b in Table 10.5).

Solution

Known See Example 10.10.

Find L

Assumptions

See Example 10.10.

Analysis   As in Example 10.10, we use properties at the average bulk mean temperature [= 0.5 (360 + 300)K = 330 K] to evaluate all of the terms in the Sieder and Tate correlation except μ_w, which is evaluated at the wall temperature T_w = 373 K. From the NIST database,

$$\mu_w(373 \text{ K}) = 282 \times 10^{-6} \text{ N}\cdot\text{s/m}^2.$$

Thus,

$$\overline{Nu}_D = 0.027 \, \text{Re}_D^{0.8} Pr^{1/3} [\mu/\mu_w]^{0.14}$$

$$= 0.027(12{,}000)^{0.8}(3.15)^{1/3}\left(\frac{489 \times 10^{-6}}{282 \times 10^{-6}}\right)^{0.14}$$

$$= 0.027(1833.8)\,1.466(1.08)$$

$$= 72.6(1.08) = 78.4,$$

and

$$\overline{h}_{\text{conv},L} = \overline{Nu}_D k/D$$

$$= 78.4(650 \times 10^{-3}/0.0125)\text{W/m}^2\cdot\text{K} = 4077 \text{ W/m}^2\cdot\text{K}.$$

Thus,

$$L_{\text{Sieder–Tate}} = L_{\text{Dittus–Boelter}}\frac{\overline{h}_{L,\text{D–B}}}{\overline{h}_{L,\text{S–T}}}$$

$$= (3.05 \text{ m})\frac{3470}{4077} = 2.6 \text{ m}.$$

Comment   From this example, we learn that the effect of the difference between the wall and bulk mean temperatures ($|T_m - T_w|$) can be substantial: The heat-transfer coefficient calculated from the Sieder and Tate correlation is about 17% greater than that from the Dittus–Boelter correlation. Second, we also see that the two correlations yield slightly different results when the effect of $|T_m - T_w|$ is ignored. To illustrate this, let us split the previous Nusselt number calculation to show the $|T_m - T_w|$ effect explicitly (i.e., $\overline{Nu}_D = 72.6 \times 1.08$). Comparing the 72.6 value here with the value of 66.7 from Example 10.10 yields a 9% difference. Adding the temperature effect contributes another 8% to produce the anticipated 17% total difference. Differences of this magnitude among correlations are not uncommon in heat-transfer calculations.

Example 10.13

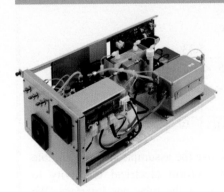

Photograph courtesy of Thermo Electron Corporation.

In the operation of a chemical analyzer that measures oxides of nitrogen (combustion-generated pollutants), sample gases flow at a rate of 3.685×10^{-5} kg/s through a 3-m-long, 4.57-mm-inside-diameter, stainless-steel tube where they are heated. The tube, insulated at its outside surface, is heated by the passage of an electrical current through the metal wall along the length of the tube as shown in the sketch. The electrical resistance of the tube and the associated voltage drop are 0.75 ohms and 2 V, respectively. Assuming the gases are composed mainly of nitrogen, determine the temperature of the gases at the tube exit when they enter at 300 K. Also determine the tube wall temperature at the exit.

Solution

Known $\dot{m}, L, D, R_{\text{elec}}, \Delta V, T_m(0)$

Find $T_m(L), T_w(L)$

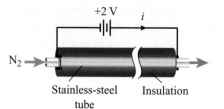

Stainless-steel tube

Insulation

Sketch

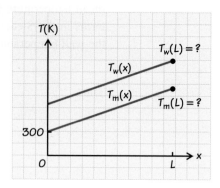

Assumptions

  i. The flow is steady state.
 ii. Entrance effects are confined to the region $x < L$.
iii. Electrical resistivity is uniform.
 iv. Heat loss through insulation is negligible.

Analysis   With assumption iv, all of the input electrical power is used to heat the gas stream. Energy conservation for a control volume cutting through the electrical wires and across the planes of the inlet and outlet flows yields

$$\dot{W}_{elec} = \dot{m}c_p[T_m(L) - T_m(0)].$$

The electrical power is expressed by (Eq. 4.11b)

$$\dot{W}_{elec} = i\Delta V,$$

or, with the aid of Ohm's law that $\Delta V = i\,R_{elec}$,

$$\dot{W}_{elec} = (\Delta V)^2/R_{elec}.$$

For the given voltage drop and electrical resistance,

$$\dot{W}_{elec} = (2)^2/0.75 = 5.33 \text{ W},$$

where we have used $1 \text{ V} \cdot \text{A} \equiv 1 \text{ W}$. Using the specific heat for N_2 at 300 K from Table F.1, we get an outlet temperature of

$$T_m(L) = T_m(0) + \frac{\dot{W}_{elec}}{\dot{m}c_p}$$

$$= 300 + \frac{5.33}{3.685 \times 10^{-5}(1041)}$$

$$= 300 + 139 = 439$$

$$[=] \frac{\text{W}}{(\text{kg/s})(\text{J/kg} \cdot \text{K})}\left[\frac{1 \text{ J/s}}{\text{W}}\right] = \text{K}.$$

To evaluate the wall temperature, we use the assumptions of negligible heat loss through the insulation and uniform electrical resistivity to conclude that a uniform heat flux exists at the tube–gas interface. We employ the definitions of the local heat-transfer coefficient (Eq. 10.28) and the heat flux (Eq. 10.25) as follows:

$$\dot{Q}'' = h_{conv,x}[T_w(L) - T_m(L)]$$

or

$$T_w(L) = T_m(L) + \frac{\dot{Q}''_w}{h_{conv,x}},$$

and

$$\dot{Q}''_w = \frac{\dot{Q}}{\pi D L} = \frac{\dot{W}_{elec}}{\pi D L}.$$

The wall heat flux is easily evaluated as

$$\dot{Q}''_w = \frac{5.33 \text{ W}}{\pi(0.00457 \text{ m})(3 \text{ m})} = 123.7 \text{ W/m}^2.$$

To evaluate $h_{conv,x}$, we apply the appropriate heat-transfer correlation. The Reynolds number is calculated to establish whether the flow is laminar or turbulent:

$$Re_D = \frac{\rho v_{avg} D}{\mu} = \frac{4\dot{m}}{\pi \mu D}$$

$$= \frac{4(3.685 \times 10^{-5})}{\pi 17.9 \times 10^{-6}(0.00457)} = 573,$$

where the viscosity for N_2 at 300 K is taken from Table F.1. (The reader should verify that this calculation yields a dimensionless quantity.) Since $Re_D < 2300$, the flow is laminar, and the appropriate Nusselt number is (Eq. 10.60)

$$Nu_D = 4.36,$$

and

$$h_{conv,x} = \frac{k}{D} Nu_D = 4.36 \, k/D$$

$$= 4.36(0.0346 \text{ W/m} \cdot \text{K})/0.00457 \text{ m}$$

$$= 33.0 \text{ W/m}^2 \cdot \text{K}.$$

The thermal conductivity value used here is for N_2 at 439 K, our previously calculated exit bulk mean temperature. The use of properties evaluated at the exit temperature yields a local value for $h_{conv,x}$ The approximate wall temperature at the exit is thus

$$T_w(L) = 439 \text{ K} + \frac{123.7 \text{ W/m}^2}{33.0 \text{ W/m}^2 \cdot \text{K}}$$

$$= 442.7 \text{ K}.$$

Comment This example combines the application of basic conservation of energy principles from earlier chapters with new concepts from the field of heat transfer.

Self Test 10.11 ✓ Redo Example 10.13 when the mass flow rate is 3×10^{-4} kg/s and the voltage drop is 5 V. The tube is smooth.

(*Answer:* $T_m(L) = 406.7 \text{ K}, T_w(L) = 413.5 \text{ K}$)

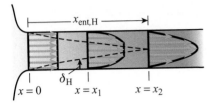

FIGURE 10.19
Flow enters a tube with a uniform velocity distribution at $x = 0$. The no-slip condition at the wall causes a boundary layer to develop. At $x = x_1$, the core of the flow is still uniform, but with a velocity greater than at the entrance, whereas at $x = x_2$, the velocity distribution is fully developed and described by Eq. 10.41, if the flow is laminar, and by Eq. 10.85, if the flow is turbulent. The entrance length $x_{ent,H}$ is defined as the distance from the entrance to the x location where the flow first becomes fully developed.

See Table 3.3 in Chapter 3 and Table 6.4 in Chapter 6.

Equation 10.90 gives the laminar flow hydrodynamic entry length.

Equation 10.91 gives the turbulent flow hydrodynamic entry length.

10.5 DEVELOPING FLOWS

In the preceding sections, we always assumed that the flow was fully developed, both hydrodynamically and thermally. In this section, we elaborate on the meaning of flow development, provide criteria for determining when a (laminar or turbulent) flow is fully developed, and provide heat-transfer correlations to deal with flows that are not fully developed.

10.5a Hydrodynamic Entry Region

Hydrodynamic Entrance Length

If the entrance to a tube or pipe is rounded with a bell mouth, the flow enters the tube with an essentially uniform velocity profile, as shown in Fig. 10.19. Because of the no-slip condition at the wall, a boundary layer develops. Since the fluid near the wall slows down, mass conservation requires that the uniform core flow speed up. Thus, as the flow proceeds downstream, the region affected by the wall grows, and the centerline velocity increases. The axisymmetric boundary layer grows until the effect of the wall reaches to the centerline (i.e., the boundary layers from opposing sides merge). The axial location at which this merger takes place is defined as the **hydrodynamic entrance length** $x_{ent,H}$, and the region beyond is called the **fully developed region**. In the fully developed region, the velocity profile no longer changes with axial location. The specific shape of the velocity profile depends on whether the flow is laminar or turbulent (with smooth or rough walls). The profiles drawn in Fig. 10.19 represent laminar flow with a parabolic distribution where the centerline velocity v_{max} is twice v_{avg}.

The magnitude of the entrance length has been established for laminar flow by solving the momentum and continuity equations (Eqs. T6.4a, T6.4c, and T3.3e) describing the developing velocity distribution $v_x(r, x)$ [23]. In dimensionless form, this result is

$$\frac{x_{ent,H}}{D} \cong 0.05\,Re_D.\tag{10.90}$$

Using the usual criterion for transition from laminar to turbulent flow ($Re_{D,crit} \approx 2300$), we see that the maximum possible hydrodynamic entrance length for laminar flow is approximately 115 tube diameters.

For turbulent flows, hydrodynamic entrance lengths have been determined empirically. The following expression is usually quoted in textbooks [26]:

$$\frac{x_{ent,H}}{D} = 4.4\,Re_D^{1/6}.\tag{10.91}$$

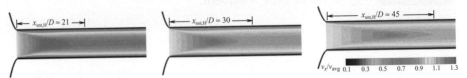

Calculated mean velocity fields show developing flows at the entrance of a tube for $Re_D = 10^4$, 10^5, and 10^6, respectively. Axial distances shortened by a scale factor of 10.

FIGURE 10.20
Control volume for integral analysis of pressure drop in developing region showing (a) forces and (b) momentum flows. The length of the control volume is $L = x_{ent,H}$.

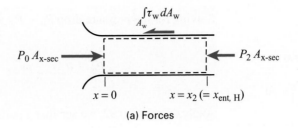

(a) Forces

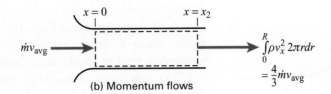

(b) Momentum flows

Entrance lengths predicted from this correlation range from 16 diameters at the lower limit ($Re_D \approx 2300$) to about 45 diameters for a Reynolds number of one million.

Increased Pressure Drop

What effect does flow development have on the pressure drop in the entrance region? To explore this, let us consider the entrance region of a laminar flow for the control volume shown in Fig. 10.20. We assume a uniform velocity profile at the entrance to the control volume and a fully developed parabolic profile at the exit. Forces acting on this control volume are the pressure forces at each end, $P_0 A_{\text{x-sec}}$ and $P_2 A_{\text{x-sec}}$, and the net viscous shear force acting on the circumferential area, $F_{\text{visc}} = \int \tau_w dA_w$, in the direction opposing the flow. The momentum flow entering the control volume is simply $\dot{m}v_{\text{avg}}$, whereas the momentum flow exiting the control volume is determined by

$$\int_0^R \rho v_x^2(r) 2\pi r dr,$$

where

$$v_x(r) = 2v_{\text{avg}}\left(1 - \frac{r^2}{R^2}\right).$$

In Chapter 6, we evaluated this integral and found that

$$\text{momentum flow out} = \frac{4}{3}\dot{m}v_{\text{avg}}.$$

See Example 6.9 in Chapter 6.

For steady flow, the fundamental conservation of momentum principle (Eq. 6.59) is expressed as

$$\left(\begin{array}{c}momentum \\ flow\ in\end{array}\right)_{x\text{-dir}} - \left(\begin{array}{c}momentum \\ flow\ out\end{array}\right)_{x\text{-dir}} + \sum F_{x\text{-dir}} = 0.$$

After isolating the forces on the left-hand side, we apply this principle to our control volume and write

$$P_0 A_{\text{x-sec}} - P_2 A_{\text{x-sec}} - \int_{A_w} \tau_w dA_w = \frac{4}{3}\dot{m}v_{\text{avg}} - \dot{m}v_{\text{avg}}.$$

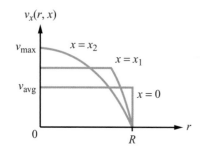

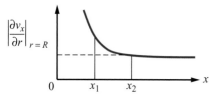

FIGURE 10.21
(a) Velocity profiles $v_x(r, x)$ are shown at the entrance to a pipe, within the developing region, and in the fully developed region (cf. Fig. 10.19). (b) Note that the magnitude of the velocity gradient $\partial v_x / \partial r$ at the wall is greatest (infinite) at $x = 0$, decreases through the developing region, and assumes a constant value beyond.

Solving for the pressure drop $P_0 - P_2$. and recognizing that $\dot{m} = \rho v_{\text{avg}} A_{\text{x-sec}}$, we obtain

$$P_0 - P_2 = \frac{1}{3}\rho v_{\text{avg}}^2 + \frac{1}{A_{\text{x-sec}}}\int_{A_{\text{w}}} \tau_{\text{w}} dA_{\text{w}}. \tag{10.92}$$

Analyzing Eq. 10.92, we see that a portion of the pressure drop results from accelerating the fluid from the initially uniform profile to the fully developed profile [i.e., the $(1/3)\rho v_{\text{avg}}^2$ term]. The acceleration of the central core fluid is evident from Figs. 10.19 and 10.21a. In the fully developed region, there is no acceleration component to the pressure drop—the momentum flow in equals the momentum flow out.

Let us now focus on the shear-stress term and compare its magnitude to that associated with fully developed flow. The velocity profiles plotted in Fig. 10.21a provide insight to make this comparison. Since $\tau_{\text{w}} = \mu|\partial v_x/\partial r|_{r=R}$, we focus on how the velocity gradient at the wall varies along the length of the control volume. At the entrance, the velocity is zero at the wall ($r = R$) and jumps to v_{avg} at $r = R^-$ (i.e., a step function); thus, the magnitude of the velocity gradient is infinite. As the boundary later develops, the magnitude of the gradient decreases until it reaches a constant value in the fully developed region (Fig. 10.21b). Thus, the average shear stress in the developing region exceeds that acting on an equivalent area in a fully developed flow, as suggested by the dashed line in Fig. 10.21b. We then conclude that

$$\left[\int_{A_{\text{w}}} \tau_{\text{w}} dA_{\text{w}}\right]_{\substack{\text{developing} \\ \text{region}}} > \left[\int_{A_{\text{w}}} \tau_{\text{w}} dA_{\text{w}}\right]_{\substack{\text{fully} \\ \text{developed}}}$$

for equal wall areas.

Thus, we see that the pressure drop per unit length is greater in the developing region for two reasons: a net outflow of momentum [$(1/3)\rho v_{\text{avg}}^2$] and a larger net viscous force retarding the flow. We discuss the engineering approach to account for this additional pressure drop in the section on minor losses later in this chapter; however, for the laminar flow situation just described, the additional pressure drop is given by [24]

$$\Delta P_{\text{entrance}} = 1.16\rho v_{\text{avg}}^2/2. \tag{10.93}$$

Example 10.14

Compare the hydrodynamic entry lengths for flows of air, water, and oil, respectively, through a 10-m-long, 20-mm-inside-diameter tube. The average velocity for all three fluids is 1 m/s. Evaluate all properties at 300 K. The air is at 100 kPa, and the water and oil are saturated liquids.

Solution

Known   Fluid, L, D, v_{avg}, T, P

Find $x_{\text{ent,H}}$

Sketch

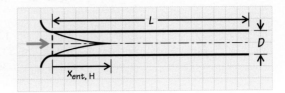

Assumptions

 i. Steady flow
 ii. Uniform velocity profile at entrance

Analysis Our procedure here is straightforward: After compiling the needed fluid properties, we calculate Re_D for each flow, establish whether the condition is laminar or turbulent, and then calculate $x_{ent,H}$ using either Eq. 10.90 (laminar) or Eq. 10.91 (turbulent).

From the appropriate appendices, we find the following properties for the given conditions:

	Air	Water	Engine Oil
$\rho\,(kg/m^3)$	1.1613	997	884.1
$\mu\,(N\cdot s/m^2)$	18.58×10^{-6}	855×10^{-6}	48.6×10^{-2}

The Reynolds number is given by

$$Re_D = \frac{\rho v_{avg} D}{\mu}.$$

For air,

$$Re_D = \frac{1.1613(1)0.02}{18.58 \times 10^{-6}} = 1250$$

$$[=] \frac{(kg/m^3)(m/s)\,m}{N \cdot s/m^2}\left[\frac{1\,N}{kg \cdot m/s^2}\right] = 1.$$

Since $1250 < 2300$, the flow is laminar and Eq. 10.90 applies:

$$x_{ent,H}/D = 0.05\,Re_D$$
$$= 0.05(1250)$$
$$= 62.5$$

and so

$$x_{ent,H} = 62.5D = 62.5(0.02\,m)$$
$$= 1.25\,m.$$

The fractional length of the tube occupied by the developing region is

$$\frac{x_{ent,H}}{L} = \frac{1.25\,m}{10\,m} = 0.125 \text{ or } 12.5\%.$$

Calculations for the other flows are summarized in the following table. Note that the air and oil flows are laminar, whereas that for water is turbulent.

Fluid	Re_D	$x_{ent,H}/D$	$x_{ent,H}/L$
Air	1250	62.5	0.125 or 12.5%
Water	23,320	23.5	0.047 or 4.7%
Oil	36.4	1.82	0.0036 or 0.36%

Comments   We see that, for the same geometrical (L, D) and flow (v_{avg}) conditions, the entrance length varies greatly with fluid type. For air, the developing region occupies a substantial portion of the tube length; for oil, the entry length is very small. Calculations of this sort can be used to estimate when entrance effects can be neglected (i.e., when $x_{ent,H} \ll L$).

Self Test 10.12 ☑ **Determine the hydrodynamic entry length for oil at 273 and 400 K for the flow described in Example 10.14.**

(Answer: $x_{ent,H}$ (273 K) = 4.67 mm, $x_{ent,H}$ (400 K) = 1.88 m)

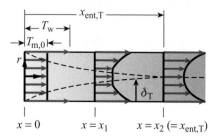

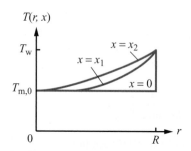

FIGURE 10.22

Thermal entry region for fluid heating $(T_w > T_{m,0})$ showing thermal boundary-layer development and thermal entrance length $x_{ent,T}$ (top). The temperature profiles at $x = 0, x_1,$ and $x_{ent,T}$ are compared in the plot of $T(r, x)$ versus r (bottom).

Equation 10.95a gives the laminar flow thermal entrance length. ▶

10.5b Thermal Entry Region

Thermal Entrance Length

When a fluid is heated or cooled after entering a pipe, a thermal entry region precedes the region of thermally fully developed flow, analogous to the hydrodynamic problem. Thermal boundary layers grow from the walls and merge on the centerline at the end of the thermal entry region to define the thermal entrance length, as illustrated in Fig. 10.22. That a fully developed thermal region should exist may seem strange since the temperature distribution within the fluid must continually change when the fluid is continually subjected to heating or cooling. This apparent contradiction is resolved when we consider that the *dimensionless* temperature distribution,

$$T^*(r, x) = \frac{T(r, x) - T_w(x)}{T_m(x) - T_w(x)}, \tag{10.94}$$

does not vary with axial position in the fully developed thermal region. (Recall that we invoked this assumption earlier in our detailed analysis of the thermal problem for fully developed, laminar tube flow.) For laminar flows, the **thermal entry length** $x_{ent,T}$ is determined by solving the energy conservation equation (Eq. T5.7b), with the result [25] that

$$\frac{x_{ent,T}}{D} \cong 0.05 \, Re_D Pr \quad \text{for} \quad Re < 2300. \tag{10.95a}$$

This result is quite similar to the corresponding expression for the hydrodynamic entrance length, except that the Prandtl number Pr now appears. For fluids with unity Pr, the thermal and hydrodynamic entry lengths are identical. For large Prandtl numbers $(Pr > 1)$, the thermal entry length exceeds the hydrodynamic entry length (water and oils fall into this category) whereas for small Prandtl numbers $(Pr < 1)$, the converse is true. This Prandtl-number dependence of the thermal entry length is analogous to the situation for flow over a flat plate (see Fig. 9.15), where the relative growth of the hydrodynamic and thermal boundary layers depends on Prandtl number. The essential physics is the same for both problems; the only significant difference is the merger of the boundary layers at the centerline for tube flows.

For turbulent flow, the Prandtl number has little effect on the thermal entry length, and the usual approximation is that

$$\frac{x_{\text{ent,T}}}{D} \approx 10 \quad \text{for} \quad Re_D > 2300. \tag{10.95b}$$

The lack of Prandtl-number dependence is a result of relatively large scale turbulent motion being the primary mechanism for transporting energy, as opposed to molecular processes.

Local Heat-Transfer Coefficients

The local heat-transfer coefficient $h_{\text{conv},x}$ falls from a very large value at the pipe entrance (theoretically infinite) to the fully developed value at some distance downstream (Fig. 10.23). That $h_{\text{conv},x}$ is large at $x = 0$ follows from the fact that the temperature gradient at the wall, $[\partial T/\partial r]_{r=R}$, is infinite at $x = 0$, as can be seen from the plot of the temperature profiles in Fig. 10.22. For laminar flow, local heat-transfer coefficients have been calculated from theory [8], and the results are shown in terms of the local Nusselt number in Fig. 10.24. The local Nusselt number for internal flow is defined by

$$Nu_D(x) \equiv \frac{h_{\text{conv},x}(x)D}{k_f}, \tag{10.96}$$

where $h_{\text{conv},x}(x)$ is the heat-transfer coefficient at location x. Note that the characteristic length is the pipe diameter, not the distance x. In addition to considering the two thermal boundary conditions (constant surface heat flux and constant surface temperature), results are shown for two entry conditions: 1. the thermal entry length alone and 2. the combined thermal and hydrodynamic entry lengths. In the former, the velocity profile is fully developed, as would be the case for a sufficiently long unheated starting length, or approximately true, for large-Prandtl-number fluids. In the combined problem, the hydrodynamic and thermal boundary layers develop simultaneously.

Local Nusselt numbers in the entry region for turbulent flow are shown in Figs. 10.25 and 10.26 for the cases indicated in Table 10.6. These figures can be used to obtain estimates of local heat-transfer coefficients for selected conditions. Note that the sharp-contraction entrance condition is actually more complicated than the simple combined entry problem since the flow separates at the sharp corner. This separation (or **vena contracta**) causes the local Nusselt number to exhibit a maximum in the region $1 < x/D < 2$ (see Fig. 10.26). The large values of local heat-transfer coefficients in the entrance region result in tube wall temperatures approaching the bulk mean temperature at the tube entrance ($x = 0$) as shown in Fig. 10.27.

Average Heat-Transfer Coefficients

To calculate the bulk mean temperature for the constant wall temperature problem, we require an average heat-transfer coefficient. This heat-transfer

---

Equation 10.95b gives the turbulent flow thermal entrance length.

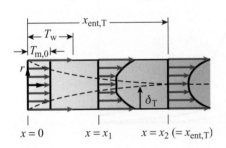

$$\frac{T - T_m}{T_w - T_m} \quad \begin{array}{ccccc} 0.00 & 0.25 & 0.50 & 0.75 & 1.00 \end{array}$$

Calculated temperature distribution in the thermal entry region for laminar tube flow. Axial distance expanded by a factor of 2. $Re_D = 500$ and $Pr = 0.1$.

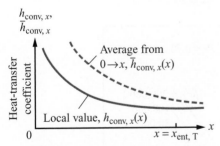

FIGURE 10.23

As the temperature profile develops in the entry region, the local heat-transfer coefficient $h_{\text{conv},x}$ drops from a very high value at the tube entrance ($x = 0$) to a constant value at the end of the thermal entry length and beyond ($x > x_{\text{ent,T}}$). The average heat-transfer coefficient $\bar{h}_{\text{conv},x}$ is always higher than the local value, which it approaches asymptotically.

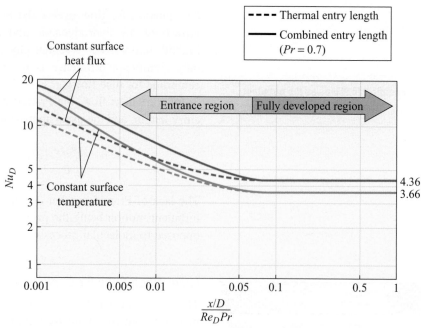

FIGURE 10.24
Local Nusselt numbers for laminar flow in a tube for thermal-entry-length and combined-entry-length problems. **Adapted from Ref. [15] with permission.**

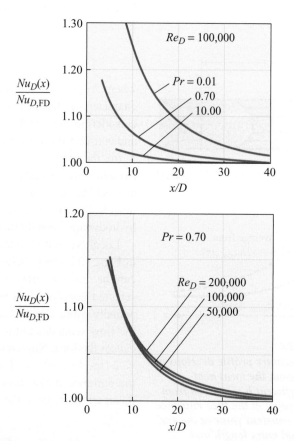

FIGURE 10.25
Local Nusselt numbers for turbulent flow in a tube for uniform wall heat flux. The effect of Prandtl number is shown at the top; the effect of Reynolds number Re_D is shown at the bottom. **Adapted from Ref. [8] with permission.**

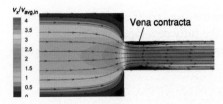

Flow through a contraction produces a vena contracta just downstream of the flow area change.

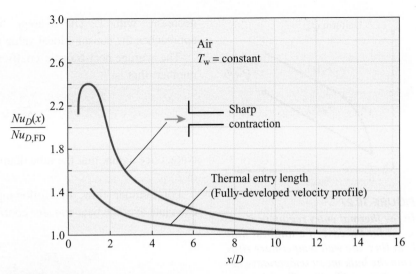

FIGURE 10.26

Local Nusselt numbers for turbulent flow of air in a tube for constant wall temperature. Shown are measurements for the thermal entry problem (with velocity profiles fully developed) and for the combined thermal-hydrodynamic problem associated with a sharp contraction. **Adapted from Ref. [8] with permission.**

coefficient is the average value from the tube entrance ($x = 0$) to an arbitrary location x and is formally defined (see Eq. 3.12) as

$$\overline{h}_{\text{conv},x} \equiv \frac{1}{x} \int_0^x h_{\text{conv},x}(x)\, dx, \tag{10.97}$$

where $h_{\text{conv},x}(x)$ is the local heat-transfer coefficient. Figure 10.23 shows a plot of both $\overline{h}_{\text{conv},x}$ and $h_{\text{conv},x}$ as functions of distance from the tube entrance. Note that at any x location, the average heat-transfer coefficient $\overline{h}_{\text{conv},x}$ is always larger than the local one, as expected for the average of a function that

Table 10.6 Summary of Heat-Transfer Relationships for Flow through Smooth Tubes: Entrance Effects

	Laminar		Turbulent	
	Local	**Average**	**Local**	**Average***
Thermal entry length				
$\dot{Q}''_w$ = constant	Fig. 10.24	—	Fig. 10.25	—
T_w = constant	Fig. 10.24	Eq. 10.99	Fig. 10.26 (air)	Eq. 10.102 $C = 1.4$ (air)
Combined hydrodynamic and thermal entry lengths				
$\dot{Q}''_w$ = constant	Fig. 10.24	—	—	—
T_w = constant	Fig. 10.24	Eq. 10.100	Fig. 10.26 (air)	Eq. 10.102 $C = 6$ (air)

* If $x/D \geq 10$, then Eqs. T10.5a–T10.5c (Table 10.5) can be used to estimate average Nusselt numbers $\overline{Nu}_D$, where properties are evaluated at the average mean temperature, that is, $\overline{T}_m \equiv [T_m(0) + T_m(x)]/2$.

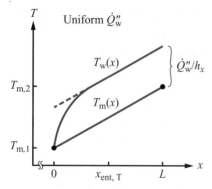

FIGURE 10.27

In the thermal entry region ($x <$ $x_{\mathrm{ent,T}}$) for the case of a uniform wall heat flux, the wall temperature rises from the bulk mean temperature at the entrance [$T_{\mathrm{w}}(0) = T_{\mathrm{m,1}}$] to a constant increment $\dot{Q}''_{\mathrm{w}}/h_{\mathrm{conv,x}}$ above the bulk mean temperature at $x =$ $x_{\mathrm{ent,T}}$ and beyond.

> The average and local Nusselt numbers have the same form: hD/k. See Eq. 10.96.

decreases with x. The average heat-transfer coefficient asymptotically approaches the constant local value in the fully developed region.

The average heat-transfer coefficient is used to define an average Nusselt number; that is,

$$\overline{Nu}_D \equiv \frac{\overline{h}_{\mathrm{conv},x} D}{k_{\mathrm{f}}}. \tag{10.98}$$

Note, once again, that the tube diameter D is the characteristic length, not x or L.

For laminar flow, the following equations [19, 25] can be used to calculate average heat-transfer coefficients for laminar flows:

$$\overline{Nu}_D = 3.66 + \frac{0.0668\,(D/x)\,Re_D Pr}{1 + 0.04\,[(D/x)\,Re_D Pr]^{2/3}} \tag{10.99}$$

for

$$T_{\mathrm{w}} = \mathrm{constant},$$
$$Re_D < 2300 \ (\mathrm{laminar\ flow}),$$
$$\mathrm{and\ thermal\ entry\ length\ alone;}$$

$$\overline{Nu}_D = 1.86 \left(\frac{Re_D Pr}{x/D} \right)^{1/3} \left(\frac{\mu}{\mu_{\mathrm{w}}} \right)^{0.14} \tag{10.100}$$

for

$$T_{\mathrm{w}} = \mathrm{constant},$$
$$Re_D < 2300 \ (\mathrm{laminar\ flow}),$$
$$\left(\frac{Re_D Pr}{x/D} \right)^{1/3} \left(\frac{\mu}{\mu_{\mathrm{w}}} \right)^{0.14} \gtrsim 2,$$
$$0.48 < Pr < 16{,}700,$$

and

$$0.0044 < \left(\frac{\mu}{\mu_{\mathrm{w}}} \right) < 9.75.$$

Except for μ_{w}, which is evaluated at the wall temperature, properties appearing in Eqs. 10.99 and 10.100 are evaluated at the average mean temperature, given by

$$\overline{T}_{\mathrm{m}} \equiv \frac{T_{\mathrm{m}}(0) + T_{\mathrm{m}}(x)}{2}. \tag{10.101}$$

For turbulent flow, the following equation can be used to provide estimates of average heat-transfer coefficients for air:

$$\frac{\overline{Nu}_D}{\overline{Nu}_{D,\mathrm{FD}}} = 1 + \frac{C}{x/D} \tag{10.102}$$

for

$$Pr \approx 0.7 \ (\mathrm{air}),$$
$$Re_D > 2300 \ (\mathrm{turbulent\ flow}),$$

where

$$C = \begin{cases} 1.4 & \text{for thermal entry length alone} \\ 6 & \text{for combined thermal and hydrodynamic entry} \\ & \text{lengths (abrupt contraction).} \end{cases}$$

Reference [28] provides additional information on thermal entrance effects for turbulent flows.

Table 10.6 summarizes heat-transfer relationships for the entry region of tubes and pipes.

As alluded to previously, the average heat-transfer coefficient appears in the integral analysis of heat transfer in a tube with a constant wall temperature. In the derivation of the bulk mean temperature distribution for this case (i.e., Eq. 10.33), we assumed that the local heat-transfer coefficient was constant; that is, we assumed that the flow was thermally fully developed. We now return to this derivation and remove this restriction by noting that $h_{conv,x}$ is a function of x and must be retained within the integrand of Eq. 10.32. Thus, Eq. 10.32 is rewritten as

$$\int_{T_m(0)}^{T_m(x)} \frac{dT_m(x)}{[T_w - T_m(x)]} = \frac{\pi D}{\dot{m}c_p} \int_0^x h_{conv,x}\, dx.$$

Multiplying and dividing the right-hand term by x yields

$$\frac{\pi D x}{\dot{m}c_p} \frac{1}{x} \int_0^x h_{conv,x}\, dx.$$

From this, we recognize the definition of the average heat-transfer coefficient $\bar{h}_{conv,x}$; thus, we rewrite this term as

$$\frac{\bar{h}_{conv,x}\, \pi D}{\dot{m}c_p} x.$$

Our revised expression for the bulk mean temperature distribution is thus identical to Eq. 10.33 except that $\bar{h}_{conv,x}$ replaces $h_{conv,x}$, that is,

$$\frac{T_m(x) - T_w}{T_m(0) - T_w} = \exp\left[-\frac{\bar{h}_{conv,x}\, \pi D}{\dot{m}c_p} x\right], \qquad (10.103a)$$

or

$$T_m(x) = T_w + (T_m(0) - T_w) \exp\left[\frac{-\bar{h}_{conv,x}\, \pi D}{\dot{m}c_p} x\right]. \qquad (10.103b)$$

The shape of the bulk mean temperature distribution $T_m(x)$ is nominally the same as that shown previously in Fig. 10.8 where entrance effects were neglected.

Example 10.15

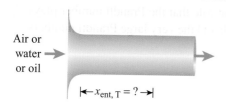

Air or water or oil

$\leftarrow x_{ent,T} = ? \rightarrow$

Consider the same three fluids and conditions presented in Example 10.14. Determine the thermal entry lengths $x_{ent,T}$ for each of the three flows.

Solution

Known   Fluids, Re_D, D, L

Find $x_{ent,T}$

Assumptions

i. Steady flow

ii. Uniform temperature profile at entrance

This we substitute into Eq. 10.103 to get

$$L = \frac{-\dot{m}c_p}{\frac{k}{D}1.86\left(\frac{Re_D Pr}{L/D}\right)^{1/3}\left(\frac{\mu}{\mu_w}\right)^{0.14}\pi D}\ln\left[\frac{T_m(L) - T_w}{T_m(0) - T_w}\right].$$

With not too much effort, we solve this expression for the unknown length L; the result is

$$L = \left\{\frac{-\dot{m}c_p}{\pi 1.86 k(Re_D Pr D)^{1/3}(\mu/\mu_w)^{0.14}}\ln\left[\frac{T_m(L) - T_w}{T_m(0) - T_w}\right]\right\}^{3/2}.$$

To independently verify our algebra, we check the dimensions of this expression:

$$[=]\left\{\frac{(\text{kg/s})(\text{J/kg}\cdot\text{K})}{(\text{W/m}\cdot\text{K})\text{m}^{1/3}}\left[\frac{1\text{ W}}{\text{J/s}}\right]\right\}^{3/2} = (\text{m}^{2/3})^{3/2} = \text{m}.$$

Noting that $\mu_w = \mu(373\text{ K}) = 0.0232\text{ N}\cdot\text{s/m}^2$, we have

$$L = \left\{\frac{-0.0576(2035)}{\pi(1.86)0.141[70.2(1205)0.0125]^{1/3}(0.0836/0.0232)^{0.14}}\right.$$
$$\left.\ln\left[\frac{360 - 373}{300 - 373}\right]\right\}^{3/2} = 90.4\text{ m}.$$

Having a value for L, we test the requirement for the use of Eq. 10.100 that

$$\left(\frac{Re_D Pr}{L/D}\right)^{1/3}\left(\frac{\mu}{\mu_w}\right)^{0.14} > 2;$$

substituting numbers on the left-hand side of this inequality, we get

$$\left(\frac{70.2(1205)}{90.4/0.0125}\right)^{1/3}\left(\frac{0.0836}{0.0232}\right)^{0.14} = 2.7.$$

Because $2.7 > 2$ we thus see that this requirement is indeed met, along with the restrictions on Pr and μ/μ_w.

Comments This length (90.4 m) is much greater than the length required to heat the same mass flow of water (3.05 m). What is responsible for this difference? The principle factor is the much greater viscosity of the oil. The large viscosity results in the flow now being laminar, rather than turbulent, as was the case for the water flow. This, in turn, produces a much smaller average convective heat-transfer coefficient (57 versus 3470 W/m²·K). The smaller heat-transfer coefficient requires a much greater length to heat the same flow, even though the specific heat of the oil is less than that of the water (2035 versus 4184 J/kg·K).

Self Test 10.14

✓ **Oil flows through a 0.1-m-inside-diameter underground pipeline at 300 K and 0.5 m/s. The pipeline then passes through a 30-m-wide lake, which is at 273 K. Determine the mean temperature of the oil in the pipe where it exits the lake.**

(Answer: $T_m(L) = 295$ K)

10.6 DUCTS OF NONCIRCULAR CROSS SECTION

10.6a Hydraulic Diameter

For ducts of circular cross section, we employ the duct diameter D as the characteristic length in correlations for friction and heat transfer. For ducts of noncircular cross section, we employ as the characteristic length the **hydraulic diameter, D_h,** defined by

$$D_h \equiv \frac{4 \times \text{flow cross-sectional area}}{\text{wetted perimeter}}. \qquad (10.104a)$$

For full flowing ducts, the wetted perimeter is the entire perimeter of the duct, and the flow cross-sectional area is the duct cross-sectional area, and so

$$D_h = \frac{4\,A_{\text{ductx-sec}}}{\mathscr{P}_{\text{duct}}}. \qquad (10.104b)$$

For a circular duct, the hydraulic diameter and the duct diameter are equal: $D_h = 4(\pi D^2/4)/(\pi D) = D$.

10.6b Laminar Flow

For fully developed, laminar flow ($Re_{D_h} < 2300$), solutions have been developed for both the friction and heat-transfer problems for a variety of duct geometries. Results for ducts of rectangular cross section are presented in Table 10.7. In this table, friction factors and Nusselt numbers are presented for various values of the duct aspect ratio b/a. Results for other geometries can be found in Refs. [15, 26]. The friction factors obtained from this table are used with Eq. 10.16, with the hydraulic diameter D_h replacing the diameter D.

10.6c Turbulent Flow

For turbulent flow ($Re_{D_h} > 2300$), solutions for circular ducts can be adapted to the problem of noncircular ducts. Specifically, we can employ the Moody chart (Fig. 10.17) and other correlations for friction (Eqs. 10.82,

Table 10.7 **Friction Factors* and Nusselt Numbers for Laminar Flow through Rectangular Ducts of Dimensions $a \times b$ [15, 26]**

b/a	$f_D Re_{D_h}$	$Nu_{D_h}(\dot{Q}''_w \text{ uniform})$	$Nu_{D_h}(T_w \text{ uniform})$
1.0	56.9	3.61	2.98
1.3$\overline{3}$	57.9	3.66†	3.06†
2.0	62.2	4.12	3.39
3.0	67.9†	4.79	3.96
4.0	72.9	5.33	4.44
8.0	82.3	6.49	5.60
∞	96.0	8.23	7.54

* Head loss is calculated as $h_L = f_D(L/D_h)v_{\text{avg}}^2/(2g)$.
† Interpolated values.

10.83, and 10.86) and heat transfer (Table 10.5) by simply substituting the hydraulic diameter D_h for the circular duct diameter D wherever it appears. For example,

$$f_D = f_D(Re_{D_h}, \varepsilon/D_h),$$

where

$$Re_{D_h} \equiv \frac{\rho v_{avg} D_h}{\mu}.$$

Similarly, the hydraulic diameter appears in the Nusselt number for the heat-transfer problem (i.e., $Nu_{D_h} \equiv h_{conv} D_h/k$).

10.7 MINOR LOSSES

In addition to the head losses associated with lengths of straight ducts, *minor losses* result for flows through values, elbows, contractions, expansions, and other common features of piping systems. To account for these minor losses, a **loss coefficient K** is defined as follows:

$$K \equiv \frac{h_{L,minor}}{v_{avg}^2/2g}. \tag{10.105}$$

The velocity appearing in Eq. 10.105 depends on the situation, and care must be exercised in employing the correct velocity. Table 10.8 presents loss coefficients for valves, elbows, and tees for U.S. standard pipe sizes, and Fig. 10.28 shows loss coefficients for entrances and sudden contractions and expansions. Note that the loss coefficient for a sudden expansion into a reservoir is unity (i.e., $K = 1$ for $d/D = 0$). Note also that the velocity appearing in Eq. 10.105 for expansions and contractions is the velocity in the smaller pipe. Additional loss coefficient data can be found in Refs. [26, 27].

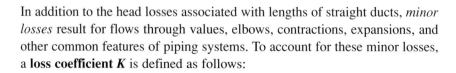

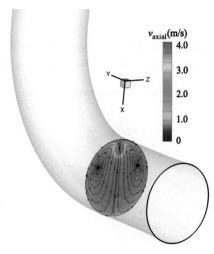

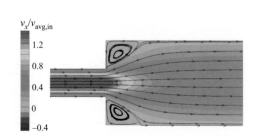

Calculations show complex velocity fields for flow through an elbow (left) and an abrupt expansion (right). Both flows exhibit recirculation zones.

Table 10.8 Loss Coeffcents* for Open Valves, Elbows, and Tees [26]

Nominal Diameter (in)	Screwed				Flanged				
	½	1	2	4	1	2	4	8	20
Valves (fully open)									
Globe	14	8.2	6.9	5.7	13	8.5	6.0	5.8	5.5
Gate	0.30	0.24	0.16	0.11	0.80	0.35	0.16	0.07	0.03
Swing check	5.1	2.9	2.1	2.0	2.0	2.0	2.0	2.0	2.0
Angle	9.0	4.7	2.0	1.0	4.5	2.4	2.0	2.0	2.0
Elbows									
45° regular	0.39	0.32	0.30	0.29					
45° long radius					0.21	0.20	0.19	0.16	0.14
90° regular	2.0	1.5	0.95	0.64	0.50	0.39	0.30	0.26	0.21
90° long radius	1.0	0.72	0.41	0.23	0.40	0.30	0.19	0.15	0.10
180° regular	2.0	1.5	0.95	0.64	0.41	0.35	0.30	0.25	0.20
180° long radius					0.40	0.30	0.21	0.15	0.10
Tee									
Line flow	0.90	0.90	0.90	0.90	0.24	0.19	0.14	0.10	0.07
Branch flow	2.4	1.8	1.4	1.1	1.0	0.80	0.64	0.58	0.41

* $K \equiv h_{L,\text{minor}}/(v_{\text{avg}}^2/2g)$, where v_{avg} is the average velocity in the connecting pipe.

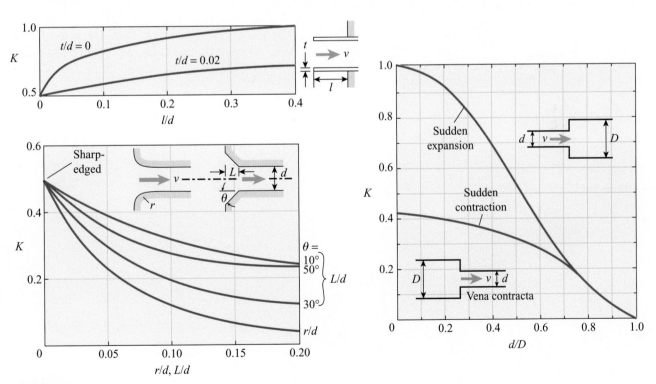

FIGURE 10.28
Loss coefficients for entrances (left) and sudden expansions and contractions (right). **Reprinted from Ref. [26] with permission.**

SUMMARY

The overarching objectives of this chapter are, first, to understand the physical phenomena that influence both isothermal and nonisothermal internal flows and, second, to be able to calculate quantities of engineering interest (e.g., τ_w, h_L, ΔP, $\dot{Q}''$, $\dot{Q}_{0-L}$, etc.) for both laminar and turbulent internal flows. You should be familiar with the concept of a critical Reynolds number applied to internal flows, the physical meaning and application of bulk mean temperature, and the idea of flow development for both the hydrodynamic and thermal problems. You should also be able to write simple force–momentum and energy balances for integral control volumes associated with the entire length of the duct or a small portion of that length. Many phenomena are represented graphically. You should be able to draw an assortment of graphs as listed in the learning objectives presented at the beginning of this chapter. Similarly, you should be able to calculate the many quantities related to friction and heat transfer in internal flows that are listed in the learning objectives.

Chapter 10
Key Concepts & Definitions Checklist[4]

10.2 The Big Picture—Integral Analyses

- ❏ Mass conservation ➤ *10.1, 10.2*
- ❏ Momentum conservation ➤ *Q10.2, Q10.3*
- ❏ Pressure drop and wall friction ➤ *Q10.2, Q10.3*
- ❏ Darcy (or Moody) friction factor ➤ *10.4*
- ❏ Mechanical energy conservation ➤ *Q10.4*
- ❏ Relationships among pressure drop, wall friction, and head loss ➤ *Q10.5*
- ❏ Energy conservation ➤ *10.8*
- ❏ Bulk mean temperature ➤ *Q10.6*
- ❏ Nonisothermal flow—uniform heat flux: $T_m(x)$, $T_w(x)$, $\dot{Q}_{0-L}$ ➤ *10.7, Q10.7*
- ❏ Nonisothermal flow—constant wall temperature: $T_m(x)$, $\dot{Q}''(x)$, $\dot{Q}_{0-L}$ ➤ *Q10.8, 10.10*

10.3 Details—Fully Developed Laminar Flow

- ❏ Hydrodynamically fully developed flow ➤ *Q10.9*
- ❏ Thermally fully developed flow ➤ *Q10.10*
- ❏ Reynolds number (Eq. 10.34) ➤ *10.13, 10.19*
- ❏ Critical Reynolds number ➤ *10.15*
- ❏ Velocity distribution ➤ *10.21*
- ❏ Average and maximum velocities ➤ *10.20*
- ❏ Wall shear stress (Eq. 10.46) ➤ *10.22*
- ❏ Pressure drop and flow rates (Eq. 10.49) ➤ *10.27*
- ❏ Friction factor–Reynolds number relationship (Eq. 10.53) ➤ *Q10.11*
- ❏ Nusselt number Nu_D: constant wall temperature and uniform wall heat flux ➤ *10.29, 10.30, 10.31*

10.4 Details—Fully Developed Turbulent Flow

- ❏ Closure problem ➤ *Q10.12*

- ❏ Universal velocity distribution for turbulent flow (Fig. 10.12) ➤ *10.44, 10.46*
- ❏ Velocity profiles: laminar versus turbulent ➤ *10.45*
- ❏ Roughness and relative roughness ➤ *10.58*
- ❏ Moody chart (Fig. 10.17) and related correlations ➤ *10.50, 10.55*
- ❏ Heat-transfer correlations (Table 10.5) ➤ *10.71*

10.5 Developing Flows

- ❏ Hydrodynamic entrance region: velocity profiles and wall shear-stress distribution ➤ *Q10.15, Q10.16*
- ❏ Hydrodynamic entrance lengths: laminar and turbulent ➤ *10.95*
- ❏ Thermal entrance region: temperature profiles and local heat-transfer coefficient distribution ➤ *Q10.18, Q10.21*
- ❏ Thermal entrance lengths: laminar and turbulent ➤ *10.99*
- ❏ Local and average heat-transfer coefficients ➤ *Q10.19, Q10.20*
- ❏ Heat-transfer correlations (Table 10.6) ➤ *10.103, 10.104*

10.6 Ducts of Noncircular Cross Section

- ❏ Hydraulic diameter ➤ *Q10.22, 10.112*
- ❏ Laminar flow correlations (Table 10.7) ➤ *10.108*
- ❏ Turbulent flows ➤ *10.107*

10.7 Minor Losses

- ❏ Loss coefficient ➤ *Q10.23*
- ❏ Calculation of minor head losses ➤ *10.121, 10.122*

---

[4] Numbers following arrows below refer to end-of-chapter Problems (e.g., 10.1) and Questions (e.g., Q10.2).

REFERENCES

1. Rouse, H., and Ince, S., *History of Hydraulics,* Iowa Institute of Hydraulics Research, State University of Iowa, Iowa City, IA, 1957.

2. Hagen, G., "Über die Bewegung des Wassers in engen zylindrischen Röhren," *Annalen der Physik und Chemie,* 46:423–442 (1839).

3. Poiseuille, J. L., "Récherches experimentelles sur le mouvement des liquids dans les tubes de très petits diamètres," *Comptes Rendus,* 11:961–967 and 1041–1048 (1840); 12:112–115 (1841).

4. Reynolds, O., "An Experimental Investigation of the Circumstances Which Determine Whether the Motion of Water Shall Be Direct or Sinuous, and the Law of Resistance in Parallel Channels," *Philosophical Transactions of the Royal Society of London,* 174:935–982, 1883.

5. Nikuradse, J., "Strömungsgesetze in rauhen Röhren," *VDI-Forschungsheft,* 361, 1933. English translation: "Laws of Flow in Rough Pipes," NACA Technical Memorandum 1292, 1950.

6. Moody, L. F., "Friction Factors for Pipe Flow," *Transactions of the ASME,* 66:671–684 (1944).

7. Layton, E. T., Jr., and Leinhard, J. H. (Eds.), *History of Heat Transfer: Essays in the Honor of the 50th Anniversary of the ASME Heat Transfer Division,* American Society of Mechanical Engineers, New York, 1988.

8. Kays, W. M., and Crawford, M. E., *Convective Heat and Mass Transfer,* 2nd ed., McGraw-Hill, New York, 1980.

9. Boussinesq, T. V., "Théorie de l'écoulement tourbillant," *Mémoires présentés, Academy of Science,* Paris, XXIII, 46 (1877).

10. Launder, B. E., and Spalding, D. B., *Lectures in Mathematical Models of Turbulence,* Academic Press, New York, 1972.

11. Prandtl, L., "Über die ausgebildete Turbulenz," *Zeitschrift für Angewandte Mathematik und Mechanik,* 5:136–139 (1925).

12. van Driest, E. R., "On Turbulent Flow near a Wall," *Journal of Aerospace Science,* 23:1007 (1956).

13. Shames, I. H., *Mechanics of Fluids,* 3rd ed., McGraw-Hill, New York, 1992.

14. Blasius, H., "Grenzschichten in Flüssigkeiten mit kleiner Reibung," *Zeitschrift für Mathematik und Physik,* 56:1–37 (1908). English translation: "Fluid Motion with Very Small Friction," NACA Technical Memorandum 1256, 1950.

15. Incropera, F. P., and DeWitt, D. P., *Fundamentals of Heat and Mass Transfer,* 5th ed., Wiley, New York, 2002.

16. Swamee, P. K., and Jain, A. K., "Explicit Equations for Pipe-Flow Problems," American Society of Civil Engineers, Hydraulics Division, *Journal,* 102:657–664 (1976).

17. Kakac, S., Shah, R. K., and Aung, W., *Handbook of Single-Phase Convective Heat-Transfer,* Wiley-Interscience, New York, 1987.

18. Dittus, F. W., and Boelter, L. M. K., "Heat Transfer in Automobile Radiators of the Tubular Type," University of California, Berkeley, Publications on Engineering, Vol. 2, pp. 443–461, 1930.

19. Sieder, E. N., and Tate, G. E., "Heat Transfer and Pressure Drop of Liquids in Tubes," *Industrial Engineering Chemistry,* 28:1429–1435 (1936).

20. Gnielinski, V., "New Equations for Heat and Mass Transfer in Turbulent Pipe and Channel Flow," *International Chemical Engineering,* 16:359–368 (1976).

21. Skupinski, E. S., Tortel, J., and Vautrey, L., "Determination des coefficients de convection d'un alliage sodium-potassium dans un tube circulaire," *International Journal of Heat and Mass Transfer,* 8:937–951 (1965).

22. Seban, R. A., and Shimazaki, T. T., "Heat Transfer to Fluid Flowing Turbulently in Smooth Pipe with Walls at Constant Temperature," *Transactions of the ASME,* 73:803–807 (1951).

23. Langhaar, H., "Steady Flow in the Transition Length of a Straight Tube," *Journal of Applied Mechanics,* 9:A55–A58 (1942).

24. Schlichting, H., *Boundary Layer Theory,* 6th ed., McGraw-Hill, New York, 1968.

25. Kays, W. M, "Numerical Solutions for Laminar-flow Heat Transfer in Circular Tubes," *Transactions of the ASME,* 77:1265–1272 (1955).

26. White, F. M., *Fluid Mechanics,* 2nd ed., McGraw-Hill, New York, 1986.

27. Fox, R. W., and McDonald, A. T., *Introduction to Fluid Mechanics,* 3rd ed., Wiley, New York, 1985.

28. Burmeister, L. C., *Convective Heat Transfer,* 2nd ed., Wiley, New York, 1993.

29. Kim, J., Moin, P., and Moser, R. D., "Turbulence Statistics in Fully Developed Channel Flow at Low Reynolds Number," *Journal of Fluid Mechanics,* 177:136–166 (1987).

Some end-of-chapter problems were adapted with permission from the following:

30. Chapman, A. J., *Fundamentals of Heat Transfer,* Macmillan, New York, 1987.

31. Pnueli, D., and Gutfinger, C., *Fluid Mechanics,* Cambridge University Press, Cambridge, England, 1992.

32. Shepherd, D. G., *Elements of Fluid Mechanics,* Harcourt, Brace & World, New York, 1965.

Nomenclature

A	Area (m^2)		t	Time (s)
c_p	Constant-pressure specific heat (J/kg·K)		T	Temperature (K)
D	Diameter (m)		T_{m}	Bulk mean temperature (K)
D_{h}	Hydraulic diameter (m)		u	Specific internal energy (J/kg)
F	Force (N)		v, V	Velocity (m/s)
f_{D}	Darcy friction factor (dimensionless)		v_x	Axial velocity (m/s)
f_{F}	Fanning friction factor (dimensionless)		v_*	Friction velocity (m/s)
g	Gravitational acceleration (m/s^2)		υ	Specific volume (m^3/kg)
h	Specific enthalpy (J/kg)		$\dot{V}$	Volume flow rate (m^3/s)
$h_{\mathrm{conv},x}$	Local convective heat-transfer coefficient (W/m^2·K or W/m^2·°C)		$\boldsymbol{V}$	Voltage (V)
$\bar{h}_{\mathrm{conv}}$	Average convective heat-transfer coefficient (W/m^2·K or W/m^2·°C)		W	Weight (N)
			$\dot{W}$	Rate of work or power (W)
h_{L}	Head loss (m)		x	Spatial coordinate in axial direction (m)
i	Electric current (A)		$x_{\mathrm{ent},H}$	Hydrodynamic entrance length (m)
k	Thermal conductivity (W/m·K or W/m·°C)		$x_{\mathrm{ent},T}$	Thermal entrance length (m)
K	Loss coefficient (dimensionless)		z	Spatial coordinate in vertical direction (m)
ℓ_{m}	Mixing length (m)			
L	Length (m)			
$\dot{m}$	Mass flow rate (kg/s)			
n	Exponent in power-law velocity distribution (dimensionless)			

GREEK

Nu_D	Nusselt number (dimensionless)		α	Thermal diffusivity (m^2/s) or kinetic energy correction factor (dimensionless)
P	Pressure (Pa)		δ_{H}	Hydrodynamic boundary-layer thickness (m)
$\mathscr{P}$	Perimeter (m)			
Pr	Prandtl number (dimensionless)		δ_{T}	Thermal boundary-layer thickness (m)
$\dot{Q}$	Heat transfer rate (W)		Δ	Difference or change
$\dot{Q}''$	Heat flux (W/m^2)		ε	Roughness height (m) or eddy viscosity (N·s/m^2)
r	Radial coordinate (m)		θ	Angle above horizontal (see Fig. 10.2)
R	Tube radius (m)		μ	Viscosity (N·s/m^2)
R_{elec}	Electrical resistance (ohm)		μ_{eff}	Effective viscosity for turbulent flow (N·s/m^2)
Re_D	Reynolds number based on diameter (dimensionless)		ν	Kinematic viscosity, μ/ρ (m^2/s)
			ρ	Density (kg/m^3)
			τ	Shear stress (N/m^2)

SUBSCRIPTS

avg	average
c	characteristic
conv	convection
crit	critical
cv	control volume
elec	electrical
f	fluid
fric	friction
FD	fully developed
H	hydrodynamic
in	into system or control volume
lam	laminar
max	maximum

out	out of system or control volume
P	pressure
T	thermal
tot	total
turb	turbulent
w	wall
1	station 1
2	station 2

SUPERSCRIPTS

*	Dimensionless quantity

OTHER NOTATION

$\overline{(\ )}$	Time-average quantity
$(\ )'$	Fluctuating component

QUESTIONS

10.1 Review the most important equations presented in this chapter (i.e., those with a reddish background). What physical principles do they express? What restrictions apply?

10.2 Sketch a control volume for flow through a long pipe that is contiguous with the inside pipe wall. Applying the conservation of momentum principle to this control volume, how do the pressure drop and the wall friction force relate for fully developed flow?

10.3 Repeat Question 10.2 for the case where the flow enters the pipe with a uniform velocity distribution and exits with a parabolic distribution.

10.4 Simplify the mechanical energy equation for steady, fully developed flow in a tube.

10.5 Use the momentum conservation principle and the mechanical energy equation to relate pressure drop, the wall shear stress, and head loss for steady, fully developed flow in a tube.

10.6 What is the bulk mean temperature? How is it used?

10.7 Consider a steady flow through a tube subjected to a constant (uniform) wall heat flux. Starting at an axial location beyond the thermal entry length, sketch both the bulk mean temperature $T_m(x)$ and tube wall temperature $T_w(x)$ as functions of downstream distance x.

10.8 A hot fluid enters a tube in which the wall temperature is fixed. The fluid cools as it flows through the tube. Sketch the bulk mean temperature as a function of downstream distance x. Show the tube wall temperature, T_w, on your graph.

10.9 What does it mean that a flow is hydrodynamically fully developed?

10.10 What does it mean that a flow is thermally fully developed?

10.11 Use the definition of the wall shear stress for a Newtonian fluid (Eq. 10.10) and the definition of the Darcy friction factor (Eq. 10.11) to show that $f_D = 64/Re_D$ for laminar flow in a circular tube where $v_x(r) = 2v_{avg}(1 - r^2/R^2)$.

10.12 What is meant by the *closure problem* in the analysis of turbulent flows?

10.13 List the similarities and differences between a uniform flow over a flat plate (external flow) and the flow in a long tube.

10.14 Define the hydrodynamic entry length $x_{ent,H}$.

10.15 Consider a steady flow entering a tube with an average velocity U and uniform velocity distribution. On v_x-r coordinates, plot $v_x(r)$ at the following locations: (a) at the tube entrance, (b) part way through the developing region, and (c) at the end of the developing region. Indicate U on your graph.

10.16 Show the velocity gradient at the wall $(r = R)$ on your plot from Question 10.15 for the three locations indicated. Draw a straight line having a slope that equals the gradient at the wall. Use your result to discuss how the wall shear stress varies from the pipe inlet $(x = 0)$ to the fully developed region and beyond. Sketch the wall shear stress τ_w versus x.

10.17 Use the results of Question 10.16 to sketch the how pressure varies with distance through a pipe, that is, plot $P(0, x)$ versus x.

10.18 Consider a cold fluid entering and flowing through a heated tube. Plot the local heat-transfer coefficient $h_{conv,x}$ as a function of distance x from the tube entrance.

10.19 Explain how the local and average heat-transfer coefficients relate in a thermally developing flow.

10.20 What is the definition of the local convective heat-transfer coefficient $h_{conv,x}$ for an internal flow? How does this definition differ from that for an external flow?

10.21 Replot $T_m(x)$ and $T_w(x)$ from Question 10.7 taking into account the thermal entrance length $x_{ent,T}$. Discuss.

10.22 Define the hydraulic diameter? How is the hydraulic diameter used to determine friction factors and heat-transfer coefficients?

10.23 Define the loss coefficient K for minor losses. Explain how loss coefficients are used to estimate total head losses in complex piping systems.

Chapter 10 Problem Subject Areas

PROBLEMS

10.1 Air enters a 30-mm-diameter tube at 25°C and 101 kPa with an average velocity of 3 m/s. The air exits the tube at 40°C and 100 kPa. Determine the exit velocity of the air.

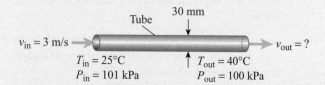

10.2 Liquid water at 200 kPa and 25°C enters a 25-mm-diameter tube with a velocity of 0.2 m/s. As the water flows through the tube it is heated and exits as saturated vapor at 198 kPa. Determine the velocity of the steam at the tube exit.

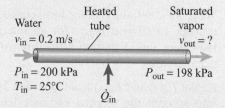

10.3 Consider water at 1 atm and 25°C flowing through a short duct. The duct has a circular cross section ($d = 50$ mm) at the entrance and a square cross section (50 mm × 50 mm) at the exit. If the water enters at 4 m/s, what is the velocity at the exit?

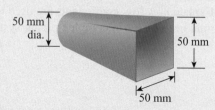

10.4 Water at 25°C and 1 atm flows through a 100-mm-diameter tube at a Reynolds number of 60,000. The Darcy friction factor is 0.02. Determine the average wall shear stress.

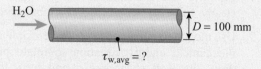

10.5 Repeat Problem 10.4 with air rather than water. Compare your results with those of Problem 10.4 and discuss.

10.6 Water enters a 40-mm-diameter tube at 25°C and a nominal pressure of 2 atm. The average velocity is 1.8 m/s. An electrical heating tape wrapped around the tube results in the heating of the water to 30°C at a downstream distance of 6 m.

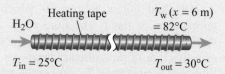

A. Determine the average heat flux at the inside wall of the tube over the 6-m length.

B. Determine the local heat-transfer coefficient at the downstream location if the wall temperature here is 82°C.

10.7 Consider a fully developed, laminar flow of water through a 10-mm-diameter tube at a flow rate of 8.4×10^{-3} kg/s. A constant heat flux of 3000 W/m² is applied to the wall. At a particular point in the flow, the bulk mean temperature of the water is 320 K.

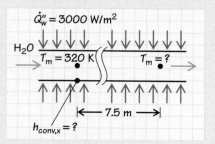

A. Estimate the convective heat-transfer coefficient and the tube wall temperature at this location.

B. Determine the bulk mean temperature of the water at a location 7.5 m downstream of the point where the water is 320 K.

10.8 Nitrogen (N_2) enters a 15-mm-diameter heated tube at 20°C and 4 atm with a velocity of 4 m/s. The N_2 exits the tube at 50°C. Determine the heat-transfer rate to the N_2.

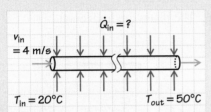

10.9 Repeat Problem 10.8 for the following fluids:

A. Water

B. Helium

C. Methane

D. Propane

Other than the fluid type, all conditions are the same. Compare and discuss your results.

10.10 Consider a thermally fully developed flow of water through a 20-mm-diameter tube at a nominal pressure of 1 atm. The tube wall is maintained at a constant wall temperature of 50°C. The mean velocity is 1.75 m/s, and the convective heat-transfer coefficient is 6900 W/m² · K. Determine the length of tube required to heat the water from 25°C to 29°C.

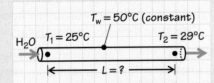

10.11 Consider thermally fully developed flow through a tube. Derive an expression equivalent to Eq. 10.33a for the case of a constant ambient temperature T_∞. Include the thermal resistance of the tube wall in your analysis for a tube with inner and outer radii of r_i and r_o, respectively, and a thermal conductivity of k_w. Assume a constant convective heat-transfer coefficient $h_{conv,o}$ at the exterior surface of the tube. Compare your result with Eq. 10.33a and discuss.

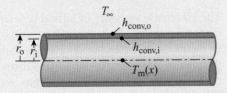

10.12 Water at 25°C flows upward through a 50-mm-diameter tube at a flow rate of 3.5 kg/s. The tube is 30 m long and inclined at an angle of 20° from the horizontal. The upper end of the tube is open to the atmosphere, and the average wall shear stress is 12 N/m². Determine the pressure at the tube entrance, assuming that the inlet and outlet velocity profiles are identical.

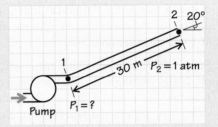

10.13 Determine the velocity required to provide a Reynolds number of 2300 for flow through a 20-mm-diameter tube for the following fluids, each at 300 K and the pressure indicated:

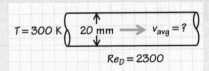

A. Water (1 atm)

B. Air (1 atm)

C. Engine oil (saturated liquid)

D. Helium (1 atm)

E. Mercury (saturated liquid)

Discuss.

10.14 Consider steady flow through a tube with an average velocity of 0.3 m/s. Determine the tube diameter required to provide a Reynolds number of 2300 for the same fluids as in Problem 10.13. Each fluid is at 300 K and the pressure indicated. Discuss.

10.15 Consider flow through a 8-mm-diameter tube with an average velocity of 2 m/s. Determine whether the flow is laminar or turbulent for the following fluids, each at 300 K and the pressure indicated:

A. Water (1 atm)

B. Air (1 atm)

C. Engine oil (saturated liquid)

D. Hydrogen (1 atm)

E. Mercury (saturated liquid)

F. Ethylene glycol (saturated liquid)

10.16 The mean flow velocity in a circular tube is 6 ft/s. What must the tube diameter be such that that the flow is most likely to be turbulent if the fluid is (a) air at 1 atm and 70 F or (b) water at 1 atm and 70 F.

10.17 Find the maximum mass flux (i.e., the mass flow rate per unit of cross-sectional area [=] kg/s·m²) for which one may expect laminar flow in a 3-cm-diameter tube if the fluid is (a) water, (b) ethylene glycol, (c) air, or (d) helium. Assume all fluids are at 1 atm pressure and 20°C.

10.18 Steam at 200 psia and 600 F flows with a velocity of 2 ft/s through a 2-in (nominal diameter), schedule-40 pipe (with an inside diameter of 2.067 in). Determine whether the flow is most likely laminar or turbulent.

10.19 Using the definition of the Reynolds number given by Eq. 10.34a, show that $Re_D = 4\dot{m}/(\pi\mu D)$ (Eq. 10.34b).

10.20 Using the velocity distribution for laminar flow given by Eq. 10.41 show that $v_{avg} = v_{max}/2$.

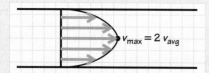

$v_{max} = 2 v_{avg}$

10.21 Starting with Eq. 10A.2, apply the boundary conditions of Eqs. 10.38a and 10.38b to derive the velocity distribution given by Eq. 10.39.

10.22 Consider a fully developed, laminar flow of water at 300 K through a 10-mm-diameter, horizontal tube. The mass flow rate is 8.4×10^{-3} kg/s and the pressure is nominally 1 atm. Determine the following:

H_2O

$\dot{m} = 8.4 \times 10^{-3}$ kg/s $D = 10$ mm

A. The average velocity

B. The maximum velocity

C. The Reynolds number

D. The wall shear stress

E. The pressure drop for a 10-m length of tube

F. The Darcy (or Moody) friction factor f_D using the results from the previous parts

G. The head loss

H. The pumping power in W required to maintain the flow

10.23 Consider a flow between two parallel plates as shown in the sketch. The top plate is externally driven at a constant velocity V_p, whereas the lower plate is stationary. The following assumptions apply to this flow:

1. The flow is steady.

2. The flow is two dimensional (i.e., there is no flow in the z-direction).

3. The flow is fully developed.

4. The fluid is incompressible and Newtonian.

5. There is no pressure gradient in the axial (i.e., x-) direction.

6. The plates are horizontal.

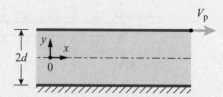

A. Simplify the x-component of the Navier–Stokes equation (linear momentum conservation)

$$\rho g_x - \frac{\partial P}{\partial x} + \mu\left(\frac{\partial^2 v_x}{\partial x^2} + \frac{\partial^2 v_x}{\partial y^2} + \frac{\partial^2 v_x}{\partial z^2}\right)$$

$$= \rho\left(\frac{\partial v_x}{\partial t} + v_x\frac{\partial v_x}{\partial x} + v_y\frac{\partial v_x}{\partial y} + v_z\frac{\partial v_x}{\partial z}\right),$$

as appropriate for this flow. Cross out terms, or otherwise simplify, indicating the assumption that allows you to perform each simplification.

B. Write the boundary conditions necessary to solve your equation from Part A [i.e., what boundary conditions are needed to find $v_x(y)$]? Use the coordinate system defined in the sketch.

C. Solve the governing equation from Part A and apply the boundary conditions from Part B to determine the velocity distribution $v_x(y)$.

10.24 Water at 25°C flows from one large tank to another at a lower elevation as shown in the sketch. The 50-m-long interconnecting tube has a diameter of 5 mm. Estimate the flow rate through the tube assuming entrance and exit effects are negligible.

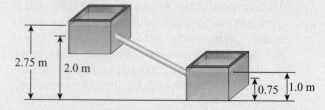

2.75 m   2.0 m

0.75 1.0 m

10.25 The kinematic viscosity of a fluid is determined by measuring the flow rate through a capillary tube fed by a tank as shown in the sketch. The fluid depth in the tank is 100 mm. The length and diameter of the capillary tube are 200 mm and 1 mm, respectively. The following flow rates were measured for various fluids at 25°C: (a) 25.3 cm³/s, (b) 1.5 cm³/s, and (c) 0.04 cm³/s. Determine the kinematic viscosity of each fluid. Assume all entrance effects associated with the tube are negligible.

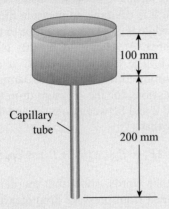

10.26 Design a laminar flow capillary flow meter (see Example 10.4) to measure a nominal 1×10^{-6} kg/s flow of air (i.e., determine D and L). The air enters the capillary tube at 25°C and 3 atm. The desired pressure drop is to be approximately 0.5 in H_2O. Constrain your design so that the tube length is at least ten times $0.05DRe_D$, the entry length associated with laminar flow.

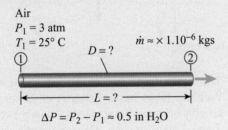

10.27 A chemiluminescent analyzer is a device used to measure the concentration of nitric oxide in combustion product gases. A 0.5-mm-diameter, 38-mm-long capillary tube is used to meter the sample flow to the analyzer. The pressure drop across the capillary is maintained at 5 in Hg by a back-pressure regulator. Determine the sample mass flow rate for a pressure and temperature at the capillary entrance of 100 kPa and 25°C, respectively. Assume the sample properties can be approximated as those of air.

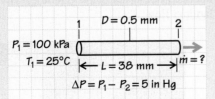

10.28 Liquid hexane flows through a 0.5-mm-diameter, 2-m-long capillary tube at 20 cm³/min at 25°C and a nominal pressure of 6 atm. Determine the pressure drop required to sustain this flow of hexane.

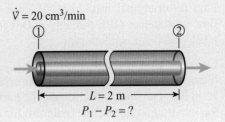

10.29 Consider laminar flow through a 10-mm-diameter tube at 27°C. A uniform heat flux is applied to the tube wall. Estimate the convective heat-transfer coefficient at the inside tube wall associated with each of the following fluids:

A. Water (1 atm)

B. Air (1 atm)

C. Helium (1 atm)

D. Mercury (saturated liquid)

E. Engine oil (saturated liquid)

Order your results from low to high and discuss.

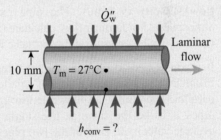

10.30 Oil flows through a 6-mm-diameter tube with a Reynolds number of 1200. The tube wall is maintained at 80°C, and the oil enters the tube at 25°C. Estimate the tube length required to heat the oil to 50°C.

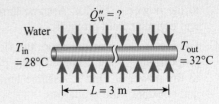

10.31 Water flows through a 3-mm-diameter tube with a Reynolds number of 1500. The water enters at 28°C. What heat flux is required at the tube wall to heat the water to 32°C in a 3-m length?

10.32 A heavy oil ($\mu = 1.5$ N·s/m²) flows in a 0.01-m-diameter pipe with an average velocity of 0.1 m/s. Determine (a) the pressure gradient and (b) the shear stress at the pipe wall. Show that the pressure drop balances the shear forces.

10.33 Find the frictional head loss (ft) for a 1 gal/min flow of water at 68 F through a 50-ft-long, 2-in-inside-diameter, smooth pipe.

10.34 Consider flow through a 6-ft-long, $1/2$-in-inside-diameter pipe having an allowable pressure loss of 15 in Hg. The flow is restricted to be laminar. Find the maximum rate of flow (gal/min) for engine oil at 300 K.

10.35 For the friction factors to be the same, what is the ratio of the volumetric flow rates of the following fluids flowing through a given smooth pipe: (a) water at 25°C and (b) air at 25°C and 1 atm?

10.36 Atmospheric-pressure air flows with a bulk mean temperature of 50°C through a 30-m-long, 60-mm-diameter tube at a mass flow rate of 0.5 kg/h. If the tube surface is maintained at 90°C, find the average surface heat-transfer coefficient and the pressure drop.

10.37 A fluid with a viscosity of 2×10^{-3} N·s/m² flows in a 25-mm-diameter, 20-m-long pipe. The average flow velocity is 1 m/s. The pipe exits to the atmosphere. For laminar flow calculate (a) the flow rate and (b) the pressure at the entrance to the pipe.

10.38 It is suggested to change the pipe described in Problem 10.37 and use a pair of smaller pipes in parallel. The combined flow rate through the two pipes is to be the same as the single pipe. The average velocity in each of the two pipes also is to be the same as the single pipe (i.e., 1 m/s).

A. Calculate the required pressure drop in these pipes for laminar flow.

B. If the pressure drop calculated in Problem 10.37 is applied to the pair of pipes, what is the average velocity and the flow rate in the two pipes?

10.39 Consider the fluid defined in Problem 10.37. The same flow rate as in Problem 10.37 must now be supplied through a 12.5-mm-diameter pipe.

A. For laminar flow calculate the pressure drop in this pipe and compare your result with that of Problem 10.37.

B. For a constant flow rate, the pressure drop is proportional to the diameter of the pipe raised to the power n. Find n. Assume laminar flow and constant fluid properties.

C. For a constant pressure drop, the flow rate is proportional to the diameter of the pipe raised to the power s. Find s. Assume laminar flow and constant fluid properties.

10.40 A fluid flows in a pipe of diameter D. The same fluid flows in the gap between two parallel, flat plates. The height of the gap is also D. The same pressure gradient exists in both systems. Calculate the ratio between the flow rate of the fluid in the

pipe and the flow rate between the plates per width equal to D. *Hint:* See Example 6.16 where $D = 2H$.

10.41 A gas turbine can operate using several different fuels. It is desirable to measure the density of the fuel while it is flowing in a tube to the combustion chamber. A scheme to do this is shown in the sketch. The fuel has viscosity μ and density ρ. Assuming fully developed, laminar flow, find expressions for the pressure drop between points 1 and 2 and between points 3 and 4. In particular, show that

$$(P_1 - P_2) + (P_4 - P_3) = 2\rho g L.$$

In other words, show that the density ρ can be obtained from pressure drop measurements, even though the viscosity and the velocity of the fluid are unknown. What assumptions are embedded in this result?

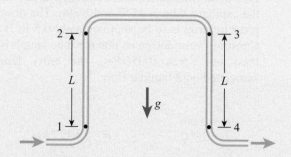

10.42 The tube shown in the sketch associated with Problem 10.41 has a diameter of 25 mm, and the lengths between points 1 and 2 and between points 3 and 4 are 2 m each. The pipe carries fuel with a viscosity of 2×10^{-3} kg/m·s and a density of 900 kg/m³ at an average velocity of 3 m/s. Assuming fully developed, laminar flow, find the pressure drops between points 1 and 2 and between points 3 and 4.

10.43 Oil flows by gravity through a 1-mile-long pipeline from a large reservoir to a lower reservoir. The difference in elevation between the reservoirs is 14 m. The discharge rate required is 1.6 m³/min. At 20°C, the oil viscosity and density are 0.9 N·s/m² and 950 kg/m³, respectively. Calculate the diameter of the steel pipe assuming fully developed, laminar flow.

10.44 Consider turbulent flow of water at 25°C through a 10-cm-diameter, horizontal, smooth pipe. The centerline velocity is 4 m/s.

    A. Use Eqs. 10.74 and 10.75 to determine the wall shear stress.

    B. Determine the pressure drop for a pipe length of 25 m.

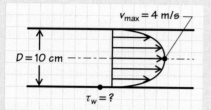

10.45 Consider the flow described in Problem 10.44. Plot the velocity distribution from centerline ($r = 0$) to the wall ($r = R$). Use linear (not logarithmic) coordinates. Superimpose on your graph the velocity distribution for a laminar flow having the same centerline velocity (see Eq. 10.41).

10.46 Consider a flow of water through a 10-cm-diameter, smooth pipe at a Reynolds number of 3.2×10^6. Plot and compare the velocity distributions predicted using the universal profile (Eq. 10.74) and the power-law approximation (Eq. 10.85). Note that you will be reproducing Fig. 10.14 for a higher Reynolds number.

10.47 Consider the flow of water at 25°C through a 30-mm-diameter, smooth tube. The tube is horizontal and 300 m long. Use the Blasius friction-factor relationship (Eq. 10.82) to develop expressions for (a) the head loss as a function of volume flow rate and (b) the pumping power requirement as a function of volume flow rate. Note that the pumping power is the equivalent power that must be supplied to a pump with the pressure difference $P_{out} - P_{in}$ equal to the pipe pressure drop. Plot the pumping power $\dot{W}_p$ versus the volume flow rate $\dot{V}$ for the range over which Eq. 10.83 applies ($4000 < Re_D < 10^5$). Discuss your results. Create two plots: one using SI units and one using units of horsepower and gallons per minute.

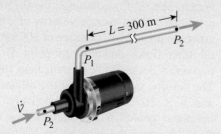

10.48 Consider a fully developed flow of water through a 10-mm-diameter, smooth, horizontal tube. The

flow rate is 0.054 kg/s and the water temperature is 300 K. Determine the following:

    A. The average velocity

    B. The Reynolds number

    C. The friction factor

    D. The pressure drop for a 50-m length of tube

    E. The pumping power in watts to maintain the flow

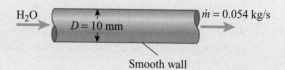

10.49 Repeat Parts C–E of Problem 10.48 for a cast-iron pipe rather than a smooth tube.

10.50 Water at 25°C is pumped from one reservoir to another through a 50-mm-diameter, 1000-m-long pipe. The flow rate is 3.4×10^{-3} m³/s, and the outlet reservoir is 10 m higher than the inlet reservoir. The surface of each reservoir is exposed to the atmosphere. Compare the pumping power requirements associated with using a plastic pipe versus a galvanized iron pipe. Neglect minor losses. Also compare the operational costs for a week of continuous operation if the pump efficiency is 75% and electricity is purchased for 8.2 cents/kW·hr.

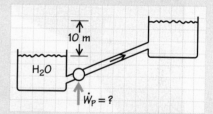

10.51 Air at atmospheric pressure flows at a volume flow rate of 1 m³/s through a 20-cm-diameter air-conditioning duct. The air and duct wall are both at 18°C. Determine the friction factor for this flow.

10.52 Superheated steam at 4 MPa and 500°C flows at 1.25 kg/s through an 8-in, schedule-80 pipe (with an inside diameter of 193.7 mm). The pipe surface temperature is 450°C. Find the pipe friction factor.

10.53 Water in a circulating chilled-water air-conditioning system enters the chilling unit at 15°C and leaves at 5°C. The tubes in the chilling unit are 3.8 cm in diameter, the surface temperature is 2°C, and the average water velocity is 0.6 m/s. Estimate the friction factor.

10.54 Air at 40 atm flows at an average velocity of 6 m/s through a 2.5-cm-diameter tube at a bulk mean temperature of 250°C. The tube surface temperature is 80°C. Calculate the friction factor.

10.55 Consider fully developed flow in a circular tube. The average flow velocity is 6 m/s, and the tube wall temperature is 150°C. Find the friction factor for the following cases:

A. Air at 90°C and 1 atm with $D = 2.5$ cm

B. Air at 90°C and 1 atm with $D = 1.25$ cm

C. Air at 90°C and 10 atm with $D = 2.5$ cm

D. Water at 90°C and 10 atm with $D = 2.5$ cm

E. Engine oil at 90°C with $D = 2.5$ cm

10.56 Water ($\rho = 1000$ kg/m³, $\nu = 10^{-6}$ m²/s) flows with an average velocity of 10 m/s in a 10-cm-diameter, smooth pipe.

A. Estimate the thickness of the laminar sublayer.

B. At what distance from the wall does the turbulent core begin?

C. Write the equations for the velocity distribution in the pipe.

D. What are the velocities at the edge of the laminar sublayer, at the beginning of the turbulent core, and at the pipe centerline?

10.57 Use the universal velocity distribution ("law of the wall") to calculate the local velocity at a point situated midway between the pipe axis and the pipe wall in a 0.5-m-diameter, smooth pipe carrying air at an average velocity of 30 m/s and a pressure of 200 kPa gage. The air temperature is 300 K and the barometric pressure is 100 kPa. Hint: Use Eq. 10.84.

10.58 Experiments show that turbulent flow in rough pipes is not affected by the roughness elements as long as the latter do not protrude outside the laminar sublayer. Mathematically, the pipe walls are said to be hydraulically smooth for

$$\varepsilon^+ = \frac{\varepsilon v_*}{\nu} \le 5,$$

where ε is the height of the roughness elements. For turbulent flow in a 5-cm-diameter pipe at a Reynolds number $Re_D = 50,000$, find the maximum height of roughness elements to consider the pipe hydraulically smooth.

10.59 Air flows at 0.3 lb$_m$/s, 92 F, and 15 psia through a 24-ft-long, 3-in-inside-diameter, smooth pipe. Determine the pressure drop attributable to friction (psi and in H$_2$O).

10.60 Calculate the pressure loss for the following conditions, assuming flow is fully developed in both cases:

A. A 45-gal/min flow of water at 10°C through a 6-m-long, smooth, 50-mm-inside-diameter pipe

B. A 0.9-kg/s flow of air at 298 K and 1.5 in Hg (gage) through a 2.5-m-long, smooth, 76.2-mm-inside-diameter pipe

10.61 Find the ratio of the pumping power required for a 6.1-gal/min flow of water at 60 F through equal lengths of (a) 1-in-inside-diameter galvanized-iron pipe and (b) ½-in-inside-diameter commercial steel pipe.

10.62 Both laminar and turbulent flow are possible in a smooth pipe at a Reynolds number of 3500 under proper conditions. Consider water at 20°C flowing at this Reynolds number through a 12.5-m-long, smooth, 25-mm-inside-diameter tube.

A. Find the ratios of the average velocity, pressure loss, and wall shear stress in turbulent flow to those in laminar flow.

B. Find the pressure loss for turbulent flow.

C. Find the velocity at the pipe centerline for laminar flow.

10.63 A major pipeline carrying crude oil drops 600 m in elevation over the last 5 miles, as shown in the sketch. The oil flows through the 0.75-m-diameter main line with an average velocity of 2 m/s. The oil arrives at point A at a pressure of 300 kPa gage. The oil then flows downhill to point B where the pressure is desired to be no more than 100 kPa gage. The density and viscosity of the oil are 860 kg/m³ and 5×10^{-3} N·s/m², respectively. What pipe diameter would you recommend for the section A–B? Assume the pipe to be hydrodynamically smooth.

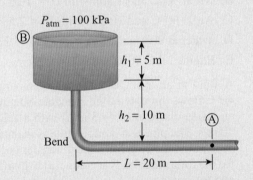

10.64 Oil flows by gravity through a 1-mile-long pipeline from a large reservoir to a lower reservoir. The difference in elevation of the two reservoirs is 14 m. The required flow rate is 1.6 m^3/min. The viscosity and density of the oil are 3×10^{-3} N·s/m^2 and 950 kg/m^3, respectively. Calculate the pipe diameter required to achieve this flow rate. Assume a hydraulically smooth pipe.

10.65 Water flows from a tank through a bend and then into the 20-m-long, 100-mm-diameter, smooth pipe, as shown in the sketch. The top of the tank is open to the atmosphere (P_{atm} = 100 kPa). The bend in the pipe presents a resistance to the flow equivalent to a 2-m length of pipe. The pressure at point A is measured to be 160 kPa absolute. Assuming the water has a density of 998 kg/m^3 and viscosity of 0.001 N·s/m^2, find the volumetric flow rate in the pipe.

27°C. Estimate the pressure drop per meter of horizontal pipe at the maximum flow rate.

10.68 A 20-cm-diameter commercial steel pipe carries water from a reservoir and discharges it as a free jet as shown in the sketch. Estimate the flow rate assuming a roughness height $\varepsilon = 4.6 \times 10^{-5}$ m.

10.69 Determine the Prandtl numbers for the following fluids at 325 K and 1 atm (except as noted) and list in order from lowest to highest:

A. Water

B. Air

C. Refrigerant 134a (saturated liquid)

D. Ethylene glycol (saturated liquid)

E. Mercury (saturated liquid)

F. Helium

G. Engine oil (saturated liquid)

10.70 Determine the thermal conductivities for the fluids in problem 10.69.

10.71 Consider the heating of a fluid as it flows through a smooth, 25-mm-diameter tube with an average velocity of 2.8 m/s. A uniform wall temperature is maintained. Determine the local heat-transfer coefficients associated with the following fluids at a bulk mean temperature of 325 K and 1 atm (except as noted):

A. Water

B. Air

C. Refrigerant 134a (saturated liquid)

D. Ethylene glycol (saturated liquid)

E. Mercury (saturated liquid)

F. Helium

G. Engine oil (saturated liquid)

Compare and discuss your results.

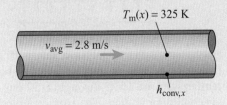

10.66 Water at 25°C flows in a smooth, 25-mm-diameter, 100-m-long pipe with an average velocity of 5 m/s. The pipe discharges to the atmosphere. Neglecting any entrance effects, estimate the pressure at the pipe entrance.

10.67 A 15-cm-diameter cast-iron pipe was installed in a water system 15 years ago. Experience indicates that the wall roughness of this type of pipe increases with time according to the relationship

$$\varepsilon = \varepsilon_{new}(1 + 0.25t).$$

Here ε_{new} (= 25×10^{-5} m) is the wall roughness of a new pipe, and ε is the roughness after t years. A new pump is installed in the system. This pump delivers a maximum flow of 0.03 m^3/s of water at

10.72 Nitrogen is heated as it flows through a 10-mm-diameter, 5-m-long, smooth, horizontal tube in which a uniform wall heat flux of 65 W/m^2 is maintained. The nitrogen enters at 300 K and 2 atm with an average velocity of 5 m/s. Assume the flow is fully developed both hydrodynamically and thermally. Estimate the following quantities:

A. The temperature of the nitrogen at the outlet

B. The tube wall temperature at the inlet and the tube wall temperature at the outlet

C. The pressure drop from inlet to outlet

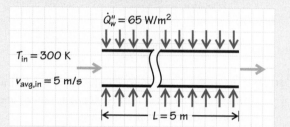

10.73 Consider the same situation described in Problem 10.72 except now, rather than imposing a constant heat flux, the tube wall temperature is maintained at 450 K. Estimate the following quantities:

A. The temperature of the nitrogen at the outlet

B. The total heat-transfer rate $\dot{Q}$ from the tube wall to the nitrogen

C. The tube wall heat flux at the inlet and the tube wall heat flux at the outlet

10.74 A solar collector for heating water consists of a 292.6-m length of black EPDM tubing. The inside diameter of the tube is 5.7 mm. Water is pumped through the tube at a volumetric flow rate of 3.9×10^{-5} m^3/s. Thermal conditions inside the collector approximate those of the application of a uniform heat flux of 725 W/m^2 to the tube wall. Water enters the collector tube at 290 K.

A. Determine the bulk mean temperature of the water at the tube exit.

B. Determine the maximum tube wall temperature.

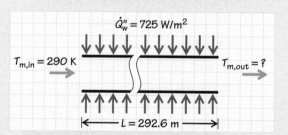

10.75 Air at atmospheric pressure flows at a volume flow rate of 1 m^3/s with an average bulk mean temperature of 15°C through a 20-cm-diameter air-conditioning duct. The surface temperature of the duct is 30°C. Determine the average convective heat-transfer coefficient at the duct wall.

10.76 Water enters a 2-in, schedule-40 pipe (with an inside diameter of 2.067 in) at 80 F and leaves

at 120 F. The flow rate is 50,000 lb$_m$/hr. The pipe surface is maintained at 140 F. Find the convective heat-transfer coefficient based on (a) the entrance conditions, (b) the exit conditions, and (c) the average of the entrance and exit conditions.

10.77 Superheated steam at 4 MPa and 500°C flows at 1.25 kg/s through an 8-in, schedule-80 pipe (with an inside diameter of 193.7 mm). The pipe surface temperature is 450°C. Find the convective heat-transfer coefficient.

10.78 Water in a circulating chilled-water air-conditioning system enters the chilling unit at 15°C and leaves at 5°C. The tubes in the chilling unit are 3.8 cm in diameter, the surface temperature is 2°C, and the average water velocity is 0.6 m/s. Estimate the convective heat-transfer coefficient.

10.79 Air at 40 atm flows at an average velocity of 6 m/s through a 2.5-cm-diameter tube at a bulk mean temperature of 250°C. The tube surface temperature is 80°C. Calculate the convective heat-transfer coefficient.

10.80 Consider fully developed flow in a circular tube. The average flow velocity is 6 m/s, and the tube wall temperature is 150°C. Find the convective heat-transfer coefficient for the following cases:

A. Air at 90°C and 1 atm with $D = 2.5$ cm

B. Air at 90°C and 1 atm with $D = 1.25$ cm

C. Air at 90°C and 10 atm with $D = 2.5$ cm

D. Water at 90°C and 10 atm with $D = 2.5$ cm

E. Engine oil at 90°C with $D = 2.5$ cm

10.81 A fluid at an average bulk mean temperature of 40°C flows through a 2-in, schedule-80 pipe (with an inside diameter of 49.3 mm) with a velocity of 10 m/s. The pipe wall temperature is maintained at 50°C. Find the convective heat-transfer coefficient for the following fluids at 1 atm: (a) helium, (b) air, and (c) water.

10.82 Water enters a 5-cm-inside-diameter pipe at 20°C and exits at 50°C. The water flows with a mean velocity of 3.5 m/s, and the pipe surface is maintained at 60°C. Find the convective heat-transfer coefficient at the pipe surface for (a) the entrance conditions, (b) the exit conditions, and (c) the average of the entrance and exit conditions.

10.83 Natural gas (methane) at 100 psia and 90 F flows through a 3/4-in-diameter pipe with an average velocity of 10 ft/s. The pipe wall temperature is 80 F. Estimate the convective heat-transfer coefficient.

10.84 Water flowing at the rate of 2.6 kg/s is heated from 7°C to 20°C as it flows through a smooth, 2-in (nominal), schedule-40 pipe (with an inside diameter of 52.5 mm). The inside pipe wall temperature is 90°C. Determine the required pipe length and the pressure drop.

10.85 Water flowing at 0.75 kg/s is to be cooled from 50°C to 20°C. Which of the following conditions offers the least pressure loss: (a) flow through a 1.25-cm-inside-diameter pipe with a wall temperature of 10°C or (b) flow through a 2.5-cm-inside-diameter pipe with a wall temperature of 15°C?

10.86 Superheated steam at 6 MPa and 400°C enters a 2.5-cm-diameter pipe with a velocity of 12 m/s. The pipe wall is maintained at 500°C. How long must the pipe be to heat the steam to 430°C? Also determine the pressure drop of the steam as it flows through the pipe.

10.87 Water at 10°C enters a 5-cm-diameter pipe flowing at a mass flow rate of 2.5 kg/s. The pipe wall is maintained at 80°C. The pipe is 5 m long. What is the outlet temperature of the water? Also determine the pressure drop through the pipe.

10.88 Superheated steam flows at 2 kg/s through a 10-m-long, 8-cm-diameter pipe. The steam enters at 4 MPa and 450°C. The pipe surface temperature is maintained at 350°C. Find the outlet temperature of the steam and the pressure drop through the pipe.

10.89 Water flowing at 1.2 kg/s is to be heated from 10°C to 30°C by passing it through a 3.5-m-long, 4-cm-diameter pipe. Estimate the required surface temperature of the pipe.

10.90 Water at 60 F enters a 1-in-diameter, 10-ft long tube. If the water flow rate is 2 lb_m/s and the tube wall is maintained at 200 F, determine the exit temperature of the water and the pressure drop.

10.91 Superheated steam flows at 7 kg/s through a 4-in-inside-diameter pipe. The steam enters the pipe at 400 psia with a bulk mean temperature of 800 F. The wall temperature of the 30-ft-long pipe is maintained at 700 F. Neglecting the pressure drop in the pipe, find (a) the outlet temperature of the steam and (b) the temperature of the steam at a point halfway along the pipe (i.e., at 15 ft from the entrance). Justify neglecting the pressure drop along the length of the pipe.

10.92 Water at 95°C flows through a copper tube (22.9-mm i.d. and 24.5-mm o.d.) with a velocity of 0.6 m/s. The outside surface of the tube is exposed to air at 20°C with a convective heat-transfer coefficient of 8.5 W/m² · °C. Determine the heat loss from a 1-m length of the tube.

10.93 Superheated steam at 400 psia and 600 F flows with a velocity of 20 ft/s through a 6-in, schedule-40, steel pipe (6.065-in i.d. and 6.625-in o.d.). The outer surface of the pipe is exposed to a stream of air (1 atm and 70°C) flowing normal to the pipe at 40 ft/s. Estimate the heat loss rate per unit length of pipe (Btu/hr · m).

10.95 Estimate the hydrodynamic entry lengths for the flows described in Problems 10.22 and 10.48.

10.96 Water enters a 4-mm-diameter, 350-mm-long tube with an average velocity of 0.32 m/s at 27°C and nominally 1 atm. The entrance to the tube is shaped such that the velocity profile at the tube entrance is essentially uniform. Estimate the pressure difference between the inlet and exit. Note that the flow is laminar.

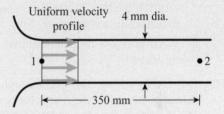

10.97 Consider the small-bore capillary-tube flowmeters discussed in Example 10.4. What is the percentage error in the mass flow rate measurement that results from neglecting entrance effects?

10.98 Allowing for entrance effects, what is the total head loss associated with the flow described in Example 10.6?

10.99 Consider the fluids and flow conditions presented in Example 10.7. Calculate the hydrodynamic and thermal entry lengths associated with each fluid at 300 and 400 K when the mean velocity is 0.172 m/s. Present your results in a table similar to that given in the example. Discuss.

10.100 Consider the flow of water through a 3-mm-diameter tube. The water enters the tube with a velocity of 0.2 m/s. Determine the hydrodynamic and thermal entry lengths for water temperatures of 300, 320, 340, and 360 K at a pressure of 1 atm. Plot your results as functions of temperature.

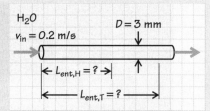

10.101 Repeat Problem 10.100 for engine oil. Assume the oil is a saturated liquid.

10.102 Determine the hydrodynamic entry length for the conditions of Example 10.8. Is neglecting entrance effects a good assumption for this flow?

10.103 Oil flows through a 10-cm-diameter, 8-m-long, horizontal tube. The oil enters with a uniform temperature of 25°C and a uniform velocity of 3 m/s. The tube wall temperature is 75°C.

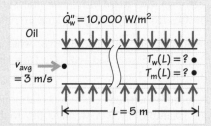

A. Determine the temperature of the oil at the tube outlet. Also determine the total heat-transfer rate to the oil.

B. What percentage error in both quantities results if the flow is assumed to be both hydro-dynamically and thermally fully developed?

C. What pressure drop is required to maintain this flow?

10.104 Oil is heated by flowing through a 12-mm-diameter, 5-m-long tube. The velocity of the oil is 3 m/s. A uniform heat flux of 10,000 W/m² is applied to the tube wall. Determine the temperature of the oil and the temperature of the tube wall at the outlet.

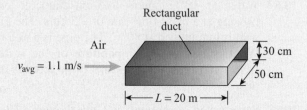

10.105 Engine oil flows with an average velocity of 0.3 m/s and an average bulk mean temperature of 160°C in a 23-mm-diameter tube. The tube wall is maintained at 140°C. Find the average convective heat-transfer coefficient and the pressure drop for (a) a 2-m-long tube and (b) a 6-m-long tube.

10.106 The cooling oil used in a transformer has properties closely resembling those of glycerin. Such an oil flows at a mean velocity of 0.5 m/s through a 5-cm-diameter, 2-m-long tube. If the oil bulk mean temperature is 50°C and the tube wall is 30°C, find the heat-transfer rate from the oil and the pressure drop through the tube.

10.107 Ventilating air for a warehouse flows through a 30-cm by 50-cm rectangular duct with an average velocity of 1.1 m/s. The air is at 20°C and nominally 1 atm. Estimate the pressure drop for a 20-m length of duct. Neglect entrance effects.

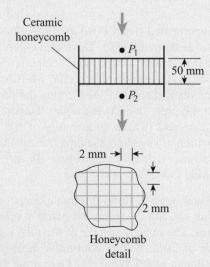

10.108 A spark-ignition engine is tested in a laboratory. To measure the air flow rate, a laminar-flow flow-meter is employed. This flowmeter consists of an upstream plenum at pressure P_1, followed by a honeycomb ceramic element that consists of 1600 square cross-sectional passages. The sides of the square passages are 2 mm. The ceramic element is 50-mm long. After passing through the honey-comb, the flow enters into another plenum in which the pressure is P_2. The temperature and pressure in the inlet plenum are 300 K and 98 kPa, respectively. Determine the volumetric air flow rate supplied to the engine when the pressure drop $P_1 - P_2$ measured by a manometer is 15 cm of water.

10.109 Air at 1 atm and 20°C enters a 2-m-long, rectan-gular duct with a cross section of 8 cm by 16 cm.

The duct wall temperature is 125°C and the mass flow rate is 0.1 kg/s. Find the heat-transfer rate to the air and the air outlet temperature.

10.110 Water flows through a 0.5-cm by 1-cm rectangular duct with a bulk mean temperature of 20°C. The duct wall temperature is 60°C. Determine the heat-transfer rate to the water per unit length.

10.111 Air at 1 atm and 30°C flows with a velocity of 5 m/s through a square (20-cm by 20-cm) air-conditioning duct. The duct wall temperature is 20°C. Estimate the convective heat-transfer coefficient.

10.112 The cross section of a duct is an equilateral triangle with sides of length b. Calculate the duct hydraulic diameter (a) when running full and (b) when running with liquid to a depth of half the altitude. One side is horizontal in both cases.

10.113 Consider a fixed flow rate through a smooth-walled rectangular duct. For a given cross-sectional area, find the ratio of the length of the sides a and b that results in the minimum friction loss per unit length of duct. Assume laminar flow.

10.114 Assuming fully developed flow, calculate the pressure loss for a 1500-ft^3/min flow of air at 150 F and 29.2 in Hg absolute through a smooth-walled, 15-in by 8-in rectangular duct. The duct length is 6 ft.

10.115 Consider the situation described in Problem 10.65. The circular cross-section pipe is now replaced with a square cross-section pipe of the same length. Find the dimensions of a square pipe that will provide the same volumetric flow rate as in Problem 10.65.

10.116 Water at 298 K flows through a smooth, square duct with sides $D = 0.1$ m. The average water velocity is 8 m/s. The duct is 100 m long. Calculate the power needed to pump the water. Compare your results with the power required to pump the same flow through a circular pipe.

10.117 Repeat Problem 10.66 for a square pipe having the same cross-sectional area as the 25-mm-diameter pipe.

10.118 Repeat Problem 10.66 for a rectangular pipe with sides in a ratio of 1:2. The rectangular pipe has the same cross-sectional area as the 25-mm-diameter pipe.

10.119 A plumber connects a 25-mm-diameter pipe (A) to a 25-mm-diameter hose (B) using a short, 13-mm-diameter connecter, as shown in the sketch. For a flow of water with an average velocity of 5 m/s in the 25-mm-diameter sections, estimate the extra pressure drop associated with the connector.

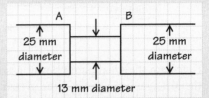

10.120 A jet of water exits from a 70-mm-diameter nozzle at a velocity of 0.335 m/s and enters an open tank as shown in the sketch. The exit of the nozzle is quite close to the top surface of the water in the tank. Water exits the tank through a 10-m length of 37-mm-diameter, galvanized iron pipe. The entrance to the pipe is sharp edged, and the pipe exits to the atmosphere. Determine the steady-state value of the height of the water above the level of the pipe (i.e., find H).

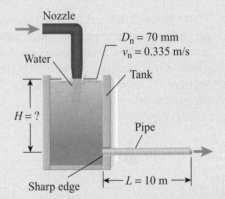

10.121 Water at 25°C is pumped through the system shown in the sketch. The flow rate is 4.3×10^{-3} m^3/s. The pipe is 2-in., schedule-40, galvanized iron (with an inside diameter of 52.5 mm). All connections are screwed (not flanged). The atmospheric pressure is 100 kPa. Determine (a) the outlet pressure of the pump and (b) the pumping power required if the water enters the pump at atmospheric pressure, and the pump efficiency is 65%. The valve head loss is nil.

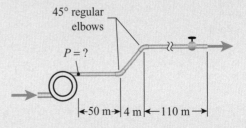

10.122 Water exits a large tank through a 20-m length of 100-mm-diameter steel pipe, followed by a 10-m length of 65-mm-diameter steel pipe. The pressure at the pipe exit is atmospheric, and the water temperature is 300 K. Estimate the mass flow rate of the water.

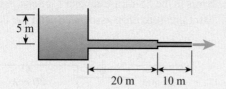

10.123 Calculate the ideal pump power associated with the arrangement shown in the sketch.

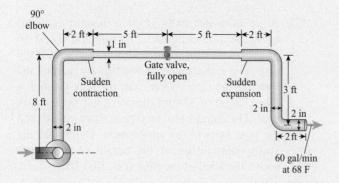

10.124 Water flows through the system shown. Neglect all friction losses except those occurring at section 2 (sudden contraction), section 3 (open globe valve), sections 5 and 6 (90° elbows), and section 7 (sudden enlargement). Find the pressure at section 8.

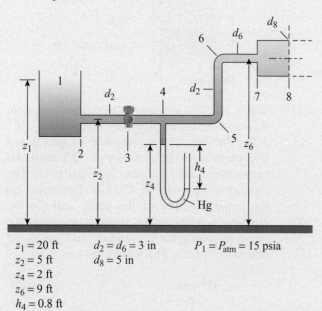

$z_1 = 20$ ft $\quad d_2 = d_6 = 3$ in $\quad P_1 = P_{atm} = 15$ psia
$z_2 = 5$ ft $\quad d_8 = 5$ in
$z_4 = 2$ ft
$z_6 = 9$ ft
$h_4 = 0.8$ ft

10.125 Water flows from a reservoir through a sharp-edged entry into a piping system as shown in the sketch. The pump maintains a flow rate of 50 gal/min. A new type of valve is being tested for pressure loss. The mercury-filled manometer indicates a reading of 6 in for the present flow rate. The piping is 1-in-inside-diameter commercial steel.

A. Find the pressure drop across the valve in feet of water and in psi.

B. Calculate the minor loss factor K_{valve} for the valve.

C. Calculate the pressure just upstream of the valve.

D. Find the total head loss through the system neglecting any losses associated with the pump.

E. Calculate the ideal pump power input.

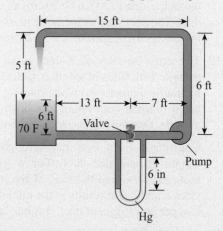

10.126 Water flows from a tank through a piping network and is discharged to the atmosphere as shown in the sketch. The 50-mm-diameter pipe used from A to E and from F to G has a total length of 60 m. The 25-mm-diameter pipe used in section E–F is 20 m long. All the pipes have a relative roughness $\varepsilon/D = 0.001$. The average velocity needed at G, when the globe valve D is fully open, is 2 m/s. Find the water level H needed to sustain such a flow of water.

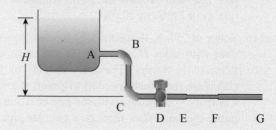

Appendix 10A
Simplifying the Governing Equations

We apply the following assumptions to simplify the governing differential equations of mass, momentum, and energy for flow through a circular tube:

1. The flow is steady state.
2. The fluid is incompressible and Newtonian.
3. All properties are constant.
4. The geometry is two dimensional (r, x).
5. The wall is impermeable.
6. The flow is fully developed.
7. Axial conduction is small compared to radial conduction.
8. Viscous dissipation is negligible.

10A.1 MASS CONSERVATION

For steady flow, Eq. T3.3f expresses mass conservation in cylindrical coordinates. This equation simplifies immediately to Eq. T3.3e as a result of assumption 4 ($v_\theta = 0$); thus, we start with

See Table 3.3 in Chapter 3.

$$\frac{1}{r}\frac{\partial}{\partial r}(\rho r v_r) + \frac{\partial}{\partial x}(\rho v_x) = 0.$$

With assumption 2, the density cancels; furthermore, the assumption of fully developed flow means that $\partial v_x/\partial x = 0$. Equation T3.3e then becomes

$$\frac{1}{r}\frac{\partial}{\partial r}(r v_r) = 0.$$

Separating and integrating this yields

$$r v_r = f(x),$$

where $f(x)$ is an unknown function of x. Since the wall is impermeable, the radial velocity v_r is zero at the wall for all x. From this, we conclude that $f(x)$ must be zero; hence,

$$v_r(r, x) = 0, \tag{10A.1}$$

that is, there is no radial velocity. This result leads to great simplification of the momentum and energy equations.

10A.2 MOMENTUM CONSERVATION

See Table 6.4 in Chapter 6.

We start with Eq. T6.4c, the x-component of momentum conservation expressed in cylindrical coordinates:

$$-\frac{\partial P}{\partial x} + \mu\left[\frac{1}{r}\frac{\partial}{\partial r}\left(r\frac{\partial v_x}{\partial r}\right) + \frac{1}{r^2}\frac{\partial^2 v_x}{\partial \theta^2} + \frac{\partial^2 v_x}{\partial x^2}\right] + \rho g_x$$

0   0

④   ⑥

$$= \rho\left(\frac{\partial v_x}{\partial t} + v_r\frac{\partial v_x}{\partial r} + \frac{v_\theta}{r}\frac{\partial v_x}{\partial \theta} + v_x\frac{\partial v_x}{\partial x}\right),$$

0   0   0   0

①   Eq. 10A.1   ④   ⑥

where the assumptions allowing each term to be eliminated are indicated by the circled numbers. Rewriting this equation, we obtain the simplified result that

$$-\frac{\partial P}{\partial x} + \mu\left[\frac{1}{r}\frac{\partial}{\partial r}\left(r\frac{\partial v_x}{\partial r}\right)\right] + \rho g_x = 0. \tag{10A.2}$$

10A.3 ENERGY CONSERVATION

See Table 5.7 in Chapter 5.

Energy conservation in cylindrical coordinates is given by Eq. T5.7b, which simplifies as follows:

$$\rho c_p\left(\frac{\partial T}{\partial t} + v_r\frac{\partial T}{\partial r} + \frac{v_\theta}{r}\frac{\partial T}{\partial \theta} + v_x\frac{\partial T}{\partial x}\right)$$

0   0   0

①   Eq. 10A.1   ④

$$= \frac{1}{r}\frac{\partial}{\partial r}\left(rk\frac{\partial T}{\partial r}\right) + \frac{1}{r^2}\frac{\partial}{\partial \theta}\left(k\frac{\partial T}{\partial \theta}\right) + \frac{\partial}{\partial x}\left(k\frac{\partial T}{\partial x}\right) + \Phi_{\text{visc}}.$$

0   0   0

④   ⑦   ⑧

Rewriting this equation yields

$$\rho c_p v_x\frac{\partial T}{\partial x} = \frac{1}{r}\frac{\partial}{\partial r}\left(rk\frac{\partial T}{\partial r}\right). \tag{10A.3}$$

With appropriate boundary conditions, Eqs. 10A.1, 10A.2, and 10A.3 can be solved for the velocity and temperature distributions, $v_x(r)$ and $T(r, x)$, respectively.

THERMAL–FLUID ANALYSIS OF STEADY–FLOW DEVICES

After studying Chapter 11, you should:

- *Appreciate the use of steady-flow devices in practical applications and have a basic understanding of their operation and the important energy inputs and outputs associated with these devices.*

- *Be able to simplify the general, steady-flow expression for energy conservation to the standard forms used to describe each steady-flow device discussed in this chapter and understand the assumptions used in these simplifications.*

- *Be able to apply the steady-flow forms of mass conservation, energy conservation, mechanical energy conservation, and linear momentum*

conservation, as appropriate, to steady-flow devices.

- *Be proficient at plotting on h–s, T–s, or other coordinates, the various thermodynamic processes associated with the various steady-flow devices.*

- *Be proficient at using the NIST database to obtain thermodynamic properties needed to analyze steady-flow devices, in particular, those involving fluids existing in both liquid and gas phases.*

- *Have an appreciation for the physical phenomena that result in loss of performance in various steady-flow devices and be able to calculate nonideal performances given isentropic efficiencies or other empirical information.*

- *Explain the meaning of choked flow, Mach number, normal shock, and stagnation conditions.*

- *Be able to explain the operation of a converging–diverging nozzle as a function of back pressure and be able to calculate the thermodynamic states and velocities through the nozzle.*

- *Be able to calculate heat-exchanger performance and to conduct preliminary design analyses of various types of heat exchangers.*

- *Be able to apply the concept of standardized enthalpies to analyze constant-pressure combustion processes.*

Chapter 11 Overview

In this chapter, we apply the basic conservation principles and other key concepts to analyze a number of important devices. Here we investigate typical components of more complex systems; these components include nozzles, diffusers, throttles, pumps, compressors, fans, turbines, heat exchangers, furnaces, and combustors. In Chapter 12, we combine these simple devices in more complex systems that include steam power plants, jet engines, other power and propulsion cycles, heat pumps, refrigeration cycles, and air conditioning and humidification systems.

This chapter should be used simultaneously with earlier chapters; it does not stand alone and does not require complete mastery of the preceding ten chapters. Appropriate entry points have been indicated in Chapters 5 and 7. The analysis of each device treated in this chapter follows the sequence of presenting mass conservation, energy conservation, and momentum conservation, where appropriate. If the reader has yet to study momentum conservation, that portion of the analysis can be ignored for the time being; however, mass and energy conservation principles are fundamental to every analysis presented here. The second law of thermodynamics is also applied when helpful. One of the purposes of this chapter is to provide an opportunity to apply the many theoretical tools developed throughout this text.

11.1 STEADY-FLOW DEVICES

In the following sections, we investigate thermal-fluid devices that are traditionally called steady-flow devices. As indicated by this designation, the underlying assumption in analyzing all of these devices is that steady state and steady flow prevail; hence, there are no time-dependent processes to consider. The particular steady-flow devices of interest are summarized in Table 11.1.

11.2 NOZZLES AND DIFFUSERS

A nozzle is a passive device having a flow area that varies in the flow direction such that the outlet velocity is higher than the inlet velocity. For subsonic flows, the area decreases in the flow direction (see Fig. 11.1). In practical applications, nozzles are purposefully used to produce a high-velocity fluid stream. Examples here include fire fighting with water jets, boring holes in relatively soft rock with high-velocity water jets, cleaning operations using solvent or water jets, drying operations using air jets, and the

Table 11.1 Steady-Flow Devices

Device	Typical Purpose(s)	Terms Usually Neglected in Energy Equation	Simplified Energy Conservation Expression*
Nozzle	Create high exit velocity	$\dot{Q}_{cv},\ \dot{W}_{cv},\ z_2 - z_1$	$h_2 - h_1 + \dfrac{\alpha_2 v_{avg,2}^2}{2} - \dfrac{\alpha_1 v_{avg,1}^2}{2} = 0$
Diffuser	Create high outlet pressure, reduce velocity	$\dot{Q}_{cv},\ \dot{W}_{cv},\ z_2 - z_1$	$h_2 - h_1 + \dfrac{\alpha_2 v_{avg,2}^2}{2} - \dfrac{\alpha_1 v_{avg,1}^2}{2} = 0$
Throttle	Reduce pressure, control flow	$\dot{Q}_{cv},\ \dot{W}_{cv},\ \dfrac{v_{avg,2}^2}{2} - \alpha_1 \dfrac{v_{avg,1}^2}{2},\ z_2 - z_1$	$h_2 - h_1 = 0$
Pump	Create flow of a liquid, increase pressure	$\dot{Q}_{cv},\ \alpha_2 \dfrac{v_{avg,2}^2}{2} - \alpha_1 \dfrac{v_{avg,1}^2}{2},\ z_2 - z_1$	$\dot{W}_{in,pump} = \dot{m}(h_2 - h_1)$
Compressor	Create flow of a gas, increase pressure	$\dot{Q}_{cv},\ \alpha_2 \dfrac{v_{avg,2}^2}{2} - \alpha_1 \dfrac{v_{avg,1}^2}{2},\ z_2 - z_1$	$\dot{W}_{in,comp} = \dot{m}(h_2 - h_1)$
Fan[†]	Create flow of a gas with minimal pressure change	$\dot{Q}_{cv},\ z_2 - z_1$, other terms depending on control volume choice	$\dot{W}_{in,fan} = \dot{m}\left(h_2 - h_1 + \alpha_2 \dfrac{v_{avg,2}^2}{2} - \alpha_1 \dfrac{v_{avg,1}^2}{2} \right)$
Turbine	Produce power by expanding a fluid	$\dot{Q}_{cv},\ \alpha_2 \dfrac{v_{avg,2}^2}{2} - \alpha_1 \dfrac{v_{avg,1}^2}{2},\ z_2 - z_1$	$-\dot{W}_{out,turbine} = \dot{m}(h_2 - h_1)$
Heat exchanger	Transfer energy from one fluid to another	$\dot{W}_{cv}$, all kinetic and potential energy terms	$\displaystyle\sum_{Inlets} \dot{m}_i h_i = \sum_{Outlets} \dot{m}_i h_i$
Furnace or boiler	Heat a load by burning a fuel	$\dot{W}_{cv}$, all kinetic and potential energy terms	$\dot{Q}_{cv,out} = \dot{m}_F h_{F,in} + \dot{m}_A h_{A,in} - \dot{m}_P h_{P,out}$
Combustor	Create a hot gas stream by burning a fuel	$\dot{Q}_{cv},\ \dot{W}_{cv}$, all kinetic and potential energy terms	$\dot{m}_P h_{P,out} = \dot{m}_F h_{F,in} + \dot{m}_A h_{A,in}$

*The usual starting point for simplification is Eq. 11.4: $\dot{Q}_{cv,net\ in} - \dot{W}_{cv,net\ out} = \dot{m}(h_2 - h_1 + \alpha_2 v_{avg,2}^2/2 - \alpha_1 v_{avg,1}^2/2 + z_2 - z_1)$, where 1 and 2 designate inlet and outlet conditions, respectively. For devices with more than a single inlet and/or outlet stream, Eq. 5.65 is the starting point.

[†]There are several definitions of fans, depending upon the application. In industrial applications, the distinction between a fan and a compressor is that, for a fan, the density increase is less than 7% [1]; in propulsion applications, fans are specialized compressors with pressure ratios of less than 1.8 [2].

Nozzles are used in many applications. Here nozzles are used in a leaf blower (top right), a fireboat (left), and a concrete sprayer (bottom right).

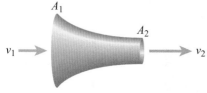

FIGURE 11.1

In a subsonic nozzle, the flow area decreases in the flow direction resulting in a high-velocity exit stream.

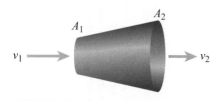

FIGURE 11.2

In a subsonic diffuser, the flow area increases in the flow direction resulting in a higher pressure and lower velocity at the exit than at the inlet.

use of liquid sprays in multitudinous applications. The thrust produced by rocket and jet engines depends critically on the high velocity generated by their exit nozzles.

A diffuser is a passive device that exchanges fluid kinetic energy for outlet flow work. Diffusers are used in applications where either high pressures or low velocities, or both, are desired. For subsonic flow through a diffuser, the flow area increases in the flow direction (Fig. 11.2). For such a device, the inlet velocity is greater than the outlet velocity, whereas the outlet pressure is greater than the inlet pressure. Applications of diffusers abound as the following examples suggest: Diffusers are frequently used at the end of wind-tunnel test sections to slow the flow and, hence, decrease pumping requirements (Fig. 11.3a). Diffusers are also used in pumps and compressors to create a high outlet pressure. Figure 11.3b shows a curved diffuser section in a radial flow pump. A diffuser section is used before the combustor in turbojet engines (Fig. 11.3c) and diffusers have been designed to improve the performance of wind turbines (Fig. 11.3d).

11.2a General Analysis

We start with a general analysis of nozzles and diffusers that applies to both incompressible and compressible flows. Because the character of incompressible and compressible flows can be quite different, separate sections dealing with each follow.

We now apply mass, energy, and momentum conservation to understand how geometric variables (inlet and outlet flow areas) affect thermodynamic and flow properties.

Mass Conservation

Consider the integral control volume shown in Fig. 11.4. Fluid enters the nozzle or diffuser at station 1 on the left and exits at station 2 on the right.

FIGURE 11.3

Various applications of diffusers include (a) decreasing the velocity after a wind-tunnel test section, (b) increasing the pressure at a pump outlet, (c) decreasing the velocity at the combustor inlet in a turbojet engine, and (d) improving the performance of a wind turbine (Vortec DAWT prototype). Drawings courtesy of Pratt & Whitney (c) and Vortec Energy, Ltd., New Zealand (d).

We assume steady state and steady flow. For this control volume, mass conservation expressed by Eq. 3.18a applies, so

$$\dot{m}_{\text{in}} = \dot{m}_{\text{out}}, \tag{11.1a}$$

or

$$\dot{m}_1 = \dot{m}_2. \tag{11.1b}$$

FIGURE 11.4

Control volume for nozzle illustrating (a) mass flows and (b) energy flows. Steady flow and steady state are assumed.

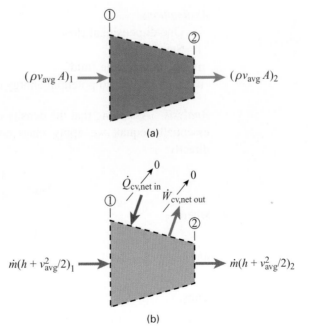

Since we are interested in the relationship between the flow velocity and flow area, we apply the definition of a flow rate (Eq. 3.15):

$$\rho_1 v_{avg,1} A_1 = \rho_2 v_{avg,2} A_2, \tag{11.2}$$

where v_{avg} is the velocity averaged over the flow cross-sectional area A (see Eq. 3.13). Solving for the exit velocity yields

$$v_{avg,2} = \frac{\rho_1 A_1}{\rho_2 A_2} v_{avg,1}. \tag{11.3}$$

From Eq. 11.3, we see that for an incompressible flow (i.e., $\rho_1 = \rho_2 = \rho$) the exit velocity equals the inlet velocity multiplied by the area ratio A_1/A_2.

Example 11.1

Water at 300 K enters a circular cross-section nozzle at an average velocity of 2 m/s. The inlet diameter is 25 mm and the exit diameter is 10 mm. Determine the exit velocity of the water.

Solution

Known $T, v_{avg,1}, D_1, D_2$

Find $v_{avg,2}$

Sketch

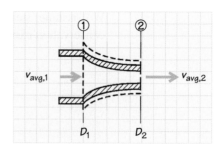

Assumptions
  i. One-dimensional flow
  ii. Steady flow
 iii. Incompressible fluid
  iv. $z_1 = z_2$ [i.e., no potential energy change]

Analysis   Assuming that the density of the water at the inlet and exit are essentially equal, we apply mass conservation expressed by Eq. 11.3 directly:

$$v_{avg,2} = \frac{A_1}{A_2} v_{avg,1},$$

where

$$A_1 = \pi D_1^2/4$$

and

$$A_2 = \pi D_2^2/4.$$

Thus,

$$v_{avg,2} = \frac{D_1^2}{D_2^2} v_{avg,1}$$

$$= \frac{(0.025)^2}{(0.010)^2} \, 2 = 12.5$$

$$[=] \frac{m^2}{m^2} \frac{m}{s} = \frac{m}{s}.$$

Comments   Since we assumed incompressible flow, no knowledge of the density at either the inlet or the exit was required to find the exit velocity.

Self Test
11.1
✓ **Calculate the volumetric flow rate at the inlet and exit of the nozzle in Example 11.1.**
(Answer: $9.817 \times 10^{-4} \, m^3/s$, $9.817 \times 10^{-4} \, m^3/s$)

Energy Conservation

Consider again the integral control volume of Fig. 11.4. For this situation of steady state and steady flow with a single inlet and a single outlet, conservation of energy expressed by the first law of thermodynamics is given by Eq. 5.63e, that is,

$$\dot{Q}_{cv,net\ in} - \dot{W}_{cv,net\ out}$$
$$= \dot{m}\left[(h_2 - h_1) + \frac{1}{2}(\alpha_2 v^2_{avg,2} - \alpha_1 v^2_{avg,1}) + g(z_2 - z_1) \right], \quad (11.4)$$

where α is the kinetic energy correction factor (Eq. 5.64) and z is the elevation. We simplify Eq. 11.4 with the following assumptions:

- The heat interaction across the control surface is zero (adiabatic) or small compared to other flows of energy (i.e., $\dot{Q}_{cv,net\ in} = 0$).
- The potential energy change is zero (horizontal nozzle) or small compared to other flows of energy (i.e., $z_2 - z_1 = 0$).
- There are no work interactions other than flow work (i.e., $\dot{W}_{cv,net\ out} = 0$).

Applying these assumptions to Eq. 11.4 yields

$$0 - 0 = \dot{m}\left[(h_2 - h_1) + \frac{1}{2}(\alpha_1 v^2_{avg,2} - \alpha_2 v^2_{avg,1}) + 0 \right],$$

Table 5.5 in Chapter 5 presents kinetic energy correction factors for turbulent flows.

or

$$(h_2 - h_1) + \frac{1}{2}(\alpha_2 v^2_{avg,2} - \alpha_1 v^2_{avg,1}) = 0. \quad (11.5)$$

Example 11.2

Estimate the pressure drop $(P_1 - P_2)$ for the nozzle and flow conditions given in Example 11.1. Assume that the flow is isothermal and that the kinetic energy correction factors are essentially unity. Neglect heat transfer.

Solution

Known v_1, v_2, T

Find $P_1 - P_2$

Sketch

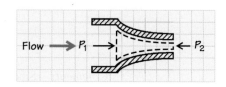

Assumptions

   i. One-dimensional flow
  ii. Steady flow
 iii. Incompressible fluid
 iv. $z_1 = z_2$ [i.e., no potential energy change]

Analysis   Although the pressure does not appear explicitly in Eq. 11.5, we recognize that it is buried in the enthalpy since $h \equiv u + P/\rho$. We again assume that the water is incompressible ($\rho_1 = \rho_2 = \rho$) and, furthermore, that $c_p = c_v = c$. Thus, we rewrite Eq. 11.5 as

$$c(T_2 - T_1) + \frac{1}{\rho}(P_2 - P_1) + \frac{1}{2}(v_{avg,2}^2 - v_{avg,1}^2) = 0.$$

Since the flow is assumed to be isothermal, $T_2 - T_1 = 0$. Solving our previous equation for the pressure drop yields

$$P_1 - P_2 = \frac{\rho}{2}(v_{avg,2}^2 - v_{avg,1}^2).$$

Approximating the density using the saturated-liquid value at 300 K (996.5 kg/m³), we substitute numerical values as follows:

$$P_1 - P_2 = \frac{996.5}{2}\left[(12.5)^2 - (2)^2\right] = 75,860$$

$$[=] \frac{\text{kg}}{\text{m}^3}\frac{\text{m}^2}{\text{s}^2}\left[\frac{1\,\text{N}}{\text{kg}\cdot\text{m/s}^2}\right] = \frac{\text{N}}{\text{m}^2}\ \text{or Pa}.$$

Comments   This is a relatively large pressure drop compared to, say, the pressure losses resulting from friction in a 100-m run of pipe. How does this result change if the kinetic energy correction factor exceeds unity (see Table 5.5)?

Example 11.3

Steam
0.13 MPa
600 K

$P_2 = 0.1$ MPa

Superheated steam enters a nozzle at 0.13 MPa and 600 K and exits at 0.1 MPa. Heat interactions and frictional effects are both negligible. The inlet and outlet diameters of the nozzle are 20 and 10 mm, respectively. Determine the mass flow rate of the steam through the nozzle. Also find the inlet and outlet velocities.

Solution

Known T_1, P_1, P_2, D_1, D_2

Find $\dot{m}, v_1, v_2$

Sketch

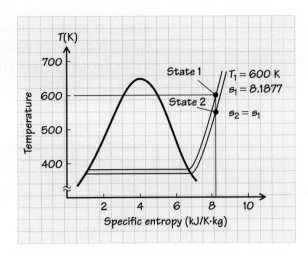

Assumptions

i. The flow is adiabatic.
ii. The flow is frictionless.
iii. Local thermodynamic equilibrium prevails through the nozzle.
iv. The velocity profiles at the inlet and outlet are uniform ($\alpha_1 = \alpha_2 = 1$).
v. $z_1 = z_2$ [i.e., no potential energy change].

Analysis Since neither the inlet nor outlet velocity is given, mass conservation alone will not allow us to find the mass flow rate. If we are able to determine the thermodynamic properties at the exit (state 2), then mass conservation (Eq. 11.2) and energy conservation (Eq. 11.5) can be combined to find the flow rate as follows: From Eq. 11.2,

$$v_2 = \frac{\dot{m}}{\rho_2 A_2} \quad \text{and} \quad v_1 = \frac{\dot{m}}{\rho_1 A_1}.$$

Rearranging the energy conservation equation (Eq. 11.5) and substituting these expressions yield

$$h_1 - h_2 = \frac{1}{2}\left(v_2^2 - v_1^2\right)$$

$$= \frac{\dot{m}^2}{2}\left[\frac{1}{(\rho_2 A_2)^2} - \frac{1}{(\rho_1 A_1)^2}\right].$$

Solving for $\dot{m}$, we obtain

$$\dot{m} = \left[\frac{2(h_1 - h_2)}{\dfrac{1}{(\rho_2 A_2)^2} - \dfrac{1}{(\rho_1 A_1)^2}}\right]^{1/2}.$$

To evaluate this expression, we require the properties h_1, h_2, ρ_1, and ρ_2. Since T_1 and P_1 are given, the thermodynamic state 1 is fully defined. We use the NIST database to find h_1 and ρ_1 given T_1 and P_1:

$$h_1 = 3128.1 \text{ kJ/kg,}$$
$$\rho_1 = 0.47070 \text{ kg/m}^3.$$

To define state 2, we assume that the flow process with negligible heat transfer and negligible friction approximates an adiabatic, reversible process. With this assumption, the process is isentropic and $s_2 = s_1$. This process is shown as a vertical line on the T–s diagram in the sketch. Having a value for s_1 ($= 8.1877$ kJ/kg$\cdot$K) defines state 2:

$$P_2 = 0.1 \text{ MPa},$$
$$s_2 = 8.1877 \text{ kJ/kg} \cdot \text{K}.$$

We again use the NIST database to find h_2 and ρ_2 given P_2 and s_2:

$$h_2 = 3057.7 \text{ kJ/kg} \cdot \text{K},$$
$$\rho_2 = 0.38461 \text{ kg/m}^3.$$

The inlet and exit areas are calculated from their respective diameters:

$$A_1 = \frac{\pi D_1^2}{4} = \frac{\pi (.020)^2}{4} \text{ m}^2 = 3.1416 \times 10^{-4} \text{ m}^2,$$

$$A_2 = \frac{\pi D_2^2}{4} = \frac{\pi (0.010)^2}{4} \text{ m}^2 = 7.854 \times 10^{-5} \text{ m}^2.$$

With numerical values now available for all of the quantities appearing in the expression derived for $\dot{m}$, we evaluate it as follows:

$$\dot{m} = \left[\frac{2(3128.1 \times 10^3 - 3057.7 \times 10^3)}{\left(\dfrac{1}{0.38461(7.854 \times 10^{-5})}\right)^2 - \left(\dfrac{1}{0.4707(3.1416 \times 10^{-4})}\right)^2} \right]^{1/2}$$

$$= 0.01158$$

$$[=] \left[\frac{\dfrac{\text{J}}{\text{kg}}}{\left(\dfrac{\text{m}^3}{\text{kg}}\dfrac{1}{\text{m}^2}\right)^2} \left[\frac{1 \text{ N}\cdot\text{m}}{\text{J}}\right]\left[\frac{1 \text{ kg}\cdot\text{m/s}^2}{\text{N}}\right] \right]^{1/2} = \text{kg/s}.$$

Now knowing the mass flow rate, we can calculate the velocities from mass conservation (Eq. 11.2):

$$v_1 = \frac{\dot{m}}{\rho_1 A_1} = \frac{0.01158}{0.4707(3.1416 \times 10^{-4})} = 78.3,$$

$$v_2 = \frac{\dot{m}}{\rho_2 A_2} = \frac{0.01158}{0.38461(7.854 \times 10^{-5})} = 383.3$$

$$[=] \frac{\text{kg/s}}{(\text{kg/m}^3)\text{m}^2} = \text{m/s}.$$

Comments   This problem brings together many thermal-fluid concepts: mass and energy conservation, the second law of thermodynamics, and thermodynamic state relations.

Self Test 11.2   ✓   **Calculate the volumetric flow rate at the inlet and exit of the nozzle in Example 11.3.**
(Answer: 2.46×10^{-2} m^3/s, 3.01×10^{-2} m^3/s)

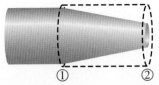

(a) Control volume selection

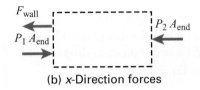

(b) x-Direction forces

(c) x-Direction momentum flows

FIGURE 11.5
Application of the principle of linear momentum conservation to a steady-flow nozzle: (a) selection of control volume, (b) identification of all x-direction forces, and (c) identification of all momentum flows having an x-component.

Linear Momentum Conservation[1]

The particular choice of a control volume now depends on which forces we wish to introduce into the analysis. One useful choice is a control volume that cuts through the solid wall at the nozzle inlet, as sketched in Fig. 11.5a. This choice exposes the tensile (or compressive) force in the pipe wall. As indicated in Fig. 11.5b, we arbitrarily assume that this force acts in the negative x-direction. If this force actually acts in the positive x-direction, our numerical result will have a negative sign in the final analysis. The control volume is deliberately chosen to be cylindrical with equal areas at stations 1 and 2, even though the flow area shown in the sketch is significantly smaller at station 2. We assume that the solid wall is thin so that $A_{\text{flow},1} \cong A_1$. As indicated in Fig. 11.5b, pressure forces act inward at the left and right faces of the control volume. The pressure forces acting on the cylindrical side surface do not act in the x-direction and, furthermore, they cancel because of symmetry. The momentum flows in and out of the nozzle are shown in Fig. 11.5c.

Having identified all of the x-direction forces and momentum flows, we can apply the conservation of linear momentum principle for steady state and steady flow expressed by Eq. 6.54 as follows:

$$\dot{m}\beta_1 v_{\text{avg},1} - \dot{m}\beta_2 v_{\text{avg},2} - F_{\text{wall}} + P_1 A_{\text{end}} - P_2 A_{\text{end}} = 0, \qquad (11.6)$$

where β is the momentum flow correction factor defined by Eq. 6.55b and $A_{\text{end}} = A_1 = A_2$. We assume that the pressure at station 2 is uniform over the entire exit area, that is, that the pressure within the emerging fluid jet equals the ambient pressure.[2] Note that we explicitly deal with the vector nature of the forces in transforming Eq. 6.54 to Eq. 11.6 by associating minus signs with F_{wall} and the left-directed pressure force $P_2 A_{\text{end}}$ to indicate that they act in the negative x-direction. Treating the force in the pipe wall as the quantity of interest, we rearrange Eq. 11.6 as follows:

$$F_{\text{wall}} = \dot{m}(\beta_1 v_{\text{avg},1} - \beta_2 v_{\text{avg},2}) + (P_1 - P_2)A_{\text{end}}. \qquad (11.7)$$

From Eq. 11.7, we see that calculating the restraining force in the pipe wall requires a knowledge of the flow rate, the inlet and exit velocities, the pressure drop, and the information required to evaluate or estimate the momentum flow correction factors. Of course, assuming the velocity profiles to be uniform makes these factors unity.

> **Table 6.1 in Chapter 6 presents values for β for laminar and turbulent flows.**

---

[1] We illustrate the application of the conservation of momentum principle to a subsonic nozzle with the fluid exiting into the atmosphere. An analysis of a diffuser would be quite similar.
[2] This assumption is generally valid for incompressible flows. The assumption, however, may break down for certain classes of compressible flows (i.e., those involving shock waves).

Example 11.4

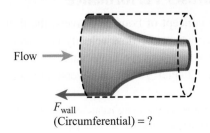

Flow

F_{wall}
(Circumferential) = ?

Estimate the force in the tube wall restraining the nozzle described in Example 11.1. Use the same flow conditions as in Example 11.1, and assume uniform inlet and exit velocity profiles.

Solution

Known $v_1, v_2, P_1 - P_2, T$

Find F_{wall}

Sketch   See Fig. 11.5.

Assumptions

  i. One-dimensional flow
 ii. Steady flow
iii. Incompressible fluid
 iv. $z_1 = z_2$ [i.e., no potential energy change]

Analysis   To calculate the restraining force, we apply Eq. 11.7 directly. From Example 11.1, we know the inlet and outlet velocities (2 and 12.5 m/s, respectively), and from Example 11.2, we know the pressure drop (75,860 Pa). The mass flow rate is calculated from Eq. 3.15 as

$$\dot{m} = \rho_1 v_{\mathrm{avg},1} A_1$$

$$= 996.5\,(2)\,\pi\,\frac{(0.025)^2}{4} = 0.978$$

$$[=]\ \frac{\mathrm{kg}}{\mathrm{m^3}}\frac{\mathrm{m}}{\mathrm{s}}\,\mathrm{m^2} = \mathrm{kg/s}.$$

With the assumption of uniform velocity profiles, $\beta_1 = \beta_2 = 1$; therefore, we now have all of the information needed to calculate F_{wall} from Eq. 11.7. Thus,

$$F_{\mathrm{wall}} = \dot{m}(\beta_1 v_{\mathrm{avg},1} - \beta_2 v_{\mathrm{avg},2}) + (P_1 - P_2)A_{\mathrm{end}}$$

$$= 0.978\,[1(2) - 1(12.5)] + 75,860\,\frac{\pi(0.025)^2}{4}$$

$$= 26.97$$

$$[=]\ \frac{\mathrm{kg}}{\mathrm{s}}\frac{\mathrm{m}}{\mathrm{s}}\left[\frac{1\,\mathrm{N}}{\mathrm{kg\cdot m/s^2}}\right] = \mathrm{N}.$$

Comment   Since our result is a positive value, the original assumption that F_{wall} is directed to the left is correct.

Self Test 11.3   ✓ **Estimate the force required to restrain the nozzle described in Example 11.3. Assume uniform inlet and exit velocity profiles.**

(Answer: 5.896 N)

The actual design and performance analysis of a diffuser is complicated by the need to deal with some complex flow phenomena in this seemingly simple device. Before proceeding, you might review Example 5.15, which considers a simple 1-D flow in a diffuser.

11.2b Flow Separation and Diffuser Performance

In Chapters 9 and 10, we discussed the concept of boundary layers, the thin region of flow adjacent to a solid surface where the flow velocity goes from zero at the wall (the no-slip condition) to relatively large values a short distance from the wall.[3] We also saw in Chapter 9 how an adverse pressure

[3] For pipe flow, the maximum boundary-layer thickness is equal to the radius of the pipe. In this instance, then, the concept of a *short distance* is relative to the length of the pipe.

FIGURE 11.6
Velocity profiles in a diffuser (a) without boundary-layer separation and (b) with boundary-layer separation. **Adapted from Ref. [3] with permission.**

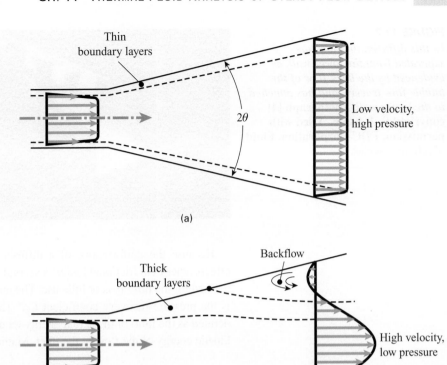

(a)

(b)

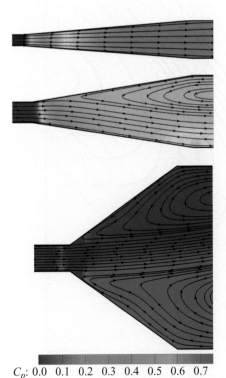

C_p: 0.0 0.1 0.2 0.3 0.4 0.5 0.6 0.7

The pressure recovery decreases as the divergence angle increases for this two-dimensional nozzle. Note the flow attached to the lower wall for $2\theta = 18°$ (middle) and the jet-like behavior with zero pressure recovery for $2\theta = 70°$ (bottom).

gradient (i.e., the case when the pressure increases in the direction of the flow) can result in the flow separating and subsequently reversing directions near the wall (see Fig. 9.24). Although the discussion in Chapter 9 pertains to external flows, an adverse pressure gradient is present in a diffuser—an internal flow—and flow separation is an important issue. Figure 11.6a illustrates an ideal flow through a diffuser in which the boundary layers remain relatively thin throughout the length; Fig. 11.6b shows the effect of separation (or stall) on the downstream velocity profile. When the flow separates in a nozzle, a relatively high velocity region exists at the exit, preventing the maximum possible recovery of pressure predicted by a one-dimensional (uniform-exit-velocity) analysis.

Several factors are important in the design and operation of a diffuser. Among these are the exit-to-inlet area ratio, the divergence angle of the diffuser, and the diffuser length-to-inlet diameter ratio. Depending on the values of these parameters, a wide range of flow behaviors can be observed. For example, as the divergence angle is increased with all other factors constant, the originally steady, unseparated flow becomes highly unsteady, with the flow separating from time to time. Increasing the divergence angle further results in a bistable flow pattern, with the flow attaching to one wall, while completely separating with a large back flow at the other wall. The flow alternates, first attaching to one wall and then the other. Figure 11.7 shows a visualization of this flow pattern. Increasing the divergence angle still further causes the flow from the inlet to penetrate into the diffuser as a jet, relatively uninfluenced by the walls. In this case, the device fails to act at all like a diffuser and produces little or no pressure recovery.

FIGURE 11.7

In this diffuser, the flow has separated from the lower wall, as evidenced by the back flow of the bubble flow tracers, and has attached to the upper wall. **Photograph [4] copyright S. J. Kline. Used with permission, EDC Corporation, Fluid Mechanics series.**

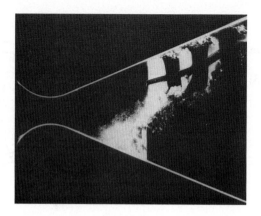

Because the performance of a diffuser is dominated by fluid mechanical effects other than frictional losses, a second-law, or isentropic, efficiency, as was defined for the nozzle, is of little use. The usual measure of diffuser performance is the pressure-recovery coefficient C_p.[4] The **pressure-recovery coefficient** is defined as the ratio of the actual energy recovered as pressure at the outlet to the kinetic energy of the flow at the inlet. Mathematically, this is expressed as

$$C_p \equiv \frac{(\dot{m}P/\rho)_2 - (\dot{m}P/\rho)_1}{\dot{m}(\alpha_1 v_1^2/2)},\qquad(11.8a)$$

FIGURE 11.8

Performance map for a conical diffuser. The inlet Reynolds and Mach numbers are 120,000 and 0.2, respectively. The inlet boundary layer effectively reduces the inlet flow area by 9%. The divergence angle 2θ is defined in Fig. 11.6. **Reprinted from Ref. [5] with permission.**

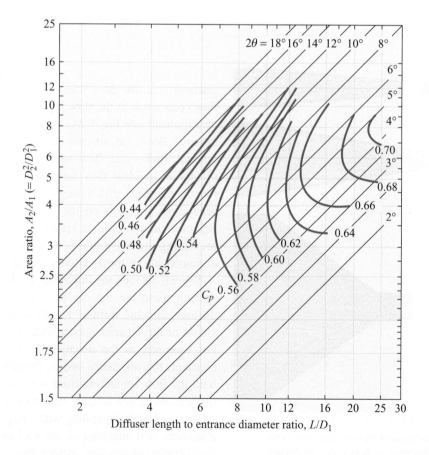

[4] Note that the pressure-recovery coefficient is represented with a capital C and should not be confused with the constant-pressure specific heat c_p (lowercase c).

See Chapter 4 for a review of flow work. ▷

where the terms in the numerator are the rates of flow work out and in, respectively, and the denominator gives the rate at which kinetic energy enters. For an incompressible flow, this simplifies to

$$C_p \equiv \frac{(P_2 - P_1)/\rho}{\frac{1}{2}\alpha_1 v_1^2}.$$
(11.8b)

Figure 11.8 presents a performance map for a conical diffuser.

Example 11.5

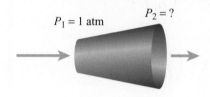

$P_1 = 1$ atm $P_2 = ?$

Consider a conical diffuser with inlet and outlet diameters of 20 and 40 mm, respectively. Water at 300 K and 1 atm enters the 0.2-m-long diffuser with a velocity of 5 m/s. Calculate the ideal (i.e., one-dimensional, frictionless flow) pressure-recovery coefficient and compare this result with that estimated from Fig. 11.8. Also determine the ideal and actual pressure increase through the diffuser.

Solution

Known $D_1, D_2, L, v_1, T_1, P_1$

Find $C_{p,\text{ideal}}, C_{p,\text{chart}}, \Delta P_{\text{ideal}}, \Delta P_{\text{actual}}$

Sketch

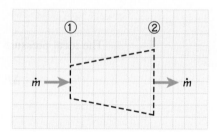

Assumptions

  i. Steady flow
  ii. Incompressible fluid
  iii. Adiabatic process
  iv. Isothermal operation
  v. $z_1 = z_2$ [i.e., no potential energy change]

Analysis   To find the ideal pressure-recovery coefficient, we combine expressions for mass conservation (Eq. 11.2 with constant ρ) and energy conservation (Eq. 11.5 with constant ρ and T) with the definition of C_p (Eq. 11.8b). Starting with mass conservation, we write

$$\rho v_1 A_1 = \rho v_2 A_2,$$

which for circular areas is

$$v_1 \pi D_1^2/4 = v_2 \pi D_2^2/4,$$

or

$$\frac{v_2}{v_1} = \left(\frac{D_1}{D_2}\right)^2.$$

Energy conservation is given by

$$h_2 - h_1 + \frac{1}{2}(v_2^2 - v_1^2) = 0,$$

$$c_p(T_2 - T_1) + \frac{1}{\rho}(P_2 - P_1) + \frac{v_1^2}{2}\left(\frac{v_2^2}{v_1^2} - 1\right) = 0.$$

Recognizing that $T_2 - T_1$ is zero for isothermal flow and substituting our result from mass conservation yields

$$\frac{1}{\rho}(P_2 - P_1) = \frac{v_1^2}{2}\left[1 - \left(\frac{D_1}{D_2}\right)^4\right],$$

or

$$\frac{(P_2 - P_1)/\rho}{\frac{1}{2}v_1^2} = 1 - \left(\frac{D_1}{D_2}\right)^4.$$

Comparing this result to Eq. 11.8b, we see that

$$C_{p,\text{ideal}} = 1 - \left(\frac{D_1}{D_2}\right)^4,$$

which we evaluate as

$$C_{p,\text{ideal}} = 1 - \left(\frac{20 \text{ mm}}{40 \text{ mm}}\right)^4 = 0.9375.$$

The ideal pressure recovery is thus

$$\Delta P_{\text{ideal}} = P_2 - P_1 = C_{p,\text{ideal}}\frac{1}{2}\rho v_1^2$$

$$= 0.9375(0.5)996.56(5)^2$$

$$= 11,700$$

$$[=] \text{kg/m}^3(\text{m/s})^2\left[\frac{1 \text{ N}}{\text{kg}\cdot\text{m/s}^2}\right] = \text{N/m}^2 \text{ or Pa},$$

where the water density is obtained from the NIST online database. To find $C_{p,\text{chart}}$ from Fig. 11.8, we determine

$$A_2/A_1 = D_2^2/D_1^2 = (40 \text{ mm}/20 \text{ mm})^2 = 4$$

and

$$L/D_1 = 0.20 \text{ m}/0.020 \text{ m} = 10.$$

Using these parameters, we find that

$$C_{p,\text{chart}} = 0.62.$$

Thus,

$$\Delta P_{\text{actual}} = C_{p,\text{chart}}\frac{1}{2}\rho v_1^2$$

$$= 0.62(0.5)996.56(5)^2 \text{ Pa}$$

$$= 7,700 \text{ Pa}$$

Comments   We see that the pressure-recovery coefficient from Fig. 11.8 is much less than the ideal value (i.e., 0.62 versus 0.9375). This indicates the importance of the actual velocity profiles and frictional effects in this diffuser. Although we did not state this directly, our use of Fig. 11.8 implicitly assumes that the Reynolds number for our flow is sufficiently close to the value of 120,000 associated with the chart. This we can easily verify by calculating $Re_{D_1} = \rho v_1 D_1 / \mu = 996.56(5)0.020/8.538 \times 10^{-4} = 116{,}720$, which we deem to be sufficiently close.

Self Test 11.4   ☑   **Using the inlet conditions given in Example 11.5, find the exit diameter, length, and maximum actual pressure-recovery coefficient of a diffuser having an ideal pressure-recovery coefficient of 0.92.**

(Answer: 37.6 mm, 0.26 m, 0.64)

11.2c Incompressible Flow

The simplification of the general conservation relationships for incompressible nozzle and diffuser flows is illustrated in Examples 11.1, 11.2, and 11.4 and Example 5.15 from Chapter 5. Application of mass conservation to an incompressible flow (see Example 11.1) is straightforward and needs no further discussion. Applying momentum conservation to an incompressible nozzle flow (see Example 11.4) is similarly straightforward. We could say the same thing for energy conservation; however, we wish to explore some subtleties that we ignored in Example 11.2 to develop a more complete understanding of nozzle performance and its connection to head loss.

We begin with the general energy conservation expression (Eq. 11.5):

$$h_2 - h_1 + \frac{1}{2}(\alpha_2 v_{\text{avg},2}^2 - \alpha_1 v_{\text{avg},1}^2) = 0.$$

Applying the definition of enthalpy yields

$$(u_2 + P_2/\rho) - (u_1 + P_1/\rho) + \frac{1}{2}(\alpha_2 v_{\text{avg},2}^2 - \alpha_1 v_{\text{avg},1}^2) = 0,$$

or

$$u_2 - u_1 + \frac{P_2 - P_1}{\rho} + \frac{1}{2}(\alpha_2 v_{\text{avg},2}^2 - \alpha_1 v_{\text{avg},1}^2) = 0.$$

The internal energy difference is related to the fluid temperature using the calorific equation of state for either an ideal gas (Eq. 2.31e) or an incompressible liquid as

$$u_2 - u_1 = c_v(T_2 - T_1),$$

where c_v is an appropriate average value of the constant-volume specific heat for the temperature range T_1 to T_2. With this substitution, energy conservation (the first law of thermodynamics) becomes

$$c_v(T_2 - T_1) + \frac{P_2 - P_1}{\rho} + \frac{1}{2}(\alpha_2 v_{\text{avg},2}^2 - \alpha_1 v_{\text{avg},1}^2) = 0. \qquad (11.9)$$

Although we discarded the term $c_v(T_2 - T_1)$ in Example 11.2, this term appears as a result of the irreversible conversion of mechanical energy (flow work and kinetic energy) to thermal energy (internal energy). Thus, if the nozzle flow is not frictionless, as all real flows must be, then $c_v(T_2 - T_1)$ is not zero. It will be small, however, in many applications.

We now retrieve the mechanical energy equation (Eq. 6.76), which was derived from the application of conservation of momentum (not energy) in Chapter 6:

$$gh_\mathrm{L} + \frac{P_2 - P_1}{\rho} + \frac{1}{2}(\alpha_2 v_{\mathrm{avg},2}^2 - \alpha_1 v_{\mathrm{avg},1}^2) = 0, \qquad (11.10)$$

> The head loss is also related to the entropy production rate (see the discussion of Eqs. 7.30 and 7.31 in Chapter 7).

where we have eliminated the control volume work and potential energy terms from Eq. 6.76 and rearranged the result to a form consistent with our energy equation (Eq. 11.9). The quantity h_L is the head loss, which we identified with frictional effects in Chapter 6. Comparing Eqs. 11.9 and 11.10 we see that

$$gh_\mathrm{L} = c_v(T_2 - T_1). \qquad (11.11)$$

In Chapter 10, we saw how the head loss h_L can be independently determined through engineering correlations for pipe flows. If we had similar correlations for nozzles, we could apply them here. Fortunately, many nozzles are short and frictional losses are small. Thus, reasonable approximations still result when they are neglected. With supersonic nozzles, however, the extreme velocities can lead to significant frictional losses. We deal with this situation in the following section by revisiting the isentropic nozzle efficiency defined in Chapter 7.

11.2d Compressible Flow

For flows of gases at high speeds, the assumption that the density is constant (i.e., the fluid is incompressible) is no longer valid. In the next section, we introduce some fundamental concepts needed to understand high-speed flows and present a criterion for determining when the effects of fluid compressibility are important.

A Few New Concepts and Definitions

Sound Speed and Mach Number   The dimensionless parameter known as the Mach number determines whether or not the effects of compressibility are important in any particular flow. The **Mach number Ma** is defined as the ratio of the local flow velocity v to the local speed of sound a; that is,

$$Ma \equiv \frac{v}{a} = \frac{\text{local flow velocity}}{\text{local speed of sound}}. \qquad (11.12)$$

A frequently applied criterion to divide "incompressible" flows from "compressible" flows is

$$\begin{aligned} Ma &< 0.3 \ \ \text{(incompressible flow)}, \\ Ma &> 0.3 \ \ \text{(compressible flow)}. \end{aligned} \qquad (11.13)$$

Capt. Charles E. Yeager and Bell X-1 supersonic research aircraft. The X-1, piloted by Yeager, was the first aircraft to fly faster than the speed of sound, reaching a Mach number of 1.06 on October 14, 1947. Photograph courtesy of U.S. Air Force.

LEVEL 2

The choice of the value 0.3 is somewhat arbitrary, but it is a typical upper limit for incompressible flows and restricts density variations to within about 5% of the mean.[5]

The speed of sound is a thermodynamic property that relates to the isentropic compressibility of a fluid (i.e., the relative change in density associated with a change in pressure for an isentropic process) as follows:

$$\frac{1}{\rho a^2} = \frac{1}{\rho}\left(\frac{\partial \rho}{\partial P}\right)_s,$$

or

$$a = \left[\left(\frac{\partial \rho}{\partial P}\right)_s\right]^{-1/2}. \qquad (11.14)$$

Values for sound speed for many common gases and liquids are available as part of the NIST database. For ideal gases, a simple relationship exists between the sound speed and temperature. Applying the ideal-gas equation of state, $P = \rho RT$ (Eq. 2.28b), together with the isentropic relationship, $T\rho^{1-\gamma} = $ constant (cf. Eq. 2.42b), yields

$$a = (\gamma RT)^{1/2}, \qquad (11.15)$$

where γ is the specific-heat ratio c_p/c_v, and R is the particular gas constant $R_u/\mathcal{M}$. From Eq. 11.15, we see that for ideal gases the sound speed depends only on the absolute temperature.

Table 11.2 presents sound speeds for a few common substances. Note how the sound speed progressively increases with decreasing compressibility of the physical state from gas to liquid to solid.

Table 11.2 Speed of Sound in Selected Gases, Liquids, and Solids

Substance	State	Sound Speed (m/s)
H_2	Gas (25°C)	1315.4*
O_2	Gas (25°C)	328.7*
N_2	Gas (25°C)	352.1*
H_2O	Liquid (25°C)	1496.7*
Hg	Liquid (25°C)	1450[†]
Al	Solid	6420[†,‡]
Cu	Solid	4760[†,‡]
Ag	Solid	3650[†,‡]

* Values from NIST online database for 1 atm.
[†] Values from *CRC Handbook of Chemistry and Physics*, 77th ed., pp. 14–36, 1996.
[‡] Velocity of plane, longitudinal wave in bulk material.

[5] Note this particular similarity in function to the Reynolds number. Here a specific value of the Mach number divides the regimes of incompressible and compressible flows, whereas a specific value for the Reynolds number divides the regimes of laminar and turbulent flow. Other dimensionless parameters perform similar regime-dividing functions.

and so

$$T_0 = 223 + 99.4 \text{ K} = 322 \text{ K}.$$

Comments   For the relatively slow moving car, the stagnation temperature is only slightly greater than the ambient value. The temperature rise is so small that we should have employed more significant figures in converting from degrees Celcius to Kelvins. For the supersonic aircraft, however, the temperature rise is quite large: 99.4 K. For this reason, the outer skins of supersonic aircraft are fabricated from materials that possess high-temperature strength (e.g., Inconel X and titanium alloys).

Self Test 11.7   ☑ Determine the stagnation pressures for the car and aircraft of Example 11.8 assuming ambient pressures of 1 and 0.26 atm, respectively.

(Answer: 101,396 Pa, 95,306 Pa)

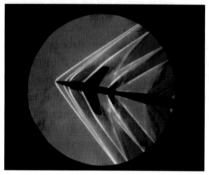

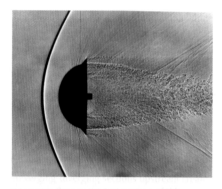

FIGURE 11.10

A bow shock wave precedes a projectile traveling at a supersonic velocity (top). At mach 1.4, several shock waves form on a model F11F-1 Tiger in the 1 × 3 ft NASA Ames supersonic wind tunnel (bottom). The shocks are made visible by schlieren photography, a flow visualization technique that is sensitive to density gradients in the flow. **Photographs courtesy of NASA.**

Choked Flow   A flow is **choked** when the mass flow rate cannot be increased by reducing the downstream pressure. Choking occurs when the Mach number is unity at the exit of a converging nozzle or at the exit of a constant-area duct. Choking also occurs when the Mach number is unity at the throat of a converging–diverging nozzle. Properties at the exit of a choked converging nozzle, or at the throat of a choked converging–diverging nozzle, are referred to as the *critical properties*[6] and are designated P^*, T^*, and ρ^*, and the area is designated A^*. The Mach number for these cases is, by definition, unity. The choked condition results in the maximum flow rate for any fixed values of upstream pressure and temperature. We examine the phenomenon of choking in more detail in our discussion of converging and converging–diverging nozzles.

Shock Waves   In high-speed flows, certain conditions result in the formation of shock waves. We define a shock wave as a region in a flow in which the thermodynamic and flow properties (P, T, ρ, and v) abruptly change. In a shock wave, pressures and temperatures greatly increase in a few molecular mean free paths. Because shock waves are so thin, we treat them as discontinuities in the flow. A familiar example of a shock wave is a sonic boom. Figure 11.10 shows a shock wave in front of a projectile traveling at a slightly supersonic velocity. Although shock waves form under a variety of circumstances, we consider only so-called **normal shocks**[7] that occur within the diverging portion of a converging–diverging nozzle.

Mach Number–Based Conservation Principles and Property Relationships

To facilitate the application of the familiar conservation principles and thermodynamic property relationships to compressible flows, it is useful to incorporate the Mach number into the relationships. We begin by rewriting the one-dimensional, steady-flow forms of mass and energy conservation

---

[6] These properties should not be confused with the liquid–vapor critical-point properties discussed in Chapter 2.

[7] *Normal* shocks derive their name from their being nominally perpendicular (normal) to the flow. Not considered here are *oblique* shocks, which form at an angle to the flow.

applied to an adiabatic control volume having one inlet and one outlet and no work interactions:

$$\rho_1 v_1 A_1 = \rho_2 v_2 A_2 \qquad \text{(mass conservation)}, \qquad (11.2)$$

$$h_1 + \frac{1}{2}v_1^2 = h_2 + \frac{1}{2}v_2^2 \qquad \text{(energy conservation)}. \qquad (11.5)$$

Other than the restrictions just listed, these expressions are quite general. We now add several restrictions typically used to analyze compressible flows:

- The fluid exhibits ideal-gas behavior.
- The calorific equation of state is simplified by the use of an average (constant) specific heat (i.e., $\Delta h = c_{p,\text{avg}}\, \Delta T$).

Since we wish to consider both shockless (isentropic) flows and flows with a normal shock, two different control volumes are required as illustrated in Fig. 11.11. We analyze control volume A first and add the following restriction:

- The flow from station 1 to station 2 is adiabatic and reversible, that is, isentropic.

Our first objective is to express energy conservation in terms of T_1, T_2, Ma_1, and Ma_2. This is easily accomplished by substituting $a_1 Ma_1$ and $a_2 Ma_2$ for v_1 and v_2, respectively, recognizing that $a = (\gamma R T)^{1/2}$, and substituting $c_{p,\text{avg}}(T_2 - T_1)$ for $h_2 - h_1$. Performing these substitutions and rearranging yield for energy conservation

$$\frac{T_1}{T_2} = \frac{\dfrac{\gamma - 1}{2} Ma_2^2 + 1}{\dfrac{\gamma - 1}{2} Ma_1^2 + 1}. \qquad (11.20)$$

For a review of ideal-gas property relationships, see Table 2.5 and related material in Chapter 2.

By applying property relationships for an isentropic process, the temperature ratio T_1/T_2 can be related to P_1/P_2 and ρ_1/ρ_2 as follows:

$$P_1/P_2 = (T_1/T_2)^{\frac{\gamma}{\gamma - 1}}, \qquad (11.21)$$

$$\rho_1/\rho_2 = (T_1/T_2)^{\frac{1}{\gamma - 1}}. \qquad (11.22)$$

Thus, if one knows the upstream state and Mach number (T_1, P_1, ρ_1, and Ma_1), the downstream state (T_2, P_2, and ρ_2) can be completely defined by

FIGURE 11.11

(a) Control volume A for analysis of isentropic flow in a nozzle or diffuser and (b) control volume B for analysis of normal shock waves (nonisentropic flow).

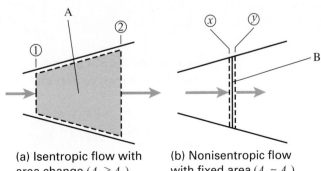

(a) Isentropic flow with area change ($A_2 \gtrless A_1$)

(b) Nonisentropic flow with fixed area ($A_y \approx A_x$)

specifying Ma_2 and applying Eqs. 11.20–11.22. Developing these relationships is left as an exercise for the reader (see Problem 11.26).

Our next objective is to employ the mass conservation expression (Eq. 11.2) to relate the downstream Mach number Ma_2 to upstream properties and the area ratio A_1/A_2. Such a relationship closes the variable-area, compressible-flow problem; that is, given upstream conditions and the downstream area, the following downstream properties can be determined: Ma_2, T_2, v_2, P_2, and ρ_2. We begin by substituting $(\gamma RT)^{1/2} Ma$ for v and use Eq. 11.22 to eliminate the densities in our simple mass conservation expression (Eq. 11.2) as follows:

$$\frac{\rho_1}{\rho_2} = \frac{a_2}{a_1} \frac{Ma_2}{Ma_1} \frac{A_2}{A_1}$$

and

$$\left(\frac{T_1}{T_2}\right)^{\frac{1}{\gamma-1}} = \left(\frac{\gamma RT_2}{\gamma RT_1}\right)^{1/2} \frac{Ma_2}{Ma_1} \frac{A_2}{A_1}.$$

Solving for T_1/T_2 and then substituting Eq. 11.20 yield our final result for mass conservation:

$$\frac{Ma_1}{Ma_2} \left[\frac{1 + \frac{1}{2}(\gamma-1)Ma_2^2}{1 + \frac{1}{2}(\gamma-1)Ma_1^2}\right]^{\frac{\gamma+1}{2(\gamma-1)}} = \frac{A_2}{A_1} \qquad \text{(control volume A).} \quad (11.23)$$

Table 11.3 summarizes the Mach number–based conservation principles and property relations that apply to control volume A (Fig. 11.11a).

To analyze control volume B (Fig. 11.11b), which applies to a normal shock, we remove the previously invoked restriction that the flow is isentropic—since a shock is a highly nonisentropic phenomenon—and add the restriction that, although the shock may be located in a nozzle having variable area, the shock is sufficiently thin that the entrance and exit areas of the control volume can be considered identical (i.e., $A_x = A_y$). Here we adopt the convention that the upstream station is denoted x and the downstream station y to avoid any confusion with the analysis of control volume A.

Conservation of energy is unaffected by the relaxation of the isentropic assumption; hence, Eq. 11.20 applies equally well to control volume B with only a change in the subscript notation as shown in the Table 11.3 summary (i.e., Eq. 11.24).

Conservation of mass follows directly from Eq. 11.2. Here the areas cancel and the density can be related to the pressure and temperature using the ideal-gas equation of state to yield the following:

$$\frac{P_x}{RT_x} v_x = \frac{P_y}{RT_y} v_y.$$

We introduce the Mach number to eliminate the velocities [i.e., $v = aMa = (\gamma RT)^{1/2} Ma$]; thus, conservation of mass becomes

$$\frac{P_y}{P_x} = \left(\frac{T_y}{T_x}\right)^{1/2} \frac{Ma_x}{Ma_y} \qquad \text{(control volume B).} \quad (11.25)$$

Table 11.3 Mach Number–Based Conservation Principles and Property Relationships for Compressible Flows of an Ideal Gas

	Control Volume A: Isentropic Flow with Variable Area*		Control Volume B: Nonisentropic Flow with Constant Area*	
Conservation of energy	$\dfrac{T_1}{T_2} = \dfrac{\dfrac{\gamma-1}{2}Ma_2^2 + 1}{\dfrac{\gamma-1}{2}Ma_1^2 + 1}$	Eq. 11.20	$\dfrac{T_y}{T_x} = \dfrac{\dfrac{\gamma-1}{2}Ma_x^2 + 1}{\dfrac{\gamma-1}{2}Ma_y^2 + 1}$	Eq. 11.24
Conservation of mass	$\dfrac{A_2}{A_1} = \dfrac{Ma_1}{Ma_2}\left[\dfrac{1+\frac{1}{2}(\gamma-1)Ma_2^2}{1+\frac{1}{2}(\gamma-1)Ma_1^2}\right]^{\frac{\gamma+1}{2(\gamma-1)}}$	Eq. 11.23	$\dfrac{P_y}{P_x} = \left(\dfrac{T_y}{T_x}\right)^{1/2}\dfrac{Ma_x}{Ma_y}$	Eq. 11.25
Conservation of momentum	—	Eq. 11.21	$\dfrac{P_y}{P_x} = \dfrac{\gamma Ma_x^2 + 1}{\gamma Ma_y^2 + 1}$ $\rho_x = \dfrac{P_x}{RT_x}$	Eq. 11.27
Property relationships	$\dfrac{P_1}{P_2} = \left(\dfrac{T_1}{T_2}\right)^{\frac{\gamma}{\gamma-1}} = \left[\dfrac{\dfrac{\gamma-1}{2}Ma_2^2 + 1}{\dfrac{\gamma-1}{2}Ma_1^2 + 1}\right]^{\frac{\gamma}{\gamma-1}}$ $\dfrac{\rho_1}{\rho_2} = \left(\dfrac{T_1}{T_2}\right)^{\frac{1}{\gamma-1}} = \left[\dfrac{\dfrac{\gamma-1}{2}Ma_2^2 + 1}{\dfrac{\gamma-1}{2}Ma_1^2 + 1}\right]^{\frac{1}{\gamma-1}}$ $s_2 - s_1 = 0$	Eq. 11.22	$\dfrac{\rho_y}{\rho_x} = \dfrac{T_x}{T_y}\dfrac{P_y}{P_x}$ $s_y - s_x = c_{p,\text{avg}}\ln(T_y/T_x) - R\ln(P_y/P_x)$	

*See Fig. 11.11 for the definition of control volumes A and B.

In our analysis of control volume A, conservation of momentum was not needed; now, however, its use is critical. Applying Eq. 6.53, an expression of momentum conservation, to control volume B yields

$$\dot{m}(v_x - v_y) + (P_x - P_y)A = 0. \tag{11.26}$$

Introducing the Mach number as before, our momentum conservation expression is transformed to

$$\frac{P_y}{P_x} = \frac{\gamma Ma_x^2 + 1}{\gamma Ma_y^2 + 1} \qquad \text{(control volume B).} \tag{11.27}$$

If we treat all upstream quantities (T_x, P_x, and Ma_x) as known, simultaneous solution of our energy, mass, and momentum equations (Eqs. 11.24, 11.25, and 11.27, respectively) yields expressions for the unknown downstream quantities T_y, P_y, and Ma_y. One particularly useful result is the following relationship between Ma_x and Ma_y:

$$Ma_y = \left[\frac{Ma_x^2(\gamma - 1) + 2}{2\gamma Ma_x^2 - \gamma + 1} \right]^{1/2}. \tag{11.28}$$

Table 11.3 again summarizes the results of our analysis and presents related results.

Converging and Converging–Diverging Nozzles

Basic Operation   Figure 11.12 illustrates a simple converging nozzle and a converging–diverging nozzle. The key to understanding the operation of these devices is being able to describe the pressure distribution through them. As shown in the figure, the nozzles are placed between two large reservoirs, each at a fixed pressure. Starting at the left (Fig. 11.12) and moving downstream in the direction of flow, we define the following important pressures:

$P_0 \equiv$ *Stagnation pressure.* This is the pressure in the upstream receiver in which the velocity is essentially zero.

$P_t \equiv$ *Throat pressure.*   This is the pressure at the minimum area. For the converging nozzle, this is also the exit plane pressure.

$P* \equiv$ *Throat pressure when the nozzle is choked, $P_{t,\text{choked}}$.* $P*$ is also known as the critical pressure.

$P_e \equiv$ *Exit plane pressure.*   Depending on conditions, this pressure may or may not equal the back pressure.

$P_b \equiv$ *Back pressure.*   This is the pressure in the large downstream receiver. This receiver may be the atmosphere or a reservoir in which the pressure is controlled by a pump, etc.

Below each nozzle schematic in Fig. 11.12 is a plot of pressure though the nozzle, normalized by the stagnation pressure [i.e., $P(x)/P_0$]. We now explore how this pressure distribution is controlled by the backpressure P_b,

FIGURE 11.12
(a) Converging nozzle (top) and pressure distributions through the nozzle for various back pressures (bottom). (b) Converging–diverging nozzle (top) and various pressure distributions (bottom).

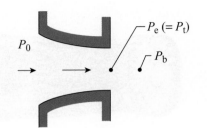

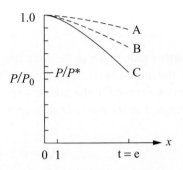

(a) Converging nozzle

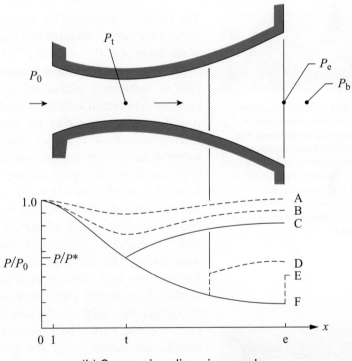

(b) Converging-diverging nozzle

a pressure that we assume is under our control or is a given quantity. When $P_b = P_0$, there is no flow. Decreasing P_b below P_0 results in flow through the nozzle. Pressure distributions denoted A and B (dashed lines in Fig. 11.12) illustrate flow conditions in which the mass flow rate for B is greater than that for A. Furthermore, the exit plane pressure equals that of the downstream receiver (i.e., $P_e = P_b$). For both the pure converging and the converging–diverging nozzle, these flows (A and B) can be modeled as isentropic with good accuracy. We also note that the diverging portion of the converging–diverging nozzle acts as a diffuser for paths A and B.

FIGURE 11.13
The F404-GE-400 engine has a hinged-flap, cam-linked, converging–diverging exhaust nozzle. The nozzle geometry can be adjusted to provide optimal performance for a wide range of flight conditions. F404-GE-400 engines power the F/A-18A/B/C/D Hornet aircraft. **Drawing courtesy of General Electric.**

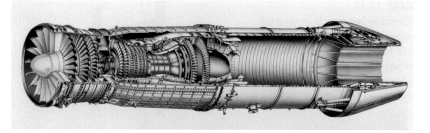

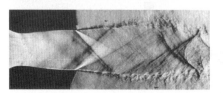

Supersonic flow from a choked converging-diverging nozzle exits into a large receiver. Mismatched pressures at the exit result in a complex pattern of shock waves corresponding to receiver pressures between E and F (top) and pressures below F (bottom). From Ref. [22] © Crown copyright 1953, reproduced by permission of the Controller of HMSO and the Queen's Printer for Scotland. Courtesy National Physical Laboratory.

FIGURE 11.14
Prototype O_2/H_2 rocket engine cutaway clearly shows the converging–diverging geometry of its nozzle. This engine was originally proposed as an attitude control thruster for the **Space Shuttle.** *The expansion ratio (A_e/A_t) is 40.*

Further reductions in the back pressure result in even greater flow rates until the pressure at the throat falls to the critical value P^*. At this condition (curve C), the nozzle becomes choked, with sonic conditions at the throat. For the pure converging nozzle, the throat and exit are identical; thus,

$$P_t\,(=P_e) = P^* = P_b. \tag{11.29}$$

For the converging–diverging nozzle,

$$P_t = P^*. \tag{11.30}$$

The diverging portion of the converging–diverging nozzle again acts as a diffuser for path C; thus, the exit plane pressure is greater than the pressure at the throat [i.e., $P_e\,(=P_b) > P^*$]. Path C is also an isentropic one.

When the back pressure is lowered to values below those associated with path C, interesting phenomena occur: For the converging nozzle, the pressure distribution within the nozzle remains unchanged; however, there is a discontinuity between the exit plane pressure and that of the exit receiver (i.e., $P_e \neq P_b$). This results in the appearance of various shock waves in the receiver. For the converging–diverging nozzle, a normal shock is present in the diverging section, the location of which depends on the value of $P_b\,(=P_e)$. Path D in Fig. 11.12 contains a shock occurring about halfway through the diverging section. A normal shock occurs just downstream of the throat when P_b is just below the pressure associated with path C. Reducing P_b causes the shock to be repositioned further downstream, with the shock ultimately standing in the exit plane (path E). Reducing the back pressure even further results in complex shock waves outside of the nozzle. Path F represents the special case of isentropic flow throughout the entire nozzle with the back pressure matched to the exit plane pressure. Further back pressure reductions result again in shock waves outside the nozzle since the pressure distribution (path F) remains unchanged with such reductions. For all of the paths below path C, the flow is supersonic upstream of the normal shock and then subsonic downstream of the shock (paths D and E). Path F is shockless.

Important applications of converging–diverging nozzles are jet engines used in high-speed aircraft and rocket engines. Figure 11.13 shows a cutaway of a jet engine that uses a sophisticated variable-geometry exhaust nozzle; Fig. 11.14 shows a rocket engine with fixed geometry.

Isentropic Flows   We now quantify the pressure distributions just discussed by applying the Mach number–based conservation principles and property relationships developed in the previous section. We begin by examining the isentropic paths, that is, all paths associated with the back pressure condition $P_0 < P_b \leq P_c$, where P_c is the back pressure for path C in Fig. 11.12. Consider the isentropic-flow relationships (Eqs. 11.20, 11.21, and 11.22) presented in Table 11.3. If we designate station 2 to be at the stagnation condition ($Ma_2 = 0$), properties through the nozzle can be related to the local Mach number and their corresponding stagnation properties as follows:

$$\frac{P}{P_0} = \left[\frac{\gamma - 1}{2} Ma^2 + 1\right]^{\frac{-\gamma}{\gamma - 1}}, \tag{11.31a}$$

$$\frac{\rho}{\rho_0} = \left[\frac{\gamma - 1}{2} Ma^2 + 1\right]^{\frac{-1}{\gamma - 1}}, \tag{11.31b}$$

and

$$\frac{T}{T_0} = \left[\frac{\gamma - 1}{2} Ma^2 + 1\right]^{-1}. \tag{11.31c}$$

We can also relate the local flow area A to the area at the throat when the flow is choked, A^*, by setting the throat Mach number to unity in our mass conservation expression (Eq. 11.23), that is,

$$\frac{A}{A^*} = \frac{1}{Ma}\left[\left(\frac{2}{\gamma + 1}\right)\left(\frac{\gamma - 1}{2} Ma^2 + 1\right)\right]^{\frac{\gamma + 1}{2(\gamma - 1)}}. \tag{11.31d}$$

To simplify the application of these relationships, tables are frequently employed. Table K.1 (Appendix K) presents values of P/P_0, ρ/ρ_0, and T/T_0 for a range of Mach numbers for air ($\gamma = 1.4$). A modern alternative to the use of tables is for the reader to program these functions using a spreadsheet or other software (see Problem 11.27). The following examples illustrate the use of these isentropic, compressible-flow functions.

Converging-diverging nozzle with an area ratio (A_e/A_t) of 1000 on test stand at NASA Glenn Research Center. Photograph courtesy of NASA.

Example 11.9

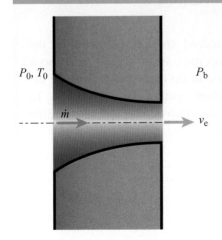

P_0, T_0 P_b

$\dot{m}$ v_e

Consider a flow of combustion products that expands through a converging nozzle. Upstream where the velocity is negligible, the pressure and temperature are 0.3 MPa and 1500 K, respectively. The nozzle exit diameter is 50 mm. The pressure is 0.1 MPa in the receiver into which the nozzle exhausts. Assume that the properties of the combustion products are essentially the same as the properties of air (i.e., $\gamma = 1.4$ and $R = 0.287$ kJ/kg·K). Determine the velocity of the combustion products at the nozzle exit and the mass flow rate through the nozzle.

Solution

Known P_0, T_0, P_b, D_e

Find $v_e, \dot{m}$

Sketch

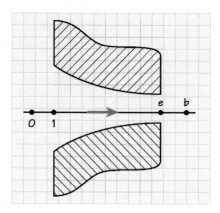

Assumptions

i. The flow is one dimensional, steady, and isentropic.
ii. The combustion products can be treated as air with constant properties.

Analysis We first determine whether the flow is choked by calculating the ratio of the back pressure to the stagnation pressure and comparing this to the critical pressure ratio $P*/P_0$:

$$\frac{P_b}{P_0} = \frac{0.1 \text{ MPa}}{0.3 \text{ MPa}} = 0.33.$$

If the flow is choked, P_b/P_0 must be equal to or less than $P*/P_0$. From Eq. 11.31a (or Table K.1), we evaluate the critical pressure ratio by setting the Mach number to unity, which gives

$$\frac{P*}{P_0} = \left[\frac{\gamma - 1}{2} Ma^2 + 1 \right]^{\frac{-\gamma}{\gamma - 1}}$$

$$= \left[\frac{1.4 - 1}{2} (1)^2 + 1 \right]^{\frac{-1.4}{1.4 - 1}} = 0.528.$$

Thus,

$$\frac{P_b}{P_0} = 0.33 < \frac{P*}{P_0} = 0.528,$$

and we conclude that the flow is indeed choked. Furthermore, the properties at the exit are the critical properties, $Ma_e = 1$, and some shock structure must exist outside of the nozzle in the downstream receiver; the flow through the nozzle, however, follows path C in Fig. 11.12a. To determine v_e, we apply the definition of the Mach number and calculate the sound speed from Eq. 11.15 as follows:

$$Ma_e = 1 = \frac{v_e}{a_e} = \frac{v_e}{(\gamma R T_e)^{1/2}},$$

or

$$v_e = (\gamma R T_e)^{1/2}.$$

The exit temperature is found from Eq. 11.31c or Table K.1:

$$\frac{T_e}{T_0} = \left[\frac{\gamma - 1}{2} Ma_e^2 + 1\right]^{-1}$$

$$= \left[\frac{1.4 - 1}{2}(1)^2 + 1\right]^{-1} = 0.833,$$

and

$$T_e = \left(\frac{T_e}{T_0}\right)T_0 = 0.833(1500\,\text{K}) = 1250\,\text{K}.$$

Thus,

$$v_e = \left[1.4(287)1250\right]^{1/2}\text{m/s} = 709\,\text{m/s}.$$

The reader should verify the units here. The mass flow rate is calculated from

$$\dot{m} = \rho_e v_e A_e,$$

where

$$\rho_e = \frac{P_e}{RT_e} = \frac{(P_e/P_0)P_0}{RT_e}$$

$$= \frac{0.528(0.3 \times 10^6)}{287(1250)}$$

$$= 0.4415$$

$$[=]\frac{\text{N/m}^2}{(\text{J/kg}\cdot\text{K})\text{K}}\left[\frac{1\,\text{J}}{\text{N}\cdot\text{m}}\right] = \text{kg/m}^3.$$

The flow rate is then

$$\dot{m} = 0.4415(709)\frac{\pi(0.05)^2}{4}$$

$$= 0.615$$

$$[=](\text{kg/m}^3)(\text{m/s})\text{m}^2 = \text{kg/s}.$$

Comment   Note the importance of ascertaining that the nozzle is choked. Remembering that the critical pressure ratio P^*/P_0 for air is just slightly greater than one-half (0.528) can be quite useful.

Self Test 11.8   Methane ($\gamma = 1.299$) is injected into a combustor operating at a chamber pressure of 1 MPa. Determine the minimum upstream methane pressure if the fuel flow is to be choked.

(Answer: 1.83 MPa)

Example 11.10

Supersonic

$Ma = 1$

Consider a flow of air through a converging–diverging nozzle. The diameter of the throat is 100 mm and the diameter at the exit is 205.8 mm. The stagnation conditions in the upstream (entrance) receiver are $P_0 = 350\,\text{kPa}$ and $T_0 = 300\,\text{K}$.

$Ma = 1$

Subsonic

A. The flow is **choked** and **supersonic** throughout the diverging portion of the nozzle. Determine the exit plane pressure P_e, the exit Mach number Ma_e, and exit velocity v_e.

B. The flow is **choked** and **subsonic** throughout the diverging portion of the nozzle. Determine the exit plane pressure P_e and the exit Mach number Ma_e.

Solution

Known P_0, T_0, D_t, D_e, choked flow

Find P_e (supersonic and subsonic), Ma_e (supersonic and subsonic), v_e (supersonic)

Sketch   See Fig. 11.12b, path F (supersonic) and path C (subsonic).

Assumptions

  i. Steady, 1-D, isentropic flow
  ii. Constant specific heats

Analysis   Because the flow is choked and, for Part A, is supersonic throughout, we know that no shocks exist; hence, path F (Fig. 11.12b) describes the flow. For the subsonic flow of Part B, path C describes the flow. Thus, the isentropic flow relationships (Eqs. 11.31a–11.31d or Table K.1) apply for both Parts A and B. We use the given diameters to determine the area ratio,

$$\frac{A_e}{A_t} = \frac{A_e}{A^*} = \frac{\pi D_e^2/4}{\pi D_t^2/4}$$

$$= \left(\frac{205.8 \text{ mm}}{100 \text{ mm}}\right)^2 = 4.235.$$

With this value of A_e/A^*, we can solve Eq. 11.31d for Ma_e. Since this equation is quadratic in Ma_e, there are two roots: a supersonic value and a subsonic value. With values for Ma_e, Eq. 11.31a can be used to find P_e/P_0. Similarly, T_e/T_0 can be evaluated from Eq. 11.31c to evaluate the exit sound speed, which is then used to calculate v_e. An alternative to this procedure is to use Table K.1 to find Ma_e. For Part A (supersonic flow), we find

$$Ma_e = 3.00.$$

Table K.1 also provides corresponding values of P_e/P_0 (Eq. 11.31a) and T_e/T_0 (Eq. 11.31c):

$$\frac{P_e}{P_0} = 0.02722,$$

$$\frac{T_e}{T_0} = 0.35714.$$

Thus,

$$P_e = \left(\frac{P_e}{P_0}\right)P_0$$

$$= 0.02722(350 \text{ kPa}) = 9.527 \text{ kPa},$$

$$T_e = \left(\frac{T_e}{T_0}\right)T_0$$

$$= 0.35714(300 \text{ K}) = 107.1 \text{ K}.$$

The exit velocity (Part A) is then

$$v_e = Ma_e a_e = Ma_e(\gamma RT_e)^{1/2}$$
$$= 3.00(1.4(287)107.1)^{1/2} \text{ m/s}$$
$$= 622.3 \text{ m/s}.$$

For Part B, we see from Table K.1 that when A_e/A^* is 4.235, Ma_e lies between 0.1 and 0.2. Solving Eq. 11.31d for the subsonic root yields

$$Ma_e = 0.1385,$$

and from Eq. 11.31a, $P_e/P_0 = 0.9867$. Thus,

$$P_b = P_e = (P_e/P_0)P_0$$
$$= 0.9867(350 \text{ kPa}) = 345.3 \text{ kPa}.$$

Comments The key to solving this problem was recognizing that paths F and C apply to Parts A and B, respectively. We also note the convenience of using the tabulated forms of Eqs. 11.31a–11.13d (i.e., Table K.1).

Self Test 11.9 A converging–diverging nozzle has an upstream stagnation pressure of 300 kPa and throat and exit diameters of 7.06 and 10 mm, respectively. Assuming no shocks exist, determine the exit Mach number when the exit pressure is 28.05 kPa for $\gamma = 1.4$.

(Answer: Ma = 2.2)

Nonisentropic Flows Normal shocks are highly irreversible and hence are nonisentropic processes. A prerequisite for a normal shock is that the flow entering the shock wave be supersonic. Shocks do not occur in subsonic flows. In a converging–diverging nozzle, supersonic flow is possible only in the diverging section. As we have already developed the basic equations relating the properties immediately upstream and downstream of a normal shock, our objective in this section is to illustrate how these results can be used. Before proceeding with specific examples, we explore some general characteristics of normal shocks described by Eqs. 11.24–11.28 and the ideal-gas property relationships shown in Table 11.3. Treating the Mach number immediately upstream of the shock (Ma_x) as a known quantity, tables can be generated and used as alternatives to solving these equations directly. Table K.2 in Appendix K is such a table (see also Problem 11.46). The following are a few selected values from this table:

Ma_x	Ma_y	P_y/P_x	T_y/T_x	ρ_y/ρ_x
1.00	1.000	1.000	1.000	1.000
1.20	0.842	1.513	1.128	1.342
2.00	0.577	4.500	1.688	2.667
5.00	0.415	29.00	5.800	5.000

What conclusions about normal shocks can we draw from this table?

- The flow enters the shock at a supersonic velocity and exits at a subsonic velocity. The greater the incoming Mach number Ma_x, the smaller the exiting Mach number Ma_y.

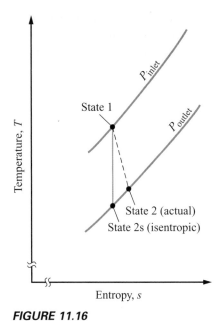

FIGURE 11.16

To define the nozzle efficiency, an ideal outlet state is defined by an isentropic expansion from the actual inlet state to the same outlet pressure as that for the actual process.

In this definition, $\dot{KE}_{act}$ is the kinetic energy flow at the nozzle outlet for the actual process, and $\dot{KE}_{isen}$ is the theoretical kinetic energy flow at the nozzle outlet for a reversible, adiabatic (isentropic) process starting at the same inlet state and ending at the same pressure as the actual process. Treating the flow as essentially one dimensional, we can express the isentropic efficiency as

$$\eta_{\text{isen,n}} = \frac{(\dot{m}v_2^2)_{\text{act}}}{(\dot{m}v_2^2)_{\text{isen}}} = \frac{(v_2^2)_{\text{act}}}{(v_2^2)_{\text{isen}}}. \tag{11.32b}$$

Figure 11.16 illustrates the actual and ideal processes on T–s coordinates. Because of irreversibilities, primarily fluid friction, the entropy of the outlet state is greater than the entropy at the inlet state. In a converging–diverging nozzle, most of the frictional losses typically occur in the usually longer diverging section where velocities are high. Assuming that there are no losses in the converging section allows the flow rates to cancel in the definition given here.

Example 11.12

Photograph courtesy of U.S. Navy.

The exit nozzle of a turbojet engine expands the flow of exhaust products from 170 to 45 kPa. The products enter the nozzle at 800 K with a velocity of 225 m/s. The isentropic efficiency of the nozzle is 97%. Determine the jet exit velocity and the temperature of the products at the nozzle exit. Also determine the entropy change from inlet to outlet.

Solution

Given $P_1, T_1, v_1, P_2, \eta_{\text{isen,n}}$

Find $v_2, T_2, s_2 - s_1$

Sketch   See Fig. 11.16.

Assumptions

  i. The flow is adiabatic and shockless.

  ii. Local thermodynamic equilibrium prevails through the nozzle.

  iii. The combustion products can be treated as an ideal gas with their properties approximated using those of air.

  iv. Constant values of c_p and γ evaluated at the average temperature, $(T_1 + T_2)/2$, can be used.

Analysis   We begin by determining the temperature at the isentropic state 2s (Fig. 11.16) using the ideal-gas relationship for an isentropic process (Eq. 11.21 or 2.41):

$$\frac{T_{2s}}{T_1} = \left(\frac{P_2}{P_1}\right)^{\frac{\gamma-1}{\gamma}}.$$

Solving for T_{2s} and evaluating using $\gamma = 1.37$ (calculated from $c_p(T_{avg}) \equiv$ $c_{p,avg} = 1070$ J/kg·K for $T_{avg} \approx 675$ K in Table C.3) yield

$$T_{2s} = T_1 \left(\frac{P_2}{P_1}\right)^{\frac{\gamma-1}{\gamma}}$$

$$= 800 \left(\frac{45}{170}\right)^{\frac{0.37}{1.37}} \text{K} = 558.7 \text{ K}.$$

We obtain the outlet velocity for the ideal (adiabatic and reversible) process by combining energy conservation (Eq. 11.5) with the ideal-gas calorific equation of state relating the enthalpy and temperature (Eq. 2.33e) as follows:

$$h_1 - h_{2s} = \frac{1}{2}\left(v_{2s}^2 - v_1^2\right)$$

and

$$c_{p,avg}(T_1 - T_{2s}) = \frac{1}{2}\left(v_{2s}^2 - v_1^2\right).$$

Solving for v_{2s} yields

$$v_{2s} = \left[2c_{p,avg}(T_1 - T_{2s}) + v_1^2\right]^{1/2}$$

$$= \left[2(1070)(800 - 558.7) + 225^2\right]^{1/2}$$

$$= 753$$

$$[=] \left[\frac{\text{J}}{\text{kg}\cdot\text{K}}\text{K}\left[\frac{\text{N}\cdot\text{m}}{1 \text{ J}}\right]\left[\frac{\text{kg}\cdot\text{m/s}^2}{1 \text{ N}}\right]\right]^{1/2} = \text{m/s}.$$

To find the actual exit velocity, we apply the definition of nozzle efficiency (Eq. 11.32b):

$$v_2 = \left(\eta_{isen,n} v_{2s}^2\right)^{1/2}$$

$$= \left[0.97(753)^2\right]^{1/2} \text{m/s} = 741.6 \text{ m/s}.$$

To find the outlet temperature T_2, we apply energy conservation (Eq. 11.5) to the actual process, that is,

$$h_1 - h_2 = c_{p,avg}(T_1 - T_2) = \frac{1}{2}\left(v_2^2 - v_1^2\right).$$

Solving for T_2 yields

$$T_2 = T_1 - \frac{1}{2c_{p,avg}}\left(v_2^2 - v_1^2\right)$$

$$= 800 - \frac{1}{2(1070)}\left(741.6^2 - 225^2\right) \text{K}$$

$$= 566.7 \text{ K}.$$

The verification of units is left as an exercise for the reader. To calculate the entropy change, we use the ideal-gas relationship derived in Chapter 2, Eq. 2.40a:

$$s_2 - s_1 = c_{p,avg} \ln\frac{T_2}{T_1} - R\ln\frac{P_2}{P_1}$$

$$= 1070 \ln\frac{566.7}{800} - 287 \ln\frac{45}{170} \text{ J/kg}\cdot\text{K}$$

$$= 12.54 \text{ J/kg}\cdot\text{K}.$$

Comments We see that the actual outlet temperature is 8 K (= 566.7 − 558.7) higher than the isentropic-process value. This increase in temperature results from the irreversible conversion of kinetic energy to thermal energy by frictional effects. The positive value for the entropy change $s_2 − s_1$ is consistent with this. Note also that our use of $T_{avg} = 675$ K was reasonable since average temperatures based on calculations of T_{2s} and T_2 are quite close to this value (i.e., 679 K and 683 K, respectively).

Self Test
11.11

 Measurements show that the exit pressure and temperature of the nozzle in Example 11.12 are actually 50 kPa and 595 K, respectively. Recalculate the isentropic efficiency for this nozzle.

(Answer: 0.918)

11.3 THROTTLES

Throttle body for spark-ignition engine.
Photograph courtesy of Marcelo Petito and FocusHacks.com.

A throttling process is used to decrease the pressure and/or control the flow rate of a flowing fluid. Any narrow constriction in a flow can be effective as a throttle. Valves, porous plugs, and fine-bore capillary tubes are all used in practice (see Fig. 11.17). Throttling devices are essential components in refrigeration systems, which we will investigate later in Chapter 12. Throttles are also used to control the load in spark-ignition engines. At idle, the intake manifold pressure is reduced by the throttle from close to atmospheric pressure to about 0.4 atm, for example.

11.3a Analysis

To analyze the steady-flow throttling process, consider the control volume shown in Fig. 11.18. The upstream and downstream boundaries are chosen so that they are sufficiently far away from any localized high-velocity regions created by the throttling device. We now apply the fundamental conservation principles.[8]

Mass Conservation

Mass conservation (Eqs. 3.18a and 3.15) is straightforwardly expressed as

$$\dot{m}_1 = \dot{m}_2,$$

or

$$\rho_1 v_{avg,1} A_1 = \rho_2 v_{avg,2} A_2.$$

Energy Conservation

To apply energy conservation, we note, first, that there is no work interaction other than flow work since no power is delivered to or from the control volume by a shaft or any other means. Other simplifications result if we assume the following:

---

[8] We will ignore conservation of linear momentum since this is usually not of interest in applications involving throttles. Note, however, that expressing momentum conservation is relatively easy to do, if needed.

FIGURE 11.17
Cutaway view of a needle valve (left) and photograph of a porous plug (right). Drawing (left) © 2002 Swagelok Company. Used with permission.

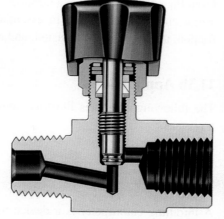

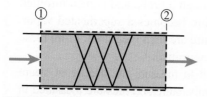

FIGURE 11.18
Control volume for analysis of steady-flow throttling process. The throttling device proper (a porous plug, a valve, or other restriction) is schematically indicated by the Xs.

- Any heat interaction across the control surface is negligible (i.e., $\dot{Q}_{cv,\text{net in}} = 0$).
- The potential energy change from inlet to outlet is negligible, (i.e., $z_2 - z_1 = 0$).
- The kinetic energy of the flow, or the change in the kinetic energy from inlet to outlet, is negligible [i.e., $(\alpha_2 v_2^2 - \alpha_1 v_1^2)/2 \approx 0$]. (See Table 11.4.)

Applying these facts and assumptions to the steady-flow energy equation (Eq. 5.63e),

$$\dot{Q}_{cv,\text{net in}} - \dot{W}_{cv,\text{net out}} = \dot{m}\left[(h_2 - h_1) + \frac{1}{2}(\alpha_2 v_2^2 - \alpha_1 v_1^2) + g(z_2 - z_1) \right],$$

yields

$$0 - 0 = \dot{m}[(h_2 - h_1) + 0 + 0],$$

or

$$h_2 - h_1 = 0. \tag{11.33}$$

Mechanical Energy Conservation

If we make the additional assumption that the flow is incompressible ($\rho = $ constant), the mechanical energy equation (Eq. 6.76) can also be simplified for a throttling process to yield

$$(P_1 - P_2)/\rho = gh_{\text{L}}. \tag{11.34}$$

Table 11.4   Conversion of Kinetic Energy to Thermal Energy for a Flow of Nitrogen

Velocity (m/s)	Specific Kinetic Energy (J/kg)	Equivalent Temperature Change* (K)
1	0.5	0.00048
10	50	0.048
20	200	0.19
50	1250	1.20
100	5000	4.8

*$\Delta T_{\text{equiv}} \equiv \Delta h/c_p = v^2/2c_p$, where $c_p = c_{p,N_2}(300 \text{ K}) = 1041$ J/kg·K.

From Eq. 11.34, we see that a throttling process is entirely dissipative; that is, useful energy is converted to essentially useless thermal energy by fluid friction, unconstrained expansion, and other irreversibilities.

11.3b Applications

The following examples illustrate some practical applications of throttling processes.

Example 11.13   Steam Power Plant Application

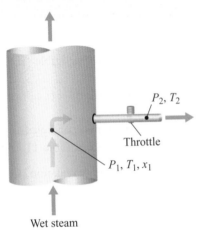

A throttling calorimeter is a device used to measure the quality of wet steam. As shown in Fig. 11.19, a saturated liquid–vapor mixture at a known saturation temperature flows through a probe and is then throttled to a lower pressure such that the mixture becomes a superheated vapor. The downstream temperature and pressure are then measured. From these data (P_1, P_2, T_2) the quality x_1 can be determined.

Consider a throttling calorimeter used to measure the quality of steam entering a reheater at 2.675 MPa. The temperature and pressure measured in the calorimeter after the throttling process are 418 K and 0.1 MPa, respectively. Determine the quality of the mixture entering the reheater. Also determine the entropy charge associated with the steam in going from state 1 to state 2.

FIGURE 11.19
A throttling calorimeter is used to measure the quality of wet steam. See Example 11.13.

Solution

Known P_1, P_2, T_2

Find $x_1, s_2 - s_1$

Sketch

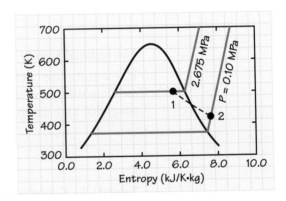

Throttling calorimeters measure steam quality.
Photograph courtesy of Cal Research, Inc.

Assumptions

  i. Steady flow through calorimeter
  ii. Adiabatic calorimeter
  iii. Negligible kinetic and potential energy changes

Analysis   We choose a control volume that cuts across the entrance to the calorimeter tube, follows the exterior surface of the tube, and cuts across the tube where P_2 and T_2 are measured (see Fig. 11.19). Applying conservation of energy to this control volume yields

$$h_1 = h_2.$$

Since we know two properties, we can determine h_2; that is, $h_2 = (T_2, P_2)$. Using the NIST database, we find

$$h_2(418 \text{ K}, 0.1 \text{ MPa}) = 2766.4 \text{ kJ/kg}$$

and

$$s_2(418 \text{ K}, 0.1 \text{ MPa}) = 7.5904 \text{ kJ/kg} \cdot \text{K}.$$

The saturation properties at state 1 ($P_1 = P_{sat}$) are also determined from the NIST database (or interpolated from Table D.2):

$$h_{f,1} = 978.83 \text{ kJ/kg}, \qquad s_{f,1} = 2.5878 \text{ kJ/kg} \cdot \text{K},$$

$$h_{g,1} = 2802.6 \text{ kJ/kg}. \qquad s_{g,1} = 6.2300 \text{ kJ/kg} \cdot \text{K}.$$

Using $h_1 = h_2$, we find the quality x_1 by applying its definition (Eq. 2.49d), that is,

$$x_1 = \frac{h_1 - h_{f,1}}{h_{g,1} - h_{f,1}}$$

$$= \frac{2766.4 - 978.83}{2802.6 - 978.83} = 0.98.$$

We now use the quality to find s_1 from

$$s_1 = (1 - x_1)s_{f,1} + x_1 s_{g,1}$$

$$= 0.02(2.5878) + 0.98(6.2300) \text{ kJ/kg} \cdot \text{K}$$

$$s_1 = 6.157 \text{ kJ/kg} \cdot \text{K}.$$

Thus, the change in entropy is

$$\Delta s = s_2 - s_1 = 7.5904 - 6.157 \text{ kJ/kg} \cdot \text{K} = 1.433 \text{ kJ/kg} \cdot \text{K},$$

which, as expected, is a positive quantity for this highly irreversible process.

Comment Throttling calorimeters are used to measure the approximate quality of steam in geothermal wells for conditions where $x < 0.995$.

Self Test 11.12   Refrigerant (R-134a) in an air-conditioning unit is throttled through several capillary tubes from a saturated-liquid condition at 0.8 MPa to a final pressure of 0.1 MPa. Determine the final quality x.

(Answer: 0.359)

Example 11.14

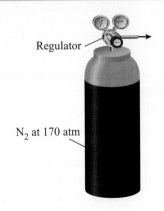

Regulator

N_2 at 170 atm

Nitrogen is contained in a steel tank at 170 atm and room temperature (298 K). A pressure regulator (throttle) is used to reduce the pressure to 2 atm as N_2 flows from the tank. Assuming the process is adiabatic, determine the temperature of the N_2 downstream of the regulator. Use real-gas properties. Compare this result to that obtained assuming ideal-gas behavior.

Solution

Known N_2, P_1, P_2, T_1

Find T_2

Sketch

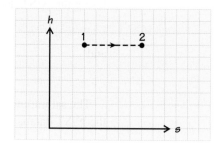

Assumptions

Adiabatic flow with negligible changes in kinetic energy

Analysis   We can use the control volume sketched in Fig. 11.18 directly, where station 1 is upstream of the regulator and station 2 downstream. With the assumptions that heat interactions and kinetic energy changes are negligible, energy conservation is given by Eq. 11.33, so

$$h_2 = h_1.$$

Since both the temperature and pressure are known at state 1, we can find the enthalpy through the calorific equation of state for N_2. The NIST database provides the needed information:

$$h_1(170 \text{ atm}, 298 \text{ K}) = 401.46 \text{ kJ/kg}.$$

From energy conservation, we know that

$$h_2(2 \text{ atm}, T_2) = h_1 = 401.46 \text{ kJ/kg}.$$

We employ the NIST database to find the state at 2 atm that has a specific enthalpy of 401.46 kJ/kg. Searching the database with a fixed pressure and incrementing the temperature, we quickly converge to the state-2 temperature:

$$T_2 = 269.69 \text{ K}.$$

For an ideal gas, the enthalpy is a function of temperature alone, independent of pressure; thus,

$$h_2(T_2) = h_1(T_1),$$

and

$$T_2 = T_1 \text{ (ideal gas)}.$$

Comments   We see that using real-gas data results in a temperature drop of 28.3 K ($= 298 - 269.69$ K) for this throttling process. As a result, we expect that some heat interaction might complicate our simple analysis. A more detailed analysis would have to consider flow rates, heat-transfer coefficients, and the specific geometry.

We also see that the 28.3-K temperature drop is not captured in the ideal-gas analysis. From this, we generalize that care must be exercised in invoking the ideal-gas approximation in throttling processes. The

Joule–Thomson coefficient, defined by

$$\mu_{\mathrm{J}} \equiv \left(\frac{\partial T}{\partial P} \right)_{h},$$

captures the effect of pressure on temperature change for a throttling process. This thermodynamic property can be positive, negative, or zero, causing the temperature downstream of a throttling device to be less than, greater than, or the same as the upstream temperature, respectively. For an ideal gas, the Joule–Thompson coefficient is zero. Joule–Thompson coefficients for selected substances are available from the NIST database.

11.4 PUMPS, COMPRESSORS, AND FANS

In this class of steady-flow devices, a power input is used to create a flow and/or increase the pressure of a fluid. You are likely to have some familiarity with these relatively common devices.

FIGURE 11.20

Various types of positive-displacement pumps: (a) reciprocating piston or plunger pump, (b) external gear pump, (c) double-screw pump, (d) sliding vane pump, (e) three-lobed pump, (f) double circumferential piston pump, and (g) flexible tube peristaltic pump. **Adapted from Ref. [3] with permission.**

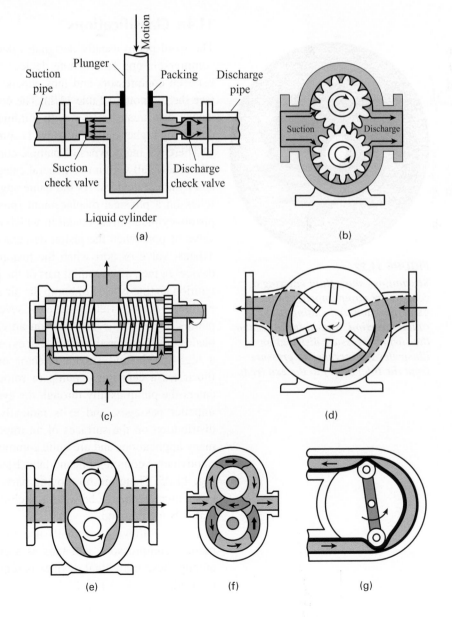

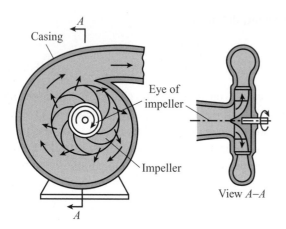
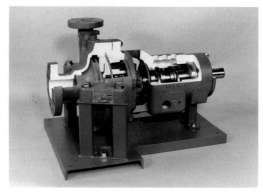

FIGURE 11.21

Centrifugal pump schematic (left) and cutaway view (right). Fluid enters the eye of the impeller, momentum is then imparted to the fluid by the impeller blades, and the fluid exits radially. The exit scroll functions as a diffuser. **Schematic adapted from Ref. [6] with permission. Photograph courtesy of Design Assistance Corporation.**

FIGURE 11.22

Streamlines show fluid entering a centrifugal pump impeller in the axial direction and exiting in the radial direction. Shown in false color, the surface pressure distribution illustrates the increase in pressure from the inlet (blue) to the exit (red).

11.4a Classifications

The word pump usually designates devices that deal with liquids, whereas compressors and fans denote devices that deal with gases. The distinction between compressors and fans depends on the specific field of application (see the footnote to Table 11.1). The common use of the word fan denotes a device that creates a flow with minimal pressure change. A fan capable of delivering higher outlet pressures is sometimes referred to as a blower.

There are many types of pumps, compressors, and fans; however, nearly all of these fall within two general categories: *positive-displacement* devices and *dynamic* devices. As the name suggests, a positive-displacement pump relies on a physical displacement (pushing) of the fluid by the device. A piston–cylinder arrangement in which a fluid enters through an open intake valve or port when the piston descends and is pushed out through an open exhaust valve or port when the piston ascends is a positive displacement device. In fact, a substantial part of the power produced by a throttled spark-ignition engine is used to pump the air and combustion products through the engine. Figure 11.20 illustrates several types of positive-displacement pumps. *Dynamic* devices rely on an exchange of momentum by spinning blades or vanes (rotodynamic devices) or on an exchange of momentum with a high-speed fluid stream (ejector or jet pump devices). Figure 11.21 illustrates a centrifugal pump (a rotodynamic device) in which the fluid enters the pump axially through the eye of the impeller, follows the curved impeller passages, and exits radically. Figure 11.22 shows the pressure distribution on the surfaces of an impeller. Centrifugal pumps are used in many applications and are quite common. Figures 11.23 and 11.24 illustrate rotodynamic compressors and fans. Ejector (or eductor) pumps are shown in Fig. 11.25. There are no moving parts in these devices, which rely on the momentum exchange between a high-pressure stream of fluid and the fluid that is being pumped.

In the following, we will use the term *pumping device* to refer to pumps, compressors, and fans as a class of devices. When a distinction among these individual devices is required, the specific device name will be used.

11.4b Analysis

Control Volume Choice

See Eq. 4.7 in Chapter 4 for the general definition of control volume power.

See Eq. 4.10 for the definition of shaft power.

In dealing with pumping devices, several choices of control volumes are possible. The particular choice depends on either the information provided as a given or on the information sought from the analysis. Three typical control volume choices are illustrated in Fig. 11.26. Let us examine these to see how they differ and the consequences of choosing one over another.

Control volume I contains only fluid; no parts of the pumping device proper are included within the volume. The control surface follows the interior surfaces of the pumping device as suggested in the sketch (Fig. 11.26a). The control volume work ($\dot{W}_{cv}$) associated with this choice is the power delivered by the solid surface of the pumping device directly to the fluid and follows from an integration of the product of the local force and velocity at the fluid–surface interface over the entire control surface. Control volume I is a useful choice when one wants to isolate inefficiencies in the fluid itself from those produced in the pumping device proper. For example, losses from friction in the bearings of the pumping device would be *excluded* in an analysis involving control volume I.

If one desires to include bearing losses, control volume II (Fig. 11.26b) may be an appropriate choice. Here the control surface cuts the inlet and exit pipes and is contiguous with the external surface of the pumping device. The control surface also cuts through the shaft of the pumping device. For this choice of control volumes, the only control volume work ($\dot{W}_{cv}$) is that associated with the spinning shaft. In choosing control volume II, inefficiencies associated with friction in the fluid being pumped and the moving parts of the pumping device are lumped together.

A third control volume choice (Fig. 11.26c) includes the device driving the pumping device. This control volume is frequently applied to electric-motor-driven pumping devices and is used to define an overall efficiency in converting electrical power to useful fluid power. We will explore this concept later in our discussion of the various efficiencies used to characterize pumping devices.

FIGURE 11.23
Multistage compressor blading for an aircraft engine. **Photograph courtesy of Pratt and Whitney.**

FIGURE 11.24
Six fans drive the flow through the 80 × 120-ft wind tunnel at NASA Ames Research Center. For a sense of scale, note the people in the photograph. **Photograph courtesy of NASA.**

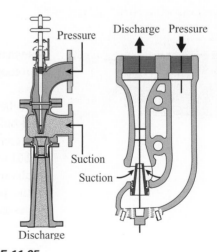

FIGURE 11.25
Cross-sectional views of eductors (i.e., liquid jet pumps). The eductor configuration shown on the right is used to pump sand and mud. **Adapted from Ref. [8] with permission.**

Control volume I—Fluid

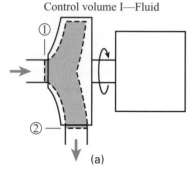

(a)

Control volume II—Fluid & Pump

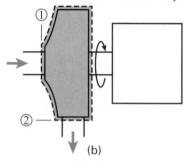

(b)

Control volume III—Fluid, Pump, & Motor

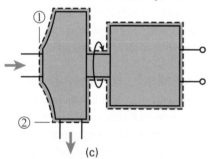

(c)

FIGURE 11.26
Various control volumes that can be used to analyze pumping devices.

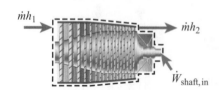

Application of Conservation Principles

We now apply the basic conservation principles to steady-flow, rotodynamic pumps and compressors.

Mass Conservation   Mass conservation is represented in the same manner as in all of our previous analyses of steady-flow devices, that is, Eq. 3.18a, which applies equally well to all control volumes of Fig. 11.26.

Energy Conservation   We begin by considering the control volume containing both the pump and its fluid (Fig. 11.26b) and invoking the following assumptions (see also Table 11.1):

- The heat interaction across the control surface, the pump external surface, is zero (adiabatic) or small compared to other flows of energy (i.e., $\dot{Q}_{cv,\,net\,in} = 0$).
- The kinetic energy of the flow, or the change in kinetic energy from inlet to outlet, is negligible compared to other flows of energy [i.e., $(\alpha_2 v_{avg,2}^2 - \alpha_1 v_{avg,1}^2)/2 \approx 0$].
- The potential energy change from inlet to outlet is negligible (i.e., $z_2 - z_1 = 0$).

Applying these assumptions to the steady-flow energy equation (Eq. 5.63e),

$$\dot{Q}_{cv,\,net\,in} - \dot{W}_{cv,\,net\,out} = \dot{m}\left[(h_2 - h_1) + \frac{1}{2}(\alpha_2 v_{avg,2}^2 - \alpha_1 v_{avg,1}^2) + g(z_2 - z_1)\right],$$

yields

$$0 - \dot{W}_{cv,\,net\,out} = \dot{m}[(h_2 - h_1) + 0 + 0].$$

Noting that power is always delivered to a pumping device (i.e., $\dot{W}_{cv,\,net\,out}$ is always negative), we define the pump shaft power as

$$\dot{W}_{shaft,in} \equiv -\dot{W}_{cv,\,net\,out}. \tag{11.35}$$

With this definition, our final simplified conservation of energy expression becomes

$$\dot{W}_{shaft,in} = \dot{m}(h_2 - h_1). \tag{11.36}$$

Note that Eq. 11.36 is quite general, subject to our assumptions, and applies to all rotodynamic pumping devices[9] for both incompressible and compressible flows.

To better understand the sources of inefficiencies, we next consider the fluid-only control volume (Fig. 11.26a). To apply energy conservation to this new control volume and still be consistent with our previous analysis (Eq. 11.36) requires that we now include a heat-transfer term, $\dot{Q}_{fluid,in}$. This term results from the transfer of thermal energy from the frictionally heated parts of the pump, such as bearings, to the fluid. Energy conservation is now expressed as

$$\dot{Q}_{fluid,in} + \dot{W}_{fluid,in} = \dot{m}(h_2 - h_1), \tag{11.37}$$

---

[9] Equation 11.36 also applies to positive-displacement devices if the pulsatile flow is treated as a time-averaged steady flow.

where $\dot{W}_{\text{fluid,in}}$ is the power input to the fluid by the moving pump (or compressor) parts. Comparing Eqs. 11.36 and 11.37, we see that

$$\dot{W}_{\text{fluid,in}} = \dot{W}_{\text{shaft,in}} - \dot{Q}_{\text{fluid,in}}. \tag{11.38}$$

Thus the actual power delivered to the fluid is less than the power delivered by the shaft, as we expect. We will return to this point later.

Momentum Conservation Generally, conservation of *linear* momentum is not particularly germane to an analysis of pumping devices. Conservation of *angular* momentum, however, is quite important to a detailed understanding of rotodynamic devices. Nevertheless, this topic reaches beyond the scope of this book and we refer the reader to other sources for further information [3, 6, 7, 9].

Mechanical Energy Conservation (Incompressible Flow) If we make the further assumption that the flow is incompressible ($\rho = $ constant), then the mechanical energy equation (Eq. 6.76),

$$\frac{P_1}{\rho} + \frac{1}{2}\alpha_1 v_{\text{avg,1}}^2 + gz_1 = \frac{P_2}{\rho} + \frac{1}{2}\alpha_2 v_{\text{avg,2}}^2 + gz_2 + \frac{\dot{W}_{\text{cv,net out}}}{\dot{m}} + gh_{\text{L}},$$

can also be applied to pumping devices. Eliminating the kinetic and potential energy terms in Eq. 6.76, rearranging, and recognizing that $\dot{W}_{\text{cv,net out}} \equiv -\dot{W}_{\text{cv,in}}$ yield

$$\frac{P_2 - P_1}{\rho} = \frac{\dot{W}_{\text{cv,in}}}{\dot{m}} - gh_{\text{L,cv}}, \tag{11.39a}$$

where h_{L} is the head loss. We retain the subscript cv to denote that both the power and head loss are specific to our choice of control volume. If we chose the fluid-only control volume, then $\dot{W}_{\text{cv,in}} = \dot{W}_{\text{fluid,in}}$ and the head loss is only associated with fluid friction. If we choose the fluid-plus-pump control volume, then $\dot{W}_{\text{cv,in}} = \dot{W}_{\text{shaft,in}}$, and the pump bearing friction losses, etc. are combined with the fluid frictional losses in the control-volume head loss. Thus, we can write the following equivalent statements of mechanical energy conservation:

$$\frac{P_2 - P_1}{\rho} = \frac{\dot{W}_{\text{shaft,in}} - \dot{Q}_{\text{fluid,in}}}{\dot{m}} - gh_{\text{L,fluid}} \tag{11.39b}$$

or

$$\frac{P_2 - P_1}{\rho} = \frac{\dot{W}_{\text{shaft,in}}}{\dot{m}} - gh_{\text{L,tot}}, \tag{11.39c}$$

where

$$h_{\text{L,tot}} \equiv h_{\text{L,fluid}} + \frac{\dot{Q}_{\text{fluid,in}}}{\dot{m}}. \tag{11.39d}$$

We now compare Eq. 11.39 with our first-law statement of energy conservation, Eq. 11.36, to which the definition of enthalpy ($h = u + P/\rho$) has been applied, that is,

$$\frac{P_2 - P_1}{\rho} = \frac{\dot{W}_{\text{shaft,in}}}{\dot{m}} - c_v(T_2 - T_1). \tag{11.40}$$

Here we see that the total head loss term in the mechanical energy equation, $gh_{\text{L,tot}}$, corresponds to the internal energy change in the first-law expression,

$c_v(T_2 - T_1)$. Friction within the control volume associated with both the fluid and moving pump parts converts energy that could have produced a higher outlet pressure into useless thermal energy.

As a useful digression, we introduce the concept of **reversible, steady-flow work.** In Chapter 4, we saw that, for a thermodynamic system undergoing a reversible process, the work performed by the system was (Eq. 4.8b)

$$W_{\text{sys,rev}} = \int_1^2 P d\mathcal{V}.$$

This follows directly from the definition of work, $\delta W = \mathbf{F} \cdot d\mathbf{s}$ (Eq. 4.5), and the concept of a quasi-equilibrium process. Similarly, the work performed in a reversible, steady-flow process can be derived by the simultaneous application of first- and second-law principles to a parcel of fluid traveling from the inlet to the outlet of a steady-flow device in a reversible manner. (See, for example, Refs. [10, 11].) The result of such a derivation is that the reversible, steady-flow work (power) delivered by the control volume can be expressed as

$$\dot{W}_{\text{SF,rev}} = -\dot{m} \int_1^2 v dP, \tag{11.41a}$$

or

$$\dot{W}_{\text{SF,rev}} = -\dot{m} \int_1^2 \frac{1}{\rho} dP, \tag{11.41b}$$

The pumping of water can often be treated as an incompressible process.

where 1 and 2 denote the inlet and outlet states, respectively, and the integral is performed following a fluid parcel from the inlet to the outlet. Furthermore, negligible kinetic and potential energy changes have been assumed. Evaluating the integral in Eq. 11.41 can be difficult because the variation of density with the pressure depends on the details of the flow process; however, for simple ideal processes the integration can be performed. For example, if the process is assumed to be incompressible, then the density is constant and can be removed from the integrand. Thus,

$$\dot{W}_{\text{SF,rev}} = -\dot{m} \frac{1}{\rho} \int_1^2 dP = -\frac{\dot{m}(P_2 - P_1)}{\rho}. \tag{11.42}$$

Note that in Eqs. 11.41 and 11.42 $\dot{W}_{\text{SF,rev}}$ denotes the power delivered *by* the control volume; for a pumping device, the reversible power is thus defined as $\dot{W}_{\text{fluid,rev}} \equiv -\dot{W}_{\text{SF,rev}}$. Using this definition, we rearrange Eq. 11.42 to look like the previously discussed energy and mechanical energy equations for incompressible flow through a pump (i.e., Eqs. 11.40 and 11.39):

$$\frac{P_2 - P_1}{\rho} = \frac{\dot{W}_{\text{fluid,rev}}}{\dot{m}}. \tag{11.43}$$

Comparing these three relationships (Eqs. 11.40, 11.39, and 11.43), we see that, to achieve a given pressure rise across a pump for a particular flow rate, the minimum power input results for a reversible process. In any real (irreversible) pumping process, a greater power input is required to compensate for the head loss (Eq. 11.39a) or, equivalently, the frictional heating of the fluid (Eq. 11.40).

Equation 11.42 can also be evaluated for ideal gased undergoing idealized processes such as constant temperature or constant entropy.

Efficiencies

The preceding discussion leads naturally to the consideration of pump efficiency. Various efficiencies can be defined. In keeping with previous analyses in this chapter, an isentropic efficiency can be defined for pumping devices (Eq. 7.28) by

$$\eta_{\text{isen,p}} \equiv \frac{(\dot{W}_{\text{fluid,in}}/\dot{m})_{\text{isen}}}{(\dot{W}_{\text{shaft,in}}/\dot{m})_{\text{act}}}, \tag{11.44}$$

where the numerator is the power input to the fluid (per unit flow rate) for the isentropic (adiabatic and reversible) process and the denominator is the actual shaft power input. Figure 11.27 illustrates the ideal (isentropic) process and the real process on h–s coordinates. Note that the real process, being irreversible, is shown as a dashed line in this sketch. Using our previous first-law statements (Eqs. 11.36 and 11.37), we can relate the isentropic efficiency to the enthalpies associated with both the real and ideal processes as follows:

$$\eta_{\text{isen,p}} = \frac{h_{2s} - h_1}{h_2 - h_1}. \tag{11.45}$$

Note that this relationship applies generally to all pumping devices, independent of whether the fluid is compressible or incompressible.

If, however, we consider an incompressible fluid, we know that, for any conditions, the minimum pumping power is that given by Eq. 11.42, which defines the steady-flow, reversible power. Furthermore, the first law reduces to Eq. 11.43. Substituting Eq. 11.43 into the definition of isentropic efficiency (Eq. 11.44) yields

$$\eta_{\text{isen,pump}} = \frac{\dot{m}(P_2 - P_1)/\rho}{(\dot{W}_{\text{shaft,in}})_{\text{act}}}. \tag{11.46}$$

This is the working definition of efficiency used in pump testing [8] and is easily calculated by measuring the inlet and outlet pressures and the shaft torque and rotational speed (see Table 4.1). We also note that Eq. 11.46 is readily interpreted as the ratio of a useful output quantity, a pressure increase,[10] to an input quantity that must be paid for, the shaft power.

Pumping devices driven by electric motors frequently employ an overall efficiency [8], defined as

$$\eta_{\text{OA,pump}} \equiv \frac{(\dot{W}_{\text{fluid,in}})_{\text{isen}}}{\dot{W}_{\text{elec,in}}}, \tag{11.47}$$

See problems 11.72 and 11.73.

Equation 7.31 in Chapter 7 provides a connection between pump isentropic efficiency and entropy production; reviewing the development of this relationship enhances the present discussion.

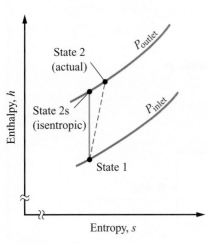

FIGURE 11.27

To define a pumping device efficiency, an ideal outlet state is defined by an isentropic compression from the actual inlet state to the same outlet pressure as that for the actual process.

---

[10] Rigorously, the useful output quantity is the net flow work (power), that is, $(\dot{m}P_2/\rho - \dot{m}P_1/\rho) = \dot{m}(P_2 - P_1)/\rho$.

where $\dot{W}_{\text{elec, in}}$ is the electrical power delivered to the motor. The overall efficiency thus includes irreversibilities within the motor, such as Joule heating, eddy currents, fan losses (windage), bearing friction, etc. (See Fig. 11.26c.) Again, if we deal with an incompressible fluid, Eq. 11.47 becomes

$$\eta_{\text{OA,pump}} = \frac{\dot{m}(P_2 - P_1)/\rho}{\dot{W}_{\text{elec,in}}}. \qquad (11.48)$$

This is a direct expression of the general idea that an efficiency is the ratio of useful energy to energy that costs.

The following examples illustrate some of the concepts discussed here related to pumps, compressors, and fans.

Example 11.15

A sales catalog presents the following data for a cast-iron centrifugal pump:

Flow rate (gal/min)	Pump Head (ft)
50	15
30	75
10	100
0	54

A 1-hp (shaft) electrical motor drives the pump. Use these data to estimate the fluid power $(\dot{W}_{\text{fluid, in}})_{\text{isen}}$ and the pump efficiency η_{pump}, for pumping water. Plot η_{pump} and the pump head as a function of flow rate. The water density is 996 kg/m^3.

Solution

Known $\dot{W}_{\text{shaft,in}}, \dot{V}, \Delta P/\rho g \ (\equiv \text{pump head}), \rho$

Find $(\dot{W}_{\text{fluid, in}})_{\text{isen}}, \eta_{\text{pump}}$

Sketch

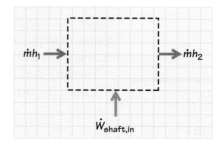

Assumptions

   i. Steady-state, steady flow
   ii. Negligible kinetic and potential energy changes

Analysis We first deal with interpreting the given data and their associated engineering units. The given *pump head* is a measure of the pressure rise across the pump, expressed as an equivalent height of a column of the pumped fluid. From our study of fluid statics in Chapter 6, we know that this conversion is given by (Eq. 6.22)

$$(\Delta P)_{\text{pump}} = \rho g (\Delta z)_{\text{equiv,pump}},$$

where $(\Delta z)_{\text{equiv,pump}}$ is the *pump head*. Converting to SI units, we calculate $\Delta P \,(= P_2 - P_1)$ for the first entry in the table as

$$(\Delta z)_{\text{pump,equiv}} = 15\ \text{ft}\,\frac{0.3048\ \text{m}}{\text{ft}} = 4.572\ \text{m}$$

and

$$\Delta P_{\text{pump}} = \rho g (\Delta z) = 996(9.807)4.572 = 44{,}700$$

$$[=]\ \frac{\text{kg}}{\text{m}^3}\,\frac{\text{m}}{\text{s}^2}\,\text{m}\left[\frac{1\ \text{N}}{\text{kg}\cdot\text{m/s}^2}\right]\left[\frac{1\ \text{Pa}}{\text{N/m}^2}\right] = \text{Pa}.$$

However, we will not need to calculate actual pressure changes since the isentropic pump work for an incompressible fluid (Eq. 11.43) can be calculated as

$$\left(\dot{W}_{\text{fluid,in}}\right)_{\text{isen}} = \dot{m}\left(\frac{P_2 - P_1}{\rho}\right) = \dot{m}g(\Delta z)_{\text{equiv,pump}}.$$

The mass flow rate is the product of the density and the volume flow rate; that is, $\dot{m} = \rho\dot{V}$ (Eq. 3.14 and 3.15). Thus,

$$\left(\dot{W}_{\text{fluid,in}}\right)_{\text{isen}} = \rho g \dot{V}(\Delta z)_{\text{equiv,pump}}.$$

The given flow data in gallons per minute are easily converted to SI units. For example,

$$50\,\frac{\text{gal}}{\text{min}}\left[\frac{3.785 \times 10^{-3}\ \text{m}^3}{\text{gal}}\right]\left[\frac{1\ \text{min}}{60\ \text{s}}\right] = 3.15 \times 10^{-3}\,\frac{\text{m}^3}{\text{s}}.$$

The fluid power for the first table entry is thus

$$\left(\dot{W}_{\text{fluid,in}}\right)_{\text{isen}} = 996(9.807)3.15 \times 10^{-3}(4.572) = 141$$

$$[=]\ \left(\frac{\text{kg}}{\text{m}^3}\right)\left(\frac{\text{m}}{\text{s}^2}\right)\left(\frac{\text{m}^3}{\text{s}}\right)(\text{m})\left[\frac{1\ \text{N}}{\text{kg}\cdot\text{m/s}^2}\right]\left[\frac{1\ \text{J}}{\text{N}\cdot\text{m}}\right] = \text{J/s or W}.$$

The pump efficiency, given by Eq. 11.46, is this fluid power divided by the shaft power. Converting the shaft power to SI units yields

$$\dot{W}_{\text{shaft,in}} = 1\ \text{hp}\left[\frac{745.7\ \text{W}}{\text{hp}}\right] = 745.7\ \text{W},$$

from which we calculate the efficiency as follows:

$$\eta_{\text{pump}} = \frac{\left(\dot{W}_{\text{fluid,in}}\right)_{\text{isen}}}{\dot{W}_{\text{shaft,in}}} = \frac{141\ \text{W}}{745.7\ \text{W}}$$

$$= 0.19\ \text{or}\ 19\%.$$

Similar calculations can be performed for the other test conditions. The following table summarizes the results; Fig. 11.28 provides the desired graphs.

FIGURE 11.28
Centrifugal pump characteristics (Example 11.15).

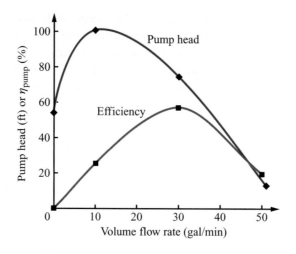

$\dot{V}$		$\dot{W}_{fluid,in}$	η_{pump}
(gal/min)	(m³/s)	(W)	(%)
50	3.15×10^{-3}	141	19
30	1.89×10^{-3}	422	57
10	0.63×10^{-3}	188	25
0	0	0	0

Comments Our first observation from this example is the common use of non-SI units in the United States. Frequently needed conversion factors are presented inside the covers of this book. The second observation concerns the actual pump characteristics plotted in Fig. 11.28. In general, the efficiency of a centrifugal pump peaks within the overall flow range. In a well-designed pumping system, the pump should operate at or near the peak efficiency. We also note that the peak pressure (pump head) is delivered at a relatively low flow rate.

Although no information was provided in the example, one might like to know the overall efficiency of the pump–motor system. Data from Ref. [12] suggest an efficiency of 83% for a 1-hp motor. (Efficiency rises with motor size; e.g., it is 90% for a 10-hp motor and 96% for a 200-hp one.) Using this value, we can estimate the peak overall efficiency (Eq. 11.47):

$$\eta_{OA, \, pump} = \frac{\dot{W}_{fluid}}{\dot{W}_{elec}} = \frac{\dot{W}_{fluid}}{\dot{W}_{shaft}/\eta_{motor}}$$

$$= \eta_{pump} \, \eta_{motor} = 0.57(0.83) = 0.47 \text{ or } 47\%.$$

Thus we conclude that, at best, less than half of the electrical energy is used to perform the desired task. The remainder heats the fluid and surroundings in a nonuseful way.

Self Test 11.13 ☑ A 0.75-kW motor-driven pump is used to raise 5 kg/s of water a total elevation of 10 m from a reservoir to a large open tank. Neglecting frictional losses in the pipe, determine the overall pump efficiency.

(Answer: 0.654)

Example 11.16 Steam Power Plant Application

The main boiler feed pump for a 250-MW electrical generating unit consists of a four-stage centrifugal pump in series with a single-stage booster pump. A steam turbine with a 6-MW shaft output drives the pump system as shown in the sketch. Liquid water enters the pump at 6.8 kPa with a flow rate of 250 kg/s and exits at 20 MPa. Determine the isentropic efficiency of the pump.

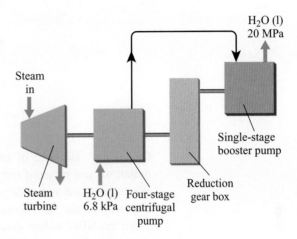

Solution

Known $\dot{W}_{\text{shaft,in}}$, P_1, P_2, $\dot{m}$

Find $\eta_{\text{isen,pump}}$

Sketch

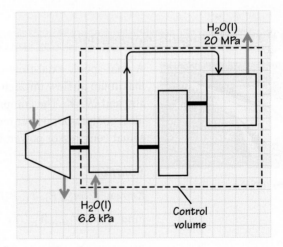

Assumptions

 i. Steady flow
 ii. Incompressible flow with $\rho = \rho(P_{\text{sat}} = P_1)$
 iii. Adiabatic process
 iv. Negligible kinetic and potential energy changes

Analysis We begin by selecting a control volume that cuts through the turbine output/pump drive shaft as shown on the sketch. This control volume includes both pumps and the interconnecting reduction gear box.

With the assumption of an incompressible fluid, Eq. 11.46 expresses the isentropic efficiency as

$$\eta_{\text{isen,pump}} = \frac{\dot{m}(P_2 - P_1)/\rho}{(\dot{W}_{\text{shaft,in}})_{\text{act}}},$$

which we evaluate using a water density of 992.75 kg/m³ [$= \rho(P_{\text{sat}} = 6.8$ kPa)] as follows:

$$\eta_{\text{isen,pump}} = \frac{250(20 \times 10^6 - 6800)/992.75}{6 \times 10^6}$$

$$= \frac{5.035 \times 10^6}{6 \times 10^6}$$

$$= 0.839$$

$$[=] \frac{(\text{kg/s})(\text{N/m}^2)/(\text{kg/m}^3)}{(\text{J/s})}\left[\frac{1 \text{ J}}{\text{N} \cdot \text{m}}\right] = 1.$$

Comments Because of the large difference in pressure from inlet to outlet, we test to see if our assumption of constant density is reasonable. We estimate the outlet density by assuming that the outlet temperature is the same as the inlet temperature [i.e., $T_2 = T_1 = T_{\text{sat}}(P_1) = 311.61$ K]. From the NIST online database ρ_2 (311.61 K, 20 MPa) = 1001.4 kg/m³; thus, the density variation from inlet to outlet is less than 0.87% ($= [(1001.4 - 992.75)/992.75] \cdot 100\%$). Since the data used in this example are from a real power plant, we have an opportunity to determine a realistic value for the ratio of the pumping power to the total power produced by the unit. This is simply $\dot{W}_{\text{pump}}/\dot{W}_{\text{elec,out}} = (6 \text{ MW})/(250 \text{ MW}) = 0.024$ or 2.4%. This is a small fraction, as expected.

Example 11.17 Jet Engine Application

Photograph courtesy of NASA.

A dual-spool, axial-flow compressor is used in a jet fighter engine. The maximum airflow through the engine is 84.4 kg/s for inlet conditions of 1 atm and 293 K. The compressor pressure ratio is 12.9:1. Determine (a) the ideal compressor power required if the isentropic efficiency is 94%, (b) the actual compressor power, and (c) the actual outlet temperature.

Solution

Known $\dot{m}_{\text{air}}, P_1, T_1, P_2/P_1, \eta_{\text{isen,p}}$

Find $\dot{W}_{\text{ideal}}, \dot{W}_{\text{act}}, T_2$

Sketch See Fig. 11.27.

Assumptions

  i. Steady flow
  ii. Adiabatic process
  iii. Negligible changes in kinetic and potential energies
  iv. Ideal-gas behavior with constant c_p and γ

Analysis Subject to our first three assumptions, conservation of energy for the compressor is expressed by Eq. 11.36. The ideal (minimum) power required to drive the compressor is for isentropic compression; thus,

$$\dot{W}_{\text{ideal}} = \dot{m}_{\text{air}}(h_{2s} - h_1).$$

Applying our fourth assumption allows us to express the enthalpy change as $c_{p,\text{avg}}(T_{2s} - T_1)$. With this substitution, the ideal power becomes

$$\dot{W}_{\text{ideal}} = \dot{m}_{\text{air}}c_{p,\text{avg}}(T_{2s} - T_1).$$

For an isentropic process (cf. Eq. 2.41),

$$\frac{T_{2s}}{T_1} = \left(\frac{P_2}{P_1}\right)^{\frac{\gamma-1}{\gamma}}.$$

Thus,

$$\dot{W}_{\text{ideal}} = \dot{m}_{\text{air}}c_{p,\text{avg}}T_1\left[(P_2/P_1)^{\frac{\gamma-1}{\gamma}} - 1\right].$$

From Table C.3, we find c_p (293 K) = 1.007 kJ/kg; and with this value, we determine $\gamma = 1.4$ [$\equiv c_p/c_v = c_p/(c_p - R)$]. The ideal power is then evaluated as

$$\dot{W}_{\text{ideal}} = 84.4(1.007)293\left[(12.9)^{\frac{1.4-1}{1.4}} - 1\right]$$

$$= 26{,}800$$

$$[=] \frac{\text{kg}}{\text{s}} \frac{\text{kJ}}{\text{kg}\cdot\text{K}}\text{K} = \frac{\text{kJ}}{\text{s}} \text{ or kW.}$$

As an aside, direct calculation of T_{2s} from Eq. 2.41 yields 608.4 K.

From the definition of the isentropic efficiency (Eq. 11.44), we find the actual power required to drive the compressor to be

$$\dot{W}_{\text{act}} = \frac{\dot{W}_{\text{ideal}}}{\eta_{\text{isen,p}}}$$

$$= \frac{26{,}800 \text{ kW}}{0.94} = 28{,}500 \text{ kW.}$$

To find the actual outlet temperature T_2, we now apply conservation of energy to the actual compressor, that is,

$$\dot{W}_{\text{act}} = \dot{m}_{\text{air}}(h_2 - h_1)$$

$$= \dot{m}_{\text{air}}c_{p,\text{avg}}(T_2 - T_1).$$

Solving for T_2 yields

$$T_2 = T_1 + \frac{\dot{W}_{\text{act}}}{\dot{m}_{\text{air}}c_{p,\text{avg}}}$$

$$= 293 + \frac{28{,}500}{84.4(1.007)} = 628.3 \text{ K.}$$

Comments We note that irreversibilities, primarily frictional effects, result in the actual outlet temperature being about 20 K ($\approx$ 628.3 K − 608.4 K) higher than that associated with the ideal isentropic compression.

Self Test
11.14

☑ The compressor of an air-conditioning unit compresses refrigerant (R-134a) from a saturated-vapor condition at 0.1 MPa to a final pressure and temperature of 0.8 MPa and 60°C. Determine the isentropic efficiency of the compressor.

(Answer: 0.693)

Example 11.18[11]

Water at 300 K is pumped from one reservoir to another at an elevation 30 m above the first as shown in the sketch. The smooth-walled, 34.3-mm-diameter interconnecting pipe is 200 m long. To maintain a flow rate of 0.921 kg/s, 358.5 W of shaft power is supplied to the pump. Neglecting all minor losses, estimate the pump isentropic efficiency.

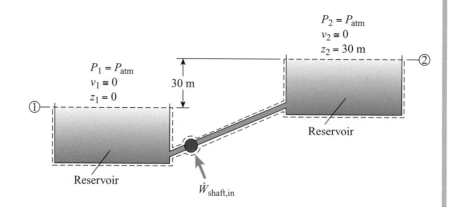

$$P_2 = P_{atm}$$
$$v_2 \cong 0$$
$$z_2 = 30 \text{ m}$$

$$P_1 = P_{atm}$$
$$v_1 \cong 0$$
$$z_1 = 0$$

30 m

Reservoir

Reservoir

$\dot{W}_{shaft,in}$

Solution

Known $\Delta z, \dot{m}, D, L, \dot{W}_{shaft,in}$

Find $\eta_{isen,pump}$

Sketch   A useful control volume is shown on the sketch with the example statement

Assumptions

i. Steady flow
ii. Incompressible fluid
iii. Large reservoirs
iv. $P_1 = P_2$
v. $v_1 = v_2 \approx 0$
vi. $\dot{Q}_{cv} = 0$ (adiabatic)
vii. Negligible minor losses
viii. $\varepsilon/D = 0$ (smooth pipe)
ix. Properties approximately those at 1 atm and 300 K.

Analysis   Our initial analysis is identical to that in Example 5.18. We apply the incompressible, steady-flow form of energy conservation (Eq. 6.76) to our control volume as follows:

$$\frac{P_1}{\rho} + \frac{1}{2}v_1^2 + gz_1 = \frac{P_2}{\rho} + \frac{1}{2}v_2^2 + gz_2 - \frac{\dot{W}_{shaft,in}}{\dot{m}} + gh_{L,tot},$$

which simplifies with assumptions iv and v to

$$\dot{W}_{shaft,in} = \dot{m}g(z_2 - z_1 + h_{L,tot}).$$

[11] This example revisits the situation discussed in Example 5.18. Before proceeding, the reader should review this Chapter 5 example.

We recognize that the total head loss can be separated into two parts: the losses associated with the pipe friction and the losses associated with the pump itself (i.e., $h_{L,tot} = h_{L,pipe} + h_{L,pump}$). For the ideal (isentropic) pump, $h_{L,pump}$ is zero; thus, we can express the ideal power input as

$$\dot{W}_{pump,ideal} = \dot{m}g(z_2 - z_1 + h_{L,pipe}).$$

The isentropic efficiency can then be obtained by applying its definition (Eq. 11.44):

$$\eta_{isen,p} = \frac{\dot{W}_{pump,ideal}}{\dot{W}_{shaft,in}} = \dot{m}g\frac{(z_2 - z_1 + h_{L,pipe})}{\dot{W}_{shaft,in}}.$$

We now apply our knowledge of internal flows from Chapter 10 to find the head loss associated with pipe friction, $h_{L,pipe}$. We begin by calculating the pipe Reynolds number:

$$Re_D = \frac{\rho v_{avg}D}{\mu} = \frac{4\dot{m}}{\pi\mu D}$$

$$= \frac{4(0.921)}{\pi 854 \times 10^{-6}(0.0343)} = 40,000$$

$$[=] \frac{kg/s}{(N \cdot s/m^2)m}\left[\frac{1\,N}{kg \cdot m/s^2}\right] = 1,$$

where the viscosity for water is found from the NIST database. With this Reynolds number, we find the friction factor from the Moody chart (Fig. 10.17) to be

$$f_D \approx 0.022.$$

The pipe head loss and friction factor are related by Eq. 10.16, that is,

$$h_{L,pipe} = f_D\left(\frac{L}{D}\right)\frac{v_{avg}^2}{2g},$$

where

$$v_{avg} = \frac{\dot{m}}{\rho A} = \frac{4\dot{m}}{\rho\pi D^2}.$$

Using the density from the NIST database, we calculate v_{avg} and then $h_{L,pipe}$:

$$v_{avg} = \frac{4(0.921)}{997\pi(0.0343)^2} = 1.0$$

$$[=] \frac{(kg/s)}{(kg/m^3)m^2} = m/s$$

and

$$h_{L,pipe} = 0.022\left(\frac{200}{0.0343}\right)\frac{(1)^2}{2(9.807)} = 6.54$$

$$[=] \frac{m}{m}\frac{(m/s)^2}{m/s^2} = m.$$

We now have all the information needed to evaluate the isentropic efficiency as expressed previously; thus,

$$\eta_{isen,p} = \frac{\dot{m}g(z_2 - z_1 + h_{L,pipe})}{\dot{W}_{shaft,in}}$$

$$= \frac{0.921(9.807)(30 - 0 + 6.54)}{358.5} = 0.92 \text{ or } 92\%$$

$$[=] \frac{(\text{kg/s})(\text{m/s}^2)\text{m}}{\text{W}} \left[\frac{1 \text{ W}}{\text{J/s}}\right]\left[\frac{1 \text{ J}}{\text{N} \cdot \text{m}}\right]\left[\frac{1 \text{ N}}{\text{kg} \cdot \text{m/s}^2}\right] = 1.$$

Comments This example illustrates how the concepts presented in Chapter 10 can be combined with those of this chapter to solve interesting engineering problems. In a real-world context, the present problem is likely to be inverted: the flow rate, elevation change, and pipe length would be given, and the engineer would have to choose a particular pump and pipe to perform the task efficiently. We also note that the minor losses associated with a sharp-edged entry and a sudden expansion are indeed small, that is,

$$h_{\text{L,minor}} = \sum K v_{\text{avg}}^2/2g = (0.5 + 1)(1)^2/[(2)9.807] \text{ m} = 0.076 \text{ m},$$

compared to the 6.54-m head loss associated with the pipe.

Table 11.5   Some Applications of Turbines

General Classification	Application
Steam turbines	Utility and industrial power plants
	Electric generator drives
	Pump drives
Gas turbines	Turbochargers for internal combustion engines (trucks, buses, locomotives, and automobiles)
	Bus and truck engines
	Aircraft propulsion (turbojet, turbofan, turboshaft, and turboprop)
	Pipeline pump drives
	Ship propulsion
	Ship electric generator drives
	Rocket engine pump drives
	Stationary power generation
Water turbines	Hydroelectric power generation
Wind turbines	Small-scale power generation
	Large-scale power generation

Table 11.6   Hydro (Water) Turbine Types

	Application		
Type	**High Head**	**Medium Head**	**Low Head**
Impulse turbines	Pelton	Cross-flow[†]	Cross-flow[†]
	Turgo*	Multijet Pelton	
		Turgo*	
Reaction tubines	—	Francis	Propeller
			Kaplan

*The water jets of a Turgo turbine enter and exit obliquely from the sides, thus preventing interference of incoming and exiting jets.

[†]Water from a rectangular nozzle strikes a bladed cylindrical drum.

11.5 TURBINES

11.5a Classifications and Applications

As shown in Table 11.5, we first classify turbines according to the working fluid (i.e., steam, gas, water, and wind). Within each working-fluid class, we can further classify turbines according to flow-passage types. For example, hydro (water) turbines are of two general types, which are then further subdivided as indicated in Table 11.6 and illustrated in Fig. 11.29. In the operation of **impulse turbines,** a free jet, or jets, strike the turbine wheel, also known as a runner (Fig. 11.29a and Fig. 11.30), whereas in reaction turbines,

Gas turbine engines play an important role in transportation systems. Here are shown the 150-mile per hour JetTrain locomotive (top) and a hydrofoil ferry (bottom). Photographs courtesy of Bombardier Transportation (top) and Peter Venema and International Hydrofoil Society (bottom).

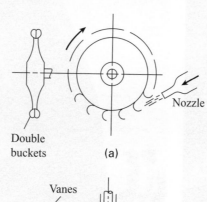

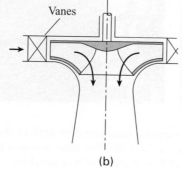

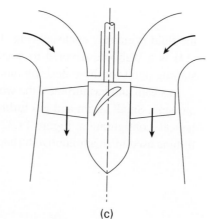

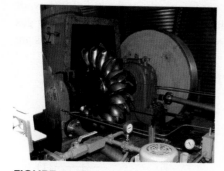

FIGURE 11.30
Open housing of Pelton-type turbine shows turbine runner. This impulse turbine converts the kinetic energy from a water jet to shaft power. (See Fig. 11.29a.) **Photograph courtesy of ABMS Consultants, Québec.**

FIGURE 11.29
Hydro (water) turbine types: the impulse-type turbine (a) utilizes a jet of water in the open air striking the buckets or blades. The double-bucket configuration is known as a Pelton type; in reaction turbines, (b) and (c), the fluid is enclosed. The fixed- or adjustable-vane radial inflow types (b) are called Francis turbines. Mixed radial- and axial-flow propeller types (c) may have fixed or variable blades. The variable-pitch type is also called a Kaplan turbine. Sketches reprinted from Ref. [13] with permission. **Photographs courtesy of General Electric Company (a), Sulzer Hydro Ltd. (b), and U.S. Army Corps of Engineers (c).**

FIGURE 11.31

The Grand Coulee Dam (left) is the largest concrete structure ever built in the United States. One of many radial inflow turbine wheels installed at the dam is shown on the right. The total installed generating capability is 6809 MW and the rated head is 330 ft. **Photographs courtesy of U.S. Bureau of Reclamation.**

the fluid is entirely enclosed (Fig. 11.29b and 11.29c and 11.31). Reference [14] is a Web site devoted to small-scale hydropower and is a interesting starting point for those desiring more information in this area.

In steam and gas turbines, the working fluid is invariably contained within a casing. Small steam and gas turbines are frequently of the radial-inflow, or helical, configuration. Figure 11.32 schematically shows a radial inflow gas turbine used in an automotive turbocharger. Larger steam and gas turbines are

FIGURE 11.32

Schematic diagram of radial-flow turbine used in an automotive turbocharger. Hot exhaust gases are accelerated through the nozzles (1–2) before entering the radial-inflow turbine rotor (2–3). **Reprinted from Ref. [15] with permission.**

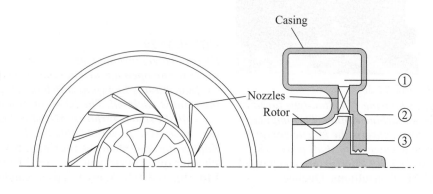

FIGURE 11.33
General Electric's 9H gas-turbine engine (480 MW) is used for stationary power generation. For a sense of scale, note the person standing on the frame. **Photograph courtesy of GE Power Systems.**

FIGURE 11.34
Aircraft turbofan engine (top) has a multistage, axial-flow, high-pressure turbine and a multistage, axial-flow, low-pressure turbine. The turboshaft engine (bottom) is configured similarly; however, the low-pressure turbine is not connected to the compressor but provides power only to the shaft. Turboshaft engines are used in helicopters, power generation, ship propulsion, and military tanks. **Drawings courtesy of Pratt and Whitney.**

generally multistage, axial-flow machines. In this configuration, the general flow path follows an axial direction, perpendicular to rows of spinning blades. Figure 1.4 in Chapter 1 shows a multistage steam turbine rotor for electrical power generation; Fig. 11.33 shows a large gas-turbine engine, also used in a power generation application. Multistage, axial-flow gas turbines are invariably employed in aircraft engines. These are illustrated in the cutaway drawings of Fig. 11.34.

Windmills have been used for centuries to provide a local source of power. In the past few decades, however, wind turbines have been developed for both small- and large-scale power production. Figure 11.35 shows a wind farm near Tehachapi, California. Wind turbines can be classified into two types depending on their axis of revolution, either horizontal or vertical (Fig. 11.36). Horizontal-axis turbines dominate the current commercial market. Installed wind energy capacity in the United States in 2004 was 6700 MW, meeting approximately 0.3% of the country's total electricity demand. Although a renewable source of energy, wind power still poses environmental concerns such as the use of large tracts of land and the effect of wind turbines on bird populations.

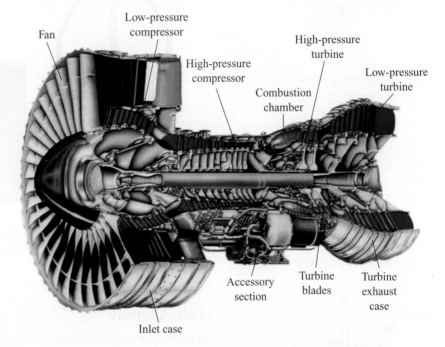

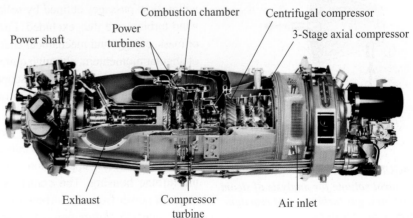

FIGURE 11.35
Wind farms at Tehachapi, California contain more than 4600 turbines with a total capacity of about 600 MW.

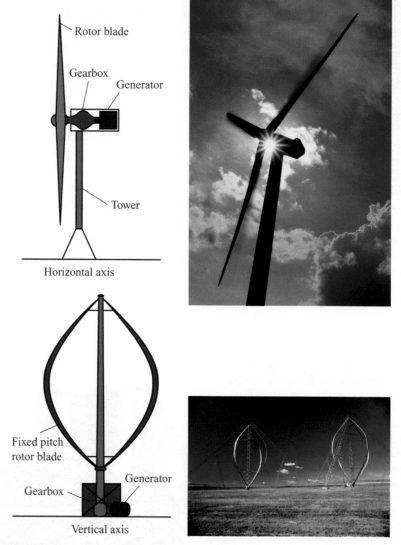

Rotor blade

Gearbox

Generator

Tower

Horizontal axis

Fixed pitch
rotor blade

Gearbox

Generator

Vertical axis

FIGURE 11.36
Wind turbines of the horizontal axis type (top) and the vertical axis type (bottom). Most commercial wind turbines are of the horizontal axis type.

≈ 0 ≈ 0

$\dot{m}(h_2 + \alpha_2 v_{avg,2}^2/2 + gz_2)$

$\dot{Q}_{cv} \approx 0$

②

$\dot{W}_{cv} = \dot{W}_{shaft,out}$

①

$\dot{m}(h_1 + \alpha_1 v_{avg,1}^2/2 + gz_1)$

≈ 0 ≈ 0

FIGURE 11.37
Control volume for analysis of steam turbines, gas turbines, and reaction-type water turbines.

11.5b Analysis

We restrict our analysis to those turbines in which the fluid is entirely contained within flow passages defined by solid boundaries; impulse hydro turbines and wind turbines are thus excluded. Furthermore, we consider only conservation of mass, energy, and mechanical energy. As discussed previously, conservation of angular momentum, although quite important to a thorough understanding of turbomachinery, is beyond the scope of this book. In Chapter 12, we will, however, include the conservation of *linear* momentum in an analysis of jet engines.

We begin by considering the generic turbine control volume shown in Fig. 11.37. The control surface cuts through any inlet and outlet pipes, if they exist in a particular application, and is contiguous with the outer surface of the turbine housing. The control volume also cuts through all shafts that transmit power from the turbine to a compressor or external load. In addition to the general requirement of steady-state and steady flow, we invoke the following simplifying assumptions:

- The heat interaction across the control surface is zero (adiabatic) or small compared to other flows of energy (i.e., $\dot{Q}_{cv} = 0$).
- The potential energy change is zero or small compared to other flows of energy (i.e., $z_2 - z_1 = 0$).
- The kinetic energy of the inlet and outlet streams is small compared to other flows of energy, or the change in kinetic energy is small (i.e., $\alpha_2 v_{avg,2}^2/2 - \alpha_1 v_{avg,1}^2/2 = 0$).

With steady flow through a control volume with a single inlet and a single outlet, mass conservation (Eqs. 3.15 and 3.18a) is expressed in the same way as for all of the previously discussed steady-flow devices:

$$\dot{m}_1 = \dot{m}_2, \tag{11.49a}$$

or

$$\rho_1 v_{avg,1} A_1 = \rho_2 v_{avg,2} A_2. \tag{11.49b}$$

With the assumptions listed, conservation of energy (Eq. 5.63e) can be simplified to yield

$$0 - \dot{W}_{cv,\,net\,out} = \dot{m}[(h_2 - h_1) + 0 + 0].$$

Identifying $\dot{W}_{cv,net\,out}$ as $\dot{W}_{shaft,out}$, this equation, upon rearrangement, becomes

$$\dot{W}_{shaft,out} = \dot{m}(h_1 - h_2). \tag{11.50}$$

For turbines utilizing incompressible fluids (e.g., water turbines), we can also apply and simplify the mechanical energy equation (Eq. 6.76), that is,

$$\dot{W}_{shaft,out} = \dot{m}\left[\frac{P_1 - P_2}{\rho} - gh_L\right]. \tag{11.51}$$

Note how the head loss h_L causes the delivered shaft power to be less than would be obtained from a turbine operating reversibly (cf. Eq. 11.42).

The isentropic efficiency for a turbine is defined by Eq. 7.27. This definition, combined with our first-law expression for the actual turbine power, yields an expression involving only enthalpies;

$$\eta_{isen,t} = \frac{\dot{W}_{act}}{\dot{W}_{isen}} = \frac{\dot{m}(h_1 - h_2)}{\dot{m}(h_1 - h_{2s})}, \tag{11.52a}$$

or

$$\eta_{isen,t} = \frac{h_1 - h_2}{h_1 - h_{2s}}. \tag{11.52b}$$

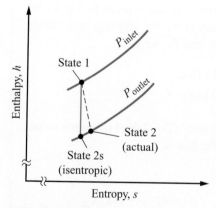

FIGURE 11.38

The maximum possible power is delivered by a turbine for the isentropic process (1–2s), whereas less power is delivered for the real process (1–2) in which irreversibilities (losses) are present. The use of a dashed line is a reminder that the real process does not follow a sequence of equilibrium states.

Figure 11.38 illustrates on h–s coordinates the states and processes associated with Eq. 11.52 (cf. Fig. 11.27 for pumping devices). Equation 11.52 applies equally well to incompressible and compressible flows.

Example 11.19 Steam Power Plant Application

Drawing courtesy of General Electric Company.

A steam turbine is used to drive a feedwater pump of a large utility boiler. A 17.78-kg/s flow of supercritical steam enters the turbine at 808.3 K and 23.26 MPa. The steam exits the turbine at 5.249 kPa with a quality of 0.9566. Determine the power produced by the turbine and the turbine isentropic efficiency.

Solution

Known $\dot{m}, P_1, T_1, P_2, x_2$

Find $\dot{W}_{act}, \eta_{isen,t}$

Sketch

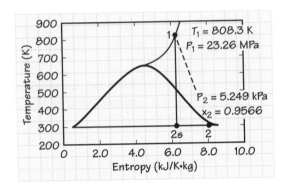

Assumptions

i. Steady state

ii. Adiabatic process ($\dot{Q}_{cv} = 0$)

iii. Negligible kinetic and potential energy changes

Analysis   With our assumptions, the turbine power is given by Eq. 11.50. To apply this, we determine the enthalpies of the entering and exiting steam from the NIST database. Given T_1 and P_1, we get

$$h_1 = 3312.1 \text{ kJ/kg},$$
$$s_1 = 6.1762 \text{ kJ/kg} \cdot \text{K}.$$

To find h_2, we apply the definition of quality (Eq. 2.49c) using the values for $h_{f,2}$ and $h_{g,2}$ at $P_2 = P_{sat} = 5.249$ kPa, that is,

$$h_2 = (1 - x_2)h_{f,2} + x_2 h_{g,2}$$
$$= (1 - 0.9566)141.38 + 0.9566(2562.3) \text{ kJ/kg}$$
$$= 2457.2 \text{ kJ/kg}.$$

The turbine power is thus

$$\dot{W}_{act} = \dot{m}(h_1 - h_2)$$
$$= 17.78(3312.1 - 2457.2)$$
$$= 15,200$$
$$[=] \frac{\text{kg}}{\text{s}} \frac{\text{kJ}}{\text{kg}} = \text{kW}.$$

To evaluate the isentropic efficiency, we apply its definition, Eq. 11.52b. Setting $s_{2s} = s_1$, we find the quality for an isentropic process using Eq. 2.49d:

$$x_{2s} = \frac{s_{2s} - s_{f,2}}{s_{g,2} - s_{f,2}}$$

$$= \frac{6.1762 - 0.48803}{8.3765 - 0.48803}$$

$$= 0.72107 \text{ (dimensionless)}.$$

Using this value, we find h_{2s}:

$$h_{2s} = (1 - x_{2s})h_{f,2} + x_{2s}h_{g,2}$$

$$= (1 - 0.72107)141.38 + 0.72107(2562.3) \text{ kJ/kg}$$

$$= 1887.0 \text{ kJ/kg}.$$

We now evaluate Eq. 11.52b:

$$\eta_{isen,t} = \frac{h_1 - h_2}{h_1 - h_{2s}}$$

$$= \frac{3312.1 - 2457.2}{3312.1 - 1887.0}$$

$$= 0.60 \text{ or } 60\%.$$

Comments   We note the importance of the steam quality in determining both the actual and ideal (isentropic) enthalpies at the exit state. We also note the relatively low value for the isentropic efficiency of this steam turbine. Since the power required to drive the feedwater pump is a small fraction of the output power of the power plant, this low efficiency has a negligible effect on the overall efficiency of the power plant. The data from this example are from an actual power plant.

**Self Test
11.15** ✓ **Helium, initially at 950 K and 800 kPa, is expanded through a turbine to a final state of 500 K and 100 kPa. Determine the isentropic efficiency of the turbine.**

(Answer: 0.839)

Example 11.20

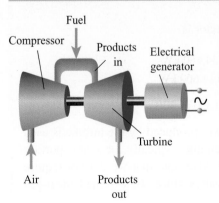

Fuel

Compressor

Products in

Electrical generator

Turbine

Air

Products out

Consider a stationary gas-turbine engine used for electrical power generation. As shown in the sketch, the turbine drives both a compressor (part of the engine) and an electrical generator (external to the engine). The shaft power supplied to the generator is 34,460 kW. Combustion products enter the turbine at 1530 K and exit at 720 K with a flow rate of 122.2 kg/s. Approximating the thermodynamic properties of the combustion products as those of air, estimate the fraction of the total turbine power used to generate electricity.

Solution

Known $T_1, T_2, \dot{m}$

Find $\dot{W}_{t,gen}/\dot{W}_{t,tot}$

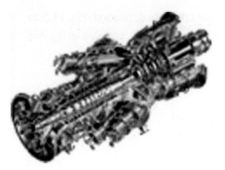

MS5002E 30 MW-class gas turbine engine.
Image courtesy of General Electric Company.

Sketch

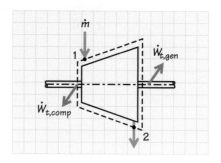

Assumptions

  i. Steady state
  ii. Air properties for combustion products
  iii. Ideal-gas behavior
  iv. Adiabatic process ($\dot{Q}_{cv} = 0$)
  v. Negligible kinetic and potential energy changes

Analysis As shown in the sketch, we choose a control volume that cuts through the shaft joining the turbine and the compressor and the shaft joining the turbine and the electrical generator. With our assumptions, conservation of energy is simply expressed by Eq. 11.50. The power in this equation is the total shaft power, given by

$$\dot{W}_{shaft,out} = \dot{W}_{t,comp} + \dot{W}_{t,gen}.$$

Thus,

$$\dot{W}_{t,comp} + \dot{W}_{t,gen} = \dot{W}_{shaft,out} = \dot{m}(h_1 - h_2).$$

The enthalpies are evaluated from Table C.2:

$$h_1(1530 \text{ K}) = 1672.37 \text{ kJ/kg},$$
$$h_2(720 \text{ K}) = 734.82 \text{ kJ/kg}.$$

The total power produced is then

$$\dot{W}_{shaft,out} = 122.2(1672.37 - 734.82)$$
$$= 114,600$$

$$[=] \frac{\text{kg}}{\text{s}} \frac{\text{kJ}}{\text{kg}} = \text{kW},$$

and the fraction delivered to the generator is

$$\frac{\dot{W}_{t,gen}}{\dot{W}_{shaft,out}} = \frac{34,460 \text{ kW}}{114,600 \text{ kW}}$$

$$= 0.300 \text{ or } 30\%.$$

Comment Note that 70% of the power produced by the turbine is used to drive the compressor. Gas-turbine engines typically use a large portion of the turbine output in this manner. Also note that we did not require values for the inlet and outlet pressures since we assumed ideal-gas behavior.

Example 11.21

Hoover Dam, located on the Colorado River at the Arizona–Nevada border, creates Lake Mead in the lower Grand Canyon. Seventeen hydro turbines generate electricity at the dam. The total rated capacity of the power plant is 2074 MW. Water enters one of two intake towers (only one of which is shown in the sketch) from which it is distributed via a 9.1-m-diameter penstock (pipe) to several smaller 3.96-m-diameter penstocks, each supplying one of the main Francis-type hydro tubines.

Consider a single main turbine with a volume flow rate of 79.4 m³/s. (a) Estimate the maximum possible power that this turbine can deliver using an average elevation change (Lake Mead surface to turbine discharge) of 158.5 m. (b) Determine the decrement in this ideal power associated with the frictional head loss of the 110.5-m-long, 3.96-m-diameter penstock. The penstock is fabricated from plate steel.

> **This example integrates topics from Chapters 5, 6, 10, and 11.**

Water intakes at Hoover Dam

Interior of a dam penstock.

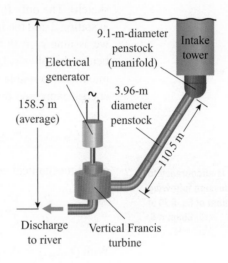

Solution

Known $z_1, z_2, \dot{V}, D_{\text{penstock}}, L_{\text{penstock}}$

Find $\dot{W}_{\text{t,ideal}}, \dot{m}gh_{\text{L,penstock}}$

Sketch

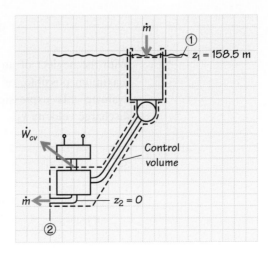

Assumptions

  i. Steady state

  ii. Incompressible flow

  iii. Adiabatic process ($\dot{Q}_{cv} = 0$)

  iv. $P_1 = P_2 = P_{atm}$

  v. $v_1 \approx 0$

  vi. Uniform discharge velocity ($\alpha_2 = 1$)

  vii. $D_2 = D_{penstock}$

  viii. Isothermal fluid with $T = 25°C$ (298 K)

  ix. Ideal (frictionless) operation of all components (turbine, penstocks, etc.) for Part a

  x. Water properties at 1 atm

Analysis   We choose a control volume that cuts through the turbine outlet shaft so that the unknown power out is included in our analysis (see sketch). The only fluid flows crossing the control surface are the discharge at station 2 and the inlet at station 1. Since the area at the inlet is very large, we assume $v_1 = 0$; furthermore, the control surface at station 1 is just a bit below the physical surface of the water so that $P_1 \approx P_{atm}$. To find the maximum possible power (ideal), we can apply either conservation of energy expressed by Eq. 5.63e,

$$\dot{Q}_{cv,\text{net in}} - \dot{W}_{cv,\text{net out}} = \dot{m}\left[h_2 - h_1 + \frac{1}{2}(\alpha_2 v_2^2 - \alpha_1 v_1^2) + g(z_2 - z_1)\right],$$

The reader is encouraged to review the discussion following the development of Eq. 6.76 in Chapter 6.

or the mechanical energy equation (Eq. 6.76),

$$\frac{P_1}{\rho} + \frac{1}{2}\alpha_1 v_1^2 + gz_1 = \frac{P_2}{\rho} + \frac{1}{2}\alpha_2 v_2^2 + gz_2 + \frac{\dot{W}_{cv,\text{out}}}{\dot{m}} + gh_{\text{L}}.$$

With $\dot{Q}_{cv,\text{net in}} = 0$, we see that these two expressions are equivalent, noting that

$$h_2 - h_1 = u_2 - u_1 + \frac{P_2 - P_1}{\rho} = gh_{\text{L}} + \frac{P_2 - P_1}{\rho}.$$

Simplifying either Eq. 5.63e or Eq. 6.76 with our assumptions that $v_1 = 0$, $P_1 = P_2$, $\alpha_2 = 1$, and $h_{\text{L}} = 0$ yields, upon rearrangement,

$$\dot{W}_{cv,\text{out}} = \dot{m}\left[g(z_1 - z_2) - \frac{1}{2}v_2^2\right].$$

With the neglect of any losses ($h_{\text{L}} = 0$), we recognize $\dot{W}_{cv,\text{out}}$ to be the maximum possible, or ideal, power produced by the hydro turbine. To evaluate this expression, we apply mass conservation to obtain the discharge velocity v_2:

$$\dot{m} = \rho v_2 A_2 = \rho \dot{V},$$

or

$$v_2 = \frac{\dot{V}}{A_2} = \frac{\dot{V}}{\pi D_2^2/4}$$

$$= \frac{79.4 \text{ m}^3/\text{s}}{\pi(3.96 \text{ m})^2/4} = 6.45 \text{ m/s}.$$

Thus, with the water density obtained from the NIST database, we get

$$\dot{W}_{\text{cv,out}} = 997.1(79.4)[9.807(158.5 - 0) - 0.5(6.45)^2]$$

$$= 1.21 \times 10^8$$

$$[=] \frac{\text{kg}}{\text{m}^3} \frac{\text{m}^3}{\text{s}} \left(\frac{\text{m}}{\text{s}^2}\right) \text{m} \left[\frac{1\,\text{N}}{\text{kg} \cdot \text{m/s}^2}\right] \left[\frac{1\,\text{W}}{\text{N} \cdot \text{m/s}}\right]$$

$$= \text{W},$$

or

$$\dot{W}_{\text{cv,out}} = 121\,\text{MW}.$$

To estimate the head loss associated with the 3.96-m-diameter penstock requires values for the pipe Reynolds number and relative roughness because

$$f_D = f_D(\text{Re}_{D_2}, \varepsilon/D_2)$$

and (Eq. 10.16)

$$h_{\text{L,penstock}} = f_D \frac{L_{\text{penstock}}}{D_2} \frac{v_2^2}{2g}.$$

Assuming that the plate steel penstock has the same roughness height as commercial steel pipe, we find from Table 10.4 that $\varepsilon = 0.046$ mm; thus,

$$\frac{\varepsilon}{D_2} = \frac{0.046 \times 10^{-3}\,\text{m}}{3.96\,\text{m}} = 1.16 \times 10^{-5}.$$

The Reynolds number is given by (Eq. 10.34)

$$Re_{D_2} = \frac{\rho v_2 D_2}{\mu}$$

$$= \frac{997.1(6.45)3.96}{893.1 \times 10^{-6}} = 2.85 \times 10^7,$$

where the viscosity is obtained from the NIST database. (The reader should verify that Re_{D_2} is dimensionless.) We use Eq. 10.86 to obtain a value for f_D:

$$f_D = \frac{0.25}{\left\{\log_{10}\left[\dfrac{\varepsilon}{3.7 D_2} + \dfrac{5.74}{Re_{D_2}^{0.9}}\right]\right\}^2}$$

$$= \frac{0.25}{\left\{\log_{10}\left[\dfrac{1.16 \times 10^{-5}}{3.96} + \dfrac{5.74}{(2.85 \times 10^7)^{0.9}}\right]\right\}^2}$$

$$= 0.00867.$$

We note that this is essentially the same as the value estimated from the Moody chart (Fig. 10.17), that is, $f_D \approx 0.0085$. The penstock head loss is thus

$$h_{\text{L,penstock}} = 0.00867\left(\frac{110.5}{3.96}\right)\frac{(6.45)^2}{2(9.807)}$$

$$= 0.51$$

$$[=] \frac{\text{m}}{\text{m}} \frac{(\text{m/s})^2}{\text{m/s}^2} = \text{m}.$$

Since the penstock L/D is relatively small ($\sim$28), the minor loss associated with the entrance to the penstock is likely to be important. From Fig. 10.28, we obtain $K = 0.5$. Employing the definition of the loss coefficient K (Eq. 10.105), we obtain the entrance head loss:

$$h_{\text{L,ent}} = K\frac{v_2^2}{2g}$$

$$= 0.5\frac{(6.45)^2}{2(9.807)} = 1.06 \text{ m}.$$

The power decrement associated with the penstock head losses can be found by retaining the head loss in our mechanical energy equation,

$$\dot{W}_{\text{cv,with loss}} = \dot{m}\left[g(z_2 - z_1) - \frac{1}{2}v_2^2 - g(h_{\text{L,penstock}} + h_{\text{L,ent}}) \right]$$

$$= \dot{W}_{\text{cv,ideal}} - \dot{m}g(h_{\text{L,penstock}} + h_{\text{L,ent}}).$$

We can express the percent power loss as

$$\%\text{loss} = \frac{\dot{m}g(h_{\text{L,penstock}} + h_{\text{L,ent}})}{\dot{W}_{\text{cv,ideal}}} \times 100\%$$

$$= \frac{79169.7(9.807)(0.51 + 1.06)}{121 \times 10^8} \cdot 100\%$$

$$= 0.01\%.$$

This result suggests that the losses associated with the penstocks are indeed small and possibly can be neglected.

Comments Note the importance of the choice of control volume in this example, as this choice results in a relatively simple analysis. Note also that the entrance loss exceeds that of the wall friction and is not really a "minor" loss in this problem.

11.6 HEAT EXCHANGERS

11.6a Classifications and Applications

The purpose of a heat exchanger is to heat or cool one fluid stream at the expense of another. A common example is an automobile radiator. In this heat exchanger, energy is removed from the hot engine coolant (a mixture of ethylene glycol and water) by a flow of relatively cool ambient air (Fig. 11.39). Heat exchangers are typically passive devices with no moving parts.

There are many types of heat exchangers. We can categorize them generally by the paths that the fluid streams take within the device, with the following being the most simple:

- parallel flow,
- counterflow, and
- cross-flow.

These basic configurations are illustrated in Fig. 11.40. Many heat exchangers combine these path types. For example, the shell-and-tube heat exchanger

FIGURE 11.39

An automobile radiator is an example of a cross-flow heat exchanger. The engine coolant is pumped through the bank of horizontal tubes, perpendicular to the flow of air. **Photograph courtesy of Delphi Automotive Systems.**

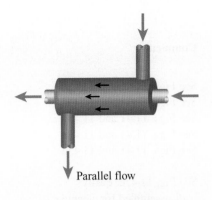

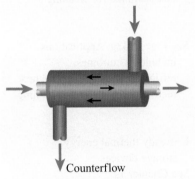

FIGURE 11.40

Three basic flow configurations for heat exchangers: parallel flow (top), counterflow (center), and cross-flow (bottom). Many heat exchangers operate with combinations of these three flow configurations. (See Fig. 11.41.)

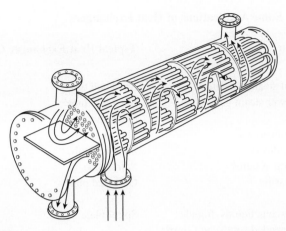

FIGURE 11.41

A U-tube shell-and-tube heat exchanger has two tube-passes. Baffles in the shell create a cross-flow component; thus the upper tube-pass is a cross-flow/parallel combination, whereas the lower tube-pass is a cross-flow/counterflow combination. **Reprinted from Ref. [23] with permission.**

shown in Fig. 11.41 exhibits elements of all three. The fluid in the first tube-pass generally runs counter to the shell fluid, whereas in the second tube-pass, the fluid generally runs parallel to the shell fluid. The baffles, however, add a cross-flow component in both cases.

We also categorize the fluid streams themselves as being *unmixed* or *mixed*. Figure 11.42 (bottom) shows a heat exchanger in which both fluid streams are unmixed (i.e., both the hot stream and the cold are subdivided into separate passages that prevent mixing); whereas Fig. 11.42 (top) shows

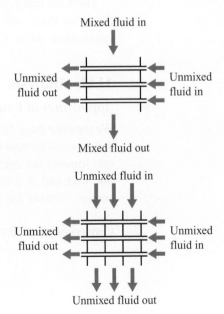

FIGURE 11.42

The meaning of mixed and unmixed fluids is illustrated in these sketches. A cross-flow heat exchanger with one mixed fluid and one unmixed fluid is shown at the top. At the bottom, both fluid streams are unmixed.

Table 11.7   Some Applications of Heat Exchangers

Application	Typical Heat Exchanger Configuration	Comment
Power plant steam condensers	Shell and tube*	See Figs. 11.41 and 11.43a
Nuclear power steam generators	Shell and tube*	See Figs. 11.41 and 11.43a
Oil coolers	Shell and tube*	See Figs. 11.41 and 11.43a
Process applications	Shell and tube*	See Figs. 11.41 and 11.43a
Chemical industry	Shell and tube*	See Figs. 11.41 and 11.43a
Refrigeration systems	Spiral tube	
Food processing	Gasketed plate	See Fig. 11.43e. Easily disassembled for cleaning
Sludges, viscous liquids, liquids with suspended solids, and slurries	Spiral plate	
Gas-to-gas exchanger	Plate-fin	See Fig. 11.43g. Applications include gas turbines, refrigeration, HVAC, waste heat recovery, and electronic cooling
Condensation and evaporation of refrigerants	Tube-fin	
Air preheaters in power plants and industrial processes	Rotary and fixed-matrix regenerators	Unsteady thermal energy storage devices
Cooling towers	Direct contact	See Chapter 12
Feedwater heaters	Shell and tube and direct contact	See Example 11.23

*More than 90% of the heat exchangers used in industry are of the shell-and-tube type [20].

a configuration in which one fluid stream is mixed and the other unmixed. In a typical automotive radiator, the flows of both the coolant and air are unmixed.

There are many, many applications of heat exchangers. Table 11.7 suggests a few of these and indicates the typical heat exchanger employed in the applications listed. Common heat exchangers are shown in Fig. 11.43.

11.6b Analysis

Application of Conservation Principles

We consider three different control volumes in our analysis of heat exchangers: 1. an overall control volume that encloses the entire device and whose surface cuts through the inlets and outlets, 2. a control volume that includes only the hot fluid, and, 3. a control volume containing only the cold fluid. These three control volumes are sketched in Fig. 11.44.

Conservation of Mass   Conservation of mass is simply expressed for each as

$$\dot{m}_{H,in} + \dot{m}_{C,in} = \dot{m}_{H,out} + \dot{m}_{C,out}, \qquad (11.53a)$$

$$\dot{m}_{H,in} = \dot{m}_{H,out} \equiv \dot{m}_H, \qquad (11.53b)$$

and

$$\dot{m}_{C,in} = \dot{m}_{C,out} \equiv \dot{m}_C, \qquad (11.53c)$$

where the subscripts H and C refer to the hot and cold streams, respectively.

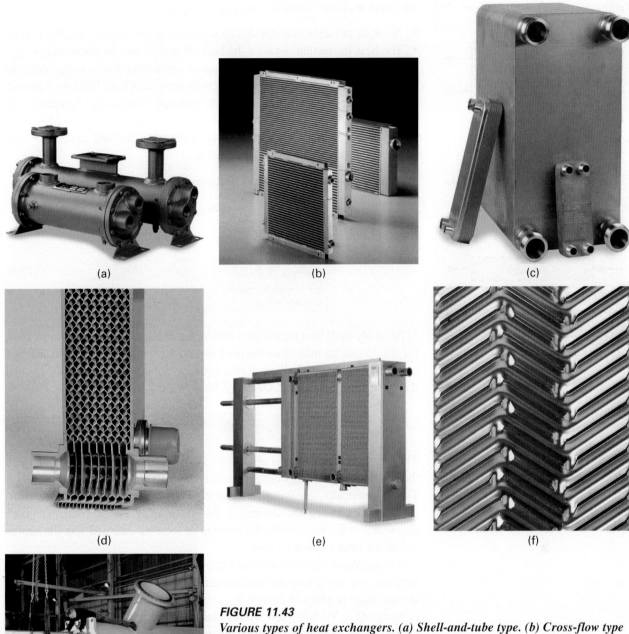

FIGURE 11.43
Various types of heat exchangers. (a) Shell-and-tube type. (b) Cross-flow type oil coolers, aftercoolers, and combination oil/aftercoolers of aluminum construction. (c) Brazed-plate type. (d) Cutaway view of flow channels in brazed-plate heat exchanger. (e) Plate-and-frame type heat exchanger for food and beverage processing. (f) Flow channel pattern for heat-transfer plate. (g) Extended-surface, plate-fin heat exchanger for compressor intercooler and aftercooler applications. **Photographs courtesy of API Heat Transfer, Inc.**

Conservation of Energy   For the overall control volume, we simplify the following steady-flow statement of conservation of energy for control volumes with multiple inlets and outlets (Eq. 5.65):

$$
\dot{Q}_{cv,net\,in} - \dot{W}_{cv,net\,out} = \sum_{k=1}^{M\,outlets} \dot{m}_{out,k}\big[h_k + \tfrac{1}{2}\alpha_k v_{avg,k}^2 + g(z_k - z_{ref})\big]
$$

$$
- \sum_{j=1}^{N\,inlets} \dot{m}_{in,j}\big[h_j + \tfrac{1}{2}\alpha_j v_{avg,j}^2 + g(z_j - z_{ref})\big].
$$

(11.54)

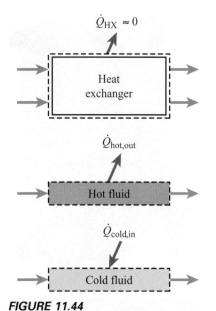

FIGURE 11.44
Control volumes for heat exchanger analysis: entire heat exchanger (top), hot fluid steam only (middle), and cold fluid stream only (bottom).

We begin with the following facts and assumptions:

- No shaft power is associated with a heat exchanger; therefore, $\dot{W}_{cv}$ is zero.
- The heat interaction between the heat exchanger and its surroundings, $\dot{Q}_{cv}$, can be neglected. For heat exchangers employing fluids hotter than the ambient surroundings, there is likely to be some heat loss. This loss, however, is usually quite small compared to the enthalpy flows on the right-hand side of Eq. 11.54; thus, we will neglect it (i.e., $\dot{Q}_{cv} \approx 0$.) (See Fig. 11.44 top.)
- Kinetic and potential energy changes between inlets and outlets for both the hot and cold streams are negligible.

With these provisions, Eq. 11.54 becomes

$$0 - 0 = \dot{m}_H h_{H,out} + \dot{m}_C h_{C,out} - \dot{m}_H h_{H,in} - \dot{m}_C h_{C,in},$$

which can be rearranged to yield

$$\underbrace{\dot{m}_H(h_{H,in} - h_{H,out})}_{\text{Rate at which hot stream loses energy}} = \underbrace{\dot{m}_C(h_{C,out} - h_{H,in})}_{\text{Rate at which cold stream gains energy}}. \tag{11.55}$$

In the analysis of heat exchangers in which a change of phase is not involved, Eq. 11.55 can be related to the inlet and outlet temperatures by assuming a constant specific heat in the calorific equation of state; that is, we assume that $\Delta h = c_{p,avg} \Delta T$ (cf. Eqs. 2.33e and 2.56b). With this assumption,

$$\dot{m}_H c_{p,H}(T_{H,in} - T_{H,out}) = \dot{m} c_{p,C}(T_{C,out} - T_{C,in}). \tag{11.56}$$

The physical interpretation of Eqs. 11.55 and 11.56 is that the rate at which the hot fluid loses energy equals the rate at which the cold fluid gains energy. This is physically satisfying, because it means that the control volume itself neither loses nor gains energy at the expense of the surroundings (i.e., $\dot{Q}_{cv} = \dot{W}_{cv} = 0$). Frequently, the product of the flow rate and specific heat is called the **heat-capacity rate C.**

Examining either the hot or cold stream in isolation (Fig. 11.44 middle and bottom), we see a heat interaction across the control surface: For the hot stream, there is a loss of energy $\dot{Q}_{H,out}$; and for the cold stream, there is a gain of energy $\dot{Q}_{C,in}$. With this new wrinkle, but retaining all of the other assumptions applied to the overall control volume, we can express energy conservation for the hot fluid by simplifying Eq. 5.63e as follows:

$$-\dot{Q}_{H,out} + 0 = \dot{m}_H(h_{H,out} - h_{H,in} + 0 + 0),$$

or

$$\dot{Q}_{H,out} = \dot{m}_H(h_{H,in} - h_{H,out}). \tag{11.57a}$$

Similarly, energy conservation for the cold stream is expressed as

$$\dot{Q}_{C,in} = \dot{m}_C(h_{C,out} - h_{C,in}). \tag{11.57b}$$

Assuming constant specific heats, we can relate the heat-transfer rates in Eqs. 11.57a and 11.57b to inlet and outlet temperatures as follows:

$$\dot{Q}_{H,out} = \dot{m} c_{p,H}(T_{H,in} - T_{H,out}) \tag{11.58a}$$

and

$$\dot{Q}_{C,in} = \dot{m}c_{p,C}(T_{C,out} - T_{C,in}).$$ (11.58b)

Comparing Eqs. 11.57a and 11.57b to our overall energy balance (Eq. 11.55), we formally write

$$\dot{Q}_{H,out} = \dot{Q}_{C,in}.$$ (11.59)

The objective in the design of a heat exchanger is to define, in detail, a device that will transfer energy at the desired rate from one fluid to another. How to do this comprises the subject of subsequent sections of this chapter. Before proceeding, however, we consider how conservation of mechanical energy impacts the analysis of a heat exchanger.

Conservation of Mechanical Energy (Incompressible Flow)   Consideration of pressure drops is important in the design and operation of heat exchangers. A major operating cost associated with heat-exchange equipment involves the energy required to push fluids through the device. Our energy analysis above does not shed any light on this. The mechanical energy equation (Eq. 6.76), a special form of momentum conservation, however, is helpful. Neglecting kinetic energy changes, we apply Eq. 6.76 to the hot-stream control volume (Fig. 11.44 middle) as follows:

$$\frac{P_{H,in} - P_{H,out}}{\rho g} = z_{H,out} - z_{H,in} + h_{L,H},$$ (11.60)

where $h_{L,H}$ is the head loss associated with flow through the hot side of the heat exchanger. We have retained the elevation change term, which might be appreciable for vertically oriented devices employing liquids. A similar expression applies for the cold stream. The difficulty in applying Eq. 11.60 is the evaluation of the head loss term. In Chapter 10, we saw methods to calculate head loss for some simple geometries. A later example illustrates the calculation of head loss in a heat exchanger.

Example 11.22

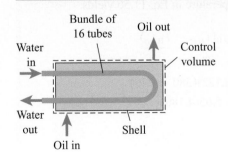

A U-tube shell-and-tube heat exchanger (Fig. 11.41) is used to cool a 5.5-kg/s flow of hot oil from 380 to 320 K. The oil flows through the shell, while water entering at 280 K flows through a bundle of 16 tubes. The average velocity of the water in the 15-mm-inside-diameter tubes is 2 m/s. The average specific heats of the oil and water can be assumed to be 2.122 and 4.180 kJ/kg·K, respectively. Determine the outlet temperature of the water.

Solution

Known $\dot{m}_H$, $c_{p,H}$, $c_{p,C}$, $T_{H,in}$, $T_{H,out}$, $T_{C,in}$, N, D_{tube}, v_{avg}

Find $T_{C,out}$

Sketch

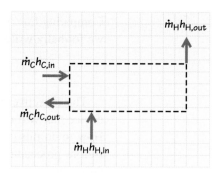

Assumptions

i. Steady flow
ii. Constant (average) properties
iii. No residual heat loss ($\dot{Q}_{cv} = 0$)
iv. Negligible kinetic and potential energy changes

Analysis   For the control volume shown in the sketch, the overall energy balance expressed by Eq. 11.56 applies. To use this equation to find $T_{C,out}$ requires that we find the mass flow rate of the water. The mass flow rate through a single tube is given by

$$\dot{m}_{1\,\text{tube}} = \rho_{\text{H}_2\text{O}} v_{\text{avg}} \pi D_{\text{tube}}^2/4,$$

and for the bundle of $N\ (=16)$ tubes,

$$
\begin{aligned}
\dot{m}_{\text{C}} = N\dot{m}_{1\,\text{tube}} &= 16\rho_{\text{H}_2\text{O}} v_{\text{avg}} \pi D_{\text{tube}}^2/4 \\
&= 16(999.1)(2.0)\pi(0.015)^2/4 \\
&= 5.65 \\
&[=] (\text{kg/m}^3)(\text{m/s})\text{m}^2 = \text{kg/s},
\end{aligned}
$$

where the water density is obtained from the NIST database for $T = 280$ K and $P = 1$ atm. (Although the pressure is unknown, using $P = 1$ atm should be reasonable since the liquid density does not vary much with pressure.) Isolating the unknown water outlet temperature in Eq. 11.56 yields

$$
\begin{aligned}
T_{\text{C,out}} = T_{\text{C,in}} + \frac{\dot{m}_{\text{H}} c_{p,\text{H}}(T_{\text{H,in}} - T_{\text{H,out}})}{\dot{m}_{\text{C}} c_{p,\text{C}}} \\
= 280 + \frac{5.5(2.122)(380 - 320)}{5.65(4.180)}\,\text{K} \\
= 310\,\text{K}.
\end{aligned}
$$

Comment   This example illustrates the application of energy conservation to find the one unknown temperature when three are given, a very common situation encountered in heat-exchanger design and analysis. We also see how the flow properties from a single tube are used to obtain the total flow rate associated with a tube bundle.

Oil flowing at 0.02 kg/s enters the inner channel of a parallel-flow heat exchanger at 400 K. Water at 290 K enters the outer channel of the heat exchanger. Neglecting any frictional effects, determine the minimum flow rate of the water such that the oil is cooled to a final temperature of 300 K.

(Answer: 0.101 kg/s)

Example 11.23 Steam Power Plant Application

Partially expanded steam is bled from a steam turbine and fed into a feedwater heater, a shell-and-tube heat exchanger in which the condensing steam on the shell side heats a flow of compressed liquid water on the tube side. The steam enters the heater at 572 K and 0.4268 MPa at a flow rate of 9.38 kg/s. Cold water enters the tubes at 386.4 K and exits at 416.3 K. The water flow rate is 189 kg/s. Assuming the pressure in the shell is uniform, determine the temperature and state of the H_2O exiting the shell.

Solution

Known $\dot{m}_H$, $\dot{m}_C$, $T_{H,in}$ $P_{H,in}(= P_{H,out})$, $T_{C,in}$, $T_{C,out}$

Find $T_{H,out}$, state (liquid, liquid–vapor mixture, or vapor)

Sketch

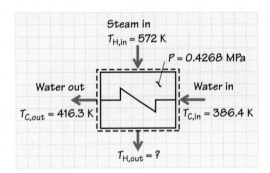

Assumptions
  i. Steady flow
  ii. No residual heat loss ($\dot{Q}_{cv} = 0$)
  iii. Negligible kinetic and potential energy changes
  iv. Specific heat for cold-side water evaluated at mean temperature and arbitrary pressure

Analysis   With our assumptions, application of conservation of energy to the control volume shown in the sketch yields (Eq. 11.55)

$$\dot{m}_H (h_{H,in} - h_{H,out}) = \dot{m}_C (h_{C,out} - h_{C,in}).$$

For the cold-water stream, we express the enthalpy difference as $c_{p,C}(T_{C,out} - T_{C,in})$, where an average value of the specific heat is employed; thus,

$$\dot{m}_H (h_{H,in} - h_{H,out}) = \dot{m}_C c_{p,C}(T_{C,out} - T_{C,in}).$$

We use this expression to determine a value of $h_{H,out}$. Because we know the pressure at the outlet of the hot stream, we have two thermodynamic properties from which all other thermodynamic properties can be obtained; specifically,

$$T_{\mathrm{H,out}} = T_{\mathrm{H,out}}(h_{\mathrm{H,out}}, P_{\mathrm{H,out}}).$$

We proceed, first, by evaluating $c_{p,\mathrm{C}}$. Since the pressure is unknown on the cold side, we will employ an arbitrary pressure of 5 atm to ensure that the cold stream enters and exits as a compressed liquid.[12] Our criterion for this choice of pressure is only that it exceed P_{sat} at the outlet temperature [i.e., $P_{\mathrm{sat}}(T_{\mathrm{C,out}}) = 3.9$ atm]. Using the NIST database, we find

$$c_p(386.4 \text{ K, 5 atm}) = 4.232 \text{ kJ/kg} \cdot \text{K},$$

$$c_p(416.3 \text{ K, 5 atm}) = 4.290 \text{ kJ/kg} \cdot \text{K},$$

the average of which is 4.261 kJ/kg·K. We also use the NIST database to evaluate $h_{\mathrm{H,in}}$:

$$h_{\mathrm{H,in}}(572 \text{ K, 0.4268 MPa}) = 3064.0 \text{ kJ/kg}.$$

The only unknown quantity left in our energy equation is $h_{\mathrm{H,out}}$. Solving for $h_{\mathrm{H,out}}$ yields

$$
\begin{aligned}
h_{\mathrm{H,out}} &= h_{\mathrm{H,in}} - \frac{\dot{m}_{\mathrm{C}} c_{p,\mathrm{C}} (T_{\mathrm{C,out}} - T_{\mathrm{C,in}})}{\dot{m}_{\mathrm{H}}} \\
&= 3064.0 - \frac{189(4.261)(416.3 - 386.4)}{9.38} \\
&= 496.9 \\
&[=] \frac{(\text{kg/s})(\text{kJ/kg} \cdot \text{K})\text{K}}{\text{kg/s}} = \text{kJ/kg}.
\end{aligned}
$$

We now create a property table for the fixed, hot-side pressure (0.4268 MPa) with temperature increments using the NIST database. A few representative values of T and h follow:

T (K)	h (kJ/kg)	State
390	490.61	compressed liquid
391	494.85	compressed liquid
392	499.09	compressed liquid
393	503.34	compressed liquid

Interpolating these data for $h_{\mathrm{H,out}} = 496.9$ yields a temperature of 391.5 K. We also see that the hot stream exits as a compressed liquid, that is, as a liquid condensate.

Comments   This example illustrates how to deal with a condensing fluid and the use of tabular data and approximations for enthalpies. The numerical values used in this example are taken from a real steam power plant.

---

[12] Note that the choice of pressure affects the specific-heat value only slightly. It is recommended that the reader verify this.

LEVEL 2

Self Test 11.17 ✓ **Heat from warm air drawn through the evaporator of an air-conditioning unit results in the evaporation of the refrigerant (R-134a) from $x = 0.359$ at 0.1 MPa to a saturated-vapor state. If the air is cooled from 27°C to 17°C at a flow rate 0.2 kg/s, determine the mass flow rate of the refrigerant.**

(Answer: 0.0146 kg/s)

Overall Heat-Transfer Coefficient

We introduced the overall heat-transfer coefficient in Chapter 5. This concept is particularly useful in our analysis of heat-exchange equipment. Figure 11.45 illustrates how we can use a thermal resistance network to express conservation of energy for a conducting wall exposed to convective environments on each side. Assuming the energy flow is one dimensional, we can express energy conservation at the fluid–solid interfaces as

$$\dot{E}_{in} = \dot{E}_{out}. \tag{11.61}$$

For this surface energy balance, we can identify $\dot{E}_{in}$ and $\dot{E}_{out}$ with the convective and conductive heat-transfer rates and write

$$\dot{Q}_{conv,i} = \dot{Q}_{wall} = \dot{Q}_{conv,o}. \tag{11.62}$$

Applying the thermal resistance concept (see Table 11.8) to Eq. 11.62 yields

$$\frac{T_i - T_{w,i}}{R_{conv,i}} = \frac{T_{w,i} - T_{w,o}}{R_{wall}} = \frac{T_{w,o} - T_o}{R_{conv,o}}. \tag{11.63}$$

Eliminating the wall temperatures $T_{w,i}$ and $T_{w,o}$ transforms Eq. 11.63 into

$$\dot{Q} = \frac{T_i - T_o}{R_{conv,i} + R_{wall} + R_{conv,o}}, \tag{11.64a}$$

where $\dot{Q}$ is the one-dimensional heat-transfer rate from the inner fluid to the outer fluid. Expressions for the thermal resistances for planar and radial systems are presented in Table 11.8.

Equation 11.64a is the starting point for defining the overall heat-transfer coefficient U_{OA}; writing

FIGURE 11.45

Electric circuit analog for convective heat transfer through (a) a plane wall and (b) a hollow cylinder with convective environments on each side.

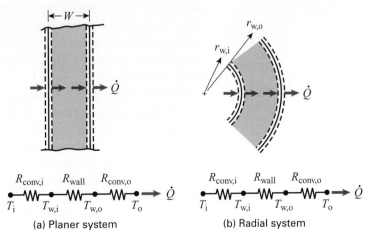

Table 11.8 Heat-Transfer Rates and Thermal Resistances Corresponding to Fig. 11.45

Situation	Heat-Transfer Rate Expression ($\dot{Q}$)	Circuit Analog Rate Expression ($\dot{Q}$)	Thermal Resistance (R_{th})
Convection at inner plane wall	$h_i A_w (T_i - T_{w,i})$	$\dfrac{T_i - T_{w,i}}{R_{conv,i}}$	$\dfrac{1}{h_{conv,i} A_w}$
Conduction through plane wall	$k_w A_w \dfrac{T_{w,i} - T_{w,o}}{W}$	$\dfrac{T_{w,i} - T_{w,o}}{R_{cond}}$	$\dfrac{W}{k_w A_w}$
Convection at outer plane wall	$h_{conv,o} A_w (T_{w,o} - T_o)$	$\dfrac{T_{w,o} - T_o}{R_{conv,o}}$	$\dfrac{1}{h_{conv,o} A_w}$
Convection at tube inner wall	$h_{conv,i} A_i (T_i - T_{w,i})$	$\dfrac{T_i - T_{w,i}}{R_{conv,i}}$	$\dfrac{1}{h_{conv,i} A_{w,i}}$
Conduction through tube wall	$\dfrac{2\pi L k_w (T_{w,i} - T_{w,o})}{\ln(r_{w,o}/r_{w,i})}$	$\dfrac{T_{w,i} - T_{w,o}}{R_{cond}}$	$\dfrac{\ln(r_{w,o}/r_{w,i})}{2\pi k_w L}$
Convection at tube outer wall	$h_{conv,o} A_o (T_{w,o} - T_o)$	$\dfrac{T_{w,o} - T_o}{R_{conv,o}}$	$\dfrac{1}{h_{conv,o} A_w}$

$$\dot{Q} = \left[\frac{1}{R_{conv,i} + R_{wall} + R_{conv,o}} \right] (T_i - T_o), \tag{11.64b}$$

we can then define

$$\dot{Q} \equiv (U_{OA} A)(T_i - T_o), \tag{11.64c}$$

where A is an appropriately defined area. From Eqs. 11.64b and 11.64c, we explicitly define the overall heat-transfer coefficient as follows:

$$U_{OA} \equiv \frac{1}{A(R_{conv,i} + R_{wall} + R_{conv,o})}, \tag{11.65a}$$

or, more generally, if other thermal resistances are involved,

$$U_{OA} \equiv \frac{1}{A \sum_j R_{thermal,j}}. \tag{11.65b}$$

Note how Eq. 11.64c has the same form as the defining relationship for a convective heat-transfer coefficient, that is, Eq. 4.18, $\dot{Q}_{conv} = \bar{h}_{conv} A (T_{fluid} - T_{surf})$. The overall heat-transfer coefficient U_{OA} plays the same role as the average convective heat-transfer coefficient $\bar{h}_{conv}$; here the temperature difference is associated with the inner and outer fluids, rather than a fluid and a solid surface.

We now need to define the area A that appears in Eqs. 11.64c, 11.65a, and 11.65b. Interestingly, the choice of this area can be made somewhat arbitrarily, so one must be careful to understand what area is meant in any particular application. For example, the inside area of the tube can be used to define a particular U_{OA} (Eq. 11.65b). Similarly, another overall heat-transfer coefficient can be defined using the outside area of the tube. From these choices, we say that the overall heat-transfer coefficient is *based on the outside area*, or *based on the inside area*, respectively. We write this explicitly as

$$(U_{OA}A)_i = U_{OA,i} A_i, \qquad (11.66a)$$

or

$$U_{OA,i} = \frac{1}{A_i(R_{conv,i} + R_{wall} + R_{conv,o})}. \qquad (11.66b)$$

For a circular-cross-section tube of length L, we substitute $A_i = 2\pi r_{w,i} L$ for the inside area and the thermal resistances from Table 11.8 into this expression to yield

$$U_{OA,i} = \frac{1}{\dfrac{1}{h_{conv,i}} + \dfrac{r_{w,i}}{k_w} \ln(r_{w,o}/r_{w,i}) + \dfrac{r_{w,i}}{r_{w,o}} \dfrac{1}{h_{conv,o}}}, \qquad (11.66c)$$

where $r_{w,i}$ and $r_{w,o}$ are the inner and outer tube radii, respectively. Similar substitutions can be performed to identify the overall heat-transfer based on the outer area:

$$(U_{OA}A)_o = U_{OA,o} A_o, \qquad (11.67a)$$

$$U_{OA,o} = \frac{1}{A_o(R_{conv,i} + R_{wall} + R_{conv,o})}, \qquad (11.67b)$$

and

$$U_{OA,o} = \frac{1}{\dfrac{r_{w,o}}{r_{w,i}} \dfrac{1}{h_{conv,i}} + \dfrac{r_{w,o}}{k_w} \ln(r_{w,o}/r_{w,i}) + \dfrac{1}{h_{conv,o}}}. \qquad (11.67c)$$

Note from the defining relationship (Eq. 11.65a) that the products of U and A, based on either area, are equivalent; that is,

$$U_{OA,i}A_i = U_{AO,o}A_o. \qquad (11.68)$$

A similar reciprocity holds for overall heat-transfer coefficients based on any area.

To evaluate an overall heat-transfer coefficient requires values for the convective heat-transfer coefficients on each side of the wall separating the hot and cold fluids. Some typical values for overall heat-transfer coefficients for shell-and-tube heat exchangers are shown in Table 11.9.

> **Chapters 9 and 10 present methods to calculate these coefficients for external and internal flows, respectively.**

Table 11.9 Typical Values for Overall Heat-Transfer Coefficients for Shell-and-Tube Heat Exchangers [19]

Hot Fluid	Cooling Fluid	Overall Heat-Transfer Coefficient* $(W/m^2 \cdot K)$
Steam	Water	1700–2800
Steam	Light oil (SAE 10)	400–570
Steam	Heavy oil	225–280
Steam	Air	170–225
Water	Water (300 K)	1560–1840
Oil (SAE 10)	Water (300 K)	400–570
Oil (SAE 30)	Water (300 K)	340–450
50% Glycol	Water	850–1020

*Upper values apply to clean heat exchangers; lower values are more appropriate for heat exchangers prone to fouling.

Example 11.24

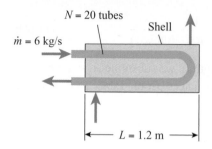

$\dot{m} = 6$ kg/s

$N = 20$ tubes

Shell

$L = 1.2$ m

Consider a shell-and-tube heat exchanger with a single shell and two tube passes of twenty tubes as shown in the sketch. Water, with a total flow rate of 6 kg/s, flows through the tubes. The length of each tube pass is 1.2 m, and the inside and outside diameters of the BWG 17 tubes are 12.9 and 15.9 mm, respectively. The tubes are fabricated from a carbon-steel alloy having a thermal conductivity of 60 W/m·K. The average heat-transfer coefficient at the outer surface of the tubes is 2000 W/m²·K. Determine the total inner and outer areas associated with the tubes and the overall heat-transfer coefficients based on the inner and outer areas, respectively. Evaluate any required properties at 300 K.

Solution

Known   Heat-exchanger configuration, N_{tubes}, M_{passes}, D_i, D_o, $\dot{m}_{tubes}$, $h_{conv,o}$, k_w

Find A_i, A_o, $U_{OA,i}$, $U_{OA,o}$

Sketch

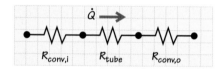

Assumptions

i. The length of tube passes are unaffected by bends.
ii. The tubes are clean.
iii. The water in the tubes is being heated (i.e., $T_w > T_m$).
iv. Entrance effects are negligible.

Analysis   The total tube area (inner or outer) is the product of the number of tubes in a tube pass, N; the number of tube passes in the heat exchanger, M; and the area (inner or outer) of a single tube of length L; that is,

$$A_{total} = NMA_{single\ tube}.$$

Thus,

$$A_{total,inner} \equiv A_i = NM(\pi D_i L)$$

$$= 20(2)\pi(0.0129\ m)1.2\ m$$

$$= 1.945\ m^2$$

and

$$A_{total,outer} \equiv A_o = NM(\pi D_o L)$$

$$= 20(2)\pi(0.0159\ m)1.2\ m$$

$$= 2.398\ m^2.$$

To calculate the overall heat-transfer coefficients $U_{OA,i}$ and $U_{OA,o}$, we employ Eqs. 11.66c and 11.67c, respectively. All quantities needed are known except $h_{conv,i}$, the average convective heat-transfer coefficient inside the tubes. As a first step in finding $h_{conv,i}$, we calculate the Reynolds number for the flow in a single tube. The mass flow rate is

$$\dot{m} \equiv \dot{m}_{\text{single tube}} = \dot{m}_{\text{total}}/N$$

$$= \frac{6\,\text{kg/s}}{20} = 0.3\,\text{kg/s},$$

and (Eq. 10.34)

$$Re_{D_i} = \frac{\rho v D_i}{\mu} = \frac{4\,\dot{m}}{\pi \mu D_i}.$$

Properties of water (Table G.1 or NIST database) at 300 K are

$$\rho = 997\,\text{kg/m}^3, \qquad\qquad Pr = 5.83,$$
$$\mu = 855 \times 10^{-6}\,\text{N}\cdot\text{s/m}^2, \qquad k = 0.613\,\text{W/m}\cdot\text{K},$$

Thus,

$$Re_D = \frac{4(0.3)}{\pi(855 \times 10^{-6})0.0129} = 34{,}630$$

$$[=] \frac{\text{kg/s}}{(\text{N}\cdot\text{s/m}^2)\text{m}}\left[\frac{1\,\text{N}}{\text{kg}\cdot\text{m/s}^2}\right] = 1.$$

Since $Re_{D_i} > 2300$, the flow is turbulent. We choose the Dittus–Boelter correlation (Eq. T.10.5a) from Table 10.5, noting that our Re and Pr values fall within the required ranges. Thus,

$$Nu_{D_i} = \frac{h_{\text{conv,i}}D_i}{k_{\text{H}_2\text{O}}} = 0.023\,Re_{D_i}^{0.8}Pr^{0.4}$$

$$= 0.023(34{,}630)^{0.8}(5.83)^{0.4}$$

$$= 199,$$

and

$$h_{\text{conv,i}} = \frac{k_{\text{H}_2\text{O}}}{D_i}Nu_{D_i} = \frac{0.613}{0.0129}199 = 9470$$

$$[=] \frac{\text{W/m}\cdot\text{K}}{\text{m}} = \text{W/m}^2\cdot\text{K}.$$

We now evaluate $U_{\text{OA,i}}$ (Eq. 11.66c) as follows:

$$U_{\text{OA,i}} = \left[\frac{1}{h_{\text{conv,i}}} + \frac{r_i}{k_{\text{tube}}}\ln\left(\frac{r_o}{r_i}\right) + \frac{r_i}{r_o}\frac{1}{h_{\text{conv,o}}}\right]^{-1}$$

$$= \left[\frac{1}{9470} + \frac{0.0129/2}{60}\ln\left(\frac{0.0159}{0.0129}\right) + \frac{0.0129}{0.0159}\frac{1}{2000}\right]^{-1}\,\text{W/m}^2\cdot\text{K}$$

$$= 1874\,\text{W/m}^2\cdot\text{K}.$$

The reader should verify the consistency of the units in each of the three right-hand-side terms. Using Eq. 11.67c, we determine the overall heat-transfer coefficient based on the outer area of the tubes to be 1520 W/m²·K, which the reader can verify for practice. We also check this result using Eq. 11.68:

$$U_{\text{OA,o}} = U_{\text{OA,i}}\frac{A_i}{A_o} = U_{\text{OA,i}}\frac{D_i}{D_o}$$

$$= 1874\frac{0.0129}{0.0159} = 1520\,\text{W/m}^2\cdot\text{K}.$$

Comments First, note the importance of understanding the particulars of the geometry to determine $A_{\text{total,i}}$ and $A_{\text{total,o}}$. We analyze the relative magnitudes of the three terms on the right-hand side of Eq. 11.66c to provide insight into the design and operation of this particular heat exchanger. We see that the thermal resistance of the tube walls makes a small contribution to U_{OA}; neglecting the $(r_i/k_{\text{tube}}) \ln (r_o/r_i)$ term results in an increase in U_{OA} of only 4%. Furthermore, we see that the outside thermal convective resistance dominates the total, as $h_{\text{conv,o}} (= 2000 \text{ W/m}^2\cdot\text{K})$ is significantly smaller than $h_{\text{conv,i}} (= 9470 \text{ W/m}^2\cdot\text{K})$. As a result, $U_{\text{OA,i}}$ is much closer to $h_{\text{conv,o}}$ than $h_{\text{conv,i}}$. Finally, we note that we did not consider any possible fouling of the heat-exchange surfaces. Dirty fluids and/or corrosion can result in fouling, which reduces the performance of a heat exchanger; that is, $U_{\text{OA,dirty}}$ will be less than $U_{\text{OA,clean}}$. For more information on fouling, the interested reader is referred to Refs. [16, 19, 21].

In the next two subsections, we present methods that utilize the overall heat-transfer coefficient to design and predict the performance of heat exchangers.

Log-Mean Temperature Difference Method

In the approach outlined in this section, the heat-transfer rate between the hot and cold fluids is expressed in a general way as

$$\dot{Q} = (UA)(\Delta T)_{\text{avg}}, \tag{11.69}$$

where $(\Delta T)_{\text{avg}}$ is an appropriately determined average temperature difference through the heat exchanger. We require an average value because the temperature difference between the hot and cold fluids varies from location to location within the heat exchanger. Note that in Eq. 11.69 the quantity UA is bracketed to emphasize that U and A are not independent; they have precise meanings only when paired (see Eq. 11.68).

Our objective in this section is to develop ways to calculate this average temperature difference, $(\Delta T)_{\text{avg}}$, using the four fluid temperatures: the hot-fluid inlet and outlet temperatures, $T_{\text{H,i}}$ and $T_{\text{H,o}}$, and the cold-fluid inlet and outlet temperatures, $T_{\text{C,i}}$ and $T_{\text{C,o}}$. We assume that the overall heat-transfer coefficient is a known quantity in this analysis. Our specific problem statement is thus the following:

Given the overall heat-transfer coefficient and the inlet temperatures and the flow rates of the hot and cold fluids, determine the heat-exchange area needed to heat (or cool) the cold (or hot) fluid to a desired temperature.

Parallel Flow We begin our solution to this problem by considering a simple parallel-flow configuration and then generalize. The situation for parallel flow is shown in Fig. 11.46. The hot fluid and the cold fluid both enter on the left. As the hot fluid passes through the heat exchanger, its temperature falls. Concomitantly, the temperature of the cold fluid rises as it passes through the heat exchanger. The temperature distributions are shown in Fig. 11.46 as functions of the heat exchange area, which varies from zero at the entrance ($x = 0$) to A_{tot} at the exit ($x = L$).

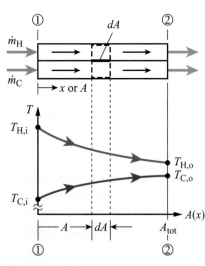

FIGURE 11.46
Schematic diagram of temperature distributions through parallel-flow heat exchanger.

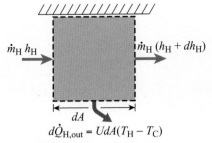

FIGURE 11.47
Differential control volume for hot fluid within a heat exchanger. The differential area dA corresponds to that shown in Fig. 11.46.

Consider now a differential control volume as suggested by the dashed lines in Fig. 11.46. The differential control volume associated with just the hot fluid is isolated and shown in Fig. 11.47. Conservation of energy for this control volume is written by simplifying Eq. 5.63, which results in

$$-d\dot{Q}_{H,out} = \dot{m}_H\left[(h_H + dh_H) - h_H\right],$$

or

$$-d\dot{Q}_{H,out} = \dot{m}_H\,dh_H.$$

Using the calorific equation of state (i.e., $dh = c_p dT$), we can reexpress this as

$$d\dot{Q}_{H,out} = -(\dot{m}c_p)_H dT_H. \tag{11.70a}$$

A similar analysis can be performed on the corresponding differential control volume for the cold stream, which yields

$$d\dot{Q}_{C,in} = (\dot{m}c_p)_C\,dT_C. \tag{11.70b}$$

The differential heat-transfer rates associated with dA can also be expressed in terms of the (local) overall heat-transfer coefficient as indicated in Fig. 11.47 for the hot fluid; that is,

$$d\dot{Q}_{H,out} = (U dA)(T_H - T_C), \tag{11.71a}$$

where T_H and T_C are the local hot and cold fluid temperatures, respectively. A similar expression applies to the cold stream:

$$d\dot{Q}_{C,in} = (U dA)(T_H - T_C). \tag{11.71b}$$

From a control volume containing both fluid streams, energy conservation yields (cf. Eq. 11.59)

$$d\dot{Q}_{H,out} = d\dot{Q}_{C,in}. \tag{11.72}$$

From this point forward, we introduce no new physics but only manipulate Eqs. 11.70 and 11.71 and integrate over the entire heat exchanger from 0 $(x = 0)$ to $A_{tot}(x = L)$. Substituting Eq. 11.71a into Eq. 11.70a and rearranging yield

$$\frac{U}{(\dot{m}c_p)_H}dA = \frac{-1}{T_H - T_C}dT_H, \tag{11.73a}$$

and performing similar operations for Eqs. 11.70b and 11.71b yields

$$\frac{U}{(\dot{m}c_p)_C}dA = \frac{1}{T_H - T_C}dT_C. \tag{11.73b}$$

Adding Eqs. 11.73a and 11.73b, we obtain

$$\left[\frac{1}{(\dot{m}c_p)_H} + \frac{1}{(\dot{m}c_p)_C}\right]U dA = \frac{dT_C - dT_H}{T_H - T_C}.$$

Recognizing that $dT_C - dT_H = d(T_C - T_H)$, we can integrate this equation from inlet to outlet as follows:

$$\left[\frac{1}{(\dot{m}c_p)_H} + \frac{1}{(\dot{m}c_p)_C}\right]U\int_{inlet}^{outlet}dA = -\int_{inlet}^{outlet}\frac{d(T_H - T_C)}{T_H - T_C},$$

where U is assumed to be a constant. Treating $(T_H - T_C)$ as a single variable that depends only on location [i.e., x or $A(x)$], we can easily evaluate this expression as

$$-\left[\frac{1}{(\dot{m}c_p)_H} + \frac{1}{(\dot{m}c_p)_C}\right]UA_{tot} = \ln\left[T_H - T_C\right]_{inlet}^{outlet}$$

$$= \ln(T_{H,o} - T_{C,o}) - \ln(T_{H,i} - T_{C,i}) \qquad (11.74)$$

$$= \ln\left[\frac{T_{H,o} - T_{C,o}}{T_{H,i} - T_{C,i}}\right].$$

Our final step is to eliminate $(\dot{m}c_p)_H$ and $(\dot{m}c_p)_C$ by applying the overall energy conservation expressions previously derived for each stream (i.e., Eqs. 11.58a and 11.58b). Rearranging these yields, respectively,

$$\frac{1}{(\dot{m}c_p)_H} = \frac{T_{H,i} - T_{H,o}}{\dot{Q}_{H,out}}$$

and

$$\frac{1}{(\dot{m}c_p)_C} = \frac{T_{C,o} - T_{C,i}}{\dot{Q}_{C,in}}.$$

Recognizing that $\dot{Q}_{H,out} = \dot{Q}_{C,in} (\equiv \dot{Q}_{PF})$ (Eq. 11.59), we substitute these expressions into Eq. 11.74 and rearrange to obtain an expression for $\dot{Q}_{PF}$, the mutual heat exchange between the two fluid streams for parallel flow:

$$\dot{Q}_{PF} = (UA_{tot})\frac{(T_{H,o} - T_{C,o}) - (T_{H,i} - T_{C,i})}{\ln\left[\dfrac{T_{H,o} - T_{C,o}}{T_{H,i} - T_{C,i}}\right]}. \qquad (11.75)$$

Comparing Eq. 11.75 with Eq. 11.69, we see that we have achieved our original objective of determining an average temperature expressed in terms of known inlet and outlet temperatures:

$$(\Delta T)_{avg,PF} = \frac{(T_{H,o} - T_{C,o}) - (T_{H,i} - T_{C,i})}{\ln\left[\dfrac{(T_{H,o} - T_{C,o})}{(T_{H,i} - T_{C,i})}\right]}.$$

This particular expression is called the **log-mean temperature difference** and is sometimes denoted as $(\Delta T)_{lm}$ or $LMTD$,

$$(\Delta T)_{lm,PF} = LMTD_{PF} \equiv \frac{(T_{H,o} - T_{C,o}) - (T_{H,i} - T_{C,i})}{\ln\left[\dfrac{(T_{H,o} - T_{C,o})}{(T_{H,i} - T_{C,i})}\right]}. \qquad (11.76)$$

Using this shorthand notation, we write the heat-transfer rate for a parallel-flow heat exchanger as

$$\dot{Q}_{PF} = (UA_{tot})(\Delta T)_{lm,PF}. \qquad (11.77)$$

Before moving on to other configurations, we note that the log-mean temperature difference can be expressed more compactly in terms of the differences in fluid temperatures at the inlet and outlet stations 1 and 2, as indicated in Fig. 11.46, that is,

$$(\Delta T)_1 \equiv T_{H,i} - T_{C,i}, \qquad (11.78a)$$

$$(\Delta T)_2 \equiv T_{H,o} - T_{C,o} \qquad (11.78b)$$

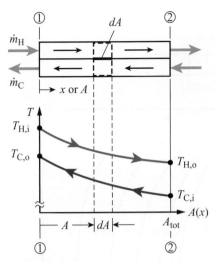

FIGURE 11.48
Schematic diagram of temperature distributions through counterflow heat exchanger.

Shell-and-tube heat exchanger with four shell passes (bottom); tube bundles for shell-and-tube heat exchangers (top). Photographs courtesy of J.F.D. Tube & Coil Products, Inc. (bottom) and Olympus NDT, Canada (top).

With this notation,

$$(\Delta T)_{\mathrm{lm,PF}} \equiv \frac{(\Delta T)_2 - (\Delta T)_1}{\ln\left(\dfrac{(\Delta T)_2}{(\Delta T)_1}\right)}. \tag{11.79}$$

Counterflow A similar analysis can be performed for a counterflow heat exchanger. Although we will present only the final result, the physical situation and the temperature distributions are shown in Fig. 11.48. Using the notation that, at station 1, the temperature difference is

$$(\Delta T)_1 \equiv T_{\mathrm{H,i}} - T_{\mathrm{C,o}}$$

and, at station 2,

$$(\Delta T)_2 \equiv T_{\mathrm{H,o}} - T_{\mathrm{C,i}},$$

we can express the heat-transfer rate for a counterflow heat exchanger, $\dot{Q}_{\mathrm{CF}}$, in an identical manner to the parallel-flow device:

$$\dot{Q}_{\mathrm{CF}} = (UA)(\Delta T)_{\mathrm{lm,CF}}, \tag{11.80}$$

where

$$(\Delta T)_{\mathrm{lm,CF}} \equiv \frac{(\Delta T)_2 - (\Delta T)_1}{\ln\left[\dfrac{(\Delta T)_2}{(\Delta T)_1}\right]}. \tag{11.81}$$

Comparing Eqs. 11.79 and 11.81, we see that the log-mean temperature difference is generalized (i.e., not specific to either parallel flow or counterflow) when the temperature differences are associated with the stations 1 and 2 and not specific inlets or outlets. Note also the mathematical equivalence

$$\frac{(\Delta T)_2 - (\Delta T)_1}{\ln\left(\dfrac{(\Delta T)_2}{(\Delta T)_1}\right)} = \frac{(\Delta T)_1 - (\Delta T)_2}{\ln\left(\dfrac{(\Delta T)_1}{(\Delta T)_2}\right)},$$

which precludes confusing stations 1 and 2 by noting that the first term in the numerator is always the top term in the logarithm.

Cross-flow and Other Configurations We now extend our analysis to cross-flow heat exchangers and other more complex geometries. Figures 11.49–11.52 illustrate these geometries schematically. Note that none of these devices are pure parallel or pure counterflow configurations; rather, they are some combination (Figs. 11.49 and 11.50) or are of a cross-flow type (Figs. 11.51 and 11.52). The problem of determining the appropriate $(\Delta T)_{\mathrm{avg}}$ for these geometries was solved by Bowman et al. [17]. The results of their calculations are expressed as a correction factor to the counterflow log-mean temperature difference, that is,

$$(\Delta T)_{\mathrm{avg}} = F(\Delta T)_{\mathrm{lm,CF}}. \tag{11.82}$$

Values for F are obtained from the charts presented in Figs. 11.49–11.52, where F is presented as a function of the two dimensionless temperature ratios, P and R, defined on each chart.

We now illustrate the use of the log-mean temperature difference method with the following example.

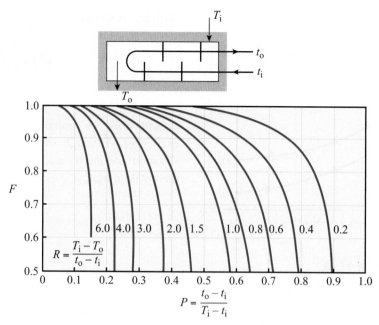

FIGURE 11.49

Correction factor for a shell-and-tube heat exchanger with one shell and any multiple of two tube-passes (i.e., two, four, etc.). **Adapted from Ref. [17] with permission.**

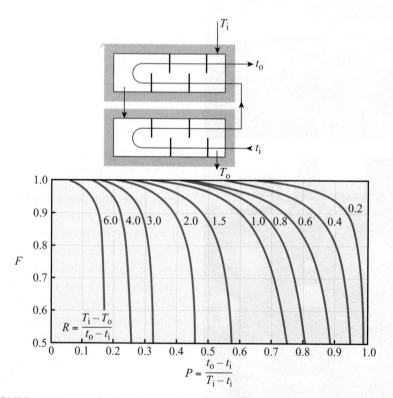

FIGURE 11.50

Correction factor for a shell-and-tube heat exchanger with two shell-passes and any multiple of four tube-passes (i.e., four, eight, etc.). **Adapted from Ref. [17] with permission.**

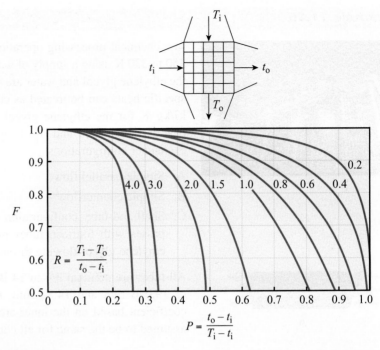

FIGURE 11.51
Correction factor for a single-pass, cross-flow heat exchanger with both fluids unmixed. **Adapted from Ref. [17] with permission.**

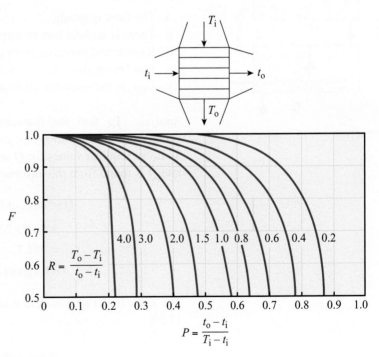

FIGURE 11.52
Correction factor for a single-pass, cross-flow heat exchanger with one fluid mixed and the other unmixed. **Adapted from Ref. [17] with permission.**

Example 11.25

In a chemical processing operation, ethylene glycol is to be cooled from 350 to 330 K using a supply of water available at 290 K. The flow rates of the ethylene glycol and water are 4.200 and 2.605 kg/s, respectively. Fluid specific heats can be treated as constants with values of 2.592 and 4.179 kJ/kg·K for the ethylene glycol and water, respectively. Determine the heat-exchange area and the length associated with the following heat-exchanger configurations:

A. Simple parallel flow

B. Simple counterflow

C. Shell-and-tube configuration with two shell passes and four tube passes with fourteen tubes in each pass (i.e., $N = 14$, $M = 4$) and ethylene glycol flowing through the shells

All tubes are nominal 3/4-in 14 BWG (with outside and inside diameters of 19.046 mm and 14.83 mm, respectively). The overall heat-transfer coefficient based on the inner area of the tube(s) is 850 W/m²·K and is assumed to be the same for all configurations.

A.

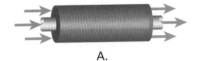

B.

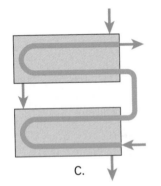

C.

Solution

Known   Configuration, $\dot{m}_\mathrm{H}$, $\dot{m}_\mathrm{C}$, $c_{p,\mathrm{H}}$, $c_{p,\mathrm{C}}$, $T_{\mathrm{H,in}}$, $T_{\mathrm{H,out}}$, $T_{\mathrm{C,in}}$, $U_{\mathrm{OA,i}}$, D_i, D_o

Find A_i, L_{HX}

Sketch   See subsequent sketches for parallel flow and counterflow.

Assumptions

   i. The flow is steady.
   ii. There is no heat loss to surroundings ($\dot{Q}_{\mathrm{cv}} = 0$).
   iii. Kinetic and potential energy changes are negligible.
   iv. Fluid properties are constant.
   v. $U_{\mathrm{OA,i}}$ is the same for all heat exchangers.

Analysis   To find the heat-exchange area for the parallel-flow and counterflow configurations we employ Eqs. 11.77 and 11.80, respectively. To do so requires values for $\dot{Q}$ and $\Delta T_{\mathrm{lm,PF}}$ and $\Delta T_{\mathrm{lm,CF}}$. The heat-transfer rate $\dot{Q}$ is found from the information given for the hot fluid (Eq. 11.58a):

$$\dot{Q} = \dot{m}_\mathrm{H} c_{p,\mathrm{H}}(T_{\mathrm{H,in}} - T_{\mathrm{H,out}})$$

$$= 4.200(2.592)(350 - 330)$$

$$= 217.7$$

$$[=]\,(\mathrm{kg/s})(\mathrm{kJ/kg \cdot K})\mathrm{K} = \mathrm{kJ/s\ or\ kW}.$$

The water outlet temperature is now found by combining Eqs. 11.58b and 11.59:

$$T_{\mathrm{C,out}} = T_{\mathrm{C,in}} + \frac{\dot{Q}}{\dot{m}_\mathrm{C} c_{p,\mathrm{C}}}$$

$$= 290 + \frac{217.7}{2.605(4.179)}\ \mathrm{K}$$

$$= 290 + 20 = 310\ \mathrm{K}.$$

We obtain $\Delta T_{\text{lm,PF}}$ (Eq. 11.79) with the aid of the following sketch:

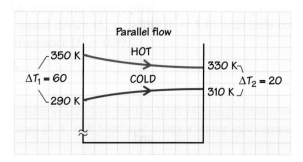

Thus,

$$\Delta T_{\text{lm,PF}} = \frac{\Delta T_2 - \Delta T_1}{\ln\left(\dfrac{\Delta T_2}{\Delta T_1}\right)} = \frac{20 - 60}{\ln\left(\dfrac{20}{60}\right)} \, K = 36.4 \, K.$$

Rearranging Eq. 11.77, we find the heat-exchange area for the parallel-flow configuration:

$$A_{\text{i,PF}} = \frac{\dot{Q}}{U_{\text{OA,i}} \Delta T_{\text{lm,PF}}}$$

$$= \frac{217{,}700}{850(36.4)} = 7.04$$

$$[=] \frac{W}{(W/m^2 \cdot K)K} = m^2.$$

The length associated with this area is given by

$$A_{\text{i,PF}} = \pi D_i L_{\text{PF}},$$

or

$$L_{\text{PF}} = \frac{A_{\text{i,PF}}}{\pi D_i} = \frac{7.04 \, m^2}{\pi (0.01483 \, m)} = 151.1 \, m.$$

For the simple counterflow configuration, $\Delta T_{\text{lm,CF}}$ is obtained with the aid of following sketch:

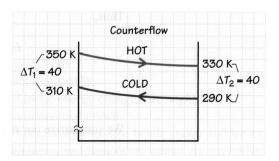

Thus,

$$\Delta T_{\text{lm,CF}} = \frac{\Delta T_2 - \Delta T_1}{\ln\dfrac{\Delta T_2}{\Delta T_1}} = \frac{40 - 40}{\ln\left(\dfrac{40}{40}\right)} = \frac{0}{0}!$$

find the ideal pumping power using Eq. 11.43, that is,

$$\dot{W}_{\text{pump,ideal}} = \dot{m}\frac{P_2 - P_1}{\rho} = \dot{m}\frac{P_2 - P_3}{\rho}.$$

The pressure drop, $P_2 - P_3$, for the tube bundle can be estimated from the head loss associated with a single tube having a length of $4L_{\text{S-T}}$, where the factor 4 is the number of tube-passes. Applying Eq. 10.17 yields

$$\Delta P = P_2 - P_3 = f_{\text{D}}\frac{(4L_{\text{S-T}})}{D_{\text{i}}}\frac{1}{2}\rho v^2,$$

where v is the average velocity in a single tube. This velocity is found by dividing the total flow rate by the number of tubes in a pass and applying mass continuity as follows:

$$v = \frac{\dot{m}_{\text{tot}}/N}{\rho\pi D_{\text{i}}^2/4} = \frac{2.605/14}{997\pi(0.01483)^2/4}$$

$$= 1.08$$

$$[=]\frac{\text{kg/s}}{(\text{kg/m}^3)\text{m}^2} = \text{m/s},$$

where the density is approximated as that of saturated liquid water at 300 K. To find the friction factor requires values for $Re_{D_{\text{i}}}$ and ε/D_{i}. The Reynolds number is

$$Re_{D_{\text{i}}} = \frac{\rho v D_{\text{i}}}{\mu} = \frac{997(1.08)0.01483}{855\times 10^{-6}}$$

$$= 18,700,$$

where the viscosity value is from Table G.2 or the NIST database. The roughness height for drawn tubing is $\varepsilon = 0.002$ mm (Table 10.4) and so

$$\frac{\varepsilon}{D_{\text{i}}} = \frac{0.002\text{ mm}}{14.83\text{ mm}} = 0.00013.$$

The friction factor (Fig. 10.17) is thus

$$f_{\text{D}} \approx 0.0265.$$

The approximate pressure drop through the four tube-passes is then

$$\Delta P = f_{\text{D}}\frac{(4L_{\text{S-T}})}{D_{\text{i}}}\frac{1}{2}\rho v^2$$

$$= 0.0265\frac{4(2.48)}{0.01483}0.5(997)(1.08)^2$$

$$= 10,300$$

$$[=]\left(\frac{\text{m}}{\text{m}}\right)\frac{\text{kg}}{\text{m}^3}\left(\frac{\text{m}}{\text{s}}\right)^2\left[\frac{1\text{ N}}{\text{kg}\cdot\text{m/s}^2}\right]\left[\frac{1\text{ Pa}}{\text{N/m}^2}\right] = \text{Pa}.$$

The minimum pumping power required is that for an ideal (adiabatic, reversible) pump (Eq. 11.43), that is,

$$\dot{W}_{\text{pump,ideal}} = \frac{\dot{m}_{\text{tot}} \Delta P}{\rho}$$

$$= \frac{2.605(10,300)}{997} = 26.9$$

$$[=] \frac{(\text{kg/s})(\text{N/m}^2)}{\text{kg/m}^3} \left[\frac{1\ \text{J}}{\text{N} \cdot \text{m}} \right] = \frac{\text{J}}{\text{s}} = \text{W}.$$

Note that the total flow rate, rather than just that of a single tube, is used to obtain the total pump power and that the pressure drop for a single tube is the same as the pressure drop for the tube bundle, as the tubes comprising the bundle are in a parallel, rather than a series, arrangement.

Comments This example illustrates how heat transfer and pumping requirements are both important aspects of heat-exchanger design and analysis. Note that the number of tubes in a pass, the number of passes, and the size of an individual tube all affect the pumping requirements; they also affect the heat-transfer performance through their influence on the overall heat-transfer coefficient. The parallel arrangement of the tubes in the shell-and-tube heat exchanger has a dramatic effect in reducing the pumping requirements associated with either the simple parallel-flow or counterflow (single-tube) heat exchangers discussed in Example 10.25. For example, the pressure drop and ideal pump power for the 151.1-m-long parallel-flow exchanger are 18.5 MPa and 48.3 kW, respectively! The huge power requirement of this device makes it as absurdly impractical as does its great length.

Effectiveness–NTU Method

The method we present here offers an alternative to the log-mean temperature approach. Why do we need another method? To use the log-mean temperature method without iteration requires that all four temperatures be known. In some situations, for example, in evaluating existing hardware at off-design conditions, only the inlet temperatures may be known. For such situations, the effectiveness–*NTU* (number of transfer units) method offers a direct (noniterative) approach to determine heat-exchanger performance. The development of this method follows.

We begin by defining a heat-exchanger effectiveness as follows:

$$\varepsilon \equiv \frac{\dot{Q}_{\text{actual}}}{\dot{Q}_{\text{max possible}}}, \tag{11.83}$$

where $\dot{Q}_{\text{actual}}$ is the heat-transfer rate between the hot and cold fluids as expressed by Eqs. 11.58a and 11.58b. For convenience in notation, we express these equations using the heat-capacity rates, which we formally define as

$$C \equiv \dot{m}c_p. \tag{11.84a}$$

Thus,

$$C_\text{C} \equiv (\dot{m}c_p)_\text{C}, \tag{11.84b}$$

$$C_\text{H} \equiv (\dot{m}c_p)_\text{H}, \tag{11.84c}$$

LEVEL 3

and

$$\dot{Q}_{\text{actual}} = \begin{cases} \dot{Q}_{\text{H,out}} = C_{\text{H}}(T_{\text{H,in}} - T_{\text{H,out}}), \\ \dot{Q}_{\text{C,in}} = C_{\text{C}}(T_{\text{C,out}} - T_{\text{C,in}}). \end{cases} \tag{11.85}$$

To define $\dot{Q}_{\text{max possible}}$, consider a counterflow heat exchanger that has sufficient heat exchange area ($A \rightarrow \infty$) such that, if the heat-capacity rate of the cold fluid is less than the heat-capacity rate of the hot fluid (i.e., $C_{\text{C}} < C_{\text{H}}$), then the outlet temperature of the cold fluid approaches the temperature of the incoming hot fluid (see Fig. 11.53 top). Alternatively, if $C_{\text{H}} < C_{\text{C}}$, then the outlet temperature of the hot fluid approaches that of the incoming cold fluid (see Fig. 11.53 bottom). In either of these two cases, the total heat-transfer rate is the maximum possible. Heating the cold fluid above the incoming temperature of the hot fluid when $C_{\text{C}} < C_{\text{H}}$, or cooling the hot fluid below the incoming temperature of the cold fluid when $C_{\text{H}} < C_{\text{C}}$, violates the second law of thermodynamics (see Table 7.4). Defining C_{min} to be the smaller of the two heat-capacity rates,

$$C_{\text{min}} \equiv \text{minimum}|C_{\text{H}}, C_{\text{C}}|, \tag{11.86}$$

we write the following general expression for the maximum possible heat-transfer rate:

$$\dot{Q}_{\text{max possible}} = C_{\text{min}}(T_{\text{H,in}} - T_{\text{C,in}}). \tag{11.87}$$

We note that the temperature difference in this expression is the maximum possible temperature difference in the heat exchanger: There is no temperature higher than that of the incoming hot fluid, and there is no temperature lower than that of the incoming cold fluid. We also note that the temperatures of the incoming streams are usually known quantities. Substituting Eqs. 11.85 and 11.87 into our definition of effectiveness (Eq. 11.83) yields

$$\varepsilon = \begin{cases} \dfrac{C_{\text{H}}(T_{\text{H,in}} - T_{\text{H,out}})}{C_{\text{min}}(T_{\text{H,in}} - T_{\text{C,in}})} \text{ (hot stream),} \\ \\ \dfrac{C_{\text{C}}(T_{\text{C,out}} - T_{\text{C,in}})}{C_{\text{min}}(T_{\text{H,in}} - T_{\text{C,in}})} \text{ (cold stream).} \end{cases} \tag{11.88}$$

This expression can be made more general by considering that, depending on the result of Eq. 11.86, either the hot or cold fluid can be designated as the *minimum fluid*; thus,

$$\varepsilon = \frac{C_{\text{min}}\Delta T_{\text{min fluid}}}{C_{\text{min}}(T_{\text{H,in}} - T_{\text{C,in}})}.$$

Canceling C_{min} from the numerator and denominator yields

$$\varepsilon = \frac{\Delta T_{\text{min fluid}}}{T_{\text{H,in}} - T_{\text{C,in}}}, \tag{11.89a}$$

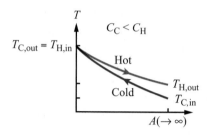

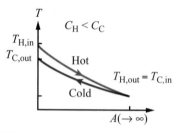

FIGURE 11.53

In a counterflow heat exchanger, the minimum fluid undergoes the greater temperature change from inlet to outlet. If the cold fluid is the minimum ($C_{\text{C}} < C_{\text{H}}$), the maximum possible heat exchange occurs when $T_{\text{C,out}} = T_{\text{H,in}}$ (top). If the hot fluid is the minimum ($C_{\text{H}} < C_{\text{C}}$), then the maximum possible heat transfer occurs when $T_{\text{H,out}} = T_{\text{C,in}}$.

or

$$\varepsilon = \frac{\Delta T_{\text{min fluid}}}{\Delta T_{\text{max possible}}}. \tag{11.89b}$$

Having defined the effectiveness solely in terms of the temperatures, we now seek to relate the effectiveness to the heat-transfer properties of the heat exchanger, namely, the product of the overall heat-transfer coefficient and the heat exchange area, (UA). For simple parallel-flow or counterflow devices, an algebraic expression can be obtained by combining $\dot{Q}_{\text{actual,PF or CF}} = (UA)\Delta T_{\text{lm,PF or CF}}$ with the relationships just derived. For a parallel-flow device, for example, we obtain the following result:

$$\varepsilon = \frac{1 - \exp\left[-\dfrac{(UA)}{C_{\text{min}}}\left(1 + \dfrac{C_{\text{min}}}{C_{\text{max}}}\right)\right]}{(1 + C_{\text{min}}/C_{\text{max}})}. \tag{11.90}$$

Similar expressions can be obtained for other heat-exchanger configurations. In all cases, the effectiveness depends only on two dimensionless parameters: $(UA)/C_{\text{min}}$ and $C_{\text{min}}/C_{\text{max}}$. Because of their importance, these two parameters are specially designated as the number of transfer units,

$$NTU = UA/C_{\text{min}}, \tag{11.91a}$$

and the heat capacity ratio,

$$CR \equiv C_{\text{min}}/C_{\text{max}}. \tag{11.91b}$$

To facilitate problem solving, Table 11.10 provides functional relationships $\varepsilon = \varepsilon(NTU, CR)$ for various heat-exchanger configurations; Table 11.11 presents the corresponding inverse relationships [i.e., $NTU = NTU\,(\varepsilon, CR)$]. Figures 11.54–11.56 present selected results in chart form.

The following example illustrates the use of the $\varepsilon - NTU$ approach.

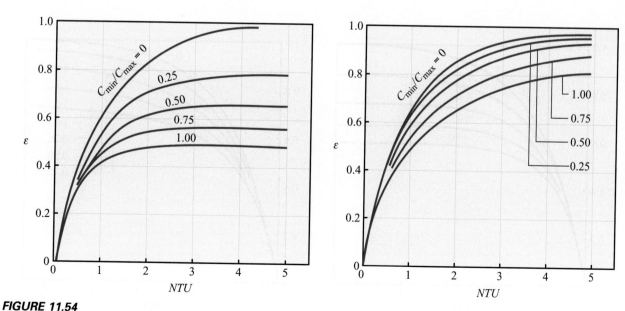

FIGURE 11.54

Effectiveness as a function of NTU for simple heat exchangers: parallel flow (left) and counterflow (right). **Adapted from Ref. [21] with permission.**

Table 11.10 Heat Exchanger Relationships of the Form $\varepsilon = \varepsilon(NTU, CR)$ (from Refs. [16, 21])

Configuration	Relationship	
Concentric Tube		
Parallel flow	$\varepsilon = \dfrac{1 - \exp[-NTU(1 + CR)]}{1 + CR}$	(T11.10a)
Counterflow	$\varepsilon = \begin{cases} \dfrac{1 - \exp[-NTU(1 - CR)]}{1 - CR\exp[-NTU(1 - CR)]} & (CR < 1) \\[2mm] \dfrac{NTU}{1 + NTU} & (CR = 1) \end{cases}$	(T11.10b)
Shell and Tube		
One shell-pass (2, 4, . . . tube-passes)	$\varepsilon_1 = 2\Big\{1 + CR + (1 + CR^2)^{1/2}$ $\qquad\qquad \times \dfrac{1 + \exp[-NTU(1 + CR^2)^{1/2}]}{1 - \exp[-NTU(1 + CR^2)^{1/2}]}\Big\}^{-1}$	(T11.10c)
n shell-passes (2n, 4n, . . . tube-passes)	$\varepsilon = \left[\left(\dfrac{1 - \varepsilon_1 CR}{1 - \varepsilon_1}\right)^n - 1\right]\left[\left(\dfrac{1 - \varepsilon_1 CR}{1 - \varepsilon_1}\right)^n - CR\right]^{-1}$	(T11.10d)
Cross-flow (Single Pass)		
Both fluids unmixed	$\varepsilon = 1 - \exp\left[\left(\dfrac{1}{CR}\right)(NTU)^{0.22}\{\exp[-CR(NTU)^{0.78}] - 1\}\right]$	(T11.10e)
C_{max} (mixed) and C_{min} (unmixed)	$\varepsilon = \left(\dfrac{1}{CR}\right)\left(1 - \exp\{-CR[1 - \exp(-NTU)]\}\right)$	(T11.10f)
C_{min} (mixed) and C_{max} (unmixed)	$\varepsilon = 1 - \exp\left(-CR^{-1}\{1 - \exp[-CR(NTU)]\}\right)$	(T11.10g)
All Exchangers ($CR = 0$)	$\varepsilon = 1 - \exp(-NTU)$	(T11.10h)

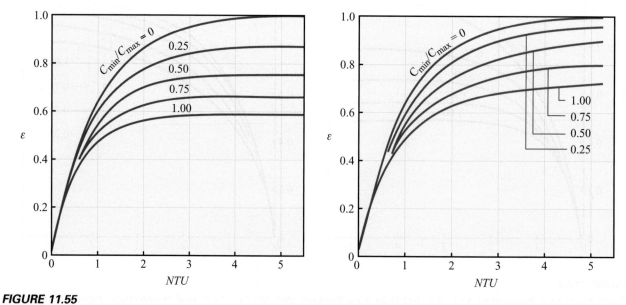

FIGURE 11.55

Effectiveness as a function of NTU for shell-and-tube heat exchangers with one (left) or two (right) shell-passes and any multiple of two tube-passes (i.e., 2, 4, 6, etc.). Adapted from Ref. [21] with permission.

Table 11.11   Heat Exchanger Relationships of the Form $NTU = NTU(\varepsilon, CR)$ (from Refs. [16, 21])

Configuration	Relationship	
Concentric Tube		
Parallel flow	$NTU = -\dfrac{\ln[1 - \varepsilon(1 + CR)]}{1 + CR}$	(T11.11a)
Counterflow	$NTU = \begin{cases} \dfrac{1}{CR - 1}\ln\left(\dfrac{\varepsilon - 1}{\varepsilon CR - 1}\right) & (CR < 1) \\[2ex] \dfrac{\varepsilon}{1 - \varepsilon} & (CR = 1) \end{cases}$	(T11.11b)
Shell and Tube		
One shell-pass (2, 4, . . . tube-passes)	$NTU = -(1 + CR^2)^{-1/2}\ln\left(\dfrac{E - 1}{E + 1}\right)$	(T11.11c)
	$E = \dfrac{2/\varepsilon_1 - (1 + CR)}{(1 + CR^2)^{1/2}}$	(T11.11d)
n shell-passes (2n, 4n,...tube-passes)	Use Eqs. T11.11c and T11.11d with $\varepsilon_1 = \dfrac{F - 1}{F - CR}, \; F = \left(\dfrac{\varepsilon CR - 1}{\varepsilon - 1}\right)^{1/n}$	(T11.11e,f)
Cross-Flow (Single Pass)		
C_{max} (mixed) and C_{mix} (unmixed)	$NTU = -\ln\left[1 + \left(\dfrac{1}{CR}\right)\ln(1 - \varepsilon CR)\right]$	(T11.11g)
C_{mix} (mixed) and C_{max} (unmixed)	$NTU = -\left(\dfrac{1}{CR}\right)\ln[CR\ln(1 - \varepsilon) + 1]$	(T11.11h)
All Exchangers ($CR = 0$)	$NTU = -\ln(1 - \varepsilon)$	(T11.11i)

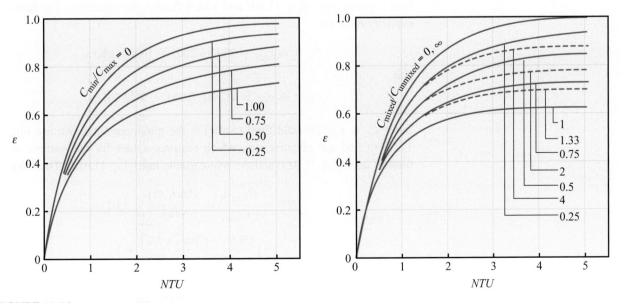

FIGURE 11.56

Effectiveness as a function of NTU for single-pass, cross-flow heat exchangers with both fluids unmixed (left) and one fluid mixed and the other unmixed (right). Adapted from Ref. [21] with permission.

Example 11.27

Photograph courtesy of Alaskan Copper Works, Seattle, WA.

Consider the shell-and-tube heat exchanger discussed in Example 11.25. Determine the effects of reducing the cold-water flow rate by a factor of 2 [i.e., $\dot{m}_C = 0.5(2.605) = 1.3025$ kg/s] assuming the overall heat-transfer coefficient is unchanged. Specifically, determine the outlet temperatures of each stream and the total heat-transfer rate.

Solution

Known $\dot{m}_C, \dot{m}_H, c_{p,H}, c_{p,C}, T_{H,in}, T_{C,in}, A_i, U_{OA,i}$

Find $T_{C,out}, T_{H,out}, \dot{Q}$

Sketch

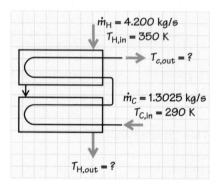

Assumptions

   i. Steady flow

   ii. Negligible $\dot{Q}_{cv}$ and changes in potential and kinetic energy

   iii. Constant c_ps

   iv. $U_{OA,i}$ unaffected by change in $\dot{m}_C$

Analysis   Since only the inlet temperatures are known, this problem is ideally suited for use of the ε–*NTU* method. We begin by determining the heat-capacity rates (Eq. 11.84) and which fluid is the minimum. The heat-capacity rates are

$$C_H = \dot{m}_H c_{p,H} = 2592(4.200) = 10{,}886 \text{ J/K} \cdot \text{s}$$

and

$$C_C = \dot{m}_C c_{p,C} = 4179(1.3025) = 5443 \text{ J/K} \cdot \text{s}.$$

Since $C_C < C_H$, the cold fluid (water) is the minimum fluid. We use Fig. 11.55 to find the effectiveness, which requires values for the number of transfer units (Eq. 11.91a) and the heat-capacity ratio (Eq. 11.91b). These are

$$NTU = \frac{U_{OA,i} A_i}{C_{min}} = \frac{850(6.47)}{5443} = 1.01$$

$$[=] \frac{(\text{W/m}^2 \cdot \text{K})\text{m}^2}{\text{J/K} \cdot \text{s}} \left[\frac{1 \text{ J/s}}{\text{W}} \right] = 1$$

and

$$CR = \frac{C_{min}}{C_{max}} = \frac{C_C}{C_H} = \frac{5433}{10{,}886} = 0.50.$$

Thus, from Fig. 11.55 we find

$$\varepsilon \approx 0.56.$$

Using the definition of ε (Eq. 11.89a), we now find the water outlet temperature:

$$\varepsilon = \frac{\Delta T_{\min}}{\Delta T_{\max \, poss}} = \frac{T_{C,out} - T_{C,in}}{T_{H,in} - T_{C,in}},$$

or

$$
\begin{aligned}
T_{C,out} &= \varepsilon(T_{H,in} - T_{C,in}) + T_{C,in} \\
&= 0.56(350 - 290) + 290 \text{ K} \\
&= 323.6 \text{ K}.
\end{aligned}
$$

The total heat-transfer rate (Eq. 11.85) is then

$$
\begin{aligned}
\dot{Q} &= C_C(T_{C,out} - T_{C,in}) \\
&= 5.443(323.6 - 290) = 182.9 \\
&[=] \left(\frac{\text{kJ}}{\text{K} \cdot \text{s}}\right) \text{K} = \text{kJ/s} = \text{kW},
\end{aligned}
$$

and the ethylene glycol outlet temperature is

$$
\begin{aligned}
T_{H,out} &= T_{H,in} - \frac{\dot{Q}}{C_H} \\
&= 350 - \frac{182.9}{10.886} \text{ K} = 333.2 \text{ K}.
\end{aligned}
$$

Comments   We first note the simplicity of the ε–*NTU* method for a situation in which only the inlet temperatures are known. As expected, reducing the water flow rate reduces the total heat-transfer rate between the fluids (182.9 versus 217.7 kW, i.e., 84% of the original rate), increases ΔT_C (33.6 versus 20 K), and decreases ΔT_H (16.8 versus 20 K). Since reducing $\dot{m}_C$ reduces $h_{conv,i}$, we might question the validity of the assumption that $U_{OA,i}$ is unchanged from Example 11.25. This is treated in Problem 11.115.

Self Test 11.19   Calculate the effectiveness of the heat exchanger described in Self Test 11.18 using the appropriate equation in Table 11.10. Check this result with Fig. 11.54.

(Answer: 0.909, 0.92)

11.7 FURNACES, BOILERS, AND COMBUSTORS

11.7a Some Applications

In this section, we consider various steady-flow combustion devices.[13] Furnaces, boilers, and like devices can be simple stand-alone units used to heat a space or to supply hot water (see Fig. 11.57), or they can be part

[13] Our requirement for steady flow thus rules out spark-ignition and diesel engines.

LEVEL 2

FIGURE 11.57

Gas-fired boilers (left) and hot-water heaters (right) are common combustion appliances found in homes. Photographs courtesy of Slant/Fin Corporation (boiler) and A. O. Smith Water Products Co. (water heater).

FIGURE 11.58

This heat-recovery steam generator is a huge heat-exchange device in which the hot products of combustion from a gas-turbine engine react with supplemental fuel to heat water and produce steam. Expansion of this steam through a steam turbine provides additional useful power. Photograph courtesy of Florida Power & Light Company.

of complex systems such as those used for electric power generation (see Fig. 11.58 and also Figs. 1.2 and 12.6). For residential and commercial applications, fuels are typically natural gas or fuel oil, whereas for large-scale power generation, coal and natural gas are the most commonly used fuels (see Table 1.1). For these devices, the useful transfer of energy occurs by heat transfer from the hot combustion products. Combustors, however, provide a stream of hot combustion products that is used directly to provide power or thrust, rather than relying on heat transfer to extract energy from the combustion products. For example, hot products expand through turbines to produce power in stationary electric power generation systems; in rocket engines, the hot products expand through a converging–diverging nozzle to produce thrust. Turbojet engines utilize both turbines and nozzles to expand a combustion product stream to useful effect. A jet engine combustor is illustrated in Fig. 11.59. (See also Figs. 1.12, 12.21, and 12.22.)

11.7b Analysis

The objective of this section is to present a brief and general analysis of these devices; additional discussions, analyses, and examples are presented in Chapter 12 where these devices are integrated into larger systems. We begin by defining an appropriate control volume and listing useful assumptions. Figure 11.60 shows a simple control volume with entering flows of fuel and oxidizer.[14] A single stream of combustion products exits the control volume.

---

[14] The oxidizer is usually air; however, other oxidizers, such as pure O_2, are employed in some applications. For example the Space Shuttle main engines burn H_2 with pure O_2.

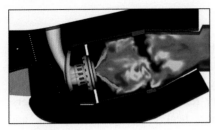

FIGURE 11.59
Modern numerical techniques are applied to simulate the complex phenomena occuring in a realistic jet engine combustor. **Image courtesy of Parviz Moin and Center for Integrated Turbulence Simulations, Stanford University.**

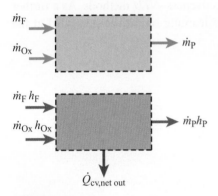

FIGURE 11.60
Control volume for analysis of steady-flow, constant-pressure combustion devices. Mass flows are shown at the top, and energy flows shown at the bottom. Subscripts F, Ox, and P refer to fuel, oxidizer, and products, respectively.

> **The reader is encouraged to consider, or review, the discussions of standardized enthalpies in Chapters 2 and 5 at this time.**

Assumptions

We employ the following assumptions to simplify our analysis:

- Steady-state, steady flow
- Constant pressure
- Uniform inlet and exit conditions
- Negligible kinetic and potential energy changes
- No work interactions other than flow work

Mass Conservation

For steady flow, the general integral expression of mass conservation (Eq. 3.18b) simplifies to

$$\dot{m}_F + \dot{m}_{Ox} = \dot{m}_P, \tag{11.92}$$

which states that the mass flow of products out of the control volume simply equals the sum of the two incoming mass flows. Note that any moisture, or other diluent, is implicitly included in the oxidizer flow rate.

Energy Conservation

The steady-flow energy equation that applies to our multistream integral control volume is Eq. 5.65:

$$\dot{Q}_{cv,in} - \dot{Q}_{cv,out} + \dot{W}_{cv,in} - \dot{W}_{cv,out}$$

$$= \sum_{k=1}^{M\text{ outlets}} \dot{m}_{out,k}\left[h_k + \frac{1}{2}\alpha_k v_{avg,k}^2 + g(z_k - z_{ref}) \right]$$

$$- \sum_{j=1}^{N\text{ inlets}} \dot{m}_{in,j}\left[h_j + \frac{1}{2}\alpha_j v_{avg,j}^2 + g(z_j - z_{ref}) \right].$$

Applying this to our specific control volume with the aforementioned assumptions yields

$$0 - \dot{Q}_{cv,out} + 0 + 0$$

$$= \dot{m}_P h_P - \left[\dot{m}_F h_F + \dot{m}_{Ox} h_{Ox} \right],$$

which can be rearranged as

$$\dot{m}_F h_F + \dot{m}_{Ox} h_{Ox} - \dot{m}_P h_P = \dot{Q}_{cv,out}. \tag{11.93}$$

It is extremely important to note that the enthalpies appearing in Eq. 11.93 are *standardized enthalpies,* that is, enthalpies that account for the making and breaking of chemical bonds in the conversion of reactants to products. The heat-transfer term in Eq. 11.93 is the energy transferred to the surroundings. For furnaces, boilers, and similar devices, this is the energy used to heat air or water or to generate steam. This term also includes any residual losses, which frequently can be neglected for well-insulated systems. For combustors, this term represents only residual heat losses, which again are usually negligible.

SUMMARY

After studying this chapter, you should have a good understanding of the operation of a wide variety of practical steady-flow devices. Specifically, you should be able to formulate, simplify, and apply the basic conservation principles and thermodynamic property relationships to nozzles, diffusers, throttles, pumps, compressors, turbines, heat exchangers, furnaces, and combustors. You should also have an appreciation for the losses and inefficiencies associated with these devices. A number of concepts related to compressible flows were presented and developed in this chapter. You should be able to ascertain when the effects of compressibility may be important in thermal-fluid devices and be able to apply simple, one-dimensional analyses to isentropic flows and to flows with normal shocks. Another extended development in this chapter is the analysis and design of heat exchangers. You should be able to perform simple design and performance calculations using both the log-mean temperature and effectiveness–*NTU* methods. As a further summary of this chapter, reviewing the learning objectives presented at the outset of this chapter is recommended.

Chapter 11
Key Concepts & Definitions Checklist[15]

11.1 Steady-Flow Devices

☐ Purpose of each device ➤ *Q11.2*

☐ Typical simplifying assumptions ➤ *Q11.3*

11.2 Nozzles and Diffusers

☐ Physical shapes ➤ *Q11.5, Q11.6*

☐ Simplified mass and energy conservation ➤ *11.1, 11.4*

☐ Linear momentum conservation ➤ *11.2*

☐ Flow separation ➤ *Q11.7*

☐ Pressure-recovery coefficient ➤ *11.8*

11.2c Incompressible Flow

☐ Relationships among friction, head loss, and internal energy ➤ *Q11.8*

11.2d Compressible Flow

☐ Speed of sound and Mach number ➤ *11.23*

☐ Criterion for incompressible flow ➤ *11.13*

☐ Supersonic nozzles and diffusers ➤ *Q11.10*

☐ Converging–diverging nozzles ➤ *Q11.11*

☐ Stagnation properties ➤ *11.42, 11.43*

☐ Choked flow ➤ *Q11.12, 11.28*

☐ Critical properties (P^*, T^*, etc.) ➤ *11.29*

☐ Isentropic flow in converging–diverging nozzles: pressure distributions (Fig. 11.12) ➤ *Q11.13*

☐ Isentropic flow Mach number relationships (Table 11.3) ➤ *11.26, 11.27*

☐ Normal shock ➤ *Q11.16*

☐ Normal shock relationships (Table 11.3) ➤ *11.46*

☐ Property changes across a shock (Fig. 11.15) ➤ *11.47*

☐ Nozzle isentropic efficiency ➤ *Q11.17*

11.3 Throttles

☐ Physical configurations ➤ *Q11.18*

☐ Throttling calorimeter ➤ *Q11.19*

☐ Simplified mass and energy conservation ➤ *11.9, 11.10*

11.4 Pumps, Compressors, and Fans

☐ Positive-displacement versus dynamic devices ➤ *Q11.20*

☐ Centrifugal and axial-flow devices ➤ *Q11.21*

☐ Simplified mass and energy conservation ➤ *11.53, 11.55*

☐ Reversible, steady-flow work ➤ *11.50*

☐ Pump head loss ➤ *11.51*

☐ Isentropic efficiency ➤ *11.57*

11.5 Turbines

☐ Applications (Table 11.5) ➤ *Q11.22*

☐ Types (Table 11.5) ➤ *Q11.22*

☐ Simplified mass and energy conservation ➤ *11.77, 11.86*

☐ Isentropic efficiency ➤ *11.78*

11.6 Heat Exchangers

☐ Parallel, counter-, and cross-flow ➤ *Q11.23*

☐ Mixed and unmixed fluids ➤ *Q11.24*

☐ Overall mass and energy conservation ➤ *11.105, 11.106*

☐ Heat-capacity rate ➤ *Q11.25*

☐ Overall heat-transfer coefficient ➤ *11.110*

☐ Log-mean temperature difference ➤ *Q11.26, 11.122*

☐ Heat-exchanger design ➤ *11.113*

☐ Heat-exchanger effectiveness ➤ *Q11.27*

☐ Effectiveness–*NTU* method ➤ *11.127*

---

[15] Numbers following arrows refer to Questions (prefaced with a Q) and Problems at the end of the chapter.

11.7 Furnaces, Boilers, and Combustors

❑ Simplified mass conservation (Eqn. 11.92)
   ➤ *11.147, 11.148*

❑ Simplified energy conservation (Eqn. 11.93)
   ➤ *11.147, 11.148*

❑ Standardized enthalpies ➤ *Q11.4*

REFERENCES

1. Baumeister, T., and Marks, L. S. (Eds.), *Standard Handbook for Mechanical Engineers*, 7th ed., McGraw-Hill, New York, 1967, Chapter 14.

2. Cumpsty, N., *Jet Propulsion*, Cambridge University Press, New York, 1997.

3. White, F. M., *Fluid Mechanics*, 4th ed., McGraw-Hill, New York, 1999.

4. Van Dyke, M., *An Album of Fluid Motion*, Parabolic Press, Stanford, CA, 1982, p. 101 (photograph from S. J. Kline motion picture *Flow Visualization*).

5. Runstadler, P. W., Jr., Dolan, F. X., and Dean, R. C., Jr., *Diffuser Data Book*, Technical Note TN-186, May 1975, Creare, Inc., Hanover, NH.

6. Roberson, J. A., and Crowe, C. T., *Engineering Fluid Mechanics*, 4th ed., Houghton Mifflin, Boston, 1990.

7. Shames, I. H., *Mechanics of Fluids*, 3rd ed., McGraw-Hill, New York, 1992.

8. Karassik, I. J., Krutzsch, W. C., Fraser, W. H., and Messina, J. P. (Eds.), *Pump Handbook*, 2nd ed., McGraw-Hill, New York, 1986.

9. Dixon, S. L., *Fluid Mechanics and Thermodynamics of Turbomachinery*, 4th ed., Butterworth-Heinemann, Woburn, MA, 1998.

10. Moran, M. J., and Shapiro, H. N., *Fundamentals of Engineering Thermodynamics*, 5th ed., Wiley, New York, 2004.

11. Howell, J. R., and Buckius, R. O., *Fundamentals of Engineering Thermodynamics*, 2nd ed., McGraw-Hill, New York, 1992.

12. Reliance Electric Co., "AC Motor Efficiency Guide," http://www.reliance.com/b7087_5/b7087_intro.htm, 1999.

13. Shepherd, D. G., *Elements of Fluid Mechanics*, Harcourt Brace & World, New York, 1965.

14. See http://www.microhydropower.net/index.php.

15. Heywood, J. B., *Internal Combustion Engine Fundamentals*, McGraw-Hill, New York, 1988.

16. Incropera, F. P., and DeWitt, D. P., *Fundamentals of Heat and Mass Transfer*, 5th ed., Wiley, New York, 2002.

17. Bowman, R. A., Mueller, A. C., and Nagle, W. M., "Mean Temperature Difference in Design," *Transactions ASME*, 62:283–294, 1940.

18. Van Dyke, M., *An Album of Fluid Motion*, Parabolic Press, Stanford, CA, 1982.

19. Kakaç, S., and Liu, H., *Heat Exchangers: Selection, Rating and Thermal Design*, 2nd. ed., CRC Press, Boca Raton, FL, 2002.

20. Kuppan, T., *Heat Exchanger Design Handbook*, Marcel Dekker, New York, 2000.

21. Kays, W. M., and London, A. L., *Compact Heat Exchangers*, 3rd ed., McGraw-Hill, New York, 1983.

22. Howarth, L. (Ed.), *Modern Developments in Fluid Dynamics, High Speed Flow*, Oxford, Oxford, UK, 1953.

23. Mills, A. F., *Heat and Mass Transfer*, Irwin, Chicago, 1995.

Some end-of-chapter problems were adapted with permission from the following:

24. Chapman, A. J., *Fundamentals of Heat Transfer*, Macmillan, New York, 1987.

25. Look, D. C., Jr., and Sauer, H. J., Jr., *Engineering Thermodynamics*, PWS, Boston, 1986.

26. Myers, G. E., *Engineering Thermodynamics*, Prentice Hall, Englewood Cliffs, NJ, 1989.

27. Pnueli, D., and Gutfinger, C., *Fluid Mechanics*, Cambridge University Press, Cambridge, England, 1992.

Nomenclature

a	Sound speed (m/s)
A	Area (m^2)
BWG	Birmingham Wire Gage
c	Specific heat (J/kg·K)
c_p	Constant-pressure specific heat (J/kg·K)
c_v	Constant-volume specific heat (J/kg·K)
C	Heat-capacity rate (J/K·s)
C_p	Diffuser pressure-recovery coefficient (dimensionless)
CR	Heat-capacity (rate) ratio (dimensionless)
D	Diameter (m)
$\dot{E}$	Energy rate (W)
f_D	Darcy friction factor (dimensionless)
F	Correction factor
g	Gravitational acceleration (m/s^2)
h	Specific enthalpy (J/kg)
h_{conv}	Convective heat-transfer coefficient (W/m^2·K or W/m^2·°C)
h_L	Head loss (m)
HX	Heat exchanger
k	Thermal conductivity (W/m·K or W/m·°C)
K	Minor loss coefficient (dimensionless)
ke	Specific kinetic energy (J/kg)
$\dot{KE}$	Kinetic energy rate (W)
L	Length (m)
LMTD	Log-mean temperature difference (K or °C)
$\dot{m}$	Mass flow rate (kg/s)
M	Number of tube-bundle passes
$\mathcal{M}$	Molecular weight
Ma	Mach number
N	Number of tubes in a bundle

Nu	Nusselt number (dimensionless)
NTU	Number of transfer units (dimensionless)
P	Pressure (Pa) or chart parameter (Figs. 11.49–11.52)
Pr	Prandtl number
pe	Specific potential energy (J/kg)
$\dot{Q}$	Heat transfer rate (W)
r	Radius (m)
R	Particular gas constant (J/kg·K), thermal resistance (K/W or °C/W), or chart parameter (Figs. 11.49–11.52)
R_u	Universal gas constant, 8314.472 (J/kmol·K)
Re	Reynolds number (dimensionless)
s	Specific entropy (J/kg·K)
T	Temperature (K)
u	Specific internal energy (J/kg)
U, U_{OA}	Overall heat-transfer coefficient (W/m^2·K or W/m^2·°C)
v	Velocity (m/s)
ν	Specific volume (m^3/kg)
$\mathcal{V}$	Volume (m^3)
$\dot{\mathcal{V}}$	Volume flowrate (m^3/s)
W	Width (m)
$\dot{W}$	Rate of work or power (W)
x	Quality (dimensionless)
z	Spatial coordinate in vertical direction (m)

GREEK

α	Kinetic energy correction factor (dimensionless)
β	Momentum flow correction factor (dimensionless)

γ Specific-heat ratio (dimensionless)

Δ Difference or change

ε Roughness height (m) or heat-exchanger effectiveness (dimensionless)

μ Viscosity $(N \cdot s/m^2)$

μ_J Joule–Thomson coefficient (K/Pa)

η Efficiency

ρ Density (kg/m^3)

SUBSCRIPTS

A   air

act   actual

avg   average

b   downstream receiver

conv   convection

C   cold fluid

CF   counterflow

e   exit plane

cv   control volume

elec   electrical

f   liquid or formation

F   fuel

g   gas or vapor

gen   generator

H   hot fluid

HX   heat exchanger

i   inner

in   into system or control volume

isen   isentropic process

j index for inlets

k index for outlets

lm   log mean

max   fluid with maximum heat capacity rate

min   fluid with minimum heat capacity rate

n   nozzle

o   outer

OA   overall

out   out of system or control volume

Ox   oxidizer

p   pump or pumping device

P   products

PF   parallel flow

ref   reference

rev   reversible

s   sensible

SF   steady flow

S-T   shell-and-tube

sys   system

t   turbine or throat

tot   total

w   wall

x upstream of shock

y downstream of shock

0   stagnation condition

1   station 1 (usually inlet)

2   station 2 (usually outlet)

SUPERSCRIPTS

$\circ$ Standard-state (e.g., $P^\circ = 1$ atm)

$*$ Throat conditions when flow is choked

QUESTIONS

11.1 Review the most important equations presented in this chapter (i.e., those with a reddish background). What physical principles do they express? What restrictions apply?

11.2 List one or more purposes for the following devices: nozzle, diffuser, throttle, pump, compressor, turbine, and heat exchanger.

11.3 What terms in the conservation of energy equation (Eqs. 5.63e or 5.65) are neglected in the thermal analysis of the following devices: nozzle, diffuser, throttle, pump, compressor, turbine, and heat exchanger?

11.4 In what way do the thermal analyses of furnaces (or boilers) and combustors differ from those of steady-flow devices that do not involve combustion?

11.5 Sketch a (incompressible or subsonic) diffuser. How does the flow area change in the flow direction? How does the velocity change in the flow direction?

11.6 Repeat Question 11.5 for a (incompressible or subsonic) nozzle.

11.7 Discuss the concept of flow separation in a diffuser. How does this phenomenon relate to the drag crisis discussed in Chapter 9?

11.8 For an incompressible flow through a pipe, how do the head loss, the heat transfer, and internal energy change relate? *Hint:* Write out the first law of thermodynamics and the mechanical energy equation.

11.9 Define the Mach number. How is the speed of sound calculated for an ideal gas?

11.10 Compare the geometries of supersonic and subsonic nozzles. Repeat for diffusers.

11.11 Explain how a converging–diverging geometry creates a nozzle.

11.12 Explain to a colleague the operation of a converging nozzle as a function of back pressure for fixed stagnation conditions.

11.13 Explain to a colleague the operation of a converging–diverging nozzle as a function of back pressure for fixed stagnation conditions.

11.14 Consider a choked converging nozzle. Describe what happens when the stagnation pressure is slowly increased while the back pressure remains fixed.

11.15 Consider an initially choked converging–diverging nozzle with supersonic flow throughout the diverging section. Describe what happens when (a) the stagnation pressure is decreased while the backpressure remains fixed and (b) the stagnation temperature is increased while both the stagnation pressure and back pressure remain fixed.

11.16 Describe a normal shock.

11.17 Define the isentropic efficiency of a nozzle.

11.18 List several devices that can act as throttles.

11.19 Explain the operation and use of a throttling calorimeter.

11.20 Distinguish between a positive-displacement pump and a dynamic (centrifugal) pump.

11.21 Create a sketch to illustrate the difference in the geometries of a centrifugal compressor and an axial-flow compressor.

11.22 List several applications of turbines. What type of turbine is typically used with the applications you list?

11.23 Distinguish among parallel-flow heat exchangers, counterflow heat exchangers, and cross-flow heat exchangers.

11.24 Distinguish between mixed and unmixed fluids in a shell-and-tube heat exchanger.

11.25 Define the heat-capacity rate.

11.26 Sketch the temperature distribution through a parallel-flow heat exchanger. Use your sketch to define the log-mean temperature difference. Repeat for a counterflow heat exchanger.

11.27 Define the effectiveness of a heat exchanger. What is the physical significance of the minimum heat capacity rate in your definition?

Chapter 11 Problem Subject Areas

11.1–11.21	**Nozzles, diffusers, and throttles**
11.22–11.48	**Compressible flows including nozzles and diffusers**
11.49–11.76	**Pumps, compressors, and fans**
11.77–11.97	**Turbines**
11.98–11.104	**Overall heat-transfer coefficients**
11.105–11.146	**Heat exchangers**
11.147–11.149	**Constant-pressure combustion devices**

PROBLEMS

11.1 Air enters a nozzle at 1.30 atm and 25°C with a velocity of 2.5 m/s. The nozzle entrance diameter is 120 mm. The air exits the nozzle at 1.24 atm with a velocity of 90 m/s. Determine the temperature of the exiting air and the nozzle exit diameter.

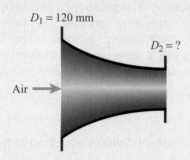

$D_1 = 120$ mm

$D_2 = ?$

Air

11.2 Water enters a nozzle with a velocity of 0.8 m/s and exits with a velocity of 5 m/s. The nozzle entrance diameter is 12 mm. Estimate the inlet pressure if the outlet pressure is 100 kPa and the water temperature is 300 K. Also determine the force required to restrain the nozzle.

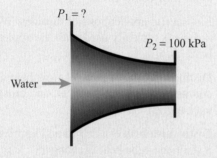

$P_1 = ?$

$P_2 = 100$ kPa

Water

11.3 Consider the garden hose nozzle described in Problem 3.32. For the flow conditions given, determine the pressure at the nozzle inlet. Assume that an ambient pressure of 100 kPa prevails at the nozzle exit.

11.4 Steam flows through a nozzle at 10^5 lb$_m$/min. The entering pressure and velocity are 250 psia and 400 ft/s, respectively; the exiting values are 1 psia and 4000 ft/s, respectively. Assuming the process is adiabatic, determine the specific enthalpy change of the steam (Btu/lb$_m$).

11.5 Steam at 100 lb$_f$/in^2 and 400 F enters a rigid, insulated nozzle with a velocity of 200 ft/s. The steam leaves at a pressure of 20 lb$_f$/in^2 and a velocity of 2000 ft/s. Assuming that the enthalpy at the entrance (h_i) is 1227.6 Btu/lb$_m$, determine the value of the enthalpy at the exit (h_e).

11.6 Air enters a diffuser at 100 kPa and 350 K with a velocity of 250 m/s. The air exits with a velocity of 30 m/s. Determine the temperature of the air at the outlet.

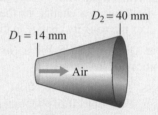

$D_2 = 40$ mm

$D_1 = 14$ mm

Air

11.7 A diffuser is an integral part of a water pump. The diameter at the entrance to the diffuser section is 30 mm and the diameter at the exit is 45 mm. Water at 300 K flows through the pump at 2.2 kg/s. Determine the pressure rise in kPa and psi associated with the diffuser.

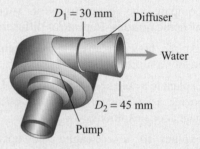

$D_1 = 30$ mm Diffuser

Water

$D_2 = 45$ mm

Pump

11.8 Consider a conical diffuser as described in Example 11.5 but now with a length of 0.1 m. Compare the ideal and actual pressure increase through this diffuser and compare your results with those of Example 11.5. Discuss why these results differ.

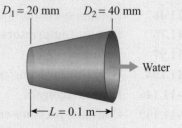

$D_1 = 20$ mm $D_2 = 40$ mm

Water

$\leftarrow L = 0.1$ m $\rightarrow$

11.9 Nitrogen flows steadily through a 23-mm-diameter sintered metal filter. The pressure upstream of the filter is 300 kPa, the downstream pressure is 150 kPa. Determine the temperature and velocity of the N_2 downstream of the filter given an inlet temperature of 320 K and an inlet velocity of 2 m/s. The inlet and exit flow areas are identical.

$P_1 = 300$ kPa

$N_2 \longrightarrow$

$P_2 = 150$ kPa
$T_2 = ?$
$v_2 = ?$

11.10 In a refrigerator, saturated liquid R-134a (a modern refrigerant) is throttled from an initial temperature of 305 K to a final pressure of 80 kPa. Determine the final temperature and specific volume of the R-134.

$T_1 = T_{sat} = 305$ K

R–134a $\longrightarrow$ •1   2•

$P_2 = 80$ kPa
$T_2 = ?$
$v_2 = ?$

11.11 Saturated liquid water at 1 MPa enters a throttling device and exits at 0.2 MPa. Determine the temperature and quality of the exiting liquid–vapor mixture.

$P_1 = 1$ MPa

$P_2 = 0.2$ MPa

$T_2 = ?$
$x_2 = ?$

Sat. liquid 1   2

11.12 Methane stored in a tank at 8 MPa and 300 K is used as a fuel for a laboratory-scale gas-turbine combustor. The methane is throttled as it passes through a pressure regulator. The regulated pressure is 120 kPa. Determine the temperature of the methane at the exit of the pressure regulator. Assume the process is adiabatic and neglect any kinetic energy changes.

$P_e = 120$ kPa
$T_e = ?$

CH$_4$
8 MPa
300 K

11.13 Water at 140°C and 10 MPa is adiabatically throttled to a pressure of 0.2 MPa. Determine the quality after throttling.

11.14 Steam flows through a nozzle from inlet conditions at 200 psia and 800 F to an exit pressure of 30 psia. The flow is reversible and adiabatic. For a flow rate of 10 lb$_m$/s, determine the exit area if the inlet velocity is negligible.

11.15 Steam at 400 psia and 600 F expands through a nozzle to 300 psia at a flow rate of 20,000 lb$_m$/hr. If the process occurs reversibly and adiabatically and the initial velocity is low, calculate (a) the velocity (ft/s) leaving the nozzle and (b) the exit area (in^2) of the nozzle.

11.16 Steam at 2 MPa and 290°C expands to 1.400 MPa and 247°C through a nozzle. If the entering velocity is 100 m/s, determine (a) the exit velocity and (b) the nozzle isentropic efficiency.

11.17 Consider a low-speed wind tunnel. At one point, the structure forms a nozzle with air inlet conditions of $v \approx 0$, $P = 14.7$ psia, and $T = 80$ F. If the nozzle isentropic efficiency is 90%, determine the air exit temperature when the exit pressure is 16 psia.

11.18 Steam enters a diffuser at 700 m/s, 200 kPa, and 200°C. It leaves the diffuser at 70 m/s. Assuming reversible adiabatic operation, determine the final pressure and temperature.

11.19 Water is throttled (constant-enthalpy process) across a valve from 20.0 MPa and 260°C to 0.143 MPa. Determine the temperature of the H_2O downstream of the valve. What is the physical state of the H_2O (i.e., superheated vapor, subcooled liquid, etc.)?

11.20 Air is throttled (constant-enthalpy process) across a valve from 20.0 MPa and 260°C to 0.143 MPa. Determine the temperature and specific volume of the air downstream of the valve. Also determine the air specific entropy change across the valve (kJ/kg·K).

11.21 In 1997, Andy Green set the world land speed record with his jet-powered vehicle, Thrust SSC, at Black Rock Desert, Nevada. With a speed of 763.035 miles/hr, Green was the first to drive a land vehicle at supersonic speeds ($Ma = 1.016$).

Photograph by Jeremy Davey
courtesy of SSC Programme, Ltd.

A. Assuming dry air, what was the temperature during Green's record run at Black Rock Desert?

B. How does the humidity affect the determination that Green broke the sound barrier?

C. Determine the stagnation temperature and pressure using your result from Part A and assuming a barometric pressure of 100 kPa.

11.22 Determine the speed of sound in the following gases at 300 K and 1 bar: (a) air, (b) helium, (c) hydrogen, and (d) a 1:1 molar mixture of helium and hydrogen.

11.23 The F-16 Fighting Falcon jet aircraft has a maximum speed of 915 miles/hr at sea level.

Photograph courtesy of U.S. Air Force.

A. Determine the Mach number of an F-16 traveling at this speed for conditions associated with a standard atmosphere at (a) sea level ($T = 288$ K) and (b) 12,000 m ($T = 216.7$ K).

B. The maximum speed for the F-16 at 12,000 m is 1320 miles/hr. What is the Mach number?

C. Discuss your results.

11.24 Determine whether or not compressible-flow effects are likely to be important for the following situations:

A. The Japanese bullet train traveling at 300 km/hr

B. An automobile speeding at 90 miles/hr across a desert when the air temperature is 100° F

C. A submerged submarine traveling at 15 knots (7.72 m/s) in water at 25°C

Photograph courtesy of U.S. Navy.

Discuss your results.

11.25 The Space Shuttle reenters the Earth's atmosphere at Mach 26 with a velocity of 28,500 km/hr.

Image courtesy of NASA.

A. Estimate the temperature rise associated with stagnating the flow.

B. Estimate the temperature of the atmosphere at reentry.

11.26 Derive expressions for 1-D, isentropic flow with variable area relating, first, the pressure ratio P_1/P_2 to Ma_1 and Ma_2, and, second, the density ratio ρ_1/ρ_2 to Ma_1 and Ma_2.

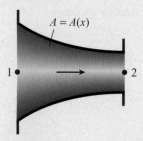

11.27 Use a spreadsheet or other software to program the isentropic compressible flow functions, Eqs. 11.31a–11.31d. Compare results from your computations with the values from Table K.1 for $Ma = 0.5, 1.0, 5$, and 10. (Save your spreadsheet as you may find it a useful alternative to Table K.1.)

11.28 Determine the mass flow rate through a converging nozzle for a flow of air with stagnation conditions $P_0 = 3$ atm, $T_0 = 300$ K. The exit diameter is 10 mm, and the nozzle exits to the atmosphere at 1 atm.

$P_0 = 3$ atm
$T_0 = 300$ K
$D_e = 10$ mm
$P_b = 1$ atm

11.29 Determine the critical pressure and temperature ratios, P^*/P_0 and T^*/T_0, respectively, for a compressible flow of the following gases: air, nitrogen, helium, hydrogen, and carbon dioxide.

11.30 Helium undergoes the following sequence of isentropic expansions:

0–1: 0.2 MPa and 300 K (stagnation conditions) to $P_1 = 0.1$ MPa,

1–2: $P_1 = 0.1$ MPa to $P_2 = 50$ kPa, and

2–3: $P_2 = 50$ kPa to $P_3 \sim 0$.

Assuming ideal-gas behavior, determine the Mach number, the local velocity, the speed of sound, and the stagnation pressure for states 1, 2, and 3. Also determine the pressure at which the Mach number is unity for the given stagnation conditions.

11.31 Find the throat and exit areas of a converging–diverging nozzle that transfers 1 kg/s air from tank A, where $P_0 = 1$ MPa and $T_0 = 300$ K, to tank B, where $P_e = 150$ kPa.

11.32 Consider the nozzle and tanks defined in Problem 11.31. The pressure at the nozzle exit P_e is now changed to 100 kPa. Find the new mass flow rate $\dot{m}$ and the pressure at the nozzle throat. Repeat your calculation for $P_e = 0.95$ MPa.

11.33 A compressor takes in air at atmospheric pressure (100 kPa) and discharges the compressed air into a large, high-pressure settling tank. To measure the mass flow rate of the compressed air as a function of the compression pressure, an engineer connects a converging nozzle to the settling tank. The exit area of the nozzle is 20×10^{-4} m². In a test series with different flow rates, the following gage pressures were measured in the settling tank: 110, 160, 250, 400, and 600 kPa. For all tests, the temperature in the settling tank was approximately 350 K. Find the corresponding mass flow rates for each pressure.

11.34 A large pressure vessel contains air ($R = 287$ J/kg·K and $\gamma = 1.4$.) at the following stagnation conditions: $P_0 = 400$ kPa and $T_0 = 420$ K. The atmospheric pressure is 100 kPa. A converging–diverging nozzle is designed to flow 1 kg/s from the vessel to the atmosphere. Determine the following:

A. The speed of the gas at the nozzle exit, v_e

B. The exit Mach number Ma_e.

C. The critical cross-sectional area A^*

D. The exit cross-sectional area A_e

11.35 The nozzle of Problem 11.34 is shortened by cutting off a portion of the diverging section. The cut occurs downstream of the critical section at point 1, where $A_1 \equiv (A^* + A_e)/2$. The shortened nozzle still connects the pressure vessel to the outside. Determine the following quantities at point 1: A_1, P_1, v_1, T_1, Ma_1, and $\dot{m}_1$.

11.36 The nozzle of problem 11.35 is further shortened, now by cutting it off before the critical section at point 2, where $A_2 \equiv A_1$. The nozzle still connects the pressure vessel to the outside. Determine the following quantities: P_2, v_2, T_2, Ma_2, and $\dot{m}_2$.

11.37 Consider a flow of air through a converging–diverging nozzle having a throat diameter of 4 mm and an exit diameter of 8 mm. The stagnation pressure and temperature are 5 atm and 300 K, respectively.

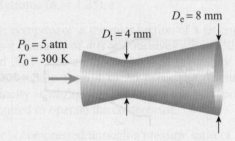

$P_0 = 5$ atm
$T_0 = 300$ K
$D_t = 4$ mm
$D_e = 8$ mm

A. Determine the maximum possible back pressure for choked flow.

B. Determine the maximum possible back pressure allowing supersonic flow throughout the diverging section and shockless operation.

C. For the conditions from Part A, determine v_t, v_e, and $\dot{m}$.

D. Discuss your results from Parts A through C.

11.38 Starting with the continuity equation, $\dot{m} = \rho_t v_t A_t$, apply Eqs. 11.31a–11.31c to derive an expression for the mass flow rate for choked flow through a nozzle. Your result should involve only the stagnation properties P_0 and T_0 and gas properties γ and R.

11.39 Following the combustion chamber of a jet engine, products of combustion and air enter the jet nozzle at low velocity at 100 psia and 1600 F. Determine the maximum velocity (ft/s) that can be obtained from the nozzle when the exit pressure is 11 psia. Assume the mixture properties are those of air. Also determine the exit diameter required to handle a flow of 23 lbm/s.

outlet, which is 10 ft higher, the velocity is 1000 ft/s. Assuming that the operation is reversible and adiabatic, determine the work produced per unit mass.

11.91   Steam flows through a turbine in a nuclear power plant at 1,230,000 lb_m/hr. The turbine inlet conditions are 500 psia and 1200 F, and the turbine outlet pressure is 5 psia. Determine the maximum turbine power output.

11.92   Air at 50 psia and 90 F flows through an expander (like a turbine) at the rate of 1.6 lb_m/s to an exit pressure of 14.7 psia. (a) What is the minimum temperature attainable at the expander exit? (b) If the inlet velocity is not to exceed 12 ft/s, what inlet diameter is required?

11.93   Steam enters a turbine as a saturated vapor at 2 MPa and exits at 101 kPa with a quality of 0.92. Determine the turbine isentropic efficiency.

11.94   For the turbine in Problem 11.93, how does the isentropic efficiency change if the exit pressure is decreased to 35 kPa?

11.95   Steam flows at the rate of 12,000 lb_m/min through a turbine from 500 psia and 700 F to an exhaust pressure of 1 psia. Determine the ideal output of the turbine (hp). If the specific entropy increases between the inlet and the outlet of the turbine by 0.1 Btu/$lb_m \cdot$R, determine the isentropic turbine efficiency.

11.96   A high-speed turbine produces 1 hp while operating on compressed air. The inlet and outlet conditions are 70 psia and 85 F and 14.7 psia and −50 F, respectively. Assume changes in kinetic and potential energies are negligible. Determine the mass flow rate.

11.97   Heat is transferred from a steam turbine at a rate of 40,000 Btu/hr while the steam mass flow rate is 10,000 lb_m/hr. Using the following data for the steam entering and leaving the turbine, find the work rate. The gravitational acceleration is g = 32.17 ft/s².

	Inlet Conditions	Outlet Conditions
Pressure	200 psia	15 psia
Temperature	700 F	—
Velocity	200 ft/s	600 ft/s
Elevation	16 ft	10 ft
Enthalpy, h	1361.2 Btu/lb_m	1150.8 Btu/lb_m

Also determine the fraction of the power contributed by each term of the first law compared to that associated with the change in enthalpy.

11.98   The following materials comprise the wall of a house: 10-cm-thick common brick (k = 0.69 W/m·K), a 1.25-cm-thick layer of Celotex (k = 0.048 W/m·K), a 9-cm-thick layer of fiberglass wool (k = 0.0414 W/m·K), and 1.25-cm-thick insulation board (k = 0.744 W/m·K). Wind blowing on the brick exterior surfaces results in an outside convective heat-transfer coefficient of 30 W/m²·K. The heat-transfer coefficient at the interior surface is 10 W/m²·K. Determine the overall heat-transfer coefficient U_{OA} for the wall.

11.99   The wall of a laboratory furnace consists of a 15-cm-thick layer of chrome brick (k = 2.4 W/m·K), a 5-cm-thick layer of fiberglass (k = 0.035 W/m·K), and a 1-cm-thick outer layer of 1%-C steel (k = 43 W/m·K). The convective heat-transfer coefficient at the interior surface is 20 W/m²·K, and the convective heat-transfer coefficient at the exterior surface is 25 W/m²·K. Determine the overall heat-transfer coefficient U_{OA} for the furnace wall.

11.100   A nominal-4-in-diameter, schedule-40, wrought-iron pipe (with inside and outside diameters of 4.026 in and 4.500 in, respectively, and k = 50 W/m·°C) is covered with 3.8-cm-thick layer of 85% magnesia insulation (k = 0.074 W/m·°C). The pipe carries superheated steam at a temperature of 400°C, and the outer insulation surface is exposed to air at 25°C. The inside and outside heat-transfer coefficients are 1400 and 10 W/m²·°C, respectively. Determine the overall heat-transfer coefficient U_{OA} of the insulated steam pipe.

11.101   A long pipe (with a 7.5-cm inside diameter, a 10-cm outside diameter, and k = 35 W/m·°C) is covered with 1.25-cm layer of insulation (k = 5 W/m·°C). The temperature of a fluid flowing inside the pipe is 20°C, and the temperature of the inner pipe wall is 45°C. The temperature of the fluid surrounding the outer surface of the insulation is 150°C, and the convective heat-transfer coefficient at that surface is 85 W/m·°C. Find the overall heat transfer coefficient based on the outside exposed surface area, $U_{OA,o}$.

11.102   Saturated steam at 10 kPa condenses on the outer surface of a 3/4-in brass condenser tube (with an inside diameter of 16.6 mm, an outside diameter of 19.1 mm, and k = 147 W/m·°C). The convective heat-transfer coefficients at the tube inner and outer surfaces are 5400 and 6800 W/m²·°C, respectively. Cooling water flows through the tube at 30°C. Find (a) the overall heat-transfer coefficient based on the outer area of the tube, $U_{OA,o}$, and (b) the steam condensation rate per unit length of tube (kg/s·m).

11.103 A water-to-water heat exchanger is made of brass tubes (with outside diameters of 1.000 in, inside diameters of 0.870 in, and $k = 147$ W/m·°C). The inside and outside convective heat-transfer coefficients are 800 and 1200 Btu/hr·ft² F, respectively. Determine the overall heat-transfer coefficient U_{OA} based on the outside area of the tube.

11.104 Water at a bulk mean temperature of 25°C flows with a velocity of 1.5 m/s through a 6-m-long, brass, steam condenser tube. The outside and inside tube diameters are 1.58 and 1.34 cm, respectively. Steam condenses on the outer tube surface where the convective heat-transfer coefficient is 12,000 W/m²·°C. Calculate the overall heat-transfer coefficient for the exchanger based on (a) the inner area of the tube and (b) the outer area of the tube.

11.105 Consider a feedwater heater from a steam power plant as shown in the sketch. Steam enters the heater at station 1 with a flow rate of 66,256 lb$_m$/hr at a pressure of 28.6 psia and an enthalpy of 1246.2 Btu/lb$_m$. The feedwater flows through the heater at 1,499,628 lb$_m$/hr, entering at station 2 with an enthalpy of 161.2 Btu/lb$_m$ and exiting at station 3 with an enthalpy of 211.0 Btu/lb$_m$. Condensate from another feedwater heater enters at station 4 with a flow rate of 69,708 lb$_m$/hr and an enthalpy of 221.3 Btu/lb$_m$ and is mixed with the condensing steam inside the heater.

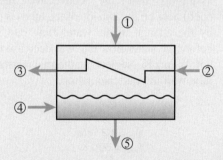

Determine the temperature of the incoming steam (station 1) in Fahrenheit and kelvins and determine the mass flow rate and enthalpy of the liquid exiting at station 5 in both U.S. customary and SI units. Also estimate the temperature of this stream assuming that the mixing within the feedwater heater occurs at constant pressure (P_1).

11.106 In the production of orange juice, it is desired to heat the juice from 4°C to 93°C in a tubular heat exchanger. To accomplish this task, saturated steam enters the outer tube of the heat exchanger at 150 kPa and exits as a saturated liquid at the same pressure. For an orange juice flow rate of 2000 kg/hr ($c_p = 3.9$ kJ/kg·K), determine the required steam flow rate.

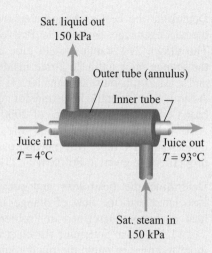

Sat. liquid out
150 kPa

Outer tube (annulus)

Inner tube

Juice in
$T = 4°C$

Juice out
$T = 93°C$

Sat. steam in
150 kPa

11.107 Steam enters the condenser of a modern power plant at a pressure of 1 psia and a quality of 0.98. The condensate leaves at 1 psia and 80 F. Determine (a) the heat rejected per pound and (b) the change in specific volume between inlet and outlet.

11.108 Water enters a heat exchanger at 180 F and 20 psia and leaves at 160 F and 19.8 psia. Air enters the heat exchanger at 70 F and 15 psia and leaves at 100 F and 14.7 psia. The mass flow rate of the water is 40 lb$_m$/min. Determine the heat-transfer rate (Btu/min) to the air and the air mass flow rate (lb$_m$/min).

11.109 Water is heated by air in a heat exchanger. The water enters at 150°C and 0.2 MPa and leaves at 300°C and 0.2 MPa. The mass flow rate of the water is 1 kg/s. The air enters at 350°C and 0.1 MPa at a rate of 2 kg/s. Determine the heat-transfer rate between the two fluids (kW) and the exit temperature (°C) of the air.

11.110 Consider the tubular heat exchanger described in Problem 11.106. The drawn 304 stainless-steel tube containing the orange juice has a 26.2-mm inside diameter and a 31.7-mm outside diameter ($1\frac{1}{4}$-in BWG 12). Assuming that the convective heat-transfer coefficient at the outside of the tube (condensing steam) has a value of 10,000 W/m²·K, estimate the overall heat-transfer coefficient based on the inside area. Orange juice properties can be assumed as follows:

$$\rho = 1000 \text{ kg/m}^3, \qquad \mu = 6 \times 10^{-4} \text{ N·s/m}^2,$$
$$c_p = 3900 \text{ J/kg·K}, \qquad k = 0.55 \text{ W/m·K}.$$

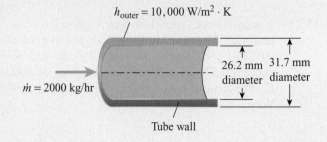

$h_{outer} = 10,000$ W/m²·K

$\dot{m} = 2000$ kg/hr

26.2 mm diameter

31.7 mm diameter

Tube wall

11.111 Determine the heat exchange area required for the heat exchanger described in Problem 11.106. The drawn 304 stainless-steel tube containing the orange juice has a 26.2-mm inside diameter and a 31.7-mm outside diameter ($1\frac{1}{4}$-in BWG). Assume an overall heat-transfer coefficient based on the tube inside area of 2000 W/m²·K. Also determine the length of the 26.2-mm-inside-diameter tube. Any needed orange juice properties are given in Problem 11.110.

11.112 Determine the head loss and pressure drop associated with the flow of orange juice in the heat exchanger described in Problems 11.106 and 11.111. Also determine the electrical power input required to pump the orange juice using a centrifugal pump having an isentropic efficiency of 75% that is driven by an electrical motor having an efficiency of 84%.

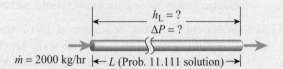

11.113 A shell-and-tube heat exchanger is to be designed to cool a 0.0736-kg/s flow of oil from 350 to 325 K using a 0.0454-kg/s flow of water entering at 285 K. The heat exchanger has a single shell with two passes of bundles of ten 10-mm-inside-diameter tubes (i.e., there are a total of twenty individual tube-passes). The water flows through the tubes. The overall heat-transfer coefficient is 220 W/m²·K based on the inner area. Determine the total heat-transfer rate, and estimate the total heat exchange area and the length of the heat exchanger.

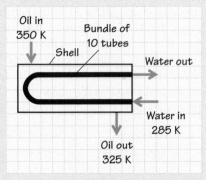

11.114 A single-pass cross-flow heat exchanger with both fluids unmixed is to be designed to cool a 0.0736-kg/s flow of oil from 350 to 325 K using a 0.539-kg/s flow of air entering at 300 K. The overall heat-transfer coefficient is 160 W/m²·K.

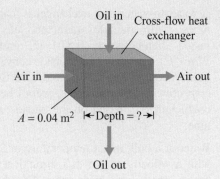

A. Estimate the heat exchange area required. Also estimate the depth of the heat exchanger given an area-to-volume ratio of 900 m²/m³ and a frontal area of 0.04 m². (Note: Heat exchangers with area-to-volume ratios of greater than 700 m²/m³ are designated *compact* heat exchangers.)

B. Physically, how does this heat exchanger compare with the shell-and-tube heat exchanger designed to perform the same task in Problem 11.113? Sketch approximately to scale the two heat exchangers to illustrate your comparison.

11.115 A shell-and-tube heat exchanger uses water to cool ethylene glycol. The geometry is described in Example 11.25 and the flow conditions are given in Example 11.27. Important features are also illustrated in the sketch. Estimate the overall heat-transfer coefficient based on the inner area, assuming that the convective heat-transfer coefficient on the outside of the tubes is unaffected by the change in the water flow rate. Note that before the water flow rate was decreased $U_{OA,i}$ was 850 W/m²·K. Neglect the thermal resistance of the tube wall in your calculation.

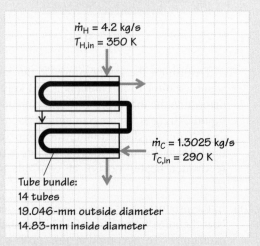

11.116 Repeat Problem 11.115 but do not neglect the thermal resistance of the tube walls. The tubes are made of 304 stainless steel.

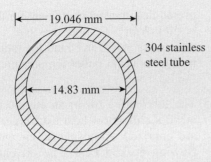

304 stainless steel tube

11.117 Consider an automotive radiator (single-pass cross-flow with both fluids unmixed) that has a heat exchange area of 3.0 m² and an overall heat-transfer coefficient of 185 W/m²·K. The air-side frontal area is 0.16 m². Ethylene glycol enters the radiator at 360 K with a flow rate of 0.845 kg/s. The velocity of the 300-K air entering the heat exchanger is 5.9 m/s and the pressure is 100 kPa. Determine the outlet temperature of the ethylene glycol and the total heat-transfer rate.

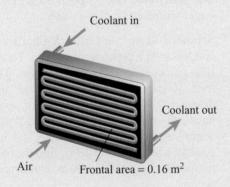

11.118 In a one-shell-pass, one-tube-pass heat exchanger, cold water ($c_{p,C} = 4.18$ kJ/kg·C) enters at 50°C with a flow rate of 900 kg/hr. Hot water ($c_{p,H} = 4.18$ kJ/kg·C) enters at 370°C with a flow rate of 1260 kg/hr. Assuming an infinitely large heat exchange area, determine (a) the maximum heat-transfer rate and (b) the outlet temperatures of the fluids for parallel flow. Repeat your calculations for counterflow.

11.119 Repeat Problem 11.118 if the cold-fluid flow rate is 1260 kg/hr and the hot-fluid flow rate is 900 kg/hr. All other conditions remain the same.

11.120 The lubricating oil cooler for a large gas turbine consists of a one-shell-pass, one-tube-pass heat exchanger. The oil ($c_p = 0.5$ Btu/lb$_m$·F) enters at 320 F flowing through the tubes at 7000 lb$_m$/hr. The water coolant enters at 80 F and flows through the shell at 9000 lb$_m$/hr. Find (a) the maximum possible heat-transfer rate (i.e., the rate for a very large heat exchanger) and (b) the outlet temperatures of the two fluids for both parallel-flow and counterflow arrangements.

11.121 Repeat Problem 11.120 if the oil flow rate is 9000 lb$_m$/hr and the water flow rate is 7000 lb$_m$/hr. All other conditions remain the same.

11.122 In a one-shell-pass, one-tube-pass heat exchanger, the hot fluid enters at 425°C and leaves at 315°C. The cold fluid enters at 40°C and leaves at 260°C. Find the log-mean temperature difference for (a) parallel-flow and (b) counterflow arrangements.

11.123 Find the mean temperature difference (i.e., the corrected *LMTD*) in a one-shell-pass, two-tube-pass heat exchanger for the temperatures noted in Problem 11.122.

11.124 The cold fluid enters a heat exchanger at 110 F and leaves at 500 F, while the hot fluid enters at 750 F and leaves at 600 F. Find the mean temperature difference (i.e., the *LMTD* corrected as necessary) for (a) parallel-flow, (b) counterflow, and (c) one-shell-pass, two-tube-pass arrangements.

11.125 In a one-shell-pass, one-tube-pass heat exchanger, the cold fluid enters at 40°C and leaves at 200°C, while the hot fluid enters at 370°C and leaves at 150°C. Find the log-mean temperature difference.

11.126 In a cross-flow heat exchanger, the cold fluid enters at 40°C and leaves at 150°C, while the hot fluid enters at 425°C and leaves at 260°C. Determine the corrected log-mean temperature difference.

11.127 The overall heat-transfer coefficient for a one-shell-pass, two-tube-pass heat exchanger is known to be 1700 W/m²·°C. The shell-side fluid ($c_p \cong 4.18$ kJ/kg·°C) enters at 260°C flowing at 9100 kg/hr. The tube-side fluid ($c_p = 3.55$ kJ/kg·°C) enters at 30°C flowing at 27,300 kg/hr. For a total surface area of 9.5 m², find the rate of heat transfer between the two fluids and their outlet temperatures (a) using the *F-LMTD* method and (b) using the ε–*NTU* method.

11.128 A one-shell-pass, one-tube-pass, counterflow heat exchanger uses 6800 kg/hr of water ($c_p = 4.18$ kJ/kg·°C) entering at 25°C to cool 18,000 kg/h of oil ($c_p = 2.9$ kJ/kg·°C) entering at 100°C. The total exchanger surface area is 14 m² and the overall coefficient is $U_{OA} = 370$ W/m²·°C. Find the heat-transfer rate and the outlet temperature of the two fluids. Use either the *F-LMTD* method or the ε–*NTU* method.

11.129 A one-shell-pass, two-tube-pass heat exchanger has a known overall heat transfer coefficient of 280 Btu/hr·ft²·F. The shell-side fluid is water ($c_p \cong 1.0$ Btu/lb$_m$·F) flowing at 18,000 lb$_m$/hr and entering at 500 F. The tube-side fluid is oil ($c_p = 0.85$ Btu/lb$_m$·F) flowing at 50,000 lb$_m$/hr and entering at 80 F. For a total surface area of 125 ft^2, find the outlet fluid temperatures (a) using the *F-LMTD* method and (b) using the ε–*NTU* method.

11.130 A one-shell-pass, one-tube-pass heat exchanger operates in counterflow. The exchanger uses 6500 kg/hr of water (c_p = 4.18 kJ/kg·°C) entering at 25°C to cool 13,000 kg/hr of oil (c_p = 2.09 kJ/kg·°C) entering at 100°C. The total surface area of the exchanger is 16 m^2 and the overall heat-transfer coefficient is 350 W/m^2·°C. Find the outlet temperatures of the two fluids using both the *F-LMTD* and the ε–*NTU* methods.

11.131 A steam condenser is made of 5-ft-long, 3/4-in, 14-gage brass tubes (with inside diameters of 0.584 in, outside diameters of 0.750 in, and k = 64.1 Btu/hr·ft·F). There are two tube-passes with 115 tubes per pass. The cooling water enters the tubes at 70 F with an average velocity of 5 ft/s. The convective heat-transfer coefficient at the inner tube wall is 1140 Btu/hr·ft^2·F, and at the outer tube surface the condensing heat-transfer coefficient is 1250 Btu/hr·ft^2·F. The steam condenses at 5 psia. Find (a) the outlet temperature of the cooling water and (b) the condensation rate of the steam (lb$_m$/hr).

11.132 A steam condenser is made of 1.8-m-long, 5/8-in, 14-gage brass tubes (with inside diameters of 11.7 mm, outside diameters of 15.9 mm, and k = 111 W/m·°C). There are two tube-passes with 125 tubes per pass. Cooling water enters the tubes at 24°C with an average velocity of 1.37 m/s. The convective heat-transfer coefficient at the inside tube wall is 6735 W/m^2·°C, and at the outer tube surface the condensing heat-transfer coefficient is 11,340 W/m^2·°C. Find the outlet water temperature and the kg/hr of steam condensed if the steam is saturated at 14 kPa.

11.133 A one-shell-pass, one-tube-pass, counterflow heat exchanger is used to cool oil (c_p = 1.811 kJ/kg·°C). The oil enters the shell at 120°C flowing at 15,000 kg/hr. Cooling water (c_p = 4.18 kJ/kg·°C) enters the tubes at 30°C flowing at 6500 kg/hr. The 1-in, 14-gage brass tubes (with inside and outside diameters of 21.2 mm and 25.4 mm, respectively) have a thermal conductivity of 111 W/m·°C. The convective heat-transfer coefficient at the inside tube surface is 700 W/m^2·°C and that at the outside tube surface is 750 W/m^2·°C. The total exposed surface area of the tubes is 20 m^2. Find the outlet temperatures of the two fluids and the heat exchange rate (W).

11.134 A one-shell-pass, two-tube-pass heat exchanger is to be used to cool ethylene glycol with water. The ethylene glycol enters at 85°C flowing at 1 kg/s; the water enters at 15°C flowing at 2 kg/s. The overall heat-transfer coefficient for the exchanger is 500 W/m^2·°C, and the total exposed tube surface area is 10 m^2. Find the total rate of heat transfer and the outlet temperatures of the two fluids.

11.135 The lubricating oil in an engine is cooled by air in a cross-flow heat exchanger; both fluids are unmixed. Atmospheric air enters at 30°C flowing at 0.55 kg/s. The oil enters at 75°C flowing at 0.025 kg/s. The overall heat-transfer coefficient based on the hot-side area (1 m^2) is 53 W/m^2·°C. Determine the exit temperature of the oil.

11.136 Water flows at an average velocity of 0.3 m/s through a 1-in, schedule-40 steel pipe (with an inside diameter of 26.6 mm, an outside diameter of 33.4 mm, and k = 43 W/m·°C). The bare, 150-m-long pipe passes horizontally through an underground service corridor, where it loses heat by free convection to the ambient air at 20°C. The water enters the pipe at 95°C. Estimate the outlet temperature of the water.

11.137 Saturated steam condenses in a horizontal steam condenser at a pressure of 14 kPa. There are two tube passes with 180 tubes per pass. The 2.4-m-long, 16-gage brass tubes (with inside and outside diameters of 12.6 mm and 15.9 mm, respectively) have a thermal conductivity of 111 W/m·°C. The water velocity through the tubes is 1 m/s. If the cooling water enters at 20°C, find the outlet water temperature and the steam condensation rate (kg/hr).

11.138 In a counterflow heat exchanger, 400,000 lb$_m$/hr of water (c_p ≅ 1 Btu/lb$_m$·F) is heated from 250 to 310 F by hot gases entering at 750 F and leaving at 500 F. The exchanger has a total surface area of 22,000 ft^2. Determine the overall heat-transfer coefficient U_{OA} using both the *F-LMTD* and ε–*NTU* methods.

11.139 A heat exchanger heats 3600 kg/hr of water (c_p ≅ 4.18 kJ/kg·°C) from 40°C to 175°C by cooling another stream of water flowing at 4550 kg/hr and entering at 315°C. The overall heat-transfer coefficient for the exchanger is 1760 W/m^2·°C. Using both the *F-LMTD* and ε–*NTU* methods, find the required surface area for (a) parallel-flow, (b) counterflow, and (c) one-shell-pass, two-tube-pass heat exchanger configurations.

11.140 A one-shell-pass, two-tube-pass heat exchanger heats 27,000 kg/hr of water from 85°C to 100°C by the condensation of saturated steam at 345 kPa. The overall heat-transfer coefficient is known to be 2800 W/m^2·°C. There are thirty 1-in-outside-diameter tubes per pass. Determine the length of a tube-pass.

11.141 A counterflow heat exchanger heats 2160 kg/hr of water (c_p = 4.18 kJ/kg·°C) from 35°C to 90°C using oil (c_p = 2.1 kJ/kg·°C), which enters at 175°C with a flow rate of 3240 kg/hr. If the overall heat-transfer coefficient is 425 W/m²·°C, find the surface area required.

11.142 A cross-flow heat exchanger is used to heat a 3-kg/s flow of water from 35°C to 85°C. The hot fluid is air (1 atm), which enters the heat exchanger at 225°C and exits at 100°C. The overall heat-transfer coefficient is 210 W/m²·°C. Determine the surface area required.

11.143 A two-shell-pass, four-tube pass heat exchanger heats 1.2 kg/s of water from 20°C to 80°C using 2.2 kg/s of oil (c_p = 2.1 kJ/kg·°C) entering at 160°C. The overall heat-transfer coefficient of the heat exchanger is 300 W/m²·°C. Find the required surface area.

11.144 The combustion air supplied to a boiler furnace is preheated in a cross-flow heat exchanger using hot exhaust gases from the furnace. The hot gases (c_p = 1.075 kJ/kg·°C) enter the exchanger at 825°C flowing at 15 kg/s, while the air (c_p = 1.007 kJ/kg·°C) enters at 25°C and flows at 10 kg/s. The overall heat-transfer coefficient is estimated to be 100 W/m²·°C. Estimate the required surface area for the heat exchanger to heat the air to 575°C.

11.145 A one-shell-pass, one-tube-pass heat exchanger is made of sixty tubes (with inside and outside diameters of 11.7 mm and 15.9 mm, respectively). Water enters the shell side at 150°C and leaves at 40°C, flowing at 9000 kg/hr. The tube water enters at 25°C and leaves at 65°C. The inside and outside tube surface heat-transfer coefficients are 1700 and 8500 W/m²·°C, respectively. The thermal resistance of the tube wall is negligible. Find the required length of the tubes.

11.146 Saturated steam at 77°C condenses within the shell of a shell-and-tube condenser. There are 150 brass tubes (with inside diameters of 14.8 mm, outside diameters of 19.1 mm, and k = 111 W/m·°C) per pass. The cooling water flows at the rate of 1.52 m/s, entering at 25°C and leaving at 50°C. The heat-transfer coefficient associated with the condensing steam is 10,200 W/m²·°C, and the convection coefficient at the inner tube surface is 6800 W/m²·°C. If the tube length is not to exceed 3 m, find the number of tube-passes and the length of the tubes.

11.147 A steam generator fired by natural gas (CH_4) is schematically shown in the sketch. The air and

the fuel both enter at 298 K and 1 atm; the products exit at 500 K and nominally 1 atm with a standardized enthalpy of −2787.0 kJ/kg. The air and fuel are supplied in stoichiometric proportions (100% theoretical air), and the fuel mass flow rate is 2.45 kg/s. Assume that the entering air is simple dry air (3.76 kmol of N_2 for each kmol of O_2). The molar mass of "simple" air is 28.85 kg/kmol. Water enters the steam generator at 443 K and 10 MPa, and steam exits at 983 K and 10 MPa. Determine (a) the mass flow rate of the stack gases (combustion products) and (b) the mass flow rate of the steam exiting the boiler tubes.

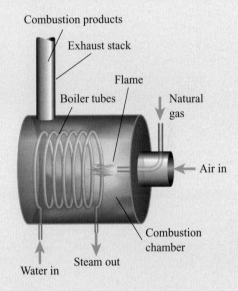

11.148 Consider an oil-fired home furnace operating steadily on a cold winter day. Fuel is supplied to the burner at 298 K at 0.01 gal/min. The liquid fuel can be treated as *n*-decane ($C_{10}H_{22}$) with the following properties: $\mathcal{M}$ = 142.284 kg/kmol, ρ = 730 kg/m³, and $\bar{h}_{f(liq)}^{\circ}$ = −289,072 kJ/mol. Air (105% theoretical) enters the burner at 298 K and 1 atm. The flue gases exit the furnace at 380 K and nominally 1 atm. Determine the rate at which energy is supplied to the home heating system.

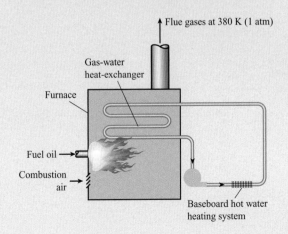

11.149 A furnace is to supply 100 MJ/hr to heat a home during the winter. The furnace completely burns natural gas (assumed to be methane) with 200% theoretical air. The methane–air mixture enters the furnace steadily at 25°C and 1 atm. The products of combustion leave at 500 K to avoid having any liquid water in the exhaust. The exhaust pressure is 1 atm. Determine the following:

A. The methane mass flow rate (kg/hr)

B. The furnace efficiency based on the higher heating value

SYSTEMS FOR POWER PRODUCTION, PROPULSION, AND HEATING AND COOLING

After studying Chapter 12, you should:

- *Understand how steady-flow devices are combined to form comprehensive systems for power production, propulsion, and heating and cooling.*

- *Be able to sketch on T–s, h–s, or other useful coordinates, the thermodynamic cycles associated with the following systems: the basic steam power plant (Rankine cycle) and its improved-efficiency variants, turbojet engines, gas-turbine power systems (Brayton cycle), and simple heat pump and refrigeration systems.*

- *Be able to perform a cycle analysis for all the various systems and calculate various performance measures such as thermal efficiency, specific thrust, and coefficient of performance.*

- *Understand the origins of the inefficiencies associated with system components and be able to include these in cycle analyses.*

- *Be able to explain using words and sketches how superheat, reheat, and regeneration improve the efficiency of a steam power plant.*

- *Be able to determine overall energy conversion efficiencies for systems that employ combustion devices.*

- *Understand the concepts of specific humidity, relative humidity, and dew point, and be able to apply them in the analysis of evaporative coolers, humidifiers, air conditioners, dehumidifiers, cooling towers, or other systems involving moist air.*

- *Be able to apply all knowledge gained in previous chapters to the analysis of complex, thermal-fluid systems.*

Chapter 12 Overview

In this chapter, we see how steady-flow devices combine to form complex systems for power production, propulsion, and heating and cooling. Not only are such systems important from an engineering perspective, but more generally, they are essential to everyday life in industrialized societies. Here we analyze these systems to understand their basic operation and to determine various performance measures. Thermodynamic cycle efficiency, first introduced in Chapter 7, provides a dominant theme for this chapter. Here we investigate in some detail the energy conversion efficiency of the various systems just listed. In some sense, this chapter is the culmination of all the preceding chapters. Chapter 12 not only provides an opportunity to integrate knowledge gained from previous chapters but also provides interesting applications that can be explored in parallel with earlier chapters.

12.1 FOSSIL-FUELED STEAM POWER PLANTS

> **The reader is encouraged to reread the appropriate material in Chapter 1 before proceeding.**

Fossil-fueled steam power plants are enormously important in providing electricity in the United States and throughout the world. In Chapter 1, we discussed this importance (Table 1.1) and introduced the basic concept of a steam power plant. In the present chapter, we build upon this introduction.

Figure 12.1 schematically illustrates the basic components and their arrangement in a fossil-fuel steam power plant. An actual power plant is usually significantly more complex than suggested by this schematic. Basic components include the following:

- a combustion chamber, or furnace, to create hot products of combustion;
- various heat exchangers (feedwater preheaters, economizers, boilers, and superheaters) to heat water and produce saturated and/or superheated steam as the working fluid for the turbines;
- pumps and pump drives (steam turbines or electric motors) to elevate the pressure of the working fluid;

Coal-fired Navajo Generating Station near Page, Arizona. Photograph courtesy of U. S. Geological Survey.

Electrical generator installation (left) and power plant control room (right)

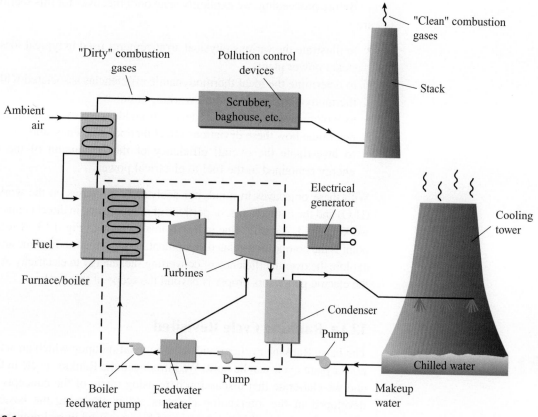

FIGURE 12.1
Schematic of utility-type fossil-fueled steam power plant showing major components.

Steam turbine assembly

- various steam turbines to extract the energy from the steam and transmit this power to electrical generators;
- electrical generators (dynamos) to convert shaft power to alternating-current electricity;
- heat-exchange equipment (condensers and cooling towers) to condense the expanded steam from the turbines;
- pollution-control equipment to remove particulate matter (baghouses), sulfur (scrubbers), and oxides of nitrogen (scrubbers and/or catalyst beds); and
- control systems of various types to ensure the integrated operation of the various subsystems.

Electrostatic precipitators control the emission of fly ash from a coal-fired power plants.
Photographs courtesy of Dr. H. Christopher Frey.

Before proceeding, we explicitly state our objectives for this section. These are

- to illustrate the various physical arrangements used in typical fossil-fueled steam power plants,
- to determine the ideal thermodynamic efficiencies associated with various thermodynamic cycles of the working fluid,
- to explore how real processes deviate from ideal processes and quantitatively determine how these deviations affect thermal efficiency, and
- to investigate the overall efficiency of the conversion of the chemical energy contained in the fuel to electrical power.

With these objectives in mind, our analysis first focuses on the working fluid (H_2O) and the components, or portions of components, in direct contact with the working fluid. A dashed line indicates this division in Fig. 12.1. A second focal point is the combustion-related equipment. Using a thermodynamic analysis, we explore the overall efficiency of converting fuel energy to electricity. Analysis of the electric generators proper is beyond the scope of this book.

12.1a Rankine Cycle Revisited

The basic Rankine cycle provides the foundation upon which an actual steam power plant cycle can be built. We introduced the Rankine cycle in Chapter 1, and we elaborate that discussion here using many of the concepts and tools developed in the intervening chapters. After exploring the basic Rankine cycle in detail, we add complexity and features used in real power plants that either increase the overall efficiency or offer other practical advantages.

Figure 12.2 shows the basic Rankine cycle components and the corresponding thermodynamic state points on a T–s diagram for the ideal cycle. To achieve this *ideal cycle*, we assume that the pump and turbine operate adiabatically and reversibility (i.e., isentropic operation) and that there are no heat losses to the surroundings associated with heat-exchange equipment, or with any of the piping interconnecting the various components. Furthermore, all of the processes are reversible (e.g., the flow through all devices and pipes is assumed to be frictionless).

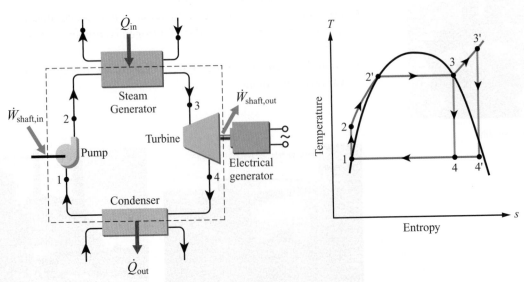

FIGURE 12.2

Components associated with a simple Rankine cycle (left) and the corresponding T–s diagram (right) for the cycle: 1–2–3–4–1. Also shown on the T–s diagram is a Rankine cycle with superheat: 1–2–3'–4'–1.

This ideal thermodynamic cycle consists of four processes:

- *Process 1–2:* A circulating pump boosts the pressure of the liquid water prior to entering the boiler. To operate the pump, a power input is required, $\dot{W}_{shaft,in}$. The ideal processes is isentropic with $s_1 = s_2$.
- *Process 2–3:* Energy is added to the water in the boiler at constant pressure, resulting, first, in an increase in the water temperature and, second, in a phase change. Heat transferred from the hot products of combustion provide this energy, $\dot{Q}_{in}$. The working fluid is all liquid at state 2 and all vapor (steam) at state 3. The phase change begins at 2′ on the T–s diagram. The ideal process is isobaric with $P_2 = P_3$.
- *Process 3–4:* Energy is removed from the high-temperature, high-pressure steam as it expands through a steam turbine. The ideal process is isentropic with $s_3 = s_4$. The shaft power of the turbine, $\dot{W}_{shaft,out}$, is delivered to an electrical generator for the production of electricity.
- *Process 4–1:* The low-pressure steam is returned to the liquid state as it flows through the condenser. The ideal process is isobaric with $P_4 = P_1$. The energy from the condensing steam, $\dot{Q}_{out}$, is transferred to the cooling water.

As outlined here, the cycle consists of two constant-pressure (isobaric) processes and two constant-specific-entropy (isentropic) processes, as shown on the T–s diagram.

We now determine the thermodynamic efficiency of this cycle. In Chapter 7, we defined the thermal efficiency of a work-producing cycle as (Eq. 7.6)

$$\eta_{th} = \frac{\text{useful work produced}}{\text{energy supplied}}. \tag{12.1}$$

For the Rankine cycle, the rate at which useful work is produced is the difference between the output shaft power of the turbine and the input shaft power of the pump, that is,

$$\dot{W}_{net,out} = \dot{W}_{turbine,out} - \dot{W}_{pump,in}.$$

The rate at which energy is supplied to the cycle is the heat-transfer rate to the water in the steam generator, $\dot{Q}_{in}$. Thus, the Rankine cycle efficiency is given by

$$\eta_{th,Rankine} = \frac{\dot{W}_{turbine,out} - \dot{W}_{pump,in}}{\dot{Q}_{in}}. \tag{12.2}$$

See Table 11.1 in Chapter 11.

From our previous component analyses, the quantities appearing in Eq. 12.2 can all be related to the thermodynamic states of the fluid at the inlets and outlets of the pump, turbine, and steam generator. Because the fluid kinetic and potential energy changes were neglected in our analyses, only the enthalpies are involved, so we have

$$\dot{W}_{turbine,out} = \dot{m}(h_3 - h_4), \tag{12.3a}$$

$$\dot{W}_{pump,in} = \dot{m}(h_2 - h_1), \tag{12.3b}$$

$$\dot{Q}_{in} = \dot{m}(h_3 - h_2). \tag{12.3c}$$

Substituting these quantities into Eq. 12.2 yields

$$\eta_{th,Rankine} = \frac{(h_3 - h_4) - (h_2 - h_1)}{h_3 - h_2},$$ (12.4a)

or, upon rearrangement,

$$\eta_{th,Rankine} = 1 - \frac{(h_4 - h_1)}{(h_3 - h_2)}.$$ (12.4b)

Since $\dot{Q}_{out} = \dot{m}(h_4 - h_1)$, Eq. 12.4b is also equivalent to

$$\eta_{th,Rankine} = 1 - \frac{\dot{Q}_{out}}{\dot{Q}_{in}}.$$

We note that Eqs. 12.3a–12.3c apply to real, as well as ideal, devices; thus, our thermal efficiency expressions (Eqs. 12.4a and 12.4b) apply equally well to real and ideal cycles.

Example 12.1

Pulp mill. Scotland.

Determine the maximum possible thermal efficiency for an industrial steam power plant operating with a high pressure of 1 MPa and a low pressure of 5 kPa. The power plant operates on a Rankine cycle with the steam entering the turbine as saturated vapor. Also determine the ratio of the pump power to the turbine power.

Solution

Known   Steam Rankine cycle with P_{low} ($= P_1 = P_4$) = 5 kPa and P_{high} ($= P_2 = P_3$) = 1 MPa.

Find η_{th} (ideal), $\dot{W}_{pump}/\dot{W}_{turbine}$

Sketch   See Fig. 12.3.

FIGURE 12.3
T–s diagram and property data summary for Example 12.1.

Region	1	2	3	4
	Saturated liquid	Compressed liquid	Saturated vapor	Liquid–vapor mixture
h (kJ/kg)	137.75	138.75	2777.1	2007.1
s (kJ/kg · K)	0.47620	0.47620	6.5850	6.5850

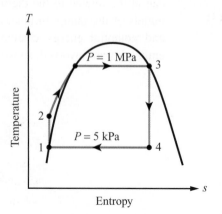

Assumptions

i. Adiabatic, reversible (i.e., isentropic) operation of pump and turbine
ii. No frictional effects in steam generator and condenser
iii. No friction or heat losses associated with any interconnecting piping

Analysis   The cycle thermal efficiency is calculated from a straightforward application of Eq. 12.1; this reduces the problem to determining the specific enthalpies at each state point in the cycle. To organize the required property data, we employ a table. The completed table is shown in Fig. 12.3; the steps required to fill the table are described in the following.

From the definition of the Rankine cycle, we can immediately identify the regions associated with states 1, 2, 3, and 4 (i.e., saturated liquid, compressed liquid, saturated vapor, and liquid–vapor mixture). Furthermore, states 1 and 3 are fully defined since they are saturated liquid ($P_{sat} = 5$ kPa) and saturated vapor ($P_{sat} = 1$ MPa) at the given pressures, respectively. From the NIST online database [1], we retrieve the enthalpies h_1 and h_3. We also retrieve the entropies s_1 and s_3, as they will be required to fully define states 2 and 4. We therefore have the following properties:

State 1 (Saturated Liquid)	State 3 (Saturated Vapor)
$P_1 = 5$ kPa (given)	$P_3 = 1$ MPa (given)
$T_1 = 306.02$ K	$T_3 = 453.03$ K
$h_1 = 137.75$ kJ/kg	$h_3 = 2777.1$ kJ/kg
$s_1 = 0.47620$ kJ/kg·K	$s_3 = 6.5850$ kJ/kg·K
$v_1 = 0.0010053$ m³/kg	

State 2 is a compressed liquid at a pressure of 1 MPa. Since the pump process is isentropic, s_2 equals s_1. Knowing two properties (P_2, s_2) allows us to find all other properties at state 2. Using the NIST database to generate a table for a compressed liquid at 1 MPa, we retrieve the following data:

State 2 (Compressed Liquid)
$P_2 = 1$ MPa (known)
$T_2 = 306.058$ K
$h_2 = 138.75$ kJ/kg
$s_2 = 0.4762$ kJ/kg·K (known)

Since we only need to find h_2 at state 2, an alternative to the use of the NIST database is to employ the incompressible approximation for the reversible pump work (Eq. 11.43):

$$\frac{\dot{W}_{pump,rev}}{\dot{m}} = h_2 - h_1 \cong v_1(P_2 - P_1),$$

or

$$h_2 = h_1 + v_1(P_2 - P_1)$$
$$= 137.75 + 0.0010053\,(1 \times 10^3 - 5)$$
$$= 138.75$$
$$[=] \frac{m^3}{kg}\frac{kN}{m^2}\left[\frac{1\ kJ}{kN \cdot m}\right] = \frac{kJ}{kg}.$$

Both approaches yield the same result to five significant digits.

With the assumption of the turbine operating adiabatically and reversibly, we know that s_4 must equal s_3. Since P_4 is given, state 4 is defined; however, to find h_4 requires the calculation of the quality x_4. Using the known entropy s_4 ($= s_3$), we calculate (Eq. 2.49d)

$$x_4 = \frac{s_4 - s_f(P_{sat} = 5\text{ kPa})}{s_g(P_{sat} = 5\text{ kPa}) - s_f(P_{sat} = 5\text{ kPa})}$$

$$= \frac{6.5850 - 0.47620}{8.3938 - 0.47620} = 0.7715,$$

where s_f and s_g are obtained from the NIST database. Knowing the quality, we perform the inverse operation of the one we just did (Eq. 2.49c) to find the unknown enthalpy h_4:

$$h_4 = h_f + x_4(h_g - h_f)$$

$$= 137.75 + 0.7715(2560.7 - 137.75)\text{ kJ/kg}$$

$$= 2007.1\text{ kJ/kg},$$

where h_f and h_g are determined in the same manner as s_f and s_g. Our property table (Fig. 12.3) is now complete, and the ideal cycle efficiency can be calculated (Eq. 12.4):

$$\eta_{th,\,Rankine} = 1 - \frac{h_4 - h_1}{h_3 - h_2}$$

$$= 1 - \frac{2007.1 - 137.75}{2777.1 - 138.75}$$

$$= 0.291,$$

or

$$\eta_{th,\,Rankine} = 29.1\%.$$

We can also calculate the ratio of the pump to turbine power:

$$\frac{\dot{W}_{pump}}{\dot{W}_{turbine}} = \frac{\dot{m}(h_2 - h_1)}{\dot{m}(h_3 - h_4)} = \frac{h_2 - h_1}{h_3 - h_4}$$

$$= \frac{138.75 - 137.75}{2771.1 - 2007.1} = \frac{1}{764} = 0.0013,$$

or

$$0.13\%.$$

Comments   First, we note the low quality of the steam exiting the turbine ($x_4 = 0.7715$). Because liquid droplets erode turbine blades, many power plants operate with qualities in excess of 0.90. We explore this in a later example (Example 12.3).

We also note that when the cycle operates with the condensed steam at near-ambient temperatures, the steam side of the condenser operates at subatmospheric conditions. For this example, $T_4 = 306.02$ K when $P_4 = 5$ kPa. This vacuum condition is one reason why the working fluid undergoes a closed cycle.

Third, we note that only 29.1% of the energy input to the working fluid is returned as useful work. If we were to consider all of the irreversibilities and their attendant losses, the thermal efficiency would be even lower than this ideal value. The direct effect of turbine losses is considered in the next example. Various means to improve the ideal efficiency comprise the subjects of the next sections.

Repeat Example 12.1 for a high pressure of 1.5 MPa. Also determine the differences in the heat added (kJ/kg), the power (kJ/kg) required by the pump, and the power produced by the turbine.

(Answer: $\eta_{\text{th,Rankine}} = 31.1\%$. The higher pressure results in a small increase in pumping power (0.5 kJ/kg) and heat input (13.4 kJ/kg) and a substantial increase in the turbine power output (57.3 kJ/kg).)

Example 12.2

Consider the same situation as described in Example 12.1, except now the turbine has an isentropic efficiency of 0.90. Calculate the effect of the turbine efficiency on the quality of the steam exiting the turbine and on the cycle efficiency.

Solution

Known $P_1, P_2, P_3, P_4, \eta_{\text{isen,t}}$

Find x_4, η_{th}

Sketch

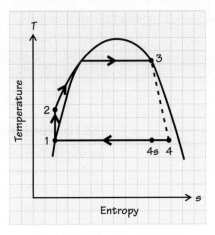

At the Nasjavellir geothermal power plant in Iceland, saturated steam from 2000-m deep bore holes enters turbines at approximately 1.2 MPa and 190°C.

Assumptions
We invoke the same assumptions applied to Example 12.1 *except* that turbine losses are considered through $\eta_{\text{isen,t}}$.

Analysis   States 1, 2, and 3 are unchanged by the addition of the turbine losses; only state 4 is affected. As indicated in the sketch, the entropy of the steam exiting the turbine (state 4) exceeds that of the corresponding isentropic-expansion state 4s. State 4s corresponds to state 4 in the previous example (i.e., $s_{4s} = 6.5850$ kJ/kg·K). We can determine a value for h_4 by applying the definition of the isentropic efficiency for the turbine (Eq. 11.52b), that is,

$$\eta_{\text{isen,t}} = \frac{h_3 - h_4}{h_3 - h_{4s}}.$$

Solving for h_4 and evaluating, we get

$$h_4 = h_3 - \eta_{\text{isen,t}}(h_3 - h_{4s})$$
$$= 2777.1 - 0.90(2777.1 - 2007.1) \text{ kJ/kg}$$
$$= 2084.1 \text{ kJ/kg},$$

where values for h_3 and h_{4s} are from Example 12.1 (see Fig. 12.3). With this value of h_4, we calculate the steam quality and the cycle efficiency as follows:

$$x_4 = \frac{h_4 - h_f(P_{sat} = 5 \text{ kPa})}{h_g(P_{sat} = 5 \text{ kPa}) - h_f(P_{sat} = 5 \text{ kPa})}$$

$$= \frac{2084.1 - 137.75}{2560.7 - 137.75} = 0.8033$$

and

$$\eta_{th} = 1 - \frac{h_4 - h_1}{h_3 - h_2} = 1 - \frac{2084.1 - 137.75}{2777.1 - 138.75}$$

$$= 0.262 \quad \text{or} \quad 26.2\%.$$

Comments As expected, the less-than-unity turbine efficiency reduces the cycle efficiency. Since the turbine was the only nonideal component considered, $\eta_{th,Rankine} = \eta_{isen,t}\eta_{th,Rankine}$ (ideal) [i.e., $\eta_{th,Rankine} = 0.90$ (0.2914) = 0.262]. Thus, there was no need to directly calculate h_4 except to find the quality x_4. Note that the turbine inefficiency has the side benefit of improving the quality of the steam ($x_4 = 0.8033$ versus 0.7715), although the steam is still quite wet.

**Self Test
12.2**

☑ **Repeat the analysis of Example 12.1 using a pump having an isentropic efficiency of 0.60. Comment on your results.**

(Answer: $\eta_{th,Rankine} = 29.1\%$. Even though there is a significant increase in the required pump work, the overall contribution of this device is insignificant to the cycle efficiency.)

12.1b Rankine Cycle with Superheat and Reheat

In this section, we discuss two modifications to the basic Rankine cycle: the addition of superheat and the addition of reheat. All but the simplest systems employ superheat, whereas more sophisticated power plants employ both superheat and reheat.

Superheat

The addition of superheat is a relatively simple means to improve the basic Rankine cycle efficiency. Rather than allowing the steam to enter the turbine

FIGURE 12.4

Thought experiment illustrating the processes involved in steam generation: At the left, cold water is heated to the saturated liquid state. The saturated liquid is then vaporized in the center vessel; and at the right, the vapor is superheated. The three vessels correspond to the economizer, the boiler, and the superheater sections used in real steam generators (see Figs. 12.5 and 12.6).

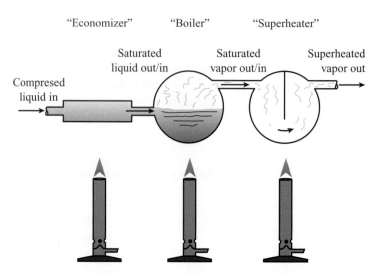

FIGURE 12.5
Basic components for a Rankine cycle with superheat showing economizer, boiler, and superheater sections in a steam generator. The state points indicated correspond to the T–s diagrams shown in Figs. 12.2 and 12.7.

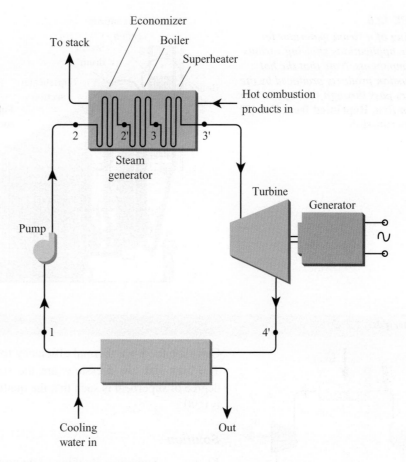

as saturated vapor, additional energy is supplied to the steam at constant pressure providing superheated vapor at the turbine inlet. A thought experiment illustrating the creation of superheated steam starting from a compressed liquid is shown in Fig. 12.4. Corresponding states are shown in the *T–s* diagram in Fig. 12.2. The liquid water is first heated from its temperature exiting the pump (state 2 in Fig. 12.2) to the saturated-liquid state (state 2′ in Fig. 12.2). This occurs in the **economizer** section of the steam generator. Energy is added to the saturated liquid in the **boiler** section of the steam generator to produce saturated vapor (state 3 in Fig. 12.2). The saturated vapor is then heated further in the **superheater** section of the boiler to state 3′ (Fig. 12.2). Figure 12.5 shows these three sections of a steam generator as part of a power-plant schematic, and Fig. 12.6 shows a drawing of an actual small-scale marine steam generator. Note that the hot products of combustion strike the superheater first, and then, sequentially, the boiler tubes and the economizer tubes. This flow path takes advantage of using the hottest gases to superheat the steam because the temperature of the combustion products decreases as these gases pass through the steam generator.

The use of superheat improves the cycle efficiency by adding energy to the working fluid at a higher mean temperature than would otherwise result at the same pressure. We saw in Chapter 7 (Eq. 7.15a) that the Carnot cycle efficiency increases with the temperature of the heat-addition reservoir. This idea carries through to the Rankine cycle as well. An additional benefit of superheat is improved steam quality at the turbine exit, thereby diminishing blade erosion from water droplets.

In the next example, we quantify these effects by adding superheat to the cycle analyzed in Example 12.1.

FIGURE 12.6

Drawing of a steam generator for marine applications showing various subcomponents. Note that the hot combustion products produced by the burners pass through the superheater section first. **Reprinted from Ref. [2] with permission.**

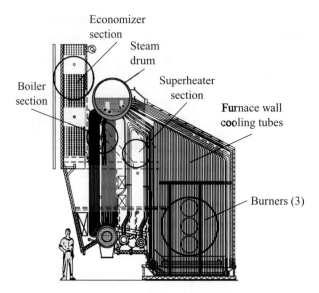

Example 12.3

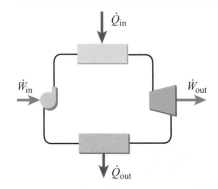

Calculate the ideal thermal efficiency for a Rankine cycle with superheat. The high and low pressures are the same as in Example 12.1, and the degree of superheat is such that the quality of the steam exiting the turbine is 0.90.

Solution

Known Superheat Rankine cycle with P_{low} ($= P_1 = P_4$) $= 5$ kPa, P_{high} ($= P_2 = P_3$) $= 1$ MPa, and $x_{4'} = 0.90$

Find η_{th} (ideal)

Sketch See T–s diagram (Fig. 12.2) for cycle 1–2–3′–4′–1. See also Fig. 12.7.

Assumptions

All processes are ideal (see assumptions in Example 12.1).

Analysis To calculate the thermal efficiency from Eq. 12.4 requires values for the enthalpy of the working fluid at states 1, 2, 3′, and 4′. States

---

FIGURE 12.7

T–s diagram for Example 12.3 drawn to scale. Note that the steam is heated well above the critical temperature.

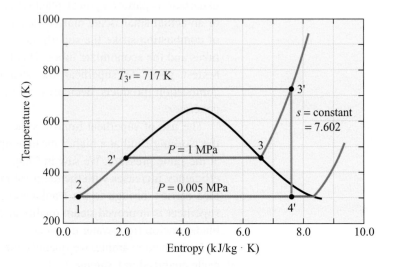

1 and 2 are unchanged from Example 12.1, so we already have those values (see Fig. 12.3):

$$h_1 = 137.75 \text{ kJ/kg},$$
$$h_2 = 138.75 \text{ kJ/kg}.$$

State 4' is defined by the quality (0.90) and the saturation pressure (5 kPa); thus, the enthalpy and entropy at 4' are calculated as

$$h_{4'} = h_f + x_{4'}(h_g - h_f)$$

and

$$s_{4'} = s_f + x_{4'}(s_g - s_f),$$

where the subscripts f and g refer to the saturated-liquid and saturated-vapor states at 5 kPa. Using the previously determined values for h_f, h_g, s_f, and s_g (see Example 12.1), $h_{4'}$ and $s_{4'}$ are determined as

$$h_{4'} = 137.75 + 0.90 (2560.7 - 137.75) \text{ kJ/kg} \cdot \text{K}$$
$$= 2318.4 \text{ kJ/kg} \cdot \text{K}$$

and

$$s_{4'} = 0.47620 + 0.90 (8.3938 - 0.47620) \text{ kJ/kg} \cdot \text{K}$$
$$= 7.602 \text{ kJ/kg} \cdot \text{K}.$$

Since we assume that the turbine operates ideally (adiabatically and reversibly), the process 3'–4' is isentropic, and so

$$s_{3'} = s_{4'} = 7.602 \text{ kJ/kg} \cdot \text{K}.$$

The state at 3' is thus defined: $s_{3'}$ and P_3 are known. Using the NIST database,[1] we then find

$$T_{3'} = 717 \text{ K},$$
$$h_{3'} = 3358.1 \text{ kJ/kg}.$$

The state points 3' and 4' are shown on the T–s diagram drawn to actual scale in Fig. 12.7. Note that the maximum superheat temperature, $T_{3'} = 717$ K, is well above the critical temperature, $T_{cr} = 647.10$ K. With values for the enthalpies at each state (1, 2, 3', 4'), we apply Eq. 12.4 to calculate the ideal cycle efficiency:

$$\eta_{th} = 1 - \frac{h_{4'} - h_1}{h_{3'} - h_2}$$

$$= 1 - \frac{2318.4 - 137.75}{3358.1 - 138.75}$$

$$= 0.323,$$

or

$$\eta_{th} = 32.3\%.$$

Comments The addition of superheat results in an increase in the ideal efficiency from 29.1% to 32.3%, a substantial improvement of 3.2 percentage points or 11%. Considering the relatively modest addition of hardware required to achieve this gain, we see why all but the smallest power plants employ superheat in practice.

---

[1] Alternatively, the property tables in Appendix D can also be used, and the desired properties can be obtained by interpolation.

We illustrated this cycle on actual T–s coordinates (Fig. 12.7) to show the true positions of the various state points. Note, in particular, that the 1-MPa compressed-liquid isobar follows the saturation line so closely that it cannot be resolved on the scale shown. As a consequence, the separation between state 1 and state 2 cannot be resolved. In Example 12.1, we found the difference between T_2 and T_1 to be only 0.04 K (= 306.06 − 306.02 K).

Self Test 12.3

☑ **Calculate the thermal efficiency and steam quality at the turbine exit for a Rankine cycle operating between the same pressures and with the same degree of superheating as in Example 12.3. The turbine has an isentropic efficiency of 0.92.**

(Answer: $x = 0.934$, $\eta_{th,Rankine} = 29.7\%$)

Example 12.4

This example combines cycle analysis with the principles of internal flows from Chapter 10. ▶

Consider the ideal Rankine cycle presented and discussed in Example 12.3. In real power plants frictional losses can be quite important. In this example, we focus on the effect of the pressure drop resulting from friction in a typical superheater. The boiler produces 22 kg/s of steam, which is superheated by flowing through forty-eight parallel, 87.7-mm-inside-diameter tubes. Each tube is 17.5 m long and connects to a receiver pipe for delivery to the turbine. The measured pressure drop ΔP across the superheater assembly is 0.1538 MPa (22.3 psi). Assume that the steam enters the superheater as saturated vapor at 1 MPa and that the heat-transfer rate to the steam per unit mass is the same as in Example 12.3. Estimate the fraction of the total superheater pressure drop that is caused by wall friction in the superheater tubes. Also determine the effect of the superheater pressure drop on an otherwise ideal cycle efficiency.

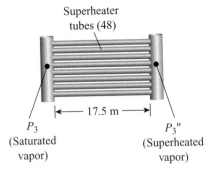

Superheater tubes (48)

|← 17.5 m →|

P_3 (Saturated vapor) P_3'' (Superheated vapor)

Solution

Known $P_3 (= P_{sat,3})$, $\Delta P (= P_3 − P_{3''})$, $\dot{m}$, N, D_i, L

Find ΔP_{tube}, η_{cycle}

Sketch

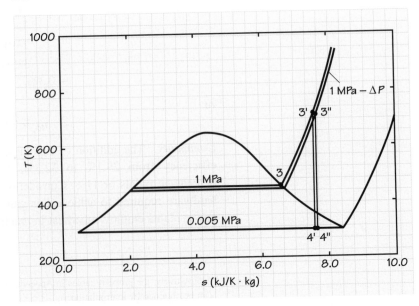

Assumptions

i. Steady flow

ii. ΔP_{tube} evaluated from incompressible flow relationships

iii. Density and viscosity evaluated at mean state: $(T_3 + T_{3''})/2$ and $(P_3 + P_{3''})/2$

Analysis The state of the steam at the superheater outlet is designated $3''$ on the sketch and lies on the 0.8462-MPa isobar. This isobar lies below the ideal 1-MPa isobar and corresponds to $P_3 - \Delta P$. The same amount of heat is transfer to the steam as in Example 12.3 (i.e., $_2\dot{Q}_{3'}/\dot{m} = {}_2\dot{Q}_{3''}/\dot{m}$); thus, $h_{3''} = h_{3'} (= 3358.1$ kJ/kg). State $3''$ is therefore defined by $P_{3''} = 0.8462$ MPa and $h_{3''} = 3358.1$ kJ/kg. Using the NIST database, we obtain the following additional properties at state $3''$:

$$T_{3''} = 716.05 \text{ K}, \qquad s_{3''} = 7.6781 \text{ kJ/kg} \cdot \text{K},$$
$$\rho_{3''} = 2.583 \text{ kg/m}^3, \qquad \mu_{3''} = 26.221 \times 10^{-6} \text{ N} \cdot \text{s/m}^2.$$

The temperature of the steam entering the superheater is 453.03 K = $T_{sat}(P_3 = 1$ MPa).

To find the pressure drop associated with the superheater tubes, we apply the methods discussed in Chapter 10. We note, however, that these methods are based on an incompressible, isothermal fluid, which is clearly not the case for the present situation. Not having any other alternative, we treat the steam as if it were incompressible and isothermal with a density and viscosity evaluated at the average temperature and pressure in the superheater, that is,

$$T_{avg} \equiv (T_3 + T_{3''})/2 = (453.03 + 716.05)/2 \text{ K} = 584.5 \text{ K}$$
$$P_{avg} = (P_3 + P_{3''})/2 = (1 + 0.8462)/2 \text{ MPa} = 0.9231 \text{ MPa}.$$

Thus, from the NIST database,

$$\rho_{avg} = 3.4954 \text{ kg/m}^3,$$
$$\mu_{avg} = 20.677 \times 10^{-6} \text{ N} \cdot \text{s/m}^2.$$

With these properties, we can determine the friction factor and hence the pressure drop associated with a single superheater tube. Since the tubes are arranged in parallel, ΔP_{tube} for a single tube is the same as that associated with all the tubes taken together. Thus,

$$\dot{m}_{1 \text{ tube}} = \frac{\dot{m}_{tot}}{N} = \frac{22 \text{ kg/s}}{48}$$
$$= 0.458 \text{ kg/s},$$

and the Reynolds number is (Eq. 10.34)

$$Re_{D_i} = \frac{\rho_{avg} v_{avg} D_i}{\mu_{avg}} = \frac{4 \dot{m}_{1 \text{ tube}}}{\pi \mu_{avg} D_i}$$

$$= \frac{4(0.458)}{\pi (20.68 \times 10^{-6}) 0.0877} = 3.21 \times 10^5$$

$$[=] \frac{\text{kg/s}}{(\text{N} \cdot \text{s/m}^2)\text{m}} \left[\frac{1 \text{ N}}{\text{kg} \cdot \text{m/s}^2} \right] = 1.$$

For drawn tubing (Table 10.4), the relative roughness is

$$\varepsilon/D_i = \frac{0.0015 \text{ mm}}{87.7 \text{ mm}} = 0.000017.$$

The friction factor can now be estimated from Fig. 10.17:

$$f_D = f_D(\varepsilon/D_i, Re_{D_i}) \approx 0.0143.$$

The pressure drop is thus (Eq. 10.17)

$$\Delta P_{\text{tube}} = f_D \frac{L}{D_i} \frac{1}{2} \rho_{\text{avg}} v_{\text{avg}}^2,$$

where

$$v_{\text{avg}} = \frac{\dot{m}_{1 \text{ tube}}}{\rho_{\text{avg}} \pi D_i^2/4}$$

$$= \frac{0.458(4)}{3.4954 \pi (0.0877)^2} = 21.7$$

$$[=] \frac{\text{kg/s}}{(\text{kg/m}^3)\text{m}^2} = \text{m/s}.$$

Compare this velocity with typical values shown in Table 3.2 in Chapter 3.

Thus,

$$\Delta P_{\text{tube}} = 0.0143 \left(\frac{17.5}{0.0877}\right) 0.5(3.4954)(21.7)^2$$

$$= 2348$$

$$[=] \left(\frac{\text{m}}{\text{m}}\right)(\text{kg/m}^3)(\text{m}^2/\text{s}^2)\left[\frac{1 \text{ N}}{\text{kg} \cdot \text{m/s}^2}\right] = \text{N/m}^2 \text{ or Pa}.$$

Because

$$\frac{\Delta P_{\text{tube}}}{\Delta P_{\text{total}}} = \frac{2348 \text{ Pa}}{0.1538 \times 10^6 \text{ Pa}} = 0.015 \text{ or } 1.5\%,$$

we see that this pressure drop from wall friction is only a small fraction of the reported total pressure drop for the superheater assembly.

To account for the superheater pressure drop in calculating the cycle efficiency, we note that only two state points are changed from those used to calculate the ideal efficiency in Example 12.3: state $3' \rightarrow$ state $3''$ and state $4' \rightarrow 4''$ (see sketch). Since $h_{3''} = h_{3'}$, the only enthalpy that we need to determine is $h_{4''}$. Assuming an isentropic process from $3''$ to $4''$, we can calculate the quality $x_{4''}$ as follows

$$s_{4''} = s_{3''} = 7.6781 \text{ kJ/kg} \cdot \text{K},$$

and, using the saturation properties at 0.005 MPa, we obtain

$$x_{4''} = \frac{s_{4''} - s_f}{s_g - s_f}$$

$$= \frac{7.6781 - 0.4762}{8.3938 - 0.4762} = 0.9096.$$

Hence,

$$h_{4''} = (1 - x_{4''})h_f + x_{4''}h_g$$

$$= 0.0904(137.75) + 0.9096(2560.7) \text{ kJ/kg}$$

$$= 2341.7 \text{ kJ/kg}.$$

The cycle efficiency is then

$$\eta_{th,nonideal} = 1 - \frac{h_{4''} - h_1}{h_{3''} - h_2}$$

$$= 1 - \frac{2341.7 - 137.75}{3358.1 - 138.75}$$

$$= 0.3154 \text{ or } 31.5\%.$$

This represents a loss of 0.76 percentage points over the ideal cycle ($\eta_{th,ideal} = 32.3\%$) in which there are no losses associated with the superheater, or, expressed as a percentage,

$$\frac{\eta_{th,nonideal}}{\eta_{th,ideal}} = \frac{0.315}{0.323} = 0.975 \text{ or } 97.5\%.$$

Comments   We note first the interesting, and perhaps surprising, result that the frictional pressure drop through the superheater tubes comprises only a small fraction of the superheater ΔP. Other losses, such as those in the headers and collectors, must dominate. We also see that the superheater loss has a relatively large influence on the cycle efficiency. In the operation of real power plants, individual percentage points of efficiency are quite important. This example also provides a quantitative illustration of how frictional effects are manifested by entropy increases: For the contant-pressure (frictionless) superheating process, the specific entropy change of the steam is $s_{3'} - s_3 = 7.602 - 6.585 \text{ kJ/kg} \cdot \text{K} = 1.017 \text{ kJ/kg} \cdot \text{K}$, whereas for the same heat addition but with friction, $s_{3''} - s_3 = 7.6781 - 6.5850 \text{ kJ/kg} \cdot \text{K} = 1.098 \text{ kJ/kg} \cdot \text{K}$, or about 8% greater.

Example 12.5

This example applies the heat-exchanger analysis developed in Chapter 11 to a real power plant superheater.

Consider the superheater and operating conditions described in Example 12.4. Hot products of combustion ($c_p = 1.28 \text{ kJ/kg} \cdot \text{K}$) enter the superheater at 1050 K at a flow rate of 33.3 kg/s. Treating the superheater as a cross-flow heat exchanger with one stream unmixed (steam) and the other (combustion products) mixed, estimate the overall heat-transfer coefficient based on the inside area of the tubes.

Solution

Known $\dot{m}_C, \dot{m}_H, T_{C,in}, T_{C,out}, T_{H,in}, c_{p,H}, N, D_i, L$

Find $U_{OA,i}$

Sketch

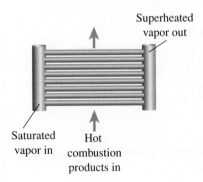

Superheated vapor out

Saturated vapor in   Hot combustion products in

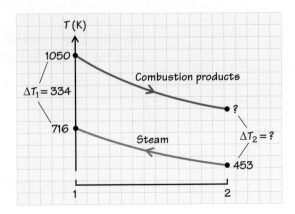

Assumptions

  i. Steady flow

  ii. No heat loss to surroundings

  iii. Negligible changes in kinetic and potential energy

  iv. Cross-flow geometry

Analysis The heat exchange between the combustion products and the steam can be modeled as (Eqs. 11.80 and 11.82)

$$\dot{Q} = U_{OA,i}A_iF(\Delta T)_{lm,CF}. \tag{A}$$

Energy balances on the combustion product stream and steam yield, respectively,

$$\dot{Q} = \dot{m}_Hc_{p,H}(T_{H,in} - T_{H,out}) \tag{B}$$

and

$$\dot{Q} = \dot{m}_{steam}(h_{3''} - h_3), \tag{C}$$

where the enthalpies are determined from the known steam states (see Examples 12.3 and 12.4). With $\dot{m}_{steam}$ given in Example 12.4, we use Eq. C to find $\dot{Q}$:

$$\dot{Q} = 22(3358.1 - 2777.1) = 12,782$$
$$[=] (kg/s)kJ/kg = kJ/s = kW.$$

The combustion products outlet temperature $T_{H,out}$ is found from Eq. B:

$$T_{H,out} = T_{H,in} - \frac{\dot{Q}}{\dot{m}_Hc_{p,H}}$$

$$= 1050 - \frac{12,782}{33.3(1.28)} = 750$$

$$[=] \frac{kW}{(kg/s)kJ/kg \cdot K}\left[\frac{1\,kJ/s}{kW}\right] = K.$$

Knowing all four temperatures, we now calculate the log-mean temperature difference for a counterflow arrangement (see sketch):

$$\Delta T_2 = 750 - 453\,K = 297\,K$$

and

$$(\Delta T)_{lm,CF} = \frac{\Delta T_2 - \Delta T_1}{\ln\left(\dfrac{\Delta T_2}{\Delta T_1}\right)}$$

$$= \frac{297 - 334}{\ln\dfrac{297}{334}}K = 315\,K.$$

We correct this counterflow mean-temperature difference for cross-flow using Fig. 11.52. The chart parameters are

$$R = \frac{T_i - T_o}{t_o - t_i} = \frac{1050 - 750}{716 - 453} = 1.14$$

and

$$P = \frac{t_o - t_i}{T_i - t_i} = \frac{716 - 453}{1050 - 453} = 0.44.$$

Using these values, we obtain

$$F \approx 0.88.$$

The total (inner) heat exchange area is simply the product of the number of tubes and the inner area for a single tube, that is,

$$A_{i,tot} = N\pi D_i L$$

$$= 48\pi (0.0877) 17.5 \text{ m}^2 = 231.4 \text{ m}^2.$$

We now apply Eq. A to find the overall heat-transfer coefficient as follows:

$$U_{OA,i} = \frac{\dot{Q}}{A_{i,tot}F(\Delta T)_{lm,CF}}$$

$$= \frac{12{,}782{,}000}{231.4(0.88)315} = 199$$

$$[=] \frac{W}{m^2(K)} \text{ or } W/m^2 \cdot K.$$

Comment This example provides some insight into how the heat-exchanger design techniques discussed in Chapter 11 apply to the design of large systems, namely, steam power plants. Note, however, that our analysis here is "reverse engineering" as the design was already established.

Tandem compound reheat steam turbine rotor displayed on half shell. Photograph courtesy of General Electric Company.

Reheat

Frequently, the simple Rankine cycle is modified with the addition of a **reheat** process. Reheat allows the use of relatively high boiler pressures while maintaining a sufficiently high value of the quality of the steam exiting the turbine. Figure 12.8a illustrates the basic component arrangement for a reheat cycle. Here we see the simple cycle modified by replacing the single turbine with two: a high-pressure turbine and a low-pressure turbine. Furthermore, the steam exiting the high-pressure turbine returns to the boiler to be reheated before entering the low-pressure turbine. A *T–s* diagram illustrating the ideal reheat cycle is shown in Fig. 12.8b. Once again, by using the word *ideal*, we

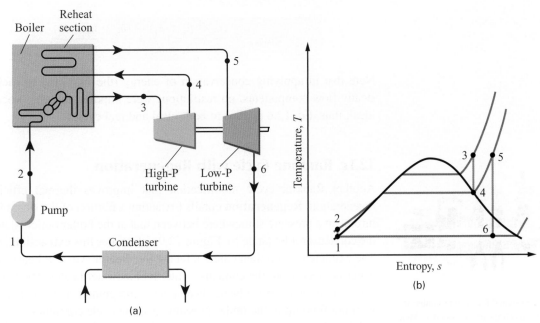

(a)

(b)

FIGURE 12.8

Rankine cycle with reheat: (a) component arrangement, (b) T–s diagram.

mean that no friction or other irreversibilities are associated with any of the components or interconnecting flow passages.

Analysis of the thermal efficiency of the reheat cycle follows that previously performed for the simple Rankine cycle with or without superheat. The primary differences are the inclusion of a second heat-addition process (state 4 to state 5) and the power delivery by two turbines (state 3 to state 4 and state 5 to state 6). To evaluate the thermal efficiency defined by Eq. 12.1, we calculate the net useful work:

$$\dot{W}_{net,out} = \dot{W}_{hi\text{-}P\ turb,out} + \dot{W}_{low\text{-}P\ turb,out} - \dot{W}_{pump,in},$$

where

$$\dot{W}_{hi\text{-}P\ turb,out} = \dot{m}(h_3 - h_4), \tag{12.5a}$$

$$\dot{W}_{low\text{-}P\ turb,out} = \dot{m}(h_5 - h_6), \tag{12.5b}$$

and

$$\dot{W}_{pump,in} = \dot{m}(h_2 - h_1). \tag{12.5c}$$

The heat added is the sum of that for the steam generation process (2–3) and the reheat process (4–5):

$$\dot{Q}_{boiler,in} = \dot{m}(h_3 - h_2) \tag{12.5d}$$

and

$$\dot{Q}_{reheat,in} = \dot{m}(h_5 - h_4). \tag{12.5e}$$

The thermal efficiency is thus expressed as

$$\eta_{th,reheat} = \frac{\dot{W}_{net,out}}{\dot{Q}_{in}} = \frac{(h_3 - h_4) + (h_5 - h_6) - (h_2 - h_1)}{(h_3 - h_2) + (h_5 - h_4)}, \tag{12.6a}$$

which can be rearranged to yield

$$\eta_{th,reheat} = \frac{(h_5 + h_3 + h_1) - (h_6 + h_4 + h_2)}{(h_5 + h_3) - (h_4 + h_2)}. \tag{12.6b}$$

Note that in applying conservation of energy (the first law) to each of the steady-flow components, no restrictions were imposed that the processes be ideal; thus, Eq. 12.6 applies to both ideal and real cycles.

12.1c Rankine Cycle with Regeneration

Another Rankine-cycle modification that improves thermal efficiency is regeneration. **Regeneration** entails extracting a portion of the steam from the turbine at a pressure somewhere between that at the boiler outlet (state 3) and the condenser inlet (state 5). Figure 12.9a illustrates this **extraction** at state 4. The extracted steam then enters a **feedwater heater** where it heats the liquid water pumped from the condenser into the feedwater heater. The two steams combine in the open feedwater heater, a second pump brings the total flow of working fluid up to the boiler pressure, and the cycle continues.

The advantage of adding regeneration is that less heat must be added in the boiler since a portion of the working fluid has been preheated in the feedwater

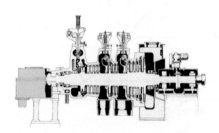

Steam turbine designed for steam to enter at 10 MPa, with extractions at 1.19 and 0.59 MPa. The steam exits at 6 kPa. Drawing courtesy of Toshiba Corporation.

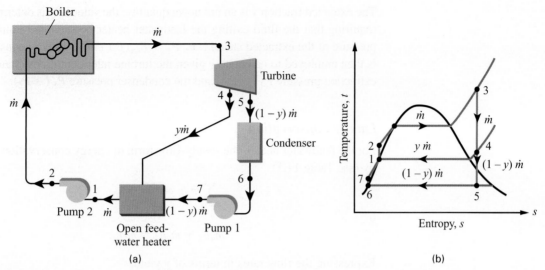

FIGURE 12.9
Regenerative Rankine cycle with a single open feedwater heater: (a) component arrangement, (b) T–s diagram for ideal components.

heater. The decreased power produced by the turbine resulting from extracting steam is more than offset by the diminished heat requirement in the boiler. In many power plants, several feedwater heaters are employed, with steam extracted at multiple locations from the turbine. Economic factors determine the optimum number of feedwater heaters for any particular power plant.

In the following, we utilize the principles of conservation of mass and energy to determine the thermal efficiency (Eq. 12.1) of a regenerative Rankine cycle employing an open feedwater heater.[2]

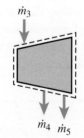

Mass Conservation

We denote the fraction of the total mass flow rate extracted from the turbine as

$$y \equiv \frac{\dot{m}_{\text{extracted}}}{\dot{m}_{\text{total}}}. \tag{12.7}$$

Applying mass conservation (Eq. 3.18b) to the turbine yields

$$\dot{m}_3 = \dot{m}_4 + \dot{m}_5, \tag{12.8a}$$

or

$$\dot{m} = y\dot{m} + (1 - y)\dot{m}, \tag{12.8b}$$

where $\dot{m} = \dot{m}_{\text{total}}(= \dot{m}_1 = \dot{m}_2 = \dot{m}_3)$. (See Fig. 12.9.) In the feedwater heater, the two streams recombine; thus, mass conservation in this device is expressed as

$$\dot{m}_4 + \dot{m}_7 = \dot{m}_1, \tag{12.9a}$$

or

$$y\dot{m} + (1 - y)\dot{m} = \dot{m}. \tag{12.9b}$$

---

[2] Two types of feedwater heaters are commonly employed: open and closed. For an analysis of closed feedwater heaters the reader is referred to Ref. [8].

The extracted fraction y is an unknown quantity; the value of y is determined by requiring that the fluid exiting the feedwater heater be saturated liquid at the pressure of the extracted steam [i.e., $P_1 = P_{sat}(T_1) = P_4$]. Energy conservation is then employed to calculate y given the turbine inlet conditions (state 3), the extraction pressure $P_4 (= P_1)$, and the condenser pressure $P_5 (= P_6)$.

Energy Conservation

For the feedwater heater, the steady-flow form of energy conservation is given by (see Table 11.1)

$$\sum_{\text{inlets}} \dot{m}_i h_i = \sum_{\text{outlets}} \dot{m}_i h_i,$$

or

$$\dot{m}_4 h_4 + \dot{m}_7 h_7 = \dot{m}_1 h_1. \qquad (12.10a)$$

Expressing the flow rates in terms of y yields

$$y\dot{m}h_4 + (1 - y)\dot{m}h_7 = \dot{m}h_1. \qquad (12.10b)$$

From this expression of energy conservation, we solve for y:

$$y = \frac{h_1 - h_7}{h_4 - h_7}. \qquad (12.11)$$

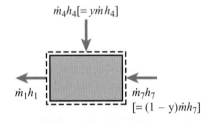

With the extracted mass fraction now determined, a thermodynamic analysis of the cycle can proceed in the same fashion as our previous analyses. The following example illustrates this procedure.

Example 12.6

Calculate the ideal cycle efficiency for a regenerative Rankine cycle employing a single open feedwater heater. A temperature–entropy diagram for the cycle is shown in the sketch. The low-pressure and high-pressure state points are the same as those in Example 12.3 (i.e., $T_3 = 717$ K and $x_5 = 0.90$). The extracted steam and the feedwater heater are at 0.2 MPa.

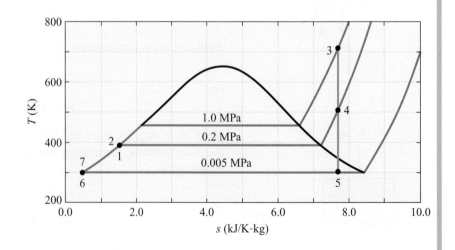

Solution

Known $P_1, P_2, P_3, P_4, P_6, P_7, T_3, x_5$

Find η_{th}

Assumptions

    i. The flow is steady.

   ii. All processes and devices are ideal.

Analysis   The cycle thermal efficiency is defined as the net useful power delivered divided by the rate at which heat is added to the working fluid. The heat added comes solely from the boiler as the feedwater heater is adiabatic with respect to the surroundings. Thus,

$$\eta_{th,regen} = \frac{\dot{W}_{net,out}}{\dot{Q}_{in}} = \frac{\dot{W}_{turbine} - \dot{W}_{pump1} - \dot{W}_{pump2}}{\dot{Q}_{boiler}}.$$

From an overall energy balance, we can also write

$$\dot{W}_{net,out} = \dot{Q}_{in} - \dot{Q}_{out}$$

$$= \dot{Q}_{boiler} - \dot{Q}_{condenser}.$$

Thus,

$$\eta_{th,regen} = \frac{\dot{Q}_{boiler} - \dot{Q}_{condenser}}{\dot{Q}_{boiler}}$$

$$= 1 - \frac{\dot{Q}_{condenser}}{\dot{Q}_{boiler}}.$$

Steady-flow first-law analyses of the boiler and condenser yield

$$\dot{Q}_{boiler} = \dot{m}(h_3 - h_2)$$

and

$$\dot{Q}_{condenser} = (1 - y)\dot{m}(h_5 - h_6).$$

Substituting these back into the thermal efficiency expression, we have

$$\eta_{th,regen} = 1 - (1 - y)\frac{h_5 - h_6}{h_3 - h_2},$$

where *y*, the extracted fraction, is evaluated from Eq. 12.11, that is,

$$y = \frac{h_1 - h_7}{h_4 - h_7}.$$

From this, we see that enthalpy values are required at each of the seven state points. States 1 and 6 are saturated-liquid states and properties are easily obtained from the NIST database; the properties for states 3 and 5 were previously evaluated in Example 12.3. These properties are summarized as follows:

Property	State 1	State 3	State 5	State 6
P (MPa)	0.20	1.00	0.005	0.005
T (K)	393.36	717	453.03	306.02
h (kJ/kg)	504.70	3358.1	2318.4	137.75
s (kJ/kg·K)	1.5302	7.602	7.602	0.4762
v (m³/kg)	0.0010605	—	—	0.0010053

The enthalpy at state 4 is determined from the NIST database using the two known properties $P_4 = 0.2$ MPa and $s_4 = s_3 = 7.602$ kJ/kg·K. This value is

$$h_4 = 2916.3 \text{ kJ/kg}.$$

The enthalpies at the pump outlets can be estimated by combining Eqs. 11.36 and 11.43:

$$h_7 \cong h_6 + v_6(P_7 - P_6)$$
$$= 137.75 + 0.0010053(200 - 5) \text{ kJ/kg} = 137.95 \text{ kJ/kg}$$

and

$$h_2 \cong h_1 + v_1(P_2 - P_1)$$
$$= 504.70 + 0.0010605(1000 - 200) \text{ kJ/kg} = 505.55 \text{ kJ/kg}.$$

With values for all seven enthalpies, we find the extracted fraction (Eq. 12.11) and the cycle thermal efficiency as follows:

$$y = \frac{h_1 - h_7}{h_4 - h_7} = \frac{504.70 - 137.95}{2916.3 - 137.95} = 0.1320$$

and

$$\eta_{\text{th,regen}} = 1 - (1 - y)\frac{h_5 - h_6}{h_3 - h_2}$$
$$= 1 - (1 - 0.1320)\frac{2318.4 - 137.75}{3358.1 - 505.55}$$
$$= 0.3365 \text{ or } 33.65\%.$$

Comment Comparing the present result ($\eta_{\text{th,regen}} = 33.65\%$) with that for the same cycle without regeneration ($\eta_{\text{th}} = 32.3\%$), we see a substantial improvement resulting from the addition of a single feedwater heater (i.e., the percentage improvement = $[(33.65 - 32.3)/32.3] \cdot 100\% = 4.2\%$).

  The net power output of the regenerative Rankine cycle of Example 12.6 is 20 MW. Determine the mass flow rate through the boiler, the power required by each pump, and the power generated by the turbine.

(Answer: $\dot{m} = 20.84$ kg/s, $\dot{W}_{\text{pump1}} = 3.62$ kW, $\dot{W}_{\text{pump 2}} = 17.71$ kW, $\dot{W}_{\text{turbine}} = 20{,}023$ kW)

12.1d Energy Input from Combustion

The reader may want to review the following topics: ideal-gas mixtures (Chapter 2), standardized enthalpies (Chapter 2), element conservation (Chapter 3), and stoichiometry (Chapter 3).

So far in our analyses of the Rankine cycle and its variants we have treated the energy input to the system as a heat-addition process. We now examine this process in greater detail. Figure 12.10 schematically shows how the hot products of combustion from the furnace section of the boiler[3] are used to heat water and to produce and superheat steam. The economizer, steam generator, and superheater are, in effect, specialized heat exchangers (see Chapter 11). Although not shown in Fig. 12.10, reheater components and their associated inlet and outlet streams may also be part of the overall

---

[3] We use the generic term *boiler* to refer to the package of components used to produce steam in a power plant.

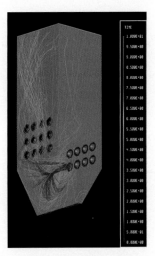

Computer simulation of coal burning in a utility boiler. Image courtesy of Natural Resources Canada, CANMET Energy Technology Centre.

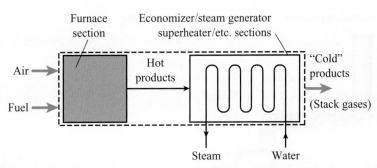

FIGURE 12.10
Schematic diagram of a steam power plant boiler. In an actual boiler, the furnace section is integrated with the boiler components as shown in Fig. 12.6.

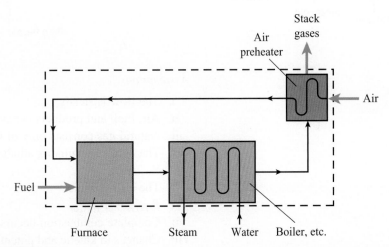

FIGURE 12.11
Schematic diagram of a steam power plant boiler with preheating of the combustion air using hot stack gases. The heat exchangers used for this purpose can be either recuperators (steady-flow devices) or regenerators (energy-storage devices). See Table 11.7.

system. Furthermore, most power plants preheat the combustion air to improve efficiency (see Fig. 12.11). Our purpose here is not to provide a detailed analysis of any particular furnace and boiler system but, rather, to show how the fundamental principles discussed in previous chapters can be applied to obtain a more complete understanding of the steam production part of a power plant. The following example illustrates this synthesis.

Example 12.7

Consider a simple, natural gas-fired steam generator as shown in the sketch. Air and fuel (assumed to be CH_4) both enter at 298 K and 1 atm, and the products of combustion exit the stack at 500 K. The fuel mass flow rate is 3 kg/s, and the air is supplied at 105% of the theoretical (stoichiometric) rate. The air can be treated as a simple mixture of 21% O_2 and 79% N_2 (i.e., 3.76 kmol of N_2 for each kmol of O_2, with a molecular weight 28.85 kg/kmol). Water enters the boiler at 400 K and 12 MPa; superheated steam exits with negligible pressure drop at 900 K. Determine the mass flow rate of the steam.

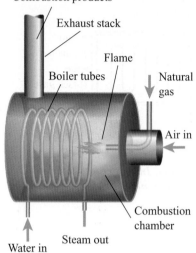

Combustion products

Exhaust stack

Flame

Boiler tubes

Natural gas

Air in

Combustion chamber

Water in   Steam out

Solution

Known $\dot{m}_F$, 105% theoretical air, T_F, T_A, P_F, P_A, T_P, $T_{H_2O,in}$, $T_{H_2O,out}$, P_{H_2O}

Find $\dot{m}_{H_2O}$

Sketch

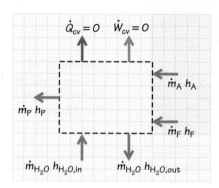

Assumptions

   i. The flow is steady.

  ii. Air, fuel, and products behave as ideal gases.

 iii. Natural gas consists only of CH_4.

 iv. The control volume is adiabatic and produces no work (i.e., $\dot{Q}_{cv} = \dot{W}_{cv} = 0$).

  v. The air is simple.

 vi. $P_{H_2O,in} = P_{H_2O,out}$.

 vii. Complete combustion occurs with negligible dissociation.

viii. Changes in kinetic and potential energy are negligible for all streams.

Analysis   We begin by applying mass conservation, recognizing that the water/steam side of the system is isolated from the combustion side; thus,

$$\dot{m}_{H_2O,in} = \dot{m}_{H_2O,out}$$

and

$$\dot{m}_A + \dot{m}_F = \dot{m}_P,$$

where A, F, and P indicate air, fuel, and products, respectively. The air flow rate can be found using the given stoichiometry, and the products flow rate can be found from our mass conservation expression.

   Before proceeding with these calculations, we apply energy conservation to our control volume to illustrate the overall approach to finding the unknown steam (water) flow rate. With our assumptions, the general steady-flow energy equation for a multistream control volume (Eq. 5.65),

$$\dot{Q}_{cv,net\,in} - \dot{W}_{cv,net\,out} = \sum_{k=1}^{M\,outlets} \dot{m}_{out,k}\left[h_k + \frac{1}{2}v_k^2 + g(z_k - z_{ref})\right]$$
$$- \sum_{j=1}^{N\,inlets} \dot{m}_{in,j}\left[h_j + \frac{1}{2}v_j^2 + g(z_j - z_{ref})\right],$$

simplifies as follows:

$$0 - 0 = \dot{m}_P(h_P + 0 + 0) + \dot{m}_{H_2O,out}(h_{H_2O,out} + 0 + 0) -$$
$$\dot{m}_A(h_A + 0 + 0) - \dot{m}_F(h_F + 0 + 0) - \dot{m}_{H_2O,in}(h_{H_2O,in} + 0 + 0),$$

or

$$\dot{m}_A h_A + \dot{m}_F h_F - \dot{m}_P h_P = \dot{m}_{H_2O}(h_{H_2O,out} - h_{H_2O,in}).$$

The unknown water flow rate is then

$$\dot{m}_{H_2O} = \frac{\dot{m}_A h_A + \dot{m}_F h_F - \dot{m}_P h_P}{h_{H_2O,out} - h_{H_2O,in}}.$$

With the big picture complete, we now focus on the details. To find $\dot{m}_A$, we combine the definitions of air–fuel ratio (Eq. 3.60) and percent stoichiometric air (Eq. 3.61):

$$\dot{m}_A = \dot{m}_F (A/F)_{stoic} \cdot \frac{\%\ stoichiometric\ air}{100\%}.$$

For the fuel composition expressed as C_xH_y, we find the stoichiometric air–fuel ratio using Eqs. 3.58–3.60:

$$(A/F)_{stoic} = \left(\frac{\dot{m}_A}{\dot{m}_F}\right)_{stoic} = \frac{4.76(x + y/4)\mathcal{M}_A}{\mathcal{M}_F}.$$

For methane (CH_4),

$$(A/F)_{stoic} = \frac{4.76(1 + 1)28.85}{16.04} = 17.1.$$

The air flow rate is then

$$\dot{m}_A = 3(17.1)\frac{105}{100} = 53.9$$

$$[=]\ kg/s(kg/kg) = kg/s,$$

and the products flow rate is

$$\dot{m}_P = 53.9 + 3\ kg/s = 56.9\ kg/s.$$

Next we find the five enthalpies involved in our energy conservation expression. The water and steam enthalpies are found using the NIST database:

$$h_{H_2O,in} = h_{H_2O}\ (400\ K,\ 12\ MPa) = 541.05\ kJ/kg,$$

$$h_{H_2O,out} = h_{H_2O}\ (900\ K,\ 12\ MPa) = 3676.2\ kJ/kg.$$

Because we are dealing with a combustion process, we must employ standardized enthalpies for the fuel, air, and products. The enthalpy of the air is expressed

$$\bar{h}_A = X_{O_2}\bar{h}_{O_2} + X_{N_2}\bar{h}_{N_2},$$

where the standardized enthalpy of each species is the sum of the enthalpy of formation and the sensible enthalpy change (Eq. 2.70):

$$\bar{h}_i(T) = \bar{h}_{f,i}^\circ + \Delta\bar{h}_{s,i}(T).$$

For O_2 and N_2 at 298 K and 1 atm, both contributions to the standardized enthalpy are zero; that,

$$\bar{h}_A(298\ K) = 0.21(0 + 0) + 0.79(0 + 0) = 0.$$

The mass-specific enthalpy is also zero:

$$h_A = \bar{h}_A/\mathcal{M}_A = 0/28.85 = 0.$$

The enthalpy of the entering fuel, found using the enthalpy of formation of methane from Table H.1, is

$$\bar{h}_F\,(298\text{ K}) = \bar{h}_{CH_4}\,(298\text{ K})$$

$$= \bar{h}^\circ_{f,CH_4} + \Delta h_{s,CH_4}$$

$$= -74{,}831 + 0\text{ kJ/kmol,}$$

and (Eq. 2.4)

$$h_F = \bar{h}_F/\mathcal{M}_F = -74{,}831/16.04 = -4665.3$$

$$[=]\,\frac{\text{kJ/kmol}}{\text{kg/kmol}} = \text{kJ/kg.}$$

We now need to determine the composition of the combustion products stream so that we can apply Eq. 2.71f to determine the products enthalpy:

$$\bar{h}_P = \sum X_i \bar{h}_i.$$

See Example 3.15 in Chapter 3 to review stoichiometry concepts.

The reaction of CH_4 with 105% stoichiometric air can be written (Eq. 3.58)

$$CH_4 + 1.05(2)(O_2 + 3.76\,N_2) \rightarrow bCO_2 + cH_2O + dO_2 + eN_2.$$

The unknown coefficients are found from atom balances as follows:

C: $1 = b$ or $b = 1$,

H: $4 = 2c$ or $c = 2$,

O: $1.05\,(2)\,2 = 2b + c + 2d$ or $d = (4.2 - 2b - c)/2 = 0.1$, and

N: $1.05\,(2)\,3.76\,(2) = 2e$ or $e = 7.896$.

The total number of moles of products (per mole of CH_4 burned) is

$$N_P = b + c + d + e$$

$$= 1 + 2 + 0.1 + 7.896 = 10.996$$

and the mole fraction of each constituent is

$$X_{CO_2} = N_{CO_2}/N_P = 1/10.996 = 0.0909,$$

$$X_{H_2O} = N_{H_2O}/N_P = 2/10.996 = 0.1819,$$

$$X_{O_2} = N_{O_2}/N_P = 0.1/10.996 = 0.0091,$$

and

$$X_{N_2} = N_{N_2}/N_P = 7.896/10.996 = 0.7181.$$

For each constituent, we calculate its molar-specific standardized enthalpy using data from Appendix B at the stack temperature, 500 K:

$$\bar{h}_{CO_2}\,(500\text{ K}) = -393{,}546 + 8301\text{ kJ/kmol} = -385{,}245\text{ kJ/kmol,}$$

$$\bar{h}_{H_2O}\,(500\text{ K}) = -241{,}845 + 6947\text{ kJ/kmol} = -234{,}898\text{ kJ/kmol,}$$

$$\bar{h}_{O_2}\,(500\text{ K}) = 0 + 6097\text{ kJ/kmol} = 6097\text{ kJ/kmol,}$$

$$\bar{h}_{N_2}\,(500\text{ K}) = 0 + 5920\text{ kJ/kmol} = 5920\text{ kJ/kmol.}$$

Using Eq. 2.71f, we find the mixture molar-specific standardized enthalpy:

$$\bar{h}_P = 0.0909(-385{,}245) + 0.1819(-234{,}898) + 0.0091(6097)$$

$$+ 0.7181(5920)\text{ kJ/kmol}$$

$$= -73{,}440\text{ kJ/kmol.}$$

Converting this to a mass-specific basis for our energy balance requires the product mixture molecular weight, which is calculated from Eq. 2.60a:

$$\mathcal{M}_P = \sum X_i \mathcal{M}_i$$
$$= 0.0909(44.011) + 0.1819(18.016)$$
$$+ 0.0091(31.999) + 0.7181(28.013) \text{ kg/kmol}$$
$$= 27.685 \text{ kg/kmol}.$$

Thus,

$$h_P = \bar{h}_P/\mathcal{M}_P$$
$$= -73,440/27.685 = -2652.7$$
$$[=] \frac{\text{kJ/kmol}}{\text{kg/kmol}} = \text{kJ/kg}.$$

Before proceeding, we summarize our intermediate results:

	$\dot{m}$ (kg/s)	h (kJ/kg)
Air	53.9	0
Fuel	3.0	−4665.3
Products	56.9	−2652.7
Water	?	541.05
Steam	?	3676.2

With this information, we can accomplish our original objective of determining the boiler water flow rate. Using the result previously derived from energy conservation, we calculate

$$\dot{m}_{H_2O} = \frac{\dot{m}_A h_A + \dot{m}_F h_F - \dot{m}_P h_P}{h_{H_2O,\,out} - h_{H_2O,\,in}}$$
$$= \frac{59.3(0) + 3.0(-4665.3) - 56.9(-2652.7)}{3676.2 - 541.05} \text{ kg/s}$$
$$= 43.68 \text{ kg/s}.$$

Comments   This example integrates many concepts and topics, among these are element, mass, and energy conservation for control volumes; stoichiometry; properties of ideal-gas mixtures; and standardized enthalpies. Application of these principles provides a quantitative and realistic understanding of how a boiler works.

This example also affords an opportunity to evaluate the cost of raising steam. Using the U.S. national average residential price of natural gas for 2001 of $9.63 per thousand cubic feet, we can estimate the price per kJ of steam energy, using the following formula:

$$\frac{\$}{E_{steam}} = \frac{\dot{E}_F}{\dot{E}_{steam}} \frac{\$}{E_F} = \frac{\dot{m}_F \text{HHV}}{\dot{m}_{steam} h_{l-v}} \frac{\$}{E_F},$$

where HHV is the higher heating value of the fuel and

$$\frac{\$}{E_F} = \frac{\$}{V_F} \frac{1}{\rho_F} \frac{1}{\text{HHV}}.$$

Thus,

$$\frac{\$}{E_{steam}} = \frac{\$}{V_F} \frac{\dot{m}_F}{\dot{m}_{steam} h_{l-v} \rho_F}.$$

Using the numbers from our example and converting to SI units yields

$$\frac{\$}{E_{\text{steam}}} = \frac{\$9.63}{1000\,\text{ft}^3}\left[\frac{1\,\text{ft}^3}{(0.3048)^3\,\text{m}^3}\right]\frac{3\,\text{kg/s}}{43.68\,\text{kg/s}\,(3135.15\,\text{kJ/kg})\,0.6559\,\text{kg/m}^3}$$

$$= 1.14 \times 10^{-5}\,\$/\text{kJ}_{\text{steam}}.$$

We can also estimate the operating cost associated with supplying fuel, which is

$$\$/\text{time} = \frac{\$}{\mathbb{V}_F}\frac{\dot{m}_F}{\rho_F}$$

$$= \frac{\$9.63}{1000\,\text{ft}^3}\left[\frac{1\,\text{ft}^3}{(0.3048)^3\,\text{m}^3}\right]\frac{3\,\text{kg/s}}{0.6559\,\text{kg/m}^3}\left[\frac{3600\,\text{s}}{\text{hr}}\right]$$

$$= \$5600/\text{hr}.$$

Clearly this is an important cost in the operation of this boiler.

Self Test 12.5 ✓ Using the same temperatures, redo Example 12.7 for stoichiometric combustion of methane and air.

(Answer: $\dot{m} = 43.84\,\text{kg/s}$)

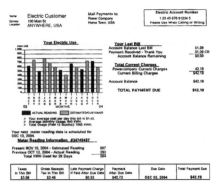

Costs of fuels and electricity are important to both energy producers and consumers

In defining HHV, the H₂O in the combustion products is assumed to be in the liquid state.

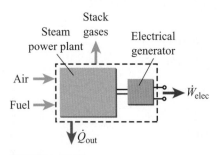

FIGURE 12.12

Control volume for evaluation of overall power-plant efficiency.

12.1e Overall Energy Utilization

Figure 12.12 shows a control volume containing the entire power plant. Energy enters this control volume with the fuel ($\dot{m}_F h_F$) and the air ($\dot{m}_A h_A$); it exits with the stack gases ($\dot{m}_P h_P$), the heat rejected in the condenser(s), and any residual heat losses combined as $\dot{Q}_{\text{out}}$, and the useful electrical power out, $\dot{W}_{\text{elec}}$. Using the general idea that an efficiency is the ratio of a desired energy output to an input energy that costs, we define an overall power-plant efficiency as

$$\eta_{\text{OA}} \equiv \frac{\text{useful electrical power produced}}{\text{supplied fuel energy rate}}$$

$$= \frac{\dot{W}_{\text{elec}}}{\dot{m}_F\,\text{HHV}}, \tag{12.12a}$$

where HHV is the higher heating value of the fuel. Recall from Chapter 2 that the HHV is the energy removed from the products of adiabatic combustion to cool them back to the initial temperature of the reactants, usually assumed to be 298 K. Consider, for example, the control volume shown in Fig. 12.12. If the stack gases exited at the same temperature as the entering fuel and air, all of the chemical energy released during combustion would go into $\dot{W}_{\text{elec}}$ and $\dot{Q}_{\text{out}}$; there would be no loss of energy associated with the stack gases. In reality, a stack loss always exists because of the limitations of heat-exchange equipment. Another way of looking at the overall efficiency of a steam power plant is to consider that the overall efficiency is the product of the efficiency of the boiler in transferring energy from the combustion products to the working fluid, η_{boiler}; the thermal efficiency of the closed steam cycle, $\eta_{\text{th,Rankine}}$; and the efficiency of the electrical generator, η_{gen}; that is,

$$\eta_{\text{OA}} = \eta_{\text{boiler}}\eta_{\text{th,Rankine}}\eta_{\text{gen}}. \tag{12.12b}$$

Typical values of boiler efficiencies range from 85% to 90%, whereas electrical generator efficiencies for large power plants range from 97% to 98%. Thermal efficiencies for Rankine cycles with reheat and regeneration are of the order of 30% to 40% in many power plants. Advanced cycles yield higher efficiencies.

Example 12.8

Determine the efficiency of the boiler operating as discussed in Example 12.7.

Solution

Known $\dot{m}_A, \dot{m}_F, \dot{m}_P, h_A, h_F, h_P$

Find η_{boiler}

Sketch

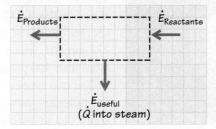

$\dot{E}_{\text{Products}}$ $\dot{E}_{\text{Reactants}}$

$\dot{E}_{\text{useful}}$
($\dot{Q}$ into steam)

Assumptions

Use the same assumptions as in Example 12.7.

Analysis   We first define the boiler efficiency to be the ratio of the energy delivered to the water to the chemical energy released by the fuel, that is,

$$\eta_{\text{boiler}} = \frac{\dot{Q}_{\text{steam}}}{\dot{m}_F \, \text{HHV}},$$

where HHV is the higher heating value of the fuel (CH_4). From Example 12.7,

$$\dot{Q}_{\text{steam}} = \dot{m}_{H_2O}(h_{H_2O,\text{out}} - h_{H_2O,\text{in}})$$
$$= \dot{m}_A h_A + \dot{m}_F h_F - \dot{m}_P h_P$$
$$= 53.9(0) + 3.0(-4665.3) - 56.9(-2652.7)$$
$$= 136{,}943$$
$$[=] (\text{kg/s})(\text{kJ/kg}) = \text{kJ/s or kW.}$$

The maximum possible thermal energy that can be extracted from the combustion process is that associated with cooling the products to the reference temperature, 298 K, and condensing the water vapor in the products, so

$$\dot{Q}_{\text{max possible}} = \dot{m}_F \, \text{HHV}$$
$$= 3.0(55{,}528) = 166{,}584$$
$$[=] (\text{kg/s})(\text{kJ/kg}) = \text{kJ/s or kW,}$$

where the higher heating value for CH_4 is from Table H.1. Applying our definition of boiler efficiency results in

$$\eta_{\text{boiler}} = \frac{136,943 \text{ kW}}{166,584 \text{ kW}} = 0.822 \text{ or } 82.2\%.$$

Comment Approximately 17.8% (= 100% − 82.2%) of the fuel energy goes up the stack, serving no useful purpose. Note, however, that the boiler efficiency is much greater than the efficiencies associated with the steam cycle. What physical factors determine the boiler efficiency?

Self Test 12.6 ☑ **Determine the efficiency of the boiler operating as discussed in Self Test 12.5.**

(Answer: 82.5%)

12.2 JET ENGINES

FIGURE 12.13
The PW4084 turbofan nominally develops 385.9 kN (86,760 lb$_f$) of thrust at takeoff. The diameter of the fan is 2.84 m (112 in). Photograph courtesy of Pratt and Whitney.

Four 363-kN-thrust (81,500 lb$_f$) engines power the Airbus A380 aircraft. The 555-seat A380 is the world's largest airliner.

Modern jet engines are amazing machines: The PW4084 turbofan engine shown in Fig. 12.13 weighs 66.7 kN (15,000 lb$_f$) and produces more than 385 kN (86,760 lb$_f$) of thrust. Two of these, or similar, engines power the Boeing 777 aircraft, which weighs more than 2446 kN (550,000 lb$_f$) and can carry more than 300 passengers.

The two types of jet engines used in aircraft applications, the turbojet and the turbofan, were introduced in Chapter 1. In the following sections, we apply our knowledge of the basic conservation principles and analyze the thermodynamic "cycle" associated with the turbojet engine to gain a firm understanding of its operation.

12.2a Basic Operation of a Turbojet Engine

A schematic of a generic, single-shaft turbojet engine is shown in Fig. 12.14; an actual engine and its application in a military aircraft are illustrated in Fig. 12.15. The propulsive thrust of the turbojet engine results from the production of a high-velocity exit jet. This is accomplished by the following sequence of processes:

1–2: Air enters the engine control volume at flight speed at station 1 and is slowed to a lower velocity at station 2. This velocity decrease from station 1 to station 2 results from the flow pattern outside the engine (1–1a) and a physical diffuser (1a–2). A pressure rise (ram effect) results from this slowing of the air.

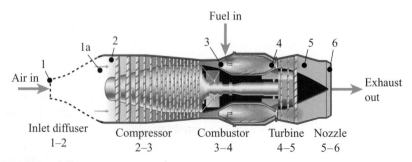

FIGURE 12.14
Schematic diagram of a turbojet engine. The gases flow through the annular space between the turbomachinery components and the engine housing.

FIGURE 12.15

Two Pratt and Whitney J52-P-408A turbojet engines power the EA-6B Prowler military aircraft. A single engine produces an intermediate thrust of 49.8 kN (11,200 lb_f), weighs 10.3 kN (2318 lb_f), has a length of 3 m (119 in), and has a maximum diameter of 0.79 m (31.1 in). **Photographs courtesy of Pratt and Whitney.**

Example 5.15 in Chapter 5 considers this diffuser section. ▶

Jet engine compressor shroud

This F100AB jet engine utilizes a variable-geometry, converging-diverging exhaust nozzle. Photograph courtesy of U. S. Air Force.

2–3: A multistage compressor compresses the air to a high pressure before entering the combustor. Typical turbojet compressors operate with pressure ratios of 10–15 or greater (see Fig. 12.16). Numerous compressor stages are required to achieve the overall pressure ratio since each stage typically operates with a pressure ratio of only 1.1 or 1.2 [3]. A diffuser section (see Fig. 12.14) is used in the transition section from the last blade row to the combustor entrance.

3–4: Fuel is injected and burned in the combustor. A portion of the high-pressure air enters the combustor, while the remaining air cools the combustor liner and dilutes the combustion products to a sufficiently low temperature to allow proper operation of the turbine. The high-temperature strength of the turbine blades is the major factor limiting the turbine inlet temperature.

4–5: A multistage gas turbine expands the high-temperature, high-pressure gas. The shaft power from the turbine drives the compressor and any auxiliary equipment. Turbojet engines typically employ one to four turbine stages.

5–6: The partially expanded hot gases complete their expansion to ambient pressure (state 6) by passing through a nozzle. For converging nozzles, the exit velocity is limited by choked conditions at the exit. High-speed aircraft employ converging–diverging nozzles to produce supersonic exit velocities.

Before investigating the details of the turbojet "cycle," we perform simple, integral control volume analyses to help understand the basic operation of this engine.

12.2b Integral Control Volume Analysis of a Turbojet

Consider a control volume that cuts across the inlet plane (station 1 in Fig. 12.14), follows the external surface of the engine housing, and cuts across the exhaust plane (station 6 in Fig. 12.14). For simplicity, we model the engine as a cylindrical tube. The control surface also cuts through the fuel feedline and through the engine mounts. Figure 12.17 shows the flows of mass, energy, and linear momentum, respectively, for this integral control volume.

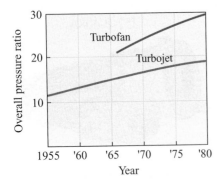

FIGURE 12.16
Overall pressure ratios increase as jet engine technology improves. **Reprinted from Ref. [4] with permission.**

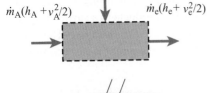

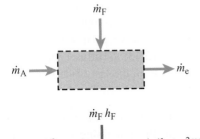

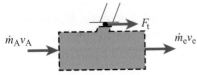

FIGURE 12.17
Integral control volumes for turbojet engine for mass conservation (top), energy conservation (middle), and axial momentum conservation (bottom).

Assumptions

Before applying the basic conservation principles, we list the assumptions used to simplify our analysis:

i. Steady-state and steady-flow conditions prevail.

ii. There are no heat interactions between the engine and the surroundings (i.e., $\dot{Q}_{cv} = 0$).

iii. Auxiliary power systems are neglected; thus, all of the turbine power is used by the compressor and $\dot{W}_{cv} = 0$.

iv. Potential energy changes are negligible (i.e., $z_1 = z_6$).

v. The inlet and exit velocity profiles are uniform (i.e., $\alpha_1 = \alpha_6 = 1$ and $\beta_1 = \beta_6 = 1$).

vi. The kinetic and potential energies of the fuel stream are negligible.

vii. All of the axial forces acting on the control volume are lumped into a single force F_t, which we designate as the total thrust.

viii. The pressure is uniform over the control surface.

Mass Conservation

Air and fuel enter the control volume as separate streams, whereas a single stream of diluted combustion products exits (Fig. 12.17 top). Conservation of mass is thus expressed by Eq. 3.18b and expanded as follows:

$$\sum_{j=1}^{N \text{ inlets}} \dot{m}_{\text{in},j} = \sum_{k=1}^{M \text{ outlets}} \dot{m}_{\text{out},k},$$

or

$$\dot{m}_A + \dot{m}_F = \dot{m}_e, \tag{12.13}$$

where $\dot{m}_A$, $\dot{m}_F$, *and* $\dot{m}_e$ are the mass flow rates of the air, fuel, and exhaust products, respectively. Rearranging Eq. 12.13 to introduce the fuel–air ratio,

$$F/A \equiv \dot{m}_F/\dot{m}_A, \tag{12.14}$$

we get

$$\dot{m}_A(1 + F/A) = \dot{m}_e. \tag{12.15}$$

Since overall fuel–air ratios of turbojet engines are of the order of 1:100, the approximation that $\dot{m}_e \approx \dot{m}_A$ is sometimes useful, depending on the issue.

Energy Conservation

The starting point for energy conservation is Eq. 5.65:

$$\dot{Q}_{cv,\text{net in}} - \dot{W}_{cv,\text{net out}} = \sum_{k=1}^{M \text{ outlets}} \dot{m}_{\text{out},k}\left[h_k + \tfrac{1}{2}\alpha_k v_{\text{avg},k}^2 + g(z_k - z_{\text{ref}})\right]$$

$$- \sum_{j=1}^{N \text{ inlets}} \dot{m}_{\text{in},j}\left[h_j + \tfrac{1}{2}\alpha_j v_{\text{avg},j}^2 + g(z_j - z_{\text{ref}})\right].$$

We simplify this statement by applying assumptions ii–vi and expanding the summations, which gives

$$0 - 0 = \dot{m}_e\left(h_e + \frac{1}{2}(1)v_e^2 + 0\right) - \dot{m}_F(h_F + 0 + 0) - \dot{m}_A\left(h_A + \frac{1}{2}(1)v_A^2 + 0\right).$$

Upon rearrangement and substitution of Eq. 12.15 for the exhaust mass flow rate, this equation becomes

$$\dot{m}_A(1 + F/A)\left(h_e + \frac{1}{2}v_e^2\right) = \dot{m}_F h_F + \dot{m}_A\left(h_A + \frac{1}{2}v_A^2\right). \qquad (12.16)$$

Anticipating that the magnitude of exhaust jet velocity is key to the performance of the engine, we isolate the kinetic energy terms in Eq. 12.16 after dividing through by $\dot{m}_A$:

$$\underbrace{(1 + F/A)\frac{1}{2}v_e^2 - \frac{1}{2}v_A^2}_{\substack{\text{Kinetic energy change} \\ \text{per mass of air}}} = \underbrace{[(F/A)h_F + h_A]}_{\substack{\text{Enthalpy of} \\ \text{reactants per mass} \\ \text{of air}}} - \underbrace{[1 + (F/A)]h_e.}_{\substack{\text{Enthalpy of} \\ \text{products per} \\ \text{mass of air}}} \qquad (12.17)$$

Assuming that the air, fuel, and products all behave as ideal gases, we interpret Eq. 12.17 using the enthalpy–temperature diagram shown in Fig. 12.18. Fixing the inlet temperature of the air (and fuel) at T_{inlet} establishes the enthalpy value of the reactants (per unit mass of air), which is shown as a point on the reactants line in Fig. 12.18. Because the chemical energy contained in the products is less than in the reactants, the products enthalpy (per unit mass of air) lies below that of the reactants. As we see in Fig. 12.18, the value of the exhaust temperature thus controls the kinetic energy change: The higher the exhaust temperature, the smaller the kinetic energy change.

FIGURE 12.18

An h–T diagram for a turbojet engine with air and fuel entering at T_{inlet} and products exiting at $T_{exhaust}$.

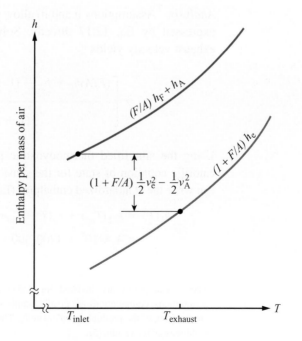

Example 12.9

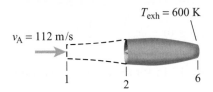

$T_{\text{exh}} = 600 \text{ K}$

$v_A = 112 \text{ m/s}$

1   2   6

Determine the exhaust jet velocity for a turbojet engine operating with a fuel–air ratio of 1:100. The air enters the engine at 112 m/s (250 miles/hr) and 300 K (station 1 in Fig. 12.14) and the fuel enters at 300 K with negligible velocity. The temperature at the exhaust plane is 600 K. Assume the following simplified thermodynamic properties[4] for the air, fuel, and products:

 i. The specific heats of the fuel, air, and products are constants and equal (i.e., $c_{p,\text{F}} = c_{p,\text{A}} = c_{p,\text{e}} = 1200 \text{ J/kg} \cdot \text{K}$).
 ii. The enthalpy of formation of the air and of the products is zero; the enthalpy of formation of the fuel is $4 \times 10^7 \text{ J/kg}$. The reference state temperature is 300 K.

Solution

Known $(F/A), v_A, T_A, T_F, T_e$

Find v_e

Sketch

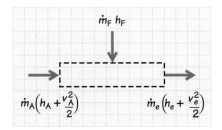

Assumptions

 i. Simplified thermodynamic properties as given
 ii. Ideal-gas behavior
 iii. $(ke)_\text{F}$ is negligible
 iv. Adiabatic process $(\dot{Q}_{\text{cv}} = 0)$

Analysis   Assumptions ii and iii allow us to use conservation of energy as expressed by Eq. 12.17 directly. Solving Eq. 12.17 for the unknown exhaust velocity yields

$$v_e = \left[\frac{(F/A)h_\text{F} + h_\text{A} - [1 + (F/A)]h_\text{e} + \frac{1}{2}v_\text{A}^2}{\frac{1}{2}(1 + F/A)} \right]^{1/2}.$$

Using the simplified thermodynamic properties given and the ideal-gas calorific equation of state for the sensible enthalpy change (Eq. 2.33), we can write the standardized enthalpies (Eq. 2.70) for the three constituents as

$$h_\text{F}(T) = h_{f,\text{F}}(T_{\text{ref}}) + c_p(T - T_{\text{ref}})$$

$$= 4 \times 10^7 + 1200(300 - 300) \text{ J/kg} = 4 \times 10^7 \text{ J/kg},$$

---

[4] These assumptions are invoked for pedagogical, not engineering, reasons. With these assumptions, computations are greatly simplified by avoiding the need to specify the detailed composition of the products to calculate h_e. The simplified properties, however, are chosen to yield reasonable results [5].

$$h_A(T) = h_{f,A}(T_{ref}) + c_p(T - T_{ref})$$
$$= 0 + 1200(300 - 300) \text{ J/kg} = 0,$$

and

$$h_e(T) = h_{f,e}(T_{ref}) + c_p(T - T_{ref})$$
$$= 0 + 1200(600 - 300) \text{ J/kg} = 360{,}000 \text{ J/kg}.$$

Substituting these values into our expression for v_e yields

$$v_e = \left[\frac{0.01(4 \times 10^7) + 0 - 1.01(360{,}000) + 0.5(112)^2}{0.5(1.01)} \right]^{1/2}$$

$$= 291$$

$$[=] \left[\frac{\text{J}}{\text{kg}} \left[\frac{1\ \text{N} \cdot \text{m}}{\text{J}} \right] \left[\frac{1\ \text{kg} \cdot \text{m/s}^2}{\text{N}} \right] \right]^{1/2} = \text{m/s}.$$

Comments   As expected, we find the exhaust velocity to be much greater than the air inlet velocity. Note how the simplified thermodynamics trivialized the computations of h_A, h_F, and h_e, while retaining the concept of standardized enthalpies necessary to deal with reacting flows.

 Self Test 12.7

(a) Determine the fuel–air ratio for stoichiometric combustion of C_7H_{14}. (b) Determine the percent stoichiometric air associated with a fuel–air ratio of 0.01.

(Answer: 0.068, 680%)

Momentum Conservation

Consider the simple control volume shown at the bottom of Fig. 12.17. (In Appendix 12A, we present a more sophisticated treatment.) The arrows sketched here represent the pertinent axial momentum flows and the axial force. With the assumption of uniform pressure, the net pressure force is zero and hence is not shown. Cutting through the engine mount exposes the thrust force F_t. The reaction to this thrust force on the aircraft side of the motor mount drives the aircraft forward (to the left).

For a control volume having a single inlet and a single outlet, both with uniform velocities, Eq. 6.53 expresses conservation of momentum:

$$\dot{m}V_{in} - \dot{m}V_{out} + \sum F_{cv} = 0.$$

The axial-direction scalar component of this equation results from the substitution of the axial momentum flows and thrust force from Fig. 12.17 as follows:

$$\dot{m}_A v_A - \dot{m}_e v_e + F_t = 0. \tag{12.18}$$

Applying mass conservation (Eq. 12.15) and solving for the thrust force yields

$$F_t = \dot{m}_A[(1 + F/A)v_e - v_A], \tag{12.19a}$$

which can be approximated by the relationship

$$F_t \cong \dot{m}_A(v_e - v_A). \tag{12.19b}$$

From Eq. 12.19a or 12.19b, we see that the thrust is directly proportional to the product of the air mass flow rate and the exhaust jet velocity. If we were to substitute our final energy conservation expression, Eq. 12.17, into Eq. 12.19, we could then identify the combustion process as the ultimate source of the thrust.

Self Test
12.8

☑ **For an air mass flow rate of 50 kg/s, determine the thrust force of the engine in Example 12.9.**

(Answer: 9.1 kN)

12.2c Turbojet Cycle Analysis

Analyzing the turbojet "cycle" allows us to determine the various thermodynamic states needed to calculate the thrust from Eq. 12.19. An ideal cycle is shown in Fig. 12.19. We also determine various measures of propulsive efficiency from this cycle analysis.

Given Conditions

We begin by defining parameters that are typically treated as known quantities:

- The inlet (atmospheric) pressure P_{amb} ($= P_1$) is given. For an aircraft at altitude, this pressure is significantly less than the standard sea-level value. The inlet (atmospheric) temperature T_1 is also known.
- The inlet velocity v_1, which is typically the flight speed of the aircraft, is given.
- The compressor pressure ratio P_3/P_2 is known.
- The turbine inlet temperature T_4 is given. This temperature is usually determined by metallurgical considerations and must be below values that lead to turbine blade failure.[5]
- The pressure at the exhaust nozzle outlet matches the ambient pressure (i.e., $P_6 = P_{amb}$).
- The isentropic efficiencies of each component are given.

Assumptions

To the given conditions, we add the following assumptions:

i. The kinetic energies of the flow can be neglected at all stations *except* at the inlet (station 1) and the outlet (station 6).
ii. Potential energy changes are negligible throughout the engine.
iii. Accessory loads to the engine are negligible; thus, the power produced by the turbine equals the power input to the compressor (i.e., $\dot{W}_{t,out} = \dot{W}_{c,in}$).

Approach

With the given conditions and assumptions, we proceed to perform a state-to-state (1–2–3–etc.), device-by-device (diffuser, compressor, combustor, etc.) analysis, ultimately concluding at state 6 at the jet exit. Defining state 6, in particular, determining the exit velocity v_6, is one of our major objectives.

---

[5] An alternative to specifying the turbine inlet temperature is to specify the fuel–air ratio supplied to the engine, as this quantity allows T_4 to be determined.

FIGURE 12.19

T–s diagram for ideal turbojet cycle. See Fig. 12.14 for definition of the diffuser (1–2).

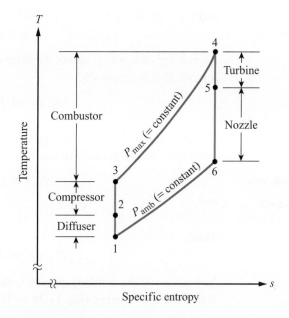

Knowing the value of this velocity allows us to determine the thrust produced by the engine (see Eq. 12.19), that is, the useful output for which the engine was designed. Complicating factors in performing this step-by-step analysis are the addition of fuel and the combustion process. In particular, the presence of combustion means we have to deal with mixtures of combustion products (CO, H_2O, CO, O_2, etc.) beginning somewhere between states 3 and 4, and continuing from state 4 through state 6. Before considering the combustion process in our analysis, we first explore what is known as the **air-standard** turbojet **cycle**. This model provides great simplification, yet it retains the essential features of the operation of a real turbojet engine:

> Various air-standard cycles can be defined for idealized power and propulsion systems (e.g., the air-standard Otto cycle and the air-standard Diesel cycle).

- The working fluid is air.
- The combustion process is replaced by a constant-pressure, heat-addition process, where the heat-addition rate is equivalent to the product of the fuel flow rate and the fuel heating value (i.e., $\dot{Q}_{added} = \dot{m}_F HV_F$).
- The exhaust and intake processes are replaced by closing the cycle with a constant-pressure heat-rejection process.

We now work through an ideal cycle (see Fig. 12.19) in which all of the isentropic efficiencies are assumed to be unity. The introduction of nonunity efficiencies into the analysis is left as an exercise for the reader.

> See Table 11.1 for a first-law analysis of each component.

Diffuser (1–2) The outlet state 2 is determined by application of energy conservation (Eq. 11.5 or Table 11.1) and the recognition that the process 1–2 is isentropic; thus,

$$h_2 = h_1 + \frac{1}{2}v_1^2 \qquad (12.20a)$$

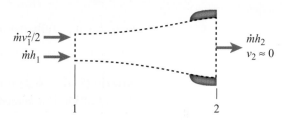

and

$$s_2 = s_1. \tag{12.20b}$$

Knowledge of h_2 and s_2 allows the determination of all other state-2 properties. Specifically, we desire P_2:

$$(h_2, s_2) \Rightarrow P_2, T_2, \text{etc.} \tag{12.20c}$$

Compressor (2–3) Since the pressure ratio P_3/P_2 is given and the process is isentropic, state 3 is easily defined:

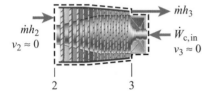

$$P_3 = P_2(P_3/P_2) \tag{12.21a}$$

and

$$s_3 = s_2. \tag{12.21b}$$

Thus,

$$(P_3, s_3) \Rightarrow h_3, T_3, \text{etc.} \tag{12.21c}$$

Now knowing the enthalpy at state 3, the compressor work can be found from conservation of energy (Eq. 11.36 or Table 11.1):

$$\dot{W}_{c,in} = \dot{m}(h_3 - h_2), \tag{12.21d}$$

or

$$\frac{\dot{W}_{c,in}}{\dot{m}} = h_3 - h_2. \tag{12.21e}$$

We will use this information later to help define state 5.

Combustor (3–4) With the assumptions that the combustor pressure is constant ($P_4 = P_3$) and that the turbine inlet temperature T_4 is given, the combustor, or air-standard heat-addition process, analysis is trivial since state 4 is defined by P_4 and T_4. Thus,

$$(P_4, T_4) \Rightarrow h_4, s_4, \text{etc.} \tag{12.22}$$

Turbine (4–5) The enthalpy at the turbine exit can be determined by equating the power produced by the turbine (Eq. 11.50 or Table 11.1) with the previously determined power input to the compressor, that is,

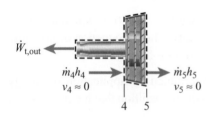

$$\dot{W}_{t,out} = \dot{W}_{c,in}, \tag{12.23a}$$

or

$$\dot{m}(h_4 - h_5) = \dot{m}(h_3 - h_2). \tag{12.23b}$$

Solving for h_5 yields

$$h_5 = h_4 + h_2 - h_3, \tag{12.23c}$$

where h_2, h_3, and h_4 are all known from our previous definitions of states 2, 3, and 4. Assuming an isentropic efficiency of unity, we also know that

$$s_5 = s_4. \tag{12.23d}$$

State 5 is thus fully defined, that is,

$$(h_5, s_5) \Rightarrow P_5, T_5, \text{etc.} \tag{12.23e}$$

Nozzle (5–6) State 6 is easy to determine since the pressure is given and the expansion is assumed to be isentropic, so

$$P_6 = P_{amb} \tag{12.24a}$$

and

$$s_6 = s_5. \tag{12.24b}$$

Thus,

$$(P_6, s_2) \Rightarrow h_6, T_6, \text{etc.} \tag{12.24c}$$

Knowing the enthalpy at state 6, the exhaust velocity is calculated from energy conservation (Eq. 11.5 or Table 11.1) as follows:

$$h_5 + 0 = h_6 + \frac{1}{2}v_6^2, \tag{12.24d}$$

or

$$v_6 = \left[2(h_5 - h_6)\right]^{1/2}. \tag{12.24e}$$

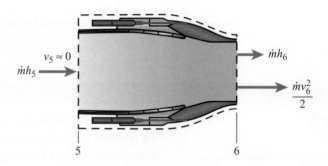

This completes the thermodynamic analysis of the cycle and allows the thrust to be computed from Eq. 12.19. Detailed implementation of this approach is presented later in Example 12.10.

12.2d Propulsive Efficiency

One measure of propulsive efficiency is the ratio of the power produced by the thrust, $\dot{W}_{\text{thrust}}$, to the rate at which energy is released in the combustor, that is,

$$\eta_{\text{prop}} \equiv \frac{\dot{W}_{\text{thrust}}}{\dot{m}(h_4 - h_3)}, \tag{12.25}$$

where the actual chemical energy release has been replaced with the air-standard-cycle heat addition, $\dot{m}(h_4 - h_3)$. The power produced by the thrust is the product of the thrust F_t and the aircraft flight speed v_{flight}:

$$\dot{W}_{\text{thrust}} \equiv F_t v_{\text{flight}}. \tag{12.26a}$$

For conditions where v_{flight} equals the inlet velocity v_1, the thrust power can be related through Eq. 12.19b to the inlet and exit velocities as follows:

$$\dot{W}_{\text{thrust}} = \dot{m}(v_6 - v_1)v_1. \tag{12.26b}$$

Substituting this expression back into the definition of propulsive efficiency yields

$$\eta_{\text{prop}} = \frac{v_1(v_6 - v_1)}{h_4 - h_3}. \tag{12.27}$$

Since all of the quantities on the right-hand side are known from our cycle analysis, the propulsive efficiency is readily calculated.

Performance checks are conducted in this jet engine test cell. Photograph courtesy of Mitsubishi Heavy Industries, Ltd.

12.2e Other Performance Measures

Two other commonly used performance measures relate engine flow rates to the production of thrust. The **specific thrust** is the ratio of the thrust produced to the mass flow rate of air through the engine:

$$\text{specific thrust} \equiv F_t / \dot{m}_A, \qquad (12.28)$$

with SI units of N/(kg/s) or m/s. Units typically used in industry are lb/lb/s or kg/kg/s. The **specific fuel consumption** is the ratio of the fuel mass flow rate to the thrust produced:

$$\text{specific fuel consumption } (SFC) \equiv \frac{\dot{m}_F}{F_t}, \qquad (12.29)$$

with SI units of (kg/s)/N or s/m. Units typically used in industry are lb/hr/lb or kg/hr/kg.

Example 12.10

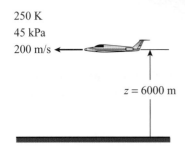

250 K
45 kPa
200 m/s

$z = 6000$ m

Consider a turbojet-powered aircraft flying 200 m/s at 6000-m altitude where the ambient temperature and pressure are 250 K and 45 kPa, respectively. The pressure ratio of the compressor is 6 and the turbine inlet temperature is 970 K. At these conditions, the fuel flow rate is 0.68 kg/s and the overall air–fuel ratio is 75:1. Estimate the thrust developed by the engine. Also determine the specific trust, the specific fuel consumption, and the propulsive efficiency. Employ an air-standard-cycle analysis with the following constant-pressure specific heats c_p and specific-heat ratios γ for each process:

	c_p (kJ/kg·K)	γ
Diffuser, 1–2	1.04	1.40
Compressor, 2–3	1.06	1.40
Combustor, 3–4	1.10	1.37
Turbine, 4–5	1.145	1.35
Nozzle, 5–6	1.098	1.37

Solution

Known $P_1, T_1, v_1, T_4, \dot{m}_F, A/F$

Find F_t, specific thrust, SFC

Sketch   The process stations are defined in Fig. 12.14, and the various processes are shown on a *T–s* diagram in Fig. 12.20.

Assumptions

i. All processes are ideal (reversible). The diffuser, compressor, turbine, and nozzle all operate adiabatically.

ii. The working fluid is air (ideal gas) with constant specific heats given separately for each process. This use of quasi-constant specific heats approximates the effects of variable specific heats.

iii. The mass flow rate is identical at each station; that is, the effect of fuel addition on the total flow is small.

FIGURE 12.20
T–s diagram and thermodynamic state summary for Example 12.10. Bold entries to the table are known or assumed values.

State	1	2	3	4	5	6
P (kPa)	**45.0**	58.3	349.8	349.8	169.1	**45.0**
T (K)	**250**	269.2	449.2	**970**	803.4	561.9
v (m/s)	**200**	~0	~0	~0	~0	728

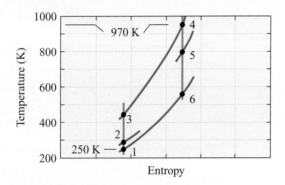

iv. The inlet and exhaust pressures are equal to the ambient pressure (i.e., $P_1 = P_6 = P_{amb}$).

v. All of the turbine output power is used to drive the compressor.

vi. The velocities at states 2, 3, 4, and 5 are negligible.

Analysis Since we assume reversible and adiabatic processes for both compression and expansion, the process sequences 1–2–3 and 4–5–6 are isentropic. With the further assumption of ideal-gas behavior, pressures and temperatures in these sequences are related by the isentropic property relationship (Eq. 2.41a)

$$\frac{P_a}{P_b} = \left(\frac{T_a}{T_b}\right)^{\frac{\gamma}{\gamma-1}}. \tag{A}$$

Furthermore, we can apply the simplified ideal-gas calorific equation of state (Eq. 2.33e)

$$\Delta h = c_p \Delta T \tag{B}$$

for any of the processes (1–2–3–4–5–6). These two relationships, together with appropriate statements of conservation of energy, are sufficient to define all of the states (1–6) and to calculate the exhaust velocity. Knowing this velocity, the thrust is calculated from momentum conservation (Eq. 12.19).

We begin by organizing the results of our calculations in a table. Known or assumed values are shown as bold entries to the table in Fig. 12.20. All other values need to be calculated.

The temperature at the diffuser outlet is determined by combining energy conservation (Eq. 12.20a) with our simplified calorific equation of state:

$$h_2 - h_1 = c_p(T_2 - T_1) = \frac{1}{2}v_1^2,$$

or

$$T_2 = \frac{v_1^2}{2c_p} + T_1$$

$$T_2 = \frac{(200)^2}{2(1040)} + 250 = 269.2$$

$$[=] \frac{(m/s)^2}{(J/kg \cdot K)}\left[\frac{1\,J}{N \cdot m}\right]\left[\frac{1\,N}{kg \cdot m/s^2}\right] = K.$$

The pressure at state 2, calculated from Eq. A, is

$$P_2 = P_1 \left(\frac{T_2}{T_1}\right)^{\frac{\gamma}{\gamma-1}}$$

$$= 45\left(\frac{269.2}{250}\right)^{\frac{1.4}{0.4}} \text{kPa} = 58.3 \text{ kPa}.$$

The compressor pressure ratio is given, so the pressure at the compressor outlet is

$$P_3 = P_2 \left(\frac{P_3}{P_2}\right)$$

$$= 58.3(6) \text{ kPa} = 349.8 \text{ kPa}.$$

The compressor outlet temperature is therefore

$$T_3 = T_2 \left(\frac{P_3}{P_2}\right)^{\frac{\gamma-1}{\gamma}}$$

$$= 269.2(6)^{\frac{0.4}{1.4}} \text{ K} = 449.2 \text{ K}.$$

From energy conservation, the compressor input power per unit mass flow can be determined. Combining this with Eq. B yields

$$\frac{\dot{W}_{c,in}}{\dot{m}} = h_3 - h_2 = c_p(T_3 - T_2)$$

$$= 1.06(449.2 - 269.2) = 190.8$$

$$[=] \frac{\text{kJ}}{\text{kg} \cdot \text{K}} (\text{K}) = \text{kJ/kg}.$$

The heat-addition (combustion) process occurs at constant pressure ($P_4 = P_3$), and the combustor outlet temperature (i.e., the turbine inlet temperature) is given ($T_4 = 970$ K). No calculation is required, and the P_4 and T_4 values are entered in the table (Fig. 12.20).

The turbine output power equals the compressor input power. Combining this knowledge with energy conservation allows us to calculate T_5 as follows:

$$\frac{\dot{W}_{c,in}}{\dot{m}} = \frac{\dot{W}_{t,out}}{\dot{m}} = h_4 - h_5 = c_p(T_4 - T_5),$$

or

$$T_5 = T_4 - \frac{\dot{W}_{c,in}/\dot{m}}{c_p}$$

$$= 970 - \frac{190.8}{1.145} = 803.4$$

$$[=] \frac{\text{kJ}}{(\text{kJ/kg} \cdot \text{K})} = \text{K}.$$

Using this turbine outlet temperature, we can calculate the outlet pressure using the isentropic process relationship (Eq. A) as follows:

$$P_5 = P_4 \left(\frac{T_5}{T_4}\right)^{\frac{\gamma}{1-\gamma}}$$

$$= 349.8\left(\frac{803.4}{970}\right)^{\frac{1.35}{0.35}} \text{kPa} = 169.1 \text{ kPa}.$$

The final process in the cycle is the expansion of the hot, high-pressure gases through the exhaust nozzle. For this isentropic expansion,

$$T_6 = T_5 \left(\frac{P_6}{P_5} \right)^{\frac{\gamma-1}{\gamma}},$$

$$= 803.4 \left(\frac{45}{169.1} \right)^{\frac{0.37}{1.37}} \text{ K} = 561.9 \text{ K},$$

where P_6 is the ambient pressure. The exhaust velocity is calculated from energy conservation (Eq. 12.24d):

$$h_5 + 0 = h_6 + \frac{1}{2} v_6^2,$$

or

$$v_6 = [2(h_5 - h_6)]^{1/2}.$$

Substituting Eq. B, we relate the velocity to the temperature difference as

$$v_6 = [2c_p(T_5 - T_6)]^{1/2}$$

$$= [2(1098)(803.4 - 561.9)]^{1/2} = 728$$

$$[=] \left(\frac{\text{J}}{\text{kg} \cdot \text{K}} \text{K} \left[\frac{\text{N} \cdot \text{m}}{1 \text{ J}} \right] \left[\frac{\text{kg} \cdot \text{m/s}^2}{1 \text{ N}} \right] \right)^{1/2} = \text{m/s}.$$

We can now calculate the thrust. Given the air–fuel ratio and the fuel flow rate, we determine the air mass flow rate (Eq. 12.14) as

$$\dot{m}_A = (A/F)\dot{m}_F,$$

$$= 75(0.68) \text{ kg/s} = 51 \text{ kg/s}.$$

Using the combined overall mass and momentum conservation relationship for the thrust (Eq. 12.19a), we calculate

$$F_t = \dot{m}_A[(1 + (F/A))v_6 - v_1],$$

$$= 51 \left[\left(1 + \frac{1}{75} \right) 728 - 200 \right] = 27,400$$

$$[=] \frac{\text{kg}}{\text{s}} \frac{\text{m}}{\text{s}} \left[\frac{1 \text{ N}}{\text{kg} \cdot \text{m/s}^2} \right] = \text{N}.$$

The specific thrust is (Eq. 12.28)

$$F_t/\dot{m}_A = \frac{27,400}{51} = 537 \text{ m/s},$$

and the specific fuel consumption (Eq. 12.29) is

$$SFC = \dot{m}_F/F_t = \frac{0.68}{27,400} \text{ s/m} = 2.48 \times 10^{-5} \text{ s/m}.$$

The reader should verify the units associated with these two calculations. Our final desired quantity is the propulsive efficiency, which we calculate directly from Eq. 12.27 as follows:

$$\eta_{prop} = \frac{v_1(v_6 - v_1)}{h_4 - h_3} = \frac{v_1(v_6 - v_1)}{c_p(T_4 - T_3)}$$

$$= \frac{200(728 - 200)}{1100(970 - 449.2)} = 0.184,$$

or expressed as a percentage,

$$\eta_{\text{prop}} = 18.4\%.$$

Comments In this example, constant specific heats are assumed to make the evaluation of thermodynamic properties nearly trivial so that attention can be focused on the cycle analysis proper. To capture some of the temperature dependence of the specific heats, different values were assigned to each process. An alternative to adopting these simplified properties is to use air tables (Appendix C).

Although we use SI units consistently throughout this book, it is useful to digress here to illustrate the use of non-SI units for specific fuel consumption. Using the awkward kilogram-force unit, which is numerically equal to a kilogram-mass in standard gravity, we convert the *SFC* to the industry-standard kg/hr/kg or lb/hr/lb units as follows:

$$F_{\text{t}} = 27{,}400 \text{ N} \frac{1 \text{ kg}_{\text{f}}}{9.807 \text{ N}} = 2794 \text{ kg}_{\text{f}}$$

and

$$\dot{m}_{\text{F}} = 0.68 \frac{\text{kg}}{\text{s}} \left[\frac{3600 \text{ s}}{\text{hr}} \right] = 2448 \text{ kg/hr};$$

therefore,

$$SFC = \frac{\dot{m}_{\text{F}}}{F_{\text{t}}} = \frac{2448}{2794} = 0.876 \text{ kg/hr/kg or lb/hr/lb}.$$

This value is consistent with *SFC* values tabulated for various engines in Ref. [6].

Note the relatively low value of 18.4% for the ideal propulsive efficiency for this particular example. Considering irreversibilities in each component would lower this value even further. We once again see how relatively inefficient some devices are in producing a useful effect from a heat input. For comparison, see Examples 12.2 and 12.3 for Rankine cycle efficiencies.

> **How would you modify this example to account for the mass flow rate through the turbine being greater than that through the compressor?**

12.2f Combustor Analysis

In our simplified analysis of the turbojet engine cycle, we replaced the actual combustion process with a constant-pressure heat-addition process. We now

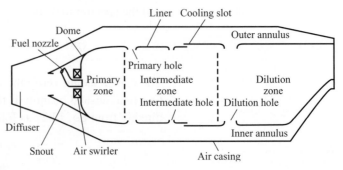

FIGURE 12.21
Schematic of annular jet-engine combustor showing various air passageways and combustion zones. **Reprinted from Ref. [7] by permission.**

FIGURE 12.22

Segment of CFM56-7 turbofan jet-engine combustor produced by CFM International (a joint company of Snecma, France, and General Electric, U.S.A.). Fuel nozzles (not shown) fit into large holes at the back (left). Note cooling air holes in louvers of combustor liner. The CFM56-7 engine is used in the Boeing 737 aircraft.

go beyond this simplification and explore how the combustion process can be treated more realistically, yet still simply.

A modern jet-engine combustor is a complex device. Multiple fuel injectors and intricate airflow passages provide the proper air–fuel mixture ratios at various locations. Air pathways also provide the necessary cooling of the combustor itself and the dilution of the hot products to temperatures safe for the turbine. Figure 12.21 schematically illustrates the various components and flow passages, and Fig. 12.22 shows a photograph of a modern turbofan combustor. Our treatment here ignores this complexity, treating the combustor as a constant-pressure, steady-flow reactor with two inlet streams, one for the air and one for the fuel, and a single outlet stream of dilute combustion products. Figure 12.23 shows the control volume for this situation.

Assumptions

We employ the following assumptions to simplify our analysis:

   i. The flow is steady and at steady state.
  ii. The pressure is constant.
 iii. Inlet and exit conditions are uniform.
 iv. The operation is adiabatic (i.e., $\dot{Q}_{cv} = 0$).
  v. Kinetic and potential energy changes are negligible.

Mass Conservation

Conservation of mass for the combustor follows that of the overall engine in that the mass flow rate of the product stream is the sum of the total air and fuel flow rates, that is,

$$\dot{m}_{A,3} + \dot{m}_F = \dot{m}_{P,4}, \tag{12.30}$$

where the subscripts A, F, and P refer to air, fuel, and products, respectively, and the numerical subscripts correspond to both the station designations shown in Fig. 12.14 and the state points of the cycle shown in Fig. 12.19. The three flow rates are also interrelated through the specification of the operating air–fuel ratio:

$$\dot{m}_{A,3} = (A/F)\dot{m}_F \tag{12.31a}$$

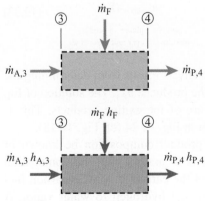

FIGURE 12.23

Control volumes used for jet-engine combustor. Mass flows for mass conservation are shown at the top, and energy flows for energy conservation at the bottom.

and

$$\dot{m}_{P,4} = [1 + (A/F)]\dot{m}_F. \tag{12.31b}$$

Energy Conservation

Applying the assumptions above to the general steady-flow energy conservation equation (Eq. 5.65),

$$\dot{Q}_{cv,\text{net in}} - \dot{W}_{cv,\text{net out}} = \sum_{k=1}^{M \text{ outlets}} \dot{m}_{\text{out},k} \left[h_k + \tfrac{1}{2}\alpha_k v_{\text{avg},k}^2 + g(z_k - z_{\text{ref}}) \right] \\ - \sum_{j=1}^{N \text{ inlets}} \dot{m}_{\text{in},j} \left[h_j + \tfrac{1}{2}\alpha_j v_{\text{avg},j}^2 + g(z_j - z_{\text{ref}}) \right],$$

yields

$$0 - 0 = \dot{m}_{P,4} h_{P,4} - \dot{m}_F h_{F,3} - \dot{m}_{A,3} h_{A,3}. \tag{12.32a}$$

Conservation of energy expressed by Eq. 12.32a is quite simple: The rate of flow of enthalpy *in* equals the rate of flow of enthalpy *out*; that is,

$$\underbrace{\dot{m}_F h_{F,3} + \dot{m}_{A,3} h_{A,3}}_{\text{Enthalpy flow in}} = \underbrace{\dot{m}_{P,4} h_{P,4}.}_{\text{Enthalpy flow out}} \tag{12.32b}$$

Dividing this expression by the air mass flow rate yields

$$\frac{\dot{m}_F h_{F,3} + \dot{m}_{A,3} h_{A,3}}{\dot{m}_{A,3}} = \frac{\dot{m}_{P,4} h_{P,4}}{\dot{m}_{A,3}},$$

or

$$(F/A)h_F + h_{A,3} = (1 + F/A)h_{P,4}. \tag{12.32c}$$

If the fuel enters at the same temperature as the air, this equation can be interpreted as follows:

$$\underset{\substack{\text{Specific enthalpy of reactant} \\ \text{mixture at } T_3 \text{ per unit mass of air}}}{h_{\text{reactants}}(T_3)} = \underset{\substack{\text{Specific enthalpy of product} \\ \text{mixture at } T_4 \text{ per unit mass of air}}}{h_{\text{products}}(T_4)} \tag{12.33}$$

From this point, any complexity in the analysis results from determining the specific enthalpies, particularly, that of the products, $h_{P,4}$. The solution of Eq. 12.33 amounts to finding the temperature of the exiting products. This is shown schematically on the h–T diagram in Fig. 12.24 (cf. Fig. 12.18).

Determining $h_{P,4}$ requires that the product composition be known or calculated. The easiest way to do this is to assume that the combustion process is "complete," that is, that all of the carbon in a hydrocarbon fuel goes to carbon dioxide, and all of the fuel hydrogen to water vapor. A second, and generally more realistic, assumption is that chemical equilibrium prevails at the combustor outlet temperature T_4 and pressure P_4 ($= P_3$).

The following example illustrates how the combustion process can be integrated into an analysis of the turbojet "cycle."

FIGURE 12.24

An h–T diagram illustrating energy conservation for the combustion process. The point labeled 3 represents the left-hand side of Eq. 12.33, and point 4 the right-hand side. For simplicity, we assume the fuel enters the combustor at the same temperature as the air, T_3.

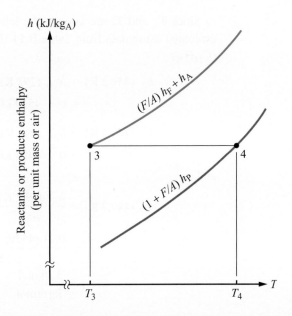

Example 12.11

Consider the combustor entrance conditions calculated in Example 12.10 (i.e., air entering at 349.8 kPa and 449.2 K). The fuel, liquid *n*-dodecane ($C_{12}H_{26}$), is sprayed into the combustor at 298 K to provide an overall air–fuel ratio of 75:1. The corresponding equivalence ratio is $\Phi = 0.1988$. Determine the temperature of the combustion products at the combustor exit. Also determine the constant-pressure specific heat of the product mixture.

Solution

Known $P, T_A, T_F, A/F$

Find $T_P, c_{p,P}$

Sketch

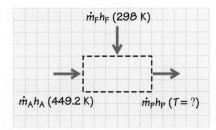

Assumptions

  i. Steady flow
 ii. Constant pressure
iii. Adiabatic process ($\dot{Q}_{cv} = 0$)
 iv. Negligible kinetic and potential energies
  v. Ideal-gas behavior
 vi. No dissociation
vii. Simple air (3.76 kmol of N_2 per kmol of O_2)

Analysis With these assumptions, we can apply conservation of energy expressed by Eq. 12.32c to find the unknown outlet temperature T_P as described in the following.

Since T_A and T_F are known, the left-hand side of Eq. 12.32c is easily evaluated using data from Table B.11 (O_2), Table B.7 (N_2), and Table H.1 ($C_{12}H_{26}$):

$$\bar{h}_{O_2}(449.2 \text{ K}) = \bar{h}^\circ_{f,O_2}(298 \text{ K}) + \Delta h_{s,O_2}(449.2 \text{ K})$$
$$= 0 + 4539 \text{ kJ/kmol} = 4539 \text{ kJ/kmol},$$

$$\bar{h}_{N_2}(449.2 \text{ K}) = \bar{h}^\circ_{f,N_2}(298 \text{ K}) + \Delta h_{s,N_2}(449.2 \text{ K})$$
$$= 0 + 4423 \text{ kJ/kmol} = 4423 \text{ kJ/kmol},$$

and

$$h_A(449.2 \text{ K}) = \frac{\bar{h}_A(449.2 \text{ K})}{\mathcal{M}_A} = \frac{X_{O_2}\bar{h}_{O_2} + X_{N_2}\bar{h}_{N_2}}{\mathcal{M}_A}$$
$$= \frac{0.21(4539) + 0.79(4423)}{28.85} = 154.2$$
$$[=] \frac{\text{kJ/kmol}}{\text{kg/kmol}} = \text{kJ/kg},$$

Since the fuel enters as a liquid at the reference-state temperature 298 K, there is no sensible contribution to the standardized enthalpy [$\Delta h_{s,F}(298 \text{ K}) = 0$]. Furthermore, the enthalpy of formation for the gaseous fuel (Table H.1) needs to be decreased by the enthalpy of vaporization as follows:

$$\mathcal{M}_F = 170.337 \text{ kg/kmol},$$
$$\bar{h}^\circ_{f,F\,(\text{vapor})} = -292,162 \text{ kJ/kmol},$$
$$h_{fg} = 256 \text{ kJ/kg},$$

where the values are taken from Table H.1. Thus,

$$h_{F\,(\text{liquid})} = \frac{\bar{h}_{f,F\,(\text{vapor})}}{\mathcal{M}_F} - h_{fg}$$
$$= \frac{-292,162}{170.337} - 256 \text{ kJ/kg} = -1971.2 \text{ kJ/kg}.$$

Rearranging Eq. 12.32c to isolate the enthalpy of the products yields

$$\frac{(F/A)}{1 + (F/A)}h_F + \frac{1}{1 + (F/A)}h_A = h_P,$$

which is evaluated as follows:

$$\frac{(1/75)}{[1 + (1/75)]}(-1971.2) + \frac{1}{[1 + (1/75)]}(154.2) \text{ kJ/kg}$$
$$= 0.01316(-1971.2) + 0.98684(154.2) \text{ kJ/kg} = 126.2 \text{ kJ/kg}.$$

Our problem now is to determine what temperature of the products yields this value of the mass-specific enthalpy. We first determine the products' mixture composition using element balances (C, H, O, N) for the given stoichiometry and then apply (Eqs. 2.71f and 2.60a)

$$h_P = \frac{\bar{h}_P}{\mathcal{M}_P} = \frac{\sum X_i \bar{h}_i(T)}{\sum X_i \mathcal{M}_i},$$

Also see Example 3.15 in Chapter 3.

where we guess a value for T to evaluate the individual species $\bar{h}_i$ using the tables in Appendix B. Iteration then provides a final result. We begin this procedure by writing a combustion reaction equation by combining Eqs.

3.58, 3.59, and 3.61b:

$$C_x H_y + \frac{x + y/4}{\Phi}(O_2 + 3.76\,N_2) \rightarrow b\,CO_2 + c\,H_2O + d\,O_2 + e\,N_2.$$

For $C_{12}H_{26}$, this becomes

$$C_{12}H_{26} + \frac{18.5}{\Phi}(O_2 + 3.76\,N_2) \rightarrow b\,CO_2 + c\,H_2O + d\,O_2 + e\,N_2.$$

Element conservation yields

C: $12 = b$ or $b = 12$,

H: $26 = 2c$ or $c = 13$,

O: $2(18.5)/\Phi = 2b + c + 2d$ or $d = 18.5\,((1 - \Phi)/\Phi)$, and

N: $[2(18.5)3.76]/\Phi = 2e$ or $e = 69.56/\Phi$.

The mole fraction associated with each product constituent can be determined from

$$X_i = \frac{N_i}{N_{tot}},$$

where $N_{tot} = b + c + d + e$. Using the given value of $\Phi = 0.1988$,[6] we evaluate the X_is:

$$X_{CO_2} = \frac{b}{N_{tot}} = \frac{12}{449.458} = 0.0267,$$

$$X_{H_2O} = \frac{c}{N_{tot}} = \frac{13}{449.458} = 0.0289,$$

$$X_{O_2} = \frac{d}{N_{tot}} = \frac{74.558}{449.458} = 0.1659,$$

and

$$X_{N_2} = \frac{349.899}{449.458} = 0.7785.$$

We begin our iteration process by guessing a product temperature of 1000 K. The following table summarizes the calculation of $\bar{h}_P$ using the Appendix B tabulations for the $\bar{h}_i$:

	X_i	$\bar{h}_i(1000\ \text{K}) = \bar{h}_{f,i}^{\circ}(298\ \text{K}) + \Delta\bar{h}_{s,i}(1000\ \text{K})$
CO_2	0.0267	$-393{,}546 + 33{,}425 = -360{,}121$
H_2O	0.0289	$-241{,}845 + 25{,}993 = -215{,}852$
O_2	0.1654	$0 + 22{,}721 = 22{,}721$
N_2	0.7785	$0 + 21{,}468 = 21{,}468$

Thus,

$$\sum X_i \bar{h}_i = \bar{h}_P(1000\ \text{K}) = 4628.9\ \text{kJ/kmol},$$

$$\mathcal{M}_P = \sum X_i \mathcal{M}_i = 28.813\ \text{kg/kmol},$$

---

[6] The reader should verify that this value is correct from the given information ($A/F = 75{:}1$ for $C_{12}H_{26}$).

and

$$h_{\mathrm{P}}(1000 \text{ K}) = \frac{4628.9}{28.813} \text{ kJ/kg} = 160.65 \text{ kJ/kg}.$$

Since $h_{\mathrm{P}}(1000 \text{ K}) > h_{\mathrm{P}} = 126.2$ kJ/kg, we need to try a lower temperature. Using $T = 900$ K yields $h_{\mathrm{P}}(900 \text{ K}) = 65.76$ kJ/kg. From this, we conclude

$$900 \text{ K} < T < 1000 \text{ K}.$$

A simple linear interpolation using these two values yields $T = 964$ K, further iteration homes in on $T = 970$ K, the value used as a given in Example 12.10.

The product mixture specific heat is now evaluated from Eq. 2.71h using $\bar{c}_{p,i}(970 \text{ K})$ values from Appendix B:

$$\begin{aligned} \bar{c}_{p,\mathrm{P}} &= \sum X_i \bar{c}_{p,i} \\ &= 0.0267(53.993) + 0.0289(40.890) \\ &\quad + 0.1659(34.791) + 0.7785(32.573) \text{ kJ/kmol} \\ &= 33.753 \text{ kJ/kmol} \end{aligned}$$

and

$$c_{p,\mathrm{P}} = \frac{\bar{c}_{p,\mathrm{P}}}{\mathcal{M}_{\mathrm{P}}} = \frac{33.753}{28.813} = 1.171 \text{ kJ/kg} \cdot \text{K}.$$

Comments The need to deal with the multicomponent product mixture complicates computations, although the principles involved (element conservation) are straightforward. This example also illustrates the importance of being able to work with both mass-specific and molar-specific properties.

Using our final result that $T = 970$ K, we can test the assumption that dissociation is negligible using software from Ref. [5]. Calculation of the equilibrium mixture composition allowing for dissociation shows that the mole fractions of all minor species (CO, H_2, OH, etc.) are less than 10^{-7}. These small values justify our original assumption of negligible dissociation.

12.3 GAS-TURBINE ENGINES

In a gas-turbine engine, air enters and is compressed by a multistage compressor, fuel is added to this compressed air and burned, the hot products of combustion expand in a turbine, and the combustion products exit the engine at low velocity. A portion of the power produced by the turbine drives the compressor, while the remainder is delivered to a load through a shaft (Fig. 12.25). Many gas-turbine engines are more complex than suggested by the basic arrangement in Fig. 12.25. For example, regeneration (using exhaust gases to heat the air before it enters the combustor), multistage compression with intercooling, and multistage expansion with reheat can be employed to improve performance and/or efficiency. Our focus, however, is the simple system of Fig. 12.25. For more information on complex systems, see Refs. [8, 9] for example.

Gas-turbine engines are frequently used in stationary electrical power generation units and range in size from tens of kilowatts to hundreds of megawatts. Gas-turbine engines are also common choices for ship propulsion and pipeline pumping systems. Their application to trucks and automobiles has been explored for many years; however, the engine characteristics and the highly transient power demands of these applications are generally not well matched.

Gas- and liquid-fired gas-turbine engines drive the pumps for the Trans-Alaska Pipeline. This above ground section of the pipeline uses a support system designed to prevent thawing of the permafrost. Photograph courtesy of DOE/NREL.

Gas turbine engines are available in a wide range of sizes. The GE 9H engine (left) has a power output of 480 MW, whereas the much smaller GE10 engine (right) produces 11.25 MW. Note the array of combustors utilized in the GE 9H engine and the single vertical can combustor used in the GE 10 engine. Photographs courtesy of General Electric Company.

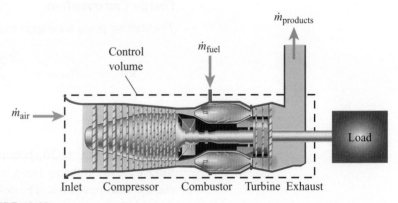

FIGURE 12.25
Schematic diagram of a gas-turbine engine. The nature of the load depends on the application. For example, the load is an electrical generator for stationary power generation; whereas for marine propulsion, the load is a gearbox that drives a propeller.

In the following sections, we analyze the gas-turbine engine using the same methods previously applied to the turbojet engine.

12.3a Integral Control Volume Analysis

As indicated by the dashed line in Fig. 12.25, we consider a control volume that crosses the air inlet plane, cuts through the exhaust duct, cuts through the turbine output shaft, and otherwise surrounds the entire engine. This control volume is reproduced in Fig. 12.26, which shows mass flows (top) and energy flows (bottom).

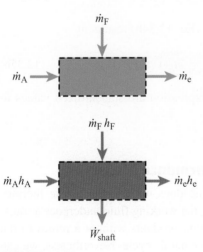

FIGURE 12.26
Integral control volumes for application of mass conservation (top) and energy conservation (bottom) to the analysis of a gas-turbine engine.

Assumptions

The assumptions invoked in the application of mass and energy conservation to this control volume are the following:

i. The flow is steady and at steady state.
ii. There are no heat interactions between the engine and the surroundings (i.e., $\dot{Q}_{cv} = 0$).
iii. The kinetic and potential energies of the air, fuel, and exhaust streams are negligible.

Mass Conservation

Air and fuel enter the control volume as separate streams, and a single stream of combustion products exits as shown in Fig. 12.26 (top). Conservation of mass is thus expressed by Eq. 3.18b as follows:

$$\dot{m}_A + \dot{m}_F = \dot{m}_e. \tag{12.34a}$$

This can also be recast by introducing the fuel–air ratio as

$$\dot{m}_A(1 + F/A) = \dot{m}_e. \tag{12.34b}$$

Similar to the operation of turbojet engines, fuel–air ratios for gas-turbine engines are small and of the order of 1:100.

Energy Conservation

The starting point for energy conservation is Eq. 5.65:

$$\dot{Q}_{cv,net\,in} - \dot{W}_{cv,net\,out} = \sum_{k=1}^{M\,outlets} \dot{m}_{out,k}\left[h_k + \tfrac{1}{2}\alpha_k v_{avg,k}^2 + g(z_k - z_{ref})\right]$$
$$- \sum_{j=1}^{N\,inlets} \dot{m}_{in,j}\left[h_j + \tfrac{1}{2}\alpha_j v_{avg,j}^2 + g(z_j - z_{ref})\right].$$

As shown in Fig. 12.26 (bottom), the only energy flows across the control surface are the enthalpy flows in (air and fuel) and out (combustion products) and the shaft power out. The only difference between this and our analysis of the turbojet engine is that now the kinetic energies of the air and exhaust streams are negligible, whereas for the turbojet, they are very important. The energy conservation equation thus simplifies to

$$0 - \dot{W}_{shaft} = \dot{m}_e h_e - \dot{m}_F h_F - \dot{m}_A h_A,$$

or

$$\dot{W}_{shaft} = \dot{m}_F h_F + \dot{m}_A h_A - \dot{m}_e h_e. \tag{12.35a}$$

Combining this with mass conservation (Eq. 12.34b) results in

Can you provide a physical interpretation of this equation? ▷

$$\dot{W}_{shaft} = \dot{m}_F h_F + \dot{m}_A h_A - \dot{m}_A(1 + F/A)h_e. \tag{12.35b}$$

Note that all of the enthalpies in this expression are standardized values to account for chemical transformations.

12.3b Cycle Analysis and Performance Measures

Like the turbojet, the gas-turbine engine does not execute a true thermo-dynamic cycle. In both of these engines, the working fluid undergoes a trans-formation from air and fuel to combustion products without a return to the original state as is required for a thermodynamic cycle. Nevertheless, we can define state points (see Fig. 12.27 top) and analyze the engine component by component (i.e., state to state) to define a "cycle." To do so, however, requires a treatment of the combustion process, a fairly lengthy procedure as we saw in Example 12.11. To gain some insight into the operation of gas-turbine

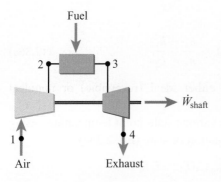

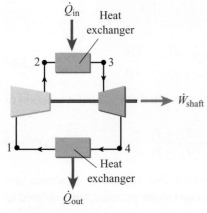

FIGURE 12.27

A gas-turbine engine (top schematic) can be idealized using an air-standard cycle (bottom schematic) in which heat-transfer processes replace the combustion process (2–3) and exhaust process (4–1).

engines, while avoiding the complexities imposed by the combustion process, we again define and analyze an air-standard cycle for this engine.

Air-Standard Brayton Cycle

Figure 12.27 illustrates how the air-standard Brayton cycle models the actual gas-turbine cycle. Using air as the working fluid, the combustion process is replaced by a heat-addition process, and the exhaust and intake processes are replaced with a loop-closing heat-rejection process. The ideal (reversible) air-standard Brayton cycle is thus defined by the following steady-flow processes and is illustrated on P–v and T–s coordinates in Fig. 12.28:

1–2: adiabatic and reversible (isentropic) compression,

2–3: constant-pressure heat addition,

3–4: adiabatic and reversible (isentropic) expansion, and

4–1: constant-pressure heat rejection.

Air-Standard Thermal Efficiency

Applying the definition of the thermal efficiency of a work-producing cycle (Eq. 12.1) to the air-standard Brayton cycle yields

$$\eta_{\text{th, Brayton}} = \frac{\dot{W}_{\text{shaft}}}{\dot{Q}_{\text{in}}} = \frac{\dot{W}_{\text{turbine,out}} - \dot{W}_{\text{compressor,in}}}{\dot{Q}_{\text{in}}}.$$

Recognizing from an overall energy balance on the engine that $\dot{W}_{\text{shaft}} = \dot{Q}_{\text{in}} - \dot{Q}_{\text{out}}$ (see Fig. 12.27 bottom), we rewrite this as

$$\eta_{\text{th, Brayton}} = \frac{\dot{Q}_{\text{in}} - \dot{Q}_{\text{out}}}{\dot{Q}_{\text{in}}} = 1 - \frac{\dot{Q}_{\text{out}}}{\dot{Q}_{\text{in}}}.$$

Neglecting changes in kinetic and potential energies across the heat-exchange devices, we can relate the $\dot{Q}$s to enthalpy changes, that is,

$$\dot{Q}_{\text{in}} = \dot{m}(h_3 - h_2)$$

and

$$\dot{Q}_{\text{out}} = \dot{m}(h_4 - h_1).$$

FIGURE 12.28

The ideal air-standard cycle for a gas-turbine engine, or ideal Brayton cycle, consists of an isentropic compression (1–2), a constant-pressure heat addition (2–3), an isentropic expansion (3–4), and a constant-pressure heat rejection (4–1).

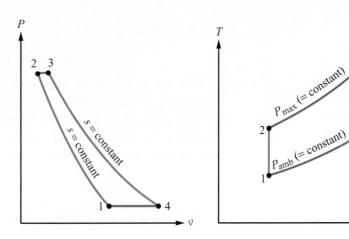

Thus,

$$\eta_{th,Brayton} = 1 - \frac{h_4 - h_1}{h_3 - h_2}. \tag{12.36a}$$

This result applies to a cycle with either ideal (reversible) or nonideal (irreversible) components.

We can also relate Eq. 12.36 to the various state-point temperatures using the approximate ideal-gas calorific equation of state (Eq. 2.33e),

$$h_4 - h_1 = c_{p,avg,1-4}(T_4 - T_1),$$

$$h_3 - h_2 = c_{p,avg,2-3}(T_3 - T_2),$$

and so

$$\eta_{th,Brayton} = 1 - \frac{c_{p,avg,1-4}(T_4 - T_1)}{c_{p,avg,2-3}(T_3 - T_2)}. \tag{12.36b}$$

The specific heats can be eliminated by assuming that c_p is not a function of temperature;[7] thus,

$$\eta_{th,Brayton} = 1 - \frac{T_4 - T_1}{T_3 - T_2}. \tag{12.36c}$$

For the ideal cycle, both the compression and expansion processes are isentropic, and the temperatures can be related to the pressure ratio, defined as

$$R_P \equiv P_2/P_1 = P_3/P_4,$$

using the second-law state relationship, Eq. 2.41, as follows:

$$T_2 = T_1\left(\frac{P_2}{P_1}\right)^{\frac{\gamma-1}{\gamma}} = T_1 R_P^{\frac{\gamma-1}{\gamma}}$$

and

$$T_4 = T_3\left(\frac{P_4}{P_3}\right)^{\frac{\gamma-1}{\gamma}} = T_3(R_P^{-1})^{\frac{\gamma-1}{\gamma}},$$

or

$$T_4 = T_3(R_P)^{-\left(\frac{\gamma-1}{\gamma}\right)}.$$

Substituting these relationships back into Eq. 12.36c yields

$$\eta_{th,Brayton} = 1 - \frac{T_3(R_P)^{-\left(\frac{\gamma-1}{\gamma}\right)} - T_1}{T_3 - T_1(R_P)^{\frac{\gamma-1}{\gamma}}},$$

which simplifies to

$$\eta_{th,Brayton} = 1 - (R_P)^{\frac{1-\gamma}{\gamma}}. \tag{12.37}$$

This interesting result shows that the ideal Brayton cycle thermal efficiency is a function only of the pressure ratio; it is independent of both the air inlet temperature T_1 and the turbine inlet temperature T_3. Figure 12.29 illustrates the dependence of thermal efficiency on pressure ratio for a specific-heat ratio γ of 1.4. Here we see that at low pressure ratios, say 2–7,

---

[7] This is technically an invalid assumption since we know indeed that $c_p = c_p(T)$; however, c_ps for air do not vary greatly over the temperature range of interest, so this is a useful approximation.

FIGURE 12.29
*Ideal air-standard Brayton cycle
efficiency depends only on the pressure
ratio $R_p = P_2/P_1$ and the specific heat
ratio γ of the working fluid. Plot
values are based on $\gamma = 1.4$.*

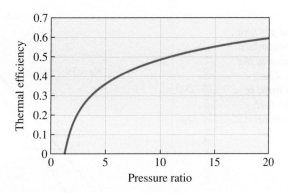

the thermal efficiency is strongly dependent on the pressure ratio, whereas the dependence is much weaker at large values of pressure ratio, say, >15.

Process Thermal Efficiency and Specific Fuel Consumption

For a real gas-turbine engine, we can define a process thermal efficiency with reference to Fig. 12.26. Starting with Eq. 12.1,

$$\eta_{process,Brayton} = \frac{\text{useful work produced}}{\text{energy supplied}},$$

and recognizing that the energy supplied is associated with the chemical energy in the fuel, we can write

See Example 2.26 for a formal
definition of HV_F and Appendix H
for HV_F values for various fuels.

$$\eta_{process,Brayton} = \frac{\dot{W}_{shaft}}{\dot{m}_F HV_F}, \qquad (12.38)$$

where HV_F is the heating value of the fuel.

A frequently used measure of engine efficiency is the **specific fuel consumption**, which we define as

$$sfc \equiv \frac{\text{rate fuel consumed}}{\text{rate work delivered}} = \frac{\dot{m}_F}{\dot{W}_{shaft}}. \qquad (12.39)$$

Comparing this definition with Eq. 12.38, we see that the specific fuel consumption is inversely proportional to the process efficiency; thus, the smaller the value of *sfc*, the higher the efficiency. Specific fuel consumption is also frequently used to quantify spark-ignition and diesel engine fuel efficiency.

Power and Size

A useful measure of engine performance is the ratio of the net power produced to the mass flow rate through the engine, $\dot{W}_{shaft}/\dot{m}$. With a value for this parameter, one can get a feel for how large a device needs to be to perform a particular task, where size enters through the cross-sectional flow area (i.e., $\dot{m} = \rho v A$). Using the air-standard Brayton cycle (Fig. 12.28) we can express $\dot{W}_{shaft}/\dot{m}$ as follows:

$$\frac{\dot{W}_{shaft}}{\dot{m}} = \frac{\dot{W}_{turbine,out} - \dot{W}_{compressor,in}}{\dot{m}}$$

$$= (h_3 - h_4) - (h_2 - h_1),$$

FIGURE 12.30
The net shaft power delivered per unit of mass flow for an ideal air-standard Brayton cycle is linearly related to the turbine inlet temperature T_3. The turbine inlet temperature, in turn, is essentially proportional to the fuel flow rate. Plot values are based on $T_1 = 300$ K, $\gamma = 1.4$, and $c_{p,\text{avg}} = 1.007$ kJ/kg·K.

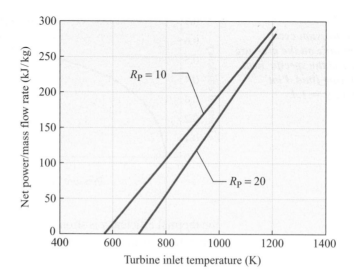

or, furthermore, if we assume a constant specific heat,

$$\frac{\dot{W}_{\text{shaft}}}{\dot{m}} = c_{p,\text{avg}}[(T_3 - T_4) - (T_2 - T_1)].$$

We now treat T_1 and T_3 as given parameters and apply the isentropic property relationships to eliminate T_2 and T_4; this yields

$$\frac{\dot{W}_{\text{shaft}}}{\dot{m}} = c_{p,\text{avg}}\left\{T_3\left[1 - R_{\text{P}}^{\frac{1-\gamma}{\gamma}}\right] + T_1\left[1 - R_{\text{P}}^{\frac{\gamma-1}{\gamma}}\right]\right\}. \qquad (12.40)$$

Figure 12.30 illustrates this relationship for two values of pressure ratio R_{P}. Here we see that a minimum turbine inlet temperature T_3 is required to produce any net power. At the zero value of $\dot{W}_{\text{shaft}}/\dot{m}$, all of the turbine power is used to drive the compressor. Above this minimum value of the turbine inlet temperature, $\dot{W}_{\text{shaft}}/\dot{m}$ increases linearly with T_3, where the slope depends on the pressure ratio R_{P}. Although this analysis is based on the air-standard cycle, we can connect it to the real engine by realizing that the temperature difference across the combustor, $T_3 - T_2$, is, to first order, proportional to the fuel flow rate. Thus we can construe Fig. 12.30 to be a plot of $\dot{W}_{\text{shaft}}/\dot{m}$ versus $\dot{m}_{\text{F}}$ (or F/A), where the minimum turbine inlet temperature to produce any net power is equivalent to the idle fuel flow rate or idle fuel–air ratio.

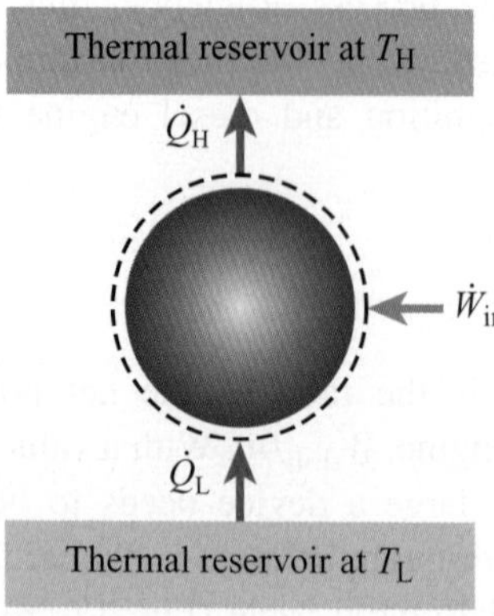

12.4 REFRIGERATORS AND HEAT PUMPS

In this section, we explore so-called reversed cycles in which a work or power input moves energy from a low-temperature thermal reservoir to high-temperature thermal reservoir (Fig. 12.31). In a refrigerator, one seeks to maintain the cold space at a desired low temperature, whereas for a heat pump, the high-temperature space is the focus. Refrigerators are common in food processing and storage—there may even be a refrigerator similar to the one pictured in Fig. 12.32 in the room in which you are reading this book. Heat pumps are becoming increasingly popular for residential and other space-heating applications. We introduced these devices in our discussion of the second law of thermodynamics in Chapter 7 without explanation of how they might accomplish their desired tasks. The objective now is to fill this void. Here we present and discuss refrigerators and heat pumps that operate on a **vapor-compression cycle**. Other types of reversed cycles exist; however,

FIGURE 12.31
A reversed cycle uses an input of work to transfer energy from a low-temperature reservoir to a high-temperature reservoir. Compare this with the power-producing cycle of Fig. 7.1.

FIGURE 12.32
Rear view of a small refrigerator suitable for use in a student dormitory room.

Refrigerated test cells or "cold rooms" are used for vehicle testing. Photograph courtesy of General Motors.

discussion of these goes beyond the scope of this book. For more information, we refer the interested reader to Refs. [8, 9].

12.4a Energy Conservation for a Reversed Cycle

Consider a steady-flow device that operates on an arbitrary reversed cycle as shown in Fig. 12.31. As indicated by the dashed line, a control surface surrounds the device. The only energy flows crossing the control surface are a power input and two heat interactions, one with the low-temperature reservoir and one with the high-temperature reservoir. We thus express conservation of energy simply as

$$\sum \dot{E}_{in} = \sum \dot{E}_{out},$$

or

$$\dot{W}_{in} + \dot{Q}_L = \dot{Q}_H. \qquad (12.41)$$

Although the cyclic device contained within our control surface may be quite complex, the overall energy conservation expression describing its operation is quite simple.

12.4b Performance Measures

As we saw in Chapter 7, the coefficient of performance is used to quantify how well a reversed-cycle device performs its job in the same way that the thermal efficiency is used to characterize the performance of a power-producing cycle; that is,

$$COP \equiv \beta \equiv \frac{\text{desired energy}}{\text{energy that costs}}. \qquad (12.42)$$

For a refrigerator, the energy removed from the low-temperature space is the desired energy; thus,

$$\beta_{refrig} = \frac{\dot{Q}_L}{\dot{W}_{in}}, \qquad (12.43a)$$

where the power in comprises the energy that costs. Applying conservation of energy (Eq. 12.41), we can also express this definition as

$$\beta_{refrig} = \frac{\dot{Q}_L}{\dot{Q}_H - \dot{Q}_L} = \frac{1}{\dot{Q}_H/\dot{Q}_L - 1}. \qquad (12.43b)$$

In refrigeration applications, the unit *ton* is frequently used to quantify the energy removal rate from the cold space. *One ton of refrigeration* equals 3.517 kW or 200 Btu/min.

Similarly, the desired energy for a heat pump is that delivered to the high-temperature space; thus,

$$\beta_{heat\ pump} = \frac{\dot{Q}_H}{\dot{W}_{in}}, \qquad (12.44a)$$

or

$$\beta_{heat\ pump} = \frac{\dot{Q}_H/\dot{Q}_L}{\dot{Q}_H/\dot{Q}_L - 1}. \qquad (12.44b)$$

See Example 7.3 in Chapter 7 to consolidate these definitions.

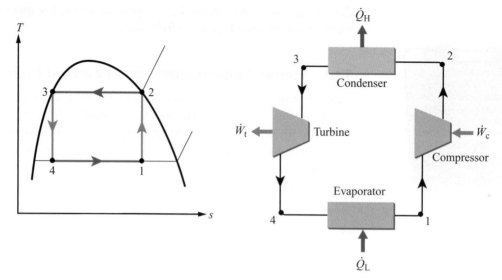

FIGURE 12.33
The reversed Carnot cycle establishes thermodynamic limits for the performance of refrigerators and heat pumps operating between two fixed temperatures.

The definitions expressed in Eqs. 12.43 and 12.44 apply to both ideal and real devices.

To get a handle on the upper limits of the coefficient of performance for a device operating between two fixed temperatures, we resurrect the ideal of the Carnot cycle, but now operating in reverse. As you may recall from Chapter 7, the Carnot cycle is an ideal cycle in which all processes are performed reversibly (i.e., there is no friction or other source of irreversibility.) Figure 12.33 illustrates a steady-flow, reversed Carnot cycle. The processes involved and the devices that accomplish them are as follows:

State Points	Process	Device
1–2	Adiabatic, reversible (isentropic) compression	Ideal compressor
2–3	Reversible heat rejection at constant temperature	Ideal condenser
3–4	Adiabatic, reversible (isentropic) expansion	Ideal turbine
4–1	Reversible heat addition at constant temperature	Ideal evaporator

Recognizing that the net power supplied is $\dot{W}_{in} = \dot{W}_c - \dot{W}_t$, we can express the coefficients of performance for the reversed-Carnot cycle refrigerator and heat pumps by Eqs. 12.43 and 12.44, respectively. We can relate the heat-transfer rates to the temperatures in the condenser and evaporator by applying the definition of entropy for a reversible, constant-temperature process. Expressing the heat-transfer rates on a per-unit-mass basis, we write

$$q_L = \dot{Q}_L/\dot{m},$$
$$q_L = T_L(s_1 - s_4)$$

and

$$q_H = \dot{Q}_H/\dot{m},$$
$$q_H = T_H(s_2 - s_3).$$

Substituting these expressions for q_L and q_H into Eq. 12.43b yields

$$\beta_{\text{refrig,Carnot}} = \frac{1}{T_H/T_L - 1}, \tag{12.45a}$$

where we recognize from Fig. 12.33 that $s_1 - s_4 \equiv s_2 - s_3$. Similarly, Eq. 12.44b is transformed to

$$\beta_{\text{heat pump,Carnot}} = \frac{T_H/T_L}{T_H/T_L - 1}. \tag{12.45b}$$

We need to point out that, although theoretically possible, the reversed Carnot cycle illustrated in Fig. 12.33 is impractical for several reasons: First, compressors do not function well with wet, liquid–vapor mixtures. In practice, the refrigerant is frequently slightly superheated upon entering the compressor. Second, the power produced by a turbine in a reversed Carnot cycle would be quite small, and the complexity and expense required to recover this small amount of energy preclude the use of a turbine. In real refrigerators and heat pumps, the working fluid expands *irreversibly* through a simple throttling device or expansion valve. The refrigerator pictured in Fig. 12.32 uses a length of capillary tube to achieve the desired expansion.

12.4c Vapor-Compression Refrigeration Cycle

The ideal vapor-compression refrigeration cycle is illustrated on *T–s* coordinates in Fig. 12.34, and a schematic diagram of the components used to effect this cycle is shown in Fig. 12.35. In the *ideal* vapor-compression refrigeration cycle, the compression process takes place adiabatically and reversibly, and hence isentropically, and the heat-transfer processes in the condenser and

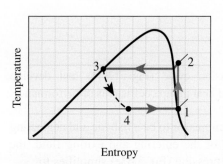

FIGURE 12.34

The ideal vapor-compression refrigeration cycle comprises an isentropic compression (1–2), a constant-pressure heat rejection in which the working fluid condenses (2–3), a constant-enthalpy expansion (3–4), and a constant-pressure heat addition in which the working fluid evaporates (4–1).

FIGURE 12.35

Schematic of components used to implement the vapor-compression refrigeration cycle.

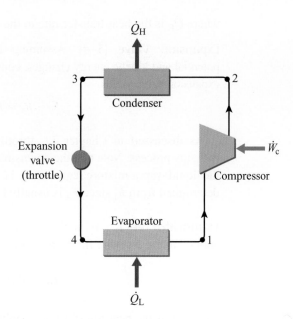

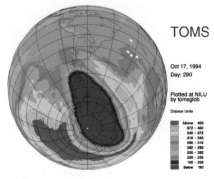

TOMS

Oct 17, 1994
Day: 290

Plotted at NILU
by tomsglob

Dobson Units

Meteor-3 TOMS (Total Ozone Monitoring
System) image shows the ozone hole (purple)
over Antarctica on October 17, 1994. TOMS
data (NASA) plotted by Norwegian Institute for
Air Research.

evaporator are assumed to be reversible. The throttling process, however, is by definition highly irreversible; thus, the ideal cycle contains an irreversible process.

A variety of working fluids are used in refrigerators and heat pumps. The NIST online database provides thermodynamic and transport properties for eight modern refrigerants. Among these, tetrafluoroethane (CH_3CH_2F)— Refrigerant 134a (R-134a)—and chlorodifluoromethane ($CHClF_2$)—Refrigerant 22 (R-22)—are commonly used in residential, commercial, and automotive systems. These refrigerants are formulated to minimize ozone destruction in the upper atmosphere. Older refrigerants (R-11 and R-12) contain chlorine, which reacts catalytically to destroy ozone. Refrigerant leakage from refrigeration systems and the improper disposal of refrigerants, combined with other source of these chlorofluorohydrocarbons (CFCs), has over several decades resulted in the "ozone hole" over Antarctica. An international treaty (Montreal Protocol, 1987) has phased out the production of CFCs and other ozone-depleting substances.

Cycle Analysis

We now analyze the vapor-compression cycle (Fig. 12.34) and develop relationships to evaluate refrigerator and heat pump coefficients of performance.

Compressor (1–2) Assuming the compressor to be adiabatic and neglecting potential and kinetic energy changes of the entering and exiting fluid, the steady-flow conservation of energy expression (Eq. 5.63e) simplifies to

$$\dot{W}_c = \dot{m}(h_2 - h_1), \tag{12.46}$$

where $\dot{W}_c$ is the power input to the compressor. Note that Eq. 12.46 applies to both reversible and irreversible compression, provided the process is adiabatic.

Condenser (2–3) Again neglecting changes in kinetic and potential energies, conservation of energy applied to the working fluid in the condenser yields

$$\dot{Q}_H = \dot{m}(h_2 - h_3), \tag{12.47}$$

where $\dot{Q}_H$ is the heat-transfer rate to the surroundings.

Expansion Valve (3–4) Assuming adiabatic operation and neglecting potential and kinetic energy changes, conservation of energy applied across the expansion valve yields

$$h_3 = h_4. \tag{12.48}$$

As discussed in Chapter 11, throttling is characterized as a constant-enthalpy process. Note that the downstream state for the expansion valve is in the liquid–vapor mixture region (Fig. 12.34). The mixture quality x_4 is readily determined from h_4 since P_4 is usually known.

Evaporator (4–1) With the usual assumptions, conservation of energy applied to the working fluid flowing through the evaporator yields

$$\dot{Q}_L = \dot{m}(h_1 - h_4), \tag{12.49}$$

where $\dot{Q}_L$ is the heat-transfer rate supplied to the refrigerant.

Coefficients of Performance

Substituting Eqs. 12.46–12.49 into the coefficient of performance definitions (Eqs. 12.43 and 12.44) yields

$$\beta_{\text{refrig}} = \frac{\dot{Q}_L}{\dot{W}_{in}} = \frac{h_1 - h_4}{h_2 - h_1} \qquad (12.50a)$$

and

$$\beta_{\text{heat pump}} = \frac{\dot{Q}_H}{\dot{W}_{in}} = \frac{h_2 - h_3}{h_2 - h_1}. \qquad (12.50b)$$

Other than the restrictions that the compression and throttling processes are adiabatic, these relationships apply to both ideal (reversible) and real (irreversible) systems. Furthermore, we note that, although the ideal cycle presented in Fig. 12.34 shows the working fluid entering the compressor as a saturated vapor, the entering fluid (state point 1) in a real compressor is likely to be slightly superheated to avoid any possibility of moisture. Similarly, state point 3 may lie in the subcooled liquid region in a real device. Because no specific designation of the state points is assumed in their derivation, Eqs. 12.50a and 12.50b still apply.

The following examples illustrate the application of the vapor-compression cycle to refrigerators and heat pumps.

Example 12.12

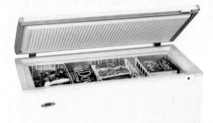

Image courtesy of Haier America.

It is desired to maintain the cold space in a freezer at 0 F. The freezer utilizes a vapor-compression cycle and operates in a room in which the air temperature is 70 F. Determine the minimum requirements for the pressures in the evaporator and condenser of this freezer (i.e., the minimum or maximum pressures required). The refrigerant is R-134a.

Solution

Known   Maximum T_L, minimum T_H, R-134a

Find   Saturation pressures corresponding to $T_{L,\max}$ and $T_{H,\min}$

Sketch

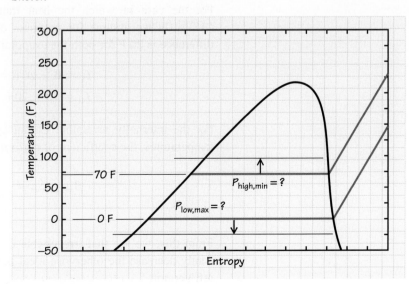

Analysis The bulk of the analysis required to solve this problem has been done in the creation of the sketch. To keep the freezer cold space at 0 F requires that energy be removed from this space. This is accomplished in the evaporator of the vapor-compression cycle. The maximum possible low temperature of the working fluid is identical to the temperature of the cold space, 0 F. This represents a reversible heat-transfer process since both the system (working fluid) and the surroundings (cold space) are at the same temperature. The corresponding pressure in the R-134a is the saturation pressure at 0 F. From the NIST database we find this to be

$$P_{\text{sat}} (T = T_{\text{sat}} = 0 \text{ F}) = 1.4406 \text{ atm.}$$

That this pressure is the *maximum possible* low pressure can be seen on the sketch. Since the real heat-transfer process in the evaporator will require that the temperature of the refrigerant be lower than the temperature of the cold space, the actual low pressure will correspond to $P_{\text{sat}} (T = T_{\text{sat}} < 0 \text{ F})$. This lower pressure is indicated by the downward-pointing arrow originating at the $P_{\text{low,max}}$ line.

Similar logic can be applied to the heat-rejection process in the condenser. Note that the real heat-transfer process will now involve a temperature difference in which the refrigerant is *hotter* than the surroundings. This is indicated by the upward-pointing arrow on the sketch. From the NIST database,

$$P_{\text{high,min}} = P_{\text{sat}} (T_{\text{sat}} = 70 \text{ F}) = 5.8387 \text{ atm.}$$

Comments This example can also be used to illustrate the effect of irreversible ($\Delta T > 0$) heat transfer on the performance of a vapor-compression refrigeration device. Using $T_L = 0$ F (255 K) and $T_H = 70$ F (294 K) as the reservoir temperatures for the operation of a reversed Carnot cycle in which all processes are both *internally* and *externally* reversible, we get a coefficient of performance for the refrigerator of

> Table 7.1 in Chapter 7 lists many irreversibilities: Second on the list is heat transfer across a finite temperature difference.

$$\beta_{\text{refrig,Carnot}} = \frac{1}{T_H/T_L - 1}$$

$$= \frac{1}{\dfrac{294}{255} - 1} = 6.54.$$

For the case in which the heat transfer is accomplished irreversibly ($\Delta T > 0$), we assume for purposes of illustration that the temperature difference is 20 F for both the evaporator and condenser; thus,

$$T_L = 0 - 20 \text{ F} = -20 \text{ F} (244 \text{ K})$$
$$T_H = 70 + 20 \text{ F} = 90 \text{ F} (305 \text{ K})$$

For the internally reversible Carnot cycle operating between these two temperatures, the coefficient of performance is

$$\beta_{\text{refrig,Carnot}} = \frac{1}{\dfrac{305}{244} - 1} = 4.0.$$

From this, we see that irreversible (real) heat-transfer processes significantly decrease the otherwise ideal performance of a vapor-compression refrigerator.

Example 12.13

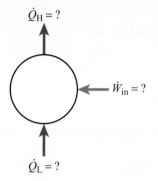

Consider an ideal vapor-compression refrigeration cycle operating between pressures of 0.14 and 0.80 MPa. The flow rate of the refrigerant (R-134a) is 0.04 kg/s. Determine the rate at which energy is removed from the cold space, the rate at which energy is rejected to the surroundings, the power input to the compressor, and the cycle coefficient of performance. Also compare the cycle *COP* with that of a reversed Carnot cycle operating between the same two pressures.

Solution

Known P_H, P_L, $\dot{m}$, R-134a

Find $\dot{Q}_L$, $\dot{Q}_H$, $\dot{W}_c$, β_{refrig}, β_{Carnot}

Sketch

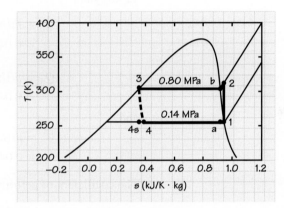

Assumptions

  i. Steady flow
 ii. Ideal vapor-compression cycle
iii. Changes in kinetic and potential energy negligible for all components

Analysis The key state points (1–2–3–4–1) are identified on the sketch for the ideal cycle. Determining the mass-specific enthalpy at each of the points allows us to find the first four desired quantities. Since state 1 lies on the saturated vapor line at 0.14 MPa, all other desired state-1 properties can be obtained from the NIST database or by another table look-up. The properties for the saturated liquid at state 3 are similarly obtained. Using the NIST database with property values based on the ASHRAE reference state,[8] we construct the following table:

State	1	2	3	4
Region	Saturated vapor	Superheated vapor	Saturated liquid	Liquid–vapor mixture
P (MPa)	0.14	0.80	0.80	0.14
T (K)	254.39	?	304.48	254.39
h (kJ/kg)	239.18	?	95.501	$h_4 = h_3$
s (kJ/kg·K)	0.9446	$s_2 = s_1$	0.3541	?

[8] Various reference states are available as options. The ASHRAE standard sets $u = s = 0$ at $-40°C$ for saturated liquid. Other examples in this chapter use the NIST default rather than the ASHRAE reference state. Care must be exercised in using property data from different sources as different reference states may be employed.

To find h_4, we recognize that the adiabatic, reversible, compression process is isentropic (i.e., $s_2 = s_1$); thus, $h_2 = h_2 (P_2, s_2)$. Again using the NIST database, we obtain

$$h_2 = h_2(0.80 \text{ MPa}, 0.9946 \text{ kJ/kg} \cdot \text{K})$$

$$= 275.39 \text{ kJ/kg},$$

and the corresponding temperature is

$$T_2 = 312.13 \text{ K}.$$

Recognizing that the enthalpy change is zero across the expansion valve, we have

$$h_4 = h_3 = 95.501 \text{ kJ/kg}.$$

First-law analyses of the individual components (Eqs. 12.49, 12.47, and 12.46) yield

$$\dot{Q}_L = \dot{m}(h_1 - h_4)$$

$$= 0.04(239.18 - 95.501) \text{ kW} = 5.747 \text{ kW},$$

$$\dot{Q}_H = \dot{m}(h_2 - h_3)$$

$$= 0.04(275.39 - 95.501) \text{ kW} = 7.196 \text{ kW},$$

and

$$\dot{W}_c = \dot{m}(h_2 - h_1)$$

$$= 0.04(275.39 - 239.18) \text{ kW} = 1.448 \text{ kW}.$$

The coefficient of performance is then (Eq. 12.43a)

$$\beta_{\text{refrig}} = \frac{\dot{Q}_L}{\dot{W}_c} = \frac{5.747}{1.448} = 3.97.$$

For a reversed Carnot cycle operating within the liquid–vapor region between the same two pressures (see the cycle a–b–3–4s–a on the sketch), the coefficient of performance is given by

$$\beta_{\text{refrig,Carnot}} = \frac{1}{\dfrac{T_H}{T_L} - 1}$$

$$= \frac{1}{\dfrac{304.48}{254.39} - 1} = 5.08.$$

Comments   We note that all of the energy rates are dictated by the refrigerant flow rate and the high- and low-pressure set points. As we saw in the previous example, the pressure set points are determined by the temperature requirements of the application. Thus, more refrigerant flow provides more cooling capacity for a given application. We also note that the coefficient of performance is much greater than unity. We also observe that the Carnot *COP* is approximately 28% higher than that of the ideal vapor-compression cycle.

Self Test
12.9

A home air conditioner operating on an ideal vapor-compression refrigeration cycle (R134-a) steadily removes heat at a rate of 500 kJ/min. If the minimum and maximum operating pressures are 0.1 and 0.9 MPa, respectively, determine the coefficient of performance and refrigerant mass flow rate of the air conditioner.

(Answer: $\beta_{\text{refrig}} = 2.89$, $\dot{m} = 0.063$ kg/s)

Example 12.14

As shown in the sketch, a heat pump is used to heat a swimming pool by extracting energy from the ambient air at 295.3 K (72 F) and transferring it to the water in the pool at 283.1 K (50 F). The heat pump operates with R-22 between 0.653 MPa absolute (~80 psig) and 1.342 MPa (~180 psig) and delivers 24.62 kW (84,000 Btu/hr) to the water. The R-22 vapor is superheated to 291.44 K (65 F) before it enters the compressor, and the R-22 liquid is subcooled to 299.78 (80 F) at the condenser outlet. The scroll compressor has an isentropic efficiency of 65%. Assuming no losses other than those associated with the compressor, plot the vapor-compression cycle on T–s coordinates and determine the cycle coefficient of performance and the mass flow rate of the refrigerant.

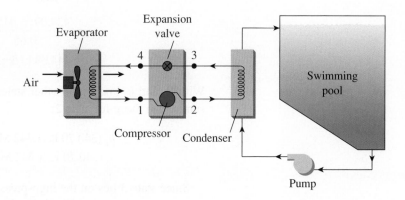

Solution

Known P_{high}, P_{low}, T_1, T_3, $\dot{Q}_{\text{H}}$, $T_{\text{H}_2\text{O}}$, T_{air}, R-22

Find T–s diagram, $\beta_{\text{heat pump}}$, $\dot{m}_{\text{R-22}}$

Assumptions

  i. Steady flow
 ii. Adiabatic (but nonideal) compression
iii. Negligible changes in potential and kinetic energy for all components
 iv. No pressure losses through coils and interconnecting plumbing

Analysis   To create a T–s diagram, we first determine the properties at the points designated 1, 2, 3, and 4 in the sketch. From the assumption that there are no pressure losses through the coils, we can write

$$P_1 = P_4 = P_{\text{low}} = 0.653 \text{ MPa}$$

and

$$P_2 = P_3 = P_{\text{high}} = 1.342 \text{ MPa}.$$

The temperature at the compressor inlet, T_1, is given. With T_1 and P_1 known, we use the NIST database to find s_1 and h_1 anticipating the use of the latter to calculate the *COP*:

For state 1.

$$s_1 (291.44 \text{ K}, 0.653 \text{ MPa}) = 1.7646 \text{ kJ/kg·K},$$
$$h_1 (291.44 \text{ K}, 0.653 \text{ MPa}) = 415.53 \text{ kJ/kg}.$$

To define state 2, we use the definition of the compressor isentropic efficiency (Eq. 11.45) as follows:

$$\eta_{isen,c} = \frac{h_{2s} - h_1}{h_2 - h_1},$$

where h_{2s} is defined as follows and evaluated using the NIST database:

$$h_{2s}(s_{2s} = s_1, P_2) = h_{2s}(1.7646 \text{ kJ/kg} \cdot \text{K}, 1.342 \text{ MPa})$$
$$= 434.09 \text{ kJ/kg}.$$

Solving the defining relationship for $\eta_{isen,c}$ for h_2 yields

$$h_2 = \frac{h_{2s} - h_1}{\eta_{isen,c}} + h_1$$

$$= \frac{434.09 - 415.53}{0.65} + 415.53 \text{ kJ/kg}$$

$$= 444.08 \text{ kJ/kg}.$$

With P_2 and h_2 known, all other state-2 properties can be determined. We summarize those for state 2:

$$h_2(340.70 \text{ K}, 1.342 \text{ MPa}) = 444.08 \text{ kJ/kg},$$
$$s_2(340.70 \text{ K}, 1.342 \text{ MPa}) = 1.7943 \text{ kJ/kg} \cdot \text{K}.$$

Since state 3 lies on the high-pressure isobar, $P_3 = 1.342$ MPa, and the temperature $T_3 = 299.78$ K is given, we obtain the subcooled liquid properties directly from the NIST database:

$$h_3(299.78 \text{ K}, 1.342 \text{ MPa}) = 232.33 \text{ kJ/kg},$$
$$s_3(299.78 \text{ K}, 1.342 \text{ MPa}) = 1.1105 \text{ kJ/kg} \cdot \text{K}.$$

The enthalpy across the expansion valve is constant (throttling process), so $h_4 = h_3$. Since $P_4 = P_{low}$, h_4 and P_4 define the state. To determine s_4, however, requires that we first determine the quality x_4, which we accomplish as follows (Eq. 2.49d):

$$x_4 = \frac{h_4 - h_{f,4}}{h_{g,4} - h_{f,4}} = \frac{232.33 - 210.21}{408.09 - 210.21} = 0.1118,$$

and (Eq. 2.49c)

$$s_4 = (1 - x_4)s_{f,4} + x_4 s_{g,4}$$
$$= (1 - 0.1118)1.0363 + 0.1118(1.7387) \text{ kJ/kg} \cdot \text{K} = 1.115 \text{ kJ/kg} \cdot \text{K}.$$

The temperature $T_4 = T_{sat}(P_{sat} = P_4 = 0.653 \text{ MPa})$; thus, the properties at state 4 are fully defined as follows:

$$T_4 = 281.76 \text{ K},$$
$$h_4(x_4 = 0.1118, 0.653 \text{ MPa}) = 232.33 \text{ kJ/kg},$$
$$s_4(x_4 = 0.1118, 0.653 \text{ MPa}) = 1.115 \text{ kJ/kg} \cdot \text{K}.$$

Using values of T and s at each state point, we create the following plot:

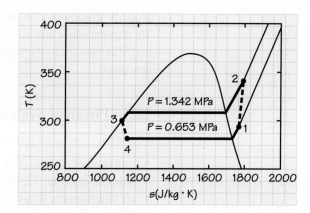

To evaluate the refrigerant flow rate, we apply the following steady-flow energy balance on the condenser:

$$\dot{Q}_H = \dot{m}_{R\text{-}22}(h_2 - h_3).$$

Solving for $\dot{m}_{R\text{-}22}$, we get

$$\dot{m}_{R\text{-}22} = \frac{\dot{Q}_H}{h_2 - h_3} = \frac{24.62}{(444.08 - 232.33)} = 0.1163$$

$$[=] \frac{kW}{kJ/kg}\left[\frac{1\ kJ/s}{kW}\right] = kg/s.$$

For a heat pump, the coefficient of performance is given by Eq. 12.50b:

$$\beta_{\text{heat pump}} = \frac{\dot{Q}_H}{\dot{W}_c} = \frac{h_2 - h_3}{h_2 - h_1}$$

$$= \frac{444.08 - 232.33}{444.08 - 415.53} = 7.4.$$

Comment Coefficients of performance for most heat pump applications are much smaller (say, 3–5) than the value obtained in this example. The reason for this difference is the relatively high temperature used in the evaporator for this application. (For a fixed temperature in the condenser, the *COP* falls with decreasing evaporator temperature, which you can easily show to be true with some simple, reversed-Carnot-cycle calculations.) In this application, air is available at a high temperature (72 F), unlike a typical heating application where the air temperature would be much less. This higher air temperature allows a higher temperature (actually pressure) to be utilized in the evaporator while maintaining an adequate temperature difference to achieve the desired heat transfer.

12.5 AIR CONDITIONING, HUMIDIFICATION, AND RELATED SYSTEMS

This section focuses primarily on devices that provide comfortable indoor climates: humidifiers, dehumidifiers, and air conditioners. We also consider cooling towers. What distinguishes these devices from others we have studied is the important role that the moisture in the air now plays. The commonplace statement, "It's not the heat, but the humidity," implies a strong connection between the amount of moisture in the air and human comfort. In this section, we apply mass and energy conservation principles together with thermodynamic property relationships for mixtures to develop an understanding of these devices. We also introduce and apply the specialized concepts and nomenclature that are normally associated with these devices: specific humidity, relative humidity, and dew point.

12.5a Physical Systems

Figures 12.36–12.40 schematically illustrate the following devices:

- evaporative coolers (Fig. 12.36),
- humidifiers (Fig. 12.36),
- air conditioners (Figs. 12.37 and 12.38),
- dehumidifiers (Figs. 12.37 and 12.39), and
- cooling towers (Fig. 12.41).

Evaporative coolers (Fig. 12.36) function by blowing warm dry air through a water spray or a porous medium saturated with liquid water. Because the warm air supplies the energy to evaporate some of the liquid water, the air is cooled. This mechanical method of cooling resembles the natural evaporation of perspiration that regulates the body temperatures of humans and other animals that perspire. Evaporative coolers, also known as "swamp coolers," work best when the moisture content of the incoming air is relatively low, as the addition of moisture can detract from the comfort associated with the cooling. Evaporative coolers are inexpensive because of their simplicity and use less energy than refrigeration-cycle-based air conditioners.

Humidifiers function in the same manner as evaporative coolers; the desired effect, however, is now the increased moisture content of the air rather than a lower temperature.

Air-conditioning systems (Figs. 12.37 and 12.38) rely on active cooling (and, sometimes, heating). Most residential and other relatively small-scale systems use a vapor-compression refrigeration cycle to provide the cooling. In Fig. 12.38, we see that the air-conditioner cooling coil is the evaporator of the core refrigeration cycle. The condenser is located outdoors, adjacent to the evaporator in the interior in a window unit, or at a distant location in central systems. The condensate from the cold side is usually led to the outside. On particularly hot and humid days, you can easily observe dripping air conditioners. Large-scale air-conditioning systems, such as those used in

Evaporative cooler. Image courtesy of Beaumark Ltd.

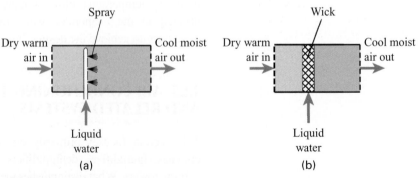

(a) (b)

FIGURE 12.36

Evaporative coolers and humidifiers both work on the same principles. The air to be cooled or moistened is drawn through a duct in which water is (a) sprayed or (b) contained in a wick or other porous material. Internal mass and energy transfers result in cool, moist air exiting the duct.

FIGURE 12.37

Air is cooled and moisture removed at the cooling coil in an air conditioner (a and b) and a dehumidifier (b). In some air conditioners the cold air may also be heated to provide a desired exit temperature (b).

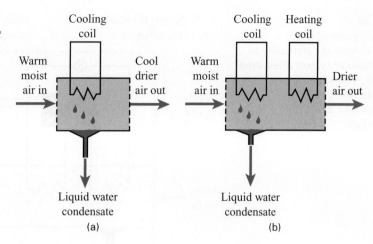

(a) (b)

office buildings, shopping malls, and other large buildings, or a complex of buildings, utilize systems that chill water rather than air, as chilled water is more easily circulated through a large building or complex of buildings than is air. A "chiller" is an integral part of such systems. Cycles other than the simple vapor-compression refrigeration cycle are also employed when efficiency is particularly important, or when particular circumstances dictate. Among these are cascade refrigeration cycles and gas-absorption systems. Discussion of these systems is beyond the scope of this book and we refer the interested reader to Refs. [8, 9].

Dehumidifiers (Fig. 12.39) operate in a manner very similar to air conditioners. The condenser (heating coil), however, is located in the air stream following the evaporator (cooling coil). The moisture condensed from the incoming air is caught in a container or led through a tube to a drain. Figure 12.40 shows a photograph of a home dehumidifier.

FIGURE 12.38

Schematic diagram of a residential air-conditioning system. At the heart of the air conditioner is the vapor-compression refrigeration cycle comprising an evaporator, compressor, condenser, and expansion valve. For window or wall units, all components are housed in a single cabinet, whereas for central systems, the evaporator is usually part of the indoor furnace assembly and the condenser is a stand-alone unit located outdoors.

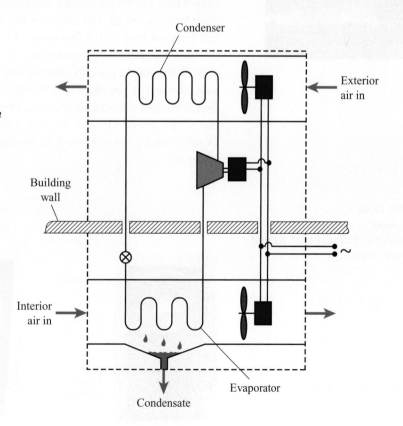

FIGURE 12.39
Schematic diagram of home dehumidifier. Note the similarity to the air-conditioner system illustrated in Fig. 12.38, with the primary difference that the condenser (heating coil) now heats the dehumidified air.

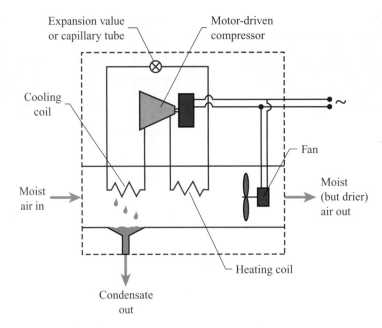

The final system we present, a cooling tower, differs significantly in both scale and application from those already discussed. Cooling towers are huge structures and frequently dominate the view of steam power generating plants. Heights range from less than 15 m (50 ft) to more than 175 m (575 ft). Cooling towers provide a semiclosed loop to the cooling side of steam condensers as shown in Fig. 12.41. The cooling water in a steam condenser exits at temperatures around 40°C (100 F). This hot water is sprayed into the upflowing air stream in the tower. The water droplets in the spray cool as they partially evaporate in the air. This process is similar to that used by the evaporative cooler previously discussed, except the desired effect is to cool the water, not the air. The cooled water is collected at the bottom of the tower and pumped back to the condenser. Make-up water is added to compensate for the evaporation. Depending on the tower design, the air stream can be pulled through the tower by fans or can flow as a result of natural convection since the density of the warm, moist air is less than that of the ambient air. So-called drift eliminators are used to prevent a large carryover of the spray out of the tower with the air. One can frequently observe a row of small clouds above a tower formed by the condensation of the moisture from the tower air.

FIGURE 12.40
Photograph of residential dehumidifier. Front view (left) and rear view (right).

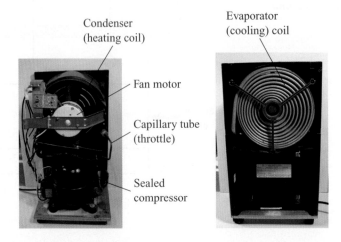

FIGURE 12.41

Schematic diagram of a cooling tower used in steam power plants. Hot water is sprayed in the cooling tower and cooled by the air that enters at the base of the tower. The cold water is then pumped to the steam condenser.

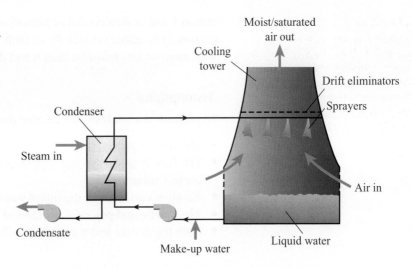

12.5b General Analysis

To develop a basic understanding of all of these devices or systems, we apply basic mass and energy conservation principles to a generic control volume in which any or all of the important processes associated with these devices occur. Figure 12.42a shows this control volume. Here moist air enters at

FIGURE 12.42

(a) General control volume for analysis of moisture-laden air streams. (b) Control volume illustrating mass flows. (c) Control volume illustrating energy flows.

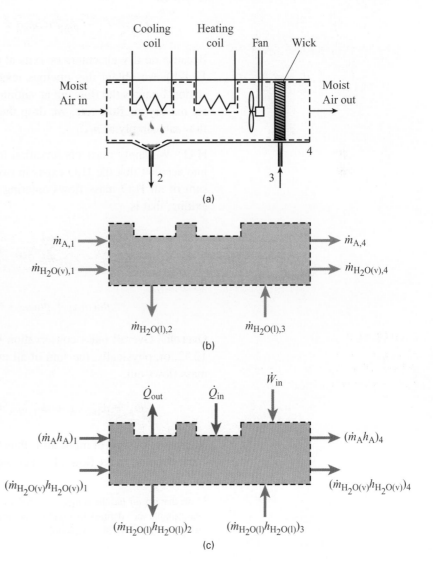

station 1 and is then cooled or heated, or both, while liquid water is removed (station 2) or added (station 3), or both. The air stream then exits at station 4 with more or less moisture than it had upon entering.

Assumptions

To perform our analysis, we assume the following:

- The flow is steady.
- The flow is at steady state with no accumulation or loss of mass within the control volume.
- No air is dissolved in the liquid water.
- The kinetic and potential energies of all streams are negligible.
- Both the dry air and water vapor behave as ideal gases.

Mass Conservation

We can write three different mass conservation expressions for the situation shown in Fig. 12.42a: one for the dry air[9] alone, one for H_2O alone, and an overall or combined relationship. Individual flows of dry air and H_2O are shown in Fig. 12.42b.

Dry Air   With the assumptions just listed, mass conservation for the dry air is written

$$\dot{m}_{A,1} = \dot{m}_{A,4} \equiv \dot{m}_A = \text{constant.} \tag{12.51}$$

Because no dry air enters or exits at any station other than 1 or 4 (see Fig. 12.42b), we obtain this obvious result that the mass flow of dry air in at station 1 equals the flow out at station 4. Because there is only a single value for the dry-air flow rate, we drop the subscripts 1 and 4 and designate this flow rate simply as $\dot{m}_A$.

H_2O   We apply mass conservation to the chemical compound H_2O, taking into account that the H_2O exists in two phases, liquid and vapor. Simply, the sum of all H_2O mass flows entering must equal the sum of all H_2O flows exiting, that is,

$$\sum^{\text{inlets}} \dot{m}_{H_2O,i} = \sum^{\text{outlets}} \dot{m}_{H_2O,i}, \tag{12.52a}$$

or

$$\dot{m}_{H_2O(v),1} + \dot{m}_{H_2O(l),3} = \dot{m}_{H_2O(v),4} + \dot{m}_{H_2O(l),2}. \tag{12.52b}$$

Overall   Overall mass conservation is formally the sum of Eqs. 12.51 and 12.52, or, physically, the sum of all mass flows in must equal the sum of all mass flows out:

$$\dot{m}_A + \dot{m}_{H_2O(v),1} + \dot{m}_{H_2O(l),3} = \dot{m}_A + \dot{m}_{H_2O(v),4} + \dot{m}_{H_2O(l),2}. \tag{12.53}$$

The mass flow rate of moist air is thus the sum of the dry-air mass flow and the corresponding mass flow of water vapor (e.g., $\dot{m}_{\text{moist air},1} = \dot{m}_A + \dot{m}_{H_2O\,(v),1}$).

---

[9] Note that *dry air* has the composition defined in Appendix C and should not be confused with the "simple air" defined to simplify the analysis of combustion processes. The molecular weight of dry air is 28.97 kg/kmol.

From (Eqs. 12.51–12.53), we see that, conceptually, mass conservation is straightforward and intuitive. Later, this result will be rewritten by expressing the amount of water vapor using the specific humidity.

Energy Conservation

With our assumptions, the general steady-flow expression of energy conservation (Eq. 5.65),

$$\sum \dot{E}_{in} = \sum \dot{E}_{out},$$

simplifies to

$$\dot{Q}_{in} + \dot{W}_{in} + \dot{m}_A h_{A,1} + (\dot{m}_{H_2O(v)} h_{H_2O(v)})_1 + (\dot{m}_{H_2O(l)} h_{H_2O(l)})_3$$
$$= \dot{Q}_{out} + (\dot{m}_{H_2O(l)} h_{H_2O(l)})_2 + \dot{m}_A h_{A,4} + (\dot{m}_{H_2O(v)} h_{H_2O(v)})_4. \tag{12.54}$$

In the analysis of the devices and systems just discussed here, the combined application of mass and energy conservation allows the determination of unknown moisture fractions or temperatures; for example, one may desire to determine the outlet temperature and relative humidity for an air-conditioner stream given the inlet conditions and cooling rate.

12.5c Some New Concepts and Definitions

In this section, we define and discuss commonly used terms related to mixtures of air and water vapor.

Psychrometry

The term **psychrometry** refers generally to the art and science of measuring the moisture content in dry air; thus, the concepts discussed in the following are topics within the field of psychrometry. The concepts of wet-bulb and dry-bulb temperatures and the use of the psychrometric chart, although generally a part of psychrometrics, are not covered in this text. References [8, 9], among many others, provide information on these topics.

Thermodynamic Treatment of Water Vapor in Dry Air

With the previously given assumption that moist air is a mixture of two ideal gases—dry air and water vapor—the total pressure of the mixture is expressed (Eq. 2.65) as

$$P = P_A + P_{H_2O(v)}. \tag{12.55}$$

The temperature, however, is uniform throughout the gas-phase mixture, and so we write

$$T = T_A = T_{H_2O(v)}. \tag{12.56}$$

Usually the total pressure (e.g., the local barometric pressure) is known, whereas the partial pressure of the water vapor depends on how much moisture is present within the mixture. For an ideal-gas mixture, the mole fraction and mass fraction of the water vapor in the dry air/water vapor mixture are defined, respectively, as

$$X_{H_2O(v)} = \frac{N_{H_2O(v)}}{N_A + N_{H_2O(v)}} = \frac{P_{H_2O(v)}}{P} \tag{12.57}$$

and

$$Y_{H_2O(v)} = \frac{M_{H_2O(v)}}{M_A + M_{H_2O(v)}}. \tag{12.58}$$

In psychrometry, however, composition is usually defined by the humidity ratio or relative humidity rather than by using mole or mass fractions. We now define and discuss these composition variables.

Humidity Ratio

The **humidity ratio** ω (also known as the **specific humidity** or the **absolute humidity**) is defined as the mass of water vapor per unit mass of dry air:

$$\omega \equiv \frac{M_{H_2O(v)}}{M_A}. \tag{12.59}$$

We can easily relate the humidity ratio to the water-vapor mass fraction as follows:

$$Y_{H_2O(v)} = \frac{M_{H_2O(v)}}{M_{H_2O(v)} + M_A} = \frac{M_{H_2O(v)}/M_A}{M_{H_2O(v)}/M_A + 1} = \frac{\omega}{\omega + 1}, \tag{12.60a}$$

or, conversely,

$$\omega = \frac{Y_{H_2O(v)}}{1 - Y_{H_2O(v)}}. \tag{12.60b}$$

Applying the relationship between mass and mole fractions (Eq. 2.59), we also obtain

$$X_{H_2O(v)} = \frac{\omega \mathcal{M}_A}{\mathcal{M}_{H_2O} + \omega \mathcal{M}_A}, \tag{12.60c}$$

or, conversely,

$$\omega = \frac{X_{H_2O(v)} \mathcal{M}_{H_2O}}{(1 - X_{H_2O(v)}) \mathcal{M}_A}. \tag{12.60d}$$

The humidity ratio can also be related to the water-vapor partial pressure. Applying the ideal-gas equation of state independently for both the water vapor and the dry air (see Eq. 2.61), we have

$$M_{H_2O(v)} = \frac{P_{H_2O(v)} \mathcal{V}_{mix}}{(R_u/\mathcal{M}_{H_2O})T}$$

and

$$M_A = \frac{P_A \mathcal{V}_{mix}}{(R_u/\mathcal{M}_A)T} = \frac{(P - P_{H_2O(v)})\mathcal{V}_{mix}}{(R_u/\mathcal{M}_A)T}.$$

Substituting these expressions into our definition of humidity ratio (Eq. 12.59) yields

$$\omega = \frac{\mathcal{M}_{H_2O}}{\mathcal{M}_A} \frac{P_{H_2O(v)}}{P - P_{H_2O(v)}}, \tag{12.61a}$$

or

$$\omega = 0.622 \frac{P_{H_2O(v)}}{P - P_{H_2O(v)}},$$ (12.61b)

where numerical values for $\mathcal{M}_{H_2O}$ (= 18.016) and $\mathcal{M}_A$ (= 28.97) have been substituted. The inverse of this relationship is sometimes useful:

$$P_{H_2O(v)} = \frac{\omega P}{0.622 + \omega}.$$ (12.61c)

Relative Humidity

The relative humidity is the fractional, or percentage, expression of the actual amount of water vapor in the air at a given temperature to the maximum possible amount that the air can hold at that same temperature. Before expressing this definition symbolically, we explore its meaning using a concrete example.

Consider a mixture of dry air and water vapor at 25°C with a total pressure of 1 atm (101.325 kPa) and a water-vapor partial pressure of 2.3393 kPa. The state of the water vapor is indicated as point A in Fig. 12.43, and the water-vapor mole fraction can be found by applying Eq. 12.57:

$$X_{H_2O(v)} = \frac{P_{H_2O(v)}}{P} = \frac{2.3393 \text{ kPa}}{101.325 \text{ kPa}} = 0.02309.$$

We now conduct a thought experiment in which we increase the amount of moisture present while maintaining the total pressure fixed and the temperature at 25°C. This process continues until the vapor pressure of the water in the air equals the saturation pressure at 25°C, point B in Fig. 12.43. Further addition

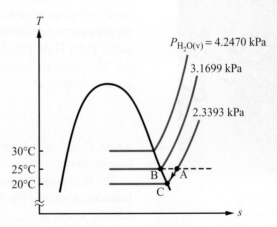

FIGURE 12.43

A T–s diagram for H₂O (not to scale) shows isobars corresponding to saturation temperatures of 20°C, 25°C, and 30°C. For air at 25°C, the maximum possible water vapor partial pressure is 3.1699 kPa. At this condition, the air is saturated with water (point B) and the relative humidity is 100%. Partial pressures less than this result in relative humidities less than 100%; for example, at point A, the relative humidity is 73.8% [= (2.3393/3.1699) 100%] and the corresponding dew-point temperature is $T_{DP} = 20°C$ (point C).

of moisture results in a two-phase system with liquid water condensing out of the air–water vapor mixture. The amount of moisture in the air at point B is thus the maximum quantity that the air can hold at 25°C, and the vapor pressure is given by $P_v = P_{sat}(25°C) = 3.1699$ kPa. The corresponding mole fraction is

$$X_{H_2O(v)max@25°C} = \frac{P_{sat}(25°C)}{P} = \frac{3.1699 \text{ kPa}}{101.325 \text{ kPa}} = 0.03128.$$

The relative humidity ϕ for this experiment is thus the ratio of the actual water vapor content $X_{H_2O(v)actual}$ to the maximum possible $X_{H_2O(v)max@25°C}$, that is,

$$\frac{X_{H_2O(v)actual}}{X_{H_2O(v)max@25°C}} \quad \frac{0.02309}{0.03128} = 0.738 \text{ or } 73.8\%.$$

We now generalize and define the **relative humidity** as

$$\phi \equiv \frac{X_{H_2O(v)actual@T}}{X_{H_2O(v)max@T}} = \frac{P_{H_2O(v)}(T)}{P_{sat}(T)}, \tag{12.62}$$

See Example 7.11 and related material in Chapter 7 to review liquid–vapor equilibrium.

which is frequently expressed on a percentage basis by multiplying by 100%. Note that using a mass basis to define relative humidity also yields the result that $\phi = P_{H_2O(v)}(T)/P_{sat}(T)$. Note also that this definition of relative humidity assumes that the H_2O liquid–vapor equilibrium is unaffected by the presence of the air. Although the air does alter this equilibrium, the effect is very small.

Dew Point

The **dew point,** or **dew-point temperature** T_{DP}, is the temperature at which moisture begins to condense out of an air–water vapor mixture upon cooling. You are most likely familiar with the appearance of liquid water on the outside of a can or glass containing a cold beverage on a hot, humid day. Here the temperature at the outside surface of the container is at or below the dew point. Point C on Fig. 12.43 indicates the dew point (20°C) associated with the 25°C mixture having a relative humidity of 73.8%. Formally, the dew point is defined as

$$T_{DP} \equiv T_{sat}(P_{sat} = P_{H_2O(v)}), \tag{12.63}$$

where $P_{H_2O(v)}$ is the partial pressure of water vapor in the moist air under consideration.

The dew point is easily related to both the humidity ratio and the relative humidity. Knowing the dew point, we can find the humidity ratio by using Eq. 12.61a, where $P_{H_2O(v)} = P_{sat}(T_{DP})$. With the additional knowledge of the mixture temperature, the relative humidity can also be determined from its definition in Eq. 12.62. We illustrate these relationships with the following example.

Example 12.15

On August 6, 2002, the local weather station at State College, Pennsylvania, reported an ambient temperature of 18°C (64 F) and a dew point of 8°C (46 F). The local barometric pressure (not corrected to sea level) was 29.16 mm Hg. Determine the relative humidity, the humidity ratio, the water-vapor mole fraction, and the water-vapor mass fraction associated with the reported measurements.

Solution

Known T, T_{DP}, P

Find $\phi, \omega, X_{H_2O(v)}, Y_{H_2O(v)}$

Sketch

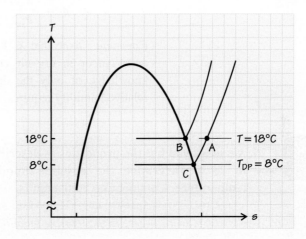

Assumptions

  i. Ideal-gas behavior of air and $H_2O(v)$
 ii. H_2O liquid–vapor equilibrium unaffected by air

Analysis   Given the dew-point temperature, we can determine the partial pressure of the water vapor in the moist air by recognizing that $P_A = P_C = P_{sat}(T_{DP})$ (see sketch). From the NIST database,

$$P_{sat}(T_{DP}) = 1.073 \text{ kPa} = P_{H_2O(v)}.$$

With this value and the total pressure ($P = 29.16 \text{ mm Hg} = 98.75 \text{ kPa}$), we can find three of the four requested quantities: For an ideal-gas mixture, the mole fraction is (Eq. 12.57)

$$X_{H_2O(v)} = \frac{P_{H_2O(v)}}{P} = \frac{1.073 \text{ kPa}}{98.75 \text{ kPa}} = 0.01087,$$

the mass fraction is (see Eqs. 2.59a and 2.60b)

$$
\begin{aligned}
Y_{H_2O(v)} &= X_{H_2O(v)} \frac{\mathcal{M}_{H_2O}}{\mathcal{M}_{mix}} \\
&= X_{H_2O(v)} \frac{\mathcal{M}_{H_2O}}{X_{H_2O(v)}\mathcal{M}_{H_2O} + (1 - X_{H_2O(v)})\mathcal{M}_A} \\
&= 0.01087 \frac{18.016}{0.01087(18.016) + (1 - .01087)28.97} \text{ kg}_{H_2O}/\text{kg}_{mix} \\
&= 0.00679 \text{ kg}_{H_2O}/\text{kg}_{mix},
\end{aligned}
$$

and the humidity ratio is (Eq. 12.61)

$$\omega = \frac{\mathcal{M}_{H_2O}}{\mathcal{M}_A} \frac{P_{H_2O(v)}}{P - P_{H_2O(v)}}$$

$$= \frac{18.016}{28.97} \frac{1.073}{98.75 - 1.073} \; kg_{H_2O}/kg_A = 0.00683 \; kg_{H_2O}/kg_A.$$

To determine the relative humidity requires finding the saturation pressure associated with the ambient temperature of 18°C (i.e., point B on the sketch). Using the NIST database, we find

$$P_B = P_{sat}(T = 18°C) = 2.0647 \; kPa,$$

and from the definition of relative humidity (Eq. 12.62),

$$\phi = \frac{P_{H_2O(v)}(T)}{P_{sat}(T)} = \frac{1.073 \; kPa}{2.0647 \; kPa} = 0.52 \text{ or } 52\%.$$

Comment Note that X, Y, ω, and ϕ are all measures of the amount of water vapor in dry air and that the first three can be found from a knowledge of only the dew point and total pressure. To evaluate the relative humidity requires, in addition, the ambient temperature.

Self Test 12.10 **Air in a room has the properties as described in Example 12.15. The air is then heated to 75 F. Determine the absolute humidity and relative humidity at this new temperature. Comment on your results.**

(Answer: $\omega = 0.00683$, $\phi = 36.3\%$. The absolute humidity is unchanged, whereas the relative humidity decreases.)

Example 12.16

Show that, for the conditions of Example 12.15, the water vapor can be treated as an ideal gas.

Solution

See also Examples 2.14 and 2.17 in Chapter 2.

We follow the method used in Example 2.16 to test the ideal-gas assumption. For an ideal gas,

$$\frac{Pv}{RT} = 1,$$

but more generally,

$$\frac{Pv}{RT} = Z,$$

where Z is the compressibility factor. For a real gas, the closer Z is to unity, the closer its P–v–T behavior is to that of an ideal gas. Using the NIST online database we determine the real-gas specific volume for the water vapor at points A, B, and C on the sketch in Example 12.15 and then evaluate Z. For example, at point C

$$P = 1.073 \; kPa,$$

$$T = 8 + 273.15 \; K = 281.15 \; K,$$

$$v = 120.83 \; m^3/kg,$$

and the gas constant is

$$R = R_u/\mathcal{M}_{H_2O} = 8314.47/18.016 \text{ J/kg·K} = 461.505 \text{ J/kg·K}.$$

Thus,

$$Z = \frac{1.073 \times 10^3(120.83)}{461.505\,(281.15)} = 0.9992.$$

Clearly, 0.9992 is close to unity. Similar calculations at points A and B yield, respectively, $Z = 0.9994$ and $Z = 0.9988$. Again, we see that ideal-gas behavior is an excellent approximation.

12.5d Recast Conservation Equations

To facilitate the use of the mass and energy conservation equations previously developed, specifically Eqs. 12.52, 12.53, and 12.54, we employ the definition of humidity ratio (i.e., $\dot{m}_{H_2O(v),i} = \dot{m}_A \omega_i$) to recast these expressions as follows:

H_2O mass conservation (12.52b):

$$\dot{m}_A \omega_1 + \dot{m}_{H_2O(l),3} = \dot{m}_A \omega_4 + \dot{m}_{H_2O(l),2}, \tag{12.64}$$

overall mass conservation (12.53):

$$\dot{m}_A(1 + \omega_1) + \dot{m}_{H_2O(l),3} = \dot{m}_A(1 + \omega_4) + \dot{m}_{H_2O(l),2}, \tag{12.65}$$

overall energy conservation (Eq. 12.54):

$$\dot{Q}_{in} + \dot{W}_{in} + \dot{m}_A(h_{A,1} + \omega_1 h_{H_2O(v),1}) + \dot{m}_{H_2O(l),3} h_{H_2O(l),3}$$
$$= \dot{Q}_{out} + \dot{m}_A(h_{A,4} + \omega_4 h_{H_2O(v),4}) + \dot{m}_{H_2O(l),2} h_{H_2O(l),2}. \tag{12.66}$$

We illustrate the use of these equations in the following examples.

Example 12.17

An evaporative cooler is used to cool a student apartment in Tempe, Arizona. Hot, moist air from outside enters the cooler at 106 F (314.2 K) with a relative humidity of 16% at a volumetric flow rate of 3000 ft³/min (1.416 m³/s). As shown in the sketch, a fan driven by a 1/8-hp (93.2-W) electric motor pulls the air through the cooler. The air exits the cooler into the apartment at 78 F (298.7 K) with an unknown humidity. The barometric pressure is 100 kPa. Determine the mass flow rate of liquid water that must be supplied to the cooler and the relative humidity of the cool air stream.

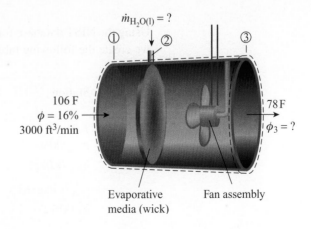

Evaporative media (wick)   Fan assembly

Solution

Known $P_{atm}, T_1, T_3, \phi_1, \dot{V}_1, \dot{W}_{elec}$

Find $\dot{m}_{H_2O(l),2}, \phi_3$

Sketch

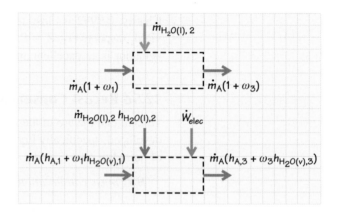

Assumptions

i. The flow is steady and at steady state.
ii. Kinetic and potential energies are negligible.
iii. The process is adiabatic ($\dot{Q}_{cv} = 0$).
iv. The pressure is essentially uniform ($P_1 = P_2 = P_3 = P_{atm}$).
v. Air and $H_2O(v)$ behave as ideal gases.
vi. Liquid water at station 2 enters at T_3 and P_{atm}.

Analysis   Our overall strategy is to write mass and energy conservation expressions using the corresponding control volume sketches shown. This results in two equations with two unknowns, ω_3 and $\dot{m}_{H_2O(l),2}$, as all other quantities are easily determined or approximated. Before writing these conservation expressions, we determine the properties of the air, water vapor, and liquid water at stations 1, 2, and 3 and the mass flow rate of the entering dry air, $\dot{m}_A$. Using the definition of relative humidity (Eq. 12.62) and the NIST database, we find $P_{H_2O(v),1}$:

$$\phi_1 = \frac{P_{H_2O(v),1}\,(314.2\ K)}{P_{sat}\,(314.2\ K)},$$

or

$$P_{H_2O(v),1} = \phi_1 P_{sat}\,(314.2\ K)$$

$$= 0.16(7.8085\ \text{kPa}) = 1.2494\ \text{kPa}.$$

Using the NIST database for H_2O properties and Tables C.2 and C.3 for air, we create the following tables to organize our calculations:

Station	1	2	3
T (K)	314.2	298.7	298.7
P_{H_2O} (kPa)	1.2494	100	?
h_{H_2O} (kJ/kg)	2577.5	107.22	2547.5 (est.)
$c_{p,A}$ (kJ/kg·K)	1.0072	—	1.007
h_A (kJ/kg)	440.34	—	424.73

The enthalpy of the water vapor at station 3 is estimated by assuming $h_{H_2O(v),3} \approx h_{sat,3}$. Since the water vapor acts essentially as an ideal gas, the enthalpy of the vapor will not be affected significantly by the pressure.

To determine $\dot{m}_A$, we combine the definitions of mass and volume flow rates (Eq. 3.14 and 3.15) and humidity ratio (Eq. 12.59) as follows:

$$\dot{m}_1 = \rho_1 \dot{V}_1 = \dot{m}_A(1 + \omega_1),$$

where (Eq. 12.61)

$$\omega_1 = 0.622 \frac{P_{H_2O(v),1}}{P_1 - P_{H_2O(v),1}}$$

$$= 0.622 \frac{1.2494}{100 - 1.2494} = 0.00787.$$

We obtain the mixture density at station 1 by applying the ideal-gas equation of state, where the mixture molecular weight is determined using the water-vapor mole fraction from Eq. 12.57, that is,

$$X_{H_2O(v),1} = \frac{P_{H_2O(v),1}}{P_1} = \frac{1.2494}{100} = 0.01249,$$

$$\mathcal{M}_1 = X_{H_2O(v),1}\mathcal{M}_{H_2O} + (1 - X_{H_2O(v),1})\mathcal{M}_A$$

$$= 0.01249(18.016) + (1 - 0.01249)28.97$$

$$= 28.83 \, \text{kg/kmol},$$

and

$$\rho_1 = \frac{P_1 \mathcal{M}_1}{R_u T_1} = \frac{100 \times 10^3 (28.83)}{8314.47(314.2)} = 1.104$$

$$[=] \frac{(\text{N/m}^2)(\text{kg/kmol})}{(\text{J/kmol} \cdot \text{K})\text{K}} \left[\frac{1 \, \text{J}}{\text{N} \cdot \text{m}} \right] = \text{kg/m}^3.$$

Therefore,

$$\dot{m}_A = \frac{\rho_1 \dot{V}_1}{1 + \omega_1} = \frac{1.104(1.416)}{1 + 0.00787}$$

$$= 1.551$$

$$[=] (\text{kg/m}^3)(\text{m}^3/\text{s}) = \text{kg/s}.$$

With these preliminary calculations out of the way, we now write mass and energy conservation using our sketches as guides. For H_2O conservation,

$$\sum^{\text{inlets}} \dot{m}_{H_2O,i} = \sum^{\text{outlets}} \dot{m}_{H_2O,i},$$

so

$$\dot{m}_{H_2O(v),1} + \dot{m}_{H_2O(l),2} = \dot{m}_{H_2O(v),3},$$

or, equivalently,

$$\omega_1 \dot{m}_A + \dot{m}_{H_2O(l),2} = \omega_3 \dot{m}_A.$$

This can be rearranged as

$$\dot{m}_{H_2O(l),2} = \dot{m}_A(\omega_3 - \omega_1). \tag{A}$$

For energy conservation,

$$\sum \dot{E}_{\text{in}} = \sum \dot{E}_{\text{out}},$$

which gives

$$\dot{m}_A(h_{A,1} + \omega_1 h_{H_2O(v),1}) + \dot{m}_{H_2O(l),2} h_{H_2O(l),2} + \dot{W}_{elec}$$
$$= \dot{m}_A(h_{A,2} + \omega_3 h_{H_2O(v),3}). \tag{B}$$

Substituting Eq. A into Eq. B and solving for the unknown ω_3 yield

$$\omega_3 = \frac{(h_{A,1} - h_{A,3}) + \omega_1(h_{H_2O(v),1} - h_{H_2O(l),2})}{h_{H_2O(v),3} - h_{H_2O(l),2}} + \frac{\dot{W}_{elec}}{\dot{m}_A(h_{H_2O(v),3} - h_{H_2O(l),2})},$$

which is evaluated as follows:

$$\omega_3 = \frac{(440.34 - 424.73) + 0.00787(2577.5 - 107.22)}{(2547.5 - 107.22)}$$
$$+ \frac{0.0932}{1.551(2547.5 - 107.22)}$$
$$= 0.01436 + 0.0000246 = 0.01438.$$

Returning to H_2O mass conservation (Eq. A), we can now find the liquid-water flow rate:

$$\dot{m}_{H_2O(l),2} = \dot{m}_A(\omega_3 - \omega_1)$$
$$= 1.551(0.01438 - 0.00787) \text{ kg/s} = 0.0101 \text{ kg/s}.$$

To find the exit relative humidity ϕ_3, we apply Eqs. 12.61b and 12.62 in succession as follows:

$$\omega_3 = 0.622 \frac{P_{H_2O(v),3}}{P_3 - P_{H_2O(v),3}},$$

or

$$P_{H_2O(v),3} = \frac{P_3}{\frac{0.622}{\omega_3} + 1} = \frac{100 \text{ kPa}}{\frac{0.622}{0.01438} + 1} = 2.260 \text{ kPa}.$$

Thus,

$$\phi_3 = \frac{P_{H_2O(v),3}}{P_{sat}(T_3)} = \frac{2.260 \text{ kPa}}{3.2754 \text{ kPa}} = 0.69 \text{ or } 69\%.$$

Comments   Using the value for $h_{H_2O(v),3}$ as the actual state of the water vapor at station 3 ($h = 2548.0$ kJ/kg at 2.260 kPa, 298.7 K) rather than our original estimate (2547.5 kJ/kg) has an insignificant effect on the determination of ω_3 (0.01437 versus 0.01438), as anticipated. We note that the contribution of $\dot{W}_{elec}$ to the final result for ω_3 is quite small (0.0000246, as revealed in our calculations).

Note that evaporative coolers work best in hot, dry climates where the addition of moisture does not significantly decrease comfort. In this example, a substantial quantity of water (0.0101 kg/s $\approx$ 9.6 gal/hr) is used to achieve the desired cooling.

Self Test
12.11

✓ **Redo Example 12.17 for an outside relative humidity of 25%.**

(Answer: $\phi_3 = 90.2\%$)

Example 12.18

Image courtesy of Sunpentown.

Consider a household dehumidifier system as pictured in Fig. 12.40 with the component arrangement as shown in Fig. 12.39. The following information is known about this system:

inlet mass flow rate (dry air + water vapor) = 0.15 kg/s,
inlet pressure = 98 kPa,
inlet temperature = 295 K (71.3 F),
inlet relative humidity = 85%,
outlet pressure = 100 kPa,
condensate flow rate = 3.284×10^{-4} kg/s (~0.3 gal/hr),
condensate temperature = 281 K (46.2 F),
total electrical power supplied (compressor and fan motor drives combined) = 688.7 W, and
heat loss from the dehumidifier system to the surroundings = 15 W.

Determine the following quantities:

A. Total mass flow rate (dry air + water vapor) at the outlet
B. Specific humidity at the outlet
C. Temperature and relative humidity at the outlet

Solution

Known $\dot{m}_1, \dot{m}_2, P_1, P_3, T_1, T_2, \phi_1$

Find $\dot{m}_3, \omega_3, T_3, \phi_3$

Sketch

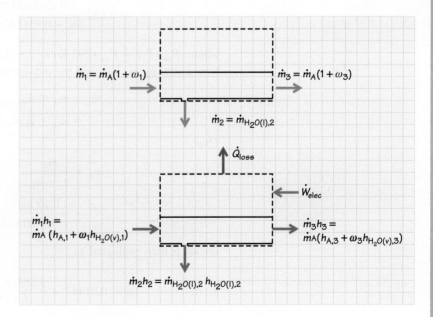

Assumptions

    i. The flow is steady and at steady state.
   ii. Kinetic and potential energies are negligible.
  iii. Air and $H_2O(v)$ behave as ideal gases.
  iv. $P_{H_2O(l),2} = P_3$.

Analysis   To find the outlet mass flow rate $\dot{m}_3$, we write the following statement of overall mass conservation:

$$\overset{\text{inlets}}{\sum \dot{m}_i} = \overset{\text{outlets}}{\sum \dot{m}_i},$$

or

$$\dot{m}_1 = \dot{m}_2 + \dot{m}_3.$$

Thus,

$$\dot{m}_3 = \dot{m}_1 - \dot{m}_2$$
$$= 0.15 - 3.284 \times 10^{-4}\,\text{kg/s} = 0.1497\,\text{kg/s}.$$

To find the specific humidity at the outlet (station 3), we apply mass conservation to the H_2O alone as follows:

$$\dot{m}_{H_2O(v),1} = \dot{m}_{H_2O(l),2} + \dot{m}_{H_2O(v),3},$$

or

$$\dot{m}_A \omega_1 = \dot{m}_{H_2O(l),2} + \dot{m}_A \omega_3.$$

Solving for ω_3 yields

$$\omega_3 = \omega_1 - \frac{\dot{m}_{H_2O(l),2}}{\dot{m}_A}.$$

The specific humidity at station 1 can be found from Eq. 12.61 combined with the definition of relative humidity (Eq. 12.62):

$$\omega_1 = 0.622\,\frac{P_{H_2O(v),1}}{P_1 - P_{H_2O(v),1}} = 0.622\,\frac{\phi_1 P_{\text{sat}}(T_1)}{P_1 - \phi_1 P_{\text{sat}}(T_1)}$$
$$= 0.622\,\frac{0.85(2.6212)}{98 - 0.85(2.6212)} = 0.01447,$$

where $P_{\text{sat}}(T_1) = P_{\text{sat}}(295\,\text{K}) = 2.6212\,\text{kPa}$ from the NIST database. Now knowing ω_1, we find the dry-air flow rate from the definition of ω:

$$\dot{m}_1 = \dot{m}_A + \dot{m}_{H_2O(v),1}$$
$$= \dot{m}_A + \omega_1 \dot{m}_A = \dot{m}_A(1 + \omega_1).$$

Thus,

$$\dot{m}_A = \frac{\dot{m}_1}{1 + \omega_1}$$
$$= \frac{0.15\,\text{kg/s}}{1 + 0.01447} = 0.1479\,\text{kg/s},$$

and ω_3 is then evaluated as

$$\omega_3 = 0.01447 - \frac{3.284 \times 10^{-4}\,\text{kg/s}}{0.1479\,\text{kg/s}} = 0.01225.$$

To find the outlet temperature T_3, we apply overall conservation of energy to our control volume using the second of our two sketches as a guide; thus,

$$\sum \dot{E}_{\text{in}} = \sum \dot{E}_{\text{out}},$$

or

$$\dot{m}_1 h_1 + \dot{W}_{\text{elec}} = \dot{m}_2 h_2 + \dot{m}_3 h_3 + \dot{Q}_{\text{loss}}.$$

Expanding the $\dot{m}h$ terms yields

$$\dot{m}_A(h_{A,1} + \omega_1 h_{H_2O(v),1}) + \dot{W}_{elec} = \dot{m}_{H_2O(l),2} h_{H_2O(l),2}$$
$$+ \dot{m}_A(h_{A,3} + \omega_3 h_{H_2O(v),3}) + \dot{Q}_{loss}.$$

Known quantities in this expression are $\dot{m}_A$, ω_1, $\dot{W}_{elec}$, $\dot{m}_{H_2O(l),2}$, ω_3, and $\dot{Q}_{loss}$, whereas the enthalpies, $h_{A,1}$, $h_{H_2O(v),1}$, and $h_{H_2O(l),2}$, are all easily evaluated from known temperatures and pressures. The remaining unknown quantities, $h_{A,3}$ and $h_{H_2O(v),3}$, are functions only of the unknown temperature if we assume negligible pressure dependence for $h_{H_2O(v),3}$ and use the approximation

$$h_{H_2O(v),3} \approx h_{H_2O(v)}(T_{sat} = T_3).$$

To proceed, we substitute $c_{p,A,avg}(T_3 - T_1)$ for $h_{A,3} - h_{A,1}$ to simplify our iterative calculation and rearrange our energy balance to separate unknown and known quantities on the left- and right-hand sides, respectively, that is,

$$\dot{m}_A c_{p,A,avg}(T_3 - T_1) + \dot{m}_A \omega_3 h_{H_2O(v),3}(T_3)$$
$$= \dot{m}_A \omega_1 h_{H_2O(v),1} - \dot{m}_{H_2O(l),2} h_{H_2O(l),2} + \dot{W}_{elec} - \dot{Q}_{loss}.$$

We obtain the following enthalpies for H_2O from the NIST database:

	Station 1	Station 2
T (K)	295	281
P_{H_2O} (kPa)	2.2280*	100
h (kJ/kg)	2541.0	33.094

* $P_{H_2O(v),1} = \phi_1 P_{sat}(T_1) = 0.85\,(2.6212) = 2.2280$ kPa.

From Table C.3, we see that for the temperature range of interest an appropriate value for $c_{p,A,avg}$ is 1.007 kJ/kg·K. Inserting numerical values into the energy balance equation yields

$$0.1479(1.007)(T_3 - 295) + 0.1479(0.01225)h_{sat\,vap}(T_3)$$
$$= 0.1479(0.01447)2541.0 - 3.284 \times 10^{-4}(33.094)$$
$$+ 0.6887 - 0.015,$$

where we recognize that each term has units of kW. Performing the indicated multiplications and rearranging and combining terms result in

$$0.1489T_3 + 0.001812\,h_{sat\,vap}(T_3) = 50.0368,$$

or

$$T_3 = 336.04 - 0.01217\,h_{sat\,vap}(T_3).$$

The form of this equation is quite amenable to iteration: We guess a value for T_3 and evaluate our expression for an improved value of T_3. The process can be repeated using this improved value to reevaluate the right-hand side. We begin by guessing $T_3 = 305$ K to obtain our first iterate:

$$T_3 = 336.04 - 0.01217\,(2558.9) \text{ K} = 304.898 \text{ K}.$$

Further iteration yields

$$T_3 = 336.04 - 0.01217\,(2558.7) \text{ K} = 304.901 \text{ K},$$

which is quite close to the input value; thus, we conclude then that, with proper rounding,

$$T_3 = 304.9 \text{ K}.$$

With knowledge now of T_3, we can determine the relative humidity ϕ_3, that is,

$$\phi_3 = \frac{P_{\text{H}_2\text{O(v)},3}}{P_{\text{sat}}(T_3)},$$

where $P_{\text{H}_2\text{O(v)},3}$ is obtained from the relationship between ω_3 and $P_{\text{H}_2\text{O(v)},3}$ (Eq. 12.61c):

$$\begin{aligned} P_{\text{H}_2\text{O(v)},3} &= \frac{\omega_3 P_3}{0.622 + \omega_3} \\ &= \frac{0.01225(100 \text{ kPa})}{0.622 + 0.01225} = 1.931 \text{ kPa}. \end{aligned}$$

Thus,

$$\phi_3 = \frac{1.931}{P_{\text{sat}}(304.9 \text{ K})} = \frac{1.931}{4.693} = 0.411 \text{ or } 41.1\%.$$

Comments   We note that the relative humidity of the outlet stream (41.1%) is much less than that of the entering stream (85%) and within the generally accepted comfort zone of 40%–60% relative humidity.

Although the calculations here are in places quite lengthy, the principles involved—mass and energy conservation—are straightforward and easy to apply once all of their constituent terms are evaluated. Drawing and labeling control volumes showing all of the mass flows and energy flows is essential to proper application of conservation principles.

SUMMARY

After studying this chapter, you should have a basic understanding of the operation of fossil-fueled steam power plants, turbojet engines, gas-turbine engines, refrigerators and heat pumps, and air conditioners and other systems in which moist air is important. Specifically, you should be able to represent the various thermodynamic cycles associated with each of these systems on appropriate thermodynamic coordinates, (e.g., T–s and P–v diagrams) and be able to model these systems to estimate measures of efficiency and performance. As a result of your study of this chapter, your ability to analyze complex thermal-fluid systems should be greatly enhanced. Specifically, your skills in selecting control volumes and in applying mass, momentum, and energy conservation principles to these control volumes should be quite well developed. These general skills should be such that you are comfortable analyzing thermal-fluid devices and systems that are not specifically discussed in this chapter. The detailed learning objectives presented at the beginning of this chapter should be reviewed at this time to complete this summary.

Chapter 12
Key Concepts & Definitions Checklist[10]

12.1 Fossil-Fueled Steam Power Plants

☐ Basic Rankine cycle ➤ *Q12.1, 12.1*

☐ Thermal efficiency ➤ *12.2*

☐ Rankine cycle with superheat ➤ *12.3*

☐ Rankine cycle with reheat ➤ *12.26*

☐ Rankine cycle with regeneration ➤ *12.31*

☐ Rankine cycle and variants: *T–s* diagrams ➤ *Q12.2*

☐ Steam extraction ➤ *Q12.3*

☐ Feedwater heater ➤ *Q12.4*

☐ Boiler efficiency ➤ *12.36*

☐ Overall efficiency ➤ *Q12.5*

12.2 Jet Engines

☐ Two major types ➤ *Q12.6*

☐ Turbojet "cycle" processes ➤ *Q12.7*

☐ Integral control volume analyses for mass, energy, and momentum ➤ *12.37*

☐ Air-standard cycle analysis ➤ *12.38*

☐ Propulsive efficiency ➤ *12.38*

☐ Specific thrust ➤ *12.38*

☐ Specific fuel consumption ➤ *12.38*

☐ Combustor operation: conservation of chemical elements, overall mass, and energy ➤ *12.42*

12.3 Gas-Turbine Engines

☐ Integral control volume analyses for mass and energy ➤ *12.81 (A&B)*

☐ Air-standard Brayton cycle ➤ *12.63*

☐ Process thermal efficiency ➤ *12.81(C)*

☐ Specific fuel consumption ➤ *12.81(D)*

12.4 Refrigerators and Heat Pumps

☐ Reversed cycles ➤ *12.82, 12.84, 12.85*

☐ Coefficients of performance ➤ *12.82, 12.84, 12.85*

☐ Vapor-compression-cycle processes ➤ *Q12.9, Q12.10, 12.89*

☐ Vapor-compression-cycle *T–s* diagram ➤ *Q12.9, Q12.10, 12.89*

☐ Cycle analysis ➤ *12.92*

12.5 Air Conditioning, Humidification, and Related Systems

☐ Evaporative cooler, humidifier, air conditioner, dehumidifier, and cooling tower ➤ *Q12.11, Q12.12, Q12.13*

☐ Integral control volume mass and energy conservation ➤ *12.142*

☐ Psychrometry ➤ *12.146*

☐ Humidity ratio (or specific humidity or absolute humidity) ➤ *12.111, 12.112*

☐ Relative humidity ➤ *12.111, 12.112*

☐ Dew point ➤ *12.111, 12.112*

☐ System performance analysis ➤ *12.143, 12.144, 12.145*

---

[10] Numbers following arrows below refer to end-of-chapter Problems (e.g., 12.1) and Questions (e.g., Q12.1).

REFERENCES

1. *Thermophysical Properties of Fluid Systems*, National Institute of Standards and Technology, Gaithersburg, MD, 2000, http://webbook.nist.gov/chemistry/fluid.

2. *Steam: Its Generation and Use*, 40th ed., S. C. Stulz and J. B. Kitto, Eds., Babock & Wilcox, Barberton, OH, 1992.

3. Mattingly, J. D., *Elements of Gas Turbine Propulsion*, McGraw-Hill, New York, 1996.

4. Hopkins, K. N., "Turbopropulsion Combustion—Trends and Challenges," AIAA Paper 80-1199, 1980.

5. Turns, S. R., *An Introduction to Combustion: Concepts and Applications*, 2nd ed., McGraw-Hill, New York, 2000.

6. St. Peter, J., *The History of Aircraft Turbine Engine Development in the United States*, International Gas Turbine Institute of the American Society of Mechanical Engineers, Atlanta, 1999.

7. Lefebvre, A. H., *Gas Turbine Combustion*, Taylor & Francis, Hemisphere, Washington, DC, 1983.

8. Moran, M. J., and Shapiro, H. N., *Fundamentals of Engineering Thermodynamics*, 3rd ed., Wiley, New York, 1995.

9. Çengel, Y. A., and Boles, M. A., *Thermodynamics: An Engineering Approach*, 4th ed., McGraw-Hill, New York, 2002.

Some end-of-chapter problems were adapted with permission from the following:

10. Look, D. C., Jr., and Sauer, H. J., Jr., *Engineering Thermodynamics*, PWS, Boston, 1986.

11. Myers, G. E., *Engineering Thermodynamics*, Prentice Hall, Englewood Cliffs, NJ, 1989.

Nomenclature

A	Area (m^2)
A/F	Air–fuel ratio (kg$_{air}$/kg$_{fuel}$)
c_p	Constant-pressure mass-specific heat (J/kg·K)
$\bar{c}_p$	Constant-pressure molar-specific heat (J/kmol·K)
COP	Coefficient of performance (dimensionless)
D	Diameter (m)
$\dot{E}$	Energy rate (W)
f_D	Darcy friction factor (dimensionless)
F	Heat exchanger correction factor (dimensionless)
$\boldsymbol{F}$	Force vector (N)
F_t	Thrust force (N)
F/A	Fuel–air ratio (kg$_{fuel}$/kg$_{air}$)
g	Gravitational acceleration (m/s^2)
h	Mass-specific enthalpy (J/kg)
$\bar{h}$	Molar-specific enthalpy (J/kmol)
HHV	Higher heating value (J/kg)
HV	Heating value (J/kg)
ke	Specific kinetic energy (J/kg)
L	Length (m)
M	Mass (kg)
$\dot{m}$	Mass flow rate (kg/s)
$\mathcal{M}$	Molecular weight (kg/kmol)
N	Number of moles or number of tubes in a bundle
P	Pressure (Pa) or chart parameter (Fig. 11.52)
pe	Specific potential energy (J/kg)
q	Heat energy per unit mass or $\dot{Q}/\dot{m}$ (J/kg)
$\dot{Q}$	Heat transfer rate (W)

R	Particular gas constant (J/kg·K) or chart parameter (Fig. 11.52)
R_u	Universal gas constant, 8314.472 (J/kmol·K)
Re	Reynolds number (dimensionless)
s	Mass-specific entropy (J/kg·K)
sfc	Specific fuel consumption for power-producing engines (kg/J)
SFC	Specific fuel consumption for thrust-producing engines [(kg/s)/N]
U_{OA}	Overall heat-transfer coefficient (W/m^2·K or W/m^2·°C)
v	Velocity (m/s)
v	Specific volume (m^3/kg)
$\boldsymbol{V}, V$	Velocity vector (m/s) and magnitude, respectively
$\mathcal{V}$	Volume (m^3)
$\dot{\mathcal{V}}$	Volume flow rate (m^3/s)
$\dot{W}$	Rate of work or power (W)
x	Steam quality (dimensionless)
X	Mole fraction (dimensionless)
y	Extracted fraction (dimensionless)
Y	Mass fraction (dimensionless)
z	Spatial coordinate in vertical direction (m)
Z	Compressibility factor (dimensionless)

GREEK

α	Kinetic energy correction factor (dimensionless)
β	Coefficient of performance or momentum flow correction factor (dimensionless)
γ	Specific-heat ratio (dimensionless)

Δ	Difference or change		fg	liquid to vapor
ε	Roughness height (m)		g	gas or vapor
μ	Viscosity ($N \cdot s/m^2$)		gen	generator
η	Efficiency		H	hot fluid
ρ	Density (kg/m^3)		i	inner
ϕ	Relative humidity (dimensionless or %)		in	into system or control volume
Φ	Equivalence ratio $(F/A)/(F/A)_{stoic}$ (dimensionless)		isen	isentropic process
			l	liquid
ω	Humidity ratio or specific humidity (kg_{H_2O}/kg_{air})		lm	log mean
			OA	overall

SUBSCRIPTS

A	air		out	out of system or control volume
amb	ambient		P	products
avg	average		prop	propulsive
b	downstream receiver		ref	reference
c	compressor		s	sensible
C	cold fluid		stoic	stoichiometric
CF	counterflow		t	turbine
e	exhaust		th	thermal
cv	control volume		v	vapor
elec	electrical			
f	liquid or formation			
F	Fuel			

SUPERSCRIPTS

$\circ$	Standard-state (e.g., pressure $P^\circ = $ 1 atm)

QUESTIONS

12.1 List the components of a simple Rankine cycle. Sketch the components and show how they interconnect.

12.2 Draw T–s diagrams for (a) the basic Rankine cycle without superheat, (b) the basic Rankine cycle with superheat, (c) the Rankine cycle with reheat, and (d) the Rankine cycle with regeneration.

12.3 Define *steam extraction*. Why is steam extracted from a turbine in a steam power plant?

12.4 What is the purpose of a feedwater heater in a steam power plant? Draw sketches of closed and open feedwater heaters.

12.5 What factors are important in determining the overall efficiency of the conversion of fuel energy to electricity in a fossil-fueled power plant?

12.6 Distinguish between turbojet and turbofan jet engines.

12.7 List the components of a turbojet engine. Sketch the components and show how they interconnect.

12.8 List the components of a stationary gas-turbine engine used for electrical power generation. Sketch the components and show how they interconnect.

12.9 List the components of a vapor-compression-cycle refrigerator. Sketch the components and show how they interconnect.

12.10 Draw a T–s diagram for an ideal vapor-compression refrigeration cycle.

12.11 Sketch an evaporative cooler. Discuss its operation.

12.12 List the components of an air conditioner that operates on a vapor-compression cycle. Sketch the components and show how they interconnect.

12.13 Explain how a cooling tower works.

Chapter 12 Problem Subject Areas

12.1–12.24	**Steam power plants with simple Rankine cycle**
12.25–12.30	**Rankine cycle with reheat**
12.31–12.36	**Rankine cycle with regeneration**
12.37–12.43	**Turbojet engines**
12.44–12.61	**Air-standard cycles**
12.62–12.81	**Gas-turbine engines**
12.82–12.110	**Refrigerators and heat pumps**
12.111–12.141	**Humidity measures and related concepts**
12.142–12.174	**Air conditioning, humidification, and related systems**

PROBLEMS

12.1 Consider an ideal Rankine cycle operating between 5 kPa and 10 MPa using H_2O as the working fluid. Steam enters the turbine as saturated vapor at a flow rate of 8 kg/s.

   A. Plot the ideal cycle on T–s coordinates.

   B. Determine the power produced by the turbine.

   C. Determine the power required to drive the pump.

   D. Determine the thermal efficiency of the cycle.

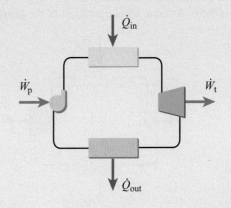

12.2 Repeat Problem 12.1 but assume now that the turbine isentropic efficiency is 95% and the pump isentropic efficiency is 85%. Also compare the steam quality at the turbine exit with the value from Problem 12.1.

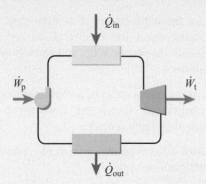

12.3 Consider an ideal Rankine cycle operating between 5 kPa and 10 MPa with a maximum cycle temperature of 750 K. The working fluid is H_2O.

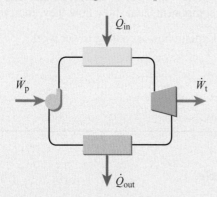

A. Plot the cycle on T–s coordinates.

B. Determine the fraction of the total energy imparted to the H_2O in the boiler that is associated with the superheater section (i.e., $\dot{Q}_{3-3'}/\dot{Q}_{2-3'}$) in Fig. 12.5.

C. Determine the cycle efficiency.

D. Compare the result from Part C to the cycle efficiency for the same cycle, but without superheat (see Problem 12.1). Discuss.

12.4 Consider a Rankine cycle operating between 5 kPa and 10 MPa with a maximum cycle temperature of 750 K. The working fluid is H_2O. Determine both the cycle thermal efficiency and the isentropic efficiency of the turbine if the steam quality at the turbine exit is 92%. Assume ideal operation of the pump.

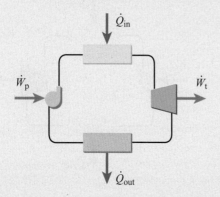

12.5 Consider the ideal Rankine cycle and operating conditions specified in Problem 12.1. River water is to be used as the cooling fluid in the steam condenser. The river water is available at 295 K, and the discharge temperature is to be no greater than 5 K above the inlet temperature. The overall heat-transfer coefficient associated with the condenser is 2500 $W/m^2 \cdot K$. Assuming primarily counterflow operation, estimate the heat exchange area required for this condenser.

Photograph courtesy of CS Energy, Ltd.

12.6 Conduct a design sensitivity study of an ideal Rankine cycle (with superheat) by performing the following tasks. The fluid is H_2O.

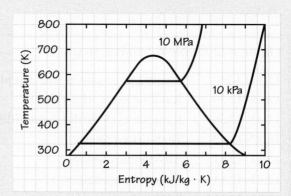

A. **Base case:** For the base case (turbine inlet temperature and pressure of 700 K and 10 MPa, respectively, and a condenser pressure of 10 kPa), calculate the following quantities:

   i.   the quality x of the steam exiting the turbine,

   ii. the work delivered by the turbine per unit mass of working fluid (i.e., $\dot{W}_{turb}/\dot{m}$),

   iii. the heat rejected in the condenser, per unit mass of working fluid (i.e., $\dot{Q}_{out,cond}/\dot{m}$),

   iv. the pump work per unit mass of working fluid, $\dot{W}_{pump}/\dot{m}$,

   v.   the heat added to the working fluid in the boiler, per unit mass of working fluid (i.e., $\dot{Q}_{in,boiler}/\dot{m}$), and

   vi. the thermal efficiency of the cycle.

   Perform all calculations without the aid of a computer; the NIST online database, however, may be useful to obtain any required thermodynamic properties.

B. **Turbine inlet temperature:** Determine the quality of the steam exiting the turbine and the cycle efficiency as a function of the turbine inlet temperature. Use the following values: 600 K, 700 K (base case), 800 K, and 900 K. All other parameters are as in the base case. Use spreadsheet software to perform your calculations (x and η) and to plot the results of your computations (x and η) as functions of turbine inlet temperature.

C. **Turbine inlet pressure:** Determine the turbine exit steam quality and the cycle thermal efficiency as a function of turbine inlet pressure. Use the following values: 5 MPa, 10 MPa (base case), 15 MPa, and 20 MPa. All other parameters are as in the base case. Use spreadsheet software and plot as in Part B.

D. **Turbine exhaust pressure:** Perform calculations and plot in the same manner as in Parts B and C. Now vary the turbine outlet pressure (condenser pressure) using the following values: 5 kPa, 10 kPa (base case), 50 kPa, and 100 kPa.

E. **Summary:** Write a few brief paragraphs discussing your results from Parts B, C, and D.

12.7 Determine the quality of the steam exiting the turbine and the cycle thermodynamic efficiency of the base-case cycle in Problem 12.6 if the turbine has an isentropic efficiency of 96% and the pump has an isentropic efficiency of 92%. Compare these results to those of the ideal cycle calculated in Part A of Problem 12.6.

12.8 In a Rankine cycle, steam leaves the boiler at 400°C and 5 MPa. The condenser pressure is 0.01 MPa. The water entering the pump is saturated. Determine the first-law thermal efficiency and compare it to that of a Carnot cycle operating between the same temperature extremes.

12.9 An engineer proposes to eliminate the condenser in an ideal Rankine cycle to form an open cycle. Lake water is available at 16°C. The lake water is pumped to 5 MPa and then heated at constant pressure to 400°C. The "spent" steam is rejected back into the lake. Determine the first-law thermal efficiency of this process and compare it with that of Problem 12.8. What are some of the practical problems with this proposed cycle?

12.10 A Rankine-cycle steam power plant operates with a turbine inlet temperature 1000 F and an exhaust pressure of 1 psia. If the turbine isentropic efficiency is 0.85, determine the maximum turbine inlet pressure that can be used if the exit quality is to be 0.90.

12.11 Consider a 50-MW, Rankine-cycle steam power plant. The steam enters the turbine at 1000 F and 1200 psia and leaves at 1 psia with a quality of 0.90. Water

leaves the condenser at 80 F and 1 psia. The power supplied to the pump may be neglected. Determine the thermal efficiency for the power plant and the mass flow rate (Mlb_m/hr) of steam in the boiler.

12.12 A solar-driven, Rankine-cycle power plant uses refrigerant R-134a as the working fluid. During operation (i.e., daytime), the turbine inlet state is 220 F and 200 psia. The air-cooled condenser operates at a pressure of 80 psia. The pump and turbine isentropic efficiencies are 0.70 and 0.80, respectively. The work output is 100 hp when the solar-collector input is 200 Btu/hr·ft². The R-134a leaves the condenser as saturated liquid. Determine (a) the thermal efficiency, (b) the refrigerant mass flow rate (klb_m/hr), and (c) the required collector surface area (ft²).

12.13 Consider a 600-MW, Rankine-cycle steam power plant. The boiler pressure is 1000 psia and the condenser pressure is 1 psia. The turbine inlet temperature is 1000 F and the pump inlet state is saturated liquid. The turbine and pump isentropic efficiencies are 100%. Determine the thermal efficiency of the power plant.

12.14 Consider a 600-MW, Rankine-cycle steam power plant. The boiler pressure is 1000 psia and the condenser pressure is 1 psia. The turbine inlet temperature is 1000 F. The condenser pressure is 1 psia and the pump inlet state is saturated liquid. The turbine isentropic efficiency is 0.90 and the pump efficiency is 0.85. Determine the thermal efficiency of the power plant.

12.15 A Rankine-cycle, steam power plant operates at a boiler pressure of 5 MPa and a condenser pressure of 0.01 MPa. The turbine inlet temperature is 500°C. The water entering the pump is saturated liquid. The mass flow rate through the boiler is 60 kg/s. The first-law thermal efficiency of the power plant is 0.30. Determine the thermal net power (MW) produced by the power plant.

12.16 Consider a 100-MW, Rankine-cycle steam power plant. The peak cycle temperature is 1000 F. The condenser pressure is 1 psia. Plot the first-law thermal efficiency and mass flow rate (Mlb_m/hr) as functions of the boiler pressure (from 1 to 1000 psia using a semilogarithic scale).

12.17 Consider a 100-MW, Rankine-cycle steam power plant. The boiler pressure is 600 psia and the condenser pressure is 1 psia. Plot the first-law thermal efficiency and the mass flow rate (lb_m/hr) as functions of the turbine inlet temperature (up to 1200 F).

12.18 Consider a Rankine-cycle steam power plant operating between a heat source at 500°C and a heat sink at 20°C. For safety reasons the boiler pressure cannot be more than 10 MPa. The mass flow rate of the steam is 50 kg/s. The pumping process occurs in the liquid region. Determine the maximum thermal efficiency and the electrical generating capacity (MW) of this power plant.

12.19 Steam leaves the boiler of a 100-MW, Rankine-cycle power plant at 700 F and 500 psia. The steam enters the turbine at 650 F and 475 psia. The turbine has an efficiency of 0.85 and exhausts at 2 psia. In the condenser, the water is subcooled to 100 F at 2 psia by lake water at 55 F. The pump efficiency is 0.75 and its discharge pressure is 550 psia. The water enters the boiler at 95 F and 530 psia. Determine the first-law thermal efficiency for this cycle and the mass flow rate (lb_m/hr) in the boiler.

12.20 Consider a 100-MW, Rankine-cycle steam power plant. Temperatures and pressures at the inlet states of the four basic components are given in the table. The turbine and the pump are both adiabatic. Determine (a) the mass flow rate (lb_m) in the boiler, (b) the heat-transfer rates across the power plant boundary, (c) the work rates across the power plant boundary, and (d) the power plant thermal efficiency.

State	Location	T (F)	P (psia)
1	Turbine inlet	1600	800
2	Condenser inlet	200	1
3	Pump inlet	100	1
4	Boiler inlet	102	800

12.21 A simple Rankine-cycle steam power plant operates with a steam flow rate of 500,000 lb_m/hr. Steam enters the turbine at 500 psia and 1000 F. The turbine exhaust (condenser inlet) conditions are 1.0 psia with a steam quality of 0.90. Determine the following:

A. Turbine output (kW)
B. Condenser heat rejection rate (Btu/hr)
C. Turbine isentropic efficiency (%)
D. Each term in the second-law entropy balance for the condensing process
E. Approximate pump work (kW)
F. Thermal efficiency of the plant (%)
G. Ideal thermal efficiency of the plant (%)

12.22 In a power plant operating on a Rankine cycle, steam at 400 kPa and quality 100% enters the turbine; the pressure is 3.5 kPa in the condenser. Determine the cycle efficiency. How is the efficiency affected if the turbine inlet conditions are changed to 4.0 MPa and 350°C?

12.23 Consider a Rankine cycle using R-134a as the working fluid. Saturated vapor leaves the boiler at 85°C and the condenser temperature is 40°C. What is the cycle efficiency?

12.24 The following data are for a simple steam power plant as shown in the sketch:

$$P_1 = 10 \text{ psia}, \qquad T_1 = 160 \text{ F},$$
$$P_2 = 500 \text{ psia},$$
$$P_3 = 480 \text{ psia}, \qquad T_3 = 800 \text{ F},$$
$$P_4 = 10 \text{ psia},$$
$$\text{steam flow rate} = 250{,}000 \text{ lb}_m/\text{hr},$$
$$\text{turbine isentropic efficiency} = 87\%.$$

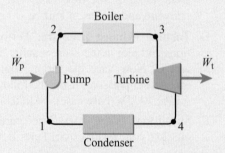

Determine the following:

A. Pipe size between condenser and pump if the velocity is not to exceed 20 ft/s
B. Minimum pump work (hp)
C. Net power output of plant (kW)
D. Condenser heat rejection rate (Btu/hr)
E. Heat input at boiler (Btu/hr)
F. Thermal efficiency (%)
G. Maximum possible thermal efficiency (%)

12.25 Add reheat to the base-case ideal Rankine cycle described in Problem 12.6. The exit pressure of the high-pressure turbine is 2.2 MPa. The steam exiting the high-pressure turbine is reheated back to 700 K before it enters the low-pressure turbine. Calculate the thermal efficiency for this ideal cycle with reheat and compare it with that of the base cycle without reheat. Discuss your results.

12.26 Consider an ideal Rankine cycle with superheat and reheat. The maximum temperature of the super-heated and reheated steam is 750 K. The maximum pressure in the cycle is 10 MPa and the minimum pressure is 5 kPa. Steam enters the high-pressure turbine at 750 K and exits as saturated vapor. The steam is then reheated to 750 K before expanding in the low-pressure turbine. The steam flow rate is 90 kg/s.

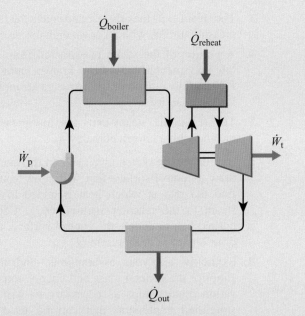

A. Plot the cycle on T–s coordinates.

B. Determine the total heat-addition rate in the boiler (economizer, steam generator, and superheater combined), the heat-addition rate in just the superheater section, and the heat-addition rate in the reheater.

C. Determine the net power produced by the cycle.

D. Determine the cycle thermal efficiency and compare with the thermal efficiency of an ideal cycle without reheat operating between the same high and low pressures (see Problem 12.3 Part C).

12.27 Consider a 100-MW, reheat-Rankine-cycle steam power plant. Superheated steam enters the high-pressure turbine at 800°C and 10 MPa. The steam enters the reheater at 700°C and 4 MPa and is reheated at constant pressure to 800°C. The steam expands through the low-pressure turbine and enters the condenser as saturated vapor at 0.01 MPa. The water enters the pump as a saturated liquid at 0.01 MPa. The turbines and pump are adiabatic. The pump is reversible. Determine the thermal efficiency of the power plant and mass flow rate (kg/s) of the steam.

12.28 Consider a 600-MW, reheat-Rankine-cycle steam power plant. The boiler pressure is 1000 psia, the reheater pressure is 200 psia, and the condenser pressure is 1 psia. Both turbine inlet temperatures are 1000 F. The water leaving the condenser is saturated liquid. Determine the thermal efficiency of the power plant and the boiler mass flow rate (lb_m/hr).

12.29 Consider a 600-MW, reheat-Rankine-cycle steam power plant. The boiler pressure is 1000 psia, the reheater pressure is 100 psia, and the condenser

pressure is 1 psia. The exit temperature of both the boiler and the reheater is 1000 F. The turbines and the pump have isentropic efficiencies of 0.90 and 0.85, respectively. The water leaving the condenser is saturated liquid. Determine the thermal efficiency of the power plant.

12.30 Consider an ideal Rankine cycle modified with the addition of a single reheat stage. The working fluid is H_2O. The maximum pressure in the cycle is 10.7 MPa, and the temperature of the steam exiting both the superheater and the reheater is 800 K. The pressure in the condenser is 0.01 MPa. A "rule of thumb" for the reheat cycle is that the optimum pressure for the reheat process is approximately one-fourth of the maximum pressure in the cycle. This problem explores the validity of this guideline.

A. Show the T–s state points for reheat cycles employing these conditions and the following reheat pressures:

   a. 4.675 MPa ($= 0.25\,P_{max} + 2$ MPa)

   b. 2.675 MPa ($= 0.25\,P_{max}$)

   c. 0.675 MPa ($= 0.25\,P_{max} - 2$ MPa)

   Show all three cycle on a single T–s diagram.

B. Calculate the thermal efficiency for the three reheat cycles given in Part A.

C. Create an engineering graph showing the effect of reheat pressure on cycle thermal efficiency by plotting $(\eta_{th}/\eta_{th,w/o\ reheat} - 1)\cdot 100\%$ versus q_{reheat} (kJ/kg).

D. Does the rule of thumb hold true? Discuss your results.

12.31 Consider an ideal regenerative Rankine cycle utilizing an open feedwater heater as shown in Fig. 12.9. The turbine inlet conditions are 10 MPa and 700 K, and the condenser pressure is 10 kPa. The mass flow rate of the steam entering the turbine is 12 kg/s. Steam is extracted between turbine stages at 2.2 MPa.

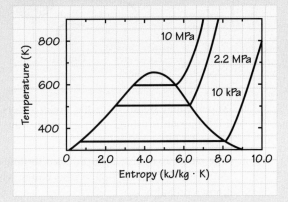

A. Determine the fraction of the total flow extracted from the turbine and the mass flow rate of the extracted steam.

B. Determine the power (kW) required to pump the condensate into the feedwater heater. Also express this pump power per unit mass of steam entering the turbine (i.e., $\dot{W}_{pump1}/\dot{m}_{tot}$).

C. Determine the rate of heat addition (kW) in the steam generator section (state 2 to state 3). Also express this on a per unit mass basis and compare it to the same quantity calculated for the base-case cycle in Problem 12.6 Part A.

D. Determine the power produced by the high-pressure portion of the turbine. Also determine the power delivered by the low-pressure turbine stages.

E. Compare the total power produced by the two turbine stages (per unit mass of total working fluid) with that produced by the turbine in the simple-cycle base case (Problem 12.6 Part A).

F. Determine the thermal efficiency and compare with that of the simple-cycle base case (Problem 12.6 Part A).

12.32 Consider the steam power plant shown in the sketch. This problem deals only with the turbines and the reheat process, with the key state points identified as follows:

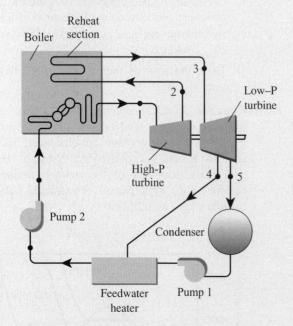

1. Superheated steam from the boiler enters the high-pressure turbine with a known flow rate $\dot{m}_1$ and a known enthalpy h_1.

2. The steam exits the high-pressure turbine and enters the reheat section of the boiler with a known enthalpy h_2.

3. The steam exits the reheater and enters the low-pressure turbine with a known enthalpy h_3.

4. A portion of the steam is extracted from the low-pressure turbine with a known enthalpy h_4. The flow rate $\dot{m}_4$ of the extracted steam is a known quantity.

5. The remaining steam exits the low-pressure turbine with a known enthalpy h_5.

A. Reproduce the sketch given and on this sketch draw a control surface that will allow you to find the rate at which heat is added to the steam in the reheater section, $\dot{Q}_{cv,in}$. Use a dashed line to denote the control surface, and draw an arrow to represent $\dot{Q}_{cv,in}$.

B. Explicitly list your assumptions, and then simplify the general mass and energy conservation relationships as necessary to write a simplified expression that can be used to determine $\dot{Q}_{cv,in}$.

C. On your sketch also draw a control surface that will allow you to find the power output, $\dot{W}_{cv,out}$, of both turbines combined. Use a dashed line to denote the control surface, and draw an arrow to represent $\dot{W}_{cv,out}$.

D. Explicitly list your assumptions, and then simplify the general mass and energy conservation relationships as necessary to write a simplified expression that can be used to determine $\dot{W}_{cv,out}$. Treat $\dot{m}_1$ and $\dot{m}_4$ as known quantities, applying mass conservation as necessary to eliminate any other flow rate(s).

12.33 Consider a 600-MW, regenerative-Rankine-cycle steam power plant with an open feedwater heater. The boiler pressure is 1000 psia, the extraction pressure is 200 psia, and the condenser pressure is 1 psia. The high-pressure turbine inlet temperature is 1000 F. The water leaving the condenser is saturated liquid. The exit state of the feedwater heater is saturated liquid. Determine the thermal efficiency of the power plant and the boiler mass flow rate (lb$_m$/hr).

12.34 Consider a 600-MW, regenerative-Rankine-cycle steam power plant. The boiler pressure is 1000 psia, the extraction pressure is 100 psia, and the condenser pressure is 1 psia. The boiler exit temperature is 1000 F. The pressure in the open feedwater heater is 100 psia. The water leaving the condenser is saturated liquid. The turbines and the pumps have isentropic efficiencies of 0.90 and 0.85, respectively. Determine the thermal efficiency of the power plant and the boiler mass flow rate.

12.35 A regenerative Rankine cycle operates with a first-stage turbine inlet state of 800°C and 5 MPa and a second-stage exhaust pressure of 0.01 MPa. A single open feedwater heater is used with an extraction pressure of 0.7 MPa. Assume the exit states of the condenser and of the feedwater heater are both saturated liquid. Turbine and pump efficiencies are 100%. Determine the first-law thermal efficiency for the cycle and the total turbine work rate per unit boiler mass flow rate (kJ/kg).

12.36 A natural gas–fired boiler raises steam for a pulp and paper manufacturing power plant. The air and natural gas enter at 298 K and 1 atm. Water enters the boiler at 305 K and 10 MPa at a flow rate of 38.7 kg/s. Steam exits the boiler/superheater at 780 K and 10 MPa. The boiler efficiency (defined in Example 12.8) is 85%.

    A. Determine the mass flow rate of the natural gas (CH_4).

    B. Determine the temperature of the products of combustion exiting the boiler. Assume stoichiometric combustion with negligible dissociation.

12.37 Air enters a turbojet engine with a velocity of 150 m/s (station 1) and a mass flow rate of 23.6 kg/s. Fuel is supplied to the engine at 0.3165 kg/s. The mass-specific standardized enthalpy of the air and fuel are 1.872 kJ/kg_A and −3210.4 kJ/kg_F, respectively, and the standardized enthalpy of the combustion products exiting the engine is −129.9 kJ/kg_e. Estimate the thrust developed by the engine.

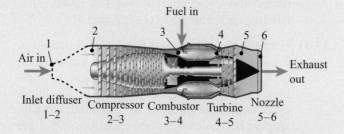

12.38 Consider the turbojet engine application presented in Example 12.10, with all conditions and properties the same, except that now the compressor and turbine have isentropic efficiencies of 94% and 96%, respectively. Repeat the cycle analysis and performance calculations, creating a table as shown at the top of Fig. 12.20. Also sketch the cycle on T–s coordinates. Use actual values of temperature, but values for specific entropy need only be qualitative. Discuss your results and compare with those of the original example.

12.39 Redo Example 12.9 using standardized enthalpies for the air, fuel, and products based on the ideal-gas properties given in Appendices B and H. Assume that the fuel is n-decane ($C_{10}H_{22}$) and that it enters the combustor at 298 K. Retain the assumption of simple air (1 kmol of O_2 per 3.76 kmol of N_2) and assume complete combustion with no dissociation.

12.40 Show that Eq. 12.33 can be expressed equivalently on (a) a per unit mass of mixture basis and (b) a per unit mass of fuel basis.

12.41 Consider the turbojet combustor described in Example 12.11. Determine the temperature of the combustion products exiting the combustor if the air–fuel ratio is reduced to 50:1.

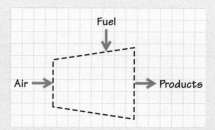

12.42 The maximum temperature that the turbine blades can withstand limits the performance of a turbojet engine. This temperature, in turn, is controlled by the temperature of the combustion products exiting the combustor. Determine the minimum air–fuel ratio that can be supplied to a turbojet engine such that the combustor outlet temperature does not exceed 1200 K (1700 F) when the air enters at 400 K. Assume the fuel is $C_{10}H_{22}$ and enters the combustor at 298 K. Also assume the usual simplified composition of air.

Photograph courtesy of NASA.

12.43 Create a spreadsheet model of a turbojet combustor in which the air temperature, fuel temperature, and air–fuel ratio are input quantities, and the combustion products temperature is an output quantity. Use iso-octane (C_8H_{10}) and simple air as the reactants.

12.44 Consider a power cycle in which air is the working fluid. The air is contained in a piston–cylinder assembly and undergoes the following processes:

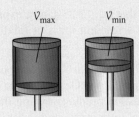

1–2: Internally reversible, adiabatic compression from the maximum volume V_{max} to the minimum volume V_{min}, where $V_{max} = 10\,V_{min}$.

2–3: Constant-volume heat addition to a specified value of $T_3 = 1000$ K.

3–4: Constant-pressure heat addition until the volume equals $3V_{min}$.

4–5: Internally reversible, adiabatic expansion from $V_4\ (= 3V_{min})$ to the maximum volume V_{max}.

5–1: Constant-volume heat rejection to return to the initial state.

A. Carefully sketch the cycle on P–V and T–s coordinates. Label the state points 1, 2, 3, etc.

B. Given the following properties for air:

$c_p = 1.008$ kJ/kg·K (a constant value),

$c_v = 0.721$ kJ/kg·K (a constant value),

$R_{air} = 0.287$ kJ/kg·K,

complete the following table:

State	1	2	3	4
P (kPa)	100		3333	
V (m³)	0.001	0.0001	0.0001	0.0003
T (K)	300		1000	
m (kg)				

C. Determine the heat added during the constant-pressure process 3–4.

12.45 An air-standard Otto cycle is often used as a very simplified model of a spark-ignition engine. The following sequence of processes applied to a fixed mass of air in a piston–cylinder assembly constitutes the Otto cycle:

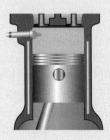

1–2: Internally reversible, adiabatic compression from the maximum volume V_{max} to the minimum volume V_{min}.

2–3: Constant-volume heat addition to a peak cycle temperature T_3.

3–4: Internally reversible, adiabatic expansion from V_{min} to V_{max}.

4–1: Constant-volume heat rejection to return to the initial state.

Sketch this cycle on P–V and T–S coordinates, labeling each state point (1, 2, 3, and 4).

12.46 Consider a fixed mass of air trapped in a piston–cylinder assembly. The air-standard diesel cycle is defined by the following sequences of processes applied to this system:

1–2: Internally reversible, adiabatic compression from the maximum volume V_{max} to the minimum volume V_{min}.

2–3: Constant-pressure heat addition to a peak cycle temperature T_3. Note that $V_{min} < V_3 < V_{max}$.

3–4: Internally reversible, adiabatic expansion to V_{max}.

4–1: Constant-volume heat rejection to return to the initial state.

Sketch this cycle on P–V and T–S coordinates, labeling each state point (1, 2, 3, and 4).

12.47 Consider the air-standard Otto cycle described in Problem 12.45 operating with a compression ratio of 8 $(= V_{max}/V_{min})$. The air just prior to compression is at 293 K and 1 atm. The maximum cycle temperature is 2500 K. Assuming constant specific heats, determine the temperature (K) and pressure (atm) at the start of each process and the thermal efficiency for the cycle.

12.48 Consider the air-standard Otto cycle described in Problem 12.45 operating with a compression ratio of 10 $(= V_{max}/V_{min})$. The air just prior to compression is at 70 F and 14.7 psia. The maximum cycle temperature is 4500 F. Assuming constant specific heats, determine the temperature and pressure at the beginning of each process and the thermal efficiency for the cycle.

12.49 An old V-8 spark-ignition engine has a 4-in-diameter cylinder and a 4-in-long stroke. The intake conditions are atmospheric (70 F and 14.7 psia). The engine has a compression ratio of 9. (See Appendix 1A to review the geometrical relationships.) The peak cycle temperature is 4000 F. Assuming constant specific heats and using the

Otto cycle as a model, determine the following quantities:

A. The net work (Btu) per cylinder per cycle

B. The pressure (psia) after compression

C. The peak pressure (psia) after combustion

D. The mean-effective pressure (psia) (i.e., net work divided by displacement volume)

E. The first-law thermal efficiency of the engine

F. The horsepower output of the engine running at 2000 rev/min, taking into account that in such a four-stroke engine, two revolutions of the crankshaft are required for each power stroke

G. The specific fuel consumption [i.e., the mass flow rate (lb_m/hr) of fuel used per horsepower], assuming the heating value of the fuel in this case is 19 $kBtu/lb_m$ of fuel

12.50 Consider the air-standard diesel cycle described in Problem 12.46 operating with a compression ratio of 15 (= $\mathcal{V}_{max}/\mathcal{V}_{min}$). The air just prior to compression is at 20°C and 0.1014 MPa. The maximum cycle temperature is 2500 K. Assuming constant specific heats, determine the temperature and pressure at the beginning of each process and the thermal efficiency for the cycle.

12.51 Determine the efficiency for the air-standard cycle indicated in the sketch. Assume the pressures and temperatures are known quantities.

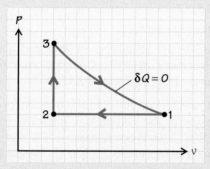

12.52 Rework Problem 12.51 for the following cycle.

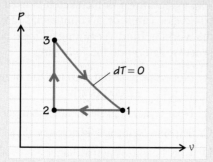

12.53 Rework Problem 12.51 for the following cycle.

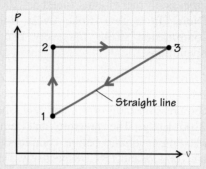

12.54 Express the thermal efficiency of a Carnot-cycle heat engine in terms of the isentropic compression ratio ($r_{isen} \equiv \mathcal{V}_{large}/\mathcal{V}_{small}$).

12.55 An inventor proposes a reversible nonflow cycle using air. The cycle consists of the following three processes:

1–2: Constant-volume compression from 101 kPa and 15°C to 700 kPa.

2–3: Constant-pressure heat addition during which the volume is tripled.

3–1: A process that appears as a straight line on a P–$\mathcal{V}$ diagram.

Draw P–$\mathcal{V}$ and T–S diagrams of the cycle. Compute the net work of the cycle in kJ/kg and Btu/lb_m.

12.56 In a compression-ignition engine, air originally at 120 F is compressed to a temperature of 980 F. Compression obeys the relationship $P\mathcal{V}^{1.34}$ = constant. Determine (a) the compression ratio required (i.e., the ratio of the volume before compression to the volume after compression), (b) the work of compression (Btu/lb_m), and (c) the heat transfer (Btu/lb_m).

12.57 In a compression-ignition engine, air originally at 50°C is compressed to a temperature of 550°C. Compression obeys the relationship $P\mathcal{V}^{1.34}$ = constant. Determine (a) the compression ratio required (i.e., the ratio of the volume before compression to the volume after compression), (b) the work of compression (kJ/kg), and (c) the heat transfer (kJ/kg).

12.58 The compression stroke for a four-stroke-cycle spark-ignition engine is approximated as a reversible, adiabatic process. Assume that the cylinder volume at bottom center is 400 in^3, the compression ratio (= $\mathcal{V}_1/\mathcal{V}_2$) is 9, and the cylinder is initially charged with air at 15 psia and 90 F. Determine (a) the temperature and pressure of the air after compression and (b) the horsepower required for this compression process if the engine speed is 2000 rev/min. (Note: In a four-stroke cycle, there is one compression stroke for every two crankshaft revolutions.)

12.59 Consider the air-standard Otto cycle described in Problem 12.45. A particular cycle operates with a compression ratio of 7:1 and has a heat input of 2100 kJ/kg. The pressure and temperature at maximum volume before compression (state 1) are 100 kPa and 15°C, respectively. Determine the net work produced by the cycle.

12.60 Consider the air-standard Otto cycle described in Problem 12.45, operating with a compression ratio of 8:1. In the constant-volume heat-addition process (state 2–state 3), 1800 kJ/kg of energy is transferred to the air. If the cycle begins with air at ambient conditions of 100 kPa and 15°C (state 1), determine the cycle efficiency. Also determine the heat rejected to the atmosphere.

12.61 The following $P-v$ and $T-s$ diagrams illustrate the so-called dual cycle, a combination of the Otto- and diesel-cycle processes. Heat addition occurs in both the constant-volume process, 2–3, and in the constant-pressure process, 3–4. Heat is rejected in the constant-volume process, 5–1. The compression and expansion processes, 1–2 and 4–5, respectively, are isentropic. Assuming that air is the working fluid, determine the cycle thermal efficiency η_{dual} as a function of the following three parameters: r_v ($\equiv v_1/v_2$), R_p ($\equiv P_3/P_2$), and β ($\equiv v_4/v_3$).

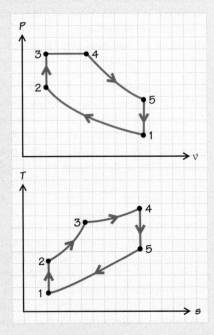

12.62 Show that Eq. 12.37 can be obtained from the simplification of the expression preceding it in the text.

12.63 Consider a gas-turbine engine operating with 30 kg/s of air entering at 300 K and a pressure ratio of 12.5. Using an ideal air-standard cycle as a model of this engine, determine the ideal thermal efficiency and the net shaft power delivered to the load. Assume the turbine inlet temperature is 1050 K.

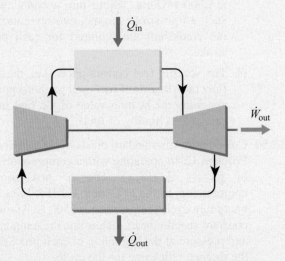

12.64 Repeat Problem 12.63, but assume a turbine inlet temperature of 1200 K.

12.65 Repeat Problem 12.63, but use a pressure ratio of 15.

12.66 Repeat Problem 12.63 assuming a turbine isentropic efficiency of 96% and a compressor efficiency of 95%. Compare your results with those from Problem 12.63.

12.67 Consider an air-standard Brayton cycle. Conditions at the compressor inlet are 100 kPa and 20°C. The pressure ratio is 7:1 and the turbine inlet temperature is 800°C. Determine (a) the compressor work (kJ/kg), (b) the heat added (kJ/kg), (c) the turbine work (kJ/kg), and (d) the cycle thermal efficiency.

12.68 The cycle shown in the sketch is used to cool the passenger cabin of a commercial aircraft. Air is the working fluid. Assuming an ideal compression process, determine the following:

A. Net work required (hp) per ton of refrigeration (1 ton = 12,000 Btu/hr)

B. Heat rejected at the heat exchanger (Btu/hr)

C. Turbine isentropic efficiency (%)

D. Coefficient of performance

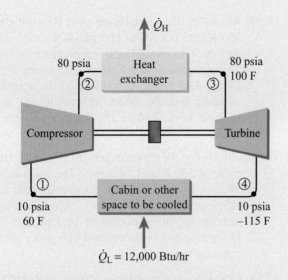

$\dot{Q}_L = 12,000$ Btu/hr

12.69 Consider a Brayton cycle using air as the working fluid. The air enters the compressor at 102 kPa and 15°C and exits at 612 kPa. If the maximum cycle temperature is 800°C, what is the ideal cycle efficiency?

12.70 Perform an air-standard cycle analysis of a gas-turbine engine that employs regeneration (i.e., some of the energy in the hot exhaust stream is transferred to the air between the compressor and the "combustor").

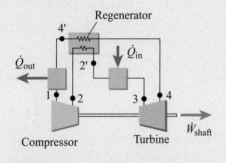

Compressor   Turbine

A. Sketch the regenerative cycle on *T–s* coordinates and label the state points.

B. From your analysis, derive a formula for the ideal cycle efficiency expressed in terms of temperatures.

C. Use your result from Part B to calculate the thermal efficiency for an ideal cycle in which the air enters at 300 K, the regenerator preheats the air 75 K, and the pressure ratio is 12.5.

D. Compare your result from Part C with the thermal efficiency calculated in Problem 12.63. Discuss.

12.71 Consider an ideal air-standard Brayton cycle operating with an inlet temperature (T_1) and pressure (P_1) of 300 K and 100 kPa, respectively. The heat added per unit mass, q_{2-3}, is 1050 kJ/kg. Assume the following constant values of the

specific heats: $c_p = 1.005$ kJ/kg·K and $c_v = 0.718$ kJ/kg·K. Investigate the effects of the compressor pressure ratio (P_2/P_1) by performing the tasks listed. Use a spreadsheet to perform all repetitive calculations; however, make sure that you check at least one calculation by hand.

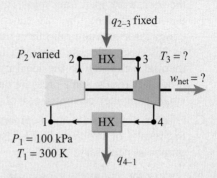

A. Sketch the cycle on *T–s* and *P–v* diagrams indicating the key state points with bold dots.

B. Calculate the turbine inlet temperature (T_3), the net work per unit mass, and the thermal efficiency as functions of the compressor pressure ratio for a range from 10 to 25. Create a single computer-generated plot of your results.

C. Discuss your results.

12.72 Consider the same situation described in Problem 12.71, but now the turbine inlet temperature is a given quantity ($T_3 = 1700$ K) and the heat added per unit mass, q_{2-3}, is unknown.

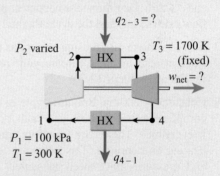

A. Calculate the heat added per unit mass, q_{2-3}, the net work per unit mass, w_{net}, and the thermal efficiency η_{th} as functions of the compressor pressure ratio for a range from 10 to 25. Create a single computer-generated plot of your results.

B. Discuss your results and compare them with the results of Problem 12.71.

12.73 Consider the situation described in Problem 12.72, but now neither the compressor nor the turbine is ideal. Assume that the isentropic efficiency of the compressor is 0.82 and the isentropic efficiency of the turbine is 0.87.

A. Sketch the cycle on T–s and P–v diagrams indicting the key state points with bold dots. Show both the isentropic and actual state points for conditions at the compressor and turbine outlets.

B. Calculate the heat added per unit mass, q_{2-3}, the net work per unit mass, w_{net}, and the thermal efficiency η_{th} as functions of the compressor pressure ratio for a range from 10 to 25. Also calculate the ratio of the cycle thermal efficiency to the ideal cycle thermal efficiency for the same pressure ratio. Create a single computer-generated plot of your results.

C. Discuss your results and compare them with the results of Problem 12.72. Use additional graphs as necessary.

12.74　Air enters the compressor of a 100-hp Brayton cycle at 530 R and 14.7 psia. The constant-pressure heating process occurs at 80 psia with a peak temperature of 2200 R. Heat is rejected at 530 R. Determine the thermal efficiency and the mass flow rate (lb$_m$/h) for the cycle.

12.75　A 75-kW Brayton cycle has a compressor inlet state at 293 K and 1 atm and a turbine inlet state at 1200 K and 4 atm. Determine the thermal efficiency and the mass flow rate (kg/hr) for both helium and air as the working fluid. Which fluid would be better? Explain.

12.76　Air enters the compressor of a Brayton cycle at 20°C and 1 atm. The turbine inlet state is at 800°C and 3.4 atm. Determine the net power per unit mass flow rate (kJ/kg) and the cycle thermal efficiency.

12.77　Solve Problem 12.76 using helium as the working fluid instead of air. Compare your results to the results of Problem 12.76.

12.78　Consider a 200-hp Brayton cycle using air. The compressor and turbine are both adiabatic but irreversible. There are pressure drops during heating and cooling. The following temperatures and pressures have been measured at the inlet states around the cycle:

State	T (R)	P (psia)
1	530	15.0
2	830	60.0
3	1960	58.8
4	1400	15.3

Assuming constant specific heats, determine (a) the thermal efficiency of the cycle and (b) the mass flow rate (lb$_m$/hr) of the air.

12.79　Air enters the compressor of a Brayton cycle at 250 K and 0.1 MPa. The pressure ratio is 4 to 1 and the maximum cycle temperature is 1300 K. The compressor and turbine efficiencies are 0.85. Using data from Appendix C.2 to account for variable specific heats, determine (a) the net specific work (kJ/kg), (b) the thermal efficiency, and (c) the ratio of compressor to turbine work.

12.80　Air enters the compressor of a 100-hp Brayton cycle at 530 R and 14.7 psia. The air is heated at 80 psia to 2200 R. The compressor and turbine efficiencies are each 0.90. After leaving the turbine, the air is cooled at 14.7 psia to 530 R. Determine the cycle thermal efficiency assuming constant specific heats.

12.81　Consider a gas-turbine engine. Air enters the compressor at 298 K and 1 atm with a flow rate of 685 kg/s. The compressor has a pressure ratio of 18 and an isentropic efficiency of 0.85. The air from the compressor enters the combustion chamber where natural gas (CH_4) at 298 K is injected, mixed, and burned with the air. The mass air–fuel ratio is 50. The hot products of combustion expand through a turbine back to 1 atm. The isentropic efficiency of the turbine is 0.92.

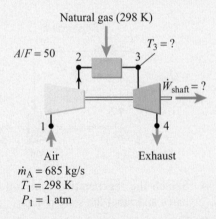

Solve Parts A and B using the following assumptions:

i. Combustion is "complete" (i.e., the only product species are CO_2, H_2O, O_2, and N_2).

ii. The air has a simplified composition with a molar O_2-to-N_2 ratio of 1:3.76 and a specific-heat ratio γ of 1.4.

iii. All species behave as ideal gases with properties obtained from Appendix B.

iv. The combustor is adiabatic and all potential and kinetic energy changes are negligible.

A. Determine the temperature of the gases at the outlet of the combustor. (Note: This is also the turbine inlet temperature.)

B. Determine the shaft power delivered by the engine.

C. Determine the engine process thermal efficiency using the product of the fuel flow rate and the fuel higher heating value as the input energy rate.

12.82 A reversible refrigerator removes energy from a thermal reservoir at 4°C and delivers energy to a thermal reservoir at 26°C. Determine the coefficient of performance for this device. Also determine the coefficient of performance for a reversible heat pump operating between the same two reservoir temperatures.

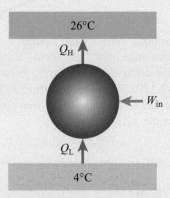

12.83 Plot the Carnot coefficient of performance for a heat pump delivering energy into a thermal reservoir at 294.2 K (70 F) as a function of the temperature of the cold reservoir. Use a range of cold reservoir temperatures from 255 K (~0 F) to 290 K (~62 F).

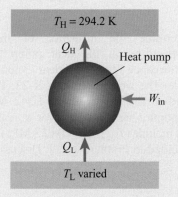

12.84 A heat pump uses the ground as a source to deliver 20,000 Btu/hr to heat a home. The heat pump coefficient of performance is 3.23. Determine (a) the electrical power required to operate the heat pump in steady state, (b) the operating cost per hour if electricity is purchased at 6.2 cents/kW·hr, and (c) the rate at which energy is removed from the ground expressed in both kW and Btu/hr.

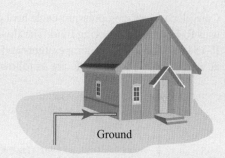

Ground

12.85 Heat leaks from the air in a kitchen through the walls of a refrigerator into the refrigerated cold space at a rate of 1.43 kW. Determine the electrical power required to maintain a steady temperature in the cold space if the refrigerator coefficient of performance is 2.6. Also determine the rate at which heat is transferred to the kitchen air from the refrigerator condenser coil and the net rate at which energy is transferred from the entire refrigerator to the air in the kitchen.

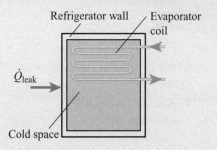

12.86 Consider an ideal vapor-compression cycle using R-134a and operating between pressures of 0.30 and 1.68 MPa. The refrigerant flow rate is 0.035 kg/s. Plot the cycle on T–s coordinates with the aid of the NIST plotting software and calculate the following quantities: $\dot{Q}_H$, $\dot{Q}_L$, $\dot{W}_{in}$, β_{refrig}, and $\beta_{heat\,pump}$.

12.87 Repeat Problem 12.86 using R-22. Compare your results with those of problem 12.86.

12.88 Consider an ideal vapor-compression cycle using R-134a and operating between pressures of 0.28 and 1.75 MPa. Determine the coefficient of performance for the cycle for application (a) in a refrigerator and (b) in a heat pump.

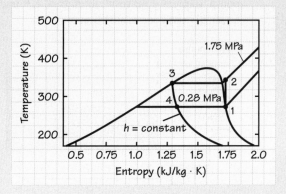

12.89 Consider a vapor-compression-cycle heat pump that uses R-134a as the working fluid. The flowate of the R-134a is 0.08 kg/s. The temperatures and pressures at various points in the cycle are as follows:

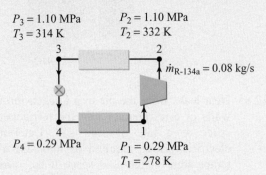

$P_3 = 1.10$ MPa $P_2 = 1.10$ MPa
$T_3 = 314$ K $T_2 = 332$ K

$\dot{m}_{R-134a} = 0.08$ kg/s

$P_4 = 0.29$ MPa $P_1 = 0.29$ MPa
                     $T_1 = 278$ K

State	Location	T (K)	P (MPa)
1	Compressor inlet/ evaporator outlet	278	0.29
2	Compressor outlet/ condenser inlet	332	1.10
3	Condenser outlet/ expansion valve inlet	314	1.10
4	Expansion valve outlet/ evaporator inlet	—	0.29

A. Sketch the cycle on a *T–s* diagram exaggerating as necessary to show how the cycle deviates from the ideal cycle.

B. Determine the power input to the compressor.

C. Determine the isentropic efficiency of the compressor.

D. Determine the heat pump coefficient of performance.

12.90 Consider the heat-pump-heated swimming pool discussed in Example 12.14. The rectangular (7.6 m × 12 m) pool is filled to an average depth of 2.3 m. Estimate the rate at which the temperature of the pool water increases for steady operation of the heat pump. Express your result in units of K/s and F/hr. Explicitly list your assumptions.

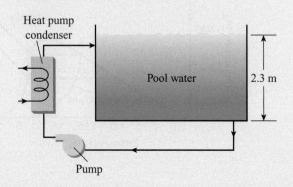

12.91 Consider the heat pump and operating conditions specified in Example 12.14. Treating the evaporator as a cross-flow heat exchanger, estimate the required product of the overall heat-transfer coefficient and the heat-exchange area $U_{OA}A$. The air flow rate though the evaporator is 2.29 kg/s. Perform a similar calculation for the condenser assuming a simple counterflow arrangement. The water flow rate through the condenser is 2.84 kg/s.

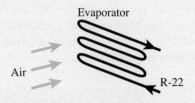

12.92 Determine the heat pump coefficient of performance for an ideal vapor-compression refrigeration cycle operating between 0.653 and 1.342 MPa. The working fluid is R-22. Compare your result with that for the real cycle described in Example 12.14.

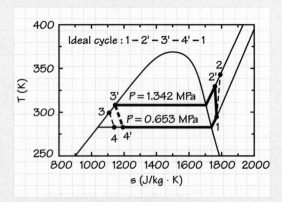

12.93 Determine the influence of the evaporator temperature on the heat pump coefficient of performance for an ideal vapor-compression cycle. The working fluid is R-22. The pressure is fixed in the condenser at 1.342 MPa. Vary the evaporator temperature over a reasonable range that includes T_{sat} ($P_{sat} = 0.653$ MPa), the condition specified in Problem 12.92. Discuss your results.

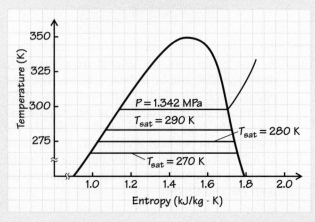

12.94 An architect wants to use an electrical work input to keep the temperature inside a building at 20°C when the outside temperature is −20°C. Determine the maximum heat-transfer rate (kW) that can be supplied to the building (per kW of electrical input) by (a) a resistance heater; (b) a vapor-compression cycle using R-134a with $T_{evap} = -20°C$ and $T_{cond} = 20°C$, saturated vapor leaving the evaporator, and saturated liquid leaving the condenser; and (c) the best possible method.

12.95 A 1-ton (200 Btu/min) vapor-compression refrigerator uses R-134a as the refrigerant. Evaporation occurs at 10 F and condensation occurs at 100 F. The compressor is reversible and adiabatic. The refrigerant leaves the evaporator at 20 F and leaves the condenser at 95 F. The heating and cooling processes are both at constant pressure. Determine the steady-state power (hp) into the compressor.

12.96 Create a T–s diagram for Problem 12.95 using the NIST software plotting capability. Your sketch should show the following:

A. The vapor dome

B. Constant-temperature lines (isotherms) for 10, 20, 95, and 100 F

C. Constant-pressure lines (isobars) for evaporation at 10 F and condensation at 100 F

D. States 1, 2, 3, and 4 with state 1 as that at the compressor inlet

12.97 Create a P–h diagram for Problem 12.95 using the NIST software plotting capability. Your sketch should show the following:

A. The vapor dome

B. Constant-temperature lines (isotherms) for 10, 20, 95, and 100 F

C. Constant-pressure lines (isobars) for evaporation at 10 F and condensation at 100 F

D. States 1, 2, 3, and 4 with state 1 at the compressor inlet

12.98 A vapor-compression refrigerator using R-134a has a maximum pressure of 120 psia and a minimum pressure of 15 psia. The R-134a leaves the condenser as a saturated liquid and leaves the evaporator as a saturated vapor. The compressor is reversible and adiabatic. Determine (a) the maximum room temperature, (b) the minimum food-compartment temperature, and (c) the coefficient of performance.

12.99 A vapor-compression heat pump circulates R-134a at 300 lb$_m$/hr. The R-134a enters the compressor at 20 F and 20 psia and leaves at 180 F and 180 psia. The refrigerant enters the expansion valve at 100 F and 160 psia and leaves the evaporator at 20 F and 20 psia. Determine (a) the heating capacity (kBtu/hr) of the heat pump and (b) the coefficient of performance.

12.100 In a vapor-compression cycle using R-134a as the working fluid, the condenser pressure is 100 psia and the evaporator pressure is 20 psia. The R-134a leaves the condenser as saturated liquid and the evaporator as saturated vapor. Determine the coefficient of performance both as a refrigerator and as a heat pump.

12.101 In a vapor-compression refrigeration cycle that circulates R-134a at a rate of 5 kg/min, the refrigerant enters the compressor at −10°C and 0.18 MPa and leaves at 70°C and 0.7 MPa. The R-134a leaves the condenser as saturated liquid at the compressor outlet pressure. Determine the refrigeration capacity (kW) and the coefficient of performance. Also suggest two modifications to this refrigeration cycle to increase its coefficient of performance. Use T–s diagrams or other means to illustrate how your suggestions will help.

12.102 The temperatures and pressures shown in the table were measured for a vapor-compression refrigerator that uses R-134a as the refrigerant.

State	T (°C)	P (MPa)
1	−30	0.08
2	60	0.60
3	15	0.60
4		0.08

The refrigerator operates between a cold region at T_C and a hot region at T_H. The positive-displacement compressor (piston–cylinder device) has a cylinder volume and operating speed such that it processes 1 m^3/min of R-134a at the inlet of the compressor.

A. Determine the minimum value of T_C and the maximum value of T_H.

B. Determine the coefficient of performance.

C. Determine the heat transfer rate from the region at T_C.

12.103 A vapor-compression refrigerator using R-134a is to provide chilled water at 40 F and 14.7 psia. Originally the water is at 70 F and 14.7 psia. The pressure in the evaporator is 40 psia and that in the condenser is 100 psia. The R-134a enters the compressor at 40 F and leaves at 120 F. The power supplied to the compressor is 1 hp. The R-134a leaves the condenser as saturated liquid. Determine the coefficient of performance for the refrigerator and the mass flow rate (lb_m/hr) of chilled water that can be produced.

12.104 In a vapor-compression cycle, evaporation occurs at $-10°C$ and condensation occurs at $40°C$. The compressor is reversible and adiabatic. R-134a exits the evaporator at $-5°C$ and exits the condenser at $38°C$. Determine the work into the compressor per unit mass of R-134a (kJ/kg).

12.105 The capacity and compressor-work characteristics for a 3-ton (600-Btu/min) vapor-compression air conditioner are given in the table, along with the seasonal house-cooling needs. If electricity costs 5 cents/kW·hr, determine the seasonal operating cost and the seasonal coefficient of performance.

B. Determine the horsepower required per ton of refrigeration (hp/ton). That is, determine $\dot{W}/\dot{Q}_e$, where $\dot{W}$ is the power supplied to the compressor (hp) and $\dot{Q}_e$ is the heat-transfer rate to the evaporator (tons). (Note: 1 ton of refrigeration = 200 Btu/min.)

12.108 A 3-ton (600-Btu/min) vapor-compression refrigeration cycle uses R-134a to maintain a freezing compartment at 0 F. Room air is at 70 F. The R-134a condenses at 80 F and evaporates at -10 F. The R-134a enters the compressor with 10 F of superheat. The compressor efficiency is 0.85. The R-134a leaves the condenser with 10 of subcooling.

A. Perform a complete thermal analysis of the cycle.

B. Determine the horsepower required per ton of refrigeration (hp/ton). That is, determine

Ambient Temperature (F)	Number of Hours	Cooling Load (Btu/hr)	Compressor Power at Capacity (kW)	Air-Conditioner Capacity (Btu/hr)
100	100	36,000	5.8	37,000
90	500	24,000	5.5	39,000
80	1200	12,000	5.2	41,000

12.106 A computer facility in the Sahara Desert is to be maintained at $15°C$ by a vapor-compression refrigeration system that uses water as the refrigerant. The water leaves the evaporator as a saturated vapor at $10°C$. The compressor is reversible and adiabatic. The pressure in the condenser is 0.01 MPa and the water is saturated liquid as it leaves the condenser. Determine (a) the coefficient of performance for the cycle and (b) the power input per unit mass flow rate of water (kJ/kg).

12.107 A 3-ton (600-Btu/min), vapor-compression refrigeration cycle using R-134a as the refrigerant maintains a freezer compartment at 0 F. The room air is at 70 F. Satisfactory operation requires a minimum temperature difference between the room air and the condensing R-134a of 10 F. Similarly, a 10-F minimum temperature difference between the evaporating R-134a and the freezing compartment is required. The R-134a enters the compressor ($\eta_{isen} = 0.85$) as a saturated vapor. The R-134a leaves the condenser as a saturated liquid.

A. Perform a complete thermal analysis of the cycle.

$\dot{W}/\dot{Q}_e$, where $\dot{W}$ is the power supplied to the compressor (hp) and $\dot{Q}_e$ is the heat-transfer rate to the evaporator (tons). (Note: 1 ton of refrigeration = 200 Btu/min.)

12.109 An old, 10-kW, vapor-compression heat pump using R-12 maintains a building at $20°C$. The inlet state to the compressor is saturated vapor at $-30°C$. The compressor exit state is at $50°C$ and 0.7 MPa. The inlet state to the expansion valve is saturated liquid at $25°C$. Perform a complete thermal analysis of the heat pump.

12.110 A Carnot refrigerator removes energy at a rate of 300 Btu/hr from a low-temperature thermal reservoir at -160 F. The high-temperature reservoir for the refrigerator is the atmosphere at 40 F. A Carnot engine operating between a reservoir at 1140 F and the atmosphere (40 F) drives this Carnot refrigerator.

A. At what rate must heat be supplied (Btu/hr) to the Carnot engine from the 1140-F reservoir?

B. What is the power input to the Carnot refrigerator (Btu/hr)?

C. What are the coefficient of performance of the Carnot refrigerator and the thermal efficiency of the Carnot engine?

12.111 On an extremely cold winter day, the air temperature and relative humidity in the living space of a home are 295.3 K (72 F) and 18%, respectively. After operating a humidifier for 8 h, the relative humidity in the house is increased to 40%, the low end of the comfort zone. The air temperature remains at 295.3 K (72 F) and the atmospheric pressure is 100 kPa. Determine the specific humidity before and after humidifying the air. Also determine the average rate at which moisture was added to the air if the volume of the humidified space is 238 m^3. Express your result in both kg/s and gal/hr.

12.112 Determine the water-vapor mole and mass fractions before and after humidifying the air as described in Problem 12.111.

12.113 For the purpose of testing air conditioners, the following indoor and outside air temperatures and relative humidities are defined for a so-called moderate climate:

Indoors: $T_A = 27°C$ (80.6 F), $\phi = 0.48$,

Outside: $T_A = 35°C$ (95 F), $\phi = 0.41$.

Determine the specific humidity and dew point associated with these two conditions.

12.114 Air at 100 F and 20 psia has a dew-point temperature of 70 F. Determine the relative humidity and the humidity ratio. Also determine the mixture volume (ft^3) containing 1 lb$_m$ of dry air.

12.115 Combustion products (10% carbon dioxide, 20% water vapor, and 70% nitrogen by volume) flow in a chimney at 94°C and 1 atm. Determine (a) the dew-point temperature (°C), (b) the humidity ratio, and (c) the relative humidity.

12.116 Combustion products (10% carbon dioxide, 20% water vapor, and 70% nitrogen by volume) flow in a pipe at 94°C and 1.36 atm. Determine (a) the dew-point temperature (°C), (b) the humidity ratio, and (c) the relative humidity.

12.117 At state 1, a piston–cylinder device contains 0.004 m^3 of moist air initially at 20°C and 0.10135 MPa with a 70% relative humidity. The volume is then decreased to 0.002 m^3 at state 2 by pushing the piston inward in a reversible, isothermal process.

A. Complete the following table:

Property	State 1	State 2
Relative humidity		
Humidity ratio		
Dew-point temperature (°C)		
Mass of dry air (g)		
Mass of water vapor (g)		
Mass of liquid water (g)		

B. Does the total internal energy of the contents of the cylinder increase, decrease, or remain constant during this process? Explain.

12.118 A dry gas mixture (60% oxygen, 40% helium by volume) is at 94°C and 3.4 atm. Moisture is then added. Properties at the final state are 94°C, 3.4 atm, and 20% relative humidity. Determine (a) the dew-point temperature (°C), (b) the molar composition of the moist gas, and (c) the humidity ratio.

12.119 Moist air initially at 150 F, 20 psia, and 60% relative humidity is compressed reversibly and isothermally in a piston–cylinder device to a pressure of 40 psia. Determine the following:

A. The dew-point temperature (F) at the initial state

B. The humidity ratio at the initial state

C. The mass fraction of water that is liquid at the final state [i.e., $M_{liquid}/(M_{liquid} + M_{vapor})$ at the final state]

12.120 A mixture [10 mol CO$_2$, 70 mol N$_2$, and 20 mol H$_2$O (liquid plus vapor)] is at 40°C and 1 atm. This temperature is below the dew point and the mixture is saturated with water vapor. Determine the number of moles of water that are vapor and the number of moles of water that are liquid.

12.121 A 10-ft by 20-ft by 8-ft room contains moist air at 90 F and 14.7 psia. The partial pressure of the water vapor is 0.3 psia. Determine the following quantities:

A. The dew-point temperature (F), relative humidity, and humidity ratio

B. The total mass (lb_m) of the water vapor in the room

C. The volume (ft^3) that would be occupied by this amount of water if it were a saturated liquid at 90 F

12.122 A dry mixture of carbon dioxide and nitrogen (30% carbon dioxide and 70% nitrogen by volume) is initially at 90 F and 14.7 psia. Moisture is then added to the mixture until it becomes saturated at 90 F and 14.7 psia. Determine the following quantities:

A. The molar mass of the original, dry mixture

B. The composition (by volume) of the saturated mixture

C. The ratio of the mixture volume when saturated to the mixture volume when dry

12.123 A 1-kg mixture of air and water vapor at 20°C, 1 atm, and 75% relative humidity is confined in a cylinder by a frictionless piston. This mixture is compressed isothermally until the pressure is 2 atm. Determine (a) the final relative humidity and humidity ratio and (b) the mass (kg) of water vapor condensed.

12.124 Combustion products (10% carbon dioxide, 20% water vapor, and 70% nitrogen by volume) enter a pipe at 200 F and 14.7 psia. The mass flow rate of products is 6 lb_m/min. The products are cooled as they flow through the pipe and leave at 100 F and 14.7 psia. Determine (a) the mass flow rate (lb_m/min) of water condensed and (b) the heat-transfer rate (Btu/min).

12.125 Moist air at 100 F, 20 psia, and 40% relative humidity is flowing in a pipe. Determine (a) the dew-point temperature (F) and (b) the humidity ratio.

12.126 Moist natural gas enters a valve at 40°C, 0.2 MPa, and 70% relative humidity with a volumetric flow rate of 20 m^3/min. An analysis of the dry gases in the natural gas shows 80% methane and 20% hydrogen by volume as the gas enters. The moist natural gas leaves the valve at 0.15 MPa. Determine the moisture condensation rate (kg/min).

12.127 A tank contains moist air at 37.7°C, 14.7 psia, and 27% relative humidity. Determine the following:

A. The dew-point temperature (°C) of the mixture

B. The temperature (°C) at which condensation will begin if the mixture is cooled down

C. The heat transfer (kJ/m^3) required to cool the contents of the tank to 10°C

12.128 A 10-ft^3 tank initially contains a moist gas (30% carbon dioxide, 50% nitrogen, and 20% water vapor by volume) at 300 F and 50 psia. Heat transfer to the atmosphere reduces the mixture temperature to 150 F. Determine the following:

A. The final pressure (psia) in the tank

B. The mass (lb_m) of water condensed

C. The volume (ft^3) of the condensed liquid water

D. The molar composition of the moist gas at the final state

E. The heat transfer (Btu)

12.129 A 20-ft × 12-ft × 8-ft room contains an air–water vapor mixture at 80 F. The barometric pressure is 14.7 psia and the measured partial pressure of the water vapor is 0.2 psia. Calculate (a) the relative humidity, (b) the humidity ratio, (c) the dew-point temperature, and (d) the mass (lb_m) of water vapor contained in the room.

12.130 One of the many methods used for drying air is to cool it below the dew-point temperature so that condensation or freezing of the moisture takes place. To what temperature must atmospheric air be cooled to have a humidity ratio of 0.00433? To what temperature must this air be cooled if the pressure is 10 atm?

12.131 One method of removing moisture from atmospheric air is to cool the air so that the moisture condenses or freezes out. A laboratory experiment requires a humidity ratio of 0.00620. To what temperature must the air be cooled at a pressure of 0.1 MPa to achieve this humidity?

12.132 A 4-m × 6-m × 2.4-m room contains an air–water vapor mixture at a total pressure of 100 kPa and a temperature of 25°C. The partial pressure of the water vapor is 1.4 kPa. Determine (a) the humidity ratio, (b) the dew point, and (c) the total mass of water vapor in the room.

12.133 Air enters an air compressor at 70 F (21.1°C) and 14.7 psia (101.3 kPa) with a 50% relative humidity. The air is compressed to 50 psia (344.7 kPa) and sent to an intercooler. If condensation of water vapor from the air is to be prevented, what is the lowest temperature to which the air can be cooled in the intercooler?

12.134 Consider a room containing moist air at 24°C, 60% relative humidity, and 1 atm. Determine the following:

A. The humidity ratio

B. The mixture enthalpy (kJ/kg)

C. The dew-point temperature (°C)

D. The mixture specific volume (m³/kg)

12.135 A 4-lb$_m$ mass of air at 80 F (26.7°C) and 50% relative humidity mixes with a 1-lb$_m$ mass of air at 60 F (15.6°C) and 50% relative humidity. The pressure is 1 atm. Determine (a) the relative humidity of the mixture and (b) the dew-point temperature of the mixture.

12.136 Air is compressed in a compressor from 30°C, 60% relative humidity, and 101 kPa to 414 kPa and then cooled in an intercooler before entering a second stage of compression. What is the minimum temperature (°C) to which the air can be cooled so that condensation does not take place?

12.137 A 0.5-m³ tank contains an air–water vapor mixture at 100 kPa and 35°C with a 70% relative humidity. The tank is cooled until the water vapor begins to condense. Determine the temperature at which condensation begins and the heat transfer for the process.

12.138 An air–water vapor mixture at 100 F (37.8°C) temperature contains 0.02 lb$_m$ water vapor per pound of dry air. The barometric pressure is 28.561 in Hg (96.7 kPa). Calculate the relative humidity and dew-point temperature.

12.139 Ethane burns with 150% stoichiometric air. Assume complete combustion with no dissociation. Determine (a) the mole percentage of each product species and (b) the dew-point temperature of the products.

12.140 Rework Problem 12.139 but use propane as the fuel.

12.141 Consider the combustion of benzene, C_6H_6, with air. Determine the dew-point temperature of the combustion products if the mass air–fuel ratio is 20:1. The pressure is 1 atm.

12.142 An evaporative cooler inducts outside air at 46°C (114.8 F) and 100 kPa having a relative humidity of 16%. The air is cooled to 29°C (84.2 F). Determine the following quantities:

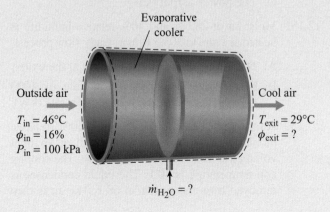

A. The mass of water added by the cooler per unit mass of dry air

B. The relative humidity of the outlet stream

C. The mass and volumetric flow rates of the added water when the volumetric flow rate of the entering moist air is 2500 ft³/min.

12.143 An air-conditioning unit inducts outside air at 46°C (114.8 F) and 100 kPa having a relative humidity of 16%. The air is cooled and dehumidified to provide an outlet temperature of 29°C (84.2 F) and relative humidity of 39%. The coefficient of performance for the vapor-compression refrigeration system contained in the air-conditioning unit is 2.92. Determine the following quantities:

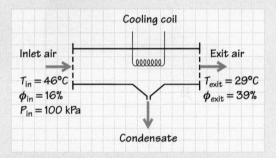

A. The mass of water removed per unit mass of dry air

B. The cooling rate per unit mass of dry air

C. The input power to the vapor-compression refrigeration system per unit mass of dry air

12.144 Consider a home dehumidifier as shown in Figs. 12.39 and 12.40. Moist air at 78 F and relative humidity of 85% enters at a volumetric flow rate of 330 ft³/min. The air exits at 92 F with a relative humidity of 50%, and the condensate exits at 52 F. Determine the electrical power supplied to the dehumidifier in kW, and determine the condensate flow rate in kg/s and gal/min. Assume that the dehumidifier is overall adiabatic and that the pressure is uniform at 100 kPa.

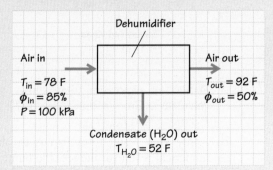

12.145 Cooling water from a power plant steam condenser enters a cooling tower at 307 K (93 F)

with a flow rate of 3500 kg/s. The water exits the tower at 295 K (71.4 F). Air enters at the base of the tower at 290 K (62.4 F) with a relative humidity of 32% and exits at the top saturated at 304 K (87.6 F). Make-up water is provided at 295 K (71.4 F). The pressure in the tower is 96 kPa. Determine the flow rate of the make-up water and the flow rate of the air (dry basis) through the tower.

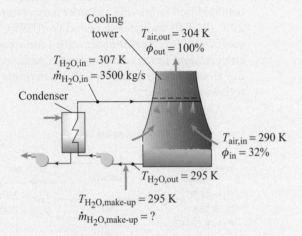

12.146 As shown in the sketch, moist air enters an insulated tube at station 1 with properties T_1, P_1, and ω_1. Moisture is added to the stream as it passes through a wick and exits at station 2 saturated with moisture (i.e., $\phi_2 = 100\%$) at a temperature T_2. The pressure at station 2 can be assumed to be the same as at station 1. The make-up water is supplied to the wick as saturated liquid also at T_2. This process is referred to as an adiabatic saturation process and the device that accomplishes this an adiabatic saturator.

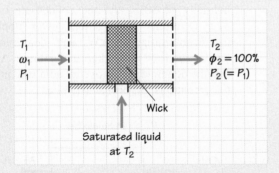

A. Derive an expression that allows you to calculate ω_1 from a knowledge of T_1, P_1, T_2, $P_2 (= P_1)$, and $P_{sat} (T_2)$.

B. Apply your result form Part A to obtain a numerical result for the specific humidity ω_1 and relative humidity ϕ_1 for the following conditions: $T_1 = 299.8$ K (80 F), $T_2 = 293.1$ K (68 F), and $P = 1$ atm.

(Note: Sling psychrometers, and similar devices used to measure humidity, provide conditions quite similar to an adiabatic saturator. In these devices, T_1 is defined as the dry-bulb temperature and T_2 is defined as the wet-bulb temperature.)

12.147 Moist air enters an air conditioner at 305.3 K, 137.9 kPa, and 80% relative humidity. The mass flow rate of dry air entering is 45.4 kg/min. The moist air leaves the air conditioner at 285.3 K, 124.1 kPa, and 100% relative humidity. The condensed moisture leaves at 285.3 K and 124.1 kPa. Determine the heat transfer rate (kW) from the moist air.

12.148 A moist gas (10% carbon dioxide, 70% nitrogen, 20% water vapor by volume) enters a reversible, adiabatic turbine at 250°C and 0.4 MPa. The turbine exit pressure is 0.10135 MPa. The inlet volumetric flow rate is 3 m³/min. Determine the following:

A. The mass flow rate (kg/min) of dry gas mixture

B. The exit temperature (°C)

C. The mass flow rate (kg/min) of liquid water at the exit

D. The power produced (kW)

12.149 An adiabatic compressor receives air from the atmosphere at 60 F, 14.7 psia, and 75% relative humidity and discharges the air at 100 psia. The compressor isentropic efficiency is 0.80. Determine the relative humidity and the humidity ratio of the discharged air.

12.150 A moist gas (10% carbon dioxide, 70% nitrogen, 20% water vapor by volume) enters a reversible, adiabatic turbine at 150°C and 0.4 MPa. The turbine exit pressure is 0.10135 MPa. The inlet volumetric flow rate is 3 m³/min. Determine the following quantities:

A. The mass flow rate (kg/min) of dry gas mixture

B. The exit temperature (°C)

C. The mass flow rate (kg/min) of liquid water at the exit

D. The power output (kW)

12.151 Moist air at 283 K and 40% relative humidity is heated at 1 atm to 305.3 K in a steady-flow process.

A. Determine the relative humidity at the exit.

B. For an inlet volumetric flow rate of 1000 ft³/min, determine the heat-transfer rate (kW).

12.152 Atmospheric air at 90 F and 60% relative humidity flows over a cooling coil at an inlet volumetric flow rate of 2000 ft³/min. The cooling coil temperature is 40 F. The liquid condensed is removed from the system at 60 F. The air is then electrically heated until a final state of 80 F and 40% relative humidity is reached.

A. Determine the mass flow rate (lb$_m$/min) of liquid water leaving the cooling section.

B. Determine the heat-transfer rate (Btu/min) in the cooling section.

C. Determine the electrical work-transfer rate (Btu/min).

12.153 A home heating system in Madison, Wisconsin, is designed to heat outside air at 4.4°C (40 F) and 100% relative humidity to 26.7°C (80 F). Assume the pressure to be 1 atm everywhere. Determine the relative humidity at 26.7°C and the heat transfer to the air (kJ/kg).

12.154 During mild exercise (e.g., walking) a person breathes air in at the rate of 20 liters/min and has a metabolism rate of 300 W. Atmospheric air is at 20°C and 30% relative humidity, and the "air" leaving the mouth is at 35°C and 80% relative humidity. Determine the following:

A. The energy loss from a person due to breathing (W)

B. The energy-transfer rate (W) to the dry air that is breathed

C. The water loss (kg/min) from a person due to breathing

12.155 Moist air at 90 F, 14.7 psia, and 80% relative humidity is removed from the top of a large room at a rate of 500 ft^3/min. The air is then cooled in an air conditioner at constant pressure to 60 F and returned to the bottom of the room. Condensed moisture leaves the air conditioner at 60 F and 14.7 psia. Determine (a) the rate of moisture removal (gal/hr) by the air conditioner and (b) the heat-transfer rate (Btu/min) rejected by the air conditioner.

12.156 Moist air enters an adiabatic humidifier at state 1 [21.1°C (70 F), 1 atm, 10% relative humidity] with a volumetric flow rate of 500 ft^3/min. The air leaves the humidifier at state 2 [23.9°C (75 F), 1 atm, 70% relative humidity]. The change in the condition of the moist air is brought about by the injection of steam at state 3 from a boiler, as shown in the sketch. The boiler is supplied with 23.3°C (74 F) water at state 4. An adiabatic valve between the boiler and the humidifier causes the steam pressure to drop from boiler pressure to the humidifier pressure of 1 atm.

A. Determine the mass flow rate (kg/min) of water at state 4.

B. Determine the heat-transfer rate to the boiler (kW).

C. Determine the pressure (kPa) in the boiler.

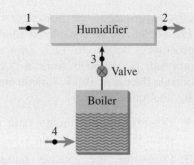

12.157 Moist air at 30°C, 1 atm, and 20% relative humidity flows steadily into an adiabatic mixing chamber. Steam at 1 atm is sprayed into the chamber to humidify the air. The air leaves the chamber completely saturated at 30°C and 1 atm.

A. Determine the temperature (°C) and/or quality of the added steam.

B. Determine the mass of steam required per mass of dry air (kg/kg).

12.158 Cold air enters an air conditioner at 4.4°C (40 F) with 50% relative humidity at a volumetric flow rate of 1000 ft^3/min. The air is then heated. Next, moisture at 21.1°C (70 F) and 1 atm is added adiabatically to bring the air to 26.7°C (80 F) with a 40% relative humidity. The entire process occurs at 1 atm. Determine (a) the rate of moisture addition to the air (kg/min), (b) the heat-transfer rate (W), and (c) the air temperature (°C) after heating.

12.159 Moist air steadily enters a room at 20°C and 1 atm with a dew-point temperature of 7°C at a rate of 15 m^3/min. As the moist air passes through the room it is heated at a rate of 7 kW by a source at 65°C. Liquid water at 20°C is also supplied to the room at a rate of 5 kg/hr and is evaporated into the moist air. Determine the temperature (°C) and relative humidity of the moist air as it leaves the room.

12.160 A stream of moist air is obtained by mixing a 1500 ft^3/min flow of air at 26.7°C (80 F) and 80% relative humidity with 500 ft^3/min flow of air at 15.6°C (60 F) and 50% relative humidity. The mixing is adiabatic and occurs at constant pressure (1 atm). Determine the temperature (°C), humidity ratio, and relative humidity of the exit stream.

12.161 An evaporative cooler for an automobile consists of an 18-in-long by 8-in-diameter cylinder that hangs on the outside of the car. At 55 mph the cooler processes 4500 lb$_m$ of dry air per hour. During desert driving the air is at 100 F with a 10% relative humidity. Inside the cooler, the air passes through a moistened burlap cloth and leaves the cooler at 80 F. This cooled air flows through a duct to the passenger compartment. The cooler can be considered adiabatic, and all processes occur at 14.7 psia. The burlap moisture is supplied from a 5-gal container outside the car at 100 F. Kinetic energy changes are negligible.

A. Determine the humidity ratio and the dew point (F) of the desert air.

B. Determine the relative humidity of the processed air entering the car.

C. If the container is full at the start, determine the time (hr) the car can be driven at these conditions before the 5-gal container runs dry.

12.162 Cooling water from an internal combustion engine is to be cooled from 150 to 110 F. The pressure is 14.7 psia. Compare the following three methods with respect to air and/or city water required (lb$_m$/1000 gallons of engine cooling water):

A. Heat transfer to city water. The city water temperature increases from 70 to 130 F.

B. Heat transfer to dry atmospheric. The air temperature increases from 90 to 120 F.

C. A cooling tower that receives moist air at 90 F with a 50% relative humidity and discharges it at 95 F with 90% relative humidity. Evaporated water is made up by city water at 70 F.

12.163 Humid air enters a dehumidifier with an enthalpy of 21.6 Btu/lb$_m$ of dry air and 1100 Btu/lb$_m$ of water vapor. There is 0.02 lb$_m$ of vapor per pound of dry air at entrance and 0.009 lb$_m$ of vapor per pound of dry air at exit. The dry air at exit has an enthalpy of 13.2 Btu/lb$_m$, and the vapor at exit has an enthalpy of 1085 Btu/lb$_m$. Condensate with an enthalpy of 22 Btu/lb$_m$ leaves the dehumidifier. The dry air flow rate is 287 lb$_m$/min. Determine (a) the amount of moisture removed from the air (lb$_m$/min) and (b) the rate at which heat must be removed.

12.164 Air is supplied to a room from the outside, where the temperature is 20 F ($-6.7°C$) and the relative humidity is 60%. The room is to be maintained at 70 F ($21.1°C$) and 50% relative humidity. What mass of water must be supplied per mass of air supplied to the room?

12.165 Saturated air at 40 F (4.4°C) is first preheated and then saturated adiabatically. This saturated air is then heated to a final condition of 105 F (40.6°C) and 28% relative humidity. To what temperature must the air initially be heated in the preheat coil?

12.166 An air–water vapor mixture enters an air-conditioning unit at a pressure of 150 kPa, a temperature of 30°C, and a relative humidity of 80%. The mass flow rate of dry air entering is 1 kg/s. The air–vapor mixture leaves the air-conditioning unit at 125 kPa, 10°C, and 100% relative humidity. The condensed moisture leaves at 10°C. Determine the heat-transfer rate (kW) for this process.

12.167 Air at 40°C and 300 kPa with a relative humidity of 35% expands in a reversible, adiabatic nozzle. To how low a pressure (kPa) can the gas be expanded if no condensation is to take place? What is the exit velocity (m/s) at this condition?

12.168 In an air-conditioning unit, air enters at 80 F, 60% relative humidity, and standard atmospheric pressure. The volumetric flow rate of the entering air (dry basis) is 71,000 ft^3/min. The moist air exits at 57 F and 90% humidity. Calculate the following:

A. The cooling capacity of the air-conditioning unit (Btu/hr)

B. The rate of water removal from the unit (lb$_m$/hr)

C. The dew point of the air leaving the conditioner

12.169 An air–water vapor mixture enters a heater–humidifier unit at 5°C, 100 kPa, and 50% relative humidity. The flow rate of dry air is 0.1 kg/s. Liquid water at 10°C is sprayed into the mixture at the rate of 0.0022 kg/s. The mixture leaves the unit at 30°C and 100 kPa. Calculate (a) the relative humidity at the outlet and (b) the rate of heat transfer to the unit.

12.170 A room is maintained at 75 F and 50% relative humidity. The outside air is available at 40 F and 50% relative humidity. Return air from the room is cooled and dehumidified by mixing it with fresh air from the outside. The air flowing into the room is 60% outdoor air and 40% return air by mass. Determine the temperature, relative humidity, and the specific humidity of the mixed air going to the room.

12.171 An air–water vapor mixture at 14.7 psia, 85 F, and 50% relative humidity is contained in a 15-ft^3 tank. At what temperature will condensation begin? If the tank and mixture are cooled an additional 15 F, how much water will condense from the mixture?

12.172 Consider 1000 ft^3/min (dry basis) of air at 14.7 psia, 90 F, and 60% relative humidity passing over a cooling coil with a mean surface temperature of 40 F. A water spray on the coil ensures that the exiting air is saturated at the coil temperature. What is the required cooling capacity of the coil in tons? (Note: 1 ton of refrigeration = 12,000 Btu/hr.)

12.173 A flow of 2832 liters/s (6000 ft³/min, dry basis) of air at 26.7°C (80 F) dry-bulb, 50% relative humidity, and standard atmospheric pressure enter an air-conditioning unit. The air exits at a 13.9°C (57 F) temperature with 90% relative humidity. Determine the following:

A. The cooling capacity of the air-conditioning unit (kW and tons)

B. The rate of water removal from the unit (kg/hr and lb_m/hr)

C. The dew point of the air leaving the conditioner (°C and F)

12.174 A gas-turbine engine burns liquid octane with 300% theoretical air in a steady-flow, constant-pressure process. The air and octane both enter the engine at 25°C. The exhaust gases leave the engine at 725°C. The overall engine is well insulated. Determine the following:

A. The power output for a fuel flow rate of 39.5 kg/hr

B. The composition of the exhaust products on a volumetric basis (dry)

C. The dew-point temperature of the exhaust gases

Appendix 12A
Turbojet Engine Analysis Revisited

This appendix provides a more rigorous analysis of momentum conservation for a turbojet engine. Earlier in the chapter we modeled the engine as a simple cylindrical tube with equal inlet and exit areas. In this appendix, we consider a more realistic geometry. Figure 12A.1 shows an engine propelling an aircraft at a steady speed v_A. The reference frame is attached to the engine. The control volume shown in Figure 12A.1 extends upstream from the physical inlet of the engine such that the air enters this control volume at a uniform pressure equal to the ambient condition and at the flight velocity. Air crossing the area A_{in} enters the engine proper, flowing between the streamlines, while air crossing the annular area $A_{an,1}$ flows around the outside of the engine. A portion of this air exits the control volume through the cylindrical side, whereas the remainder flows out through the annular area $A_{an,2}$ at the exit plane. Mass conservation requires that the mass flow rate

FIGURE 12A.1

To analyze a turbojet engine, we employ a control volume that extends beyond the physical boundaries of the engine. The air that enters the engine proper flows between the streamlines.

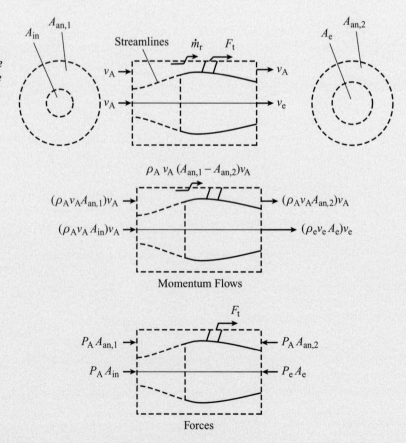

Momentum Flows

Forces

through the cylindrical side $\dot{m}_r$ be the difference between that entering $A_{an,1}$ and that exiting at $A_{an,2}$; that is,

$$\dot{m}_r = \rho_A v_A (A_{an,1} - A_{an,2}). \tag{12A.1}$$

The center panel of Fig. 12A.1 shows all of the x-direction momentum flows associated with our large control volume. Similarly, the lower panel shows all of the forces acting in the x-direction. Here we allow the pressure at the engine exit to be different from the ambient pressure. Assuming uniform velocities at the inlets and outlets, we apply the conservation of momentum principle expressed by Eq. 6.56 to yield

$$F_t = (\dot{m}_e v_e - \dot{m}_A v_A) + (P_e - P_A) A_e. \tag{12A.2}$$

We note that, if the pressure in the exit plane of the engine P_e equals the ambient pressure P_A, Eq. 12A.2 to yields the same result given by Eq. 12.18 for our simple analysis.

Lifetime	Key People	Date	Developments in Thermal-Fluid Sciences	Date	World Events
287–212 BCE	Archimedes	250 BCE	Laws of buoyancy formulated	264–241	First Punic War
1452–1519	Leonardo da Vinci	ca. 1506–1510	Formulated 1-D, steady, incompressible continuity equation	1492	Columbus first European to sail to Caribbean islands
1548–1620	Simon Stevin	1586	Solved hydrostatic paradox	1584	Sir Walter Raleigh lands expedition in Roanoke and names land Virginia
				1603	Shakespeare's *Hamlet* written
1642–1727	Isaac Newton	1686	Established principles of momentum conservation and formulated stress–strain relationship for "Newtonian" fluids; *Principia* published	1689	Peter the Great becomes Czar of Russia and attempts westernization
1663–1729	Thomas Newcomen	1712	Coal-burning steam engine/pump put in service	1707	United Kingdom of Britain formed by union of England, Scotland, and Wales
1700–1782	Daniel Bernoulli	1738	Published *Hydrodynamica*		
1707–1783	Leonhard Euler	1750	Formulated integral and differential forms of continuity equation and derived "Euler equation" for inviscid fluid (ca. 1750); provided modern form of Bernoulli equation	1755	Samuel Johnson's *Dictionary* published
1717–1783	Jean le Rond d'Alembert	1752	Published paper containing famous paradox		
1724–1792	John Smeaton	1759	First (?) use of scale-model experiments; conducted systematic parametric testing of water wheels; founded the British Society of Civil Engineers		
1736–1819	James Watt	1765	Improved steam engine using separate condenser		
1718–1798	Antoine Chézy	1770	Measured channel and pipe flow resistances; developed first resistance formula ca. 1775	1770	Boston Massacre
1743–1794	Antoine Lavoisier	1777	Formulated concept of element conservation	1775–1783	American Revolution
1734–1809	Pierre Louis Georges Du Buat	1786	French military engineer; measured resistance in pipes and channels; helped resolved d'Alembert's paradox	1787	U.S. Constitution

1753–1814	Benjamin Thompson (Count Rumford)	Cannon boring experiments show motion and heat are related	1798	French Revolution
			1789–1795	Malthus publishes *Essay on the Principles of Population*
1773–1829	Thomas Young	Defined energy as capacity to do work	1801	Napoleonic Wars: 1796–1815
1768–1830	Jean Baptiste Joseph Fourier	First to formulate theory of dimensional analysis ca. 1807–1822	1807	
1785–1836	Louis Marie Henri Navier	Extends Euler's formulation of momentum conservation to viscous fluids—"Navier–Stokes equation"	1822	Schubert's Symphony No. 8 (*The Unfinished*) composed
1796–1832	Sadi Carnot	Formulated second law of thermodynamics and Carnot cycle	1824	Beethoven composes Symphony No. 9 (*Choral*)
1797–1884	Gotthilf Heinrich Ludwig Hagen	Developed empirical law that flow rate is proportional to the fourth-power of the tube diameter for laminar flow	1839	Charles Goodyear discovers vulcanization process for rubber
1799–1869	Jean Louis Poiseuille	Obtained same result as Hagen in nearly coincident studies	1839	Opium War between Britain and China 1839–1842
1814–1878	Julius Robert Mayer	Conservation of energy principle stated	1842	Ether first used as surgical anesthesia
1797–1886	Claude Barre de Saint-Venant	Developed generalized form of momentum conservation for fluids	1843	Charles Dickens' *A Christmas Carol* published
1789–1857	Augustin Louis de Cauchy	Also extended Euler's formulation of momentum conservation to viscous fluids	1845	*The Raven and Other Poems* of Edgar Allen Poe published
1781–1840	Simeon Denis Poisson	Also extended Euler's formulation of momentum conservation to viscous fluids	1845	Annexation of Texas to the United States
1819–1903	George Gabriel Stokes	Also extended Euler's formulation of momentum conservation to viscous fluids—"Navier–Stokes equation"	1845	
1821–1894	Hermann Helmholtz	Extended conservation of energy principle	1847	Liberia established as independent republic
1824–1907	William Thomson (Lord Kelvin)	Establishes concept of absolute temperature and defines "Kelvin" scale	1848	Marks and Engels publish *Communist Manifesto*
1818–1889	James Prescott Joule	Measured mechanical equivalent of heat	1849	British annex Punjab
1822–1888	Rudolf Clausius	Stated that the energy of the universe is constant and created the word *entropy*	1850	Invention of the Bunsen gas burner
1798–1895	Franz Neumann	Developed theory for Hagen–Poiseuille flow (independently from Hagenbach)	1856	Treaty of Paris ends Crimean War
1833–1910	Eduard Hagenbach	Developed theory for Hagen–Poiseuille flow (independently from Neumann)	1856	Neanderthal skull discovered

(continued)

Lifetime	Key People	Date	Developments in Thermal-Fluid Sciences	Date	World Events
1803–1858	Henry Philibert Gaspard Darcy	1857	Performed experimental study that conclusively showed the importance of wall roughness		Financial panic in United States and Europe
				1861–1865	U.S. Civil War
1810–1879	William Froude	1868	Proposed drag laws for ship hulls from model testing		Louisa Alcott's *Little Women* and Dostoyevsky's *The Idiot* published
1842–1912	Osborne Reynolds	1874	Developed analogy among mass, heat, and momentum transfer (i.e., the Reynolds analogy)	1870–1871	Franco–Prussian War
1839–1903	Josiah Willard Gibbs	1876	First to develop generalized thermodynamic relationships		Invention of telephone (Bell) and internal combustion engine (Otto)
1844–1906	Ludwig Boltzmann	1877	Pioneer of statistical mechanics—derived microscopic interpretation of entropy		Invention of phonograph (Edison)
1842–1919	John William Strutt (Lord Rayleigh)	1877	Published "Method of Dimensions"		Edison's invention of electric light bulb
1842–1912	Osborne Reynolds	1883	Defined critical parameter (i.e., Reynolds number) for transition from laminar to turbulent flow		Eruption of Krakatoa volcano in Indonesia
1857–1899	Aimeé Vaschy	1892	Developed concepts of dimensional analysis and similitude; published "Sur les lois similitude en physique"		Tchaikowsky's *Nutcracker;* Diesel engine patented
1842–1912	Osborne Reynolds	1895	Analyzed turbulent flows by decomposing variables into mean and fluctuating components (i.e., the Reynolds decomposition)	1894–1895	Sino–Japanese War
1858–1947	Max Planck	1900	Advanced understanding of second law and entropy from a macroscopic viewpoint; won Nobel Prize for quantum theory of energy	1903	Wright brothers first powered flight at Kittyhawk, NC
1875–1953	Ludwig Prandtl	1904	"Inventor" of boundary layers and separation; "Father of Modern Fluid Mechanics"	1904–1905	Russo–Japanese War
1882–1962	Dimitri Riabouchinsky	1911	Engineering applications of dimensional analysis and similitude		Chinese Republic replaces Manchu Dynasty
1881–1963	Theodor von Kármán	1911	Numerous contributions to fluid mechanics; published studies of vortex street that bears his name		Roald Amundsen reaches South Pole
1883–1970	Paul Heinrich Blasius	1913	First to plot friction coefficient as a function of Reynolds number and relative roughness; student of Prandtl		Niels Bohr formulates theory of atomic structure
1867–1940	Edgar Buckingham	1914	Engineering applications of dimensional analysis and similitude	1914–1918	World War One

Dates	Person	Year	Contribution
1871–1951	Moritz Weber	1919	Gave modern form to principles of similitude; gave names to Reynolds number and Froude number
1882–1961	Percy Bridgeman	1922	Author of classic monograph *Dimensional Analysis* (1922) and Nobel Prize winner for work and invention in high-pressure physics (1946)
1894–1979	Johann Nikuradse	1933	Performed meticulous experiments to quantify wall roughness effects in pipe flows using various sized sand grains; student of Prandtl
1859–1958	William Frederick Durand	1934	Developed dimensionless analysis for aeronautics; published six-volume reference, *Aerodynamic Theory*
1886–1975	Geoffrey Ingram Taylor	1935	Developed statistical theories of turbulence; pioneer in the study of hydrodynamic stability
1900–1977	Joseph Keenan	1941	Modern codifier of thermodynamics and author of classical textbook
1880–1953	Lewis Ferry Moody	1944	Presents paper containing useful summary of friction data for internal flows (Moody chart)
1886–1984	Jerome Clarke Hunsaker	1947	First formal control volume analyses presented in *Engineering Applications of Fluid Mechanics* with B. G. Rightmire; developed modern wind tunnel (1914)

Year	World Event
1917–1920	Russian Revolution
1927	Economic collapse in Germany
1929	"Black Friday" world economic crisis begins; Spain becomes a republic
1932	Aldous Huxley's *Brave New World* published
	Famine in USSR and Roosevelt's "New Deal" in U.S.; Hitler appointed chancellor in Germany
	Mao Zedong begins "Long March" north; German plebiscite elects Hitler as Führer
	Persia renamed Iran
1938	Orson Welles radio broadcast of *War of the Worlds*
1939	Germany invades Poland; World War Two begins
	Japan attacks Pearl Harbor; Manhattan Project begins
	Tennessee Williams's *The Glass Menagerie* presented
1945	World War Two ends; United Nations created
1947	Jackie Robinson joins Brooklyn Dodgers and breaks color barrier; Chuck Yeager breaks the sound barrier
1947–1951	Marshall Plan for European recovery

Appendix B
Thermodynamic Properties of Ideal Gases and Carbon

Tables B.1–B.13 Present values for $\bar{c}_p(T)$, $\bar{h}^\circ(T) - \bar{h}^\circ_{f,\text{ref}}$, $\bar{h}^\circ_f(T)$, $\bar{s}^\circ(T)$, and $\Delta\bar{g}^\circ_f(T)$ at standard reference state ($T = 298.15$ K, P = 1 atm) for various species of the C–H–O–N system (with ideal-gas values for gaseous species).

Note that enthalpy of formation and Gibbs function of formation for compounds are calculated from the elements as follows:

$$\bar{h}^\circ_{f,i}(T) = \bar{h}^\circ_i(T) - \sum_{j\,\text{elements}} v'_j \bar{h}^\circ_j(T),$$

$$\bar{g}^\circ_{f,i}(T) = \bar{g}^\circ_i(T) - \sum_{j\,\text{elements}} v'_j \bar{g}^\circ_j(T)$$

$$= \bar{h}^\circ_{f,i}(T) - T\bar{s}^\circ_i(T) - \sum_{j\,\text{elements}} v'_j[-T\bar{s}^\circ_j(T)].$$

Sources: Tables B.1–B.12 are from Key, R. J., Rupley, F. M., and Miller, J. A., "The Chemkin Thermodynamic Data Base," Sandia Report, SAND87-8215B, March 1991. Table B.13 is from Myers, G. E., *Engineering Thermodynamics,* Prentice-Hall, Englewood Cliffs, NJ, 1989. Table B.14 is from Key, R. J., et al., ibid.

Table B.1 CO (Molecular Weight = 28.010, Enthalpy of Formation at 298 K = −110,541 kJ/kmol)

T (K)	$\bar{c}_p$ (kJ/kmol·K)	$\bar{h}°(T) - \bar{h}°_f(298)$ (kJ/kmol)	$\bar{h}°_f(T)$ (kJ/kmol)	$\bar{s}°(T)$ (kJ/kmol·K)	$\Delta\bar{g}°_f(T)$ (kJ/kmol)
200	28.687	−2835	−111,308	186.018	−128,532
298	29.072	0	−110,541	197.548	−137,163
300	29.078	54	−110,530	197.728	−137,328
400	29.433	2979	−110,121	206.141	−146,332
500	29.857	5943	−110,017	212.752	−155,403
600	30.407	8955	−110,156	218.242	−164,470
700	31.089	12,029	−110,477	222.979	−173,499
800	31.860	15,176	−110,924	227.180	−182,473
900	32.629	18,401	−111,450	230.978	−191,386
1000	33.255	21,697	−112,022	234.450	−200,238
1100	33.725	25,046	−112,619	237.642	−209,030
1200	34.148	28,440	−113,240	240.595	−217,768
1300	34.530	31,874	−113,881	243.344	−226,453
1400	34.872	35,345	−114,543	245.915	−235,087
1500	35.178	38,847	−115,225	248.332	−243,674
1600	35.451	42,379	−115,925	250.611	−252,214
1700	35.694	45,937	−116,644	252.768	−260,711
1800	35.910	49,517	−117,380	254.814	−269,164
1900	36.101	53,118	−118,132	256.761	−277,576
2000	36.271	56,737	−118,902	258.617	−285948
2100	36.421	60,371	−119,687	260.391	−294,281
2200	36.553	64,020	−120,488	262.088	−302,576
2300	36.670	67,682	−121,305	263.715	−310,835
2400	36.774	71,354	−122,137	265.278	−319,057
2500	36.867	75,036	−122,984	266.781	−327,245
2600	36.950	78,727	−123,847	268.229	−335,399
2700	37.025	82,426	−124,724	269.625	−343,519
2800	37.093	86,132	−125,616	270.973	−351,606
2900	37.155	89,844	−126,523	272.275	−359,661
3000	37.213	93,562	−127,446	273.536	−367,684
3100	37.268	97,287	−128,383	274.757	−375,677
3200	37.321	101,016	−129,335	275.941	−383,639
3300	37.372	104,751	−130,303	277.090	−391,571
3400	37.422	108,490	−131,285	278.207	−399,474
3500	37.471	112,235	−132,283	279.292	−407,347
3600	37.521	115,985	−133,295	280.349	−415,192
3700	37.570	119,739	−134,323	281.377	−423,008
3800	37.619	123,499	−135,366	282.380	−430,796
3900	37.667	127,263	−136,424	283.358	−438,557
4000	37.716	131,032	−137,497	284.312	−446,291
4100	37.764	134,806	−138,585	285.244	−453,997
4200	37.810	138,585	−139,687	286.154	−461,677
4300	37.855	142,368	−140,804	287.045	−469,330
4400	37.897	146,156	−141,935	287.915	−476,957
4500	37.936	149,948	−143,079	288.768	−484,558
4600	37.970	153,743	−144,236	289.602	−492,134
4700	37.998	157,541	−145,407	290.419	−499,684
4800	38.019	161,342	−146,589	291.219	−507,210
4900	38.031	165,145	−147,783	292.003	−514,710
5000	38.033	168,948	−148,987	292.771	−522,186

Table B.2   CO₂ (Molecular Weight = 44.011, Enthalpy of Formation at 298 K = −393,546 kJ/kmol)

T (K)	$\bar{c}_p$ (kJ/kmol·K)	$\bar{h}°(T) - \bar{h}_f°(298)$ (kJ/kmol)	$\bar{h}_f°(T)$ (kJ/kmol)	$\bar{s}°(T)$ (kJ/kmol·K)	$\Delta\bar{g}_f°(T)$ (kJ/kmol)
200	32.387	−3423	−393,483	199.876	−394,126
298	37.198	0	−393,546	213.736	−394,428
300	37.280	69	−393,547	213.966	−394,433
400	41.276	4003	−393,617	225.257	−394,718
500	44.569	8301	−393,712	234.833	−394,983
600	47.313	12,899	−393,844	243.209	−395,226
700	49.617	17,749	−394,013	250.680	−395,443
800	51.550	22,810	−394,213	257.436	−395,635
900	53.136	28,047	−394433	263.603	−395,799
1000	54.360	33,425	−394,659	269.268	−395,939
1100	55.333	38,911	−394,875	274.495	−396,056
1200	56.205	44,488	−395,083	279.348	−396,155
1300	56.984	50,149	−395,287	283.878	−396,236
1400	57.677	55,882	−395,88	288.127	−396,301
1500	58.292	61,681	−395,691	292.128	−396,352
1600	58.836	67,538	−395,897	295.908	−396,389
1700	59.316	73,446	−396,110	299.489	−396,414
1800	59.738	79,399	−396,332	302.892	−396,425
1900	60.108	85,392	−396,564	306.132	−396,424
2000	60.433	91,420	−396,808	309.223	−396,410
2100	60.717	97,477	−397,065	312.179	−396,384
2200	60.966	103,562	−397,338	315.009	−396,346
2300	61.185	109,670	−397,626	317.724	−396,294
2400	61.378	115,798	−397,931	320.333	−396,230
2500	61.548	121,944	−398,253	322.842	−396,152
2600	61.701	128,107	−398,594	325.259	−396,061
2700	61.839	134,284	−398,952	327.590	−395,957
2800	61.965	140,474	−399,329	329.841	−395,840
2900	62.083	146,677	−399,725	332.018	−395,708
3000	62.194	152,891	−400,140	334.124	−395,562
3100	62.301	159,116	−400,573	336.165	−395,403
3200	62.406	165,351	−401,025	338.145	−395,229
3300	62.510	171,597	−401,495	340.067	−395,041
3400	62.614	177,853	−401,983	341.935	−394,838
3500	62.718	184,120	−402,489	343.751	−394,620
3600	62.825	190,397	−403,013	345.519	−394,388
3700	62.932	196,685	−403,553	347.242	−394,141
3800	63.041	202,983	−404,110	348.922	−393,879
3900	63.151	209,293	−404,684	350.561	−393,602
4000	63.261	215,613	−405,273	352.161	−393,311
4100	63.369	221,945	−405,878	353.725	−393,004
4200	63.474	228,287	−406,499	355.253	−392,683
4300	63.575	234,640	−407,135	356.748	−392,346
4400	63.669	241,002	−407,785	358.210	−391,995
4500	63.753	247,373	−408,451	359.642	−391,629
4600	63.825	253,752	−409,132	361.044	−391,247
4700	63.881	260,138	−409,828	362.417	−390,851
4800	63.918	266,528	−410,539	363.763	−390,440
4900	63.932	272,920	−411,267	365.081	−390,014
5000	63.919	279,313	−412,010	366.372	−389,572

Table B.3 H$_2$ (Molecular Weight = 2.016, Enthalpy of Formation at 298 K = 0 kJ/kmol)

T (K)	$\bar{c}_p$ (kJ/kmol·K)	$\bar{h}°(T) - \bar{h}_f°(298)$ (kJ/kmol)	$\bar{h}_f°(T)$ (kJ/kmol)	$\bar{s}°(T)$ (kJ/kmol·K)	$\Delta\bar{g}_f°(T)$ (kJ/kmol)
200	28.522	−2818	0	119.137	0
298	28.871	0	0	130.595	0
300	28.877	53	0	130.773	0
400	29.120	2954	0	139.116	0
500	29.275	5874	0	145.632	0
600	29.375	8807	0	150.979	0
700	29.461	11,749	0	155.514	0
800	29.581	14,701	0	159.455	0
900	29.792	17,668	0	162.950	0
1000	30.160	20,664	0	166.106	0
1100	30.625	23,704	0	169.003	0
1200	31.077	26,789	0	171.687	0
1300	31.516	29,919	0	174.192	0
1400	31.943	33,092	0	176.543	0
1500	32.356	36,307	0	178.761	0
1600	32.758	39,562	0	180.862	0
1700	33.146	42,858	0	182.860	0
1800	33.522	46,191	0	184.765	0
1900	33.885	49,562	0	186.587	0
2000	34.236	52,968	0	188.334	0
2100	34.575	56,408	0	190.013	0
2200	34.901	59,882	0	191.629	0
2300	35.216	63,388	0	193.187	0
2400	35.519	66,925	0	194.692	0
2500	35.811	70,492	0	196.148	0
2600	36.091	74,087	0	197.558	0
2700	36.361	77,710	0	198.926	0
2800	36.621	81,359	0	200.253	0
2900	36.871	85,033	0	201.542	0
3000	37.112	88,733	0	202.796	0
3100	37.343	92,455	0	204.017	0
3200	37.566	96,201	0	205.206	0
3300	37.781	99,968	0	206.365	0
3400	37.989	103,757	0	207.496	0
3500	38.190	107,566	0	208.600	0
3600	38.385	111,395	0	209.679	0
3700	38.574	115,243	0	210.733	0
3800	38.759	119,109	0	211.764	0
3900	38.939	122,994	0	212.774	0
4000	39.116	126,897	0	213.762	0
4100	39.291	130,817	0	214.730	0
4200	39.464	134,755	0	215.679	0
4300	39.636	138,710	0	216.609	0
4400	39.808	142,682	0	217.522	0
4500	39.981	146,672	0	218.419	0
4600	40.156	150,679	0	219.300	0
4700	40.334	154,703	0	220.165	0
4800	40.516	158,746	0	221.016	0
4900	40.702	162,806	0	221.853	0
5000	40.895	166,886	0	222.678	0

Table B.4 H (Molecular Weight = 1.008, Enthalpy of Formation at 298 K = 217,979 kJ/kmol)

T (K)	$\bar{c}_p$ (kJ/kmol·K)	$\bar{h}°(T) - \bar{h}_f°(298)$ (kJ/kmol)	$\bar{h}_f°(T)$ (kJ/kmol)	$\bar{s}°(T)$ (kJ/kmol·K)	$\Delta\bar{g}_f°(T)$ (kJ/kmol)
200	20.786	−2040	217,346	106.305	207,999
298	20.786	0	217,977	114.605	203,276
300	20.786	38	217,989	114.733	203,185
400	20.786	2117	218,617	120.713	198,155
500	20.786	4196	219,236	125.351	192,968
600	20.786	6274	219,848	129.141	187,657
700	20.786	8353	220,456	132.345	182,244
800	20.786	10,431	221,059	135.121	176,744
900	20.786	12,510	221,653	137.569	171,169
1000	20.786	14,589	222,234	139.759	165,528
1100	20.786	16,667	222,793	141.740	159,830
1200	20.786	18,746	223,329	143.549	154,082
1300	20.786	20,824	223,843	145.213	148,291
1400	20.786	22,903	224,335	146.753	142,461
1500	20.786	24,982	224,806	148.187	136,596
1600	20.786	27,060	225,256	149.528	130,700
1700	20.786	29,139	225,687	150.789	124,777
1800	20.786	31,217	226,099	151.977	118,830
1900	20.786	33,296	226,493	153.101	112,859
2000	20.786	35,375	226,868	154.167	106,869
2100	20.786	37,453	227,226	155.181	100,860
2200	20.786	39,532	227,568	156.148	94,834
2300	20.786	41,610	227,894	157.072	88,794
2400	20.786	43,689	228,204	157.956	82,739
2500	20.786	45,768	228,499	158.805	76,672
2600	20.786	47,846	228,780	159.620	70,593
2700	20.786	49,925	229,047	160.405	64,504
2800	20.786	52,003	229,301	161.161	58,405
2900	20.786	54,082	229,543	161.890	52,298
3000	20.786	56,161	229,772	162.595	46,182
3100	20.786	58,239	229,989	163.276	40,058
3200	20.786	60,318	230,195	163.936	33,928
3300	20.786	62,396	230,390	164.576	27,792
3400	20.786	64,475	230,574	165.196	21,650
3500	20.786	66,554	230,748	165.799	15,502
3600	20.786	68,632	230,912	166.384	9350
3700	20.786	70,711	231,067	166.954	3194
3800	20.786	72,789	231,212	167.508	−2967
3900	20.786	74,868	231,348	168.048	−9132
4000	20.786	76,947	231,475	168.575	−15,299
4100	20.786	79,025	231,594	169.088	−21,470
4200	20.786	81,104	231,704	169.589	−27,644
4300	20.786	83,182	231,805	170.078	−33,820
4400	20.786	85,261	231,897	170.556	−39,998
4500	20.786	87,340	231,981	171.023	−46,179
4600	20.786	89,418	232,056	171.480	−52,361
4700	20.786	91,497	232,123	171.927	−58,545
4800	20.786	93,575	232,180	172.364	−64,730
4900	20.786	95,654	232,228	172.793	−70,916
5000	20.786	97,733	232,267	173.213	−77,103

Table B.5   OH (Molecular Weight = 17.007, Enthalpy of Formation at 298 K = 38,986 kJ/kmol)

T (K)	$\bar{c}_p$ (kJ/kmol·K)	$\bar{h}°(T) - \bar{h}_f°(298)$ (kJ/kmol)	$\bar{h}_f°(T)$ (kJ/kmol)	$\bar{s}°(T)$ (kJ/kmol·K)	$\Delta\bar{g}_f°(T)$ (kJ/kmol)
200	30.140	−2948	38,864	171.607	35,808
298	29.932	0	38,985	183.604	34,279
300	29.928	55	38,987	183.789	34,250
400	29.718	3037	39,030	192.369	32,662
500	29.570	6001	39,000	198.983	31,072
600	29.527	8955	38,909	204.369	29,494
700	29.615	11,911	38,770	208.925	27,935
800	29.844	14,883	38,599	212.893	26,399
900	30.208	17,884	38,410	216.428	24,885
1000	30.682	20,928	38,220	219.635	23,392
1100	31.186	24,022	38,039	222.583	21,918
1200	31.662	27,164	37,867	225.317	20,460
1300	32.114	30,353	37,704	227.869	19,017
1400	32.540	33,586	37,548	230.265	17,585
1500	32.943	36,860	37,397	232.524	16,164
1600	33.323	40,174	37,252	234.662	14,753
1700	33.682	43,524	37,109	236.693	13,352
1800	34.019	46,910	36,969	238.628	11,958
1900	34.337	50,328	36,831	240.476	10,573
2000	34.635	53,776	36,693	242.245	9194
2100	34.915	57,254	36,555	243.942	7823
2200	35.178	60,759	36,416	245.572	6458
2300	35.425	64,289	36,276	247.141	5099
2400	35.656	67,843	36,133	248.654	3746
2500	35.872	71,420	35,986	250.114	2400
2600	36.074	75,017	35,836	251.525	1060
2700	36.263	78,634	35,682	252.890	−275
2800	36.439	82,269	35,524	254.212	−1604
2900	36.604	85,922	35,360	255.493	−2927
3000	36.759	89,590	35,191	256.737	−4245
3100	36.903	93,273	35,016	257.945	−5556
3200	37.039	96,970	34,835	259.118	−6862
3300	37.166	100,681	34,648	260.260	−8162
3400	37.285	104,403	34,454	261.371	−9457
3500	37.398	108,137	34,253	262.454	−10,745
3600	37.504	111,882	34,046	263.509	−12,028
3700	37.605	115,638	33,831	264.538	−13,305
3800	37.701	119,403	33,610	265.542	−14,576
3900	37.793	123,178	33,381	266.522	−15,841
4000	37.882	126,962	33,146	267.480	−17,100
4100	37.968	130,754	32,903	268.417	−18,353
4200	38.052	134,555	32,654	269.333	−19,600
4300	38.135	138,365	32,397	270.229	−20,841
4400	38.217	142,182	32,134	271.107	−22,076
4500	38.300	146,008	31,864	271.967	−23,306
4600	38.382	149,842	31,588	272.809	−24,528
4700	38.466	153,685	31,305	273.636	−25,745
4800	38.552	157,536	31,017	274.446	−26,956
4900	38.640	161,395	30,722	275.242	−28,161
5000	38.732	165,264	30,422	276.024	−29,360

Table B.6 H$_2$O (Molecular Weight = 18.016, Enthalpy of Formation at 298 K = −241,845 kJ/kmol, Enthalpy of Vaporization = 44,010 kJ/kmol)

T (K)	$\bar{c}_p$ (kJ/kmol·K)	$\bar{h}^\circ(T) - \bar{h}_f^\circ(298)$ (kJ/kmol)	$\bar{h}_f^\circ(T)$ (kJ/kmol)	$\bar{s}^\circ(T)$ (kJ/kmol·K)	$\Delta\bar{g}_f^\circ(T)$ (kJ/kmol)
200	32.255	−3227	−240,838	175.602	−232,779
298	33.448	0	−241,847	188.715	−228,608
300	33.468	62	−241,865	188.922	−228,526
400	34.437	3458	−242,858	198.686	−223,929
500	35.337	6947	−243,822	206.467	−219,085
600	36.288	10,528	−244,753	212.992	−214,049
700	37.364	14,209	−245,638	218.665	−208,861
800	38.587	18,005	−246,461	223.733	−203,550
900	39.930	21,930	−247,209	228.354	−198,141
1000	41.315	25,993	−247,879	232.633	−192,652
1100	42.638	30,191	−248,475	236.634	−187,100
1200	43.874	34,518	−249,005	240.397	−181,497
1300	45.027	38,963	−249,477	243.955	−175,852
1400	46.102	43,520	−249,895	247.332	−170,172
1500	47.103	48,181	−250,267	250.547	−164,464
1600	48.035	52,939	−250,597	253.617	−158,733
1700	48.901	57,786	−250,890	256.556	−152,983
1800	49.705	62,717	−251,151	259.374	−147,216
1900	50.451	67,725	−251,384	262.081	−141,435
2000	51.143	72,805	−251,594	264.687	−135,643
2100	51.784	77,952	−251,783	267.198	−129,841
2200	52.378	83,160	−251,955	269.621	−124,030
2300	52.927	88,426	−252,113	271.961	−118,211
2400	53.435	93,744	−252,261	274.225	−112,386
2500	53.905	99,112	−252,399	276.416	−106,555
2600	54.340	104,524	−252,532	278.539	−100,719
2700	54.742	109,979	−252,659	280.597	−94,878
2800	55.115	115,472	−252,785	282.595	−89,031
2900	55.459	121,001	−252,909	284.535	−83,181
3000	55.779	126,563	−253,034	286.420	−77,326
3100	56.076	132,156	−253,161	288.254	−71,467
3200	56.353	137,777	−253,290	290.039	−65,604
3300	56.610	143,426	−253,423	291.777	−59,737
3400	56.851	149,099	−253,561	293.471	−53,865
3500	57.076	154,795	−253,704	295.122	−47,990
3600	57.288	160,514	−253,852	296.733	−42,110
3700	57.488	166,252	−254,007	298.305	−36,226
3800	57.676	172,011	−254,169	299.841	−30,338
3900	57.856	177,787	−254,338	301.341	−24,446
4000	58.026	183,582	−254,515	302.808	−18,549
4100	58.190	189,392	−254,699	304.243	−12,648
4200	58.346	195,219	−254,892	305.647	−6742
4300	58.496	201,061	−255,093	307.022	−831
4400	58.641	206,918	−255,303	308.368	5085
4500	58.781	212,790	−255,522	309.688	11,005
4600	58.916	218,674	−255,751	310.981	16,930
4700	59.047	224,573	−255,990	312.250	22,861
4800	59.173	230,484	−256,239	313.494	28,796
4900	59.295	236,407	−256,501	314.716	34,737
5000	59.412	242,343	−256,774	315.915	40,684

Table B.7 N_2 (Molecular Weight = 28.013, Enthalpy of Formation at 298 K = 0 kJ/kmol)

T (K)	$\bar{c}_p$ (kJ/kmol·K)	$\bar{h}°(T) - \bar{h}_f°(298)$ (kJ/kmol)	$\bar{h}_f°(T)$ (kJ/kmol)	$\bar{s}°(T)$ (kJ/kmol·K)	$\Delta\bar{g}_f°(T)$ (kJ/kmol)
200	28.793	−2841	0	179.959	0
298	29.071	0	0	191.511	0
300	29.075	54	0	191.691	0
400	29.319	2973	0	200.088	0
500	29.636	5920	0	206.662	0
600	30.086	8905	0	212.103	0
700	30.684	11,942	0	216.784	0
800	31.394	15,046	0	220.927	0
900	32.131	18,222	0	224.667	0
1000	32.762	21,468	0	228.087	0
1100	33.258	24,770	0	231.233	0
1200	33.707	28,118	0	234.146	0
1300	34.113	31,510	0	236.861	0
1400	34.477	34,939	0	239.402	0
1500	34.805	38,404	0	241.792	0
1600	35.099	41,899	0	244.048	0
1700	35.361	45,423	0	246.184	0
1800	35.595	48,971	0	248.212	0
1900	35.803	52,541	0	250.142	0
2000	35.988	56,130	0	251.983	0
2100	36.152	59,738	0	253.743	0
2200	36.298	63,360	0	255.429	0
2300	36.428	66,997	0	257.045	0
2400	36.543	70,645	0	258.598	0
2500	36.645	74,305	0	260.092	0
2600	36.737	77,974	0	261.531	0
2700	36.820	81,652	0	262.919	0
2800	36.895	85,338	0	264.259	0
2900	36.964	89,031	0	265.555	0
3000	37.028	92,730	0	266.810	0
3100	37.088	96,436	0	268.025	0
3200	37.144	100,148	0	269.203	0
3300	37.198	103,865	0	270.347	0
3400	37.251	107,587	0	271.458	0
3500	37.302	111,315	0	272.539	0
3600	37.352	115,048	0	273.590	0
3700	37.402	118,786	0	274.614	0
3800	37.452	122,528	0	275.612	0
3900	37.501	126,276	0	276.586	0
4000	37.549	130,028	0	277.536	0
4100	37.597	133,786	0	278.464	0
4200	37.643	137,548	0	279.370	0
4300	37.688	141,314	0	280.257	0
4400	37.730	145,085	0	281.123	0
4500	37.768	148,860	0	281.972	0
4600	37.803	152,639	0	282.802	0
4700	37.832	156,420	0	283.616	0
4800	37.854	160,205	0	284.412	0
4900	37.868	163,991	0	285.193	0
5000	37.873	167,778	0	285.958	0

Table B.8 N (Molecular Weight = 14.007, Enthalpy of Formation at 298 K = 472,629 kJ/kmol)

T (K)	$\bar{c}_p$ (kJ/kmol·K)	$\bar{h}^\circ(T) - \bar{h}_f^\circ(298)$ (kJ/kmol)	$\bar{h}_f^\circ(T)$ (kJ/kmol)	$\bar{s}^\circ(T)$ (kJ/kmol·K)	$\Delta\bar{g}_f^\circ(T)$ (kJ/kmol)
200	20.790	−2040	472,008	144.889	461,026
298	20.786	0	472,628	153.189	455,504
300	20.786	38	472,640	153.317	455,398
400	20.786	2117	473,258	159.297	449,557
500	20.786	4196	473,864	163.935	443,562
600	20.786	6274	474,450	167.725	437,446
700	20.786	8353	475,010	170.929	431,234
800	20.786	10,431	475,537	173.705	424,944
900	20.786	12,510	476,027	176.153	418,590
1000	20.786	14,589	476,483	178.343	412,183
1100	20.792	16,668	476,911	180.325	405,732
1200	20.795	18,747	477,316	182.134	399,243
1300	20.795	20,826	477,700	183.798	392,721
1400	20.793	22,906	478,064	185.339	386,171
1500	20.790	24,985	478,411	186.774	379,595
1600	20.786	27,064	478,742	188.115	372,996
1700	20.782	29,142	479,059	189.375	366,377
1800	20.779	31,220	479,363	190.563	359,740
1900	20.777	33,298	479,656	191.687	353,086
2000	20.776	35,376	479,939	192.752	346,417
2100	20.778	37,453	480,213	193.766	339,735
2200	20.783	39,531	480,479	194.733	333,039
2300	20.791	41,610	480,740	195.657	326,331
2400	20.802	43,690	480,995	196.542	319,612
2500	20.818	45,771	481,246	197.391	312,883
2600	20.838	47,853	481,494	198.208	306,143
2700	20.864	49,938	481,740	198.995	299,394
2800	20.895	52,026	481,985	199.754	292,636
2900	20.931	54,118	482,230	200.488	285,870
3000	20.974	56,213	482,476	201.199	279,094
3100	21.024	58,313	482,723	201.887	272,311
3200	21.080	60,418	482,972	202.555	265,519
3300	21.143	62,529	483,224	203.205	258,720
3400	21.214	64,647	483,481	203.837	251,913
3500	21.292	66,772	483,742	204.453	245,099
3600	21.378	68,905	484,009	205.054	238,276
3700	21.472	71,048	484,283	205.641	231,447
3800	21.575	73,200	484,564	206.215	224,610
3900	21.686	75,363	484,853	206.777	217,765
4000	21.805	77,537	485,151	207.328	210,913
4100	21.934	79,724	485,459	207.868	204,053
4200	22.071	81,924	485,779	208.398	197,186
4300	22.217	84,139	486,110	208.919	190,310
4400	22.372	86,368	486,453	209.431	183,427
4500	22.536	88,613	486,811	209.936	176,536
4600	22.709	90,875	487,184	210.433	169,637
4700	22.891	93,155	487,573	210.923	162,730
4800	23.082	95,454	487,979	211.407	155,814
4900	23.282	97,772	488,405	211.885	148,890
5000	23.491	100,111	488,850	212.358	141,956

Table B.9 NO (Molecular Weight = 30.006, Enthalpy of Formation at 298 K = 90,297 kJ/kmol)

T (K)	$\bar{c}_p$ (kJ/kmol·K)	$\bar{h}°(T) - \bar{h}_f°(298)$ (kJ/kmol)	$\bar{h}_f°(T)$ (kJ/kmol)	$\bar{s}°(T)$ (kJ/kmol·K)	$\Delta\bar{g}_f°(T)$ (kJ/kmol)
200	29.374	−2901	90,234	198.856	87,811
298	29.728	0	90,297	210.652	86,607
300	29.735	55	90,298	210.836	86,584
400	30.103	3046	90,341	219.439	85,340
500	30.570	6079	90,367	226.204	84,086
600	31.174	9165	90,382	231.829	82,828
700	31.908	12,318	90,393	236.688	81,568
800	32.715	15,549	90,405	241.001	80,307
900	33.489	18,860	90,421	244.900	79,043
1000	34.076	22,241	90,443	248.462	77,778
1100	34.483	25,669	90,465	251.729	76,510
1200	34.850	29,136	90,486	254.745	75,241
1300	35.180	32,638	90,505	257.548	73,970
1400	35.474	36,171	90,520	260.166	72,697
1500	35.737	39,732	90,532	262.623	71,423
1600	35.972	43,317	90,538	264.937	70,149
1700	36.180	46,925	90,539	267.124	68,875
1800	36.364	50,552	90,534	269.197	67,601
1900	36.527	54,197	90,523	271.168	66,327
2000	36.671	57,857	90,505	273.045	65,054
2100	36.797	61,531	90,479	274.838	63,782
2200	36.909	65,216	90,447	276.552	62,511
2300	37.008	68,912	90,406	278.195	61,243
2400	37.095	72,617	90,358	279.772	59,976
2500	37.173	76,331	90,303	281.288	58,711
2600	37.242	80,052	90,239	282.747	57,448
2700	37.305	83,779	90,168	284.154	56,188
2800	37.362	87,513	90,089	285.512	54,931
2900	37.415	91,251	90,003	286.824	53,677
3000	37.464	94,995	89,909	288.093	52,426
3100	37.511	98,744	89,809	289.322	51,178
3200	37.556	102,498	89,701	290.514	49,934
3300	37.600	106,255	89,586	291.670	48,693
3400	37.643	110,018	89,465	292.793	47,456
3500	37.686	113,784	89,337	293.885	46,222
3600	37.729	117,555	89,203	294.947	44,992
3700	37.771	121,330	89,063	295.981	43,766
3800	37.815	125,109	88,918	296.989	42,543
3900	37.858	128,893	88,767	297.972	41,325
4000	37.900	132,680	88,611	298.931	40,110
4100	37.943	136,473	88,449	299.867	38,900
4200	37.984	140,269	88,283	300.782	37,693
4300	38.023	144,069	88,112	301.677	36,491
4400	38.060	147,873	87,936	302.551	35,292
4500	38.093	151,681	87,755	303.407	34,098
4600	38.122	155,492	87,569	304.244	32,908
4700	38.146	159,305	87,379	305.064	31,721
4800	38.162	163,121	87,184	305.868	30,539
4900	38.171	166,938	86,984	306.655	29,361
5000	38.170	170,755	86,779	307.426	28,187

Table B.10 NO$_2$ (Molecular Weight = 46.006, Enthalpy of Formation at 298 K = 33,098 kJ/kmol)

T (K)	$\bar{c}_p$ (kJ/kmol·K)	$\bar{h}°(T) - \bar{h}_f°(298)$ (kJ/kmol)	$\bar{h}_f°(T)$ (kJ/kmol)	$\bar{s}°(T)$ (kJ/kmol·K)	$\Delta\bar{g}_f°(T)$ (kJ/kmol)
200	32.936	−3432	33,961	226.016	45,453
298	36.881	0	33,098	239.925	51,291
300	36.949	68	33,085	240.153	51,403
400	40.331	3937	32,521	251.259	57,602
500	43.227	8118	32,173	260.578	63,916
600	45.737	12,569	31,974	268.686	70,285
700	47.913	17,255	31,885	275.904	76,679
800	49.762	22,141	31,880	282.427	83,079
900	51.243	27,195	31,938	288.377	89,476
1000	52.271	32,375	32,035	293.834	95,864
1100	52.989	37,638	32,146	298.850	102,242
1200	53.625	42,970	32,267	303.489	108,609
1300	54.186	48,361	32,392	307.804	114,966
1400	54.679	53,805	32,519	311.838	121,313
1500	55.109	59,295	32,643	315.625	127,651
1600	55.483	64,825	32,762	319.194	133,981
1700	55.805	70,390	32,873	322.568	140,303
1800	56.082	75,984	32,973	325.765	146,620
1900	56.318	81,605	33,061	328.804	152,931
2000	56.517	87,247	33,134	331.698	159,238
2100	56.685	92,907	33,192	334.460	165,542
2200	56.826	98,583	33,233	337.100	171,843
2300	56.943	104,271	33,256	339.629	178,143
2400	57.040	109,971	33,262	342.054	184,442
2500	57.121	115,679	33,248	344.384	190,742
2600	57.188	121,394	33,216	346.626	197,042
2700	57.244	127,116	33,165	348.785	203,344
2800	57.291	132,843	33,095	350.868	209,648
2900	57.333	138,574	33,007	352.879	215,955
3000	57.371	144,309	32,900	354.824	222,265
3100	57.406	150,048	32,776	356.705	228,579
3200	57.440	155,791	32,634	358.529	234,898
3300	57.474	161,536	32,476	360.297	241,221
3400	57.509	167,285	32,302	362.013	247,549
3500	57.546	173,038	32,113	363.680	253,883
3600	57.584	178,795	31,908	365.302	260,222
3700	57.624	184,555	31,689	366.880	266,567
3800	57.665	190,319	31,456	368.418	272,918
3900	57.708	196,088	31,210	369.916	279,276
4000	57.750	201,861	30,951	371.378	285,639
4100	57.792	207,638	30,678	372.804	292,010
4200	57.831	213,419	30,393	374.197	298,387
4300	57.866	219,204	30,095	375.559	304,772
4400	57.895	224,992	29,783	376.889	311,163
4500	57.915	230,783	29,457	378.190	317,562
4600	57.925	236,575	29,117	379.464	323,968
4700	57.922	242,367	28,761	380.709	330,381
4800	57.902	248,159	28,389	381.929	336,803
4900	57.862	253,947	27,998	383.122	343,232
5000	57.798	259,730	27,586	384.290	349,670

Table B.11 O$_2$ (Molecular Weight = 31.999, Enthalpy of Formation at 298 K = 0 kJ/kmol)

T (K)	$\bar{c}_p$ (kJ/kmol·K)	$\bar{h}°(T) - \bar{h}°_f(298)$ (kJ/kmol)	$\bar{h}°_f(T)$ (kJ/kmol)	$\bar{s}°(T)$ (kJ/kmol·K)	$\Delta\bar{g}°_f(T)$ (kJ/kmol)
200	28.473	−2836	0	193.518	0
298	29.315	0	0	205.043	0
300	29.331	54	0	205.224	0
400	30.210	3031	0	213.782	0
500	31.114	6097	0	220.620	0
600	32.030	9254	0	226.374	0
700	32.927	12,503	0	231.379	0
800	33.757	15,838	0	235.831	0
900	34.454	19,250	0	239.849	0
1000	34.936	22,721	0	243.507	0
1100	35.270	26,232	0	246.852	0
1200	35.593	29,775	0	249.935	0
1300	35.903	33,350	0	252.796	0
1400	36.202	36,955	0	255.468	0
1500	36.490	40,590	0	257.976	0
1600	36.768	44,253	0	260.339	0
1700	37.036	47,943	0	262.577	0
1800	37.296	51,660	0	264.701	0
1900	37.546	55,402	0	266.724	0
2000	37.788	59,169	0	268.656	0
2100	38.023	62,959	0	270.506	0
2200	38.250	66,773	0	272.280	0
2300	38.470	70,609	0	273.985	0
2400	38.684	74,467	0	275.627	0
2500	38.891	78,346	0	277.210	0
2600	39.093	82,245	0	278.739	0
2700	39.289	86,164	0	280.218	0
2800	39.480	90,103	0	281.651	0
2900	39.665	94,060	0	283.039	0
3000	39.846	98,036	0	284.387	0
3100	40.023	102,029	0	285.697	0
3200	40.195	106,040	0	286.970	0
3300	40.362	110,068	0	288.209	0
3400	40.526	114,112	0	289.417	0
3500	40.686	118,173	0	290.594	0
3600	40.842	122,249	0	291.742	0
3700	40.994	126,341	0	292.863	0
3800	41.143	130,448	0	293.959	0
3900	41.287	134,570	0	295.029	0
4000	41.429	138,705	0	296.076	0
4100	41.566	142,855	0	297.101	0
4200	41.700	147,019	0	298.104	0
4300	41.830	151,195	0	299.087	0
4400	41.957	155,384	0	300.050	0
4500	42.079	159,586	0	300.994	0
4600	42.197	163,800	0	301.921	0
4700	42.312	168,026	0	302.829	0
4800	42.421	172,262	0	303.721	0
4900	42.527	176,510	0	304.597	0
5000	42.627	180,767	0	305.457	0

Table B.12 O (Molecular Weight = 16.000, Enthalpy of Formation at 298 K = 249,197 kJ/kmol)

T (K)	$\bar{c}_p$ (kJ/kmol·K)	$\bar{h}°(T) - \bar{h}_f°(298)$ (kJ/kmol)	$\bar{h}_f°(T)$ (kJ/kmol)	$\bar{s}°(T)$ (kJ/kmol·K)	$\Delta\bar{g}_f°(T)$ (kJ/kmol)
200	22.477	−2176	248,439	152.085	237,374
298	21.899	0	249,197	160.945	231,778
300	21.890	41	249,211	161.080	231,670
400	21.500	2209	249,890	167.320	225,719
500	21.256	4345	250,494	172.089	219,605
600	21.113	6463	251,033	175.951	213,375
700	21.033	8570	251,516	179.199	207,060
800	20.986	10,671	251,949	182.004	200,679
900	20.952	12,768	252,340	184.474	194,246
1000	20.915	14,861	252,698	186.679	187,772
1100	20.898	16,952	253,033	188.672	181,263
1200	20.882	19,041	253,350	190.490	174,724
1300	20.867	21,128	253,650	192.160	168,159
1400	20.854	23,214	253,934	193.706	161,572
1500	20.843	25,299	254,201	195.145	154,966
1600	20.834	27,383	254,454	196.490	148,342
1700	20.827	29,466	254,692	197.753	141,702
1800	20.822	31,548	254,916	198.943	135,049
1900	20.820	33,630	255,127	200.069	128,384
2000	20.819	35,712	255,325	201.136	121,709
2100	20.821	37,794	255,512	202.152	115,023
2200	20.825	39,877	255,687	203.121	108,329
2300	20.831	41,959	255,852	204.047	101,627
2400	20.840	44,043	256,007	204.933	94,918
2500	20.851	46,127	256,152	205.784	88,203
2600	20.865	48,213	256,288	206.602	81,483
2700	20.881	50,300	256,416	207.390	74,757
2800	20.899	52,389	256,535	208.150	68,027
2900	20.920	54,480	256,648	208.884	61,292
3000	20.944	56,574	256,753	209.593	54,554
3100	20.970	58,669	256,852	210.280	47,812
3200	20.998	60,768	256,945	210.947	41,068
3300	21.028	62,869	257,032	211.593	34,320
3400	21.061	64,973	257,114	212.221	27,570
3500	21.095	67,081	257,192	212.832	20,818
3600	21.132	69,192	257,265	213.427	14,063
3700	21.171	71,308	257,334	214.007	7307
3800	21.212	73,427	257,400	214.572	548
3900	21.254	75,550	257,462	215.123	−6212
4000	21.299	77,678	257,522	215.662	−12,974
4100	21.345	79,810	257,579	216.189	−19,737
4200	21.392	81,947	257,635	216.703	−26,501
4300	21.441	84,088	257,688	217.207	−33,267
4400	21.490	86,235	257,740	217.701	−40,034
4500	21.541	88,386	257,790	218.184	−46,802
4600	21.593	90,543	257,840	218.658	−53,571
4700	21.646	92,705	257,889	219.123	−60,342
4800	21.699	94,872	257,938	219.580	−67,113
4900	21.752	97,045	257,987	220.028	−73,886
5000	21.805	99,223	258,036	220.468	−80,659

Table B.13   C(s) (Graphite, Molecular Weight = 12.011, Enthalpy of Formation at 298 K = 0 kJ/kmol)

T (K)	$\bar{c}_p$ (kJ/kmol·K)	$\bar{h}°(T) - \bar{h}_f°(298)$ (kJ/kmol)	$\bar{h}_f°(T)$ (kJ/kmol)	$\bar{s}°(T)$ (kJ/kmol·K)	$\Delta\bar{g}_f°(T)$ (kJ/kmol)
100	1.65	−1000	0	0.88	0
200	5.03	−670	0	3.01	0
298	8.53	0	0	5.69	0
400	11.93	1050	0	8.68	0
500	14.63	2380	0	11.65	0
600	16.89	3960	0	14.52	0
700	18.58	5740	0	17.26	0
800	19.83	7660	0	19.83	0
900	20.79	9700	0	22.22	0
1000	21.54	11,820	0	24.45	0
1100	22.19	14,000	0	26.53	0
1200	22.72	16,250	0	28.49	0
1300	23.12	18,540	0	30.33	0
1400	23.45	20,870	0	32.05	0
1500	23.72	23,230	0	33.68	0
1600	23.94	25,610	0	35.22	0
1700	24.12	28,020	0	36.67	0
1800	24.28	30,440	0	38.06	0
1900	24.42	32,870	0	39.38	0
2000	24.54	35,320	0	40.63	0
2100	24.65	37,780	0	41.83	0
2200	24.74	40,250	0	42.98	0
2300	24.84	42,730	0	44.08	0
2400	24.92	45,220	0	45.14	0
2500	25.00	47,710	0	46.16	0
2600	25.07	50,220	0	47.14	0
2700	25.14	52,730	0	48.09	0
2800	25.21	55,240	0	49.00	0
2900	25.28	57,770	0	49.89	0
3000	25.34	60,300	0	50.75	0
3100	25.41	62,840	0	51.58	0
3200	25.47	65,380	$\bar{h}_f°(T)$	52.39	$\Delta\bar{g}_f°(T)$
3300	25.53	66,880	0	53.17	0
3400	25.60	70,490	0	53.94	0
3500	25.66	73,050	0	54.68	0
3600	25.73	75,620	0	55.40	0
3700	25.79	78,200	0	56.11	0
3800	25.86	80,780	0	56.80	0
3900	25.93	83,370	0	57.47	0
4000	26.00	85,960	0	58.13	0
4100	26.07	88,570	0	58.77	0
4200	26.14	91,180	0	59.40	0
4300	26.21	93,800	0	60.02	0
4400	26.28	96,420	0	60.62	0
4500	26.36	99,050	0	61.21	0
4600	26.43	101,690	0	61.79	0
4700	26.51	104,340	0	62.36	0
4800	26.59	106,990	0	62.92	0
4900	26.66	109,650	0	63.47	0
5000	26.74	112,320	0	64.01	0

Table B.14 Curve-Fit Coefficients for Thermodynamic Properties (C–H–O–N System):

$$\bar{c}_p/R_u = a_1 + a_2T + a_3T^2 + a_4T^3 + a_5T^4,$$

$$\bar{h}^\circ/R_uT = a_1 + \frac{a_2}{2}T + \frac{a_3}{3}T^2 + \frac{a_4}{4}T^3 + \frac{a_5}{5}T^4 + \frac{a_6}{T},$$

$$\bar{s}^\circ/R_u = a_1\ln T + a_2T + \frac{a_3}{2}T^2 + \frac{a_4}{3}T^3 + \frac{a_5}{4}T^4 + a_7$$

Species	T (K)	a_1	a_2	a_3	a_4	a_5	a_6	a_7
CO	1000–5000	0.03025078E+02	0.14426885E-02	-0.05630827E-05	0.10185813E-09	-0.06910951E-13	-0.14268350E+05	0.06108217E+02
	300–1000	0.03262451E+02	0.15119409E-02	-0.03881755E-04	0.05581944E-07	-0.02474951E-10	-0.14310539E+05	0.04848897E+02
CO$_2$	1000–5000	0.04453623E+02	0.03140168E-01	-0.12784105E-05	0.02393996E-08	-0.16690333E-13	-0.04896696E+06	-0.09553959E+01
	300–1000	0.02275724E+02	0.09922072E-01	-0.10409113E-04	0.06866686E-07	-0.02117280E-10	-0.04837314E+06	0.10188488E+02
H$_2$	1000–5000	0.02991423E+02	0.07000644E-02	-0.05633828E-06	-0.09231578E-10	0.15827519E-14	-0.08350340E+04	-0.13551101E+01
	300–1000	0.03298124E+02	0.08249441E-02	-0.08143015E-05	0.09475434E-09	0.04134872E-11	-0.10125209E+04	-0.03294094E+02
H	1000–5000	0.02500000E+02	0.00000000E+00	0.00000000E+00	0.00000000E+00	0.00000000E+00	0.02547162E+06	-0.04601176E+01
	300–1000	0.02500000E+02	0.00000000E+00	0.00000000E+00	0.00000000E+00	0.00000000E+00	0.02547162E+06	-0.04601176E+01
OH	1000–5000	0.02882730E+02	0.10139743E-02	-0.02276877E-05	0.02174683E-09	-0.05126305E-14	0.03886888E+05	0.05595712E+02
	300–1000	0.03637266E+02	0.01859010E-02	-0.16761646E-05	0.02387202E-07	-0.08431442E-11	0.03606781E+05	0.13588605E+01
H$_2$O	1000–5000	0.02672145E+02	0.03056293E-01	-0.08730260E-05	0.12009964E-09	-0.06391618E-13	-0.02989921E+06	0.06862817E+02
	300–1000	0.03386842E+02	0.03474982E-01	-0.06354696E-04	0.06968581E-07	-0.02506588E-10	-0.03020811E+06	0.02590232E+02
N$_2$	1000–5000	0.02926640E+02	0.14879768E-02	-0.05684760E-05	0.10097038E-09	-0.06753351E-13	-0.09227977E+04	0.05980528E+02
	300–1000	0.03298677E+02	0.14082404E-02	-0.03963222E-04	0.05641515E-07	-0.02444854E-10	-0.10208999E+04	0.03950372E+02
N	1000–5000	0.02450268E+02	0.10661458E-03	-0.07465337E-06	0.01879652E-09	-0.10259839E-14	0.05611604E+06	0.04448758E+02
	300–1000	0.02503071E+02	-0.02180018E-03	0.05420529E-06	-0.05647560E-09	0.02099904E-12	0.05609890E+06	0.04167566E+02
NO	1000–5000	0.03245435E+02	0.12691383E-02	-0.05015890E-05	0.09169283E-09	-0.06275419E-13	0.09800840E+05	0.06417293E+02
	300–1000	0.03376541E+02	0.12530634E-02	-0.03302750E-04	0.05217810E-07	-0.02446262E-10	0.09817961E+05	0.05829590E+02
NO$_2$	1000–5000	0.04682859E+02	0.02462429E-01	-0.10422585E-05	0.01976902E-08	-0.13917168E-13	0.02261292E+05	0.09885985E+01
	300–1000	0.02670600E+02	0.07838500E-01	-0.08063864E-04	0.06161714E-07	-0.02320150E-10	0.02896290E+05	0.11612071E+02
O$_2$	1000–5000	0.03697578E+02	0.06135197E-02	-0.12588420E-06	0.01775281E-09	-0.11364354E-14	-0.12339301E+04	0.03189165E+02
	300–1000	0.03212936E+02	0.11274864E-02	-0.05756150E-05	0.13138773E-08	-0.08768554E-11	-0.10052490E+04	0.06034737E+02
O	1000–5000	0.02542059E+02	-0.02755061E-03	-0.03102803E-07	0.04551067E-10	-0.04368051E-14	0.02923080E+06	0.04920308E+02
	300–1000	0.02946428E+02	-0.16381665E-02	0.02421031E-04	-0.16028431E-08	0.03890696E-11	0.02914764E+06	0.02963995E+02

Appendix C
Thermodynamic and Thermo-Physical Properties of Air

Table C.1 Approximate Composition, Apparent Molecular Weight, and Gas Constant for Dry Air

Constituent	Mole %
N_2	78.08
O_2	20.95
Ar	0.93
CO_2	0.036
Ne, He, CH_4, others	0.003

$$\mathcal{M}_{air} = 28.97 \text{ kg/kmol} \times \frac{1 \text{ kmol}}{1000 \text{ mol}}$$
$$R_{air} = 287.0 \text{ J/kg} \cdot \text{K}$$

Table C.2 Thermodynamic Properties of Air at 1 atm*

T (K)	h (kJ/kg)	u (kJ/kg)	$s°$ (kJ/kg·K)	c_p (kJ/kg·K)	c_v (kJ/kg·K)	$\gamma \, (= c_p/c_v)$
200	325.42	268.14	3.4764	1.007	0.716	1.406
210	335.49	275.32	3.5255	1.007	0.716	1.405
220	345.55	282.49	3.5723	1.006	0.716	1.405
230	355.62	289.67	3.6171	1.006	0.716	1.404
240	365.68	296.85	3.6599	1.006	0.716	1.404
250	375.73	304.02	3.7009	1.006	0.717	1.404
260	385.79	311.20	3.7404	1.006	0.717	1.403
270	395.85	318.38	3.7783	1.006	0.717	1.403
280	405.91	325.56	3.8149	1.006	0.717	1.403
290	415.97	332.74	3.8502	1.006	0.718	1.402
300	426.04	339.93	3.8844	1.007	0.718	1.402
310	436.11	347.12	3.9174	1.007	0.719	1.401
320	446.18	354.31	3.9494	1.008	0.719	1.401
330	456.26	361.51	3.9804	1.008	0.720	1.400
340	466.34	368.72	4.0105	1.009	0.721	1.400
350	476.43	375.94	4.0397	1.009	0.721	1.399
360	486.53	383.16	4.0682	1.010	0.722	1.399
370	496.64	390.39	4.0959	1.011	0.723	1.398
380	506.75	397.63	4.1228	1.012	0.724	1.398
390	516.88	404.89	4.1491	1.013	0.725	1.397

*Property values generated from NIST Database 23: REFPROP Version 7.0 (August 2002). Reference state: h (78.903 K) = 0.0; s (78.903 K) = 0.0.

(continued)

Table C.2 (continued)

T (K)	h (kJ/kg)	u (kJ/kg)	$s°$ (kJ/kg·K)	c_p (kJ/kg·K)	c_v (kJ/kg·K)	$\gamma\ (= c_p/c_v)$
400	527.02	412.15	4.1748	1.014	0.727	1.396
410	537.17	419.42	4.1999	1.016	0.728	1.395
420	547.33	426.71	4.2244	1.017	0.729	1.395
430	557.51	434.01	4.2483	1.018	0.731	1.394
440	567.70	441.33	4.2717	1.020	0.732	1.393
450	577.90	448.66	4.2947	1.021	0.734	1.392
460	588.13	456.01	4.3171	1.023	0.735	1.391
470	598.36	463.38	4.3392	1.025	0.737	1.390
480	608.62	470.76	4.3608	1.026	0.739	1.389
490	618.89	478.16	4.3819	1.028	0.741	1.388
500	629.18	485.58	4.4027	1.030	0.743	1.387
510	639.50	493.01	4.4231	1.032	0.745	1.386
520	649.83	500.47	4.4432	1.034	0.747	1.385
530	660.18	507.95	4.4629	1.036	0.749	1.384
540	670.55	515.45	4.4823	1.038	0.751	1.383
550	680.94	522.97	4.5014	1.040	0.753	1.382
560	691.35	530.51	4.5201	1.042	0.755	1.381
570	701.79	538.07	4.5386	1.045	0.757	1.380
580	712.25	545.65	4.5568	1.047	0.759	1.378
590	722.73	553.26	4.5747	1.049	0.762	1.377
600	733.23	560.89	4.5924	1.051	0.764	1.376
610	743.76	568.55	4.6098	1.054	0.766	1.375
620	754.30	576.22	4.6269	1.056	0.769	1.374
630	764.88	583.92	4.6438	1.058	0.771	1.373
640	775.47	591.65	4.6605	1.061	0.773	1.371
650	786.09	599.40	4.6770	1.063	0.776	1.370
660	796.74	607.17	4.6932	1.066	0.778	1.369
670	807.41	614.96	4.7093	1.068	0.781	1.368
680	818.10	622.78	4.7251	1.070	0.783	1.367
690	828.81	630.63	4.7408	1.073	0.786	1.366
700	839.55	638.50	4.7562	1.075	0.788	1.365
710	850.32	646.39	4.7715	1.078	0.790	1.364
720	861.11	654.30	4.7866	1.080	0.793	1.362
730	871.92	662.24	4.8015	1.082	0.795	1.361
740	882.76	670.21	4.8162	1.085	0.798	1.360
750	893.62	678.20	4.8308	1.087	0.800	1.359
760	904.50	686.21	4.8452	1.090	0.802	1.358
770	915.41	694.24	4.8595	1.092	0.805	1.357
780	926.34	702.30	4.8736	1.094	0.807	1.356
790	937.29	710.39	4.8875	1.097	0.809	1.355
800	948.27	718.49	4.9014	1.099	0.812	1.354
820	970.30	734.77	4.9285	1.104	0.816	1.352
840	992.41	751.15	4.9552	1.108	0.821	1.350
860	1014.62	767.61	4.9813	1.113	0.825	1.348
880	1036.91	784.16	5.0069	1.117	0.830	1.346
900	1059.29	800.80	5.0321	1.121	0.834	1.344
920	1081.76	817.52	5.0568	1.125	0.838	1.343
940	1104.30	834.32	5.0810	1.129	0.842	1.341
960	1126.93	851.21	5.1048	1.133	0.846	1.339
980	1149.64	868.18	5.1283	1.137	0.850	1.338

T (K)	h (kJ/kg)	u (kJ/kg)	$s°$ (kJ/kg·K)	c_p (kJ/kg·K)	c_v (kJ/kg·K)	γ ($= c_p/c_v$)
1000	1172.43	885.22	5.1513	1.141	0.854	1.336
1020	1195.29	902.34	5.1739	1.145	0.858	1.335
1040	1218.23	919.53	5.1962	1.149	0.861	1.333
1060	1241.23	936.80	5.2181	1.152	0.865	1.332
1080	1264.31	954.13	5.2397	1.156	0.868	1.331
1100	1287.46	971.53	5.2609	1.159	0.872	1.329
1120	1310.67	989.00	5.2818	1.162	0.875	1.328
1140	1333.95	1006.54	5.3024	1.165	0.878	1.327
1160	1357.29	1024.13	5.3227	1.169	0.881	1.326
1180	1380.69	1041.79	5.3427	1.172	0.884	1.325
1200	1404.15	1059.51	5.3624	1.175	0.887	1.324
1220	1427.67	1077.29	5.3819	1.177	0.890	1.323
1240	1451.25	1095.12	5.4010	1.180	0.893	1.322
1260	1474.88	1113.01	5.4199	1.183	0.896	1.321
1280	1498.56	1130.95	5.4386	1.186	0.898	1.320
1300	1522.30	1148.95	5.4570	1.188	0.901	1.319
1320	1546.09	1166.99	5.4751	1.191	0.903	1.318
1340	1569.92	1185.08	5.4931	1.193	0.906	1.317
1360	1593.81	1203.23	5.5108	1.195	0.908	1.316
1380	1617.74	1221.42	5.5282	1.198	0.911	1.315
1400	1641.71	1239.65	5.5455	1.200	0.913	1.315
1420	1665.74	1257.93	5.5625	1.202	0.915	1.314
1440	1689.80	1276.25	5.5793	1.204	0.917	1.313
1460	1713.91	1294.61	5.5960	1.206	0.919	1.312
1480	1738.05	1313.02	5.6124	1.208	0.921	1.312
1500	1762.24	1331.46	5.6286	1.210	0.923	1.311
1520	1786.46	1349.94	5.6447	1.212	0.925	1.310
1540	1810.73	1368.46	5.6605	1.214·	0.927	1.310
1560	1835.03	1387.02	5.6762	1.216	0.929	1.309
1580	1859.36	1405.61	5.6917	1.218	0.931	1.309
1600	1883.73	1424.24	5.7070	1.219	0.932	1.308
1620	1908.14	1442.90	5.7222	1.221	0.934	1.307
1640	1932.58	1461.60	5.7372	1.223	0.936	1.307
1660	1957.05	1480.33	5.7520	1.224	0.937	1.306
1680	1981.55	1499.09	5.7667	1.226	0.939	1.306
1700	2006.08	1517.88	5.7812	1.227	0.940	1.305
1750	2067.54	1564.99	5.8168	1.231	0.944	1.304
1800	2129.19	1612.27	5.8516	1.235	0.947	1.303
1850	2190.99	1659.72	5.8854	1.238	0.951	1.302
1900	2252.96	1707.33	5.9185	1.241	0.954	1.301
1950	2315.08	1755.09	5.9508	1.244	0.957	1.300
2000	2377.34	1803.00	5.9823	1.247	0.959	1.299
2050	2439.73	1851.04	6.0131	1.249	0.962	1.298
2100	2502.26	1899.21	6.0432	1.252	0.965	1.298
2150	2564.90	1947.50	6.0727	1.254	0.967	1.297
2200	2627.67	1995.91	6.1016	1.256	0.969	1.296
2250	2690.54	2044.43	6.1298	1.259	0.971	1.296
2300	2753.53	2093.05	6.1575	1.261	0.974	1.295
2350	2816.61	2141.78	6.1846	1.263	0.976	1.294

Table C.3A   Thermo-Physical Properties of Air (100–1000 K at 1 atm)*

T (K)	ρ (kg/m³)	c_p (kJ/kg·K)	μ (μPa·s)	ν (m²/s)	k (W/m·K)	α (m²/s)	Pr
100	3.6043	1.0356	7.1551	1.985E-06	0.010116	2.710E-06	0.73245
120	2.9772	1.0211	8.4995	2.855E-06	0.011996	3.946E-06	0.72349
140	2.5403	1.0142	9.7899	3.854E-06	0.013802	5.357E-06	0.71940
160	2.2169	1.0105	11.029	4.975E-06	0.015540	6.936E-06	0.71723
180	1.9674	1.0084	12.221	6.212E-06	0.017213	8.677E-06	0.71593
200	1.7688	1.0071	13.370	7.559E-06	0.018829	1.057E-05	0.71508
220	1.6068	1.0063	14.479	9.011E-06	0.020392	1.261E-05	0.71449
240	1.4721	1.0059	15.552	1.056E-05	0.021908	1.480E-05	0.71407
260	1.3584	1.0058	16.592	1.221E-05	0.023381	1.711E-05	0.71376
280	1.2610	1.0061	17.601	1.396E-05	0.024817	1.956E-05	0.71354
300	1.1767	1.0066	18.582	1.579E-05	0.026220	2.214E-05	0.71339
320	1.1030	1.0075	19.536	1.771E-05	0.027594	2.483E-05	0.71330
340	1.0380	1.0087	20.465	1.972E-05	0.028944	2.764E-05	0.71324
360	0.98022	1.0103	21.372	2.180E-05	0.030272	3.057E-05	0.71323
380	0.92856	1.0122	22.256	2.397E-05	0.031583	3.361E-05	0.71326
400	0.88208	1.0144	23.121	2.621E-05	0.032880	3.675E-05	0.71331
420	0.84004	1.0169	23.966	2.853E-05	0.034164	3.999E-05	0.71340
440	0.80183	1.0198	24.794	3.092E-05	0.035437	4.334E-05	0.71352
460	0.76695	1.0230	25.605	3.339E-05	0.036702	4.678E-05	0.71366
480	0.73497	1.0264	26.400	3.592E-05	0.037960	5.032E-05	0.71384
500	0.70556	1.0301	27.180	3.852E-05	0.039212	5.395E-05	0.71403
520	0.67842	1.0340	27.946	4.119E-05	0.040458	5.767E-05	0.71425
540	0.65329	1.0382	28.698	4.393E-05	0.041699	6.148E-05	0.71449
560	0.62995	1.0424	29.438	4.673E-05	0.042935	6.538E-05	0.71475
580	0.60823	1.0469	30.166	4.960E-05	0.044167	6.936E-05	0.71503
600	0.58795	1.0514	30.883	5.253E-05	0.045395	7.343E-05	0.71532
620	0.56898	1.0561	31.589	5.552E-05	0.046618	7.758E-05	0.71562
640	0.55120	1.0608	32.284	5.857E-05	0.047837	8.181E-05	0.71593
660	0.53450	1.0656	32.970	6.168E-05	0.049050	8.612E-05	0.71626
680	0.51878	1.0704	33.646	6.486E-05	0.050259	9.051E-05	0.71659
700	0.50396	1.0752	34.313	6.809E-05	0.051462	9.497E-05	0.71693
720	0.48996	1.0800	34.972	7.138E-05	0.052659	9.951E-05	0.71727
740	0.47672	1.0848	35.623	7.473E-05	0.053851	1.041E-04	0.71761
760	0.46417	1.0896	36.266	7.813E-05	0.055036	1.088E-04	0.71796
780	0.45227	1.0943	36.901	8.159E-05	0.056215	1.136E-04	0.71831
800	0.44097	1.0989	37.529	8.511E-05	0.057388	1.184E-04	0.71866
820	0.43022	1.1035	38.151	8.868E-05	0.058553	1.233E-04	0.71901
840	0.41997	1.1081	38.765	9.230E-05	0.059712	1.283E-04	0.71936
860	0.41021	1.1125	39.374	9.599E-05	0.060863	1.334E-04	0.71971
880	0.40089	1.1169	39.976	9.972E-05	0.062007	1.385E-04	0.72006
900	0.39198	1.1212	40.573	1.035E-04	0.063143	1.437E-04	0.72041
920	0.38346	1.1254	41.164	1.074E-04	0.064271	1.489E-04	0.72075
940	0.37530	1.1295	41.749	1.112E-04	0.065392	1.543E-04	0.72109
960	0.36749	1.1335	42.329	1.152E-04	0.066505	1.597E-04	0.72143
980	0.35999	1.1374	42.904	1.192E-04	0.067610	1.651E-04	0.72176
1000	0.35279	1.1412	43.474	1.232E-04	0.068708	1.707E-04	0.72210

*Values generated from NIST Database 23: REFPROP Version 7.0 (August 2002).

Table C.3B   Thermo-Physical Properties of Air (1000–2300 K at 1 atm)*

T (K)	ρ (kg/m³)	c_p (kJ/kg·K)	μ (μPa·s)	ν (m²/s)	k (W/m·K)	α (m²/s)	Pr
1000	0.35281	1.1412	43.474	1.23E-04	0.068708	6.871E-06	0.72210
1100	0.32074	1.1590	46.258	1.44E-04	0.074077	7.408E-06	0.72372
1200	0.29402	1.1745	48.941	1.66E-04	0.079254	7.925E-06	0.72530
1300	0.27141	1.1881	51.539	1.90E-04	0.084248	8.425E-06	0.72683
1400	0.25202	1.1999	54.063	2.15E-04	0.089070	8.907E-06	0.72835
1500	0.23523	1.2103	56.524	2.40E-04	0.093733	9.373E-06	0.72986
1600	0.22053	1.2194	58.930	2.67E-04	0.098251	9.825E-06	0.73138
1700	0.20756	1.2274	61.286	2.95E-04	0.10263	1.026E-05	0.73293
1800	0.19603	1.2345	63.600	3.24E-04	0.10690	1.069E-05	0.73451
1900	0.18571	1.2409	65.877	3.55E-04	0.11105	1.111E-05	0.73614
2000	0.17643	1.2466	68.119	3.86E-04	0.11509	1.151E-05	0.73781
2100	0.16803	1.2517	70.332	4.19E-04	0.11904	1.190E-05	0.73954
2200	0.16039	1.2564	72.518	4.52E-04	0.12290	1.229E-05	0.74134
2300	0.15342	1.2607	74.681	4.87E-04	0.12668	1.267E-05	0.74320

*Values generated from NIST Database 23: REFPROP Version 7.0 (August 2002).

Appendix D
Thermodynamic Properties of H₂O

Table D.1   Saturation Properties of Water and Steam—Temperature Increments

T (K)	P (kPa)	v (m³/kg) sat. liquid	v (m³/kg) sat. vapor	u (kJ/kg) sat. liquid	u (kJ/kg) sat. vapor	h (kJ/kg) sat. liquid	h (kJ/kg) sat. vapor	s (kJ/kg·K) sat. liquid	s (kJ/kg·K) sat. vapor
273.16	0.611650	0.0010002	205.99	0	2374.9	0.00061	2500.9	0	9.1555
274	0.650030	0.0010002	194.43	3.5435	2376.1	3.5442	2502.5	0.012952	9.1331
275	0.698460	0.0010001	181.60	7.7590	2377.5	7.7597	2504.3	0.028309	9.1066
276	0.750070	0.0010001	169.71	11.971	2378.8	11.972	2506.1	0.043600	9.0804
277	0.805020	0.0010001	158.70	16.181	2380.2	16.182	2508.0	0.058825	9.0544
278	0.863500	0.0010001	148.48	20.388	2381.6	20.389	2509.8	0.073985	9.0287
279	0.925700	0.0010001	139.00	24.593	2383.0	24.594	2511.6	0.089083	9.0031
280	0.991830	0.0010001	130.19	28.795	2384.3	28.796	2513.4	0.10412	8.9779
281	1.062200	0.0010002	122.01	32.996	2385.7	32.997	2515.3	0.11909	8.9528
282	1.136800	0.0010003	114.40	37.194	2387.1	37.195	2517.1	0.13401	8.9280
283	1.216000	0.0010003	107.32	41.391	2388.4	41.392	2518.9	0.14886	8.9034
284	1.300000	0.0010004	100.74	45.586	2389.8	45.587	2520.8	0.16366	8.8791
285	1.389100	0.0010005	94.602	49.779	2391.2	49.780	2522.6	0.17840	8.8549
286	1.483600	0.0010006	88.887	53.971	2392.6	53.973	2524.4	0.19308	8.8310
287	1.583600	0.0010008	83.560	58.162	2393.9	58.163	2526.2	0.20771	8.8073
288	1.689500	0.0010009	78.592	62.351	2395.3	62.353	2528.1	0.22228	8.7838
289	1.801600	0.0010011	73.955	66.540	2396.7	66.542	2529.9	0.23680	8.7605
290	1.920100	0.0010012	69.625	70.727	2398.0	70.729	2531.7	0.25126	8.7374
291	2.045400	0.0010014	65.581	74.914	2399.4	74.916	2533.5	0.26568	8.7145
292	2.177900	0.0010016	61.801	79.099	2400.8	79.101	2535.3	0.28003	8.6918
293	2.317800	0.0010018	58.267	83.284	2402.1	83.286	2537.2	0.29434	8.6693
294	2.465500	0.0010020	54.960	87.468	2403.5	87.471	2539.0	0.30860	8.6471
295	2.621300	0.0010022	51.865	91.652	2404.8	91.654	2540.8	0.32280	8.6250
296	2.785700	0.0010025	48.966	95.835	2406.2	95.837	2542.6	0.33696	8.6031
297	2.959100	0.0010027	46.251	100.02	2407.6	100.02	2544.4	0.35106	8.5814
298	3.141800	0.0010030	43.705	104.20	2408.9	104.20	2546.2	0.36512	8.5599
299	3.334300	0.0010032	41.318	108.38	2410.3	108.38	2548.0	0.37913	8.5385
300	3.536900	0.0010035	39.078	112.56	2411.6	112.56	2549.9	0.39309	8.5174
301	3.750200	0.0010038	36.976	116.74	2413.0	116.75	2551.7	0.40700	8.4964
302	3.974600	0.0010041	35.002	120.92	2414.4	120.93	2553.5	0.42087	8.4756
303	4.210600	0.0010044	33.147	125.10	2415.7	125.11	2555.3	0.43469	8.4550
304	4.458700	0.0010047	31.403	129.28	2417.1	129.29	2557.1	0.44846	8.4346
305	4.719400	0.0010050	29.764	133.46	2418.4	133.47	2558.9	0.46219	8.4144
306	4.993299	0.0010053	28.222	137.64	2419.8	137.65	2560.7	0.47587	8.3943
307	5.280799	0.0010057	26.770	141.82	2421.1	141.83	2562.5	0.48950	8.3744
308	5.582599	0.0010060	25.403	146.00	2422.5	146.01	2564.3	0.50310	8.3546
309	5.899199	0.0010063	24.116	150.18	2423.8	150.19	2566.1	0.51664	8.3351
310	6.231199	0.0010067	22.903	154.36	2425.2	154.37	2567.9	0.53015	8.3156
311	6.579299	0.0010071	21.759	158.54	2426.5	158.55	2569.7	0.54361	8.2964
312	6.944099	0.0010074	20.680	162.72	2427.8	162.73	2571.5	0.55702	8.2773
313	7.326199	0.0010078	19.663	166.90	2429.2	166.91	2573.2	0.57040	8.2584
314	7.726299	0.0010082	18.702	171.08	2430.5	171.09	2575.0	0.58373	8.2396
315	8.145199	0.0010086	17.795	175.26	2431.9	175.27	2576.8	0.59702	8.2210
316	8.583499	0.0010090	16.938	179.44	2433.2	179.45	2578.6	0.61027	8.2025
317	9.041899	0.0010094	16.129	183.62	2434.5	183.63	2580.4	0.62348	8.1842
318	9.521299	0.0010099	15.363	187.80	2435.9	187.81	2582.2	0.63664	8.1660
319	10.022989	0.0010103	14.639	191.98	2437.2	191.99	2583.9	0.64977	8.1480
320	10.546989	0.0010107	13.954	196.16	2438.5	196.17	2585.7	0.66285	8.1302

(continued)

T (K)	P (MPa)	ν (m³/kg) sat. liquid	ν (m³/kg) sat. vapor	u (kJ/kg) sat. liquid	u (kJ/kg) sat. vapor	h (kJ/kg) sat. liquid	h (kJ/kg) sat. vapor	s (kJ/kg·K) sat. liquid	s (kJ/kg·K) sat. vapor
320	0.010547	0.0010107	13.954	196.16	2438.5	196.17	2585.7	0.66285	8.1302
325	0.013532	0.0010130	11.039	217.07	2445.2	217.08	2594.6	0.72768	8.0430
330	0.017214	0.0010155	8.8050	237.98	2451.8	238.00	2603.3	0.79154	7.9592
335	0.021719	0.0010181	7.0788	258.9	2458.3	258.93	2612.1	0.85447	7.8787
340	0.027189	0.0010209	5.7339	279.84	2464.8	279.87	2620.7	0.91650	7.8013
345	0.033784	0.0010239	4.6776	300.79	2471.2	300.82	2629.3	0.97766	7.7267
350	0.041683	0.0010270	3.8419	321.75	2477.6	321.79	2637.7	1.0380	7.6549
355	0.051081	0.0010303	3.1759	342.73	2483.9	342.78	2646.1	1.0975	7.5857
360	0.062195	0.0010337	2.6414	363.73	2490.1	363.79	2654.4	1.1562	7.5190
365	0.075261	0.0010373	2.2098	384.75	2496.2	384.82	2662.5	1.2142	7.4545
370	0.090536	0.0010410	1.8590	405.79	2502.3	405.88	2670.6	1.2715	7.3923
375	0.108310	0.0010449	1.5722	426.86	2508.2	426.97	2678.5	1.3281	7.3321
380	0.128860	0.0010490	1.3364	447.96	2514.1	448.09	2686.2	1.3839	7.2738
385	0.152529	0.0010532	1.1414	469.09	2519.8	469.25	2693.9	1.4392	7.2174
390	0.179649	0.0010575	0.97928	490.25	2525.4	490.44	2701.3	1.4938	7.1627
395	0.210609	0.0010620	0.84389	511.45	2530.9	511.67	2708.6	1.5478	7.1097
400	0.245779	0.0010667	0.73024	532.69	2536.2	532.95	2715.7	1.6013	7.0581
405	0.285589	0.0010715	0.63441	553.98	2541.4	554.28	2722.6	1.6541	7.0081
410	0.330459	0.0010765	0.55323	575.31	2546.5	575.66	2729.3	1.7065	6.9593
415	0.380879	0.0010817	0.48418	596.69	2551.4	597.10	2735.8	1.7583	6.9119
420	0.437309	0.0010870	0.42520	618.13	2556.2	618.60	2742.1	1.8097	6.8656
425	0.500259	0.0010926	0.37463	639.62	2560.7	640.17	2748.1	1.8606	6.8205
430	0.570269	0.0010983	0.33110	661.18	2565.1	661.80	2753.9	1.9110	6.7764
435	0.647879	0.0011042	0.29350	682.8	2569.3	683.52	2759.5	1.9610	6.7333
440	0.733679	0.0011103	0.26090	704.5	2573.3	705.31	2764.7	2.0106	6.6911
445	0.828249	0.0011166	0.23255	726.26	2577.1	727.19	2769.7	2.0598	6.6498
450	0.932209	0.0011232	0.20781	748.11	2580.7	749.16	2774.4	2.1087	6.6092
455	1.046289	0.0011299	0.18616	770.05	2584.0	771.23	2778.8	2.1571	6.5694
460	1.170988	0.0011369	0.16715	792.07	2587.2	793.41	2782.9	2.2053	6.5303
465	1.306987	0.0011442	0.15041	814.2	2590.1	815.69	2786.6	2.2532	6.4917
470	1.455187	0.0011517	0.13564	836.42	2592.7	838.09	2790.0	2.3007	6.4538
475	1.616086	0.0011594	0.12255	858.75	2595.0	860.62	2793.1	2.3480	6.4164
480	1.790586	0.0011675	0.11094	881.19	2597.1	883.28	2795.8	2.3950	6.3794
485	1.979285	0.0011758	0.10061	903.76	2598.9	906.09	2798.1	2.4418	6.3428
490	2.183184	0.0011845	0.091390	926.45	2600.5	929.04	2800.0	2.4884	6.3066
495	2.402884	0.0011935	0.083149	949.28	2601.7	952.15	2801.4	2.5348	6.2708
500	2.639283	0.0012029	0.075764	972.26	2602.5	975.43	2802.5	2.5810	6.2351
505	2.893182	0.0012127	0.069131	995.38	2603.0	998.89	2803.1	2.6271	6.1997
510	3.165582	0.0012228	0.063161	1018.7	2603.2	1022.5	2803.2	2.6731	6.1645
515	3.457181	0.0012334	0.057776	1042.1	2603.0	1046.4	2802.7	2.7189	6.1293
520	3.769080	0.0012445	0.052910	1065.8	2602.4	1070.5	2801.8	2.7647	6.0942
525	4.101980	0.0012561	0.048503	1089.6	2601.4	1094.8	2800.3	2.8104	6.0591
530	4.456979	0.0012682	0.044503	1113.7	2599.9	1119.3	2798.2	2.8561	6.0239
535	4.834978	0.0012809	0.040868	1138	2597.9	1144.2	2795.5	2.9019	5.9885
540	5.236977	0.0012942	0.037556	1162.5	2595.5	1169.3	2792.2	2.9476	5.9530
545	5.664076	0.0013083	0.034535	1187.3	2592.5	1194.7	2788.1	2.9935	5.9171
550	6.117276	0.0013231	0.031772	1212.4	2588.9	1220.5	2783.3	3.0394	5.8809
555	6.597675	0.0013387	0.029242	1237.8	2584.8	1246.6	2777.7	3.0855	5.8443
560	7.106274	0.0013553	0.026920	1263.5	2579.9	1273.1	2771.2	3.1319	5.8071
565	7.644473	0.0013729	0.024786	1289.6	2574.4	1300.1	2763.9	3.1785	5.7693
570	8.213272	0.0013917	0.022820	1316	2568.0	1327.5	2755.5	3.2254	5.7307
575	8.814071	0.0014118	0.021005	1343	2560.9	1355.4	2746.0	3.2727	5.6912
580	9.448070	0.0014334	0.019328	1370.4	2552.7	1383.9	2735.3	3.3205	5.6506
585	10.117686	0.0014567	0.017773	1398.4	2543.5	1413.1	2723.3	3.3690	5.6087
590	10.821674	0.0014820	0.016329	1426.9	2533.2	1443.0	2709.9	3.4181	5.5654
595	11.563662	0.0015095	0.014984	1456.2	2521.5	1473.7	2694.8	3.4680	5.5203
600	12.345649	0.0015399	0.013728	1486.4	2508.3	1505.4	2677.8	3.5190	5.4731
605	13.167635	0.0015735	0.012552	1517.4	2493.4	1538.1	2658.7	3.5713	5.4234
610	14.033620	0.0016112	0.011446	1549.6	2476.4	1572.2	2637.0	3.6252	5.3707
615	14.943604	0.0016541	0.010400	1583.2	2456.9	1608.0	2612.3	3.6811	5.3142
620	15.901586	0.0017039	0.0094067	1618.6	2434.3	1645.7	2583.9	3.7396	5.2528
625	16.908567	0.0017634	0.0084538	1656.5	2407.8	1686.3	2550.7	3.8019	5.1851
630	17.969544	0.0018374	0.0075279	1697.7	2375.8	1730.7	2511.1	3.8698	5.1084
635	19.086516	0.0019353	0.0066074	1744.3	2335.7	1781.2	2461.8	3.9463	5.0181
640	20.265481	0.0020767	0.0056451	1799.7	2281.1	1841.8	2395.5	4.0375	4.9027
645	21.515425	0.0023527	0.0044553	1880.5	2185.2	1931.1	2281.0	4.1722	4.7147
647.096	22.064000	0.0031056	0.0031056	2015.7	2015.7	2084.3	2084.3	4.4070	4.4070

Table D.2   Saturation Properties of Water and Steam—Pressure Increments

P (kPa)	T (K)	ν (m³/kg) sat. liquid	ν (m³/kg) sat. vapor	u (kJ/kg) sat. liquid	u (kJ/kg) sat. vapor	h (kJ/kg) sat. liquid	h (kJ/kg) sat. vapor	s (kJ/kg·K) sat. liquid	s (kJ/kg·K) sat. vapor
2.0	290.64	0.0010014	66.987	73.426	2398.9	73.428	2532.9	0.26056	8.7226
4.0	302.11	0.0010041	34.791	121.38	2414.5	121.39	2553.7	0.42239	8.4734
6.0	309.31	0.0010065	23.733	151.47	2424.2	151.48	2566.6	0.52082	8.3290
8.0	314.66	0.0010085	18.099	173.83	2431.4	173.84	2576.2	0.59249	8.2273
10.0	318.96	0.0010103	14.670	191.80	2437.2	191.81	2583.9	0.6492	8.1488
12.0	322.57	0.0010119	12.358	206.90	2442.0	206.91	2590.3	0.69628	8.0849
14.0	325.70	0.0010134	10.691	219.98	2446.1	219.99	2595.8	0.73664	8.0311
16.0	328.46	0.0010147	9.4306	231.55	2449.8	231.57	2600.6	0.77201	7.9846
18.0	330.95	0.0010160	8.4431	241.95	2453.0	241.96	2605.0	0.80355	7.9437
20.0	333.21	0.0010172	7.6480	251.40	2456.0	251.42	2608.9	0.83202	7.9072
22.0	335.28	0.0010183	6.9936	260.09	2458.7	260.11	2612.5	0.85800	7.8743
24.0	337.20	0.0010193	6.4453	268.13	2461.2	268.15	2615.9	0.88191	7.8442
26.0	338.99	0.0010203	5.9792	275.62	2463.5	275.64	2619.0	0.90407	7.8167
28.0	340.67	0.0010213	5.5778	282.64	2465.7	282.66	2621.8	0.92472	7.7912
30.0	342.25	0.0010222	5.2284	289.24	2467.7	289.27	2624.5	0.94407	7.7675
32.0	343.74	0.0010231	4.9215	295.49	2469.6	295.52	2627.1	0.96228	7.7453
34.0	345.15	0.0010240	4.6497	301.41	2471.4	301.45	2629.5	0.97948	7.7246
36.0	346.50	0.0010248	4.4072	307.05	2473.1	307.09	2631.8	0.99579	7.7050
38.0	347.78	0.0010256	4.1895	312.43	2474.8	312.47	2634.0	1.0113	7.6865
40.0	349.01	0.0010264	3.9930	317.58	2476.3	317.62	2636.1	1.0261	7.6690
42.0	350.18	0.0010271	3.8146	322.52	2477.8	322.56	2638.0	1.0402	7.6524
44.0	351.32	0.0010279	3.6520	327.27	2479.3	327.31	2639.9	1.0537	7.6365
46.0	352.40	0.0010286	3.5031	331.83	2480.6	331.88	2641.8	1.0667	7.6214
48.0	353.45	0.0010293	3.3663	336.24	2481.9	336.29	2643.5	1.0792	7.6069
50.0	354.47	0.0010299	3.2400	340.49	2483.2	340.54	2645.2	1.0912	7.5930
52.0	355.45	0.0010306	3.1232	344.60	2484.4	344.66	2646.8	1.1028	7.5797
54.0	356.40	0.0010312	3.0148	348.59	2485.6	348.64	2648.4	1.1140	7.5669
56.0	357.32	0.0010319	2.9139	352.45	2486.8	352.51	2649.9	1.1248	7.5545
58.0	358.21	0.0010325	2.8198	356.20	2487.9	356.26	2651.4	1.1353	7.5426
60.0	359.08	0.0010331	2.7317	359.84	2489.0	359.91	2652.9	1.1454	7.5311
62.0	359.92	0.0010337	2.6492	363.39	2490.0	363.45	2654.2	1.1553	7.5200
64.0	360.74	0.0010342	2.5716	366.84	2491.0	366.91	2655.6	1.1649	7.5093
66.0	361.54	0.0010348	2.4986	370.20	2492.0	370.27	2656.9	1.1742	7.4989
68.0	362.32	0.0010354	2.4298	373.48	2493.0	373.55	2658.2	1.1833	7.4888
70.0	363.08	0.0010359	2.3648	376.68	2493.9	376.75	2659.4	1.1921	7.4790
72.0	363.82	0.0010364	2.3033	379.80	2494.8	379.88	2660.6	1.2007	7.4695
74.0	364.55	0.0010370	2.2450	382.86	2495.7	382.94	2661.8	1.2091	7.4602
76.0	365.26	0.0010375	2.1897	385.84	2496.5	385.92	2663.0	1.2172	7.4512
78.0	365.96	0.0010380	2.1371	388.77	2497.4	388.85	2664.1	1.2252	7.4425
80.0	366.64	0.0010385	2.0871	391.63	2498.2	391.71	2665.2	1.2330	7.4339
82.0	367.30	0.0010390	2.0394	394.43	2499.0	394.51	2666.3	1.2407	7.4256
84.0	367.95	0.0010395	1.9940	397.18	2499.8	397.26	2667.3	1.2482	7.4175
86.0	368.59	0.001040	1.9506	399.87	2500.6	399.96	2668.3	1.2555	7.4096
88.0	369.22	0.0010404	1.9091	402.51	2501.3	402.60	2669.3	1.2626	7.4018
90.0	369.84	0.0010409	1.8694	405.10	2502.1	405.20	2670.3	1.2696	7.3943
92.0	370.44	0.0010414	1.8313	407.65	2502.8	407.75	2671.3	1.2765	7.3869
94.0	371.04	0.0010418	1.7949	410.15	2503.5	410.25	2672.2	1.2833	7.3796
96.0	371.62	0.0010423	1.7599	412.61	2504.2	412.71	2673.1	1.2899	7.3726
98.0	372.19	0.0010427	1.7262	415.02	2504.9	415.13	2674.1	1.2964	7.3656
100.0	372.76	0.0010432	1.6939	417.40	2505.6	417.50	2674.9	1.3028	7.3588

P (MPa)	T (K)	ν (m³/kg) sat. liquid	ν (m³/kg) sat. vapor	u (kJ/kg) sat. liquid	u (kJ/kg) sat. vapor	h (kJ/kg) sat. liquid	h (kJ/kg) sat. vapor	s (kJ/kg·K) sat. liquid	s (kJ/kg·K) sat. vapor
0.10	372.76	0.0010432	1.6939	417.40	2505.6	417.50	2674.9	1.3028	7.3588
0.20	393.36	0.0010605	0.88568	504.49	2529.1	504.70	2706.2	1.5302	7.1269
0.30	406.67	0.0010732	0.60576	561.10	2543.2	561.43	2724.9	1.6717	6.9916
0.40	416.76	0.0010836	0.46238	604.22	2553.1	604.65	2738.1	1.7765	6.8955
0.50	424.98	0.0010925	0.37481	639.54	2560.7	640.09	2748.1	1.8604	6.8207
0.60	431.98	0.0011006	0.31558	669.72	2566.8	670.38	2756.1	1.9308	6.7592
0.70	438.10	0.0011080	0.27277	696.23	2571.8	697.00	2762.8	1.9918	6.7071
0.80	443.56	0.0011148	0.24034	719.97	2576.0	720.86	2768.3	2.0457	6.6616
0.90	448.50	0.0011212	0.21489	741.55	2579.6	742.56	2773.0	2.0940	6.6213
1.00	453.03	0.0011272	0.19436	761.39	2582.7	762.52	2777.1	2.1381	6.5850
1.10	457.21	0.0011330	0.17745	779.78	2585.5	781.03	2780.6	2.1785	6.5520
1.20	461.11	0.0011385	0.16326	796.96	2587.8	798.33	2783.7	2.2159	6.5217
1.30	464.75	0.0011438	0.15119	813.11	2589.9	814.60	2786.5	2.2508	6.4936
1.40	468.19	0.0011489	0.14078	828.36	2591.8	829.97	2788.8	2.2835	6.4675
1.50	471.44	0.0011539	0.13171	842.83	2593.4	844.56	2791.0	2.3143	6.4430
1.60	474.52	0.0011587	0.12374	856.60	2594.8	858.46	2792.8	2.3435	6.4199
1.70	477.46	0.0011634	0.11667	869.76	2596.1	871.74	2794.5	2.3711	6.3981
1.80	480.26	0.0011679	0.11037	882.37	2597.2	884.47	2795.9	2.3975	6.3775
1.90	482.95	0.0011724	0.10470	894.48	2598.2	896.71	2797.2	2.4227	6.3578
2.00	485.53	0.0011767	0.099585	906.14	2599.1	908.50	2798.3	2.4468	6.3390
2.10	488.01	0.0011810	0.094938	917.39	2599.9	919.87	2799.3	2.4699	6.3210
2.20	490.4	0.0011852	0.090698	928.27	2600.6	930.87	2800.1	2.4921	6.3038
2.30	492.71	0.0011894	0.086815	938.79	2601.1	941.53	2800.8	2.5136	6.2872
2.40	494.94	0.0011934	0.083244	949.00	2601.6	951.87	2801.4	2.5343	6.2712
2.50	497.10	0.0011974	0.079949	958.91	2602.1	961.91	2801.9	2.5543	6.2558
2.60	499.20	0.0012014	0.076899	968.55	2602.4	971.67	2802.3	2.5736	6.2409
2.70	501.23	0.0012053	0.074066	977.93	2602.7	981.18	2802.7	2.5924	6.2264
2.80	503.21	0.0012091	0.071429	987.07	2602.9	990.46	2802.9	2.6106	6.2124
2.90	505.13	0.0012129	0.068968	995.99	2603.1	999.51	2803.1	2.6283	6.1988
3.00	507.00	0.0012167	0.066664	1004.7	2603.2	1008.3	2803.2	2.6455	6.1856
3.10	508.83	0.0012204	0.064504	1013.2	2603.2	1017.0	2803.2	2.6623	6.1727
3.20	510.61	0.0012241	0.062475	1021.5	2603.2	1025.4	2803.1	2.6787	6.1602
3.30	512.35	0.0012278	0.060564	1029.7	2603.2	1033.7	2803.0	2.6946	6.1479
3.40	514.05	0.0012314	0.058761	1037.7	2603.1	1041.8	2802.9	2.7102	6.1360
3.50	515.71	0.0012350	0.057058	1045.5	2602.9	1049.8	2802.6	2.7254	6.1243
3.60	517.33	0.0012385	0.055446	1053.1	2602.8	1057.6	2802.4	2.7403	6.1129
3.70	518.92	0.0012421	0.053918	1060.7	2602.6	1065.3	2802.1	2.7549	6.1018
3.80	520.48	0.0012456	0.052467	1068.1	2602.3	1072.8	2801.7	2.7691	6.0908
3.90	522.01	0.0012491	0.051089	1075.3	2602.0	1080.2	2801.3	2.7831	6.0801
4.00	523.50	0.0012526	0.049776	1082.5	2601.7	1087.5	2800.8	2.7968	6.0696
4.10	524.97	0.0012560	0.048525	1089.5	2601.4	1094.7	2800.3	2.8102	6.0592
4.20	526.41	0.0012594	0.047332	1096.4	2601.0	1101.7	2799.8	2.8234	6.0491
4.30	527.83	0.0012629	0.046192	1103.2	2600.6	1108.7	2799.2	2.8363	6.0391
4.40	529.22	0.0012663	0.045102	1109.9	2600.1	1115.5	2798.6	2.8490	6.0293
4.50	530.59	0.0012696	0.044059	1116.5	2599.7	1122.2	2797.9	2.8615	6.0197
4.60	531.93	0.0012730	0.043059	1123.0	2599.2	1128.9	2797.3	2.8738	6.0102
4.70	533.25	0.0012764	0.042100	1129.5	2598.7	1135.5	2796.5	2.8859	6.0009
4.80	534.55	0.0012797	0.041180	1135.8	2598.1	1141.9	2795.8	2.8978	5.9917
4.90	535.83	0.0012831	0.040296	1142.0	2597.6	1148.3	2795.0	2.9095	5.9826
5.00	537.09	0.0012864	0.039446	1148.2	2597.0	1154.6	2794.2	2.9210	5.9737

(*continued*)

Table D.2 (continued)

P (MPa)	T (K)	v (m³/kg) sat. liquid	v (m³/kg) sat. vapor	u (kJ/kg) sat. liquid	u (kJ/kg) sat. vapor	h (kJ/kg) sat. liquid	h (kJ/kg) sat. vapor	s (kJ/kg·K) sat. liquid	s (kJ/kg·K) sat. vapor
5.0	537.09	0.0012864	0.039446	1148.2	2597.0	1154.6	2794.2	2.9210	5.9737
5.5	543.12	0.0013029	0.035642	1177.9	2593.7	1185.1	2789.7	2.9762	5.9307
6.0	548.73	0.0013193	0.032448	1206.0	2589.9	1213.9	2784.6	3.0278	5.8901
6.5	554.01	0.0013356	0.029727	1232.7	2585.7	1241.4	2778.9	3.0764	5.8516
7.0	558.98	0.0013519	0.027378	1258.2	2581	1267.7	2772.6	3.1224	5.8148
7.5	563.69	0.0013682	0.025330	1282.7	2575.9	1292.9	2765.9	3.1662	5.7793
8.0	568.16	0.0013847	0.023526	1306.2	2570.5	1317.3	2758.7	3.2081	5.7450
8.5	572.42	0.0014013	0.021923	1329.0	2564.7	1340.9	2751.0	3.2483	5.7117
9.0	576.49	0.0014181	0.020490	1351.1	2558.5	1363.9	2742.9	3.2870	5.6791
9.5	580.40	0.0014352	0.019199	1372.6	2552.0	1386.2	2734.4	3.3244	5.6473
10.0	584.15	0.0014526	0.018030	1393.5	2545.2	1408.1	2725.5	3.3606	5.6160
10.5	587.75	0.0014703	0.016965	1414.0	2538.0	1429.4	2716.1	3.3959	5.5851
11.0	591.23	0.0014885	0.015990	1434.1	2530.5	1450.4	2706.3	3.4303	5.5545
11.5	594.58	0.0015071	0.015093	1453.8	2522.6	1471.1	2696.1	3.4638	5.5241
12.0	597.83	0.0015263	0.014264	1473.1	2514.3	1491.5	2685.4	3.4967	5.4939
12.5	600.96	0.0015461	0.013496	1492.3	2505.6	1511.6	2674.3	3.5290	5.4638
13.0	604.00	0.0015665	0.012780	1511.1	2496.5	1531.5	2662.7	3.5608	5.4336
13.5	606.95	0.0015877	0.012112	1529.9	2487.0	1551.3	2650.5	3.5921	5.4032
14.0	609.82	0.0016097	0.011485	1548.4	2477.1	1571.0	2637.9	3.6232	5.3727
14.5	612.60	0.0016328	0.010895	1566.9	2466.6	1590.6	2624.6	3.6539	5.3418
15.0	615.31	0.0016570	0.010338	1585.3	2455.6	1610.2	2610.7	3.6846	5.3106
15.5	617.94	0.0016824	0.0098106	1603.8	2444.1	1629.9	2596.1	3.7151	5.2788
16.0	620.50	0.0017094	0.0093088	1622.3	2431.8	1649.7	2580.8	3.7457	5.2463
16.5	623.00	0.0017383	0.0088299	1641.0	2418.9	1669.7	2564.6	3.7765	5.2130
17.0	625.44	0.0017693	0.0083709	1659.9	2405.2	1690.0	2547.5	3.8077	5.1787
17.5	627.82	0.0018029	0.0079292	1679.2	2390.5	1710.8	2529.3	3.8394	5.1431
18.0	630.14	0.0018398	0.0075017	1699.0	2374.8	1732.1	2509.8	3.8718	5.1061
18.5	632.41	0.0018807	0.0070856	1719.3	2357.8	1754.1	2488.8	3.9053	5.0670
19.0	634.62	0.0019268	0.0066773	1740.5	2339.1	1777.2	2466.0	3.9401	5.0256
19.5	636.79	0.0019792	0.0062725	1762.8	2318.5	1801.4	2440.8	3.9767	4.9808
20.0	638.90	0.002040	0.0058652	1786.4	2295.0	1827.2	2412.3	4.0156	4.9314
20.5	640.96	0.0021126	0.0054457	1812.0	2267.6	1855.3	2379.2	4.0579	4.8753
21.0	642.98	0.0022055	0.0049961	1841.2	2233.7	1887.6	2338.6	4.1064	4.8079
21.5	644.94	0.0023468	0.0044734	1879.1	2186.9	1929.5	2283.1	4.1698	4.7181
22.0	646.86	0.0027044	0.0036475	1951.8	2092.8	2011.3	2173.1	4.2945	4.5446
22.064	647.096	0.0031056	0.0031056	2015.7	2015.7	2084.3	2084.3	4.4070	4.4070

Table D.3 Superheated Vapor (Steam)*

Table D.3A Isobaric Data for $P = 0.006$ MPa

T (K)	P (MPa)	ρ (kg/m³)	v (m³/kg)	u (kJ/kg)	h (kJ/kg)	s (kJ/kg·K)
309.31	0.006	0.042135	23.733	2424.2	2566.6	8.3290
320	0.006	0.040708	24.565	2439.7	2587.1	8.3940
340	0.006	0.038291	26.116	2468.4	2625.1	8.5092
360	0.006	0.036151	27.662	2497.0	2663.0	8.6176
380	0.006	0.034239	29.206	2525.7	2701.0	8.7202
400	0.006	0.032522	30.748	2554.6	2739.0	8.8179
420	0.006	0.030969	32.290	2583.5	2777.3	8.9112
440	0.006	0.029559	33.831	2612.7	2815.7	9.0005
460	0.006	0.028272	35.371	2642.1	2854.3	9.0863
480	0.006	0.027092	36.911	2671.6	2893.1	9.1689
500	0.006	0.026008	38.450	2701.4	2932.1	9.2485
520	0.006	0.025006	39.990	2731.4	2971.4	9.3255
540	0.006	0.024079	41.529	2761.7	3010.9	9.4000
560	0.006	0.023219	43.068	2792.2	3050.6	9.4722
580	0.006	0.022418	44.607	2822.9	3090.6	9.5424
600	0.006	0.02167	46.146	2853.9	3130.8	9.6105
620	0.006	0.020971	47.685	2885.1	3171.2	9.6769
640	0.006	0.020315	49.224	2916.6	3211.9	9.7415
660	0.006	0.019699	50.763	2948.3	3252.9	9.8046
680	0.006	0.01912	52.301	2980.4	3294.2	9.8661
700	0.006	0.018573	53.840	3012.6	3335.7	9.9263
720	0.006	0.018057	55.379	3045.2	3377.4	9.9851
740	0.006	0.017569	56.917	3078.0	3419.5	10.043
760	0.006	0.017107	58.456	3111.0	3461.8	10.099
780	0.006	0.016668	59.995	3144.4	3504.3	10.154
800	0.006	0.016251	61.533	3178.0	3547.2	10.209

*Property values generated from NIST Database 23: REFPROP Version 7.0 (August 2002).

Table D.3B Isobaric Data for $P = 0.035$ MPa

T (K)	P (MPa)	ρ (kg/m³)	v (m³/kg)	u (kJ/kg)	h (kJ/kg)	s (kJ/kg·K)
345.83	0.035	0.22099	4.5251	2472.3	2630.7	7.7146
360	0.035	0.21197	4.7176	2493.5	2658.6	7.7939
380	0.035	0.20052	4.9871	2523.1	2697.7	7.8994
400	0.035	0.1903	5.2549	2552.5	2736.5	7.9989
420	0.035	0.1811	5.5218	2581.9	2775.2	8.0934
440	0.035	0.17278	5.7879	2611.4	2813.9	8.1835
460	0.035	0.16519	6.0535	2640.9	2852.8	8.2699
480	0.035	0.15826	6.3187	2670.7	2891.8	8.353
500	0.035	0.15189	6.5837	2700.6	2931.0	8.433
520	0.035	0.14602	6.8484	2730.7	2970.4	8.5102
540	0.035	0.14059	7.113	2761.1	3010.0	8.5849
560	0.035	0.13555	7.3775	2791.6	3049.8	8.6573
580	0.035	0.13086	7.6419	2822.4	3089.9	8.7276
600	0.035	0.12648	7.9062	2853.4	3130.1	8.7958
620	0.035	0.12239	8.1704	2884.7	3170.7	8.8623
640	0.035	0.11856	8.4346	2916.2	3211.4	8.927
660	0.035	0.11496	8.6987	2948.0	3252.5	8.9901
680	0.035	0.11157	8.9627	2980.0	3293.7	9.0517
700	0.035	0.10838	9.2268	3012.3	3335.3	9.1119
720	0.035	0.10537	9.4908	3044.9	3377.1	9.1708
740	0.035	0.10251	9.7547	3077.7	3419.1	9.2284
760	0.035	0.099813	10.019	3110.8	3461.4	9.2848
780	0.035	0.097251	10.283	3144.2	3504.0	9.3402
800	0.035	0.094818	10.547	3177.8	3546.9	9.3944
820	0.035	0.092503	10.81	3211.7	3590.1	9.4477
840	0.035	0.090299	11.074	3245.9	3633.5	9.5

Table D.3C Isobaric Data for $P = 0.070$ MPa

T (K)	P (MPa)	ρ (kg/m^3)	v (m^3/kg)	u (kJ/kg)	h (kJ/kg)	s (kJ/kg·K)
363.08	0.07	0.42287	2.3648	2493.9	2659.4	7.479
380	0.07	0.40301	2.4813	2519.9	2693.5	7.5709
400	0.07	0.38205	2.6175	2550.0	2733.2	7.6727
420	0.07	0.3633	2.7525	2579.9	2772.6	7.7687
440	0.07	0.3464	2.8868	2609.7	2811.8	7.8599
460	0.07	0.33106	3.0206	2639.6	2851.0	7.9471
480	0.07	0.31706	3.154	2669.5	2890.3	8.0307
500	0.07	0.30422	3.2871	2699.6	2929.7	8.1111
520	0.07	0.2924	3.42	2729.9	2969.3	8.1886
540	0.07	0.28147	3.5528	2760.3	3009.0	8.2636
560	0.07	0.27134	3.6854	2790.9	3048.9	8.3362
580	0.07	0.26193	3.8179	2821.8	3089.0	8.4066
600	0.07	0.25314	3.9503	2852.9	3129.4	8.475
620	0.07	0.24494	4.0827	2884.2	3170.0	8.5416
640	0.07	0.23725	4.215	2915.8	3210.8	8.6064
660	0.07	0.23003	4.3472	2947.6	3251.9	8.6696
680	0.07	0.22324	4.4794	2979.6	3293.2	8.7312
700	0.07	0.21685	4.6116	3012.0	3334.8	8.7915
720	0.07	0.2108	4.7437	3044.5	3376.6	8.8504
740	0.07	0.20509	4.8758	3077.4	3418.7	8.9081
760	0.07	0.19968	5.0079	3110.5	3461.1	8.9645
780	0.07	0.19455	5.14	3143.9	3503.7	9.0199
800	0.07	0.18968	5.2721	3177.5	3546.6	9.0742
820	0.07	0.18504	5.4041	3211.5	3589.7	9.1275
840	0.07	0.18063	5.5361	3245.7	3633.2	9.1798
860	0.07	0.17643	5.6681	3280.1	3676.9	9.2313

Table D.3D Isobaric Data for $P = 0.100$ MPa

T (K)	P (MPa)	ρ (kg/m^3)	v (m^3/kg)	u (kJ/kg)	h (kJ/kg)	s (kJ/kg·K)
372.76	0.1	0.59034	1.6939	2505.6	2674.9	7.3588
380	0.1	0.57824	1.7294	2517.0	2689.9	7.3986
400	0.1	0.54761	1.8261	2547.8	2730.4	7.5025
420	0.1	0.52038	1.9217	2578.2	2770.3	7.5999
440	0.1	0.49592	2.0165	2608.3	2810.0	7.6921
460	0.1	0.47378	2.1107	2638.4	2849.5	7.7799
480	0.1	0.45361	2.2045	2668.5	2889.0	7.864
500	0.1	0.43514	2.2981	2698.7	2928.6	7.9447
520	0.1	0.41815	2.3915	2729.1	2968.2	8.0226
540	0.1	0.40247	2.4847	2759.6	3008.1	8.0978
560	0.1	0.38794	2.5777	2790.3	3048.1	8.1705
580	0.1	0.37444	2.6707	2821.3	3088.3	8.2411
600	0.1	0.36185	2.7635	2852.4	3128.8	8.3096
620	0.1	0.3501	2.8563	2883.8	3169.4	8.3763
640	0.1	0.33909	2.9491	2915.4	3210.3	8.4411
660	0.1	0.32876	3.0418	2947.2	3251.4	8.5044
680	0.1	0.31904	3.1344	2979.3	3292.8	8.5661
700	0.1	0.30988	3.227	3011.7	3334.4	8.6264
720	0.1	0.30124	3.3196	3044.3	3376.2	8.6854
740	0.1	0.29307	3.4122	3077.1	3418.3	8.7431
760	0.1	0.28533	3.5047	3110.3	3460.7	8.7996
780	0.1	0.27799	3.5972	3143.6	3503.4	8.855
800	0.1	0.27102	3.6897	3177.3	3546.3	8.9093
820	0.1	0.2644	3.7822	3211.3	3589.5	8.9626
840	0.1	0.25809	3.8747	3245.5	3632.9	9.015
860	0.1	0.25207	3.9671	3280.0	3676.7	9.0665
880	0.1	0.24633	4.0595	3314.7	3720.7	9.1171

Table D.3E   Isobaric Data for _P_ = 0.150 MPa

_T_ (K)	_P_ (MPa)	_ρ_ (kg/m³)	_v_ (m³/kg)	_u_ (kJ/kg)	_h_ (kJ/kg)	_s_ (kJ/kg·K)
384.5	0.15	0.8626	1.1593	2519.2	2693.1	7.223
400	0.15	0.82612	1.2105	2544.0	2725.6	7.3058
420	0.15	0.78408	1.2754	2575.2	2766.5	7.4057
440	0.15	0.74658	1.3394	2605.9	2806.8	7.4995
460	0.15	0.71278	1.403	2636.4	2846.9	7.5884
480	0.15	0.6821	1.4661	2666.9	2886.8	7.6733
500	0.15	0.65407	1.5289	2697.3	2926.6	7.7547
520	0.15	0.62835	1.5915	2727.8	2966.6	7.833
540	0.15	0.60463	1.6539	2758.5	3006.6	7.9086
560	0.15	0.58268	1.7162	2789.4	3046.8	7.9816
580	0.15	0.5623	1.7784	2820.4	3087.1	8.0524
600	0.15	0.54333	1.8405	2851.6	3127.7	8.1212
620	0.15	0.52562	1.9025	2883.1	3168.4	8.188
640	0.15	0.50903	1.9645	2914.7	3209.4	8.253
660	0.15	0.49348	2.0264	2946.6	3250.6	8.3164
680	0.15	0.47886	2.0883	2978.8	3292.0	8.3782
700	0.15	0.46508	2.1502	3011.1	3333.7	8.4386
720	0.15	0.45209	2.212	3043.8	3375.6	8.4976
740	0.15	0.4398	2.2738	3076.7	3417.7	8.5554
760	0.15	0.42817	2.3355	3109.8	3460.2	8.6119
780	0.15	0.41714	2.3973	3143.3	3502.9	8.6674
800	0.15	0.40667	2.459	3177.0	3545.8	8.7217
820	0.15	0.39671	2.5207	3210.9	3589.0	8.7751
840	0.15	0.38724	2.5824	3245.1	3632.5	8.8275
860	0.15	0.3782	2.6441	3279.7	3676.3	8.879
880	0.15	0.36958	2.7058	3314.4	3720.3	8.9296
900	0.15	0.36135	2.7674	3349.5	3764.6	8.9794

Table D.3F   Isobaric Data for _P_ = 0.300 MPa

_T_ (K)	_P_ (MPa)	_ρ_ (kg/m³)	_v_ (m³/kg)	_u_ (kJ/kg)	_h_ (kJ/kg)	_s_ (kJ/kg·K)
406.67	0.3	1.6508	0.60576	2543.2	2724.9	6.9916
420	0.3	1.5906	0.62868	2565.8	2754.4	7.063
440	0.3	1.5101	0.6622	2598.4	2797.1	7.1623
460	0.3	1.4387	0.69507	2630.3	2838.8	7.255
480	0.3	1.3746	0.72749	2661.7	2879.9	7.3426
500	0.3	1.3165	0.75958	2692.9	2920.8	7.426
520	0.3	1.2635	0.79144	2724.0	2961.5	7.5057
540	0.3	1.2149	0.82312	2755.2	3002.1	7.5824
560	0.3	1.17	0.85467	2786.4	3042.8	7.6564
580	0.3	1.1285	0.8861	2817.7	3083.6	7.7279
600	0.3	1.09	0.91744	2849.2	3124.4	7.7973
620	0.3	1.0541	0.94871	2880.9	3165.5	7.8646
640	0.3	1.0205	0.97992	2912.7	3206.7	7.93
660	0.3	0.98904	1.0111	2944.8	3248.1	7.9937
680	0.3	0.95951	1.0422	2977.1	3289.7	8.0558
700	0.3	0.93173	1.0733	3009.6	3331.6	8.1165
720	0.3	0.90553	1.1043	3042.4	3373.6	8.1757
740	0.3	0.88079	1.1353	3075.3	3416.0	8.2337
760	0.3	0.85738	1.1663	3108.6	3458.5	8.2904
780	0.3	0.8352	1.1973	3142.1	3501.3	8.346
800	0.3	0.81415	1.2283	3175.9	3544.3	8.4005
820	0.3	0.79414	1.2592	3209.9	3587.6	8.4539
840	0.3	0.7751	1.2901	3244.2	3631.2	8.5064
860	0.3	0.75696	1.3211	3278.7	3675.1	8.558
880	0.3	0.73966	1.352	3313.6	3719.2	8.6087
900	0.3	0.72314	1.3829	3348.7	3763.5	8.6586
920	0.3	0.70734	1.4138	3384.1	3808.2	8.7076

Table D.3G Isobaric Data for $P = 0.50$ MPa

T (K)	P (MPa)	ρ (kg/m^3)	v (m^3/kg)	u (kJ/kg)	h (kJ/kg)	s (kJ/kg·K)
424.98	0.5	2.668	0.37481	2560.7	2748.1	6.8207
440	0.5	2.5579	0.39095	2587.6	2783.1	6.9015
460	0.5	2.429	0.41169	2621.6	2827.4	7.0001
480	0.5	2.3153	0.43191	2654.5	2870.5	7.0917
500	0.5	2.2135	0.45176	2686.8	2912.7	7.1779
520	0.5	2.1215	0.47136	2718.8	2954.5	7.2599
540	0.5	2.0376	0.491	2750.6	2996.0	7.3382
560	0.5	1.9607	0.510	2782.4	3037.4	7.4134
580	0.5	1.8898	0.529	2814.1	3078.7	7.486
600	0.5	1.8242	0.548	2846.0	3120.1	7.5561
620	0.5	1.7631	0.567	2878.0	3161.5	7.6241
640	0.5	1.7063	0.586	2910.1	3203.1	7.6901
660	0.5	1.6531	0.605	2942.4	3244.8	7.7542
680	0.5	1.6032	0.624	2974.8	3286.7	7.8168
700	0.5	1.5564	0.643	3007.5	3328.8	7.8777
720	0.5	1.5123	0.661	3040.4	3371.1	7.9373
740	0.5	1.4706	0.680	3073.6	3413.5	7.9955
760	0.5	1.4313	0.699	3106.9	3456.3	8.0524
780	0.5	1.394	0.717	3140.5	3499.2	8.1082
800	0.5	1.3587	0.736	3174.4	3542.4	8.1629
820	0.5	1.3252	0.755	3208.5	3585.8	8.2165
840	0.5	1.2932	0.77325	3242.9	3629.5	8.2691
860	0.5	1.2629	0.79186	3277.5	3673.4	8.3208
880	0.5	1.2339	0.81045	3312.4	3717.6	8.3716
900	0.5	1.2062	0.82904	3347.6	3762.1	8.4216
920	0.5	1.1798	0.84762	3383.0	3806.8	8.4708
940	0.5	1.1545	0.86619	3418.7	3851.8	8.5191

Table D.3H Isobaric Data for $P = 0.70$ MPa

T (K)	P (MPa)	ρ (kg/m^3)	v (m^3/kg)	u (kJ/kg)	h (kJ/kg)	s (kJ/kg·K)
438.1	0.7	3.666	0.27277	2571.8	2762.8	6.7071
440	0.7	3.6453	0.27433	2575.5	2767.6	6.718
460	0.7	3.4477	0.29005	2612.3	2815.3	6.8242
480	0.7	3.2775	0.30511	2647.0	2860.5	6.9204
500	0.7	3.1274	0.31976	2680.5	2904.3	7.0099
520	0.7	2.9929	0.33413	2713.4	2947.3	7.0941
540	0.7	2.8712	0.34828	2745.9	2989.7	7.1742
560	0.7	2.7603	0.36228	2778.3	3031.9	7.2508
580	0.7	2.6585	0.37616	2810.5	3073.8	7.3244
600	0.7	2.5645	0.38994	2842.7	3115.7	7.3953
620	0.7	2.4775	0.40364	2875.0	3157.6	7.464
640	0.7	2.3965	0.41728	2907.4	3199.5	7.5306
660	0.7	2.3209	0.43086	2939.9	3241.5	7.5952
680	0.7	2.2502	0.4444	2972.6	3283.7	7.6582
700	0.7	2.1839	0.4579	3005.4	3326.0	7.7195
720	0.7	2.1215	0.47137	3038.5	3368.5	7.7793
740	0.7	2.0626	0.48481	3071.8	3411.1	7.8378
760	0.7	2.0071	0.49823	3105.3	3454.0	7.895
780	0.7	1.9545	0.51163	3139.0	3497.1	7.9509
800	0.7	1.9047	0.52501	3172.9	3540.4	8.0058
820	0.7	1.8575	0.53837	3207.1	3584.0	8.0595
840	0.7	1.8125	0.55172	3241.6	3627.8	8.1123
860	0.7	1.7697	0.56505	3276.3	3671.8	8.1641
880	0.7	1.729	0.57838	3311.3	3716.1	8.215
900	0.7	1.6901	0.59169	3346.5	3760.7	8.2651
920	0.7	1.6529	0.60499	3382.0	3805.5	8.3143
940	0.7	1.6174	0.61829	3417.7	3850.5	8.3628

Table D.3I   Isobaric Data for $P = 1.0$ MPa

T (K)	P (MPa)	ρ (kg/m³)	v (m³/kg)	u (kJ/kg)	h (kJ/kg)	s (kJ/kg·K)
453.03	1	5.145	0.19436	2582.7	2777.1	6.585
460	1	5.0376	0.19851	2597.0	2795.5	6.6253
480	1	4.7658	0.20983	2634.9	2844.7	6.7301
500	1	4.5323	0.22064	2670.6	2891.2	6.825
520	1	4.3268	0.23112	2705.0	2936.1	6.9131
540	1	4.1431	0.24137	2738.7	2980.1	6.996
560	1	3.9771	0.25144	2772.0	3023.4	7.0748
580	1	3.8258	0.26138	2804.9	3066.3	7.1501
600	1	3.6871	0.27122	2837.7	3109.0	7.2224
620	1	3.5591	0.28097	2870.5	3151.5	7.2921
640	1	3.4404	0.29066	2903.3	3194.0	7.3596
660	1	3.33	0.3003	2936.2	3236.5	7.425
680	1	3.227	0.30988	2969.2	3279.1	7.4885
700	1	3.1305	0.31943	3002.3	3321.7	7.5504
720	1	3.04	0.32895	3035.6	3364.5	7.6107
740	1	2.9547	0.33844	3069.1	3407.5	7.6695
760	1	2.8744	0.3479	3102.7	3450.6	7.727
780	1	2.7984	0.35734	3136.6	3494.0	7.7833
800	1	2.7265	0.36677	3170.7	3537.5	7.8384
820	1	2.6583	0.37618	3205.1	3581.2	7.8924
840	1	2.5936	0.38557	3239.6	3625.2	7.9454
860	1	2.532	0.39495	3274.4	3669.4	7.9974
880	1	2.4733	0.40432	3309.5	3713.8	8.0484
900	1	2.4174	0.41367	3344.8	3758.5	8.0986
920	1	2.3639	0.42302	3380.4	3803.4	8.148
940	1	2.3129	0.43236	3416.3	3848.6	8.1966
960	1	2.264	0.4417	3452.4	3894.1	8.2444

Table D.3J   Isobaric Data for $P = 1.5$ MPa

T (K)	P (MPa)	ρ (kg/m³)	v (m³/kg)	u (kJ/kg)	h (kJ/kg)	s (kJ/kg·K)
471.44	1.5	7.5924	0.13171	2593.4	2791.0	6.443
480	1.5	7.3885	0.13534	2612.3	2815.3	6.4942
500	1.5	6.9775	0.14332	2652.6	2867.6	6.6009
520	1.5	6.629	0.15085	2690.1	2916.4	6.6967
540	1.5	6.3249	0.1581	2726.1	2963.2	6.7851
560	1.5	6.0549	0.16515	2761.0	3008.8	6.8678
580	1.5	5.8121	0.17206	2795.3	3053.4	6.9462
600	1.5	5.5915	0.17884	2829.2	3097.5	7.0209
620	1.5	5.3897	0.18554	2862.9	3141.2	7.0925
640	1.5	5.2039	0.19216	2896.4	3184.7	7.1616
660	1.5	5.0319	0.19873	2929.9	3228.0	7.2282
680	1.5	4.8721	0.20525	2963.4	3271.3	7.2929
700	1.5	4.7231	0.21173	2997.0	3314.6	7.3556
720	1.5	4.5836	0.21817	3030.7	3358.0	7.4167
740	1.5	4.4527	0.22458	3064.5	3401.4	7.4762
760	1.5	4.3295	0.23097	3098.5	3445.0	7.5343
780	1.5	4.2133	0.23734	3132.7	3488.7	7.5911
800	1.5	4.1036	0.24369	3167.0	3532.6	7.6466
820	1.5	3.9997	0.25002	3201.6	3576.6	7.701
840	1.5	3.9011	0.25634	3236.4	3620.9	7.7543
860	1.5	3.8075	0.26264	3271.4	3665.3	7.8066
880	1.5	3.7184	0.26893	3306.6	3710.0	7.858
900	1.5	3.6335	0.27522	3342.1	3754.9	7.9084
920	1.5	3.5525	0.28149	3377.8	3800.0	7.958
940	1.5	3.4752	0.28775	3413.8	3845.4	8.0068
960	1.5	3.4013	0.29401	3450.0	3891.0	8.0548
980	1.5	3.3305	0.30026	3486.5	3936.9	8.1021

Table D.3K   Isobaric Data for P = 2.0 MPa

T (K)	P (MPa)	ρ (kg/m³)	ν (m³/kg)	u (kJ/kg)	h (kJ/kg)	s (kJ/kg·K)
485.53	2	10.042	0.099585	2599.1	2798.3	6.339
500	2	9.5781	0.10441	2632.6	2841.4	6.4265
520	2	9.0447	0.11056	2674.0	2895.1	6.5319
540	2	8.5934	0.11637	2712.6	2945.4	6.6267
560	2	8.2008	0.12194	2749.5	2993.4	6.7141
580	2	7.853	0.12734	2785.3	3040.0	6.7959
600	2	7.5406	0.13262	2820.4	3085.6	6.8732
620	2	7.2573	0.13779	2855.0	3130.6	6.947
640	2	6.9983	0.14289	2889.4	3175.1	7.0176
660	2	6.7599	0.14793	2923.5	3219.4	7.0857
680	2	6.5395	0.15292	2957.6	3263.4	7.1514
700	2	6.3346	0.15786	2991.6	3307.4	7.2151
720	2	6.1436	0.16277	3025.8	3351.3	7.277
740	2	5.9648	0.16765	3059.9	3395.2	7.3372
760	2	5.7969	0.1725	3094.2	3439.3	7.3959
780	2	5.639	0.17734	3128.7	3483.4	7.4532
800	2	5.4901	0.18215	3163.3	3527.6	7.5092
820	2	5.3493	0.18694	3198.1	3572.0	7.564
840	2	5.2159	0.19172	3233.1	3616.5	7.6176
860	2	5.0894	0.19649	3268.3	3661.2	7.6702
880	2	4.9691	0.20124	3303.7	3706.1	7.7219
900	2	4.8547	0.20599	3339.3	3751.3	7.7726
920	2	4.7456	0.21072	3375.2	3796.6	7.8224
940	2	4.6415	0.21545	3411.3	3842.2	7.8714
960	2	4.5421	0.22016	3447.6	3887.9	7.9196
980	2	4.4469	0.22488	3484.2	3934.0	7.967
1000	2	4.3558	0.22958	3521.1	3980.2	8.0137

Table D.3L   Isobaric Data for P = 3.0 MPa

T (K)	P (MPa)	ρ (kg/m³)	ν (m³/kg)	u (kJ/kg)	h (kJ/kg)	s (kJ/kg·K)
507	3	15.001	0.066664	2603.2	2803.2	6.1856
520	3	14.309	0.069888	2637.1	2846.7	6.2704
540	3	13.442	0.074391	2682.9	2906.1	6.3825
560	3	12.729	0.078561	2724.7	2960.4	6.4812
580	3	12.119	0.082512	2764.1	3011.6	6.5712
600	3	11.587	0.086307	2801.9	3060.8	6.6546
620	3	11.113	0.089986	2838.7	3108.6	6.733
640	3	10.687	0.093576	2874.7	3155.5	6.8073
660	3	10.299	0.097096	2910.3	3201.6	6.8783
680	3	9.9443	0.10056	2945.6	3247.3	6.9465
700	3	9.6174	0.10398	2980.7	3292.6	7.0122
720	3	9.3147	0.10736	3015.7	3337.7	7.0758
740	3	9.033	0.1107	3050.6	3382.7	7.1374
760	3	8.77	0.11403	3085.6	3427.7	7.1973
780	3	8.5236	0.11732	3120.6	3472.6	7.2557
800	3	8.292	0.1206	3155.8	3517.6	7.3126
820	3	8.0738	0.12386	3191.0	3562.6	7.3682
840	3	7.8678	0.1271	3226.4	3607.7	7.4226
860	3	7.6729	0.13033	3262.0	3653.0	7.4758
880	3	7.488	0.13355	3297.8	3698.4	7.528
900	3	7.3124	0.13675	3333.7	3744.0	7.5792
920	3	7.1454	0.13995	3369.9	3789.7	7.6295
940	3	6.9863	0.14314	3406.2	3835.7	7.6789
960	3	6.8344	0.14632	3442.8	3881.8	7.7275
980	3	6.6893	0.14949	3479.7	3928.1	7.7752
1000	3	6.5506	0.15266	3516.7	3974.7	7.8223
1020	3	6.4177	0.15582	3554.0	4021.5	7.8686

Table D.3M Isobaric Data for P = 4.0 MPa

T (K)	P (MPa)	ρ (kg/m³)	ν (m³/kg)	u (kJ/kg)	h (kJ/kg)	s (kJ/kg·K)
523.5	4	20.09	0.049776	2601.7	2800.8	6.0696
540	4	18.831	0.053103	2648.4	2860.8	6.1824
560	4	17.642	0.056683	2696.9	2923.7	6.2968
580	4	16.675	0.05997	2740.9	2980.8	6.397
600	4	15.857	0.063063	2782.1	3034.3	6.4878
620	4	15.147	0.066018	2821.4	3085.4	6.5716
640	4	14.52	0.069	2859.4	3134.9	6.6501
660	4	13.958	0.072	2896.6	3183.2	6.7244
680	4	13.449	0.074	2933.2	3230.6	6.7953
700	4	12.985	0.077	2969.4	3277.5	6.8632
720	4	12.558	0.080	3005.4	3323.9	6.9285
740	4	12.163	0.082	3041.1	3370.0	6.9917
760	4	11.796	0.085	3076.8	3415.9	7.0529
780	4	11.454	0.087	3112.5	3461.7	7.1124
800	4	11.134	0.090	3148.2	3507.4	7.1702
820	4	10.833	0.092	3183.9	3553.1	7.2267
840	4	10.55	0.095	3219.7	3598.9	7.2818
860	4	10.283	0.097	3255.7	3644.7	7.3357
880	4	10.03	0.100	3291.8	3690.6	7.3885
900	4	9.791	0.102	3328.1	3736.6	7.4402
920	4	9.5636	0.105	3364.5	3782.8	7.4909
940	4	9.3473	0.10698	3401.2	3829.1	7.5407
960	4	9.1412	0.10939	3438.0	3875.6	7.5897
980	4	8.9446	0.1118	3475.1	3922.3	7.6378
1000	4	8.7568	0.1142	3512.4	3969.1	7.6851
1020	4	8.5771	0.11659	3549.9	4016.2	7.7317
1040	4	8.405	0.11898	3587.6	4063.5	7.7776

Table D.3N Isobaric Data for P = 6.0 MPa

T (K)	P (MPa)	ρ (kg/m³)	ν (m³/kg)	u (kJ/kg)	h (kJ/kg)	s (kJ/kg·K)
548.73	6	30.818	0.032448	2589.9	2784.6	5.8901
560	6	29.166	0.034287	2629.1	2834.8	5.9807
580	6	26.947	0.03711	2687.2	2909.8	6.1125
600	6	25.247	0.039608	2737.6	2975.2	6.2233
620	6	23.865	0.041903	2783.5	3034.9	6.3211
640	6	22.697	0.044058	2826.5	3090.8	6.4099
660	6	21.686	0.046112	2867.5	3144.2	6.4921
680	6	20.795	0.048089	2907.2	3195.7	6.569
700	6	19.998	0.050006	2945.9	3246.0	6.6418
720	6	19.277	0.051875	2984.0	3295.2	6.7112
740	6	18.62	0.053706	3021.5	3343.8	6.7777
760	6	18.017	0.055504	3058.7	3391.8	6.8417
780	6	17.46	0.057274	3095.7	3439.4	6.9036
800	6	16.943	0.059022	3132.6	3486.7	6.9635
820	6	16.461	0.06075	3169.4	3533.9	7.0217
840	6	16.01	0.062461	3206.1	3580.9	7.0784
860	6	15.587	0.064157	3242.9	3627.9	7.1336
880	6	15.188	0.06584	3279.8	3674.8	7.1876
900	6	14.812	0.067512	3316.7	3721.8	7.2404
920	6	14.457	0.069173	3353.8	3768.8	7.2921
940	6	14.119	0.070825	3391.0	3815.9	7.3427
960	6	13.799	0.072469	3428.4	3863.2	7.3924
980	6	13.494	0.074105	3465.9	3910.5	7.4413
1000	6	13.204	0.075735	3503.6	3958.0	7.4892
1020	6	12.927	0.077358	3541.5	4005.6	7.5364
1040	6	12.662	0.078977	3579.6	4053.4	7.5828
1060	6	12.409	0.08059	3617.9	4101.4	7.6285

Table D.3O   Isobaric Data for $P = 8.0$ MPa

T (K)	P (MPa)	ρ (kg/m³)	v (m³/kg)	u (kJ/kg)	h (kJ/kg)	s (kJ/kg·K)
568.16	8	42.507	0.023526	2570.5	2758.7	5.745
580	8	39.647	0.025223	2619.0	2820.7	5.8531
600	8	36.218	0.027611	2684.7	2905.6	5.9971
620	8	33.704	0.02967	2740.2	2977.6	6.1151
640	8	31.713	0.031532	2789.9	3042.1	6.2176
660	8	30.065	0.033261	2835.8	3101.9	6.3096
680	8	28.658	0.034894	2879.3	3158.4	6.394
700	8	27.431	0.036455	2921.0	3212.6	6.4726
720	8	26.343	0.037961	2961.5	3265.2	6.5466
740	8	25.367	0.039422	3001.1	3316.5	6.6168
760	8	24.482	0.040847	3040.0	3366.8	6.6839
780	8	23.673	0.042242	3078.5	3416.4	6.7484
800	8	22.929	0.043612	3116.6	3465.5	6.8105
820	8	22.242	0.044961	3154.5	3514.2	6.8706
840	8	21.602	0.046292	3192.3	3562.6	6.9289
860	8	21.006	0.047606	3229.9	3610.8	6.9856
880	8	20.447	0.048907	3267.6	3658.8	7.0409
900	8	19.922	0.050196	3305.2	3706.8	7.0948
920	8	19.427	0.051475	3342.9	3754.7	7.1474
940	8	18.96	0.052744	3380.7	3802.6	7.199
960	8	18.517	0.054004	3418.6	3850.6	7.2495
980	8	18.097	0.055257	3456.6	3898.6	7.299
1000	8	17.698	0.056503	3494.7	3946.8	7.3476
1020	8	17.318	0.057742	3533.1	3995.0	7.3953
1040	8	16.956	0.058976	3571.5	4043.3	7.4423
1060	8	16.61	0.060205	3610.2	4091.8	7.4885
1080	8	16.279	0.06143	3649.0	4140.5	7.5339

Table D.3P   Isobaric Data for $P = 10.0$ MPa

T (K)	P (MPa)	ρ (kg/m³)	v (m³/kg)	u (kJ/kg)	h (kJ/kg)	s (kJ/kg·K)
584.15	10	55.463	0.01803	2545.2	2725.5	5.616
600	10	49.773	0.020091	2619.1	2820.0	5.7756
620	10	45.151	0.022148	2689.7	2911.2	5.9253
640	10	41.84	0.0239	2748.7	2987.7	6.0468
660	10	39.259	0.025472	2801.1	3055.8	6.1516
680	10	37.145	0.026922	2849.2	3118.4	6.2451
700	10	35.355	0.028285	2894.5	3177.4	6.3305
720	10	33.804	0.029582	2937.8	3233.7	6.4098
740	10	32.437	0.030829	2979.7	3288.0	6.4843
760	10	31.215	0.032036	3020.6	3341.0	6.5549
780	10	30.112	0.033209	3060.7	3392.8	6.6222
800	10	29.107	0.034356	3100.2	3443.7	6.6867
820	10	28.185	0.035479	3139.3	3494.1	6.7489
840	10	27.335	0.036584	3178.1	3543.9	6.8089
860	10	26.546	0.037671	3216.7	3593.4	6.8671
880	10	25.81	0.038744	3255.1	3642.6	6.9237
900	10	25.123	0.039804	3293.5	3691.6	6.9787
920	10	24.478	0.040854	3331.9	3740.4	7.0324
940	10	23.87	0.041893	3370.3	3789.2	7.0849
960	10	23.297	0.042924	3408.7	3837.9	7.1362
980	10	22.755	0.043947	3447.2	3886.7	7.1864
1000	10	22.241	0.044963	3485.8	3935.5	7.2357
1020	10	21.752	0.045972	3524.6	3984.3	7.284
1040	10	21.287	0.046976	3563.4	4033.2	7.3315
1060	10	20.844	0.047975	3602.5	4082.2	7.3782
1080	10	20.421	0.048969	3641.6	4131.3	7.4241
1100	10	20.017	0.049959	3681.0	4180.6	7.4693

Table D.3Q   Isobaric Data for P = 12.0 MPa

T (K)	P (MPa)	ρ (kg/m³)	ν (m³/kg)	u (kJ/kg)	h (kJ/kg)	s (kJ/kg·K)
597.83	12	70.106	0.014264	2514.3	2685.4	5.4939
600	12	68.549	0.014588	2528.8	2703.8	5.5247
620	12	59.113	0.016917	2628.8	2831.8	5.7346
640	12	53.502	0.018691	2701.7	2926.0	5.8843
660	12	49.499	0.020202	2762.6	3005.0	6.0059
680	12	46.392	0.021555	2816.6	3075.3	6.1108
700	12	43.857	0.022801	2866.3	3139.9	6.2045
720	12	41.718	0.023971	2912.9	3200.5	6.2899
740	12	39.87	0.025082	2957.4	3258.4	6.3692
760	12	38.245	0.026147	3000.4	3314.2	6.4436
780	12	36.796	0.027177	3042.3	3368.4	6.514
800	12	35.49	0.028177	3083.3	3421.4	6.5811
820	12	34.303	0.029152	3123.7	3473.5	6.6454
840	12	33.215	0.030107	3163.6	3524.9	6.7073
860	12	32.213	0.031044	3203.2	3575.7	6.7671
880	12	31.284	0.031966	3242.5	3626.1	6.8251
900	12	30.419	0.032874	3281.7	3676.2	6.8813
920	12	29.611	0.033771	3320.7	3726.0	6.9361
940	12	28.853	0.034658	3359.7	3775.6	6.9895
960	12	28.14	0.035536	3398.7	3825.2	7.0416
980	12	27.468	0.036406	3437.8	3874.6	7.0926
1000	12	26.832	0.037269	3476.9	3924.1	7.1425
1020	12	26.229	0.038126	3516.0	3973.5	7.1915
1040	12	25.657	0.038976	3555.3	4023.0	7.2395
1060	12	25.112	0.039821	3594.7	4072.5	7.2867
1080	12	24.593	0.040662	3634.2	4122.1	7.3331
1100	12	24.098	0.041498	3673.9	4171.8	7.3787

Table D.3R   Isobaric Data for P = 14.0 MPa

T (K)	P (MPa)	ρ (kg/m³)	ν (m³/kg)	u (kJ/kg)	h (kJ/kg)	s (kJ/kg·K)
609.82	14	87.069	0.011485	2477.1	2637.9	5.3727
620	14	77.66	0.012877	2549.9	2730.1	5.5229
640	14	67.43	0.01483	2646.6	2854.2	5.72
660	14	61.128	0.016359	2719.6	2948.6	5.8653
680	14	56.586	0.017672	2781.1	3028.5	5.9847
700	14	53.047	0.018851	2836.0	3099.9	6.0882
720	14	50.153	0.019939	2886.5	3165.7	6.1808
740	14	47.711	0.020959	2934.1	3227.5	6.2655
760	14	45.602	0.021929	2979.5	3286.5	6.3441
780	14	43.747	0.022859	3023.3	3343.3	6.418
800	14	42.094	0.023756	3066.0	3398.5	6.4879
820	14	40.605	0.024628	3107.7	3452.5	6.5545
840	14	39.252	0.025477	3148.8	3505.5	6.6184
860	14	38.012	0.026307	3189.5	3557.8	6.6799
880	14	36.871	0.027122	3229.7	3609.4	6.7392
900	14	35.813	0.027923	3269.7	3660.6	6.7967
920	14	34.829	0.028712	3309.5	3711.4	6.8526
940	14	33.91	0.02949	3349.1	3762.0	6.9069
960	14	33.048	0.030259	3388.7	3812.3	6.96
980	14	32.237	0.03102	3428.2	3862.5	7.0117
1000	14	31.472	0.031774	3467.8	3912.6	7.0623
1020	14	30.749	0.032521	3507.4	3962.7	7.1119
1040	14	30.064	0.033262	3547.1	4012.8	7.1605
1060	14	29.414	0.033998	3586.8	4062.8	7.2082
1080	14	28.795	0.034729	3626.7	4112.9	7.255
1100	14	28.205	0.035455	3666.7	4163.1	7.301
1120	14	27.641	0.036178	3706.8	4213.3	7.3463

Table D.3S Isobaric Data for $P = 16.0$ MPa

T (K)	P (MPa)	ρ (kg/m³)	v (m³/kg)	u (kJ/kg)	h (kJ/kg)	s (kJ/kg·K)
620.5	16	107.42	0.009309	2431.8	2580.8	5.2463
640	16	85.058	0.011757	2579.1	2767.2	5.5427
660	16	74.683	0.01339	2670.5	2884.8	5.7237
680	16	67.984	0.014709	2742.1	2977.5	5.8622
700	16	63.065	0.015857	2803.5	3057.2	5.9778
720	16	59.195	0.016893	2858.6	3128.9	6.0788
740	16	56.014	0.017853	2909.6	3195.3	6.1697
760	16	53.32	0.018755	2957.7	3257.8	6.253
780	16	50.988	0.019613	3003.7	3317.5	6.3306
800	16	48.935	0.020435	3048.1	3375.1	6.4035
820	16	47.103	0.021230	3091.4	3431.1	6.4726
840	16	45.452	0.022001	3133.8	3485.8	6.5386
860	16	43.951	0.022753	3175.5	3539.5	6.6018
880	16	42.576	0.023488	3216.7	3592.5	6.6627
900	16	41.308	0.024208	3257.5	3644.8	6.7215
920	16	40.134	0.024916	3298.0	3696.7	6.7785
940	16	39.042	0.025614	3338.3	3748.2	6.8338
960	16	38.021	0.026301	3378.5	3799.4	6.8877
980	16	37.063	0.026981	3418.6	3850.3	6.9403
1000	16	36.163	0.027653	3458.7	3901.1	6.9916
1020	16	35.313	0.028318	3498.8	3951.8	7.0418
1040	16	34.51	0.028977	3538.8	4002.5	7.091
1060	16	33.749	0.029631	3579.0	4053.1	7.1391
1080	16	33.026	0.030279	3619.2	4103.7	7.1864
1100	16	32.338	0.030924	3659.5	4154.3	7.2329
1120	16	31.682	0.031564	3700.0	4205.0	7.2785
1140	16	31.056	0.0322	3740.5	4255.7	7.3235

Table D.3T Isobaric Data for $P = 18.0$ MPa

T (K)	P (MPa)	ρ (kg/m³)	v (m³/kg)	u (kJ/kg)	h (kJ/kg)	s (kJ/kg·K)
630.14	18	133.3	0.0075017	2374.8	2509.8	5.1061
640	18	110	0.0090911	2489.2	2652.8	5.3314
660	18	91.08	0.010979	2613.3	2810.9	5.5749
680	18	80.955	0.012353	2698.9	2921.2	5.7397
700	18	74.095	0.013496	2768.4	3011.3	5.8704
720	18	68.943	0.014505	2829.0	3090.1	5.9813
740	18	64.839	0.015423	2883.9	3161.6	6.0793
760	18	61.439	0.016276	2935.0	3228.0	6.1679
780	18	58.544	0.017081	2983.4	3290.9	6.2495
800	18	56.029	0.017848	3029.8	3351.0	6.3257
820	18	53.809	0.018584	3074.7	3409.2	6.3975
840	18	51.825	0.019296	3118.4	3465.7	6.4656
860	18	50.034	0.019986	3161.3	3521.0	6.5307
880	18	48.403	0.02066	3203.5	3575.3	6.5931
900	18	46.908	0.021318	3245.2	3628.9	6.6533
920	18	45.529	0.021964	3286.5	3681.8	6.7115
940	18	44.251	0.022598	3327.5	3734.3	6.7679
960	18	43.06	0.023223	3368.3	3786.3	6.8227
980	18	41.948	0.023839	3409.0	3838.1	6.876
1000	18	40.904	0.024448	3449.5	3889.6	6.9281
1020	18	39.922	0.025049	3490.0	3940.9	6.9789
1040	18	38.995	0.025645	3530.6	3992.2	7.0286
1060	18	38.118	0.026234	3571.1	4043.3	7.0774
1080	18	37.287	0.026819	3611.7	4094.4	7.1251
1100	18	36.497	0.0274	3652.3	4145.5	7.172
1120	18	35.745	0.027976	3693.1	4196.6	7.2181
1140	18	35.029	0.028548	3733.9	4247.8	7.2633

Table D.3U Isobaric Data for P = 20.0 MPa

T (K)	P (MPa)	ρ (kg/m³)	v (m³/kg)	u (kJ/kg)	h (kJ/kg)	s (kJ/kg·K)
638.9	20	170.5	0.005865	2295.0	2412.3	4.9314
640	20	160.5	0.006231	2328.4	2453.0	4.995
660	20	112.09	0.008921	2543.7	2722.1	5.4104
680	20	96.059	0.01041	2650.1	2858.3	5.6139
700	20	86.38	0.011577	2730.2	2961.8	5.7639
720	20	79.525	0.012575	2797.4	3048.9	5.8867
740	20	74.257	0.013467	2857.0	3126.3	5.9927
760	20	70.002	0.014285	2911.5	3197.2	6.0872
780	20	66.444	0.01505	2962.5	3263.5	6.1733
800	20	63.396	0.015774	3010.9	3326.4	6.253
820	20	60.736	0.016465	3057.5	3386.8	6.3276
840	20	58.379	0.017129	3102.7	3445.3	6.398
860	20	56.268	0.017772	3146.8	3502.2	6.465
880	20	54.357	0.018397	3190.1	3558.0	6.5291
900	20	52.615	0.019006	3232.7	3612.8	6.5907
920	20	51.015	0.019602	3274.8	3666.8	6.6501
940	20	49.538	0.020186	3316.5	3720.2	6.7075
960	20	48.168	0.020761	3358.0	3773.2	6.7633
980	20	46.891	0.021326	3399.2	3825.7	6.8174
1000	20	45.696	0.021884	3440.3	3878.0	6.8702
1020	20	44.574	0.022435	3481.3	3930.0	6.9217
1040	20	43.518	0.022979	3522.2	3981.8	6.972
1060	20	42.521	0.023518	3563.2	4033.5	7.0213
1080	20	41.577	0.024052	3604.1	4085.1	7.0695
1100	20	40.682	0.024581	3645.1	4136.7	7.1168
1120	20	39.831	0.025106	3686.1	4188.3	7.1633
1140	20	39.021	0.025627	3727.3	4239.8	7.2089

Table D.3V Isobaric Data for P = 24.0 MPa (Supercritical)

T (K)	P (MPa)	ρ (kg/m³)	v (m³/kg)	u (kJ/kg)	h (kJ/kg)	s (kJ/kg·K)
660	26	365.88	0.002733	2010.9	2082.0	4.386
680	26	167.08	0.005985	2448.8	2604.4	5.1692
700	26	134.79	0.007419	2591.2	2784.1	5.43
720	26	118.02	0.008473	2688.8	2909.1	5.6062
740	26	106.98	0.009348	2767.2	3010.2	5.7447
760	26	98.863	0.010115	2834.6	3097.6	5.8613
780	26	92.512	0.010809	2895.3	3176.3	5.9635
800	26	87.325	0.011451	2951.2	3248.9	6.0554
820	26	82.963	0.012054	3003.7	3317.1	6.1396
840	26	79.211	0.012625	3053.7	3382.0	6.2178
860	26	75.927	0.01317	3101.9	3444.3	6.2912
880	26	73.015	0.013696	3148.6	3504.7	6.3606
900	26	70.403	0.014204	3194.2	3563.5	6.4267
920	26	68.039	0.014697	3239.0	3621.1	6.4899
940	26	65.882	0.015179	3283.0	3677.6	6.5507
960	26	63.902	0.015649	3326.5	3733.3	6.6094
980	26	62.074	0.01611	3369.5	3788.4	6.6661
1000	26	60.378	0.016562	3412.2	3842.9	6.7211
1020	26	58.797	0.017008	3454.7	3896.9	6.7747
1040	26	57.318	0.017447	3497.0	3950.6	6.8268
1060	26	55.93	0.01788	3539.2	4004.0	6.8777
1080	26	54.623	0.018307	3581.2	4057.2	6.9274
1100	26	53.389	0.01873	3623.3	4110.2	6.976
1120	26	52.222	0.019149	3665.3	4163.1	7.0237
1140	26	51.115	0.019564	3707.3	4216.0	7.0704
1160	26	50.062	0.019975	3749.4	4268.7	7.1163
1180	26	49.06	0.020383	3791.5	4321.4	7.1614
1200	26	48.104	0.020788	3833.7	4374.2	7.2057

Table D.3W Isobaric Data for $P = 28.0$ MPa (Supercritical)

T (K)	P (MPa)	ρ (kg/m³)	v (m³/kg)	u (kJ/kg)	h (kJ/kg)	s (kJ/kg·K)
660	28	455.47	0.0021955	1897.4	1958.9	4.1923
680	28	210.73	0.0047453	2347.1	2480.0	4.9705
700	28	157.04	0.0063676	2533.7	2712.0	5.3072
720	28	133.92	0.0074672	2647.0	2856.1	5.5104
740	28	119.74	0.0083511	2733.9	2967.7	5.6634
760	28	109.74	0.0091127	2806.9	3062.0	5.7892
780	28	102.1	0.0097942	2871.3	3145.6	5.8977
800	28	95.978	0.010419	2930.1	3221.9	5.9943
820	28	90.896	0.011002	2984.9	3293.0	6.0821
840	28	86.571	0.011551	3036.8	3360.2	6.1632
860	28	82.818	0.012075	3086.5	3424.6	6.2389
880	28	79.511	0.012577	3134.5	3486.6	6.3102
900	28	76.563	0.013061	3181.1	3546.8	6.3779
920	28	73.906	0.013531	3226.8	3605.6	6.4425
940	28	71.494	0.013987	3271.6	3663.2	6.5044
960	28	69.286	0.014433	3315.8	3719.9	6.5641
980	28	67.254	0.014869	3359.5	3775.8	6.6217
1000	28	65.374	0.015297	3402.8	3831.1	6.6775
1020	28	63.626	0.015717	3445.8	3885.8	6.7318
1040	28	61.994	0.016131	3488.5	3940.2	6.7845
1060	28	60.465	0.016538	3531.1	3994.2	6.8359
1080	28	59.029	0.016941	3573.6	4047.9	6.8862
1100	28	57.675	0.017339	3615.9	4101.4	6.9353
1120	28	56.395	0.017732	3658.3	4154.8	6.9833
1140	28	55.183	0.018121	3700.6	4208.0	7.0304
1160	28	54.033	0.018507	3743.0	4261.2	7.0767
1180	28	52.938	0.01889	3785.3	4314.3	7.122
1200	28	51.895	0.01927	3827.8	4367.3	7.1666

Table D.3X Isobaric Data for $P = 32.0$ MPa (Supercritical)

T (K)	P (MPa)	ρ (kg/m³)	v (m³/kg)	u (kJ/kg)	h (kJ/kg)	s (kJ/kg·K)
660	32	516.64	0.001936	1823.6	1885.6	4.0689
680	32	351.66	0.002844	2100.2	2191.2	4.5242
700	32	217.87	0.00459	2395.5	2542.3	5.0338
720	32	172.38	0.005801	2553.3	2739.0	5.3111
740	32	148.9	0.006716	2661.8	2876.7	5.4999
760	32	133.76	0.007476	2747.9	2987.1	5.6471
780	32	122.84	0.00814	2821.2	3081.7	5.77
800	32	114.42	0.00874	2886.5	3166.2	5.877
820	32	107.62	0.009292	2946.2	3243.6	5.9726
840	32	101.96	0.009808	3002.0	3315.9	6.0597
860	32	97.13	0.010295	3054.9	3384.4	6.1403
880	32	92.934	0.01076	3105.6	3449.9	6.2156
900	32	89.235	0.011206	3154.5	3513.1	6.2866
920	32	85.935	0.011637	3202.1	3574.4	6.354
940	32	82.962	0.012054	3248.6	3634.3	6.4184
960	32	80.262	0.012459	3294.3	3692.9	6.4802
980	32	77.791	0.012855	3339.3	3750.6	6.5396
1000	32	75.517	0.013242	3383.7	3807.5	6.597
1020	32	73.413	0.013622	3427.7	3863.6	6.6527
1040	32	71.458	0.013994	3471.5	3919.3	6.7067
1060	32	69.633	0.014361	3514.9	3974.4	6.7592
1080	32	67.924	0.014722	3558.1	4029.3	6.8105
1100	32	66.318	0.015079	3601.2	4083.8	6.8605
1120	32	64.804	0.015431	3644.3	4138.0	6.9094
1140	32	63.374	0.015779	3687.2	4192.1	6.9572
1160	32	62.019	0.016124	3730.1	4246.1	7.0041
1180	32	60.734	0.016465	3773.0	4299.9	7.0502
1200	32	59.511	0.016804	3816.0	4353.7	7.0953

Table D.4 Compressed Liquid (Water)*

Table D.4A Isobaric Data for P = 5.0 MPa

T (K)	P (MPa)	ρ (kg/m³)	v (m³/kg)	u (kJ/kg)	h (kJ/kg)	s (kJ/kg·K)
273.15	5	1002.3	0.000998	0.044068	5.0	0.0001
280	5	1002.3	0.000998	28.731	33.7	0.1039
300	5	998.74	0.001001	112.15	117.2	0.3917
320	5	991.56	0.001009	195.5	200.5	0.66066
340	5	981.68	0.001019	278.9	284.0	0.91364
360	5	969.62	0.001031	362.5	367.7	1.1528
380	5	955.65	0.001046	446.5	451.7	1.38
400	5	939.91	0.001064	530.9	536.2	1.5968
420	5	922.46	0.001084	616.1	621.5	1.8047
440	5	903.27	0.001107	702.2	707.7	2.0053
460	5	882.22	0.001134	789.6	795.2	2.1998
480	5	859.08	0.001164	878.6	884.5	2.3897
500	5	833.51	0.001200	970.0	976.0	2.5764
520	5	804.92	0.001242	1064.3	1070.5	2.7618
537.09	5	777.37	0.001286	1148.2	1154.6	2.921

*Property values generated from NIST Database 23: REFPROP Version 7.0 (August 2002).

Table D.4B Isobaric Data for P = 10.0 MPa

T (K)	P (MPa)	ρ (kg/m³)	v (m³/kg)	u (kJ/kg)	h (kJ/kg)	s (kJ/kg·K)
273.15	10	1004.8	0.0009952	0.1171	10.069	0.0003376
280	10	1004.7	0.0009954	28.659	38.613	0.10355
300	10	1001.0	0.0009991	111.74	121.73	0.39029
320	10	993.7	0.0010063	194.78	204.84	0.65846
340	10	983.84	0.0010164	277.92	288.08	0.91079
360	10	971.85	0.001029	361.27	371.56	1.1493
380	10	957.99	0.0010439	444.93	455.37	1.3759
400	10	942.42	0.0010611	529.06	539.67	1.5921
420	10	925.19	0.0010809	613.84	624.65	1.7994
440	10	906.28	0.0011034	699.52	710.55	1.9992
460	10	885.59	0.0011292	786.38	797.67	2.1928
480	10	862.94	0.0011588	874.8	886.39	2.3816
500	10	838.02	0.0011933	965.25	977.18	2.5669
520	10	810.36	0.001234	1058.4	1070.7	2.7504
540	10	779.15	0.0012835	1155.3	1168.1	2.9341
560	10	743.0	0.0013459	1257.6	1271.1	3.1213
580	10	699.05	0.0014305	1368.8	1383.1	3.3177
584.15	10	688.42	0.0014526	1393.5	1408.1	3.3606

Table D.4C   Isobaric Data for _P_ = 15.0 MPa

_T_ (K)	_P_ (MPa)	_ρ_ (kg/m³)	_v_ (m³/kg)	_u_ (kJ/kg)	_h_ (kJ/kg)	_s_ (kJ/kg·K)
273.15	15	1007.3	0.000993	0.17746	15.069	0.00044686
280	15	1007.0	0.000993	28.581	43.476	0.10316
300	15	1003.1	0.000997	111.34	126.29	0.38886
320	15	995.83	0.001004	194.1	209.16	0.65627
340	15	985.98	0.001014	276.99	292.2	0.90797
360	15	974.05	0.001027	360.07	375.47	1.1459
380	15	960.3	0.001041	443.45	459.07	1.3719
400	15	944.88	0.001058	527.26	543.13	1.5875
420	15	927.86	0.001078	611.68	627.85	1.7942
440	15	909.22	0.0011	696.94	713.44	1.9933
460	15	888.87	0.001125	783.3	800.18	2.186
480	15	866.68	0.001154	871.1	888.4	2.3738
500	15	842.36	0.001187	960.75	978.56	2.5578
520	15	815.52	0.001226	1052.8	1071.2	2.7395
540	15	785.52	0.001273	1148.2	1167.3	2.9208
560	15	751.28	0.001331	1248.3	1268.2	3.1042
580	15	710.78	0.001407	1355.3	1376.4	3.294
600	15	659.41	0.001517	1474.9	1497.7	3.4994
615.31	15	603.52	0.001657	1585.3	1610.2	3.6846

Table D.4D   Isobaric Data for _P_ = 20.0 MPa

_T_ (K)	_P_ (MPa)	_ρ_ (kg/m³)	_v_ (m³/kg)	_u_ (kJ/kg)	_h_ (kJ/kg)	_s_ (kJ/kg·K)
273.15	20	1009.7	0.00099	0.22569	20.033	0.00046962
280	20	1009.4	0.000991	28.496	48.31	0.10271
300	20	1005.3	0.000995	110.94	130.84	0.38741
320	20	997.93	0.001002	193.44	213.48	0.65408
340	20	988.09	0.001012	276.07	296.31	0.90516
360	20	976.22	0.001024	358.89	379.38	1.1426
380	20	962.58	0.001039	442.0	462.77	1.368
400	20	947.31	0.001056	525.5	546.62	1.583
420	20	930.48	0.001075	609.58	631.08	1.7891
440	20	912.09	0.001096	694.44	716.37	1.9874
460	20	892.08	0.001121	780.32	802.74	2.1794
480	20	870.3	0.001149	867.53	890.51	2.3662
500	20	846.53	0.001181	956.44	980.07	2.5489
520	20	820.44	0.001219	1047.6	1071.9	2.7291
540	20	791.5	0.001263	1141.6	1166.9	2.9082
560	20	758.86	0.001318	1239.7	1266.0	3.0885
580	20	721.04	0.001387	1343.5	1371.3	3.2731
600	20	675.11	0.001481	1456.8	1486.4	3.4682
620	20	613.23	0.001631	1588.6	1621.2	3.6891
638.9	20	490.19	0.00204	1786.4	1827.2	4.0156

Table D.4E   Isobaric Data for $P = 30.0$ MPa

T (K)	P (MPa)	ρ (kg/m³)	v (m³/kg)	$\widehat{u}$ (kJ/kg)	h (kJ/kg)	s (kJ/kg·K)
273.15	30	1014.5	0.0009857	0.28791	29.858	0.0002688
280	30	1014.0	0.0009862	28.307	57.893	0.10164
300	30	1009.6	0.0009905	110.16	139.87	0.38444
320	30	1002.1	0.0009979	192.15	222.08	0.64972
340	30	992.24	0.0010078	274.29	304.52	0.89961
360	30	980.48	0.0010199	356.62	387.21	1.1359
380	30	967.04	0.0010341	439.19	470.21	1.3603
400	30	952.05	0.0010504	522.11	553.62	1.5742
420	30	935.59	0.0010688	605.53	637.6	1.7791
440	30	917.67	0.0010897	689.63	722.33	1.9761
460	30	898.25	0.0011133	774.62	808.02	2.1666
480	30	877.23	0.00114	860.75	894.95	2.3516
500	30	854.45	0.0011703	948.32	983.43	2.5322
520	30	829.66	0.0012053	1037.7	1073.9	2.7095
540	30	802.5	0.0012461	1129.5	1166.9	2.885
560	30	772.42	0.0012946	1224.4	1263.2	3.0601
580	30	738.56	0.001354	1323.4	1364.0	3.237
600	30	699.47	0.0014296	1428.5	1471.4	3.4189
620	30	652.41	0.0015328	1542.9	1588.9	3.6116
640	30	591.01	0.001692	1674.8	1725.5	3.8283

Table D.4F   Isobaric Data for $P = 50.0$ MPa

T (K)	P (MPa)	ρ (kg/m³)	v (m³/kg)	u (kJ/kg)	h (kJ/kg)	s (kJ/kg·K)
273.15	50	1023.8	0.000977	0.28922	49.126	−0.0010315
280	50	1022.9	0.000978	27.862	76.742	0.098824
300	50	1017.8	0.000982	108.63	157.76	0.37828
320	50	1010.1	0.00099	189.68	239.18	0.64102
340	50	1000.3	0.001	270.92	320.9	0.88875
360	50	988.72	0.001011	352.32	402.89	1.123
380	50	975.62	0.001025	433.9	485.15	1.3454
400	50	961.12	0.00104	515.76	567.78	1.5573
420	50	945.31	0.001058	597.99	650.88	1.7601
440	50	928.2	0.001077	680.73	734.6	1.9548
460	50	909.8	0.001099	764.15	819.1	2.1426
480	50	890.05	0.001124	848.42	904.59	2.3245
500	50	868.88	0.001151	933.75	991.29	2.5015
520	50	846.15	0.001182	1020.4	1079.5	2.6744
540	50	821.66	0.001217	1108.6	1169.5	2.8442
560	50	795.16	0.001258	1198.9	1261.8	3.012
580	50	766.3	0.001305	1291.7	1356.9	3.1789
600	50	734.55	0.001361	1387.6	1455.7	3.3463
620	50	699.21	0.00143	1487.7	1559.2	3.516
640	50	659.19	0.001517	1593.4	1669.3	3.6907

Table D.5   Vapor Properties: Saturated Solid (Ice)–Vapor (Sublimation Line: 200–273.16 K)*

T (K)	P (kPa)	ρ (kg/m³)	v (m³/kg)	u (kJ/kg)	h (kJ/kg)	s (kJ/kg·K)
200	0.000162	1.758E-06	568840	2273.5	2365.8	12.3800
205	0.000343	3.628E-06	275640	2280.4	2375.1	12.08
210	0.000701	7.231E-06	138290	2287.4	2384.3	11.795
215	0.001385	1.395E-05	71663	2294.3	2393.6	11.524
220	0.002653	2.613E-05	38277	2301.3	2402.8	11.267
225	0.004938	4.755E-05	21031.000	2308.2	2412.1	11.021
230	0.008947	8.429E-05	11864.000	2315.2	2421.3	10.788
235	0.015806	0.0001457	6861.300	2322.1	2430.6	10.565
240	0.027271	0.0002462	4061.300	2329.1	2439.8	10.352
245	0.046015	0.000407	2457.100	2336.0	2449.1	10.149
250	0.076029	0.000659	1517.400	2343.0	2458.3	9.9545
255	0.12316	0.0010467	955.360	2349.9	2467.6	9.7685
260	0.19583	0.0016324	612.590	2356.8	2476.8	9.5903
265	0.30594	0.0025024	399.610	2363.7	2486.0	9.4195
270	0.47008	0.0037742	264.960	2370.6	2495.1	9.2556
273.15	0.61115	0.0048508	206.15	2374.9	2500.9	9.1558
273.16	0.61166	0.0048546	205.990	2374.9	2500.9	9.1555

*Property values generated from NIST Database 23: REFPROP Version 7.0 (August 2002).

Appendix E
Various Thermodynamic Data

Table E.1 Critical Constants and Specific Heats for Selected Gases*

Substance	$\mathcal{M}$ (kg/kmol)	T_c (K)	P_c (10^5 Pa)	$\bar{v}_c$ (m³/kmol)	Z_c	c_v (kJ/kg·K)	c_p (kJ/kg·K)
Acetylene (C_2H_2)	26.04	309	62.4	0.112	0.272	1.37	1.69
Air (equivalent)	28.97	133	37.7	0.0829	0.284	0.718	1.005
Ammonia (NH_3)	17.04	406	112.8	0.0723	0.242	1.66	2.15
Benzene (C_6H_6)	78.11	562	48.3	0.256	0.274	0.67	0.775
n-Butane (C_4H_{10})	58.12	425.2	37.9	0.257	0.274	1.56	1.71
Carbon dioxide (CO_2)	44.01	304.2	73.9	0.0941	0.276	0.657	0.846
Carbon monoxide (CO)	28.01	133	35.0	0.0928	0.294	0.744	1.04
Refrigerant 134a ($C_2F_4H_2$)	102.03	374.3	40.6	0.200	0.262	0.76	0.85
Ethane (C_2H_6)	30.07	305.4	48.8	0.148	0.285	1.48	1.75
Ethylene (C_2H_4)	28.05	283	51.2	0.128	0.279	1.23	1.53
Helium (He)	4.003	5.2	2.3	0.0579	0.300	3.12	5.19
Hydrogen (H_2)	2.016	33.2	13.0	0.0648	0.304	10.2	14.3
Methane (CH_4)	16.04	190.7	46.4	0.0991	0.290	1.70	2.22
Nitrogen (N_2)	28.01	126.2	33.9	0.0897	0.291	0.743	1.04
Oxygen (O_2)	32.00	154.4	50.5	0.0741	0.290	0.658	0.918
Propane (C_3H_8)	44.09	370	42.5	0.200	0.278	1.48	1.67
Sulfur dioxide (SO_2)	64.06	431	78.7	0.124	0.268	0.471	0.601
Water (H_2O)	18.02	647.1	220.6	0.0558	0.230	1.40	1.86

*Adapted from Wark, K., Jr., and Richards, D. E., *Thermodynamics,* 6th ed., McGraw-Hill, New York, 1999.

Table E.2 Van der Waals Constants for Selected Gases*

Substance	a [10^5 Pa·(m³/kmol)²]	b (m³/kmol)	Substance	a [10^5 Pa·(m³/kmol)²]	b (m³/kmol)
Acetylene (C_2H_2)	4.410	0.0510	Ethylene (C_2H_4)	4.563	0.0574
Air (equivalent)	1.358	0.0364	Helium (He)	0.0341	0.0234
Ammonia (NH_3)	4.223	0.0373	Hydrogen (H_2)	0.247	0.0265
Benzene (C_6H_6)	18.63	0.1181	Methane (CH_4)	2.285	0.0427
n-Butane (C_4H_{10})	13.80	0.1196	Nitrogen (N_2)	1.361	0.0385
Carbon dioxide (CO_2)	3.643	0.0427	Oxygen (O_2)	1.369	0.0315
Carbon monoxide (CO)	1.463	0.0394	Propane (C_3H_8)	9.315	0.0900
Refrigerant 134a ($C_2F_4H_2$)	10.05	0.0957	Sulfur dioxide (SO_2)	6.837	0.0568
Ethane (C_2H_6)	5.575	0.0650	Water (H_2O)	5.507	0.0304

*Adapted from Wark, K., Jr., and Richards, D. E., *Thermodynamics,* 6th ed., McGraw-Hill, New York, 1999.

Appendix F
Thermo-Physical Properties of Selected Gases at 1 atm

Table F.1A   Ammonia (NH$_3$)*

T (K)	ρ (kg/m^3)	c_p (kJ/kg·K)	μ (μPa·s)	ν (m^2/s)	k (W/m·K)	α (m^2/s)	Pr
239.824	0.8895	2.297	8.054	9.054E-06	0.0210	1.026E-05	0.8822
240	0.8888	2.296	8.059	9.068E-06	0.0210	1.028E-05	0.8820
260	0.8135	2.207	8.734	1.074E-05	0.0220	1.228E-05	0.8744
280	0.7515	2.172	9.436	1.256E-05	0.0234	1.435E-05	0.8748
300	0.6990	2.165	10.160	1.454E-05	0.0251	1.659E-05	0.8762
320	0.6538	2.174	10.902	1.668E-05	0.0271	1.904E-05	0.8759
340	0.6143	2.193	11.657	1.898E-05	0.0293	2.173E-05	0.8734
360	0.5795	2.219	12.422	2.144E-05	0.0317	2.467E-05	0.8691
380	0.5485	2.249	13.195	2.406E-05	0.0344	2.786E-05	0.8634
400	0.5207	2.283	13.971	2.683E-05	0.0372	3.130E-05	0.8572
420	0.4956	2.320	14.751	2.976E-05	0.0402	3.498E-05	0.8510
440	0.4728	2.358	15.531	3.285E-05	0.0433	3.886E-05	0.8453
460	0.4521	2.399	16.310	3.607E-05	0.0465	4.292E-05	0.8405
480	0.4331	2.440	17.088	3.945E-05	0.0498	4.714E-05	0.8369
500	0.4157	2.483	17.863	4.297E-05	0.0531	5.147E-05	0.8349
520	0.3996	2.526	18.635	4.663E-05	0.0564	5.587E-05	0.8346
540	0.3848	2.570	19.403	5.043E-05	0.0596	6.030E-05	0.8362
560	0.3710	2.615	20.167	5.436E-05	0.0628	6.471E-05	0.8401
580	0.3581	2.660	20.927	5.843E-05	0.0658	6.904E-05	0.8463
600	0.3462	2.706	21.682	6.264E-05	0.0686	7.324E-05	0.8552

*Property values generated from NIST Database 23: REFPROP Version 7.0 (August 2002).

Table F.1B   Carbon Dioxide (CO_2)*

T (K)	ρ (kg/m³)	c_p (kJ/kg·K)	μ (μPa·s)	ν (m²/s)	k (W/m·K)	α (m²/s)	Pr
220	2.472	0.781	11.06	4.475E-06	0.0109	5.647E-06	0.792
240	2.258	0.796	12.07	5.344E-06	0.0122	6.808E-06	0.785
260	2.079	0.814	13.06	6.282E-06	0.0137	8.079E-06	0.778
280	1.927	0.833	14.05	7.288E-06	0.0152	9.465E-06	0.770
300	1.797	0.853	15.02	8.361E-06	0.0168	1.096E-05	0.763
320	1.683	0.872	15.98	9.499E-06	0.0184	1.256E-05	0.756
340	1.583	0.890	16.93	1.070E-05	0.0201	1.426E-05	0.750
360	1.494	0.908	17.87	1.196E-05	0.0218	1.605E-05	0.745
380	1.415	0.925	18.79	1.328E-05	0.0235	1.792E-05	0.741
400	1.343	0.942	19.70	1.466E-05	0.0251	1.988E-05	0.738
420	1.279	0.958	20.59	1.610E-05	0.0268	2.190E-05	0.735
440	1.220	0.973	21.47	1.759E-05	0.0285	2.401E-05	0.733
460	1.167	0.988	22.33	1.913E-05	0.0302	2.618E-05	0.731
480	1.118	1.002	23.18	2.073E-05	0.0318	2.842E-05	0.729
500	1.073	1.015	24.02	2.237E-05	0.0335	3.072E-05	0.728
520	1.032	1.029	24.84	2.407E-05	0.0351	3.309E-05	0.727
540	0.994	1.041	25.65	2.581E-05	0.0368	3.553E-05	0.727
560	0.958	1.053	26.44	2.760E-05	0.0384	3.802E-05	0.726
580	0.925	1.065	27.23	2.944E-05	0.0400	4.057E-05	0.726
600	0.894	1.076	28.00	3.131E-05	0.0416	4.318E-05	0.725
620	0.865	1.087	28.76	3.324E-05	0.0431	4.585E-05	0.725
640	0.838	1.098	29.50	3.520E-05	0.0447	4.858E-05	0.725
660	0.813	1.108	30.24	3.721E-05	0.0462	5.136E-05	0.724
680	0.789	1.117	30.96	3.925E-05	0.0478	5.420E-05	0.724
700	0.766	1.127	31.68	4.134E-05	0.0493	5.709E-05	0.724
720	0.745	1.136	32.38	4.347E-05	0.0508	6.004E-05	0.724
740	0.725	1.145	33.07	4.563E-05	0.0523	6.304E-05	0.724
760	0.706	1.153	33.75	4.783E-05	0.0538	6.609E-05	0.724
780	0.688	1.161	34.43	5.007E-05	0.0553	6.920E-05	0.724
800	0.670	1.169	35.09	5.234E-05	0.0567	7.236E-05	0.723

*Property values generated from NIST Database 23: REFPROP Version 7.0 (August 2002).

Table F.1C   Carbon Monoxide (CO)*

T (K)	ρ (kg/m^3)	c_p (kJ/kg·K)	μ (µPa·s)	ν (m^2/s)	k (W/m·K)	α (m^2/s)	Pr
200	1.7112	1.0443	12.8977	7.537E-06	0.01923	1.076E-05	0.701
220	1.5544	1.0430	13.9377	8.967E-06	0.02080	1.283E-05	0.699
240	1.4241	1.0420	14.9369	1.049E-05	0.02232	1.504E-05	0.697
260	1.3140	1.0412	15.8998	1.210E-05	0.02378	1.738E-05	0.696
280	1.2198	1.0407	16.8304	1.380E-05	0.02520	1.985E-05	0.695
300	1.1382	1.0402	17.7315	1.558E-05	0.02656	2.243E-05	0.694
320	1.0669	1.0399	18.6057	1.744E-05	0.02789	2.514E-05	0.694
340	1.0040	1.0396	19.4549	1.938E-05	0.02918	2.795E-05	0.693
360	0.9481	1.0394	20.2811	2.139E-05	0.03043	3.088E-05	0.693
380	0.8982	1.0393	21.0859	2.348E-05	0.03165	3.391E-05	0.692
400	0.8532	1.0392	21.8707	2.563E-05	0.03285	3.705E-05	0.692
450	0.7583	1.0392	23.7543	3.133E-05	0.03571	4.532E-05	0.691
500	0.6824	1.0395	25.5404	3.743E-05	0.03844	5.419E-05	0.691
550	0.6204	1.0400	27.2444	4.392E-05	0.04105	6.363E-05	0.690
600	0.5687	1.0408	28.8791	5.078E-05	0.04357	7.362E-05	0.690
650	0.5249	1.0418	30.4548	5.802E-05	0.04601	8.414E-05	0.690
700	0.4874	1.0429	31.9798	6.561E-05	0.04838	9.518E-05	0.689
750	0.4549	1.0443	33.4606	7.355E-05	0.05070	1.067E-04	0.689
800	0.4265	1.0457	34.9026	8.184E-05	0.05297	1.188E-04	0.689

*Property values generated from NIST Database 12: NIST Pure Fluids Version 5.0 (September 2002).

Table F.1D   Helium (He)*

T (K)	ρ (kg/m³)	c_p (kJ/kg·K)	μ (μPa·s)	ν (m²/s)	k (W/m·K)	α (m²/s)	Pr
100	0.487	5.149	9.78	2.008E-05	0.0737	2.914E-05	0.689
120	0.4060	5.1938	10.79	2.659E-05	0.0833	3.952E-05	0.673
140	0.3481	5.1935	11.94	3.432E-05	0.0925	5.117E-05	0.671
160	0.3046	5.1933	13.05	4.284E-05	0.1013	6.404E-05	0.669
180	0.2708	5.1932	14.11	5.212E-05	0.1098	7.806E-05	0.668
200	0.2437	5.1931	15.14	6.213E-05	0.1180	9.322E-05	0.667
220	0.2216	5.1931	16.14	7.286E-05	0.1260	1.095E-04	0.666
240	0.2031	5.1930	17.12	8.429E-05	0.1337	1.268E-04	0.665
260	0.1875	5.1930	18.08	9.641E-05	0.1413	1.451E-04	0.664
280	0.1741	5.1930	19.01	1.092E-04	0.1487	1.645E-04	0.664
300	0.1625	5.1930	19.93	1.226E-04	0.1560	1.848E-04	0.664
320	0.1524	5.1930	20.83	1.367E-04	0.1631	2.061E-04	0.663
340	0.1434	5.1930	21.72	1.514E-04	0.1701	2.284E-04	0.663
360	0.1354	5.1930	22.59	1.668E-04	0.1770	2.516E-04	0.663
380	0.1283	5.1930	23.45	1.827E-04	0.1837	2.757E-04	0.663
400	0.1219	5.1930	24.29	1.993E-04	0.1904	3.007E-04	0.663
420	0.1161	5.1930	25.13	2.164E-04	0.1969	3.266E-04	0.663
440	0.1108	5.1930	25.95	2.342E-04	0.2034	3.534E-04	0.663
460	0.1060	5.1930	26.76	2.525E-04	0.2098	3.811E-04	0.663
480	0.1016	5.1930	27.57	2.714E-04	0.2161	4.096E-04	0.663
500	0.0975	5.1930	28.36	2.908E-04	0.2223	4.389E-04	0.663
550	0.0887	5.1930	30.31	3.419E-04	0.2376	5.159E-04	0.663
600	0.0813	5.1930	32.22	3.963E-04	0.2524	5.980E-04	0.663
650	0.0750	5.1930	34.07	4.541E-04	0.2669	6.850E-04	0.663
700	0.0697	5.1930	35.89	5.152E-04	0.2811	7.768E-04	0.663
750	0.0650	5.1930	37.68	5.794E-04	0.2949	8.733E-04	0.663
800	0.0610	5.1930	39.43	6.468E-04	0.3085	9.745E-04	0.664
850	0.0574	5.1930	41.15	7.172E-04	0.3219	1.080E-03	0.664
900	0.0542	5.1930	42.85	7.907E-04	0.3350	1.190E-03	0.664
950	0.0513	5.1930	44.52	8.671E-04	0.3479	1.305E-03	0.664
1000	0.0488	5.1930	46.16	9.464E-04	0.3606	1.424E-03	0.665

*Property values generated from NIST Database 12: NIST Pure Fluids Version 5.0 (September 2002).

Table F.1E   Hydrogen (H₂)*

T (K)	ρ (kg/m³)	c_p (kJ/kg·K)	μ (μPa·s)	ν (m²/s)	k (W/m·K)	α (m²/s)	Pr
100	0.2457	11.23	4.190	1.705E-05	0.0683	2.477E-05	0.688
150	0.1637	12.61	5.561	3.397E-05	0.1010	4.894E-05	0.694
200	0.1228	13.54	6.780	5.523E-05	0.1324	7.970E-05	0.693
250	0.09820	14.05	7.903	8.047E-05	0.1606	1.164E-04	0.691
300	0.08184	14.31	8.953	1.094E-04	0.1858	1.586E-04	0.690
350	0.07016	14.43	9.946	1.418E-04	0.2103	2.077E-04	0.682
400	0.06139	14.47	10.89	1.774E-04	0.2341	2.634E-04	0.674
450	0.05457	14.49	11.80	2.162E-04	0.2570	3.249E-04	0.665
500	0.04912	14.51	12.67	2.580E-04	0.2805	3.936E-04	0.656
550	0.04465	14.53	13.52	3.027E-04	0.3042	4.689E-04	0.646
600	0.04093	14.54	14.34	3.503E-04	0.3281	5.514E-04	0.635

*Property values generated from NIST Database 12: NIST Pure Fluids Version 5.0 (September 2000).

Table F.1F   Nitrogen (N₂)*

T (K)	ρ (kg/m³)	c_p (kJ/kg·K)	μ (μPa·s)	ν (m²/s)	k (W/m·K)	α (m²/s)	Pr
100	3.4831	1.0718	6.97	2.000E-06	0.0099	2.644E-06	0.756
150	2.2893	1.0486	10.10	4.410E-06	0.0145	6.061E-06	0.728
200	1.7107	1.0435	12.92	7.555E-06	0.0187	1.045E-05	0.723
250	1.3666	1.0418	15.51	1.135E-05	0.0224	1.573E-05	0.721
300	1.1382	1.0414	17.90	1.572E-05	0.0259	2.182E-05	0.721
350	0.9753	1.0423	20.12	2.063E-05	0.0291	2.863E-05	0.721
400	0.8532	1.0450	22.22	2.604E-05	0.0322	3.612E-05	0.721
450	0.7584	1.0497	24.20	3.191E-05	0.0352	4.423E-05	0.721
500	0.6825	1.0564	26.08	3.821E-05	0.0381	5.290E-05	0.722
550	0.6204	1.0650	27.88	4.493E-05	0.0410	6.211E-05	0.723
600	0.5687	1.0751	29.60	5.205E-05	0.0439	7.182E-05	0.725
700	0.4875	1.0981	32.87	6.742E-05	0.0496	9.267E-05	0.728
800	0.4266	1.1223	35.93	8.424E-05	0.0552	1.153E-04	0.731
900	0.3792	1.1457	38.83	1.024E-04	0.0607	1.396E-04	0.733
1000	0.3413	1.1674	41.60	1.219E-04	0.0660	1.656E-04	0.736
1100	0.3103	1.1868	44.25	1.426E-04	0.0712	1.933E-04	0.738
1200	0.2844	1.2040	46.81	1.646E-04	0.0762	2.224E-04	0.740
1300	0.2625	1.2191	49.29	1.878E-04	0.0810	2.532E-04	0.742
1400	0.2438	1.2324	51.70	2.121E-04	0.0858	2.855E-04	0.743
1500	0.2275	1.2439	54.06	2.376E-04	0.0904	3.193E-04	0.744
1600	0.2133	1.2541	56.36	2.642E-04	0.0948	3.545E-04	0.745
1700	0.2008	1.2630	58.61	2.919E-04	0.0992	3.913E-04	0.746
1800	0.1896	1.2708	60.83	3.208E-04	0.1035	4.295E-04	0.747
1900	0.1796	1.2778	63.01	3.508E-04	0.1077	4.692E-04	0.748
2000	0.1707	1.2841	65.16	3.818E-04	0.1118	5.104E-04	0.748

*Property values generated from NIST Database 23: REFPROP Version 7.0 (August 2002).

Table F.1G   Oxygen (O₂)*

T (K)	ρ (kg/m³)	c_p (kJ/kg·K)	μ (µPa·s)	ν (m²/s)	k (W/m·K)	α (m²/s)	Pr
100	3.995	0.9356	7.74835	1.940E-06	0.00931	2.491E-06	0.779
150	2.619	0.9198	11.34	4.329E-06	0.01399	5.807E-06	0.745
200	1.956	0.9146	14.65	7.491E-06	0.01840	1.029E-05	0.728
250	1.562	0.9150	17.71	1.134E-05	0.02260	1.581E-05	0.717
300	1.301	0.9199	20.56	1.581E-05	0.02666	2.228E-05	0.710
350	1.114	0.9291	23.23	2.085E-05	0.03067	2.962E-05	0.704
400	0.9749	0.9417	25.75	2.641E-05	0.03469	3.779E-05	0.699
450	0.8665	0.9564	28.14	3.247E-05	0.03872	4.672E-05	0.695
500	0.7798	0.9722	30.41	3.900E-05	0.04275	5.639E-05	0.692
550	0.7089	0.9880	32.58	4.597E-05	0.04676	6.677E-05	0.688
600	0.6498	1.003	34.67	5.336E-05	0.05074	7.783E-05	0.686
700	0.5569	1.031	38.62	6.935E-05	0.05850	1.019E-04	0.681
800	0.4873	1.054	42.32	8.685E-05	0.06593	1.283E-04	0.677
900	0.4332	1.074	45.82	1.058E-04	0.07300	1.569E-04	0.674
1000	0.3899	1.090	49.15	1.261E-04	0.07968	1.875E-04	0.672
1100	0.3544	1.103	52.33	1.477E-04	0.08601	2.200E-04	0.671
1200	0.3249	1.114	55.40	1.705E-04	0.09198	2.541E-04	0.671
1300	0.2999	1.123	58.36	1.946E-04	0.09762	2.897E-04	0.672
1400	0.2785	1.131	61.23	2.199E-04	0.1030	3.267E-04	0.673
1500	0.2599	1.138	64.02	2.463E-04	0.1080	3.650E-04	0.675

*Property values generated from NIST Database 23: REFPROP Version 7.0 (August 2002).

Table F.1H   Water Vapor (H₂O)*

T (K)	ρ (kg/m³)	c_p (kJ/kg·K)	μ (µPa·s)	ν (m²/s)	k (W/m·K)	α (m²/s)	Pr
373.124	0.5977	2.080	12.27	2.053E-05	0.02509	2.019E-05	1.017
400	0.5549	2.009	13.28	2.394E-05	0.02702	2.423E-05	0.988
450	0.4910	1.976	15.25	3.105E-05	0.03117	3.213E-05	0.966
500	0.4409	1.982	17.27	3.917E-05	0.03586	4.105E-05	0.954
550	0.4003	2.001	19.33	4.828E-05	0.04096	5.112E-05	0.944
600	0.3667	2.027	21.41	5.838E-05	0.04637	6.239E-05	0.936
650	0.3383	2.056	23.49	6.944E-05	0.05205	7.485E-05	0.928
700	0.3140	2.087	25.56	8.142E-05	0.05796	8.847E-05	0.920
750	0.2930	2.119	27.63	9.429E-05	0.06408	1.032E-04	0.914
800	0.2746	2.153	29.67	1.080E-04	0.07039	1.191E-04	0.907
850	0.2584	2.187	31.69	1.226E-04	0.07685	1.360E-04	0.902
900	0.2440	2.222	33.69	1.380E-04	0.08347	1.539E-04	0.897
950	0.2312	2.257	35.65	1.542E-04	0.09022	1.729E-04	0.892
1000	0.2196	2.292	37.59	1.712E-04	0.09709	1.929E-04	0.888

*Property values generated from NIST Database 23: REFPROP Version 7.0 (August 2002).

Appendix G
Thermo-Physical Properties of Selected Liquids

Table G.1 Thermo-Physical Properties of Saturated Water*

Temperature (K)	Pressure (MPa)	Liquid Density (kg/m³)	Vapor Density (kg/m³)	Liqid c_p (kJ/kg·K)	Vapor c_p (kJ/kg·K)	Liquid Viscosity (μPa·s)	Vapor Viscosity (μPa·s)	Liquid Therm. Cond. (W/m·K)	Vapor Therm. Cond. (W/m·K)	Liquid Prandtl	Vapor Prandtl	Surface Tension (N/m)	Liquid Expansion Coef. β (1/K)	Vapor Expansion Coef. β (1/K)	T (K)
273.16	0.000612	999.79	0.0048546	4.2199	1.8844	1791.2	9.2163	0.56104	0.017071	13.472	1.0173	0.075646	-0.000067965	0.0036807	273.16
280	0.000992	999.86	0.0076812	4.2014	1.8913	1433.7	9.3815	0.57404	0.017442	10.493	1.0173	0.074677	0.000043569	0.0035962	280
285	0.001389	999.47	0.010571	4.1927	1.8967	1239.3	9.509	0.58348	0.017729	8.9052	1.0173	0.073951	0.00011191	0.0035375	285
290	0.00192	998.76	0.014363	4.1869	1.9023	1084	9.6414	0.59273	0.018031	7.6573	1.0172	0.07321	0.0001721	0.0034814	290
295	0.002621	997.76	0.019281	4.1832	1.9081	957.87	9.7784	0.60169	0.018345	6.6594	1.017	0.072455	0.00022593	0.0034277	295
300	0.003537	996.51	0.02559	4.1809	1.9141	853.84	9.9195	0.61028	0.018673	5.8495	1.0168	0.071686	0.00027471	0.0033765	300
305	0.004719	995.03	0.033598	4.1798	1.9204	766.95	10.064	0.61841	0.019014	5.1837	1.0165	0.070903	0.00031942	0.0033276	305
310	0.006231	993.34	0.043663	4.1795	1.927	693.54	10.213	0.62605	0.019369	4.6301	1.0161	0.070106	0.00036081	0.0032811	310
315	0.008145	991.46	0.056195	4.1798	1.9341	630.91	10.364	0.63315	0.019736	4.1651	1.0156	0.069295	0.00039947	0.0032369	315
320	0.010546	989.39	0.071662	4.1807	1.9417	577.02	10.518	0.63971	0.020117	3.7711	1.0152	0.06847	0.00043586	0.0031951	320
325	0.013531	987.15	0.09059	4.1821	1.9499	530.29	10.675	0.64571	0.020512	3.4346	1.0147	0.067632	0.00047035	0.0031557	325
330	0.017213	984.75	0.11357	4.1838	1.9587	489.49	10.833	0.65118	0.020922	3.145	1.0143	0.066781	0.00050326	0.0031186	330
335	0.021718	982.2	0.14127	4.186	1.9684	453.64	10.994	0.65611	0.021345	2.8942	1.0139	0.065917	0.00053484	0.0030839	335
340	0.027188	979.5	0.1744	4.1885	1.979	421.97	11.157	0.66055	0.021784	2.6757	1.0136	0.06504	0.00056531	0.0030516	340
345	0.033783	976.67	0.21378	4.1913	1.9906	393.85	11.321	0.6645	0.022238	2.4842	1.0135	0.06415	0.00059485	0.0030218	345
350	0.041682	973.7	0.26029	4.1946	2.0033	368.77	11.487	0.668	0.022707	2.3156	1.0135	0.063248	0.00062362	0.0029945	350
355	0.05108	970.61	0.31487	4.1983	2.0173	346.3	11.654	0.67108	0.023193	2.1665	1.0137	0.062333	0.00065178	0.0029697	355
360	0.062194	967.39	0.37858	4.2024	2.0326	326.1	11.823	0.67376	0.023695	2.034	1.0142	0.061406	0.00067944	0.0029476	360
365	0.07526	964.05	0.45253	4.207	2.0493	307.87	11.992	0.67606	0.024213	1.9158	1.0149	0.060467	0.00070671	0.0029281	365
370	0.090535	960.59	0.53792	4.2122	2.0676	291.36	12.162	0.67802	0.02475	1.81	1.016	0.059517	0.00073371	0.0029114	370
373.15	0.10142	958.35	0.59817	4.2157	2.08	281.74	12.269	0.67909	0.025096	1.749	1.0169	0.058912	0.00075062	0.0029023	373.15
375	0.1083	957.01	0.63605	4.2178	2.0877	276.36	12.332	0.67966	0.025303	1.715	1.0175	0.058555	0.00076053	0.0028975	375
380	0.12885	953.33	0.7483	4.2241	2.1096	262.69	12.504	0.681	0.025875	1.6294	1.0194	0.057581	0.00078726	0.0028865	380
385	0.15252	949.53	0.87615	4.2309	2.1334	250.21	12.675	0.68205	0.026465	1.5521	1.0218	0.056596	0.00081398	0.0028786	385
390	0.17964	945.62	1.0212	4.2384	2.1594	238.77	12.848	0.68283	0.027074	1.4821	1.0247	0.055601	0.00084079	0.0028737	390

(continued)

Table G.1 (continued)

Temperature (K)	Pressure (MPa)	Liquid Density (kg/m³)	Vapor Density (kg/m³)	Liqid c_p (kJ/kg·K)	Vapor c_p (kJ/kg·K)	Liquid Viscosity (μPa·s)	Vapor Viscosity (μPa·s)	Liquid Therm. Cond. (W/m·K)	Vapor Therm. Cond. (W/m·K)	Liquid Prandtl	Vapor Prandtl	Surface Tension (N/m)	Liquid Expansion Coef. β (1/K)	Vapor Expansion Coef. β (1/K)	T (K)
395	0.2106	941.61	1.185	4.2466	2.1877	228.27	13.02	0.68335	0.027701	1.4185	1.0282	0.054595	0.00086777	0.0028719	395
400	0.24577	937.49	1.3694	4.2555	2.2183	218.8	13.192	0.68364	0.028347	1.3607	1.0324	0.053578	0.00089499	0.0028735	400
405	0.28558	933.26	1.5763	4.2652	2.2514	209.68	13.365	0.68369	0.029013	1.308	1.0371	0.052551	0.00092254	0.0028784	405
410	0.33045	928.92	1.8076	4.2756	2.2871	201.43	13.538	0.68352	0.029699	1.26	1.0425	0.051514	0.0009505	0.0028868	410
415	0.38087	924.48	2.0654	4.2868	2.3254	193.78	13.711	0.68313	0.030404	1.216	1.0487	0.050468	0.00097895	0.0028987	415
420	0.4373	919.93	2.3518	4.299	2.3666	186.68	13.883	0.68253	0.031128	1.1758	1.0555	0.049411	0.001008	0.0029142	420
425	0.50025	915.27	2.6693	4.312	2.4105	180.07	14.056	0.68172	0.031873	1.139	1.063	0.048346	0.0010377	0.0029334	425
430	0.57026	910.51	3.0202	4.326	2.4573	173.91	14.228	0.6807	0.032638	1.1052	1.0712	0.047272	0.0010681	0.0029564	430
435	0.64787	905.63	3.4072	4.341	2.507	168.16	14.401	0.67948	0.033424	1.0743	1.0801	0.046189	0.0010994	0.0029834	435
440	0.73367	900.65	3.8329	4.3571	2.5597	162.77	14.573	0.67805	0.03423	1.0459	1.0898	0.045098	0.0011317	0.0030143	440
445	0.82824	895.55	4.3001	4.3743	2.6154	157.72	14.745	0.67642	0.035056	1.02	1.1001	0.043999	0.001165	0.0030495	445
450	0.9322	890.34	4.812	4.3927	2.6742	152.98	14.917	0.67459	0.035904	0.99615	1.1111	0.042891	0.0011995	0.0030889	450
455	1.0462	885.01	5.3717	4.4124	2.7362	148.52	15.089	0.67254	0.036773	0.97438	1.1228	0.041777	0.0012354	0.0031328	455
460	1.1709	879.57	5.9826	4.4334	2.8014	144.31	15.261	0.67028	0.037663	0.9545	1.1352	0.040655	0.0012727	0.0031814	460
465	1.3069	874	6.6484	4.4559	2.8701	140.34	15.434	0.66781	0.038576	0.93639	1.1483	0.039527	0.0013116	0.003235	465
470	1.4551	868.31	7.3727	4.4799	2.9422	136.58	15.606	0.66512	0.039512	0.91994	1.1621	0.038392	0.0013522	0.0032937	470
475	1.616	862.49	8.1598	4.5055	3.0181	133.02	15.779	0.66221	0.040471	0.90506	1.1767	0.037252	0.0013948	0.003358	475
480	1.7905	856.54	9.0139	4.533	3.0979	129.64	15.952	0.65907	0.041455	0.89167	1.1921	0.036105	0.0014396	0.0034282	480
485	1.9792	850.45	9.9397	4.5623	3.1819	126.43	16.126	0.65569	0.042464	0.87972	1.2083	0.034954	0.0014867	0.0035048	485
490	2.1831	844.22	10.942	4.5937	3.2705	123.37	16.3	0.65206	0.043502	0.86914	1.2255	0.033797	0.0015365	0.0035882	490
495	2.4028	837.84	12.027	4.6274	3.3641	120.45	16.476	0.64819	0.044568	0.85989	1.2436	0.032637	0.0015891	0.0036791	495
500	2.6392	831.31	13.199	4.6635	3.4631	117.66	16.653	0.64405	0.045666	0.85195	1.2628	0.031472	0.001645	0.0037781	500
505	2.8931	824.63	14.465	4.7022	3.568	114.98	16.831	0.63964	0.046799	0.84529	1.2832	0.030304	0.0017045	0.0038861	505
510	3.1655	817.77	15.833	4.744	3.6796	112.42	17.011	0.63495	0.047969	0.83991	1.3049	0.029133	0.001768	0.004004	510
515	3.4571	810.74	17.308	4.7889	3.7986	109.95	17.193	0.62997	0.049182	0.83581	1.3279	0.027959	0.001836	0.0041328	515
520	3.769	803.53	18.9	4.8375	3.9257	107.57	17.377	0.62468	0.050442	0.83301	1.3524	0.026784	0.001909	0.0042738	520
525	4.1019	796.13	20.617	4.8901	4.0622	105.27	17.564	0.61908	0.051756	0.83154	1.3786	0.025608	0.0019876	0.0044285	525
530	4.4569	788.53	22.47	4.9471	4.209	103.05	17.755	0.61315	0.05313	0.83143	1.4066	0.02443	0.0020727	0.0045986	530
535	4.8349	780.71	24.469	5.0092	4.3677	100.89	17.949	0.60688	0.054575	0.83275	1.4365	0.023253	0.0021652	0.0047862	535
540	5.2369	772.66	26.627	5.077	4.54	98.792	18.149	0.60026	0.056102	0.83558	1.4688	0.022077	0.0022659	0.0049936	540
545	5.664	764.36	28.956	5.1513	4.7277	96.746	18.353	0.59329	0.057723	0.84001	1.5031	0.020902	0.0023764	0.0052239	545
550	6.1172	755.81	31.474	5.2331	4.9332	94.746	18.563	0.58595	0.059456	0.84616	1.5402	0.01973	0.002498	0.0054805	550
555	6.5976	746.97	34.198	5.3235	5.1594	92.785	18.781	0.57826	0.061321	0.85418	1.5802	0.018561	0.0026326	0.0057678	555
560	7.1062	737.83	37.147	5.4239	5.4099	90.857	19.007	0.57021	0.063341	0.86425	1.6234	0.017396	0.0027826	0.0060909	560
565	7.6444	728.36	40.346	5.5361	5.6889	88.956	19.242	0.56181	0.065549	0.87658	1.67	0.016236	0.0029508	0.0064566	565

570	8.2132	718.53	43.822	5.6624	6.002	87.074	19.489	0.55308	0.067981	0.89146	1.7207	0.015082	0.0031409	0.0068731	570
575	8.814	708.3	47.607	5.8055	6.356	85.206	19.749	0.54405	0.070685	0.90923	1.7758	0.013937	0.0033575	0.0073509	575
580	9.448	697.64	51.739	5.9691	6.7598	83.342	20.024	0.53474	0.073721	0.93033	1.8361	0.0128	0.0036067	0.007904	580
585	10.117	686.48	56.265	6.1579	7.2252	81.477	20.318	0.52519	0.077163	0.95534	1.9025	0.011673	0.0038965	0.0085503	585
590	10.821	674.78	61.242	6.3784	7.7679	79.6	20.634	0.51543	0.081108	0.98504	1.9762	0.010559	0.0042377	0.0093144	590
595	11.563	662.45	66.738	6.6393	8.4096	77.703	20.976	0.50551	0.085682	1.0205	2.0588	0.0094591	0.0046454	0.01023	595
600	12.345	649.41	72.842	6.9532	9.1809	75.773	21.35	0.49546	0.091052	1.0634	2.1528	0.0083756	0.0051415	0.011345	600
605	13.167	635.53	79.669	7.3391	10.127	73.798	21.765	0.4853	0.097442	1.116	2.2619	0.0073112	0.0057587	0.012731	605
610	14.033	620.65	87.369	7.8268	11.315	71.759	22.229	0.47503	0.10517	1.1823	2.3917	0.006269	0.0065497	0.014496	610
615	14.943	604.55	96.15	8.4674	12.857	69.632	22.759	0.46465	0.11468	1.2689	2.5517	0.0052528	0.0076048	0.016815	615
620	15.901	586.88	106.31	9.3541	14.945	67.382	23.374	0.4541	0.12666	1.388	2.758	0.0042676	0.0090924	0.019997	620
625	16.908	567.09	118.29	10.673	17.944	64.951	24.109	0.44338	0.14223	1.5635	3.0417	0.0033194	0.011355	0.024628	625
630	17.969	544.25	132.84	12.827	22.658	62.244	25.018	0.43251	0.16344	1.8459	3.4683	0.0024169	0.015162	0.032006	630
635	19.086	516.71	151.35	16.795	31.271	59.101	26.208	0.42189	0.19479	2.3528	4.2073	0.0015728	0.022437	0.045668	635
640	20.265	481.53	177.15	25.942	52.586	55.247	27.938	0.41493	0.25001	3.4542	5.8764	0.00080882	0.039706	0.07995	640
645	21.515	425.05	224.45	93.35	191.32	49.357	31.348	0.46136	0.42459	9.9868	14.125	0.00017569	0.171	0.30797	645
647.1	22.064	322	322	—	—	39.43	39.43	0.19748	0.19748	—	—	0	—	—	647.1

*Property values generated from NIST Database 23: REFPROP Version 7.0 (August 2002).

Table G2.A   R-134a (1,1,1,2-Tetrafluoroethane)—Saturated*

T (K)	ρ (kg/m^3)	c_p (kJ/kg·K)	μ (μPa·s)	ν (m^2/s)	k (W/m·K)	α (m^2/s)	Pr	β (1/K)
170	1590.7	1.1838	2139.7	1.345E-06	0.14515	7.708E-08	17.45	0.001658
180	1564.2	1.1871	1479.1	9.456E-07	0.1391	7.492E-08	12.62	0.0017
190	1537.5	1.1950	1106.2	7.195E-07	0.1333	7.257E-08	9.91	0.001748
200	1510.5	1.2058	867.3	5.742E-07	0.1277	7.014E-08	8.19	0.001802
210	1483.1	1.2186	702.3	4.735E-07	0.1224	6.771E-08	6.99	0.001864
220	1455.2	1.2332	582.2	4.001E-07	0.1172	6.529E-08	6.13	0.001934
230	1426.8	1.2492	491.2	3.443E-07	0.1121	6.292E-08	5.47	0.002017
240	1397.7	1.2669	420.2	3.006E-07	0.1073	6.058E-08	4.96	0.002113
250	1367.9	1.2865	363.3	2.656E-07	0.1025	5.827E-08	4.56	0.002226
260	1337.1	1.3082	316.6	2.368E-07	0.0979	5.598E-08	4.23	0.002361
270	1305.1	1.3326	277.5	2.127E-07	0.0934	5.371E-08	3.96	0.002525
280	1271.8	1.3606	244.3	1.927E-07	0.0890	5.143E-08	3.74	0.002726
290	1236.8	1.3933	215.6	1.744E-07	0.0846	4.912E-08	3.55	0.002979
300	1199.7	1.4324	190.5	1.588E-07	0.0803	4.675E-08	3.40	0.003303
310	1159.9	1.4807	168.0	1.449E-07	0.0761	4.429E-08	3.27	0.003733
320	1116.8	1.5426	147.8	1.323E-07	0.0718	4.167E-08	3.18	0.004326
330	1069.1	1.6267	129.2	1.209E-07	0.0675	3.879E-08	3.12	0.005191
340	1015.0	1.7507	111.8	1.102E-07	0.0631	3.549E-08	3.10	0.006567
350	951.32	1.9614	95.1	9.996E-08	0.0586	3.14E-08	3.18	0.009102
360	870.11	2.4368	78.1	8.981E-08	0.0541	2.55E-08	3.52	0.015393
370	740.32	5.1048	58.0	7.829E-08	0.0518	1.37E-08	5.72	0.055237

*Property values generated from NIST Database 23: REFPROP Version 7.0 (August 2002).

Table G2.B   Engine Oil (Unused)—Saturated*

T (K)	ρ (kg/m^3)	c_p (kJ/kg·K)	μ (Pa·s)	ν (m^2/s)	k (W/m·K)	α (m^2/s)	Pr	β (1/K)
273	899.1	1.796	3.85	0.00428	0.147	9.10E-08	47,000	0.0007
280	895.3	1.827	2.17	0.00243	0.144	8.80E-08	27,500	0.0007
290	890	1.868	0.999	0.00112	0.145	8.72E-08	12,900	0.0007
300	884.1	1.909	0.486	0.00055	0.145	8.59E-08	6,400	0.0007
310	877.9	1.951	0.253	0.000288	0.145	8.47E-08	3,400	0.0007
320	871.8	1.993	0.141	0.000161	0.143	8.23E-08	1,965	0.0007
330	865.8	2.035	0.0836	0.0000966	0.141	8.00E-08	1,205	0.0007
340	859.9	2.076	0.0531	0.0000617	0.139	7.79E-08	793	0.0007
350	853.9	2.118	0.0356	0.0000417	0.138	7.63E-08	546	0.0007
360	847.8	2.161	0.0252	0.0000297	0.138	7.53E-08	395	0.0007
370	841.8	2.206	0.0186	0.000022	0.137	7.38E-08	300	0.0007
380	836.0	2.250	0.0141	0.0000169	0.136	7.23E-08	233	0.0007
390	830.6	2.294	0.0110	0.0000133	0.135	7.09E-08	187	0.0007
400	825.1	2.337	0.00874	0.0000106	0.134	6.95E-08	152	0.0007
410	818.9	2.381	0.00698	0.00000852	0.133	6.82E-08	125	0.0007
420	812.1	2.427	0.00564	0.00000694	0.133	6.75E-08	103	0.0007
430	806.5	2.471	0.0047	0.00000583	0.132	6.62E-08	88	0.0007

*Property values from Incropera, F. P., and DeWitt, D. P., *Fundamentals of Heat and Mass Transfer*, 3rd ed., Wiley, New York, 1990.

Table G2.C   Ethylene Glycol (C$_2$H$_4$(OH)$_2$)—Saturated*

T (K)	ρ (kg/m³)	c_p (kJ/kg·K)	μ (Pa·s)	ν (m²/s)	k (W/m·K)	α (m²/s)	Pr	β (1/K)
273	1130.8	2.294	0.0651	0.0000576	0.242	9.33E-08	617	0.00065
280	1125.8	2.323	0.0420	0.0000373	0.244	9.33E-08	400	0.00065
290	1118.8	2.368	0.0247	0.0000221	0.248	9.36E-08	236	0.00065
300	1114.4	2.415	0.0157	0.0000141	0.252	9.39E-08	151	0.00065
310	1103.7	2.460	0.0107	0.00000965	0.255	9.39E-08	103	0.00065
320	1096.2	2.505	0.00757	0.00000691	0.258	9.40E-08	73.5	0.00065
330	1089.5	2.549	0.00561	0.00000515	0.260	9.36E-08	55.0	0.00065
340	1083.8	2.592	0.00431	0.00000398	0.261	9.29E-08	42.8	0.00065
350	1079.0	2.637	0.00342	0.00000317	0.261	9.17E-08	34.6	0.00065
360	1074.0	2.682	0.00278	0.00000259	0.261	9.06E-08	28.6	0.00065
370	1066.7	2.728	0.00228	0.00000214	0.262	9.00E-08	23.7	0.00065
373	1058.5	2.742	0.00215	0.00000203	0.263	9.06E-08	22.4	0.00065

*Property values from Incropera, F. P., and DeWitt, D. P., *Fundamentals of Heat and Mass Transfer*, 3rd ed., Wiley, New York, 1990.

Table G2.D   Glycerin (C$_3$H$_5$(OH)$_3$)—Saturated*

T (K)	ρ (kg/m³)	c_p (kJ/kg·K)	μ (Pa·s)	ν (m²/s)	k (W/m·K)	α (m²/s)	Pr	β (1/K)
273	1276.0	2.261	10.6	0.00831	0.282	9.77E-08	85,000	0.00047
280	1271.9	2.298	5.34	0.00420	0.284	9.72E-08	43,200	0.00047
290	1265.8	2.367	1.85	0.00146	0.286	9.55E-08	15,300	0.00048
300	1259.9	2.427	0.799	0.000634	0.286	9.35E-08	6,780	0.00048
310	1253.9	2.490	0.352	0.000281	0.286	9.16E-08	3,060	0.00049
320	1247.2	2.564	0.210	0.000168	0.287	8.97E-08	1,870	0.00050

*Property values from Incropera, F. P., and DeWitt, D. P., *Fundamentals of Heat and Mass Transfer*, 3rd ed., Wiley, New York, 1990.

Table G2.E   Mercury (Hg)—Saturated*

T (K)	ρ (kg/m³)	c_p (kJ/kg·K)	μ (Pa·s)	ν (m²/s)	k (W/m·K)	α (m²/s)	Pr	β (1/K)
273	13,595	0.1404	0.001688	1.240E-07	8.18	4.285E-06	0.0290	0.000181
300	13,529	0.1393	0.001523	1.125E-07	8.54	4.530E-06	0.0248	0.000181
350	13,407	0.1377	0.001309	9.76E-08	9.18	4.975E-06	0.0196	0.000181
400	13,287	0.1365	0.001171	8.82E-08	9.80	5.405E-06	0.0163	0.000181
450	13,167	0.1357	0.001075	8.16E-08	10.40	5.810E-06	0.0140	0.000181
500	13,048	0.1353	0.001007	7.71E-08	10.95	6.190E-06	0.0125	0.000182
550	12,929	0.1352	0.000953	7.37E-08	11.45	6.555E-06	0.0112	0.000184
600	12,809	0.1355	0.000911	7.11E-08	11.95	6.880E-06	0.0103	0.000187

*Property values from Incropera, F. P., and DeWitt, D. P., *Fundamentals of Heat and Mass Transfer*, 3rd ed., Wiley, New York, 1990.

Appendix H
Thermo-Physical Properties of Hydrocarbon Fuels

Table H.1 Selected Properties of Hydrocarbon Fuels: Enthalpy of Formation,[a] Gibbs Function of Formation,[a] Entropy,[a] and Higher and Lower Heating Values All at 298.15 K and 1 atm; Boiling Points[b] and Latent Heat of Vaporization[c] at 1 atm; Constant-Pressure Adiabatic Flame Temperature at 1 atm;[d] Liquid Density[c]

Formula	Fuel	Molecular Weight (kg/kmol)	$\bar{h}_f^\circ$ (kJ/kmol)	$\Delta \bar{g}_f^\circ$ (kJ/kmol)	$\bar{s}^\circ$ (kJ/kmol·K)	HHV* (kJ/kg)	LHV* (kJ/kg)	Boiling Pt. (°C)	h_{fg} (kJ/kg)	$T_{ad}^\dagger$ (K)	$\rho_{liq}^\ddagger$ (kg/m³)
CH_4	Methane	16.043	−74,831	−50,794	186.188	55,528	50,016	−164	509	2226	300
C_2H_2	Acetylene	26.038	226,748	209,200	200.819	49,923	48,225	−84	—	2539	—
C_2H_4	Ethene	28.054	52,283	68,124	219.827	50,313	47,161	−103.7	—	2369	—
C_2H_6	Ethane	30.069	−84,667	−32,886	229.492	51,901	47,489	−88.6	488	2259	370
C_3H_6	Propene	42.080	20,414	62,718	266.939	48,936	45,784	−47.4	437	2334	514
C_3H_8	Propane	44.096	−103,847	−23,489	269.910	50,368	46,357	−42.1	425	2267	500
C_4H_8	1-Butene	56.107	1172	72,036	307.440	48,471	45,319	−63	391	2322	595
C_4H_{10}	n-Butane	58.123	−124,733	−15,707	310.034	49,546	45,742	−0.5	386	2270	579
C_5H_{10}	1-Pentene	70.134	−20,920	78,605	347.607	48,152	45,000	30	358	2314	641
C_5H_{12}	n-Pentane	72.150	−146,440	−8201	348.402	49,032	45,355	36.1	358	2272	626
C_6H_6	Benzene	78.113	82,927	129,658	269.199	42,277	40,579	80.1	393	2342	879
C_6H_{12}	1-Hexene	84.161	−41,673	87,027	385.974	47,955	44,803	63.4	335	2308	673
C_6H_{14}	n-Hexane	86.177	−167,193	209	386.811	48,696	45,105	69	335	2273	659
C_7H_{14}	1-Heptene	98.188	−62,132	95,563	424.383	47,817	44,665	93.6	—	2305	—
C_7H_{16}	n-Heptane	100.203	−187,820	8745	425.262	48,456	44,926	98.4	316	2274	684
C_8H_{16}	1-Octene	112.214	−82,927	104,140	462.792	47,712	44,560	121.3	—	2302	—
C_8H_{18}	n-Octane	114.230	−208,447	17,322	463.671	48,275	44,791	125.7	300	2275	703
C_9H_{18}	1-Nonene	126.241	−103,512	112,717	501.243	47,631	44,478	—	—	2300	—
C_9H_{20}	n-Nonane	128.257	−229,032	25,857	502.080	48,134	44,686	150.8	295	2276	718
$C_{10}H_{20}$	1-Decene	140.268	−124,139	121,294	539.652	47,565	44,413	170.6	—	2298	730
$C_{10}H_{22}$	n-Decane	142.284	−249,659	34,434	540.531	48,020	44,602	174.1	277	2277	730
$C_{11}H_{22}$	1-Undecene	154.295	−144,766	129,830	578.061	47,512	44,360	—	—	2296	—
$C_{11}H_{24}$	n-Undecane	156.311	−270,286	43,012	578.940	47,926	44,532	195.9	265	2277	740
$C_{12}H_{24}$	1-Dodecene	168.322	−165,352	138,407	616.471	47,468	44,316	213.4	—	2295	—
$C_{12}H_{26}$	n-Dodecane	170.337	−292,162	—	—	47,841	44,467	216.3	256	2277	749

*Based on gaseous fuel.

† For stoichiometric combustion with air (79% N_2, 21% O_2).

‡ For liquids at 20°C or for gases at the boiling point of the liquefied gas.

Sources:

[a] Rossini, F. D., et al., *Selected Values of Physical and Thermodynamic Properties of Hydrocarbons and Related Compounds*, Carnegie Press, Pittsburgh, PA, 1953.

[b] Weast, R. C. (Ed.), *Handbook of Chemistry and Physics*, 56th ed., CRC Press, Cleveland, OH, 1976.

[c] Obert, E. F., *Internal Combustion Engines and Air Pollution*, Harper & Row, New York, 1973.

[d] Turns, S. R., *An Introduction to Combustion*, 2nd ed., McGraw-Hill, New York, 2000.

Table H.2   Curve-Fit Coefficients for Fuel Specific Heat and Enthalpy[a] for Reference State of Zero Enthalpy of the Elements at 298.15 K and 1 atm:

$$\bar{c}_p \text{ (kJ/kmol} \cdot \text{K)} = 4.184(a_1 + a_2\theta + a_3\theta^2 + a_4\theta^3 + a_5\theta^{-2}),$$
$$\bar{h}°\text{(kJ/kmol)} = 4184(a_1\theta + a_2\theta^2/2 + a_3\theta^3/3 + a_4\theta^4/4 - a_5\theta^{-1} + a_6),$$
$$\text{where } \theta \equiv T \text{ (K)}/1000$$

Formula	Fuel	Molecular Weight	a_1	a_2	a_3	a_4	a_5	a_6	a_8^*
CH_4	Methane	16.043	−0.29149	26.327	−10.610	1.5656	0.16573	−18.331	4.300
C_3H_8	Propane	44.096	−1.4867	74.339	−39.065	8.0543	0.01219	−27.313	8.852
C_6H_{14}	Hexane	86.177	−20.777	210.48	−164.125	52.832	0.56635	−39.836	15.611
C_8H_{18}	Isooctane	114.230	−0.55313	181.62	−97.787	20.402	−0.03095	−60.751	20.232
CH_3OH	Methanol	32.040	−2.7059	44.168	−27.501	7.2193	0.20299	−48.288	5.3375
C_2H_5OH	Ethanol	46.07	6.990	39.741	−11.926	0	0	−60.214	7.6135
$C_{8.26}H_{15.5}$	Gasoline	114.8	−24.078	256.63	−201.68	64.750	0.5808	−27.562	17.792
$C_{7.76}H_{13.1}$		106.4	−22.501	227.99	−177.26	56.048	0.4845	−17.578	15.232
$C_{10.8}H_{18.7}$	Diesel	148.6	−9.1063	246.97	−143.74	32.329	0.0518	−50.128	23.514

*To obtain 0 K reference state for enthalpy, add a_8 to a_6.
[a]*Source*: From Heywood, J. B., *Internal Combustion Engine Fundamentals*, McGraw-Hill, New York, 1988, by permission of McGraw-Hill, Inc.

Table H.3 Curve-Fit Coefficients for Fuel Vapor Thermal Conductivity, Viscosity, and Specific Heat:[a]

$$\left.\begin{array}{l} k\ (\text{W/m}\cdot\text{K}) \\ \mu\ (\text{N}\cdot\text{s/m}^2)\times 10^{6} \\ c_p\ (\text{J/kg}\cdot\text{K}) \end{array}\right\} = a_1 + a_2 T + a_3 T^2 + a_4 T^3 + a_5 T^4 + a_6 T^5 + a_7 T^6$$

Formula	Fuel	T range (K)	Property	a_1	a_2	a_3	a_4	a_5	a_6	a_7
CH_4	Methane	100–1000	k	−1.34014990E−2	3.66307060E−4	−1.82248608E−6	5.93987998E−9	−9.14055050E−12	−6.78968890E−15	−1.95048736E−18
		70–1000	μ	2.96826700E−1	3.71120100E−1	1.21829800E−5	−7.02426000E−8	7.54326900E−11	−2.72371660E−14	0
			c_p	See Table B.2						
C_3H_8	Propane	200–500	k	−1.07682209E−2	8.38590325E−5	4.22059864E−8	0	0	0	0
		270–600	μ	−3.54371100E−1	3.08009600E−2	−6.99723000E−6	0	0	0	0
			c_p	See Table B.2						
C_6H_{14}	n-Hexane	150–1000	k	1.28775700E−3	−2.00499443E−5	2.37858831E−7	−1.60944555E−10	7.71027290E−14	0	0
		270–900	μ	1.54541200E+0	1.15080900E−2	2.72216500E−5	−3.26900000E−8	1.24545900E−11	0	0
			c_p	See Table B.2						
C_7H_{16}	n-Heptane	250–1000	k	−4.60614700E−2	5.95652224E−4	−2.98893153E−6	8.44612876E−9	−1.22927E−11	9.0127E−15	−2.62961E−18
		270–580	μ	1.54009700E+0	1.09515700E−2	1.80066400E−5	−1.36379000E−8	0	0	0
		300–755	c_p	9.46260000E+1	5.86099700E+1	−1.98231320E−3	−6.88699300E−8	−1.93795260E−10	0	0
		755–1365	c_p	−7.40308000E+2	1.08935370E+2	−1.26512400E−2	9.84376300E−6	−4.32282960E−9	7.86365000E−13	0
C_8H_{18}	n-Octane	250–500	k	−4.01391940E−3	3.38796092E−5	8.19291819E−8	0	0	0	0
		300–650	μ	8.32435400E−1	1.40045000E−2	8.79376500E−6	−6.84030000E−9	0	0	0
		275–755	c_p	2.14419800E+2	5.35690500E+0	−1.17497000E−3	−6.99115500E−7	0	0	0
		755–1365	c_p	2.43596860E+3	−4.46819470E+0	−1.66843290E−2	−1.78856050E−5	8.64282020E−9	−1.61426500E−12	0
$C_{10}H_{22}$	n-Decane	250–500	k	−5.88274000E−3	3.72449646E−5	7.55109624E−8	0	0	0	0
			μ	Not available						
		300–700	c_p	2.40717800E+2	5.09965000E+0	−6.29026000E−4	−1.07155000E−6	0	0	0
		700–1365	c_p	−1.35345890E+4	9.14879000E+1	−2.20700000E−1	2.91406000E−4	−2.15307400E−7	8.38600000E−11	−1.34404000E−14
CH_3OH	Methanol	300–550	k	−2.02986750E−2	1.21910927E−4	−2.23748473E−8	0	0	0	0
		250–650	μ	1.19790000E+0	2.45028000E−2	1.86162740E−5	−1.30674820E−8	0	0	0
			c_p	See Table B.2						
C_2H_5OH	Ethanol	250–550	k	−2.46663000E−2	1.55892550E−4	−8.22954822E−8	0	0	0	0
		270–600	μ	−6.33595000E−2	3.20713470E−2	−6.25079576E−6	0	0	0	0
			c_p	See Table B.2						

[a]Source: Andrews, J. R., and Biblarz, O., "Temperature Dependence of Gas Properties in Polynomial Form," Naval Postgraduate School, NPS67-81-001, January 1981.

Appendix 1
Thermo-Physical Properties of Selected Solids

Table I.1   Thermo-Physical Properties of Selected Metallic Solids[a]

Composition	Melting Point (K)	Properties at 300 K ρ (kg/m³)	c_p (J/kg·K)	k (W/m·K)	$\alpha \cdot 10^6$ (m²/s)	k (W/m·K) and c_p (J/kg·K) at Various Temperatures (K) 100	200	400	600	800	1000	1200	1500	2000	2500
Aluminum															
Pure	933	2702	903	237	97.1	302	237	240	231	218					
						482	798	949	1033	1146					
Alloy 2024-T6 (4.5% Cu, 1.5% Mg, 0.6% Mn)	775	2770	875	177	73.0	65	163	186	186						
						473	787	925	1042						
Alloy 195, Cast (4.5% Cu)		2790	883	168	68.2			174	185						
								—	—						
Beryllium	1550	1850	1825	200	59.2	990	301	161	126	106	90.8	78.7			
						203	1114	2191	2604	2823	3018	3227	3519		
Bismuth	545	9780	122	7.86	6.59	16.5	9.69	7.04							
						112	120	127							
Boron	2573	2500	1107	27.0	9.76	190	55.5	16.8	10.6	9.60	9.85				
						128	600	1463	1892	2160	2338				
Cadmium	594	8650	231	96.8	48.4	203	99.3	94.7							
						198	222	242							
Chromium	2118	7160	449	93.7	29.1	159	111	90.9	80.7	71.3	65.4	61.9	57.2	49.4	
						192	384	484	542	581	616	682	779	937	
Cobalt	1769	8862	421	99.2	26.6	167	122	85.4	67.4	58.2	52.1	49.3	42.5		
						236	379	450	503	550	628	733	674		
Copper															
Pure	1358	8933	385	401	117	482	413	393	379	366	352	339			
						252	356	397	417	433	451	480			
Commercial bronze (90% Cu, 10% Al)	1293	8800	420	52	14		42	52	59						
							785	460	545						

Source: [a]Adapted from Incropera, F. P., and DeWitt, D. P., *Fundamentals of Heat and Mass Transfer*, 3rd ed., Wiley, New York, 1990, with permission. Data for wrought iron, cast iron, and various carbon steels adapted from Chapman, A. J., *Fundamentals of Heat Transfer*, Macmillan, New York, 1987.

(continued)

Table I.1 (continued)

Composition	Melting Point (K)	Properties at 300 K				k (W/m·K) and c_p (J/kg·K) at Various Temperatures (K)									
		ρ (kg/m³)	c_p (J/kg·K)	k (W/m·K)	$\alpha \cdot 10^6$ (m²/s)	100	200	400	600	800	1000	1200	1500	2000	2500
Phosphor gear bronze (89% Cu, 11% Sn)	1104	8780	355	54	17		41	65	74						
							—	—	—						
Cartridge brass (70% Cu, 30% Zn)	1188	8530	380	110	33.9	75	95	137	149						
							360	395	425						
Constantan (55% Cu, 45% Ni)	1493	8920	384	23	6.71	17	19								
						237	362								
Germanium	1211	5360	322	59.9	34.7	232	96.8	43.2	27.3	19.8	17.4	17.4			
						190	290	337	348	357	375	395			
Gold	1336	19300	129	317	127	327	323	311	298	284	270	255			
						109	124	131	135	140	145	155			
Iridium	2720	22500	130	147	50.3	172	153	144	138	132	126	120	111		
						90	122	133	138	144	153	161	172		
Iron — Pure	1810	7870	447	80.2	23.1	134	94.0	69.5	54.7	43.3	32.8	28.3	32.1		
						216	384	490	574	680	975	609	654		
Armco (99.75% pure)		7870	447	72.7	20.7	95.6	80.6	65.7	53.1	42.2	32.3	28.7	31.4		
						215	384	490	574	680	975	609	654		
Iron — Wrought iron* (C < 0.5%)		7849	460	59	16.3	83	60	56	47	39	34	33	33		
Cast iron* (C ≈ 4%)		7272	420	52	17										
Carbon steels — Carbon steel* (C ≈ 0.5%)		7833	465	54	14.7		57	51	44	38	32	30	31		
Carbon steel* (C ≈ 1.0%)		7801	473	43	11.7		43	43	39	34	30	28	29		
Carbon steel* (C ≈ 1.5%)		7753	486	36	9.7		36	36	34	32	29	28	29		
Carbon steels — Plain carbon (Mn ≤ 1%, Si ≤ 0.1%)		7854	434	60.5	17.7			56.7	48.0	39.2	30.0				
								487	559	685	1169				
AISI 1010		7832	434	63.9	18.8			58.7	48.8	39.2	31.3				
								487	559	685	1168				

Composition	Melting Point (K)	ρ (kg/m³)	c_p *	k *	$\alpha \cdot 10^6$ *										
Carbon–silicon (Mn ≤ 1%, 0.1% < Si ≤ 0.6%)		7817	446	51.9	14.9			49.8 / 501	44.0 / 582	37.4 / 699	29.3 / 971				
Carbon–manganese–silicon (1% < Mn ≤ 1.65%, 0.1% < Si ≤ 0.6%)		8131	434	41.0	11.6			42.2 / 487	39.7 / 559	35.0 / 685	27.6 / 1090				
Chromium (low) steels $\frac{1}{2}$ Cr–$\frac{1}{4}$ Mo–Si (0.18% C, 0.65% Cr, 0.23% Mo, 0.6% Si)		7822	444	37.7	10.9			38.2 / 492	36.7 / 575	33.3 / 688	26.9 / 969				
1 Cr–$\frac{1}{2}$ Mo (0.16% C, 1% Cr, 0.54% Mo, 0.39% Si)		7858	442	42.3	12.2			42.0 / 492	39.1 / 575	34.5 / 688	27.4 / 969				
1 Cr–V (0.2% C, 1.02% Cr, 0.15% V)		7836	443	48.9	14.1			46.8 / 492	42.1 / 575	36.3 / 688	28.2 / 969				
Stainless steels AISI 302		8055	480	15.1	3.91			17.3 / 512	20.0 / 559	22.8 / 585	25.4 / 606				
AISI 304	1670	7900	477	14.9	3.95	9.2 / 272	12.6 / 402	16.6 / 515	19.8 / 557	22.6 / 582	25.4 / 611	28.0 / 640	31.7 / 682		
AISI 316		8238	468	13.4	3.48			15.2 / 504	18.3 / 550	21.3 / 576	24.2 / 602				
AISI 347		7978	480	14.2	3.71			15.8 / 513	18.9 / 559	21.9 / 585	24.7 / 606				
Lead	601	11340	129	35.3	24.1	39.7 / 118	36.7 / 125	34.0 / 132	31.4 / 142						
Magnesium	923	1740	1024	156	87.6	169 / 649	159 / 934	153 / 1074	149 / 1170	146 / 1267					
Molybdenum	2894	10240	251	138	53.7	179 / 141	143 / 224	134 / 261	126 / 275	118 / 285	112 / 295	105 / 308	98 / 330	90 / 380	86 / 459

*Properties at 293 K.

(continued)

Table I.1 (continued)

page **1134**

Composition	Melting Point (K)	Properties at 300 K				k (W/m·K) and c_p (J/kg·K) at Various Temperatures (K)									
		ρ (kg/m³)	c_p (J/kg·K)	k (W/m·K)	$\alpha \cdot 10^6$ (m²/s)	100	200	400	600	800	1000	1200	1500	2000	2500
Nickel Pure	1728	8900	444	90.7	23.0	164 / 232	107 / 383	80.2 / 485	65.6 / 592	67.6 / 530	71.8 / 562	76.2 / 594	82.6 / 616		
Nichrome (80% Ni, 20% Cr)	1672	8400	420	12	3.4			14 / 480	16 / 525	21 / 545					
Inconel X-750 (73% Ni, 15% Cr, 6.7% Fe)	1665	8510	439	11.7	3.1	8.7 / —	10.3 / 372	13.5 / 473	17.0 / 510	20.5 / 546	24.0 / 626	27.6 / —	33.0 / —		
Niobium	2741	8570	265	53.7	23.6	55.2 / 188	52.6 / 249	55.2 / 274	58.2 / 283	61.3 / 292	64.4 / 301	67.5 / 310	72.1 / 324	79.1 / 347	
Palladium	1827	12020	244	71.8	24.5	76.5 / 168	71.6 / 227	73.6 / 251	79.7 / 261	86.9 / 271	94.2 / 281	102 / 291	110 / 307		
Platinum Pure	2045	21450	133	71.6	25.1	77.5 / 100	72.6 / 125	71.8 / 136	73.2 / 141	75.6 / 146	78.7 / 152	82.6 / 157	89.5 / 165	99.4 / 179	
Alloy 60Pt-40Rh (60% Pt, 40% Rh)	1800	16630	162	47	17.4			52 / —	59 / —	65 / —	69 / —	73 / —	76 / —		
Rhenium	3453	21100	136	47.9	16.7	58.9 / 97	51.0 / 127	46.1 / 139	44.2 / 145	44.1 / 151	44.6 / 156	45.7 / 162	47.8 / 171	51.9 / 186	
Rhodium	2236	12450	243	150	49.6	186 / 147	154 / 220	146 / 253	136 / 274	127 / 293	121 / 311	116 / 327	110 / 349	112 / 376	
Silicon	1685	2330	712	148	89.2	884 / 259	264 / 556	98.9 / 790	61.9 / 867	42.2 / 913	31.2 / 946	25.7 / 967	22.7 / 992		
Silver	1235	10500	235	429	174	444 / 187	430 / 225	425 / 239	412 / 250	396 / 262	379 / 277	361 / 292			
Tantalum	3269	16600	140	57.5	24.7	59.2 / 110	57.5 / 133	57.8 / 144	58.6 / 146	59.4 / 149	60.2 / 152	61.0 / 155	62.2 / 160	64.1 / 172	65.6 / 189

Element															
Thorium	2023	11700	118	54.0	39.1	59.8/99	54.6/112	54.5/124	55.8/134	56.9/145	56.9/156	58.7/167			
Tin	505	7310	227	66.6	40.1	85.2/188	73.3/215	62.2/243							
Titanium	1953	4500	522	21.9	9.32	30.5/300	24.5/465	20.4/551	19.4/591	19.7/633	20.7/675	22.0/620	24.5/686		
Tungsten	3660	19300	132	174	68.3	208/87	186/122	159/137	137/142	125/145	118/148	113/152	107/157	100/167	95/176
Uranium	1406	19070	116	27.6	12.5	21.7/94	25.1/108	29.6/125	34.0/146	38.8/176	43.9/180	49.0/161			
Vanadium	2192	6100	489	30.7	10.3	35.8/258	31.3/430	31.3/515	33.3/540	35.7/563	38.2/597	40.8/645	44.6/714	50.9/867	
Zinc	693	7140	389	116	41.8	117/297	118/367	111/402	103/436						
Zirconium	2125	6570	278	22.7	12.4	33.2/205	25.2/264	21.6/300	20.7/322	21.6/342	23.7/362	26.0/344	28.8/344	33.0/344	

Table I.2 Thermo-Physical Properties of Selected Nonmetallic Solids[a]

Composition	Melting Point (K)	Properties at 300 K ρ (kg/m³)	c_p (J/kg·K)	k (W/m·K)	$\alpha \cdot 10^6$ (m²/s)	k (W/m·K) and c_p (J/kg·K) at Various Temperatures (K) 100	200	400	600	800	1000	1200	1500	2000	2500
Aluminum oxide, sapphire	2323	3970	765	46	15.1	450	82	32.4	18.9	13.0	10.5				
						—	—	940	1110	1180	1225				
Aluminum oxide, polycrystalline	2323	3970	765	36.0	11.9	133	55	26.4	15.8	10.4	7.85	6.55	5.66	6.00	
						—	—	940	1110	1180	1225	—	—	—	
Beryllium oxide	2725	3000	1030	272	88.0			196	111	70	47	33	21.5	15	
								1350	1690	1865	1975	2055	2145	2750	
Boron	2573	2500	1105	27.6	9.99	190	52.5	18.7	11.3	8.1	6.3	5.2			
						—	—	1490	1880	2135	2350	2555			
Boron fiber epoxy (30% vol) composite	590	2080													
k, ‖ to fibers				2.29		2.10	2.23	2.28							
k, ⊥ to fibers				0.59		0.37	0.49	0.60							
c_p			1122			364	757	1431							
Carbon															
Amorphous	1500	1950	—	1.60	—	0.67	1.18	1.89	2.19	2.37	2.53	2.84	3.48		
						—	—	—	—	—	—	—	—		
Diamond, type IIa insulator	—	3500	509	2300		10000	4000	1540							
						21	194	853							
Graphite, pyrolytic	2273	2210													
k, ‖ to layers				1950		4970	3230	1390	892	667	534	448	357	262	
k, ⊥ to layers				5.70		16.8	9.23	4.09	2.68	2.01	1.60	1.34	1.08	0.81	
c_p			709			136	411	992	1406	1650	1793	1890	1974	2043	
Graphite fiber epoxy (25% vol) composite	450	1400													
k, heat flow ‖ to fibers				11.1		5.7	8.7	13.0							
k, heat flow ⊥ to fibers				0.87		0.46	0.68	1.1							
c_p			935			337	642	1216							

Composition	Melting Point (K)	ρ (kg/m³)	c_p (J/kg·K)	k (W/m·K)	$\alpha \cdot 10^6$ (m²/s)	100 K	200 K	400 K	600 K	800 K	1000 K	1200 K	1500 K
Pyroceram, Corning 9606	1623	2600	808	3.98	1.89	5.25 / —	4.78 / —	3.64 / 908	3.28 / 1038	3.08 / 1122	2.96 / 1197	2.87 / 1264	2.79 / 1498
Silicon carbide	3100	3160	675	490	230	— / —	— / —	— / 880	— / 1050	— / 1135	87 / 1195	58 / 1243	30 / 1310
Silicon dioxide, crystalline (quartz)	1883	2650											
k, ∥ to c axis				10.4		39	16.4	7.6	5.0	4.2			
k, ⊥ to c axis				6.21		20.8	9.5	4.70	3.4	3.1			
c_p			745					885	1075	1250			
Silicon dioxide, polycrystalline (fused silica)	1883	2220	745	1.38	0.834	0.69 / —	1.14 / —	1.51 / 905	1.75 / 1040	2.17 / 1105	2.87 / 1155	4.00 / 1195	
Silicon nitride	2173	2400	691	16.0	9.65	— / —	— / 578	13.9 / 778	11.3 / 937	9.88 / 1063	8.76 / 1155	8.00 / 1226	7.16 / 1306
Sulfur	392	2070	708	0.206	0.141	0.165 / 403	0.185 / 606						
Thorium dioxide	3573	9110	235	13	6.1			10.2 / 255	6.6 / 274	4.7 / 285	3.68 / 295	3.12 / 303	2.73 / 315
Titanium dioxide, polycrystalline	2133	4157	710	8.4	2.8			7.01 / 805	5.02 / 880	3.94 / 910	3.46 / 930	3.28 / 945	

Source: [a]Adapted from Incropera, F. P., and Dewitt, D. P., *Fundamentals of Heat and Mass Transfer*, 3rd ed., Wiley, New York, 1990, with permission.

Table I.3 Thermo-Physical Properties of Common Materials[a]

Description/ Composition	Temperature (K)	Density, ρ (kg/m³)	Thermal Conductivity, k (W/m·K)	Specific Heat, c_p (J/kg·K)
Asphalt	300	2115	0.062	920
Bakelite	300	1300	1.4	1465
Brick, refractory				
Carborundum	872	—	18.5	—
	1672	—	11.0	—
Chrome brick	473	3010	2.3	835
	823		2.5	
	1173		2.0	
Diatomaceous	478	—	0.25	
silica, fired	1145	—	0.30	—
Fire clay, burnt 1600 K	773	2050	1.0	960
	1073	—	1.1	
	1373	—	1.1	
Fire clay, burnt 1725 K	773	2325	1.3	960
	1073		1.4	
	1373		1.4	
Fire clay brick	478	2645	1.0	960
	922		1.5	
	1478		1.8	
Magnesite	478	—	3.8	1130
	922	—	2.8	
	1478		1.9	
Clay	300	1460	1.3	880
Coal, anthracite	300	1350	0.26	1260
Concrete (stone mix)	300	2300	1.4	880
Cotton	300	80	0.06	1300
Foodstuffs				
Banana (75.7% water content)	300	980	0.481	3350
Apple, red (75% water content)	300	840	0.513	3600
Cake, batter	300	720	0.223	—
Cake, fully baked	300	280	0.121	—
Chicken meat, white	198	—	1.60	—
(74.4% water content)	233	—	1.49	
	253		1.35	
	263		1.20	
	273		0.476	
	283		0.480	
	293		0.489	
Glass				
Plate (soda lime)	300	2500	1.4	750
Pyrex	300	2225	1.4	835

(continued)

Description/ Composition	Temperature (K)	Density, ρ (kg/m³)	Thermal Conductivity, k (W/m·K)	Specific Heat, c_p (J/kg·K)
Ice	273	920	1.88	2040
	253	—	2.03	1945
Leather (sole)	300	998	0.159	—
Paper	300	930	0.180	1340
Paraffin	300	900	0.240	2890
Rock				
Granite, Barre	300	2630	2.79	775
Limestone, Salem	300	2320	2.15	810
Marble, Halston	300	2680	2.80	830
Quartzite, Sioux	300	2640	5.38	1105
Sandstone, Berea	300	2150	2.90	745
Rubber, vulcanized				
Soft	300	1100	0.13	2010
Hard	300	1190	0.16	—
Sand	300	1515	0.27	800
Soil	300	2050	0.52	1840
Snow	273	110	0.049	—
		500	0.190	—
Teflon	300	2200	0.35	—
	400		0.45	—
Tissue, human				
Skin	300	—	0.37	—
Fat layer (adipose)	300	—	0.2	—
Muscle	300	—	0.41	—
Wood, cross grain				
Balsa	300	140	0.055	—
Cypress	300	465	0.097	—
Fir	300	415	0.11	2720
Oak	300	545	0.17	2385
Yellow pine	300	640	0.15	2805
White pine	300	435	0.11	—
Wood, radial				
Oak	300	545	0.19	2385
Fir	300	420	0.14	2720

Source: [a]Adapted from Incropera, F. P., and DeWitt, D. P., *Fundamentals of Heat and Mass Transfer,* 3rd ed., Wiley, New York, 1990, with permission.

Table I.4 Thermo-Physical Properties of Structural Building Materials[a]

Description/ Composition	Typical Properties at 300 K		
	Density, ρ (kg/m³)	Thermal Conductivity, k (W/m·K)	Specific Heat, c_p (J/kg·K)
Building boards			
Asbestos–cement board	1920	0.58	—
Gypsum or plaster board	800	0.17	—
Plywood	545	0.12	1215
Sheathing, regular density	290	0.055	1300
Acoustic tile	290	0.058	1340
Hardboard, siding	640	0.094	1170
Hardboard, high density	1010	0.15	1380
Particle board, low density	590	0.078	1300
Particle board, high density	1000	0.170	1300
Woods			
Hardwoods (oak, maple)	720	0.16	1255
Softwoods (fir, pine)	510	0.12	1380
Masonry materials			
Cement mortar	1860	0.72	780
Brick, common	1920	0.72	835
Brick, face	2083	1.3	—
Clay tile, hollow			
1 cell deep, 10 cm thick	—	0.52	—
3 cells deep, 30 cm thick	—	0.69	—
Concrete block, 3 oval cores			
sand/gravel, 20 cm thick	—	1.0	—
cinder aggregate, 20 cm thick	—	0.67	—
Concrete block, rectangular core			
2 cores, 20 cm thick, 16 kg	—	1.1	—
same with filled cores	—	0.60	—
Plastering materials			
Cement plaster, sand aggregate	1860	0.72	—
Gypsum plaster, sand aggregate	1680	0.22	1085
Gypsum plaster, vermiculite aggregate	720	0.25	—
Blanket and batt			
Glass fiber, paper faced	16	0.046	—
	28	0.038	—
	40	0.035	—
Glass fiber, coated; duct liner	32	0.038	835
Board and slab			
Cellular glass	145	0.058	1000
Glass fiber, organic bonded	105	0.036	795
Polystyrene, expanded			
extruded (R-12)	55	0.027	1210
molded beads	16	0.040	1210
Mineral fiberboard; roofing material	265	0.049	—
Wood, shredded/cemented	350	0.087	1590
Cork	120	0.039	1800

(continued)

Description/ Composition	Typical Properties at 300 K		
	Density, ρ (kg/m^3)	Thermal Conductivity, k (W/m·K)	Specific Heat, c_p (J/kg·K)
Loose fill			
Cork, granulated	160	0.045	—
Diatomaceous silica, coarse	350	0.069	—
powder	400	0.091	—
Diatomaceous silica, fine powder	200	0.052	—
	275	0.061	—
Glass, fiber, poured or blown	16	0.043	835
Vermiculite, flakes	80	0.068	835
	160	0.063	1000
Formed/foamed-in-place			
Mineral wool granules with asbestos/inorganic binders, sprayed	190	0.046	—
Polyvinyl acetate cork mastic; sprayed or troweled	—	0.100	—
Urethane, two-part mixture; rigid foam	70	0.026	1045
Reflective			
Aluminum foil separating fluffy glass mats; 10–12 layers; evacuated; for cryogenic applications (150 K)	40	0.00016	—
Aluminum foil and glass paper laminate; 75–150 layers; evacuated; for cryogenic application (150 K)	120	0.000017	—
Typical silica powder, evacuated	160	0.0017	—

Source: [a]Adapted from Incropera, F. P., and DeWitt, D. P., *Fundamentals of Heat and Mass Transfer*, 3rd ed., Wiley, New York, 1990, with permission.

Table I.5 Thermo-Physical Properties of Industrial Insulation[a]

Description/Composition	Max Service Temp (K)	Typical Density (kg/m³)	\[Typical Thermal Conductivity k (W/m·K) at Various Temperatures (K)\] 200	215	230	240	255	270	285	300	310	365	420	530	645	750
Blankets																
Blanket, mineral fiber, metal reinforced	920	96–192									0.038	0.046	0.056	0.078		
	815	40–96									0.035	0.045	0.058	0.088		
Blanket, mineral fiber, glass; fine fiber, organic bonded	450	10				0.036		0.040	0.043	0.048	0.052	0.076				
		12				0.035	0.036	0.039	0.042	0.046	0.049	0.069				
		16				0.033	0.035	0.036	0.039	0.042	0.046	0.062				
		24				0.030	0.032	0.033	0.036	0.039	0.040	0.053				
		32				0.029	0.030	0.032	0.033	0.036	0.038	0.048				
		48				0.027	0.029	0.030	0.032	0.033	0.035	0.045				
Blanket, alumina–silica fiber	1530	48												0.071	0.105	0.150
		64												0.059	0.087	0.125
		96												0.052	0.076	0.100
		128												0.049	0.068	0.091
Felt, semirigid; organic bonded	480	50–125					0.036	0.035	0.036	0.038	0.039	0.051	0.063			
	730	50	0.023	0.025	0.026	0.027	0.029	0.030	0.032	0.033	0.035	0.051	0.079			
Felt, laminated; no binder	920	120											0.051	0.065	0.087	
Blocks, boards, and pipe insulations																
Asbestos paper, laminated and corrugated																
4-ply	420	190								0.078	0.082	0.098				
6-ply	420	255								0.071	0.074	0.085				
8-ply	420	300								0.068	0.071	0.082				
Magnesia, 85%	590	185									0.051	0.055	0.061			
Calcium silicate	920	190									0.055	0.059	0.063	0.075	0.089	0.104
Cellular glass	700	145			0.046	0.048	0.051	0.052	0.055	0.058	0.062	0.069	0.079			
Diatomaceous silica	1145	345												0.092	0.098	0.104
	1310	385												0.101	0.100	0.115

Description/Composition										
Polystyrene, rigid										
Extruded (R-12)	350	56	0.023	0.023	0.022	0.023	0.025	0.026	0.027	0.029
Extruded (R-12)	350	35	0.023	0.023	0.023	0.025	0.026	0.027	0.029	
Molded beads	350	16	0.026	0.029	0.030	0.033	0.035	0.036	0.038	0.040
Rubber, rigid foamed	340	70	0.029	0.030	0.032	0.033				
Insulating cement										
Mineral fiber										
(rock, slag, or glass)										
with clay binder	1255	430	0.071	0.079	0.088	0.105	0.123			
with hydraulic										
setting binder	922	560	0.108	0.115	0.123	0.137				
Loose fill										
Cellulose, wood,										
or paper pulp	—	45	0.036	0.039	0.038	0.039	0.042	0.043		
Perlite, expanded	—	105	0.042	0.046	0.049	0.051	0.053	0.056		
Vermiculite, expanded	—	122	0.056	0.061	0.063	0.065	0.068	0.071		
		80	0.049	0.055	0.058	0.061	0.063	0.066		

Source: [a]Adapted from Incropera, F. P., and DeWitt, D. P., *Fundamentals of Heat and Mass Transfer*, 3rd ed., Wiley, New York, 1990, with permission.

Appendix J
Radiation Properties of Selected Materials and Substances

Table J.1 Total, Normal (n), or Hemispherical (h) Emissivity of Selected Surfaces: Metallic Solids and Their Oxides[a]

Description/Composition		100	200	300	400	600	800	1000	1200	1500	2000	2500
Aluminum												
Highly polished, film	(h)	0.02	0.03	0.04	0.05	0.06						
Foil, bright	(h)	0.06	0.06	0.07								
Anodized	(h)			0.82	0.76							
Chromium												
Polished or plated	(n)	0.05	0.07	0.10	0.12	0.14						
Copper												
Highly polished	(h)			0.03	0.03	0.04	0.04	0.04				
Stably oxidized	(h)					0.50	0.58	0.80				
Gold												
Highly polished or film	(h)	0.01	0.02	0.03	0.03	0.04	0.05	0.06				
Foil, bright	(h)	0.06	0.07	0.07								
Molybdenum												
Polished	(h)					0.06	0.08	0.10	0.12	0.15	0.21	0.26
Shot-blasted, rough	(h)					0.25	0.28	0.31	0.35	0.42		
Stably oxidized	(h)					0.80	0.82					
Nickel												
Polished	(h)					0.09	0.11	0.14	0.17			
Stably oxidized	(h)					0.40	0.49	0.57				
Platinum												
Polished	(h)							0.10	0.13	0.15	0.18	
Silver												
Polished	(h)			0.02	0.02	0.03	0.05	0.08				
Stainless steels												
Typical, polished	(n)			0.17	0.17	0.19	0.23	0.30				
Typical, cleaned	(n)			0.22	0.22	0.24	0.28	0.35				
Typical, lightly oxidized	(n)						0.33	0.40				
Typical, highly oxidized	(n)						0.67	0.70	0.76			
AISI 347, stably oxidized	(n)					0.87	0.88	0.89	0.90			
Tantalum												
Polished	(h)								0.11	0.17	0.23	0.28
Tungsten												
Polished	(h)							0.10	0.13	0.18	0.25	0.29

Source: [a]Adapted from Incropera, F. P., and DeWitt, D. P., *Fundamentals of Heat and Mass Transfer*, 3rd ed., Wiley, New York, 1990, with permission.

Table J.2 Total, Normal (n), or Hemispherical (h) Emissivity of Selected Surfaces: Nonmetallic Substances

Description/Composition		Temperature (K)	Emissivity ε
Aluminum oxide	(n)	600	0.69
		1000	0.55
		1500	0.41
Asphalt pavement	(h)	300	0.85–0.93
Building materials			
Asbestos sheet	(h)	300	0.93–0.96
Brick, red	(h)	300	0.93–0.96
Gypsum or plaster board	(h)	300	0.90–0.92
Wood	(h)	300	0.82–0.92
Cloth	(h)	300	0.75–0.90
Concrete	(h)	300	0.88–0.93
Glass, window	(h)	300	0.90–0.95
Ice	(h)	273	0.95–0.98
Paints			
Black (Parsons)	(h)	300	0.98
White, acrylic	(h)	300	0.90
White, zinc oxide	(h)	300	0.92
Paper, white	(h)	300	0.92–0.97
Pyrex	(n)	300	0.82
		600	0.80
		1000	0.71
		1200	0.62
Pyroceram	(n)	300	0.85
		600	0.78
		1000	0.69
		1500	0.57
Refractories (furnace liners)			
Alumina brick	(n)	800	0.40
		1000	0.33
		1400	0.28
		1600	0.33
Magnesia brick	(n)	800	0.45
		1000	0.36
		1400	0.31
		1600	0.40
Kaolin insulating brick	(n)	800	0.70
		1200	0.57
		1400	0.47
		1600	0.53
Sand	(h)	300	0.90
Silicon carbide	(n)	600	0.87
		1000	0.87
		1500	0.85
Skin	(h)	300	0.95
Snow	(h)	273	0.82–0.90
Soil	(h)	300	0.93–0.96
Rocks	(h)	300	0.88–0.95
Teflon	(h)	300	0.85
		400	0.87
		500	0.92
Vegetation	(h)	300	0.92–0.96
Water	(h)	300	0.96

Source: ªAdapted from Incropera, F. P., and DeWitt, D. P., *Fundamentals of Heat and Mass Transfer*, 3rd ed., Wiley, New York, 1990, with permission.

Appendix K
Mach Number Relationships for Compressible Flow

Table K.1 One-Dimensional, Isentropic, Variable-Area Flow of Air with Constant Properties ($\gamma = 1.4$)

Ma	$\dfrac{A}{A^*}$	$\dfrac{P}{P_0}$	$\dfrac{\rho}{\rho_0}$	$\dfrac{T}{T_0}$
0	∞	1.00000	1.00000	1.00000
0.10	5.8218	0.99303	0.99502	0.99800
0.20	2.9635	0.97250	0.98027	0.99206
0.30	2.0351	0.93947	0.95638	0.98232
0.40	1.5901	0.89562	0.92428	0.96899
0.50	1.3398	0.84302	0.88517	0.95238
0.60	1.1882	0.78400	0.84045	0.93284
0.70	1.09437	0.72092	0.79158	0.91075
0.80	1.03823	0.65602	0.74000	0.88652
0.90	1.00886	0.59126	0.68704	0.86058
1.00	1.00000	0.52828	0.63394	0.83333
1.10	1.00793	0.46835	0.58169	0.80515
1.20	1.03044	0.41238	0.53114	0.77640
1.30	1.06631	0.36092	0.48291	0.74738
1.40	1.1149	0.31424	0.43742	0.71839
1.50	1.1762	0.27240	0.39498	0.68965
1.60	1.2502	0.23527	0.35573	0.66138
1.70	1.3376	0.20259	0.31969	0.63372
1.80	1.4390	0.17404	0.28682	0.60680
1.90	1.5552	0.14924	0.25699	0.58072
2.00	1.6875	0.12780	0.23005	0.55556
2.10	1.8369	0.10935	0.20580	0.53135
2.20	2.0050	0.09352	0.18405	0.50813
2.30	2.1931	0.07997	0.16458	0.48591
2.40	2.4031	0.06840	0.14720	0.46468
2.50	2.6367	0.05853	0.13169	0.44444
2.60	2.8960	0.05012	0.11787	0.42517
2.70	3.1830	0.04295	0.10557	0.40684
2.80	3.5001	0.03685	0.09462	0.38941
2.90	3.8498	0.03165	0.08489	0.37286
3.00	4.2346	0.02722	0.07623	0.35714
3.50	6.7896	0.01311	0.04523	0.28986
4.00	10.719	0.00658	0.02766	0.23810
4.50	16.562	0.00346	0.01745	0.19802
5.00	25.000	$189(10)^{-5}$	0.01134	0.16667
6.00	53.180	$633(10)^{-6}$	0.00519	0.12195
7.00	104.143	$242(10)^{-6}$	0.00261	0.09259
8.00	190.109	$102(10)^{-6}$	0.00141	0.07246
9.00	327.189	$474(10)^{-7}$	0.000815	0.05814
10.00	535.938	$236(10)^{-7}$	0.000495	0.04762
∞	∞	0	0	0

Table K.2   One-Dimensional Normal-Shock Functions for Air with Constant Properties ($\gamma = 1.4$)

Ma_x	Ma_y	$\dfrac{P_y}{P_x}$	$\dfrac{\rho_y}{\rho_x}$	$\dfrac{T_y}{T_x}$	$\dfrac{P_{0y}}{P_{0x}}$	$\dfrac{P_{0y}}{P_x}$
1.00	1.00000	1.0000	1.0000	1.0000	1.00000	1.8929
1.10	0.91177	1.2450	1.1691	1.0649	0.99892	2.1328
1.20	0.84217	1.5133	1.3416	1.1280	0.99280	2.4075
1.30	0.78596	1.8050	1.5157	1.1909	0.97935	2.7135
1.40	0.73971	2.1200	1.6896	1.2547	0.95819	3.0493
1.50	0.70109	2.4583	1.8621	1.3202	0.92978	3.4133
1.60	0.66844	2.8201	2.0317	1.3880	0.89520	3.8049
1.70	0.64055	3.2050	2.1977	1.4583	0.85573	4.2238
1.80	0.61650	3.6133	2.3592	1.5316	0.81268	4.6695
1.90	0.59562	4.0450	2.5157	1.6079	0.76735	5.1417
2.00	0.57735	4.5000	2.6666	1.6875	0.72088	5.6405
2.10	0.56128	4.9784	2.8119	1.7704	0.67422	6.1655
2.20	0.54706	5.4800	2.9512	1.8569	0.62812	6.7163
2.30	0.53441	6.0050	3.0846	1.9468	0.58331	7.2937
2.40	0.52312	6.5533	3.2119	2.0403	0.54015	7.8969
2.50	0.51299	7.1250	3.3333	2.1375	0.49902	8.5262
2.60	0.50387	7.7200	3.4489	2.2383	0.46012	9.1813
2.70	0.49563	8.3383	3.5590	2.3429	0.42359	9.8625
2.80	0.48817	8.9800	3.6635	2.4512	0.38946	10.569
2.90	0.48138	9.6450	3.7629	2.5632	0.35773	11.302
3.00	0.47519	10.333	3.8571	2.6790	0.32834	12.061
4.00	0.43496	18.500	4.5714	4.0469	0.13876	21.068
5.00	0.41523	29.000	5.0000	5.8000	0.06172	32.654
10.00	0.38757	116.50	5.7143	20.388	0.00304	129.217
∞	0.37796	∞	6.000	∞	0	∞

Answers to Selected Problems

Chapter 1

1.17

Quantity (Units)	Mass (kg)	Precision (#)	Standard (#)	Substandard (#)	Value (k$)
Inflow	520	0	0	0	0
Produced	0	3000	5000	2000	30
Outflow	295	1500	2000	2000	13.5
Stored	225	1500	3000	0	16.5
Destroyed	0	0	0	0	0

1.19 A.

Quantity (Units)	Mass (lb_m/hr)	Oranges (#/hr)	Juice (cans/hr)	P & P (lb_m/hr)	Money ($/hr)
Inflow	500	1000	0	0	160
Produced	0	0	140	70	0
Outflow	350	0	150	50	250
Stored	150	300	−10	20	−90
Destroyed	0	700	0	0	0

B. No. The plant will either run out of cans or money or become filled with mass, oranges, or peels and pulp.

C. No. It is impossible to create an orange from juice, peels, and pulp.

1.21

Quantity (Units)	Mass (lb_m)	Apples (#)	Sauce (lb_m)
Inflow	460	100	0
Produced	0	0	23
Outflow	419	18	0
Stored	41	36	23
Destroyed	0	46	0

1.32 0.006375 m³, 259.5 kW, 580 N·m, 0.402 km, 14.8 s, 42.5 m/s

1.34 9.36×10^{-11} lb_m/ft^3

1.35 (a) 179.8 N, 40.4 lb_f; (b) 1088 N, 244.7 lb_f; (c) 244.7 lb_m

1.38 0.352 or 35.2%

1.42 128.7 lb_m

1.46 98.70 ft/s² up, 158.7 ft/s² down

1.47 1920 kg/m³, 0.294 kW/m·K, 1.20 kW/m²·K, 733 J/kg·K, 0.020 N·s/m², 3700 N·s/m², 0.28 m²/s, 5.67×10^{-8} W/m²·K⁴, 3.66 m/s²

Chapter 2

2.2 3.342×10^{25}

2.4 5490.2 Pa, 0.7963 psia, 41.18 mm Hg

2.7 2.7 in Hg, 1.33 psia, 0.0902 atm

2.10 152 kg

2.11 579.67 R, 48.89°C, 322.04 K

2.14 532 R, 22.2°C, 295 K

2.17 0.362 kg/m³, 3128.75 kJ/kg, 51,385.99 kJ/kmol, 56,364.43 kJ/kmol

2.20 32.704 kJ/kmol·K, 1.167 kJ/kg·K

2.24 0.993 kg/m³

2.27 8720 kg

2.30 119 m³/kg

2.33 39.3 MPa

2.37 4.843×10^{-3} kmol (all gases), 0.136 kg N_2, 0.193 kg Ar, 0.0194 kg He

2.41 433 kJ/kg

2.43 746.4 kJ/kg, 751.7 kJ/kg

2.47 7.73 Btu/lb_m, 1.13 ft³/lb_m

2.51 1.2669 kJ/kg·K

2.53 198.0 kJ/kg, 142.0 kJ/kg, 0.0360 kJ/kg·K

2.57 2.82 kg/m^3 for $\gamma = 1.4$

2.60 716.59 K, 79.62 kPa; 1012.21 K, 112.47 kPa

2.63 423°C, 1.86 MPa, 405 kJ/kg

2.67 −0.029 kJ/kg·K

2.70 605.2 kJ/kg, −5%; 1.060 kJ/kg, −4%

2.73 0.10441 m^3/kg, 0.1154 m^3/kg (ideal gas)

2.75 168.7 kg/m^3 (ideal gas), 286.12 kg/m^3 (van der Waals), 365.73 kg/m^3 (NIST)

2.81 A. 952.37 kPa, 138.13 psia
      B. 0.0020322 m^3/kg, 0.04836 m^3/kg, 23.83
      C. 335.25 kJ/kg

2.84 2390 kJ/kg

2.87

Inlet	*Outlet*
$v = 0.0010363$ m^3/kg	$v = 0.074387$ m^3/kg
$h = 403.29$ kJ/kg	$h = 4250.5$ kJ/kg
$u = 396.14$ kJ/kg	$u = 3737.2$ kJ/kg
$s = 1.2453$ kJ/kg·K	$s = 7.7013$ kJ/kg·K
Region: compressed liquid	Region: superheated vapor

2.90 125.2 kJ/kg, 0.4203 kJ/kg·K

2.92 A. 0.826, B. 214°C, C. 0.994, D. 0.799, E. 249°C

2.96 0.0947 kg

2.99 6.95 × 10^{-4} m^3, 0.628 kg; 0.565 m^3, 2.07 kg

2.102 411 K, 0.34008 MPa, 0.48478 m^3/kg

2.105 7.84 × 10^{-4} m^3, 0.962 kg; 0.00112 m^3, 0.0382 kg

2.109 liquid: 0.00425 m^3, 4.21 kg; vapor: 84.996 m^3, 5.79 kg

2.112 457 K

2.115 533.23 kJ/kg, 533.13 kJ/kg (NIST); 937.47 kg/m^3, 937.62 kg/m^3 (NIST)

2.117 −732 kJ

2.120 28.85 kg/kmol, 0.210 kmol/kmol, 0.233 kg/kg

2.123 0.55, 0.3, 0.15, 325 kg, 32.6 kg/kmol

2.126 1.51 atm, 0.0011 kmol, 0.023 kg

2.130 N_2: 5.56 kg; CO_2: 8.73 kg

2.132 70,492 kJ/kmol, −142,235 kJ/kmol, 110,405 kJ/kmol; 184.622 kJ/kmol·K, 264.890 kJ/kmol·K, 238.588 kJ/kmol·K

2.136 37.4 kJ/K

2.140 zero

2.141 A. −802,409 kJ, B. 27.62 kg/kmol, C. −2761.6 kJ/kg$_{mix}$, D. −50,025.5 kJ/kg$_{CH_4}$

Chapter 3

3.2 6.55 × 10^{-5} kg

3.3 1.6929 kg, 0.0214

3.5 1685 lb$_m$

3.8 A. 0.0487 kg
     B. 0.0254 kg, 0.0560 m^3
     C. 0.0233 kg, 2.41 × 10^{-5} m^3

3.10 33.6% liquid, 66.4% vapor

3.14 A. 0.6 MPa, 20°C, 0.140 m^3/kg; 0.0714 kg

3.19 3.596 kg/s

3.22 0.111 m

3.26 1421 lb$_m$/hr

3.30 0.367 m^3/s

3.33 9.238 × 10^{-3} kg/s, 0.2422 m/s

3.36 2.1 m/s

3.38 10.72 m/s

3.41 4 m/s, 16 m/s

3.44 2121 lb$_m$/hr, 33.8 ft/s

3.47 2 kg/s

3.50 14.7 ft/s, 5.3 in

3.53 2.03 hr, 30.3 hr

3.56 2190 s or 36.5 min

3.58 −20 kg/s

3.63 1800 kg/s·m^2, 36,700 kg/s·m^2

3.67 A. Yes, B. No, C. No, D. No

3.73 0, Yes

3.77 A. 8.1 m/s; B. 1.92 m/s, −0.546 m/s, 0.888 m/s

3.78 Compressible ($Ma = 0.71$)

3.82 0.779, 128.4%, 28.4%

3.85 17.35, 111.4%, 11.4%, 0.898

3.88 A. 0.205; B. $X_{CO_2} = 0.0291$, $X_{H_2O} = 0.0267$, $X_{O_2} = 0.165$, $X_{N_2} = 0.779$

3.90 24.0, $X_{CO_2} = 0.0756$, $X_{H_2O} = 0.1134$, $X_{O_2} = 0.0662$, $X_{N_2} = 0.7448$

3.94 18.9, 110.1%, 10.1%

3.104 A. 84.45% C, 15.55% H; B. 125.5%; C. 0.7968

3.108 14 kmol

Chapter 4

4.1 233,812 J

4.5 442.4 m/s

4.8 0.60

4.11 A. 9.41 kJ
  B. 1.29 kJ
  C. 8.13 kJ
  D. Inside pressure force will equal sum of atmospheric and connecting-rod forces.
  E. Work transfer to connecting rod

4.13 A. 12.6 Btu, B. 5.77 Btu

4.16 6.096 Btu, 6.096 Btu

4.20 ac: $3RT_1/2$ abc: RT_1

4.24 589 kJ, 559 Btu

4.26 54 W

4.28 291 kW

4.33 B. 6818 K/m
  C. $-170,450$ W/m^2, i.e., directed left
  D. 0.011 m, 462.5 K
  E. Slab to surroundings

4.37 72,380 W/m^2, 723.8 W/m$^2 \cdot$K

4.40 6.67 W/m$^2 \cdot$K

4.43 18.19 W/m$^2 \cdot$K; Velocity is not needed to obtain solution, although the velocity does influence the value of the heat-transfer coefficient.

4.47 217.1 W/m^2, 173.7 W/m^2

Chapter 5

5.3 14.3 kJ, 37.0 kJ, 51.2 kJ, 51.2 kJ

5.7 0, -2013.09 kJ/kg, -2250.06 kJ/kg, -2013.09 kJ/kg

5.9 370.8 K

5.11 71,928 K/s

5.16 201 W/m$^2 \cdot$K

5.20 (a) 797 kJ, 1116 kJ, 797 kJ, 0 kJ
  (b) 797 kJ, 1116 kJ, 1116 kJ, 319 kJ

5.24 2938 kJ

5.27 ab: $9RT_1$ adb: $17 RT_1/2$

5.29 A. 240 MJ, B. zero

5.30 619 Btu

5.35 6.18 psia, 125.1 Btu

5.37 357 Btu

5.41 A. 2.425 kJ, B. 2.425 kJ, C. 20.06°C, D. 2.425 kJ

5.44 -62.5 cm

5.48 A. 100 psia, B. 538 F, C. 5.83 ft^3, D. 1191 Btu

5.50 470°C, 342 kJ

5.54 1483 K (using specific heats at 1200 K), 105.923 kJ

5.57 2979 K (using specific heats at 1200 K)

5.60 A. 18,300 cm^3
  B. 8.8044 g (liquid), 0.1818 g (vapor)
  C. 1.006 kJ into contents
  D. -258 kJ
  E. 259 kJ

5.65 1266.7 W

5.69 370°C

5.71 0.050 W/m $\cdot$ K

5.75 73.7°C, 548 W/m^2

5.80 0.418 m

5.84 No. The thermal conductivity must be a function of temperature since the heat flux is more than doubled when the temperature difference is doubled.

5.86 90.9 W/m^2

5.92 630 W/m$^2 \cdot$K, 6300 W/m^2

5.94 412.4 K

5.98 262.37 W/m; neglecting pipe wall, 262.42 W/m

5.101 289.4°C

5.105 342.2 W/m

5.109 18.95 mm

5.111 1.565 m (impractical)

5.115 69.7°C, 59.0°C

5.119 67,565 W

5.122 254.4°C, 112.8 mm^2

5.126 316 K

5.131 2649 kW or 3552 hp

5.133 3473 ft/s

5.137 3.447 kJ/kg

5.141 0.152

5.144 60.84 m/s

5.148 229.8 MW

5.151 A. 92,300 Btu/hr, B. 1.959

5.154 115.8 kg/kg

5.157 41.3 hp

5.159 A. −44.99 kJ/kg, B. −11.7°C

5.162 A. 0.357 Btu/lb$_m$
       B. 133.7 ft/s
       C. Viscous friction converts some KE to internal energy; some heat transfer to surroundings; some KE remains with water flowing away.

5.166 0.3 kg/s, 329.5°C

5.169 49,528 kJ/kg, 45,742 kJ/kg

5.172 1.30 MW

5.175 2664 K

5.177 2908 K

5.180 A. 891.3 MJ/kmol or 55.56 MJ/kg
       B. 673.5 MJ/kmol or 41.98 MJ/kg

5.182 1356 s

5.184 0.0072 kg

5.189 A. 700 W, B. 3700 W, C. 3000 W

5.190

State	$\dot{m}$ (kg/s)	h (J/kg)	Device	$\dot{Q}$ (W)	$\dot{W}$ (W)
1	30	15	A	150	0
2	25	13	B	30	25
3	25	9	C	100	0
4	30	10	D	0	5
5	5	14			

5.192 460 Btu/s

Chapter 6

6.2   0.00049 or 0.049%

6.5   137.098 kPa, 1.3531 atm, 1.37098 bar, 19.884 psia

6.8   564.1 m

6.11   598.2 kg/m^3

6.14   18.47 kPa

6.17   1.89 kN

6.23   A. 0.59 lb$_f$/in^2 or 0.0401 atm
       B. 14.92 lb$_f$/in^2 or 1.015 atm
       C. 1.207 in Hg

6.25   A. 110.9 ft H$_2$O, B. 127.0 in Hg

6.26   122.95 kPa

6.29   316.8 kPa

6.32   100.048 Pa

6.35   (a) 124.427 kPa
       (b) 24.431 kN, 24.431 kN, 97.725 kN
       (c) 4796.2 N, 11,191 N, 11,191 N

6.37   8334 N, −0.0203 m from CG, 169.2 N·m

6.38   207.2 N, 2659 N

6.45   A. 62,130 N, on centerline at depth of 5.51 m,
       B. 54.73°

6.50   0.086 N

6.55   A. positive x-direction, B. right-hand corner, C. 4906 Pa (gage)

6.60   Constant-P lines run upward at 45° left to right; 108,335 Pa

6.65   11.3°

6.68   0, 0.225 m/s, 0.2009 N/m^2 or Pa

6.72   400 N/m^2 (magnitude), 800 N/m^2

6.76   5.044 kg/s, 10.324 N

6.78   25.45 N

6.83   298.1 kN

6.86   4277 N (tension)

6.91   −2500 N (i.e., to the left)

6.96   −1178 N (i.e., to the left)

6.99   9425 N for $\rho = 1.2$ kg/m^3

6.102 16.6 m, 56.3%

6.108 A. 2.24×10^{-3} m^3/s, B. 5.617×10^{-3} m^3/s

6.110 38.9 m/s

6.112 3.96 m/s, 2.80×10^{-5} m^3/s

6.120 27,930 Pa

6.123 0.129 m

6.133 6.25 N·s/m^2

Chapter 7

7.1   449.4 cycles/min, 16 kW

7.4   3.32 kW, 4.03

7.7   B. 1–2: 1434.7 kJ/kg (in), 1434.7 kJ/kg (out); 2–3: 0 kJ/kg, 3744 kJ/kg (in); 3–4: 7173.5 kJ/kg (out), 7173.5 kJ/kg (in); 4–1: 0 kJ/kg, 3744 kJ/kg (out)
       C. 0.526

7.10 1.26 kW, 0.84 kW

7.13 B. 45.3%, 0.6766, 0.3243

C. Difficulty in operating pump and turbine with two-phase mixtures, etc.

7.16 0.198, 0.369 (reversible)

7.20 $\eta_{A+B} = 1 - (1 - \eta_A)(1 - \eta_B)$

7.24 457 Btu, 0.543

7.28 A. 315°C

B. 241°C

7.29 Possible, but unlikely, since reversible COP is 7.51.

7.31 No. The claimed efficiency is greater than the reversible value.

7.34 21.1 kW

7.38 342.9 Btu/min or 8.08 hp, 311.4 F

7.42 5.06, 151.4 kJ

7.45 250 Btu/min

7.49 9.54, 9.54 kJ, 10.54, kJ

7.53 127.06 J/K

7.55 0.164 J/K·s

7.57 −0.0487 Btu/lb$_m$·R, 46.8 Btu/lb$_m$, 42.9 Btu/lb$_m$

7.59 No. It violates the second law.

7.64 0.296 kJ/kg·K

7.66 $\Delta s = a[\ln(T_2/T_1) - b(T_2 - T_1)]$

7.70 5889 kg/hr, 4553 kJ/K·hr, 35,730 kJ/K·hr, 43,408 kJ/K·hr, 3125 kJ/K·hr

7.75 8.368 MW, 422.7 K

7.77 2.524 kJ/K·s

7.81 −168 kJ/kg, −171 kJ/kg

7.84 A. 0.92

B. 460 kW·hr

C. −0.5 kW·hr/K, 0.5 kW·hr/K

D. −0.333 kW·hr/K, 0.5 kW·hr/K

E. zero, 0.1667 kW·hr/K

7.87 A. 564 K for $\gamma = 1.4$

B. 1.16 m³/kg

C. −377 kJ

D. −527 kJ/kg

E. 527 kJ/kg

7.90 192 MW, 0.79

7.93 301 kJ/kg

7.95 A. 419 kJ/kg, B. 0.194 kJ/kg·K

7.97 300°C, 0.837

7.99 341 kW

7.103 A. 13.7 kW, B. 12.6 kW, C. 0.916

7.106 367 K for $\gamma = 1.288$

7.110 *1 atm*: 1.3815×10^{-5} atm, 0.999986 atm; 1.3815×10^{-5}, 0.999986

0.1 atm: 4.3688×10^{-6} atm, 0.0999956 atm; 4.3688×10^{-5}, 0.999956

7.113 8.347×10^{-6}, dimensionless

7.118 $X_{CO} = 0.1115$, $X_{H_2O} = 0.4448$, $X_{CO_2} = 0.2219$, $X_{H_2} = 0.2219$

7.120 $X_{H_2O} = 0.0211$, $X_{N_2} = 0.8787$, $X_{H_2} = 0.0668$, $X_{O_2} = 0.0334$

7.123 $K_{p,2} = K_{p,1}^{1/2}$

Chapter 8

8.5 $\partial T^*/\partial t^* = (\alpha t_c/L_c^2)\nabla^2 T^*$, *Fourier number* $\equiv \alpha t_c/L_c^2$

8.6 1.59 times

8.7 A. 7.69×10^5, B. 536, C. 15,800, D. 0.743

8.14 32.1 W/m²·K, 58.7 W/m²·K

8.18 15,500 W/m², 6650 W/m²

8.22 A. 1.72 m/s, B. 22,900 N

8.26 11.5 m/s, 1.29×10^9, 6.70×10^6

8.29 $\mu/(\rho VL)$, $F_D/(V^2\rho L^2)$, D/L

8.34 $\mu_a/(\rho_a VD)$, $D\rho_1 V^2/\sigma$, $D^{5/2}(\rho_1/\sigma)^{3/2}gV$, ρ_a/ρ_1, μ_1/μ_a

8.36 W/d, $dfA\rho/\mu$, $f^2A^2/(gd)$, A/d, where d = depth, W = width, f = frequency, and A = amplitude (m)

Chapter 9

9.2 minimum [78.4 m/s (air), 4.26 m/s (H$_2$O), 2748 m/s (oil)] = 4.26 m/s; 3.03×10^{-3} m (air), 7.07×10^{-4} m (H$_2$O), 0.018 m (oil)

9.5 (a) turbulent, $Re = 5.27 \times 10^7$; laminar, $Re = 2.90 \times 10^5$; turbulent, $Re = 1.31 \times 10^6$

9.8 0.093 N

9.11 0.293

9.18 4.33 N (both sides), 5.11 N (both sides)

9.21 0.283 N

9.24 $3.464Re^{-1/2}$, $0.577Re^{-1/2}$

9.26 35.86 W

9.31 24.76 W/m²·K, 2.38 W

9.35 (a) 3410 W/m²·K, 2410 W/m²·K, 1970 W/m²·K; (b) 4060 W/m²·K

9.36 A. 17.6 W, B. 79.5 W

9.40   2250 W, 51.9 N

9.43   (a) 18.2 N, (b) 3750 W

9.46   (a) 0.0125 lb$_f$, 687 Btu/hr; (b) 0.00816 lb$_f$, 448 Btu/hr

9.48   46.7 kN, 65.3 kN

9.50   4090 W/m

9.52   15.9 kW or 21.4 hp, 23.7%, 76% reduction with 8:1 aspect-ratio fairings

9.54   *head on:* 131.3 kN, 1182 kN, 13,130 kN; *45°:* 141.5 kN, 1274 kN, 14,150 kN

9.56 $Nu_\theta(\theta = 0) = 0.1674 Re_D^{0.6574}$

9.57   2042 W/m^2

9.62   106 W/m$^2 \cdot$K, 11.9 N/m

9.65   0.071 m/s

9.70   2.37 m

9.73   646 W or 0.866 hp

9.75   400 W/m$^2 \cdot$K for a terminal velocity of 26.9 m/s

9.76   19 W

9.79   2(121.2 W) + 202 W = 444.4 W

9.81   490

9.82   Yes, marginally. 0.42 W

9.89   11.9 W, 36.5 W, 115 W, 179 W

9.93   275 F, 73.6 F

9.97   429 W

9.100   5.48 W/m$^2 \cdot$K

9.103   18.3 W/m$^2 \cdot$K, 1370 W/m$^2 \cdot$K

9.106   623 W/m

9.109   A. 69.5°C, B. 2725°C (destructively hot)

9.111   0.768 W

9.113   (a) 1.46 W, (b) 173 W, (c) 52.5 W

Chapter 10

10.2   178.3 m/s

10.4   0.718 N/m^2

10.8   103 W

10.10   0.921 m

10.13   A. 0.10 m/s, B 1.2 m/s, C 63.22 m/s, D. 14.08 m/s, E. 0.01 m/s

10.16   0.754 in, 0.0484 in

10.18   Likely turbulent

10.22   A. 0.107 m/s, B. 0.214 m/s, C. 1250, D. 0.733 N/m^2, E. 293.1 Pa, F. 0.051, G. 0.03 m, H. 2.47 × 10^{-3} W

10.24   5.88 × 10^{-3} kg/s

10.28   64.9 kPa

10.30   1014 m

10.31   1656 W/m^2

10.36   1.71 W/m$^2 \cdot$K, 0.235 Pa

10.43   0.331 m

10.44   20.39 N/m^2, 20.4 kPa

10.48   A. 0.69 m/s, B. 8055, C. 0.0334, D. 39.6 kPa, E. 2.15 W

10.49   A. 0.69 m/s, B. 8055, C. 0.058, D. 68.8 kPa, E. 3.73 W

10.50   2884 W, 4239 W, $39.72/week, $58.40/week

10.53   0.0275

10.57   32.4 m/s

10.64   0.168 m

10.68   0.261 m^3/s

10.72   A. 311 K; B. 305.8 K, 316.7 K; C. 142.6 Pa

10.73   A. 420 K; B. 111 W; C. 1422 W/m^2, 287 W/m^2

10.76   Using Sieder & Tate correlation: (a) 1720 Btu/hr·ft^2·F, (b) 2050 Btu/hr·ft^2·F, (c) 1890 Btu/hr·ft^2·F

10.79   553 W/m$^2 \cdot$K

10.82   Using Sieder & Tate correlation: (a) 10,700 W/m$^2 \cdot$K, (b) 13,600 W/m$^2 \cdot$K, (c) 12,200 W/m$^2 \cdot$K

10.86   1.01 m, 893 Pa

10.89   79.2°C

10.95   0.313 m, 0.427 m

10.96   86.3 Pa

10.98   0.06 m

10.102 0.175, yes

10.104 28.3°C, 218°C

10.108 8.96 × 10^{-4} m^3/s

10.111 18.9 W/m$^2 \cdot$K

10.116 30.2 kW, 26.2 kW

10.120 0.714 m

10.121 (a) 275 kPa, (b) 1.16 kW

Chapter 11

11.1 20.98°C, 0.0203 m

11.2 112.1 kPa

11.6 380.5 K

11.7 3.9 kPa, 0.566 psi

11.11 393.36 K, 0.117

11.16 412.3 m/s, 0.954

11.19 110°C, liquid–vapor mix ($x = 0.301$)

11.20 260°C, 1.07 m^3/kg, 1.42 kJ/kg·K

11.23 A. 1.20, 1.39; B. 2.0

11.25 A. 29,845 K, B. 230.8 K

11.31 23.3 mm, 29 mm

11.32 1 kg/s, 0.528 MPa, 0.69 kg/s, 0.867 MPa

11.37 A. 4.9 atm; B. 0.149 atm; C. 316.9 m/s, 56.6 m/s, 0.0148 kg/s

11.39 3404 ft/s, 0.564 ft

11.44 304.2 K, 349.7 m/s, 456 kPa, 0.0045 m^2, 633 kPa

11.47 6.5 mm, 517.6 m/s, 194.1 m/s

11.50 1079 W

11.51 1564 W, 23.9 m

11.55 136.8 kJ/kg, 134.7 kW

11.56 0.956

11.60 (a) −104.5 Btu/lb$_m$; 0 (b) −76.8 Btu/lb$_m$, −76.8 Btu/lb$_m$; (c) −95 Btu/lb$_m$, −28.7 Btu/lb$_m$

11.63 0.146

11.68 1.27 kJ/kg

11.71 0.463

11.73 5.61 kW, 5.2 kW

11.77 9229 kW

11.78 19.5 kg/s, 20.3 kg/s

11.80 125 MW

11.84 810.7 K

11.85 26.85 MW, 830 K

11.86 986.7 J/kg, 1.013 × 10^5 kg/s

11.90 504 Btu/lb$_m$

11.93 0.604

11.96 79.3 lb$_m$/hr

11.98 0.347 W/m^2·K

11.104 $U_0 = 3652$ W/m^2·K, $U_i = 4280$ W/m^2·K

11.105 415.15 F, 486.0 K, 135,964 lb$_m$/hr, 17.13 kg/s, 171.47 Btu/lb$_m$, 203.17 F

11.106 0.0866 kg/s

11.109 303 kW, 204°C

11.113 3801 W, 0.471 m^2, 0.75 m

11.114 0.783 m^2, 0.0218 m

11.122 parallel: 170°C; counter flow: 215°C

11.123 parallel: 154°C; counter flow: 196°C

11.126 240°C

11.127 1640 kW, $T_{H,o} = 105$°C, $T_{C,o} = 91.0$°C

11.132 42°C, 2090 kg/hr

11.135 46.4°C

11.140 0.777 m

11.145 8.78 m

11.147 44.4 kg/s, 35.4 kg/s

11.149 2.36 kg/hr, 0.762

Chapter 12

12.1 B. 8.12 MW, C. 80.2 kW, D. 0.39

12.2 B. 7.71 MW, C. 94.4 kW, D. 0.37, E. $x = 0.67$

12.5 1007 m^2

12.11 0.342, 0.342 Mlb$_m$/hr

12.14 0.3635

12.15 58.3 MW

12.20 (a) 0.489 Mlb$_m$/hr; (b) 870 MBtu/hr (boiler), 529 MBtu/hr (condenser); (c) 343 MBtu/hr (turbine), 2.00 MBtu/hr (pump); (d) 0.392

12.22 0.257, 0.376

12.26 B. 342.86 MW, 53.02 MW, 57.84 MW
C. 146.4 MW
D. 0.4269 versus 0.42 (a 1.64% improvement)

12.27 0.423, 56.9 kg/s

12.35 0.447, 1540 kJ/kg

12.37 7.187 kN

12.41 1200 K

12.47

State	T (K)	P (atm)
1	293	1.00
2	673.1	18.38
3	2500	68.26
4	1088.2	3.71
$\eta = 0.565$		

12.54 $\eta = 1 - r_{\text{isen}}^{1-\gamma}$

12.58 (a) 1324 R, 325.1 psia; (b) 2.26 Btu/compression or 53.3 hp

12.63 0.514, 6.72 MW

12.67 219 kJ/kg, 566 kJ/kg, 461 kJ/kg, 0.426

12.71

P_2/P_1	T_3 (K)	W_{net}/M (kJ/kg)	η_{th}
10	1624	506.2	0.482
15	1695.1	565.6	0.539
20	1750.8	603.9	0.575

12.74 0.384, 2053 lb_m/hr

12.76 197 kJ/kg, 0.298

12.80 0.302

12.82 12.6, 13.6

12.84 1.815 kW, \$0.11/hr, 4.046 kW, 13,806 Btu/hr

12.86 5.16 kW, 3.90 kW, 1.25 kW, 3.12, 4.12

12.88 2.83, 3.83

12.92 10.38

12.109 Compressor work, 2.63 kW; condenser heat transfer, 10.0 kW; evaporator heat transfer, 7.37 kW; *COP*, 3.80; flowrate, 0.0641 kg/s

12.111 0.0030, 0.0067, 3.56×10^{-5} kg/s, 0.034 gal/hr

12.113 Indoors: 0.0107, 15.06°C; Outdoors: 0.0145, 19.78°C

12.117 A.

Property	State 1	State 2
Relative humidity	70	100
Humidity ratio	0.01021	0.00726
Dew point (°C)	14.4	20
Mass of dry air (g)	4.74	4.74
Mass of water vapor (g)	0.0484	0.0344
Mass of water liquid (g)	0	0.0140

B. Decreases

12.119 A. 130 F, B. 0.0782, C. 0.184

12.123 100%, 0.00726, 0.00366 kg

12.130 1.89°C, 39.0°C

12.132 0.00883, 12.0°C, 0.586 kg

12.137 28.2°C, 2.77 kJ

12.142 A. 0.0073
B. 91.5%
C. 0.0092 kg/s, $9.233 \cdot 10^{-6}$ m^3/s

12.144 0.743 kW, 0.0002 kg/s, 0.00317 gal/min

12.148 A. 6.62 kg/min, B. 88.2°C, C. 0, D. 23.7 kW

12.149 0.166%, 0.00824

12.152 A. 1.34 lb_m/min, B. 2470 Btu/min, C. 685 Btu/min

12.157 A. 100°C, B. 0.0219 kg/kg

12.162 A. 5460 lb_m
B. 45,300 lb_m
C. 15,900 lb_m (air), 279 lb_m (water)

12.165 101 F

12.166 −40 kW

12.169 89%, 8.05 kW

12.172 9.3 tons

12.174 A. 132 hp
B. CO_2: 4.6%, O_2: 14.4%, N_2: 81.0%
C. 32.8°C

ILLUSTRATION CREDITS

CHAPTER 1: Opener: Courtesy of Scott/AGE fotostock; p. 4, left to right: Courtesy JERRY MASON/SCIENCE PHOTO LIBRARY, Courtesy of Delespinasse/AGE fotostock; p. 5, counterclockwise from top right corner: Courtesy of Raga/AGE fotostock, Courtesy of Carroll/AGE fotostock, Courtesy of Lehn/AGE fotostock, Courtesy of Turner/AGE fotostock, Courtesy of Poelking/AGE fotostock, Courtesy of Siteman/AGE fotostock, Courtesy of AGE fotostock; p. 6, top left: Courtesy of Cheadle/AGE fotostock; p. 6, middle left: Courtesy of Rugner/AGE fotostock; Fig. 1.3: Florida Power and Light Co.; Fig. 1.4: Smithsonian Institution Neg. # EMP21.035, from the Science Service Historical Image Collection, courtesy Westinghouse; p. 10, top to bottom: Courtesy of AGE fotostock, Courtesy of Lasalle/AGE fotostock; Fig. 1.7: © PITCHAL FREDERIC/CORBIS SYGMA; p. 11, top to bottom: Courtesy MARTIN BOND/SCIENCE PHOTO LIBRARY, Courtesy of AGE fotostock; Fig. 1.8: © 2003 CSU/Photography and Digital Imaging; p. 13, top to bottom: Courtesy of Borchardt/AGE fotostock, Courtesy of Khornak/AGE fotostock, Courtesy of Greenberg/AGE fotostock; p. 14: Courtesy of Pobereskin/AGE fotostock; p. 16, top left: Courtesy of Wassell/AGE fotostock; Figure 1.13: Courtesy Pratt and Whitney; Fig. 1.14, left to right: Courtesy of AGE fotostock, Courtesy of Pitcher/AGE fotostock, Courtesy of Forbes/AGE fotostock, Courtesy of Román/AGE fotostock; p. 18, top left: Courtesy of Ocaña/AGE fotostock; p. 18, left middle: Courtesy of Gold/AGE fotostock; Fig. 1.18b: Courtesy of Andrade/AGE fotostock; Fig. 1.21b, Fig. 1.21c: Courtesy of Silva/AGE fotostock, Photo courtesy NASA; p. 21: Courtesy of Schön/AGE fotostock; p. 24, top to bottom: Courtesy of Iwasaki/AGE fotostock, Courtesy of AGE fotostock, Courtesy of Logan/AGE fotostock; p. 25: US Air Force photo; p. 26: Courtesy of Amengual/AGE fotostock; Fig. 1.26: Courtesy Stirling Tech Co.; p. 28: Courtesy MEHAU KULYK/SCIENCE PHOTO LIBRARY; Fig. 1.30: Courtesy of AGE fotostock; p. 31, top to bottom: *A Gallery of Fluid Motion*, ed. M. Samimy et al. Published by Cambridge University Press. © Cambridge University Press 2003, Courtesy of Gay/AGE fotostock, Courtesy of Gay/AGE fotostock, Courtesy of Forbes/AGE fotostock; p. 34: Courtesy of Gual/AGE fotostock; Pr. 1.8: Courtesy of AGE fotostock; Pr. 1.10: Courtesy of Logan/AGE fotostock; Pr. 1.11: Courtesy of Cary/AGE fotostock; Pr. 1.16: Courtesy of Moellers/AGE fotostock; Pr. 1.31: Courtesy DOE/NREL; Pr. 1.32: General Motors Corporation. Used with permission, GM Media Archive; Pr. 1.33: Courtesy of Kearney/AGE fotostock; Pr. 1.35: Photo courtesy NASA; Pr. 1.36: Courtesy of AGE fotostock; Pr. 1.40: Photo courtesy NASA; Pr. 1.44: Courtesy of AGE footstock. **CHAPTER 2:** Opener: Courtesy of Foxx/AGE fotostock; p. 48: Courtesy MATT MEADOWS/SCIENCE PHOTO LIBRARY; p. 49, top to bottom: Courtesy of Brown/AGE fotostock, Courtesy of Rugner/AGE fotostock; p. 54: Courtesy of Larrea/AGE fotostock; Ex. 2.1: Courtesy of Mata/AGE fotostock, Courtesy of Powietrzynski/AGE fotostock; Ex. 2.2, top to bottom: Courtesy of López/AGE fotostock, Courtesy of Miller/AGE fotostock; Ex. 2.3: Courtesy of Bibikow/AGE fotostock; p. 60: Courtesy LAGUNA DESIGN/SCIENCE PHOTO LIBRARY; Ex. 2.4: Photo courtesy NASA; p. 65, top to bottom: Courtesy CLIVE FREEMAN/BIOSYM TECHNOLOGIES/SCIENCE PHOTO LIBRARY, Photo E. Lange, Heidelberg, Courtesy AIP Emilio Segrè Visual Archives; p. 66: Courtesy of ARL (Penn State) Computational Mechanics; p. 69: Photo courtesy Air Liquide Group; Ex. 2.6: Courtesy of AGE fotostock; p. 71: Courtesy of ARL (Penn State) Computational Mechanics; Ex. 2.8: Courtesy CLIVE FREEMAN, THE ROYAL INSTITUTION/SCIENCE PHOTO LIBRARY; p. 80: Courtesy SCIENCE, INDUSTRY & BUSINESS LIBRARY/NEW YORK PUBLIC LIBRARY/SCIENCE PHOTO LIBRARY; p. 84: Courtesy of AGE fotostock; Ex. 2.12: Courtesy of Gold/AGE fotostock; p. 87: Image courtesy of Eugene Kung and Daniel Haworth; p. 103: Courtesy of Nowitz/AGE fotostock; p. 126: Courtesy of Morandi/AGE fotostock; p. 131: Photo courtesy Dr. Kenneth Libbrecht; p. 132, top to bottom: Courtesy of Gold/AGE fotostock, Courtesy of AGE fotostock; p. 135: Courtesy of Millich/AGE fotostock; p. 144: Courtesy of AGE fotostock; p. 146, clockwise starting from top left: Courtesy of Grill/AGE fotostock, Courtesy of Larrea/AGE fotostock, Courtesy of Nebot/AGE fotostock, Courtesy of AGE fotostock, Courtesy of Cary/AGE fotostock, Courtesy of AGE fotostock; p. 149: Courtesy of Grill/AGE fotostock; Pr. 2.4: Photo courtesy of VACUUBRAND GMBH + CO KG, Germany; Pr. 2.11: Courtesy of Khornak/AGE fotostock; Pr. 2.12: Courtesy of Morsch/AGE fotostock; Pr. 2.24b: Courtesy Barry Howe; Pr. 2.29: Courtesy of Pitamitz/AGE fotostock; Pr. 2.47: Courtesy of Carroll/AGE fotostock; Pr. 2.117: Courtesy of AGE fotostock; Pr. 2.121: Courtesy of Bott/AGE fotostock. **CHAPTER 3:** Opener: Courtesy of Janes/AGE fotostock; p. 174: Courtesy of AGE fotostock; p. 175: Photo courtesy of Panopticon Lavoisier/Institute and Museum of History of Science-Florence; Ex. 3.4: Courtesy Compact Radial Engines, Inc.; Fig. 3.3: Courtesy of ARL (Penn State) Computational Mechanics; Ex. 3.7: Courtesy of ARL (Penn State) Computational Mechanics; Ex. 3.8: Courtesy of ARL (Penn State) Computational Mechanics; Ex. 3.10: Building design and solar system by architect Sture Larsen, www.solarsen.com; p. 200, left top to bottom: Photo courtesy NASA, Courtesy of Hatter/AGE fotostock; p. 200, bottom right: Courtesy of MacDonald/AGE fotostock; Fig. 3.10: Courtesy of the author; p. 206: Photo courtesy NASA; p. 208: Courtesy of Silva/AGE fotostock; p. 209: Courtesy of ARL (Penn State) Computational Mechanics; p. 213, left to right: Courtesy of ARL (Penn State) Computational Mechanics; p. 214: Courtesy US DEPT. OF ENERGY/SCIENCE PHOTO LIBRARY; p. 216: Courtesy of AGE fotostock; p. 217: Courtesy of Larrea/AGE fotostock; p. 218: Courtesy of AGE fotostock; p. 221: Courtesy of Moss Motors, Ltd.; Pr. 3.5: Courtesy of Gual/AGE fotostock; Pr. 3.8: Courtesy of de Gail/AGE fotostock; Pr. 3.24: Courtesy of Firebaugh/AGE fotostock; Pr. 3.70: Courtesy of AGE fotostock; Pr. 3.81: Courtesy of AGE fotostock; Pr. 3.82: Courtesy of Delespinasse/AGE fotostock; Pr. 3.87: Courtesy of Gold/AGE fotostock; Pr. 3.88: Courtesy of Lauritz/AGE fotostock; Pr. 3.110a: Courtesy of AGE fotostock; Pr. 3.110c: Courtesy of Bibikow/AGE fotostock. **CHAPTER 4:** Opener: Photo courtesy NASA/JPL-Caltech; p. 244: AIP Emilio Segrè Visual Archives; Fig. 4.1a: Courtesy of Toomey/AGE fotostock; Fig. 4.1b: Courtesy of AGE fotostock; Fig. 4.1c: Photo courtesy NASA; p. 246, bottom left: Courtesy of Zyvex; p. 247, top to bottom: Courtesy Sandia National Laboratories, SUMMiT™ Technologies, www.mems.sandia.gov, Courtesy COLIN CUTHBERT/SCIENCE PHOTO LIBRARY; p. 248, top to bottom: Courtesy of AGE fotostock, Courtesy of Cary/AGE fotostock, Courtesy of Coll/AGE fotostock; p. 250: Courtesy of Paynter/AGE fotostock; p. 258: Courtesy of AGE fotostock; Ex. 4.3: Courtesy of AGE fotostock; Fig. 4.6a: Courtesy of GE Energy; Fig. 4.6b: Courtesy of Scania; Fig. 4.6c: Photo courtesy NASA; p. 263, bottom left: Courtesy of Larrea/AGE fotostock; p. 264, left middle: Courtesy of Arici/AGE fotostock; Fig. 4.8, left to right: Courtesy of AGE fotostock; Ex. 4.4: Courtesy of Warren/AGE fotostock; p. 267, top to bottom: Image provided courtesy of Flowserve Corporation, Courtesy DOE/NREL, Courtesy of Allen/AGE fotostock; p. 268, p. 271: Courtesy DOE/NREL; p. 273: Courtesy of Algarra/AGE fotostock; Ex. 4.8: Courtesy of Banús/AGE fotostock; p. 278, top left: Courtesy of ARL (Penn State) Computational Mechanics; Fig. 4.16 a, c, d: Courtesy of Paul Ruby; Fig. 4.16b: Courtesy of Sibtosh Pal; Fig. 4.16e: Courtesy of Gary Settles; p. 279, top to bottom: Courtesy of Cancalosi/AGE fotostock, Courtesy of Amengual/AGE fotostock; p. 281: Courtesy of Williams/AGE fotostock; Ex. 4.11: Courtesy of Mike Yip; p. 288: Courtesy of Kunst & Scheidulin/AGE fotostock, Courtesy NASA; Ex. 4.13: "Reproduced with the permission of Omega Engineering, Inc., Stamford, CT 06907" www.omega.com. This image is a registered trademark of Omega Engineering, Inc.; Ex. 4.15: Courtesy of AGE fotostock. **CHAPTER 5:** Opener: Courtesy of Reede/AGE fotostock; p. 310, left to right: Burndy Library, courtesy AIP Emilio Segrè Visual Archives; Photo E. Lange, Heidelberg, courtesy AIP Emilio Segrè Visual Archives; Courtesy of AIP Emilio Segrè Visual Archives, Physics Today Collection; Ex. 5.1: Photo courtesy NASA Glenn Research Center; p. 315: Photo courtesy NASA; Ex. 5.6: Courtesy of Datene/AGE fotostock; p. 326: Courtesy of Rugner/AGE fotostock; p. 337: Courtesy of Forbes/AGE fotostock; Ex. 5.10: Courtesy of Satushek/AGE fotostock; p. 340, left to right: Courtesy of AGE fotostock, Courtesy NASA, Courtesy of Beauregard/AGE fotostock; p. 341, top to bottom: Courtesy NASA Glenn Research Center, Courtesy DOE/NREL; Fig. 5.8: Courtesy of Jim Wiedenhoefer and Rolf Reitz, Engine Research Center, University of Wisconsin-Madison; Fig. 5.9: Courtesy of IBM Corporate Archives; p. 344: Courtesy DOE/NREL; Ex. 5.12: Photo courtesy of photostocker.com; p. 348: Courtesy of Wells/AGE fotostock; Ex. 5.13: Courtesy DOE/NREL; Ex. 5.14: Courtesy of Yuba Heat Transfer division of Connell LP; Ex. 5.16: Courtesy of Lasalle/AGE fotostock; Ex. 5.17: Courtesy of Yuba Heat Transfer division of Connell LP; p. 364: Courtesy of Kearney/AGE fotostock; Ex. 5.18, p. 366, p. 367: Courtesy of AGE fotostock; p. 368: Courtesy of Forbes/AGE fotostock; Fig. 5.18, bottom: Photo courtesy NASA; p. 375: Courtesy of ARL (Penn State) Computational Mechanics; p. 378: Image courtesy of Daniel C. Haworth; Pr. 5.11: © Minister of Natural Resources Canada 2001, Reproduced with permission; Pr. 5.14: Courtesy of Banús/AGE fotostock; Pr. 5.29: Courtesy of Greenberg/AGE fotostock; Pr. 5.72: Courtesy of Bott/AGE fotostock; Pr. 5.74: Courtesy of Paras/AGE fotostock; Pr. 5.86: Courtesy of de Alda/AGE fotostock; Pr. 5.93: General Motors Corporation. Used with permission, GM Media Archive; Pr. 5.94: Photo courtesy CLASIC CZ Ltd.; Pr. 5.95: Courtesy of Larrea/AGE fotostock; Pr. 5.123: US Air Force photo; Pr. 5.124: Courtesy of AGE fotostock; Pr. 5.127: Courtesy of Buss/AGE fotostock; Pr. 5.131: Courtesy of Peebles/AGE fotostock; Pr. 5.133: Courtesy of Scott/AGE fotostock; Pr. 5.137: Courtesy of AGE fotostock; Pr. 5.162: Courtesy of Wall/AGE fotostock. **CHAPTER 6:** Opener: Courtesy of Sanford/AGE fotostock; p. 408, all: Courtesy of AGE fotostock; p. 409: Courtesy of AIP Emilio Segrè Visual Archives, Lande Collection; p. 410: Courtesy of Silva/AGE fotostock; p. 412, top to bottom: Courtesy of Cary/AGE fotostock, Courtesy of Steeger/AGE fotostock; p. 413: Courtesy of AGE fotostock; p. 418: Courtesy of Bellurget/AGE fotostock; p. 421: Courtesy NASA Glenn Research Center; p. 423, top to bottom: Courtesy of Raga/AGE fotostock, © Bettmann/Corbis; p. 426: Courtesy of GE Energy; p. 428: US Navy photo by Photographer's Mate 3rd Class Danielle M. Sosa; p. 431: Courtesy of Milchanowski/AGE fotostock; p. 432: Courtesy of Bowater/AGE fotostock; Ex. 6.6: General Motors Corp. Used with permission, GM Media Archive; p. 438: Courtesy of Delespinasse/AGE fotostock; Ex. 6.9: Courtesy of Welsh/AGE fotostock; p. 443: Courtesy of Gold/AGE fotostock; p. 444: US Air Force photo; p. 446: Courtesy NASA; p. 448, top to bottom: Courtesy of Andersson/AGE fotostock, Courtesy of Grill/AGE fotostock; p. 450: Courtesy NASA; p. 461, all: Courtesy of ARL (Penn State) Computational Mechanics; p. 464, all: Images courtesy of Daniel Haworth and Eugene Kung; Ex. 6.13: Courtesy of ARL (Penn State) Computational Mechanics; p. 472: Courtesy of Moon/AGE fotostock; p. 473: Courtesy of Kearney/AGE fotostock; p. 474: Courtesy of Beauregard/AGE fotostock; Ex. 6.18: Courtesy NASA; p. 478, top to bottom: Courtesy of ARL (Penn State) Computational Mechanics; Pr. 6.3: Courtesy of Seale/AGE fotostock; Pr. 6.24: Courtesy of Wyle/AGE fotostock; Pr. 6.33: Courtesy of Mora/AGE fotostock; Pr. 6.56: Courtesy of AGE fotostock; Pr. 6.99: Courtesy of Kunst & Scheidulin/AGE fotostock. **CHAPTER 7:** Opener: Courtesy of AGE fotostock; p. 514: A.R. Thayer, from a portrait by Bailly, courtesy AIP Emilio Segrè Visual Archives; p. 515, top to bottom: AIP Emilio Segrè Visual Archives, Physics Today Collection; AIP Emilio Segrè Visual Archives, Segrè Collection; p. 518, left to right: Courtesy of Peebles/AGE fotostock, Courtesy of Arnold/AGE fotostock, Courtesy of AGE fotostock; p. 519: Courtesy of Allen/AGE fotostock; Fig. 7.3a: Courtesy DOE/NREL; Fig. 7.3b: Courtesy of Holmes/AGE fotostock; Ex. 7.2: Schematic diagram courtesy of Kockums AB,

Index